Selected Works in Probability and Statistics

For further volumes:
http://www.springer.com/series/8556

Javier Rojo
Editor

Selected Works of E. L. Lehmann

Springer

Editor
Javier Rojo
Rice University
Houston, TX 77005, USA
jrojo052@gmail.com

ISBN 978-1-4614-1411-7 e-ISBN 978-1-4614-1412-4
DOI 10.1007/978-1-4614-1412-4
Springer New York Dordrecht Heidelberg London

Library of Congress Control Number: 2011941600

Printed on acid-free paper

Springer is part of Springer Science+Business Media (www.springer.com)

Preface to the Series

Springer's Selected Works in Probability and Statistics series offers scientists and scholars the opportunity of assembling and commenting upon major classical works in statistics, and honors the work of distinguished scholars in probability and statistics. Each volume contains the original papers, original commentary by experts on the subject's papers, and relevant biographies and bibliographies.

Springer is committed to maintaining the volumes in the series with free access of SpringerLink, as well as to the distribution of print volumes. The full text of the volumes is available on SpringerLink with the exception of a small number of articles for which links to their original publisher is included instead. These publishers have graciously agreed to make the articles freely available on their websites. The goal is maximum dissemination of this material.

The subjects of the volumes have been selected by an editorial board consisting of Anirban DasGupta, Peter Hall, Jim Pitman, Michael Sörensen, and Jon Wellner.

Preface

Erich L. Lehmann's legacy is profound and extensive. His scholarly contributions, as partially presented and commented upon in this volume of selected works, clearly demonstrate his broad interests and the widespread impact that his "system building" life-long work had on the discipline. Stephen Stigler (2009), writing about Erich's honorary degree from the University of Chicago, states that the citation on Erich's degree read as follows:

> *Your research on the application of decision theory to statistical problems has helped create and organize modern mathematical statistics; your elegant treatises have guided the curricula of a majority of the nation's graduate programs and given shape to the discipline, and your teaching has inspired a generation of scholars.*

The present collection of selected papers shows Erich's multifaceted contributions. His work on concepts of dependence in 1966 has created a substantial literature and continues to have a large impact on applied probability. His joint work with Joseph L. Hodges on the efficiency of rank tests, and the corresponding work on efficiency of estimators derived from rank tests, helped to assuage the fears of the nonbelievers of nonparametric methodologies. The Lehmann alternatives, with its implicit connection to the proportional hazards model of Cox, and its spillover into the modeling of ROC curves literature, has been used both in the applied and theoretical statistical camps. Erich's work on unifying concepts, ideas, and results in decision theory under an overarching structure hinged on his use of concepts of unbiasedness and invariance applied in the context of exponential and group families of distributions, a literature to which he contributed substantially and extensively. In this regard, Erich's role was one of a system builder. But he also found himself as the originator of ideas in a very substantive way. One example is a concept that serves as one of the main pillars of the statistical decision theory of Wald. Wald, in his book Statistical Decision Functions (1950), on page 29 credits Erich with the concept of a complete class. Here is the exact text:

> *The concept of a complete class of decision functions was introduced by Lehmann, and the first result regarding such classes is due to him [30]. . . .*

Wald then proceeds to discuss the first result on complete classes proved by Erich. The citation [30] refers to Erich's "On Families of Admissible Tests" published in the the *Ann. Math. Stat.* in 1947.

Erich's books, translated into several languages, continue to have a large impact on the education of the next generations of statisticians. His enjoyment for teaching one-on-one is reflected in the large number of Ph.D. students that he mentored.

Erich also served the broader statistical community well. He was Editor (1953-1956), and Associate Editor (1956-1968) of the Annals of Statistics. He served as President of the Institute of Mathematical Statistics (1960-1961) and chaired the Department of Statistics at Berkeley (1973-1976). He was a member of committees of visitors to the department of statistics at Harvard (1974-1980) and Princeton (1975-1980) Universities.

Many honors were bestowed upon him. Among others, there are the following: Guggenheim Fellow (1955-1956, 1966-1967, 1980-1981); Research Professor, Miller institute for Basic Research (1962-1963, 1972-1973); IMS Wald Memorial Lecturer (1964); Elected to the American Academy of Arts and Sciences (1975) and to the National Academy of Sciences (1978); He received honorary degrees from the University of Leiden (1985) – this was the first honorary degree awarded by Leiden to a mathematician since 1884 when Stieltjes received it. An honorary degree from the University of Chicago followed in 1991; He was elected Honorary Fellow of the Royal Statistical Society (1986) and was the COPSS R. A. Fisher Memorial Lecturer (1988); He received the American Statistical Association (ASA) Wilks Memorial Award, and he was the first ASA Goffried Noether Awardee and lecturer. Posthumously, he received the 2009 best paper award from the Journal of Nonparametric Statistics.

In addition, Erich's gentle and unselfish spirit and mentorship did not go unnoticed by his students who organized several meetings in his honor and published collections of works related to Erich's interests. These include: A Festschrift for Erich L. Lehmann (1983), and four Erich L. Lehmann Symposia on optimality (2002, 2004, 2007 and 2011).

In planning this volume, Erich and I met several times to decide on the organization of the papers into the various chapters, and who the contributors would be. Clearly, the chapters cannot be defined sharply as there are papers that could very easily have been placed in a different group. The careful reader will notice two surprises in the collection. The first one is easy to detect. Sections 3.10 - 3.12 of *Testing Statistical Hypotheses* 1959, are being included here. This is happening thanks to Lawrence D. Brown who suggested, and Erich agreed, that they should be reproduced here.

The second pleasant surprise is the inclusion of Erich's work on the history of confidence statements (1958). Stephen Stigler was kind enough to point out this paper to us.

I would like to thank Jim Pitman for his patience and encouragement during the preparation of this volume. John Kimmel and Eric Strauss have been very supportive of this project. Thanks also go to the contributors to this work. The requests for contributions, as it always happens, came at a busy time for everyone. The contributors' special efforts in helping to bring this project to fruition is greatly appreciated.

My deep gratitude goes to Juliet Shaffer. Julie was a great source of information during the preparation of this work. In addition, the many invitations to join her and Erich for lunch or dinner, were great opportunities to get to know them better. Erich's and Julie's concern for my family will always be appreciated.

Erich Leo Lehmann passed away in the early morning of September 12th, 2009. I first learned of his death as I waited for my wife at the Oakland airport so that, together, we could visit Erich – but it was not to be. I and my family were very fortunate to get to know Erich and enjoy his and Julie's company. Erich will always be in our thoughts.

Houston, TX Javier Rojo

References

1. A. Wald. *Statistical Decision Functions*. John Wiley, New York, 1950.
2. E. L. Lehmann. *Some Early Instances of Confidence Statements*. Statistical Laboratory, University of California, Berkeley. ONR 5 Technical Report to the Office of Naval Research. Unpublished. September 1958.
3. E. L. Lehmann. *Testing Statistical Hypotheses*, John Wiley and Sons, New York, pp. xiii + 369, 1959.
4. P. J. Bickel, K. Doksum, and J. L. Hodges. *A Festschrift For Erich L. Lehmann*, Wadsworth Statistics/Probability Series, 1983.
5. J. Rojo and V. Perez-Abreu, *The First Erich L. Lehmann Symposium: Optimality*. IMS Lecture Notes–Monograph Series, Volume 44, 2004.
6. J. Rojo. *Optimality: The Second Erich L. Lehmann Symposium*. IMS Lecture Notes–Monograph Series, Volume 49, 2007.
7. J. Rojo. *Optimality: The Third Erich L. Lehmann Symposium*. IMS Lecture Notes–Monograph Series, Volume 49, 2007.
8. S. Stigler. The Honorable Erich L. Lehmann. In *The Third Erich L. Lehmann Symposium – Optimality*, (J. Rojo, Ed.) IMS Lecture Notes-Monograph Series, Vol. 57, pp. 10-17, 2009.

Acknowledgements

This series of selected works is possible only because of the efforts and cooperation of many people, societies, and publishers. The series editors originated the series and directed its development. The volume editors spent a great deal of time organizing the volume and compiling the previously published material. The contributors provided comments on the significance of the papers. The societies and publishers who own the copyright to the original material made the volume possible and affordable by their generous cooperation:

American Statistical Association
Indian Statistical Institute
Institute of Mathematical Statistics
International Statistical Institute
John Wiley and Sons
Karolinum Press
National Academy of Sciences
Oxford University Press
Royal Statistical Society
Springer Science+Business Media
Taylor & Francis
University of California Press

Contents

Part III Estimation

Part IV Nonparametrics I

Part XI Historical work

Part XII Philosophical work

Part XIII Books

Contributors

George Casella Distinguished Professor, Department of Statistics, University of Florida, P.O. Box 118545, Gainesville, FL 32611-8545

David Draper Department of Applied Mathematics and Statistics, Baskin School of Engineering, University of California, 1156 High Street , Santa Cruz CA 95064 USA

Hannu Oja Tampere School of Public Health, FIN-33014, University of Tampere, FINLAND

Walter Piegorsch Department of Mathematics, The University of Arizona, 617 N. Santa Rita Ave., P.O. Box 210089, Tucson, AZ 857210089 USA

Calyampudi R. Rao Professor Emeritus of Statistics, Eberly Professor Emeritus, Penn State University, 326 Thomas Building, University Park, PA 16802

Javier Rojo Professor of Statistics, Department of Statistics, Rice University, 6100 Main St., Houston, TX 77005

Joseph P. Romano Professor of Statistics, Department of Statistics, Sequoia Hall, 390 Serra Mall, Stanford University, Stanford, CA 94305-4065

Marco Scarsini Dipartimento di Scienze Economiche e Aziendali, LUISS, Viale Romania 32, 00197 Roma, ITALY

Juliet Shaffer Senior Lecturer Emerita, Department of Statistics, 367 Evans Hall, University of California, Berkeley, CA. 94720

Moshe Shaked Department of Mathematics, The University of Arizona, 617 N. Santa Rita Ave., P.O. Box 210089, Tucson, AZ 857210089 USA

Willem van Zwet Mathematical statistics and probability, Mathematical Institute, University of Leiden, P.O.Box 9512, 2300 RA Leiden, The Netherlands

Douglas A. Wolfe Chair, Department of Statistics, The Ohio State University, 1958 Neil Avenue, 404 Cockins Hall, Columbus, OH 43210-1247

Erich Leo Lehmann

Erich Leo Lehmann was born in Strasbourg on November 20th, 1917. The year of 1917 was a turbulent year. The Mexican revolution was coming to a close while the Bolshevik revolution had just started, and World War I, although ending short of one year after Erich's birth, was still ravaging Europe. Strasbourg was part of Germany at that time, and Erich's father was stationed there as part of the German army to which he had been drafted and now served in his capacity of Captain. At two months of age, the family moved to Frankfurt. Erich was raised in Germany but with the coming of the Nazis to power in 1933, the family moved to Switzerland. Erich stated in DeGroot (1986) that: "After some wondering, because I had a brother who was dying and needed to be cared for first, we settled in Switzerland." One aspect of his education in Germany, however, would come to play a very important role in his future. Erich felt that his strong respect for authority was a direct consequence of his being raised in Germany. Thus, although his dream was to study German literature, given the conditions in Germany at the time, his father suggested mathematics for which Erich had shown affinity. Erich accepted the advice without much resistance. After consulting with the mathematician Edmund Landau, and Landau having assessed, and been satisfied with, Erich's mathematical ability, Erich studied mathematics for five years in Zürich. In 1938, feeling that Switzerland was not a good place for a German refugee, and seeking Landau's advice once more, Erich enrolled in Trinity College in Cambridge. After two years in Trinity College, and not having adjusted well to the environment there, together with the threat of a German invasion of England, Erich traveled to the United States on a French visa under the French quota. Traveling under the French quota was possible since the Treaty of Versailles had returned Strasbourg to France in 1919. The 1924 Immigration Act, also known as The Johnson-Reed Act, provided for a quota of two percent of the total number of people of each nationality in the United States as of the 1890 national census. The Act was revised by Congress in 1952. Erich arrived in New York in November of 1940. He had with him a letter of introduction for Richard Courant from Landau's wife. Courant recommended Berkeley as a good place, a growing and upcoming university.

Erich arrived in Berkeley on January 2nd of 1941. Not having completed a B. A. in Europe, he was admitted as a probationary graduate student in the mathematics department by Griffith C. Evans who was the chair of the mathematics department at the time. Some time later, Evans advised Erich that it was perhaps best to look for an area with more application to the war effort and he suggested physics or statistics as a better choice than mathematics. Erich chose statistics over physics and, after speaking with Neyman, started taking classes in statistics. However, the course material was not particularly to Erich's liking and he decided to go back to mathematics. Before he was able to gather the courage to speak to Neyman and Evans about his plans, Neyman offered him a job, and he accepted. This stage of Erich's life is vividly and amusingly recounted in DeGroot (1988): "So three times: First, German literature – my father says do mathematics. Second, I try to do mathematics and Evans says do something applied. The third time, I want to go back to pure mathematics and Neyman says here's a job for you". One may also want to include a fourth occasion when Erich followed advice. Namely when Landau recommended Trinity College in Cambridge.

In 1942, Erich was granted the M. A. degree from the University of California at Berkeley. After a hiatus of roughly a year, from the spring of 1944 to August of 1945 when Erich was stationed in Guam working as an Operations Analyst for the Air Force, Erich returned to Berkeley and started work on his thesis. In preparing to write up the results, a citation of related work led to other works that, painfully, led to the realization that Erich's work was already part of Markov's, and other Russian probabilists' work. This created a difficult situation for Erich, specially since Neyman was going away as a supervisor in the Greek elections. Neyman then asked Pao-Lu Hsu, who had been his student in London in 1938 and who was visiting Berkeley for the 1945 fall term, to give Erich a new thesis topic. Hsu proposed a problem on which he had been working and for which he had already obtained independent results. Erich learnt about Hsu's generosity some time later and wanted to thank him personally after Hsu, as Neyman was hoping, returned to Berkeley from his one-semester visit to Columbia. But Hsu never returned to Berkeley having gone from Columbia to North Carolina and then back to China. Erich, however, always remained grateful for Hsu's generosity. With both Neyman and Hsu gone, Neyman asked George Polya at Stanford to supervise Erich's thesis. Weekly or biweekly meetings with Polya, commuting between Berkeley and Stanford, eventually led to Polya urging Lehmann to write up the results in preparation for finishing. It turned out that Neyman was able to return in time to be part of the examination.

Erich was offered a faculty position in 1946 at Berkeley and, despite offers from Stanford in 1951 and from the Eidgenössische Technische Hochschule (ETH), the Swiss Federal Institute of Technology, in 1959, Erich never left Berkeley. Some notable exceptions were the visits to Columbia and Princeton during the 1950-1951 academic year, a one year visit to Stanford during the 1951-1952 academic year to allow more time for the unrest created by the Loyalty Oath to die out, and a stay in Princeton from 1995 to 1997 after Julie accepted a position as principal research scientist and coordinating director of the Large Scale Assessment Group, and Erich served as distinguished research scientist at the Educational Testing Service.

Erich returned to Berkeley from Stanford in 1952 as an Associate Professor of Mathematics, and in 1954 was promoted to Professor of Mathematics. In 1955, after

Neyman succeeded in creating a separate department of statistics, Erich's title changed to Professor of Statistics. In 1988, Erich retired and became Professor Emeritus.

Throughout his prolific career, Erich, who as a teenager had dreams of becoming a writer, wrote several influential textbooks. Some of these have been translated to several languages. The books, *Testing statistical hypotheses* (first edition, 1959; second edition, 1986; third edition, with Joseph P. Romano, 2005 – the first edition has been translated into Russian, Polish, and Japanese), and *Theory of Point Estimation*, (first edition, 1983; second edition with George Casella, 1998 – the first edition was translated into Russian, and the second into Chinese), continue to have impact in preparing the next generations of professional statisticians. Together with *Nonparametrics: Statistical Methods Based on Ranks* – first published in 1975, then in paperback in 1998, and then by Springer Science + Business Media, LLC in 2006, and translated into Japanese, these three books have been cited more than 7,500 times according to the ISI Web of knowledge. And the number of citations keeps increasing. Other books include: *Basic Concepts of Probability and Statistics*, with Joseph L. Hodges; *Elements of Finite Probability*, with Joseph L. Hodges; *Elements of Large-Sample Theory*; and *Reminiscences of a Statistician: The Company I Kept*.

In addition, Lehmann was a *Special Editor* for the second edition of *Statistics: A Guide to the Unknown*, edited by Judith M. Tanur. For the third edition, published in 1989, Judith M. Tanur was the editor while Lehmann continued as a co-editor. Two more works, *Statistics: a guide to the study of the biological and health sciences* and *Statistics: a guide to political and social issues*, both published in 1977, were a result of these collaborative efforts. Lehmann was a *Special Editor* for the latter and a co-editor for the former.

Among his many influential papers, Erich seemed to have a special liking for "The power of rank tests", *Ann. Math. Statist.*, Vol. 24, pp. 23-43, 1953, and for "The efficiency of some nonparametric competitors of the t-test", *Ann. Math. Statist.*, Vol. 27, pp. 324-335, 1956. In the former, Erich introduced a set of alternatives for which calculating the power of rank tests against the usual alternatives in the two-sample problem is simplified. This set of alternatives is now known as *The Lehmann alternatives*. In the latter paper, co-authored with Joseph L. Hodges, it is shown that the asymptotic relative efficiency (ARE) of the two-sample Wilcoxon test, relative to the t-test, is never under .864. This assuaged the anxiety of the doubtful who feared that nonparametric techniques had very little power. These results were later extended to the problem of estimating a location parameter in "Estimates of location based on rank tests", *Ann. Math. Statist.*, Vol. 34, pp. 593-611, 1963, where it was shown that the ARE of the estimates obtained from rank tests to the classical linear estimates agrees precisely with the ARE of the rank tests to the corresponding t-tests. Another paper that continues to have impact on applied probability is the paper "Some concepts of dependence", *Ann. Math. Statist.*, Vol. 37, pp. 1137-1153, 1966. According to the ISI web of knowledge, the paper has been cited at least 540 times, with at least 20 of these citations occurring in the first six months of 2011.

Erich chaired the statistics department at Berkeley from 1973-1976, and also served the broader statistical community as Editor (1953-1955) and Associate Editor (1955-1968) of the Annals of Mathematical Statistics, President of the Institute of Mathematical Statistics 1960-1961, and as a member of committees of visitors to the departments of statistics at Harvard (1974-1980) and Princeton (1975-1980) Universities.

During his lifetime, Erich received many accolades for his contributions. He was a three-time Guggenheim Fellow (1995-1956, 1966-1967, and 1980-1981) and a two-time Miller Institute for Basic Research Professor (1962-1963 and 1972-1973). He gave the IMS Wald Lectures in 1964, and the COPSS R. A. Fisher Memorial Lecture in 1988. Surprisingly, he was never invited to deliver the IMS Neyman Lecture. In 2000, he became the first recipient of the ASA Goffried Noether Award, and he received the Journal of Nonparametric Statistics Award for best paper in 2009. He was elected to the American Academy of Arts and Sciences in 1975 and, in 1978, became an elected member of the National Academy of Sciences. He received Dr. *honoris causa* degrees from the University of Leiden in 1985 and from the University of Chicago in 1991. The former was the first Dr. h. c. awarded to a mathematician at Leiden in a century, Stieltjes having been the last mathematician to receive it in 1884. Erich wrote about this experience in Lehmann (2008). In addition, there was a *Festschrift for Erich L. Lehmann* organized by Bickel, Doksum, and Hodges in 1982, and a series of *Lehmann Symposia*, organized by Rojo and Perez-Abreu in 2002, and Rojo in 2004, 2007, and 2011. Although he did not like teaching large classes, he acknowledged that the one-on-one interaction with his Ph.D. students, of whom he had 43+, was the most rewarding aspect of teaching.

Erich Leo Lehmann died on the morning of September 12th, 2009. At the time of his death, he was working on the manuscript for the book *Fisher, Neyman, and the Creation of Classical Statistics*, soon to be published by Springer-Verlag. After his death, Erich's life and work were celebrated with a memorial service on the Berkeley campus on November 9, 2009. In addition, the statistical community he appreciated so much honored him with a memorial session, organized by Peter Bickel, at the joint statistical meetings in Vancouver, Canada on August of 2010; a lecture by Peter Bickel at the IX Latin American Congress in Probability and Mathematical Statistics meeting in Venezuela in November of 2009; and a memorial session, organized by Willem van Zwet, during the IMS meeting in Sweden during August of 2010.

Several obituaries have been published that shed light on Erich's work and life. See, for example, Brillinger (2010), and Bickel (2009). The papers by Stigler (2009) and Rojo (2009) provide accounts of some of Erich's experiences.

Outside statistics, Erich loved music and Franz Schubert was one his favorite composers. He also enjoyed playing the piano and entertaining visitors at his home. He derived enormous pleasure from family life. While working with Erich on the planning of this collection of works, I once asked him what were the aspects of his life that he found the most rewarding. Immediately, and without hesitation, he responded "my children, grand-children, and great-grand-children and having met Julie".

Houston, TX — Javier Rojo

References

1. Bickel, P. J., Doksum, K., and Hodges, J. L. (1983). *A Festschrift For Erich L. Lehmann*, Wadsworth Statistics/Probability Series.
2. Bickel, P. J. (2009). Erich L. Lehmann, 1917-2009, *IMS Bulletin* Vol. 38, Issue 9. pp. 10.
3. Brillinger, D. R. (2010). Erich Leo Lehmann, 1917-2009. *J. R. Statist. Soc. A*, Vol. 173, Part 3, pp. 683-686.
4. DeGroot, M. H. (1986). A Conversation with Erich L. Lehmann. *Statist. Sci.*, Vol. I, 243-258.
5. Lehmann, E. L. (1959). *Testing Statistical Hypotheses*. John Wiley & Sons, New York.
6. Lehmann, E. L., and Hodges, J. L. (1964). *Basic Concepts of Probability and Statistics*. Holden-Day, Inc., San Francisco, London, Amsterdam. Reprinted in 2005 as part of the SIAM series Classics in Applied Mathematics.
7. Lehmann, E. L. (1975). Nonparametrics: Statistical Methods Based on Ranks, with the assistance of H. J. M. D'Abrera. First published as a hardcover by Holden-Day in 1975. In paperback by Prentice-Hall, Inc., Simon & Schuster in 1998, and then by Springer Science + Business Media, LLC in 2006.
8. Lehmann, E. L. (1983). *Theory of Point Estimation*. Wiley & Sons, New York. Transferred to Wadsworth & Brooks/Cole 1991.
9. Lehmann, E. L. (1986). *Testing Statistical Hypotheses*, second edition. John Wiley & Sons, New York. Transferred to Wadsworth & Brooks/Cole, 1991.
10. Lehmann, E. L., and Romano, J. P. (2005). *Testing Statistical Hypotheses,* third edition. Springer-Verlag, New York.
11. Lehmann, E. L. (2008). *Reminiscences of a Statistician: The Company I Kept*. Springer-Verlag, New York.
12. Lehmann, E. L. (2011). *Fisher, Neyman, and the Creation of Classical Statistics*. In press. Springer, 2011.
13. *The First Erich L. Lehmann Symposium: Optimality*. IMS Lecture Notes–Monograph Series, Volume 44, J. Rojo and V. Perez-Abreu, Eds, 2004.
14. *Optimality: The Second Erich L. Lehmann Symposium*. IMS Lecture Notes–Monograph Series, Vol. 49, J. Rojo Ed., 2006.
15. *Optimality: The Third Erich L. Lehmann Symposium* IMS Lecture Notes–Monograph Series, Vol. 57, J. Rojo Ed., 2009.
16. Rojo, J. (2009). Erich L. Lehmann, the Lehmann Symposia, and November 20th 1917. *Optimality: The Third Erich L. Lehmann Symposium*, IMS Lecture Notes-Monograph Series, Vol. 57, (J. Rojo, Ed.), pp. 1-7.
17. Stigler, S. (2009). The Honorable Erich L. Lehmann. *Optimality: The Third Erich L. Lehmann Symposium*, IMS Lecture Notes-Monograph Series, Vol. 57, (J. Rojo, Ed.), pp. 8-10.
18. Tanur, J. M., Mosteller, F., Kruskal, W. H., Lehmann, E. L., Link, R. F., Pieters, R. S. and Rising, G. R. Co-editors (1977). *Statistics: a guide to the study of the biological and health sciences*. Holden-Day, Inc., Publishers, San Francisco, 140 pp.
19. Tanur, J. M. (Editor), Lehmann, E. L. (Special Editor), Mosteller, F., Kruskal, W. H., Link, R. F., Pieters, R. S. and Rising, G. R. Co-editors (1977). *Statistics: a guide to political and social issues*. Holden-Day, Inc., Publishers, San Francisco, 141 pp.
20. Tanur, J. M. (Editor), Lehmann, E. L. (Special Editor), Mosteller, F., Kruskal, W. H., Link, R. F., Pieters, R. S. and Rising, G. R. Co-editors (1978). *Statistics: A Guide to the Unknown, Second Edition*, Holden-Day, San Francisco.
21. Tanur, J. M., Mosteller, F., Kruskal, W. H., Lehmann, E. L., Link, R. F., Pieters, R. S. and Rising, G. R. Co-editors (1989). *Statistics: A Guide to the Unknown, Third Edition*. Wadsworth.

Bibliography

Erich L. Lehmann

Articles

1. E. L. Lehmann. Une propriété optimale de certains ensembles critiques du type A1. C. R. Acad. Sci. Paris, Vol. 223, pp. 567–569, 1946.
2. E. L. Lehmann. On families of admissible tests. Ann. Math. Statist., Vol. 18, pp. 97–104, 1947.
3. E. L. Lehmann. On optimum tests of composite hypotheses with one constraint. Ann. Math. Statist., Vol. 18, pp. 473–495, 1947.
4. E. L. Lehmann and H. Scheffé. On the problem of similar regions. Proc. of the Natl. Acad. Sci. U S A, pp. 382–386, 1947.
5. E. L. Lehmann and C. Stein. Most powerful tests of composite hypotheses. I. Normal Distributions. Ann. Math. Statist., Vol. 19, No. 4, pp. 495–516, 1948.
6. E. L. Lehmann. Some comments on large sample tests. Proc. Berkeley Symp. Math. Statist. Prob., University of California Press, pp. 451–457, 1949.
7. E. L. Lehmann and C. Stein. On the theory of some non-parametric hypotheses. Ann. Math. Statist., Vol. 20, pp. 28–45, 1949.
8. E. L. Lehmann. Some principles of the theory of testing hypotheses. Ann. Math. Statist., Vol. 21, pp. 1–26, 1950.
9. J. L. Hodges, Jr. and E. L. Lehmann. Some problems in minimax point estimation, Ann. Math. Statist., Vol. 21, pp. 182–197, 1950.
10. E. L. Lehmann and C. Stein. Completeness in the sequential case. Ann. Math. Statist., Vol. 21, pp. 376–385, 1950.
11. E. L. Lehmann and H. Scheffé. Completeness, similar regions and unbiased estimation. Part I. Sankhya, Vol. 10, pp. 305–340, 1950.
12. J. L. Hodges, Jr. and E. L. Lehmann. Some applications of the Cramér-Rao inequality. Proc. Second Berkeley Symp. Math. Statist. Prob., University of California Press, pp. 13–22, 1951.
13. E. L. Lehmann. Consistency and unbiasedness of certain nonparametric tests. Ann. Math. Statist., Vol. 22, pp. 165–179, 1951.

14. E. L. Lehmann. A general concept of unbiasedness. Ann. Math. Statist., Vol. 22, pp. 587–591, 1951.

15. J. L. Hodges, Jr. and E. L. Lehmann. The use of previous experience in reaching statistical decisions. Ann. Math. Statist., Vol. 23, pp. 396–407, 1952.

16. E. L. Lehmann. On the existence of least favorable distributions. Ann. Math. Statist., Vol. 23, pp. 408–416, 1952.

17. E. L. Lehmann. Testing multiparameter hypotheses. Ann. Math. Statist., Vol. 23, pp. 541–552, 1952.

18. E. L. Lehmann. The power of rank tests. Ann. Math. Statist., Vol. 24, pp. 23–43, 1953.

19. E. L. Lehmann and C. Stein. The admissibility of certain invariant statistical tests involving a translation parameter. Ann. Math. Statist., Vol. 24, pp. 473–479, 1953.

20. H. Chernoff and E. L. Lehmann. The use of maximum likelihood estimates in χ^2 tests for goodness of fit. Ann. Math. Statist., Vol. 25, pp. 579–586, 1954.

21. J. L. Hodges, Jr. and E. L. Lehmann. Matching in paired comparisons. Ann. Math. Statist., Vol. 25, pp. 787–791, 1954.

22. J. L. Hodges, Jr. and E. L. Lehmann. Testing the approximate validity of statistical hypothe-ses. J. Roy. Statist. Soc. Ser. B, Vol. 26, pp. 261–268, 1954.

23. T. W. Anderson, H. Cramér, J. L. Hodges, Jr. H. A. Freeman, E. L. Lehmann, A. M. Mood, C. Stein. The life of Abraham Wald. In Selected Papers in Statistics and Probability. New York: McGraw-Hill, 1955.

24. R. R. Bahadur and E. L. Lehmann. Two comments on sufficiency and statistical decision functions. Ann. Math. Statist., Vol. 26, pp. 139–142, 1955.

25. E. L. Lehmann and H. Scheffé. Completeness, similar regions and unbiased estimation, Part II. Sankhya, Vol. 15, pp. 219–236, 1955.

26. E. L. Lehmann. Ordered families of distributions. Ann. Math. Statist., Vol. 26, pp. 399–419, 1955.

27. J. L. Hodges, Jr. and E. L. Lehmann. Two approximations to the Robbins-Monro process. Proc. Third Berkeley Symp. Math. Statist. Prob., Vol. I, pp. 95–104, 1956.

28. J. L. Hodges, Jr. and E. L. Lehmann. The efficiency of some nonparametric competitors of the t-test. Ann. Math. Statist., Vol. 27, pp. 324–335, 1956.

29. E. L. Lehmann. A theory of some multiple decision problems. I. Ann. Math. Statist., Vol. 28, pp. 1–25, 1957.

30. E. L. Lehmann. A theory of some multiple decision problems. II. Ann. Math. Statist., Vol. 28, pp. 547–572, 1957.

31. E. L. Lehmann. Significance level and power. Ann. Math. Statist., Vol. 29, pp. 1167–1176, 1958.

32. E. Fix, J. L. Hodges, Jr. and E. L. Lehmann. The restricted chi-square test. Probability and Statistics, The Harald cramér Volume, (U. Grenander, Ed.), pp. 92–107, John Wiley and Sons, 1959.

33. E. L. Lehmann. Optimum invariant tests. Ann. Math. Statist., Vol. 30, pp. 881–884, 1959.

34. J. L. Hodges, Jr. and E. L. Lehmann. Comparison of the normal scores and Wilcoxon tests. Proc. Fourth Berkeley Symp. Math. Prob., Vol. I, pp. 307–319, 1961.

35. E. L. Lehmann. Some model I problems of selection. Ann. Math. Statist., Vol. 32, pp. 990–1012, 1961.

36. J. L. Hodges, Jr. and E. L. Lehmann. Rank methods for combination of independent experi-ments in analysis of variance. Ann. Math. Statist., Vol. 33, pp. 482–497, 1962.

37. J. L. Hodges, Jr. and E. L. Lehmann. Probabilities of rankings of two widely separated normal distributions. Studies in Math. Analysis and Related Topics: Essays in Honor of George Polya. Stanford University Press, pp. 146–151, 1962.

38. J. L. Hodges, Jr. and E. L. Lehmann. Estimates of location based on rank tests. Ann. Math. Statist., Vol. 34, pp. 593–611, 1963.

39. E. L. Lehmann. Robust estimation in analysis of variance. Ann. Math. Statist., Vol. 34, pp. 957–966, 1963.

40. E. L. Lehmann. A class of selection procedures based on ranks. Math. Annalen, Vol. 150, pp. 268–275, 1963.

41. E. L. Lehmann. Asymptotically nonparametric inference: an alternative approach to linear models. Ann. Math. Statist., Vol. 34, pp. 1494–1506, 1963.

42. E. L. Lehmann. Nonparametric confidence intervals for a shift parameter. Ann. Math. Statist., Vol. 34, pp. 1507–1512, 1963.

43. E. L. Lehmann. Asymptotically nonparametric inference in some linear models with one observation per cell. Ann. Math. Statist., Vol. 35, pp. 726–734, 1964.

44. E. L. Lehmann. On the non-verifiability of certain parametric functions. Teoria Veroyatnostey i Primenem, Vol. 10, pp. 758–760, 1965.

45. E. L. Lehmann. On a theorem of Bahadur and Goodman. Ann. Math. Statist., Vol. 37, pp. 1–6, 1966.

46. E. L. Lehmann. Some concepts of dependence. Ann. Math. Statist., Vol. 37, pp. 1137–1153, 1966.

47. E. L. Lehmann. Moments of chi and power of t. Proc. Fifth Berkeley Symp. Math. Statist. Prob., University of California Press, Vol. I, pp. 187–201, 1967.

48. J. L. Hodges, Jr. and E. L. Lehmann. On medians and quasi-medians. J. Amer. Statist. Assoc., Vol. 62, pp. 926–931, 1967.

49. J. L. Hodges, Jr. and E. L. Lehmann. A compact table for power of the t-test. Ann. Math. Statist., Vol. 39, pp. 1629–1637, 1968.

50. E. L. Lehmann. Hypothesis testing. In International Encyclopedia of the Social Sciences, Vol. 7, David L. Sills, Ed. New York: Macmillan, pp. 40–47, 1968.

51. P. J. Bickel and E. L. Lehmann. Unbiased estimation in convex families. Ann. Math. Statist., Vol. 40, pp. 1523–1535, 1969.

52. J. L. Hodges, Jr. and E. L. Lehmann. Deficiency. Ann. Math. Statist., Vol. 41, pp. 783–801, 1970.

53. J. L. Hodges, Jr. and E. L. Lehmann. Wilcoxon and t-tests for matched pairs of typed subjects. J. Amer. Statist. Assoc., pp. 151–158, 1973.

54. L. Le Cam and E. L. Lehmann. J. Neyman: On the occasion of his 80th birthday. Ann. Statist., Vol. 2, vii-xiii, 1974.

55. P. J. Bickel and E. L. Lehmann. Measures of location and scale. In Proc. Prague Symposium on Asymptotic Statistics, J. Hajek, Ed., Charles University, Prague, Vol. 1, pp. 25–36, 1974.

56. P. J. Bickel and E. L. Lehmann. Descriptive statistics for nonparametric models. I. Introduction. Ann. Statist., Vol. 3, pp. 1038–1044, 1975.

57. L. L'o Kam and E. L. Lehmann. Professor Jerzy Neyman. On the occasion of his 80th birthday. (Bulgarian) Translated from the English (Ann. Statist. 2 (1974), vii-xiii). Fiz.-Mat. Spis. B"lgar. Akad. Nauk., 18(51), no. 2, pp. 152–156, 1975.

58. P. J. Bickel and E. L. Lehmann. Descriptive statistics for nonparametric models II. Location. Ann. Statist., Vol. 4, pp. 1045–1069, 1976.

59. P. J. Bickel and E. L. Lehmann. Descriptive statistics for nonparametric models. III. Dispersion. Ann. Statist., Vol. 4, pp. 1139–1158, 1976.

60. E. L. Lehmann and J. P. Shaffer. On a fundamental theorem in multiple comparisons. J. Amer. Statist. Assoc., Vol. 72, pp. 576–578, 1977.

61. E. L. Lehmann. Henry Scheffé 1907–1977. Internat. Statist. Rev., 46, 126, 1978.

62. E. L. Lehmann and J. P. Shaffer. Optimum significance levels for multistage comparison procedures. Ann. Statist., Vol. 7, pp. 27–45, 1979.

63. P. J. Bickel and E. L. Lehmann. Descriptive statistics for nonparametric models. IV. Spread. In Contributions to Statistics, Jaroslav Hajek Memorial Volume, Prague, 1979.

64. T. W. Anderson, K. L. Chung, and E. L. Lehmann. Pao-Lu Hsu, 1909–1970. Ann. Statist., Vol. 7, pp. 467–470, 1979.

65. E. L. Lehmann. Hsu's work on inference. Ann. Statist., Vol. 7, pp. 471–473, 1979.

66. C. Daniel and E. L. Lehmann. Henry Scheffé, 1907–1977. Ann. Statist., Vol. 7, pp. 1149–1161, 1979.

67. E. L. Lehmann. Efficient likelihood estimators. Amer. Statist., Vol. 34, pp. 233–235, 1980.

68. E. L. Lehmann. The work of Pao-Lu Hsu on statistical inference. (Chinese) Knowledge Practice Math., No. 3, pp. 6–8, 1980.

69. T. W. Anderson, K. L. Chung, and E. L. Lehmann. Pao-Lu Hsu: 1910–1970. (Chinese) Knowledge Practice Math., No. 3, pp. 3–5, 1980.

70. E. L. Lehmann. An interpretation of completeness and Basu's theorem. J. Amer. Statist. Assoc., Vol. 76, pp. 335–340, 1981.

71. E. L. Lehmann. A minimax property of the sample mean in finite populations. Ann. Statist., Vol. 8, pp. 1119–1122, 1981.

72. J. L. Hodges, Jr. and E. L. Lehmann. Minimax estimation in simple random sampling. In Statistics and Probability: Essays in honor of C. R. Rao, (Kallianpur et al, Eds.), North Holland, 1982.

73. E. L. Lehmann and C. Reid. In memoriam: Jerzy Neyman (1894–1981). Amer. Statist., Vol. 36, pp. 161–162, 1982.

74. E. L. Lehmann. Classical estimation. In Encycl. Statist. Sci., Vol. 2, pp. 2079–2087, 1982.

75. J. L. Hodges and E. L. Lehmann. Hodges-Lehmann Estimators. In Encycl. Statist. Sci., Vol. 3, pp. 3180–3183, 1983.

76. E. L. Lehmann. Estimation with inadequate information. J. Amer. Statist. Assoc., Vol. 78, pp. 624–627, 1983.

77. E. L. Lehmann. Least informative distributions. In Recent Advances in Statistics, (Rizvi et al, Eds.), pp. 593–599, Academic Press, 1983.

78. E. L. Lehmann. Comparison of experiments for some multivariate normal situations. In Studies in Econometrics, Time Series, and Multivariate Statistics, (Karlin et al, Eds.). Academic Press, 1983.

79. E. L. Lehmann. Specification problems in the Neyman-Pearson-Wald theory. In Statistics: an appraisal. Proceedings 50th Anniversary Conference. Iowa State University Press, (H. A. David and H. T. David Eds.), pp. 425–436, 1984.

80. E. L. Lehmann. The Neyman-Pearson Lemma. In Encycl. Statist. Sci., Vol. 6, pp. 224–230, 1985.

81. E. L. Lehmann. The Neyman-Pearson Theory After Fifty Years. In Proceedings of the Berkeley Conference in Honor of Jerzy Neyman and Jack Kiefer, Vol. I. (L. Le Cam and R. A. Olshen, Eds), pp. 1–14, Wadsworth 1985.

82. E. L. Lehmann. Unbiasedness. Encycl. Statist. Sci., Vol. 9, pp. 386–391, 1988.

83. E. L. Lehmann. Statistics - An Overview. Encycl. Statist. Sci., Vol. 8, pp. 683–702, 1988.

84. E. L. Lehmann. Comparing location experiments. Ann. Statist. Vol. 16, pp. 521–533, 1988.

85. E. L. Lehmann and J. Shaffer. Inverted distributions. Am. Statist., Vol. 42, pp. 191–194, 1988.

86. E. L. Lehmann. Group families. In Encycl. Statist. Sci. Suppl., pp. 70–71, 1989.

87. P. Diaconis and E. L. Lehmann. Contributions to mathematical statistics. In A Statistical model: Frederick Mosteller's contributions to statistics, science, and public policy, (S. E. Fienberg, Ed.), pp. 59–80. Springer 1990.

88. E. L. Lehmann. Verifiability and strong verifiability. In Zapiski Nauchnykh Seminarov Leningradskogo Otdeleniya Matematicheskogo Instituta imeni V. A. Steklova Akademii Nauk SSSR (LOMI), Vol. 184, (Ibragimov et al, Eds.), pp. 182–188, 1990.

89. E. L. Lehmann. Model specification: The views of Fisher and Neyman, and later developments. Statist. Sci., Vol. 5, pp. 160–168, 1990.

90. E. L. Lehmann and W-Y. Loh. Pointwise vs. uniform robustness of some large-sample tests and confidence intervals. Scand. J. Statist., Vol. 17, pp. 177–187, 1990.

91. E. L. Lehmann. Comment on Lindley: The present position in Bayesian statistics. Statist. Sci., Vol. 5, pp. 82–83, 1990.

92. E. L. Lehmann. Henry Scheffé. In Dictionary of Scientific Biography, Vol. 18, Supplement II, 1990.

93. E. L. Lehmann. Jerzy Neyman. In Dictionary of Scientific Biography, Vol. 18, Supplement II, 1990. (Reproduced as: Neyman, Jerzy. Complete Dictionary of Scientific Biography. Vol. 18. Detroit: Charles Scribner's Sons, 2008. 669–675. Gale Virtual Reference Library. Web. 14 June 2011).

94. E. L. Lehmann. Introduction to Student (1908): The probable error of a mean. In Breakthroughs in Statistics, Vol. II, (S. Kotz and N. L. Johnson Eds.). New York, Springer, pp. 29–32, 1991.

95. E. L. Lehmann. Introduction to Neyman, J. and Pearson, E. S. (1933): On the problem of the most efficient tests of statistical hypotheses. In Breakthroughs in Statistics, Vol. I, (S. Kotz and N. L. Johnson Eds.). New York, Springer, pp. 67–72, 1991.

96. E. L. Lehmann and F. Scholz. Ancillarity. In Current Issues in Statistical Inference: Essays in Honor of D. Basu, IMS Lecture Notes - Monograph Series, Vol. 17, pp. 32–51, 1992.

97. E. L. Lehmann and J. Rojo. Invariant directional orderings. Ann. Statist., Vol. 20, pp. 2100–2110, 1992.

98. E. L. Lehmann. The Fisher, Neyman-Pearson theories of testing hypotheses: One theory or two? J. Amer. Statist. Assoc., Vol. 88, pp. 1242–1249, 1993.

99. E. L. Lehmann. The Bertrand-Borel Debate and the origins of the Neyman-Pearson theory. In Statistics and Probability. A Raghu Raj Bahadur Festschrift, (Ghosh et al Eds.), pp. 371–380. Wiley Eastern Ltd. New Delhi, 1993.

100. E. L. Lehmann. Mentors and early collaborators: reminiscences from the years 1940–1956 with an epilogue. Statist. Sci., Vol. 8, pp. 331–341, 1993.

101. E. L. Lehmann. Jerzy Neyman (1894–1981). In Biographical Memoirs of the Nat. Acad. of Sci., Vol. 63, pp. 395–420. The National Academies Press, 1994.

102. E. L. Lehmann. English translation of Verifiability and strong verifiability. (Originally published in Russian in Zap. Nauchn. Sem. Leningrad. Otdel. Mat. Inst. Steklov. (LOMI) 184 (1990), Issled. po Mat. Statist. 9, 182–188, 323.) J. Math. Sci., Vol. 68, Number 4, pp. 526–530, 1994.

103. E. L. Lehmann. Foundational issues in statistics: theory and practice. Foundations of Science, Vol. 1, pp. 45–49, 1995.

104. E. L. Lehmann. Neyman's statistical philosophy. Dedicated to the memory of Jerzy Neyman. Probability and Math. Statistics, Vol. 15, pp. 29–36, 1995.

105. E. L. Lehmann. The creation and early history of the Berkeley statistics department. In Festschrift for David Blackwell, IMS Lecture Notes - Monograph Series, Vol. 30, (T. Ferguson, Ed.), pp. 139–146, 1996.

106. E. L. Lehmann. Le Cam at Berkeley. In Festschrift for Lucien Le Cam, pp. 297–304, Springer Verlag, 1997.

107. E. L. Lehmann. Testing statistical hypotheses: The story of a book. Statist. Sci. Vol. 12, pp. 48–52, 1997.

108. E. L. Lehmann. 'Student' and small-sample theory. Statist. Sci., Vol. 14, pp. 418–426, 1999.

109. P. J. Bickel and E. L. Lehmann. Frequentist inference. In International Encyclopedia of the Social and Behavioral Sciences, pp. 5789–5796. Pergamon, Oxford, 2001.

110. P. J. Bickel and E. L. Lehmann. Frequentist interpretation of probability. In International Encyclopedia of the Social and Behavioral Sciences, N. J. Smelser and Paul B. Baltes (Eds.), pp. 5796–5798. Pergamon, Oxford, 2001.

111. E. L. Lehmann. Optimality and symposia: Some history. In The First Erich L. Lehmann Symposium - Optimality. (V. Perez-Abreu and J. Rojo, Eds.), IMS Lecture Notes-Monograph Series, Vol. 44, pp. 1–10, 2004.

112. E. L. Lehmann, J. P. Romano, and J. P. Shaffer. On optimality of step-down and step-up multiple test procedures. Ann. Statist., Vol. 33, pp. 1084–1108, 2005.

113. E. L. Lehmann and J. P. Romano. Generalizations of the family-wise error rate. Ann. Statist., Vol. 33, pp. 1138–1154, 2005.

114. K. J. Arrow and E. L. Lehmann. Harold Hotelling. In Biographical Memoirs, National Academy of Sciences, Vol. 87, pp. 221–233. The National Academies Press, 2005.

115. E. L. Lehmann. Hodges, Joseph Lawson, Jr. In Encyclopedia of Statistical Science, 2nd Ed., pp. 3179–3180, 2006.

116. E. L. Lehmann. Scheffé, Henry. In Encyclopedia of Statistical Science, 2nd Ed., pp. 7472–7474, 2006.

117. E. L. Lehmann. On likelihood ratio tests. In The Second Erich L. Lehmann Symposium Optimality, (J. Rojo, Ed.), IMS Lecture Notes-Monograph Series Vol. 49, pp. 1–9, 2006.

118. E. L. Lehmann. On the history and use of some standard statistical models. In Probability and Statistics: Essays in Honor of David A. Freedman, (D. Nolan and T. Speed, Eds), IMS Collections Vol. 2, pp. 114–126, 2008.

119. P. Diaconis and E. L. Lehmann. Comment on: Zabell, S. L., On Student's 1908 article: The probable error of a mean. J. Amer. Statist. Assoc., Vol. 103, pp. 16–19, 2008.

120. E. L. Lehmann. Parametric versus nonparametrics: two alternative methodologies. Invited paper with discussion. J. Nonpar. Statist., Vol. 21, pp. 397–405, 2009.

121. E. L. Lehmann. Rejoinder to "Parametric versus nonparametrics: two alternative methodologies", with discussion. J. Nonpar. Statist., Vol. 21, pp. 425–426, 2009.

122. E. L. Lehmann. Some history of optimality. In The Third Erich L. Lehmann Symposium Optimality, (J. Rojo, Ed.) IMS Lecture Notes-Monograph Series, Vol. 57, pp. 10–17, 2009.

Technical Reports

1. H. Chernoff and E. L. Lehmann. Stanford University Technical Report No. 18 for Office of Naval Research (mimeographed), pp. 1–16, 1953.

2. E. L. Lehmann. Optimum tests of a certain class of hypotheses specifying the value of a correlation coefficient. Thesis (Ph.D.)-Univ. of California, Berkeley, June 1946.

3. E. L. Lehmann. Notes on the theory of estimation. Chapters I to V. Lectures by E. L. Lehmann, 1949–50, University of California. Notes recorded by Colin Blyth. Berkeley, Calif., Associated Students Store, University of California, 1950.

4. E. L. Lehmann. Theory of testing hypotheses. Lectures by E. L. Lehmann. Notes recorded by Colin Blyth. Berkeley, Calif., Associated Students Store, University of California, 1950.

5. E. L. Lehmann. Discriminatory analysis. On the simultaneous classification of several individuals. U.S.A.F. School of Aviation Medicine, Report No. 6, Project 21-49-004, pp. 1–27, 1951.

6. E. L. Lehmann. Discriminatory analysis. On the theory of selection. U.S.A.F. School of Aviation Medicine, Report No. 7, Project 21-49-004, pp. 1–10, 1951.

7. E. L. Lehmann. Some Early Instances of Confidence Statements. Statistical Laboratory, University of California, Berkeley. ONR 5 Technical Report to the Office of Naval Research. Unpublished. September 1958.

8. E. L. Lehmann. Statistics: an overview. Technical report, No. 72, University of California, Berkeley. Dept. of Statistics, 1986.

9. E. L. Lehmann. Jerzy Neyman, 1894–1981. Technical report, No. 155, University of California, Berkeley. Dept. of Statistics, 1988.

10. E. L. Lehmann and J. Rojo. Invariant directional orderings. Technical report, No. 278, University of California, Berkeley. Dept. of Statistics, Nov. 1990, revised No. 1991.

11. E. L. Lehmann, Joseph P. Romano. Generalizations of the family-wise error rate. Technical report no. 2003–37, Stanford University. Dept. of Statistics.

12. E. L. Lehmann, Joseph P. Romano, Juliet Popper Shaffer. On optimality of stepdown and stepup procedures. Technical report, no. 2003–12, Stanford University. Dept. of Statistics.

Books and their Translations

1. J. L. Hodges, Jr. and Erich L. Lehmann. Basic concepts of Probability and Statistics, Holden-Day, Inc. San Francisco, pp. xiii + 375, 1964.

2. J. L. Hodges, Jr. and Erich L. Lehmann. Elements of Finite Probability. Holden-Day series in probability and statistics, 230 pp., San Francisco, Holden-Day, 1965.

3. J. L. Hodges, Jr. and E. L. Lehmann. Grundbegreger i sandsynhghedsregning og statistik, Nyt Nordisk Forlag. Danish translation of Basic concepts of Probability and Statistics, Copen-hagen, 1969.

4. J. L. Hodges, Jr. and E. L. Lehmann. Basic concepts of Probability and Statistics, Second Edition. Holden-Day, Inc., San Francisco, 441 pp., 1970.

5. J. L. Hodges, Jr. and Erich L. Lehmann. Elements of Finite Probability, Second Edition. Holden-Day series in probability and statistics, 258 pp., San Francisco, Holden-Day, 1970.

6. J. L. Hodges, Jr. and E. L. Lehmann. Italian translation of Basic concepts of Probability and Statistics, two volumes. Societa Editrice il Mulino, Bologna, Italy, 1971.

7. J. L. Hodges, Jr. and E. L. Lehmann. Hebrew translation of Basic concepts of Probability and Statistics, Second Edition, 1972.

8. J. L. Hodges, Jr. and E. L. Lehmann. Farsi translation of Basic concepts of Probability and Statistics, 1994.

9. E. L. Lehmann. Testing Statistical Hypotheses, John Wiley and Sons, New York, pp. xiii + 369,1959.

10. E. L. Lehmann. Russian translation of Testing Statistical Hypotheses, Moscow, 1964.

11. E. L. Lehmann. Tesowanie Hipotez Statvstycznych, Polish translation of Testing Statistical Hypotheses, Warsaw. Panstwowe Wydawnicto Naukowe, 1968.

12. E. L. Lehmann. Japanese translation of Testing Statistical Hypotheses, 418 pp., Iwanami Shoten, Publishers, Tokyo. (Year of publication - unavailable).

13. E. L. Lehmann. Nonparametrics: Statistical Methods Based on Ranks, Holden-Day, Inc., Publishers, San Francisco, 457 pp., 1975.

14. E. L. Lehmann. Japanese translation of Nonparametrics: Statistical Methods Based on Ranks, Tokyo, 1978.

15. E. L. Lehmann. Theory of Point Estimation. Wiley, 1983.

16. E. L. Lehmann. Testing Statistical Hypotheses, Second Edition, John Wiley, 1986.

17. E. L. Lehmann. Russian Translation of Theory of Point Estimation, 1991.

18. E. L. Lehmann and G. Casella. Theory of Point Estimation, 2nd Edition. Springer Verlag, 1998.

19. E. L. Lehmann. Elements of Large Sample Theory. Springer-Verlag, 1999.

20. E. L. Lehmann and G. Casella. Chinese translation of Theory of Point Estimation, 2nd ed. China Statistics Press, (ISBN: 5037-4557-6), 2004.

21. E. L. Lehmann and J. P. Romano. Testing Statistical Hypotheses, 3rd Edition. Springer Verlag, xiii + 765, 2005.

22. E. L. Lehmann. Reminiscences of a Statistician: The Company I Kept, Springer, 2008.

23. E. L. Lehmann. Fisher, Neyman, and the Creation of Classical Statistics. Springer, 2011.

24. J. M. Tanur, F. Mosteller, W. H. Kruskal, E. L. Lehmann, R. F. Link, R. S. Pieters, and G. R. Rising, Co-editors. Statistics: a guide to the study of the biological and health sciences. Holden-Day, Inc., Publishers, San Francisco, 140 pp., 1977.

25. J. M. Tanur (Editor), E. L. Lehmann (Special Editor). F. Mosteller, W. H. Kruskal, R. F. Link, R. S. Pieters, and G. R. Rising, Co-editors. Statistics: a guide to political and social issues. Holden-Day, Inc., Publishers, San Francisco, 141 pp., 1977.

26. J. M. Tanur (Editor), E. L. Lehmann (Special Editor). F. Mosteller, W. H. Kruskal, R. F. Link, R. S. Pieters, and G. R. Rising, Co-editors. Statistics: A Guide to the Unknown, Second Edition, Holden-Day, San Francisco, 1978.

27. J. M. Tanur (Editor), E. L. Lehmann (Special Editor). F. Mosteller, W. H. Kruskal, R. F. Link, R. S. Pieters, and G. R. Rising, Co-editors. Reprint of

Statistics: A Guide to the Unknown, Second Edition, Wadsworth & Brooks/ Cole, 1985.

28. J. M. Tanur, F. Mosteller, W. H. Krusal, E. L. Lehmann, R. F. Link, R. S. Pieters, G. R. Rising, Co-editors. Statistics: A Guide to the Unknown, Third Edition. Wadsworth, 1989.

29. J. M. Tanur (Editor), E. L. Lehmann (Special Editor). F. Mosteller, W. H. Kruskal, R. F. Link, R. S. Pieters, and G. R. Rising, Co-editors. Chinese Translation of Statistics: A Guide to the Unknown, Second Edition, 1990.

30. J. M. Tanur (Editor), E. L. Lehmann (Special Editor). F. Mosteller, W. H. Kruskal, R. F. Link, R. S. Pieters, and G. R. Rising, Co-editors. Spanish translation of Statistics: A Guide to the Unknown, Second Edition. Alianza Editorial, S. A., Spain, ISBN 842069651X, 1992.

Interviews, Recorded Lectures, and Biographies/Obituaries

1. AMSTAT NEWS http://magazine.amstat.org/blog/2009/11/01/obitnov09/

2. N. Asimov. Erich L. Lehmann - Berkeley professor - dies October 16, 2009. S. F. Chronicle, http://articles.sfgate.com/2009-10-16/bay-area/17184544 1 uc-berkeley-testing-statistical-hypotheses-classical-statistics

3. P. J. Bickel. Erich L. Lehmann, 1917–2009. http://isi.cbs.nl/bnews/09b/bn 06.html

4. P. J. Bickel. Erich L. Lehmann, 1917–2009, IMS Bulletin Vol. 38, Issue 9. pp. 10, 2009.

5. D. R. Brillinger. Erich Leo Lehmann, 1917–2009. J. R. Statist. Soc. A, Vol. 173, Part 3, pp. 683–686, 2010.

6. M. H. DeGroot. A Conversation with Erich L. Lehmann. Statist. Sci., Vol. I, 243–258, 1986.

7. P. Diaconis and A. Herzberg. Teachers, Colleagues and Students. A conversation with Erich Lehmann. Filmed at the University of Connecticut, April 1992.

8. E. L. Lehmann. Two centuries of hypothesis testing. Lecture filmed at the University of Connecticut for the Amer. Statist. Assoc., April 1992.

9. E. L. Lehmann. The Early History of the Berkeley Statistics Department [QuickTime, 29MB]. Jerzy Neyman Room. Recorded by the Statistics Graduate Student Association. http://www. stat.berkeley.edu/videos/lehmann-history-mov. November 7, 2005.

10. J. Rojo. Erich L. Lehmann, the Lehmann Symposia, and November 20th 1917. Optimality: The Third Erich L. Lehmann Symposium, IMS Lecture Notes-Monograph Series, Vol. 57, (J. Rojo, Ed.), pp. 1–7, 2009.

11. S. Stigler. The Honorable Erich L. Lehmann. Optimality: The Third Erich L. Lehmann Symposium, IMS Lecture Notes-Monograph Series, Vol. 57, (J. Rojo, Ed.), pp. 8–10, 2009.

12. Webcast of The Second Erich L. Lehmann Symposium. Free public access. (J. Rojo, Ed.), http://edtech.rice.edu/www/?option=com iwebcast&action=details &event=408, Wednesday, May 19, 2004 - Friday, May 21, 2004.

13. Webcast of The Third Erich L. Lehmann Symposium. Free public access. (J. Rojo, Ed.), http://edtech.rice.edu/www/?option=com iwebcast&task=webcast &action=details&event= 1057, Wednesday, May 16, 2007 - Saturday, May 19, 2007.

Reviews

1. E. L. Lehmann. Review of Elementary Statistical Analysis, by S. S. Wilks. Amer. Math. Monthly, Vol. 56, no. 6, pp. 429–430, 1949.

2. E. L. Lehmann. Review of An Introduction to Scientific Research, by E. Bright Wilson. Journal of the American Statistical Association, Vol. 48, No. 262, pp. 358–359, 1953.

3. E. L. Lehmann. Review of Asymptotic Techniques for Use in Statistics by O. E. Barndorff-Nielsen and D. R. Cox. Short Book Reviews, Vol. 9, pp. 25, 1989.

4. E. L. Lehmann. Review of Error and the Growth of Experimental Knowledge by Deborah G. Mayo. J. Amer. Statist. Assoc., Vol. 92, pp. 789, 1997.

5. E. L. Lehmann. Review of Leading personalities in the statistical sciences. Johnson, Norman L. and Kotz, Samuel (Eds.). J. Amer. Statist. Assoc., Vol. 93, pp. 398, 1998.

6. E. L. Lehmann. Review of Statistical Inference in Science by D. A. Sprott. Springer, 2000. Short Book Reviews, Vol. 21, pp. 7, 2001.

Miscellaneous

1. P. J. Bickel, K. Doksum, and J. L. Hodges. A Festschrift For Erich L. Lehmann, Wadsworth Statistics/Probability Series, 1983.

2. P. J. Bickel and E. L. Lehmann. Joseph Lawson Hodges, Jr. IMS Bulletin, Vol. 4 (2000), pp. 411–412, 2000. Also in ISI Newsletter, Vol. 24, 10–11, 2000.

3. D. Blackwell, L. Le Cam, E. L. Lehman, H. Lewy, E. L. Scott. Jerzy Neyman, Statistics: Berkeley 1894–1981. University of California: In Memoriam, pp 313–315, 1985. http://content.cdlib.org/view?docId=hb4d5nb20m&chunk. id=div00120&brand=calisphere& doc.view=entire_text

4. D. Blackwell, R. N. Bellah, P. J. Bickel, W. W. Borah, E. L. Lehmann. Edward William Barankin, Statistics: Berkeley 1920–1985. University of California: In Memoriam, pp. 8–9, 1986. http://content.cdlib.org/view?docId=hb767nb3z6&chunk.id=div00005&brand=calisphere& doc.view=entire_text

5. D. Blackwell, E. Colson, S. Ervin-Tripp, L. Le Cam, E. L. Lehmann, L. Nader. Elizabeth Leonard Scott, Statistics: Berkeley 1917–1988. University of California: In Memoriam, pp 186–190, 1991. http://content.cdlib.org/view?docId=hb4t1nb2bd&chunk.id=div00061&brand=calisphere&doc.view=entire_text

6. D. Blackwell, A. H. Bowker, and E. L. Lehmann. Joseph L. Hodges, Jr. In Memoriam, University of California, pp. 83–85, 2000.

7. E. L. Lehmann, D. Blackwell, J. L. Hodges. Henry Scheffé, Statistics; Engineering: Berkeley and Los Angeles, 1907–1977. University of California: In Memoriam, September, pp 96–98, 1979. http://content.cdlib.org/view?docId=hb4q2nb2px&chunk.id=div00039&brand=calisphere&doc.view=entire_text

8. E. L. Lehmann, D. R. Brillinger, and I. M. Singer. Jack Carl Kiefer, Statistics and Mathematics: Berkeley, 1924–1981. University of California: In Memoriam, pp. 215–216, 1985. http://content.cdlib.org/view docId=hb4d5nb20m&chunk.id=div00081&brand=calisphere& doc.view=entire_text

9. E. L. Lehmann. Parametrics versus nonparametrics: Two alternative methodologies. Noether Award Lecture printed by Noether Award Committee, 2000.

10. Erich L. Lehmann received the Journal of Nonparametric Statistics Best Paper Award in 2009 for his discussion paper, 'Parametric versus Nonparametrics: Two Alternative Methodologies.'

11. J. Neyman, J. L. Hodges, Jr., L. Le Cam, E. L. Lehmann, E. L. Scott. Evelyn Fix, Statistics: Berkeley University of California: In Memoriam, June pp 36–38, 1967. http://content.cdlib.org/view?docId=hb629006vt&chunk.id=div00009&brand=calisphere&doc.view=entire_text

12. Optimality: The Second Erich L. Lehmann Symposium. IMS Lecture Notes-Monograph Series, Vol. 49, J. Rojo Ed., 2006.

13. Optimality: The Third Erich L. Lehmann Symposium IMS Lecture Notes-Monograph Series, Vol. 57, J. Rojo Ed., 2009.

14. The First Erich L. Lehmann Symposium: Optimality. IMS Lecture Notes-Monograph Series, Volume 44, J. Rojo and V. Perez-Abreu, Eds, 2004.

Erich L. Lehmann's Ph.D. Students

In 1992, during one of my summer visits to the Berkeley Statistics Department, I approached Juliet Shaffer with a plan to celebrate Erich's 75th birthday. I proposed to organize a technical meeting. Julie thought that Erich would not be receptive to the idea since he felt that the Lehmann Festschrift organized by Bickel, Doksum, and Hodges (1983) was enough celebration. Instead, Julie suggested that a collection of reminiscences from his former Ph.D. students would be something that Erich would like. As a result, I contacted all of Erich's former Ph.D. students and asked that they provide a couple of pages on their past and current, professional and social, interactions with Erich. I was amazed at the uniformity of the good experiences reported by all of Erich's students. And this was not an accident. Erich truly enjoyed the individual interaction with his students. In Lehmann (2008), he writes: *While I eschewed very large courses, I loved the teaching that occurred at the other end of the spectrum. Working on a one-on-one basis with Ph.D. students was, for me, the most enjoyable and rewarding aspect of teaching.*

The list below contains the names, year of degree, and title of dissertations for forty three of Erich's Ph.D. students. There are two names with an asterisk. These students earned their degree from the Statistics Department at Columbia and they were Abraham Wald's students at the time of Wald's untimely death. Erich happened to be visiting Columbia at the time and, at the students' request, agreed to serve as their advisor.

As an appreciation for Erich's commitment and support, his students have reciprocated by honoring him with the following:

- *A Festschrift for Erich L. Lehmann*, Bickel, Doksum, and Hodges (1983).
- *The First Lehmann Symposium on Optimality*, at CIMAT, May 2002. Rojo and Perez-Abreu.
- *The Second Lehmann Symposium on Optimality*, at Rice University, May 2004. Rojo.
- *The Third Lehmann Symposium on Optimality*, at Rice University, May 2007. Rojo.
- *The Fourth Lehmann Symposium on Optimality*, at Rice University, May 2011. Rojo.

In the preface to the volume of reminiscences prepared for Erich's 75th birthday, I summarized the Ph.D. students' contributions to the volume as follows:

The reader of this volume of reminiscences will correctly infer that inside Erich's prolific and prodigious scientific mind, there is room for a gentle, kind, responsible, and fatherly human being who has not only directed over forty Ph.D. dissertations but has also provided his students with advice on classical music, has shown a genuine concern for the well-being of his students and their families, and has even given away one of his students as a bride in the absence of her father. I think all of Erich's Ph.D. students would agree that Erich has been, throughout the years, more than just a Ph.D. thesis advisor to each of them.

Erich's impact went beyond the immediate impact on his Ph.D. students. The students' families were also affected by Erich's caring ways. On the title page of the same volume, Maria Luisa Rojo wrote:

You have touched our lives in ways you will never know.

List of Ph.D. Students

1. **COLIN ROSS BLYTH 1950**
   I. Contribution to the Statistical Theory of the Geiger-Muller Counter; II. On Minimax Statistical Decision Procedures and Their Admissibility.
2. **FRED CHARLES ANDREWS 1953**
   Asymptotic Behavior of Some Rank Tests for Analysis of Variance.
3. **JACK LADERMAN* 1953**
   On Statistical Decision Functions for Selecting one of k Populations.
4. **HENDRIK SALOMOM KONIJN 1954**
   On the Power of Some Tests for Independence.
5. **ALLAN BIRNBAUM* 1955**
   Characterizations of Complete Classes of Tests of Some Multiparametric Hypotheses, with Applications to Likelihood Ratio Tests.
6. **BALKRISHNA V. SUKHATME 1955**
   Testing the Hypothesis that Two Populations Differ Only in Location.
7. **V. J. CHACKO 1959**
   Testing Homogeneity Against Ordered Alternatives.
8. **PIOTR WITOLD MIKULSKI 1961**
   Some Problems in the Asymptotic Theory of Testing Statistical Hypotheses.
9. **MADAN LAL PURI 1962**
   Asymptotic Efficiency of a Class of C-Sample Tests.
10. **KRISHEN LAL MEHRA 1962**
    Rank Tests for Incomplete Block Designs. Paired-Comparison Case.
11. **SUBHA BHUCHONGKUL SUTCHRITPONGSA 1962**
    Class of Non-Parametric Tests for Independence in Bivariate Populations.

12. **Shishirkumar Shreedhar Jogdeo 1962**
 Nonparametric Tests for Regression Models.
13. **Peter J. Bickel 1963**
 Asymptotically Nonparametric Statistical Inference in the Multivariate Case.
14. **Arnljot Høyland 1963**
 Some Problems in Robust Point Estimation.
15. **Milan Kumar Gupta 1964**
 An Asymptotically Nonparametric Test of Symmetry.
16. **Madabhushi Raghavachari 1964**
 The Two-Sample Scale Problem When Locations are Unknown.
17. **Ponnapalli Venkata Ramachandramurty 1964**
 On Some Nonparametric Estimates and Tests in The Behrens-Fisher Situation.
18. **Vida Greenberg 1964**
 Robust Inference in Some Experimental Designs.
19. **Kjell Andreas Doksum 1965**
 Asymptotically Minimax Distribution-Free Procedures.
20. **William Harvey Lawton 1965**
 Concentration of Random Quotients.
21. **Shulamith Gross 1966**
 Nonparametric Tests When Nuisance Parameters Are Present.
22. **Bruce Hoadley 1966**
 The Theory of Large Deviations with Statistical Applications.
23. **Gouri Kanta Bhattacharyya 1966**
 Multivariate Two-Sample Normal Scores Test for Shift.
24. **James Nwoye Adichie 1966**
 Nonparametric Inference in Linear Regression.
25. **Dattaprabhakar V. Gokhale 1966**
 Some Problems in Independence and Dependence.
26. **Frank Rudolph Hampel 1968**
 Contributions to the Theory of Robust Estimation.
27. **Wilhelmine von Turk Stefansky 1969**
 On the Rejection of Outliers by Maximum Normed Residual.
28. **Neil H. Timm 1969** – Co-advisors Erich Leo Lehman & Leonard Marascuilo
 Estimating variance-covariance and correlation matrices from incomplete data.
29. **Louis Jaeckel 1969**
 Robust Estimates of Location.
30. **Friedrich Wilhelm Scholz 1971**
 Comparison of Optimal Location Estimators.
31. **Dan Anbar 1971**
 On Optimal Estimation Methods Using Stochastic Approximation Procedures.
32. **Michael Denis Stuart 1972**
 Components of 2 for Testing Normality Against Certain Restricted Alternatives.
33. **Claude L. Guillier 1972**
 Asymptotic Relative Efficiencies of Rank Tests for Trend Alternatives.

34. **Sherali Mavjibhai Makani 1972**
Admissibility of Linear Functions for Estimating Sums and Differences of Exponential Parameters.
35. **Howard J. M. D'Abrera 1973**
Rank Tests for Ordered Alternatives.
36. **Hyun-Ju Yoo Jin 1974**
Robust Measures of Shift.
37. **Amy Poon Davis 1977**
Robust Measures of Association.
38. **Jan F. Bjørnstad 1978**
On Optimal Subset Selection Procedures.
39. **William Paul Carmichael 1981**
The Rate of Weak Convergence of a Vector of U-Statistics Generated by a Single Sample.
40. **David Draper 1981**
Rank-Based Robust Analysis of Linear Models.
41. **Wei-Yin Loh 1982**
Tail-Orderings on Symmetric Distributions with Statistical Applications.
42. **Marc J. Sobel 1983**
Admissibility in Exponential Families.
43. **Javier Rojo 1984**
On Lehmann's General Concept of Unbiasedness and the Existence of L-unbiased Estimators.

Houston, TX Javier Rojo

References

1. P. J. Bickel, K. Doksum, and J. L. Hodges. *A Festschrift for Erich L. Lehmann*, Wadsworth Statistics/Probability Series, 1983.
2. J. Rojo. To Erich Leo Lehmann. *On his Seventy-fifth Birthday from his PH.D. Students*, November 1992.
3. *The First Erich L. Lehmann Symposium: Optimality*. IMS Lecture Notes–Monograph Series, Volume 44, J. Rojo and V. Perez-Abreu, Eds, 2004.
4. *Optimality: The Second Erich L. Lehmann Symposium*. IMS Lecture Notes–Monograph Series, Vol. 49, J. Rojo Ed., 2006.
5. *E. L. Lehmann. Reminiscences of a Statistician: The Company I Kept*, Springer, 2008.
6. *Optimality: The Third Erich L. Lehmann Symposium*. IMS Lecture Notes–Monograph Series, Vol. 57, J. Rojo Ed., 2009.

Part I
Decision Theory

Erich L. Lehmann's Work on Decision Theory

Javier Rojo

This chapter collects some of Erich's work on areas that attracted his attention during the early years in his professional career. Thus, the papers discuss the concepts of completeness (minimal complete families and complete sufficient statistics), minimal sufficiency, admissibility, invariance, unbiasedness, and minimaxity.

In Lehmann (1947) Erich argued that, in the absence of a uniformly most powerful (UMP) test, it is then a difficult choice to select a "best" test. Since there may be too many choices, it was proposed to reduce the totality of possible tests, by restricting attention to a class of tests $\mathcal{F}$ such that for any test ϕ of a hypothesis $\mathcal{H}$ of a given significance level, there is a corresponding test ϕ^* in $\mathcal{F}$ with power uniformly larger than or equal to the power of ϕ; and if $\phi_1, \phi_2 \in \mathcal{F}$, then the power of ϕ_1 is not uniformly larger than the power of ϕ_2 and vice versa. That is, in the absence of a UMP test, $\mathcal{F}$ is the smallest class from which to select a test when the selection is made based on power. Although there are other important ideas in the paper, this concept became a central part of Wald's statistical decision theory. In Lehmann (2008), Erich writes that

> *Wald at that time had resumed work on the general theory of statistical inference he had outlined in his 1939 paper. He found that my suggestion fit well into his general framework, and he magically transformed it into a theorem of great beauty and generality, which became one of the principal pillars of his decision theory.*

Motivated by the lack of specific examples of minimax estimators, and hence the lack of knowledge about how well minimax estimators behave in particular cases, Hodges and Lehmann (1950, 1951, 1952) embarked on a program to work out minimax estimators in several examples. In particular, these works are concerned with the admissibility of minimax estimators and connections with Bayes estimators. In the first of these papers, Hodges and Lehmann set out to find a minimax estimator for the probability of success p in a binomial experiment. Using the beta conjugate prior for p with parameters α and β, the Bayes estimator turns out to be $(\alpha + X)/(\alpha + \beta + n)$,

J. Rojo
Professor of Statistics, Department of Statistics, Rice University, Houston, TX 77005
e-mail: jrojo@rice.edu

J. Rojo (ed.), *Selected Works of E. L. Lehmann*, Selected Works in Probability and Statistics, DOI 10.1007/978-1-4614-1412-4_1, © Springer Science+Business Media, LLC 2012

where X is the number of successes in n Bernoulli trials. The risk function of this estimator is easily computed and, for $\alpha = \beta = \sqrt{n}/2$, has constant risk. Thus, the estimator defined by $(\bar{X}\sqrt{n} + 1/2)/(1 + \sqrt{n})$, where $\bar{X} = X/n$, is minimax. Admissibility of the minimax estimator then follows from the uniqueness of the Bayes estimator. Other strategies for finding minimax estimators in the nonparametric and prediction settings are also discussed. For example, in the nonparametric case, the paper presents an earlier version of the following result: Let $X \sim F$, with $F \in \mathcal{F}$. Suppose that $\delta(X)$ is minimax for $g(F)$ when $F \in \mathcal{F}_I$ and $\mathcal{F}_I \subset \mathcal{F}$. If in addition,

$$\sup_{F \in \mathcal{F}_I} R(F, \delta(X)) = \sup_{F \in \mathcal{F}} R(F, \delta(X)), \tag{1}$$

then $\delta(X)$ is also minimax over $\mathcal{F}$. (See Lehmann and Casella (1998)). Assuming that $X_1, X_2, \ldots, X_n$ are independent with a joint distribution in some family $\mathcal{F}$, and if the family $\mathcal{F}$ contains $\mathcal{F}_I$, the family of joint distributions of n independent and identically distributed Bernoulli random variables with probability of success $p, 0 < p < 1$, then as discussed above, the minimax estimator for p when restricting attention to $\mathcal{F}_I$ is $(\bar{X}\sqrt{n} + 1/2)/(1 + \sqrt{n})$. That this estimator remains minimax for the larger class of distributions $\mathcal{F}$, follows after showing that (1) holds. The concept of complete classes having been formalized by Wald in (1950), the paper also shows that, for convex loss functions, the class of nonandomized estimators is essentially complete.

Minimax estimators are not typically Bayes estimators. Rather, they arise as limits of Bayes estimators and, hence, the trick of finding a prior distribution for which the Bayes estimator has constant risk does not usually work. Therefore, in Hodges and Lehmann (1951) a different approach is used to find minimax and admissible estimators. Using the Cramer-Rao lower bound, results are provided to obtain admissible/minimax estimators with respect to the loss function $L(\delta, \theta) = h(\theta)(\delta - \theta)^2$, where $h \geq 0$. The method requires the solution of a differential inequality involving the lower bound for the Mean Squared Error. Extensions were also presented to certain sequential problems.

Motivated by the observation that Bayes estimators are computed assuming complete knowledge of a prior distribution, and minimax estimators correspond to Bayes estimators with respect to a least favorable (or a sequence of least favorable) distribution(s), Hodges and Lehmann (1952) considered, instead, an approach based on *restricted Bayes solutions*. An estimator δ is said to be a *restricted Bayes estimator* with respect to a prior λ if it minimizes $\int R_\delta(\theta) d\lambda(\theta)$ subject to $R_\delta(\theta) \leq C_0$, for all θ, where $C_0 > C^*$, and C^* represents the maximum of the risk of the minimax estimator, and $R_\delta(\theta) = \int R(\delta(X), \theta) dF_\theta(x)$ represents the risk function of δ. It is seen that this is equivalent to choosing a δ^* that minimizes, instead, $\rho \int R_\delta(\theta) d\lambda(\theta) + (1 - \rho) \sup_\theta R_\delta(\theta)$ for some $0 \leq \rho \leq 1$. Conditions are discussed for the existence of restricted Bayes estimators and several examples are provided that illustrate the method.

In Lehmann (1951), in the spirit of Wald's decision theory, Erich proposed a general concept of unbiasedness that is associated with the loss function. Thus, an estimator δ is *L-unbiased*, with respect to the loss function $L(\delta, \theta)$, provided that

$E_\theta(L(\delta, \theta)) \leq E_\theta(L(\delta, \theta^*))$ for all $\theta \neq \theta^*$. It turns out that this reduces, when the loss is squared error, to the usual (mean) unbiasedness concept of David and Neyman (1938). When the loss is absolute error, L-unbiasedness is equivalent to the usual median-unbiasedness concept of Brown (1947). And, for appropriate loss functions, this concept reduces to the usual concepts of unbiased tests and unbiased confidence sets. Rojo (1984) studied the existence of L-unbiased estimators for many other types of loss functions.

Continuing with his pursuit of results dealing with minimaxity and admissibility, Lehmann (1952) addresses the issue of the existence of least favorable distributions. Wald (1950) had provided conditions – albeit too restrictive – for the existence of least favorable distributions that required compactness of the parameter space. In this paper, Erich did away with compactness of the parameter space in the case of hypothesis testing and, more generally, in the case where only a finite number of decisions are available. In Lehmann and Stein (1953), the most powerful invariant test for testing one location parameter family against another is shown to be admissible.

In Bahadur and Lehmann (1955), the relationship between the existence of a necessary and sufficient field and the existence of a necessary and sufficient statistic is studied. In these early papers necessary and sufficient meant minimally sufficient, and a field meant a $\sigma-$field. The concepts of minimal sufficient σ-fields and minimal sufficient statistics first appeared in work of Halmos and Savage (1949) and Bahadur (1954). Detailed methods for constructing minimally sufficient statistics can be found in Lehmann and Scheffé (1950), and in Bahadur (1954). But the question, for example, first raised in Bahadur (1954) of whether the existence of a minimally sufficient $\sigma-$field always implies the existence of a minimally sufficient statistic had not been dealt with. In Bahadur and Lehmann (1955), some light is shed on this problem. The paper provides an example of a minimally sufficient $\sigma-$field that cannot be induced by a statistic. However, the question still remained unanswered. Additional work by Pitcher (1957) provided an example of a family of probability measures for which no minimally sufficient $\sigma-$field and no minimally sufficient statistic exists. The question remained: Does the existence of a minimally sufficient $\sigma-$field always implies the existence of a minimally sufficient statistic or vice versa? The first answer to this question appeared in Landers and Rogge (1974): The existence of a minimally sufficient statistic is neither necessary nor sufficient for the existence of a minimally sufficient $\sigma-$field.

Lehmann (1966) provides an alternative proof to a theorem of Bahadur and Goodman (1952) that shows that the "natural" rule for selecting the best population, out of k populations, is optimal. Other optimality properties of various other procedures proposed in the literature are obtained as a consequence of the new proof. Thus, for example, as a consequence of the results provided in this paper, the optimality of the selection procedures in Bechhofer (1954), Bechhofer *et al.* (1959), and Bechhofer and Milton (1954) are seen to be optimal, when the best population is sufficiently better than the second best, in the sense of maximizing the minimum probability of a correct decision.

References

[1] G. W. Brown. On Small-Sample Estimation. *Ann. Math. Statist.*, Vol. 18, No. 4, pp. 582-585, 1947.
[2] R. R. Bahadur and L. A. Goodman. Impartial Decision Rules and Sufficient Statistics. *Ann. Math. Statist.*, Vol. 23, No. 4, pp. 553-562, 1952.
[3] R. R. Bahadur. Sufficiency and Statistical Decision Functions. *Ann. Math. Statist.*, Vol. 25, No. 3, pp. 423-462, 1954.
[4] R. R. Bahadur and E. L. Lehmann. Two comments on sufficiency and statistical decision functions. *Ann. Math. Statist.*, Vol. 26, pp. 139-142, 1955.
[5] R. E. Bechhofer. A Single-Sample Multiple Decision Procedure for Ranking Means of Normal Populations with known Variances. *Ann. Math. Statist.*, Vol. 25, No. 1, pp. 16-39, 1954.
[6] R. E. Bechhofer and M. Sobel. A Single-Sample Multiple Decision Procedure for Ranking Variances of Normal Populations. *Ann. Math. Statist.*, Vol. 25, No. 2, pp. 273-289, 1954.
[7] R. E. Bechhofer, S. Elmaghraby, and N. Morse. A Single-Sample Multiple-Decision Procedure for Selecting the Multinomial Event Which Has the Highest Probability. *Ann. Math. Statist.*, Vol. 30, No. 1, pp. 102-119, 1959.
[8] F. N. David and J. Neyman. Extension of the Markoff Theorem on Least Squares. *Statistical Research Memoirs*, Vol. II, London, pp. 105-116, 1938.
[9] P. R. Halmos and L. J. Savage. Application of the Radon-Nikodym Theorem to the Theory of Sufficient Statistics. *Ann. Math. Statist.*, Vol. 20, No. 2, pp. 225-241, 1949.
[10] J. L. Hodges, Jr. and E. L. Lehmann. Some problems in minimax point estimation. *Ann. Math. Statist.*, Vol. 21, pp. 182-197, 1950.
[11] J. L. Hodges, Jr. and E. L. Lehmann. Some applications of the Cramer-Rao inequality. *Proc. Second Berkeley Symp. Math. Statist. Prob.*, University of California Press, pp. 13-22, 1951.
[12] J. L. Hodges, Jr. and E. L. Lehmann. The use of previous experience in reaching statistical decisions. *Ann. Math. Statist.*, Vol. 23, pp. 396-407, 1952.
[13] D. Landers and L. Rogge. Minimal Sufficient σ-Fields and Minimal Sufficient Statistics. Two Counterexamples. *Ann. Math. Statist.*, Vol. 43, No. 6, pp. 2045-2049, 1972.
[14] E. L. Lehmann. On families of admissible tests. *Ann. Math. Statist.*, Vol. 18, pp. 97-104, 1947.
[15] E. L. Lehmann and H. Scheffé. Completeness, similar regions and unbiased estimation. Part I. *Sankhya*, Vol. 10, pp. 305-340, 1950.
[16] E. L. Lehmann. A general concept of unbiasedness. *Ann. Math. Statist.*, Vol. 22, pp. 587-591, 1951.
[17] E. L. Lehmann. On the existence of least favorable distributions. *Ann. Math. Statist.*, Vol. 23, pp. 408-416, 1952.
[18] E. L. Lehmann and C. Stein. The admissibility of certain invariant statistical tests involving a translation parameter. *Ann. Math. Statist.*, Vol. 24, pp. 473-479, 1953.
[19] E. L. Lehmann. On a theorem of Bahadur and Goodman. *Ann. Math. Statist.*, Vol. 37, pp. 1-6, 1966.
[20] E. L. Lehmann and G. Casella. *Theory of Point Estimation,* 2nd Edition. Springer-Verlag, 1998.
[21] E. L. Lehmann. *Reminiscences of a Statistician: The Company I Kept*, Springer, 2008.
[22] T. S. Pitcher. Sets of Measures not Admitting Necessity and Sufficient Statistics or Subfields. *Ann. Math. Statist.*, Vol. 28, No. 1, pp. 267-268, 1957.
[23] J. Rojo. *On Lehmann's General Concept of Unbiasedness and the Existence of L-unbiased Estimators*. Ph.D. Thesis, Department of Statistics, University of California at Berkeley, 1984.
[24] A. Wald. *Statistical Decision Functions*. John Wiley & Sons, 1950.

ON FAMILIES OF ADMISSIBLE TESTS

By E. L. LEHMANN

University of California, Berkeley

1. Summary. For each hypothesis H of a certain class of simple hypotheses, a family F of tests is determined such that

(a) given any test w of H there exists a test w' belonging to F which has power uniformly greater than or equal to that of w.

(b) no member of F has power uniformly greater than or equal to that of any other member of F.

The effect on F of various assumptions about the set of alternatives are considered. As an application an optimum property of the known type A_1 tests is proved, and a result is obtained concerning the most stringent tests of the hypotheses considered.

2. Introduction. In the theory of testing simple hypotheses, if a uniformly most powerful test exists, it is the most desirable test to use. If, as is generally the case, such a test does not exist, the choice between tests none of which is "altogether better" than all the others, has to be based on information not contained in the general formulation of the testing problem. If no such additional information is available, the choice must of necessity be somewhat arbitrary.

Now although a single uniformly most powerful test exists only in exceptional cases, there will always exist a family F of tests such that

(a) given any test w of the hypotheses H under consideration and of prescribed level of significance, there exists a test w' belonging to F which has power uniformly greater than or equal to that of w.

(b) no member of F has power uniformly greater than or equal to that of any other member of F.

The family F is essentially unique. Arbitrariness occurs only since a test region is not uniquely determined by its power function. But since two tests with the same power function are equivalent for testing purposes, it is from the present point of view immaterial which one is included in F.

With the same restriction F is essentially the family of admissible tests, a test w being admissible if there is no test of the same level of significance which has power uniformly greater than or equal to but not identically equal to that of w. This definition differs only trivially from the one given by Wald [1, p. 15] who defines a test w to be non-admissible if there exists a w' with power everywhere greater than that of w (except at the hypothetical point).

F naturally depends on the class of alternatives considered. A restriction in the class of alternatives may (although it will not necessarily) diminish F. The family F may also be decreased by other additional information: For instance a probability distribution may be assumed for the set of alternatives, and some properties of this distribution may be presupposed.

J. Rojo (ed.), *Selected Works of E. L. Lehmann*, Selected Works in Probability and Statistics, DOI 10.1007/978-1-4614-1412-4_2, © Springer Science+Business Media, LLC 2012

The determination of the family F, (and a description of the power functions of the tests in F) might be considered a solution of the testing problem. The solution is not unique and hence does not provide a basis for action. This reflects the fact that additional information is needed to make possible the unique choice of a best test. On the basis of the available information, F represents the furthest reduction of the problem that seems possible. On the one hand, if the choice of test is to be made from the point of view of power, the only contestants for "best test" are the members of F. On the other hand, the available information does not give preference to any one member of F over any other unless additional principles (such as unbiasedness for instance) are introduced.

It is the purpose of the present paper to illustrate the above notions by determining F for a very simple case.

3. Determination of the family F.

Let the random variable

$$E = (X_1, X_2, \cdots; X_n)$$

have a probability density function

$$p_\theta \tag{1}$$

depending on parameter θ. Concerning (1) we shall make the assumptions under which Neyman [2, 3] has shown the existence of the type A_1 test of the hypothesis

$$H: \theta = \theta_0 . \tag{2}$$

ASSUMPTIONS:

(*a*) *Conditions of regularity:*

The integral

$$\int_w p_\theta(e)\, de \qquad e = (x_1, \cdots, x_n) \quad de = dx_1 \cdots dx_n \tag{3}$$

extended over any region w in the sample space, admits of two successive derivatives with respect to θ under the integral sign, i.e.

$$\frac{d^k}{d\theta^k} \int_w p_\theta(e)\, de = \int_w \frac{\partial^k}{\partial\theta^k} p_\theta(e)\, de \quad \text{for} \quad k = 1, 2. \tag{4}$$

(*b*) *A differential equation:*

If

$$\varphi_\theta(e) = \frac{\partial}{\partial\theta} \log p_\theta(e) \tag{5}$$

$$\varphi'_\theta(e) = \frac{\partial}{\partial\theta} \varphi_\theta(e),$$

φ_{θ_0} is not identically zero, and there exist functions of θ (but independent of e), A and B, such that

$$\varphi_\theta' = A + B\varphi_\theta . \tag{6}$$

Under these assumptions Neyman has shown

A. that the probability density function p_θ is of the form

$$p_\theta(e) = \exp \{P(\theta) + T(e)\cdot Q(\theta) + R(e)\} \tag{7}$$

where Q is a monotone function with $\frac{d}{d\theta} Q(\theta) \mid_{\theta=\theta_0} \neq 0$ (without loss of generality we shall assume Q monotonely increasing) and

B. that the type A_1 test of the hypothesis H exists, and is given by

$$T(e) \leq c_1 , \qquad T(e) \geq c_2 \tag{8}$$

for suitable choice of c_1 and c_2 .

In what follows we shall assume that the permissible first kind error in testing H is fixed throughout and has the value ϵ. By a test w of H we shall always mean a test of level of significance ϵ, i.e. satisfying

$$\int_w p_{\theta_0}(e)\, de = \epsilon. \tag{9}$$

Let us consider the family of tests

$$w(k)\colon T(e) \leq k, T(e) \geq f(k); k < f(k) \tag{10}$$

where $f(k)$ is determined by (9). It easily follows from (9) that k can take on all values from $-\infty$ to k_0 , say, where k_0 is such that

$$f(k_0) = +\infty . \tag{11}$$

For the family F of tests $\{w(k)\}$, $-\infty \leq k \leq k_0$ we now state

THEOREM 1. *All members of* F *are admissible, and if* w *is any admissible test not in* F, *there exists a member of* F *which has power identical with that of* w.

We first prove the

LEMMA. *Let* β_w *denote the powerfunction of a test* w. *Then if* $k_1 < k_2$

$$\begin{aligned} \beta_{w(k_1)}(\theta) &< \beta_{w(k_2)}(\theta) && \text{if } \theta < \theta_0 \\ \beta_{w(k_1)}(\theta) &> \beta_{w(k_2)}(\theta) && \text{if } \theta > \theta_0 . \end{aligned} \tag{12}$$

PROOF: Let $\bar{w}$ denote the complement of a region w. Consider the intervals

$$\begin{aligned} I &= w(k_1) \cdot \overline{w(k_2)} \\ J &= \overline{w(k_1)} \cdot w(k_2). \end{aligned} \tag{13}$$

I lies entirely to the right of J. Let $\theta > \theta_0$. Then

$$\frac{p_\theta(e)}{p_{\theta_0}(e)} = C(\theta) \exp \{T(e)[Q(\theta) - Q(\theta_0)]\} \tag{14}$$

is a strictly increasing function of T since Q is increasing. Therefore there exists a constant C such that

$$\begin{aligned} \frac{p_\theta(e)}{p_{\theta_0}(e)} &\leq C \quad \text{if} \quad T(e) \text{ is in } J \\ C &\leq \frac{p_\theta(e)}{p_{\theta_0}(e)} \quad \text{if} \quad T(e) \text{ is in } I. \end{aligned} \tag{15}$$

Since

$$\int_{w(k_1)} p_{\theta_0}(e)\, de = \int_{w(k_2)} p_{\theta_0}(e)\, de \tag{16}$$

we have

$$\int_{T(e)\epsilon I} p_{\theta_0}(e)\, de = \int_{T(e)\epsilon J} p_{\theta_0}(e)\, de \tag{17}$$

and therefore

$$\int_J p_\theta(e)\, de < C \cdot \int_J p_{\theta_0}(e)\, de = C \cdot \int_I p_{\theta_0}(e)\, de < \int_I p_\theta(e)\, de \tag{18}$$

from which it follows that

$$\int_{w(k_2)} p_\theta(e)\, de < \int_{w(k_1)} p_\theta(e)\, de \tag{19}$$

which is the desired result.

Proof of Theorem 1. The proof consists of several parts.

I. Let m be any real number, and assume that there exists a value of k such that

$$\beta_w(\theta_0) = \epsilon \tag{20}$$

$$\frac{d}{d\theta} \beta_w(\theta) \big|_{\theta=\theta_0} = m \tag{21}$$

for $w = w(k)$. Then $w(k)$ has power uniformly greater than or equal to that of any other test satisfying (20) and (21).

For $m = 0$ this becomes Neyman's theorem stating that the type A test is also of type A_1. The proof of the theorem however is independent of the value of

$$\frac{d}{d\theta} \beta_w(\theta) \big|_{\theta=\theta_0} \tag{23}$$

and hence carries over to arbitrary m.

II. If there exists any test satisfying (20) and (21) then there exists a number k for which $w(k)$ also satisfies (20) and (21).

To prove this let us determine, of all tests satisfying (20), the one which maximizes

$$\frac{d}{d\theta}\beta_w(\theta)\Big|_{\theta=\theta_0} = \int_w \frac{\partial}{\partial\theta} p_\theta(e)\Big|_{\theta=\theta_0} de. \tag{24}$$

This can be done by means of the lemma of Neyman and Pearson [4, p. 11] which gives sufficient conditions for a region w, subject to restrictions

$$\int_w f_i(e)\, de = a_i, \qquad (i = 1, \cdots, p), \tag{25}$$

to maximize an integral

$$\int_w g(e)\, de. \tag{26}$$

According to this lemma the desired test is of the form

$$\frac{\partial}{\partial\theta} p_\theta(e)\Big|_{\theta=\theta_0} \geq a \cdot p_{\theta_0}(e) \tag{25}$$

provided a value of a exists for which this test satisfies (20). (25) is equivalent to

$$P'(\theta_0) + T(e) \cdot Q'(\theta_0) \geq a \qquad \text{from (7)} \tag{26}$$

or, since $Q'(\theta_0) > 0$, to

$$T(e) \geq b. \tag{27}$$

Thus, if a number b exists such that the test (27) satisfies (20), this test is the one maximizing (24). But such a number does exist, namely $f(-\infty)$. Therefore $w(-\infty)$ is the desired test.

Similarly it is easy to show that of all tests satisfying (20), $w(k_0)$ minimizes (24). But

$$\frac{d}{d\theta}\beta_{w(k)}(\theta)\Big|_{\theta=\theta_0} = \int_{T<k,\, T\geq f(k)} \frac{\partial}{\partial\theta} p_\theta(e)\Big|_{\theta=\theta_0} de \tag{28}$$

is a continuous function of k, and therefore takes on all intermediate values, which establishes II.

III. From I. and II. we conclude that given any test w there exists a member of F which has power uniformly greater than or equal to that of w. For let w be any test of H. From the condition of regularity it follows that its power-function has a derivative at θ_0. By II. there exists a value of k such that the powerfunction of $w(k)$ has the same slope at θ_0, and from I. it follows that $w(k)$ is uniformly more powerful than w.

But from the lemma we see that none of the tests $w(k)$ is uniformly more powerful than any other. Hence all members of F are admissible, and the theorem is proved.

From the lemma and Theorem 1 we can conclude for all members of F the following optimum property:

COROLLARY 1: *Let* w *be any test, and let* w_0 *be any member of* F. *Then at least one of the two statements*

$$(29)\qquad \begin{aligned} \beta_w(\theta) &\leq \beta_{w_0}(\theta) && \text{for all } \theta < \theta_0 \\ \beta_w(\theta) &\leq \beta_{w_0}(\theta) && \text{for all } \theta > \theta_0 \end{aligned}$$

must hold.

The lemma and Theorem 1 also give the following result concerning most stringent tests, defined by Wald [1, p. 33].

COROLLARY 2: *There exists a uniformly most powerful of all most stringent tests. It is that unique member* w_0 *of* F *for which*

$$\underset{\theta<\theta_0}{\text{l.u.b.}}\left[\underset{w}{\text{l.u.b.}}\ \beta_w(\theta) - \beta_{w_0}(\theta)\right] = \underset{\theta>\theta_0}{\text{l.u.b.}}\left[\underset{w}{\text{l.u.b.}}\ \beta_w(\theta) - \beta_{w_0}(\theta)\right].$$

4. The effect on F of assumptions about the alternatives. Let us next consider how a restriction in the set of alternatives effects the family F. From the lemma it follows that there is no change as long as the set of alternatives contains values of θ both greater and less than θ_0. On the other hand, if the alternatives are restricted to values of θ greater than θ_0, say, the family F for testing H against these alternatives consists of only a single member, the test $w(-\infty)$, (and similarly for the other onesided case). This follows from

THEOREM 2: *Under conditions a. and b. the test* $w(-\infty)$ *is uniformly most powerful against the alternatives* $\theta > \theta_0$, *the test* $w(k_0)$ *is uniformly most powerful against the alternatives* $\theta < \theta_0$.

PROOF: Let w be any test. By Theorem 1 there exists a number k such that

$$(30)\qquad \beta_w(\theta) \leq \beta_{w(k)}(\theta) \qquad \text{for all } \theta.$$

From the lemma it follows that

$$(31)\qquad \begin{aligned} \beta_{w(k)}(\theta) &\leq \beta_{w(k_0)}(\theta) && \text{if } \theta < \theta_0 \\ \beta_{w(k)}(\theta) &\leq \beta_{w(-\infty)}(\theta) && \text{if } \theta > \theta_0. \end{aligned}$$

Combining (30) and (31) we have the desired result.

(It is also easy to prove Theorem 2 directly from the Neyman-Pearson lemma.)

In order to illustrate how the assumption of an a priori distribution of θ together with some information about this distribution affects F, let us consider a special case of the class of hypotheses discussed so far.

Let

$$(32)\qquad p_\theta(x_1, \cdots, x_n) = c \cdot e^{-\frac{1}{2}\Sigma(x_i-\theta)^2}$$

so that $E = (X_1, X_2, \cdots, X_n)$ is a sample from a normal distribution with unit variance and unknown mean. We want to test the hypothesis

$$(33)\qquad H\colon \theta = 0.$$

We shall show that if θ has a probability density function g which is symmetric about the origin, then the family F for testing H consists, as might be expected, of a single member, the type A_1 test.

Our problem is to find the test w satisfying

$$\int_w p_0(x_1, \cdots, x_n)\, dx_1 \cdots dx_n = \epsilon \tag{34}$$

and which maximizes

$$\int_{-\infty}^{\infty} g(\theta) \int_w p_\theta(x_1, \cdots, x_n)\, dx_1 \cdots dx_n \cdot d\theta. \tag{35}$$

Inverting the order of integration, which is permissible in this case, the Neyman-Pearson lemma shows the desired test to be of the form

$$\int_{-\infty}^{\infty} g(\theta) p_\theta(x_1, \cdots, x_n)\, d\theta \geq a \cdot p_0(x_1, \cdots, x_n) \tag{36}$$

provided a value of a exists for which (36) satisfies (34). Substituting from (32), (36) becomes

$$f(\bar{x}) = \int_{-\infty}^{\infty} g(\theta) e^{-\frac{1}{2}\theta^2 + n\theta\bar{x}}\, d\theta \geq a \tag{37}$$

where

$$\bar{x} = \frac{1}{n} \sum_{i=1}^{n} x_i . \tag{38}$$

Since

$$\frac{d^2}{d\bar{x}^2} f(\bar{x}) > 0 \tag{39}$$

the region (37) is either empty, which would contradict (34), or else can be described by inequalities

$$\bar{x} \leq a_1 , \qquad \bar{x} \geq a_2 \tag{40}$$

where

$$f(a_1) = f(a_2) \tag{41}$$

the latter equation becoming, on substitution from (37)

$$\int_{-\infty}^{\infty} g(\theta) e^{-(n/2)\theta^2} (e^{na_2\theta} - e^{na_1\theta})\, d\theta = 0. \tag{42}$$

If g is an even function, (42) is certainly satisfied when $a_1 = -a_2$. Our test then becomes

$$\bar{x} \leq -a_2 , \qquad \bar{x} \geq a_2 \tag{43}$$

which for proper choice of a_2 satisfies (34) and is the well known type A_1 test.

5. Concluding remarks. Let us consider once more a probability density function satisfying a. and b. We have seen that the family F for testing H against the alternatives $\theta \neq \theta_0$ contains an infinity of elements unless we make some additional assumptions. On the other hand, if the principle of unbiasedness is accepted, F shrinks to a single element: the type A_1 test.

But unbiasedness does not insure power. Thus conceivably some other test might be more powerful than the test chosen, everywhere except in a small one sided neighbourhood of θ_0. That this is not so is shown by Corollary 1 to Theorem 1. This remark illustrates how intuitively appealing principles and a knowledge of the family F may be used in conjunction to arrive at a choice of a satisfactory test, when not enough information is available to make the choice compelling.

Finally, it should be pointed out that although we restricted our considerations to simple hypotheses, the notions developed also apply to composite hypotheses.

REFERENCES

[1] A. Wald, "On the principles of statistical inference". *Notre Dame Mathematical Lectures.* Number 1.

[2] J. Neyman, "L'estimation statistique traitée comme un problème classique de probabilité". *Conférences internationales des sciences mathématiques à Genève: Colloque d'Octobre 1937, sur le Calcul des Probabilités*, Paris, 1938.

[3] J. Neyman and E. S. Pearson, "Contributions to the theory of testing statistical hypotheses, Part II". *Statistical Research Memoirs*, Vol. II, London, 1938.

[4] J. Neyman and E. S. Pearson, "Contributions to the theory of testing statistical hypotheses, Part I". *Statistical Research Memoirs*, Vol. I, London, 1936.

SOME PROBLEMS IN MINIMAX POINT ESTIMATION

By J. L. Hodges, Jr.,[1] and E. L. Lehmann

University of California, Berkeley

1. Summary. In the present paper the problem of point estimation is considered in terms of risk functions, without the customary restriction to unbiased estimates. It is shown that, whenever the loss is a convex function of the estimate, it suffices from the risk viewpoint to consider only nonrandomized estimates. For a number of specific problems the minimax estimates are found explicitly, using the squared error as loss. Certain minimax prediction problems are also solved.

2. Introduction. The principles most commonly applied in the selection of a point estimate are the principles of maximum likelihood (R. A. Fisher) and of minimum variance unbiased estimation (Markoff).[2] Both of these principles are intuitively appealing, but neither of them can be justified very well in a systematic development of statistics. This holds also for some modifications of these principles proposed by G. W. Brown [1], as the author himself points out.

In an important early paper [2], Wald indicated a more systematic approach to the problem, which he later developed into his general theory of statistical decision problems [3, 4, 5]. Consider a random variable X distributed over a space $\mathfrak{X}$ according to a distribution P_θ^X with $\theta \epsilon \Omega$. It is desired to estimate some $g(\theta)$. If the value x of X is observed one makes an estimate, say $f(x)$, and thereby incurs a loss of $W[g(\theta), f(x)]$ when θ is the true value of the parameter. We shall assume that the loss function is nonnegative. It then follows that the expectation of the loss will always exist (although it may be infinite). The risk associated with the estimate f is defined to be the expected loss, as given by

$$R_f(\theta) = E_\theta W[g(\theta), f(x)] = \int_{\mathfrak{X}} W[g(\theta), f(x)]\, dP_\theta^X(x). \tag{2.1}$$

The choice of estimate should then be made according to the risk function. As a particular possibility Wald suggests the use of minimax estimates, i.e. estimates which minimize $\sup_\theta R_f(\theta)$.

The main purpose of the present paper is to obtain minimax estimates for a number of specific problems. Only few such problems have been worked out so far, the emphasis in Wald's work having been on the general theory. In [2] Wald obtained the minimax estimate of an unknown location parameter. Stein and Wald [6] treated the sequential problem of estimating the mean of a normal dis-

[1] This work was supported in part by the Office of Naval Research.

[2] Actually, the principle of minimum variance unbiased estimation goes back to Gauss. For discussions of the history of these ideas, see E. Czuber's *Theorie der Beobachtungsfehler*, Leipzig, 1891, and R. L. Plackett, "A historical note on the method of least squares", *Biometrika*, Vol. 36 (1950), p. 458.

tribution with known variance, and in his forthcoming book Wald considers as an example the sequential problem of estimating the mean of a random variable distributed uniformly over an interval of length 1.

It seems worthwhile to consider further special problems both because one may obtain estimates that in some cases are preferable to the conventional ones, and because these examples throw some light on the general desirability of the minimax principle. As we shall see below, it does not seem possible to reach any definite conclusions on this latter point, and to obtain a generally valid comparison between the minimax estimate and, for example, the unbiased estimate with uniformly smallest variance (when such an estimate exists).

Consider, for example, the problem of estimating the probability of success from a number of independent trials each of which may be a success or a failure, when the loss-function is the squared error. If the number of trials is one, the minimax estimate (as is shown below) is given by $f(X) = \frac{1}{2}X + \frac{1}{4}$, where X is 1 or 0 as the trial is a success or failure. As is easily seen, this estimate has smaller risk than the usual estimate $f^*(X) = X$ whenever $0.07 \leqq p \leqq 0.93$. On the other hand, when the number of trials is large the standard estimate $\bar{X}$ has smaller risk than the minimax estimate nearly everywhere. The minimax estimate is only slightly better in a small interval centered at $p = \frac{1}{2}$, whose length tends to zero as the number of trials tends to infinity, and is worse everywhere else.

For our purpose it is convenient to formulate the problem of point estimation as follows (see in this connection [7]). A random variable X is distributed over a space $\mathfrak{X}$ according to a distribution P belonging to a family $\mathfrak{F}$. We wish to estimate $g(P)$ where g is a function whose domain is $\mathfrak{F}$ and whose range is contained in some space $\mathcal{Y}$ (in any example $\mathcal{Y}$ is usually a Euclidean space, mostly even a one dimensional Euclidean space). An estimate is a statistic $f(X)$ taking on values in $\mathcal{Y}$. We denote by $W[g(P), f(x)]$ the loss which results from making the estimate $f(x)$ when P is the true distribution, and we define the risk function of the estimate f by

$$R_f(P) = E_P W[g(P), f(X)]. \tag{2.2}$$

The problem is to determine f so as to minimize $\sup_{P \epsilon \mathfrak{F}} R_f(P)$.

Our principal tool will be the following theorem, which is essentially contained in Wald's work but which is not stated there explicitly. The theorem is a slight modification of one used for the theory of testing in [8].

THEOREM 2.1. *Let* $\{P_\theta\}$, $\theta \epsilon \omega$ *(where* ω *is a subset of a Euclidean space), be a parametric subfamily of* $\mathfrak{F}$, *and let* λ *be a probability measure over* ω. *Suppose that* f *minimizes*

$$\int_\omega E_\theta W[g(P_\theta), f(X)] \, d\lambda(\theta) \tag{2.3}$$

and that

(i) $E_\theta W[g(P_\theta), f(X)]$ *is constant (say* c*) for all* $\theta \epsilon \omega$,

(ii) $E_P W[g(P), f(X)] \leq c$ *for all* P *in* $\mathfrak{F}$.

Then f *is a minimax estimate for estimating* g.

PROOF. Let f^* be any other estimate of g. Then

$$
\begin{aligned}
\sup_{P \in \mathfrak{F}} E_P W[g(P), f(X)] &= \int_\omega E_\theta W[g(P_\theta), f(X)] \, d\lambda(\theta) \\
&\leqq \int_\omega E_\theta W[g(P_\theta), f^*(X)] \, d\lambda(\theta) \qquad (2.4) \\
&\leqq \sup_{P \in \mathfrak{F}} E_P W[g(P), f^*(X)].
\end{aligned}
$$

We note that if f is the unique function minimizing (2.3), then the first inequality in (2.4) becomes strict, and hence f is the unique minimax estimate of g.

Following Wald we shall call the function f that minimizes (2.3) the Bayes estimate of g associated with the a priori distribution λ. As a corollary to theorem 2.1, we note that a Bayes estimate whose risk function is constant, is a minimax estimate.

3. Randomization. In the formulation of the problem of point estimation given above, the estimate $f(x)$ is assumed to be completely determined by the observed value x of the random variable X. In the present section a broader formulation of the problem will be considered, in which the estimate corresponding to x may itself be a random variable, say T_x. This extension is a special case of the notion of randomized decision function introduced by Wald in his general decision theory. We associate with each x in $\mathfrak{X}$ a probability distribution F_x, with the convention that when X is observed to have the value x, we estimate $g(P)$ by means of a random variable T_x which is distributed according to F_x. Estimates of this latter kind we shall call *randomized*, and the fixed estimates $f(x)$ *nonrandomized*.

The motivation behind the admission of randomized estimates (or more generally of randomized statistical decision funtions) is that in some problems of statistical inference the performance of the decision function is considerably improved by randomization. It is clear however that the randomized functions are more complicated, and hence that it is useful to know when their consideration is not necessary. Before investigating this question we give the following definition, which makes precise a sense in which certain estimates may be omitted from consideration. (See Wald [9]).

DEFINITION. For a given estimation problem a class C of estimates will be said to be essentially complete with respect to a class D of estimates, if for every estimate g in D there exists an estimate f in C such that $R_f(P) \leqq R_g(P)$ for all P in $\mathfrak{F}$. If D is the class of all randomized estimates we simply say that C is essentially complete for the given problem.

It is clear that if one adopts the risk function point of view, one loses nothing by restricting consideration to an essentially complete class of estimates. In the present section we find conditions under which the totality of nonrandomized estimates forms an essentially complete class.

For this purpose we need the notion of convexity. A set S in a k-dimensional Euclidean space is said to be convex if, whenever P and Q are in S, then all points on the line segment from P to Q are also in S. A real valued function ψ defined over a k-dimensional Euclidean space is said to be convex, if for any points $(x_1, \cdots, x_k)$ and $(y_1, \cdots, y_k)$ of the space, and any number $0 < \alpha < 1$ we have

$$\begin{aligned} &\alpha\psi(x_1, \cdots, x_k) + (1 - \alpha)\psi(y_1, \cdots, y_k) \geqq \\ &\qquad \psi(\alpha x_1 + (1 - \alpha)y_1, \cdots, \alpha x_k + (1 - \alpha)y_k). \end{aligned} \tag{3.1}$$

We use the following notation for conditional expectation. If U and V are two random variables which have a joint distribution, then $E(U|v)$ denotes the conditional expectation of U given that $V = v$; $E(U|S)$ denotes the conditional expectation of U given that V is in S. Let $\Phi(v) = E(U|v)$; then for $\Phi(V)$ we write $E(U \mid V)$.

LEMMA 3.1. *Let U, V be two random variables with a joint distribution, such that U is distributed in a k-dimensional space and $E(U)$ is finite. Let ψ be a real-valued convex function defined over this space and bounded from below. Then*

$$E\{\psi\{E(U \mid V)\}\} \leqq E\{\psi(U)\}.$$

PROOF. The proof is immediate in the special case that, for almost all v, there exists a determination of the conditional probability distribution of U given v which is a measure. We then know, from the convexity of ψ, that for almost all values v of V, $\psi\{E(U \mid v)\} \leqq E\{\psi(U) \mid v\}$. Replacing v by V and taking expectations of both sides, we obtain the desired result.

If we do not assume the existence of conditional measures, the proof is more complicated. Since $E(U)$ is finite, there exists a function $E(U \mid v)$ such that for any set S, $E(U \mid S) = E\{E(U \mid V) \mid S\}$; see [10], p. 47. Since ψ is convex it is measurable, and since ψ is bounded from below $E\{\psi(U)\}$ exists. Excluding the trivial case $E\{\psi(U)\} = \infty$, we know there exists a function $E\{\psi(U) \mid v\}$ such that for any set S, $E\{\psi(U) \mid S\} = E\{E\{\psi(U) \mid V\} \mid S\}$.

If the lemma were false, we should have $E\{E\{\psi(U) \mid V\}\} < E\{\psi\{E(U \mid V)\}\}$, and could find an $\epsilon > 0$ and a set A of positive V measure such that for every $v \in A$, $E\{\psi(U) \mid v\} + 2\epsilon < \psi\{E(U \mid v)\}$. This implies the existence of a number d and a set B of positive V measure such that for every $v \in B$, $E\{\psi(U) \mid v\} \leqq d$ and $d + \epsilon \leqq \psi\{E(U \mid v)\}$. Since ψ is convex, the domain D of points P for which $\psi(P) < d + \epsilon$ is convex, and we may find a subset C of B, of positive V measure, for which the set of points $E(U \mid v)$, $v \in C$, lies in a convex domain E disjoint of D. It follows that $E(U \mid C)$ lies in E, and hence that $\psi\{E(U \mid C)\} \geqq d + \epsilon$. Clearly $d \geqq E\{\psi(U) \mid C\}$. Thus we have the contradiction $E\{\psi(U) \mid C\} \geq \psi\{E(U \mid C)\}$.

DEFINITION. A loss function W will be called convex if for every $u \in \mathcal{Y}$, $W(u, v)$ is a convex function of the estimate v.

An example of a convex loss function is provided by the Markoff principle of estimation. The variance of an unbiased estimate may be considered as a risk function if we take the loss function to be the squared error, i.e. the square of the difference between the true value $g(P)$ and the estimated value $f(x)$ or T_x; and this loss function is clearly convex.

THEOREM 3.2. *If the loss function W is convex, if $\mathcal{Y}$ is in a Euclidean space, and if we consider only estimates having finite expectation, then the class of nonrandomized estimates is essentially complete.*

PROOF. Let T_X be any randomized estimate such that $E(T_X)$ exists and is finite. Applying lemma 3.1 we see that $E(T_X \mid X)$, which as a function of X only is a nonrandomized estimate, has a risk never greater than that of T_X.

The restriction in theorem 3.2 to estimates having finite expectation may be replaced by the requirement that for each $u \epsilon \mathcal{Y}$ there exist a number M_u such that if $|v - u| = M_u$ then $W(u, v) > W(u, u)$. With this requirement and the convexity assumption, it follows that the risk associated with T_X is infinite whenever $E(T_X)$ is infinite.

Theorem 3.2 is related to a generalization of a theorem of Blackwell. If Y is a sufficient statistic for $g(P)$, and if for almost all y the conditional distribution of X given y exists in the sense of measure, we may regard estimation of $g(P)$ based on X as randomized estimation of $g(P)$ based on Y; and if the assumptions of theorem 3.1 are satisfied, we may apply this theorem to conclude the essential completeness of the class of nonrandomized estimates based on Y. In the general case we may resort again to lemma 3.1 to prove the following theorem; the proof is the same as that of theorem 3.2 if X is replaced by Y throughout.

THEOREM 3.3. *If the loss function W is convex, if $\mathcal{Y}$ is in a Euclidean space, if we consider only estimates having a finite expectation, and if Y is a sufficient statistic for $\mathcal{F}$, then the class of nonrandomized estimates which are functions of Y only is essentially complete.*

Blackwell [11] proved that if U is a sufficient statistic for a real-valued parameter θ, and if T is an unbiased estimate for θ, then $E(T \mid U)$, which is a function of U only and also an unbiased estimate for θ, has a variance which never exceeds that of T. Observing that the theorems above hold true when we restrict attention to unbiased estimates, Blackwell's result may be obtained from theorem 3.3 by letting $\mathcal{Y}$ be one-dimensional, letting W be the squared error, and restricting ourselves to unbiased estimates. In a similar manner we can get from theorem 3.3 an extension of Blackwell's theorem given by Barankin [12], who treated the case in which $W(\theta, t) = |\theta - t|^s$, $s \geq 1$. It is clear that these loss functions are convex.

If the convexity assumption is removed, theorems 3.2 and 3.3 cease to be true. For example, if $\mathcal{X}$ has only n points, if $\mathcal{Y}$ is a finite line segment of length greater than $2n\alpha$, and if the loss is 0 whenever $|g(P) - f(x)| \leq \alpha$, and 1 otherwise, then the minimax risk among nonrandomized estimates is 1. By admitting randomization, however, the maximum risk can be brought below 1 without using X at all; if our estimate T is uniformly distributed over $\mathcal{Y}$, then the maximum risk will be $1 - \alpha/(\text{length of } \mathcal{Y})$.

The example just given may seem inappropriate, in that with the specified loss function the problem would customarily be considered one of interval estimation rather than point estimation. This objection does not apply however to the loss functions considered in the following theorem.

THEOREM 3.4. *Let* $\mathfrak{X} = \{0, 1, \cdots, n\}$, $n \geq 1$. *Let* $\mathfrak{F}$ *be the set of binomial distributions* P_p *defined by* $P_p(X = x) = \binom{n}{x} p^x(1-p)^{n-x}$, $0 \leq p \leq 1$, *Let* $\mathfrak{Y}$ *be the closed interval* $[0, 1]$ *and* $g(P_p) = p$. *Let* $W(p, t) = |p - t|^s$, $0 < s < 1$. *Then no minimax estimate can be nonrandomized, and the class of nonrandomized estimates is not essentially complete.*

PROOF. For any nonrandomized estimate f, $R_f(p)$, being a sum of products of continuous functions of p, is itself a continuous function of p. The nonrandomized minimax risk is less than 1, as may be shown by considering any estimate of the following kind: $f(0) = 0$, $f(n) = 1$, and $0 \leq f(x) \leq 1$ for all x. Here $R_f(0) = R_f(1) = 0$, while if $0 < p < 1$, $R_f(p) \leq \max_x |p - f(x)|^s < 1$. By continuity $\sup_{0 \leq p \leq 1} R_f(p) < 1$.

It is easy to see that there exists among the nonrandomized estimates a minimax estimate, say h. Let the corresponding minimax risk be denoted by M. We know that $M = \sup_{0 \leq p \leq 1} R_h(p) < 1$; it is obvious that $M > 0$. Observe that $h(0) < 1$, since $h(0) \geq 1$ leads to the contradiction $R_h(0) = |h(0)|^s \geq 1$. We can write

$$R_h(p) = \sum_{h(x)=h(0)} P_p(X = x) \cdot |p - h(x)|^s + \sum_{h(x)\neq h(0)} P_p(X = x) \cdot |p - h(x)|^s.$$

The second sum has a finite derivative with respect to p at $p = h(0)$, while the first sum increases with infinite speed as p is moved away from $h(0)$. Therefore $R_h\{h(0)\} < M$; and by an exactly symmetrical argument, $0 < h(n)$ and $R_h\{h(n)\} < M$. Using the continuity of R_h, we can find a positive number ω so small that $R_h(p) < M$ whenever $|p - h(0)| < \omega$ or $|p - h(n)| < \omega$.

Consider now the randomized estimate T_X defined by $T_x = h(x)$ if $0 < x < n$, and by $T_x = h(x) + \alpha Y$ otherwise, where Y is a random variable independent of X and taking on the values 1 and -1 each with probability $\frac{1}{2}$, and where $0 < \alpha < \omega$. Observe

$$R_{T_X}(p) - R_h(p) = (1-p)^n[\tfrac{1}{2}\{|p - h(0) + \alpha|^s + |p - h(0) - \alpha|^s\} - |p - h(0)|^s] + p^n[\tfrac{1}{2}\{|p - h(n) + \alpha|^s + |p - h(n) - \alpha|^s\} - |p - h(n)|^s].$$

By the concavity of the functions involved, the first square bracketted term is negative whenever $|p - h(0)| \geq \alpha$, and the second is negative whenever $|p - h(n)| \geq \alpha$. We can choose α so small that whenever either $|p - h(0)|$ or $|p - h(n)|$ is less than α, $R_{T_X}(p) - R_h(p) < \omega$. A continuity argument now shows that $\sup_{0 \leq p \leq 1} R_{T_X}(p) < M$. But this proves that no minimax estimate, with randomization permitted, can be nonrandomized. It is also now obvious that the class of nonrandomized estimates is not essentially complete: every nonrandomized estimate must have a risk function which somewhere exceeds $\sup_{0 \leq p \leq 1} R_{T_X}(p)$.

4. General properties of minimax estimation. Whether a principle such as the minimax principle is a desirable one has to be decided mainly on two criteria:

(i) its general properties, and

(ii) its performance in many particular instances.

It has already been remarked that in the second respect the minimax principle does not seem entirely satisfactory. With regard to the former, one great advantage of this principle is that when there is a unique minimax estimate, it is admissible. Here an estimate f is said to be admissible (see [3]) if there exists no other estimate f^* such that $R_{f^*}(P) \leq R_f(P)$ for all P in $\mathcal{F}$ with strict inequality holding for some P. It is interesting that, as we shall show below, this admissibility property is not shared by either the principle of unbiasedness or the maximum likelihood principle.

In this connection we begin by proving another theorem concerning essentially complete classes.

THEOREM 4.1. *Suppose that the space $\mathcal{Y}$ is a finite interval $[a, b]$ on the real line, and that for each $u \in \mathcal{Y}$, $W(u, v)$ is a non-decreasing function of v when $v > u$ and a non-increasing function of v when $v < u$. Then the class of estimates whose range is contained in $\mathcal{Y}$ is essentially complete with respect to the class of all real valued estimates.*

PROOF. If T is any real-valued estimate, define T^* by

$$T^* = \begin{cases} T & \text{if} \quad T \in \mathcal{Y}, \\ a & \text{if} \quad T < a, \\ b & \text{if} \quad t > b. \end{cases} \tag{4.1}$$

It is clear that $R_{T^*}(P) \leq R_T(P)$ for every $P \in \mathcal{F}$.

Halmos [7] has provided an example in which the minimum variance unbiased estimate takes on, with positive probability, values outside the range of the parameter. It can be shown from the proof of theorem 4.1 that in this case any unbiased estimate is inadmissible, provided the loss function is of the kind described in theorem 4.1.

That the maximum likelihood principle may also lead to inadmissible estimates is easy to show, since this is the case in many familiar situations. The following example may be of interest in that here the maximum likelihood estimate is uniformly worst among all estimates which one would consider using.

Example. Let X be a random variable with only 0 and 1 as possible values, and let $P(X = 1) = p$. Assume it to be known that $\frac{1}{3} \leq p \leq \frac{2}{3}$. Then the maximum likelihood estimate for p is easily seen to be $\frac{1}{3}(X + 1)$, and, if the loss function is the squared error, the associated risk function is $\frac{1}{3}(p - \frac{1}{2})^2 + \frac{1}{36}$. This risk function is, for every possible value of p, greater than that of any estimate $f(x)$ satisfying: $\frac{1}{3} \leq f(0) \leq f(1) = 1 - f(0) \leq \frac{2}{3}$.

The selection of loss function in any problem should in theory be governed by metastatistical considerations, but in fact the circumstances of statistical problems do not usually offer compelling reasons for using one loss function rather than another. Considerations of mathematical facility are often determining. Thus, various classical unbiased estimates become minimax estimates when the loss function is judiciously chosen. For, if we take as loss function the ratio of squared error to the variance of the unbiased estimate, the risk becomes constant, and we can easily obtain the classical estimates as minimax estimates in the familiar binomial, Poisson, and rectangular problems, and in some of the nonparametric problems considered in section 6.

However, this approach seems to be somewhat artificial, and hereafter we shall restrict ourselves to a single loss function, namely the squared error. There are two reasons for this choice. With squared error for the loss, the mathematical problems are rather simple. And as was remarked above, squared error (if one restricts oneself to unbiased estimates) is the traditional loss function. Fortunately, the squared error loss function is convex, and hence theorem 3.2 permits us to avoid considering randomized estimates.

When the loss function is squared error, we have the following obvious linearity property, which for later reference we state as

THEOREM 4.2. *If $f(X)$ is the minimax estimate for $g(P)$, then $af(X) + b$ is the minimax estimate for $a \cdot g(P) + b$.*

However, as we shall show by an example in the next section, it need not be true that if $X_1, \cdots, X_n$ are independent and $f_i(X_i)$ is the minimax estimate for $g_i(P_i)$, $i = 1, \cdots, n$, then $\sum_{i=1}^n a_i f_i(X_i)$ is the minimax estimate for $\sum_{i=1}^n a_i g_i(P_i)$. This is a definite disadvantage of the minimax principle as compared with the Markoff principle which does possess the linearity property mentioned.

We conclude this section with an explicit solution of the Bayes problem in the squared error case. If the distribution P is itself a random variable distributed over $\mathcal{F}$ according to some distribution λ, we may compare estimates f by means of their expected loss $Q(f) = E[g(P) - f(X)]^2$. Since $Q(f) = E\{E[g(P) - f(X)]^2 \mid X\}$, it is well known that $Q(f)$ is minimized by using the estimate $f(x) = E[g(P) \mid x]$, provided the conditional measures exist. In fact, this result holds even without this assumption.

THEOREM 4.3. *$E[g(P) - f(X)]^2$ is minimized by $f(x) = E[g(P) \mid x]$.*

PROOF.
$$E[g(P) - f(X)]^2 - E\{g(P) - E[g(P) \mid X]\}^2 = E\{E[g(P) \mid X] - f(X)\}^2$$
$$+ 2E[E\{g(P) - E[g(P) \mid X]\}\{E[g(P) \mid X] - f(X)\} \mid X] \geq 0.$$

In applications it is convenient to write $E[g(P) \mid X]$ more explicitly. Suppose that with respect to some measure μ over $\mathfrak{X}$, each distribution $P \in \mathcal{F}$ has a generalized probability density p_P, so that for any A, the probability that $X \in A$ computed for P, is given by

$$\int_A p_P(x)\, d\mu(x)\,.$$

Minimizing a quadratic expression shows that

$$\frac{\int_{\mathfrak{F}} g(P)p_P(x)\, d\lambda(P)}{\int_{\mathfrak{F}} p_P(x)\, d\lambda(P)} \tag{4.2}$$

is a Bayes solution.

5. Binomial and hypergeometric distributions. In the present section we shall consider three discrete minimax problems.

PROBLEM 1. (Binomial.) Let X be a binomial random variable witd parameter p, $0 \leq p \leq 1$, so that $P(X = k) = \binom{n}{k}p^k(1 - p)^{n-k}$. We shall show that the minimax estimate for p is

$$\frac{X}{n} \cdot \frac{\sqrt{n}}{(\sqrt{n} + 1)} + \frac{1}{2(\sqrt{n} + 1)}. \tag{5.1}$$

Consider any linear estimate $\alpha X + \beta$. The risk $E_p(\alpha X + \beta - p)^2$ is a quadratic function of p which is constantly equal to β^2 when $\alpha = \frac{1}{\sqrt{n}\,(1 + \sqrt{n})}$ and $\beta = \frac{1}{2(1 + \sqrt{n})}$. Hence (5.1) is a constant risk estimate of p. Since it is easily seen that

$$\frac{\int_0^1 p \cdot p^k q^{n-k} \cdot p^{a-1} q^{b-1}\, dp}{\int_0^1 p^k q^{n-k} \cdot p^{a-1} q^{b-1}\, dp} = \frac{a + k}{a + b + n}, \qquad (q = 1 - p),$$

it follows that (5.1) is the Bayes estimate when p is distributed with probability density $C(pq)^{(\sqrt{n}/2)-1}$, and hence by Theorem 2.1 we conclude that (5.1) is the minimax estimate of p.

After obtaining this result we were informed that it had been obtained earlier by H. Rubin, to whom, therefore, the priority belongs.

It is interesting to compare the risk of the above estimate with that of the standard unbiased estimate X/n. We have

$$E\left(\frac{X}{n} - p\right)^2 = \frac{pq}{n},$$

$$E\left[\frac{1}{1 + \sqrt{n}}\left(\frac{X}{\sqrt{n}} + \frac{1}{2}\right) - p\right]^2 = \frac{1}{4(1 + \sqrt{n})^2}.$$

As is easily seen, $\frac{pq}{n} \leq \frac{1}{4(1 + \sqrt{n})^2}$ if and only if

$$\left|p - \frac{1}{2}\right| \geq \frac{\sqrt{1 + 2\sqrt{n}}}{2(1 + \sqrt{n})}.$$

Thus the standard estimate is better than the minimax estimate outside an interval around $p = \frac{1}{2}$ whose length decreases with increasing n, tending to 0 as n tends to infinity. However, for very small values of n the minimax estimate has the smaller risk over nearly the whole range.

PROBLEM 2. (Difference of binomials.) Let X and Y be independent binomial random variables, where $P(X = k) = \binom{n}{k} p_1^k (1 - p_1)^{n-k}$ and $P(Y = l) = \binom{n}{l} p_2^l(1 - p_2)^{n-l}$. By use of theorem 2.1 we shall show that the minimax estimate for $p_1 - p_2$ is $\frac{\sqrt{2n}}{1 + \sqrt{2n}}\left(\frac{X}{n} - \frac{Y}{n}\right)$. For the set ω of theorem 2.1 we take $p_1 = p$, $p_2 = 1 - p$, $0 \leq p \leq 1$, and we let $Z = X + n - Y$. Applying the result of Problem 1 to Z, we find the minimax estimate of p to be $\alpha_{2n} \cdot Z + \beta_{2n}$, and by Theorem 4.2 the minimax estimate based on Z for $p_1 - p_2 = 2p - 1$, is $\frac{\sqrt{2n}}{1 + \sqrt{2n}}\left(\frac{X}{n} - \frac{Y}{n}\right)$, and the risk of this estimate is constant over ω.

To prove that this is also the minimax estimate of $p_1 - p_2$ for the original problem, we consider the risk as a function of p_1 and p_2. It is easy to show that $(1 + \sqrt{2n})^2 R(p_1, p_2) = 2\cdot[p_1(1 - p_1) + p_2(1 - p_2)] + (p_1 - p_2)^2$. Finally it can be shown that $p_1(1 - p_1) + p_2(1 - p_2)$ is maximized, subject to the condition that $p_1 - p_2$ be constant, when $p_1 + p_2 = 1$.

PROBLEM 3. (Hypergeometric.) We finally consider the problem of estimating the number of defectives in a lot from a sample drawn from this lot at random. We denote by N and n the number of elements in lot and sample respectively, and by D and X the corresponding number of defectives. For later reference we note

$$P(X = k) = \frac{\binom{D}{k}\binom{n - D}{n - k}}{\binom{N}{n}},$$

$$E(X) = n\frac{D}{N},$$

$$\sigma_x^2 = \frac{nD(N - n)(N - D)}{N^2(N - 1)}.$$

As in Problem 1 we easily find a linear function of X whose risk is constant. In fact

$$E_D(\alpha X + \beta - D)^2 \equiv \beta^2$$

when

$$\alpha = \frac{N}{n + \sqrt{\frac{n(N - n)}{N - 1}}}, \beta = \frac{N}{2}\left(1 - \frac{\alpha n}{N}\right).$$

To prove that $\alpha X + \beta$ is the minimax estimate of D we shall show that it is the Bayes estimate corresponding to

$$P(D = d) = \int_0^1 \binom{N}{d} p^d q^{N-d} \cdot C\, p^{a-1} q^{b-1}\, dp, \tag{5.2}$$

where $a, b > 0$, and

$$C = \frac{\Gamma(a+b)}{\Gamma(a)\,\Gamma(b)}.$$

In this connection it is useful to notice that since (5.2) is a distribution

$$\sum_{d=0}^{N} \binom{N}{d} \frac{\Gamma(a+d)\,\Gamma(N+b-d)}{\Gamma(N+a+b)} \cdot \frac{\Gamma(a+b)}{\Gamma(a)\,\Gamma(b)} = 1. \tag{5.3}$$

Using theorem 4.3, we find the Bayes estimate associated with (5.2) to be

$$f(k) = \frac{\sum_{d=k}^{N-n+k} d \binom{d}{k}\binom{N-d}{n-k}\binom{N}{d}\Gamma(a+d)\,\Gamma(N+b-d)}{\sum_{d=k}^{N-n+k} \binom{d}{k}\binom{N-d}{n-k}\binom{N}{d}\Gamma(a+d)\,\Gamma(N+b-d)}.$$

Replacing d by $(d-a)+a$, and using the relation

$$\binom{d}{k}\binom{N-d}{n-k}\binom{N}{d} = \binom{N-n}{d-k} \cdot \text{(terms not involving } d\text{)},$$

we find:

$$f(k) = \frac{\sum_{i=0}^{N-n} \binom{N-n}{i} \Gamma(d+a+1)\,\Gamma(N+b-d)}{\sum_{i=0}^{N-n} \binom{N-n}{i} \Gamma(d+a)\,\Gamma(N+b-d)} - a.$$

Now apply (5.3) to numerator and denominator separately; then

$$f(k) = k\frac{a+b+N}{a+b+n} + \frac{a(N-n)}{a+b+n}.$$

Putting $\dfrac{a+b+N}{a+b+n} = \alpha$, $\dfrac{a(N-n)}{a+b+n} = \beta$ one obtains easily

$$a = \frac{\beta}{\alpha - 1}, \quad b = \frac{N - \alpha n - \beta}{\alpha - 1}.$$

Substituting the values of α and β one finds that $\beta > 0$, $N > \alpha n + \beta$ and that $\alpha > 1$ provided $N > n + 1$. In the special case $N = n$ the result is immediate, while if $N = n + 1$, the result is obtained by giving to D a binomial distribution with $p = \frac{1}{2}$.

6. Non parametric problems. We shall in this section consider estimation problems in which the functional form of the distribution of X is not assumed known. Restrictions will be imposed on the variables only to insure the existence of estimates with bounded risk. The problem will be treated under two different such restrictions: (i) that the variables are bounded with known bounds, (ii) that the variables have bounded variances.

In the first of these cases we can assume without loss of generality that the variables are distributed over the interval [0, 1], and then obtain

THEOREM 6.1. *Let $X_1, \cdots, X_n$ be independently distributed over* [0, 1] *according to a joint distribution belonging to a family $\mathcal{F}$. Suppose that $\mathcal{F}$ contains the subfamily $\mathcal{F}_0$ according to which $X_1, \cdots, X_n$ are independently and identically distributed with $P(X_i = 1) = p$, $P(X_i = 0) = 1 - p$, $0 \leq p \leq 1$. Let $E(X_i) = \mu_i$, $\frac{1}{n}\sum_{i=1}^{n} \mu_i = \bar{\mu}$. Then the minimax estimate of $\bar{\mu}$ is*

$$\frac{1}{1+\sqrt{n}}(\sqrt{n}\,\bar{X} + \tfrac{1}{2}). \tag{6.1}$$

PROOF. Since (6.1) is the minimax estimate of $\bar{\mu} = p$ when the distribution of the X's is known to belong to $\mathcal{F}_0$, we only need to show that its risk is largest for the distributions of $\mathcal{F}_0$. But

$$E(A\bar{X} + B - \bar{\mu})^2 = A^2\sigma_{\bar{X}}^2 + [B + (A-1)\bar{\mu}]^2 = \frac{A^2}{n^2}\sum_{i=1}^{n}\sigma_{x_i}^2 + [B + (A-1)\bar{\mu}]^2$$

and

$$\Sigma\sigma_{x_i}^2 = \Sigma E(X_i^2) - \Sigma\mu_i^2 \leqq \Sigma\mu_i - \Sigma\mu_i^2 = n\bar{\mu} - \Sigma(\mu_i - \bar{\mu})^2 - n\bar{\mu}^2 \leqq n\bar{\mu}(1-\bar{\mu})$$

where equality holds for the distributions in $\mathcal{F}_0$.

COROLLARY 6.2. *Let $X_1, \cdots, X_n$ be a sample from an unknown univariate distribution over* [0, 1]. *Then the minimax estimate of $E(X_i) = \mu$ is given by* (6.1).

COROLLARY 6.3. *Let $X_1, \cdots, X_n$ be a sample from an unknown absolutely continuous univariate distribution over* [0, 1]. *Then the minimax estimate of $E(X_i) = \mu$ is given by* (6.1).

Corollary 6.3 follows from the fact that any risk function that can be obtained for binomial distribution can be approximated by means of absolutely continuous distributions.

Theorem 6.1 can be extended to include variables that are negatively correlated. Namely if $X_1, \cdots, X_n$ are distributed over [0, 1] according to a joint distribution belonging to some family $\mathcal{F}$, if for each distribution of $\mathcal{F}$ the correlation coefficient ρ_{ij} of X_i, X_j is $\leqq 0$ for all i, j, and if $\mathcal{F}$ contains the family $\mathcal{F}_0$ of theorem 6.1, then the conclusion of this theorem remains valid. This result can be used for example in the following situation. Suppose a sample of n is taken from a lot of unknown size, and suppose it is desired to estimate the proportion p of defectives in the lot. If k is the number of defectives in the sample, it follows from the above remarks that the minimax estimate of p is $\frac{1}{1+\sqrt{n}}\left(\frac{k}{\sqrt{n}} + \frac{1}{2}\right)$.

It should be pointed out that this result holds only if no upper bound is assumed known for the lot size. If it is known that the number of items in the lot is $\leq N_0$, then the minimax estimate is that found in section 5 for the case of a hypergeometric distribution with $N = N_0$.

Next let us consider estimating the difference of the average means in two groups of variables.

THEOREM 6.4. *Let* $X_1, \cdots, X_n; Y_1, \cdots, Y_n$ *be independently distributed over the interval* $[0, 1]$ *according to a joint distribution belonging to a family* $\mathfrak{F}$. *Suppose that* $\mathfrak{F}$ *contains the subfamily* $\mathfrak{F}_1$, *according to which* $X_1, \cdots, X_n; Y_1, \cdots, Y_n$ *are two samples with* $P(X_i = 1) = p_1$, $P(X_i = 0) = 1 - p_1$; $P(Y_i = 1) = p_2$, $P(Y_i = 0) = 1 - p_2$, $0 \leq p_1, p_2 \leq 1$. *If* $E(X_i) = \mu_i$, $E(Y_i) = \nu_i$, $\frac{1}{n}\mu_i = \bar{\mu}$, $\frac{1}{n}\nu_i = \bar{\nu}$, *then the minimax estimate of* $\bar{\mu} - \bar{\nu}$ *is*

$$\frac{\sqrt{2n}}{1 + \sqrt{2n}}(\bar{X} - \bar{Y}). \tag{6.2}$$

PROOF. Again, since (6.2) is the minimax estimate in the binomial case (Problem 2 of section 5), we need only verify that its risk is a maximum in $\mathfrak{F}_1$. But

$$\begin{aligned} &E[A(\bar{X} - \bar{Y}) - (\bar{\mu} - \bar{\nu})]^2 \\ &= E[A(\bar{X} - \bar{\mu}) - A(\bar{Y} - \bar{\nu}) + (A - 1)(\bar{\mu} - \bar{\nu})]^2 \\ &= A^2(\sigma_{\bar{X}}^2 + \sigma_{\bar{Y}}^2) + (A - 1)^2(\bar{\mu} - \bar{\nu})^2, \end{aligned}$$

of which we already have shown that it is maximized in the binomial case.

Up to now we assumed the variables to be bounded. Let us now suppose instead that the variances are bounded. With this assumption we can give an analogue of the classical Markoff theorem on least squares.

THEOREM 6.5. *Suppose that* $X_1, \cdots, X_n$ *are independently distributed according to a joint distribution belonging to some family* $\mathfrak{F}$, *which contains the subfamily* f_0 *where the* X*'s are normal with variance* M^2. *Suppose that for all distributions in* $\mathfrak{F}$, $E(X_i) = \sum_{j=1}^{s} a_{ij}\theta_j$ *and* $\sigma_{X_i}^2 \leqq M^2$. *We assume the matrix* (a_{ij}) *to be known and of rank* $s \leqq n$. *Then the estimate* $[f_1(X), \cdots, f_s(X)]$ *of* $(\theta_1, \cdots, \theta_s)$ *which minimizes* $\sup E \sum_i [f_i(X) - \theta_i]^2$, *is the Markoff estimate.*

PROOF. Consider first the subfamily $\mathfrak{F}_0$. Then there exists an orthogonal transformation to $Y_1, \cdots, Y_n$ such that $E(Y_i) = k_i\theta_i$ for $i = 1, \cdots, s$, where $k_i > 0$; $E(Y_i) = 0$ for $i = s + 1, \cdots, n$; and $\sigma_{Y_i}^2 \leqq M^2$ for $i = 1, \cdots, n$. Then $(Y_1, \cdots, Y_s)$ is a sufficient statistic for $(\theta_1, \cdots, \theta_s)$, and it is easily shown, using the methods of [6], that $\left(\frac{Y_1}{k_1}, \cdots, \frac{Y_s}{k_s}\right)$ is the minimax estimate for $(\theta_1, \cdots, \theta_s)$. But this is the Markoff estimate. In order to complete the proof we must show that the risk of this estimate takes on in $\mathfrak{F}_0$ its supremum over $\mathfrak{F}$. But this is immediate; for $E\sum_{i=1}^{s}[f_i(X) - \theta_i]^2 = E\sum_{i=1}^{s}\left(\frac{Y_i}{k_i} - \theta_i\right)^2 \leqq M^2\sum_{i=1}^{s}\frac{1}{k_i^2}$.

In a similar manner it is easily shown that the least squares estimate for a linear function of one or more of the θ's, is the minimax estimate.

Theorem 6.5 gives a justification of the least squares estimate different from that of the Markoff theorem. In the Markoff theorem, it is shown that the least squares estimate has *uniformly* smallest risk among all linear unbiased estimates; here it is shown that the least squares estimate minimizes the maximum risk among all estimates. (The assumptions concerning variances also differ.)

7. Prediction problems. Frequently one is interested in estimating the value of a random variable rather than that of a parameter. A customary method for this is to estimate the expectation of the random variable (a parameter) and then to "identify" the variable and its expectation; i.e., to use the estimate of the expectation as a prediction for the variable. As we shall see below one is led to this procedure if one adopts the point of view of unbiased estimation, so that from this point of view prediction poses no new problem. This however is no longer true when one employs the minimax principle.

Consider a pair X, Y of random variables having a joint distribution P belonging to a family $\mathcal{F}$ of distributions. It is desired to use the observed X to predict, say, $g(Y)$. We are interested in minimax predictions; i.e., functions $f(X)$ which minimize $\sup_{P \in \mathcal{F}} E_P W[g(Y), f(X)]$. To obtain minimax predictions we need the following analogue of Theorem 2.1.

THEOREM 7.1. *Let* $\{P_\theta\}$, $\theta \in \omega$ *be a parametric subfamily of* $\mathcal{F}$, *and let* λ *be a probability measure over* ω. *Suppose that* f *is such that* $\int E_\theta W[g(Y), f(X)]\, d\lambda(\theta)$ *is minimum, and that*

(i) $E_\theta W[g(Y), f(X)]$ *is constant, say* $= c$, *for all* $\theta \in \omega$,

(ii) $E_P W[g(Y), f(X)] \leqq c$ *for all* $P \in \mathcal{F}$.

Then f *is a minimax prediction for* $g(Y)$.

The proof is the exact analogue of that of theorem 2.1.

COROLLARY 7.2. *A constant risk Bayes prediction is a minimax prediction.*

Suppose now that X and Y are independent and that $W[g(y), f(x)] = [g(y) - f(x)]^2$. Consider the problem first from the point of view of unbiasedness. A prediction could reasonably be called unbiased if $E_P f(X) = E_P g(Y)$. Subject to unbiasedness, the risk is given by $E_P[g(Y) - f(X)]^2 = \sigma_P^2 f(X) + \sigma_P^2\, g(Y)$. But $\sigma_P^2 g(Y)$ is a known function of P, and hence the problem of minimizing (for a particular P) the expected squared error reduces to that of finding an unbiased estimate of $E_P g(Y)$ with minimum variance at P. In a similar way one sees, without any restriction to unbiased predictions, that the Bayes prediction for $g(Y)$ is the same as the Bayes estimate for $E_P g(Y)$, and hence that formula (4.2), with $g(P)$ replaced by $E_P g(Y)$, may be used if the assumptions there made are valid.

One might expect that as in the unbiased theory the prediction will coincide with the estimate. This however is not the case since the λ's that give constant risk in the two cases will usually be distinct. In fact the two problems are rather different in that the "least favorable" λ for the prediction problem must not only take into account the difficulty of finding the correct value of θ for various a priori distributions but also the difficulty of predicting $g(Y)$ when θ is known.

As a first example consider the prediction analogue of problem 1 of section 5. Let X, Y be independent binomial variables such that $P(X = k) = \binom{m}{k} p^k(1-p)^{m-k}$ and $P(Y = l) = \binom{n}{l} p^l(1-p)^{m-l}$. We shall obtain the minimax prediction of Y in a manner quite analogous to the one in which we determined the minimax estimate of p. Actually, the present problem is a generalization of the earlier one, to which it can be reduced by letting $n \to \infty$. First it is easily seen that

$$E\left(\alpha \frac{X}{m} + \beta - \frac{Y}{n}\right)^2$$

is a quadratic function of p, which when $m > 1$ is constant for

$$\alpha = \frac{m}{m-1}\left[1 - \sqrt{\frac{1}{m} + \frac{1}{n} - \frac{1}{mn}}\right],$$

$$\beta = \frac{1-\alpha}{2}.$$

But we have already seen that $\alpha \frac{X}{m} + \beta$ is the Bayes solution corresponding to $Cp^{a-1}q^{b-1}$ where $\alpha = \frac{m}{m+a+b}$, $\beta = \frac{a}{m+a+b}$. Clearly $\beta = \frac{1-\alpha}{2}$ when $a = b$, and $a > 0$ provided $0 < \alpha < 1$, which is easily verified when $m, n > 1$. We note that as $n \to \infty$, the values of α, β tend to those of the minimax estimate of P.

When $m = 1$, $E\left(\alpha \frac{X}{m} + \beta - \frac{Y}{n}\right)^2$ is constant for $\alpha = \frac{n-1}{2n}$, $\beta = \frac{1-\alpha}{2}$, and again $\alpha \frac{X}{m} + \beta$ is the Bayes estimate of a beta distribution when $n > 1$, and hence minimax.

Finally in the case $n = 1$, the situation degenerates. Since $E(\frac{1}{2} - Y)^2 = \frac{1}{4}$, the prediction $f(X) = \frac{1}{2}$ has constant risk. In addition it is the Bayes prediction corresponding to the distribution which assigns probability 1 to $p = \frac{1}{2}$. Hence in this case, regardless of the value of X one would predict for Y the value $\frac{1}{2}$.

It is interesting that the above prediction problem can be interpreted also as an estimation problem in the following manner. Suppose a lot of size $N = m + n$ is such that the number of defectives follow a binomial distribution; this is the case when the items making up the lot are produced by a manufacturing process that is in statistical control. It is desired to estimate from a sample of size m taken from this lot, the proportion of defectives in the remainder. That this is equivalent to the prediction problem treated above follows from a remark of Mood [13] that in such a lot the number of defectives in the sample and in the remainder are independently distributed according binomial distributions with common p.

We can again use the binomial results to obtain the solutions of certain non-parametric problems. For example, let $X_1, \cdots, X_m$ be independently and identically distributed on [0, 1] and let $Y_1, \cdots, Y_n$ be another sample from the same distribution. Then the minimax prediction for $\bar{Y}$ is given by $\alpha\bar{X} + \beta$ with $\alpha = \frac{m}{m-1}\left[1 - \sqrt{\frac{1}{m} + \frac{1}{n} - \frac{1}{mn}}\right]$, $\beta = \frac{1-\alpha}{2}$. This follows from the fact that

$$\begin{aligned} E(\alpha\bar{X} + \beta - \bar{Y})^2 &= E[\alpha(\bar{X} - \bar{\mu}) - (\bar{Y} - \mu) + (\beta + (\alpha - 1)\mu)]^2 \\ &= \alpha^2\left(\frac{1}{m} + \frac{1}{n}\right)\sigma^2 + [\beta + (\alpha - 1)\mu]^2 \\ &\leq \alpha^2\left(\frac{1}{m} + \frac{1}{n}\right)\mu(1 - \mu) + [\beta + (\alpha - 1)\mu]^2. \end{aligned}$$

An analogous modification clearly is possible for theorem 6.4.

For the situation considered in 6.5, the prediction problem gives the same result as the estimation problem. For consider first two samples $X_1, \cdots, X_m$; $Y_1, \cdots, Y_n$ from a normal distribution with known variance σ^2. Here

$$E_\theta[f(X_1, \cdots, X_n) - \bar{Y}]^2 = E_\theta[f(X_1, \cdots, X_m) - \theta]^2 + \frac{\sigma^2}{n},$$

and hence the risk differs from that of the estimation problem only by a constant. Thus $\bar{X}$ is the minimax prediction of $\bar{Y}$, and it is then seen immediately that it is also the minimax prediction for $\bar{Y}$ when of the underlying common distribution of the X's and Y's it is assumed only that the variance is bounded.

REFERENCES

[1] G. W. Brown, "On small sample estimation," *Annals of Math. Stat.*, Vol. 18 (1949), p. 514.

[2] A. Wald, "Contributions to the theory of statistical estimation and testing hypotheses," *Annals of Math. Stat.*, Vol. 10 (1939), p. 299.

[3] A. Wald, *On the Principles of Statistical Inference*, Notre Dame Math. Lectures, No. 1 (1942).

[4] A. Wald, "Statistical decision functions which minimize the maximum risk," *Annals of Math.*, Vol. 46 (1945), p. 265.

[5] A. Wald, "Statistical decision functions," *Annals of Math. Stat.*, Vol. 20 (1949), p. 165.

[6] C. Stein and A. Wald, "Sequential confidence intervals for the mean of a normal distribution with known variance," *Annals of Math. Stat.*, Vol. 18 (1947), p. 427.

[7] P. R. Halmos, "The theory of unbiased estimation," *Annals of Math. Stat.*, Vol. 17 (1946), p. 34.

[8] E. L. Lehmann and C. Stein, "Most powerful tests of composite hypotheses. I. Normal distributions," *Annals of Math. Stat.*, Vol. 19 (1948), p. 495.

[9] A. Wald, "An essentially complete class of admissible decision functions," *Annals of Math. Stat.*, Vol. 18 (1947), p. 549.

[10] A. Kolmogoroff, *Grundbegriffe der Wahrscheinlichkeitsrechnung*, Berlin, 1933.

[11] D. Blackwell, "Conditional expectation and unbiased sequential estimation," *Annals of Math. Stat.*, Vol. 18 (1947), p. 105.

[12] E. W. Barankin, "Extension of a theorem of Blackwell," *Annals of Math. Stat.*, Vol. 21 (1950), p. 280.

[13] A. M. Mood, "On the dependence of sampling inspection plans upon population distributions," *Annals of Math. Stat.*, Vol. 14 (1943), p. 145.

SOME APPLICATIONS OF THE CRAMÉR-RAO INEQUALITY

J. L. HODGES, JR. AND E. L. LEHMANN
UNIVERSITY OF CALIFORNIA

1. Summary and introduction

In 1945 and 1946, Cramér [1] and Rao [2] independently investigated the problem of obtaining a simple lower bound to the variance of point estimates. In 1947 Wolfowitz [3] simplified the conditions under which Cramér had obtained this bound and extended the result to sequential estimates. In the present paper, use is made of the Cramér-Rao result, in Wolfowitz's form, to investigate some problems of the minimax theory of estimation.

The Bayes method for obtaining minimax estimates developed by Wald since 1939 [4], [5], is completely satisfactory whenever the minimax estimate is the Bayes solution for some *a priori* distribution of the parameter. However, frequently minimax estimates are not Bayes solutions, but only limits of Bayes solutions. When this occurs, the possibility is left open that the minimax estimate is not admissible; that is, that there exists some other minimax estimate whose risk is never greater and is for some parameter value less than that of the given estimate.

In section 2 we consider certain estimation problems in which the loss is proportional to the square of the error of estimate, and use the Cramér-Rao bound to establish directly that certain estimates, which can be shown to be minimax by the Bayes method, are in addition admissible. In section 3 we consider several problems of sequential estimation, for some of which previously no minimax estimates have been known. In all of these cases it turns out that there are minimax estimates based on samples of fixed size.

Problems similar to those treated in the present paper were considered simultaneously by Girshick and Savage [6], the scope of whose work is much larger than ours. Portions of both papers were presented at the joint colloquium of the Stanford and California statistical groups, resulting in a fruitful exchange of ideas. The method introduced here has been employed in [6] to obtain extensions of some of our results.

2. Estimates based on samples of fixed size

Let X be a random variable with distribution P_θ, $\theta \in D$, so that the probability that X falls in a set A is given by

$$P_\theta(A) = \int_A p_\theta(x)\, d\mu(x)\,. \tag{2.1}$$

Let $f(X)$ be any estimate of θ and let $b_f(\theta) = E_\theta[f(X)] - \theta$ be its bias. Then the Cramér-Rao inequality states that the variance $\sigma_f^2(\theta)$ of $f(X)$ satisfies

$$\sigma_f^2(\theta) \geqq \frac{[1 + b_f'(\theta)]^2}{E_\theta\left[\left(\frac{\partial}{\partial\theta}\log p_\theta(X)\right)^2\right]}. \tag{2.2}$$

We shall now prove a theorem which will essentially reduce the problem of proving that certain estimates are admissible and minimax, to proving that there is a unique solution to a differential inequality related to (2.2). It will be convenient to associate with each bias function $b(\theta)$ the function $C_b(\theta)$ defined by

$$C_b(\theta) = b^2(\theta) + \frac{[1 + b'(\theta)]^2}{E_\theta\left[\frac{\partial}{\partial\theta}\log p_\theta(X)\right]^2}. \tag{2.3}$$

If the loss is defined to be the square of the error of estimation, (2.3) has the significance of a lower bound on the risk of an estimate whose bias function is $b(\theta)$. Suppose now that $g(X)$ is an estimate for which the risk everywhere attains this lower bound. We may then substitute, in a proof of the admissibility of $g(X)$, the bound (2.3) for the actual risk.

THEOREM 1. *If the loss is squared error, if $g(X)$ is an estimate for which* (2.2) *becomes an equality, if the inequality* (2.2) *is satisfied for all estimates, and if, for every bias function $b(\theta)$,*

$$C_b(\theta) \leqq C_{b_g}(\theta) \tag{2.4}$$

for all $\theta \in D$ implies $b(\theta) \equiv b_g(\theta)$, then $g(X)$ is admissible.

PROOF. Since loss is squared error, $R_f(\theta) = b_f^2(\theta) + \sigma_f^2(\theta) \geqq C_{b_f}(\theta)$. Suppose for some estimate $f(X)$, $R_g(\theta) \geqq R_f(\theta)$ for all $\theta \in D$. Since by assumption $C_{b_g}(\theta) = R_g(\theta)$, we have $C_{b_g}(\theta) \geqq C_{b_f}(\theta)$ and from (2.4) conclude $b_g(\theta) = b_f(\theta)$. From this follows $b_g'(\theta) = b_f'(\theta)$, $C_{b_g}(\theta) = C_{b_f}(\theta)$, $R_f(\theta) \geqq C_{b_f}(\theta) = C_{b_g}(\theta) = R_g(\theta)$, and hence $R_f(\theta) = R_g(\theta)$.

COROLLARY 1. *Theorem 1 remains valid if the loss function is squared error divided by a function $q(\theta)$ which is everywhere positive and finite.*

PROOF. Admissibility is not affected when the risk function is divided by such a function.

COROLLARY 2. *If in addition to the assumptions of theorem 1 we assume that $g(X)$ is a constant risk estimate, then $g(X)$ is an admissible minimax estimate.*

PROOF. Any constant risk admissible estimate is minimax.

Remarks. (i) A statistical problem is not completely specified until the loss function has been stated. Squared error is the classical loss function for estimation, primarily for reasons of convenience [7, p. 516]. An alternative loss function is obtained if we divide squared error by the variance of X, thus measuring the seriousness of errors in terms of the difficulty of estimation as reflected by σ_X^2 as a function of θ. This alternative approach is particularly desirable in those problems for which, when the loss function is squared error itself, the minimax risk is infinite. For, when this happens, every estimate is minimax and the minimax principle provides no basis for choice.

Those loss functions obtained by dividing squared error by a function of θ have been termed "quadratic loss functions" by Girshick and Savage [6].

(ii) In all of the problems considered below, the family of distributions is complete in the sense of [8], and hence every estimate is uniquely determined by its bias function. Since in the proof of theorem 1 we have established the uniqueness of the admissible minimax bias function, it follows that the estimates shown below to be admissible minimax estimates are in fact the unique minimax estimates.

(iii) In the statistical applications which we shall make of this theorem, we sometimes replace a sample $X_1, X_2, \ldots, X_n$ by a single sufficient statistic, say X. It is known that nothing is lost by this simplification, since from the risk point of view one may restrict oneself to estimates which are nonrandomized functions of a sufficient statistic [9].

Actually, it is not necessary to work with the single sufficient statistic, since the Cramér-Rao inequality may be applied directly to a sample, but the regularity conditions are easier to check when dealing with a single variable.

As an application of theorem 1 we shall now consider five specific problems, showing in each case that a given estimate is admissible and minimax. To apply theorem 1 we must check the validity of (2.2) for all estimates. By the method of proof given by Wolfowitz [3], (2.2) can be shown to be valid under the following assumptions:

(i) The parameter θ lies in an open interval D of the real line, which may be infinite or semi-infinite;

(ii) For almost all x, $\dfrac{\partial p_\theta(x)}{\partial \theta}$ exists for all $\theta \in D$;

(iii) The expression $\int p_\theta(x)\, d\mu(x)$ may be differentiated under the integral sign;

(iv) For every $\theta \in D$, $E_\theta\left[\dfrac{\partial \log p_\theta(X)}{\partial \theta}\right]^2 > 0$;

(v) The expression $\int f(x)\, p_\theta(x)\, d\mu(x)$ may be differentiated under the integral sign.

The problems we treat concern the binomial, Poisson, normal, and chi-square distributions, and we now check the validity of (2.2) for these distributions.

Lemma 1. *If $p_\theta(x)$ is any of the following:*

(a) $\binom{n}{x} \theta^x (1-\theta)^{n-x}$, $x = 0, 1, \ldots, n$; $0 < \theta < 1$; $\mu =$ *counting measure;*

(b) $\dfrac{\theta^x}{x!} e^{-\theta}$, $x = 0, 1, \ldots$; $0 < \theta < \infty$; $\mu =$ *counting measure;*

(c) $\dfrac{1}{\sqrt{2\pi}\sigma} e^{-1/2(x-\theta)^2}$, $-\infty < x < \infty$; $-\infty < \theta < \infty$; $\mu =$ *Lebesgue measure;*

(d) $\dfrac{x^{\nu/2-1} e^{-x/2\theta}}{2^{\nu/2}\Gamma\left(\frac{\nu}{2}\right) \theta^{\nu/2}}$, $0 < x < \infty$; $0 < \theta < \infty$; $\mu =$ *Lebesgue measure;*

then (2.2) *is satisfied.*

Proof. Conditions (i)–(iv), none of which involve the estimate $f(X)$, are obviously satisfied. In checking condition (v), there is no loss of generality in assuming that $f(X)$ has finite variance, since otherwise (2.2) certainly holds.

For distribution (a), condition (v) is obvious, since we are dealing with a finite sum. In cases (c) and (d) the result follows immediately from well known properties of the bilateral and unilateral Laplace transform, respectively. In case (b) our assumption guarantees the absolute convergence of the power series $\sum_{x=0}^{\infty} f(x)\,\theta^x/x!$ in the open interval $0 < \theta < \infty$, and hence the series may be differentiated term by term in that interval.

In the examples below we need in each case only check (2.4), the remaining conditions of theorem 1 obviously being satisfied.

Problem 1. Let $X_1, X_2, \ldots, X_n$ be a sample from the normal distribution with unknown expectation θ and known variance which we may without loss of generality take to be 1. Let the loss be squared error. It has been known for some time that $\bar{X} = \sum X_i/n$ is a minimax estimate for θ. This result was obtained by Stein and Wald [10] for a different loss function for the much harder sequential problem, and was proved explicitly for the loss function here employed by Wolfowitz [11]. We shall now use theorem 1 to prove both the admissibility and minimaxity of $\bar{X}$.

Since $\bar{X}$ is sufficient we need only consider estimates of the form $f(\bar{X})$, and since $\bar{X}$ is normally distributed we may apply (c) of lemma 1. We need only check (2.4) which now becomes

$$b^2(\theta) + \frac{1}{n}[1 + b'(\theta)]^2 \leqq \frac{1}{n} \tag{2.5}$$

for every $\theta \in D$ implies $b(\theta) \equiv 0$. Since neither term on left side of (2.5) can be negative, $|b(\theta)|$ is bounded and $b'(\theta)$ is never positive. Consequently there exists a sequence $\{\theta_i\}$ for which $b'(\theta_i)$ approaches 0 as $|\theta_i|$ approaches ∞, and hence by the hypothesis of (2.5), $b(\theta_i)$ does likewise. But since $b(\theta)$ is monotone, it must always be 0.

It is interesting to observe that if we assume certain additional information about θ, the estimate $\bar{X}$ may continue to be minimax without any longer being admissible. This is the case, for example, if we assume it known that $\theta > \theta_0$. For, $b^2(\theta) + [1 + b'(\theta)]^2/n$ is still a lower bound for the risk, and by an argument analogous to the one just given it is easily seen that $\sup_{\theta > \theta_0} b^2(\theta) + [1 + b'(\theta)]^2/n \geqq 1/n$. Hence the minimax risk is still 1 and $\bar{X}$ is minimax; its inadmissibility follows from the fact that $P(\bar{X} < \theta_0) > 0$, [9].

Problem 2. Let $X_1, \ldots, X_n$ be a sample from a Poisson distribution with unknown mean θ. For the loss we take squared error divided by θ; see remark (i) above. Since $X = \sum_{i=1}^{n} X_i$ is sufficient for θ we may restrict consideration to estimates $f(X)$. Taking $g(X) = X/n$, we shall prove admissibility of X/n by checking (2.4), which now becomes the condition

$$\frac{n\,b^2(\theta)}{\theta} + [1 + b'(\theta)]^2 \leqq 1 \quad \text{for all } \theta \in D \tag{2.6}$$

implies $b(\theta) = 0$ for all $\theta \in D$.

Since neither term on the left can be negative, $|b(\theta)|$ is bounded by $\sqrt{\theta/n}$ and $b'(\theta)$ satisfies the inequality

$$b'(\theta) \leqq \sqrt{1 - \frac{nb^2(\theta)}{\theta}} - 1 \leqq -\frac{nb^2(\theta)}{2\theta}.$$

Thus $\lim_{\theta \to 0} b(\theta) = 0, b'(\theta) \leqq 0$, and hence $b(\theta) \leqq 0$ for all θ. But if for some θ_0, $b(\theta_0)$ were negative, it would thereafter always be less than or equal to the function $c(\theta)$ for which $c'(\theta) = -nc^2(\theta)/(2\theta)$ and $c(\theta_0) = b(\theta_0)$. This latter function may be obtained explicitly by solving the differential equation, and is easily seen not to be absolutely bounded by $\sqrt{\theta/n}$.

We justified the choice of loss functions for the present problem in part by the remark that there exists no estimate with bounded risk function when the loss is squared error. That this is so is easily seen. For let $f(X)$ be such an estimate. Then

$$c_{b_f}(\theta) = b_f^2(\theta) + \frac{\theta}{n}[1 + b_{f'}(\theta)]^2$$

is bounded. But boundedness of the second term of $c_{b_f}(\theta)$ implies that $b_f'(\theta) \leqq -1 + \epsilon$ for all sufficiently large θ, and hence the unboundedness of the first term. An analogous remark applies to the χ^2-problem which we treat next and, in general, whenever the range of θ and the function $E_\theta\left[\frac{\partial}{\partial\theta}\log p_\theta(X)\right]^2$ are both unbounded and (2.2) holds for all estimates.

Problem 3. We next consider the estimation of the parameter in the chi-square distribution. This problem arises, for example, if we have a sample $X_1, X_2, \ldots, X_n$ from a normal distribution of known expectation but unknown variance. Then the statistic $\sum_{i=1}^{n}[X_i - E(X_i)]^2$ is sufficient for the variance and has a chi-square distribution. Suppose now that X has a chi-square distribution of n degrees of freedom with expectation θ, and take the loss to be squared error divided by θ^2. This loss function is chosen according to the principle discussed in remark (i) above. We shall now show, by means of theorem 1, that the estimate $g(X) = X/(n + 2)$ is the unique admissible minimax estimate for θ. It is interesting that this estimate is biassed, while the minimum variance unbiassed estimate has constant risk but is neither minimax nor admissible.

Condition (2.4) of theorem 1 now becomes the condition

$$\frac{b^2(\theta)}{\theta^2} + \frac{2}{n}[1 + b'(\theta)]^2 \leqq \frac{2}{n+2} \quad \text{for every} \quad \theta \in D \tag{2.7}$$

implies $b(\theta) \equiv -2\theta/(n + 2)$.

Since neither term on the left of the hypothesis of (2.7) can be negative, we have $b'(\theta) < 0$ and $|b(\theta)| < \theta$. It follows that $b(0+) = 0$. If for any 0, $b'(\theta) = b(\theta)/\theta$, their common value must be $-2/(n + 2)$, since the expression $r^2 + 2[1 + n]^2/n$ has a minimum of $2/(n + 2)$ when $r = -2/(n + 2)$.

We next observe that $b'(\theta) \leqq b(\theta)/\theta$. For, suppose that for some θ, $b'(\theta) > b(\theta)/\theta$. Then we should have

$$\left[\frac{b(\theta)}{\theta}\right]^2 + \frac{2}{n}[1 + b'(\theta)]^2 > \left[\frac{b(\theta)}{\theta}\right]^2 + \frac{2}{n}\left[1 + \frac{b(\theta)}{\theta}\right]^2,$$

which, by the previous paragraph, is not less than $2/(n+2)$. But this contradicts (2.7). Observing $\theta^2[b(\theta)/\theta]' = \theta b'[\theta - b(\theta)]$, we conclude that $b(\theta)/\theta$ is a nonincreasing function of θ.

We shall now prove that $b'(\theta)$ is not bounded away from $b(\theta)/\theta$ for large θ. For suppose $b'(\theta) \leqq b(\theta)/\theta - \epsilon$ for all $\theta \geqq \theta_0$. Then for $\theta \geqq \theta_0$, $b(\theta)$ will lie below that function $c(\theta)$ for which $c(\theta_0) = b(\theta_0)$ and $c'(\theta) = c(\theta)/\theta - \epsilon$. But $c(\theta) = \epsilon \cdot \theta \log \theta + k \cdot \theta$, which violates $-\theta < b(\theta) \leqq c(\theta)$. Analogously, $b'(\theta)$ cannot be bounded away from $b(\theta)/\theta$ as $\theta \to 0$. For otherwise $b(\theta)/\theta \geqq c(\theta)/\theta > 0$ for θ sufficiently small, while we know that $b(\theta) \leqq 0$.

We next see that if for some sequence $\{\theta_i\}$, $b'(\theta) - b(\theta_i)/\theta_i \to 0$ then $b(\theta_i)/\theta_i \to -2/(n+2)$. For, the hypothesis of (2.7) may be written

$$\left\{\left[\frac{b(\theta)}{\theta}\right]^2 + \frac{2}{n}\left[1 + \frac{b(\theta)}{\theta}\right]^2\right\} + \frac{2}{n}\left\{\left[b'(\theta) - \frac{b(\theta)}{\theta}\right]^2 - 2\left[b'(\theta) - \frac{b(\theta)}{\theta}\right]\left[1 + \frac{b(\theta)}{\theta}\right]\right\} \leqq \frac{2}{n+2} \tag{2.8}$$

and our hypothesis implies that the second term on the left side of (2.8) approaches 0; consequently the first term must approach its minimum which implies our statement.

Combining the results of the two preceding paragraphs, and using the monotoneness of $b(\theta)/\theta$, we see that $b(\theta)/\theta \to -2/(n+2)$ as $\theta \to 0$ or ∞, whence our result (2.7) follows.

Problem 4. Suppose that X has the binomial distribution (a) of lemma 1, and that the loss is squared error divided by $\theta(1-\theta)$. Condition (2.4) now becomes the condition

$$b^2(\theta) + \frac{\theta(1-\theta)}{n}[1 + b'(\theta)]^2 \leqq \frac{\theta(1-\theta)}{n} \text{ implies } b(\theta) \equiv 0. \tag{2.9}$$

Letting θ tend to 0 and 1 yields $b(0) = b(1) = 0$, while $b'(\theta) \leqq 0$ since $b^2(\theta)$ is nonnegative.

Problem 5. As a final example, we consider the previous problem, using however the classical squared error loss function. It is already known [9] that in this case $\frac{X}{\sqrt{n}(\sqrt{n}+1)} + \frac{1}{2(\sqrt{n}+1)}$ is an admissible minimax estimate for θ, but we shall now establish this fact as a consequence of theorem 1. The verification of (2.4) involved will in any case be needed when considering the sequential problem in the next section.

Condition (2.4) now becomes

$$b^2(\theta) + \frac{\theta(1-\theta)}{n}[1 + b'(\theta)]^2 \leqq \frac{1}{4(\sqrt{n}+1)^2} \text{ implies } b(\theta) \equiv \frac{\frac{1}{2} - \theta}{\sqrt{n}+1}. \tag{2.10}$$

Proof. Since the second term on the left side of the hypothesis of (2.10) cannot be negative, we have $|b(\theta)| \leqq \frac{1}{2(\sqrt{n}+1)}$. We next observe that $b(\theta) \geqq \frac{\frac{1}{2} - \theta}{\sqrt{n}+1}$ for $\frac{1}{2} \leqq \theta < 1$. For if on the contrary we had, for some $\frac{1}{2} \leqq \theta_0 < 1$,

$b(\theta_0) < \frac{\frac{1}{2} - \theta}{\sqrt{n}+1}$, we should be able to find a point $\theta_0 < \theta_1 < 1$ at which $b'(\theta_1) = \frac{b(1) - b(\theta_0)}{1 - \theta_0} > -\frac{1}{\sqrt{n}+1}$, and at which $b(\theta_1) < \frac{\frac{1}{2} - \theta}{\sqrt{n}+1}$. It is clear from the identical satisfaction of (2.10) by $b(\theta) = \frac{\frac{1}{2} - \theta}{\sqrt{n}+1}$ that this would imply the violation of (2.10) at θ_1.

By the symmetrical argument in the interval $0 < \theta \leqq \frac{1}{2}$, we find that in that interval $b(\theta) \leqq \frac{\frac{1}{2} - \theta}{\sqrt{n}+1}$, and hence that $b(\frac{1}{2}) = 0$. It also follows that $b'(\frac{1}{2}) \geqq -\frac{1}{\sqrt{n}+1}$; but on substituting $\frac{1}{2}$ for θ in the hypothesis of (2.10) and using $b(\frac{1}{2}) = 0$, we find $b'(\frac{1}{2}) = -\frac{1}{\sqrt{n}+1}$.

We can now conclude that (2.10) is satisfied. For suppose $b(\theta)$ satisfies the hypothesis of (2.10). By symmetry we need only consider the interval $\frac{1}{2} \leqq \theta < 1$, and need only show that $b(\theta_0) > \frac{\frac{1}{2} - \theta}{\sqrt{n}+1}$ for $\frac{1}{2} < \theta_0 < 1$ leads to a contradiction. Consider the function $c(\theta) = b(\theta) - \frac{\frac{1}{2} - \theta}{\sqrt{n}+1}$. $c(\theta)$ is continuous and has a continuous derivative, is nonnegative for $\frac{1}{2} \leqq \theta < 1$, and $c(\frac{1}{2}) = c'(\frac{1}{2}) = 0$. Hence for every $\epsilon > 0$ and every $k > 0$ we can find $\frac{1}{2} < \theta_1 < \theta_0$ for which $|c'(\theta_1)| < \epsilon$ and $c'(\theta_1) > kc(\theta_1)$. This is easily seen by considering $[\log c(\theta)]' = c'(\theta)/c(\theta)$, and using the continuity of $c'(\theta)$.

Since $b(\theta)$ is assumed to satisfy the hypothesis of (2.10) we can subtract to obtain $c(\theta)\left[b(\theta) + \frac{\frac{1}{2} - \theta}{\sqrt{n}+1}\right] + \frac{\theta(1-\theta)}{n} c'(\theta)\left[2 + b'(\theta) - \frac{1}{\sqrt{n}+1}\right] \leqq 0.$

Take now $\epsilon < 2 - \frac{2}{\sqrt{n}+1}$ and $k > \frac{-n}{2\theta_0(1-\theta_0)(\sqrt{n}-\epsilon)}$ to obtain a contradiction.

3. Estimates based on sequential procedures

In the previous section we have considered only the class of estimates based on a sample of fixed size, and have shown that certain estimates are optimum within this class. However, as is well known [12], the efficiency of statistical procedures can often be improved by taking the observations sequentially. Various definitions of optimum sequential procedures are possible within the minimax theory. For example, one may try to minimize the maximum expectation of a linear combination of loss and cost, measuring the latter by the number of observations. Alternatively, one may place a bound on the expected number of observations and try to minimize the maximum expected loss. Both of these formulations have been considered in the literature [5], [10], [11]. We have applied the method of the present paper to obtain optimum sequential procedures only under the second definition; it seems doubtful that our method would give easy results under the first definition. Although the first definition of an optimum estimate has theoretical advantages, in practical applications the second is sometimes more reasonable. This may happen, for example, when cost and loss cannot be measured on a common scale of value, or when budgetary considerations compel one to place a separate bound on the average cost of experimentation.

It is interesting that in a number of problems it turns out that a procedure of fixed sample size n is optimum among all sequential procedures for which the expected value of the number of observations N never exceeds n. For example, this is the case in problems treated by Stein and Wald and by Wolfowitz. We shall now show that the same holds true for the five problems treated in section 2.

The basis of our results in the present section is the extention by Wolfowitz [3] of the Cramér-Rao inequality to the sequential case. Wolfowitz proved under certain regularity conditions that

$$\sigma_f^2(\theta) \geq \frac{[1+b_f'(\theta)]^2}{E_\theta(N)E_\theta\left[\frac{\partial}{\partial\theta}\log p_\theta(X)\right]^2} \tag{3.1}$$

where $p_\theta(x)$ is the density of an individual observation and N is the (random) number of observations taken.

It is clear that theorem 1, with the obvious modifications, remains valid for the class of all sequential estimates, if we replace inequality (2.2) by inequality (3.1). Further, if we consider only those sequential procedures for which, for some integer n,

$$E_\theta(N) \leq n \quad \text{for all} \quad \theta \in D\,, \tag{3.2}$$

theorem 1 will be valid if in (3.1) we replace $E_\theta(N)$ by n. To extend the results of section 2 to the sequential case, we must verify the satisfaction of (2.4) and of the regularity conditions under which Wolfowitz proved (3.1). We carry out these checks not for all sequential procedures satisfying (3.2), but only for bounded procedures; that is, for procedures such that

$$P(N \leq m) = 1 \quad \text{for all} \quad \theta \in D \tag{3.3}$$

for some finite number m.

Since our results will be independent of the value of the bound m, provided only that it is sufficiently large, the restriction (3.3) is not serious from a practical point of view: any actual experiment does have a bound on the number of observations. However, the restriction is theoretically undesirable. We shall show below that the solutions obtained retain their minimax character when the restriction is removed. On the other hand, our argument does not establish the admissibility of the estimate within the unrestricted class of sequential procedures.

The Wolfowitz regularity conditions are contained in section 3 of his paper [3]. If the sequential procedure is bounded, and if the density is one of those considered in our lemma 1, all of these conditions are trivial, except for his (3.4). An examination of the proof shows that this condition is used only to permit a certain differentiation under the integral sign. We shall assure the applicability of the inequality by checking this differentiability directly.

LEMMA 3. *If the sequential procedure is bounded, and if the density is one of those considered in lemma* 1, *then* (3.1) *holds.*

PROOF. Let R_j be the set of points $(x_1, x_2, \ldots, x_j)$ for which $N = j$. Then

$$E_\theta[f(X_1, X_2, \ldots, X_N)] = \sum_{j=1}^{m}\int_{R_j} f(x_1, x_2, \ldots, x_j)\, p_\theta(x_1)\, p_\theta(x_2)\ldots \times p_\theta(x_j)\, d\mu(x_1)\, d\mu(x_2)\ldots d\mu(x_j)\,. \tag{3.4}$$

In view of the remarks just made we need only check that the right side of (3.4) may be differentiated under the integral sign. For density (a) of lemma 1 there is no difficulty, since we have simply a polynomial in θ. With (b) we have a convergent multiple series of nonnegative terms. This can be rearranged as a convergent series of powers of θ, which may be differentiated termwise. The normal cases are somewhat more involved. We may assume $E_\theta|f(X_1, X_2, \ldots, X_N)| < \infty$ for all $\theta \in D$, and hence the finiteness of each integral on the right of (3.4). Let $\phi_j(x_1, x_2, \ldots, x_j)$ be the characteristic function on R_j; we must show the differentiability under the integral of

$$(3.5) \quad \int_{-\infty}^{\infty}\!\!\cdots\!\int \phi_j(x_1, x_2, \ldots, x_j)\, |\, f(x_1, x_2, \ldots, x_j)\, |\, p_\theta(x_1)\, p_\theta(x_2) \ldots \times p_\theta(x_j)\, d(x_1)\, d(x_2) \ldots d(x_j).$$

Recalling $p_\theta(x) = \frac{1}{\sqrt{2\pi}} e^{-1/2(x-\theta)^2}$, collecting the exponents, and making an orthogonal transformation with $y_1 = x_1 + x_2 + \ldots + x_j$, we see that (3.5) may be rewritten as

$$(3.6) \qquad e^{-(j/2)\theta^2} \int_{-\infty}^{\infty}\!\!\cdots\!\int K(y_1, y_2, \ldots, y_j)\, e^{\theta y_1} dy_1 dy_2 \ldots dy_j.$$

Using the Fubini theorem, we see that the integral in (3.6) is a convergent Laplace transform and may therefore be differentiated.

A similar argument applies to the chi-square situation, using a unilateral instead of a bilateral Laplace transform.

We can now conclude that the estimates found to be admissible minimax estimates in problems 1–5 of the preceding section continue to have this property in the class of all estimates based on sequential procedures satisfying (3.2) and (3.3). We shall not have to recheck the differential inequalities which result from (2.4), since they are in each case unchanged.

Finally, we observe that condition (3.3) may be removed in all of these problems without effecting the conclusion that all estimates considered are minimax. For, if there were a sequential estimate δ not satisfying (3.3) having a maximum risk r less by $\epsilon > 0$ than the minimax risk for bounded sequential procedures, then we could construct a bounded sequential estimate with maximum risk $< r + \frac{\epsilon}{2}$. To see this, notice that in each of the cases treated there exists an estimate δ_0 of θ, based on a single observation, whose risk is bounded, say by a constant k. Since $E(N) \leq n$, $P_\theta(N_\delta > m) \to 0$ uniformly in θ. Let the estimate δ be defined as follows. If $N_\delta \leqq m$, let δ' agree with δ. If $N_\delta > m$, take an $(m+1)$-st observation and let δ' agree with $\delta_0(x_{n+1})$. It is clear that $P(N_{\delta'} \leqq m + 1) = 1$, that $E(N_\delta) \leqq n$, and that $\sup_{\theta \in D} R_\theta(\delta')$ can be made less than $r + \frac{\epsilon}{2}$ by taking m sufficiently large.

REFERENCES

[1] H. Cramér, "A contribution to the theory of statistical estimation," *Skandinavisk Aktuarietidskrift*, Vol. 29 (1946), pp. 85–94.

[2] C. R. Rao, "Information and the accuracy attainable in the estimation of statistical parameters," *Bull. Calcutta Math. Soc.*, Vol. 37, No. 3 (1945), pp. 81–91.

[3] J. Wolfowitz, "The efficiency of sequential estimates and Wald's equation for sequential processes," *Annals of Math. Stat.*, Vol. 18 (1947), pp. 215–230.

[4] A. Wald, "Contributions to the theory of statistical estimation and testing hypothesis," *Annals of Math. Stat.*, Vol. 10 (1939), pp. 299–326.

[5] ———, *Statistical Decision Functions*, Wiley, New York, 1950.

[6] M. A. Girshick and L. J. Savage, "Bayes and minimax estimates arising from quadratic risk functions," *Proceedings of the Second Berkeley Symposium on Mathematical Statistics and Probability*, University of California Press, Berkeley, 1951, pp. 53–73.

[7] C. F. Gauss, *Abhandlungen zur Methode der Kleinsten Quadrate*, Berlin, 1887.

[8] E.L. Lehmann and H. Scheffé, "Completeness, similar regions, and unbiased estimation —Part I," Sankhyā, Vol. 10 (1950), pp. 305–340.

[9] J. L. Hodges, Jr. and E. L. Lehmann, "Some problems in minimax point estimation," *Annals of Math. Stat.*, Vol. 21 (1950), pp. 182–197.

[10] C. Stein and A. Wald, "Sequential confidence intervals for the mean of a normal distribution with known variance," *Annals of Math. Stat.*, Vol. 18 (1947), pp. 427–433.

[11] J. Wolfowitz, "Minimax estimates of the mean of a normal distribution with known variance," *Annals of Math. Stat.*, Vol. 21 (1950), pp. 218–230.

[12] A. Wald, *Sequential Analysis*, Wiley, New York, 1947.

NOTES

A GENERAL CONCEPT OF UNBIASEDNESS

By E. L. Lehmann

University of California, Berkeley, and Princeton University

The term unbiasedness was introduced by Neyman and Pearson [1] in connection with hypothesis testing. A test of the hypothesis $\theta \varepsilon \omega$ against the alternatives $\theta \varepsilon \Omega - \omega$ is said to be unbiased at level α if its power function β satisfies

$$\beta(\theta) \leq \alpha \text{ for } \theta \varepsilon \omega, \quad \beta(\theta) \geq \alpha \text{ for } \theta \varepsilon \Omega - \omega. \tag{1}$$

In 1937 Neyman [2] developed a theory of estimation by confidence sets. He established a duality with the theory of hypothesis testing, so that to each notion of one theory corresponds an analogous one in the other. In particular, he defined a family of confidence sets $A(x)$ to be unbiased if

$$P_\theta(A(X) \supset \theta') \leq P_\theta(A(X) \supset \theta) \text{ for all } \theta \text{ and } \theta'. \tag{2}$$

While the above two definitions are closely related, a third use of the term unbiasedness was made in a rather different context. In presenting their version of the Gauss-Markov theorem on least squares David and Neyman [3] defined a point estimate $\delta(X)$ of $g(\theta)$ to be unbiased if its expectation coincides with the estimated value, that is, if

$$E_\theta \delta(X) \equiv g(\theta). \tag{3}$$

It was pointed out later by Brown [4] that one obtains other analogous definitions by postulating that some central value of the distribution of $\delta(X)$ other than the mean coincides with the estimated value. Using the median as an example he defined $\delta(X)$ to be median-unbiased if

$$P_\theta(\delta(X) > g(\theta)) = P_\theta(\delta(X) < g(\theta)) \text{ for all } \theta. \tag{4}$$

In view of Wald's theory of decision functions [5] it seems tempting to try to give a definition of unbiasedness at the level of generality of this theory. Suppose we are concerned with a decision problem where the loss resulting from a decision $\delta(X)$ is $W(\theta, \delta(X))$ when the true parameter value is θ. In analogy with (2) we shall say that a decision procedure $\delta(X)$ is unbiased if for each θ

$$E_\theta W(\theta', \delta(X)) = \min \text{ when } \theta' = \theta. \tag{5}$$

This clearly reduces to Neyman's definition for confidence sets if one uses for loss function,

$$W(\theta, \delta(x)) = \begin{cases} 0 \text{ if the confidence set } \delta(x) \text{ covers } \theta, \\ 1 \text{ otherwise.} \end{cases} \tag{6}$$

J. Rojo (ed.), *Selected Works of E. L. Lehmann*, Selected Works in Probability and Statistics, DOI 10.1007/978-1-4614-1412-4_5, © Springer Science+Business Media, LLC 2012

In order to obtain an interpretation of condition (5), let us consider the case that for each parameter value θ there exists a unique "correct" decision d and that each d is correct for at least some θ. This is the case for example in hypothesis testing and in point estimation. Here a correct decision may be defined by the condition $W(\theta, d) = 0$. Let us say that two parameter values θ, θ' are equivalent, $\theta \sim \theta'$, if the correct decision is the same for both of them, and let us suppose that for any decision d'

$$W(\theta_1, d') = W(\theta_2, d') \text{ whenever } \theta_1 \sim \theta_2 . \tag{7}$$

Then the loss $W(\theta, d')$ depends only on the actual decision taken, say d', and the decision d that would have been correct, and we may write for it $W(d, d')$. The loss $W(d, d')$ is a measure of how far the two decisions d and d' are apart, and (5) states that a decision function $\delta(X)$ is unbiased if on the average it comes closer to the correct decision than to any incorrect decision.

Let us now apply this notion to some particular examples. Let the decision to accept and reject the hypothesis $H: \theta \varepsilon \omega$ be denoted by d_0 and d_1, respectively. Since in the Neyman-Pearson theory of hypothesis testing one is concerned only with the probabilities of the two types of error, the natural associated loss function is of the form

$$\begin{aligned} W(\theta, d_0) &= \begin{cases} a \text{ if } \theta \epsilon \Omega - \omega, \\ 0 \text{ if } \theta \epsilon \omega; \end{cases} \\ W(\theta, d_1) &= \begin{cases} b \text{ if } \theta \epsilon \omega, \\ 0 \text{ if } \theta \epsilon \Omega - \omega. \end{cases} \end{aligned} \tag{8}$$

It is easy to see that in this case (5) becomes

$$\begin{aligned} P_\theta(d_1) &\leq \frac{a}{a+b} \text{ for } \theta \epsilon \omega, \\ P_\theta(d_1) &\geq \frac{a}{a+b} \text{ for } \theta \epsilon \Omega - \omega, \end{aligned} \tag{9}$$

where $P_\theta(d)$ denotes the probability that $\delta(X) = d$ when θ is the true parameter value. This is exactly the usual definition (1) with $\alpha = a/(a + b)$.

Let us next consider point estimation where the loss is taken as the square of the error. If the function to be estimated is $g(\theta)$, condition (5) becomes

$$E_\theta[\delta(X) - g(\theta')]^2 \geq E_\theta[\delta(X) - g(\theta)]^2 \text{ for all } \theta, \theta'. \tag{10}$$

Let $E_\theta\delta(X) = h(\theta)$. In the usual case that $h(\theta)$ is one of the possible values of the function g, the left-hand side of (10) is minimized for $g(\theta') = h(\theta)$. Thus the inequality holds for all θ' if and only if $g(\theta) = h(\theta)$, which is equivalent to (3). So again (5) reduces just to the usual definition.

Even if $h(\theta)$ is not one of the possible values of g, it is easily seen that (10) is equivalent to

$$| h(\theta) - g(\theta) | = \min_{\theta'} | h(\theta) - g(\theta') | .$$

Then, if for example Ω is a real interval and g is continuous and strictly monotone, there can exist at most two values of θ for which $g(\theta) \neq h(\theta)$. If further, as is usually the case, $h(\theta)$ is continuous for all estimates δ, we must have $h(\theta) \equiv g(\theta)$.

Quite analogously one sees that if $W(\theta, \delta(x)) = | \delta(x) - g(\theta) |$, definition (5) reduces to Brown's notion of median-unbiasedness.

While the definition given here seems satisfactory in that it does reduce under reasonable assumptions to the usual concepts, it is somewhat more restrictive than appears at first sight. If for example there exists for each θ a unique correct decision d and if the loss function is of the form

$$W(\theta, d') = f(\theta)V(d, d'),$$

then, with the trivial exception of procedures for which $E_\theta V(d, \delta(x)) = 0$ for some d and some value of θ in ω_d, no unbiased procedure can exist unless $f(\theta)$ is constant on each ω_d. For let $\theta, \theta' \in \omega_d$. On substituting in (5) we see that unbiasedness implies $f(\theta') \geq f(\theta)$ and hence by symmetry $f(\theta') = f(\theta)$. In hypothesis testing for example if the loss is zero for a correct decision, it follows, again with trivial exceptions, that unbiased tests can exist only if the loss function is given by (8).

It is perhaps worth pointing out certain connections between the principle of unbiasedness and that of invariance. Consider for example the problem of estimating θ from a sample $X_1, \cdots, X_n$ where the X's are uniformly distributed on $(0, \theta)$. If one takes as loss function

$$W(\theta, \delta(x_1, \cdots, x_n)) = [\delta(x_1, \cdots, x_n) - \theta]^2/\theta^2 \tag{11}$$

the problem transforms in an obvious manner under a change of scale, and one may wish to consider only estimates having the invariance property

$$\delta(cX_1, \cdots, cX_n) = c\delta(X_1, \cdots, X_n) \text{ for all } c > 0. \tag{12}$$

If $Y = \max(X_1, \cdots, X_n)$, it is easily seen that among all invariant estimates the one that uniformly minimizes the expected loss is

$$\frac{n+2}{n+1} Y. \tag{13}$$

This estimate does not have the usual unbiasedness property since

$$E_\theta \left[\frac{n+2}{n+1} Y\right] = \frac{n(n+2)}{(n+1)^2}\,\theta.$$

However a simple computation shows that (13) is unbiased in the sense of (5) with respect to the invariant loss function (11).

More generally, let $\mathcal{G}$ be a group of measurable 1:1 transformations on the sample space. Let gX be the random variable that takes on the value gx when $X = x$, and suppose that when X has a distribution p_θ, $\theta \in \Omega$, then gX has a distribution $p_{\theta'}$, $\theta' \in \Omega$. Denote this θ' by $\bar{g}\theta$ and suppose that $\bar{g}\theta$ defines a 1:1 transformation on Ω. Let $\bar{\mathcal{G}}$ be the group of transformations $\bar{g}$ and assume that there exists a group $\mathcal{G}^*$ of 1:1 transformations on the decision space D such that $\mathcal{G}^*$ is homomorphic to $\bar{\mathcal{G}}$ and

$$W(\bar{g}\theta, g^*d) = W(\theta, d) \text{ for all } \theta \varepsilon \Omega, d \varepsilon D. \tag{14}$$

Then a decision function δ is said to be invariant if

$$\delta(gX) = g^*\delta(X). \tag{15}$$

This is a natural generalization of the definition of invariance given by Hunt and Stein [6, 7], and is essentially the definition used by Peisakoff [8]. Further, δ is said to be almost invariant if (15) holds except on a set N_g of measure 0.

Whenever among all unbiased procedures there exists a unique[1] one that uniformly minimizes the risk, then it is almost invariant. This follows easily from the fact that if $\delta(X)$ is unbiased $g^*\delta(g^{-1}X)$ is also unbiased. It is not in general true that conversely an optimum invariant test is necessarily unbiased. However, this result does hold under certain restrictions.[2] If

(i) $\mathcal{G}$ is transitive, i.e., given any θ, θ' there exists $\bar{g}$ such that $\theta = \bar{g}\theta'$,

(ii) $\mathcal{G}^*$ is commutative,

and if among all invariant (or almost invariant) procedures there exists one that uniformly minimizes the risk, then it is unbiased.

To see this, let δ be invariant and such that for any other invariant procedure δ'

$$E_\theta W(\theta, \delta'(X)) \geq E_\theta W(\theta, \delta(X)).$$

Let $\theta' \neq \theta$, $\theta = \bar{g}\theta'$, say. Then

$$E_\theta W(\theta', \delta(X)) = E_\theta W(\theta, g^*\delta(X)) \geq E_\theta W(\theta, \delta(X)).$$

Here the inequality follows since by (ii) the invariance of $\delta(X)$ implies that $g^*\delta(X)$ is also invariant.

While assumptions (i) and (ii) are satisfied in many estimation problems, (i) will in general not hold in a problem of hypothesis testing because of the asymmetry of d_0 and d_1. Here the result in question follows when the loss function is given by (8) from the fact that if a test is unbiased so is any test that is uniformly better, together with the unbiasedness and invariance of the test $\varphi(x) \equiv a/(a + b)$ (i.e., the test that rejects the hypothesis with probability $a/(a + b)$ regardless of the observations).

That the result is not true in general if we drop either one of the two conditions (i) or (ii) can be seen from the following example. For estimating the mean ξ of a normal variable with unknown variance σ^2 when the loss function is $[(\delta(x) - \xi)/\sigma]^2$, the best invariant estimate is X both with respect to the group

$$\mathcal{G}_1: gx = x + b, \qquad -\infty < b < \infty$$

---

[1] Throughout, this is understood to mean unique up to a set of measure zero.

[2] I am grateful to the referee for pointing out an error in my original statement of this result.

and with respect to

$$\mathcal{G}_2: gx = ax + b, \qquad 0 < a < \infty, -\infty < b < \infty.$$

For this problem an unbiased estimate in the sense of (5) does not exist, and it is seen that $\mathcal{G}_1$ satisfies (ii) but not (i) while $\mathcal{G}_2$ satisfies (i) but not (ii).

The notion of unbiasedness in many cases leads to reasonable decision procedures and this seems to be in general the value of such concepts. On the other hand there is no guarantee that an optimum unbiased procedure is necessarily satisfactory. As an example (for another example see [9]) consider a Poisson variable X which is observed only if $X \neq 0$, so that the distribution of X is given by

$$P(X = K) = \frac{\lambda^K}{K!} e^{-\lambda}(1 - e^{-\lambda})^{-1}, \qquad K = 1, 2, \cdots. \tag{15}$$

It is desired to estimate the probability $e^{-\lambda}$ of X being zero, and the loss function is squared error. The condition of unbiasedness gives

$$\sum_{K=1}^{\infty} \delta(K) \frac{\lambda^K}{K!} \equiv 1 - e^{-\lambda} \equiv \sum_{K=1}^{\infty} (-1)^{K+1} \frac{\lambda^K}{K!}, \tag{16}$$

so that $\delta(K) = (-1)^{K+1}$. Thus the estimate takes on only impossible values and instead of decreasing with K as one would expect, it does not depend on the order of magnitude of K at all.

As a final remark we mention, without going into details, the following extension of the notion of unbiasedness. Instead of comparing $E_\theta W(\theta', \delta(X))$ only with $E_\theta W(\theta, \delta(X))$ we may ask that $E_\theta W(\theta', \delta(X))$ be a nondecreasing function of $v(\theta, \theta')$, where $v(\theta, \theta')$ in some sense measures the distance between θ and θ'. This notion is a generalization of one used by P. L. Hsu [10] in the theory of hypothesis testing. It is also closely connected with the principle of invariance. In fact if there exists a group of transformations leaving the problem invariant then with a suitable definition of $v(\theta, \theta')$ it is easy to see under weak assumptions on the loss function that Theorem 7.1 of [7] generalizes to the present case. This theorem states essentially that the totality of procedures for which $E_\theta W(\theta', \delta(X))$ depends only on $v(\theta, \theta')$ coincides with the totality of invariant procedures.

REFERENCES

[1] J. Neyman and E. S. Pearson, "Contributions to the theory of testing statistical hypotheses. I. Unbiased critical regions of type A and type A_1," *Stat. Res. Mem.*, Vol. 1 (1936), pp. 1–37.

[2] J. Neyman, "Outline of a theory of statistical estimation based on the classical theory of probability," *Phil. Trans. Roy. Soc. London*, Series A, Vol. 236 (1937), pp. 333–380.

[3] F. N. David and J. Neyman, "Extension of the Markoff theorem on least squares," *Stat. Res. Mem.*, Vol. 2 (1938), pp. 105–116.

[4] G. W. Brown, "On small sample estimation," *Annals of Math. Stat.*, Vol. 18 (1947), pp. 582–585.

[5] A WALD, *Statistical Decision Functions*, John Wiley and Sons, 1950.
[6] G. HUNT AND C. STEIN, "Most stringent tests of statistical hypotheses," unpublished.
[7] E. L. LEHMANN, "Some principles of the theory of testing hypotheses," *Annals of Math. Stat.*, Vol. 21 (1950), pp. 1–26.
[8] M. PEISAKOFF, "Transformation parameters," unpublished thesis, Princeton University, 1950.
[9] P. R. HALMOS, "The theory of unbiased estimation," *Annals of Math. Stat.*, Vol. 17 (1946), pp. 34–43.
[10] P. L. HSU, "Analysis of variance from the power function standpoint," *Biometrika*, Vol. 32 (1941), pp. 62–69.

ONE-SIDED CONFIDENCE CONTOURS FOR PROBABILITY DISTRIBUTION FUNCTIONS[1]

BY Z. W. BIRNBAUM AND FRED H. TINGEY[2]

University of Washington

Summary. Let $F(x)$ be the continuous distribution function of a random variable X, and $F_n(x)$ the empirical distribution function determined by a sample $X_1, X_2, \cdots, X_n$. It is well known that the probability $P_n(\epsilon)$ of $F(x)$ being everywhere majorized by $F_n(x) + \epsilon$ is independent of $F(x)$. The present paper contains the derivation of an explicit expression for $P_n(\epsilon)$, and a tabulation of the 10%, 5%, 1%, and 0.1% points of $P_n(\epsilon)$ for $n = 5, 8, 10, 20, 40, 50$. For $n = 50$ these values agree closely with those obtained from an asymptotic expression due to N. Smirnov.

1. Introduction. Let X be a random variable with the continuous probability distribution function $F(x) = \text{Prob. } \{X \le x\}$. An ordered sample $X_1 \le X_2 \le \cdots \le X_n$ of X determines the empirical distribution function

$$F_n(x) = \begin{cases} 0 & \text{for } x < X_1, \\ \dfrac{k}{n} & \text{for } X_k \le x < X_{k+1}, \qquad k = 1, 2, \cdots, n-1, \\ 1 & \text{for } X_n \le x. \end{cases}$$

The function

$$F^+_{n,\epsilon}(x) = \min\,[F_n(x) + \epsilon, 1],$$

also determined by the sample, will be called an *upper confidence contour*. It is well known [2] that the probability

$$P_n(\epsilon) = \text{Prob. } \{F(x) \le F^+_{n,\epsilon}(x) \text{ for all } x\}$$

of $F(x)$ being everywhere majorized by $F^+_{n,\epsilon}(x)$ is independent of the distribution $F(x)$. An expression for $P_n(\epsilon)$ in determinant form was given by A. Wald and

[1] Presented to the American Mathematical Society on April 28, 1951.
[2] Research under the sponsorship of the Office of Naval Research.

THE USE OF PREVIOUS EXPERIENCE IN REACHING STATISTICAL DECISIONS[1]

By J. L. Hodges, Jr., and E. L. Lehmann

University of California, Berkeley, and University of Chicago and Stanford University

1. Summary. Instead of minimizing the maximum risk it is proposed to restrict attention to decision procedures whose maximum risk does not exceed the minimax risk by more than a given amount. Subject to this restriction one may wish to minimize the average risk with respect to some guessed a priori distribution suggested by previous experience. It is shown how Wald's minimax theory can be modified to yield analogous results concerning such restricted Bayes solutions. A number of examples are discussed, and some extensions of the above criterion are briefly considered.

2. Introduction. Among various possible approaches to the problem of defining a best decision procedure we may mention the following two extremes.

(i) *The Bayes solution.* If the unknown parameter θ is a random variable distributed according to a known probability distribution λ, and if $R_\delta(\theta)$ denotes the risk function of the decision procedure δ, we simply minimize with respect to δ the average risk $\int R_\delta(\theta)\, d\lambda(\theta)$.

(ii) *The minimax principle.* Here one focuses attention on the maximum of the risk function and wishes to minimize $\sup_\theta R_\delta(\theta)$. The reader may consult Wald [1] for definitions and examples of these terms.

Of these two methods of treating the problem the first one assumes complete knowledge of the a priori distribution, an assumption that is usually not satisfied in practice. Even if extensive past experience is available, it will in most cases be difficult to exclude the possibility of some change in conditions. On the other hand, the minimax principle forces us to act as if θ were following a particular a priori distribution, the one least favorable to us, even though we may feel pretty sure that actually θ is distributed in quite a different manner. Thus it would seem that the minimax principle is suitable, if at all, only in situations characterized by a complete absence of past experience or other sources of knowledge concerning θ.

The situations occurring in practice usually lie between the two extremes just described. On the one hand, one does frequently have a good idea as to the range of θ, and as to which values in this range are more or less likely. On the other hand, such information cannot be expected to be either sufficiently precise or sufficiently reliable to justify complete trust in the Bayes approach.

The purpose of the present paper is to discuss an approach to the problem of

[1] Work sponsored in part by the Office of Naval Research.

optimal decisions that utilizes the available information but at the same time provides a safeguard in case this information is not correct.

Suppose that the maximum risk of the minimax procedure is $\underline{C}$. This is the smallest possible value for the maximum of a risk function. But we may be willing to tolerate a somewhat bigger maximum risk $C_0 > \underline{C}$ if, in case the guess at λ has been a good one, there results a substantial decrease in the average risk. This leads to the

DEFINITION. The procedure δ_0 is said to be a *restricted Bayes solution* with respect to the a priori distribution λ and subject to the restriction

$$R_\delta(\theta) \leqq C_0 \text{ for all } \theta, \tag{2.1}$$

if it minimizes $\int R_\delta(\theta)\, d\lambda(\theta)$ among all procedures satisfying (2.1).

This definition takes into account two aspects of the risk function, its supremum and its average with result to the distribution λ. Within a certain range of values each of these can be improved at the expense of the other. The proper balance between the two, that is, the value we select for C_0, depends on the confidence we have in θ's actually following a distribution close to λ and on the decrease in $\int R_\delta\, d\lambda$ that can be achieved by further increasing C_0.

To the above approach can also be given the following slightly different form. Instead of setting an upper bound for the risk, we specify a constant $0 \leqq \rho_0 \leqq 1$ and minimize

$$\rho_0 \int R_\delta\,(\theta)\, d\lambda(\theta) + (1 - \rho_0) \sup_\theta R_\delta\,(\theta). \tag{2.2}$$

Here it is ρ_0 that indicates the confidence we have in λ. The two principles are clearly equivalent; for if δ_0 minimizes (2.2) and if $\sup R_{\delta_0}(\theta) = C_0$, then δ_0 is a restricted Bayes solution, and the converse also holds.

The formulations given here may be applicable also to games played against an opponent rather than against Nature. This would be the case if one believed from past experience that the opponent is likely to make certain mistakes. One could then take advantage of these and still protect oneself in case the opponent has improved.

3. Restricted Bayes solutions. The principal aim of the present section is to obtain sufficient conditions for a decision procedure to be a restricted Bayes solution. For this purpose the modified problem mentioned at the end of the last section turns out to be the more natural one to consider; the results concerning restricted Bayes solutions follow as immediate corollaries. All of the theorems of this section will be simple generalizations of the corresponding results in Wald's minimax theory.

THEOREM 1. *Let ν_0 be a distribution for which there exists a constant $0 < \rho_0 \leqq 1$*

and a distribution μ_0 *such that*

$$\nu_0 = \rho_0\lambda + (1 - \rho_0)\mu_0 . \tag{3.1}$$

Then if the Bayes solution δ_0 *of* ν_0 *satisfies*

$$\int R_{\delta_0}(\theta)\, d\mu_0(\theta) = \sup_\theta R_{\delta_0}(\theta), \tag{3.2}$$

the procedure δ_0 *minimizes*

$$\rho_0 \int R_\delta(\theta)\, d\lambda(\theta) + (1 - \rho_0) \sup R_\delta(\theta). \tag{3.3}$$

PROOF. Let δ be any procedure. Then

$$\rho_0 \int R_\delta\, d\lambda + (1 - \rho_0) \sup R_\delta(\theta) \geqq \rho_0 \int R_\delta\, d\lambda + (1 - \rho_0) \int R_\delta\, d\mu_0$$

$$\geqq \rho_0 \int R_{\delta_0}\, d\lambda + (1 - \rho_0) \int R_{\delta_0}\, d\mu_0 = \rho_0 \int R_{\delta_0}\, d\lambda + (1 - \rho_0) \sup R_{\delta_0}(\theta).$$

COROLLARY. *Under the assumptions of Theorem* 1 *suppose that* $\sup R_{\delta_0}(\theta) = C_0$. *Then* δ_0 *minimizes* $\int R_\delta\, d\lambda$ *among all procedures satisfying* $R_\delta(\theta) \leqq C_0$ *for all* θ.

Since a distribution ν_0 with the required properties does not always exist, we state the following generalization of Theorem 1. *If* ν_i *is a sequence of a priori distributions with* $\nu_i = \rho_0\lambda + (1 - \rho_0)\mu_i$, *and if* δ_i *are the associated Bayes solutions, then*

$$\lim_{i\to\infty} \int R_{\delta_i}\, d\mu_i = \sup R_{\delta_0}(\theta) \tag{3.4}$$

is a sufficient condition for δ_0 *to minimize* (3.3).

Before proceeding with this development let us discuss briefly the decomposition (3.1). For a given pair of distributions λ, ν consider the totality of numbers $0 \leqq \rho \leqq 1$ such that $\nu = \rho\lambda + (1 - \rho)\mu$ for some μ. It is easily seen that this set is a closed interval $0 \leqq \rho \leqq \rho_\lambda$, and we shall call ρ_λ the λ-*component* of ν. It is of interest to note that under the conditions of Theorem 1 we have $\rho_0 = \rho_\lambda$ unless C_0 is the minimax risk. For if ρ_0 is not the λ-component of ν_0 it follows that μ_0 must have a positive λ-component. Thus any point of increase of λ is also one of μ_0 and we have $\int R_{\delta_0}\, d\lambda = C_0$. But by Theorem 1 this is possible only if C_0 is the minimax risk.

As in the minimax theory the distribution ν_0 of Theorem 1 is "least favorable" for the statistician, in a sense made precise in

THEOREM 2. *Under the assumptions of Theorem* 1, *the distribution* ν_0 *maximizes the Bayes risk among all a priori distributions* ν *that permit a representation*

$$\nu = \rho\lambda + (1 - \rho)\mu \textit{ with } \rho \geqq \rho_0 .$$

PROOF. Let $\nu = \rho\lambda + (1 - \rho)\mu$ be any distribution with $\rho \geqq \rho_0$. Then

$$\int R_{\delta_0}\, d\nu_0 = \rho_0 \int R_{\delta_0}\, d\lambda + (1 - \rho_0) \sup R_{\delta_0}(\theta) \geqq \rho \int R_{\delta_0}\, d\lambda$$

$$+ (1 - \rho) \sup R_{\delta_0}(\theta) \geqq \rho \int R_{\delta_0}\, d\lambda + (1 - \rho) \int R_{\delta_0}\, d\mu$$

$$\geqq \int R_{\delta_\nu}(\theta)\, d\nu(\theta).$$

Thus δ_0 is a maximum solution for Nature (in Wald's interpretation of a decision problem as a two-person zero sum game whose players are Nature and the statistician), if she is restricted to a priori distributions whose λ-component is at least ρ_0.

From the proof of Theorem 2 it is seen that ν_0 not only maximizes the Bayes risk but also—and this is a slightly stronger result—the restricted Bayes risk, that is, the quantity $\inf \int R_\delta\, d\lambda$ when δ is restricted by the condition $R_\delta(\theta) \leqq C_0$ for all θ.

In Theorem 1 we gave sufficient conditions for a procedure δ_0 to be a restricted Bayes solution, and the distribution ν_0 in terms of which these conditions were formulated was further characterized in Theorem 2. We must still prove the existence of a distribution with the desired properties, at least for some class of decision problems. This is easily done, along the lines of Wald's proof of Theorem 3.10 of [1], under the following assumptions.

ASSUMPTION 1.

$$\inf_\delta \sup_\mu \left[\rho_0 \int R_\delta\, d\lambda + (1 - \rho_0) \int R_\delta\, d\mu\right]$$

$$= \sup_\mu \inf_\delta \left[\rho_0 \int R_\delta\, d\lambda + (1 - \rho_0) \int R_\delta\, d\mu\right].$$

This states that the decision problem is strictly determined when Nature is restricted to distributions with λ-components $\geqq \rho_0$.

ASSUMPTION 2. *There exists a least favorable distribution μ_0, that is a distribution that maximizes*

$$\inf_\delta \left[\rho_0 \int R_\delta\, d\lambda + (1 - \rho_0) \int R_\delta\, d\mu\right].$$

ASSUMPTION 3. *There exists a decision procedure δ_0 that minimizes*

$$\sup_\mu \left[\rho_0 \int R_\delta\, d\lambda + (1 - \rho_0) \int R_\delta\, d\mu\right].$$

It follows from Wald's work that 1 and 3 hold, for example, when Conditions 3.1–3.6 of [1] are satisfied, while in general 2 requires the stronger Condition

(3.7). All these conditions are satisfied, for example, when we are dealing with discrete or absolutely continuous distributions, the problem is nonsequential, the loss function is bounded and the parameter space is compact.

To prove our result, we note that

$$\sup_{\mu}\left[\rho_0\int R_\delta\,d\lambda + (1-\rho_0)\int R_\delta\,d\mu\right] = \rho_0\int R_\delta\,d\lambda + (1-\rho_0)\sup_\theta R_\delta\,(\theta).$$

Therefore, we have by assumption

$$\rho_0\int R_{\delta_0}\,d\lambda + (1-\rho_0)\sup R_{\delta_0}(\theta) = \inf_\delta\left[\rho_0\int R_\delta\,d\lambda + (1-\rho_0)\int R_\delta\,d\mu_0\right],$$

and hence

$$\rho_0\int R_{\delta_0}\,d\lambda + (1-\rho_0)\sup R_{\delta_0}(\theta) = \rho_0\int R_{\delta_0}\,d\lambda + (1-\rho_0)\int R_{\delta_0}\,d\mu_0.$$

But this implies, when $\rho_0 < 1$, that

$$\sup R_{\delta_0}\,(\theta) = \int R_{\delta_0}\,d\mu_0.$$

On the other hand, the result is true vacuously when $\rho_0 = 1$.

4. A continuity theorem. We shall next consider the dependence of the restricted Bayes solution on the restricting quantity C_0. The main result will be a continuity theorem that permits us, at least in some cases, to characterize the topological structure of the restricted Bayes solutions.

Throughout this section we shall make the following two assumptions whose validity was proved by Wald (Theorems 3.1 and 3.2a of [1]) under very general conditions.

ASSUMPTION A. *The space of decision functions is compact, that is, every sequence δ_i of decision procedures possesses a subsequence δ_{i_j} that converges to some decision function δ^*. Here convergence means what Wald calls "regular convergence"* (see pp. 65–66 of [1]).

ASSUMPTION B. *If $\delta_i \to \delta^*$, then for every distribution ν*

$$\int R_{\delta_i}\,d\nu \to \int R_{\delta^*}\,d\nu.$$

This convergence of integrals implies the pointwise convergence of the risk functions, as may be seen by letting the distributions ν degenerate at single points θ. In some cases (for example, if the loss function is bounded) the two notions of convergence are equivalent.

Let $\bar{C} = \sup R_{\delta_\lambda}(\theta)$ be the maximum risk of the unrestricted Bayes solution (for simplicity assume that δ_λ is unique; otherwise we would put $\bar{C} = \inf_{\delta_\lambda} \sup_\theta R_{\delta_\lambda}(\theta)$, the inf taken with respect to all Bayes solutions), and let $\underline{C}$ be the maximum risk of the minimax procedure. Then $\underline{C} \leqq \bar{C}$ and we may exclude the case $\underline{C} = \bar{C}$ as trivial. For any $\underline{C} \leqq C \leqq \bar{C}$ let $\delta_\lambda(C)$ be an associated restricted Bayes solution, and $r_\lambda(C) = \int R_{\delta_\lambda(C)}\, d\lambda$ the corresponding restricted Bayes risk. Concerning $r_\lambda(C)$ we have the

LEMMA. *The function $r_\lambda(C)$ is convex, continuous, and strictly decreasing.*

PROOF.

(i) $r_\lambda(C)$ is obviously nonincreasing. If $C' < C''$, $0 < \gamma < 1$ and

$$\delta = \gamma\delta_\lambda(C') + (1 - \gamma)\delta_\lambda(C''),$$

then

$$\int R_\delta\, d\lambda = \gamma r_\lambda(C') + (1 - \gamma)\, r_\lambda(C'')$$

and

$$\sup R_\delta(\theta) \leqq \gamma C' + (1 - \gamma)C'',$$

from which the convexity of $r_\lambda(C)$ follows.

(ii) From well known properties of convex functions, the continuity of $r_\lambda(C)$ is now obvious except at the point $C = \underline{C}$. But let $C_i \downarrow \underline{C}$, let δ_i be the corresponding restricted Bayes solutions and δ_{i_j} the convergent subsequence guaranteed by Assumption A. If $\delta_{i_j} \to \delta^*$, we have $r_\lambda(C_i) = \int R_{\delta_{i_j}}\, d\lambda \to \int R_{\delta^*}\, d\lambda$. By monotonicity $r_\lambda(\underline{C}) \geqq \int R_{\delta^*}\, d\lambda$. Also, since $R_{\delta_{i_j}}(\theta) \to R_{\delta^*}(\theta)$ for each θ, we have that $\sup R_{\delta^*}(\theta) \leqq \lim C_{i_j} = \underline{C}$, so that $\sup R_{\delta^*}(\theta) = \underline{C}$. But this implies $r_\lambda(\underline{C}) \leqq \int R_{\delta^*}\, d\lambda$ and hence $r_\lambda(\underline{C}) = \int R_{\delta^*}\, d\lambda = \lim r_\lambda(C_{i_j})$.

(iii) Before proving that $r_\lambda(C)$ is strictly decreasing we shall now show that for any $\underline{C} < C < \bar{C}$ we have

$$\sup_\theta R_{\delta_\lambda(C)}(\theta) = C.$$

Suppose that $\sup R_{\delta_\lambda}(\theta) < C$. If $\bar{C} < \infty$ there exists $0 < \gamma < 1$ so that $\delta = \gamma\delta_\gamma + (1 - \gamma)\delta_\lambda(C)$ still satisfies $\sup R_\delta(\theta) \leqq C$, but we would have $\int R_\delta\, d_\lambda < \int R_{\delta_\lambda(C)}\, d\lambda$. If $\bar{C} = \infty$, then $C_i \to \bar{C}$ implies that $\int R_{\delta_\lambda(C_i)}\, d\lambda \to \int R_{\delta_\lambda(\bar{C})}\, d\lambda$ and the same argument applies.

(iv) Strict monotonicity of $r_\lambda(C)$ is an obvious consequence of (iii). For let $C' < C''$ and suppose that $r_\lambda(C') = r_\lambda(C'')$. Then $\delta_\lambda(C')$ would be a solution not only corresponding to C' but also to C'' in contradiction to (iii).

We can now state a closure and continuity theorem.

THEOREM 3. *If $\{\delta_i\}$ is a sequence of restricted Bayes solutions converging regularly to δ^*, then δ^* is a restricted Bayes solution, and $\sup_\theta R_{\delta^*}(\theta) = \lim_{i\to\infty} \sup_\theta R_{\delta_i}(\theta)$.*

PROOF. Let r_i, r^* be the Bayes risks of δ_i, δ^* respectively. By Assumption

B, $r_i \to r^*$. By the lemma, we can conclude that the sequence $C_i = \sup_\theta R_{\delta_i}(\theta)$ is convergent; denote $\lim C_i$ by C_0. Clearly $C^* = \sup R_{\delta^*}(\theta) \geqq C_0$, and since for each θ $R_{\delta_i^*}(\theta) = \lim R_{\delta_i}(\theta) \leqq C_0$, we conclude $C^* \leqq C_0 = C^*$. The lemma now assures δ^* to be a restricted Bayes solution.

COROLLARY. *The set of risk functions corresponding to restricted Bayes solutions is closed with respect to the convergence of Assumption B.*

PROOF. Let R_i be the risk functions corresponding to restricted Bayes solutions δ_i, and let $R_i \to R^*$. By the compactness Assumption A, we may extract a subsequence δ_{i_j} which converges regularly to some δ^*. By Theorem 3, δ^* is a restricted Bayes solution, whose risk function R^* is the limit of $R_{\delta_{i_j}}$, and hence of R_{δ_i}.

5. Extensions. We shall now mention briefly some extensions of the notion of a restricted Bayes solution. To give a first simple example, it may happen that we have ideas concerning the range of the parameter but not concerning a possible distribution over this range. Thus it may be known that $\theta \in \Omega$ and may further be indicated by other considerations that actually $\theta \in \omega$ where $\omega \subset \Omega$. We may then, subject to the condition $R_\delta(\theta) \leqq C_0$ for all $\theta \in \Omega$, wish to minimize $\sup_{\theta \in \omega} R_\delta(\theta)$. For example, when testing the hypothesis that three means θ_1, θ_2, θ_3 are equal we may believe that the most likely alternatives are such that $\theta_1 < \theta_2 < \theta_3$ without however definitely being able to exclude the other possibilities.

We can get further refinements as follows. Let $\Lambda_0 \subset \Lambda_1 \subset \cdots \subset \Lambda_{r-1} \subset \Lambda_r$ be a nested set of families of distributions. We may be certain that the true distribution of θ is an element of Λ_r (Λ_r may of course contain all distributions that assign probability 1 to a single point), fairly sure that it lies in Λ_{r-1}, still believe that is in Λ_{r-2}, etc. Accordingly we could select a decreasing sequence of constants $C_r > C_{r-1} > \cdots > C_1$, each C_i greater than the restricted minimax risk at this stage, and minimize $\sup_{\lambda \in \Lambda_i} \int R_\delta \, d\lambda$ subject to the condition $\sup_{\lambda \in \Lambda_i} \int R_\delta \, d\lambda \leqq C_i$ for $i = 1, \cdots, r$.

There is an extension of Theorem 1 to the present case that we give as an example of how the theory of the earlier sections generalizes.

THEOREM 4. *Suppose there exists $\rho_0 > 0$ and a distribution*

$$\nu_0 = \rho_0 \lambda_0 + \sum_{i=1}^{r} \rho_i \lambda_i, \qquad \rho_i \geqq 0, \qquad \sum_{i=0}^{r} \rho_i = 1,$$

with $\lambda_i \in \Lambda_i$ and such that the Bayes solution δ_0 of ν_0 satisfies

$$\sup_{\lambda \in \Lambda_i} \int R_{\delta_0} \, d\lambda = C_i = \int R_{\delta_0} \, d\lambda_i \qquad \textit{for } i = 1, \cdots, r \tag{1}$$

and

$$\sup_{\lambda \in \Lambda_0} \int R_{\delta_0} \, d\lambda = \int R_{\delta_0} \, d_{\lambda_0}. \tag{2}$$

Then δ_0 *minimizes* $\sup_{\lambda\epsilon\Lambda_0} \int R_\delta \, d\lambda$ *subject to*

$$\sup_{\lambda\epsilon\Lambda_i} \int R_\delta \, d\lambda \leqq C_i \qquad \textit{for } i = 1, \cdots, r. \tag{3}$$

PROOF. Let δ be any procedure satisfying (3). Then

$$\sum_{i=0}^{r} \rho_i \sup_{\lambda\epsilon\Lambda_i} \int R_\delta \, d\lambda \geqq \sum \rho_i \int R_\delta \, d\lambda_i \geqq \sum \rho_i \int R_{\delta_0} \, d\lambda_i = \sum \rho_i \sup_{\lambda\epsilon\Lambda_i} \int R_\delta \, d\lambda.$$

But for $i = 1, \cdots, r$ we have

$$\sup_{\lambda\epsilon\Lambda_i} \int R_\delta \, d\lambda \leqq C_i = \sup_{\lambda\epsilon\Lambda_i} \int R_{\delta_0} \, d\lambda$$

and the conclusion follows.

Of the conditions, (2) of course becomes vacuous in case Λ_0 contains only a single distribution.

6. Examples. We conclude the paper by discussing a number of examples which serve to illustrate the ideas and theorems. As might be expected, it is more difficult to obtain explicit results with the restricted Bayes approach than with the logically simpler minimax or Bayes principles. In fact, the minimax principle has been successfully applied, in most cases, by guessing the answer and then verifying it through use of the specialized form of Theorem 1. It appears that the restricted Bayes solutions are often much less simple in their mathematical structure than the minimax solutions, and will accordingly be much harder to guess. We suspect that the widespread application of the restricted Bayes approach would require numerical methods in combination with theoretical results of the kind given above.

EXAMPLE 1. Let $X = (X_1, X_2, \cdots, X_n)$ be random variables having the joint density function $p_\theta(x) = c(\theta)b(x)e^{t(x)q(\theta)}$, where θ is an unknown real parameter and $q(\theta)$ is a monotonely increasing function of θ. This form includes many of the distributions frequently used in statistics, such as the Poisson, binomial, and normal with known mean or variance. Consider the problem of testing, at a given level of significance, the hypothesis $H: \theta = \theta_0$ against the alternatives $\theta \neq \theta_0$. We are indifferent to alternatives $\theta_1 \leqq \theta \leqq \theta_2$, where $\theta_1 < \theta_0 < \theta_2$. As risk function, we shall use the probability of false acceptance. Thus, the risk is $1 - \beta(\theta)$, where $\beta(\theta)$ is the power function. Suppose finally that past experience or other considerations suggest that, should the hypothesis $\theta = \theta_0$ be false, then one of the alternatives $\theta \geqq \theta_2$ is true.

It is known [2] that an essentially complete class of tests consists in those of the form $w(k)$: reject H if $t(x) \leqq k$ or $t(x) \geqq f(k)$, where $f(k)$ is determined by the given level of significance. We may restrict attention to these tests. Denote the power of $w(k)$ at alternative θ by $\beta_k(\theta)$. If we had complete confidence in the presumption that either $\theta = \theta_0$ or $\theta \geqq \theta_2$, we should seek to maximize the power against the latter alternatives, with no regard to the power against the alternatives $\theta \leqq \theta_1$.

This leads to the known uniformly most powerful one-sided test $w(-\infty)$. This test is obtainable as the Bayes test corresponding to any a priori distribution for θ which assigns all of the probability to values $\theta \geqq \theta_2$. However, the power of $w(-\infty)$ tends to 0 as θ tends to its lower limit, so we may get very poor performance if our presumption is not correct. At the other extreme, if we placed no reliance on the presumption, we might seek the minimax test. It is easy to see that the power functions all have unique minima, and are continuous in k for fixed θ, whence the minimax test will be $w(\underline{k})$, where $\underline{k}$ is determined by the condition $\beta_{\underline{k}}(\theta_1) = \beta_{\underline{k}}(\theta_2)$. This test then gives no better protection against large alternatives than against small, and thus makes no use at all of our presumption that large alternatives are the ones to fear.

The restricted Bayes approach suggests a compromise, under which we would seek to maximize $\inf_{\theta \geq \theta_2} \beta(\theta)$ subject to $\inf_\theta \beta(\theta) \geqq C$. It can be shown that, for $-\infty < k \leqq \underline{k}$, $\inf_{\theta \geq \theta_2} \beta_k(\theta) = \beta_k(\theta_2)$, while $\inf_\theta \beta_k(\theta)$ is attained at a finite value of $\theta \leqq \theta_1$. From these facts the Lemma of [2] enables us to conclude, for $k' < k < k''$, $k \leqq \underline{k}$, that $\inf_{\theta \geq \theta_2} \beta_k(\theta) > \inf_{\theta \geq \theta_2} \beta_{k''}(\theta)$, while $\inf_\theta \beta_k(\theta) > \inf_\theta \beta_{k'}(\theta)$. It follows that for every $-\infty < k \leqq \underline{k}$, $w(k)$ is a restricted Bayes solution in the sense given. The same result holds if we assign to θ, any a priori distribution under which $P(\theta \geqq \theta_2) = 1$, and apply the restricted Bayes principle in the narrower sense of Section 2.

It is easily seen that $C(k) = \inf_\theta \beta_k(\theta)$ is a continuous, monotonely increasing function of k in the interval $-\infty \leqq k \leqq \underline{k}$. The tests $w(k)$, $k \leqq \underline{k}$, thus provide essentially unique restricted Bayes solutions for all possible values of C. In this example, our principle provides nothing essentially new, since the admissible tests already form only a one-parameter family.

Example 2. Consider next the binomial random variable X = number of successes on n independent trials, each having the probability p of success, and the problem of estimating p from X. Take as loss function the square of the error of estimate. Suppose that a theoretical examination of the experimental situation reveals considerations of symmetry which suggest that $p = \frac{1}{2}$. We wish to design an estimate which will take advantage of this theory, in the sense of providing small risk at $p = \frac{1}{2}$, but will not place complete reliance in the theory, in the sense of giving some control over the risk for all values of p. This problem falls within the framework of the restricted Bayes approach, if we give to p the a priori distribution λ which assigns all of its probability to the value $\frac{1}{2}$. Our objective is then to minimize the risk at $\frac{1}{2}$, subject to a given maximum C for the risk over $0 \leqq p \leqq 1$.

Denote the estimate corresponding to a value x of X by $\delta(x)$, and the risk involved in using this estimate by $R_\delta(p) = E[\delta(X) - p]^2$. The ordinary Bayes estimate is $\bar{\delta}(x) = \frac{1}{2}$ for every x, since this estimate reduces the risk at $\frac{1}{2}$ to 0, while every other estimate has positive risk at $\frac{1}{2}$. The maximum risk of $\bar{\delta}$ is $\bar{C} = 1/4$, attained at $p = 0$ and $p = 1$. The minimax estimate for this problem is known (see [3]) to be $\underline{\delta}(x) = \dfrac{\sqrt{n}}{\sqrt{n}+1}\dfrac{X}{n} + \dfrac{1}{2(\sqrt{n}+1)}$. This estimate has the constant risk $\underline{C} = 1/4(\sqrt{n}+1)^2$.

In getting the restricted Bayes estimates, we first observe that we need only consider those estimates which possess the symmetry property $\delta(x) + \delta(n - x) = 1$. For, if $\delta(x)$ is any estimate, the estimate $\delta'(x) = 1 - \delta(n - x)$ will have the same maximum risk and the same Bayes risk as δ, for any a priori distribution which is symmetric about $p = 1/2$. Thus, the estimate $\frac{1}{2}(\delta(x) + \delta'(x))$, which possesses the symmetry property cited, will have the same Bayes risk as δ and no larger maximum risk.

Consider the distribution μ_1 which assigns probability 1/2 to each of the points 0 and 1. The Bayes solution δ_1 corresponding to the a priori distribution $\rho\lambda + (1 - \rho)\mu_1$ is easily found to be of the form: $\delta_1(0) = 1 - \delta_1(n)$, $\delta_1(x) = 1/2$ for $0 < x < n$. By virtue of Theorem 2, δ_1 is a restricted Bayes estimate if it achieves

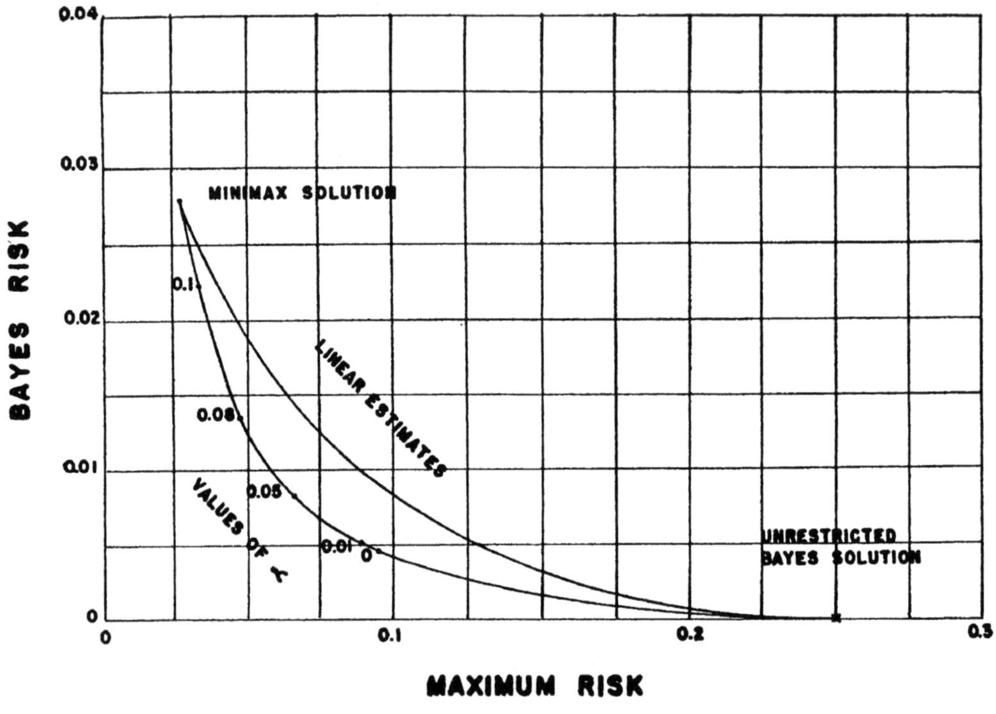

its maximum risk only at 0 and 1. But this is true for $\bar{R}(p) = (p - \frac{1}{2})^2$, and hence by continuity (Theorem 3) must hold true for some interval $(C_1, \bar{C})$ of values of C. In fact, C_1 may be calculated from the condition $R'_1(0) = 0$. Again using the continuity of the variation of the risk function with C, we see that for some interval (C_2, C_1), the distribution μ_2 will consist of probability 1/2 at points α and $1 - \alpha$, the value of α varying continuously with C. At $C = C_2$, the risk function will develop further maxima, requiring a modification of the form of the distribution μ, and so forth. But in any case, the distribution μ must have only finitely many points of positive probability, since the risk function must achieve its maximum at all such points, is a polynomial in p, and is not constant except for $C = \underline{C}$.

The computing program outlined in the preceding paragraph was carried out numerically for the case $n = 4$, leading to the relation between C and $R(1/2)$

shown in the graph. The restricted Bayes estimates are (except for the extremes of minimax and Bayes estimates) not linear functions of X. For comparison, the graph also shows the minimum risk at $p = 1/2$ attainable for given maximum risk using only linear estimates. We see that, over a wide range of values of C, the use of nonlinear estimates permits nearly a 50 per cent reduction in the risk at $p = 1/2$.

EXAMPLE 3. As an illustration of the application of our principle to a finite decision problem, consider the following genetic situation. We are concerned with the inheritance of a simple Mendelian trait. Two individuals F and G are crossed to produce n progeny, each of whom is then crossed with a known hybrid to produce a single offspring. Of the n third-generation individuals, R are found to be recessive, while the remaining $n - R$ are either dominant or hybrid. Suppose it is known from other evidence that F and G are neither both dominant nor both recessive. Then the number k of recessive genes which they possess between them is either 1, 2, or 3. Our problem is to infer the value of k from the observed value of R. As our risk, we take the probability of wrong inference.

It is easily seen that R has the binomial distribution corresponding to $p = k/8$. The complete class of inference procedures is obtainable as the class of Bayes procedures corresponding to a priori distributions λ over the three possible values of p. If $P(p = i/8) = \lambda_i$, $i = 1, 2, 3$, then the Bayes solution, when R is observed to have the value r, is to decide for that value of k for which

$$\lambda_k\binom{n}{k}(k/8)^r(1 - k/8)^{n-r}$$

is greatest. These solutions have the following structure. There exist two numbers a and b, such that our decision is for $k = 1$ if $R < a$, for $k = 2$ if $a < R < b$, for $k = 3$ if $b < R$, while if $R = a$, we choose between $k = 1$ and $k = 2$, and if $R = b$, we choose between $k = 2$ and $k = 3$ according to certain probabilities. Let $P(\text{choosing } k = 1 \mid R = a) = \pi_1$, and $P(\text{choosing } k = 2 \mid R = b) = \pi_2$.

Suppose the proportions of dominant and recessive genes in the population at large are known to be μ and $1 - \mu$, respectively. Then, if F and G are bred under panmixia (i.e., chosen independently and randomly from the population), we should have

$$P(k = 1):P(k = 2):P(k = 3) = 2(1 - \mu)^2:3\mu(1 - \mu):2\mu^2.$$

The Bayes solution corresponding to this distribution for k would be the reasonable decision procedure to use if we were sure that F and G had been bred under panmixia. If we placed no reliance at all in the panmixia hypothesis, we might prefer to employ the minimax solution, which is characterized by equal probabilities of error corresponding to the three possible values of k. If we placed some reliance in the panmixia hypothesis, but not complete reliance, we might impose a limit on the permissible probability of error for any value of k, and subject to this limit seek to minimize the average probability of error under panmixia.

The minimax, Bayes, and restricted Bayes solutions are easy to obtain numerically. For illustration, suppose $n = 20$ and $\mu = 0.8$. The three solutions are as follows.

Solution	*a*	*b*	π_1	π_2
Minimax	3	7	0.482	0.142
Bayes	0.5	4.5	...	...
Restricted Bayes, maximum probability of error = 1/2	2	5	0.868	0.828

The performance characteristic properties of the three solutions may be conveniently compared in tabular form.

Solution	*Maximum probability of error*	*Average probability of error under panmixia*
Minimax	0.354	0.354
Restricted Bayes	0.500	0.239
Bayes	0.931	0.234

The restricted Bayes solution loses little efficiency as compared with the Bayes solution if panmixia holds, but gives considerably better protection if it does not hold.

It may be remarked that, in any finite decision problem, the restricted Bayes solution may be found by means of a finite number of applications of the Neyman-Pearson fundamental lemma. Suppose there are m decisions, so that the risks associated with any decision procedure may be represented as a vector $(r_1, r_2, \cdots, r_m)$ in m-dimensional Euclidean space. Except in the trivial situation in which the Bayes solution itself satisfies the restrictive condition that $r_i \leqq C$ for every $i = 1, 2, \cdots, m$, we may conclude from part (iii) of the proof of the Lemma of Section 3, that the restricted Bayes solution will have $r_i = C$ for at least one i. Let $I(C)$ be the set of i such that $r_i = C$. The restricted Bayes solution corresponding to C may then be obtained by minimizing $\sum \lambda_j r_j$, subject to $r_i = C$ for all i in $I(C)$. If we compute these minimizing solutions for all sets I, using the Neyman-Pearson lemma, we may select the restricted Bayes solution from among them by inspection.

REFERENCES

[1] A. Wald, *Statistical Decision Functions*, John Wiley and Sons, 1950.

[2] E. L. Lehmann, "On families of admissible tests," *Annals of Math. Stat.*, Vol. 18 (1947), pp. 97–104.

[3] J. L. Hodges, Jr. and E. L. Lehman, "Some problems in minimax point estimation," *Annals of Math. Stat.*, Vol. 21 (1950), pp. 182–197.

ON THE EXISTENCE OF LEAST FAVORABLE DISTRIBUTIONS[1]

By E. L. Lehmann

Stanford University and University of California, Berkeley

1. Summary. Sufficient conditions for the existence of a least favorable distribution were given by Wald in his work on general decision theory. It is shown here that for problems of hypothesis testing and more generally for multiple decision problems involving a finite number of decisions, the result holds under a much weaker restriction than Wald's assumption of a compact parameter space.

2. Introduction. In his general theory of decision functions Wald has given conditions under which there exists a least favorable distribution for Nature. One of these, which is rather restrictive and not required for many parts of the theory, is condition (3.7) of [1], a compactness assumption on the parameter space. This condition can, of course, in general not be dispensed with; there are many examples of simple statistical decision problems for which a least favorable distribution does not exist. However, as we shall show below, for certain special classes of problems the compactness assumption is not required.

The following is a typical example. Suppose that θ is a location parameter and that we are interested in testing $H:\theta \leqq \theta_0$ against $K:\theta \geqq \theta_1$. If a least favorable distribution does not exist this means essentially that positive probability is assigned to at least one of the points $\theta = \pm\infty$. But it is intuitively clear that by doing this, Nature would be playing into the hands of the statistician, since for sufficiently large $|\theta|$ it is very easy to determine whether or not H is true. In fact one would expect Nature to do better if the above strategy S were replaced by the (conditional) distribution for θ obtained from S, given that no probability is placed on $\theta = \pm\infty$.

In the next section we shall make this argument precise and use it to establish the existence of a least favorable distribution for certain hypothesis testing and multiple decision problems. In particular our condition is satisfied for many problems of testing a composite hypothesis against a simple alternative. This shows that a method given in [2] for determining the most powerful test against a simple alternative is applicable in most problems of the kind usually considered.

3. Least favorable distributions for problems of hypothesis testing. In the present section we shall consider the hypothesis H that the probability density of a random variable X is $f(x)$ with $f \varepsilon \mathfrak{F}$, where $\mathfrak{F}$ is some given class of densities, and the class of alternatives K that the density of X is $g_\theta(x)$ with $\theta \varepsilon \Omega$. We shall assume that the sample space $\mathfrak{X}$ is Euclidean and that $\theta = (\theta_1, \cdots, \theta_s)$, where Ω is a Borel set in R_s. Since throughout we shall be concerned with dis-

[1] Work sponsored in part by the office of Naval Research.

tributions over Ω we may, by defining these distributions to be zero outside of Ω, assume without loss of generality that Ω is the full Euclidean space R_s. We assume further that for almost all x, $g_\theta(x)$ is continuous in θ and measurable in $\Omega \times \mathfrak{X}$.

For the present we consider the usual asymmetric formulation of the problem in which a level of significance α is assigned and it is desired to maximize the minimum power of the test over K. In this situation, as is pointed out on page 133 of [1], assumptions (3.1) – (3.6) of [1] hold, and we have, as a simple consequence of Theorem 3.12 of [1],

LEMMA 3.1. *There exists a level-α test ϕ of H that maximizes the minimum power against K, and a sequence of distributions λ_i over Ω such that*

$$\lim_{i\to\infty} \int E_\theta \phi(X)\, d\lambda_i(\theta) = \inf_{\theta\epsilon\Omega} E_\theta \phi(X) \tag{3.1}$$

and

$$\lim_{i\to\infty} \beta_i = \inf_{\theta\epsilon\Omega} E_\theta \phi(X), \tag{3.2}$$

where β_i is the power of the most powerful level-α test for testing H against the simple alternative

$$\int g_\theta(x)\, d\lambda_i(\theta).$$

Now by Helly's compactness theorem there exists a subsequence of the sequence $\{\lambda_i\}$, which without loss of generality we shall take to be the original sequence, such that

$$\lambda_i \to \gamma\mu \tag{3.3}$$

for all continuity intervals of μ, where μ is again a distribution and $0 \leqq \gamma \leqq 1$. If here we could assume that $\gamma = 1$ it would follow that μ is least favorable (see the proof of Theorem 3.14 of [1]). We shall now show that μ is least favorable by proving that either $\gamma = 1$ or $\lim \beta_i = 1$ under the following

ASSUMPTION A. *Given any $\epsilon > 0$ and any closed bounded set $\omega \subset \Omega$ there exists a Borel set S in the sample space such that*

$$\begin{aligned} P(\bar{S} \mid f) &\leqq \epsilon \qquad \text{for all } f \ \epsilon\ \mathfrak{F}, \\ P(\bar{S} \mid g_\theta) &\leqq \epsilon \qquad \text{for all } \theta\ \epsilon\ \omega, \end{aligned} \tag{3.4}$$

and

$$P(S \mid g_\theta) \to 0 \text{ uniformly as } \theta_1^2 + \cdots + \theta_s^2 \to \infty. \tag{3.5}$$

Clearly Assumption A is satisfied if both $\mathfrak{F}$ and Ω are compact, but as we shall show later the assumption also holds in many cases in which (3.7) of [1] is not satisfied.

The proof of our result is based on the following two lemmas.

LEMMA 3.2. *Let ω be any closed bounded set in Ω and let $\epsilon > 0$. Then there exists a Borel set S in $\mathfrak{X}$ and an integer i_0 such that*

$$\int_{\bar{\omega}} P(\bar{S}) \mid \theta) \, d\lambda_i(\theta) \geqq 1 - \gamma - \epsilon \qquad \textit{for } i \geqq i_0, \tag{3.6}$$

and even

$$\int_{\bar{\omega}} E_\theta[\phi(X) c_{\bar{S}}(X)] \, d\lambda_i(\theta) \geqq 1 - \gamma - 2\epsilon \qquad \textit{for } i \geqq i_0, \tag{3.7}$$

where ϕ is the test referred to in Lemma 3.1 and where $c_{\bar{S}}(x)$ is the characteristic set function of $\bar{S}$.

Here and in what follows the notation $P(\bar{S} \mid \theta)$ and E_θ indicates that the probability or expectation is computed with respect to the probability density $g_\theta(x)$.

LEMMA 3.3. *For any $\epsilon > 0$ there exists a continuity interval ω of μ such that for any test ϕ*

$$\int_{\omega} E_\theta \phi(X) \, d\lambda_i(\theta) \geqq \gamma \int_{\Omega} E_\theta \phi(X) \, d\mu(\theta) - \epsilon \tag{3.8}$$

for sufficiently large i.

Before proving these lemmas we shall show that they imply $\gamma = 1$ or $\lim \beta_i = 1$ and hence the existence of a least favorable distribution. Let ϕ be the test referred to in Lemma 3.1 and let

$$\lim_{i \to \infty} \int E_\theta \phi(X) \, d\lambda_i(\theta) = \beta. \tag{3.9}$$

Given any $\epsilon > 0$ let ω be a bounded subset of Ω for which (3.8) holds. Then

$$\int_{\Omega} E_\theta \phi(X) \, d\lambda_i(\theta) = \int_{\omega} E_\theta \phi(X) \, d\lambda_i(\theta) + \int_{\bar{\omega}} E_\theta \phi(X) \, d\lambda_i(\theta),$$

where for sufficiently large i the first term is $\geqq \gamma \int_{\Omega} E_\theta \phi(X) \, d\mu(\theta) - \epsilon$ by (3.8) and the second term is $\geqq 1 - \gamma - 2\epsilon$ by (3.7), so that

$$\beta \geqq (1 - \gamma) + \gamma \int_{\Omega} E_\theta \phi(X) \, d\mu(\theta) \geqq \int_{\Omega} E_\theta \phi(X) \, d\mu(\theta). \tag{3.10}$$

Here the last inequality is strict and therefore contradicts (3.1), unless either

$$\gamma = 1 \text{ or } \int_{\Omega} E_\theta \phi(X) \, d\mu(\theta) = 1.$$

But in the latter case $\beta = 1$ by (3.10), so that

$$\int_{\Omega} E_\theta \phi(X) \, d\mu(\theta) = \inf E_\theta \phi(X)$$

and μ would be least favorable.

PROOF OF LEMMA 3.2. Let S be a set guaranteed by Assumption A for which

$$P(\bar{S} \mid \theta) \leqq \epsilon/2 \quad \text{for all } \theta \in \omega, \tag{3.11}$$

$$P(\bar{S} \mid f) \leqq \epsilon \quad \text{for all } f \in \mathfrak{F}, \tag{3.12}$$

and such that

$$P(S \mid \theta) \to 0 \text{ uniformly as } \theta_1^2 + \cdots + \theta_s^2 \to \infty. \tag{3.13}$$

From (3.3) and (3.13) it follows that

$$\int_\Omega P(\bar{S} \mid \theta)\, d\lambda_i(\theta) \geqq 1 - \gamma - \epsilon/2 \quad \text{for } i \geqq i_0, \tag{3.14}$$

and (3.6) is an immediate consequence of (3.14) and (3.11).

Suppose now that (3.7) did not hold. Then there would exist a subsequence of $\{\lambda_i\}$, which we may again take to be the full sequence such that

$$\lim_{i\to\infty} \int_{\bar{\omega}} E_\theta[\phi(X)c_S(X)]\, d\lambda_i(\theta) = \delta < 1 - \gamma. \tag{3.15}$$

We define a new test ϕ^* by

$$\phi^*(x) = \begin{cases} 1 & \text{if } x \in \bar{S}, \\ \dfrac{\alpha - \epsilon}{\alpha}\, \phi(x) & \text{if } x \in S. \end{cases} \tag{3.16}$$

Then for any $f \in \mathfrak{F}$ we have, because of (3.12)

$$E(\phi^*(X) \mid f) \leqq \frac{d - \epsilon}{\alpha}\, \alpha + P(\bar{S} \mid f) \leqq \alpha, \tag{3.17}$$

so that ϕ^* is again a level-α test. Also, for all x,

$$\phi^*(x) \geqq \frac{\alpha - \epsilon}{\alpha}\, \phi(x) + c_S(x)[1 - \phi(x)], \tag{3.18}$$

and hence

$$\begin{aligned} \int E_\theta\phi^*(X)\, d\lambda_i(\theta) \geqq{}& \frac{\alpha - \epsilon}{\alpha} \int E_\theta\phi(X)\, d\lambda_i(\theta) \\ &+ \int E_\theta\{c_S(X)[1 - \phi(X)]\}\, d\lambda_i(\theta). \end{aligned} \tag{3.19}$$

The first term on the right-hand side of (3.19) tends to $\beta(\alpha - \epsilon)/\alpha$ as $i \to \infty$. On the other hand the second term is greater than or equal to

$$\int_{\bar{\omega}} P_\theta(\bar{S})\, d\lambda_i(\theta) - \int_{\bar{\omega}} E_\theta[c_S(X)\phi(X)]\, d\lambda_i(\theta),$$

which for sufficiently large i exceeds $(1 - \gamma - \epsilon) - (\delta + \epsilon)$ by (3.15) and (3.6). This shows that for sufficiently large i

$$\int E_\theta \phi^*(X)\, d\lambda_i(\theta) > \beta + \frac{1-\gamma-\delta}{2}$$

and thus contradicts (3.2), since it implies that for sufficiently large i ϕ^* is more powerful than the most powerful test for testing H against $\int g_\theta(x)\, d\lambda_i(\theta)$.

Proof of Lemma 3.3. Let ω be a continuity interval of μ, and so large that

$$\mu(\bar{\omega}) \leqq \epsilon/2. \tag{3.20}$$

Then by the Helly-Bray theorem, as $i \to \infty$

$$\int_\omega g_\theta(x)\, d\lambda_i(\theta) \to \gamma \int_\omega g_\theta(x)\, d\mu(\theta),$$

and hence by Fatou's lemma

$$\liminf_i \int \left[\phi(x) \int_\omega g_\theta(x)\, d\lambda_i(\theta)\right] dx \geqq \gamma \int \left[\phi(x) \int_\omega g_\theta(x)\, d\mu(\theta)\right] dx.$$

Interchanging the order of integration we then have for i sufficiently large

$$\int_\omega E_\theta \phi(X)\, d\lambda_i(\theta) \geqq \gamma \int_\omega E_\theta \phi(X)\, d\mu(\theta) - \epsilon/2 \geqq \gamma \int_\Omega E_\theta \phi(X)\, d\mu(\theta) - \epsilon$$

by (3.20).

We can sum up the work of this section in

Theorem 3.1. *If $g_\theta(x)$ is continuous in θ and if Assumption A holds, then for the problem of testing H against K there exists a least favorable distribution for θ.*

As an illustration, let $X_1, \cdots, X_n$ be a sample from a distribution with density $f(x - \theta)$ and consider testing $H: \theta \leqq \theta_0$ against $\theta \geqq \theta_1$. Here we could clearly take for S a set of the form $X_1 \leqq c$, so that Assumption A is satisfied.

4. Testing against a simple alternative. As an application of Theorem 3.1 we now consider the problem of testing a composite hypothesis against a simple alternative. Let the hypothesis state that the density of X is $f_\theta(x)$, $\theta \in \Omega$, and let the alternative be $g(x)$, where we assume as before that the sample space is Euclidean and that $\theta = (\theta_1, \cdots, \theta_s)$ with Ω being the full Euclidean space R_s.

Assumption B. *For every closed bounded set S in the sample space, $P(S \mid f_\theta) \to 0$ uniformly as $\theta_1^2 + \cdots + \theta_s^2 \to \infty$.*

Theorem 4.1. *If $f_\theta(x)$ is a continuous function of θ for almost all x, then under Assumption B there exists a least favorable distribution over Ω.*

Proof. Theorem 3.1 clearly applies if we interchange H, α with K, β. We therefore only need to show that Assumption B implies Assumption A with f and g interchanged. That is, we need to prove that Assumption B implies the existence of a set S such that

$$P(\bar{S} \mid g) \leqq \epsilon, \tag{4.1}$$

$$P(\bar{S} \mid f_\theta) \leqq \epsilon \qquad \text{for all } \theta \in \omega, \tag{4.2}$$

and

(4.3) $$P(S \mid f_\theta) \to 0 \text{ as } \theta_1^2 + \cdots + \theta_s^2 \to \infty.$$

It was pointed out by Scheffe [3] that when $g_\theta(x)$ is continuous in θ, the convergence $\theta_i \to \theta$ implies

(4.4) $$\int_S g_{\theta_i}(x)\, dx \to \int_S g_\theta(x)\, dx \text{ uniformly for all Borel sets } S.$$

Therefore if ω is bounded and closed, the associated set of distributions is compact with respect to the convergence definition (4.4). Let $\epsilon > 0$. Then we can find a finite number of points in ω, say $\theta_1, \cdots, \theta_r$, such that every $\theta \varepsilon \omega$ is within $\frac{1}{2}\epsilon$ of one of these θ_i in the sense that $| P(S \mid \theta) - P(S \mid \theta_i) | \leqq \frac{1}{2}\epsilon$ for all Borel sets S. It follows that there exists a closed bounded set S for which (4.1) and (4.2) hold and (4.3) then follows from Assumption B.

Assumption C. *There exist s statistics $Y_i = h_i(X), i = 1, \cdots, s$, such that the h_i are continuous, the marginal distribution of Y_i depends only on θ_i, and Y_i tends to $+\infty$ or $-\infty$ as $\theta_i \to +\infty$ or $-\infty$.*

Lemma 4.1. *A sufficient condition for Assumption B to hold is Assumption C.*

Proof. Let S be a closed bounded set, and let $a_i = \min_S h_i(x), b_i = \max_S h_i(x)$. Then $P_\theta(X \varepsilon S) \leqq P_{\theta_i}(a_i \leqq Y_i \leqq b_i)$ for $i = 1, \cdots, s$. Now if $\theta_1^2 + \cdots + \theta_s^2 > C$ we must have $\theta_i^2 > C/s$ for at least one value of i. Given $\epsilon > 0$, let C be so large that

$$P_{\theta_i}(a_i \leqq Y_i \leqq b_i) < \epsilon \qquad \text{if} \qquad \theta_i^2 > \frac{C}{s} \qquad \text{for all } i.$$

Then $P_\theta(X \varepsilon S) \leqq \epsilon$ provided $\theta_1^2 + \cdots + \theta_s^2 > C$.

As an example let $X_1, \cdots, X_n$ be a sample from a distribution with unknown location and scale parameter ξ and η. Let $\theta_1 = \xi$, $\theta_2 = \log \eta$, and take $Y_1 = X_1/| X_2 - X_1 |$, $Y_2 = \log | X_2 - X_1 |$. Then Assumption C is seen to be satisfied, so that for testing a hypothesis of this type against a simple alternative, a least favorable distribution exists.

Theorem 4.1 answers a problem raised by [2], where in Theorem 1 a method is given for proving that a test is most powerful for testing a composite hypothesis against a simple alternative. Combining Theorem 4.1 with Theorem 3.10 of [1], we see that this method is always applicable when Assumption B is satisfied.

While it is quite likely that Assumption B can be weakened somewhat, it is perhaps of interest to point out that some condition of this type is required, which restricts the behavior of the distributions as $\theta_1^2 + \cdots + \theta_s^2 \to \infty$. As an example consider the following situation in which Ω is denumerable and the densities are given by

$$H: f_i(x) = \frac{1}{2\sqrt{2\pi}} [e^{-\frac{1}{2}(x-i)^2} + e^{-\frac{1}{2}(x-(1/i))^2}], \qquad i = m, m+1, \cdots,$$

$$K: g(x) = \frac{1}{\sqrt{2\pi}} e^{-\frac{1}{2}x^2}.$$

We shall show that if m is sufficiently large the most powerful level-α test rejects when $x \leqq b$, where b is determined by

$$\lim_{i\to\infty} \int_{-\infty}^{b} f_i(x)\, dx = \tfrac{1}{2} \int_{-\infty}^{b} g(x)\, dx = \alpha. \tag{4.5}$$

First one checks easily that for large i $\alpha(i) = \int_{-\infty}^{b} f_i(x)\, dx$ is an increasing function of i. Let m be such that $\alpha(i)$ is increasing for $i \geqq m$. Then the proposed test is of size α and power 2α by (4.5). But if β is the power of the most powerful test ϕ we clearly have

$$\beta \leqq 2 \limsup_{i\to\infty} \int f_i(x)\, \phi(x)\, dx \leqq 2\, \alpha,$$

which proves that the region of rejection $x \leqq b$ is most powerful. If a least favorable distribution λ over Ω existed, the test $x \leqq b$ would be a Bayes solution for λ (Theorem 3.9 of [1]), that is, would be most powerful for testing $\int f_i(x)\, d\lambda(i)$ against $g(x)$. But an application of the Neyman-Pearson fundamental lemma shows immediately that no Bayes solution is of this form.

It may be worth pointing out that the conditions of the present paper are really of a quite different nature from those given by Wald in his general theory. Thus they do not even imply that the parameter space is weakly compact in the sense of (3.1) of [1]. As an example, consider a random variable X the distribution of which depends on an unknown location parameter θ. Let H denote the hypothesis $\theta \leqq \theta_0$ and suppose that the simple alternative is $\theta_1 > \theta_0$. Let $\{\lambda_i\}$ be any sequence of distributions over H, and let $\phi(x)$ be any test. Then weak compactness of H would imply the existence of a subsequence $\{\lambda_{i_j}\}$ and a distribution λ_0 over H such that

$$\limsup_{j\to\infty} \int E_\theta\, \phi(x)\, d\lambda_{i_j}(\theta) \leqq \int E_\theta\, \phi(x)\, d\lambda_0(\theta). \tag{4.6}$$

But let $\phi(x)$ be the characteristic set function of the set $x \leqq C$ and let $\lambda_i(\theta)$ assign probability 1 to the point $\theta = -i$. Then the left-hand side of (4.6) is 1, while for any distribution λ_0 the right-hand side is <1, so that H is not weakly compact. On the other hand, Assumption B is satisfied.

5. Some extensions. We shall now indicate briefly how the results of Section 3 may be extended to somewhat more general decision problems. As a first generalization consider the problem of hypothesis testing, formulated as in Section 3 except that the loss function need no longer be simple. Suppose that acceptance of the hypothesis when the true distribution is given by $g_\theta(x)$, $\theta \in \Omega$ results in a loss $W(\theta)$, and that $V(f)$ is the loss resulting from rejection of H when f is the true density. We shall assume that V and W are bounded and that we are interested in minimizing $\sup_\theta W(\theta) E_\theta[1 - \phi(x)]$ among all tests ϕ for which $V(f)E[\phi(x) \mid f] \leqq \alpha$ for all $f \in \mathfrak{F}$.

Let Assumption A′ be obtained from Assumption A by replacing $P(\bar{S} \mid g_\theta)$ and $P(S \mid g_\theta)$ in (3.4) and (3.5) by $W(\theta)P(\bar{S} \mid g_\theta)$ and $W(\theta)P(S \mid g_\theta)$ respectively, and $P(\bar{S} \mid f)$ in (3.4) by $V(f)P(\bar{S} \mid f)$. Then Assumption A′ implies the existence of a least favorable distribution over Ω. Only the obvious changes are required in the proof of Theorem 3.1.

As an immediate further generalization one next obtains an analogous result for the following two-decision problem. The parameter space is partitioned into two set $\omega_1 \cup \omega_2$. The loss is zero if $\theta \varepsilon \omega_i$ and decision d_i is taken; it is $W_i(\theta)$ if

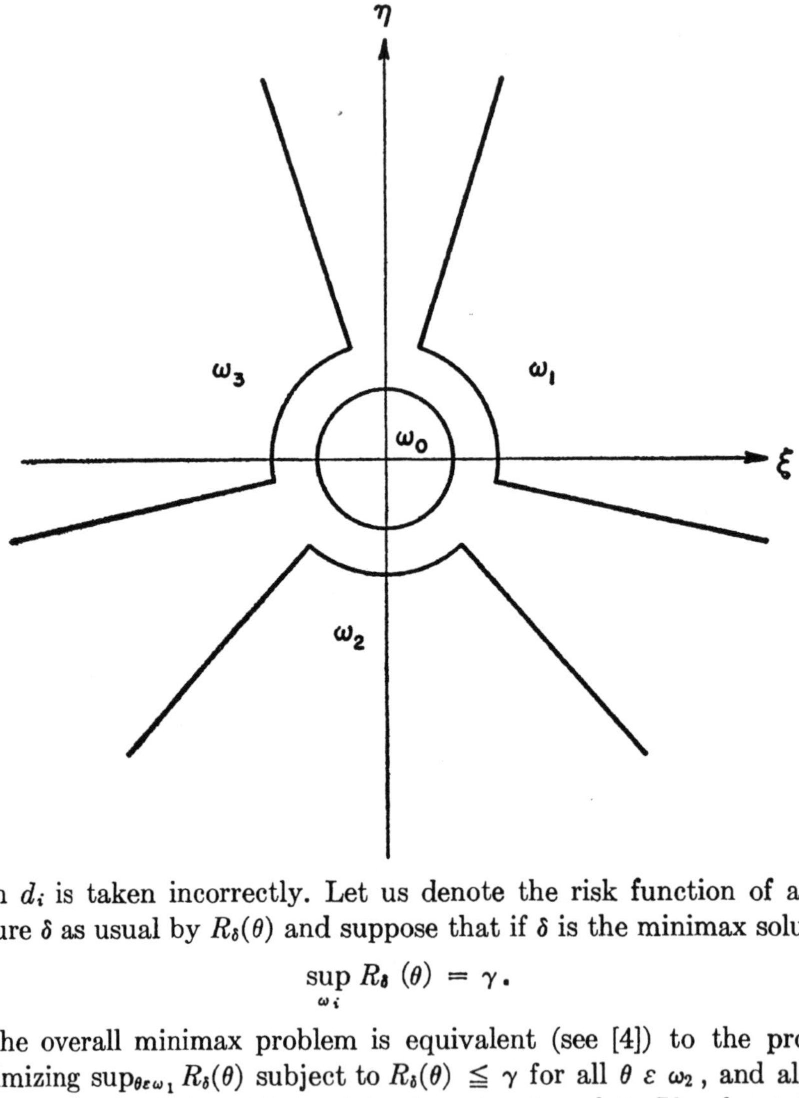

decision d_i is taken incorrectly. Let us denote the risk function of a decision procedure δ as usual by $R_\delta(\theta)$ and suppose that if δ is the minimax solution

$$\sup_{\omega_i} R_\delta(\theta) = \gamma.$$

Then the overall minimax problem is equivalent (see [4]) to the problem Π_1 of minimizing $\sup_{\theta \varepsilon \omega_1} R_\delta(\theta)$ subject to $R_\delta(\theta) \leqq \gamma$ for all $\theta \varepsilon \omega_2$, and also to the problem Π_2 obtained from Π_1 by interchanging 1 and 2. If a least favorable distribution over ω_i exists for Π_i $(i = 1, 2)$, then one exists for the overall problem. But each of the partial problems is of the type discussed at the beginning of this section.

The whole argument extends without much difficulty to multiple decision problems involving a finite number of decisions and having a bounded loss function. We shall not give the details but simply state the result. Let $\omega_1, \cdots, \omega_K$ be the regions in the parameter space in which the decisions $d_1, \cdots, d_K$ are appropriate so that the loss $W(\theta, d_i) = 0$ if $\theta \varepsilon \omega_i$. Then we consider the two-decision problems Π_i $(i = 1, \cdots, K)$, where the decision lies only between $\theta \varepsilon \omega_i$ and $\theta \varepsilon \Omega - \omega_i$. A least favorable distribution exists over Ω provided each of the problems Π_i satisfies Assumption A′.

As an example consider the case that $X_1, \cdots, X_m$; $Y_1, \cdots, Y_n$ are $m + n$ independent random variables, and that the density of the X's is $f(x - \xi)$ while that of the Y's is $g(y - \eta)$. Suppose that the partition of Ω is as shown in the diagram and that the loss is 0 or 1 as the correct or an incorrect decision is taken. Then a least favorable distribution exists. To show that Π_3 satisfies Assumption A′ suppose that the bounding rays l_1 and l_2 of ω_3 are given by $\eta = m_1\xi$ and $\eta = m_2\xi$. Then for $\bar{S}$ we may take a set in the sample space of the form

$$x_1^2 + y_1^2 \geqq R, \qquad m_1 + c_1 \leqq y_1 \leqq m_2x_1 + c_2,$$

where R, c_1 and c_2 must be chosen sufficiently large.

REFERENCES

[1] A. Wald, *Statistical Decision Functions*, John Wiley & Sons, 1950.

[2] E. L. Lehmann and C. Stein, "Most powerful tests of composite hypotheses. I. Normal distributions," *Annals of Math. Stat.*, Vol. 19 (1948), pp. 495–516.

[3] H. Scheffé, "A useful convergence theorem for probability distributions," *Annals of Math. Stat.*, Vol. 18 (1947), pp. 434–438.

[4] C. Blyth, "On minimax statistical decision procedures and their admissibility," *Annals of Math. Stat.*, Vol. 22 (1951), pp. 22–42.

$$\frac{1}{B\left(\frac{1}{2}, \frac{\nu}{2}\right)} \int_{x_\alpha}^{1} p^{-\frac{1}{2}}(1 - p)^{(\nu-2)/2} \, dp = \alpha.$$

Using the approximation of ordinate over abscissa for the cumulative normal for extreme abscissa we find that z is the abscissa of a cumulative normal which is approximately equal to the power of the t-test for alternative δ. In a similar manner the normal approximation to the binomial yields $z = \delta\sqrt{r+1}$ for the sign test. A fixed value of N and α determines r, α, x_α and we may solve for ν.

REFERENCES

[1] National Bureau of Standards, *Tables of the Binomial Probability Distribution*, U. S. Government Printing Office, Washington, D. C.

[2] K. Pearson, *Tables of the Incomplete Beta Function*, Cambridge University Press, 1934.

[3] Burton H. Camp, "Approximation to the point binomial," *Ann. Math. Stat.*, Vol. 22 (1951), pp. 130–131.

[4] W. G. Cochran, "The efficiencies of the binomial test of significance of a mean and of a correlation coefficient," *J. Roy. Stat. Soc.*, Vol. 100 (1937), pp. 69–73.

[5] W. J. Dixon and A. M. Mood, "The statistical sign test," *J. Amer. Stat. Assn.*, Vol. 41 (1946), pp. 557–566.

[6] Murray F. Freeman and John W. Tukey, "Transformations related to the angular and the square root," *Ann. Math. Stat.*, Vol. 21 (1950), pp. 607–611.

[7] T. A. Jeeves and Robert Richards, "A note on the power of the sign test," *Ann. Math. Stat.*, Vol. 21 (1950), p. 618.

[8] W. L. Nicholson, "A computing formula for the power of the analysis of variance test," submitted to *Ann. Math. Stat.*

[9] John E. Walsh, "On the power function of the sign test for slippage of means," *Ann. Math. Stat.*, Vol. 17 (1946), pp. 358–362.

[10] John E. Walsh, "On the asymptotic power efficiency of the sign test for slippage of means," Douglas Aircraft Co., Inc. Report.

[11] John E. Walsh, "Some comments on the efficiency of statistical tests," *Human Biology*, Vol. 21 (1949), pp. 205–217.

[12] John E. Walsh, "Some significance tests for the median which are valid under very general conditions," *Ann. Math. Stat.*, Vol. 20, (1949), pp. 64–81.

[13] John E. Walsh, "Some bounded significance level properties of the equal-tail sign test," *Ann. Math. Stat.*, Vol 22 (1951), pp. 408–417.

THE ADMISSIBILITY OF CERTAIN INVARIANT STATISTICAL TESTS INVOLVING A TRANSLATION PARAMETER

By E. L. Lehmann[1] and C. M. Stein

University of California, Berkeley, and University of Chicago

1. Introduction. The notion of invariance (or symmetry) has such strong intuitive appeal that many current statistical procedures have the invariance property and are in fact the best invariant procedures although they were pro-

Received 2/10/53.

[1] This work is supported in part by the Office of Naval Research.

posed long before a general discussion of invariance was available. Hotelling [1], [2] and Pitman [3], [4] emphasized the invariant nature of certain tests and estimates. A general definition of the notion for the problem of testing hypotheses was given by Hunt and Stein who showed that in this case under severe restrictions on the group of transformations an optimum invariant test is most stringent or more generally minimax with respect to an invariant loss function (see [5]). This result has been extended to more general decision problems and more general groups by Peisakoff [6]. However, these results do not prove admissibility of the procedures in question unless the group of transformations is compact.

The problem of admissibility in the case of point estimation of a location parameter was treated in the normal case by Blyth [7] and by Hodges and Lehmann [8] and for a general class of location parameter-problems by Blackwell [9]. In the latter paper the surprising fact was brought to light that even in the location parameter problem the best invariant estimate may, under certain circumstances, be inadmissible.

In the present note we prove under conditions which are presumably unnecessarily restrictive the admissibility of the most powerful invariant test for testing one location parameter family against another. As an example, consider the problem in which $Z_1, \cdots, Z_n$ are normally distributed with unknown mean ζ and variance σ^2. If we wish to test $H: \zeta \leqq 0$ against the alternatives $K: \zeta > 0$, it was already pointed out in ([5], p. 15) that Student's t-test is admissible for this problem. This result is quite elementary and rests on the fact that unbiasedness in this case implies that the probability of rejection equals the level of significance for all points (ζ, σ) with $\zeta = 0$. However, this argument breaks down if we introduce an indifference zone and restrict our class of alternatives to $K': \zeta/\sigma \geqq \delta$ where δ is some specified positive number.

Consider now the general problem in which one observes a random point (X, Y) where X ranges over an arbitrary set, Y over the real line. There are two hypotheses H_i according to which the distribution of $(X, Y - \eta)$ is $F_i (= 1, 2)$ where η is unspecified. The problem discussed above is an example of this, if we take H_i to be $\zeta/\sigma = \delta_i$, $X = \sum Z_i/\sqrt{\sum Z_j^2}$, $Y = \log \sum Z_j^2$ and $\eta = \log \sigma$. As another example let $(Z_1 - \eta, Z_2 - \eta, \cdots, Z_n - \eta)$ have distribution F_i under H_i. Then we can take for X the set of differences $X = (Z_1 - Z_n, \cdots, Z_{n-1} - Z_n)$ and for Y the mean $\bar{Z}$ or the observation Z_n, or any of a number of other statistics.

2. The principal theorem. Let $\mathfrak{X}$ be a set (which for all practical purposes may be taken to be a Euclidean space), $\mathfrak{A}$ a σ-algebra of subsets of $\mathfrak{X}$ (say, the ordinary Borel sets if $\mathfrak{X}$ is Euclidean), $\mathfrak{R}$ the real line, $\mathfrak{B}$ the set of all ordinary Borel subsets of $\mathfrak{R}$, λ_1, λ_2 probability measures on $\mathfrak{A}$ and for each x, let F_{1x}, F_{2x} be probability measures on $\mathfrak{B}$ such that for each $B \varepsilon \mathfrak{B}$, real k, and $i = 1, 2$ $\{x \mid F_{ix}(B) \leqq k\} \varepsilon \mathfrak{A}$. We suppose that the distribution of the random point (X, Y) ranging over $\mathfrak{X} \times \mathfrak{R}$ is, for some real η, with $i = 1$ or 2

$$P_{i\eta}((X, Y) \varepsilon C) = \int_C d\lambda_i(x) \int dF_{ix}(y - \eta). \tag{1}$$

A test for the hypothesis H_1 that it is $P_{1\eta}$ (with η unspecified) is a function φ on $\mathfrak{X} \times \mathfrak{R}$ to $[0, 1]$, $\mathfrak{A}\mathfrak{B}$ measurable. The test φ is said to be better than φ_0 if for all η

$$\begin{aligned} &E_{1\eta}\varphi(X, Y) \leqq E_{1\eta}\varphi_0(X, Y) \\ &E_{2\eta}\varphi(X, Y) \geqq E_{2\eta}\varphi_0(X, Y), \end{aligned} \tag{2}$$

strictly better if (2) holds with strict inequality for some η. φ_0 is admissible if there exists no φ strictly better than φ_0.

THEOREM 1. *If* $E_{10} \mid Y \mid, E_{20} \mid Y \mid < \infty, 0 < c < 1,$

$$\lambda_1 \left\{ x \,\middle|\, \frac{d\lambda_2}{d(\lambda_1 + \lambda_2)}(x) = c \right\} = 0,$$

φ_0 is the test defined by

$$\varphi_0(x, y) = \begin{cases} 1 \text{ if } \dfrac{d\lambda_2}{d(\lambda_1 + \lambda_2)}(x) \geqq c \\[2ex] 0 \text{ if } \dfrac{d\lambda_2}{d(\lambda_1 + \lambda_2)}(x) < c, \end{cases} \tag{3}$$

and φ is better than φ_0, then $\varphi - \varphi_0 = 0$ a.e. $(\lambda_1 + \lambda_2)\mu$ where μ is ordinary Lebesgue measure on the real line.

COROLLARY. *If in addition all F_{ix} are absolutely continuous with respect to μ, then φ_0 is admissible.*

The corollary is an immediate consequence of Theorem 1.

PROOF OF THEOREM 1. Putting $\lambda = \lambda_1 + \lambda_2$, $f(x) = d\lambda_2/d(\lambda_1 + \lambda_2)(x)$ we can rewrite the condition (2) that φ be better than φ_0

$$\begin{aligned} &\int_{f(x)<c} (1 - f(x))\, d\lambda(x) \int (\varphi - \varphi_0)(x, y)\, dF_{1x}(y - \eta) \\ &\qquad - \int_{f(x)\geqq c} (1 - f(x))\, d\lambda(x) \int (\varphi_0 - \varphi)(x, y)\, dF_{1x}(y - \eta) \leqq 0 \end{aligned} \tag{4}$$

$$\begin{aligned} &- \int_{f(x)<c} f(x)\, d\lambda(x) \int (\varphi - \varphi_0)(x, y)\, dF_{2x}(y - \eta) \\ &\qquad + \int_{f(x)\geqq c} f(x)\, d\lambda(x) \int (\varphi_0 - \varphi)(x, y)\, dF_{2x}(y - \eta) \leqq 0. \end{aligned} \tag{5}$$

Multiplying (4) by c and (5) by $1 - c$ and adding we obtain

$$\begin{aligned} &\int_{f(x)<c} c(1 - f(x))\, d\lambda(x) \int (\varphi - \varphi_0)(x, y)\, dF_{1x}(y - \eta) \\ &\qquad + \int_{f(x)\geqq c} (1 - c)f(x)\, d\lambda(x) \int (\varphi_0 - \varphi)(x, y)\, dF_{2x}(y - \eta) \\ &\qquad \leqq \int_{f(x)<c} (1 - c)f(x)\, d\lambda(x) \int (\varphi - \varphi_0)(x, y)\, dF_{2x}(y - \eta) \\ &\qquad + \int_{f(x)\geqq c} c(1 - f(x))\, d\lambda(x) \int (\varphi_0 - \varphi)(x, y)\, dF_{1x}(y - \eta). \end{aligned} \tag{6}$$

In order to derive the conclusion of Theorem 1 from (6) we shall need the

Lemma. *If* $\mathfrak{X}$, $\mathfrak{A}$, $\mathfrak{R}$, $\mathfrak{B}$ *are as before,* ρ *a probability measure on* $\mathfrak{A}$, h_1, h_2 $\mathfrak{A}$-*measurable functions on* $\mathfrak{X}$ *to* $[0, 1]$ *with* $h_1 - h_2 > 0$ a.e. (ρ), ψ *an* $\mathfrak{A}\mathfrak{B}$-*measurable function on* $\mathfrak{X} \times \mathfrak{R}$ *to* $[0, 1]$, *and for each* x, H_{1x}, H_{2x} *probability measures on* $\mathfrak{B}$ *such that*

$$\text{(7)} \qquad \text{for each } B \in \mathfrak{B} \text{ real } k \text{ and } i = 1, 2, \{x \mid H_{ix}(B) \leq k\} \in \mathfrak{A}$$

$$\text{(8)} \qquad \int h_i(x)\, d\rho(x) \int |y|\, dH_{ix}(y) < \infty.$$

$$\text{(9)} \qquad \begin{aligned} &\int h_1(x)\, d\rho(x) \int \psi(x, y)\, dH_{1x}(y - \eta) \\ &\qquad \leq \int h_2(x)\, d\rho(x) \int \psi(x, y)\, dH_{2x}(y - \eta) \qquad \text{for all real } \eta \end{aligned}$$

then $\psi = 0$ a.e. $(\rho\mu)$.

Proof of Lemma. We can rewrite (9)

$$\text{(10)} \quad \int h_1(x)\, d\rho(x) \int \psi(x, y + \eta)\, dH_{1x}(y) \leq \int h_2(x)\, d\rho(x) \int \psi(x, y + \eta)\, dH_{2x}(y).$$

Now

$$\text{(11)} \qquad \begin{aligned} &\int_{-n}^{n} d\eta \int \psi(x, y + \eta)\, dH_{2x}(y) - \int_{-n}^{n} \psi(x, \eta)\, d\eta \\ &\qquad = \int dH_{2x}(y) \left[\int_{-n+y}^{n+y} \psi(x, \eta)\, d\eta - \int_{-n}^{n} \psi(x, \eta)\, d\eta \right] \\ &\qquad \leq \int_{y \geq 0} dH_{2x}(y) \int_{n}^{n+y} \psi(x, \eta)\, d\eta + \int_{y \leq 0} dH_{2x}(y) \int_{-n+y}^{-n} \psi(x, \eta)\, d\eta \\ &\qquad \leq \int |y|\, dH_{2x}(y) \end{aligned}$$

and

$$\text{(12)} \qquad \begin{aligned} &\int_{-n}^{n} d\eta \int \psi(x, y + \eta)\, dH_{1x}(y) - \int_{-n}^{n} \psi(x, \eta)\, d\eta \\ &\qquad = \int dH_{1x}(y) \left[\int_{-n+y}^{n+y} \psi(x, \eta)\, d\eta - \int_{-n}^{n} \psi(x, \eta)\, d\eta \right] \\ &\qquad \geq -\int_{y \geq 0} dH_{1x}(y) \int_{-n}^{-n+y} \psi(x, \eta)\, d\eta - \int_{y \leq 0} dH_{1x}(y) \int_{n+y}^{n} \psi(x, \eta)\, d\eta \\ &\qquad \geq -\int |y|\, dH_{1x}(y). \end{aligned}$$

Integrating (10) with respect to η from $-n$ to n and using the final forms of (11), (12) we obtain

$$\int [h_1(x) - h_2(x)]\, d\rho(x) \int_{-n}^{n} \psi(x, \eta)\, d\eta \tag{13}$$
$$\leqq \int h_1(x)\, d\rho(x) \int |y|\, dH_{1x}(y) + \int h_2(x)\, d\rho(x) \int |y|\, dH_{2x}(y).$$

Consequently,

$$\int [h_1(x) - h_2(x)]\, d\rho(x) \int_{-\infty}^{\infty} \psi(x, \eta)\, d\eta < \infty \tag{14}$$

and for every $\delta > 0$ there exists n such that

$$\int [h_1(x) - h_2(x)]\, d\rho(x) \int_{|\eta| \geqq n} \psi(x, \eta)\, d\eta \leqq \delta. \tag{15}$$

If instead of using the final forms of (11) and (12) for all x, we use them only in the range $h_1(x) - h_2(x) < \epsilon$ and use the next to final forms when $h_1(x) - h_2(x) \geqq \epsilon$ we obtain instead of (13).

$$\int [h_1(x) - h_2(x)]\, d\rho(x) \int_{-n}^{n} \psi(x, \eta)\, d\eta$$
$$\leqq \int_{h_1(x)-h_2(x)<\epsilon} d\rho(x) \left[h_1(x) \int |y|\, dH_{1x}(y) + h_2(x) \int |y|\, dH_{2x}(y) \right]$$
$$+ \int_{h_1(x)-h_2(x)\geqq\epsilon} d\rho(x) \left[h_1(x) \left\{ \int_{y\geqq 0} dH_{1x}(y) \int_{-n}^{-n+y} \psi(x, \eta)\, d\eta \right.\right. \tag{16}$$
$$\left. + \int_{y\leqq 0} dH_{1x}(y) \int_{n+y}^{n} \psi(x, \eta)\, d\eta \right\}$$
$$\left. + h_2(x) \left\{ \int_{y\geqq 0} dH_{2x}(y) \int_{n}^{n+y} \psi(x, \eta)\, d\eta + \int_{y\leqq 0} dH_{2x}(y) \int_{-n+y}^{-n} \psi(x, \eta)\, d\eta \right\} \right].$$

The first term on the right-hand side can be made arbitrarily small by taking ϵ sufficiently small since $h_1(x) - h_2(x) > 0$ a.e. (ρ), $0 \leqq h_i(x) \leqq 1$ (using (8)). For, given $\epsilon > 0$, the second half of the last term can be made arbitrarily small by choosing $n \geqq n(\epsilon)$ sufficiently large since by (15)

$$\int_{h_1(x)-h_2(x)\geqq\epsilon} d\rho(x) \int_{|\eta|\geqq n} \psi(x, \eta)\, d\eta \leqq \delta/\epsilon.$$

Also

$$\int_{h_1(x)-h_2(x)\geqq\epsilon} d\rho(x) h_1(x) \int_{n/2\leqq y} dH_{1x}(y) \int_{-n}^{-n+y} \psi(x, \eta)\, d\eta \tag{17}$$
$$\leqq \int_{h_1(x)-h_2(x)\geqq\epsilon} d\rho(x)\, h_1(x) \int_{n/2\leqq y} y dH_{1x}(y).$$

Again, for fixed ϵ this can be made arbitrarily small by choosing $n \geqq n(\epsilon)$ sufficiently large. Finally

$$\int_{h_1(x)-h_2(x)\geqq\epsilon} d\rho(x)\, h_1(x) \int_0^{n/2} dH_{1x}(y) \int_{-n}^{-n+y} \psi(x, \eta)\, d\eta$$

$$\leqq \int_{h_1(x)-h_2(x)\geqq\epsilon} d\rho(x)\, h_1(x) \int_{-\infty}^{-n/2} \psi(x, \eta)\, d\eta,$$

which is disposed of in the same way as the second half of the last term. The remaining integral with $y \leqq 0$ is analogous. Then, since the right-hand side of (16) is arbitrarily small for sufficiently large n, $\psi = 0$ a.e. $(\lambda_1 + \lambda_2)\mu$. This completes the proof of the Lemma.

To apply the Lemma to (6) we make the following identifications:

(i) If $f(x) < c$,

$$h_1(x) = c(1 - f(x)) \qquad h_2(x) = (1 - c)f(x)$$

$$H_{1x} = F_{1x} \qquad H_{2x} = F_{2x}.$$

$$\psi = \varphi - \varphi_0$$

(ii) If $f(x) \geqq c$

$$h_1(x) = (1 - c)f(x) \qquad h_2(x) = c(1 - f(x))$$

$$H_{1x} = F_{2x} \qquad H_{2x} = F_{1x}.$$

$$\psi = \varphi_0 - \varphi$$

In any case $\rho = \lambda/2$. The reader will readily verify that (7), (8), (9) are satisfied so that the theorem follows.

A moment's reflection shows that the origin of $\mathcal{R}$ for given x is arbitrary so that the hypotheses $E_{i0} \mid Y \mid < \infty$ could be replaced by: There exists an $\mathcal{A}$-measurable real-valued function τ on $\mathfrak{X}$ such that $E_{i0} \mid Y - \tau(X) \mid < \infty$.

It is seen that the admissibility of the noncentral t-test for testing $\zeta/\sigma = \delta_0$ against $\zeta/\sigma = \delta$, (central in case $\delta_0 = 0$) follows immediately from the theorem since

$$E \mid \log \sum Z_1^2 \mid < \infty$$

and $P(\sum Z_i = c\sqrt{\sum Z_i^2}) = 0$.

Another example is that of testing for the same random variables $\sigma = \sigma_0$ against $\sigma = \sigma_1$. Here we may take $X = \sum(Z_i - \bar{Z})^2$ and $Y = \sum Z_i$. Actually in this case the result can be proved quite easily by other means. Instead of taking for ζ the usual least favorable sequence of a priori distributions which in the limit is invariant, we may, if $\sigma_0 < \sigma_1$ take in H the a priori distribution $P(\zeta = a) = 1$ where a is any constant, and in K a normal distribution with mean a and variance $n(1/\sigma_1^2 - 1/\sigma_0^2)$. The Bayes solution is seen to be the F-test which is therefore admissible. (For details see [10]).

We can also consider the general linear hypothesis with no unknown means as nuisance parameters. For brevity we use the terminology of [5]. In the canonical form we have $U_1 \cdots U_m$, $V_1 \cdots V_n$ independently normally distributed with $EU_i = \nu_i$, $EV_j = 0$, $E(U_i - \nu_i)^2 = EV_j^2 = \sigma^2$ where σ^2, ψ_i are unknown and we want to test the hypothesis that all $\nu_i = 0$ say, against $\sum \nu_i^2 \geqq \gamma\sigma^2$. A sufficient statistic is $(U_1 \cdots U_m, \sum V_j^2)$. The problem is invariant under rotation of the vector $U_1, \cdots, U_m$ and multiplication of all U_i, V_j by the same constant c. Since the rotation group G_0 possesses a finite invariant measure, any test invariant under G_0 and admissible among all tests invariant under G_0 is admissible. Thus, in proving the usual F-test admissible we may restrict our attention to tests depending only on $(\sum U_i^2, \sum V_j^2)$. Under multiplication by c this goes into $(c^2 \sum U_i^2, c^2 \sum V_j^2)$. Taking $X = \sum U_i^2 / \sum V_j^2$ and $Y = \log \sum V_j^2$, applying Theorem 1 and the optimum property of the F-test among all those based only on $\sum U_i^2 / \sum V_j^2$, we obtain the admissibility of the usual test: Reject H_0 if $\sum U_i^2 \geqq k \sum V_j^2$. The same argument applies to the problem of testing $H_0 : \sum \nu_i^2 \leqq \gamma_1 \sigma^2$ against $H_1 : \sum \nu_i^2 \geqq \gamma_2 \sigma^2$ with $\gamma_2 > \gamma_1$.

REFERENCES

[1] H. Hotelling, "Analysis of a complex of statistical variables into principal components," *J. Educational Psychology*, Vol. 24 (1933), pp. 417–441 and pp. 498–520.

[2] H. Hotelling, "Relations between two sets of variates," *Biometrika*, Vol. 28 (1936), pp. 321–377.

[3] E. J. G. Pitman, "The estimation of the location and scale parameters of a continuous population of any given form," *Biometrika*, Vol. 30 (1938), pp. 391–421.

[4] E. J. G. Pitman, "Tests of hypotheses concerning location and scale parameters, *Biometrika*, Vol. 31 (1939), pp. 200–215.

[5] E. L. Lehmann, "Some principles of the theory of testing hypotheses," *Ann. Math. Stat.*, Vol. 21 (1950), pp. 1–26.

[6] M. Peisakoff, Unpublished Ph.D. thesis, Princeton University, 1950.

[7] C. Blyth, "On minimax statistical decision procedures and their admissibility," *Ann. Math. Stat.*, Vol. 22 (1951), pp. 22–42.

[8] J. L. Hodges, Jr. and E. L. Lehmann, "Some applications of the Cramér-Rao inequality," *Proceedings of the Second Berkeley Symposium on Mathematical Statistics and Probability*, University of California Press, 1951.

[9] D. Blackwell, "On the translation parameter problem for discrete variables," *Ann. Math. Stat.*, Vol. 22 (1951), pp. 393–399.

[10] E. L. Lehmann and C. Stein, "Most powerful tests of composite hypotheses, I. Normal distributions," *Ann. Math. Stat.*, Vol. 19 (1948), pp. 495–516.

follow. It may be that the use of decision limits which depend upon the value of n will produce a better test than (2), which uses the customary constant limits for the sequential probability ratio test.

REFERENCES

[1] P. G. Hoel, "A sequential test for linear hypotheses," *Ann. Math. Stat.*, Vol. 22 (1951), p. 483, (Abstract).

[2] N. L. Johnson, "Some notes on the application of sequential methods in the analysis of variance," *Ann. Math. Stat.*, Vol. 24 (1953), pp. 614–623.

[3] D. R. Cox, "Sequential tests of composite hypotheses," *Proc. Camb. Philos. Soc.*, Vol. 48 (1952), pp. 290–299.

[4] A. Wald, *Sequential Analysis*, John Wiley and Sons, 1947, pp. 81–84, 203–207.

TWO COMMENTS ON "SUFFICIENCY AND STATISTICAL DECISION FUNCTIONS"

By R. R. Bahadur and E. L. Lehmann

Columbia University and *University of California, Berkeley*

In the following comments we employ the notation and definitions of [1]. The first comment answers a question raised in [1] by giving an example of a necessary and sufficient subfield which cannot be induced by a statistic. The second remark clarifies this example somewhat by discussing the connection between statistics and subfields in general. It was hoped that this connection would be so close as to provide the answer to another question raised in [1]: whether the existence of a necessary and sufficient subfield implies that of a necessary and sufficient statistic. However, an example given at the end of the second comment shows that such a result cannot be proved without making deeper use of sufficiency.

1. A counter example. The following result was communicated to us by David Blackwell.

Lemma 1. (*Blackwell*). *Let* $\mathbf{S}_0$ *be a proper subfield of* $\mathbf{S}$ *and suppose that for each* x *the set* $\{x\}$ *consisting of the single point* x *is in* $\mathbf{S}_0$. *Then* $\mathbf{S}_0$ *cannot be induced by a statistic.*

Proof. Suppose there exists such a statistic, say T, and let $\mathbf{T}$ be the field of sets B in the range of T such that $T^{-1}(B) \in \mathbf{S}$. Since $\{x\} \in \mathbf{S}_0$, there exists $B \in \mathbf{T}$ such that $T^{-1}(B) = \{x\}$, and, by definition of $\mathbf{T}$, a set $A \in \mathbf{S}$ such that $T(A) = B$. We therefore have $T^{-1}[T(A)] = \{x\}$, and since always $T^{-1}[T(A)] \supseteq A$, we have that $T^{-1}[T(x)] = x$ for all x. Therefore, if A is any set in $\mathbf{S}$, we see that $T^{-1}[T(A)] = A$ so that $A \in \mathbf{S}_0$ and hence our assumption that $\mathbf{S}_0$ is induced by T implies that $\mathbf{S}_0 = \mathbf{S}$.

We now give an example of a necessary and sufficient subfield that cannot be

Received July 24, 1953, revised May 24, 1954.

induced by a statistic. A trivial such example would be the case of two normal distributions defined over the Lebesgue sets S of the real line. Then S_0, the class of Borel sets, is a necessary and sufficient subfield, and this cannot, by Lemma 1, be induced by a statistic. The example is, however, uninteresting since $S_0 = S[S, P]$ and S is induced by the necessary and sufficient statistic $T(x) = x$. In this case, therefore, the necessary and sufficient statistic is equivalent to the necessary and sufficient subfield.

An example in which this is not the case can be based on a problem discussed by Halmos ([2], p. 71, prob. 2). Let X be the interval (0, 1), let S_0 be the class of Lebesgue sets of X, and let p denote Lebesgue measure on S_0. Let $M \subset X$ be a fixed set which is not Lebesgue-measurable, and let the complement of M be denoted by $\bar{M}$. Define S to be the class of all sets of the form $EM + F\bar{M}$, with E and F in S_0. Then S is a field containing S_0. We take (X, S) to be the sample space. To define P, let G and H be fixed sets in S_0 such that

$$G \subset M \subset H \text{ and } p(G) = p_*(M),\ p^*(M) = p(H),$$

where p_* and p^* denote inner and outer Lebesgue measure. Let $D = H - G$, and for any set $N = EM + F\bar{M}$ in S define

$$\lambda(N) = p[(EM + F\bar{M})\bar{D}] + \tfrac{1}{2}p(ED) + \tfrac{1}{2}p(FD).$$

Halmos shows that λ is a measure on S such that $\lambda = p$ on S_0. Let the set P of probability measures on S consist of the two measures λ and μ, where μ is defined by $d\mu = 2x\, d\lambda$.

It is easily seen that in this case (i) S_0 is a necessary and sufficient subfield (cf. Theorem 6.2 of [1]), (ii) $T_0(x) = x$ is a necessary and sufficient statistic (cf. Theorem 6.3 of [3]), and (iii) P is a completed set of measures on S, that is, if a set A is S-P-null, then every $B \subset A$ is in S (and therefore S-P-null). We shall show that no necessary and sufficient subfield is equivalent to a statistic; a fortiori, no such subfield is inducible by a statistic.

Suppose to the contrary that S_* is a necessary and sufficient subfield, and that $S_* = S_T[S, P]$, where S_T is the subfield induced by a statistic T. It follows from the essential uniqueness of the necessary and sufficient subfield (cf. Corollary 6.2 of [1]) that S_T is necessary and sufficient. Consequently, by (i), S_T is equivalent to S_0. The sufficiency of S_T means that T is a sufficient statistic. Hence, by (ii), T_0 is essentially a function of T. More precisely, there exists a function F on the range of T into X, and an S-P-null set N, such that $T_0(x) = F(T(x))$ on $X - N$. This, together with (iii), implies that the subfield induced by T_0 is essentially a subfield of S_T. However, T_0 induces S itself, so that S_T must be equivalent to S.

Thus S_T is equivalent to S_0 and also to S. We conclude that S_0 is equivalent to S, that is, $S_0 = S[S, P]$. Since $\bar{M}$ is in S, and P contains λ, this conclusion implies that there exists a set in S_0, E_0 say, such that the symmetric difference of E_0 and $\bar{M}$ is of λ-measure zero. This is, however, a contradiction, since the symmetric difference in question is $E_0M + \bar{E}_0\bar{M}$, and its λ-measure is not less than $\tfrac{1}{2}p(E_0D) + \tfrac{1}{2}p(\bar{E}_0D) = \tfrac{1}{2}p(D)$, and $p(D) = p^*(M) - p_*(M) > 0$.

2. The connection between subfields and statistics. The above Lemma of Blackwell provides a necessary condition for a subfield to be inducible by a statistic. We shall now obtain a necessary and sufficient condition. As is pointed out in [1], any subfield $\mathbf{S}_0$ of $\mathbf{S}$ induces a partition π if we put $x \sim x'$, provided for all $A_0 \varepsilon \mathbf{S}_0$ we have $x \varepsilon A_0 \Leftrightarrow x' \varepsilon A_0$. Let E_x denote the set of π containing x. Then we may characterize E_x as the largest set containing x and such that for all $A_0 \varepsilon \mathbf{S}_0$ either $E_x \subseteq A_0$ or $E_x \subseteq \bar{A}_0$.

Not every partition π can be induced by a subfield of $\mathbf{S}$, and if it can there may be more than one subfield inducing it. Let us denote by C_π the (possibly empty) class of all subfields of $\mathbf{S}$ that induce π. We then have

LEMMA 2. *If C_π is nonempty, it contains a largest member $\mathbf{S}_\pi$ which is given by*

$$\mathbf{S}_\pi = \{A : A \varepsilon \mathbf{S}, x \varepsilon A \Rightarrow E_x \subseteq \mathrm{A} \text{ for all } x \varepsilon X\}.$$

PROOF. Since C_π is nonempty, there exists a subfield $\mathbf{S}_1$ of $\mathbf{S}$ which induces π. For any $A_1 \varepsilon \mathbf{S}_1$ and any $x \varepsilon X$ it follows that $x \varepsilon A_1 \Rightarrow E_x \subseteq A_1$ so that $\mathbf{S}_1 \subseteq \mathbf{S}_\pi$. Therefore, if π' is the partition induced by $\mathbf{S}_\pi$, we see that π' is a refinement of π. (In general, if $\mathbf{S}_i$ induces π_i for $i = 0$ or 1 and if $\mathbf{S}_1 \subseteq \mathbf{S}_0$, then π_0 is a refinement of π_1).

We shall now prove that, conversely, π is a refinement of π'. This will show that $\pi' = \pi$ and hence that $\mathbf{S}_\pi$ induces π. Since we have already shown that $\mathbf{S}_\pi$ contains any $\mathbf{S}_1$ that induces π, this will establish that $\mathbf{S}_\pi$ is the largest member of C_π.

Let E'_x be the set of π' containing x. Then E'_x is the intersection of all $A \varepsilon \mathbf{S}_\pi$ that contain x. Since by definition of $\mathbf{S}_\pi$ all of these sets also contain E_x it follows that $E_x \subseteq E'_x$, as was to be proved.

We can now state

LEMMA 3. *Let C_π be nonempty. Then one and only one of the subfields constituting C_π, can be induced by a statistic, namely $\mathbf{S}_\pi$.*

Hence: A necessary and sufficient condition for a subfield $\mathbf{S}_0$ to be inducible by a statistic is that, $\mathbf{S}_0 = \mathbf{S}_\pi$, if $\mathbf{S}_0$ induces π.

PROOF. We remark first that if T is a statistic, then the subfield $\mathbf{S}_0$ induced by it is the class of all $A_0 \varepsilon \mathbf{S}$ for which $T^{-1}[T(A)] = A$. Now let E_x be defined relative to π as before and let $T(x) = E_x$. Then $T^{-1}[T(x)] = E_x$ and we see that $A_0 \varepsilon \mathbf{S}_0$ (the subfield induced by T) if and only if $A_0 \varepsilon \mathbf{S}$ and $x \varepsilon A_0 \Rightarrow E_x \subseteq A_0$. This shows that $\mathbf{S}_0 = \mathbf{S}_\pi$ and hence that $\mathbf{S}_\pi$ can be induced by a statistic.

On the other hand, let T be any statistic whose subfield $\mathbf{S}_0$ induces π and let $F_x = \{x' : T(x') = T(x)\}$. Then

$$x \varepsilon A_0 \Rightarrow F_x \subseteq A_0, \qquad x \varepsilon \bar{A}_0 \Rightarrow F_x \subseteq \bar{A}_0.$$

It follows from the characterization of E_x given earlier that $F_x \subseteq E_x$. Therefore, $A \varepsilon \mathbf{S}_\pi, x \varepsilon A \Rightarrow F_x \subseteq \mathrm{A}$, and hence $A \varepsilon \mathbf{S}_0$. It follows that $\mathbf{S}_0 = \mathbf{S}_\pi$.

One might hope that Lemma 3 would establish the existence of a necessary and sufficient statistic as $\mathbf{S}_\pi$, where π is the partition induced by a necessary and sufficient subfield. Unfortunately, however, the notion of statistic is not invariant under equivalence [S, P]. A subfield equivalent to a statistic need not itself be a statistic. In an attempt to avoid this difficulty, one may define a *pseudo-statistic* as any subfield equivalent to a statistic. If Lemma 3 remained valid for pseudo-statistics in the sense that a member of C_π is a pseudo-statistic if and only if it is equivalent to $\mathbf{S}_\pi$, this would establish the desired result.

The following example shows that this stronger version of Lemma 3 is not correct. Let $\mathbf{S}_\pi$ be the class of all Lebesgue sets on the real line and $\mathbf{S}_0$ the class of all Lebesgue sets differing only by a set 0 from a set symmetric with respect to the origin. Clearly, $\{x\} \varepsilon \mathbf{S}_0$ for all x so that $\mathbf{S}_0 \varepsilon C_\pi$. Also $\mathbf{S}_0$ is a pseudo-statistic since it is equivalent to the subfield induced by $T(x) = |x|$. But clearly $\mathbf{S}_0$ and $\mathbf{S}_\pi$ are not equivalent.

REFERENCES

[1] R. R. Bahadur, "Sufficiency and statistical decision functions," *Ann. Math. Stat.*, Vol. 25 (1954), pp. 423–462.

[2] P. R. Halmos, *Measure Theory*, D. Van Nostrand Company, Inc., New York, 1950.

[3] E. L. Lehmann and H. Scheffé, "Completeness, similar regions, and unbiased estimation. Part I," *Sankhyā*, Vol. 10 (1950), pp. 305–340.

A NOTE ON CONFIDENCE SETS FOR RANDOM VARIABLES

Lionel Weiss

University of Virginia

Suppose the chance variables $X_1, \cdots, X_m, Y_1, \cdots, Y_n$ have a joint probability distribution depending on the unknown parameters $\theta_1, \cdots, \theta_k$, but otherwise of known form. We assume that there is a set of sufficient statistics for

$$\theta_1, \cdots, \theta_k, \quad \text{denoted by} \quad T_1(X_1, \cdots, X_m, Y_1, \cdots, Y_n), \cdots,$$

$T_r(X_1, \cdots, X_m, Y_1, \cdots, Y_n)$. We shall let X denote the vector $(X_1, \cdots, X_m)$, Y the vector $(Y_1, \cdots, Y_n)$, θ the vector $(\theta_1, \cdots, \theta_k)$, and $T(X, Y)$ the vector $(T_1(X, Y), \cdots, T_r(X, Y))$. $P_\theta(A)$ shall denote the probability of A when the vector of parameters equals θ, and $P_\theta(A \mid B)$ shall denote the conditional probability of A given B when the vector of parameters equals θ.

Given a number α between 0 and 1, if for each vector X we can find a set $S(X)$ in n-dimensional Euclidean space such that $P_\theta(Y \text{ in } S(X)) = \alpha$ identically in θ, then the system of sets $S(X)$ is called a "parameter-free confidence set of level α for the random vector Y."

Since $T(X, Y)$ is a set of sufficient statistics for θ, the joint conditional distribution of Y given that $T(X, Y) = t = (t_1, \cdots, t_r)$ is independent of θ. But then for any given vector t, it is possible to construct a region $S'(t)$ in n-dimensional Euclidean space such that $P_\theta(Y \text{ in } S'(t) \mid T(X, Y) = t) = \alpha$ identically in θ

Received April 7, 1954, revised November 10, 1954.

ON A THEOREM OF BAHADUR AND GOODMAN[1]

By E. L. Lehmann

University of California, Berkeley

1. Introduction. This note is concerned with the problem of selecting the best one (or any other specified number) of several populations. It is restricted to the symmetric case where typically the observations consist of samples of equal size from the different populations. For certain families of distributions, Bahadur (1950) and Bahadur and Goodman (1952) have proved that the natural selection procedure uniformly minimizes the risk among all symmetric procedures for a large class of loss functions. In Section 2 we give an alternative proof of this theorem, and in Section 3 show that the theorem implies many other optimum properties including one obtained in a different manner by Hall (1959).

The problem of selecting the best one of s populations is a finite decision problem with s possible decisions. Let us more generally consider any finite decision problem with possible decisions $d_1, \cdots, d_s$. A (randomized) decision procedure is a vector $\varphi = (\varphi_1, \cdots, \varphi_s)$ where $\varphi_i(x)$ denotes the probability of taking decision d_i when the value of the random observable X is x, and where $\sum \varphi_i = 1$ for all x. We suppose that the distribution P_θ of X depends on the parameter θ and that the loss resulting from decision d_i when θ is the true parameter value is $L(\theta, d_i) = L_i(\theta)$.

Corresponding to the symmetry assumed for the selection problem, we shall assume that the problem is invariant under the finite transformation group $G = \{g_1, \cdots, g_N\}$: if the distribution of X is $d[X] = P_\theta$, the random variable g_iX has distribution $d[g_iX] = P_{\bar{g}_i\theta}$ where g_i and $\bar{g}_i$ are 1:1 mappings respectively of the sample space and of the parameter space onto themselves; furthermore there exist transformations $g_1^*, \cdots, g_N^*$ of the decision space (i.e. permutations of $d_1, \cdots, d_s$) such that for any i, j and θ

$$L(\bar{g}_i\theta, g_i^* d_j) = L(\theta, d_j). \tag{1}$$

A procedure φ is then said to be invariant if

$$g^*\varphi(x) = \varphi(gx) \text{ for all } x \text{ and } g. \tag{2}$$

The procedure taking on the value $\varphi(gx)$ at the point x will be denoted by φg, and (2) can then be written as $g^* \varphi g^{-1} = \varphi$.

To prove that their procedure uniformly minimizes the risk among all invariant procedures, Bahadur and Goodman first characterize the totality of invariant procedures. An altenative proof can be based on the following lemma concerning general finite invariant decision problems.

Received 12 August 1965.

[1] With the partial support of the Office of Naval Research under contract number NONR-222(43).

LEMMA 1. *A necessary and sufficient condition for an invariant procedure* $\varphi^{(0)}$ *to minimize the risk* $R(\theta, \varphi)$ *among all invariant procedures is that it minimizes the average risk*

$$r(\theta, \varphi) = \sum_{i=1}^{N} R(\bar{g}_i \theta, \varphi)/N \tag{3}$$

among all procedures.

PROOF. (i) Let $\varphi^{(0)}$ be an invariant procedure which minimizes $r(\theta, \varphi)$ among all procedures. If φ' is any other invariant procedure, it follows from the fact that the risk function of any invariant procedure is constant over each orbit $\{\bar{g}_i\theta: i = 1, \cdots, N\}$ that

$$R(\theta, \varphi') = r(\theta, \varphi') \geqq r(\theta, \varphi^{(0)}) = R(\theta, \varphi^{(0)})$$

Hence $\varphi^{(0)}$ minimizes $R(\theta, \varphi)$ among all invariant procedures.

(ii) Suppose conversely that $\varphi^{(0)}$ minimizes $R(\theta, \varphi)$ among all invariant procedures and let φ' be any procedure. Then there exists an invariant procedure φ'' such that $r(\theta, \varphi') = r(\theta, \varphi'')$. (This follows from Lemma 2 in Section 3, or can be verified directly by taking $\varphi'' = \sum g_i^* \varphi' g_i^{-1}/N$). We now have

$$r(\theta, \varphi') = r(\theta, \varphi'') = R(\theta, \varphi'') \geqq R(\theta, \varphi^{(0)}) = r(\theta, \varphi^{(0)}),$$

2. The Bahadur-Goodman theorem. Let the random observable X have a density of the form

$$h_\theta(t) = C(\theta) f_{\theta_1}(t_1) \cdots f_{\theta_s}(t_s) \tag{4}$$

with respect to some σ-finite measure μ, where $\theta = (\theta_1, \cdots, \theta_s)$, $t = (t_1, \cdots, t_s)$ and where $t_i = t_i(x)$, $i = 1, \cdots, s$, are real-valued statistics. Suppose that f_{θ_i} has monotone (non-decreasing) likelihood ratio in t_i for each $i = 1, \cdots, s$. Consider the problem of selecting the largest among $\theta_1, \cdots, \theta_s$ and let d_i denote the selection of θ_i.

The following are some typical examples of (4).

(i) X_{ij} ($j = 1, \cdots, n; i = 1, \cdots, s$) are independently normally distributed with mean θ_i and variance σ^2.

(ii) X_i are independently distributed according to the binomial distribution $b(\theta_i, n)$.

(iii) X_i are independently distributed according to the Poisson distribution with mean θ_i.

(iv) $(X_1, \cdots, X_s)$ have a multinomial distribution with success probabilities $(\theta_1, \cdots, \theta_s)$.

(v) A population consists of N items of s different types with $r_i = \theta_i N$ items of type i. A sample of n items is drawn at random. If X_i denotes the number of items of type i in the sample, then the joint distribution of $(X_1, \cdots, X_s)$ is the multiple hypergeometric distributon

$$P(X_1 = x_1, \cdots, X_s = x_s) = \binom{r_1}{x_1} \cdots \binom{r_s}{x_s} \Big/ \binom{N}{n}, \qquad \left(\sum r_i = N, \qquad \sum x_i = n\right).$$

In all these examples, the distribution is of form (4) with $T_i = \sum X_{ij}$ in (i), and $T_i = X_i$ in the other cases.

Invariance can be introduced by assuming that $x = (x_1, \cdots, x_N)$ and that there exist permutations of the x's which leave μ invariant. It is perhaps simpler instead to reduce the problem first to the sufficient statistics $t = (t_1, \cdots, t_s)$ whose joint density is given by (4) with respect to a σ-finite measure ν in t-space which we assume to be invariant under the group G of all permutations of $(t_1, \cdots, t_s)$. (For a discussion of the relation between these approaches, see [9]; for the particular problem at hand it is also discussed in [2]).

THEOREM (Bahadur-Goodman). *Let the distribution of the sufficient statistics* $T = (T_1, \cdots, T_s)$ *have density* (4) *with respect to a σ-finite measure ν which is invariant under G. For any permutation g of* $(t_1, \cdots, t_s)$ *define $\bar{g}$ and g^* as the same permutation of* $(\theta_1, \cdots, \theta_s)$ *and* $(d_1, \cdots, d_s)$ *respectively, and suppose that the loss function L satisfies* (1) *and*

$$\theta_i < \theta_j \Rightarrow L_i(\theta) \geqq L_j(\theta). \tag{5}$$

Let $\varphi^{(0)}$ be the procedure which takes decision d_i when t_i is the unique largest among $(t_1, \cdots, t_s)$, *and which takes decisions $d_{i_1}, \cdots, d_{i_r}$ each with probability $1/r$ if* $(t_{i_1}, \cdots, t_{i_r})$ *is the set of t-values equal to* $\max t_j$ (i.e. *which breaks ties at random*). *Then $\varphi^{(0)}$ uniformly minimizes the risk among all procedures based on t which are invariant under G.*

PROOF. To prove this theorem, it is by Lemma 1 enough to show that the procedure $\varphi^{(0)}$, which is clearly invariant under G, uniformly minimizes $r(\theta, \varphi)$ among all procedures. Now

$$r(\theta, \varphi) = \sum_{i=1}^{s} \int A_i \varphi_i(t)\, d\nu(t) \tag{6}$$

where

$$A_i = \sum_{j=1}^{N} L_i(\bar{g}_j \theta) h_{\bar{g}_j \theta}(t). \tag{7}$$

Suppose without loss of generality that $\theta_1 < \cdots < \theta_s$ and consider the difference $A_i - A_k$ for $i < k$. For any permutation g of $(1, \cdots, s)$, say $g = (j_1, \cdots j_s)$, with $j_i < j_k$ let g' denote the permutation obtained from g by interchanging j_i and j_k. The contribution to $A_i - A_k$ from the terms (in (7)) corresponding to g and g' is

$$[L_i(\bar{g}\theta) - L_k(\bar{g}\theta)] h_{\bar{g}\theta}(t) + [L_i(\bar{g}'\theta) - L_k(\bar{g}'\theta)] h_{\bar{g}'\theta}(t). \tag{8}$$

Since the loss function is invariant, we have

$$L_i(\bar{g}\theta) = L_k(\bar{g}'\theta) \quad \text{and} \quad L_k(\bar{g}\theta) = L_i(\bar{g}'\theta)$$

so that (8) reduces to

$$[L_i(\bar{g}\theta) - L_k(\bar{g}\theta)][h_{\bar{g}\theta}(t) - h_{\bar{g}'\theta}(t)]. \tag{9}$$

The second factor in (9) is $\geqq$ or $\leqq 0$ as

$$f_{\theta_{j_i}}(t_i) f_{\theta_{j_k}}(t_k) \quad \text{is} \quad \geqq \quad \text{or} \quad \leqq \quad f_{\theta_{j_k}}(t_i) f_{\theta_{j_i}}(t_k)$$

and hence, since $\theta_{j_i} < \theta_{j_k}$ and f_θ has monotone likelihood ratio, as t_i is $\leqq$ or $\geqq t_k$.

Hence, for a given point t, the quantity A_i is a minimum for all values of i for which $t_i = \max t_j$. The average risk $r(\theta, \varphi)$ is therefore minimized for any procedure φ which puts $\varphi_i(t) = 0$ whenever $t_i < \max t_j$. Since Lemma 2 requires the procedure to be invariant under G, a minimum risk procedure is obtained by selecting a population with maximum t-value, where ties between maxima are broken at random.

We conclude this section with a discussion of two extensions of the theorem.

(i) One extension is indicated by the example of samples X_{ij} ($j = 1, \cdots, n$; $i = 1, \cdots, s$) from the normal distributions $N(\xi_i, \sigma_i^2)$. Suppose we wish to select the population with the smallest variance, with a loss function independent of the ξ's, satisfying (1), and

$$\sigma_i^2 < \sigma_j^2 \Rightarrow L_i(\sigma_1^2, \cdots, \sigma_s^2) \leqq L_j(\sigma_1^2, \cdots, \sigma_s^2). \tag{10}$$

If the ξ_i are known, the theorem applies with $t_i = \sum (x_{ij} - \xi_i)^2$. If they are unknown, the problem remains invariant not only under permutations of the samples but also under the translations

$$X'_{ij} = X_{ij} + c_i. \tag{11}$$

These induce in the parameter space the transformations

$$\xi_{i'} = \xi_i + c_i, \qquad \sigma_{i'} = \sigma_i$$

and leave the decision space unchanged (i.e. $d_{i'} = d_i$).

A necessary and sufficient condition for a procedure to be invariant under these translations is that it only depend on the differences $X_{ij} - X_{ik}$ within samples, and for these differences the statistics $T_i = \sum (X_{ij} - X_{i\cdot})^2$, $i = 1, \cdots, s$, form a set of sufficient statistics. Application of the Bahadur-Goodman theorem to the distribution of the T's now proves the selection of the population for which $T_i = \min T_j$, to uniformly minimize the risk among all procedures which are invariant both under the translations (11) and under permutation of the T_i (or of the samples).

(ii) The theorem extends easily to the ordered or unordered selection of a fixed number of populations greater than one. The situation is sufficiently well illustrated with the selection of two populations. A selection procedure may now be denoted by $\varphi = \{\varphi_{ij}\}$, the loss resulting from the selection of the ith and jth population by $L_{ij}(\theta)$. In the unordered case, (in which we do not specify which of the two selected populations is considered the best and which the second best) condition (5) is replaced by

$$\theta_i < \theta_j \Rightarrow L_{ik}(\theta) \geqq L_{jk}(\theta) \quad \text{and} \quad L_{ki}(\theta) \geqq L_{kj}(\theta). \tag{12}$$

The optimum procedure selects the two populations with the largest t-values, with ties being broken at random. This procedure uniformly minimizes the risk among all procedures that are invariant under G. If in the proof of the original theorem, A_i in Equations (6) and (7) is replaced by A_{ik} we must show that for fixed t, $A_{ik} = A_{ik}(t)$ is minimized by taking for i and k the subscripts of the two

two largest t's. But this follows from the earlier argument by first holding i fixed and minimizing with respect to $k \neq i$ and then, having determined k, minimizing with respect to i.

In the ordered case, in which d_{ij} corresponds to the selection of θ_i as the largest and θ_j as the second largest value, we must add to (12) the condition

$$\theta_i < \theta_j \Rightarrow L_{ij}(\theta) \geqq L_{ji}(\theta). \tag{13}$$

Then it is seen as in the original proof that

$$A_{ik} \leqq A_{ki} \quad \text{if} \quad t_i \geqq t_k,$$

and this, in conjunction with the result for the unordered case, establishes the optimum property of the obvious procedure in the ordered case.

3. Other optimum properties of $\varphi^{(0)}$. The fact that $\varphi^{(0)}$ uniformly minimizes the risk among all invariant procedures entails a number of other optimum properties. The approach which is essentially that of [10], is most simply given in terms of the general finite decision problem discussed in Section 1. In such a problem, which is assumed to remain invariant under a finite group G, suppose an ordering $\leqq$ is introduced among the procedures (read $\varphi \leqq \varphi'$ as φ' is at least as good as φ).

LEMMA 2. (a) *If the ordering is such that*

(i) $\varphi \leqq \varphi' \Rightarrow g^*\varphi g^{-1} \leqq g^*\varphi' g^{-1}$, *and*

(ii) $\varphi \leqq \varphi^{(i)}, i = 1, \cdots, r \Rightarrow \varphi \leqq \sum_{i=1}^{r} \varphi^{(i)}/r$

then given any procedure φ there exists an invariant procedure φ' such that $\varphi \leqq \varphi'$.

(b) *Suppose that in addition to* (i) *and* (ii) *the ordering satisfies*

(iii) $R(\theta, \varphi') \leqq R(\theta, \varphi)$ *for all* $\theta \Rightarrow \varphi \leqq \varphi'$.

Then, if there exists a procedure $\varphi^{(0)}$ that uniformly minimizes the risk among all invariant procedures, $\varphi^{(0)}$ is optimum with respect to the ordering $\leqq$, i.e.

$$\varphi \leqq \varphi^{(0)} \quad \textit{for all} \quad \varphi.$$

PROOF. (a) The argument is familiar from many special cases, for example Theorem 8.6.4. of Blackwell and Girshick (1954), and consists in noting that

$$\varphi' = \sum_{i=1}^{r} g_i^* \varphi g_i^{-1}/r$$

is invariant and, by (i) and (ii), at least as good as φ.

(b) This is an immediate consequence of (a).

Before applying the lemma we note that it remains valid if instead of considering all procedures φ, attention is restricted to some class $\mathcal{C}$ of procedures which is closed under convex combinations and such that

$$\varphi \in \mathcal{C} \Rightarrow g^* \varphi g^{-1} \in \mathcal{C} \quad \text{for all} \quad g \in G.$$

The following are examples of some orderings satisfying (i)–(iii).

EXAMPLE 1. *Minimax.* $\varphi \leqq \varphi'$ if $\sup R(\theta, \varphi') \leqq \sup R(\theta, \varphi)$

In this case, Lemma 2 reduces to Theorem 8.6.4 of Blackwell and Girshick (1954).

EXAMPLE 2. *Minimax regret*: For a finite decision problem this ordering is given by $\varphi \leqq \varphi'$ if

$$\sup [R(\theta, \varphi') - \min_i L_i(\theta)] \leqq \sup [R(\theta, \varphi) - \min_i L_i(\theta)]$$

EXAMPLE 3. *Average risk*. $\varphi \leqq \varphi'$ if for some specified θ_0

$$\sum R(\bar{g}_i\theta_0, \varphi')/N \leqq \sum R(\bar{g}_i\theta_0, \varphi)/N.$$

Lemma 2, in conjunction with the Bahadur-Goodman theorem proves, for example, that the procedure $\varphi^{(0)}$ of that theorem maximizes the minimum probability of a correct decision whenever the best population is sufficiently much better than the second best (and hence establishes an optimum property of the procedures given in [4], [5], [6] and [11]). This result was proved by the method of least favorable distributions by Hall (1959). It also follows that $\varphi^{(0)}$ minimizes, for example, $\sup [\max \theta_i - \sum \theta_j E_\theta \varphi_j(X)]$, a measure of regret considered in [1] and [3].

REFERENCES

[1] BAHADUR, RAGHU RAJ (1950). On a problem in the theory of k populations. *Ann. Math. Statist.* **20** 362–375.

[2] BAHADUR, RAGHU RAJ and GOODMAN, LEO A. (1952). Impartial decision rules and sufficient statistics. *Ann. Math. Statist.* **23** 553–562.

[3] BAHADUR, RAGHU RAJ and ROBBINS, HERBERT (1950). The problem of the greater mean. *Ann. Math. Statist.* **21** 469–487.

[4] BECHHOFER, ROBERT E. (1954). A single-sample multiple decision procedure for ranking means of normal populations with known variances. *Ann. Math. Statist.* **25** 16–39.

[5] BECHHOFER, ROBERT E., ELENAGHRABI, SALAH and MORSE, NORMAN (1959). A single-sample multiple-decision procedure for selecting the multinomial event which has the highest probability. *Ann. Math. Statist.* **30** 102–119.

[6] BECHHOFER, ROBERT E. and SOBEL, MILTON (1954). A single-sample multiple decision procedure for ranking variances of normal populations. *Ann. Math. Statist.* **25** 273–289.

[7] BLACKWELL, DAVID and GIRSHICK, M. A. (1954). *Theory of Games and Statistical Decisions*. Wiley, New York.

[8] HALL, WM. JACKSON (1959). The most-economical character of some Bechhofer and Sobel decision rules. *Ann. Math. Statist.* **30** 964–969.

[9] HALL, W. J., WIJSMAN, R. A., and GHOSH, J. K. (1965). The relationship between sufficiency and invariance with applications in sequential analysis. *Ann. Math. Statist.* **36** 575–614.

[10] LEHMANN, E. L. (1959). Optimum invariant tests. *Ann. Math. Statist.* **30** 881–884.

[11] SOBEL, MILTON and HUYETT, MARYLIN J. (1957). Selecting the best one of several binomial populations. *Bell System Tech. J.* **36** 537–576.

Part II
Hypothesis Testing

On Testing of Hypotheses

Javier Rojo

Despite the fact that Erich L. Lehmann contributed substantially to several other areas of statistics, his name is prominently linked to hypothesis testing. This is not only because the tremendous impact of his 1959 book *Testing Statistical Hypotheses*, TSH, now in its third edition – Lehmann and Romano (2005), but it is also due to Erich's life-long pioneering and substantial contributions to the area. The chapter on Lehmann's books in the second volume of these selected works contains a fascinating account by Joseph P. Romano on the history of the writing of the substantially expanded third edition of *Testing Statistical Hypotheses*.

This chapter examines and summarizes some of Erich's early work on hypothesis testing. Much of this early work was incorporated into the books on hypothesis testing. The first edition of TSH contained three sections on sequential analysis, but these sections were not included in subsequent editions. In the preface to the second edition, Erich explains the decision to leave these sections out:

> *They are outdated and have been deleted, since it was not possible to do justice to the extensive and technically demanding expansion of this area. This is consistent with the decision not to include the theory of experimental design. Together with sequential analysis and survey sampling, this topic should be treated in a separate book.*

Lawrence D. Brown suggested their inclusion in this volume. Here's Lawrence's motivation:

> *The first edition of TSH has a beautiful treatment of the SPRT in Sections 3.10 - 3.12. This treatment was taken out of subsequent editions of TSH. (At the time, I tried to persuade Erich to leave this in, but he held his motivation for removing it, feeling that the theory was incomplete.) So far as I know no one else has written nearly as elegant a treatment since. It is still useful for contemporary students and academics to understand the fundamental ideas in this topic, even if the modern applications usually involve group sequential trials rather than ordinary one-by-one decisions, and composite hypotheses rather than only the simple vs simple special case treated here. It would thus be of interest - and useful - to reprint these 14 pages. I believe they can be read on their own, without reference to any earlier material in the text.*

J. Rojo
Professor of Statistics, Department of Statistics, Rice University, Houston, TX 77005
e-mail: jrojo@rice.edu

J. Rojo (ed.), *Selected Works of E. L. Lehmann*, Selected Works in Probability and Statistics, DOI 10.1007/978-1-4614-1412-4_11, © Springer Science+Business Media, LLC 2012

Erich and I agreed. These three sections, sections 3.10 - 3.12, of TSH 1959, have been reproduced here.

The papers Lehmann (1947) and Lehmann and Stein (1948) deal with the same theme of testing a composite (null) hypothesis. The 1947 paper extends the work of Scheffé (1942). Let Θ denote a k−dimensional parameter space. Let Θ_0 denote the subset of Θ given by $\{\bar{\theta} \in \Theta : \theta_i = \theta_i^0\}$, for one $i = 1, \ldots, k$. Then the null hypothesis $H_0 : \bar{\theta} \in \Theta_0$ is an example of a composite (null) hypothesis with one constraint. Thus, in fact, the parameters $\theta_j, j \neq i$ are nuisance parameters. The case of more than one constraint is less fruitful, but results of Hsu (1945) are useful in this regard. Neyman (1935) had provided Type B regions for the case of a single nuisance parameter. These results were extended by Scheffé to the case of several nuisance parameters (under H_0), and Scheffé provided sufficient conditions for these Type B regions to also be Type B_1 (uniformly most powerful unbiased) regions. Of special interest, in Lehmann (1947), is the use of Neyman and Pearson's (1933) and Hsu's (1945) methods for finding the totality of similar regions as these regions play a central role in the existence of uniformly most powerful (UMP) or UMP one-sided regions in the case of composite hypotheses with one (or several) constraints. In Lehmann (1947), Erich extended Scheffé's results using the Neyman-Pearson method to obtain uniformly most powerful tests against one-sided alternatives and an example for testing for circular serial correlation in a normal population is presented; Hsu's method is also used to obtain UMP regions in cases, e.g. location & scale exponential and uniform distributions, where the method of Neyman and Pearson does not apply. This is achieved only after obtaining results that characterize the totality of similar regions for a large class of probability distributions that admit a sufficient statistic.

In Lehmann and Stein (1948) the problem of testing a composite hypothesis against a single alternative is addressed. Prior to this work, the approach to handle these problems utilized Neyman and Pearson's idea of restricting attention to similar regions. To fix ideas, let $\mathcal{F}$ denote a family of probability density functions and let g be a probability density function, $g \notin \mathcal{F}$. The test of interest is $H_0 : f \in \mathcal{F}$ against the alternative $H_1 : f = g$. For a level of significance α, Neyman and Pearson restricted attention to critical regions $\mathcal{W}$ such that $\int_{\mathcal{W}} f(x)dx = \alpha$ for all $f \in \mathcal{F}$. Such regions $\mathcal{W}$ are called *similar* regions. Hence a (similar) region is optimal of size α if it maximizes the power $\int_{\mathcal{W}} g(x)dx$ subject to $\mathcal{W}$ being *similar*. In this paper, Lehmann and Stein relaxed the condition of similarity to one requiring only that $\int_{\mathcal{W}} f(x)dx \leq \alpha$ or all $f \in \mathcal{F}$. By adapting the Neyman-Pearson lemma to apply in the present context, they derived sufficient conditions for most powerful tests. The newly developed theory is applied to some examples involving the normal family. Of special interest was Student's problem for which the composite null hypothesis consisted of the normal family with mean 0 and unspecified variance, while the simple alternative hypothesis consisted of the normal distribution with specified mean and variance. The result was somewhat surprising. Here is Erich's account as given in Lehmann (2008):

> *Of our four joint papers, I shall mention only one: "Most Powerful Tests of Composite Hypotheses" (1948). As the title indicates, the problem was to determine a level α test that would maximize the power against a specific alternative. Its purpose was to fill a gap in the classical Neyman-Pearson theory. This theory showed that many standard tests, for example*

the t-test, maximize the power among all unbiased tests. (A test is unbiased if its power against all alternatives is greater or equal to α.) But does this optimum property still hold when the restriction to unbiased tests is dropped and all level α tests are permitted to compete? We developed some general theory for this problem that naturally suggested itself as an adaptation from decision theory, and followed this by examining a number of classical testing problems. In some examples, the standard test retained its optimality against this wider competition; in others it did not. What was needed for these results was the construction of a "least favorable" weighted average of distributions in the hypothesis H as close to the alternative as possible. This least favorable distribution was often suggested by intuition, and then all went swimmingly. However, this turned out not to be the case when we came to the most interesting example, that of the t-test. We conjectured that the least favorable distribution would concentrate all its probability on a single point, but then intuition deserted us. We saw no way of determining this point and were quite frustrated. A few days later, Charles told me that he had solved the problem. Although not determining the point explicitly, he showed by a careful analysis that for $\alpha < 1/2$ such a point exists and gives the right test, which is quite different from the t-test. It has better power in the neighborhood of the specific alternative for which it was designed but lower power elsewhere. For $\alpha \geq 1/2$, the situation turned out to be much easier, and in that case the t-test cannot be improved.

The starting point for Lehmann (1950), Lehmann (1959), and Lehmann (2006) is the same – the likelihood ratio principle for testing. The three papers start by recognizing and lauding the intuition behind the likelihood ratio principle and the "reasonable" tests, in the sense that likelihood ratio tests usually agree with tests derived under optimality criteria, it produces. The three papers then proceed to examine different aspects of the testing problem motivated by the optimality of the likelihood ratio test in some cases, and its total failure in other cases.

In Lehmann (1959), Erich considers the case when the testing problem remains invariant under a group of transformations G. Furthermore, consider a class of invariant tests $\mathcal{F}$ endowed with an order that satisfies certain properties. Erich then proceeds to show that in this case, the optimality properties enjoyed by the likelihood ratio test follow directly from the fact that the test is uniformly most powerful invariant. The paper is also successful in unifying optimality results found in Kiefer (1958), who proved optimality results for symmetrical non-randomized designs, and results found in Wald (1942), who obtained optimality results for the analysis of variance test for the general univariate linear hypothesis. This unification of results under the same umbrella, is another clear example of Erich's system-building, rather than problem-solving, interests.

In Lehmann (2006) and Lehmann (1950), a closer examination of an ever-growing collection of examples where the likelihood ratio test becomes, for all practical purposes, useless, quickly changes the tone and each paper proceeds to examine, against this backdrop, the properties of tests produced by other approaches. When the testing problem remains invariant with respect to a transitive group of transformations, Lehmann (2006) proposed a testing approach – *the likelihood averaged or integrated with respect to an invariant measure approach.* In these cases the resulting test turns out to be uniformly at least as powerful as the corresponding likelihood ratio test, with the former being strictly better except when the two coincide. Moreover, even in the absence of the invariance property, the proposed approach continues to improve on the likelihood ratio test for many cases. In Lehmann (1950), Erich embarked on a survey, and provided some extensions and modifications, of the existing theory. In particular, for example, the *Rao-Blackwellization* version of testing is provided. As a consequence, only tests which are functions of sufficient statistics need to be considered. When the problem remains invariant under certain groups of transformations, using the Hunt-Stein ideas, Erich discussed most stringent tests. Other approaches, such as unbiased tests, are also discussed.

The issue of "statistical significance" versus "practical significance" is addressed in Hodges and Lehmann (1954). Consider, for example, the problem of testing for the mean θ of a normal distribution. Suppose we consider the simple null hypothesis $H_0 : \theta = 0$ vs the alternative $H_a : \theta \neq 0$. Any reasonable test will have power increasing to 1 for every θ^* under the alternative, regardless of how close θ^* is to 0, as the sample size increases to ∞. Motivated by remarks in Berkson (1938), the paper proposes testing instead the null hypothesis that $\theta \in \Theta_0$, where Θ_0 is some neighborhood of 0 – an indifference zone –, and where the choice of Θ_0 is guided by the user's idea of practical significance for the problem at hand. The size of the test is then $\sup_{\theta \in \Theta_0} P_\theta(\textit{rejection})$. Several examples are discussed. Following is Erich's discussion of this paper as it appears in Lehmann (2008):

> *In the classical Neyman-Pearson theory, the hypothesis H: $\theta = \theta_0$ completely specifies the value of the parameter being tested. Frequently, it is more reasonable to consider instead the hypothesis H': $|\theta - \theta_0| < \Delta$, for some $\Delta > 0$. We worked out a number of examples, the most interesting (and rather complicated) being the case of a normal mean. The paper did not find much resonance, but the problem was later revived. It became known as "testing for bio-equivalence", however with the hypothesis and alternatives interchanged so that the hypothesis being tested was: H': $|\theta - \theta_0| > \Delta$.*

In the expository paper Lehmann (1993), Erich discusses the issue of the use of p-values versus the use of a predetermined significance level α, and the Fisher-Neyman feud over this issue. While acknowledging that there are other approaches to testing (Bayesian, fiducial) the paper focuses only on the $p-$value *vs.* predetermined significance level α debate. Erich's approach is, as it was in the case of the *data analytic vs. Bayesian vs. Frequentist* debate – see the chapter on Erich's philosophical work in this volume – to find common ground where both approaches complement, rather than contradict, each other. Erich had stated in various works and interviews that he did not think too much about foundational issues. See, for example, his comments included in the chapter on philosophical work. I commented there that his work demonstrated otherwise. Thus, it is interesting that in this paper he reaffirms this by stating that "I am not a philosopher, and this article is written from a statistical, not a philosophical, point of view". And while this is true, it is also true that the issue at hand is a foundational issue, although the topic might be considered of historical interest, an aspect that Erich was willing to accept. Thus in Lehmann (2008) he writes:

> *During the 1990's, I became interested in the history of statistics and found occasions to write about some aspects of Fisher's work. The first such instance was the surprising selection of me as the Fisher Lecturer for 1988 – surprising in view of the hostility between Neyman and Fisher and my close association with Neyman.*

And,

> *What has been emphasized by many authors is the difference between the calculation of p-values and the use of a predetermined significance level α, the former attributed to Fisher and the latter to Neyman and Pearson. However, in practice most users combine the two by calculating a p-value and then rejecting or accepting the hypothesis as the p-value is below or above α. This is in fact Fisher's own frequent practice. In 1993, I presented an account along these lines in an expository paper,* The Fisher, Neyman-Pearson Theories of Testing Hypotheses: One Theory or Two?

Erich concluded the paper by stating that an approach that takes into account, $p-$values, predetermined significance levels, power considerations, and conditioning, is feasible, while clearly giving conditioning a prominent place in the discussion. In connection with this last issue see the work of Kiefer (1975, 1976, 1977) and chapter 10 of TSH, 3rd edition. This issue of conditioning also appears in an earlier paper – Lehmann (1958). In it, Erich discusses conditional tests in the exponential family setting, where conditioning is with respect to the complete sufficient statistic T. Noticing that, in some testing situations, the conditional power $\beta^*(t) = P_{\theta_1}\{rejecting \mid T = t\}$ at the alternative θ_1 depends on t, Erich then asks the question:

> *Suppose that $\beta^*(t)$ is quite small for the observed t, or quite high; is this value not more relevant to the case in hand than the average $\beta(\theta)$?*

Rather than answering the question directly, Erich then goes on to justify the use of $\beta^*(t)$ in a different way. Since T is complete and sufficient, then $\beta^*(T)$ becomes the best unbiased estimator for $\beta(\theta)$.

References

[1] J. Berkson. Some Difficulties of Interpretation Encountered in the Application of the Chi-Square Test. *J. Am. Statist. Assoc.*, Vol. 33, No. 203, pp. 526-536, 1938.

[2] J. L. Hodges, Jr. and E. L. Lehmann. Testing the approximate validity of statistical hypotheses. *J. Roy. Statist. Soc. Ser. B*, Vol. 26, pp. 261-268, 1954.

[3] P. L. Hsu. On the Power Functions of the E^2-Test and the T^2-Test. *Ann. Math. Statist.*, Vol. 16, No. 3, pp. 278-286, 1945.

[4] J. Kiefer. Conditional confidence approach in multi-decision problems. In *Multivariate Analysis IV*, (P. R. Krishnaiah, ed.), pp. 143-158. Academic Press, New York, 1975.

[5] J. Kiefer. Admissibility of Conditional Confidence Procedures. *Ann. Statist.*, Vol. 4, No. 5, pp. 836-865, 1976.

[6] J. Kiefer. Conditional Confidence Statements and Confidence Estimators. *J. Am. Statist. Assoc.*, Vol. 72, No. 360, pp. 789-808, 1977.

[7] J. Neyman. Sur la vérification des hypothèses statistiques composées. *Bull. Soc. Math. France*, Vol. 63, pp. 1, 1935.

[8] J. Neyman and E. S. Pearson. On the Problem of the Most Efficient Tests of Statistical Hypotheses. *Phil. Trans. R. Soc. Lond. A*, 231, pp. 289-337, 1933.

[9] E. L. Lehmann. On optimum tests of composite hypotheses with one constraint. *Ann. Math. Statist.*, Vol. 18, pp. 473-495, 1947.

[10] E. L. Lehmann and C. Stein. Most powerful tests of composite hypotheses. I. Normal Distributions. *Ann. Math. Statist.*, Vol. 19, No. 4, pp. 495-516, 1948.

[11] E. L. Lehmann. Some principles of the theory of testing hypotheses. *Ann. Math. Statist.*, Vol. 21, pp. 1-26, 1950.

[12] E. L. Lehmann. Significance Level and Power. *Ann. Math. Statist.*, Vol. 29, No. 4, pp. 1167-1176, 1958.
[13] E. L. Lehmann. Optimum invariant tests. *Ann. Math. Statist.*, Vol. 30, pp. 881-884, 1959.
[14] E. L. Lehmann. *Testing Statistical Hypotheses*, John Wiley and Sons, New York, pp. xiii + 369, 1959.
[15] E. L. Lehmann. The Fisher, Neyman-Pearson theories of testing hypotheses: One theory or two? *J. Amer. Statist. Assoc.*, Vol. 88, pp. 1242-1249, 1993.
[16] E. L. Lehmann and J. P. Romano. *Testing Statistical Hypotheses, 3rd Edition*, Springer, 2005.
[17] E. L. Lehmann. On likelihood ratio tests. In *The Second Erich L. Lehmann Symposium Optimality*, (J. Rojo, Ed.), IMS Lecture Notes-Monograph Series Vol. 49, pp. 1-9, 2006.
[18] E. L. Lehmann. *Reminiscences of a Statistician: The Company I Kept*, Springer, 2008.
[19] H. Scheffé. On the Theory of Testing Composite Hypotheses with One Constraint *Ann. Math. Statist.*, Vol. 13, No. 3, pp. 280-293, 1942.

ON OPTIMUM TESTS OF COMPOSITE HYPOTHESES WITH ONE CONSTRAINT[1]

By E. L. Lehmann

University of California, Berkeley

Summary. This paper is concerned with optimum tests of certain composite hypotheses. In section 2 various aspects of a theorem of Scheffé concerning type B_1 tests are discussed. It is pointed out that the theorem can be extended to cover uniformly most powerful tests against a one-sided set of alternatives. It is also shown that the method for determining explicitly the optimum test region may in certain cases be reduced to a simple formal procedure. These results are used in section 3 to obtain optimum tests for the composite hypothesis specifying the value of the circular serial correlation coefficient in a normal distribution. A surprising feature of this example is the fact that for the simple hypothesis obtained by specifying values for the nuisance parameters no test with the corresponding optimum properties exists.

In section 4 the totality of similar regions is obtained for a large class of probability laws which admit a sufficient statistic. Some composite hypotheses concerning exponential and rectangular distributions are treated in section 5. It is proved that the likelihood ratio tests of these hypotheses have various optimum properties.

1. Introduction. In developing tests for a class of hypotheses three phases may be distinguished. First, tests are obtained which are intuitively appealing; next, it is shown that these tests have certain attractive features; finally, it is proved that they are "best possible" tests.

In dealing with parametric hypotheses, the likelihood ratio principle is frequently used to obtain a reasonable test. For many of the tests so derived for normal and exponential distributions, the question of bias has been investigated. In most cases unbiasedness has been established; in the other cases, usually a test based on the same criterion but with the boundaries shifted, can be proved to be unbiased. Other desirable properties which likelihood ratio tests have been shown to possess, relate to the asymptotic behaviour of these tests as the sample sizes tend to infinity. An interesting problem which does not seem to have been treated is the question of admissibility of likelihood ratio tests, a test being admissible if its power can not be improved upon uniformly by any other test of the same level of significance.

Investigations of optimum tests of composite hypotheses have been carried through for many hypotheses concerning normal distributions. When the hypothesis specifies the value of one parameter (hypothesis with one constraint), uniformly most powerful one-sided and type B_1 (uniformly most powerful un-

[1] Presented at a meeting of the Institute of Mathematical Statistics in San Diego, June, 1947.

J. Rojo (ed.), *Selected Works of E. L. Lehmann*, Selected Works in Probability and Statistics, DOI 10.1007/978-1-4614-1412-4_12, © Springer Science+Business Media, LLC 2012

biased) tests have been obtained. When the number of constraints is larger than one, not so much can be expected. It has been shown for some of the tests in this class that they have maximum average power uniformly over a family of surfaces in the parameter space, or that they are uniformly most powerful with respect to the subclass of tests whose power depends only on some function of the parameters. (All optimum properties mentioned are relative only to the class of all similar regions. This will be so throughout the paper and will usually not be stated explicitly).

Two methods for finding uniformly most powerful or uniformly most powerful one-sided regions and type B_1 tests, if they exist are known. Neyman and Pearson [1] developed a method for determining all similar regions, and applied it to obtain uniformly most powerful one-sided tests of certain hypotheses. Neyman [2, 3] extended the method to obtain, for certain hypotheses, the class of all bisimilar (unbiased similar) regions, and Scheffé [4], developing the method further, proved the existence of type B_1 tests for an important class of hypotheses.

A different method for obtaining all similar and bisimilar regions was devised by P. L. Hsu and was used by him and other writers to prove various optimum properties of the likelihood ratio tests for the general linear hypothesis, of Hotelling's T^2 and of other tests [5, 6, 7, 8].

In the present paper we are concerned with applications of these two methods to composite hypotheses with one constraint. However, the applicability is not so restricted. In fact, the second method has been used mainly in connection with composite hypotheses with many constraints, and the author believes it to be suitable also for deriving optimum classification procedures. An essential restriction of both methods seems to be that a set of sufficient statistics must exist with respect to the parameters involved: with respect to the nuisance parameters so that all similar regions can be found, with respect to the parameters specified by the hypothesis so that there exists a best of all similar regions.

Extensions of the existing theory based on the first method are obtained in section 2, and the theory is applied in section 3 to a hypothesis concerning a multivariate normal distribution. Sections 4 and 5 are concerned with applications of the second method to problems to most of which the earlier method is not applicable, in particular to hypotheses concerning exponential and rectangular distributions, hitherto only treated from the likelihood ratio point of view.

2. On the theory of optimum tests.

2.1 *One-sided tests.* In an interesting paper [4], Scheffé determined the type B and type B_1 tests of a certain class of composite hypotheses specifying the value θ_0 of a parameter θ in the presence of nuisance parameters.

Scheffé's results can, in an obvious way, be extended to cover one-sided sets of alternatives. To show this, consider the method used in [4]. Under certain assumptions all tests[2] are found which satisfy the two conditions:

[2] The terms "the test w" and "the region [of rejection] w" will be used interchangeably.

(a) The power function β_w at θ_0 has a preassigned value ϵ (the level of significance), independent of the nuisance parameters;

(b) the power function at θ_0 has derivative 0. (Condition of unbiasedness). Then that test w_0 is determined for which, of all those satisfying (a) and (b),

(c) the second derivative at θ_0, $\beta''_w(\theta_0)$, is as large as possible.

By definition w_0 is a type B test. Under a certain additional assumption (this is the convexity assumption $\frac{\partial^2 g}{\partial y_1^2} > 0$ of Scheffé's Theorem 2) it is shown that of all tests satisfying (a) and (b), w_0 has maximum power against all alternatives, i.e. is of type B_1.

If now we want to maximize the power against only the one-sided set of alternatives, $\theta > \theta_0$, we determine that test w_1 of all those satisfying (a), for which

(d) the first derivative at θ_0, $\beta'_w(\theta_0)$, is as large as possible.

Under a certain additional assumption (in Scheffé's notation this would be the monotonicity assumption $\frac{\partial g}{\partial y_1} > 0$) it can then be shown that of all tests satisfying (a), w_1 has maximum power against all alternatives $\theta > \theta_0$, (it also has minimum power against all alternatives $\theta < \theta_0$), i.e. w_1 is uniformly most powerful against alternatives $\theta > \theta_0$. We shall not carry through the discussion in detail since Scheffé's argument applies step by step, with only the obvious changes.

2.2 *Determination of the boundaries.* Let $X_1, \cdots, X_n$ be n random variables with a joint probability density function p, depending on parameters θ_1 and $\theta = (\theta_2, \cdots, \theta_l)$. We shall denote the probability density function of a set of random variables $X_1, \cdots, X_n$ whose distribution depends on a parameter θ by $p(x_1, \cdots, x_n \mid \theta)$ or simply by $p(x_1, \cdots, x_n)$ when the dependence on θ is clear from the context. The set of points $(x_1, \cdots, x_n)$ for which

$$p(x_1, \cdots, x_n \mid \theta)$$

is positive we shall denote by $W_+(\theta)$.

Let

$$\varphi_i(x_1, \cdots, x_n) = \frac{\partial}{\partial\theta_i} \log p(x_1, \cdots, x_n \mid \theta_1, \theta) \big|_{\theta_1 = \theta_1^0}, \qquad (i = 1, \cdots, l), \tag{2.1}$$

and let the random variable Φ_i be defined by

$$\Phi_i = \varphi_i(X_1, \cdots, X_n). \tag{2.2}$$

Then for testing the hypothesis $H\colon \theta_1 = \theta_1^0$, under the assumptions stated by Scheffé, the type B_1 test w_0 is defined by the inequalities

$$\varphi_1 < k_1, > k_2 \qquad (k_1 < k_2) \tag{2.3}$$

where k_1, k_2 depend on $\theta_1^0, \theta, \varphi_2, \cdots, \varphi_l$ and are determined by the two equations[3]

$$\int_{k_1}^{k_2} \varphi_1^s\, p(\varphi_1, \cdots, \varphi_l)\, d\varphi_1 = (1 - \epsilon) \int_{-\infty}^{\infty} \text{same} \qquad (s = 0, 1). \tag{2.4}$$

[3] Although k_1 and k_2 may depend on θ, w_0 is independent of θ, as was shown in [4].

The equations (2.3) and (2.4) are not suitable for the determination of the boundary of w_0. The variables have to be transformed so as to obtain for w_0 an expression from which the calculation of the boundaries becomes feasible, (cf. [9]). This part of the work may be formalized in the following theorem.

THEOREM 1. *Let*

$$U = f(\Phi_1, \Phi_2, \cdots, \Phi_l)$$
$$V_i = g_i(\Phi_2, \cdots, \Phi_l), \qquad (i = 2, \cdots, l), \tag{2.5}$$

be a system of functions, continuously differentiable and with non-vanishing Jacobian almost everywhere, and such that

(i) *U is a linear function of Φ_1*

$$U = a\Phi_1 + b \tag{2.6}$$

with coefficients which may depend on $\Phi_2, \cdots, \Phi_l$ and such that[4] $a(\Phi_2, \cdots, \Phi_l) > 0$,

(ii) *it is possible to solve for $\Phi_2, \cdots, \Phi_l$ in terms of the V's,*

(iii) *under the hypothesis H, U is distributed independently of*

$$V = (V_2, \cdots, V_l).$$

Then the region w_0 is equivalent to the region

$$u < c_1, > c_2 \qquad (c_1 < c_2) \tag{2.7}$$

where c_1, c_2 are determined by

$$\int_{c_1}^{c_2} u^s p(u)\, du = (1 - \epsilon) \int_{-\infty}^{\infty} u^s p(u)\, du \qquad (s = 0, 1). \tag{2.8}$$

PROOF.

$$\begin{aligned} p(\varphi_1, \varphi_2, \cdots, \varphi_l) &= p(u, v_2, \cdots, v_l) \cdot \left| \frac{\partial(u, v_2, \cdots v_l)}{\partial(\varphi_1, \cdots, \varphi_l)} \right| \\ &= p(u) \cdot p(v_2, \cdots, v_l) \frac{\partial u}{\partial \varphi_1} \cdot \left| \frac{\partial(v_2, \cdots, v_l)}{\partial(\varphi_2, \cdots, \varphi_l)} \right|. \end{aligned} \tag{2.9}$$

But

$$\begin{aligned} u &= a(\varphi_2, \cdots, \varphi_l) \cdot \varphi_1 + b(\varphi_2, \cdots, \varphi_l) \\ &= \alpha(v_2, \cdots, v_l) \cdot \varphi_1 + \beta(v_2, \cdots, v_l) \end{aligned} \tag{2.10}$$

so that (2.4) reduces to

$$\begin{aligned} &\int_{c_1(v_2,\cdots,v_l)}^{c_2(v_2,\cdots,v_l)} \left(\frac{u - \beta}{\alpha}\right)^s p(u)p(v_2, \cdots, v_l)\, du \\ &\qquad = (1 - \epsilon) \int_{-\infty}^{\infty} \text{same} \qquad (s = 0, 1) \end{aligned} \tag{2.11}$$

[4] A similar theorem holds when we assume $\alpha(\Phi_2, \ldots, \Phi_l) < 0$.

and hence to

$$\int_{c_1(v_2,\cdots,v_l)}^{c_2(v_2,\cdots,v_l)} u^s p(u)\,du = (1-\epsilon)\int_{-\infty}^{\infty} \text{same} \qquad (s = 0, 1) \tag{2.12}$$

which shows c_1 and c_2 to be independent of the v's. Also obviously (2.3) transforms into (2.7) which completes the proof.

If U is such that its distribution (when $\theta_1 = \theta_1^0$) is independent of θ, c_1 and c_2 of theorem 1 will depend only on the data of the problem: ϵ, n, θ_1^0. However, the existence of constants c_1 and c_2 satisfying (2.8) still has to be proved. We may show more generally the existence of k_1 and k_2 satisfying (2.4). A proof is immediately supplied by an argument which was used by Neyman [10] and Wald [11] to prove the existence of type A tests, and which may be stated in the following

LEMMA. *Let* $0 < \alpha < 1$, *let* $f(x) \geq 0$ *and* $\int_{-\infty}^{\infty} x^s f(x)\,dx < \infty$ *for* $s = 0, 1$. *Then there exist* A, B *such that*

$$\int_A^B x^s f(x)\,dx = \alpha \int_{-\infty}^{\infty} x^s f(x)\,dx \qquad (s = 0, 1). \tag{2.13}$$

3. Testing for circular serial correlation in a normal population. We now apply the results of the previous section to obtain the optimum tests (i.e. uniformly most powerful against the one-sided set of alternatives, type B_1 in the two-sided case) for the hypothesis specifying the value of the circular serial correlation coefficient in the normal population considered by Dixon [12]. (For the literature on testing for non-circular serial correlation in normal populations cf. [12]).

We assume

$$p(x_1, \cdots, x_n) = \frac{1-\delta^n}{(\sqrt{2\pi}\sigma)^n}\exp\left[-\frac{1}{2\sigma^2}\sum_{i=1}^n [(x_i - \xi) - \delta(x_{i+1} - \xi)]^2\right] \tag{3.1}$$

where $x_{n+1} = x_1$ and $|\delta| < 1$, and we test the hypothesis $\delta = \delta_0$. For testing purposes only the value $\delta_0 = 0$ is of interest presumably, however, the family of tests for arbitrary δ_0 is required for estimating δ by means of confidence intervals, and therefore the more general hypothesis is considered.

Making a transformation in one of the parameters we write

$$\begin{aligned} &p(x_1, \cdots, x_n) \\ &\quad = C(\delta, \alpha)\exp\left[\alpha\left[(1+\delta^2)\sum_{i=1}^n (x_i - \xi)^2 - 2\delta\sum_{i=1}^n (x_i - \xi)(x_{i+1} - \xi)\right]\right] \end{aligned} \tag{3.2}$$

where in the notation of the previous section $\theta_1 = \delta$, $\theta_2 = \alpha$, $\theta_3 = \xi$.

THEOREM 2. *For testing the hypothesis* $\delta = \delta_0$ *for the distribution* (3.2)

(a) *the type* B_1 *test exists and is given by*

$$r < r_1,\ > r_2 \tag{3.3}$$

where

$$(3.4)\qquad r = \frac{\sum_{i=1}^{n} (x_i - \bar{x})(x_{i+1} - \bar{x})}{\sum_{i=1}^{n} (x_i - \bar{x})^2}; \qquad \bar{x} = \frac{1}{n}\sum_{i=1}^{n} x_i$$

and where r_1 *and* r_2 *are determined by*

$$(3.5)\qquad \int_{r_1}^{r_2} \left(\frac{r}{1 + \delta_0^2 - 2\delta_0 r}\right)^s p(r)\, dr = (1 - \epsilon) \int_{-\infty}^{\infty} \text{same} \qquad (s = 0, 1).$$

(b) *the uniformly most powerful similar region for testing H against the alternatives* $\delta > \delta_0$ *exists and is given by*

$$(3.6)\qquad r > r'$$

where r' *is determined by*

$$(3.7)\qquad \int_{-\infty}^{r'} p(r)\, dr = (1 - \epsilon) \int_{-\infty}^{\infty} p(r)\, dr.$$[5]

Proof. We compute

$$(3.8)\qquad \begin{aligned} \varphi_1 &= C_1(\delta_0, \alpha) + 2\alpha[\delta_0 \Sigma(x_i - \xi)^2 - \Sigma(x_i - \xi)(x_{i+1} - \xi)] \\ \varphi_2 &= C_2(\delta_0, \alpha) + (1 + \delta_0^2)\Sigma(x_i - \xi)^2 - 2\delta_0 \Sigma(x_i - \xi)(x_{i+1} - \xi) \\ \varphi_3 &= -2n\alpha(1 - \delta_0^2)(\bar{x} - \xi). \end{aligned}$$

There is no difficulty in checking the conditions of Scheffé's theorems [4].

Next we apply Theorem 1 of the previous section, and define

$$(3.9)\qquad \begin{aligned} V_2 &= (1 + \delta_0^2)\Sigma(X_i - \bar{X})^2 - 2\delta_0 \Sigma(X_i - \bar{X})(X_{i+1} - \bar{X}) \\ V_3 &= \bar{X} - \xi \\ U &= \frac{\Sigma(X_i - \bar{X})(X_{i+1} - \bar{X})}{V_2}. \end{aligned}$$

Conditions (i) and (ii) of Theorem 1 are easily seen to be satisfied. To show that U is independent of $V = (V_2, V_3)$ we employ arguments which have recently been used by various authors in a number of similar problems (cf. [13, 14, 15]).

It is seen that an orthonormal transformation exists:

$$X_1, \cdots, X_n \to Y_1, \cdots, Y_n$$

such that

$$(3.10)\qquad \begin{aligned} \sqrt{n}\,\bar{X} &= Y_1 \\ \sum_{i=1}^{n} (X_i - \bar{X})(X_{i+1} - \bar{X}) &= \sum_{i=2}^{n} \lambda_i Y_i^2 \\ \sum_{i=1}^{n} (X_i - \bar{X})^2 &= \sum_{i=2}^{n} Y_i^2. \end{aligned}$$

[5] A corresponding result holds for the other one-sided case.

Under H the Y's are distributed with probability density

$$p(y_1, \cdots, y_n) = C(\delta_0, \alpha) \exp\left[\alpha\left[k(y_1 - \sqrt{n}\xi)^2 + \sum_{i=2}^{n} \mu_i y_i^2\right]\right] \tag{3.11}$$

where $k, \mu_2, \cdots, \mu_n$ depend on δ_0 and where the μ's are all positive. Introducing new variables

$$Z_i = \sqrt{\mu_i}\, Y_i, \qquad (i = 2, \cdots, n), \tag{3.12}$$

and, then, generalized polar coordinates in the space of the Z's,

$$R = \sqrt{\sum_{i=2}^{n} Z_i^2}, \qquad \Psi_1, \cdots, \Psi_{u-2} \tag{3.13}$$

we see that Y_1, R and $\Psi_1, \cdots, \Psi_{n-2}$ are completely independent. Also

$$V_2 = R^2, \qquad V_3 = \frac{1}{\sqrt{n}}(Y_1 - \xi)$$

while U, being homogeneous of degree 0 in the Z's, is a function of the Ψ's only. This proves that U, V_2 and V_3 are completely independent. The type B_1 test of H is therefore given by

$$u = \frac{\sum_{i=1}^{n} (x_i - \bar{x})(x_{i+1} - \bar{x})}{(1 + \delta_0^2)\sum_{i=1}^{n}(x_i - \bar{x})^2 - 2\delta_0 \sum_{i=1}^{n}(x_i - \bar{x})(x_{i+1} - \bar{x})} < c_1, > c_2 \tag{3.14}$$

where c_1 and c_2 are determined by

$$\int_{c_1}^{c_2} u^s p(u)\, du = (1 - \epsilon)\int_{-\infty}^{\infty} u^s pu()\, du \qquad (s = 0, 1). \tag{3.15}$$

We still have to show that this test is equivalent to the one defined by (3.3) and (3.5). For $\delta_0 = 0$ this is trivial. Let us assume $\delta_0 < 0$. (The other case goes through similarly.) The inequality $u < c_1$ is equivalent to

$$(1 + 2\delta_0 c_1)\Sigma(x_i - \bar{x})(x_{i+1} - \bar{x}) < (1 + \delta_0^2)\Sigma(x_i - \bar{x})^2 \tag{3.16}$$

and hence to

$$\frac{\Sigma(x_i - \bar{x})(x_{i+1} - \bar{x})}{\Sigma(x_i - \bar{x})^2} < w_1 \tag{3.17}$$

provided $1 + 2c_1\delta_0 > 0$. Suppose $1 + 2c_1\delta_0 \leq 0$, i.e. $c_1 \geq -\frac{1}{2\delta_0}$. Then[6]

$$P\{U < c_1\} \geq P\left\{U < -\frac{1}{2\delta_0}\right\} = P\{0 < \Sigma(X_i - \bar{X})^2\} = 1 \tag{3.18}$$

[6] We denote the probability of an event A by $P\left\{A\right\}$.

i.e. $P\{U < c_1\} = 1$ which would contradict (3.15). Similarly if $1 + 2c_2\delta_0 \leq 0$ we would have $P\{U > c_2\} = 0$ and hence our test would be one-sided and therefore not unbiased. The inequalities $u < c_1, > c_2$ are thus equivalent to the inequalities $r < r_1, > r_2$ and since

$$u = \frac{r}{1 + \delta_0^2 - 2\delta_0 r},$$

(3.5) also follows.

The existence of type B_1 and uniformly most powerful one-sided tests of the hypothesis H is rather surprising. For when α and ξ are assumed known, neither the type A_1 test nor the uniformly most powerful one-sided test of the simple hypothesis H': $\delta = \delta_0$ exists. This is easily seen by determining the most powerful and the most powerful unbiased test against a specific alternative δ_1 for the hypothesis H' in the population

$$p(x_1, \cdots, x_n) = \frac{1 - \delta^n}{(\sqrt{2\pi})^n} \exp\left[-\tfrac{1}{2}[(1 + \delta^2)\Sigma x_i^2 - 2\delta\Sigma x_i x_{i+1}]\right]. \tag{3.19}$$

The distribution of the criterion R was obtained by R. L. Anderson [16] (see also [17]) for the case $\delta = 0$. Madow [15] using Anderson's result found the distribution for arbitrary δ. (Approximations to the distribution have been studied by various authors; for the literature on this cf. [18]. Recently Hsu [19] obtained an asymptotic expansion.) A direct derivation for arbitrary δ may be based on the following theorem of Cramér, which was communicated to the author by Dr. P. L. Hsu.

THEOREM 3. *(Cramér)*[7]. *If X, Y are two random variables, (not necessarily independent), $Y > 0$, then*

$$P\left\{\frac{X}{Y} \leq x\right\} = -\frac{1}{2\pi}\int_{-\infty}^{\infty} \frac{\varphi_x(t) - \psi(t)}{it}\, dt \tag{3.20}$$

where φ_x and ψ are the characteristic functions of $X - xY$ and Y respectively, provided

$$\int_{-\infty}^{\infty} \left|\frac{\varphi_x(t) - \psi(t)}{t}\right| dt < \infty. \tag{3.21}$$

THEOREM 4. *If*

$$p(x_1, \cdots, x_n) = \frac{1 - \delta^n}{(\sqrt{2\pi}\,\sigma)^n} \exp\left[-\frac{1}{2\sigma^2}\sum_{i=1}^{n}[(x_i - \xi) - \delta(x_{i+1} - \xi)]^2\right], \qquad (x_{n+1} = x_1) \tag{3.22}$$

[7] Differentiated forms of the theorem were given by R. C. Geary [*Jour. Roy. Stat. Soc.* Vol. 107 (1944) p. 56] and H. Cramér [Exercise 6 on p. 317 of *Mathematical Methods of Statistics*. Princeton Univ. Press (1946)].

and if

$$R = \frac{\sum_{i=1}^{n} (X_i - \bar{X})(X_{i+1} - \bar{X})}{\sum_{i=1}^{n} (X_i - X^2)}, \tag{3.23}$$

then

$$P\{R > r\} = \frac{2^{n+1/2}}{n} \frac{1 - \delta^n}{(1 - \delta)(1 + \delta^2 - 2\delta r)} \cdot \sum_j \frac{(-1)^{j+1}(\lambda_j - r)^{n-3/2} \sin \frac{j\pi}{n} \sin \frac{2j\pi}{n}}{1 + \delta^2 - 2\delta\lambda_j} \tag{3.24}$$

where the summation is extended over all integer j, $1 \leq j \leq \frac{n}{2}$, for which $\lambda_j > r$, and where

$$\lambda_j = 2 \cos \frac{2j\pi}{n}. \tag{3.25}$$

The proof of this theorem from Theorem 3 is straightforward and only will be indicated here. If X and Y denote the numerator and denominator of R respectively, the characteristic functions of Y and $X - rY$ may be obtained by the method of circulants (cf. [12, 17]). The integral on the right hand side of (3.20) is then easily evaluated by the theory of residues when n is odd. In the case that n is even, the integrand has two branchpoints, one in the lower and one in the upper half plane. These may be separated, and then again the method of residues may be applied.

4. Similar regions. The problem of finding all regions similar to the sample space with respect to a parameter θ was solved by Neyman and Pearson [1] for a certain class of probability laws. In a later paper Neyman proved ([20] proposition IX) that if there exists a sufficient statistic T for a parameter θ, then w is similar with respect to θ if it has the following structure: For the intersection $w(t)$ of w with the surface $T = t$, the relative probability of $w(t)$ given $T = t$ has a constant value independent of t. We shall show in this section that for a large class of probability laws which admit a sufficient statistic for θ the regions with the above structure are the only ones that are similar with respect to θ.

We consider samples from a univariate distribution and we distinguish three cases as one, both or neither of the extremes of the range of the distribution depend on the parameter θ. For the first of these cases (cf. Pitman [21]) we consider samples from a distribution with probability density

$$p(x) = \frac{f(x)}{g(\theta)}, \qquad k(\theta) \le x \le c, \tag{4.1}$$

where $k(\theta)$ is a strictly monotone continuous function of θ and where c may be infinite. Introducing a new parameter $\delta = k(\theta)$ the distribution of a sample from (4.1) is given by

$$p(x_1, \cdots, x_n) = \frac{f(x_1) \cdots f(x_n)}{b(\delta)}, \qquad \delta \le x_i \le c. \tag{4.2}$$

To obtain the totality of regions w similar with respect to δ let us denote by $W_1, \cdots, W_n$ the portions of the sample space where the smallest of the x's is $x_1, \cdots, x_n$ respectively. For any region w denote by w_k' the intersection of w with W_k. Consider a transformation carrying $W_2, \cdots, W_n$ into W_1, letting $y_1 = \min(x_1, \cdots, x_n)$ and letting in W_k:

$$y_2 = x_1,\ y_3 = x_2,\ \cdots,\ y_k = x_{k-1},\ y_{k+1} = x_{k+1},\ \cdots,\ y_n = x_n. \tag{4.3}$$

Denote by w_k the image of w_k' under this transformation. The condition that w be similar with respect to δ,

$$\int_w \frac{f(x_1) \cdots f(x_n)}{b(\delta)}\, dx_1 \cdots dx_n \overset{(\delta)}{\equiv} \epsilon, \tag{4.4}$$

may be written in the form

$$\begin{aligned} &\int_\delta^c \frac{f(y_1)}{b(\delta)} \left\{ \sum_{k=1}^n \int_{w_k(y_1)} f(y_2) \cdots f(y_n)\, dy_2 \cdots dy_n \right\} dy_1 \\ &\qquad \overset{(\delta)}{\equiv} n\epsilon \int_\delta^c \frac{f(y_1)}{b(\delta)} \left\{ \int_{(W(y_1))} f(y_2) \cdots f(y_n)\, dy_2 \cdots dy_n \right\} dy_1 \end{aligned} \tag{4.5}$$

where $W(y_1)$ denotes the region $y_1 \le y_i \le c$, $(i = 2, \cdots, n)$, that is, the region of variation of $y_2, \cdots, y_n$ given y_1, and where $w_k(y_1)$ denotes the region of variation of $y_2, \cdots, y_n$ given y_1 and w_k. From (4.5) we obtain

$$\frac{1}{b(\delta)} \int_\delta^c f(y_1)\psi(y_1)\, dy_1 \overset{(\delta)}{\equiv} 0 \tag{4.6}$$

where

$$\begin{aligned} \psi(y_1) = &\sum_{k=1}^n \int_{w_k(y_1)} f(y_2) \cdots f(y_n)\, dy_2 \cdots dy_n \\ &- n\epsilon \int_{y_1}^c \cdots \int_{y_1}^c f(y_2) \cdots f(y_n)\, dy_2 \cdots dy_n. \end{aligned} \tag{4.7}$$

But (4.6) implies

$$\psi(y_1) = 0 \text{ almost everywhere} \tag{4.8}$$

and since we can only determine w up to a set of measure 0, we may omit the qualification in (4.8). Therefore a necessary and sufficient condition for w to be similar is

$$\frac{1}{n\left[\int_{y_1}^{c} f(y)\,dy\right]^{n-1}} \sum_{k=1}^{n} \int_{w_k(y_1)} f(y_2) \cdots f(y_n)\,dy_2 \cdots dy_n = \epsilon \tag{4.9}$$

for all y_1.

To see more clearly the structure of these regions, let us take $n = 2$. Equation (4.9) states that on each of the broken lines of Fig. 1 the relative probability of $w = w_1' + w_2'$ given $Y_1 = y_1$ is ϵ, where the decomposition of this probability into its two components may vary with y_1.

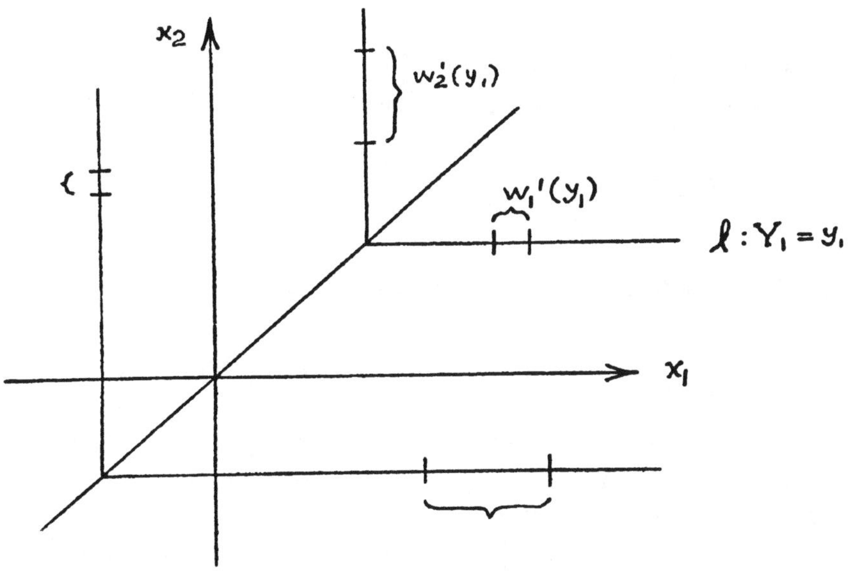

FIG. 1

In general equation (4.9) states that on each hyperplane $Y_1 = y_1$ the relative probability of w is independent of y_1. Since $Y_1 = \min(X_1, \cdots, X_n)$ is a sufficient statistic for θ, Neyman's theorem in this case does give all similar regions.

Next let us consider the case where both extremes of the range of the distribution depend on the parameter. We shall assume (cf. [21]) that $X_1, \cdots, X_n$ are distributed with probability density

$$p(x) = \frac{f(x)}{g(\theta)} \quad \text{in} \quad \theta \leq x \leq b(\theta) \tag{4.10}$$

where b is a strictly decreasing continuous function over an interval $[-\infty, b(-\infty)]$ and where $b[b(-\infty)] = -\infty$. These assumptions insure that there exists a unique number a, $-\infty < a < b(-\infty)$, such that $b(a) = a$.

Denote by W_{ij}, $(i, j = 1, \cdots, n; i \neq j)$, the portion of the sample space where the smallest and the largest of the x's are x_i and x_j respectively. Denote by W_{ij1} and W_{ij2} those portions of W_{ij} where x_i is greater than and less than $b^{-1}(x_j)$ respectively. For any region w denote by w'_{ijk} the intersection of w with

W_{ijk}. Consider a transformation carrying the sample-space into W_{1n}, letting $y_1 = \min(x_1, \cdots, x_n)$, $y_n = \max(x_1, \cdots, x_n)$ and in W_{ij} letting $y_2, \cdots, y_{n-1}$ denote the remaining x's in the order of their subscripts. Next make a transformation carrying W_{1n} into W_{1n1}, letting $z_1 = \max[y_1, b^{-1}(y_n)]$, $z_n = \min[y_1, b^{-1}(y_n)]$ and $z_k = y_k$ for $k = 2, \cdots, n-1$. Denote by w_{ijk} the image of w'_{ijk} in W_{1n1}.

Then Z_n is a sufficient statistic for θ (cf. [21]) and there exist functions f_1, g_1 such that the density of Z_n is given by

$$(4.11) \qquad p(z_n) = \frac{f_1(z_n)}{g_1(\theta)} \quad \text{in} \quad \theta \leq z_n \leq a$$

while the distribution of the remaining Z's given Z_n is independent of θ.

The condition that w be a similar region may now be written, analogously to (4.5), in the form

$$(4.12) \quad \int_a^\theta \frac{f_1(z_n)}{g_1(\theta)} \sum_{i,j,k} \int_{w_{ijk}(z_n)} p(z_1, \cdots, z_{n-1} \mid z_n)\, dz_1 \cdots dz_{n-1}\, dz_n \equiv \epsilon \int_\theta^a \frac{f_1(z_n)}{g_1(\theta)}\, dz_n$$

and hence by the argument which led to (4.6), as

$$(4.13) \qquad \sum_{i,j,k} \int_{w_{ijk}(z_n)} p(z_1, \cdots, z_{n-1} \mid z_n)\, dz_1 \cdots dz_{n-1} = \epsilon \qquad \text{for all } z_n.$$

Thus in this case also Neyman's theorem gives the most general similar region.

For the case that neither extreme of the range of the distribution depends on the parameter θ, it has been shown by various authors [22, 21, 23] under slightly varying assumptions concerning the regularity of the distribution function, that the existence of a sufficient statistic implies

$$(4.14) \qquad p(x \mid \theta) = \exp[P(\theta) + T(x)Q(\theta) + R(x)].$$

This (cf. [10]) is a special case of that for which Neyman and Pearson determined the totality of similar regions, however under the restriction that the moments of $\Phi = \frac{\partial}{\partial\theta} \sum_{i=1}^n \log p(X_i)$ uniquely determine the distribution of Φ. We shall briefly indicate how this assumption may be avoided.

Let $X_1, \cdots, X_n$ be a sample from (4.14), or, more generally, (this is the case considered by Neyman and Pearson), let $X_1, \cdots, X_n$ be distributed with probability density

$$(4.15) \qquad \begin{aligned} &p(x_1, \cdots, x_n) \\ &\quad = \exp[P(\theta) + u(x_1, \cdots, x_n)Q(\theta) + v(x_1, \cdots, x_n)] \end{aligned}$$

in a sample space W_+ which is independent of θ. We shall assume that the set of values which Q takes on contains at least some interval. Introducing $\delta =$

$-Q(\theta)$ as a new parameter, we shall obtain all regions similar to δ (where the set of values of δ contains an interval) for the distribution[8]

$$(4.16)\quad \begin{aligned} &p(x_1, \cdots, x_n) \\ &\quad = \exp\,[p_1(\delta) - \delta \cdot u(x_1, \cdots, x_n) + v(x_1, \cdots, x_n)] \end{aligned}$$

under the assumption that $\sum_{i=1}^{n} \left(\frac{\partial u}{\partial x_i}\right)^2 \neq 0$ except possibly on a set of measure 0.

Let us for a moment assume that there exist functions $f_i(x_i, \cdots, x_n)$, $(i = 2, \cdots, n)$, with continuous partial derivatives almost everywhere and such that the transformation

$$(4.17)\quad y_1 = u(x_1, \cdots, x_n); \quad y_i = f_i(x_1, \cdots, x_n), \qquad (i = 2, \cdots, n),$$

is one to one on W_+ except possibly on a subset of measure 0. Applying this transformation we may write the condition of similarity in the form

$$(4.18)\quad \begin{aligned} &\int_{-\infty}^{\infty} e^{P_1(\delta)-\delta y_1} \int_{w(y_1)} f(y_1, \cdots, y_n)\, dy_2 \cdots dy_n \cdot dy_1 \\ &\quad = \epsilon \int_{-\infty}^{\infty} e^{P_1(\delta)-\delta y_1} \int_{W(y_1)} f(y_1, \cdots, y_n)\, dy_2 \cdots dy_n \cdot dy_1 \end{aligned}$$

where $W(y_1)$ denotes the region of variation of $y_2, \cdots, y_n$ given y_1, and where $w(y_1)$ denotes the region of variation of $y_2, \cdots, y_n$ given y_1 and w. Furthermore $f(y_1, \cdots, y_n)$ is independent of δ. From the theory of bilateral Laplace transforms it is known that (4.18) implies that

$$(4.19)\quad \int_{w(y_1)} f(y_1, \cdots, y_n)\, dy_2 \cdots dy_n = \epsilon \int_{W(y_1)} f(y_1, \cdots, y_n)\, dy_2 \cdots dy_n$$

which is the desired result.

More generally it may be shown that our assumption concerning $u(x_1, \cdots, x_n)$ insures the existence of functions f_i, $(i = 2, \cdots, n)$, such that under the transformation (4.17) no point $(y_1, \cdots, y_n)$ has more than a denumerable infinity of counter images in x-space. Our proof can be modified to cover this case. The argument is similar to that used to obtain equations (4.9) and (4.13) which were also arrived at through many to one transformations.

5. Testing exponential and rectangular distributions.

In their fundamental 1928 paper [24] on likelihood ratio tests, Neyman and Pearson discussed various hypotheses relating to normal, exponential and rectangular distributions. Later they and other authors developed a theory of similar and bisimilar regions which made it possible to obtain optimum tests of many composite hypotheses with

[8] An assumption that we can solve for θ as a function of δ is not needed since we can determine $P_1(\delta)$ by integrating the density (4.16) over W_+.

one constraint concerning normal populations. This theory however is not applicable to most hypotheses concerning exponential or rectangular distributions. We shall in this section obtain optimum tests of some hypotheses relating to these latter distributions, using the method of the previous section.

Let us first consider a sample $X_1, \cdots, X_n$ from an exponential population, the probability density of the sample being.

$$(5.1) \quad p(x_1, \cdots, x_n) = \frac{1}{a^n} \exp\left[-\frac{1}{a}\sum_{i=1}^{n}(x_i - b)\right] \quad \text{if } x_i > b, \quad (i = 1, \cdots, n)$$

and let us consider the two hypotheses $H_1{:}a =_0$, $H_2{:}b = b_0$ where, without loss of generality, we shall take $a_0 = 1$, $b_0 = 0$. The likelihood ratio tests of both these hypotheses were shown to be completely unbiased by Paulson [25]. We shall prove

THEOREM 5. *The likelihood ratio tests of H_1 and H_2 are type B_1 and uniformly most powerful, respectively. The one-sided tests based on the likelihood ratio criterion for H_1 are the uniformly most powerful one-sided similar regions for testing this hypothesis.*

PROOF. In order to simplify the argument we shall give a detailed proof only for the restricted class of tests which are symmetric in the variables $X_1, \cdots, X_n$.

For testing H_1 let us make the following transformation introduced by Sukhatme [26]:

$$(5.2) \quad \begin{aligned} Z_1 &= nY_1 \\ Z_i &= (n - i + 1)(Y_i - Y_{i-1}), \qquad (i = 2, \cdots, n), \end{aligned}$$

where Y_i is the ith of the X's in order of magnitude. Then

$$(5.3) \quad \begin{aligned} p(z_1, \cdots, z_n) &= \frac{1}{a^n} \exp\left[-\frac{1}{a}(z_1 - nb) - \frac{1}{a}\sum_{i=2}^{n} z_i\right] \\ &\text{if} \quad z_1 \geq nb; z_i \geq 0 \qquad (i = 2, \cdots, n). \end{aligned}$$

We want to determine all regions w which under H are similar to the sample space with respect to b, i.e. all regions w satisfying

$$(5.4) \quad \begin{aligned} &\int_w e^{-(z_1 - nb)} \exp\left[-\sum_{i=2}^{n} z_i\right] dz_2 \cdots dz_n \, dz_1 \\ &\quad = \int_{nb}^{\infty} e^{-(z_1-nb)} \left\{\int_{w(z_1)} \exp\left[-\sum_{i=2}^{n} z_i\right] dz_2 \cdots dz_n\right\} dz_1 \\ &\quad \overset{(b)}{\equiv} \epsilon \overset{(b)}{\equiv} \epsilon \int_{nb}^{\infty} e^{-(z_1-nb)} \, dz_1 \end{aligned}$$

where $w(z_1)$ denotes the intersection of w with the hyperplane $Z_1 = z_1$. Now (5.4) is equivalent to

$$e^{nb} \int_{nb}^{\infty} e^{-z_1} f(z_1)\, dz_1 \overset{(b)}{\equiv} 0 \tag{5.5}$$

where

$$f(z_1) = \int_{w(z_1)} \exp\left[-\sum_{i=2}^{n} z_i \right] dz_2 \cdots dz_n - \epsilon \tag{5.6}$$

and this in turn is equivalent to

$$f(z_1) = 0 \text{ for all } z_1 . \tag{5.7}$$

Of all the regions w satisfying (5.7) we want to determine the one which against a specific alternative, say a_1, has maximum power, i.e. for which

$$\int_{nb}^{\infty} e^{-(1/a_1)(z_1-nb)} \int_{w(z_1)} \exp\left[-\frac{1}{a_1} \sum_{i=2}^{n} z_i \right] dz_2 \cdots dz_n\, dz_1 \tag{5.8}$$

is as large as possible. We thus see that w will have the desired properties if $w(z_1)$ is determined according to the two conditions

$$\int_{w(z_1)} \exp\left[-\sum_{i=2}^{n} z_i \right] dz_2 \cdots dz_n = \epsilon \tag{5.9}$$

and

$$\int_{w(z_1)} \exp\left[-\frac{1}{a_1} \sum_{i=2}^{n} z_i \right] dz_2 \cdots dz_n = \max. \tag{5.10}$$

Hence by the Neyman-Pearson fundamental lemma $w(z_1)$ is the set of points satisfying

$$\exp\left[\left(-\frac{1}{a_1} \sum_{i=2}^{n} z_i + \sum_{i=2}^{n} z_i \right)\right] \geq C(a_1, z_1) \tag{5.11}$$

and therefore according as a_1 is greater or less than 1, $w(z_1)$ is determined by

$$\begin{aligned} \sum_{i=2}^{n} z_i &= \sum_{i=1}^{n} [x_i - \min (x_1, \cdots, x_n)] \geq k(a_1, z_1), \text{ or} \\ \sum_{i=2}^{n} z_i &= \sum_{i=1}^{n} [x_i - \min (x_1, \cdots, x_n)] \leq k'(a_1, z_1). \end{aligned} \tag{5.12}$$

But $\sum_{i=2}^{n} Z_i$ is independently distributed of Z_1 and under H the distribution of $\sum_{i=2}^{n} Z_i$ does not depend on a_1, in fact it is a chi-square distribution with $2n - 2$ degrees of freedom. Thus k and k', as determined by (5.9) are independent of a_1 and the two tests (5.12) are uniformly most powerful one-sided.

Next we consider the more restricted class of unbiased similar regions. For w to be unbiased we must have

$$\frac{d}{da}\left\{\frac{1}{a^n}\int_w \exp\left[-\frac{z_1-nb}{a}\right]\exp\left[-\frac{1}{a}\sum_{i=2}^n z_i\right]dz_1\cdots dz_n\right\}\Bigg|_{a=1}$$

$$(5.13)\quad = \int_{nb}^{\infty}(z_1-nb-n)\exp\left[-(z_1-nb)\right]\int_{w(z_1)}\exp\left[-\sum_{i=2}^n z_i\right]dz_2\cdots dz_n\,dz_1$$

$$+\int_{nb}^{\infty}\exp\left[-(z_1-nb)\right]\int_{w(z_1)}\left(\sum_{i=2}^n z_i\right)\exp\left[-\sum_{i=2}^n z_i\right]dz_2\cdots dz_n\,dz_1 = 0.$$

The first of the integrals in the middle member equals

$$(5.14)\quad \begin{aligned}&\int_0^{\infty}(z-n)\,e^{-z}\int_{w(z+nb)}\exp\left[-\sum_{i=2}^n z_i\right]dz_2\cdots dz_n\,dz\\ &=\epsilon\int_0^{\infty}(z-n)\,e^{-z}\,dz = -(n-1)\epsilon.\end{aligned}$$

Therefore

$$(5.15)\quad \begin{aligned}&\int_{nb}^{\infty}e^{-(z_1-nb)}\int_{w(z_1)}\left(\sum_{i=2}^n z_i\right)\exp\left[-\sum_{i=2}^n z_i\right]dz_2\cdots dz_n\,dz_1\\ &=(n-1)\epsilon = (n-1)\epsilon\int_{nb}^{\infty}e^{-(z_1-nb)}\,dz_1\end{aligned}$$

or

$$(5.16)\quad \int_{nb}^{\infty}e^{-z_1}\,g(z_1)\,dz_1 \overset{(b)}{\equiv} 0$$

where

$$(5.17)\quad g(z_1) = \int_{w(z_1)}\left(\sum_{i=2}^n z_i\right)\exp\left[-\sum_{i=2}^n z_i\right]dz_2\cdots dz_n - (n-1)\epsilon.$$

Thus finally the condition of unbiasedness reduces to

$$(5.18)\quad \int_{w(z_1)}\left(\sum_{i=2}^n z_i\right)\exp\left[-\sum_{i=2}^n z_i\right]dz_2\cdots dz_n = (n-1)\epsilon$$

and we seek the region $w(z_1)$ which satisfies (5.9), (5.10) and (5.18).

By the fundamental lemma $w(z_1)$ is given by

$$(5.19)\quad \exp\left[-\frac{1}{a_1}\sum_{i=2}^n z_i\right] \geq \left[C_1(a_1, z_1)\sum_{i=2}^n z_i + C_2(a_1, z_1)\right]\cdot\exp\left[-\sum_{i=2}^n z_i\right]$$

which is equivalent to

$$(5.20)\quad \sum_{i=2}^n z_i \leq k_1(a_1, z_1),\ \geq k_2(a_1, z_1)$$

where k_1 and k_2 are determined by (5.9) and (5.18), and are therefore independent of z_1 and a. Thus the region (5.20) which of all unbiased similar regions maximizes the power against the alternative $a = a_1$ is independent of a_1 and hence is a region of type B_1. This completes the proof since it is easily verified that (5.10) is equivalent to the likelihood ratio test.

The proof for regions which are not necessarily symmetric in the variables follows similarly if instead of the transformation (5.2) one uses a transformation $U_i = f_i(X_1, \cdots, X_n)$ which is one to one and such that $U_1 = Z_1$, $U_2 = \sum_{i=2}^{n} Z_i$. The distribution of $U_3, \cdots, U_n$ is then independent of a and b since U_1, U_2 are a pair of sufficient statistics for these parameters, and the proof carries over step by step.

Next we consider the hypothesis $H_2 : b = 0$, and again we restrict ourselves to regions which are symmetric in the variables, although as before the proof can be modified to cover also nonsymmetric regions.

We first make the transformation to $Z_1, \cdots, Z_n$ given by (5.2). In the $n - 1$ dimensional space of $Z_2, \cdots, Z_n$, we then transform to new variables $U, \Psi_1, \cdots, \Psi_{n-2}$ where $U = \sum_{i=2}^{n} Z_i$ and where the Ψ's are the generalized polar angles. Obviously the distribution of the Ψ's does not depend on a, since they are homogeneous of degree 0 in the Z's. Furthermore the Ψ's are independently distributed of U since the probability density of the Z's is constant over the hyperplanes $U = u$. Thus

$$p(z_1, u, \psi_1, \cdots, \psi_{n-2}) = \frac{K}{a^n} \exp\left[-\frac{z_1 - nb}{a}\right] u^{n-2} e^{-u/a} p(\psi_1, \cdots, \psi_{n-2}). \tag{5.21}$$

We next introduce new variables

$$V = Z_1 + U \text{ and } T = \frac{Z_1}{Z_1 + U} \tag{5.22}$$

and find

$$p(v, t, \psi_1, \cdots, \psi_{n-2}) = \frac{K}{a^n} \exp\left[-\frac{v - nb}{a}\right] v^{n-1} (1 - t)^{n-2} p(\psi_1, \cdots, \psi_{n-2}) \quad \text{for } v \geq nb,\ \frac{nb}{v} \leq t \leq 1. \tag{5.23}$$

For w under H_2 to be similar with respect to a, we must have

$$\int_0^\infty \frac{K}{a^n} \exp\left[-\frac{v}{a}\right] v^{n-1} \int_{w_0(v)} (1 - t)^{n-2} p(\psi_1, \cdots, \psi_{n-2})\, dt\, d\psi_1 \cdots d\psi_{n-2} \cdot dv = \epsilon \int_0^\infty \frac{K}{a^n} \exp\left[-\frac{v}{a}\right] v^{n-1}\, dv \tag{5.24}$$

where $w(v)$ designates the intersection of w with the hyperplane $V = v$, and where $w_0(v)$ denotes the part of $w(v)$ lying between the hyperplanes $t = 0$ and $t = 1$.

Hence the condition of similarity may be written as

$$\int_0^\infty \exp\left[-\frac{v}{a}\right] v^{n-1} f(v)\, dv = 0 \qquad \text{for all } a > 0 \tag{5.25}$$

where

$$f(v) = \int_{w_0(v)} (1 - t)^{n-2} p(\psi_1, \cdots \psi_{n-2})\, dt\, d\psi_1 \cdots d\psi_{n-2} - \epsilon. \tag{5.26}$$

By the uniqueness theorem for Laplace transforms, (5.25) implies $f(v) = 0$ for all $v > 0$, so that the condition of similarity finally reduces to

$$\int_{w_0(v)} (1 - t)^{n-2} p(\psi_1, \cdots, \psi_{n-2})\, dt\, d\psi_1 \cdots d\psi_{n-2} = \epsilon. \tag{5.27}$$

Of all similar regions, let us find the one which has maximum power. Obviously we want to include in $w(v)$ all points for which $t < 0$. In addition we want to choose $w_0(v)$ such that

$$\int_{w_b(v)} (1 - t)^{n-2} p(\psi_1, \cdots, \psi_{n-2})\, dt\, d\psi_1 \cdots d\psi_{n-2} = \max \tag{5.28}$$

where $w_b(v)$ is that part of $w(v)$ in which $\max\left(0, \frac{nb}{v}\right) \leq t$.

If, for some alternative b, $w_0(v)$ is contained in $\frac{nb}{v} < t < 1$, then $w_b(v)$ and $w_0(v)$ coincide and hence (5.28) attains its maximum value ϵ whatever the position of $w_0(v)$ in $\frac{nb}{v} \leq t \leq 1$. If on the other hand $\frac{nb}{v}$ is so close to 1 that $\frac{nb}{v} \leq t \leq 1$ is too small to contain $w_0(v)$, then (5.28) attains its maximum for any $w_0(v)$ containing $\frac{nb}{v} \leq t \leq 1$. There exists therefore a unique $w_0(v)$ which maximizes (5.28) for all values of b and v, namely the region defined by

$$C(v) \leq t \leq 1 \tag{5.29}$$

where C is determined by (5.27).

Since under H_2, the statistics V and T are independent, C does not depend on v. The test

$$t \leq 0, \qquad \geq C \tag{5.30}$$

which we have just shown to be uniformly most powerful, is also the likelihood ratio test which completes the proof of the theorem.

We shall finally consider an example of an optimum test in connection with a rectangular distribution. Let $X_1, \cdots, X_n$ be independently and uniformly distributed over $(a, a + \theta)$, where θ is positive. *For testing the hypothesis* $H: a = a_0$, *the test*

$$\frac{Y_1 - a_0}{Y_n - Y_1} \leq 0, \qquad \geq C \tag{5.31}$$

where Y_1 and Y_n are the smallest and the largest of the X's respectively, is the uniformly most powerful of all similar regions.

The proof of this goes through very much like that for H_2 in Theorem 5. Without loss of generality we take $a_0 = 0$. Also again, to simplify the proof, we restrict ourselves to regions which are symmetric in the variables. We need the following lemma.

LEMMA. *Let $X_1, \cdots, X_n$ be independently and uniformly distributed over $(a, a + \theta)$. Let Y_i denote the ith X in order of magnitude, and let*

$$T_n = Y_n,\ T_k = \frac{Y_k}{Y_{k+1}}, \qquad (= 1, \cdots, n - 1). \tag{5.32}$$

Then for $a > 0$

$$p(t_1, \cdots, t_n) = \frac{n!}{\theta^n} t_n^{n-1} t_{n-1}^{n-2} \cdots t_2 \tag{5.33}$$

when

$$a \leq t_n \leq a + \theta, \frac{a}{t_n \cdot t_{n-1} \cdots t_{k+1}} \leq t_k \leq 1, \qquad (k = 1, \cdots, n - 1).$$

This is easily seen by applying the usual method of Jacobians. The inequalities describing the sample space of the T's are equivalent to the following more convenient ones:

$$a \leq t_n \leq a + \theta, \frac{a}{t_n} \leq t_1 t_2 \cdots t_{n-1} \leq 1; t_k \leq 1, \qquad (k = 1, \cdots, n - 1). \tag{5.34}$$

Let us denote by $w(t_n)$ the intersection of a region w with the hyperplane $T_n = t_n$, and by $w_0(t_n)$ that part of $w(t_n)$ contained in the cylinder $0 \leq t_k \leq 1$, $(k = 1, \cdots, n - 1)$; then we find as a necessary and sufficient condition for w to be similar with respect to θ (assuming H)

$$(n - 1)! \int_{w_0(t_n)} t_{n-1}^{n-2} t_{n-2}^{n-3} \cdots t_2 \, dt_{n-1} \cdots dt_1 = \epsilon. \tag{5.35}$$

Of all regions satisfying (5.35) we want to find the most powerful one. Let us first consider alternatives $a > 0$. If $w_a(t_n)$ denotes the common part of $w_0(t_n)$ and the region

$$\frac{a}{t_n} \leq t_{n-1} t_{n-2} \cdots t_1 \leq 1, \tag{5.36}$$

we must choose $w_a(t_n)$ such that

$$\int_{w_a(t_n)} t_{n-1}^{n-2} \cdot t_{n-2}^{n-3} \cdots t_2 \, dt_{n-1} \cdots dt_1 = \max. \tag{5.37}$$

From this it follows easily that against alternatives $a > 0$ the uniformly best choice for $w_0(t_n)$ is

$$t_1 t_2 \cdots t_{n-1} = \frac{y_1}{y_n} \geq C'(t_n), \tag{5.38}$$

and since under H, $\dfrac{Y_1}{Y_n}$ is independently distributed of T_n, $C'(t_n)$ does not depend on t_n.

Consider next alternatives $a < 0$. We include in the region of rejection all points for which $Y_1 \leq 0$. To determine $w_0(t_n)$ we notice that, given $Y_1 > 0$, the X's are uniformly distributed between 0 and $a + \theta$. (Provided $a + \theta > 0$; the case $a + \theta \leq 0$ is trivial). Hence the probability distribution of the T's given $Y_1 > 0$ is

$$p(t_1, \cdots, t_n \mid Y_1 > 0) = \frac{n!}{(a + \theta)^n} t_n^{n-1} \cdots t_2 \tag{5.39}$$

when

$$0 \leq t_n \leq a + \theta, \qquad 0 \leq t_k \leq 1 \qquad \text{for } k = 1, \cdots, n-1.$$

Thus

$$\frac{p(t_1, \cdots, t_{n-1} \mid t_n, a < 0, Y_1 > 0)}{p(t_1, \cdots, t_{n-1} \mid t_n, a = 0)} \tag{5.40}$$

is independent of $t_1, \cdots, t_{n-1}$ and hence the power of w against alternatives $a < 0$ is independent of the choice of $w_0(t_n)$. Therefore the region

$$y_1 \leq 0, \; \frac{y_1}{y_n} \geq C' \tag{5.41}$$

is uniformly most powerful against all alternatives. But (5.41) is equivalent to

$$\frac{y_1}{y_n - y_1} \leq 0, \; \geq C. \tag{5.42}$$

It is interesting to compare this result with that for the corresponding simple hypothesis. Let H' be the hypothesis: $a = 0$ when the X's are assumed independently and uniformly distributed over $(a, a + 1)$. There exists no uniformly most powerful test of H'; instead the two uniformly most powerful one-sided tests exist. By analogy with the normal case one might then expect for H' that of all tests with symmetric power-functions, there be a uniformly most powerful one. This however is not so: there exist infinitely many admissible tests with symmetric powerfunction.

In this and the previous section we restricted ourselves to problems involving only one nuisance parameter. However, the method applies also to problems involving several nuisance parameters.

In the usual way (cf. [20, 9]) the results of this section may be translated to give optimum sets of confidence intervals for estimating the parameters in question. In this connection it is an open question whether the confidence regions based on the type B_1 tests discussed in section 2 will always be intervals; one would expect this to be the case.

The author wishes to acknowledge his indebtedness to Professor P. L. Hsu for many helpful suggestions.

Added in proof: In a joint paper by Professor Henry Scheffé and the present author which has been submitted to the Proceedings of the National Academy of Sciences, a result is given concerning the existence of certain 1:1 transformations. This result bears on Section 4 of the present paper where a question arises concerning the existence of a 1:1 transformation. The existence of such a transformation is now assured and, as a consequence, the last paragraph of Section 4 has become superfluous.

REFERENCES

[1] J. Neyman and E. S. Pearson, "On the problem of the most efficient tests of statistical hypotheses", *Roy. Soc. London Phil. Trans.*, Ser. A, Vol. 231 (1933), p. 289.

[2] J. Neyman, "Sur la vérification des hypothèses statistiques composées", *Bull. Soc. Math. France*, Vol. 63 (1935), p. 1.

[3] J. Neyman, "On a statistical problem arising in routine analysis and in sampling inspection of mass production", *Annals of Math. Stat.*, Vol. 12 (1941), p. 46.

[4] H. Scheffé, "On the theory of testing composite hypotheses with one constraint", *Annals of Math. Stat.*, Vol. 13 (1942), p. 280.

[5] P. L. Hsu, "Analysis of variance from the power function standpoint", *Biometrika*, Vol. 32 (1941), p. 62.

[6] J. B. Simaika, "On an optimum property of two important statistical tests", *Biometrika*, Vol. 32 (1941), p. 70.

[7] A. Wald, "On the power function of the analysis of variance test", *Annals of Math. Stat.*, Vol. 13 (1942), p. 434.

[8] P. L. Hsu, "On the power function of the E^2-test and the T^2-test", *Annals of Math. Stat.*, Vol. 16 (1945), p. 278.

[9] H. Scheffé, "On the ratio of the variances of two normal populations", *Annals of Math. Stat.*, Vol. 13 (1942), p. 371.

[10] J. Neyman, "L'estimation statistique traitée comme un problème classique de probabilité", *Actualités Sci. et Ind. No. 739*, Hermann et Cie, Paris, 1938.

[11] A. Wald, "Notes on the theory of statistical estimation and testing hypotheses", Columbia University.

[12] W. J. Dixon, "Further contributions to the problem of serial correlation", *Annals of Math. Stat.*, Vol. 15 (1944), p. 119.

[13] J. von Neumann, "Distribution of the ratio of the mean square successive difference to the variance", *Annals of Math. Stat.*, Vol. 12 (1941), p. 367.

[14] M. Kac, "A remark on independence of linear and quadratic forms involving independent Gaussian variables", *Annals of Math. Stat.*, Vol. 16 (1945), p. 400.

[15] W. G. Madow, "Note on the distribution of the serial correlation coefficient", *Annals of Math. Stat.*, Vol. 16 (1945), p. 308.

[16] R. L. ANDERSON, "Distribution of the serial correlation coefficient", *Annals of Math. Stat.*, Vol. 13 (1942), p. 1.
[17] T. KOOPMANS, "Serial correlation and quadratic forms in normal variables", *Annals of Math. Stat.*, Vol. 13 (1942), p. 14.
[18] R. B. LEIPNIK, "Distribution of the serial correlation coefficient in a circularly correlated universe", *Annals of Math. Stat.*, Vol. 18 (1947), p. 80.
[19] P. L. HSU, "On the asymptotic distributions of certain statistics used in testing the independence between successive observations from a normal population", *Annals of Math. Stat.*, Vol. 17 (1946), p. 350.
[20] J. NEYMAN, "Outline of a theory of statistical estimation based on the classical theory of probability", *Roy. Soc. London Phil. Trans.*, Ser. A, Vol. 236 (1937), p. 333.
[21] E. J. G. PITMAN, "Sufficient statistics and intrinsic accuracy", *Proc. Camb. Phil. Soc.*, Vol. 32 (1936), p. 567.
[22] B. O. KOOPMAN, "On distributions admitting a sufficient statistic", *Trans. Amer. Math. Soc.*, Vol. 39 (1936), p. 399.
[23] D. DUGUÉ, "Application des propriétés de la limite au sens du calcul des probabilités à l'étude des diverses questions d'estimation", *J. École Poly.*, Vol. 3 (1937), p. 305.
[24] J. NEYMAN AND E. S. PEARSON, "On the use and interpretation of certain test criteria for purposes of statistical inference", *Biometrika*, Vol. 20A (1928), p. 175.
[25] E. PAULSON, "On certain likelihood ratio tests associated with the exponential distribution", *Annals of Math Stat.*, Vol. 12 (1941), p. 301.
[26] P. V. SUKHATME, "Tests of significance for samples of the χ^2 population with two degrees of freedom", *Annals of Eugenics*, Vol. 8 (1937), p. 52.

MOST POWERFUL TESTS OF COMPOSITE HYPOTHESES. I. NORMAL DISTRIBUTIONS

By E. L. Lehmann and C. Stein

University of California, Berkeley

Summary. For testing a composite hypothesis, critical regions are determined which are most powerful against a particular alternative at a given level of significance. Here a region is said to have level of significance ϵ if the probability of the region under the hypothesis tested is bounded above by ϵ. These problems have been considered by Neyman, Pearson and others, subject to the condition that the critical region be similar. In testing the hypothesis specifying the value of the variance of a normal distribution with unknown mean against an alternative with larger variance, and in some other problems, the best similar region is also most powerful in the sense of this paper. However, in the analogous problem when the variance under the alternative hypothesis is less than that under the hypothesis tested, in the case of Student's hypothesis when the level of significance is less than $\frac{1}{2}$, and in some other cases, the best similar region is not most powerful in the sense of this paper. There exist most powerful tests which are quite good against certain alternatives in some cases where no proper similar region exists. These results indicate that in some practical cases the standard test is not best if the class of alternatives is sufficiently restricted.

1. Introduction. The problem to be discussed in this paper is that of testing a composite hypothesis against a simple alternative. More specifically let $\mathcal{F} = \{f\}$ be a family of probability density functions defined over a Euclidean space R_n and let g be a probability density function not in $\mathcal{F}$. We wish to test the hypothesis H_0 that the random variable $X = (X_1, \cdots, X_n)$ is distributed according to a density f of $\mathcal{F}$ against the alternative H_1 that X is distributed according to g. By a test we mean a region of rejection, w in R_n.

Neyman and Pearson, in the fundamental paper [1] which laid the groundwork of the theory of optimum tests, restricted their considerations to similar regions. They considered a region (set) w to be optimum for the given level of significance ϵ if it maximizes the power

$$\int_w g(x)\,dx \tag{1}$$

subject to the restriction

$$\int_w f(x)\,dx = \epsilon \quad \text{for all} \quad f \text{ in } \mathcal{F}. \tag{2}$$

As Neyman, Wald and others have pointed out, it is more natural to replace the condition of similarity (2) by the weaker restriction

$$\int_w f(x)\,dx \leq \epsilon \quad \text{for all} \quad f \text{ in } \mathcal{F}. \tag{3}$$

J. Rojo (ed.), *Selected Works of E. L. Lehmann*, Selected Works in Probability and Statistics, DOI 10.1007/978-1-4614-1412-4_13, © Springer Science+Business Media, LLC 2012

A region w maximizing (1) subject to (3) is called most powerful against the alternative g at the level of significance. Here and throughout the paper, all functions and sets are assumed to be Borel measurable.

In the present paper we shall consider certain composite hypotheses, and derive tests for them which are most powerful against a simple alternative. For the cases in which these tests coincide with the standard similar regions it will thus be established that no further increase in power is possible with tests of fixed sample sizes. In the more usual situation where the most powerful test depends strongly on the specific alternative chosen, no such absolute justification of the standard test is possible. In these cases, any justification must take account of the fact that it is desired to obtain good power against a large class of alternatives. This can be done, for instance, by using Wald's definition of a most stringent test [2] or his concept of minimizing the maximum risk.[1] If, on the other hand, the class of alternatives is sufficiently restricted, the results of the present paper indicate that for small samples there may exist a test which is appreciably better than the standard test.

Frequently the probability of an error of the first kind is an analytic function of a nuisance parameter for every choice of critical region. Hence, if it is known that some nuisance parameter θ lies, say in a certain finite interval I, then any test which is similar for θ in I will be similar for all θ. Consequently, the knowledge concerning θ cannot be used to find a more powerful test. On the other hand, as is indicated at the end of section 5, restrictions of the nuisance parameters may, for small samples, lead to considerably more powerful tests if the condition of similarity is replaced by the weaker condition (3).

There is one class of problems to which it may be desirable to apply the method of the present paper regardless of sample size; namely, if no similar region exists. Suppose, for instance, that $X_1, \cdots, X_n$ are known to be normally and independently distributed, X_i having unknown mean and variance ξ_i and σ_i^2 for $i = 1, \cdots, n$. For testing the hypothesis

$$H_0 : \sigma_i = 1, \qquad (i = 1, \cdots, n)$$

no similar region exists, while it is easy to see that against any simple alternative

$$H_1 : \sigma_i = \sigma_{i1} < 1, \qquad \xi_i = \xi_{i1},$$

there exists a test which satisfies condition (3) and which has good power against H_1 provided the σ_{i1} are sufficiently small.

The present first part of this paper is restricted to hypotheses concerning normal distributions. It is intended to extend the considerations to exponential

[1] In an unpublished paper, it is shown by G. Hunt and C. Stein that the traditional test is most stringent in several cases, including the (univariate) linear hypothesis and the hypothesis specifying the ratio of the variances of two normal distributions. These results can be extended to analogous problems for distributions other than the normal, and similar results can be proved regarding minimization of the maximum risk if the weight function has a certain type of symmetry.

and rectangular distributions, to consider non-parametric problems and possibly also more complicated problems connected with normal distributions, in later parts of the paper.

2. Sufficient conditions for a most powerful test. The method which will be used in this paper to obtain most powerful tests is an adaptation of the fundamental lemma of Neyman and Pearson [1]. At the same time it is essentially a special case of much more general results of Wald [3, 4], although the exact conditions of Wald's investigation are not satisfied in most of our problems.

Let h and g be two functions defined over R_n, let k be a constant and let w be a region in R_n such that

$$g(x) \geq k\ h(x) \text{ in } w; \tag{4}$$
$$g(x) \leq k\ h(x) \text{ in } R_n - w.$$

Then if w' is such that

$$\int_{w'} h(x)\, dx \leq \int_{w} h(x)\, dx, \tag{5}$$

it follows as in the fundamental lemma where in (5) equality is assumed instead of inequality, that

$$\int_{w'} g(x)\, dx \leq \int_{w} g(x)\, dx. \tag{6}$$

Throughout the present paper we shall be concerned with the special case in which $\mathcal{F}$ is an s-parameter family. We may denote the members of $\mathcal{F}$ by f_θ and we shall obtain all members of $\mathcal{F}$ as θ ranges over a set ω in an s-dimensional Euclidean space. In the theorem which we shall now state, we shall be concerned with point functions λ defined over ω. We shall assume that $\lambda = c\mu$ where c is a positive constant and μ a cumulative distribution function.[2] Also we suppose that $f_\theta(x)$ is a measurable function of x and θ jointly. However, the theorem is also valid if ω is an abstract space and λ a (finite) non-negative additive set function (measure) over ω. Such more general interpretation may be required when applying the theory to non-parametric problems.

THEOREM 1. *Let H_0 be the hypothesis that the random variable X is distributed according to a density function f_θ with θ in ω, and let H_1 denote the alternative that X is distributed according to a density g. Let λ be a function defined over ω and such that*

$$\lambda = c\mu, \tag{7}$$

[2] The introduction of the distribution μ is simply a mathematical device and does not imply that θ is a random variable (see Wald [16] p. 282).

where c is a positive constant and μ a cumulative distribution function. Let k be a constant and let w be a region in R_n such that

$$(8)\qquad \begin{aligned} g(x) &\geq k \int_w f_\theta(x)\, d\lambda(\theta) \quad \text{in} \quad w; \\ g(x) &\leq k \int_w f_\theta(x)\, d\lambda(\theta) \quad \text{in} \quad R_n - w. \end{aligned}$$

Suppose that w is of level of significance ϵ for testing H_0 against H_1, that is that

$$(9)\qquad \int_w f_\theta(x)\, dx \leq \epsilon \quad \text{for all} \quad \theta \text{ in } \omega,$$

and suppose that the subset of ω for which

$$(10)\qquad \int_w f_\theta(x)\, dx < \epsilon$$

has λ-measure zero. Then w is most powerful for testing H_0 against H_1 at level of significance ϵ.

PROOF. Without loss of generality we shall assume $c = 1$. Let w' be any test of level of significance ϵ. Then

$$(11)\qquad \int_{w'} f_\theta(x)\, dx \leq \epsilon \quad \text{for all} \quad \theta \text{ in } \omega,$$

and because of (7)

$$(12)\qquad \int_\omega \left\{ \int_{w'} f_\theta(x)\, dx \right\} d\lambda(\theta) \leq \epsilon \int_\omega d\lambda(\theta) = \epsilon.$$

Since λ is of bounded variation we may interchange the order of integration in (12) and obtain

$$(13)\qquad \int_{w'} h(x)\, dx \leq \epsilon,$$

where

$$(14)\qquad h(x) = \int_\omega f_\theta(x)\, d\lambda(\theta).$$

From (9) and the condition surrounding (10) it follows that

$$(15)\qquad \int_\omega \left\{ \int_w f_\theta(x)\, dx \right\} d\lambda(\theta) = \epsilon,$$

and therefore that

$$(16)\qquad \int_w h(x)\, dx = \epsilon.$$

Thus w and w' satisfy conditions (4) and (5), and hence also (6) which completes the proof.

It is useful to notice that, the assumptions of theorem 1 will be satisfied provided

$$\int_w f_\theta(x)\,dx$$

attains its maximum ϵ at all points of increase of λ, and therefore in particular whenever w is a similar region of size ϵ.

We shall in many problems exhibit a function λ which satisfies the conditions of theorem 1 without giving the reasons which led us to this function. However the following comments concerning the tentative process that we used, may be helpful. One may first examine the known most powerful similar region. If there exists a cumulative distribution function λ such that (8) is the most powerful similar region, the problem is solved. If the most powerful similar region cannot even be approximated by (8) with a sequence of λ's, it is reasonable to conclude that the most powerful test is not similar. Because the probability (under the null hypothesis) of any test is in all the problems considered here an analytic function of the parameter, this implies that the probability (under the null hypothesis) of the most powerful test attains its maximum at an at most denumerable (in some cases finite) set of points. In all the cases of this kind which we considered in the present part I, it was then possible to prove the existence of a function λ with a single point of increase, which satisfied the conditions of theorem 1.

A theorem analogous to theorem 1 holds for most powerful similar regions. Let H_0 and H_1 be as before and let λ be a function of bounded variation not necessarily non-decreasing. Let w be a region in R_n such that

$$\begin{aligned} g(x) &\geq k \int_\omega f_\theta(x)\,d\lambda(\theta) \quad \text{in} \quad w; \\ g(x) &\leq k \int_\omega f_\theta(x)\,d\lambda(\theta) \quad \text{in} \quad R_n - w. \end{aligned} \tag{17}$$

Let w be a similar region of level of significance ϵ for testing H_0 against H_1, that is, let

$$\int_w f_\theta(x)\,dx = \epsilon \quad \text{for all} \quad \theta \text{ in } \omega; \tag{18}$$

then w is a most powerful similar region for testing H_0 against H_1.

For all the problems considered in this paper we shall prove the existence of functions λ satisfying the conditions of theorem 1, but we have not investigated the corresponding existence problem in general. On the other hand one verifies easily that for many of the cases treated here in which the most powerful test is not similar, the method for obtaining most powerful similar regions does not apply. However, for all the problems considered in the present paper the most powerful similar tests can be obtained easily by other methods [1, 5, 6, 7, 8]. For most of the problems the corresponding derivations have been carried out in the literature.

Although we restrict ourselves in the present paper to the problem of maximizing the power at a single alternative, theorem 1 clearly also applies to the more general problem of maximizing the average power over surfaces in a space of alternatives. Such problems have been considered from the point of view of similar regions by Wald, Hsu and others [9, 10, 11].

3. Testing the values of one or several variances. Let $X_1, \cdots, X_n$ be a sample from a normal population with mean ξ and variance σ^2, both unknown. We want to test the hypothesis H_0 that $\sigma = \sigma_0$ against the simple alternative that $\sigma = \sigma_1$, $\xi = \xi_1$. We shall show that the most powerful test for H_0 against H_1 is

$$\Sigma(x_i - \xi_1)^2 \leq k \quad \text{when} \quad \sigma_1 < \sigma_0; \tag{19}$$

$$\Sigma(x_i - \bar{x})^2 \geq c \quad \text{when} \quad \sigma_1 > \sigma_0, \tag{20}$$

where k and c are determined by the level of significance. Thus the best similar region is most powerful if the variance under the alternative is greater than that under the null hypothesis, while the most powerful tests against the other alternatives are not similar. That the region $\Sigma(x_i - \bar{x})^2 \geq c$ ($\leq c'$) is most powerful of all similar regions against $\sigma_1 > \sigma_0$ ($\sigma_1 < \sigma_0$) was shown by Neyman and Pearson [1].

We consider first the case $\sigma_1 < \sigma_0$, and apply theorem 1 with λ a stepfunction having a single jump at ξ_1, that is,

$$\lambda(\xi) = \begin{matrix} 0 & \text{if} & \xi < \xi_1; \\ 1 & \text{if} & \xi \geq \xi_1. \end{matrix} \tag{21}$$

The region w given by (8) thus becomes

$$\frac{\exp\left[-\frac{1}{2\sigma_1^2}\Sigma(x_i - \xi_1)^2\right]}{\exp\left[-\frac{1}{2\sigma_0^2}\Sigma(x_i - \xi_1)^2\right]} \geq k', \tag{22}$$

which is equivalent to

$$\Sigma(x_i - \xi_1)^2 \leq k, \tag{23}$$

since $\sigma_1 < \sigma_0$. The size of the region (23), that is, its probability under the null hypothesis is a function of ξ and clearly attains its maximum when $\xi = \xi_1$. Thus all conditions of theorem 1 are satisfied provided we choose k so that the maximum size of (23) equals ϵ.

Before considering the case $\sigma_1 > \sigma_0$ we state for later reference the following:

LEMMA 1. *If $\sigma_1 > \sigma_0$ there exists an absolutely continuous non-decreasing function λ of bounded variation such that*

$$\int_{-\infty}^{\infty} \exp\left[-\frac{1}{2\sigma_0^2}(t - \xi)^2\right] d\lambda(\xi) = C \exp\left[-\frac{1}{2\sigma_1^2}(t - \xi_1)^2\right]. \tag{24}$$

This follows immediately from the well known representation of $\exp\left(-\frac{1}{a^2}t^2\right)$ as a Laplace transform by applying a translation, and is easily verified directly by substituting

$$\lambda'(\xi) = \exp\left[-\frac{1}{2(\sigma_1^2 - \sigma_0^2)}(\xi - \xi_1)^2\right]. \tag{25}$$

Now let $\sigma_1 > \sigma_0$ and $n > 1$. The region w given by (8) can be expressed in the form

$$\frac{\exp\left[-\frac{1}{2\sigma_1^2}\Sigma(x_i - \bar{x})^2\right]}{\exp\left[-\frac{1}{2\sigma_0^2}\Sigma(x_i - \bar{x})^2\right]} \cdot \frac{\exp\left[-\frac{n}{2\sigma_1^2}(\bar{x} - \xi_1)^2\right]}{\int \exp\left[-\frac{n}{2\sigma_0^2}(\bar{x} - \xi)^2\right]d\lambda(\xi)} \geq k'. \tag{26}$$

By lemma 1 there exists an absolutely continuous function λ for which the second factor is constant. For this λ (26) is equivalent to

$$\Sigma(x_i - \bar{x})^2 \geq c, \tag{27}$$

and since this is a similar region, the conditions of theorem 1 are satisfied provided c is chosen so as to give the correct level of significance.

We next consider the problem in which the random variables X_i $(i = 1, \cdots, n)$ are independently normally distributed with unknown means ξ_i and unknown variances σ_i^2. We wish to test the hypothesis $H_0 : \sigma_i = \sigma_{i0}$ for $i = 1, \cdots, n$ against the alternative $H_1 : \sigma_i = \sigma_{i1}$, $\xi_i = \xi_{i1}$. Feller [12] showed that there exist no similar regions for this problem. However, as we shall show now, when the critical regions are not required to be similar, non-trivial tests against H_1 do exist provided $\sigma_{i1} < \sigma_{i0}$ for at least one value of i.

Let us assume without loss of generality that $\sigma_{i1} < \sigma_{i0}$ for $i = 1, \cdots, m$; $\sigma_{i1} > \sigma_{i0}$ for $i = m + 1, \cdots, n$ where $n - m$ may be zero but where for the moment we shall assume $m > 0$. With $\lambda(\xi_1, \cdots, \xi_n) = \prod_{i=1}^{n} \lambda_i(\xi_i)$, the region (8) becomes

$$\prod_{i=1}^{m} \frac{\exp\left[-\frac{1}{2\sigma_{i1}^2}(x_i - \xi_{i1})^2\right]}{\int_{-\infty}^{\infty} \exp\left[-\frac{1}{2\sigma_{i0}^2}(x_i - \xi_i)^2\right]d\lambda_i(\xi_i)} \cdot \prod_{j=m+1}^{n} \frac{\exp\left[-\frac{1}{2\sigma_{j1}^2}(x_j - \xi_{j1})^2\right]}{\int_{-\infty}^{\infty} \exp\left[-\frac{1}{2\sigma_{j0}^2}(x_j - \xi_j)^2\right]d\lambda_j(\xi_j)} \geq k. \tag{28}$$

For $\lambda_i(i = 1, \cdots, m)$ we take step functions with a single jump at ξ_{i1}, while for the remaining λ's we choose the absolutely continuous functions which make the second factor constant and whose existence is guaranteed by lemma 1. The region (28) thus reduces to

$$\sum_{i=1}^{m}\left(\frac{1}{\sigma_{i1}^2}-\frac{1}{\sigma_{i0}^2}\right)(x_i-\xi_{i1})^2 \leq c. \tag{29}$$

Since the probability of the region (29) is independent of $\xi_{m+1}, \cdots, \xi_n$ and with varying $\xi_1, \cdots, \xi_m$ takes on its maximum when $\xi_i = \xi_{i1}$ it follows from theorem 1 that this region is most powerful for testing H_0 against H_1.

We still have to consider the case $m = 0$, that is, the case in which $\sigma_{i1} > \sigma_{i0}$ for all i. To treat this problem we adjoin to the variables $X_1, \cdots, X_n$ a random variable Y uniformly distributed between 0 and 1, that is, essentially a table of random numbers. In the space of $n + 1$ random variables we determine a region w according to (8), letting $\lambda(\xi_1, \cdots, \xi_n) = \prod_{i=1}^{n} \lambda_i(\xi_i)$ and choosing the λ's so as to make the left hand side of (8) equal to the right hand side. This is possible by lemma 1 and with this choice of the λ's the inequalities (9) become

$$\begin{aligned} k &\geq k \text{ in } w; \\ k &\leq k \text{ in } R_{n+1} - w, \end{aligned} \tag{30}$$

and hence they impose no restrictions on w. Thus any similar region of the correct size will satisfy the conditions of theorem 1. It follows that the region

$$w: 0 \leq y \leq \epsilon, \tag{31}$$

being a similar region of size ϵ, is most powerful. This result means that we do not use the observations $x_1, \cdots, x_n$ at all but consult a table of random numbers.

The situation just described occurs in other problems to which the same method of proof can be applied. It is therefore convenient for later reference to formulate the following

THEOREM 2. *Let H_0 be the hypothesis that the random variable X is distributed according to a probability density function f_θ with θ in ω, and let H_1 denote the alternative that X is distributed according to the density function g. Let Y be a random variable known to be uniformly distributed over the interval* [0, 1]. *If there exists a real valued function λ satisfying* (7) *for which*

$$g(x) = k\int_\omega f_\theta(x)\, d\lambda(\theta), \tag{32}$$

then the critical region $0 \leq y \leq \epsilon$ is most powerful for testing H_0 against H_1 at level of significance ϵ.

4. Testing equality of variances and the value of the circular serial correlation coefficient. For each $i = 1, \cdots, m$ let $X_{ij} (j = 1, \cdots, n_i)$ be a sample from a normal distribution with $E(X_{ij}) = \xi_i$ and $E(X_{ij} - \xi_i)^2 = \sigma_i^2$. We are concerned with the hypothesis H_0 that $\sigma_1 = \sigma_2 = \cdots = \sigma_m$, where first we shall assume the ξ's to be known, so that without loss of generality we may assume them equal to 0. The alternative hypothesis specifies $\sigma_i = \sigma_{i1}$, $i = 1 \cdots m$. Let σ^2 denote the unknown common variance under H_0 and let $\lambda(\sigma)$ be a step function with a single jump at a point σ_0 to be determined later. With $k = \prod_{i=1}^{m} \left(\frac{\sigma_0}{\sigma_{i1}}\right)^{n_i}$, the test (8) takes on the form

$$\frac{\exp\left[-\frac{1}{2}\sum_{i,j}\frac{x_{ij}^2}{\sigma_{i1}^2}\right]}{\exp\left[-\frac{1}{2\sigma_0^2}\sum_{i,j} x_{ij}^2\right]} \geq 1, \tag{33}$$

or equivalently

$$\frac{\sum_{i,j} x_{ij}^2}{\sum_{i,j} \frac{1}{\sigma_{i1}^2} x_{ij}^2} \geq \sigma_0^2. \tag{34}$$

Since the function on the left hand side is homogeneous of degree 0 in the x's, this is a similar region and the conditions of theorem 1 are therefore satisfied provided the region has the correct size. This can be achieved for any level of significance ϵ by proper choice of σ_0^2.

As stated earlier, the conditions of theorem 1 imply that the size of the critical region is equal to ϵ at all points of increase of λ. As a consequence, if the size equals ϵ at only a finite number of points of ω, λ must be a step function. Also if each point of a certain interval is a point of increase of λ, the critical region must be similar over that interval (and, if the functions involved are analytic, the region must be similar over ω). However, the last problem shows that the converse of neither of these two statements is correct. For the region (34) is a similar region although the corresponding λ has only a single point of increase.

Next we consider the hypothesis of equality of variances without assuming the means to be known. For the case $m = 2$ the most powerful similar region was obtained by Neyman and Pearson [1]. We assume first that $n_i > 1$ for all i, and we take $\lambda(\sigma, \xi_1, \cdots, \xi_m) = \lambda_0(\sigma)\prod_{i=1}^{m}\lambda_i(\xi_i)$, with $\lambda_0(\sigma)$ as before a step function with a single jump at a point σ_0 to be determined later. Suppose now that $\sigma_0 > \sigma_{i1}$ for $i = 1, \cdots, s$; $\quad \sigma_0 < \sigma_{i1}$ for $i = s + 1, \cdots, m$, $\sigma_{11} \leq \sigma_{21} \leq \cdots$ where $0 \leq s \leq m$ and s depends on σ_0. Then define

$$\lambda_i(\xi_i) = \begin{cases} 0 & \text{if } \xi_i < \xi_{i1}; \\ 1 & \text{if } \xi_i \geq \xi_{i1} \end{cases} \tag{35}$$

for $i = 1, \cdots, s$ and use lemma 1 for $i = s + 1, \cdots, m$.

For proper choice of k the critical region will then be determined by the inequality

$$(36)\qquad \sum_{i=s+1}^{m}\left(\frac{1}{\sigma_0^2}-\frac{1}{\sigma_{i1}^2}\right)\sum_{j=1}^{n_i}(x_{ij}-\bar{x}_i)^2-\sum_{i=1}^{s}\left(\frac{1}{\sigma_{i1}^2}-\frac{1}{\sigma_0^2}\right)\sum_{j=1}^{n_i}(x_{ij}-\xi_{i1})^2\geq 0.$$

The probability of this region computed under H_0, is independent of $\xi_{s+1}, \cdots, \xi_n$ and for any σ attains its maximum when $\xi_i = \xi_{i1}$ $(i = 1, \cdots, s)$. Since the probability of the region is independent of σ when $\xi_i = \xi_{i1}$ for $i = 1, \cdots, s$, the conditions of theorem 1 are again established. That for $\xi_i = \xi_{i1}$ the size of (36) goes continuously from 0 to 1 with decreasing σ_0 is easily checked since at the only doubtful points $\sigma_0 = \sigma_{i1}$ (where the value of s changes), the corresponding coefficient $\frac{1}{\sigma_0^2} - \frac{1}{\sigma_{i1}^2}$ passes through 0.

We still have to consider the case that some of the n_i are equal to 1. If $n_i = 1$ for some $i \leq s$ there is no change whatever, while if $n_i = 1$ for some $i > s$, the corresponding term in (36) vanishes. It follows easily that if $n_i > 1$ for at least one value of $i > 1$ the solution (36) is valid. On the other hand, if $n_i = 1$ for all $i > 1$, we can apply theorem 2 by taking $\sigma_0 = \sigma_{i1}$, $\lambda_1(\xi_1)$ as a step function with a single jump at ξ_{11} and the remaining $\lambda_i(\xi_i)$ according to lemma 1. It thus follows that for this problem no non-trivial test exists.

The following problem can be reduced to the hypothesis of equality of variances with means assumed known: Under the null hypothesis $X_1, \cdots, X_n$ have a joint multivariate normal distribution with density $C \exp\left[-\frac{1}{2\sigma^2}\Sigma\, a_{ij}x_ix_j\right]$ where the a's are known and where σ is an unknown scale factor. Under H_1 the X's have a joint multivariate normal distribution with density $C' \exp\left[-\frac{1}{2}\Sigma b_{ij}x_ix_j\right]$. A number of hypotheses specifying the value of one or several correlation coefficients have this form. The most powerful test of H_0 against H_1 is given by

$$(36)\qquad \frac{\Sigma b_{ij}x_ix_j}{\Sigma a_{ij}x_ix_j}$$

as is easily shown by applying a non-singular linear transformation which reduces $\Sigma b_{ij}x_ix_j$ to diagonal form and $\Sigma a_{ij}x_ix_j$ to a sum of squares, or by applying directly the method of proof of the earlier problem.

A corresponding reduction when the X's have a common but unknown mean is usually impossible. One problem of this kind for which the solution is simple is the hypothesis specifying the value of a serial correlation coefficient in a circular population. The most powerful similar region for testing this hypothesis was obtained in [7]. Consider the probability density function

$$(37)\qquad C\exp\left[-\alpha\left\{\sum_{i=1}^{n}(x_i-\xi)-\delta(x_{i+1}-\xi)\right\}^2\right],\qquad (x_{n+1}=x_1),\qquad |\delta|<1,$$

and let H_0 specify $\delta = \delta_0$ while H_1 assigns to the parameters the values α_1, ξ_1, δ_1. Then the most powerful test of H_0 against H_1 is

$$(38)\qquad \begin{aligned} &\frac{\Sigma(x_i - \bar{x})(x_{i+1} - \bar{x})}{\Sigma(x_i - \bar{x})^2} \geq k \quad \text{if} \quad \delta_1 > \delta_0;\\ &\frac{\Sigma(x_i - \xi_1)(x_{i+1} - \xi_1)}{\Sigma(x_i - \xi_1)^2} \leq k' \quad \text{if} \quad \delta_1 < \delta_0. \end{aligned}$$

We shall omit the proof of this result, since the method is the same as in the other problems considered in this section.

5. Student's hypothesis and some generalizations. As the principal result of the present section we shall prove that for testing Student's hypothesis against a simple alternative the most powerful test is a non-similar region of the form

$$(39)\qquad \Sigma(X_i - \eta)^2 \leq k,$$

if the level of significance ϵ is less than or equal to $\frac{1}{2}$. Here η and k depend on ϵ and on the alternative, and they will not be determined explicitly. It will be shown also that if ϵ is greater than or equal to $\frac{1}{2}$, Student's test is most powerful. These results will be extended rather easily to the general univariate linear hypothesis. The corresponding investigation for similar regions was carried through for Student's hypothesis by Neyman and Pearson [1] while the extension to a general linear hypothesis is contained in a paper by Hsu [13].

The proof of the main result mentioned above is rather lengthy. We shall begin by proving two lemmas.

LEMMA 2. *Let $Y_1, \cdots, Y_n$ be n independent random variables, normally distributed with 0 mean and unit variance, and let*

$$(40)\qquad \begin{aligned} &P(a, k) = P\left\{\sum_{i=1}^{n} (Y_i - a)^2 \leq (n - k)a^2\right\};\\ &\varphi(k) = \sup_a P(a, k) \quad \textit{for} \quad 0 < k < n, \quad 0 < a. \end{aligned}$$

Then for each k there exists $a(k)$ such that

$$(41)\qquad P(a(k), k) = \varphi(k).$$

PROOF. If $Z_i = Y_i/a$, $(i = 1, \cdots, n)$ the Z's are independently normally distributed with zero mean and variance $1/a^2$ and (40) may be written as

$$(42)\qquad P(a, k) = P\{\Sigma (Z_i - 1)^2 \leq n - k\}.$$

Hence it is seen that for any k, $P(a, k)$ tends to zero as a tends to either zero or infinity. This proves the lemma since for any k, $P(a, k)$ is a continuous function of a.

LEMMA 3. *Given any ϵ, $0 < \epsilon < \frac{1}{2}$ there exists $k(\epsilon)$ between zero and n such that $\varphi(k(\epsilon)) = \epsilon$.*

PROOF. The proof will be given in a number of steps.

(i) $\varphi(k) \to \frac{1}{2}$ as $k \to 0$.

Clearly $P(a, k)$ never exceeds $\frac{1}{2}$. The result will therefore follow if we exhibit a sequence a_k such that $P(a_k, k) \to \frac{1}{2}$ as $k \to 0$. Let $a_k = 1/\sqrt{k}$. Then

$$P(a_k, k) = P\{\sqrt{k}\,\Sigma Y_i^2 - 2\Sigma Y_i + \sqrt{k} \le 0\}. \tag{43}$$

The right hand side is a continuous function of k and therefore tends to

$$P\{\Sigma Y_i \ge 0\} = \tfrac{1}{2}, \tag{44}$$

as k tends to zero.

(ii) $\varphi(k) \to 0$ as $k \to n$.

Consider $P(a, k)$ as in (42). Written as an integral of the probability density of the Z's, the region of integration is independent of a and its volume tends to 0 as k tends to n. On the other hand the probability density depends on a but is uniformly bounded over the region of integration if $k > 0$, and hence the result follows.

(iii) If $0 < k_0$, $P(a, k)$ tends to zero uniformly for k in the interval $k_0 \le k \le n$ as a tends to zero or infinity.

This follows from the fact that $0 \le P(a, k) \le P(a, k_0)$ since $P(a, k_0)$ tends to 0 as a tends to zero or infinity.

(iv) Given k_0 and k_1 there exist numbers a_0 and a_1 with $0 < a_0 < a_1 < \infty$ such that $0 < k_0 \le k \le k_1 < n$ implies $a_0 \le a(k) \le a_1$.

If this were not true there would exist a sequence $k^{(i)}$ with $k_0 \le k^{(i)} \le k_1$ and $a(k^{(i)})$ tending to infinity or zero. Then $\varphi(a(k^{(i)}))$ would tend to zero by (iii). On the other hand consider $P(1, k)$ for $k_0 \le k \le k_1$. This is a continuous non-vanishing function of k and hence attains its lower bound m for some k in $k_0 \le k \le k_1$. Therefore m is positive and we have a contradiction.

(v) Given any k_0, k_1 with $0 < k_0 < k_1 < n$, $\varphi(k)$ is continuous on the interval $[k_0, k_1]$.

To see this, select a_0 and a_1 in accordance with (iv). Then $P(a, k)$ is uniformly continuous in the rectangle $a_0 \le a \le a_1$, $k_0 \le k \le k_1$. Given $\eta > 0$ let δ be such that $|k' - k''| < \delta$ implies $|P(a, k') - P(a, k'')| < \eta$. Then $\varphi(k') \ge P(a(k''), k') \ge P(a(k''), k'') - \eta = \varphi(k'') - \eta$, and by symmetry $\varphi(k'') \ge \varphi(k') - \eta$, which establishes the continuity of φ.

The proof of the lemma is now immediate. For let $0 < \epsilon < \frac{1}{2}$. It follows from (i) and (ii) that there exist k_0 and k_1 such that

$$\varphi(k_0) \le \epsilon/2, \qquad \varphi(k_1) \ge \epsilon + \tfrac{1}{2}(\tfrac{1}{2} - \epsilon),$$

and hence by (v) there exists $k(\epsilon)$ for which $\varphi(k(\epsilon)) = \epsilon$.

Let us now consider Student's hypothesis. The random variables $X_1, \cdots, X_n$ are a sample from a normal distribution which under H_0 has mean 0 and unknown variance σ^2, while under H_1 the mean is ξ_1 and the variance σ_1^2. Without loss of generality we shall assume $\xi_1 > 0$. Applying theorem 1 with λ a step-function having a single jump at a point $\sigma_0 > \sigma_1$ to be determined later, we obtain the critical region in the form

$$\left(\frac{1}{\sigma_1^2} - \frac{1}{\sigma_0^2}\right) \Sigma X_i^2 - 2 \frac{\xi_1}{\sigma_1^2} \Sigma X_i \leq c. \tag{45}$$

Let $Y_i = X_i/\sigma$ so that under H_0 the Y's are distributed with zero mean and unit variance. Then (45) becomes

$$\Sigma Y_i^2 - 2 \frac{\xi_1}{\sigma(1 - \sigma_1^2/\sigma_0^2)} \Sigma Y_i \leq \frac{c}{\sigma^2}, \tag{46}$$

which may be written as

$$\Sigma(Y_i - a)^2 \leq (n - k)a^2, \tag{47}$$

where

$$a = \frac{\xi_1}{\sigma(1 - \sigma_1^2/\sigma_0^2)}, \qquad k = \frac{-c}{\xi_1^2}\left(1 - \frac{\sigma_1^2}{\sigma_0^2}\right)^2. \tag{48}$$

As σ varies from 0 to ∞, a goes from ∞ to 0. Let $P(a, k)$, $\varphi(k)$ and $a(k)$ be defined as in lemma 2. Given the level of significance ϵ $(0 < \epsilon < \frac{1}{2})$, let k^* and a^* be determined according to lemma 2 and 3 so that

$$\varphi(k^*) = \epsilon \text{ and } P(a^*, k^*) = \varphi(k^*). \tag{49}$$

We now select $\sigma_0 > \sigma_1$ and c so that

$$a^* = \frac{\xi_1}{(1 - \sigma_1^2/\sigma_0^2)\sigma_0} \quad \text{and} \quad k^* = \frac{c}{\xi_1^2}\left(1 - \frac{\sigma_1^2}{\sigma_0^2}\right). \tag{50}$$

We have to show that for this choice of σ_0 and c the size of the critical region attains its maximum when $\sigma = \sigma_0$ and that this maximum size is ϵ. Substituting from (50) we express the region (47) in the form

$$\Sigma\left(Y_i - \frac{\sigma_0}{\sigma} a^*\right)^2 \leq (n - k^*) \frac{\sigma_0^2}{\sigma^2} a^{*2}. \tag{51}$$

Thus the probability of the region is

$$P\left(\frac{\sigma_0}{\sigma} a^*, k^*\right). \tag{52}$$

As σ varies, (52) attains its maximum when $\frac{\sigma_0}{\sigma}a^* = a(k^*) = a^*$, that is, when $\sigma = \sigma_0$ and the maximum value of (52) is $\varphi(k^*) = \epsilon$.

This derivation is valid even when $n = 1$, i.e., when the hypothesis $\xi = 0$ is to be tested by observing only a single random variable X, known to be normally distributed but whose mean ξ and variance are unknown. For this problem no similar region exists. However, critical regions of the form $0 < \xi_1 - a < x < \xi_1 + b$ will give any level of significance $< \frac{1}{2}$ for proper choice of a and b, while the power of such regions will tend to 1 as σ_1 tends to 0. Therefore, the power of the most powerful test will be close to 1 if σ_1 is sufficiently small.

Having completed the discussion of the case $\epsilon < \frac{1}{2}$ let us next suppose that $\epsilon \geq \frac{1}{2}$. We shall need the following

LEMMA 4. *Let c and α_1 be positive constants. Then there exists a function f such that $f(\alpha) = 0$ when $\alpha < \alpha_1$ and such that for all $w > 0$*

$$\int_0^\infty e^{-\alpha w} f(\alpha)\, d\alpha = k e^{-\alpha_1 w - c\sqrt{w}}. \tag{53}$$

This follows from the well known representation of $e^{-c\sqrt{w}}$ as a Laplace transform by applying a translation. (53) can be checked directly by substituting

$$f(\alpha) = \frac{ce^{-(c^2/4(\alpha-\alpha_1))}}{(\alpha - \alpha_1)^{3/2}} \quad \text{for} \quad \alpha \geq \alpha_1. \tag{54}$$

Applying theorem 1 to Student's hypothesis, where again we shall assume ξ_1 to be positive, for proper choice of k we obtain from (9)

$$\frac{\exp\left[-\frac{1}{2\sigma_1^2}\Sigma X_i^2 + \frac{\xi_1}{\sigma_1^2}\Sigma X_i\right]}{\int_0^\infty \exp\left[-\frac{1}{2\sigma^2}\Sigma X_i^2\right]\frac{1}{\sigma^n}\, d\lambda(\sigma)} \geq 1. \tag{55}$$

It follows from lemma 4 that for any positive c there exists a non-decreasing function λ of bounded variation with $\lambda(\sigma)$ constant for $\sigma > \sigma_1$, such that

$$\int_0^\infty \exp\left[-\frac{1}{2\sigma^2}\Sigma X_i^2\right]\frac{1}{\sigma^n}\, d\lambda(\sigma) = \exp\left[-\frac{1}{2\sigma_1^2}\Sigma X_i^2 - c\sqrt{\Sigma x_i^2}\right]. \tag{56}$$

For this choice of λ, (55) reduces to

$$\exp\left[\frac{\xi_1}{\sigma_1^2}\Sigma x_i\right] \geq \exp\left[-c\sqrt{\Sigma x_i^2}\right], \tag{57}$$

and hence to

$$\frac{\Sigma x_i}{\sqrt{\Sigma x_i^2}} \geq c'. \tag{58}$$

This is a similar region and therefore most powerful for testing Student's hypothesis against H_1. By adjusting c, the size of the region can be made equal to any $\epsilon \geq \frac{1}{2}$.

The argument for $\epsilon > \frac{1}{2}$ must be modified slightly in the case $n = 1$, that is, when we want to test Student's hypothesis on the basis of a single observation. Let us adjoin to the variable X a random variable Y known to be uniformly distributed over the interval [0, 1]. Using the same λ and k as before, (58) becomes

(59)

For $c' = -1$ the critical region includes all points (x, y) for which x is positive while (59) places no restriction on which of the remaining points to include in the critical region. The similar region

$$x \geq 0; \quad x < 0, \quad 0 < y < 2(\epsilon - \tfrac{1}{2}) \tag{60}$$

therefore satisfies all conditions of theorem 1 and hence is most powerful

In extending these results to the general linear hypothesis, we shall assume the hypothesis reduced to canonical form [14, 15]. We shall therefore assume that $X_1, \cdots, X_n$ are normally distributed with common variance which is unknown under H_0 and has the value σ_1^2 under H_1. Furthermore, under H_0, $E(X_i) = 0$ for $i = 1, \cdots, s, s+1, \cdots, m$; $E(X_i)$ unknown for $i = m+1, \cdots, n$ while under H_1 $E(X_i) = 0$ for $i = s+1, \cdots, m$; $E(X_i) = \xi_{i1}$ for the remaining values of i.

For $\epsilon < \frac{1}{2}$ we shall consider critical regions of the form

$$\frac{\exp\left\{-\frac{1}{2\sigma_1^2}\left[\sum_{i=1}^{s}(x_i - \xi_{i1})^2 + \sum_{i=s+1}^{m} x_i^2 + \sum_{i=m+1}^{n}(x_i - \xi_{i1})^2\right]\right\}}{\exp\left\{-\frac{1}{2\sigma_1^2}\left[\sum_{i=1}^{s} x_i^2 + \sum_{i=s+1}^{m} x_i^2 + \sum_{i=m+1}^{n}(x_i - \xi_{i1})^2\right]\right\}} \geq k, \tag{61}$$

which are obtained from (8) by substituting for λ a step-function with a single jump at the parameter point $(\sigma_i, \xi_{m+1,1}, \cdots, \xi_{n,1})$. Making an orthonormal transformation from $x_s, \cdots, x_s$ to $y_1, \cdots, y_s$ such that $y_1 = \dfrac{\sum_{i=1}^{s} \zeta_{i1} x_i}{\sqrt{\sum \zeta_{i1}^2}}$ and letting $y_i = x_i$ for $i = s+1, \cdots, m$; $y_i = x_i - \xi_{i1}$ for $i = m+1, \cdots, n$, (61) reduces to

$$\frac{\exp\left\{-\frac{1}{2\sigma_1^2}\left[\sum_{i=1}^{n} y_i^2 - 2y_1\sqrt{\sum_{i=1}^{s}\xi_{i1}^2}\right]\right\}}{\exp\left\{-\frac{1}{2\sigma_0^2}\sum_{i=1}^{n} y_i^2\right\}} \geq c. \tag{62}$$

For $\sigma_0 > \sigma_1$ we can rewrite (62) as

$$\sum_{i=m+1}^{n} y_i^2 \leq \frac{1}{\sigma_1^2 - \sigma_0^2}\left[c + \frac{2y_1\sqrt{\sum \xi_{i1}^2}}{\sigma_1^2}\right] - \sum_{i=1}^{m} y_i^2, \tag{63}$$

and we see that under H_0 for any σ the size of this region considered as a function of the unknown means of $Y_{m+1}, \cdots, Y_n$ takes on its maximum when these means are zero, i.e. when $\xi_i = \xi_{i1}$ for $i = m+1, \cdots, n$. For these maximizing values of the means the existence of a suitable σ_0 and c follows from the corresponding result in connection with Student's hypothesis.

Thus the most powerful test for testing H_0 against H_1 at level of significance $\epsilon = \frac{1}{2}$ has the form

$$\sum_{i=1}^{s}\left[x_i - \frac{\xi_{i1}}{1 - \sigma_1^2/\sigma_0^2}\right]^2 + \sum_{i=s+1}^{m} x_i^2 + \sum_{i=m+1}^{n}(x_i - \xi_{i1})^2 \leq c. \tag{64}$$

It is interesting that the variables $X_i(i = m + 1, \cdots, n)$ which may be discarded when considerations are restricted to similar regions [18], do contribute to the power when similarity is not required. The same phenomenon also occurs in certain problems considered earlier in this paper.

For the case $\epsilon \geq \frac{1}{2}$, let us take

$$\lambda(\sigma, \xi_{m+1}, \cdots, \xi_n) = \lambda(\sigma) \prod_{i=m+1}^{n} \lambda_i(\xi_i \mid \sigma). \tag{65}$$

We shall select $\lambda(\sigma)$ such that $\lambda(\sigma)$ is constant when $\sigma \geq \sigma_1$. Hence it is enough to define $\lambda_i(\xi_i \mid \sigma)$ for $\sigma < \sigma_1$. For any $\sigma < \sigma_1$ there exists by lemma 1 a function $\lambda_i(\xi_i \mid \sigma)$ such that

$$\int_{-\infty}^{\infty} \exp\left[-\frac{1}{2\sigma^2}(x_i - \xi_i)^2\right] d\lambda_i(\xi_i \mid \sigma) = k \exp\left\{-\frac{1}{2\sigma_1^2}(x_i - \xi_{i1})^2\right\}. \tag{66}$$

For this choice of the λ_i, (9) becomes

$$\frac{\exp\left\{-\frac{1}{2\sigma_1^2}\left[\sum_{i=1}^{s}(x_i - \xi_{i1})^2 + \sum_{i=s+1}^{m} x_i^2\right]\right\}}{\int_0^{\infty} \exp\left\{-\frac{1}{2\sigma^2}\sum_{i=1}^{m} x_i^2\right\} d\lambda\ (\sigma)} \geq k'. \tag{67}$$

Next we chose $\lambda(\sigma)$ according to lemma 4 such that

$$\int_0^{\infty} \exp\left[-\frac{1}{2\sigma^2}\sum_{i=1}^{m} x_i^2\right] \frac{1}{\sigma^m}\, d\lambda(\sigma) = \exp\left[-\frac{1}{2\sigma_1^2}\sum_{i=1}^{m} x_i^2 - c\sqrt{\sum_{i=1}^{m} x_i^2}\right], \tag{68}$$

thus, by proper choice of k', reducing (67) to

$$\frac{\sum_{i=1}^{s} \xi_{i1}\, x_i}{\sqrt{\sum_{i=1}^{m} x_i^2}} \geq -c. \tag{69}$$

The probability of this region under H_0 is independent of $\xi_{m+1}, \cdots, \xi_n$ and σ, and hence (69) is most powerful for testing H_0 against H_1.

Let us return once more to the problem of testing Student's hypothesis against a simple alternative $\xi = \xi_1$, $\sigma = 1$ and let us assume as known that $\sigma \leq 1$. No use can be made of this knowledge if consideration is restricted to similar regions. For the probability of first kind error is an analytic function of σ, and consequently, if a test is similar with respect to all values of σ which are ≤ 1, it is similar with respect to all values of σ. Let us now consider this problem without the restriction of similarity. If $\epsilon \geq \frac{1}{2}$, the knowledge concerning σ does not enable us to find a test which is more powerful than that given by (58), since the function $\lambda(\sigma)$ on which (58) was based had all its points of increase for $\sigma \leq 1$.

On the other hand we may expect improvement for $\epsilon < \frac{1}{2}$ since the most powerful test in this case was based on a function λ with a single point of increase

$\sigma_0 > 1$ which is no longer admitted as a possible value of σ. If, instead, we take for λ the step function with a single jump at $\sigma = 1$ we obtain the critical region

$$\frac{\exp\left[-\frac{1}{2}\sum(x_i - \xi_1)^2\right]}{\exp\left[-\frac{1}{2}\sum x_i^2\right]}, \tag{70}$$

which is equivalent to

$$\bar{x} \geq c. \tag{71}$$

Here $c > 0$ since $\epsilon < \frac{1}{2}$, and therefore, when $\xi = 0$ the probability of (71) is an increasing function of σ and hence takes on its maximum at $\sigma = 1$. It follows from theorem 1 that (71) is most powerful under the conditions stated.

In the opposite problem in which it is known that $\sigma \geq 1$, the situation is reversed. For $\epsilon \leq \frac{1}{2}$ no improvement over (45) is possible while for $\epsilon > \frac{1}{2}$ we can use for λ the step function with a single step at $\sigma = 1$ thus obtaining the critical region (70) but this time with $c < 0$. When $\xi = 0$ the probability of this region is a decreasing function of σ and it follows that (70) is most powerful in this case.

Similar remarks apply to other problems. We mention as one further example a modification of the Behrens-Fisher problem. Let $X_1, \cdots, X_n$ and $Y_1, \cdots, Y_m$ be independently normally distributed, the X's with mean ξ and variance σ^2, the Y's with mean η and variance τ^2, all four parameters being unknown. We wish to test, at level of significance $\epsilon \leq \frac{1}{2}$, the hypothesis $\xi = \eta$ against the simple alternative $\xi = \xi_1, \eta = \eta_1, \sigma = 1, \tau = 1$, where $\xi_1 \neq \eta_1$ and we assume it known that $\sigma \leq 1, \tau \leq 1$. Basing the test on a step function λ with a single jump at $\sigma = 1, \tau = 1, \xi = \dfrac{n\xi_1 + m\eta_1}{m+n}$ we obtain for w the region

$$\frac{\exp\left[-\frac{1}{2}\sum(x_i - \xi_1)^2 - \frac{1}{2}\sum(y_i - \eta_1)^2\right]}{\exp\left[-\frac{1}{2}\sum\left(x_i - \frac{n\xi_1 + m\eta_1}{n+m}\right)^2 - \frac{1}{2}\sum\left(y_i - \frac{n\xi_1 + m\eta_1}{n+m}\right)^2\right]} \geq k, \tag{73}$$

which is equivalent to

$$\bar{y} - \bar{x} \geq c \qquad (c > 0), \tag{74}$$

if we assume, as we may without loss of generality, that $\eta_1 > \xi_1$. When $\eta = \xi_2$, $\bar{Y} - \bar{X}$ is normally distributed with zero mean and variance $\dfrac{\sigma^2}{n} + \dfrac{\tau^2}{m}$. Therefore the probability of (74) is an increasing function of $\dfrac{\sigma^2}{n} + \dfrac{\tau^2}{m}$ and hence attains its maximum when $\sigma = \tau = 1$. It follows from theorem 1 that the region (74) is most powerful for the problem under consideration.

6. Admissibility. The general problem to be considered in this paper has been formulated in section 1: To obtain a region w

$$\text{maximizing} \quad \int_w g(x)\,dx \tag{75}$$

subject to the restriction

$$\int_w f_\theta(x)\,dx \leq \epsilon \qquad \textit{for all } \theta \in \omega. \tag{76}$$

Since for any particular such problem there may exist several essentially different regions satisfying these conditions, it may happen that there exists a region w' such that

$$\int_{w'} g(x)\,dx = \int_w g(x)\,dx, \tag{77}$$

and

$$\int_{w'} f_\theta(x)\,dx \leq \int_w f_\theta(x)\,dx \qquad \textit{for all } \theta \in \omega, \tag{78}$$

with inequality holding for some θ. Clearly w' is preferable to w. In this case, following the definition of Wald [4], we say that w is not admissible. We shall rule out this possibility for a large class of problems by proving

THEOREM 3. *If w satisfies the conditions of theorem* 1, *and if the set of points x for which equality holds in* (8) *has measure zero, then any region satisfying* (75) *and* (76) *differs from w only on a set of measure zero.*

PROOF. Without loss of generality we shall assume λ of theorem 1 to be a distribution function. Then

$$h(x) = \int_\omega f_\theta(x)\,d\lambda(\theta)$$

is a completely specified probability density function, and w is the unique[3]—up to a set of measure zero—most powerful test for testing the simple hypothesis $H_0':h$ against the simple alternative $H_1:g$. Suppose now that w' satisfies (75) and (76). Then

$$\int_{w'} h(x)\,dx \leq \epsilon, \tag{79}$$

and w' is most powerful for testing H_0' against H_1. It follows that w' differs from w at most by a null set.

Earlier we enlarged the problem of testing by adjoining to the original random variable X a random variable with a known distribution. This is equivalent to the following modification of the original problem. Instead of defining a test to be a critical region (of rejection) in the space of x, we define it to be a critical

[3] One sees this easily from Neyman and Pearson's proof of the fundamental lemma [1], by using the assumption that the set of points for which equality holds in (8), has measure zero.

function φ $(0 \leq \varphi(x) \leq 1)$ which with every point x associates a probability of rejection $\varphi(x)$. If x is observed, the hypothesis is rejected with probability $\varphi(x)$ according to a table of random numbers. In the case where random numbers are not employed, φ merely becomes the characteristic function of the set w.

We shall now state a theorem which will prove admissibility for all but one of those problems treated in sections two to five, to which theorem 3 does not apply.

THEOREM 4. *Suppose* $\omega = \{\theta\}$ *is a subset of an s-dimensional Euclidean space, and that for any measurable function* φ *and for any set* S *which has positive measure and is contained in* ω

$$\int \varphi(x) f_\theta(x)\, dx = c \qquad \textit{for } \theta \in S \tag{80}$$

implies

$$\int \varphi(x) f_\theta(x)\, dx = c \qquad \textit{for } \theta \in \omega. \tag{81}$$

(Here and in all that follows whenever a region of integration is not indicated, the integral extends over the whole x *space). Suppose further that* φ *is a critical function satisfying the conditions of theorem* 1 *and that the set* S_0 *of points of increase of* λ *has positive measure. Then* φ *is admissible.*

PROOF. If φ were not admissible there would exist φ_1 with

$$\int \varphi_1(x)\, g(x)\, dx = \int \varphi(x) g(x)\, dx; \tag{82}$$

$$\int \varphi_1(x) f_\theta(x)\, dx \leq \int \varphi(x) f_\theta(x)\, dx \qquad \textit{for all } \theta \in \omega; \tag{83}$$

$$\int \varphi_1(x) f_\theta(x)\, dx < \int \varphi(x) f_\theta(x)\, dx \qquad \textit{for some } \theta \in \omega. \tag{84}$$

The set T of points θ for which (84) holds, differs from ω at most by a null set. For

$$\int [\varphi_1(x) - \varphi(x)] f_\theta(x)\, dx = 0 \qquad \textit{for } \theta \in \omega - T, \tag{85}$$

and if $\omega - T$ had positive measure, (85) would hold for all $\theta \in \omega$.

Let h and H_0' be defined as in the proof of theorem 3. Since S has positive measure, it follows that

$$\epsilon = \int \varphi(x) h(x)\, dx > \int \varphi_1(x) h(x)\, dx = \eta, \qquad \textit{say}. \tag{86}$$

Let $\varphi_2(x) = \min\left[1, \varphi_1(x) + \epsilon - \eta\right]$. Then

$$\int \varphi_2(x) h(x)\, dx \leq \epsilon \tag{87}$$

and

$$\int \varphi_2(x)g(x)\,dx > \int \varphi_1(x)g(x)\,dx. \tag{88}$$

But φ_1 is most powerful for testing H_0' against H_1 and we have a contradiction.

By applying theorems 3 and 4 one can easily show for all but one of the problems treated in sections three to five that the tests obtained there are admissible. The one exception occurs when testing equality of variances. Simplifying the notation, since we are now concerned with a special case, we shall assume that $X_i(i = 1, \cdots, n)$, $Y_1, \cdots, Y_r$ are independently and normally distributed, the X's with mean ξ_0 and variance σ_0^2, Y_i with mean ξ_i and variance σ_i^2, all parameters being unknown. We wish to test the hypothesis of equality of variances against the simple alternative

$$H_1 : \xi_i = \xi_{i1}, \qquad \sigma_i = \sigma_{i1} \qquad (i = 0, \cdots, r),$$

with

$$\sigma_{01} < \sigma_{11} < \cdots < \sigma_{r1}.$$

We shall first consider the case $n = 1$, and prove admissibility of the critical function

$$\varphi(x, y_1, \cdots, y_r) = \epsilon \tag{89}$$

by using a different distribution function for the parameters from the one used earlier. With some specialization of the distribution function, (8) becomes for our problem

$$\frac{\exp\left\{-\frac{1}{2\sigma_{01}}(x-\xi_{01})^2 - \frac{1}{2}\sum_{i=1}^{r}\frac{1}{\sigma_{i1}^2}(y_i-\xi_{i1})^2\right\}}{\int \frac{1}{\sigma^{r+1}}\left\{\int \exp\left[-\frac{1}{2\sigma^2}(x-\xi_0)^2\,d\lambda_\sigma^{(0)}(\xi_0)\right] \cdot \prod_{i=1}^{r}\int \exp\left[-\frac{1}{2\sigma^2}(y_i-\xi_i)^2\right]d\lambda_\sigma^{(i)}(\xi_i)\right\} d\,\mu(\sigma)} \geq k \tag{90}$$

For any $\sigma < \sigma_{01}$ we select the $\lambda_\sigma^{(i)}(\xi_i)$ according to lemma 1. If we then take for μ the uniform distribution over $(\sigma_{01} - 1, \sigma_{01})$ the left hand side of (90) will reduce to k. Admissibility of the critical function (89) then follows from theorem 4.

That a constant critical function is not admissible in the case $n > 1$ is easily seen if one compares it for instance with the critical region

$$\left|\frac{\bar{x} - \xi_{01}}{\sqrt{\Sigma(x_1 - \bar{x})^2}}\right| \leq c. \tag{91}$$

We shall not obtain a complete family of admissible tests (cf. [4]) for the case $n > 1$ but we shall show that this problem is equivalent to the following one: To find a complete class of unbiased admissible tests for the hypothesis specifying the mean and variance of a normal distribution on the basis of a sample from this distribution, the class of alternatives being the totality of univariate normal distributions.

Let $n > 1$ and let φ be any most powerful critical function for testing the hypothesis of equality of variances against H_1. If φ corresponds to the level of significance ϵ and if β_φ denotes the power of φ, we have

$$\beta_\varphi(\sigma, \sigma, \cdots, \sigma, \xi_0, \xi_1, \cdots, \xi_r) \leq \epsilon \tag{92}$$

for all admissible values of the arguments. It also follows from section 4 that

$$\beta_\varphi(\sigma_{01}, \sigma_{11} \cdots, \sigma_{r1}, \xi_{01}, \xi_{11}, \cdots, \xi_{r1}) = \epsilon. \tag{93}$$

Consider for a moment the hypothesis $H_0': \sigma_i = \sigma_{01} (i = 0, \cdots, r)$, $\xi_0 = \xi_{01}$, ξ_i unspecified for $i = 1, \cdots, r$. It is easily seen that the maximum power for testing H_0' against H_1 is ϵ. Therefore any most powerful test for testing H_0 against H_1 is also most powerful for testing H_0' against H_1, and in particular this holds for φ. Furthermore, it follows easily from theorem 4 that for any most powerful test of H_0' against H_1 the probability of an error of the first kind must be identically equal to ϵ. Therefore

$$\beta_\varphi(\sigma_{01}, \cdots, \sigma_{01}, \xi_{01}, \xi_1, \cdots, \xi_r) = \epsilon \text{ for all } \xi_1, \cdots, \xi_r. \tag{94}$$

But (94) is equivalent to the condition that φ is similar with respect to $\xi_1, \cdots, \xi_r$, and it follows [12] that φ is a function of $x_1, \cdots, x_n$ only. The problem is therefore reduced to that of finding all admissible critical functions $\varphi(x_1, \cdots, x_n)$ satisfying

$$\beta_\varphi(\sigma_{01}, \xi_{01}) = \epsilon; \qquad \beta_\varphi(\sigma_0, \xi_0) \leq \epsilon \text{ for all } \sigma_0, \xi_0. \tag{95}$$

That this problem in turn is equivalent to the one stated above is immediate when one considers the complementary critical functions $1 - \varphi$.

REFERENCES

[1] J. Neyman and E. S. Pearson, "On the problem of the most efficient tests of statistical hypotheses," *Roy. Soc. Phil. Trans.*, Ser. A, Vol. 231 (1933), p. 289.

[2] A. Wald, "Test of statistical hypotheses concerning several parameters when the number of observations is large," *Am. Math. Soc. Trans.*, Vol. 54 (1943), p. 426.

[3] A. Wald, "Statistical decision functions which minimize the maximum risk," *Annals of Math.*, Vol. 46 (1945), p. 265.

[4] A. Wald, "An essentially complete class of admissible decision-functions," *Annals of Math. Stat.*, Vol. 18 (1947), p. 549.

[5] J. Neyman, "On a statistical problem arising in routine analysis and in sampling inspection of mass production," *Annals of Math. Stat.*, Vol. 12 (1941), p. 46.

[6] H. Scheffé, "On the theory of testing composite hypotheses with one constraint," *Annals of Math. Stat.*, Vol. 13 (1942), p. 280.

[7] E. Lehmann, "On optimum tests of composite hypotheses with one constraint," *Annals of Math. Stat.*, Vol. 18 (1947), p. 473.

[8] E. Lehmann and H. Scheffé, "On the problem of similar regions," *Proc. Nat. Acad. Sci.*, Vol. 33 (1947), p. 382.

[9] A. Wald, "On the power function of the analysis of variance test," *Annals of Math. Stat.*, Vol. 13 (1942), p. 434.
[10] P. L. Hsu, "On the power function of the E^2-test and the T^2-test," *Annals of Math. Stat.*, Vol. 16 (1945), p. 278.
[11] H. K. Nandi, "On the average power of test criteria," *Sankhyā*, Vol. 8 (1946), p. 67.
[12] W. Feller, "Note on regions similar to the sample sapce," *Stat. Res. Memoirs*, Vol. 2 (1938), p. 117.
[13] P. L. Hsu, "Analysis of variance from the power function standpoint," *Biometrika*, Vol. 32 (1941), p. 62.
[14] S. Kolodzieczyk, "On an important class of statistical hypotheses," *Biometrika*, Vol. 27 (1935), p. 161.
[15] P. C. Tang, "The power function of the analysis of variance tests with tables and illustrations of their use," *Stat. Res. Memoirs*, Vol. 2 (1938), p. 126.
[16] A. Wald, "Foundations of a general theory of sequential decision functions," *Econometrica*, Vol. 15 (1947) p. 279.

SOME PRINCIPLES OF THE THEORY OF TESTING HYPOTHESES[1]

By E. L. Lehmann

University of California, Berkeley

Introduction:

1. The likelihood ratio principle. The development of a theory of hypothesis testing (as contrasted with the consideration of particular cases), may be said to have begun with the 1928 paper of Neyman and Pearson [16]. For in this paper the fundamental fact is pointed out that in selecting a suitable test one must take into account not only the hypothesis but also the alternatives against which the hypothesis is to be tested, and on this basis the likelihood ratio principle is proposed as a generally applicable criterion. This principle has proved extremely successful; nearly all tests now in use for testing parametric hypotheses are likelihood ratio tests, (for an extension to the non-parametric case see [33]), and many of them have been shown to possess various optimum properties.

At least in the parametric case the likelihood ratio test has a number of desirable properties. Among these we mention:

(i) Frequently it is easy to apply and leads to a definite and reasonable test.

(ii) If the sample size is large, and if certain regularity conditions are satisfied an approximate solution can be given for the distribution problems that arise in the determination of size and power of the test (Wilks [32], Wald [25]). In fact, if the likelihood ratio is denoted by λ, $-2 \log \lambda$ approximately has a central χ^2-distribution under the hypothesis, a non-central χ^2-distribution under the alternatives. The number of degrees of freedom in these distributions equal the number of constraints imposed by the hypothesis.

(iii) As was shown by Wald [25], under certain restrictions the likelihood ratio test possesses various pleasant large sample properties.

In view of this, one may feel that the likelihood ratio principle, although perhaps not always leading to the optimum test, is completely satisfactory, and that a more systematic study of the problem of test selection is not necessary. Unfortunately, against the pleasant properties just mentioned there stands a very unpleasant one. Cases exist, in which the likelihood ratio test is not only unsatisfactory but worse than useless, and hence the likelihood ratio principle is not reliable. Examples of this kind were constructed independently by H. Rubin and C. Stein; the following is Stein's example.

---

[1] Parts of this paper were presented in an invited address at the meeting of the Institute of Mathematical Statistics on Dec. 30, 1948, in Cleveland, Ohio.

J. Rojo (ed.), *Selected Works of E. L. Lehmann*, Selected Works in Probability and Statistics, DOI 10.1007/978-1-4614-1412-4_14, © Springer Science+Business Media, LLC 2012

Let X be a random variable capable of taking on the values 0, ± 1, ± 2 with probabilities as indicated:

	-2	2	-1	1	0
Hypothesis H:	$\frac{\alpha}{2}$	$\frac{\alpha}{2}$	$\frac{1}{2} - \alpha$	$\frac{1}{2} - \alpha$	α
Alternatives:	pC	$(1 - p)C$	$\frac{1 - C}{1 - \alpha}\left(\frac{1}{2} - \alpha\right)$	$\frac{1 - C}{1 - \alpha}\left(\frac{1}{2} - \alpha\right)$	$\alpha \frac{1 - C}{1 - \alpha}$

Here α, C are constants, $0 < \alpha \leqq \frac{1}{2}$, $\frac{\alpha}{2-\alpha} < C < \alpha$, and p ranges over the interval [0, 1].

It is desired to test the hypothesis H at significance level α. The likelihood ratio test rejects when $X = \pm 2$, and hence its power is C against each alternative. Since $C < \alpha$, this test is literally worse than useless, for a test with power α can be obtained without observing X at all, simply by the use of a table of random numbers. It is worth noting that the test, which rejects H when $X = 0$, has power $\alpha \frac{1 - C}{1 - \alpha} > \alpha$, so that a reasonable test of the hypothesis in question does exist.

The existence of such examples gives added importance to the problem of developing a systematic theory of hypothesis testing. It is the purpose of the present paper to give a brief survey of the work done on some aspects of such a theory and to indicate certain extensions and modifications of the existing theory. Some examples and applications will be considered. These will be restricted to parametric problems. For applications to testing non-parametric hypotheses see [12].

The results of sections 5 and 8 were obtained jointly by Gilbert Hunt and Charles Stein in 1945. They have not been published and were communicated to me by Professor Stein. I should like to express to him my gratitude for acquainting me with this material and for giving me permission to include it in this paper. I should also like to acknowledge my indebtedness to Professor Henry Scheffé who read the manuscript and made many helpful suggestions.

2. Formulation of the problem. The problem of testing a statistical hypothesis was formulated by Neyman and Pearson [18] as follows.

A random variable X is known to be distributed over a space $\mathfrak{X}$ according to some member of a family of probability distributions $\{P_\theta^X\}$, $\theta \in \Omega$. It will be assumed here that there is specified an additive class $\mathfrak{B}$ of sets in $\mathfrak{X}$, and that the probability distributions P_θ^X are probability measures defined over $\mathfrak{B}$. All sets or real valued functions mentioned in this paper will be assumed measurable $\mathfrak{B}$ unless otherwise stated. If $B \in \mathfrak{B}$, we shall write for the measure assigned to B by P_θ^X interchangeably $P_\theta^X(X \in B)$, $P_\theta^X(B)$, and if there is no possibility of confusion, $P_\theta(B)$. Throughout most of the paper it will be assumed that the probability measures P_θ^X are absolutely continuous with respect to a given sigma finite measure μ defined over $\mathfrak{B}$, so that there exist non-negative functions f_θ such that

$$P_\theta(B) = \int_B f_\theta(x)\, d\mu(x). \tag{2.1}$$

We shall then say that $f_\theta(x)$ is a generalized probability density *w.r.* to μ.

A statistical hypothesis H specifies a subset ω of Ω, and states that the distribution of X is some P_θ^X with $\theta \in \omega$. A test of H is any subset w of $\mathfrak{X}$, the convention being that H is rejected if the observed value x of X is in w, and that in the contrary case H is accepted. The selection of w is to be made as follows. A number α is given, $0 < \alpha < 1$, the level of significance, and w must be such that

$$P_\theta(w) = \alpha \text{ for all } \theta \in \omega. \tag{2.2}$$

Subject to this restriction it is desired to maximize $P_\theta(w)$ for θ in $\Omega - \omega$. The interpretation of these conditions is immediate. Since $P_\theta(w)$ is the probability of rejecting H computed under the assumption that P_θ^X is the distribution of X, equation (2.2) states that the probability of rejecting H is to be α (usually some small number such as .01 or .05) whenever H is true. Similarly the second condition expresses the fact that H is to be rejected with high probability when θ is in $\Omega - \omega$.

Naturally the second condition is not to be taken literally but rather as a loosely stated principle of choice. For in general there will exist a unique set w maximizing $P_{\theta_1}(w)$ for any given $\theta_1 \in \Omega - \omega$, but this w will change with θ_1. The condition has a clear meaning only in the case that the set $\Omega - \omega$ contains only a single point, and in a few special problems in which the same set w maximizes $P_\theta(w)$ for all $\theta \in \Omega - \omega$. In the general case there are available two main methods for making the condition precise. One may restrict consideration to some class of "nice" tests, so that within this class the maximization of $P_\theta(w)$ can be achieved uniformly for $\theta \in \Omega - \omega$. Alternatively, instead of asking that a local optimum property hold uniformly, one may look for a test whose power function possesses some optimum property in the large. Both of these approaches have an element of arbitrariness. In the first, the selection of a class of nice tests, in the second, the choice of an appropriate optimum property. Fortunately, in a number of important special cases, both methods, for various reasonable definitions, lead to the same test.

Before proceeding with this development, we shall modify the formulation of the problem slightly. First, as has been pointed out by many writers, it seems more natural to replace (2.2) by

$$P_\theta(w) \leqq \alpha \text{ for all } \theta \in \omega. \tag{2.3}$$

Secondly, we shall permit "randomized" tests (see [11, 29]), that is, instead of demanding that the statistician decide for each value of x whether to accept or to reject H, we shall allow the possibility that for certain x the decision be reached by means of some chance device such as a table of random numbers. By a test of H we shall therefore mean a function ϕ from $\mathfrak{X}$ to the interval [0, 1], with the convention that when x is the observed value of X some chance experiment with two possible outcomes R, $\bar{R}$ will be performed such that $P(R) = \phi(x)$, and that H will be rejected when the outcome is R and will otherwise be accepted. The case of a non-randomized test w clearly is obtained as a special case by taking for ϕ the characteristic function of the set w.

For a test ϕ the probability of rejection is given by

$$\int_{\mathfrak{X}} \phi(x)\, d\, P_\theta^X(x) = E_\theta \phi(X) \tag{2.4}$$

where E_θ denotes expectation computed with respect to the probability distribution P_θ^X. We therefore obtain the following formulation of the problem: To determine a test function ϕ $(0 \leqq \phi(x) \leqq 1)$ which maximizes $E_\theta\, \phi(X)$, the *power* of ϕ against the alternative θ, for θ in $\Omega - \omega$ subject to the condition

$$E_\theta \phi(X) \leqq \alpha \text{ for all } \theta \in \omega. \tag{2.5}$$

In this connection it is convenient to use the term *"level of significance"* for the preassigned number α, and to define the *size* of the test ϕ as

$$\sup_{\theta \in \omega} E_\theta \phi(X). \tag{2.6}$$

Except in the trivial case that there exists a test of size $< \alpha$ whose power is 1 against all alternatives, the size of any optimum test (in fact, of any admissible test) equals the level of significance.

3. Testing against a simple alternative. A complete solution of the problem formulated in the last section is available only in the case that ω and $\Omega - \omega$ each contains only a single point, that is, in the case that both the hypothesis and the alternative are simple. The solution is then given by the fundamental lemma of Neyman and Pearson [18], which we may state in the following slightly more complete form.

THEOREM 3.1. *Let*

$$P_\theta(A) = \int_A f_\theta(x)\, d\mu(x). \tag{3.1}$$

(a) *For testing the hypothesis H: $\theta = \theta_0$ against the alternative $\theta = \theta_1$ at level of significance α, there exists a number k and a test ϕ of size α such that*

$$\begin{aligned} \phi(x) &= 1 \quad \textit{when} \quad f_{\theta_1}(x) > k f_{\theta_0}(x), \\ \phi(x) &= 0 \quad \textit{when} \quad f_{\theta_1}(x) < k f_{\theta_0}(x). \end{aligned} \tag{3.2}$$

(b) *If $f_{\theta_0}(x)$ and $f_{\theta_1}(x)$ are $\neq 0$ for all x in $\mathfrak{X}$, then a test ϕ is most powerful for testing H against $\theta = \theta_1$ if and only if it satisfies* (3.2) *except possibly on a set of $\mu-$ measure* 0[2]. (Note that the number k of (3.2) is essentially unique).

[2] Throughout the paper we shall consider two tests as equal if they differ only on a set of μ-measure 0.

The second half of the theorem may be paraphrased by saying that under the conditions stated the most powerful test is uniquely determined by (3.2) except on the set on which

$$f_{\theta_1}(x) = k f_{\theta_0}(x). \tag{3.3}$$

On this set the value of ϕ may be assigned arbitrarily provided the resulting test has size α. If in particular the set on which (3.3) holds has measure 0, the most powerful test is unique.

It should be mentioned that (3.1) is no restriction since any two probability measures P_1, P_2 defined over a common additive class can be represented in this form with $\mu = P_1 + P_2$. If the assumption of (b) is not satisfied, the theorem is still true in essence but some trivial modifications are necessary.

No such complete solution is available for the problem of testing a composite hypothesis against a simple alternative. However, as was shown in [11], this problem may in many cases be reduced to the one just considered. Let the hypothesis state that θ is an element of ω, and consider the simple alternative $\theta = \theta_1$. Suppose that an additive class of sets has been defined on ω (in most of the applications ω is a subset of Euclidean space, and the additive class is formed by the Borel sets contained in ω). Then for any probability distribution λ over ω,

$$h_\lambda(x) = \int_\omega f_\theta(x)\, d\lambda(\theta) \tag{3.4}$$

is a probability density function with respect to μ.

Under certain conditions to be stated below, the most powerful test ϕ_λ for testing the simple hypothesis H_λ that X is distributed with probability density h_λ against the alternative f_{θ_1} is also most powerful for testing the original hypothesis H against the same alternative. This is essentially the Bayes approach developed by Wald for his general decision theory, and in fact, under the conditions which we shall state, λ is a least favorable distribution over ω in the following sense. Let β_λ be the power of ϕ_λ against f_{θ_1}, and for any distribution λ^* over ω denote by H_{λ^*}, ϕ_{λ^*}, β_{λ^*} the associated hypothesis, the most powerful test for testing it against f_{θ_1}, and the power of this test respectively. Then λ is said to be least favorable if for all λ^*

$$\beta_\lambda \leqq \beta_{\lambda^*}. \tag{3.5}$$

THEOREM 3.2. *Suppose there exists a probability distribution λ over ω such that the most powerful test ϕ_λ of size α for testing H_λ against f_{θ_1} is of size α also with respect to the original hypothesis H. Then*

(i) *ϕ_λ is most powerful for testing H against f_{θ_1};*

(ii) *λ is a least favorable distribution.*

Also, if ϕ_λ is the unique most powerful test for testing H_λ against f_{θ_1}, it is the unique most powerful test for testing H against f_{θ_1}.

These results are essentially contained in Wald's work (see for example theorem 4.8 of [26]).

There are many trivial applications of this theorem to finding most powerful tests of one-sided hypotheses concerning a single real-valued parameter, such as testing $H: p \leqq p_0$ against $p = p_1(p_0 < p_1)$ when X has a binomial distribution with parameter p. As is well known, it turns out in a number of these cases that the most powerful tests are in fact uniformly most powerful against the one-sided class of alternatives.

In [11] Theorem 3.2 was used to determine most powerful tests of certain hypotheses concerning normal distributions. As an example consider the case that $X_1, \cdots, X_n$ are independently normally distributed with common mean ξ and variance σ^2. Denote by H_1 and H_2 the hypotheses $\sigma = 1$ and $\xi = 0$ respectively, and let the alternative be: $\xi = \xi_1$, $\sigma^2 = \sigma_1^2$. Then the most powerful test of H_1 rejects if

$$(3.6) \qquad \begin{aligned} \Sigma(x_i - \xi_1)^2 < k_1 \quad &\text{when} \quad \sigma_1 < 1, \\ \Sigma(x_i - \bar{x})^2 > c_1 \quad &\text{when} \quad \sigma_1 > 1, \end{aligned}$$

and accepts otherwise. Here k_1 and c_1 depend only on the level of significance, that is, are independent of ξ_1, σ_1. If $\xi_1 > 0$, the most powerful test for testing H_2 rejects if

$$(3.7) \qquad \begin{aligned} \Sigma(x_i - b)^2 \leqq k_2 b^2 \quad &\text{when} \quad \alpha < \tfrac{1}{2}, \\ \frac{\bar{x}}{\sqrt{\Sigma(x_i - \bar{x})^2}} \leqq c_2 \quad &\text{when} \quad \alpha \geqq \tfrac{1}{2}, \end{aligned}$$

and accepts H_2 otherwise. Here k_2 and c_2 depend only on α, while b depends on ξ_1, σ_1 and α.

These results indicate that even when the class of alternatives is larger than in the above problems, some improvement over the standard tests may be possible provided good power is desired only against a narrow class of alternatives.

4. Sufficient statistics. Before treating the problem of composite alternatives, we shall consider an important simplification that can be obtained by making use of sufficient statistics. This notion was introduced by R. A. Fisher, and was further developed by J. Neyman [13] and in [2] and [10]. Consider any measurable partition of $\mathfrak{X}$. For any point x in $\mathfrak{X}$, let $t(x)$ be that set of the partition in which x lies. A set in the range of t is said to be measurable if the corresponding set of points x is an element of $\mathfrak{B}$. Denote the class of measurable t-sets by $\mathfrak{A}$. Then the *statistic* $T = t(X)$ is a random variable defined over $\mathfrak{A}$. Kolmogoroff has shown how for any $B \in \mathfrak{B}$ one can define the conditional probability $P(B \mid t)$ of B given $T = t$ uniquely up to a set of measure zero by the equation

$$(4.1) \qquad P(B \cap t^{-1}(A)) = \int_A P(B \mid t)\, dP^T(t) \quad \text{for all} \quad A \in \mathfrak{A}.$$

Suppose now that we are given a class $\mathfrak{F}$ of probability distributions for X, $\mathfrak{F} = \{P_\theta^X\}$, $\theta \in \Omega$. Denote by $P_\theta(B \mid t)$ the conditional probability of B given

$T = t$ computed for the distribution P_θ^X. The statistic T is said to be a *sufficient statistic* for $\mathfrak{F}$ (or for θ) if for every $B \in \mathfrak{B}$ there exists a determination of $P_\theta(B \mid t)$ that is independent of θ.

According to the above definition of statistic, $t(x)$ is an element of a measurable partition. However, one may consider instead any function t^* for which $t^*(x) = t^*(x')$ if and only if $t(x) = t(x')$, that is, any function that leads to this partition; the values that the function takes on are really immaterial. It will be convenient here to use this wider definition of statistic. For a rigorous treatment of some of the problems that will be referred to one needs to define an equivalence of statistics and to include in this definition the appropriate nullset considerations. A detailed account of these matters is given in [2] and [10].

From our present point of view tests are compared solely in terms of their power functions. On this basis two tests ϕ_1 and ϕ_2 may be considered equivalent if they have identical power, that is, if

$$E_\theta\phi_1(X) = E_\theta\phi_2(X) \text{ for all } \theta \in \Omega. \tag{4.2}$$

We can then state

THEOREM 4.1. *If T is a sufficient statistic for θ and $\phi(X)$ any test of a hypothesis concerning θ then there exists an equivalent test that is a function of T only.*

The proof of this theorem is immediate since

$$\psi(T) = E[\phi(X) \mid T] \tag{4.3}$$

is such a test.

It follows from Theorem 4.1 that we lose nothing by restricting consideration to tests based on a sufficient statistic.[3] The problem of determining whether or not some statistic is sufficient for a given family of distributions is simplified through the use of a criterion for sufficiency that can be checked on sight. This criterion is due to Neyman [13] who proved it in a somewhat special setting, and was recently proved in a very general form by Halmos and Savage [2]. It states that if $\mathfrak{F} = \{p_\theta\}$, $\theta \in \Omega$ is a family of generalized probability densities for X, then under certain mild restrictions a necessary and sufficient condition for $T = t(X)$ to be a sufficient statistic for $\mathfrak{F}$ is that $p_\theta(x)$ factors into one factor depending on θ but on x only through $t(x)$ and a second factor depending only on x.

The question arises as to which of various sufficient statistics to use. Since the purpose of introducing sufficient statistics is to reduce the complexity of a given statistical problem, one is led to seek a sufficient statistic that reduces the problem as far as possible and hence to the notion of a *minimal sufficient statistic*, a sufficient statistic T being *minimal* if it is a function of every other sufficient statistic (see [10]). It can be shown under fairly general conditions that a minimal sufficient statistic exists, and one can give an explicit construction for it.

[3] A justification for the use of sufficient statistics in the general statistical decision problem was given in [2].

As one would expect it turns out that the sufficient statistics commonly associated with various families of distributions are actually minimal. Thus for example, if $X_1, \cdots, X_n$ are independently normally distributed with common mean ξ and variance σ^2, the statistic $(\bar{X}, \Sigma(X_i - \bar{X})^2)$ is a minimal sufficient statistic for $\theta = (\xi, \sigma^2)$. If $X_1, \cdots, X_n$ are independently uniformly distributed over $(0, \theta)$, $\max(X_1, \cdots, X_n)$ is the minimal sufficient statistic for θ. If $\mathfrak{F}$ is the family of distributions according to which $X_1, \cdots, X_n$ are identically independently distributed according to an arbitrary univariate distribution (or according to an arbitrary probability density with respect to a fixed univariate measure), then the minimal sufficient statistic is obtained by defining for each point $x = (x_1, \cdots, x_n)$ the set $t(x)$ as the set of points obtainable from x by permutation of coordinates. Alternatively one can define it by $t(x_1, \cdots, x_n) = (\Sigma x_i, \Sigma x_i^2, \cdots, \Sigma x_i^n)$.

5. The principle of invariance. The notion of invariance was introduced into the statistical literature in the writings of R. A. Fisher, Hotelling, Pitman [20] and others, in connection with various special problems. A general formulation was given by Hunt and Stein who, in an unpublished paper [5], utilized this notion to find most stringent tests, and who obtained the examples of uniformly most powerful invariant tests that will be given below. The point of view in the present section is different from theirs however, since here invariance will only be considered as an intuitively appealing restriction that one may wish to impose on statistical tests.

We shall begin by considering an example. Suppose it were known that the height of people is distributed about a known mean, which for convenience we shall take to be zero, either according to a normal or to a Cauchy distribution, with unknown scale factor so that either

$$f_\theta(x) = \frac{1}{\sqrt{2\pi}\,\theta} \exp\left(-\frac{x^2}{2\theta^2},\right) \qquad 0 < \theta < \infty \tag{5.1}$$

or

$$f_\theta(x) = \frac{\theta}{\pi} \frac{1}{\theta^2 + x^2}, \qquad 0 < \theta < \infty. \tag{5.2}$$

Suppose we wish to test from a sample $X_1, \cdots, X_n$ the hypothesis H that the true probability density belongs to the first of these classes against the alternative that it belongs to the second. Then it seems desirable that the decision of whether or not to accept H should be independent of the scale adopted for measuring the heights. For otherwise one worker expressing his data in feet might reject H while another worker using the same data but expressing them in inches would reach the contrary decision (In this connection see for example [34], p. 104). A "nice" test function ϕ therefore would be independent of the choice of scale, i.e., it would satisfy the condition

(5.3) $\phi(cx_1, \cdots, cx_n) = \phi(x_1, \cdots, x_n)$ for all $c > 0$ and for all $(x_1, \cdots, x_n)$

except possibly on a set N, independent of c and of measure zero.

On analyzing this problem one is led to the following observation. Multiplying each of the random variables $X_1, \cdots, X_n$ by the same constant leaves both ω and $\Omega - \omega$ invariant, i.e., if the X's are normally distributed with zero mean and arbitrary scale so are $cX_1, \cdots, cX_n$, and analogously for the Cauchy distributions. It is this fact that makes it so desirable to have ϕ invariant under multiplication of the x's by a common constant.

More generally consider measurable 1:1 transformations g of $\mathfrak{X}$ into itself, and let $Y = gX$. Suppose that when X is distributed according to $\theta \in \omega$, Y is distributed according to $\theta' \in \omega$—we shall then write $\theta' = \bar{g}\theta$—and that as θ ranges over ω so does θ'. Suppose that the analogous condition is satisfied for $\Omega - \omega$, so that the problem of testing ω against $\Omega - \omega$ is left invariant under g. Now whether one expresses the observations in terms of X or in terms of Y is essentially a matter of choice of coordinates. The principle of invariance asks that if such a change of coordinates leaves the problem invariant, then it should also leave the test invariant, i.e., if G is a group of measurable 1:1 transformations of $\mathfrak{X}$ such that

$$\bar{g}\omega = \omega \text{ and } \bar{g}(\Omega - \omega) = \Omega - \omega \text{ for all } g \in G, \tag{5.4}$$

then ϕ should satisfy the condition

$$\phi(gx) = \phi(x) \text{ for all } g \in G, \tag{5.5}$$

and for all x except on a set N independent of g and such that $\mu(N) = 0$. If this condition were not satisfied, two workers, using the same data but expressing them in different coordinate systems might arrive at contrary conclusions.

As an example consider the general linear univariate hypothesis. In canonical form $X_1, \cdots, X_r; X_{r+1}, \cdots, X_s; X_{s+1}, \cdots, X_n$ are independently normally distributed with common variance. The means of the first s variables are unknown, the means of the last n-s variables are known to be zero. The hypothesis states that the first r means are zero. Adding arbitrary constants to each of the variables of the middle group leaves ω and $\Omega - \omega$ invariant. So does any orthogonal transformation of the first r variables, and any orthogonal transformation of the last $n - s$ variables. Finally, the problem is also left invariant when all of the variables are multiplied by the same constant. It is easy to see that a function ϕ is invariant under these transformations if and only if it is a function of

$$\sum_{i=1}^{r} x_i^2 \Big/ \sum_{i=s+1}^{n} x_i^2 .$$

But, as is well known and easy to show, among all tests based on this statistic there is a uniformly most powerful one, namely the test that rejects H when

$$\sum_{i=1}^{r} x_i^2 \Big/ \sum_{i=s+1}^{n} x_i^2$$

is too large. Therefore, among all tests satisfying the condition of invariance the standard test is uniformly most powerful.

To formulate a corresponding reduction procedure in general, we define a function h on $\mathfrak{X}$ to be maximal invariant (under G) if it is invariant and if $h(x') = h(x)$ implies the existence of $g \in G$ such that $x' = gx$. Then a function φ on $\mathfrak{X}$ is invariant under G if and only if it depends on x only through $h(x)$, that is, if there exists a function ψ such that $\varphi(x) = \psi[h(x)]$. Hence a necessary and sufficient condition for a test to be invariant under G is that it be based on the statistic $Y = h(X)$. The principle of invariance therefore reduces the problem from X to $Y = h(X)$. To determine the resulting statistical reduction, that is, the simplification of the parameter space, one may consider the group $\bar{G}$ of transformations over Ω induced by G. If $v(\theta)$ is a maximal invariant function under $\bar{G}$, it is easily shown that the distribution of Y depends only on $v(\theta)$. Hence under the principle of invariance any two θ-values with common $v(\theta)$ (that is, such that each can be obtained from the other by a transformation of $\bar{G}$) are identified. If in particular $v(\theta)$ is constant over ω, the hypothesis H, when expressed for Y, becomes simple, and there may even exist a uniformly most powerful invariant test.

Besides for the example already mentioned this is the case for Hotelling's T^2-problem and for the hypothesis specifying the value of a multiple correlation coefficient. Another example is obtained when $X_1, \cdots, X_n$ are independently identically distributed, each with probability density $p_\theta(x)$ where under H_i $p_\theta(x) = f_i(x - \theta)$, $(i = 0, 1)$, and where it is desired to test H_0 against H_1. One may also in this example replace the location parameter by a scale parameter or have both parameters present.

It may be worth noting that the likelihood ratio test is invariant under any transformation leaving the statistical problem invariant. In the problems concerning normal distributions mentioned above, when there exists a uniformly most powerful invariant test, it coincides with the likelihood ratio test. That this is not so in general can be seen from Stein's example given in section 1. There the problem remains invariant under multiplication of X by -1, and there exists a uniformly most powerful invariant test. However, the likelihood ratio test is instead uniformly least powerful.

For certain applications it is more useful to consider a somewhat weaker definition of invariance. We shall say that a function φ is *almost invariant* under a group G of transformations if for each $g \in G$, $\varphi(gx) = \varphi(x)$ for all x except on a set N_g such that $\mu(N_g) = 0$. This definition differs from the previous one in that the null set N_g is now permitted to depend on g. It was shown by Hunt and Stein that under certain conditions on G, which are satisfied for the problems mentioned above, any almost invariant test is invariant.

We have indicated how for certain hypotheses one can find a group of transformations leaving the problem invariant, such that among all tests invariant under this group there exists a uniformly most powerful one. The question may be raised whether this approach is consistent, or whether there may exist some other group of transformations also leaving the problem invariant but leading to a different test. Also in problems where among all invariant tests there does not exist a uniformly most powerful one, the question arises whether one is using the totality of transformations leaving the problem invariant, or whether perhaps one can reduce the problem further. It therefore seems of interest to determine the totality of transformations leaving a given problem invariant. This was carried out for a few simple problems in [8].

We finally mention a connection between the notions of invariance and sufficiency. Consider any problem in which the variables $X_1, \cdots, X_n$ are independently identically distributed under all distributions of Ω. Such a problem clearly is left invariant under any permutation of the variables. Actually, these transformations leave not only ω and $\Omega - \omega$ invariant but each point of Ω individually. No essential reduction of the problem is obtained since the maximal invariant statistic is a sufficient statistic. It is easily seen that this will always be the case when the transformations leave Ω pointwise invariant, but that in this way one does not obtain all sufficient statistics. These can be obtained, however, by considering more general transformations, where each point x of $\mathfrak{X}$ is transformed into the points of $\mathfrak{X}$ according to a probability distribution P_x.

6. The principle of unbiasedness. As a second principle of reduction we shall consider the principle of unbiasedness proposed by Neyman and Pearson. A test is said to be unbiased [19] if

$$P_\theta \text{ (rejecting } H) \geq \alpha \text{ for all } \theta \in \Omega - \omega.$$

This seems a desirable property for a test to have since it assures that there do not exist θ_0 in ω and θ_1 in $\Omega - \omega$, for which

$$P_{\theta_0} \text{ (rejecting } H) > P_{\theta_1} \text{ (rejecting } H).$$

We shall therefore be concerned in this section with the totality of tests ϕ for which

$$\begin{aligned} E_\theta \phi(X) &\leq \alpha \quad \text{for all} \quad \theta \in \omega \\ E_\theta \phi(X) &\geq \alpha \quad \text{for all} \quad \theta \in \Omega - \omega. \end{aligned} \tag{6.1}$$

For a number of important special cases there exists, among all tests satisfying (6.1), one that is uniformly most powerful in $\Omega - \omega$ and uniformly least powerful in ω. (The latter property is of course very desirable since when H is true one wants to reject it as rarely as possible.) This follows immediately from well known results concerning best similar tests since for the problems in question Ω is a subset of a Euclidean space and for any test ϕ, $E_\theta\phi(X)$ is a continuous function of θ. If then Λ is the set of points that are boundary points both of ω and of $\Omega - \omega$, it follows from (6.1) that

$$E_\theta\phi(X) = \alpha \text{ for all } \theta \in \Lambda, \tag{6.2}$$

i.e., that ϕ is similar for θ in Λ. But if among all tests satisfying (6.2) there exists one that is uniformly most powerful in $\Omega - \omega$ and uniformly least powerful in ω, it automatically satisfies (6.1) as is seen by comparison with the test $\phi(X) \equiv \alpha$.

As an example suppose that $X_1, \cdots, X_n$ are independently normally distributed with common mean ξ and common variance σ^2. If the hypothesis is $H_1 : \sigma \leq 1$ and the alternatives are $\sigma > 1$, the set Λ becomes the line $\sigma = 1$. As was shown by Neyman and Pearson [18], among all tests satisfying (6.2) with this Λ, the test that rejects H_1 when $\Sigma(x_i - \bar{x})^2 \leq k$ (where k is an appropriately chosen constant) is uniformly most powerful for θ in $\Omega - \omega$, and uniformly least powerful for θ in ω.

If instead we consider testing the hypothesis $H_2 : \sigma = 1$ against the alternatives $\sigma \neq 1$, we find that $\Lambda = \omega$, and our problem reduces to that of finding the best test among all those that are similar in ω and unbiased. As is well known, it turns out that rejecting when $\Sigma(x_i - \bar{x})^2 \leq k_1$ and when $\Sigma(x_i - \bar{x})^2 \geq k_2$ (where $k_1 < k_2$ are two appropriately chosen constants) is uniformly most powerful among all similar unbiased tests.

A third hypothesis concerning σ that might be of interest is $H_3 : \sigma_1 \leq \sigma \leq \sigma_2$. Here Λ consists of the two lines $\sigma = \sigma_1$ and $\sigma = \sigma_2$ and it is easy to show that the test that is uniformly most powerful in $\Omega - \omega$ and uniformly least powerful in ω rejects H_3 if and only if $\Sigma(x_i - \bar{x})^2 \leq c_1$ or $\Sigma(x_i - \bar{x})^2 \geq c_2$ where again $c_1 < c_2$ are two appropriately selected constants.

The question arises as to the connection of the principles of invariance and unbiasedness. Clearly if there exists a unique test ϕ that is uniformly most powerful unbiased, this test is invariant under any group G leaving the problem invariant. If then in addition there exists a uniformly most powerful invariant (under G) test, this must coincide with ϕ. Thus, if both principles lead to a unique optimum solution, these solutions coincide.

We have seen that frequently optimum unbiased tests can be obtained through a study of tests that are similar over certain sets in the parameter space. The totality of similar tests was obtained for a number of important problems by Neyman and Pearson. In his 1937 paper on confidence intervals [15] Neyman gave a general method for constructing similar regions with the help of sufficient statistics. Let T be a sufficient statistic for $\theta \in \Lambda$. The condition for ϕ to be similar with respect to Λ and of size α, is that

$$E_\theta\phi(X) = E_\theta E[\phi(X) \mid T] = \alpha \text{ for all } \theta \in \Lambda, \tag{6.3}$$

i.e., that

$$E_\theta\{E[\phi(X) \mid T] - \alpha\} = 0 \text{ for all } \theta \in \Lambda. \tag{6.4}$$

Clearly any test ϕ for which

$$E[\phi(X) \mid t] = \alpha \text{ for almost all } t \tag{6.5}$$

is similar. This is the construction given by Neyman, and we shall say that a test ϕ satisfying (6.5) has the Neyman structure with respect to T. The question whether this exhausts the totality of similar tests is easily reduced to an analytic problem the solution of which is known in many special cases. This method was first employed by P. L. Hsu [3] for some problems concerning normal distributions, and was extended to other cases in [7]. The present general formulation was given by H. Scheffé and the author in [9] and [10]. We shall say that a family of distributions $\{P_\theta^T\}$, $\theta \in \Lambda$, is boundedly complete if

(i) $f(t)$ is bounded,

(ii) $E_\theta f(T) = 0$ for all $\theta \in \Lambda$

imply $f(t) = 0$ except on a set N with $P_\theta(N) = 0$ for all $\theta \in \Lambda$. Then we can state

THEOREM 6.1. *A necessary and sufficient condition for the totality of tests similar for* Λ *to have Neyman structure with respect to a sufficient statistic* T *is that* $\{P_\theta^T\}$, $\theta \in \Lambda$, *be boundedly complete.*

7. Tests whose power increases with the distance from the hypothesis. Frequently, even among the unbiased tests, there does not exist a uniformly most powerful one. The general univariate linear hypothesis with more than one constraint is an example of this situation. The following extension of the idea of unbiasedness may then be used to reduce the class of tests still further. Unbiasedness distinguishes between values of θ as they belong to ω or $\Omega - \omega$. However, one may further classify the points of $\Omega - \omega$ according to their "distance" from ω, and then ask of a test φ that the further be θ from ω the larger be the power $\beta_\varphi(\theta)$.

One possible such ordering of the alternatives is that induced by the envelope power function. Here the envelope power at θ (Wald [24]) is defined by

$$\beta_\alpha^*(\theta) = \sup_{\varphi \in \mathfrak{F}(\alpha)} \beta_\varphi(\theta) \tag{7.1}$$

where $\mathfrak{F}(\alpha)$ is the class of all tests φ with $E_\theta\varphi(X) \leq \alpha$ for all $\theta \in \omega$. Of two points θ_1, θ_2 one may then say that θ_1 is closer to ω than θ_2, equally close or less close, as $\beta_\alpha^*(\theta_1)$ is less than, equal to or greater than $\beta_\alpha^*(\theta_2)$. The distance of θ from ω is thus measured by the ease with which one can detect that the hypothesis is false when θ is the true parameter value.

When θ lies in a Euclidean space and $\beta_\varphi(\theta)$ is a continuous function of θ for all θ, as is the case in most applications, the condition that the power increase with β_α^* will usually imply that $\beta_\varphi(\theta_1) = \beta_\varphi(\theta_2)$ whenever $\beta_\alpha^*(\theta_1) = \beta_\alpha^*(\theta_2)$. In the case of the general linear hypothesis considered in section 5, for example, one would obtain the condition that the power be a function only of $\sum_{i=1}^{r} \xi_i^2/\sigma^2$ where $\xi_i = E(X_i)$. As was shown by P. L. Hsu [3], the standard (likelihood ratio) test is uniformly most powerful among all tests satisfying this condition. Analogous remarks apply to Hotelling's T^2- problem, and to the hypothesis specifying the value of the multiple correlation coefficient. The corresponding optimum properties in these cases were proved by Simaika [21].

It is interesting to compare the above condition with that of invariance.

This comparison yields nothing of interest if the totality of tests is considered. We may, however, restrict our attention to tests depending only on a sufficient statistic T. We already know that $\varphi(X)$ and $E[\varphi(X) \mid T]$ have identical power. In order to validate the comparison we wish to make, we state the following

LEMMA. *Let T be a sufficient statistic for $\theta \in \Omega$, and let G be a group of 1:1 transformations g on X leaving Ω invariant. Then if $\varphi(x)$ is invariant under G, $E[\varphi(X) \mid t]$ is almost invariant under G.*

We can now state the desired comparison in the following

THEOREM 7.1. *Let G be a group of 1:1 transformations on X, let $\bar{G}$ be the induced group of transformations on Ω, let $v(\theta)$ be maximal invariant under $\bar{G}$, and suppose that $\bar{G}$ leaves ω and $\Omega - \omega$ invariant. Suppose further that T is a sufficient statistic for Ω, and that $\{P_\theta^T\}$, $\theta \in \Omega$, is boundedly complete. Then a necessary and sufficient condition that the power of a test $\psi(T)$ be a function only of $v(\theta)$, is that $\psi(t)$ be almost invariant under G.*

This theorem is an immediate extension of some results of Wolfowitz [35].

Theorem 7.1 together with the results of section 5 proves that the standard tests of the general linear hypothesis, Hotelling's T^2-problem and the hypothesis concerning the multiple correlation coefficient possess the optimum property that was obtained for these problems by Hsu and Simaika, respectively. The method of proof indicated here is due to Wolfowitz [35].

8. Most stringent tests. We shall now turn to the third aspect of the theory: Optimum properties defined with reference to the whole class of alternatives, and attainable with no restrictions imposed on the class of tests. In the present section we shall consider the property of stringency. Wald [25] defines a test φ to be most stringent if it minimizes

$$\sup_{\theta \in \Omega - \omega} [\beta_\alpha^*(\theta) - \beta_\varphi(\theta)], \tag{8.1}$$

where β_α^* again denotes the envelope power, and β_φ the power of φ. The rationale of this definition is clear. The difference $\beta_\alpha^*(\theta) - \beta_\varphi(\theta)$ measures the amount by which the test falls short at the alternative θ of the power that could be attained against this particular alternative. A test φ is therefore most stringent if it minimizes its maximum shortcoming.

A theory of most stringent tests was developed by Hunt and Stein [5], who based it on the notion of invariance. Consider, as in section 5, a group G of measurable 1:1 transformations on $\mathfrak{X}$ leaving the problem invariant. Hunt and Stein obtained their results in connection with the following groups of transformations.

(i) $gx = x + c$, $-\infty < c < \infty$, x a real variable;
(ii) $gx = ax$, $0 < a$, x a real variable;
(iii) $gx = ax + c$, $0 < a$, $-\infty < c < \infty$, x a real variable;
(iv) the group of orthogonal transformations on a Euclidean space;
(v) any finite group.

THEOREM 8.1. (Hunt and Stein). *If G is the direct product of a finite number of groups of types* (i)–(v), *and if G leaves the problem invariant, that is, if G satisfies* (5.4), *then there exists a most stringent test invariant under G.*

Actually, it is not necessary here to require that G be a direct product. The result holds also if the factoring of G is according to normal subgroups, where the normal subgroup at each stage and the final factor group are of the types mentioned. In the light of this one may omit type (iii) from the list since it has a normal subgroup of type (i) with factor group of type (ii).

The proof of Theorem 8.1 is based on the following lemma, which has applications to many related problems.

LEMMA (Hunt and Stein). *If G is a direct product of a finite number of groups of types* (i)–(v) *then given any function f over $\mathfrak{X}$* $(0 \leq f(x) \leq 1)$ *there exists a function F* $(0 \leq F(x) \leq 1)$ *such that F is invariant under G, and*

$$\inf_{g \in G} \int f(gx)\varphi(x)\, d\mu(x) \leq \int F(x)\varphi(x)\, d\mu(x) \leq \sup_{g \in G} \int f(gx)\varphi(x)\, d\mu(x) \tag{8.2}$$

for all φ that are integrable μ.

It follows from Theorem 8.1 that if there exists a uniformly most powerful invariant test, this test is most stringent. In this way Hunt and Stein show, for example, (see in this connection section 5), that the likelihood ratio test of the general univariate linear hypothesis is most stringent. A question that is left open is the uniqueness of such a most stringent test.

In general, the possibility therefore remains that there might exist another most stringent test uniformly more powerful than the invariant one. In certain particular cases this possibility can be ruled out by the following consideration. Suppose that Ω is a subset of a Euclidean space and that every point of ω is a limit point of $\Omega - \omega$. Suppose further that for any test ϕ, $E_\theta\phi(X)$ is continuous in θ. Then clearly, if ϕ_1 is similar of size α for testing ω and ϕ_2 is of size $\leqq \alpha$ but not similar, ϕ_2 can not be uniformly as powerful as ϕ_1. Hence any test that is admissible among all similar tests of size α is also admissible among the totality of tests of size $\leqq \alpha$. Now admissibility among all similar tests is sometimes not too difficult to prove. For the likelihood ratio test of the general linear univariate hypothesis, for example, it is an immediate consequence of the properties of this test proved by Wald [23] and Hsu [4].

The following alternative method for obtaining most stringent tests is also mentioned by Hunt and Stein.

THEOREM 8.2. (Hunt and Stein). *Let $\Omega - \omega$ be partitioned into disjoint subsets Ω_δ such that $\beta^*_\alpha(\theta)$ is constant on each Ω_δ, and let φ_δ be the test that maximizes $\inf_{\theta \in \Omega_\delta} \beta_{\varphi_\delta}(\theta)$. Then if $\varphi_\delta = \varphi$ is independent of δ, φ is most stringent.*

This result may be supplemented by the following method for finding tests that maximize $\inf_{\theta \in \omega_1} \beta_\varphi(\theta)$ over a given set of alternatives ω_1 (not necessarily satisfying the conditions imposed above on the Ω_δ's).

THEOREM 8.3. *Suppose additive classes of sets have been defined over ω and ω_1, and consider probability measures λ and λ_1 over ω and ω_1. Let the functions $f_\theta(x)$ be generalized probability densities with respect to μ, so that $h(x) = \int_\omega f_\theta(x)\, d\lambda(\theta)$ and $h_1(x) = \int_{\omega_1} f_\theta(x)\, d\lambda_1(\theta)$ are again probability densities with respect to μ. Let φ be the most powerful test of size α for testing the simple hypothesis $H{:}h$ against the simple alternative h_1, and suppose that the power of φ against h_1 is β. Then if*

$$
\begin{aligned}
E_\theta\varphi(x) &\le \alpha \quad \text{for all} \quad \theta \in \omega, \\
E_\theta\varphi(x) &\ge \beta \quad \text{for all} \quad \theta \in \omega_1,
\end{aligned} \tag{8.3}
$$

φ maximizes $\inf_{\theta \in \omega_1} \beta_\varphi(\theta)$ at level of significance α.

This method, when applicable, has the advantage of giving the totality of most stringent tests (see in this connection Theorem 3.1) and hence of settling the question of admissibility. However, in many applications probability measures λ, λ_1 with the desired properties do not exist but instead only sequences $\lambda^{(n)}$, $\lambda_1^{(n)}$, which satisfy the conditions in the limit. In this case again only the weak conclusion is possible: The test obtained is most stringent but has not been proved admissible. (For an example in which the analogous method has been carried through in detail for an estimation problem, see [22]).

Actually, the two methods are closely related, as can be seen from the proof of the main lemma. In those cases in which there exists a group G giving the maximum possible reduction, the group $\bar{G}$ induces a partition of Ω (through the equivalence: $\theta_1 \sim \theta_2$ if there exists $\bar{g}$ such that $\theta_2 = \bar{g}\theta_1$), just into ω and the sets Ω_δ. (This is so mainly because, as was shown by Hunt and Stein, the envelope power remains invariant under any transformations that leave the problem invariant.) Then the measures λ, λ_δ over ω, Ω_δ respectively, which figure in the application of Theorems 8.2 and 8.3, become invariant measures over $\bar{G}$ through the obvious 1:1 mapping from ω and the Ω_δ's respectively to $\bar{G}$. Thus the second method will allow the strong conclusion when the group $\bar{G}$ involved in the first method possesses a finite invariant measure [types (iv) and (v)] but not if any of its factors are of type (i)–(iii).

To conclude this section we shall give an example where the method of invariance leads only to a partial reduction but where the solution may be completed by certain additional considerations. Suppose that $(X_1, \cdots, X_n)$ is a sample from a normal distribution with mean ξ and variance σ^2, both unknown, and that we wish to find the most stringent test of the hypothesis H: $\sigma = 1$ against the alternatives $\sigma \neq 1$. Theorem 8.1 reduces the problem to the statistic $Y = \Sigma(X_i - \bar{X})^2$, but among the tests of H based on this statistic there does not exist a uniformly most powerful one. It may also be shown [8] that no further reduction is possible by means of the method of invariance.

However, one may now consider the problem of finding the most stringent test based on Y. (The envelope power function $\beta^*(\xi, \sigma)$ that must be used naturally is not the one for Y but that for the original problem.) From an argument given in [6] it follows that this test is of the form

$$\varphi_{k_1,k_2}\text{: reject when } Y < k_1 \text{ or } > k_2 ,$$

where k_1, k_2 are determined by the two conditions

(i) $P(\text{rejection} \mid \sigma = 1) = \alpha$,

(ii) $\sup_{\sigma<1} [\beta_\alpha^*(\xi, \sigma) - \beta_{\varphi_{k_1,k_2}}(\sigma)] = \sup_{\sigma>1} [\beta_\alpha^*(\xi, \sigma) - \beta_{\varphi_{k_1,k_2}}(\sigma)]$.

Here $\beta_\alpha^*(\xi, \sigma)$ is independent of ξ and can be obtained from a table of the χ^2-distribution (with n degrees of freedom for $\sigma < 1$ and n-1 degrees of freedom for $\sigma > 1$ as can be seen from (3.6)). Hence k_1 and k_2 can be computed fairly easily.

Another problem that may be treated in this way is the hypothesis of equality of variances for two normal samples. If the two samples are of equal size, there exists a uniformly most powerful invariant test for a suitable group of transformations. However, if the sample sizes are different the method of invariance reduces the problem only to $\Sigma(X_i - \bar{X})^2 / \Sigma(Y_i - \bar{Y})^2$, and the cut off points giving the most stringent test may be determined by an argument analogous to that given above.

This method may be extended to allow determination of most stringent test of hypotheses such as H: $\sigma_1 \leqq \sigma \leqq \sigma_2$. This requires a certain modification of Theorem 1 of [6], which is easily obtained. One finds agains that one may restrict consideration to a one-parameter family of tests (determined by a somewhat different condition than above), and that among these the most stringent test is obtained by the analogue of condition (ii) above.

If should be mentioned that the results of [6] apply also to the hypothesis specifying the value of the parameter in a binomial or Poisson distribution. This is easily seen since in either case the distributions of Ω are absolutely continuous with respect to a common sigma finite measure and since for the appropriate choice of this measure the generalised density is of the form assumed for the density in [6]. Hence in both the binomial and the Poisson case the most stringent test is determined by conditions analogous to (i) and (ii) above.

9. Tests that minimize the maximum loss. In the Neyman-Pearson theory one classifies the errors into two kinds: Rejecting the hypothesis when it is true, accepting it when it is false. One may however analyze the situation further and distinguish, say, between accepting when one or some other alternative is true. Thus one is led to introduce the losses that result in a given situation from the various possible errors, and to look for a test that, in an appropriate sense, minimizes the expected loss. This possibility was mentioned by Neyman and Pearson [17], and was taken as the starting point of his general theory by Wald (see for example [24]).

In order to stay within the framework of this exposition we shall here introduce losses only for the errors of accepting the hypothesis when it is false, while still demanding that the probability of rejection when the hypothesis is true should not exceed α. Actually, there are many cases where this seems to be a reasonable formulation. For it frequently happens that the two types of error entail consequences of such completely different nature that the resulting losses cannot be measured on a common scale while usually the different errors of the same type are comparable.

We shall therefore assume that for each $\theta \in \Omega - \omega$ there is defined a $W(\theta)$, which measures the loss resulting from acceptance of H when θ is true. The risk which one runs by using a test φ, when $\theta \in \Omega - \omega$ is the true parameter value is given by the expected loss $R_\varphi(\theta) = W(\theta)\, E_\theta[1 - \varphi(X)]$. When a uniformly most powerful test exists for the hypothesis in question, this test also minimizes the expected loss uniformly for θ in $\Omega - \omega$. In the contrary case one may again restrict the class of tests in some way, so that within the restricted class there exists a uniformly most powerful test, and hence a test that uniformly minimizes the expected loss. Alternatively we may again consider some optimum property of the risk function $R_\varphi(\theta)$ as a whole. We shall here consider the minimax principle introduced by Wald, and seek a test, which, subject to $E_\theta\varphi(X) \leq \alpha$ for all $\theta \in \omega$, minimizes

$$\sup_{\theta \in \Omega - \omega} W(\theta)\, E_\theta[1 - \varphi(X)],$$

the maximum risk.

If one introduces losses also for the other type of error it is easy to see that for a suitably chosen loss function the definition of minimax expected loss coincides with that of stringency. It is therefore not surprising that the methods of the previous section can be extended to cover the problems considered in the present one. (They are actually much more general, and may be applied also, for example, to the problem of point estimation, and in fact to the general decision problem).

From the lemma of Hunt and Stein stated in the previous section we immediately obtain the following extension of Theorem 8.1.

THEOREM 9.1. *If G is a group of transformations leaving the hypothesis and the class of alternatives invariant, if G can be factored by normal subgroups into factors of types* (i)–(v), *and if the loss function $W(\theta)$ is invariant under $\bar{G}$, then there exists a test φ invariant under G and minimizing*

$$\sup_{\theta \in \Omega - \omega} W(\theta)\, E_\theta[1 - \varphi(X)]. \tag{9.1}$$

It follows that when a uniformly most powerful invariant test exists, this test has the property of minimizing the maximum expected loss with respect to any invariant loss function. Thus Student's test, for example, minimizes the maximum risk for any loss function that depends only on $|\xi|/\sigma$.

Clearly the second method mentioned in section 8 can be extended in an analogous manner if in Theorem 8.2 one replaces the sets Ω_δ by sets over which $W(\theta)$ is constant.

Again it may happen that the method of invariance does not reduce the problem sufficiently far but that the solution may be completed by other considerations. Let us once more consider the hypothesis $H: \sigma = 1$ of the previous section, and let us suppose that the loss function has the necessary invariance property, so that it is a function only of σ but not of the unknown mean. It follows from Theorem 9.1 that there exists a test minimizing the maximum risk, which is a function only of $Y = \Sigma(X_i - \bar{X})^2$. From [6] it is easily seen that a test φ_{k_1,k_2} which rejects when $Y < k_1$ or $> k_2$, has the desired property if its size is α and if in addition

$$\sup_{\sigma<1} W(\sigma)E_\sigma[1 - \varphi(Y)] = \sup_{\sigma>1} W(\sigma)E_\sigma[1 - \varphi(Y)]. \tag{9.2}$$

It follows that depending on the choice of $W(\sigma)$ the solution may be any member of the one-parameter family of tests φ_{k_1,k_2} of size α.

Under the conditions of Theorem 9.1, when a uniformly most powerful invariant test exists, this also maximizes the average power for a large class of weight functions. If there exists a common finite invariant measure over the sets Ω_δ in the sense indicated in section 8, the uniformly most powerful invariant test will maximize the average power with this measure as weight function, over Ω_δ for all δ. It follows that it maximizes the average power over $\Omega - \omega$ with respect to any weight function for which the conditional distribution over each Ω_δ is the above invariant measure. If the invariant measure over the Ω_δ's is not finite one can obtain analogous results with respect to a sequence of weight functions invariant in the limit. The results indicated here are much weaker than those obtained for the general linear univariate hypothesis by Wald [23] and Hsu [4] under the restriction to similar regions. However their results are no longer valid when this restriction is omitted.

10. Applications to sequential analysis. So far we have restricted consideration to the case that the hypothesis is to be tested on the basis of a preassigned experiment. However, frequently there is available for this purpose a large class of experiments, and the selection of an optimum experiment out of this class is part of the problem. We shall consider here only the following situation, which has recently been studied extensively (see Wald [28, 29]). There is given a sequence of random variables $X_1, X_2, \cdots$ whose joint distribution is known to belong to some family $\mathfrak{F} = \{P_\theta\}$, $\theta \in \Omega$; the hypothesis specifies some subfamily: $\theta \in \omega$. The X's are observed one by one, and the decision, whether or not to continue experimentation at any given stage, is allowed to depend on the observations taken up to that point. Thus the number n of observations that will be taken is a random variable whose distribution depends on θ. Usually, by an appropriate choice of stopping rule, there may be effected a considerable saving in the expectation of the number of observations necessary to achieve a given discrimination between hypothesis and alternatives. The problem is to determine the stopping rule and test that minimizes this expectation.

As we have seen in the previous sections the principal methods for obtaining optimum tests consist in reducing the problem to that of testing a simple hypothesis against a simple alternative. This basic problem was solved in the non-sequential case by Neyman and Pearson (Theorem 3.1). The solution of the much more difficult corresponding sequential problem was obtained for a large class of cases by Wald and Wolfowitz [31] in the following

THEOREM 10.1. *Let $X_1, X_2, \cdots$ be identically and independently distributed. It is desired to test the hypothesis that the common probability density of the X's is $f(x)$ against the alternative that it is $g(x)$. Given two numbers $0 < \alpha < \beta < 1$, there exists a test which, subject to the condition*

$$\begin{aligned} &P\,(\text{rejection} \mid f) \leq \alpha \\ &P\,(\text{rejection} \mid g) \geq \beta, \end{aligned} \tag{10.1}$$

minimizes simultaneously $E_f(n)$ and $E_g(n)$, the expected number of observations computed for the distributions f and g. This test is given in terms of two numbers A and B by the following rule. After m observations have been taken,

$$\textit{reject if } \frac{g(x_1)\cdots g(x_m)}{f(x_1)\cdots f(x_m)} > A,$$

$$\textit{accept if } \frac{g(x_1)\cdots g(x_m)}{f(x_1)\cdots f(x_m)} < B,$$

$$\textit{take another observation if } B < \frac{g(x_1)\cdots g(x_m)}{f(x_1)\cdots f(x_m)} < A.$$

Here A and B are determined so that condition (10.1) *holds with the inequality signs replaced by equality.*

So as to be able to treat the various problems considered non-sequentially in the previous sections one would have to extend this theorem at least to the case that the variables $X_1, X_2, \cdots$ form a set of equivalent variables in the sense of de Finetti [1]. Instead, we shall here restrict ourselves to a few problems that can be solved on the basis of Theorem 10.1. All of the tests discussed below were derived from various points of view and some of their properties were discussed by Girshick in his important "Contributions to the theory of sequential analysis" *Annals of Math. Stat.*, vol. 17 (1946) pp. 123–143 and 282–298, and by Wald in his basic book on the subject [28].

It is convenient here to modify slightly the formulation of the problem of hypothesis testing. Let the parameter space Ω be divided into three sets, the set ω_0 specified by the hypothesis, the class of alternatives ω_1, and a region of indifference $\Omega - \omega_0 - \omega_1$ where we do not much care whether the hypothesis is accepted or rejected (see [28]). Let us denote the sequential random variable $(X_1, \cdots, X_n)$ by X. Then we wish to determine a sequential test φ, which, subject to

$$\begin{aligned} &E_\theta\varphi(X) \leq \alpha \text{ for } \theta \in \omega_0 \\ &E_\theta\varphi(X) \geq \beta \text{ for } \theta \in \omega_1, \end{aligned} \tag{10.2}$$

minimizes $\sup_{\theta \epsilon \omega_0 + \omega_1} E_\theta(n)$. (Actually, this is a rather artificial formulation. The natural requirement is the minimization of $\sup_{\theta \epsilon \Omega} E_\theta(n)$ but this is a much more difficult problem.) The reduction to the problem of testing a simple hypothesis against a simple alternative is achieved by the following obvious extension of Theorem 8.3.

THEOREM 10.2. *Let* λ_0, λ_1 *be distributions over* ω_0, ω_1 *respectively, and let* φ *be a test, which subject to*

$$\int_{\omega_0} E_\theta \varphi(X) d\lambda_0(\theta) \leq \alpha \tag{10.3}$$

$$\int_{\omega_1} E_\theta \varphi(X) d\lambda_1(\theta) \geq \beta,$$

minimizes $\sup_{i \epsilon \{0,1\}} \int E_\theta(n)\, d\lambda_i(\theta)$. *Then if* φ *satisfies* (10.2) *and*

$$E_\theta(n) \leq \sup_{i \epsilon \{0,1\}} \int E_\theta(n)\, d\lambda_i(\theta) \text{ for all } \theta \epsilon \omega_0 + \omega_1, \tag{10.4}$$

φ *minimizes* $\sup_{\omega_1 + \omega_2} E_\theta(n)$ *subject to* (10.2).

As in section 3 we can make certain trivial applications to problems concerning a single real parameter such as testing the hypothesis $H: p \leq p_0$ against the alternatives $p \geq p_1$ $(p_0 < p_1)$, where p is the probability of success in a binomial sequence of trials. In this example condition (10.2) of Theorem 10.2 obviously is satisfied when λ_0 and λ_1 assign probability 1 to p_0 and p_1 respectively. Hence the probability ratio test for testing $p = p_0$ against $p = p_1$ has the desired properties, whenever (10.4) holds, that is, whenever $E_p(n)$ attains its maximum between p_0 and p_1.

The following is another example that may be solved in this manner. Let $X_1, X_2, \cdots; Y_1, Y_2, \cdots$ be independently normally distributed, all with unit variance and means $E(X_i) = \xi$, $E(Y_i) = \eta$. In order to test the hypothesis $H: \xi \geq \eta$ against the alternatives $\eta - \xi \geq \delta$ where $\delta > 0$ is given, a pair (X_1, Y_1) is observed. If after this observation experimentation continues another pair (X_2, Y_2) is observed, etc. In this case we may take for λ_0, λ_1 the distributions that assign probability 1 to the parameter points $(\xi, \eta) = (0, 0)$ and $\left(-\frac{\delta}{2}, \frac{\delta}{2}\right)$ respectively. Then the probability ratio after m observations is given by

$$\frac{\exp\left[-\frac{1}{2}\sum_{i=1}^{m}\left(x_i + \frac{\delta}{2}\right)^2 - \frac{1}{2}\sum_{i=1}^{m}\left(y_i - \frac{\delta}{2}\right)^2\right]}{e^{-\frac{1}{2}\Sigma x_i^2 - \frac{1}{2}\Sigma y_i^2}} = e^{-(m\delta^2/4) + \delta[\Sigma y_i - \Sigma x_i]} \tag{10.5}$$

Since the distribution of $Y - X$ depends only on $\eta - \xi$, it is easily seen that condition (10.2) is satisfied.

Some further results can be obtained through extension to the sequential case of Theorems 8.1 and 9.1.

THEOREM 10.3. *Suppose that G is of the type described in Theorem 9.1, let $Y = f(X_1, X_2, \cdots)$ be maximal invariant under G, let $v(\theta)$ be maximal invariant under $\bar{G}$, and let the set of values of $v(\theta)$ corresponding to ω_0 and ω_1 be $\bar{\omega}_0$ and $\bar{\omega}_1$, respectively. If among all tests of $\bar{\omega}_0$ against $\bar{\omega}_1$ based on Y, the test φ minimizes $\sup_{v(\theta)\epsilon\bar{\omega}_0+\bar{\omega}_1} E_\theta(n)$ subject to*

$$E_\theta\varphi(Y) \leq \alpha \text{ if } v(\theta)\epsilon\bar{\omega}_0 \qquad (10.6)$$

$$E_\theta\varphi(Y) \geq \beta \text{ if } v(\theta)\epsilon\bar{\omega}_1,$$

then φ also minimizes $\sup_{\omega_0+\omega_1} E_\theta(n)$ among all tests based on the X's and which satisfy (10.2).

As an example consider the problem of testing the hypothesis $\sigma \leq \sigma_0$ against the alternatives $\sigma \geq \sigma_1$ $(\sigma_0 < \sigma_1)$ when the X's are identically, independently normally distributed with unknown mean and variance. Since the problem remains invariant under a common translation of the X's we can take for Y of the theorem $Y = (X_2 - X_1, X_3 - X_1, \cdots)$. Equivalently we may take as our new sequence of variables $(Y_1, Y_2, \cdots)$ where

$$Y_k = \frac{kX_{k+1} - (X_1 + \cdots + X_k)}{\sqrt{k(k+1)}}. \qquad (10.7)$$

Then $Y_1, Y_2, \cdots$ are independently normally distributed with zero mean and the same variance as the X's. Hence the problem reduces to a type which we have already considered. The optimum test is based on

$$\sum_{i=1}^{m} Y_i^2 = \sum_{i=1}^{m+1}\left(X_i - \frac{X_1 + \cdots + X_{m+1}}{m+1}\right)^2.$$

It may be worth pointing out that Theorems 3.2, 8.3, 10.2 all are special cases of simple results in the general theory of statistical decision functions, of which the following is the prototype. (For a detailed treatment of this theory see, for example, [30]). Let $\{P_\theta\}$, $\theta\epsilon\Omega$, be the family of possible distributions of a random variable X, and let $\{\delta\}$ be a family of decision functions. The loss resulting from the use of $\delta(x)$ when P_θ is the true distribution is $W[\theta, \delta(x)]$ and the risk function associated with δ is $R_\delta(\theta) = E_\theta W[\theta, \delta(X)]$. Let λ be a probability measure over Ω, and let δ_λ be a decision function that minimizes $\int R_\delta(\theta)\, d\lambda(\theta)$. Then if λ is such that

$$R_{\delta_\lambda}(\theta) \leqq \int R_{\delta_\lambda}(\zeta)\, d\lambda(\zeta) \text{ for all } \theta \epsilon \Omega, \qquad (10.8)$$

δ_λ minimizes $\sup_\theta R_\delta(\theta)$.

PROOF. Let δ^* be any other decision function. Then

$$\sup_\theta R_{\delta_\lambda}(\theta) \leq \int R_{\delta_\lambda}(\theta)\, d\lambda(\theta) \leqq \int R_{\delta^*}(\theta)\, d\lambda(\theta) \leqq \sup_\theta R_{\delta^*}(\theta).$$

In an analogous manner one can give an extension of Theorems 8.1, 9.1, 10.3.

11. Two sided tests considered as 3-decision problems. In a number of important special problems the hypothesis specifies the value of a real valued parameter or states that this parameter lies in a certain interval, and it is desired to test this hypothesis against the obvious two-sided class of alternatives. It seems that in nearly any problem of this kind that would arise in practice one would want to decide when rejecting the hypothesis, whether the true parameter value lies below or above the hypothetical ones. If for example one rejects the hypothesis that the means of two normal populations are equal, one usually wants to decide which of the two is larger. It would therefore seem most natural to formulate such problems as 3-decision problems.

Problems of this kind, as all problems of hypothesis testing, naturally are special cases of the general decision problem formulated by Wald. We shall here consider the case that upper bounds are given for the probabilities of certain types of errors and thereby obtain a formulation, which is closely analogous to the classical formulation of hypothesis testing discussed in this paper, and which will allow immediate application of a large portion of the theory discussed here.

Consider the case that Ω is partitioned into 3 parts, ω, ω_1, ω_2 where in a certain sense ω lies between ω_1 and ω_2. We wish to test the hypothesis $H: \theta \in \omega$. When we reject the hypothesis, we shall reach either decision D_1 that $\theta \in \omega_1$ or decision D_2 that $\theta \in \omega_2$. Correspondingly we prescribe two positive numbers α_1, α_2 and impose the restriction that

$$
\begin{aligned}
&P_\theta(D_1) \leq \alpha_1 \text{ if } \theta \in \omega + \omega_2 \\
&P_\theta(D_2) \leq \alpha_2 \text{ if } \theta \in \omega + \omega_1 .
\end{aligned}
\tag{11.1}
$$

Subject to this condition it is desired to maximize

$$
\begin{aligned}
&P_\theta(D_1) \text{ for } \theta \in \omega_1 \\
&P_\theta(D_2) \text{ for } \theta \in \omega_2 .
\end{aligned}
\tag{11.2}
$$

A test will now consist of two non-negative functions ϕ_1 and ϕ_2 satisfying

$$
\phi_1(x) + \phi_2(x) \leq 1, \tag{11.3}
$$

with the convention that when $X = x$ the decision D_i will be taken with probability $\phi_i(x)$ $(i = 1, 2)$.

There is no difficulty concerning the extension of the notions of invariance or sufficient statistic, in fact these notions obviously apply to the general decision problem. The notion of unbiasedness is extended in the obvious way by the condition

$$
\begin{aligned}
&P_\theta(D_1) \geq \alpha_1 \text{ for } \theta \in \omega_1 \\
&P_\theta(D_2) \geq \alpha_2 \text{ for } \theta \in \omega_2 .
\end{aligned}
\tag{11.4}
$$

One then obtains the following

THEOREM 11.1. *Suppose that for testing the hypothesis H_1: $\theta \in \omega + \omega_2$ against the alternatives $\theta \in \omega_1$ at level of significance α_1, the test ϕ_1 among all unbiased tests* *is uniformly most powerful in* $\omega + \omega_2$ *and uniformly least powerful in* ω_1, *and that* ϕ_2 *has the analogous property for testing* $H_2 : \theta \epsilon \omega + \omega_1$ *against* $\theta \epsilon \omega_2$ *at significance level* α_2. *If* $\phi_1(x) + \phi_2(x) \leqq 1$ *for all x, then among all procedures satisfying* (11.1) *and* (11.4), *the procedure* (ϕ_1, ϕ_2) *uniformly maximizes the probability of a correct decision.* (If the tests ϕ_1, ϕ_2 take on only the values 0 and 1, the condition $\phi_1(x) + \phi_2(x) \leqq 1$ states that the rejection region of each of the two hypotheses is contained in the acceptance region of the other.)

As an example consider the case that $X_1, \cdots, X_n$ are independently, normally distributed with common mean ξ and variance σ^2. Suppose we wish to test the hypothesis that $\sigma_1 \leqq \sigma \leqq \sigma_2$ where σ_1 may equal σ_2. Then it follows from Theorem 11.1 that among all unbiased procedures of level (α_1, α_2), there exists one that maximizes the probability of a correct decision uniformly in ξ, σ. This is the procedure under which decision D_1 or D_2 is taken as $\Sigma(x_1 - \bar{x})^2 \leqq k_1$ or $\geqq k_2$ and the hypothesis is accepted otherwise. Here the k's are determined by

$$P(\Sigma(x_i - \bar{x})^2 \leqq k_1 \mid \sigma_1) = \alpha_1 \tag{11.5}$$
$$P(\Sigma(x_i - \bar{x})^2 \geqq k_2 \mid \sigma_2) = \alpha_2 .$$

REFERENCES

[1] B. de Finetti, "La prévision: ses lois logiques, ses sources subjectives," *Annales de l'Institut Henri Poincaré*, Vol. 7 (1937), p. 1.

[2] P. R. Halmos and L. J. Savage, "Applications of the Radon-Nikodym theorem to the theory of sufficient statistics," *Annals of Math. Stat.*, Vol. 20 (1949), p. 225.

[3] P. L. Hsu, "Analysis of variance from the power function standpoint," *Biometrika*, Vol. 32 (1941), p. 62.

[4] P. L. Hsu, "On the power function of the E^2-test and the T^2-test," *Annals of Math. Stat.*, Vol. 16 (1945), p. 278.

[5] G. Hunt and C. Stein, "Most stringent tests of statistical hypotheses," unpublished.

[6] E. L. Lehmann, "On families of admissible tests," *Annals of Math. Stat.*, Vol. 18 (1947), p. 97.

[7] E. L. Lehmann, "On the theory of testing composite hypotheses with one constraint," *Annals of Math. Stat.*, Vol. 18 (1947), p. 473.

[8] E. L. Lehmann, "On the totality of transformations leaving a statistical problem invariant," unpublished.

[9] E. L. Lehmann and H. Scheffé, "On the problem of similar regions," *Proc. Nat. Acad. Sci.*, Vol. 33 (1947), p. 382.

[10] E. L. Lehmann and H. Scheffé, "Completeness, similar regions and unbiased estimation," unpublished.

[11] E. L. Lehmann and C. Stein, "Most powerful tests of composite hypotheses. I. Normal distributions," *Annals of Math. Stat.*, Vol. 19 (1948), p. 495.

[12] E. L. Lehmann and C. Stein, "On the theory of some non-parametric hypotheses," *Annals of Math. Stat.*, Vol. 20 (1949), p. 28.

[13] J. Neyman, "Su un teorema concernente le cosidette statistiche sufficienti," *Giorn. Ist. Ital. Att.*, Vol. 6 (1935), p. 320.

[14] J. Neyman, "Sur la vérification des hypothèses statistiques composées," *Bull. Soc. Math. France*, Vol. 63 (1935), p. 1.

[15] J. Neyman, "Outline of a theory of statistical estimation based on the classical theory of probability," *Phil. Trans. Roy. Soc. London*, Series A, Vol. 236 (1937), p. 333.

[16] J. Neyman and E. S. Pearson, "On the use and interpretation of certain test criteria for purposes of statistical inference," *Biometrika*, Vol. 20A (1928), p. 175 and p. 263.

[17] J. NEYMAN AND E. S. PEARSON, "On the testing of statistical hypotheses in relation to probability a priori," *Proc. Camb. Phil. Soc.*, Vol. 29 (1933), p. 492.

[18] J. NEYMAN AND E. S. PEARSON, "On the problem of the most efficient tests of statistical hypotheses," *Phil. Trans. Roy. Soc. London*, Series A, Vol. 231 (1933), p. 289.

[19] J. NEYMAN AND E. S. PEARSON, "Contributions to the theory of testing statistical hypotheses. I. Unbiased critical regions of type A and type A_1," *Stat. Res. Mem.*, Vol. 1 (1936), p. 1.

[20] E. J. G. PITMAN, "Tests of hypotheses concerning location and scale parameters," *Biometrika*, Vol. 31 (1939), p. 200.

[21] J. B. SIMAIKA, "On an optimum property of two important statistical tests," *Biometrika*, Vol. 32 (1941), p. 70.

[22] C. STEIN AND A. WALD, "Sequential confidence intervals for the mean of a normal distribution with known variance," *Annals of Math. Stat.*, Vol. 18 (1947), p. 427.

[23] A. WALD, "On the power function of the analysis of variance test," *Annals of Math. Stat.*, Vol. 13 (1942), p. 434.

[24] A. WALD, "On the principles of statistical inference," *Notre Dame Math. Lectures*, No. 1 (1942).

[25] A. WALD, "Tests of statistical hypotheses concerning several parameters when the number of observations is large," *Trans. Am. Math. Soc.*, Vol. 54 (1943), p. 426.

[26] A. WALD, "Statistical decision functions which minimize the maximum risk," *Annals of Math.*, Vol. 46 (1945), p. 265.

[27] A. WALD, "An essentially complete class of admissible decision functions," *Annals of Math. Stat.*, Vol. 18 (1947), p. 549.

[28] A. WALD, *Sequential analysis*, John Wiley and Sons, 1947.

[29] A. WALD, "Foundations of a general theory of sequential decision functions," *Econometrica*, Vol. 15 (1947), p. 279.

[30] A. WALD, "Statistical decision functions," *Annals of Math. Stat.*, Vol. 20 (1949), p. 165.

[31] A. WALD AND J. WOLFOWITZ, "Optimum character of the sequential probability ratio test," *Annals of Math. Stat.*, Vol. 19 (1948), p. 326.

[32] S. S. WILKS, "The large-sample distribution of the likelihood ratio for testing composite hypotheses," *Annals of Math. Stat.*, Vol. 9 (1938), p. 60.

[33] J. WOLFOWITZ, "Additive partition functions and a class of statistical hypotheses," *Annals of Math. Stat.*, Vol. 13 (1942), p. 247.

[34] J. WOLFOWITZ, "Non-parametric statistical inference," *Proceedings of the Berkeley Symposium on mathematical statistics and probability* (1949), p. 93.

[35] J. WOLFOWITZ, "The power of the classical tests associated with the normal distribution," *Annals of Math. Stat.*, Vol. 20 (1949), p. 540.

Some related papers not referred to in the text.

[36] T. W. ANDERSON, "On the theory of testing serial correlation," *Skand. Aktuarietidskrift*, (1948), p. 88.

[37] R. A. FISHER, *The design of experiments*, Oliver and Boyd, 1935.

[38] M. N. GHOSH, "On the problem of similar regions," *Sankhya*, Vol. 8 (1948), p. 329.

[39] P. G. HOEL, "Testing the homogeneity of Poisson frequencies," *Annals of Math. Stat.*, Vol. 16 (1945), p. 362.

[40] P. G. HOEL, "Discriminating between binomial distributions," *Annals of Math. Stat.*, Vol. 18 (1947), p. 556.

[41] P. G. HOEL, "On the uniqueness of similar regions," *Annals of Math. Stat.*, Vol. 19 (1948), p. 66.

[42] E. L. LEHMANN, "Some comments on large sample tests," *Proceedings of the Berkeley Symposium on mathematical statistics and probability* (1949), p. 451.

[43] H. B. MANN AND A. WALD, "On the choice of the number of intervals in the application of the χ^2-test," *Annals of Math. Stat.*, Vol. 13 (1942), p. 306.

[44] J. NEYMAN, *Lectures and conferences on mathematical statistics*, Graduate School of the U. S. Dept. of Agriculture, 1938.

[45] J. NEYMAN, "Basic ideas and some recent results of the theory of testing statistical hypotheses," *Journal Roy. Stat. Soc.*, Vol. cv (1942), p. 292.

[46] J. NEYMAN, "On a statistical problem arising in routine analysis and in sampling inspection of mass production," *Annals of Math. Stat.*, Vol. 12 (1941), p. 46.

[47] S. N. ROY, "Notes on testing composite hypotheses, I, II," *Sankhya*, Vol. 8 (1947), p. 257 and Vol. 9 (1948), p. 19.

[48] H. SCHEFFÉ, "On the theory of testing composite hypotheses with one constraint," *Annals of Math. Stat.*, Vol. 13 (1942), p. 280.

[49] H. SCHEFFÉ, "On the ratio of the variances of two normal samples," *Annals of Math. Stat.*, Vol. 13 (1942), p. 371.

[50.] C. STEIN, "A two sample test for a linear hypothesis whose power is independent of the variance," *Annals of Math. Stat.*, Vol. 16 (1945), p. 243.

[51] A. WALD, "Asymptotically most powerful tests of statistical hypotheses," *Annals of* Math. Stat., Vol. 12 (1941), p. 1.

[52] A. WALD, "Some examples of asymptotically most powerful tests," *Annals of Math. Stat.*, Vol. 12 (1941), p. 396.

[53] A. WALD, "On the efficient design of statistical investigations," *Annals of Math. Stat.*, Vol. 14 (1943), p. 134.

Testing the Approximate Validity of Statistical Hypotheses

By J. L. Hodges, Jr. and E. L. Lehmann*

University of California, Berkeley

[Received March, 1954]

SUMMARY

The distinction between statistical significance and material significance in hypotheses testing is discussed. Modifications of the customary tests, in order to test for the absence of material significance, are derived for several parametric problems, for the chi-square test of goodness of fit, and for Student's hypothesis. The latter permits one to test the hypothesis that the means of two normal populations of equal variance, do not differ by more than a stated amount.

1. *Introduction*

When testing statistical hypotheses, we usually do not wish to take the action of rejection unless the hypothesis being tested is false to an extent sufficient to matter. For example, we may formulate the hypothesis that a population is normally distributed, but we realize that no natural population is ever exactly normal. We would want to reject normality only if the departure of the actual distribution from the normal form were great enough to be material for our investigation. Again, when we formulate the hypothesis that the sex ratio is the same in two populations, we do not really believe that it could be exactly the same, and would only wish to reject equality if they are sufficiently different. Further examples of the phenomenon will occur to the reader.

In practice, this imprecision in the formulation does not usually cause much trouble, since the tests employed are sufficiently lacking in power that we do not run serious risk of rejecting the hypothesis unless it is false to a considerable extent. But whenever the available data are extensive, the tests may become embarrassingly powerful, leading to a paradox enunciated by Berkson (1938):

"I believe that an observant statistician who has had any considerable experience with applying the chi-square test repeatedly will agree with my statement that, as a matter of observation, when the numbers in the data are quite large, the P's tend to come out small. Having observed this, and on reflection, I make the following dogmatic statement, referring for illustration to the normal curve: 'If the normal curve is fitted to a body of data representing any real observations whatever of quantities in the physical world, then if the number of observations is extremely large—for instance, on the order of 200,000—the chi-square P will be small beyond any usual limit of significance.'

"This dogmatic statement is made on the basis of an extrapolation of the observation referred to and can also be defended as a prediction from *a priori* considerations. For we may assume that it is practically certain that any series of real observations does not actually follow a normal curve *with absolute exactitude* in all respects, and no matter how small the discrepancy between the normal curve and the true curve of observations, the chi-square P will be small if the sample has a sufficiently large number of observations in it.

"If this be so, then we have something here that is apt to trouble the conscience of a reflective statistician using the chi-square test. For I suppose it would be agreed by statisticians that a large sample is always better than a small sample. If, then, we know in advance the P that will result from an application of a chi-square test to a large sample there would seem to be no use in doing it on a smaller one. But since the result of the former test is known, it is no test at all!"

It seems to us that this difficulty can be avoided by making a clear distinction, in the formulation of the problem, between "statistical significance" and what might be called "material significance".

* This paper was prepared with the partial support of the Office of Naval Research, U.S.A.

In the space of the parameters we may distinguish a set H_0 of values, representing the idealized hypothesis as it is customarily formulated. About the set H_0 we may then distinguish a larger set H_1 of values, representing situations close enough to H_0 that the difference is not materially significant in the problem at hand. If we knew that some distribution θ in H_1 were the true one, we should still wish to accept the hypothesis idealized by H_0. The size of the test should be the maximum value of the power function $\beta(\theta)$ in H_1, not H_0. Outside of H_1, however, we want to reject, so that consistency outside of H_1 is no longer undesirable. We reject as soon as there is statistically significant evidence that the departure from H_0 is materially significant.

It might be objected that there is nothing novel in the point of view just presented. We may just forget about H_0 and let H_1 play the role of the hypothesis in the classical formulation. There is however often a practical advantage in keeping H_0, as it is mathematically simple and corresponds to the idea underlying the situation to be tested. The boundaries of H_1 will be less precise in the experimenter's mind. It will usually be best to introduce into the space of parameters a measure, say $\Delta(\theta)$ of the "distance" of θ from H_0 on a scale reflecting at least roughly the materiality of departures from H_0, and then define H_1 as the set of those θ for which $\Delta(\theta)$ does not exceed a specified value Δ_0. The choice of Δ_0 will present problems similar to those encountered in choosing the alternative at which specified power is to be obtained.

If such a formulation is adopted, it is necessary to modify the customary tests, so as to adjust them to the new situation. This will be done briefly in section 2, for a few simple parametric situations which require only trivial modifications of the standard tests. In section 3 a modification of Student's problem is treated, and in section 4 a discussion from the present point of view is given of the χ^2-test for goodness of fit.

2. *Some Parametric Problems*

The simplest situation of the kind described in the introduction arises when a single parameter is involved, as for example in the case of a sample from a binomial or Poisson variable. We then have a (vector-valued) random variable X, with generalized probability density p_θ. For testing $H : \theta = \theta_0$ we have available a test statistic T such that

$$P_\theta(a < T < b) = G_{a,b}(\theta) \qquad (2.1)$$

is a unimodal function of θ. For testing $H : \theta = \theta_0$ we would adjust a and b such that $G_{a,b}(\theta)$ takes on its maximum at $\theta = \theta_0$ and that the value of this maximum is $1 - \alpha$. If now instead we test $H^* : \theta_1 \leqq \theta \leqq \theta_2$ we adjust a, b so that $G_{a,b}(\theta_1) = G_{a,b}(\theta_2) = 1 - \alpha$. It then follows that the power function $\beta(\theta) = 1 - G_{a,b}(\theta)$ satisfies $\beta(\theta) \leqq \alpha$ for $\theta_1 \leqq \theta \leqq \theta_2$ and $\beta(\theta) > \alpha$ for $\theta < \theta_1$ and $\theta_2 < \theta$.

The following are some examples of this approach :

(i) X has a binomial distribution with probability θ of success. We can then take $T = X$.

(ii) $X_1, \ldots, X_n$ is a sample from a Poisson distribution with $E(X_i) = \theta$. We take $T = \Sigma X_i$.

(iii) $X_1, \ldots, X_n$ is a sample from a normal distribution, mean ξ, variance σ^2. We wish to test $H^* : \sigma_1 \leqq \sigma \leqq \sigma_2$, and take $T = \Sigma(X_i - \bar{X})^2$.

(iv) $X_1, \ldots, X_n$ is a sample from a normal distribution, mean ξ, variance σ^2, and we wish to test $H^* : \delta_0 \leqq \xi/\sigma \leqq \delta_1$. We may then take

$$T = \frac{\bar{X}}{\sqrt{\{\Sigma(X_i - \bar{X})^2\}}}.$$

This is a special case of the general linear hypothesis which we consider later in this section.

Certain two-sample problems immediately reduce to situations of the kind just considered. In particular, let $X_1, \ldots, X_m$; $Y_1, \ldots, Y_n$ be samples from two normal populations with means ξ, η and variances σ^2 and τ^2 respectively. Then we can test

(v) the hypothesis $H^* : a \leqq \sigma^2/\tau^2 \leqq b$ by putting $T = \Sigma(X_i - \bar{X})^2/\Sigma(Y_j - \bar{Y})^2$.

If we assume in addition that $\sigma = \tau$ we can base a test of the hypothesis

(vi) $H^* : \delta_0 \leqq (\eta - \xi)/\sigma \leqq \delta_1$ on the statistic

$$T = \frac{\bar{Y} - \bar{X}}{\sqrt{\{\Sigma(X_i - \bar{X})^2 + \Sigma(Y_j - \bar{Y})^2\}}}.$$

The case of two Poisson or binomial populations also can be reduced to a one-parameter problem of the kind already considered.

(vii) Suppose that X, Y are two independent Poisson variables with means λ and μ respectively, and that we wish to test H^*: $a \leq \lambda/\mu \leq b$. As is well known and easily checked, the conditional distribution of Y given $X + Y = m$ is the binomial distribution of m trials and success probability $p = \mu/(\lambda + \mu) = 1/(1 + \lambda/\mu)$. Thus we obtain a test of H^* by making on each line $X + Y = m$ a conditional binomial test of the hypothesis $1/(1 + b) \leq p \leq 1/(1 + a)$.

In a completely analogous manner we can test $H^* : a \leq (p_1/q_1)/(p_2/q_2) \leq b$ where X, Y are independent binomial variables with probabilities p_1, p_2 of success respectively, and where $q_i = 1 - p_i$, since again the conditional distribution of Y given $X + Y = m$ depends on p_1, p_2 only through $\theta = (p_1/q_1)/(p_2/q_2)$.

As a last example consider the general univariate linear hypothesis, which we shall take in the canonical form, according to which $Y_1, \ldots, Y_r; \; Y_{r+1}, \ldots, Y_s; \; Y_{s+1}, \ldots, Y_n$ are independently distributed with common (unknown) variance σ^2 and means

$$E(Y_i) = \eta_i \quad i = 1, \ldots, s; \quad E(Y_i) = 0 \quad i = s + 1, \ldots, n.$$

The usual hypothesis $H : \eta_1 = \ldots = \eta_r = 0$ is tested by rejecting when

$$W = \frac{\sum_{i=1}^{r} Y^2_i/r}{\sum_{i=s+1}^{n} Y^2_i/(n - s)} \quad \quad (2.2)$$

is too large. Here W has a noncentral F-distribution with noncentrality parameter $\lambda = \sum_{i=1}^{r} \eta^2_i/\sigma^2$ and in fact the probability $P(W \geq C)$ is a strictly increasing function of λ. From this it is obvious how to test $H^* : \Sigma\eta^2_i/\sigma^2 \leq \lambda_0$. We again reject when W exceeds a constant C, but instead of determining C from the central F-distribution, we now determine it from the equation

$$P(W \geq C \mid \lambda = \lambda_0) = \alpha.$$

It is interesting to note that this is the uniformly most powerful test for testing H^* among all those based on W (or equivalently that it is the most powerful invariant test for testing H^*; see, for example, Lehmann (1950)). To obtain the test having this property we must consider the probability ratio test based on W for testing $\Sigma\eta^2_i/\sigma^2 = \lambda_0$ against the alternative $\Sigma\eta^2_i/\sigma^2 = \lambda_1$, which rejects when

$$\frac{p_{\lambda_1}(w)}{p_{\lambda_0}(w)} \quad \quad (2.3)$$

is too large. (Here p_λ denotes the density of W for parameter value λ). As we shall show below, (2.3) is an increasing function of w and this test is therefore equivalent to rejecting when $w \geq C$. From this it is then seen that the probability $P\{p_{\lambda_1}(W)/p_{\lambda_0}(W) \geq k\}$ is an increasing function of the true parameter value λ, and the result follows.

It only remains to show that the ratio (2.3) is an increasing function of w. (This fact has been known for some time. The proof which follows is essentially the same as that given by Paul L. Meyer in *An Application of the Invariance Principle to the Student Hypothesis*, Technical Report No. 24, Department of Statistics, Stanford University, California.) We have

$$p_\lambda(w) = e^{-\lambda^2/2} \sum_{k=0}^{\infty} c_k \frac{(\lambda^2/2)^k}{k!} \frac{w^{r/2-1+k}}{(1 + w)^{(r+n-s)/2+k}}$$

where $0 < w$ and the c_k are positive constants. Hence, if we put $z = w/2(1 + w)$, $a_k = c_k \lambda_0^{2k}/k!$ and $b_k = c_k \lambda_1^{2k}/k!$, we have

$$\frac{p_{\lambda_1}(w)}{p_{\lambda_0}(w)} = \frac{\Sigma b_k z^k}{\Sigma a_k z^k} = f(z), \text{ say.}$$

It is easy to show that the derivative of $f(z)$ is given by

$$(\Sigma a_k z^k)^2 f'(z) = \sum_{k<n} (n-k)(a_k b_n - a_n b_k) z^{k+n-1},$$

which is positive since $b_k/a_k < b_n/a_n$ for $k < n$. It follows that $p_{\lambda_1}(w)/p_{\lambda_0}(w)$ is an increasing function of z and hence of w.

3. *Student's Problem*

We next consider the modification of the classical Student testing problem in which absolute units are used, rather than σ-units as was done in example (iv) above. Let X have a normal distribution with $E(X) = \xi$, $V(X) = \sigma^2$, and let S, independent of X, be such that S/σ has the chi distribution with m degrees of freedom. The problem of testing the hypothesis that ξ equals a specified value ξ_0, with σ playing the role of a nuisance parameter, is often formulated. For example, it arises when testing whether two normal populations, of the same but unknown variance, have the same mean. In line with the arguments advanced above, we feel that in many applications it would be more realistic to formulate as our hypothesis the assertion that ξ does not differ from ξ_0 by more than a specified quantity chosen in the light of the material problem. For instance, instead of testing that two normal populations have the same mean, we would test that their means do not differ by more than an amount specified to represent the smallest difference of practical interest.

One possible test of the modified hypothesis that $\xi_1 \leq \xi \leq \xi_2$, which uses only existing tables, consists in performing separate one-tailed t-tests of the two one-sided hypotheses:

$$H: \xi \leq \xi_2 \text{ against alternative } \xi > \xi_2$$

and

$$H: \xi \geq \xi_1 \text{ against alternative } \xi < \xi_1.$$

We then reject if either of these separate tests rejects. The size of this composite t-test is the sum of their separate sizes, as is seen by letting σ tend to ∞. (By the size of a test we mean as usual the upper bound of its probability of first kind error.) Thus, we would obtain a test of size α if each of the one-sided t-tests was carried out at level $\alpha/2$. The main objection to the composite t-test solution of our problem lies in the fact that, when σ is small, the test has the power only of the t-test with level $\alpha/2$, although we are paying for a test of size α. The obvious remedy is to seek an unbiased test. Such a test will of necessity have power identically α when $\xi = \xi_1$ or $\xi = \xi_2$; that is, it will be "similar on the boundary". For simplicity we shall assume throughout that $\alpha \leq \frac{1}{2}$, and make a scale change to insure $\xi_1 = -1$, $\xi_2 = 1$.

In the present paragraph we consider the right boundary, that is, we assume $\xi = 1$. We may specify the location of the sample point (x, s) by the polar coordinates (r, θ) defined by $r^2 = (x-1)^2 + s^2$, $\sin\theta = s/r$. It is well known that the corresponding random variables R, Θ are independent and that $P[\Theta \leq \theta] = \frac{1}{2} I_{\sin^2\theta}(\frac{1}{2}m, \frac{1}{2})$ for $0 \leq \theta \leq \pi/2$, where I is the Incomplete Beta function. The distribution of Θ is further symmetric about $\pi/2$. It is known (Lehmann and Scheffé, 1950) that $(X-1)^2 + S^2 = R^2$ is a complete sufficient statistic for σ, so that the only tests of size α which are similar for $\xi = 1$ are those whose critical regions intersect each semicircle $R = r$ in a set of points whose conditional probability is α.

As the analogous argument goes through for $\xi = -1$, we have a double condition on the test region. We also impose the intuitively reasonable requirements that the test region be symmetric about the s-axis, and that for given values of R we want to reject for extreme values of θ. The problem now formulated is to construct a test region satisfying these conditions. We shall carry through this construction in detail in the simple case $m = 1$, where an explicit solution can be given.

When $m = 1$, Θ has the uniform distribution over $(0, \pi)$, so that we must take from each semicircle an arc of size $\pi\alpha$. The construction is made inductively on increasing values of r. From semicircles of radius $r \leq 2 = r_1$, we take arcs of size $\pi\alpha$ adjacent to the positive x-axis (see Fig. 1). This gives the segment P_0P_1 as part of the boundary of the rejection region. By the symmetry requirement, we must also reject at points below the segment Q_0Q_1, obtained by reflecting P_0P_1

about the s-axis. Let r_2 denote the distance from P_0 to Q_1, and next consider the semicircles with radii $r_1 \leq r \leq r_2$. From these we have already taken as rejection points the arcs below Q_0Q_1, which are the same size as the arcs below the segment R_0R_1 obtained by translating P_0P_1 two units to the right. To bring the rejection arcs up to size $\pi\alpha$, we cut off the additional rejection arcs below P_1P_2. The notable fact, obvious from the symmetry of the picture, is that the boundary portion P_1P_2 is a horizontal straight line segment. The construction proceeds inductively, producing a boundary of the rejection region consisting of straight line segments of length 2, alternately horizontal and of slope $\pi\alpha$, together of course with its reflection about the s-axis.

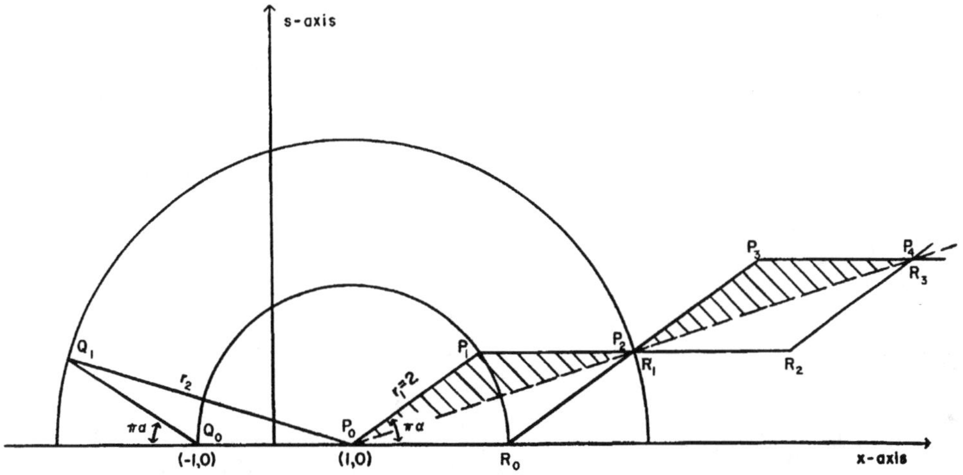

FIG. 1.—Construction of the modified Student test when $m = 1$.

The test we have obtained is by its construction similar on the boundary $\xi = \pm 1$; to verify that it is in addition unbiased, we need only observe that for each given value of s, we accept for x in an interval centered at 0. Therefore, the conditional probability of rejection, given s, is a monotonely increasing function of $|\xi|$, whatever s may be. The same must then hold for the power function as a function of $|\xi|$ for any given value of σ.

For comparison, we show as the dashed line of the Figure the boundary of the composite t-test of size α. The shaded region represents the additional rejection points which the new test affords without increase in size. It is these points which provide our test with its greater power for finite σ.

So far we have been considering only the case $m = 1$, where the problem is simple and we can give an explicit solution. When we turn to the general case things are not so easy. The inductive construction may still be employed, alternately translating boundary segments to the right by two units and then generating new segments by the requirement that a conditional probability α be taken on each arc. This leads to the relation

$$\left.\begin{aligned} (r')^2 &= 4r\cos\theta + r^2 + 4 \\ I(\sin^2\theta') &= 2\alpha - I\left[\frac{r^2\sin^2\theta}{r'^2}\right] \end{aligned}\right\} \quad . \quad . \quad . \quad . \quad . \quad (3.1)$$

where (r, θ) and (r', θ') are two boundary points, and we write $I(x)$ for $I_x(\frac{1}{2}m, \frac{1}{2})$. These equations, together with the initial values $\theta = \arcsin\sqrt{\{I^{-1}(2\alpha)\}}$ for $0 \leq r \leq 2$, permit the effective computation of a boundary curve C. It appears from the numerical computations underlying the chart that C has positive slope everywhere, but we have not found a proof of this. From the assumption that C has positive slope it would follow (as for $m = 1$) that the test region with boundary C is unbiased, and at least approximate unbiasedness is in any case guaranteed by the numerical results.

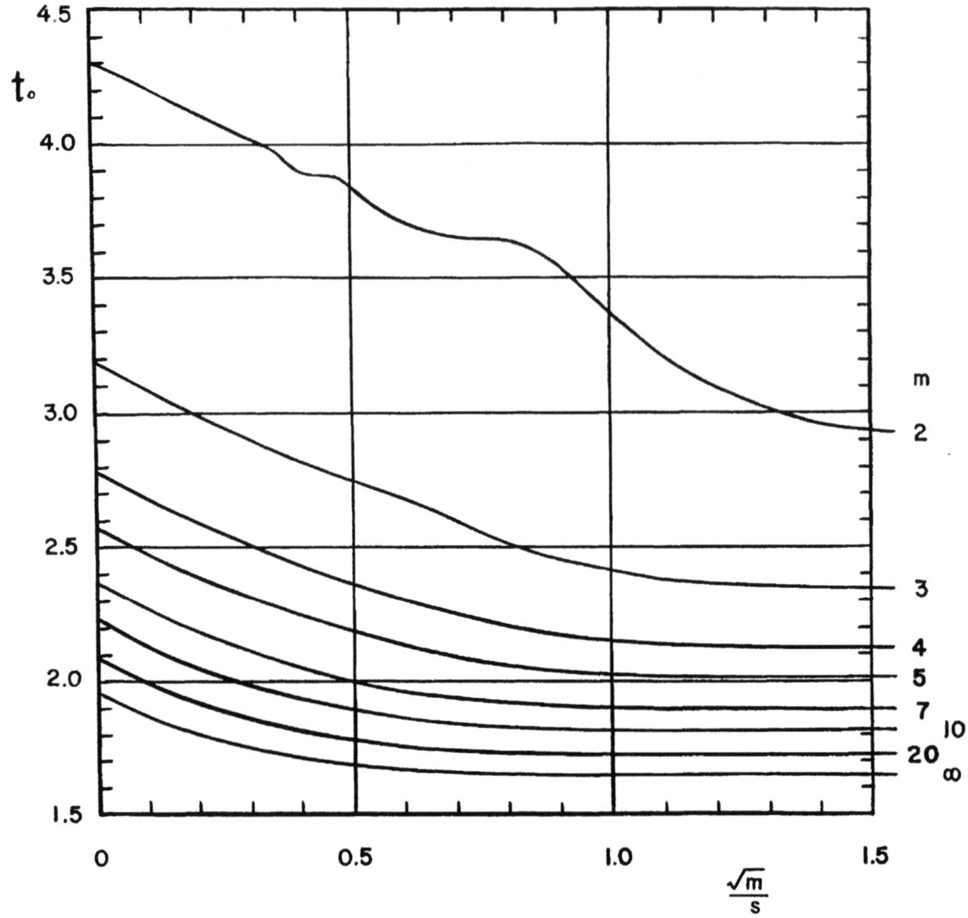

FIG. 2.—5 per cent. critical values for the modified Student test.

Fig. 2 shows critical values of the 5 per cent. test for various degrees of freedom m. Suppose X is normal, $E(X) = \xi$, $V(X) = \sigma^2$, S^2/σ^2 has the chi-square distribution of m degrees of freedom, X and S are independent, and we wish to test $|\xi| \leq 1$. If $|X| \leq 1$, accept. Otherwise, let t be $(X - 1)\sqrt{m}/S$ if $X > 1$, or $(-1 - X)\sqrt{m}/S$ if $X < -1$. Read the critical value t_0 from Fig. 2 corresponding to the value of m and of $\sqrt{m}/S$. Reject if $t > t_0$.

To illustrate the test, suppose $Y_1, \ldots, Y_k$ is a sample from the normal population of expectation η and variance τ^2, while $Z_1, \ldots, Z_n$ is a sample from the normal population of expectation ζ and variance τ^2. We wish to test the hypothesis that the absolute difference $|\eta - \zeta|$ of the population means does not exceed Δ. If we let $k\bar{Y} = \Sigma Y_i$, $n\bar{Z} = \Sigma Z_i$, $X = (\bar{Y} - \bar{Z})/\Delta$, $\xi = (\eta - \zeta)/\Delta$, $\sigma^2 = (k + n)\tau^2/kn\Delta^2$, $S^2 = (k + n)\{\Sigma(Y_i - \bar{Y})^2 + \Sigma(Z_i - \bar{Z})^2\}/kn\,\Delta^2$, and $m = k + n - 2$, then the assumptions of the test are satisfied.

In the computations for the chart good use was made of the approximate value $t_{\alpha/2} - \sqrt{m}/S$ for $t_0(\sqrt{m}/S)$. This approximation rests on the fact that as $r \to \infty$, the distance of the point (r, θ) in the first quadrant from the line $y = x\sqrt{m}/t_{\alpha/2}$ tends to 0. The recursion formulae (3.1) were used until the error of this approximation was less than 0.1, and thereafter the error was interpolated quadratically.

Inspection of Fig. 2 shows that the test has an intuitively reasonable form. When $s/\sqrt{m}$ is very small, it is plausible that σ is small, and the two sides do not interfere with each other. We may then safely use separate one-tailed t-tests each of level α; this corresponds to the flat portions of the curves. When $s/\sqrt{m}$ is large, our hypothesis is not essentially different from the classical Student hypothesis, and the two-tailed t-test of size α is appropriate; this is given in the left margin of the chart. When m is very large, we may safely take the consistent estimate $s/\sqrt{m}$ as if it were σ, and use the obvious normal test available when σ is known; this is the lowest curve of the chart. For any given finite value of m, the critical value of t changes in a more or less regular way, as $s/\sqrt{m}$ is increased, from one extreme to the other. The waves, visible for $m = 2$ and to a lesser extent for $m = 3$, are to be expected in the light of the situation for $m = 1$.

We conclude by remarking that our test also provides confidence intervals in the usual way. For instance, it will give an upper confidence limit for the absolute amount by which the mean of a normal population differs from a given quantity, or for the absolute difference of the means of two normal populations of the same unknown variance.

4. *Testing for Goodness of Fit*

It is in connection with testing for goodness of fit that the existence of the problem under consideration has been pointed out by Berkson and others. We are concerned with a multinomial distribution with, say, r classes and with the hypothesis that the point $p = (p_1, \ldots, p_r)$, where p_i is the probability of the ith class, lies on a specified surface $\mathscr{S}$. The simplest case is that in which the hypothesis specifies the p's completely so that $\mathscr{S}$ consists of a single point. Let $\Delta(p, p')$ be a distance function in the sense that $\Delta(p, p') \geqq 0$ for all p, p' and $\Delta(p, p') = 0$ if and only if $p = p'$. A typical example is ordinary Euclidean distance or more generally $\Delta(p, p') = \Sigma\, w_i(p'_i - p_i)^2$ where the weights w_i may be functions of p, p'. This includes in particular the usual χ^2-measure of distance. Let $d(p)$ denote the smallest distance of a point p from $\mathscr{S}$ so that

$$d(p) = \inf_{p' \varepsilon S} \Delta\,(p, p'). \quad \ldots \quad (4.1)$$

We shall then test instead of H the hypothesis H^* that the true point p lies within a given distance of $\mathscr{S}$, that is, the hypothesis

$$H^* : d(p) \leqq c. \quad \ldots \quad (4.2)$$

In order to obtain a test of H^* we shall assume that $\mathscr{S}$ and Δ are such that the function $d(p)$ possesses continuous first and second partial derivatives. We shall not discuss conditions on $\mathscr{S}$ and Δ which would insure this since such conditions do not appear to be particularly simple and since in applications it will always be necessary to obtain the function $d(p)$, and the regularity assumption is then easily checked directly. Of course in the particular case that $\mathscr{S}$ consists of a single point, say p^0, we simply have $d(p) = \Delta\,(p, p^0)$ and the condition on d immediately reduces to one on Δ.

Under these assumptions we propose the following procedure for testing H^*. Let $q_i\ (i = 1, \ldots, r)$ denote the relative frequency in the ith class and let $q = (q_1, \ldots, q_r)$. Then if $d(q) \leqq c$ accept H^*. If $d(q) > c$ test the hypothesis

$$H' : d(p) = c \quad \ldots \quad (4.3)$$

by means of minimum χ^2 (or some asymptotically equivalent test) in the usual manner. For example, we may reject when the minimum (modified) χ^2

$$n \sum_{i=1}^{r} \frac{(q_i - \hat{p}_i)^2}{q_i} > K \quad \ldots \quad (4.4)$$

where n is the sample size, where the $\hat{p}_i$ are BAN estimates (Neyman, 1949) of the p_i subject to H', and where K is determined from the distribution of χ_1^2, that is, χ^2 with 1 degree of freedom (See Neyman (1949), p. 267. The one degree of freedom arises because the only restriction on $\check{p}_i$ is $d(\hat{p}_i) = 0$). However, since we are subjecting the rejection region to the additional restriction $d(q) > c$, the cut-off point K is determined so that

$$P(\chi_1^2 > K) = 2\alpha \quad \ldots \quad (4.5)$$

where α is the desired level of significance. This differs from the usual procedure in which the right-hand side of (4.5) is taken to be α.

We shall now prove that the power function $\beta_n(p)$ of this test has the following property:

$$\lim_{n \to \infty} \beta_n(p) = \begin{matrix} 0 \\ \alpha \\ 1 \end{matrix} \text{ as } d(p) \begin{matrix} < \\ = \\ > \end{matrix} c. \qquad (4.6)$$

This result follows easily from the theory of minimum-χ^2 tests. (For details see Cramér (1946) and Neyman (1949).) In particular, if $d(p) < c$, it follows from the continuity of d and the fact that $q \to p$ in probability that $d(q) < c$ with probability tending to one, so that $\beta_n(p) \to 0$. Suppose next that $d(p) > c$. Then it follows similarly that $d(q) > c$ with probability tending to one. Also it follows from the consistency of the χ^2 test that the probability of (4.4) tends to one. Hence $\beta_n(p)$ is the probability of the simultaneous occurrence of two events, say A_n and B_n, for which $\lim P(A_n) = \lim P(B_n) = 1$. But this implies that $\lim \beta_n(p) = \lim P(A_n \cap B_n) = 1$.

Let us finally consider the case in which the true probability point, say p^0, lies on the surface $\mathscr{S}': d(p) = c$. Then it is known that $n\Sigma\{(q_i - \hat{p}_i)^2/q_i\}$ has the same limiting distribution as $[\Sigma\gamma_i\sqrt{n}\,(q_i - p_i^0)]^2$ where the γ's are a set of coefficients such that

(i) The equation of the tangent plane of the surface $\mathscr{S}'$ at p^0 is $\Sigma\gamma_i\,(p_i - p_i^0) = 0$, and

(ii) $\Sigma p_i^0\,\gamma_i = 0$, $\Sigma p_i^0\,\gamma_i^2 = 1$.

Therefore, $d(q) = d(p^0) + \Sigma\,\gamma_i(q_i - p_i^0) + 0_p\,(1/n)$ and hence, since $d(p^0) = c$,

$$d(q) - c \geqq 0 \text{ implies } \sqrt{n}\,\Sigma\gamma_i\,(q_i - p_i^0) + 0_p\,(1/\sqrt{n}) > 0.$$

Therefore

$$\beta_n(p^0) = P\{n\Sigma\{(q_i - \hat{p}_i)^2/q_i\} > K,\ d(q) > c\}$$

becomes in the limit equal to

$$P\{[\Sigma\,\gamma_i\,\sqrt{n}(q_i - p_i^0)]^2 > K,\ \Sigma\gamma_i\,\sqrt{n}(q_i - p_i^0) > 0\}.$$

But if we let $Y = \Sigma\gamma_i\,\sqrt{n}(q_i - p_i^0)$, then Y has a limiting normal distribution with zero mean and unit variance, and the result follows.

The test is of course very simple to carry out since in the minimization of χ^2 we have only the single condition $d(p) = c$. If, following Neyman, we replace this by the asymptotically equivalent condition $d(q) + \Sigma a_i(q)\,(p_i - q_i) = c$ where $a_i = a_i(q) = \frac{\partial}{\partial q_i}\,d(q)$, we get the solution

$$q_i - \hat{p}_i = (a_i - \bar{a})\,q_i\,[d(q) - c]/\sigma_a^2 \qquad (4.7)$$

where $\bar{a} = \Sigma a_i q_i$, $\sigma_a^2 = \Sigma q_i(a_i - \bar{a})^2$.

As an example, suppose that we wish to test that all r cell probabilities are equal, and that we set $\Delta(p, p') = \Sigma(p_i' - p_i)^2$. Then the modified hypothesis becomes $H^*: \sum_{i=1}^{r} (p_i - 1/r)^2 \leqq c$, and we must determine the $\hat{p}_i$ subject to the condition $\Sigma(p_i - 1/r)^2 = c$. We then have $a_i = 2(q_i - 1/r)$ and can compute $q_i - \hat{p}_i$ from (4.7).

References

BERKSON, J. (1938), *J. Amer. Statist. Ass.*, **33**, 526–542.
CRAMÉR, H. (1946), *Mathematical Methods of Statistics*. Princeton University Press.
LEHMANN, E. L. (1950), *Ann. Math. Statist.*, **21**, 1–26.
—— & SCHEFFÉ, H. (1950), *Sankhyā*, **10**, 305–340.
NEYMAN, J. (1949), *Proceedings of the Berkeley Symposium on Mathematical Statistics and Probability*. Berkeley and Los Angeles: University of Calif. Press.

SIGNIFICANCE LEVEL AND POWER[1]

By E. L. Lehmann

University of California, Berkeley

1. Summary and introduction. Significance testing, as described in most textbooks, consists in fixing a standard significance level α such as .01 or .05 and rejecting the hypothesis $\theta = \theta_0$ if a suitable statistic Y exceeds C where $P_{\theta_0}\{Y > C\} = \alpha$. Such a procedure controls the probability of false rejection (error of the first kind) at the desired level α but leaves the power of the test and hence the probability of an error of the second kind to the mercy of the experiment. It seems more natural when deciding on a significance level (and this suggestion is certainly not new) to take into account also what power can be achieved with the given experiment. In Section 3 a specific suggestion will be made as to how to balance α against the power β obtainable against the alternatives of interest.

The adoption of this or some similar rule for choosing a significance level has important consequences for the theory of testing composite hypotheses, where nuisance parameters are present. Since the quantity α is then potentially a function of the nuisance parameter ϑ, the classical rule of a fixed significance level leads to the condition that the tests be *exact* or *similar*, that is, that $\alpha(\vartheta)$ equal the preassigned value α for all ϑ. On the other hand, the power β that can be attained against any alternative $\theta = \theta_1$ frequently depends on ϑ. The requirement that $\alpha(\vartheta)$ and $\beta(\vartheta)$ be in a certain balance thus leads to tests which are not similar and hence do not agree with the standard solutions.

To obtain a suitable setting for this discussion, we consider first a minimal complete class of tests for testing the hypothesis $H:\theta \leqq \theta_0$ in a multiparameter exponential family (Section 2). The proposed α, β-relation is discussed in Section 3, and in Section 4 is applied to the exponential family. Section 5 gives some illustrations of the theory.

2. A complete class theorem. Many standard testing problems concern an exponential family of distributions, which has probability densities of the form

$$p_{\theta,\vartheta}(x) = C(\theta, \vartheta) \exp\left[\theta U(x) + \sum_{i=1}^{r} \vartheta_i T_i(x)\right] h(x) \tag{1}$$

with respect to a σ-finite measure μ, where θ, U, the ϑ_i and T_i are real-valued and where $\vartheta = (\vartheta_1, \cdots, \vartheta_r)$. In this family, the statistics U and $T = (T_1, \cdots, T_r)$ constitute a set of sufficient statistics for (θ, ϑ).

The problem of testing the hypothesis $H:\theta \leqq \theta_0$ against the one-sided al-

Received January 17, 1958; revised July 22, 1958.

[1] This paper was prepared with the partial support of the Office of Naval Research. This paper in whole or in part may be reproduced for any purpose of the United States Government.

ternatives $\theta > \theta_0$ has been treated by many authors (usually in the formulation $\theta = \theta_0$ against $\theta > \theta_0$). The solution of this testing problem according to the Neyman-Pearson theory is the uniformly most powerful unbiased test; this depends only on U and T and is given by the critical function[2]

$$\phi(u, t) = \begin{cases} 1 \text{ if } u > C(t), \\ \gamma(t) \text{ if } u = C(t), \\ 0 \text{ if } u < C(t), \end{cases} \tag{2}$$

where the functions C and γ are determined by the conditions $E_{\theta_0}[\phi(U, T) \mid T = t] = \alpha$ and $E_{\theta_0}[U\phi(U, T) \mid T = t] = \alpha\, E_{\theta_0}[U \mid T = t]$ for all t. The condition of unbiasedness

$$E_{\theta,\vartheta}\phi(U, T) \lesseqqgtr \alpha \qquad \text{as} \qquad \theta \lesseqqgtr \theta_0,$$

and that of similarity

$$E_{\theta_0,\vartheta}\phi(U, T) = \alpha \qquad \text{for all } \vartheta$$

which it implies and which by itself is sufficient to justify the test, are not inherent in the problem but are imposed, at least in part, to facilitate the solution. Before proposing an alternative approach, it is interesting to see how far the problem can be reduced without the introduction of extraneous principles. This can be done by viewing it within the framework of decision theory.

Let d_0 and d_1 denote the decisions of accepting and rejecting the hypothesis H, and denote by $L_i(\theta, \vartheta)$ the loss resulting from decision d_i when (θ, ϑ) are the true parameter values. Then for fixed ϑ, the function $L_0(\theta, \vartheta)$ typically will be zero for $\theta \leqq \theta_0$ and increasing for $\theta \geqq \theta_0$, while $L_1(\theta, \vartheta)$ will be decreasing for $\theta \leqq \theta_0$ and zero for $\theta \geqq \theta_0$. In particular, the difference then satisfies

$$L_1(\theta, \vartheta) - L_0(\theta, \vartheta) \gtreqqless 0 \qquad \text{as} \qquad \theta \lesseqqgtr \theta_0. \tag{3}$$

The risk function of a test ϕ, which is the expected loss resulting from its use considered as a function of the parameters, is

$$\begin{aligned} R_\varphi(\theta, \vartheta) = \int \{&\varphi(U(x), T(x))L_1(\theta, \vartheta) \\ &+ [1 - \varphi(U(x), T(x))]L_0(\theta, \vartheta)\}p_{\theta,\vartheta}(x)\, d\mu(x). \end{aligned} \tag{4}$$

Let $\mathcal{C}$ be the class of all tests satisfying (2) for some functions C and γ. For all loss functions satisfying (3) it was shown by Truax [13] that $\mathcal{C}$ is essentially complete; that is, given any φ there exists $\varphi' \varepsilon\, \mathcal{C}$ such that

$$R_{\varphi'}(\theta, \vartheta) \leqq R_\varphi(\theta, \vartheta) \qquad \text{for all } (\theta, \vartheta). \tag{5}$$

We shall now prove that among essentially complete classes, $\mathcal{C}$ is minimal in the sense that if (5) holds for two tests φ, φ' in $\mathcal{C}$, then $\varphi = \varphi'$ a.e. μ.*

---

[2] See for example [7].

* Recently I learned that this result has been obtained also by D. L. Burkholder. His results are sketched in Abstract 18, *Ann. Math. Stat.*, Vol. 29 (1958), p. 616.

Let φ and φ' belong to $\mathcal{C}$ and let

$$(6) \qquad \alpha(\vartheta) = E_{\theta_0,\vartheta}\,\varphi(U, T), \qquad \alpha'(\vartheta) = E_{\theta_0,\vartheta}\,\varphi'(U, T).$$

(i) If the functions α and α' do not agree for all ϑ, suppose without loss of generality that there exists ϑ_0 such that $\alpha(\vartheta_0) < \alpha'(\vartheta_0)$. Since for $\vartheta = \vartheta_0$, the expected values of φ and φ' are continuous functions of θ, there exist $\theta_1 < \theta_0 < \theta_2$ such that

$$(7) \qquad E_{\theta,\vartheta_0}\varphi(U, T) < E_{\theta,\vartheta_0})\varphi'(U, T) \qquad \text{for } \theta = \theta_1 \text{ and } \theta = \theta_2 .$$

Then $R_\phi(\theta_1, \vartheta_0) < R_{\phi'}(\theta_1, \vartheta_0)$ and $R_\phi(\theta_2, \vartheta_0) > R_{\phi'}(\theta_2, \vartheta_0)$, and hence neither of the procedures ϕ and ϕ' is uniformly better than the other. (ii) Suppose on the other hand that $\alpha(\vartheta) \equiv \alpha'(\vartheta)$. The standard proof showing a similar test satisfying (2) to be uniformly most powerful similar also shows that a test ϕ_0 satisfying (2) and

$$(8) \qquad E_{\theta_0,\vartheta}\,\phi_0(U, T) = \alpha(\vartheta) \qquad \text{for all } \vartheta$$

is uniformly most powerful among all tests satisfying (8). The tests ϕ and ϕ' are therefore both uniformly most powerful within this class and hence

$$E_{\theta,\vartheta}\,\phi(U, T) = E_{\theta,\vartheta}\,\phi'(U, T) \qquad \text{for all } \theta > \theta_0 \text{ and all } \vartheta.$$

Since the family of distributions of the sufficient statistics (U, T) is complete, it follows that $\phi(u, t) = \phi'(u, t)$ a.e., as was to be proved.

3. Significance level and power. It follows from the result of the preceding section that the class $\mathcal{C}$ of tests (2) represents the maximum reduction that can be achieved by comparing only tests of which one has a uniformly better risk function than the other. The selection of a specific test from $\mathcal{C}$, involves two difficulties. It requires the adoption of some principle (Bayes, minimax, etc.) leading to a definite choice;[3] in addition, it requires knowledge of the loss functions L_0 and L_1. An alternative approach, utilizing the fortunate circumstance that the complete class is independent of the actual loss functions (subject only to their satisfying (3)), consists in making the choice by some simple rule of thumb, which does not require (the usually unavailable) knowledge of these losses.

Consider the simplest case of the family (1) with $r = 0$, which involves no nuisance parameters. The family of tests (2) is then a one-parameter family, one test corresponding to each value of

$$\alpha_0 = E_{\theta_0}\phi(X), \qquad 0 \leqq \alpha_0 \leqq 1.$$

A simple method of choice consists in specifying a value of α_0 and selecting the test corresponding to this value. This need not be a purely formal or arbitrary

[3] Particular proposals of this kind that have been made in the literature include those of Jeffreys [5] involving considerations of *a priori* probabilities, and of Lindley [8] based on his concept of unlikelihood.

procedure since α_0 as the maximum probability of false rejection is of course an important quantity in its own right.

Nevertheless, as was pointed out in Section 1, the above rule appears to neglect too many aspects of the problem. In particular, suppose that the alternatives of primary interest, for which it is important to reject the hypothesis, are those satisfying $\theta \geqq \theta_1$ ($\theta_0 < \theta_1$). Since the power function of any test (2) is increasing in θ, the probability β_1 of rejection when $\theta = \theta_1$ is the minimum power against these alternatives. It seems then reasonable that the choice of test should involve at least β_1 in addition to α_0.

The quantities α_0 and $\alpha_1 = 1 - \beta_1$ are the error probabilities associated with the problem of testing the simple hypothesis $\theta = \theta_0$ against the simple alternative $\theta = \theta_1$. The attainable pairs (α_0, α_1) form a convex set, the lower boundary of which corresponds to the admissible tests (2). This lower boundary is a convex curve S connecting the points (0, 1) and (1, 0), and what is needed is a reasonable way of selecting a point on each such curve. One possible approach to this question is in terms of indifference curves. Suppose that a system of curves could be specified in the (α_0, α_1)-plane such that any two points lying on the same curve are equally desirable, with the curves closer to the origin being more desirable than those further away. The optimum test would then be given by that point of S lying on the indifference curve closest to the origin (Fig. 1).

It seems likely that even this approach is too complex for most applications. To obtain an even simpler formulation, consider once more the rule of fixing the significance level without regard to power. If the level is α, this means restricting attention to the points (α_0, α_1) lying on the vertical line segment $L: \alpha_0 = \alpha$, $0 \leqq \alpha_1 \leqq 1 - \alpha$. The test then corresponds to the point (α_0, α_1), which is the

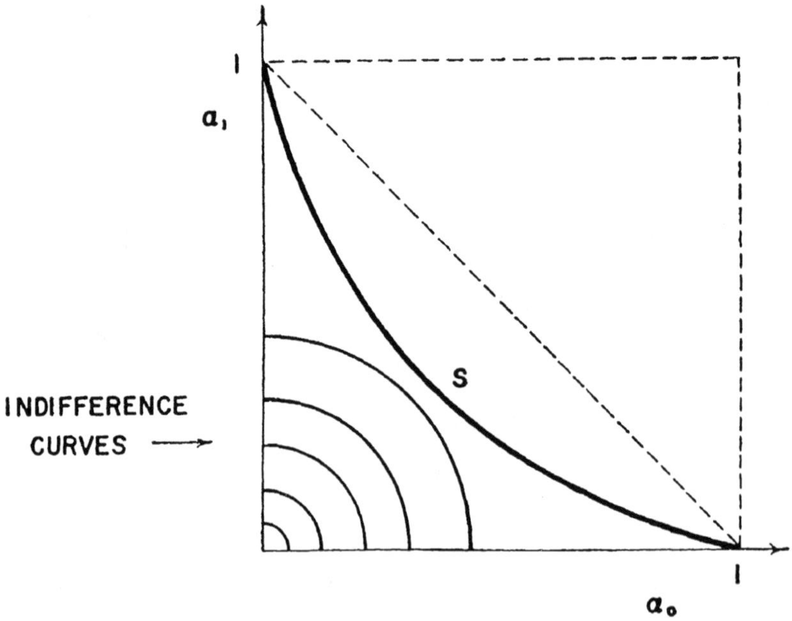

FIG. 1

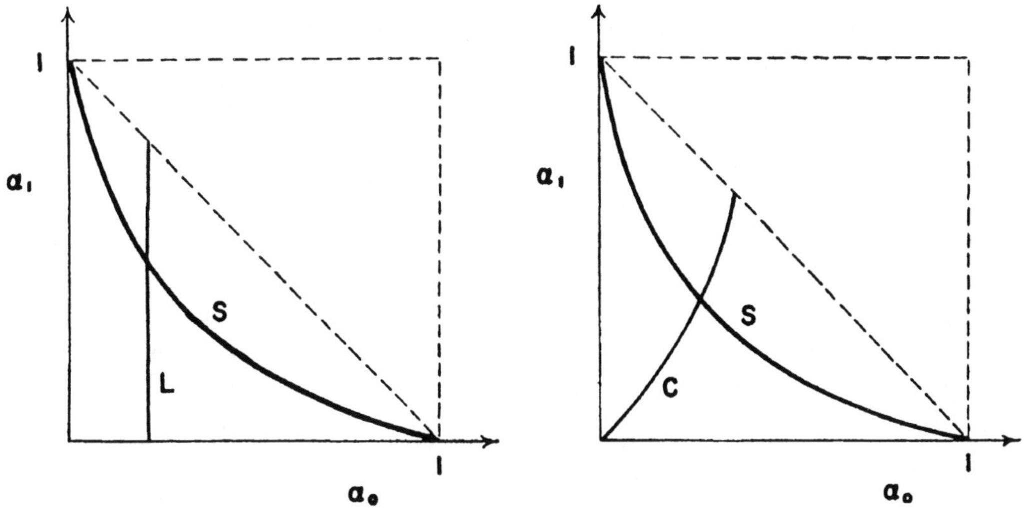

FIG. 2

intersection of S and L. This procedure is commonly justified on the grounds that the error of the first kind is of a higher order of importance, and should therefore be controlled at the prescribed level. However, if the curve S is sufficiently close to the α_0- and α_1-axis, as will always be the case if the sample size is sufficiently large, then α_1 is much smaller than α_0, which is inconsistent with the assumed relative importance of the two errors.

A more reasonable solution is obtained if one replaces the vertical line segment L by a curve $C: \alpha_1 = f(\alpha_0)$ where f is a continuous strictly increasing function with $f(0) = 0$. A particularly simple choice for f is a linear function

$$\alpha_1 = k\alpha_0 . \tag{9}$$

Since $\alpha_0 \leqq 1 - \alpha_1$ for all admissible tests, one has $\alpha_0 \leqq 1/(k + 1)$ so that $1/(k + 1)$ is an upper bound for α_0. As an example, consider (9) with $k = 9$. If $\beta_1 = 1 - \alpha_1$ denotes the power of a test against the alternative θ_1, some typical pairs of values of (α_0, β_1) are

α_0	.1	.05	.04	.03	.02	.01	.005
β_1	.1	.55	.64	.73	.82	.91	.955

with .1 being an upper bound for α_0.

One would of course hope to avoid cases such as $\alpha_0 = .1$, $\beta_1 = .1$ or even $\alpha_0 = .05$, $\beta_1 = .55$. When no nuisance parameters are present, this can be achieved by taking a sample of sufficient size. In the composite case, on the other hand, it can frequently not be achieved by samples of fixed size no matter how large, but only by resorting to sequential experimentation.

To avoid misunderstandings, it should be emphasized that (9) is not being proposed as a logically convincing rule, nor as one fitting all occasions. Actually, it seems clear that no rule satisfying these requirements exists, except the Bayes solution when sufficient knowledge concerning losses and *a priori* probabilities is available. In the absence of this knowledge it may be convenient to employ a simple rule of thumb. Such a rule is in fact being used in much of present practice: It consists in choosing α to be .05 or .01 depending on the seriousness attached to the committing of an error of the first kind. To this, (9) is suggested as an alternative which appears to be more reasonable in many cases.

It so happens that (9) is the minimax solution if the loss for rejecting $H: \theta \leqq \theta_0$ is a_0 when H is true, and the loss is a_1 for accepting H when $\theta \geqq \theta_1$, where the constant k of (9) is then given by $k = a_0/a_1$. However, this is not the basis for the present suggestion of (9), and the minimax property does not carry over to the application to be made in the next section to composite hypotheses.

4. Conditional tests. We return now to the composite case of the exponential family (1) with $r > 0$. The minimal complete class $\mathcal{C}$ is then more complex than in the preceding section, its members being characterized by the function $\alpha(\vartheta)$ instead of the single number α_0. Given any function $\alpha(\vartheta)$, which is the expectation of some critical function ϕ, there exists a unique member of $\mathcal{C}$ whose expectation function for $\theta = \theta_0$ is also $\alpha(\vartheta)$. This uniformly minimizes the risk (and maximizes the power) among all critical functions having this expectation.

If the alternatives of interest are as before those satisfying $\theta \geqq \theta_1$, let $\beta(\vartheta)$ denote the power function of a test against the alternative (θ_1, ϑ). The proposal made in the preceding section suggests selecting that member of $\mathcal{C}$ which satisfies

$$1 - \beta(\vartheta) = k\alpha(\vartheta) \qquad \text{for all } \vartheta. \tag{10}$$

However, this relationship depends on the particular parametrization chosen, and we shall not discuss it here. Instead an alternative approach will be proposed in which this difficulty does not arise.

Consider once more the case of the similar test with $\alpha(\vartheta) \equiv \alpha$. Since T is a complete sufficient statistic for ϑ when $\theta = \theta_0$, the functions C and γ of (2) are determined by the requirement that the conditional probability of rejection

$$\alpha^*(t) = P_{\theta_0}\{U > C(t) \mid t\} + \gamma(t)P_{\theta_0}\{U = C(t) \mid t\}$$

be equal to α for all t.[4] However, the conditional power $\beta^*(t) = P_{\theta_1}\{\text{rejecting } H \mid t\}$ of the test against the alternative $\theta = \theta_1$, typically depends on t. The question then arises: Suppose that $\beta^*(t)$ is quite small for the observed t, or quite high; is this value not more relevant to the case in hand than the average value $\beta(\vartheta)$?

Without entering into the difficulties raised by this question, there is an alternative and simpler justification for considering $\beta^*(t)$. The actual power β against the alternative $\theta = \theta_1$ generally depends on the nuisance parameter ϑ and is therefore unknown. It can however be estimated from the observations,

[4] This method of constructing exact tests was originated by Bartlett [1] and Neyman [9]. That in the present case it provides the totality of such tests has been noted by many authors. For a recent discussion and references see [7].

and $\beta^*(T)$ is the unbiased estimate with (uniformly) minimum variance. That it is unbiased is clear since $\beta(\vartheta) = E_{\theta_1,\vartheta}\beta^*(T)$. The minimum variance property is an immediate consequence of the completeness of the sufficient statistic T for (θ_1, ϑ) and of Theorem 5.1 of [7].

Analogous remarks apply in the more general case, in which the tests are not required to be exact. If the relevant frame of reference is obtained by considering t as fixed, the error probabilities of interest are the conditional probabilities $\alpha_0^*(t) = P_{\theta_0}$ (rejecting $H \mid t$) and $\alpha_1^*(t) = P_{\theta_1}$ (accepting $H \mid t$), and the quantities $C(t)$ and $\gamma(t)$ can therefore be determined from the relation

$$\alpha_1^*(t) = k\alpha_0^*(t). \tag{11}$$

The resulting test will of course not be similar. However, since $\alpha_0^*(t) \leqq 1/(k+1)$ for all t, the quantity $1/(k+1)$ is an upper bound also for the average probability $\alpha_0(\vartheta)$ of an error of the first kind.

The above discussion applies only to problems in which the parameter of interest is one of the "natural" parameters of the exponential distribution (1). As was pointed out in [7], any parameter of the form $\theta + \sum a_i\vartheta_i$ is natural for a suitable definition of U, the T's and ϑ's. When the parameter of interest is not of this form, related methods may be applicable as is indicated by the following example.

If $X_1, \cdots, X_n$ are a sample from a normal distribution $N(\xi, \sigma^2)$, neither the parameter ξ nor ξ/σ are of this form. The problem of testing $\xi/\sigma \leqq \delta_0$ against $\xi/\sigma \geqq \delta_1$ can be reduced by invariance considerations to the statistic $\bar{X}/[\sum(X_i - X)^2]^{\frac{1}{2}}$, the distribution of which depends on the single parameter $\delta = \xi/\sigma$. If $\alpha_i = P_{\delta_i}\{X > C[\sum(X_i - \bar{X})^2]^{\frac{1}{2}}\}$, the quantity C can be determined so that $\alpha_1 = k\alpha_0$. The problem of testing $\xi \leqq \xi_0$ against $\xi \geqq \xi_1$ appears to be more difficult; a possible approach may be that of [4], Section 3.

5. Examples. We shall now briefly indicate some examples in which the natural parameter θ is the relevant one so that the method of the preceding sections is applicable. Of these, Examples 1, 2, 3 have been treated by the same method (but from a different point of view) by Tocher [12], and Examples 2, 3 by Sverdrup [11].

Example 1. Let X, Y be independent Poisson variables with $E(X) = \lambda$, $E(Y) = \mu$, and consider the problem of testing $\mu/\lambda \leqq a_0$ against $\mu/\lambda \geqq a_1$. The joint distribution of X, Y forms an exponential family with $T = X + Y$, $U = Y$, $\theta = \log(\mu/\lambda)$ and $\vartheta = \log \lambda$. The conditional distribution of Y given $X + Y = t$ is a binomial distribution corresponding to the success probability $p = \mu/(\lambda + \mu)$ and number of trials equal to t. In terms of p, the hypothesis and class of alternatives becomes $p \leqq a_0/(a_0 + 1)$ and $p \geqq a_1/(a_1 + 1)$ so that the test satisfying (2) and (11) can be determined from a table of the binomial distribution.

Example 2. If X, Y are independent variables with binomial distributions $b(p_1, m)$ and $b(p_2, n)$, their joint distribution has the exponential form (1) with

$T = X + Y$, $U = Y$, $\theta = \log (p_2/q_2 \div p_1/q_1)$ and $\vartheta = \log (p_1/q_1)$. The method is therefore applicable to the problem of testing $p_2/q_2 \leqq a_0(p_1/q_1)$, and in particular $p_2 \leqq p_1$ by letting $a_0 = 1$, against the alternatives $p_2/q_2 \geqq a_1(p_1/q_1)$. Putting $\rho = (p_2/q_2) \div (p_1/q_1)$, the conditional distribution of Y given t is

$$(12) \qquad P_\rho\{Y = y \mid X + Y = t\} = C_t(\rho) \binom{m}{t-y} \binom{n}{y} \rho^y, \qquad y = 0, 1, \cdots, t,$$

which for $\rho = 1$ reduces to the hypergeometric distribution.

EXAMPLE 3. In a 2×2 table representing the results of classifying a sample of size s according to two characteristics A and B, the joint distribution of the numbers X, Y, Y' in the

	A	$\tilde{A}$	
B	X	X'	M
$\tilde{B}$	Y	Y'	N
	T	T'	S

categories AB, $A\tilde{B}$ and $\tilde{A}\tilde{B}$ constitute an exponential family with $U = Y$, $T_1 = X + Y$, $T_2 = Y + Y'$ and $\theta = \log (p_{A\tilde{B}}p_{\tilde{A}B}/p_{AB}p_{\tilde{A}\tilde{B}})$. Putting $\Delta = (p_{A\tilde{B}}p_{\tilde{A}B}/p_{AB}p_{\tilde{A}\tilde{B}})$ one finds

$$p_{AB} = p_A p_B + \frac{1-\Delta}{\Delta} p_{\tilde{A}B}p_{A\tilde{B}}; \qquad p_{\tilde{A}\tilde{B}} = p_{\tilde{A}}p_{\tilde{B}} + \frac{1-\Delta}{\Delta} p_{\tilde{A}B}p_{A\tilde{B}}$$

$$p_{A\tilde{B}} = p_A p_{\tilde{B}} - \frac{1-\Delta}{\Delta} p_{\tilde{A}B}p_{A\tilde{B}}; \qquad p_{\tilde{A}B} = p_{\tilde{A}}p_B - \frac{1-\Delta}{\Delta} p_{\tilde{A}B}p_{A\tilde{B}}$$

where p_{AB} denotes the probability of having the characteristics A and B, $p_A = p_{AB} + p_{A\tilde{B}}$ the probability of having the characteristic A, etc. The quantity Δ is therefore a measure of the degree of dependence,[5] $\Delta = 1$ corresponding to independence, $\Delta < 1$ to negative and $\Delta > 1$ to positive dependence. The method of the preceding section is applicable to testing $\Delta \leqq 1$ or more generally $\Delta \leqq \Delta_0$ against the alternatives $\Delta \geqq \Delta_1$. The conditional distribution of Y given $X + Y = t$, $Y + Y' = n$ is given by (12) with Δ in place of ρ.

EXAMPLE 4. Consider a number of paired comparisons (U_k, V_k) where only the sign of the differences $W_k = V_k - U_k$ are observed for each pair $k = 1, \cdots, n$. If the probability of a positive, negative and zero observation are p_+, p_- and p_0 in each case and if the comparisons are independent, the joint distribution of the numbers X, Y and Z of positive, negative and zero cases is the multinomial distribution

$$\frac{n!}{x!y!z!} p_+^x p_-^y p_0^z.$$

[5] Δ is equivalent to Yule's measure of association, which is $Q = (1 - \Delta)/(1 + \Delta)$. For a discussion of this and related measures, see [2].

This is an exponential family with $U = Y$, $T = Z$, $\theta = \log (p_+/p_-)$ and $\vartheta = \log (p_0/p_-)$. The test of $p_+ \leqq p_-$ (or $p_+ \leqq a_0p_-$) against $p_+ \geqq a_1p_-$ is therefore performed conditionally given $Z = t$. Since the conditional distribution of Y given $Z = t$ is the binomial distribution $b(p_+/(p_+ + p_-), n - t)$, the constants $C(t)$ and $\gamma(t)$ for which the test satisfies (2) and (11) can be obtained from the binomial tables.[6]

EXAMPLE 5. Let $Y_1, \cdots, Y_N$ be independently distributed according to the binomial distributions $b(p_i, n_i)$ $i = 1, \cdots, N$ where

$$p_i = 1/[1 + e^{-(\alpha+\beta x_i)}]$$

This is the model frequently assumed in bioassay, where x_i denotes the dose or some function of the dose such as its logarithm, of a drug given to n_i experimental subjects and where Y_i is the number among these subjects which respond to the drug at level x_i. Here the x_i are known, and α and β are unknown parameters. The joint distribution of the Y's is

$$e^{\alpha\Sigma y_i+\beta\Sigma x_i y_i} \prod_{i=1}^{N} \binom{n_i}{y_i} \left[\frac{e^{-(\alpha+\beta x_i)}}{1 + e^{-(\alpha+\beta x_i)}}\right]^{n_i}, \tag{13}$$

which is an exponential family with the parameters α, β and sufficient statistics $\sum Y_i$, $\sum x_i Y_i$. The method is therefore applicable to testing $\alpha \leqq \alpha_0$ against $\alpha \geqq \alpha_1$ or $\beta \leqq \beta_0$ against $\beta \geqq \beta_1$. It is interesting to note that for the particular case $x_i = ic$ and $H: \beta \leqq 0$, the conditional test given $Y = t$ is a form of the Wilcoxon test in a setting similar to that discussed by Haldane and Smith [3].

As a last example we mention without going into details the comparison of two distributions of type (13). If the parameters in these are α, β and α', β' the differences $\alpha' - \alpha$ and $\beta' - \beta$ are natural parameters of the resulting exponential families, and can therefore be tested by the method discussed here.

REFERENCES

[1] M. S. BARTLETT, "Properties of sufficiency and statistical tests," *Proc. Roy. Soc. London*, Vol. 160 (1937), pp. 268–281.

[2] LEO A. GOODMAN AND WILLIAM H. KRUSKAL, "Measures of association for mass classifications," *J. Amer. Stat. Assn.*, Vol. 49 (1954), pp. 732–764.

[3] J. B. S. HALDANE AND CEDRIC A. B. SMITH, "A simple exact test for birth order effect," *Ann. Eugenics*, Vol. 14 (1948), pp. 117–124.

[4] J. L. HODGES, JR. AND E. L. LEHMANN, "Testing the approximate validity of statistical hypotheses," *J. Roy. Stat. Soc., Ser. B.*, Vol. 16 (1954), pp. 261–268.

[5] H. JEFFREYS, "Theory of probability (2d ed.), Oxford: Clarendon Press, 1948.

[6] SAMUEL KARLIN AND HERMAN RUBIN, "The theory of decision procedures for distribu-

[6] This problem has been considered previously in [10]. The statement made there that the test satisfying (2) and $\alpha^*(t) \equiv \alpha$ is uniformly most powerful is too strong. The test is however uniformly most powerful among all similar (and hence all unbiased) tests. An analogous remark applies to Example 2, which is among those considered in [12]. As was proved there, the test which is conditionally unbiased at fixed level α for each t, is uniformly unbiased; it is however not uniformly most powerful without this restriction as is claimed in [12] for all the cases treated there.

tions with monotone likelihood ratio," *Ann. Math. Stat.*, Vol. 27 (1956), pp. 272–299.

[7] E. L. LEHMANN AND HENRY SCHEFFÉ, "Completeness, similar regions and unbiased estimation," *Sankhya*, Vol. 10 (1950), pp. 305–340 and Vol. 15 (1955), pp. 219–236.

[8] D. V. LINDLEY, "Statistical inference," *J. Roy. Stat. Soc., Ser. B.*, Vol. 15 (1953), pp. 30–76.

[9] J. NEYMAN, "Outline of a theory of statistical estimation based on the classical theory of probability," *Philos. Trans. Roy. Soc. London*, Vol. 236 (1937), pp. 333–380.

[10] JOSEPH PUTTER, "The treatment of ties in some nonparametric tests," *Ann. Math. Stat.*, Vol. 26 (1955), pp. 368–386.

[11] ERLING SVERDRUP, "Similarity, unbiasedness, minimaxibility, and admissibility of statistical test procedures," *Skand. Aktuarietids.*, Vol. 36 (1953), pp. 64–86.

[12] K. D. TOCHER, "Extension of Neyman-Pearson theory of tests to discontinuous variates," *Biometrika*, Vol. 37 (1950), pp. 130–144.

[13] DONALD R. TRUAX, "Multidecision problems for the multivariate exponential family," Stanford Technical Report No. 32 (1955) (to be published).

OPTIMUM INVARIANT TESTS[1]

By E. L. Lehmann

University of California, Berkeley

Summary. The standard (likelihood ratio) test of the general linear hypothesis has been shown to possess numerous different optimum properties. A brief survey of these was included in a recent paper by Kiefer [2]. In the present note it is shown that all of these, and in fact a wide class of optimum properties of which the above are special cases, are consequences of the fact that the test is uniformly most powerful invariant.

1. Order relations among tests. Let X be a random variable with possible distributions $\mathcal{P} = \{P_\theta, \theta \varepsilon \Omega\}$ and consider the hypothesis $H: \theta \varepsilon \omega$ where ω is a subset of Ω. Suppose that the problem of testing H against the alternatives $K: \theta \varepsilon \Omega - \omega$ remains invariant under a group G of transformations of the sample space. Let $\mathfrak{I}$ be a class of tests φ of H, for example the class of all level α tests or of all unbiased level α tests, which is invariant under G in the sense that $\varphi \in \mathfrak{I}$ implies $\varphi g \varepsilon \mathfrak{I}$ for all $g \varepsilon G$. Here φg denotes the critical function defined by

$$\varphi g(x) = \varphi(gx).$$

Suppose that a relation $\precsim$ has been defined among the tests of $\mathfrak{I}$ such that every pair $\varphi, \varphi' \varepsilon \mathfrak{I}$ satisfies either $\varphi \precsim \varphi'$ or $\varphi' \precsim \varphi$. When both of these relations hold, we write $\varphi \approx \varphi'$. Let the (weak) ordering $\precsim$ satisfy the following conditions:

(i) If φ' is uniformly at least as powerful as φ, then $\varphi \precsim \varphi'$.

(ii) If $\varphi_\gamma, \gamma \varepsilon \Gamma$ is any family of tests belonging to $\mathfrak{I}$ and ν any probability measure over the label space Γ, then $\varphi \precsim \varphi_\gamma$ for all $\gamma \varepsilon \Gamma$ implies $\varphi \precsim \int \varphi_\gamma \, d\nu(\gamma)$.

(iii) If $\varphi_0 \precsim \varphi_n$ for $n = 1, 2, \cdots$ and if φ is a critical function such that the power-functions $\beta_{\varphi_n}(\theta) \to \beta_\varphi(\theta)$ for all $\theta \varepsilon \Omega$ as $n \to \infty$, then $\varphi_0 \precsim \varphi$.

(iv) If $\varphi \precsim \varphi'$ then $\varphi g \precsim \varphi' g$ for all $g \varepsilon G$.

A test $\varphi_0 \varepsilon \mathfrak{I}$ will be called *optimum* within $\mathfrak{I}$ according to this ordering if $\varphi \precsim \varphi_0$ for all $\varphi \varepsilon \mathfrak{I}$.

The following are some examples of such orderings, which have been considered in the literature. Throughout, β_φ denotes the power function of φ.

Example 1. Let $a(\theta) \geqq 0$ and $b(\theta)$ be functions which are invariant under the transformations $\bar{G}$ induced by G in the parameter space, and let $\varphi \precsim \varphi'$ if

$$\inf_{\Omega-\omega} [a(\theta)\beta_\varphi(\theta) + b(\theta)] \leqq \inf_{\Omega-\omega} [a(\theta)\beta_{\varphi'}(\theta) + b(\theta)].$$

Then conditions (i) to (iv) are clearly satisfied. A particular case is obtained by putting $b(\theta) = 0$; $a(\theta) = 1$ if $\theta \varepsilon \omega'$ and $a(\theta) = 0$ otherwise, where ω' is

Received September 15, 1958.

[1] This paper was prepared with the partial support of the Office of Naval Research (Nonr-222-43). This paper in whole or in part may be reproduced for any purpose of the United States Government.

an invariant subset of $\Omega - \omega$. Then $\varphi \precsim \varphi'$ if

$$\inf_{\omega'} \beta_\varphi(\theta) \leqq \inf_{\omega'} \beta_{\varphi'}(\theta).$$

A test is optimum according to this ordering if it maximizes the minimum power over ω'.

Example 2. Let the tests be ordered according to $-s(\varphi)$ where $s(\varphi)$ is the stringency of φ defined by

$$s(\varphi) = \sup_{\Omega-\omega} [\beta^*(\theta) - \beta_\varphi(\theta)]$$

with β^* denoting the envelope power function. Then $\varphi \precsim \varphi'$ if $s(\varphi) \geqq s(\varphi')$ and the four conditions are again easily verified.

Example 3. Let ω' be an invariant subset of $\Omega - \omega$ and suppose that there exists a probability distribution λ over ω' which is invariant under the group $\bar{G}$ induced by G in the parameters face. Then the relation $\varphi \precsim \varphi'$ if

$$\int_{\omega'} \beta_\varphi(\theta) d\lambda(\theta) \leqq \int_{\omega'} \beta_{\varphi'}(\theta)\, d\lambda(\theta)$$

also satisfies conditions (i) to (iv).

Example 4. Suppose that $\theta = (\theta_1, \cdots, \theta_r)$ and that ω consists of the single point $\theta^0 = (\theta_1^0, \cdots, \theta_r^0)$. We shall assume that the power function $\beta_\varphi(\theta)$ of any test φ possesses continuous second derivatives $\partial^2\beta/\partial\theta_i\theta_j$ for all i and j at θ^0. Let $\mathfrak{J}$ be the class of all level α tests that are strictly unbiased in the neighborhood of θ^0 and let $\Delta(\varphi)$ denote the Gaussian curvature of the power surface at θ^0, which is given by the determinant of the positive definite matrix $(\partial^2\beta/\partial\theta_i\partial\theta_j)\,|_{\theta^0}$. The relation $\varphi \precsim \varphi'$ if $\Delta(\varphi) \leqq \Delta(\varphi')$ clearly satisfies (i) and (iii). It follows from a remark of Isaacson [1] that the relation is invariant provided the transformations $\bar{g}$ of the parameter space possess continuous second partial derivatives at θ^0, which (under this restriction) verifies (iv). Condition (ii), finally, is easily verified. Optimum tests according to the present ordering correspond to the type D tests of Isaacson.

2. Consequences of the Hunt-Stein theorem. Under the assumptions of the preceding section we shall now show that if G satisfies the conditions of the Hunt-Stein theorem (cf. [3], p. 336) and if there exists test ψ_0 which is optimum according to the ordering $\precsim$, then there exists an almost invariant test which is optimum. Here we require of $\mathfrak{J}$ that it be closed under convex combinations and under weak limits.

The proof is completely analogous to and essentially follows from that of the Hunt-Stein theorem, and can be indicated very briefly. If ν_n is the sequence of almost invariant probability measures over G postulated in the theorem, consider the sequence of tests

$$\psi_n = \int \psi_0 g\, d\nu_n(g)$$

Let ψ be the weak limit of a subsequence ψ_{n_i}. Then it is shown in the proof of the Hunt-Stein theorem that ψ is almost invariant, and it remains only to show that ψ is optimum. By conditions (iv) and (ii) it follows for any $\varphi \varepsilon \mathfrak{J}$ that $\varphi \precsim \psi_n$ for all n. Hence by condition (iii) also $\varphi \precsim \psi$ for all $\varphi \varepsilon \mathfrak{J}$ as was to be proved.

Under the above assumptions, whenever there exists a UMP almost invariant test, this will be optimum with respect to any ordering $\precsim$ satisfying conditions (i)–(iv). This explains the great variety of optimum properties possessed by certain tests and makes it unnecessary to prove each of them separately.

3. Applications. Consider a sequence of n independent trials and let $X_i = 1$ or 0 as the ith trial is or is not successful. Let $P(X_i = 1) = p_i$ and consider the hypothesis $H: p_1 = \cdots = p_n = \frac{1}{2}$ against the alternatives

$$p_i > \tfrac{1}{2} \ (i = 1, \cdots, n).$$

The problem is invariant under any permutation of the variables and the sign test, which rejects when $\sum X_i > C$, is uniformly most powerful almost invariant (cf. [3], p. 219). This test therefore maximizes the minimum power over the alternatives ω': $\min p_i \geqq \frac{1}{2} + \Delta$ or ω: $\max p_i \geqq \frac{1}{2} + \Delta$ for any $\Delta > 0$; it is most stringent and of type D.

As a second application, consider the general univariate linear hypothesis in the canonical form according to which the variables $X_1, \cdots, X_r; Y_1, \cdots, Y_s; Z_1, \cdots, Z_m$ are independently normally distributed with common variance σ^2 and means $E(X_i) = \xi_i$, $E(Y_j) = \eta_j$, $E(Z_k) = 0$. The hypothesis to be tested is $H: \xi_1 = \cdots = \xi_r = 0$. This problem remains invariant under the three groups

G_1: $Y_j' = Y_j + c_j(-\infty < c_j < \infty)$; $X_i' = X_i$; $Z_k' = Z_k$.

G_2: Orthogonal transformations of $X_1, \cdots, X_r$; $Y_j' = Y_j$; $Z_k' = Z_k$.

G_3: $X_i' = aX_i$; $Y_j' = aY_j$; $Z_k' = aZ_k$ $(a \neq 0)$.

The standard test has the following two basic optimum properties:

(a) It is uniformly most powerful among level α tests which are almost invariant with respect to G_1, G_2, G_3.

(b) It is uniformly most powerful among all unbiased (or similar) level α tests which are almost invariant with respect to G_2.

The first of these is well known; the second is easily shown by a standard argument.

Since the groups $G_1 - G_3$ satisfy the conditions of the Hunt-Stein theorem, it follows from (a), for example that the standard test is most stringent and that it maximizes the minimum power against the class of alternatives

$$\omega': \sum \xi_i^2/\sigma^2 \geqq \Delta.$$

To apply (b), consider fixed values of $\eta_1, \cdots, \eta_s$ and σ, so that the power becomes a function only of $\xi_1, \cdots, \xi_r$. It then follows that for any $\eta_1, \cdots, \eta_s$ and σ the standard test maximizes (among all unbiased level α tests), for example the minimum power over the sets $\omega'(\eta_1, \cdots, \eta_s, \sigma): \sum \xi_i^2 \geq \Delta$ and the average power over the spheres $\sum \xi_i^2 = \Delta$. This was first proved by Wald [4]. It follows further that the test maximizes the Gaussian curvature of the power surface, considered for fixed $\eta_1, \cdots, \eta_s, \sigma$ as a function of the ξ's, and hence is of Isaacson's type E. This has been shown previously by Kiefer [2], who deduced it as a consequence of the test maximizing the average power over the spheres $\sum \xi_i^2 = \Delta$.

REFERENCES

[1] Stanley L. Isaacson, "On the theory of unbiased tests of simple statistical hypotheses specifying the values of two or more parameters," *Ann. Math. Stat.*, Vol. 22 (1951), pp. 217–234.

[2] Jack Kiefer, "On the nonrandomized optimality and randomized nonoptimality of symmetrical designs," *Ann. Math. Stat.*, Vol. 29 (1958), pp. 675–699.

[3] E. L. Lehmann, *Testing Statistical Hypotheses*, New York, John Wiley and Sons, 1959.

[4] Abraham Wald, "On the power function of the analysis of variance test," *Ann. Math. Stat.*, Vol. 13 (1942), pp. 434–439.

Since the distribution of $\Sigma(X_i - \bar{X})^2/\sigma^2$ does not depend on ξ or σ, the probability $P\{\Sigma(X_i - \bar{X})^2 \geqq C|\xi, \sigma\}$ is independent of ξ and increases with σ, so that the conditions of Corollary 5 are satisfied. The test (32), being independent of ξ_1 and σ_1, is UMP for testing $\sigma \leqq \sigma_0$ against $\sigma > \sigma_0$. It is also seen to coincide with the likelihood ratio test (29). On the other hand, the most powerful test (31) for testing $\sigma \geqq \sigma_0$ against $\sigma < \sigma_0$ does depend on the value ξ_1 of ξ under the alternative.

It was tacitly assumed so far that $n > 1$. If $n = 1$, the argument applies without change with respect to H_1, leading to (31) with $n = 1$. However, in the discussion of H_2 the statistic U now drops out, and Y coincides with the single observation X. Using the same λ as before one sees that X has the same distribution under H_λ as under K, and the test ϕ_λ therefore becomes $\phi_\lambda(x) \equiv \alpha$. This satisfies the conditions of Corollary 5 and is therefore the most powerful test for the given problem. It follows that a single observation is of no value for testing the hypothesis H_2 as seems intuitively obvious, but that it could be used to test H_1 if the class of alternatives were sufficiently restricted.

The corresponding derivation for the hypothesis $\xi \leqq \xi_0$ is less straightforward. It turns out* that Student's test given by (30) is most powerful if the level of significance α is $\geqq 1/2$, regardless of the alternative $\xi_1 > \xi_0, \sigma_1$. This test is therefore UMP for $\alpha \geqq 1/2$. On the other hand, when $\alpha < 1/2$ the most powerful test of H rejects when $\Sigma(x_i - a)^2 \leqq b$, where the constants a and b depend on the alternative (ξ_1, σ_1) and on α. Thus for the significance levels that are of interest, a UMP test of H does not exist. No new problem arises for the hypothesis $\xi \geqq \xi_0$ since this reduces to the case just considered through the transformation $Y_i = \xi_0 - (X_i - \xi_0)$.

10. SEQUENTIAL PROBABILITY RATIO TESTS

According to the Neyman-Pearson fundamental lemma, the best procedure for testing the simple hypothesis H that the probability density of X is p_0 against the simple alternative that it is p_1 accepts or rejects H as

$$\frac{p_{1n}}{p_{0n}} = \frac{p_1(x_1) \cdots p_1(x_n)}{p_0(x_1) \cdots p_0(x_n)}$$

is less or greater than a suitable constant C. However, further improvement is possible if the sample size is not fixed in advance but is permitted to depend on the observations. The best procedure, in a certain sense, is then the following *sequential probability ratio test*. Let $A_0 < A_1$ be

* See Lehmann and Stein, "Most powerful tests of composite hypotheses. I. Normal distributions," *Ann. Math. Stat.*, Vol. 19 (1948), pp. 495–516.

J. Rojo (ed.), *Selected Works of E. L. Lehmann*, Selected Works in Probability and Statistics, DOI 10.1007/978-1-4614-1412-4_18, © Springer Science+Business Media, LLC 2012

two given constants and suppose that observation is continued as long as the probability ratio p_{1n}/p_{0n} satisfies the inequality

$$A_0 < \frac{p_{1n}}{p_{0n}} < A_1. \tag{33}$$

The hypothesis H is accepted or rejected at the first violation of (33) as $p_{1n}/p_{0n} \leqq A_0$ or $\geqq A_1$.

The usual measures of the performance of such a procedure are the probabilities, say α_0 and α_1, of rejecting H when $p = p_0$ and of accepting it when $p = p_1$ and the expected number of observations $E_i(N)$ when $p = p_i$ $(i = 0, 1)$.

Theorem 8. *Among all tests (sequential or not) for which*

$$P_0 \text{ (rejecting } H) \leqq \alpha_0, \qquad P_1 \text{ (accepting } H) \leqq \alpha_1$$

and for which $E_0(N)$ and $E_1(N)$ are finite, the sequential probability ratio test with error probabilities α_0 and α_1 minimizes both $E_0(N)$ and $E_1(N)$.

In particular, the sequential probability ratio test therefore requires on the average fewer observations than the fixed sample size test which controls the errors at the same levels. The proof of this result will be deferred to Section 12. In this and the following sections some of the basic properties of sequential probability ratio tests will be sketched.

Because of the difficulty of determining exactly the boundaries A_0 and A_1 for which α_0 and α_1 take on preassigned values, the following inequalities are useful. Let R_n be the part of n-space defined by the inequalities

$$A_0 < \frac{p_{1k}}{p_{0k}} < A_1 \quad \text{for} \quad k = 1, \cdots, n-1 \qquad \text{and} \qquad A_1 \leqq \frac{p_{1n}}{p_{0n}}.$$

This is the set of points $(x_1, \cdots, x_n)$ for which the procedure stops with $N = n$ observations and rejects H. Then

$$\alpha_0 = \sum_{n=1}^{\infty} \int_{R_n} p_{0n} \leqq \frac{1}{A_1} \sum_{n=1}^{\infty} \int_{R_n} p_{1n} = \frac{1-\alpha_1}{A_1}.$$

Similarly, if S_n denotes the part of n-space in which $N = n$ and H is accepted, one has

$$1 - \alpha_0 = \sum_{n=1}^{\infty} \int_{S_n} p_{0n} \geqq \frac{\alpha_1}{A_0}.$$

Here it has been tacitly assumed that

$$\sum_{n=1}^{\infty} P_i\{N = n\} = \sum_{n=1}^{\infty} \int_{R_n \cup S_n} p_{in} = 1 \quad \text{for} \quad i = 0, 1,$$

that is, that the probability is 0 of the procedure continuing indefinitely. For a proof of this fact see Problems 34 and 35. The inequalities

(34) $$A_0 \geqq \frac{\alpha_1}{1-\alpha_0}, \qquad A_1 \leqq \frac{1-\alpha_1}{\alpha_0}$$

suggest the possibility of approximating the boundaries A_0 and A_1 that would yield the desired α_0 and α_1 by

$$A_0' = \frac{\alpha_1}{1-\alpha_0}, \qquad A_1' = \frac{1-\alpha_1}{\alpha_0}.$$

By (34) the error probabilities of the approximate procedure then satisfy

$$\frac{\alpha_1'}{1-\alpha_0'} \leqq A_0' = \frac{\alpha_1}{1-\alpha_0} \quad \text{and} \quad \frac{1-\alpha_1'}{\alpha_0'} \geqq A_1' = \frac{1-\alpha_1}{\alpha_0}$$

and hence

$$\alpha_0' \leqq \frac{\alpha_0}{1-\alpha_1} \quad \text{and} \quad \alpha_1' \leqq \frac{\alpha_1}{1-\alpha_0}.$$

If typically α_0 and α_1 are of the order .01 to .1, the amount by which α_i' can exceed α_i $(i = 1, 0)$ is negligible so that the probabilities of the two kinds of error are very nearly bounded above by the specified α_0 and α_1. This conclusion is strengthened by the fact that $\alpha_0' + \alpha_1' \leqq \alpha_0 + \alpha_1$, as is seen by adding the inequalities $\alpha_1'(1-\alpha_0) \leqq \alpha_1(1-\alpha_0')$ and $\alpha_0'(1-\alpha_1') \leqq \alpha_0(1-\alpha_1')$.

The only serious risk in using the approximate boundaries A_0', A_1' is therefore that α_0' and α_1' are much smaller than required, which would lead to an excessive number of observations. There is some reason to hope that this effect is also moderate. For let

(35) $$z_i = \log\,[p_1(x_i)/p_0(x_i)].$$

Then (33) becomes

$$\log A_0 < \sum_{i=1}^{n} z_i < \log A_1,$$

and when H is rejected the z's satisfy

$$z_1 + \cdots + z_{n-1} < \log A_1 \leqq z_1 + \cdots + z_n.$$

The approximation consists in replacing $z_1 + \cdots + z_n$ by $\log A_1$. The error will usually be moderate since after $n - 1$ observations Σz_i is still $< A_1$ and the excess has therefore had no possibility to accumulate, but is due to a single observation. An analogous argument applies to the other boundary.

Example 9. Consider a sequence of binomial trials with constant probability p of success, and the problem of testing $p = p_0$ against $p = p_1 (p_0 < p_1)$. Then

$$\frac{p_{1n}}{p_{0n}} = \frac{p_1^{\Sigma x_i}(1 - p_1)^{n - \Sigma x_i}}{p_0^{\Sigma x_i}(1 - p_0)^{n - \Sigma x_i}} = \left(\frac{p_1 q_0}{p_0 q_1}\right)^{\Sigma x_i}\left(\frac{q_1}{q_0}\right)^n.$$

In the case that $\log (p_1 p_0^{-1})/\log (q_0 q_1^{-1})$ is rational, exact formulas have been obtained† for the error probabilities and expected sample size which make it possible to compute the effects involved in the approximation of A_0, A_1 by A_0', A_1'. As an illustration,‡ suppose that $p_0 = .05$, $p_1 = .17$, $\alpha_0 = .05$, $\alpha_1 = .10$. It then turns out that $\alpha_0' = .031$, $\alpha_1' = .099$, and that the expectations of the sample size for the approximate procedure are $E_0'(N) = 31.4$, $E_1'(N) = 30.0$. There is an alternate plan, determined by trial and error, with $\alpha_0^* = .046$, $\alpha_1^* = .097$, $E_0^*(N) = 30.5$, $E_1^*(N) = 26.1$. On the other hand, the fixed sample size procedure with error probabilities .05 and .10 requires 57 observations.

In order to be specific, we assumed in the definition of a sequential probability ratio test that observation continues only as long as the probability ratio is strictly between A_0 and A_1. The discussion applies equally well to the rule of continuing as long as $A_0 < p_{1n}/p_{0n} < A_1$, coming to the indicated conclusion the first time that $p_{1n}/p_{0n} < A_0$ or $> A_1$, and deciding on the boundaries according to any fixed probabilities. The term *sequential probability ratio test* is applied also to this more general procedure. If the probability ratio $p_1(X)/p_0(X)$ has a continuous distribution, all these procedures are equivalent. However, in case of discrete probability ratios the possibility of randomization on the boundary is necessary to achieve preassigned error probabilities. If randomization is permitted also between taking at least one observation or reaching a decision without taking any observations, it can be shown that actually any preassigned error probabilities can be achieved.§

11. POWER AND EXPECTED SAMPLE SIZE OF SEQUENTIAL PROBABILITY RATIO TESTS

The preceding section is somewhat misleading in that it discusses the problem in a setting, that of testing a simple hypothesis against a simple alternative, which is interesting mainly because of its implications for the more realistic situation of a continuous parameter family of distributions.

† Girshick, "Contributions to the theory of sequential analysis, II, III," *Ann. Math. Stat.*, Vol. 17 (1946), pp. 282–298, and Polya, "Exact formulas in the sequential analysis of attributes," *Univ. Calif. Publs. Mathematics*, New Series, Vol. 1 (1948), pp. 229–240.

‡ Taken from Robinson, "A note on exact sequential analysis," *Univ. Calif. Publs. Mathematics*, New Series, Vol. 1 (1948), pp. 241–246.

§ This result is contained in an as yet unpublished paper by Stein, "Existence of sequential probability ratio tests." See also the abstract by Wijsman, "On the existence of Wald's sequential test," *Ann. Math. Stat.*, Vol. 29 (1958), pp. 938–939.

Unfortunately, the property of being uniformly most powerful, which the fixed sample size probability ratio test possesses for families with monotone likelihood ratio (Theorem 2), does not extend to the sequential case. More specifically, consider the sequential probability ratio test for testing $H: \theta_0$ against $K: \theta_1$, and let its power function be $\beta(\theta) = P_\theta$ (rejecting H). Then if θ_2 is some other alternative, the sequential probability ratio test for testing θ_0 against θ_2 with error probabilities α_0 and α_1 does not in general coincide with the original test, which therefore does not minimize $E_{\theta_2}(N)$. It seems in fact likely that from an over-all point of view the sequential probability ratio test is not the best sequential procedure in the continuous parameter case, although it is usually better than the best competitive test with fixed sample size.

When the probability density depends on a real parameter θ and one is testing the hypothesis $\theta \leqq \theta_0$, one is usually not concerned with the power of the test against alternatives θ close to θ_0, but would like to be able to control the probability of detecting alternatives sufficiently far away. The test should therefore satisfy

$$\begin{aligned} &\beta(\theta) \leqq \alpha \quad \text{for} \quad \theta \leqq \theta_0 \\ &\beta(\theta) \geqq \beta \quad \text{for} \quad \theta \geqq \theta_1 \end{aligned} \qquad (\theta_0 < \theta_1), \tag{36}$$

which it will do in particular if

$$\beta(\theta_0) = \alpha, \qquad \beta(\theta_1) = \beta,$$

and if $\beta(\theta)$ is a nondecreasing function of θ. The sequential probability ratio test for testing θ_0 against θ_1 with error probabilities $\alpha_0 = \alpha, \alpha_1 = 1 - \beta$ thus is a solution of the stated problem provided its power function is nondecreasing.

Lemma 4. *Let $X_1, X_2, \cdots$ be independently distributed with probability density $p_\theta(x)$, and suppose that the densities $p_\theta(x)$ have monotone likelihood ratio in $T(x)$. Then any sequential probability ratio test for testing θ_0 against θ_1 $(\theta_0 < \theta_1)$ has a nondecreasing power function.*

Proof. Let $Z_i = \log [p_{\theta_1}(X_i)/p_{\theta_0}(X_i)] = h(T_i)$, where h is nondecreasing, and let $\theta < \theta'$. By Lemma 2, the cumulative distribution function $F_\theta(t)$ of T_i satisfies $F_{\theta'}(t) \leqq F_\theta(t)$ for all t, and by Lemma 1 there exists therefore a random variable V_i and functions f and f' such that $f(v) \leqq f'(v)$ for all v and that the distributions of $f(V_i)$ and $f'(V_i)$ are F_θ and $F_{\theta'}$ respectively. The sequential test under consideration has the following graphical representation in the $(n, \sum_{i=1}^n h(t_i))$ plane. Observation is continued as long as the sample points fall inside the band formed by the parallel straight lines

$$\sum_{i=1}^{n} h(t_i) = \log A_j, \qquad j = 0, 1.$$

The hypothesis is rejected if the path formed by the points $(1, h(t_1))$, $(2, h(t_1) + h(t_2)), \cdots, (N, h(t_1) + \cdots + h(t_N))$ leaves the band through the upper boundary. The probability of this event is therefore the probability of rejection, for θ when each T_i is replaced by $f(V_i)$ and for θ' when T_i is replaced by $f'(V_i)$. Since $f(V_i) \leqq f'(V_i)$ for all i, the path generated by the $f'(V_i)$ leaves the band through the upper boundary whenever this is true for the path generated by the $f(V_i)$. Hence $\beta(\theta) \leqq \beta(\theta')$, as was to be proved.

In the case of monotone likelihood ratios, the sequential probability ratio test with error probabilities $\alpha_0 = \alpha$, $\alpha_1 = 1 - \beta$ therefore satisfies (36). It follows from the optimum property stated in Section 10 that among all tests satisfying (36) the sequential probability ratio test minimizes the expected sample size for $\theta = \theta_0$ and $\theta = \theta_1$. However, one is now concerned with $E_\theta(N)$ for all values of θ. Typically, the function $E_\theta(N)$ has a maximum at a point between θ_0 and θ_1, and decreases as θ moves away from this point in either direction. It frequently turns out that the maximum is $< n_0$, the smallest fixed sample size for which there exists a test satisfying (36). On the other hand, this is not always the case. Thus, in Example 9 for $p_0 = .4$, $p_1 = .6$, $\alpha_0 = \alpha_1 = .005$ for example, the fixed sample size n_0 is 160, and $E_p(N)$, while below this for most values of p, equals 170 for $p = 1/2$. The important problem of determining the test that minimizes $\sup E_\theta(N)$ subject to (36) is still unsolved.

An exact evaluation of the power function $\beta(\theta)$ and the expected sample size $E_\theta(N)$ of a sequential probability ratio test is in general extremely difficult. However, a simple approximation is available provided the equation

$$E_\theta \{[p_{\theta_1}(X)/p_{\theta_0}(X)]^h\} = 1 \tag{37}$$

has a nonzero solution $h = h(\theta)$, as is the case under mild assumptions. (See Problem 38.) Then

$$p_\theta^*(x) = \left[\frac{p_{\theta_1}(x)}{p_{\theta_0}(x)}\right]^h p_\theta(x)$$

is again a probability density. Suppose now that $h > 0$—the other case can be treated similarly—and consider the sequential probability ratio test with boundaries A_0^h, A_1^h for testing p_θ against p_θ^*. With this procedure observation is continued as long as

$$A_0^h < \frac{p_\theta^*(x_1)}{p_\theta(x_1)} \cdots \frac{p_\theta^*(x_n)}{p_\theta(x_n)} < A_1^h .$$

If α_0^* and $1 - \alpha_1^*$ denote the probability of rejection when p_θ and p_θ^* are the true densities, it is seen from (34) that the boundaries are given approximately by

$$A_0^h \sim \frac{\alpha_1^*}{1 - \alpha_0^*}, \qquad A_1^h \sim \frac{1 - \alpha_1^*}{\alpha_0^*} .$$

However, the test under consideration is exactly the same as the sequential probability ratio test with error probabilities $\alpha_0 = \alpha$, $\alpha_1 = 1 - \beta$ for testing θ_0 against θ_1. Hence α_0^* and $\beta(\theta)$, the probability of rejection for the two tests when p_θ is the true density, must be equal. Solving for α_0^* from the above two approximate equations one therefore finds

$$\beta(\theta) \sim \frac{1 - A_0^h}{A_1^h - A_0^h} . \tag{38}$$

An approximation for $E_\theta(N)$ can be based on *Wald's equation*

$$E_\theta(Z_1 + \cdots + Z_N) = E_\theta(N) E_\theta(Z), \tag{39}$$

which is valid whenever the Z's are identically and independently distributed and the procedure is such that the expected sample size $E_\theta(N)$ is finite. For a proof of this equation see Problem 37. If the Z's are defined by (35) and the procedure is a sequential probability ratio test, $Z_1 + \cdots + Z_N$ can be approximated as before by $\log A_1$ and $\log A_0$ when H is rejected and accepted respectively, so that from (39) one obtains

$$E_\theta(N) \sim \frac{\beta(\theta) \log A_1 + [1 - \beta(\theta)] \log A_0}{E_\theta(Z)} \tag{40}$$

provided $E_\theta(Z) \neq 0$.

Example 10. In the binomial problem of Example 9, equation (37) becomes

$$p\left(\frac{p_1}{p_0}\right)^h + q\left(\frac{q_1}{q_0}\right)^h = 1. \tag{41}$$

Since the left-hand side is a convex function of h which is 1 for $h = 0$, it is seen that the equation has a unique nonzero solution except when $p = \log(q_0/q_1)/\log(p_1q_0/p_0q_1)$, in which case the left-hand side has its minimum at $h = 0$. Equations (38) and (41) provide a parametric representation of the approximate power function, which can now be computed by giving different values to h and obtaining the associated values p and β from (38) and (41). (For $h = 0$,

β can be obtained by continuity.) The following is a comparison of the approximate with the exact values of $\beta(p)$ and $E_p(N)$ in the numerical case considered in Example 9, with $p_0 = .05$, $p = .099$, $p_1 = .17$:*

$\beta(p_0)$	$\beta(p)$	$\beta(p_1)$	$E_{p_0}(N)$	$E_p(N)$	$E_{p_1}(N)$	
.05	.44	.90	30	39	25	Approx.
.031	.409	.901	31.4	46.8	30.0	Exact

12. OPTIMUM PROPERTY OF SEQUENTIAL PROBABILITY RATIO TESTS†

The main part of the proof of Theorem 8 is contained in the solution of the following auxiliary problem. For testing the hypothesis H that p_0 is the true probability density of X against the alternative that it is p_1, let the losses resulting from false rejection and acceptance of H be w_0 and w_1, and let the cost of each observation be c. The risk (expected loss plus expected cost) of a sequential procedure is then

$$\alpha_i w_i + cE_i(N)$$

when p_i is the true density, where

$$\alpha_0 = P_0 \text{ (rejecting } H), \qquad \alpha_1 = P_1 \text{ (accepting } H)$$

are the two probabilities of error. If one supposes that the subscript i of the probability density is itself a random variable, which takes on the values 0 and 1 with probability π and $1 - \pi$ respectively, the total average risk of a procedure δ is

$$r(\pi, \delta) = \pi[\alpha_0 w_0 + cE_0(N)] + (1 - \pi)[\alpha_1 w_1 + cE_1(N)]. \tag{42}$$

We shall now determine the Bayes procedure for this problem, that is, the procedure that minimizes (42). Here the interpretation of (42) as a Bayes risk is helpful for an understanding of the proof and gives the auxiliary problem independent interest. However, from the point of view of Theorem 8, the introduction of the w's, c, and π is only a mathematical device, and the problem is simply that of minimizing the formal expression (42).

The Bayes solutions involve two numbers $\pi' \leq \pi''$ which are uniquely determined by w_0, w_1, and c through equations (44) and (45) below, and which are independent of π. It will be sufficient to restrict attention to the case that $0 < \pi' < \pi'' < 1$ and to a priori probabilities π satisfying $\pi' \leq \pi \leq \pi''$.

* Taken from Robinson, *loc. cit.*, where a number of further examples are given.

† This section treats a special topic to which no reference is made in the remainder of the book.

Lemma 5. *Let π', π'' satisfy the equations* (44). *If $0 < \pi' < \pi'' < 1$, then for all $\pi' \leqq \pi \leqq \pi''$ the Bayes risk* (42) *is minimized by any sequential probability ratio test with boundaries*

$$A_0 = \frac{\pi}{1-\pi} \cdot \frac{1-\pi''}{\pi''}, \qquad A_1 = \frac{\pi}{1-\pi} \cdot \frac{1-\pi'}{\pi'}. \tag{43}$$

Proof. (1) We begin by investigating whether at least one observation should be taken, in which case the resulting risk will be at least c, or whether it is better to come to a decision immediately. Let δ_0 denote the procedure that rejects H without taking any observations, and δ_1 the corresponding procedure that accepts H, so that

$$r(\pi, \delta_0) = \pi w_0 \qquad \text{and} \qquad r(\pi, \delta_1) = (1-\pi)w_1.$$

Let

$$\rho(\pi) = \inf_{\delta \in \mathscr{C}} r(\pi, \delta)$$

where $\mathscr{C}$ is the class of all procedures requiring at least one observation. Then for any $0 < \lambda < 1$ and any π_0, π_1,

$$\begin{aligned}\rho[\lambda\pi_0 + (1-\lambda)\pi_1] &= \inf_{\delta \in \mathscr{C}} [\lambda r(\pi_0, \delta) + (1-\lambda) r(\pi_1, \delta)] \\ &\geqq \lambda\rho(\pi_0) + (1-\lambda)\rho(\pi_1).\end{aligned}$$

Hence ρ is concave, and since it is bounded below by zero it is continuous in the interval (0, 1).* If

$$\rho\left(\frac{w_1}{w_0 + w_1}\right) < \frac{w_0 w_1}{w_0 + w_1},$$

define π' and π'' by

$$r(\pi', \delta_0) = \rho(\pi') \qquad \text{and} \qquad r(\pi'', \delta_1) = \rho(\pi''). \tag{44}$$

(See Figure 3.) Otherwise let

$$\pi' = \pi'' = \frac{w_1}{w_0 + w_1}. \tag{45}$$

In the case $0 < \pi' < \pi'' < 1$ with which we are concerned, δ_0 minimizes (42) if and only if $\pi \leqq \pi'$, and δ_1 minimizes (42) if and only if $\pi \geqq \pi''$. This establishes the following uniquely as an optimum first step for $\pi \neq \pi', \pi''$: If $\pi < \pi'$ or $> \pi''$, no observation is taken and H is rejected or accepted respectively; if $\pi' < \pi < \pi''$ the variable X_1 is observed.

(2) The proof is now completed by induction. Suppose that $\pi' < \pi < \pi''$ and that n observations have been taken with outcomes $X_1 = x_1, \cdots, X_n = x_n$, and that one is faced with the alternatives of not taking another

* See, for example, section 3.18 of Hardy, Littlewood, Pólya, *Inequalities*, Cambridge Univ. Press, 1934.

observation and rejecting or accepting H with losses w_0, w_1 for possible wrong decisions, or of going on to observe X_{n+1}. The situation is very similar to the one analyzed in part (1). An unlimited supply of observations $X_{n+1}, X_{n+2}, \cdots$ is available. The fact that one has already incurred the expense of nc units does not affect the problem, since once this loss has been sustained no future action can retrieve it. The procedure is therefore as before: No further observation is taken if the probability of H

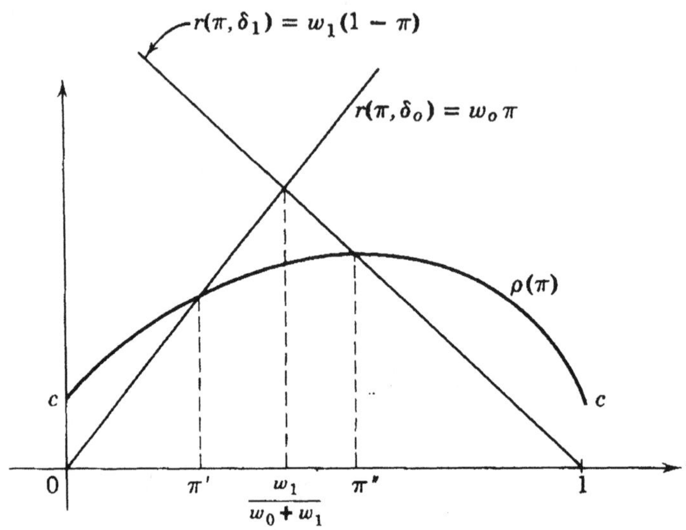

Figure 3.

being true is $< \pi'$ or $> \pi''$, whereas X_{n+1} is observed if this probability is strictly between π' and π''.

One aspect of the situation has changed as a result of observing $x_1, \cdots, x_n$. The probability of H being true is no longer π but has become

$$\pi(x_1, \cdots, x_n) = \frac{\pi p_{0n}}{\pi p_{0n} + (1 - \pi)p_{1n}},$$

the conditional (a posteriori) probability of H given $X_1 = x_1, \cdots, X_n = x_n$. A complete procedure therefore consists in continuing as long as

$$\pi' < \pi(x_1, \cdots, x_n) < \pi''$$

or equivalently as long as

$$A_0 = \frac{\pi}{1 - \pi} \cdot \frac{1 - \pi''}{\pi''} < \frac{p_{1n}}{p_{0n}} < \frac{\pi}{1 - \pi} \cdot \frac{1 - \pi'}{\pi'} = A_1.$$

H is accepted if, at the first violation of these inequalities, p_{1n}/p_{0n} is $< A_0$ and rejected if it is $> A_1$.

(3) In part (1) of this proof the first step of the procedure was uniquely determined as δ_0 for $\pi < \pi'$, as δ_1 for $\pi > \pi''$, and as taking at least one observation when $\pi' < \pi < \pi''$. For $\pi = \pi'$, the procedure δ_0 still minimizes (42) but it is no longer unique, that is, there also exists a procedure $\delta \in \mathscr{C}$ for which $r(\pi', \delta) = \rho(\pi')$. In order to belong to $\mathscr{C}$, such a procedure must require at least one observation. Once X_1 has been observed, it follows from part (2) that the best procedure in $\mathscr{C}$ is obtained by continuing observation as long as $\pi' < \pi(x_1, \cdots x_n) < \pi''$. At the first step it is therefore immaterial whether on the boundary experimentation is continued or the indicated decision is taken. The same is then true at the subsequent steps. This establishes in particular that for $\pi' \leqq \pi \leqq \pi''$ the procedure of taking a first observation and then following the sequential probability ratio test with boundaries (43) is Bayes.

The required connection between the auxiliary problem and the original one is established by the following lemma.

Lemma 6. *Given any* $0 < \pi_0' < \pi_0'' < 1$, *there exist numbers* $0 < w < 1$, $0 < c$ *such that the Bayes solution of the auxiliary problem defined by* $w_0 = 1 - w$, $w_1 = w$, c, *and an a priori probability* π *satisfying* $\pi_0' < \pi < \pi_0''$ *is a sequential probability ratio test with boundaries*

$$A_0 = \frac{\pi}{1-\pi} \cdot \frac{1-\pi_0''}{\pi_0''}, \qquad A_1 = \frac{\pi}{1-\pi} \cdot \frac{1-\pi_0'}{\pi_0'}.$$

*Proof.** (1) By Lemma 5, the quantities π' and π'' are functions of w and c, and it is therefore sufficient to find w and c such that $\pi'(w, c) = \pi_0'$, $\pi''(w, c) = \pi_0''$. For fixed w, let $\pi'(c) = \pi'(w, c)$ and $\pi''(c) = \pi''(w, c)$. If c_0 is the smallest value of c such that $\pi'(c_0) = \pi''(c_0)$, then for $0 < c < c_0$ the quantities $\pi'(c)$ and $\pi''(c)$ are determined by the equations

$$(1 - w)\pi' = \rho(\pi', c), \qquad (1 - \pi'')w = \rho(\pi'', c),$$

where $\rho(\pi, c)$ stands for the quantity previously denoted by $\rho(\pi)$. The function $\rho(\pi', c)$ considered as a function of c for fixed π' has the following properties. (i) It is continuous. This follows as before from its being concave. (ii) It is strictly increasing, since for any $\delta \in \mathscr{C}$ the risk $r(\delta, \pi')$ increases strictly with c and since the minimum risk $\rho(\pi', c)$ is taken on by a procedure $\delta \in \mathscr{C}$. (iii) As c tends to zero, so do $\rho(\pi', c)$ and $\rho(\pi'', c)$. This follows from the fact that for n sufficiently large there exists a test of fixed sample size n for which the two error probabilities are arbitrarily small.

* This proof was communicated to me by L. LeCam.

These properties of the function ρ imply that for $0 < c < c_0$ the functions π' and π'' are also continuous, strictly increasing and decreasing respectively, and that $\pi'(c) \to 0$, $\pi''(c) \to 1$ as $c \to 0$. On the other hand, as $c \to c_0$, $\pi''(c) - \pi'(c) \to 0$ so that both quantities tend to the solution $\pi' = \pi'' = w$ of the equation $\pi'(1 - w) = (1 - \pi')w$. It follows from these properties that for fixed w

$$\lambda(c) = \frac{\pi'(c)}{1 - \pi'(c)} \cdot \frac{1 - \pi''(c)}{\pi''(c)}$$

is a continuous, strictly increasing function of c, which increases from 0 to 1 as c varies from 0 to $c_0 = c_0(w)$.

(2) Let

$$\lambda(w, c) = \frac{\pi'(w, c)}{1 - \pi'(w, c)} \cdot \frac{1 - \pi''(w, c)}{\pi''(w, c)}, \qquad \gamma(w, c) = \frac{\pi''(w, c)}{1 - \pi''(w, c)} .$$

Instead of working with the variables π' and π'', it is equivalent and more convenient to work with λ and γ, and to prove the existence of w, c such that

$$\lambda(w, c) = \frac{\pi_0'}{1 - \pi_0'} \cdot \frac{1 - \pi_0''}{\pi_0''} = \lambda_0, \qquad \gamma(w, c) = \frac{\pi_0''}{1 - \pi_0''} = \gamma_0.$$

For any w, there exists by part (1) a unique cost $c = c(w)$ such that $\lambda(w, c) = \lambda_0$. It will be shown below that $\gamma(w) = \gamma[w, c(w)]$ is a 1:1 mapping of the interval $0 < w < 1$ onto $0 < \gamma < \infty$, and hence that there exists a unique value w such that $\gamma(w) = \gamma_0$. This will complete the proof of the lemma.

(3) For the auxiliary problem defined by w, $c = c(w)$, and $\pi = \pi'[w, c(w)]$ there exists by Lemma 5 a Bayes solution δ' which is a sequential probability ratio test with boundaries

$$A_0' = \frac{\pi'[w, c(w)]}{1 - \pi'[w, c(w)]} \cdot \frac{1 - \pi''[w, c(w)]}{\pi''[w, c(w)]} = \lambda[w, c(w)] = \lambda_0, \qquad A_1' = 1.$$

Let δ'' be the corresponding solution of the problem defined by w, $c = c(w)$, and $\pi = \pi''[w, c(w)]$, so that its boundaries are

$$A_0'' = 1, \qquad A_1'' = \frac{\pi''[w, c(w)]}{1 - \pi''[w, c(w)]} \cdot \frac{1 - \pi'[w, c(w)]}{\pi'[w, c(w)]} = \frac{1}{\lambda_0} .$$

Then the error probabilities and the expectations of the sample size α_0', α_1', $E_0'(N)$, $E_1'(N)$ of δ' and α_0'', α_1'', $E_0''(N)$, $E_1''(N)$ of δ'' depend on w and c only through λ_0 and not through γ, so that for fixed λ_0 they are fixed numbers. The Bayes risks for $\pi = \pi'[w, c(w)]$ and $\pi = \pi''[w, c(w)]$ are given by

$$\rho(\pi') = r(\pi', \delta') \qquad \text{and} \qquad \rho(\pi'') = r(\pi'', \delta'')$$

and it follows from (44) that

$$r(\pi', \delta_0) = r(\pi', \delta') \qquad \text{and} \qquad r(\pi'', \delta_1) = r(\pi'', \delta'').$$

These equations can be written more explicitly as

$$\pi'(1 - w) = \pi'[\alpha_0'(1 - w) + cE_0'(N)] + (1 - \pi')[\alpha_1' w + cE_1'(N)]$$

and

$$(1 - \pi'')w = \pi''[\alpha_0''(1 - w) + cE_0''(N)] + (1 - \pi'')[\alpha_1'' w + cE_1''(N)].$$

If one substitutes $\lambda_0\gamma$ for $\pi'/(1 - \pi')$ and γ for $\pi''/(1 - \pi'')$ and eliminates c, this reduces to a single equation connecting γ and w:

$$\{\lambda_0\gamma(1 - \alpha_0') - w[\lambda_0\gamma(1 - \alpha_0') + \alpha_1']\}\{\gamma E_0''(N) + E_1''(N)\}$$
$$= \{-\gamma\alpha_0'' + w[(1 - \alpha_1'') + \gamma\alpha_0'']\}\{\lambda_0\gamma E_0'(N) + E_1'(N)\}.$$

This is linear in w and for any $\gamma > 0$ has a solution $0 < w < 1$. As a function of γ it is quadratic, and the coefficients of the constant and quadratic terms have opposite signs provided $0 < w < 1$. In this case there exists therefore a unique positive solution γ, which establishes the required 1:1 relation between γ and w.

To complete the proof of Theorem 8, consider now any sequential probability ratio test with $A_0 < 1 < A_1$, and any constant $0 < \pi < 1$. Let

$$\pi' = \frac{\pi}{A_1(1 - \pi) + \pi}, \qquad \pi'' = \frac{\pi}{A_0(1 - \pi) + \pi}.$$

These values satisfy (43) and $0 < \pi' < \pi < \pi'' < 1$, and by Lemma 6 there exist therefore constants $0 < w < 1$ and $c > 0$ such that the given test is a Bayes solution for the auxiliary problem with an a priori probability π of p_0 being the true density, with losses $w_0 = 1 - w$ and $w_1 = w$, and cost c. Let the error probabilities and expectations of the sample size be α_0, α_1, $E_0(N)$, $E_1(N)$ for the given test, and consider any competitive procedure δ^*, with error probabilities $\alpha_i^* \leqq \alpha_i$ and expectations of sample size $E_i^*(N) < \infty$ $(i = 0, 1)$. Since the given test minimizes the Bayes risk, it satisfies

$$\pi[(1 - w)\alpha_0 + cE_0(N)] + (1 - \pi)[w\alpha_1 + cE_1(N)]$$
$$\leqq \pi[(1 - w)\alpha_0^* + cE_0^*(N)] + (1 - \pi)[w\alpha_1^* + cE_1^*(N)]$$

and hence

$$\pi E_0(N) + (1 - \pi)E_1(N) \leqq \pi E_0^*(N) + (1 - \pi)E_1^*(N).$$

The validity of this inequality for all $0 < \pi < 1$ implies

$$E_0(N) \leqq E_0^*(N) \quad \text{and} \quad E_1(N) \leqq E_1^*(N),$$

as was to be proved.

The Fisher, Neyman–Pearson Theories of Testing Hypotheses: One Theory or Two?

E. L. LEHMANN*

The Fisher and Neyman–Pearson approaches to testing statistical hypotheses are compared with respect to their attitudes to the interpretation of the outcome, to power, to conditioning, and to the use of fixed significance levels. It is argued that despite basic philosophical differences, in their main practical aspects the two theories are complementary rather than contradictory and that a unified approach is possible that combines the best features of both. As applications, the controversies about the Behrens–Fisher problem and the comparison of two binomials (2 × 2 tables) are considered from the present point of view.

KEY WORDS: Behrens–Fisher problem; Conditioning; Power; p-value; Significance level.

1. INTRODUCTION

The formulation and philosophy of hypothesis testing as we know it today was largely created in the period 1915–1933 by three men: R. A. Fisher (1890–1962), J. Neyman (1894–1981), and E. S. Pearson (1895–1980). Since then it has expanded into one of the most widely used quantitative methodologies, and has found its way into nearly all areas of human endeavor. It is a fairly commonly held view that the theories due to Fisher on the one hand, and to Neyman and Pearson on the other, are quite distinct. This is reflected in the fact that separate terms are often used (although somewhat inconsistently) to designate the two approaches: significance testing for Fisher and hypothesis testing for Neyman and Pearson. (Since both are concerned with the testing of hypotheses, it is convenient here to ignore this terminological distinction and to use the term "hypothesis testing" regardless of whether the testing is carried out in a Fisherian or Neyman–Pearsonian mode.)

There clearly are important differences, both in philosophy and in the treatment of specific problems. These were fiercely debated by Fisher and Neyman in a way described by Zabell (1992) as "a battle which had a largely destructive effect on the statistical profession." I believe that the ferocity of the rhetoric has created an exaggerated impression of irreconcilability. The purpose of this article is to see whether there exists a common ground that permits a resolution of some of the principal differences and a basis for rational discussion of the remaining ones.

Some of the Fisher-Neyman debate is concerned with issues studied in depth by philosophers of science (see, for example, Braithwaite 1953; Hacking 1965; Kyburg 1974; and Seidenfeld 1979). I am not a philosopher, and this article is written from a statistical, not a philosophical, point of view.

Section 2 presents some historical background for the two points of view. Section 3 discusses the basic philosophical difference between Fisher and Neyman. (Although the main substantive papers [NP 1928 and 1933a] were joint by Neyman and Pearson, their collaboration stopped soon after Neyman left Pearson's Department to set up his own program in Berkeley. After that, the debate was carried on primarily by Fisher and Neyman.) Sections 4, 5, and 6 discuss three specific issues on which the two schools differ (fixed levels versus p values, power, and conditioning). Section 7 illustrates the effect of these differences on the treatment of two statistical problems, the 2 × 2 table and the Behrens–Fisher problem, that have become focal points of the controversy. Finally, Section 8 suggests a unified point of view that does not resolve all questions but provides a common basis for discussing the remaining issues.

For the sake of completeness, it should be said that in addition to the Fisher and Neyman–Pearson theories there exist other philosophies of testing, of which we shall mention only two. There is Bayesian hypothesis testing, which, on the basis of stronger assumptions, permits assigning probabilities to the various hypotheses being considered. All three authors were very hostile to this formulation and were in fact motivated in their work by a desire to rid hypothesis testing of the need to assume a prior distribution over the available hypotheses.

Finally, in certain important situations tests can be obtained by an approach also due to Fisher for which he used the term *fiducial*. Most comparisons of Fisher's work on hypothesis testing with that of Neyman and Pearson (see, for example, Barnett 1982; Carlson 1976; Morrison and Henkel 1970; Spielman 1974, 1978; Steger 1971) do not include a discussion of the fiducial argument, which most statisticians have found difficult to follow. Although Fisher himself viewed fiducial considerations to be a very important part of his statistical thinking, this topic can be split off from other aspects of his work, and here I shall consider neither the fiducial nor the Bayesian approach any further.

Critical discussion of the issues considered in this article with references to the extensive literature, in a wider context and from viewpoints differing from that presented here, can be found in, for example, Oakes (1986) and Gigerenzer et al. (1989).

2. TESTING STATISTICAL HYPOTHESES

The modern theory of testing hypotheses began with Student's discovery of the t test in 1908. This was followed by

* E. L. Lehmann is Professor Emeritus, Department of Statistics, University of California, Berkeley, CA 94720. This research was supported by National Science Foundation Grant DMS-8908670. The author thanks the referees for helpful suggestions, Sandy Zabell for suggesting improvement to an early version of the article, and his wife Juliet Shaffer for discussions and critical comments at all stages.

Journal of the American Statistical Association
December 1993, Vol. 88, No. 424, Theory and Methods

Fisher with a series of papers culminating in his book *Statistical Methods for Research Workers* (1925), in which he created a new paradigm for hypothesis testing. He greatly extended the applicability of the t test (to the two-sample problem and the testing of regression coefficients) and generalized it to the testing of hypotheses in the analysis of variance. He advocated 5% as the standard level (with 1% as a more stringent alternative); through applying this new methodology to a variety of practical examples, he established it as a highly popular statistical approach for many fields of science.

A question that Fisher did not raise was the origin of his test statistics: Why these rather than some others? This is the question that Neyman and Pearson considered and which (after some preliminary work in Neyman and Pearson 1928) they later answered (Neyman and Pearson 1933a). Their solution involved not only the hypothesis but also a class of possible alternatives and the probabilities of two kinds of error: false rejection (Error I) and false acceptance (Error II). The "best" test was one that minimized P_A (Error II) subject to a bound on P_H (Error I), the latter being the significance level of the test. They completely solved this problem for the case of testing a simple (i.e., single distribution) hypothesis against a simple alternative by means of the Neyman–Pearson lemma. For more complex situations, the theory required additional concepts, and working out the details of this program was an important concern of mathematical statistics in the following decades.

The Neyman–Pearson introduction to the two kinds of error contained a brief statement that was to become the focus of much later debate. "Without hoping to know whether each separate hypothesis is true or false", the authors wrote, "we may search for rules to govern our behavior with regard to them, in following which we insure that, in the long run of experience, we shall not be too often wrong." And in this and the following paragraph they refer to a test (i.e., a rule to reject or accept the hypothesis) as "a rule of behavior".

3. INDUCTIVE INFERENCE VERSUS INDUCTIVE BEHAVIOR

Fisher considered statistics, the science of uncertain inference, able to provide a key to the long-debated problem of induction. He started one paper (Fisher 1932, p. 257) with the statement "Logicians have long distinguished two modes of human reasoning, under the respective names of deductive and inductive reasoning. . . . In inductive reasoning we attempt to argue from the particular, which is typically a body of observational material, to the general, which is typically a theory applicable to future experience." He developed his ideas in more detail in a later paper (Fisher 1935a, p. 39)

> . . . everyone who does habitually attempt the difficult task of making sense of figures is, in fact, essaying a logical process of the kind we call inductive, in that he is attempting to draw inferences from the particular to the general. Such inferences we recognize to be uncertain inferences. . . .

He continued in the next paragraph:

> Although some uncertain inferences can be rigorously expressed in terms of mathematical probability, it does not follow that mathematical probability is an adequate concept for the rigorous expression of uncertain inferences of every kind. . . . The inferences of the classical theory of probability are all deductive in character. They are statements about the behaviour of individuals, or samples, or sequences of samples, drawn from populations which are fully known. . . . More generally, however, a mathematical quantity of a different kind, which I have termed mathematical likelihood, appears to take its place [i.e., the place of probability] as a measure of rational belief when we are reasoning from the sample to the population.

Neyman did not believe in the need for a special inductive logic but felt that the usual processes of deductive thinking should suffice. More specifically, he had no use for Fisher's idea of likelihood. In his discussion of Fisher's 1935 paper (Neyman, 1935, p. 74, 75) he expressed the thought that it should be possible "to construct a theory of mathematical statistics . . . based solely upon the theory of probability," and went on to suggest that the basis for such a theory can be provided by "the conception of frequency of errors in judgment." This was the approach that he and Pearson had earlier described as "inductive behavior"; in the case of hypothesis testing, the behavior consisted of either rejecting the hypothesis or (provisionally) accepting it.

Both Neyman and Fisher considered the distinction between "inductive behavior" and "inductive inference" to lie at the center of their disagreement. In fact, in writing retrospectively about the dispute, Neyman (1961, p. 142) said that "the subject of the dispute may be symbolized by the opposing terms "inductive reasoning" and "inductive behavior." How strongly Fisher felt about this distinction is indicated by his statement in Fisher (1973, p. 7) that "there is something horrifying in the ideological movement represented by the doctrine that reasoning, properly speaking, cannot be applied to empirical data to lead to inferences valid in the real world."

4. FIXED LEVELS VERSUS p VALUES

A distinction frequently made between the approaches of Fisher and Neyman–Pearson is that in the latter the test is carried out at a fixed level, whereas the principal outcome of the former is the statement of a p value that may or may not be followed by a pronouncement concerning significance of the result.

The history of this distinction is curious. Throughout the 19th century, testing was carried out rather informally. It was roughly equivalent to calculating an (approximate) p value and rejecting the hypothesis if this value appeared to be sufficiently small. These early approximate methods required only a table of the normal distribution. With the advent of exact small-sample tests, tables of χ^2, t, F, . . . were also required. Fisher, in his 1925 book and later, greatly reduced the needed tabulations by providing tables not of the distributions themselves but of selected quantiles. (For an explanation of this very influential decision by Fisher see Kendall [1963]. On the other hand Cowles and Davis [1982] argue that conventional levels of three probable errors or two standard deviations, both roughly equivalent [in the normal case] to 5% were already in place before Fisher.) These tables allow the calculation only of ranges for the p values; however, they are exactly suited for determining the critical values at which the statistic under consideration becomes significant at a given level. As Fisher wrote in explaining the use of his χ^2 table (1946, p. 80):

> In preparing this table we have borne in mind that in practice we do not want to know the exact value of P for any observed χ^2, but, in the first place, whether or not the observed value is open to suspicion. If P is between .1 and .9, there is certainly no reason to suspect the hypothesis tested. If it is below .02, it is strongly indicated that the hypothesis fails to account for the whole of the facts. We shall not often be astray if we draw a conventional line at .05 and consider that higher values of χ^2 indicate a real discrepancy.

Similarly, he also wrote (1935, p. 13) that "it is usual and convenient for experimenters to take 5 percent as a standard level of significance, in the sense that they are prepared to ignore all results which fail to reach this standard . . ."

Fisher's views and those of some of his contemporaries are discussed in more detail by Hall and Selinger (1986).

Neyman and Pearson followed Fisher's adoption of a fixed level. In fact, Pearson (1962, p. 395) acknowledged that they were influenced by "[Fisher's] tables of 5 and 1% significance levels which lent themselves to the idea of choice, in advance of experiment, of the risk of the 'first kind of error' which the experimenter was prepared to take." He was even more outspoken in a letter to Neyman of April 28, 1978 (unpublished; in the Neyman collection of the Bancroft Library, University of California, Berkeley): "If there had not been these % tables available when you and I started work on testing statistical hypotheses in 1926, or when you were starting to talk on confidence intervals, say in 1928, how much more difficult it would have been for us! The concept of the control of 1st kind of error would not have come so readily nor your idea of following a rule of behaviour. . . . Anyway, you and I must be grateful for those two tables in the 1925 Statistical Methods for Research Workers." (For an idea of what the Neyman–Pearson theory might have looked like had it been based on p values instead of fixed levels, see Schweder 1988.)

It is interesting to note that unlike Fisher, Neyman and Pearson (1933a, p. 296) did not recommend a standard level but suggested that "how the balance [between the two kinds of error] should be struck must be left to the investigator," and (1933b, p. 497) "we attempt to adjust the balance between the risks P_{I} and P_{II} to meet the type of problem before us."

It is thus surprising that in SMSI Fisher (1973, p. 44–45) criticized the NP use of a fixed conventional level. He objected that

> the attempts that have been made to explain the cogency of tests of significance in scientific research, by reference to supposed frequencies of possible statements, based on them, being right or wrong, thus seem to miss the essential nature of such tests. A man who 'rejects' a hypothesis provisionally, as a matter of habitual practice, when the significance is 1% or higher, will certainly be mistaken in not more than 1% of such decisions. . . . However, the calculation is absurdly academic, for in fact no scientific worker has a fixed level of significance at which from year to year, and in all circumstances, he rejects hypotheses; he rather gives his mind to each particular case in the light of his evidence and his ideas.

The difference between the reporting of a p value or that of a statement of acceptance or rejection of the hypothesis was linked by Fisher in Fisher (1973, pp. 79–80), to the distinction between drawing conclusions or making decisions.

> The conclusions drawn from such tests constitute the steps by which the research worker gains a better understanding of his experimental material, and of the problems which it presents. . . . More recently, indeed, a considerable body of doctrine has attempted to explain, or rather to reinterpret, these tests on the basis of quite a different model, namely as means to making decisions in an acceptance procedure.

Responding to earlier versions of these and related objections by Fisher to the Neyman–Pearson formulation, Pearson (1955, p. 206) admitted that the terms "acceptance" and "rejection" were perhaps unfortunately chosen, but of his joint work with Neyman he said that "from the start we shared Professor Fisher's view that in scientific inquiry, a statistical test is 'a means of learning' " and "I would agree that some of our wording may have been chosen inadequately, but I do not think that our position in some respects was or is so very different from that which Professor Fisher himself has now reached."

The distinctions under discussion are of course related to the argument about "inductive inference" vs. "inductive behavior," but in this debate Pearson refused to participate. He concludes his response to Fisher's 1955 attack with: "Professor Fisher's final criticism concerns the use of the term 'inductive behavior'; this is Professor Neyman's field rather than mine."

5. POWER

As was mentioned in Section 2, a central consideration of the Neyman–Pearson theory is that one must specify not only the hypothesis H but also the alternatives against which it is to be tested. In terms of the alternatives, one can then define the type II error (false acceptance) and the power of the test (the rejection probability as a function of the alternative). This idea is now fairly generally accepted for its importance in assessing the chance of detecting an effect (i.e., a departure from H) when it exists, determining the sample size required to raise this chance to an acceptable level, and providing a criterion on which to base the choice of an appropriate test.

Fisher never wavered in his strong opposition to these ideas. Following are some of his objections:

1. A type II error consists in falsely accepting H, and Fisher (1935b, p.) emphasized that there is no reason for "believing that a hypothesis has been proved to be true merely because it is not contradicted by the available facts." This is of course correct, but it does not diminish the usefulness of power calculations.

2. A second point Fisher raised is, in modern terminology, that the power cannot be calculated because it depends on the unknown alternative. For example (Fisher 1955, p. 73), he wrote:

> The frequency of the 1st class [type I error] . . . is calculable and therefore controllable simply from the specification of the null hypothesis. The frequency of the 2nd kind must depend . . . greatly on how closely they [rival hypotheses] resemble the null

> hypothesis. Such errors are therefore incalculable . . . merely from the specification of the null hypothesis, and would never have came into consideration in the theory only of tests of significance, had the logic of such tests not been confused with that of acceptance procedures. (He discussed the same point in Fisher 1947, p. 16–17.)

Fisher was of course aware of the importance of power, as is clear from the following remarks (1947, p. 24): "With respect to the refinements of technique, we have seen above that these contribute nothing to the validity of the experiment and of the test of significance by which we determine its result. They may, however, be important, and even essential, in permitting the phenomenon under test to manifest itself." The section in which this statement appears is tellingly entitled "Qualitative Methods of Increasing Sensitiveness." Fisher accepted the importance of the concept but denied the possibility of assessing it quantitatively.

Later in the same book Fisher made a very similar distinction regarding the choice of test. Under the heading "Multiplicity of Tests of the Same Hypothesis," he devoted a section (sec. 61) to this topic. Here again, without using the term, he referred to alternatives when he wrote (Fisher 1947, p. 182) that "we may now observe that the same data may contradict the hypothesis in any of a number of different ways." After illustrating how different tests would be appropriate for different alternatives, he continued (p. 185):

> The notion that different tests of significance are appropriate to test different features of the same null hypothesis presents no difficulty to workers engaged in practical experimentation but has been the occasion of much theoretical discussion among statisticians. The reason for this diversity of view-point is perhaps that the experimenter is thinking in terms of observational values, and is aware of what observational discrepancy it is which interests him, and which he thinks may be statistically significant, before he inquires what test of significance, if any, is available appropriate to his needs. He is, therefore, not usually concerned with the question: To what observational feature should a test of significance be applied?

The idea that there is no need for a theory of test choice, because an experienced experimenter knows what is the appropriate test, is expressed more strongly in a letter to W. E. Hick of October 1951 (Bennett 1990, p. 144), who, in asking about "one-tail" vs. "two-tail" in χ^2, had referred to his lack of knowledge concerning "the theory of critical regions, power, etc.":

> I am a little sorry that you have been worrying yourself at all with that unnecessarily portentous approach to tests of significance represented by the Neyman and Pearson critical regions, etc. In fact, I and my pupils throughout the world would never think of using them. If I am asked to give an explicit reason for this I should say that they approach the problem entirely from the wrong end, i.e., not from the point of view of a research worker, with a basis of well grounded knowledge on which a very fluctuating population of conjectures and incoherent observations is continually under examination. In these circumstances the experimenter does know what observation it is that attracts his attention. What he needs is a confident answer to the question "ought I to take any notice of that?" This question can, of course, and for refinement of thought should, be framed as "Is this particular hypothesis overthrown, and if so at what level of significance, by this particular body of observations?" It can be put in this form unequivocally only because the genuine experimenter already has the answers to all the questions that the followers of Neyman and Pearson attempt, I think vainly, to answer by merely mathematical consideration.

6. CONDITIONAL INFERENCE

While Fisher's approach to testing included no detailed consideration of power, the Neyman–Pearson approach failed to pay attention to an important concern raised by Fisher. To discuss this issue, we must begin by considering briefly the different meanings that Fisher and Neyman attach to probability.

For Neyman, the idea of probability is fairly straightforward: It represents an idealization of long-run frequency in a long sequence of repetitions under constant conditions (see, for example, Neyman 1952, p. 27; 1957, p. 9). Later (Neyman 1977), he pointed out that by the law of large numbers, this idea permits an extension: If a sequence of independent events is observed, each with probability p of success, then the long-run success frequency will be approximately p even if the events are not identical. This property adds greatly to the appeal and applicability of a frequentist probability. In particular, it is the way in which Neyman came to interpret the value of a significance level.

On the other hand, the meaning of probability is a problem with which Fisher grappled throughout his life. Not surprisingly, his views too underwent some changes. The concept at which he eventually arrived is much broader than Neyman's: "In a statement of probability, the predicand, which may be conceived as an object, as an event, or as a proposition, is asserted to be one of a set of a number, however large, of like entities of which a known proportion, P, have some relevant characteristic, not possessed by the remainder. It is further asserted that no subset of the entire set, having a different proportion, can be recognized" (Fisher 1973, p. 113). It is this last requirement, Fisher's version of the "requirement of total evidence" (Carnap 1962, sec. 45), which is particularly important to the present discussion.

Example 1 (Cox 1958). Suppose that we are concerned with the probability $P(X \leq x)$, where X is normally distributed as $N(\mu, 1)$ or $N(\mu, 4)$, depending on whether the spin of a fair coin results in heads (H) or tails (T). Here the set of cases in which the coin falls heads is a recognizable subset; therefore, Fisher would not admit the statement

$$P(X \leq x) = \frac{1}{2}\,\Phi(x - \mu) + \frac{1}{2}\,\Phi\left(\frac{x - \mu}{2}\right) \qquad (1)$$

as legitimate. Instead, he would have required $P(X \leq x)$ to be evaluated conditionally as

$$P(X \leq x \mid \mathrm{H}) = \Phi(x - \mu) \quad \text{or}$$

$$P(X \leq x \mid \mathrm{T}) = \Phi\left(\frac{x - \mu}{2}\right), \qquad (2)$$

depending on the outcome of the spin.

On the other hand, Neyman would have taken (1) to provide the natural assessment of $P(X \leq x)$. Despite this preference, there is nothing in the Neyman–Pearson (frequentist) approach to prevent consideration of the conditional probabilities (2). The critical issue from a frequentist viewpoint is what to consider as the relevant replications of the experiment: a sequence of observations from the same normal distribution or a sequence of coin tosses, each followed by an observation from the appropriate normal distribution.

Consider now the problem of testing $H\colon \mu = 0$ against the simple alternative $\mu = 1$ on the basis of a sample $X_1, \ldots, X_n$ from the distribution (1). The Neyman–Pearson lemma would tell us to reject H when

$$\frac{1}{2}\cdot\frac{1}{\sqrt{2\pi}} e^{-\Sigma(x_i-1)^2/2} + \frac{1}{2}\frac{1}{2\sqrt{2\pi}} e^{-\Sigma(x_i-1)^2/8}$$

$$\geq K\left[\frac{1}{2}\frac{1}{\sqrt{2\pi}} e^{-\Sigma x_i^2/2} + \frac{1}{2}\frac{1}{2\sqrt{2\pi}} e^{-\Sigma x_i^2/8}\right], \quad (3)$$

where K is determined so that the probability of (3) when $\mu = 0$ is equal to the specified level α.

On the other hand, a Fisherian approach would adjust the test to whether the coin falls H or T and would use the rejection region

$$\frac{1}{\sqrt{2\pi}} e^{-\Sigma(x_i-1)^2/2} \geq K_1 \frac{1}{\sqrt{2\pi}} e^{-\Sigma x_i^2/2}$$

$$\text{when the coin falls H} \quad (4)$$

and

$$\frac{1}{2\sqrt{2\pi}} e^{-\Sigma(x_i-1)^2/8} \geq K_2 \frac{1}{2\sqrt{2\pi}} e^{-\Sigma x_i^2/8}$$

$$\text{when the coin falls T}, \quad (5)$$

where K_1 and K_2 are determined so that the null probability of both (4) and (5) is equal to α. It is easily seen that these two tests are not equivalent. Which one should we prefer?

Test (3) has the advantage of being more powerful in the sense that when the full experiment of spinning a coin and then taking n observations on X is repeated many times, and when $\mu = 1$, this test will reject the hypothesis more frequently.

The second test has the advantage that its conditional level given the outcome of the spin is α both when the outcome is H and when it is T. [The conditional level of the first test will be $<\alpha$ for one of the two outcomes and $>\alpha$ for the other.]

Which of these considerations is more important depends on the circumstances. Echoing Fisher, we might say that we prefer (1) in an acceptance sampling situation where interest focuses not on the individual cases but on the long-run frequency of errors, but that we would prefer the second test in a scientific situation where long-run considerations are irrelevant and only the circumstances at hand (i.e., H or T) matter. As Fisher put it (1973, p. 101–102), referring to a different but similar situation: "It is then obvious at the time that the judgment of significance has been decided not by the evidence of the sample, but by the throw of a coin. It is not obvious how the research worker is to be made to forget this circumstance, and it is certain that he ought not to forget it, if he is concerned to assess the weight only of objective observational facts against the hypothesis in question."

The present example is of course artificial, but the same issue arises whenever there exists an ancillary statistic (see, for example, Cox and Hinkley 1974; Lehmann 1986), and it seems to lie at the heart of the cases in which the two theories disagree on specific tests. The two most prominent of these cases are discussed in the next section.

7. TWO EXAMPLES

For many problems, a pure Fisherian or Neymann–Pearsonian approach will lead to the same test. Suppose in particular that the observations X follow a distribution from an exponential family with density

$$p_{\theta,\vartheta}(x) = C(\theta, \vartheta)e^{\theta U(x) + \Sigma_{i=1}^{k}\vartheta_i T_i(x)} \quad (6)$$

and consider testing the hypothesis

$$H\colon \theta = \theta_0 \quad (7)$$

against the one-sided alternatives $\theta > \theta_0$. Then Fisher would condition on $T = (T_1, \ldots, T_k)$ and would in the conditional model consider it natural to calculate the p value as the conditional probability of $U \geq u$, where u is the observed value of U. At a given level α, the result would be declared significant if $U \geq C(t)$, where $C(t)$ is determined by

$$P[U > C(t) \mid T = t] = \alpha. \quad (8)$$

A Neyman–Pearson viewpoint would lead to the same test as being uniformly most powerful among all similar tests.

But as we have seen in Example 1, the two approaches do not always lead to the same result. We next consider the two examples that have engendered the most controversy.

Example 2: The 2 × 2 table with one fixed margin. Let X, Y be two independent binomial variables with success probabilities p_1 and p_2 and corresponding to m and n trials. The problem of testing $H\colon p_2 = p_1$ against the alternatives $p_2 > p_1$ is of the form given by (6) and (7) with $\theta = \log[(p_2/q_2)/(p_1/q_1)]$, $T = X + Y$ and $U = Y$. Basically, there is therefore no conflict between the two approaches. However, because of the discreteness of the conditional distribution of U given t, condition (8) typically cannot be satisfied. Fisher's exact test then chooses $C(t)$ to be the largest constant for which

$$P[U > C(t) \mid T = t] \leq \alpha. \quad (9)$$

For small values of t, this may lead to conditional levels substantially less than α; for small m and n, the same may be true for the unconditional level. For this reason, Fisher's exact test has been criticized as being too conservative. Many alternatives have been proposed for which the unconditional level (which is a function of $p_1 = p_2$) is much closer to α. Upton (1982) lists 22; for other surveys, see Yates (1984) and Agresti (1992).

The issues are similar to those encountered in Example 1. If conditioning is considered appropriate (and in the present case it typically is), and if control of type I error at level α is considered essential, then the only sensible test available is of the form $U > C(t)$, where $C(t)$ is the largest value satisfying (9). If, on the other hand, only the unconditional performance is considered relevant, then we may allow the conditional level of the region $U > C(t)$ to exceed α for some values of t in such a way that the unconditional level (which is the expected value of the conditional level) gets closer to

α while remaining $\leq\alpha$ for all values of $p_2 = p_1$. This is essentially what the alternatives to Fisher's exact tests try to achieve. (The same issues arise also when analyzing 2×2 tables in which none of the margins are fixed.)

Example 3: Behrens–Fisher problem. Here we are dealing with independent samples $X_1, \ldots, X_m$ and $Y_1, \ldots, Y_n$ from normal distributions $N(\xi, \sigma^2)$ and $N(\eta, \tau^2)$ and we wish to test the hypothesis H: $\eta = \xi$. Against the two-sided alternatives $\eta \neq \xi$, there is general agreement that the rejection region should be of the form

$$\frac{|\bar{Y} - \bar{X}|}{\sqrt{\dfrac{S_X^2}{m} + \dfrac{S_Y^2}{n}}} \geq g\left(\frac{S_Y^2}{S_X^2}\right), \tag{10}$$

where S_Y^2 and S_X^2 are the usual estimators of σ^2 and τ^2.

Suppose that we consider it appropriate, as Fisher does, to carry out the analysis conditionally on the value of S_Y^2/S_X^2. If the conditional distribution of the left side of (10) given $S_Y^2/S_X^2 = c$ were independent of the parameters and hence known, there would be no problem. Everyone would agree to calculate g so that the conditional level is α for each c, which would then also result in an unconditional level identically equal to α. Unfortunately, the conditional distribution depends on the unknown variances. Two principal ways out of this difficulty have been proposed.

1. From a Neyman–Pearson point of view, the attempt has been to construct a function g for which the probability of (10) is $\equiv\alpha$ under H for all σ and τ (it actually depends only on the ratio $\theta = \tau^2/\sigma^2$). After much effort in this direction, it became clear that an acceptable function g satisfying this condition does not exist. But Welch and Aspin have produced tests whose level differs from α so little over the entire range of θ that, for all practical purposes, they can be viewed as solutions to the problem. (For a discussion and references see, for example, Stuart and Ord 1991, sec. 20.33.)

2. These tests are unacceptable to Fisher, however, because they admit recognizable subsets. In particular, Fisher (1956) produced an example for which the conditional level given $S_Y^2/S_X^2 = 1$ is always $>\alpha + \varepsilon$ for some positive ε. Fisher's own solution to the problem is the so-called Behrens–Fisher test, which he derived by means of a fiducial argument. Although it does not follow from this derivation, numerical evidence (Robinson 1976) strongly suggests that this test is conservative; that is, its unconditional level is always $<\alpha$. But a proof of this fact for all m, n, and θ is not yet available.

Let us call a set C in the sample space for which there exists $\varepsilon > 0$ such that

$$P_H[\text{rejecting} \mid X \in C] > \alpha + \varepsilon \quad \text{for all distributions in } H,$$

a liberally biased relevant subset. (The corresponding concept for confidence intervals is called negatively biased.) Robinson (1976) showed that no such subsets exist for the Behrens–Fisher test. (Because of this test's conservative nature, this is perhaps not too surprising.) He proposed calling a test conservative if its unconditional level is always $\leq\alpha$ and if it does not admit a liberally biased relevant subset, and expressed the hope that "perhaps the Behrens–Fisher test is optimal in some sense among the class of procedures which are conservative" (Robinson 1976, p. 970). This conjecture seems to have been disproved by Linssen (1991).

8. ONE THEORY OR TWO?

From the preceding sections it is clear that considerable differences exist between the viewpoints of Fisher and Neyman–Pearson. Are these sufficiently contradictory to preclude a unified theory that would combine the best features of both?

A first difference, discussed in Section 4, concerns the reporting of the conclusions of the analysis. Should this consist merely of a statement of significance or nonsignificance at a given level, or should a p value be reported? The original reason for fixed, standardized levels—unavailability of more detailed tables—no longer applies, and in any case reporting the p value provides more information. On the other hand, definite decisions or conclusions are often required. Additionally, in view of the enormously widespread use of testing at many different levels of sophistication, some statisticians (and journal editors) see an advantage in standardization; fortunately, this is a case where you can have your cake and eat it too. One should routinely report the p value and, where desired, combine this with a statement on significance at any stated level. (This was in fact common practice throughout the 19th Century and is the procedure frequently used by Fisher.) Two other principal differences, considered in Sections 5 and 6, are the omissions of power (by Fisher) and of conditioning (by Neyman–Pearson). It seems clear that a unified approach needs to incorporate both of these ideas.

For some problems this will cause no difficulty, because both approaches will lead to the same test, as illustrated at the beginning of Section 7. But the principles of conditioning on the one hand and of maximizing the unconditional power on the other may be in conflict, as is seen from Examples 1–3. This conflict disappears when it is realized that in such cases priority must be given to deciding on the appropriate frame of reference; that is, the real or hypothetical sequence of events that determine the meaning of any probability statement. Only after this has been settled do probabilistic concepts such as level and power acquire meaning, and it is only then that the problem of maximizing power comes into play.

This leaves the combined theory with its most difficult issue: What is the relevant frame of reference? It seems clear that even in the simplest situations (such as Ex. 1), no universal answer is possible. In any specific case, the solution will depend on contextual considerations that cannot easily be captured by a general theory.

That conflicting considerations argue for different solutions in specific cases is not an indictment of a theory, provided that the theory furnishes a basis for discussing the issues. Although Neyman and Pearson never seem to have raised the problem of just what constitutes a replication of an experiment, this question is as important for a frequentist as it is for an adherent of Fisherian probability. This was recognized, for example, by Bartlett (1984, p. 453), who stated "I regard the 'frequence requirement of repeated sampling' as including conditional inferences." A common basis for the discussion of various conditioning concepts, such as ancillaries and relevant subsets, thus exists. The proper choice of framework is a problem needing further study.

We conclude by considering some more detailed issues and by reviewing Examples 2 and 3 from the present point of view.

1. Both Neyman–Pearson and Fisher would give at most lukewarm support to standard significance levels such as 5% or 1%. Fisher, although originally recommending the use of such levels, later strongly attacked any standard choice. Neyman–Pearson, in their original formulation of 1933, recommended a balance between the two kinds of error (i.e., between level and power). For a disucssion of how to achieve such a balance, see, for example, Sanathanan (1974). Both level and power should of course be considered conditionally whenever conditioning is deemed appropriate. Unfortunately, this is not possible at the planning stage.

2. A second point on which there appears to be no conflict between the two approaches is "truth in advertising." Even if a particular nominal level α, say 5%, is the target, when it cannot be achieved because of discreteness the test should not just be described as conservative or liberal relative to the nominal level; instead, the actual (conditional or unconditional) level should be stated. If this level is not known because it depends on unknown parameters, at least its range should be given and, if possible, also an estimated value.

3. In both the 2×2 example and the Behrens–Fisher problems, the conflict between the solutions proposed by the two schools is often discussed as that of a desire for a similar test (i.e., one for which the unconditional level is $\equiv\alpha$) versus a suitable conditional test. The issue becomes clearer if one asks for the reason that Neyman–Pearson proposed the condition of similarity. The explanation begins with the case of a simple hypothesis where these authors take it for granted that in order to maximize the power, one would want the attained level to be equal to rather than less than α. For a composite hypothesis H, they therefore stated that the level should equal α for each of the simple hypotheses making up H. The requirement for similarity thus has its origin in the desire to maximize power, the issue discussed in Section 5.

In the light of (1) and (2), a unified theory less concerned with standard nominal levels might jettison not only the demand for similarity but also that of conservatism relative to a nominal level.

When similarity cannot be achieved and conservation is not required, various compromise solutions may be available. Thus in the 2×2 case of Example 2, one could, for example, select for each t the conditional level closest to α. If this seems too permissive, then the rule could be modified by adding a cap on the conditional level beyond which one would not go. Tests with a variable conditional level that will sometimes be $<\alpha$ and sometimes $>\alpha$ have been discussed by Barnard (1989) under the name "flexible Fisher." Alternatively, one might give up on a nominal level altogether and instead for each t adjust the level to the attainable (conditional) power.

The situation is much more complicated for the Behrens–Fisher problem. On the one hand, the arguments for conditioning an S_Y^2/S_X^2 seems less compelling; on the other hand, even if this conditioning requirement is accepted, the conditional distribution depends on unknown parameters, and thus it is less clear how to control the conditional level. Robinson's formulation, mentioned in Section 7, provides an interesting possibility but requires much further investigation. But such work can be carried out from the present point of view by combining considerations of both conditioning and power.

To summarize, p values, fixed-level significance statements, conditioning, and power considerations can be combined into a unified approach. When long-term power and conditioning are in conflict, specification of the appropriate frame of reference takes priority, because it determines the meaning of the probability statements. A fundamental gap in the theory is the lack of clear principles for selecting the appropriate framework. Additional work in this area will have to come to terms with the fact that the decision in any particular situation must be based not only on abstract principles but also on contextual aspects.

[Received January 1992. Revised February 1993.]

REFERENCES

Agresti, A. (1992), "A Survey of Exact Inference for Contingency Tables" (with discussion), *Statistical Science*, 7, 131–177.

Barnard, G. A. (1989), "On Alleged Gains in Power from Lower p Values," *Statistics in Medicine*, 8, 1469–1477.

Barnett, V. (1982), *Comparative Statistical Inference* (2nd ed.), New York: John Wiley.

Bartlett, M. S. (1984), "Discussion of 'Tests of Significance for 2 × 2 Contingency Tables,' by F. Yates." *Journal of the Royal Statistical Society*, Ser. A, 147, 453.

Bennett, J. H. (1990), *Statistical Inference and Analysis (Selected Correspondence of R. A. Fisher)*, Oxford, U.K.: Clarendon Press.

Braithwaite, R. B. (1953), *Scientific Explanation*, Cambridge, U.K.: Cambridge University Press.

Brown, L. (1967), "The Conditional Level of Student's t Test," *Annals of Mathematical Statistics*, 38, 1068–1071.

Carlson, R. (1976), "The Logic of Tests of Significance" (with discussion), *Philosophy of Science*, 43, 116–128.

Carnap, R. (1962), *Logical Foundations of Probability* (2nd ed.), Chicago: the University of Chicago Press.

Cowles, M., and Davis, C. (1982), "On the Origins of the .05 Level of Statistical Significance," *American Psychologist*, 37, 553–558.

Cox, D. R. (1958), "Some Problems Connected With Statistical Inference," *Annals of Mathematical Statistics*, 29, 357–372.

Cox, D. R., and Hinkley, D. V. (1974), *Theoretical Statistics*, London: Chapman and Hall.

Fisher, R. A. (1925) (10th ed., 1946), *Statistical Methods for Research Workers*, Edinburgh: Oliver & Boyd.

——— (1932), "Inverse Probability and the Use of Likelihood," *Proceedings of the Cambridge Philosophical Society*, 28, 257–261.

——— (1935a), "The Logic of Inductive Inference," *Journal of the Royal Statistical Society*, 98, 39–54.

——— (1935b), "Statistical Tests," *Nature*, 136, 474.

——— (1935c) (4th ed., 1947), *The Design of Experiments*, Edinburgh: Oliver & Boyd.

——— (1939), "Student," *Annals of Engenics*, 9, 1–9.

——— (1947), *The Design of Experiments* (4th ed.), New York: Hafner Press.

——— (1955), "Statistical Methods and Scientific Induction," *Journal of the Royal Statistical Society*, Ser. B, 17, 69–78.

——— (1956), "On a Test of Significance in Pearson's Biometrika Tables (No. 11), *Journal of the Royal Statistical Society*, Ser. B, 18, 56–60.

——— (1958), "The Nature of Probability," *Centennial Review*, 2, 261–274.

——— (1959), "Mathematical Probability in the Natural Sciences," *Technometrics*, 1, 21–29.
——— (1960), "Scientific Thought and the Refinement of Human Reason," *Journal of the Operations Research Society of Japan*, 3, 1–10.
——— (1973), *Statistical Methods and Scientific Inference*, (3rd ed.) London: Collins Macmillan.
Gigerenzer, G., et al. (1989), *The Empire of Chance*, New York: Cambridge University Press.
Hacking, I. (1965), *Logic of Statistical Inference*, New York: Cambridge University Press.
Hall, P., and Selinger, B. (1986), "Statistical Significance: Balancing Evidence Against Doubt," *Australian Journal of Statistics*, 28, 354–370.
Hedges, L., and Olkin, I. (1985), *Statistical Methods for Meta-Analysis*, Orlando, FL: Academic Press.
Hockberg, Y., and Tamhane, A. C. (1987), *Multiple Comparison Procedures*, New York: John Wiley.
Kendall, M. G. (1963), "Ronald Aylmer Fisher, 1890–1962," *Biometrika*, 50, 1–15.
Kyburg, H. E., Jr. (1974), *The Logical Foundations of Statistical Inference*, Boston: D. Reidel.
Linhart, H., and Zucchini, W. (1986), *Model Selection*, New York: John Wiley.
Linssen, H. N. (1991), "A Table for Solving the Behrens–Fisher Problem," *Statistics and Probability Letters*, 11, 359–363.
Miller, R. G. (1981), *Simultaneous Statistical Inference*, (2nd ed.), New York: Springer-Verlag.
Morrison, D. E., and Henkel, R. E. (1970), *The Significance Test Controversy*, Chicago: Aldine.
Neyman, J. (1935), "Discussion of Fisher (1935a)." *Journal of the Royal Statistical Society*, 98, 74–75.
——— (1938), "L'Estimation Statistique Traitée Comme un Problème Classique de Probabilité," *Actualités Scientifiques et Industrielles*, 739, 25–57.
——— (1952), *Lectures and Conferences on Mathematical Statistics and Probability* (2nd ed.), Graduate School, Washington, D.C.: U.S. Dept. of Agriculture.
——— (1955), "The Problem of Inductive Inference," *Communications in Pure and Applied Mathematics*, 8, 13–46.
——— (1956), "Note on an Article by Sir Ronald Fisher," *Journal of the Royal Statistical Society*, Ser. B, 18, 288–294.
——— (1957), " 'Inductive behavior' as a Basic Concept of Philosophy of Science," *Review of the International Statistical Institute*, 25, 7–22.
——— (1961), "Silver Jubilee of My Dispute With Fisher," *Journal of the Operations Research Society of Japan*, 3, 145–154.
——— (1966), "Behavioristic Points of View on Mathematical Statistics," in *On Political Economy and Econometrics: Essays in Honour of Oscar Lange*, Warsaw: Polish Scientific Publishers, pp. 445–462.
——— (1976), "Tests of Statistical Hypotheses and Their Use in Studies of Natural Phenomena," *Communications in Statistics, Part A—Theory and Methods*, 5, 737–751.
——— (1977), "Frequentist Probability and Frequentist Statistics," *Synthèse*, 36, 97–131.
Neyman, J., and Pearson, E. S. (1928), "On the Use and Interpretation of Certain Test Criteria for Purposes of Statistical Inference," *Biometrika*, 20A, 175–240, 263–294.
——— (1933a), "On the Problem of the Most Efficient Tests of Statistical Hypotheses," *Philosophical Transactions of the Royal Society of London*, Ser. A, 231, 289–337.
——— (1933b), "The Testing of Statistical Hypotheses in Relation to Probabilities A Priori," *Proceedings of the Cambridge Philosophical Society*, 29, 492–510.
Oakes, M. (1986), *Statistical Inference: A Comment for the Social and Behavioral Sciences*, New York: John Wiley.
Pearson, E. S. (1955), "Statistical Concepts in Their Relation to Reality," *Journal of the Royal Statistical Society*, Ser. B, 17, 204–207.
——— (1962), "Some Thoughts on Statistical Inference," *Annals of Mathematical Statistics*, 33, 394–403.
——— (1974), "Memories of the Impact of Fisher's Work in the 1920's," *International Statistical Review*, 42, 5–8.
Pearson, E. S., and Hartley, H. O. (1954), *Biometrika Tables for Statisticians (Table No. 11)*, New York: Cambridge University Press.
Pedersen, J. G. (1978), "Fiducial Inference," *International Statistical Review*, 46, 147–170.
Robinson, G. K. (1976), "Properties of Student's t and of the Behrens–Fisher Solution to the Two-Means Problem," *The Annals of Statistics*, 4, 963–971.
——— (1982), "Behrens–Fisher Problem," in Encyclopedia of Statistical Sciences (Vol. 1, eds. S. Kotz and N. L. Johnson), New York: John Wiley, pp. 205–209.
Savage, L. J. (1976), "On Rereading R. A. Fisher" (with discussion), *Annals of Statistics*, 4, 441–500.
Schweder, T. (1988), "A Significance Version of the Basic Neyman–Pearson Theory for Scientific Hypothesis Testing," *Scandanavian Journal of Statistics*, 15, 225–242.
Seidenfeld, T. (1979), *Philosophical Problems of Statistical Inference*, Boston: D. Reidel.
Spielman, S. (1974), "The Logic of Tests of Significance," *Philosophy of Science*, 41, 211–226.
——— (1978), "Statistical Dogma and the Logic of Significance Testing," *Philosophy of Science*, 45, 120–135.
Steger, J. A. (ed.) (1971), *Readings in Statistics for the Behavioral Scientist*, New York: Holt, Rinehart and Winston.
Stuart, A., and Ord, J. K. (1991), *Kendall's Advanced Theory of Statistics, Vol. II* (5th ed.), New York: Oxford University Press.
Tukey, J. W. (1960), "Conclusions vs. Decisions," *Technometrics*, 2, 424–432.
Upton, G. J. G. (1982), "A Comparison of Alternative Tests for the 2×2 Comparative Trial," *Journal of the Royal Statistical Society*, Ser. A, 145, 86–105.
Wallace, D. L. (1980), "The Behrens–Fisher and Fieller–Creasy Problems," in *R. A. Fisher: An Application*, eds. S. E. Fienberg and D. V. Hinkley, New York: Springer-Verlag, pp. 119–147.
Yates, F. (1984), "Tests of Significance for 2×2 Contingency Tables" (with discussion), *Journal of the Royal Statistical Society*, Ser. A, 147, 426–463.
Zabell, S. L. (1992), "R. A. Fisher and the Fiducial Argument," *Statistical Science*, 7, 369–387.

IMS Lecture Notes–Monograph Series
2nd Lehmann Symposium – Optimality
Vol. 49 (2006) 1–8

DOI: 10.1214/074921706000000356

On likelihood ratio tests

Erich L. Lehmann[1]

University of California at Berkeley

Abstract: Likelihood ratio tests are intuitively appealing. Nevertheless, a number of examples are known in which they perform very poorly. The present paper discusses a large class of situations in which this is the case, and analyzes just how intuition misleads us; it also presents an alternative approach which in these situations is optimal.

1. The popularity of likelihood ratio tests

Faced with a new testing problem, the most common approach is the likelihood ratio (LR) test. Introduced by Neyman and Pearson in 1928, it compares the maximum likelihood under the alternatives with that under the hypothesis. It owes its popularity to a number of facts.

(i) It is intuitively appealing. The likelihood of θ,

$$L_x(\theta) = p_\theta(x)$$

i.e. the probability density (or probability) of x considered as a function of θ, is widely considered a (relative) measure of support that the observation x gives to the parameter θ. (See for example Royall [8]). Then the likelihood ratio

$$\sup_{alt}[p_\theta(x)] / \sup_{hyp}[p_\theta(x)] \tag{1.1}$$

compares the best explanation the data provide for the alternatives with the best explanations for the hypothesis. This seems quite persuasive.

(ii) In many standard problems, the LR test agrees with tests obtained from other principles (for example it is UMP unbiased or UMP invariant). Generally it seems to lead to satisfactory tests. However, counter-examples are also known in which the test is quite unsatisfactory; see for example Perlman and Wu [7] and Menéndez, Rueda, and Salvador [6].

(iii) The LR test, under suitable conditons, has good asymptotic properties.

None of these three reasons are convincing.

(iii) tells us little about small samples.

(i) has no strong logical grounding.

(ii) is the most persuasive, but in these standard problems (in which there typically exist a complete set of sufficient statistics) all principles typically lead to tests that are the same or differ only by little.

[1]Department of Statistics, 367 Evans Hall, University of California, Berkeley, CA 94720-3860, e-mail: shaffer@stat.berkeley.edu

AMS 2000 subject classifications: 62F03.

Keywords and phrases: likelihood ratio tests, average likelihood, invariance.

In view of lacking theoretical support and many counterexamples, it would be good to investigate LR tests systematically for small samples, a suggestion also made by Perlman and Wu [7]. The present paper attempts a first small step in this endeavor.

2. The case of two alternatives

The simplest testing situation is that of testing a simple hypothesis against a simple alternative. Here the Neyman-Pearson Lemma completely vindicates the LR-test, which always provides the most powerful test. Note however that in this case no maximization is involved in either the numerator or denominator of (1.1), and as we shall see, it is just these maximizations that are questionable.

The next simple situation is that of a simple hypothesis and two alternatives, and this is the case we shall now consider.

Let $\underline{X}=(X_1,\ldots,X_n)$ where the X's are iid. Without loss of generality suppose that under H the X's are uniformly distributed on $(0,1)$. Consider two alternatives f,g on $(0,1)$. To simplify further, we shall assume that the alternatives are symmetric, i.e. that

$$\begin{aligned} p_1(\underline{x}) &= f(x_1)\cdots f(x_n) \\ p_2(\underline{x}) &= f(1-x_1)\cdots f(1-x_n). \end{aligned} \tag{2.1}$$

Then it is natural to restrict attention to symmetric tests (that is the invariance principle) i.e. to rejection regions R satisfying

$$(x_1,\ldots,x_n)\in R \quad \text{if and only if } (1-x_1,\ldots,1-x_n)\in R. \tag{2.2}$$

The following result shows that under these assumptions there exists a uniformly most powerful (UMP) invariant test, i.e. a test that among all invariant tests maximizes the power against both p_1 and p_2.

Theorem 2.1. *For testing H against the alternatives* (2.1) *there exists among all level α rejection regions R satisfying* (2.2) *one that maximizes the power against both p_1 and p_2 and it rejects H when*

$$\frac{1}{2}[p_1(\underline{x})+p_2(\underline{x})] \quad \text{is sufficiently large.} \tag{2.3}$$

We shall call the test (2.3) the *average likelihood ratio test* and from now on shall refer to (1.1) as the *maximum likelihood ratio test.*

Proof. If R satisfies (2.2), its power against p_1 and p_2 must be the same. Hence

$$\int_R p_1 = \int_R p_2 = \int_R \frac{1}{2}(p_1+p_2). \tag{2.4}$$

□

By the Neyman–Pearson Lemma, the most powerful test of H against $\frac{1}{2}[p_1+p_2]$ rejects when (2.3) holds.

Corollary 2.1. *Under the assumptions of Theorem* 2.1, *the average LR test has power greater than or equal to that of the maximum likelihood ratio test against both p_1 and p_2.*

Proof. The maximum LR test rejects when

$$\max(p_1(\underline{x}),\, p_2(\underline{x})) \quad \text{is sufficiently large.} \tag{2.5}$$

Since this test satisfies (2.2), the result follows. □

The Corollary leaves open the possibility that the average and maximum LR tests have the same power; in particular they may coincide. To explore this possibility consider the case $n = 1$ and suppose that f is increasing. Then the likelihood ratio will be

$$f(x) \quad \text{if } x > \tfrac{1}{2} \quad \text{and } f(1-x) \text{ if } x < \frac{1}{2}. \tag{2.6}$$

The maximum LR test will therefore reject when

$$\left| x - \frac{1}{2} \right| \quad \text{is sufficiently large} \tag{2.7}$$

i.e. when x is close to either 0 or 1.

It turns out that the average LR test will depend on the shape of f and we shall consider two cases: (a) f is convex; (b) f is concave.

Theorem 2.2. *Under the assumptions of Theorem* 2.1 *and with* $n = 1$,

(i) (a) *if f is convex, the average LR test rejects when* (2.7) *holds;*
(b) *if f is concave, the average LR test rejects when*

$$\left| x - \frac{1}{2} \right| \quad \textit{is sufficiently small.} \tag{2.8}$$

(ii) (a) *if f is convex, the maximum LR test coincides with the average LR test, and hence is UMP among all tests satisfying* (2.2) *for n=1.*
(b) *if f is concave, the maximum LR test uniformly* minimizes *the power among all tests satisfying* (2.2) *for n=1, and therefore has power* $< \alpha$.

Proof. This is an immediate consequence of the fact that if $x < x' < y' < y$ then

$$\frac{f(x) + f(y)}{2} \text{ is } \gtrless \frac{f(x') + f(y')}{2} \text{ if } f \text{ is } \begin{matrix}\text{convex.}\\ \text{concave.}\end{matrix} \tag{2.9}$$

It is clear from the argument that the superiority of the average over the likelihood ratio test in the concave case will hold even if p_1 and p_2 are not exactly symmetric. Furthermore it also holds if the two alternatives p_1 and p_2 are replaced by the family $\theta p_1 + (1-\theta)p_2$, $0 \le \theta \le 1$. □

3. A finite number of alternatives

The comparison of maximum and average likelihood ratio tests discussed in Section 2 for the case of two alternatives obtains much more generally. In the present section we shall sketch the corresponding result for the case of a simple hypothesis against a finite number of alternatives which exhibit a symmetry generalizing (2.1).

Suppose the densities of the simple hypothesis and the s alternatives are denoted by $p_0, p_1, \ldots, p_s$ and that there exists a group G of transformations of the sample which leaves invariant both p_0 and the set $\{p_1, \ldots, p_s\}$ (i.e. each transformation results in a permutation of $p_1, \dots, p_s$). Let $\bar{G}$ denote the set of these permutations and suppose that it is transitive over the set $\{p_1, \dots, p_s\}$ i.e. that given any i and j there exists a transformation in $\bar{G}$ taking p_i into p_j. A rejection region R is said to be invariant under $\bar{G}$ if

$$x \in R \quad \text{if and only if} \quad g(x) \in R \text{ for all } g \text{ in } G. \tag{3.1}$$

Theorem 3.1. *Under these assumptions there exists a uniformly most powerful invariant test and it rejects when*

$$\frac{\sum_{i=1}^{s} p_i(x)/s}{p_0(x)} \quad \text{is sufficiently large.} \tag{3.2}$$

In generalization of the terminology of Theorem 2.1 we shall call (3.2) the average likelihood ratio test. The proof of Theorem 3.1 exactly parallels that of Theorem 2.1.

The Theorem extends to the case where G is a compact group. The average in the numerator of (3.2) is then replaced by the integral with respect to the (unique) invariant probability measure over $\bar{G}$. For details see Eaton ([3], Chapter 4). A further extension is to the case where not only the alternatives but also the hypothesis is composite.

To illustrate Theorem 3.1, let us extend the case considered in Section 2. Let (X, Y) have a bivariate distribution over the unit square which is uniform under H. Let f be a density for (X, Y) which is strictly increasing in both variables and consider the four alternatives

$$p_1 = f(x, y),\ p_2 = f(1-x, y),\ p_3 = f(x, 1-y),\ p_4 = f(1-x, 1-y).$$

The group G consists of the four transformations

$$g_1(x, y) = (x, y),\ g_2(x, y) = (1-x, y),\ g_3(x, y) = (x, 1-y),$$
$$\text{and } g_4(x, y) = (1-x, 1-y).$$

They induce in the space of $(p_1, \dots, p_4)$ the transformations:

$\bar{g}_1 =$ the identity
$\bar{g}_2$: $p_1 \to p_2$, $p_2 \to p_1$, $p_3 \to p_4$, $p_4 \to p_3$
$\bar{g}_3$: $p_1 \to p_3$, $p_3 \to p_1$, $p_2 \to p_4$, $p_4 \to p_2$
$\bar{g}_4$: $p_1 \to p_4$, $p_4 \to p_1$, $p_2 \to p_3$, $p_3 \to p_2$.

This is clearly transitive, so that Theorem 3.1 applies. The uniformly most powerful invariant test, which rejects when

$$\sum_{i=1}^{4} p_i(x, y) \quad \text{is large}$$

is therefore uniformly at least as powerful as the maximum likelihood ratio test which rejects when

$$\max\left[p_1(x, y), p_2(x, y), p_3(x, y), p_4(x, y)\right]$$

is large.

4. Location-scale families

In the present section we shall consider some more classical problems in which the symmetries are represented by infinite groups which are not compact. As a simple example let the hypothesis H and the alternatives K be specified respectively by

$$H: \ f(x_1-\theta,\ldots,x_n-\theta) \text{ and } K: \ g(x_1-\theta,\ldots,x_n-\theta) \tag{4.1}$$

where f and g are given densities and θ is an unknown location parameter. We might for example want to test a normal distribution with unknown mean against a logistic or Cauchy distribution with unknown center.

The symmetry in this problem is characterized by the invariance of H and K under the transformations

$$X_i' \ = \ X_i + c \ \ (i=1,\ldots,n). \tag{4.2}$$

It can be shown that there exists a uniformly most powerful invariant test which rejects H when

$$\frac{\int_{-\infty}^{\infty} g(x_1-\theta,\ldots,x_n-\theta)d\theta}{\int_{-\infty}^{\infty} f(x_1-\theta,\ldots,x_n-\theta)d\theta} \quad \text{is large.} \tag{4.3}$$

The method of proof used for Theorem 2.1 and which also works for Theorem 3.1 no longer works in the present case since the numerator (and denominator) no longer are averages. For the same reason the term average likelihood ratio is no longer appropriate and is replaced by integrated likelihood. However an easy alternative proof is given for example in Lehmann ([5], Section 6.3).

In contrast to (4.2), the maximum likelihood ratio test rejects when

$$\frac{g(x_1-\hat{\theta}_1,\ldots,x_n-\hat{\theta}_1)}{f(x_1-\hat{\theta}_0,\ldots,x_n-\hat{\theta}_0)} \quad \text{is large,} \tag{4.4}$$

where $\hat{\theta}_1$and $\hat{\theta}_0$ are the maximum likelihood estimators of θ under g and f respectively. Since (4.4) is also invariant under the transformations (4.2), it follows that the test (4.3) is uniformly at least as powerful as (4.4), and in fact more powerful unless the two tests coincide which will happen only in special cases.

The situation is quite similar for scale instead of location families. The problem (4.1) is now replaced by

$$H: \ \frac{1}{\tau^n} f(\frac{x_1}{\tau},\ldots,\frac{x_n}{\tau}) \ \text{ and } \ K: \ \frac{1}{\tau^n} g(\frac{x_1}{\tau},\ldots,\frac{x_n}{\tau}) \tag{4.5}$$

where either the x's are all positive or f and g and symmetric about 0 in each variable.

This problem remains invariant under the transformations

$$X_i' = cX_i, \ c > 0. \tag{4.6}$$

It can be shown that a uniformly most powerful invariant test exists and rejects H when

$$\frac{\int_0^{\infty} \nu^{n-1} g(\nu x_1,\ldots,\nu x_n)d\nu}{\int_0^{\infty} \nu^{n-1} f(\nu x_1,\ldots,\nu x_n)d\nu} \quad \text{is large.} \tag{4.7}$$

On the other hand, the maximum likelihood ratio test rejects when

$$\frac{g(\frac{x_1}{\hat{\tau}_1}, \ldots, \frac{x_n}{\hat{\tau}_n})/\hat{\tau}_1^n}{f(\frac{x_1}{\hat{\tau}_0}, \ldots, \frac{X_n}{\hat{\tau}_0})/\hat{\tau}_0^n} \quad \text{is large} \tag{4.8}$$

where $\hat{\tau}_1$ and $\hat{\tau}_0$ are the maximum likelihood estimators of τ under g and f respectively. Since it is invariant under the transformations (4.6), the test (4.8) is less powerful than (4.7) unless they coincide.

As in (4.3), the test (4.7) involves an integrated likelihood, but while in (4.3) the parameter θ was integrated with respect to Lebesgue measure, the nuisance parameter in (4.6) is integrated with respect to $\nu^{n-1}d\nu$. A crucial feature which all the examples of Sections 2–4 have in common is that the group of transformations that leave H and K invariant is transitive i.e. that there exists a transformation which for any two members of H (or of K) takes one into the other. A general theory of this case is given in Eaton ([3], Sections 6.7 and 6.4).

Elimination of nuisance parameters through integrated likelihood is recommended very generally by Berger, Liseo and Wolpert [1]. For the case that invariance considerations do not apply, they propose integration with respect to non-informative priors over the nuisance parameters. (For a review of such prior distributions, see Kass and Wasserman [4]).

5. The failure of intuition

The examples of the previous sections show that the intuitive appeal of maximum likelihood ratio tests can be misleading. (For related findings see Berger and Wolpert ([2], pp. 125–135)). To understand just how intuition can fail, consider a family of densities p_θ and the hypothesis $H\colon \theta = 0$. The Neyman–Pearson lemma tells us that when testing p_0 against a specific p_θ, we should reject y in preference to x when

$$\frac{p_\theta(x)}{p_0(x)} < \frac{p_\theta(y)}{p_0(y)} \tag{5.1}$$

the best test therefore rejects for large values of $p_\theta(x)/p_0(x)$, i.e. is the maximum likelihood ratio test.

However, when more than one value of θ is possible, consideration of only large values of $p_\theta(x)/p_0(x)$ (as is done by the maximum likelihood ratio test) may no longer be the right strategy. Values of x for which the ratio $p_\theta(x)/p_0(x)$ is small now also become important; they may have to be included in the rejection region because $p_{\theta'}(x)/p_0(x)$ is large for some other value θ'.

This is clearly seen in the situation of Theorem 2 with f increasing and g decreasing, as illustrated in Fig. 1.

For the values of x for which f is large, g is small, and vice versa. The behavior of the test therefore depends crucially on values of x for which $f(x)$ or $g(x)$ is small, a fact that is completely ignored by the maximum likelihood ratio test.

Note however that this same phenomenon does not arise when all the alternative densities $f, g, \ldots$ are increasing. When $n = 1$, there then exists a uniformly most powerful test and it is the maximum likelihood ratio test. This is no longer true when $n > 1$, but even then all reasonable tests, including the maximum likelihood ratio test, will reject the hypothesis in a region where all the observations are large.

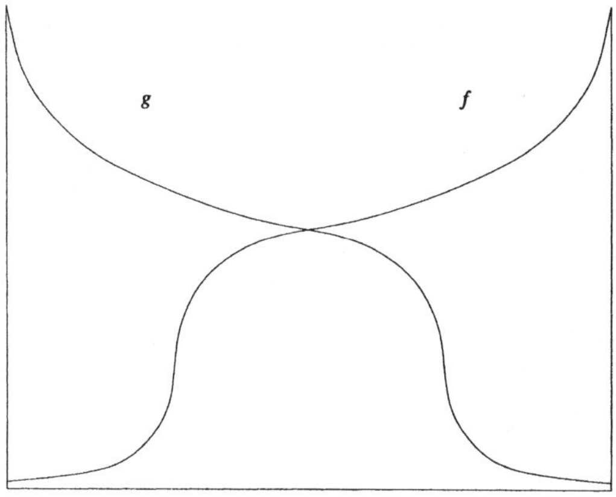

FIG 1.

6. Conclusions

For the reasons indicated in Section 1, maximum likelihood ratio tests are so widely accepted that they almost automatically are taken as solutions to new testing problems. In many situations they turn out to be very satisfactory, but gradually a collection of examples has been building up and is augmented by those of the present paper, in which this is not the case.

In particular when the problem remains invariant under a transitive group of transformations, a different principle (likelihood averaged or integrated with respect to an invariant measure) provides a test which is uniformly at least as good as the maximum likelihood ratio test and is better unless the two coincide. From the argument in Section 2 it is seen that this superiority is not restricted to invariant situations but persists in many other cases. A similar conclusion was reached from another point of view by Berger, Liseo and Wolpert [1].

The integrated likelihood approach without invariance has the disadvantage of not being uniquely defined; it requires the choice of a measure with respect to which to integrate. Typically it will also lead to more complicated test statistics. Nevertheless: In view of the superiority of integrated over maximum likelihood for large classes of problems, and the considerable unreliability of maximum likelihood ratio tests, further comparative studies of the two approaches would seem highly desirable.

References

[1] BERGER, J. O., LISEO, B. AND WOLPERT, R. L. (1999). Integrated likelihood methods for eliminating nuisance parameters (with discussion). *Statist. Sci.* **14**, 1–28. MR1702200

[2] BERGER, J. O. AND WOLPERT, R. L. (1984). *The Likelihood Principle.* IMS Lecture Notes, Vol. 6. Institue of Mathematical Statistics, Haywood, CA. MR0773665

[3] EATON, M. L. (1989). *Group Invariance Applications in Statistics.* Institute of Mathematical Statistics, Haywood, CA. MR1089423

[4] KASS, R. E. AND WASSERMAN, L. (1996). The selection of prior distributions by formal rules. *J. Amer. Statist. Assoc.* **91**, 1343–1370. MR1478684

[5] Lehmann, E. L. (1986). *Testing Statistical Hypotheses*, 2nd Edition. Springer-Verlag, New York. MR0852406

[6] Menéndez, J. A., Rueda, C. and Salvador, B. (1992). Dominance of likelihood ratio tests under cone constraints. *Amer. Statist.* **20**, 2087, 2099. MR1193327

[7] Perlman, M. D. and Wu, L. (1999). The Emperor's new tests (with discussion). *Statist. Sci.* **14**, 355–381. MR1765215

[8] Royall, R. (1997). *Statistical Evidence.* Chapman and Hall, Boca Raton. MR1629481

Part III
Estimation

Professor Erich Lehmann's Contributions to Testing of Hypotheses and Estimation

C.R. Rao

Professor Erich Lehmann, a major figure in the field of statistics, made fundamental contributions to several areas of statistical theory, especially in testing of hypotheses and estimation of unknown parameters. Although the foundations of these statistical methods were laid by Fisher, Neyman and Wald during the first half of the last century, no critical evaluation of these techniques was available. Lehmann by himself and in collaboration with some of his colleagues at the statistics department of the University of California, Berkeley, made a systematic study of these problems and provided a coherent theoretical structure. His contributions are reflected in his two great books. Testing of Hypotheses (1959) and Theory of Point Estimation (1983), which continue to be center pieces of graduate statistical education and have been translated into several languages.

The concepts of similar regions and sufficient statistics were introduced at an early stage in the development of statistical theory and both have proved extremely fruitful. A comprehensive discussion of these concepts was given in papers [4] and [6], making appropriate modifications and extending their use to a wide variety of problems in testing of hypotheses and estimation of parameters. The concept of minimal sufficient statistic was introduced to ensure optimality of certain procedures. A justification was given for the restriction to functions of sufficient statistic in testing of hypotheses similar to Rao-Blackwellization in estimators. The theory of completeness, similar regions and unbiased estimation was developed in greater generality than is customary in statistical papers. In [5], the uniqueness of unbiased estimators with uniformly minimum variance in sequential sampling was explored with special reference to Poisson, rectangular and binomial distributions. These three papers constitute a final authority on various issues related to hypotheses testing and unbiased estimation.

C.R. Rao
Professor Emeritus of Statistics, Eberly Professor Emeritus, Department of Statistics, Penn State University, University Park, PA 1602
e-mail: crr1@psu.edu

J. Rojo (ed.), *Selected Works of E. L. Lehmann*, Selected Works in Probability and Statistics, DOI 10.1007/978-1-4614-1412-4_21, © Springer Science+Business Media, LLC 2012

Situations where unbiased estimators do not exist or do not seem to be reasonable were discussed in [9]. Several examples were given where unbiased estimators do not exist or seem to be unsatisfactory. The same situation arises with maximum likelihood (ML) estimation when the sample does not contain sufficient information. In such cases ML estimators may be subject to large bias which make them not useful for practical applications. It is an open problem whether alternative methods of estimation exist in cases where reasonable minimum variance unbiased estimators or ML estimators do not exist.

Two problems of interest on unbiased estimation in convex families of distributions were considered in [2]. Let $X_1, \ldots, X_n$ be an i.i.d. sample from some distribution $F \in \mathbb{F}$, a family of distributions. Two problems of interest discussed in the earlier literature are as follows:

Problem 1: If a functional q is given on $\mathbb{F}$, does there exist a statistic $\boldsymbol{\delta}(X_1, \ldots, X_n)$ such that $E(\boldsymbol{\delta}|F) = q(F)$ for all $F \in \mathbb{F}$ or in other words, when does there exist an unbiased estimate of $q(F)$ based on a sample of n observations?

Problem 2: The degree of a functional q is defined as the smallest sample size $n \leq \infty$ such that an unbiased estimator $\boldsymbol{\delta}$ exists for q.

Satisfactory solutions were obtained for these two problems when $\mathbb{F}$ is closed under finite mixtures (convex combinations). Applications of these results were considered in a number of cases.

A minimax property of the sample mean in finite populations was proved in [3], under some conditions. Let $A = (a_1, \ldots, a_n)$ be the population and $(y_1, \ldots, y_n)$ be an estimate of $\bar{a}$, where $y_1, \ldots, y_n$ are i.i.d. observations from A. Further let $\tau^2 = (a_i - \bar{a})^2$. In an earlier paper, [1], it was proved that

$$\sup E(\delta - \bar{a})^2/\tau^2$$

is minimized when $\delta = \bar{y} = (y_1 + \cdots + y_n)/n$. It was proved in [3] that

$$\sup E(\delta - \bar{a})^2$$

is a minimum when $\delta = \bar{y}$ under the conditions $\tau^2 \leq M$. It was further shown that the same result holds if the sample includes information on the labels of the sample units, and also under stratified sampling.

A comprehensive discussion of ancillary and sufficient statistics, and the concepts of completeness and bounded completeness was given in [8]. Basu proved that when a sufficient statistic T is boundedly complete, then all ancillary statistics are independent of T. The converse of Basu's theorem that independence of all ancillary statistics from a sufficient statistic implies that T is boundedly complete is not true. Some alternative definitions of ancillarity and completeness were given in [8] which imply the converse of Basu's theorem.

Nonparametric tests were introduced quite early in the development of statistics by Pitman, Willcoxon, Kruskal, Wallis and others. In [7], an appropriate methodology was developed to obtain point estimates of location or shift parameters of a distribution using rank test statistics. This seems to be an attractive area where further research is necessary.

References

[1] Aggarwal, O. P. (1955). Bayes and minimax procedures in sampling from finite and infinite populations–I. *Ann. Math. Statist.,* **30**, 206–218.

[2] Bickel, P. J. and Lehmann, E. L. (1969). Unbiased estimation in convex families. *The Annals of Mathematical Statistics*, **40**, 1523–1535.

[3] Bickel, P. J. and Lehmann, E. L. (1981). A minimax property of the sample mean in finite populations. *The Annals of Statistics*, **9**, 119–122.

[4] Hodges, Jr. J. L. and Lehmann, E. L. (1963). Estimation of location based on rank tests. *The Annals of Mathematical Statistics*, **34**, 598–611.

[5] Lehmann, E. L. and Scheffe, H. (1950). Completeness, similar regions and unbiased estimation part 1. *Sankhya*, **10**, 305–340.

[6] Lehmann, E. L. and Scheffe, H. (1956). Completeness, similar regions and unbiased estimation part 2. *Sankhya*, **15**, 219–236, 17, 250.

[7] Lehmann, E. L. and Stein, C. (1950). Completeness in the sequential case. *The Annals of Statistics*, **21**, 376–385.

[8] Lehmann, E. L. (1981). An interpretation of completeness and Basu's theorem. *Journal of the American Statistical Association*, **76**, 335–340.

[9] Lehmann, E. L. (1983). Estimation with inadequate information. *Journal of the American Statistical Association*, **78**, 624–627.

COMPLETENESS IN THE SEQUENTIAL CASE

By E. L. Lehmann and Charles Stein

University of California, Berkeley

1. Summary. Recently, in a series of papers, Girshick, Mosteller, Savage and Wolfowitz have considered the uniqueness of unbiased estimates depending only on an appropriate sufficient statistic for sequential sampling schemes of binomial variables. A complete solution was obtained under the restriction to bounded estimates. This work, which has immediate consequences with respect to the existence of unbiased estimates with uniformly minimum variance, is extended here in two directions. A general necessary condition for uniqueness is found, and this is applied to obtain a complete solution of the uniqueness problem when the random variables have a Poisson or rectangular distribution. Necessary and sufficient conditions are also found in the binomial case without the restriction to bounded estimates. This permits the statement of a somewhat stronger optimum property for the estimates, and is applicable to the estimation of unbounded functions of the unknown probability.

2. Introduction. The notions of completeness and bounded completeness of a family of distributions were introduced in [1, 2] in connection with the problems of similar regions and unbiased estimation. The question of whether either of these two properties pertains to various families of distributions that are of interest in statistics was discussed in [2] under the assumption of fixed sample size. The only sequential problems of this kind that have been treated in the literature (with quite different terminology) refer to the binomial case. For this case Girshick, Mosteller and Savage [3] found necessary (and also certain sufficient) conditions on the sequential sampling scheme for completeness, while Wolfowitz [4] and Savage [5] gave necessary and sufficient conditions for bounded completeness.

If T is a random variable distributed over an additive class of sets in some space according to a distribution P_θ^T with θ in some set ω, then the family $\mathcal{P}^T = \{P_\theta^T \mid \theta \in \omega\}$ of possible distributions of T is said to be complete if

$$\int f(t)\, dP_\theta^T(t) = 0, \qquad \text{for all} \quad \theta \in \omega, \tag{1}$$

implies

$$f(t) = 0, \qquad \text{a.e. } \mathcal{P}^T, \tag{2}$$

that is, for all t except possibly in a set N for which $P_\theta^T(N) = 0$ for all $\theta \in \omega$. The family $\mathcal{P}^T$ is said to be boundedly complete if this implication holds under the assumption that f is bounded.

The relation of these concepts to the problem of unbiased estimation is an

J. Rojo (ed.), *Selected Works of E. L. Lehmann*, Selected Works in Probability and Statistics, DOI 10.1007/978-1-4614-1412-4_22, © Springer Science+Business Media, LLC 2012

immediate consequence of a theorem of Blackwell [6]. Let X be a random variable with distribution P_θ^X, $\theta \epsilon \omega$, and let T be a sufficient statistic for θ. Denote by P_θ^T the distribution of T, and suppose that $\mathcal{P}^T$ is complete. Then every function $g(\theta)$ for which there exists an unbiased estimate, that is, a function ϕ such that

$$E_\theta \phi(X) = g(\theta), \qquad \text{for all} \quad \theta \epsilon \omega,$$

possesses an unbiased estimate with uniformly minimum variance. One can say furthermore that if $\phi(X)$ is any unbiased or bounded unbiased estimate of $g(\theta)$, then the optimum estimate guaranteed by the above statements is the conditional espectation of $\phi(X)$ given T.

The aim of the present paper is to obtain certain results concerning completeness in sequential sampling schemes. Some necessary conditions for completeness are given in section 3, and these are used to obtain necessary and sufficient conditions for completeness when the random variable being sampled has a Poisson or rectangular distribution. In section 4 it is shown that certain necessary conditions given in [3] for the binomial case are also sufficient.

3. A necessary condition for completeness. The sequential sampling schemes with which we are concerned are of the following nature. There is given a sequence of real valued random variables $X_1, X_2, \cdots$ with a joint distribution depending on a real parameter θ, which ranges over a set ω. We shall assume that for each m the set of variables $X_1, \cdots, X_m$ admits a real valued sufficient statistic $T_m = t_m(X_1, \cdots, X_m)$ for θ, and that for each m the family $\mathcal{P}^{T_m}$ of distributions of T_m is complete. We next suppose that there is given a stopping rule, which is such that after m observations have been taken, the decision of whether or not to take an $m+1st$ observation depends only on the value of $t_m(X_1, \cdots, X_m)$. It follows (see [6]) that if the total number of observations is n (a random variable which may be infinite), then (T_n, n) is a sufficient statistic for θ. We shall say that the sequential procedure is complete if the family of distributions of (T_n, n) is complete. Throughout, we shall assume that all sequential procedures in question are closed, i.e. that for each $\theta \epsilon \omega$, n is finite with probability 1.

Let Y be a random variable distributed over a Euclidean space according to a distribution P_θ^Y with θ in ω. We shall say that a point y lies in the positive sample space of Y if there exists $\theta \epsilon \omega$ such that every open set containing y has positive probability for this θ, and that y is an impossible point if it lies in the complement of the positive sample space. Consider now a sequential sampling scheme as described above. For any integers $m < p$ we shall denote by W_p^m the positive sample space of T_p given the first m steps of the stopping rule, that is, given for $i = 1, \cdots, m$ the set S_i of values of T_i for which sampling is discontinued after the ith observation. Since all the T's are real valued, the sets W_p^m are sets of real numbers satisfying the obvious condition $W_p^{m-1} \geqq W_p^m$. The union $\bigcup S_m$ (S_m is the set of points of W_m^{m-1} for which no $m+1st$ observation is taken) will be called the set of stopping or boundary points, the points belonging to some $W_m^{m-1} - S_m$ are the continuation points.

We need the following

LEMMA 1. *A necessary condition for a sequential procedure of the type described above to be complete is that every procedure obtained from the given one by truncation be complete.*[1]

This is an immediate consequence of the following more general

LEMMA 2. *Let $X_1, X_2, \cdots$ be as before a sequence of random variables such that for each m the set $X_1, \cdots, X_m$ admits a real valued sufficient statistic $T_m = t_m(X_1, \cdots, X_m)$. Let $\Sigma_1, \Sigma_2, \cdots, \Sigma_r$ each be a complete, closed, sequential procedure based on these sufficient statistics. Let $\Sigma_1 \cup \Sigma_2 \cup \cdots \cup \Sigma_r$ denote the sequential procedure according to which we continue taking observations until at least one of the stopping rules $\Sigma_1, \cdots, \Sigma_r$ tells us to stop. Then the procedure $\Sigma_1 \cup \cdots \cup \Sigma_r$ is complete.*

This clearly implies Lemma 1. For if one takes for Σ_1 any closed, complete sequential procedure and for Σ_2 a procedure of fixed sample size, then $\Sigma_1 \cup \Sigma_2$ is the associated truncated procedure.

PROOF OF LEMMA 2. It is sufficient to prove the result for the case $r = 2$. Let n_1, n_2, n denote the number of observations taken under $\Sigma_1, \Sigma_2, \Sigma_1 \cup \Sigma_2$ respectively. Then $n = n_1$ if $n_1 \leqq n_2$, $n = n_2$ if $n_1 \geqq n_2$. Let f be any function on $\Sigma_1 \cup \Sigma_2$ such that

$$E_\theta f(T_n, n) = 0 \quad \text{for all} \quad \theta \in \omega.$$

Then

$$\left.\begin{aligned} E_\theta\, E[f(T_n, n) \mid T_{n_1}, n_1] &= 0 \\ E_\theta\, E[f(T_n, n) \mid T_{n_2}, n_2] &= 0 \end{aligned}\right\} \quad \text{for all} \quad \theta \in \omega.$$

Since Σ_1 and Σ_2 are complete it follows that

$$E[f(T_n, n) \mid T_{n_1} = t_1, n_1 = \gamma_1] = E[f(T_n, n) \mid T_{n_2} = t_2, n_2 = \gamma_2] = 0, \text{ a.e.}$$

Hence

$$\begin{aligned} 0 &= P(n_1 \leqq n_2 \mid T_{n_1} = t_1, n_1 = \gamma_1) f(t_1, \gamma_1) \\ &\quad + P(n_1 > n_2 \mid T_{n_1} = t_1, n_1 = \gamma_1) E[f(T_{n_2}, n_2) \mid T_{n_1} = t_1, n_1 = \gamma_1, n_1 > n_2], \end{aligned} \tag{3}$$

and the analogous condition holds with the subscripts 1 and 2 interchanged.

We shall prove that $f(T_n, n) = 0$, a.e., by induction over the possible values of n. Suppose, therefore, that for some integer m

$$P_\theta(n \leqq m, f(T_n', n) \neq 0) = 0.$$

(This is certainly true for $m = 0$.) It then follows that if we take $\gamma_1 = m + 1$ in (3) the second term of the right hand side vanishes, so that

$$0 = P(n = n_1 \mid T_{n_1} = t_1, n_1 = m + 1) f(t_1, m + 1).$$

[1] The authors would like to thank Mr. E. Fay for pointing out an error in the original proof of this Lemma.

Hence,

$$P_\theta(n = n_1 = m + 1, f(T_{n_1}, n_1) \neq 0)$$
$$\leqq P_\theta(n = n_1 = m + 1, P(n = n_1 \mid T_{n_1}, n_1) = 0) = 0.$$

Analogously we see that

$$P_\theta(n = n_2 = m + 1, f(T_{n_2}, n_2) \neq 0) = 0$$

and, adding, that

$$P_\theta(n = m + 1, f(T_n, n) \neq 0) = 0.$$

This completes the induction.

We need further the notion of strong completeness. Consider a random variable $W = (U, V)$, suppose that the distribution of W depends on θ, and that U is a sufficient statistic for θ. Let P_u^V be the conditional distribution of V given $U = u$—this is independent of θ since U is a sufficient statistic for θ—and let $\mathcal{P}^{V*} = \{\mathcal{P}_u^V\}$. We say that the pair $\mathcal{P}^W, \mathcal{P}^{V*}$ is strongly complete if the conditions
(i) $E_\theta f(V)$ exists for all θ,
(ii) $E(f(V) \mid U = u) = 0$ for almost all u,
imply

$$f(v) = 0, \qquad \text{a.e. } \mathcal{P}^V.$$

For brevity, we shall then usually say that $\{\mathcal{P}_u^V\}$ is strongly complete.

We can now state the following necessary condition for completeness.

THEOREM. *If a closed sequential procedure of the type considered above is complete, then*

(i) *S_m is almost empty for every m for which $W_{m+1}^{m-1} - W_{m+1}^m$ is almost empty,*

(ii) *for each m for which S_m is not almost empty, the family of conditional distributions of T_m given $T_{m+1} = t$ (as t ranges over $W_{m+1}^{m-1} - W_{m+1}^m$) is strongly complete.*

PROOF. For any $t \in W_{m+1}^{m-1} - W_{m+1}^m$ the positive sample space of T_m given $T_{m+1} = t$ is clearly contained in S_m. Suppose first that (ii) is violated and consider the sequential procedure obtained from the given one by truncation after $m + 1$ observations. By the lemma it will be enough to show that the truncated procedure is not complete. For this purpose let us assume that regardless of the stopping rule all $m + 1$ variables $X_1, \cdots, X_{m+1}$ are observed. We want to construct an estimate of zero based on the sufficient statistic for the truncated procedure. This estimate must be a function of T_1 for $T_1 \in S_1$, of T_2 for $T_2 \in S_2$, etc. That is, although we may imagine that the full sample of size $m + 1$ is taken, we must be careful not to use observations that are impossible when the stopping rule is followed.

We shall now show that there exists an unbiased estimate of zero which is zero over $S_1, \cdots, S_{m-1}$, equal to $f(T_m)$ on S_m and $g(T_{m+1})$ on W_{m+1}^m where f and g will be defined below. Since expectation equals expectation of conditional expectation, a statistic is an unbiased estimate of zero if its expectation exists and its conditional expectation given $T_{m+1} = t$ is zero for almost all t. In our case this condition is equivalent to

$$(4) \qquad \int_{S_m} f(u)\, dP_m(u \mid T_{m+1} = t) + g(t) \int_{W^{m-1}_m - S_m} dP_m(u \mid T_{m+1} = t) = 0$$

for almost all $t \in W^m_{m+1}$,

$$(5) \qquad \int_{S_m} f(u)\, dP_m(u \mid T_{m+1} = t) = 0$$

for almost all $t \notin W^m_{m+1}$, i.e. for almost all $t \in W^{m-1}_{m+1} - W^m_{m+1}$, since $t \notin W^{m-1}_{m+1}$ implies $P(S_m \mid T_{m+1} = t) = 0$,

together with the existence of $E_\theta(f(T_m) \mid n = m)$ and $E_\theta(g(T_{m+1}) \mid n = m + 1)$. Since (ii) does not hold there exists f not vanishing a.e. such that $E_\theta(f(T_m) \mid n = m)$ exists and (5) is satisfied. If g is defined by (4), $E_\theta(g(T_{m+1}) \mid n = m + 1)$ exists, and this completes the proof of the necessity of (ii).

The necessity of (i) is now obvious. For if (i) is violated, then (5) is satisfied vacuously, and we can take f to be an arbitrary positive valued function (for example) and (4) will then be satisfied.

As immediate consequences of this theorem we shall obtain two conditions, which are easier to apply than condition (ii).

COROLLARY 1. *A necessary condition for completeness is that for no m there exists a subset A of S_m such that*

$$P_\theta(A) > 0 \quad \textit{for some} \quad \theta$$

and

$$P(A \mid T_{m+1} = t) = 0 \quad \textit{for almost all} \quad t \in W^{m-1}_{m+1} - W^m_{m+1}.$$

COROLLARY 2. *Suppose that the sequence of X's is such that in the non-sequential case for all m, p with $m < p$ the positive sample space of T_m given $T_p = t$ is the intersection of the unconditional positive sample space of T_m with the interval* $[0, t]$. *Then a necessary condition for a sequential procedure to be complete is that each S_m differ from a half-open interval (possibly empty) $[a_m, b_m)$ with $a_m \leqq b_m$, $a_1 = 0$, $a_{m+1} = b_m$, by a set of probability* 0.

PROOF. Let r be the first value of m for which this condition is not satisfied. Then there exists $c > b_{r-1}$ such that the sets $S_r \cap [c, \infty)$ and $\bar{S}_r \cap [b_{r-1}, c)$ both have positive probability. The result now follows from Corollary 1 if one puts $A = S_r \cap [c, \infty)$.

Next we consider some examples.

EXAMPLE 1. Let $X_1, X_2, \cdots$ be independently normally distributed with known variance and unknown mean θ. In this case $T_m = \sum_{i=1}^m X_i$, and since the positive sample space of T_{m+1} is the infinite interval regardless of the values of $T_1, \cdots, T_m$ it follows from condition (i) of the theorem that no sequential procedure is complete, with the trivial exception of the procedures with fixed sample size.

EXAMPLE 2. Let $X_1, X_2, \cdots$ be independently uniformly distributed over the interval $(0, \theta)$, $0 < \theta < \infty$. Then $T_m = \max(X_1, \cdots, X_m)$ and Corollary 2 gives a necessary condition for completeness. If the procedure is truncated we can deduce sufficiency of this condition from (5). However, this proof does not apply to the general case. The following proof of sufficiency is similar to some of the proofs in [3, 4, 5].

Suppose $S_1, S_2, \cdots$ form a set of adjoining intervals (some of them possibly empty), $S_m = [a_m, b_m)$, and suppose there is a non-zero unbiased estimate of zero, $\Phi = \phi(T_n, n)$. Let m be the smallest integer for which ϕ is not zero almost everywhere on S_m. Then

$$E_\theta(\Phi) = P_\theta(n = m)E_\theta(\Phi \mid n = m) + \sum_{j=m+1}^{\infty} P_\theta(n = j)E_\theta(\Phi \mid n = j) \overset{(\theta)}{\equiv} 0,$$

and hence

$$P_\theta(n = m)E_\theta(\Phi \mid n = m) \overset{(\theta)}{\equiv} -\sum_{j=m+1}^{\infty} P_\theta(n = j)E_\theta(\Phi \mid n = j). \tag{6}$$

Now the right hand side of (6) is zero when $\theta \leqq b_m$, since it is then impossible that $T_j \in S_j$ for any $j > m$. Hence

$$E_\theta[\phi(T_m, m) \mid a_m \leqq T_m < b_m] = 0 \quad \text{for all} \quad \theta \leqq b_m,$$

and therefore

$$\int_{a_m}^{\theta} \phi(x, m)x^{m-1}\, dx = 0 \quad \text{for all} \quad \theta \text{ in } [a_m, b_m].$$

But this implies $\phi(x, m) = 0$ almost everywhere in S_m, which is a contradiction.

EXAMPLE 3. Let $X_1, X_2, \cdots$ be independently distributed according to a Poisson distribution with mean θ. Then $T_m = \sum_{i=1}^{m} X_i$ and again we can apply Corollary 2. To prove sufficiency we proceed as in example 2. If the condition of Corollary 2 is satisfied we may write without ambiguity $\psi(T_n)$ for $\phi(T_n, n)$. Let c be the smallest value of T_n for which $\psi(T_n) \neq 0$. Then if the probability of $T_n = j$ is $k(j)\theta^j e^{-\theta m_j}$, the identity $E_\theta(\Phi) \overset{(\theta)}{\equiv} 0$ implies

$$\phi(c)k(c)\theta^c \rho \overset{(\theta)}{\equiv} \sum_{j=c+1}^{\infty} \phi(j)k(j)\phi^j \cdot \varphi(c)k(c)\theta^c e^{-\theta m_c} \equiv \sum_{j=c+1}^{\infty} \varphi(j)k(j)\theta^j e^{-\theta m_j}.$$

Dividing this equation by θ^c and letting θ tend to zero we see that the right hand side tends to zero, which implies $\phi(c) = 0$ and hence a contradiction.

4. The binomial case. As was mentioned in section 1, the problem of bounded completeness was solved for the binomial case in [3, 4, 5]. Since presumably one is unwilling to estimate the bounded parameter p by means of an unbounded estimate, further work here may seem unnecessary. However, the problem of completeness seems to be of interest for two reasons. If the procedure is boundedly complete without being complete then, even though one may be reluctant to use such an estimate, there may exist an unbounded unbiased estimate of p, which for some values of p has smaller variance than the minimum variance bounded estimate. (An example of this is given in [2]). Since this possibility is ruled out when the procedure is complete it is seen that completeness permits statement of a stronger optimum property. Apart from this one may be interested in estimating some unbounded function of p such as $1/p$. In this case bounded completeness does not permit any statements concerning existence of optimum estimates.

In the present section we shall change our notation somewhat. We are concerned with a sequence of independent trials with constant probability p of success. On the basis of m trials the total number y of successes is a sufficient statistic for p. Instead of representing the sufficient statistic for the sequential procedure by (y, n), we shall use the representation (x, y) where x is the total number of failures, so that $x + y = n$. The couples (x, y) may be thought of as making up the points with integral-valued coordinates of the first quadrant of an xy-plane, and as before may be classified as boundary points, continuation points, and impossible points. Adopting the terminology of [3], we shall call the value of $x + y$ the index of the point (x, y), so that the points of index m lie on the line $x + y = \text{m}$.

Girshick, Mosteller and Savage defined a sequential procedure to be simple if for each m the continuation points of index m form an interval. They proved that a necessary and sufficient condition for a bounded procedure to be complete is that it be simple. (A procedure is said to be bounded if there exists N so that the number of observations is $\leqq N$.) They also showed that in general simplicity is not sufficient for completeness. However, it was shown later [4, 5] that simplicity is sufficient for bounded completeness.

A sequential procedure is said to be closed if the probability of termination is unity for every p with $0 < p < 1$. It was proved by Girshick, Mosteller and Savage that a necessary condition for completeness of a closed sequential procedure is that no procedure obtained from the given one by removing a boundary point be closed. (Removing a boundary point here means converting it into a continuation point.) We shall prove below that this condition together with simplicity is also sufficient for completeness. An interesting question is whether these two conditions are sufficient for completeness for the general sequential schemes considered in section 2, when simplicity is replaced by the condition that every procedure obtained from the given one by truncation is complete, and when the second condition is modified by the appropriate null set qualifications. It is easily seen that both of these conditions are necessary.

The following definitions will be needed below. A boundary point (a, b) is a lower (upper) boundary point if for some $x < 0$ (>0) the point $(a + x, b - x)$ is a continuation point. An impossible point (a, b) is a lower (upper) impossible point if for some $x < 0$ (>0) the point $(\text{a} + x, b - x)$ is either a continuation point or a boundary point.

If the procedure is unbounded every boundary point is either a lower or an upper boundary point. If it is simple, no point can be both an upper and a lower boundary point. The same remarks apply to impossible points.

THEOREM. *A necessary and sufficient condition for completeness of a closed procedure in the binomial case is that*
(i) *the procedure is simple,*
and
(ii) *the removal of any boundary point destroys closure.*

PROOF. Necessity was proved in [3] as was sufficiency for bounded procedures. Sufficiency for unbounded procedures will follow from the following two facts, which we shall prove below.

I. Suppose (i) holds and there exist numbers a, $M > 0$ such that for all boundary points (x, y) of index $m \geqq M$ the ratio $y/x \geqq a$. Let $f(x, y)$ be a non-zero unbiased estimate of zero defined over the set B of boundary points, and let m_0 be the smallest index for which there are points with $f(x, y) \neq 0$. Then $f(x, y) = 0$ for all lower boundary points of index m_0.

II. If (i) holds and if for every positive number a there exist infinitely many boundary points (x, y) with $y/x \leqq a$, then one may remove any lower boundary point without destroying closure.

Suppose now that a sequential procedure satisfies (i) and (ii). Then, since no lower boundary point can be removed without destroying closure, it follows from II. that there exist a and M such that $y/x \geqq a$ for all boundary points of index $\geqq M$. Hence if $f(x, y)$ is an unbiased estimate of zero, and if m_0 is defined as in I., $f(x, y) = 0$ for all lower boundary points of index m_0. Because of symmetry the statements concerning upper boundary points analogous to I. and II. also hold. It then follows analogously that $f(x, y) = 0$ for all upper boundary points of index m_0. But for a simple unbounded procedure every boundary point is either an upper or a lower boundary point, and hence we obtain a contradiction with the definition of m_0.

Before proving I. and II. we state the following corollary, which generalises an example given in [3].

COROLLARY. *A sequential procedure that is not bounded and that has a finite non-zero number of lower boundary points is not complete. The analogous result holds for upper boundary points.*

PROOF OF COROLLARY. This follows easily from II., since if a procedure of this type is to be closed there must exist for each $a > 0$ infinitely many upper boundary points (x, y) with $y/x \leqq a$.

In the remainder of the paper we are concerned with the proofs of I. and II.

PROOF OF I. Assume I. to be false, and let (x_0, y_0) be the lowest boundary point of index m_0 for which $f(x_0, y_0) \neq 0$. Then $y > y_0$ for all other boundary points (x, y) for which $f(x, y) \neq 0$. Hence if the probability of a point (x, y) is $c(x, y)p^y q^x$ and if $k(x, y) = c(x, y)f(x, y)$,

$$k(x_0, y_0)p^{y_0}q^{x_0} \equiv -\Sigma k(x, y)p^y q^x,$$

where the summation extends over all boundary points of index $\geqq m_0$ for which $y > y_0$. Dividing both sides by p^{y_0} we see that

$$k(x_0, y_0)q^{x_0} = -p\Sigma k(x, y)p^{y-y_0-1}q^x.$$

If we can show that the expression multiplying $-p$ on the right hand side remains bounded as p tends to zero, we have a contradiction. For letting p tend to zero, we would then see that the right hand side tends to zero and the left hand side to $k(x_0, y_0)$, and hence that $f(x_0, y_0) = 0$.

To prove this, note that

$$| \Sigma k(x, y)p^{y-y_0-1}q^x | \leqq \Sigma | k(x, y) | p^{y-y_0-1}.$$

The right hand side is a power series in p. We shall show that this series converges for some $p_0 > 0$. This implies uniform convergence for $| p | < p_0$, and therefore the series remains bounded at $p = 0$. By assumption there exist numbers a and M' such that $y/x \geqq a$ for all boundary points with $y > M'$. From now on we shall consider all series as being summed over the set of boundary points for which $y > M'$ and hence $q^x \geqq q^{y/a}$. Since only a finite number of terms are omitted this does not affect any convergence properties.

Let $0 < p_1 < 1$. Then, since f is an unbiased estimate of zero, the series

$$\Sigma k(x, y)p_1^y q_1^x$$

converges absolutely. Hence, so does

$$\Sigma | k(x, y) | p_1^{y-y_0-1}q_1^{x-\frac{1}{a}(y_0+1)} \geqq \Sigma | k(x, y) | (q_1 p_1^{\frac{1}{a}})^{y-y_0-1} = \Sigma | k(x, y) | p_0^{y-y_0-1},$$

and consequently the last series is convergent.

Proof of II. Let R be any closed simple procedure satisfying the conditions of II., and let (x_0, y_0) be any lower boundary point of R. We denote by R^* the procedure obtained from R by taking (x_0, y_0) to be a continuation point and by n^* the number of observations for R^*.

We first prove that any upper impossible point of R is also an impossible point of R^*. The negation of this would imply that one can get from a lower boundary point to an upper impossible point going only through impossible points. This would require at least one step of either of the following kinds:

Lower impossible point $\rightarrow$ upper impossible point;

Lower boundary point $\rightarrow$ upper impossible point.

One can easily convince oneself with the aid of a diagram that any procedure under which such steps are permitted cannot be simple.

Let $0 < p, \pi < 1$, and let a be such that $0 < a < p/q$. If p is the true probability of success, y/x tends in probability to p/q, and hence there exists N such that

$$P(y/x \geqq a \mid p) > \pi$$

whenever the index of (x, y) exceeds N. By assumption there exists $N_1 > N$ and a boundary point (x_1, y_1) of R^* of index N_1 such that $y_1/x_1 \leqq a$. Then the probability exceeds π that the random point (x, y) of index N_1 will lie above (x_1, y_1). Since (x_1, y_1) is a boundary point, the probability is therefore greater than π that the point (x, y) of index N is either an upper impossible point for R and hence impossible for R^*, or a stopping or continuation point for R. We have therefore proved that the probability is $>\pi$ that either $n^* \leqq N_1$ or the point (x, y) of index N_1 is a continuation point of R.

But given that one has reached a continuation point (a, b) of R, there exists N_2 such that

$$P(n^* \leqq N_2 \mid p, (a, b)) \geqq \pi.$$

For

$$P(n^* > N_2 \mid (a, b)) = P(n > N_2 \mid (a, b)) \rightarrow 0 \quad \text{as} \quad N_2 \rightarrow \infty.$$

Since there are only a finite number of continuation points of index N_1, it is now clear that there exists N_0 such that

$$P(n^* \leqq N_0 \mid p) \geqq \pi + \pi^2 - 1,$$

which can be made arbitrary close to 1 by proper choice of π. Therefore R^* is closed.

REFERENCES

[1] E. L. Lehmann and H. Scheffé, "On the problem of similar regions," *Proc. Nat. Acad. Sci.*, Vol. 33 (1947), pp. 382–386.

[2] E. L. Lehmann and H. Scheffé, "Completeness, similar regions and unbiased estimation," unpublished.

[3] M. A. Girshick, Frederick Mosteller and L. J. Savage, "Unbiased estimates for certain binomial sampling problems, with applications," *Annals of Math. Stat.*, Vol. 17 (1946), pp. 13–23.

[4] J. Wolfowitz, "On sequential binomial estimation," *Annals of Math. Stat.*, Vol. 17 (1946), pp. 489–493.

[5] L. J. Savage, "A uniqueness theorem for unbiased sequential binomial estimation," *Annals of Math. Stat.*, Vol. 18 (1947), pp. 295–297.

[6] D. Blackwell, "Conditional expectation and unbiased sequential estimation," *Annals of Math. Stat.*, Vol. 18 (1947), pp. 105–110.

SANKHYĀ

THE INDIAN JOURNAL OF STATISTICS

Edited By : P. C. MAHALANOBIS

VOL. 10 PART 4 1950

COMPLETENESS, SIMILAR REGIONS, AND UNBIASED ESTIMATION—PART I

By E. L. LEHMANN

University of California

AND

HENRY SCHEFFÉ

Columbia University

1. INTRODUCTION

The aim of this paper is the study of two classical problems of mathematical statistics, the problems of similar regions and of unbiased estimation. The reason for studying these two problems together is that both are concerned with a family of measures and that essentially the same condition on this family insures a very simple solution of both.

The concepts of similar region and unbiased estimate were introduced at an early stage in the development of statistical theory, and both have proved extremely fruitful. On the other hand it seems rather difficult to justify either in a completely systematic development of statistics.

Similar regions were defined by Neyman and Pearson (1933) in connection with the problem of testing composite hypotheses. Suppose one wishes to test the hypothesis H that a random variable X, taking on values in a "sample space" of points x, is distributed according to some probability distribution P_θ^x of a family $\mathfrak{P}^x = \{P_\theta^x \mid \theta\epsilon\omega\}$. If the hypothesis is to be tested at level of significance α, the critical region A must satisfy the condition

$$P_\theta \; (A) \leqslant \alpha \qquad \text{for all } \theta \text{ in } \omega. \qquad \ldots \quad (1.1)$$

Neyman and Pearson replaced this by the much stronger condition of similarity

$$P_\theta^x \; (A) = \alpha \qquad \text{for all } \theta \text{ in } \omega. \qquad \ldots \quad (1.2)$$

In a number of important cases the problem reduces by this device to that of testing a simple hypothesis, and furthermore there exists among the similar regions one which is either uniformly most powerful or uniformly most powerful unbiased (Scheffé, 1942; Lehmann, 1947). On the other hand, in the same cases the most powerful test (among all those satisfying (1.1)) frequently depends very strongly on the specific alternative at which the power is maximized (Lehmann and Stein, 1948). In these situations therefore the simplifying value of the restriction (1.2) is considerable.

Somewhat analogous remarks hold for the problem of point estimation. Given again a family of measures $\mathfrak{P}^x = \{P_\theta^x \mid \theta \epsilon \omega\}$, it is desired to estimate some real-valued function $g(\theta)$. A reasonable estimate T would seem to be one which minimizes the risk for some given weight function; say, which minimizes

$$E_\theta[T - g(\theta)]^2 \qquad \dots \quad (1.3)$$

where E_θ denotes the expected value calculated with P_θ^x. Unfortunately, this estimate will depend on the value of θ for which (1.3) is minimized. If this value is θ_0 one clearly chooses $T = g(\theta_0)$. One way of avoiding this difficulty would be to replace (1.3) by

$$\sup_\theta E_\theta[T - g(\theta)]^2. \qquad \dots \quad (1.4)$$

Another approach consists in restricting the class of estimates by the intuitively rather appealing condition of unbiasedness

$$E_\theta(T) = g(\theta) \qquad \text{for all } \theta \text{ in } \omega. \qquad \dots \quad (1.5)$$

It appears that among the unbiased estimates there frequently exists one minimizing (1.3)—which now becomes the variance of T—uniformly in θ. (This clearly cannot happen if we omit the condition of unbiasedness unless $g(\theta)$ is constant).

In order to obtain the results alluded to above, we introduce the notion of completeness of a family of measures $\mathfrak{M}^x = \{M_\theta^x \mid \theta \epsilon \omega\}$. The family $\mathfrak{M}^x$ is said to be complete if

$$\int f(x)\, d\, M_\theta^x\,(x) = 0 \quad \text{for all } \theta \text{ in } \omega \qquad \dots \quad (1.6)$$

implies $f(x) = 0$ except on a set N with $M_\theta^x(N) = 0$ for all θ in ω. The family is said to be boundedly complete if this implication holds for all bounded functions f. Now let X be a random variable with distribution P_θ^x in the family $\mathfrak{P}^x$, let T be a sufficient statistic (not necessarily real-valued) for $\mathfrak{P}^x$, and denote by P_θ^t the distribution of T. We shall show that if $\mathfrak{P}^t$ is complete, then for any real valued estimable function $g(\theta)$ there exists an unbiased estimate with uniformly smallest variance, and that this is the only unbiased estimate of $g(\theta)$ which is a function of T only. This result is an immediate consequence of a theorem of Rao (1945) and Blackwell (1947) on unbiased estimation. We shall show also that whenever $\mathfrak{P}^t$ is boundedly complete, all similar regions A have a very simple structure (roughly speaking, that the conditional probability of X falling in A is independent of t), which was first described by Neyman

(1937), and that as a result the original composite hypothesis in these cases is reduced essentially to a simple one. The first to employ this method for finding the totality of similar regions was P. L. Hsu (1941).

The applicability of the above remarks hinges on the existence of a sufficient statistic T such that $\mathfrak{P}^t$ is complete (or boundedly complete). Now in general there are many different sufficient statistics, and the question arises, as to which has the best chance of satisfying the completeness condition. Actually, the question as to which is the appropriate sufficient statistic arises also in other statistical problems. Speaking intuitively, one makes use of sufficient statistics in order to reduce the complexity of a statistical problem without losing information of value. The latter condition being guaranteed by sufficiency, one is led to seek that sufficient statistic which reduces the statistical problem as far as possible, and hence to the definition (to be stated more precisely later with appropriate null set qualifications): A sufficient statistic T is said to be minimal if T is a function of all other sufficient statistics. Using this definition we prove that whenever a minimal sufficient statistic exists, a sufficient statistic can satisfy the condition of completeness only if it is equivalent to the minimal sufficient statistic in a certain sense. We also establish the existence of the minimal sufficient statistic when the sample space is Euclidean and the distribution of the sample possesses a generalized probablity density (defined in section 6), and we give a method of constructing it, which we show to be valid in this case. We remark that the result of our construction of the minimal sufficient statistic is essentially equivalent to the definition of sufficient statistic adopted by Koopman (1936) in a more special setting.

Some results for the problem of unbiased estimation are also found in the case where the minimal sufficient statistic is not complete: A formal theory is obtained which characterizes those estimable functions possessing unbiased estimates of uniformly minimum variance and these estimates. Finally a justification is given (at the end of section 4) for the restriction to sufficient statistics when testing hypotheses. (A justification in the case of point estimation was given by Rao and Blackwell). By permitting randomised decision functions, we show that given any test and any sufficient statistic there exists a test based only on the sufficient statistic and having identically the same power function as the given test.

In this paper (with the exception of section 6) the theory is developed in greater generality than is customary in statistical papers. The reason for not limiting ourselves to Euclidean spaces is that these are insufficient both for problems that have already arisen in sequential analysis and for statistical problems that may be expected to arise in connection with stochastic processes. Many of the difficulties encountered in the paper are associated with exceptional null sets. We believe that most of these difficulties are inherent in the nature of the problems treated, and that they would arise even if considerations were limited to Euclidean spaces.

It came to our attention while this paper was in proof that some of our results on minimum variance estimates were obtained earlier in papers by Rao (1947, 1949) and in an abstract by Seth (1949).

The material outlined above forms part I of the present study. In part II the general theory is applied to a number of more special problems. As the main application some theorems on similar regions (Neyman, 1941), type B_1 tests (Scheffé, 1942), and uniformly most powerful one-sided tests (Lehmann, 1947) are simplified and extended. These results are obtained by solving the differential equations introduced by Neyman and Scheffé. The theory of part I is then applied to the families of probability densities which are solutions of these equations, and also to some more general families of probability densities. Some of these results were summarized in a previous publication (Lehmann and Scheffé, 1947), and of these some were obtained independently and have been published since then by Ghosh (1948). Applications are also made to some non-parametric problems of estimation and testing. For example, a very simple proof is given of Scheffé's theorem (1943) concerning similar regions in the non-parametric case, this result is generalized, and the solution is given of a problem formulated by Halmos (1946) concerning point estimation.

2. TERMINOLOGY AND NOTATION

Unfortunately a considerable complexity of terminology and notation seems unavoidable; to minimize this we adopt the following conventions: Several spaces, to be denoted by W^x, W^t, etc., will have to be considered; here W^x is the whole space of points x, W^t is the space of t, etc. In each space there will be a fixed countably additive family of sets, $\mathfrak{F}^x$ in W^x, $\mathfrak{F}^t$ in W^t, etc. Here a family $\mathfrak{F}^x$ is said to be *countably additive* if it contains W^x, and with any set A in $\mathfrak{F}^x$ its complement $W^x—A$, and with any countable (i.e., finite or denumerable) number of sets in $\mathfrak{F}^x$ also their union. The sets in $\mathfrak{F}^x$ will be called *measurable* ($\mathfrak{F}^x$). We shall need to define *measurable functions* only for the case of real-valued functions: A function $f(x)$ defined on W^x is said to be measurable ($\mathfrak{F}^x$) if for every real c the set $\{x \mid f(x)<c\}$ is in $\mathfrak{F}^x$. A non-negative set function M^x defined for all A in $\mathfrak{F}^x$ is said to be a *measure* on $\mathfrak{F}^x$ it it is countably additive, that is, if for any disjoint sets $A_1, A_2, \ldots$ in $\mathfrak{F}^x$, $M^x(\cup_i A_i) = \Sigma_i M^x(A_i)$. A *probability measure* on $\mathfrak{F}^x$ is a measure M^x on $\mathfrak{F}^x$ for which $M^x(W^x) = 1$. Probability measures will usually be denoted by P^x, P^t, etc.

A family of measures M^x on $\mathfrak{F}^x$ will be denoted by $\mathfrak{M}^x$. It is convenient to index the members of the family by a subscript θ that takes on values in an abstract space ω, $\mathfrak{M}^x = \{M_\theta^x \mid \theta\epsilon\omega\}$. Similarly, we may write $\mathfrak{P}^x = \{P_\theta^x \mid \theta\epsilon\omega\}$ if the measures are probability measures. A set A in $\mathfrak{F}^x$ for which $M_\theta^x(A) = 0$ will be called a *null set for the measure* M_θ^x. A set will be called a *null set for the family* $\mathfrak{M}^x$ if it is a null set for every measure in the family. If a statement about the points of W^x is true for all x in $W^x—N$, we shall say it is valid *almost everywhere* (M_θ^x) if N is a null set for M_θ^x, we shall say it is valid *almost everywhere* ($\mathfrak{M}^x$) if N is a null set for $\mathfrak{M}^x$, and we shall abbreviate this by writing (*a.e.* M_θ^x) or (*a.e.*$\mathfrak{M}^x$) after the statement. An arbitrary function $t(x)$ from W^x to a space W^t generates a countably additive family $\mathfrak{F}^t$ of sets in W^t and a family $\mathfrak{M}^t$ of measures on W^t : the family $\mathfrak{F}^t$ consists of all those sets in W^t whose pre-images ("complete counter images") are in $\mathfrak{F}^x$, while the measure

M_θ^t on $\mathfrak{F}^t$ corresponding to M_θ^x on $\mathfrak{F}^x$ is defined for B in $\mathfrak{F}^t$ by $M_\theta^t(B) = M_\theta^x(t^{-1}(B))$, where $t^{-1}(B) = \{x \mid t(x) \epsilon B\}$ is the pre-image of B. It is easily seen that with this definition of $\mathfrak{F}^t$, a real-valued function $f(t)$ defined on W^t is measurable ($\mathfrak{F}^t$ if and only if the function $f(t(x))$ is measurable ($\mathfrak{F}^x$).

A family $\mathfrak{D}$ of sets D which are disjoint and cover W^x will be called a *deocmposition* of W^x; if all the elements D in $\mathfrak{D}$ are measurable ($\mathfrak{F}^x$) we will say that $\mathfrak{D}$ is a *measurable* ($\mathfrak{F}^x$) *decomposition*. If $t(x)$ is a function defined on W^x, and x^0 is any point of W^x, we shall say the set $\{x \mid t(x) = t(x^0)\}$ is a *contour* of $t(x)$ through x^0. Any function $t(x)$ defined on W^x determines a decomposition of W^x which may be denoted by $\mathfrak{D}_t$, the elements of $\mathfrak{D}_t$ being the contours of $t(x)$. Conversely, given any decomposition $\mathfrak{D}$ of W^x, a function $t(x)$ can be found (it is not unique, but the ranges of any two such functions are in 1:1 correspondence) such that $\mathfrak{D} = \mathfrak{D}_t$: For example, $t(x)$ may be taken to be the set-valued function whose value at x is the element of $\mathfrak{D}$ containing x. If for this function $t(x)$ we form according to the above description the countably additive family $\mathfrak{F}^t$, it turns out that it consists precisely of those sets of elements D of $\mathfrak{D}$ whose union is in $\mathfrak{F}^x$. This countably additive subfamily of $\mathfrak{F}^{x|\mathfrak{D}}$ generated by a decomposition $\mathfrak{D}$ or an associated function $t(x)$ will be denoted by $\mathfrak{F}^{x|\mathfrak{D}}$ or $\mathfrak{F}^{x|t}$. We say that a function $t(x)$ and decomposition $\mathfrak{D}$ are *associated* if $\mathfrak{D} = \mathfrak{D}_t$. For many purposes in probability and statistics only the decomposition associated with a function is of importance, it being immateral which of the different functions associated with the same decomposition is chosen.

We will encounter two kinds of equivalence for decompositions: Two decompositions $\mathfrak{D}$ and $\mathfrak{D}'$ of W^x are *equivalent* ($\mathfrak{P}^x$) *in the strict sense* if there exists a null set N for $\mathfrak{P}^x$ such that on $W^x—N$ the decompositions $\mathfrak{D}$ and $\mathfrak{D}'$ coincide; $\mathfrak{D}$ and $\mathfrak{D}'$ are *equivalent* ($\mathfrak{P}^x$) *in the weak sense* if for every set A in $\mathfrak{F}^{x|\mathfrak{D}}$ there exists an A' in $\mathfrak{F}^{x|\mathfrak{D}'}$ and for every A′ in $\mathfrak{F}^{x|\mathfrak{D}'}$ there exists an A in $\mathfrak{F}^{x|\mathfrak{D}}$ such that A and A' differ by a null set for $\mathfrak{P}^x$, that is, $(A-A') \cup (A'-A)$ is a null set for $\mathfrak{P}^x$.

When one of the spaces being considered is the sample space of a statistical problem we shall always denote it by W^x. The family of measures on $\mathfrak{F}^x$ will then be a family $\mathfrak{P}^x$ of probability distributions. By saying that the random variable X (the "sample") is distributed according to P_θ^x we mean, as usual, that for any set A in $\mathfrak{F}^x$ the probability of X falling in A is P_θ^x (A). The term *statistic* will be used in this paper to mean any function of X, that is, if $t(x)$ is an arbitrary function of x defined on W^x, $T = t(X)$ will be called a statistic[1]. If $\mathfrak{F}^t$ and $\mathfrak{P}^t$ are defined as above then the probability of T falling in any set B in $\mathfrak{F}^t$ is $P_\theta^t(B)$. If $\mathfrak{D}_t$ is the decomposition of W^x associated with $t(x)$, it will be convenient to say also that it is associated with the statistic $T = t(X)$. We shall say that two statistics are *equivalent* ($\mathfrak{P}^x$) *in the strict sense*

[1] While for some purposes it may be convenient to restrict the definition of statistic, for example to functions whose associated decompositions are measurable ($\mathfrak{F}^x$), nothing is gained by such a restriction in this paper. We remark however that our construction of the minimal sufficient statistic in Section 6 does give a statistic satisfying this restriction.

or *in the weak sense* if their associated decompositions are. The reader may want to consider the statistical implications of these two kinds of equivalence of statistics.

Fundamental for this study is the notion of sufficient statistic. We base it on the definition given by Kolmogoroff (1933, p.41) of the conditional probability $P_\theta(A|t)$ of any set A in $\mathfrak{F}^x$ given that the statistic T has the value t: For each fixed θ and A, $P_\theta(A|t)$ is a real-valued point function of t, measurable ($\mathfrak{F}^t$), and defined implicitly by the equation

$$P_\theta^x(A \cap t^{-1}(B)) = \int_B P_\theta(A|t) dP_\theta^t, \qquad \dots \quad (2.1)$$

where B is any set in $\mathfrak{F}^t$, and (2.1) is regarded as an identity in B. For fixed θ and A, $P_\theta(A|t)$ is not defined uniquely, but if $f_{\theta, A}(t)$ and $g_{\theta, A}(t)$ are two determinations of $P_\theta(A/t)$, the set $N_{\theta, A}$ where they are unequal is a null set for P_θ^t. A statistic T is said to be a *sufficient statistic for* $\mathfrak{P}^x$ if there exists a determination of $P_\theta(A|t)$ indedent of θ, that is, if there exists a function $P(A|t)$ not depending on θ, measurable ($\mathfrak{F}^t$) for each fixed A, and such that for every θ in ω, every A in $\mathfrak{F}^x$, and any determination $P_\theta(A|t)$, $P(A|t) = P_\theta(A|t)$ except for t in a null set $B_{\theta, A}$ for P_θ^t. We note that two determinations of $P(A|t)$ must be equal except on a null set N_A for $\mathfrak{P}^t$. By putting $B = W^t$ in (2.1), we see that if T is a sufficient statistic for $\mathfrak{P}^x$, then for all A in $\mathfrak{F}^x$.

$$P_\theta(A) = \int_{W^t} P(A|t) dP_\theta^t. \qquad \dots \quad (2.2)$$

We shall have need also of the concept of conditional expectation as defined by Kolmogoroff (1933, p. 46). Suppose $\phi(x)$ is a real-valued measurable ($\mathfrak{F}^x$) function of x such that

$$E_\theta(\Phi) = \int_{W^x} \phi(x) d P_\theta, \qquad \dots \quad (2.3)$$

the expected value of the statistic $\Phi = \phi(X)$, calculated under the probability distribution P_θ^x, is finite. If $T = t(X)$ is a statistic (in general, not real-valued), then the conditional expected value of Φ, given $T = t$, calculated under P_θ^x, and to be denoted by $E_\theta(\Phi|t)$, is a point function of t, integrable ($\mathfrak{F}^t$, P_θ^t), defined implicitly by

$$\int_{t^{-1}(B)} \phi(x) d P_\theta^x = \int_B E_\theta(\Phi|t) d P_\theta^t, \qquad \dots \quad (2.4)$$

where B is in $\mathfrak{F}^t$, and (2.4) is regarded as an identity in B. For fixed θ, two determinations of $E_\theta(\Phi|t)$ are equal except on a null set N_θ of P_θ^t. By comparing the definitions (2.1) and (2.4) it is seen that if $\phi = \phi_A(x)$ is the characteristic function of a set A in $\mathfrak{F}^x$, then

$$P_\theta(A|t) = E_\theta(\Phi_A|t) \qquad (\text{a.e. } P_\theta^t). \qquad \dots \quad (2.5)$$

Returning to the general case of Φ with finite $E_\theta(\Phi)$, we remark that $E_\theta(\Phi|t)$ can also be calculated from the conditional probability $P_\theta(A|t)$ if this is known for all sets A in $\mathfrak{F}^x$, as was proved by Kolmogoroff (1933, p. 48). It then follows from the above definition of sufficient statistic that if T is a sufficient statistic for $\mathfrak{P}^x$, and if $E_\theta(\Phi)$ is finite for all θ in ω, there exists a measureable ($\mathfrak{F}^t$) point function of t, independent of θ, which we shall denote by $E(\Phi|t)$ and which has the property that for every θ in ω, $E(\Phi|t) = E_\theta(\Phi|t)$ except on a null set N_θ for P_θ^t.

We shall have use for the following three formulae from the calculus of conditional probabilities: Suppose T is a sufficient statistic for $\mathfrak{P}^x$. (*i*) Then if $E_\theta(\Phi)$ is finite

$$E_\theta(\Phi) = \int_{W^t} E(\Phi|t)\, d\, P_\theta^t. \qquad \dots \quad (2.6)$$

This follows from (2.4) by taking $B = W^t$. (*ii*) If $f(t)$ is a real-valued measurable ($\mathfrak{F}^t$) function of t, and if $E_\theta(f(T)\Phi)$ and $E_\theta(\Phi)$ are finite for all θ in ω, then

$$E(f(T)\Phi|t) = f(t)E(\Phi|t) \qquad \text{(a.e. } \mathfrak{P}^t). \qquad \dots \quad (2.7)$$

This formula is proved by Kolmogoroff (1933, p. 50), differently by Blackwell (1947, p. 105). (*iii*) It follows easily from (2.4) that if $c_1 \leqslant \phi(x) \leqslant c_2$ then

$$c_1 \leqslant E(\Phi|t) \leqslant c_2 \qquad \text{(a.e. } \mathfrak{P}^t). \qquad \dots \quad (2.8)$$

We remark finally that the values of conditional probabilities and expectations, given a statistic, depend only on the decomposition associated with the statistic.

Of two statistics $T = t(X)$ and $T' = t'(X)$ we shall say *T is a function of T'* (a.e. $\mathfrak{P}^x$) if there exists a function $t = \psi(t')$ on $W^{t'}$ to W^t such that $t(x) = \psi(t'(x))$ (a.e. $\mathfrak{P}^x$). In terms of the decompositions $\mathfrak{D}_t$ and $\mathfrak{D}_{t'}$ associated with the functions $t(x)$ and $t'(x)$ this means there exists a null set N for $\mathfrak{P}^x$ such that on $W^x - N$ the decomposition $\mathfrak{D}_{t'}$ subdivides the decomposition $\mathfrak{D}_t$, that is every element of $\mathfrak{D}_{t'}$ is contained in an element of $\mathfrak{D}_t$. A sufficient statistic T for $\mathfrak{P}^x$ will be called a *minimal sufficient statistic* for $\mathfrak{P}^x$ if, for any other sufficient statistic T' for $\mathfrak{P}^x$, T is a function of T' (a.e. $\mathfrak{P}^x$). Sufficient conditions for the existence of minimal sufficient statistic, and a method of finding it, are given in section 6.

3. Completeness of a Family of Measures

Given a family $\mathfrak{M}^x = \{M_\theta^x|\theta\epsilon\omega\}$ of measures on the additive family $\mathfrak{F}^x$ of sets in W^x, consider integrals of the form

$$\int_{W^x} f(x)\, d\, M_\theta^x, \qquad \dots \quad (3.1)$$

where $f(x)$ is real-valued and measurable ($\mathfrak{F}^x$). The value (if any) of this integral will in general depend on which measure M_θ^x of the family $\mathfrak{M}^x$ is used; (3.1) is a function of θ, which we may regard as a transform from a function f(x) defined on W^x to a function of θ defined on a part of ω. Under this transformation the function that is everywhere zero on W^x goes into the function that is everywhere zero on ω. Completeness means roughly that the zero function on W^x is the only function going into the zero function on ω; it is a unicity property of the transform. The exact definition is the following: The family $\mathfrak{M}^x$ of measures is *complete* if

$$\int_{W^x} f(x)\, d\, M_\theta^x = 0 \quad \text{for all } \theta \text{ in } \omega \qquad \dots \quad (3.2)$$

implies $f(x) = 0$ (*a.e.* $\mathfrak{M}^x$). This definition of completeness is appropriate for the problem of unbiased estimation; for the problem of similar regions we require the weaker property of bounded completeness: The family $\mathfrak{M}^x$ is called *boundedly complete*[2] if the condition (3.2) and the condition that $f(x)$ is bounded jointly imply that $f(x) = 0$ (*a.e.* $\mathfrak{M}^x$).

We note that if $\mathfrak{M}^x$ is complete it is boundedly complete. The following simple example of a family of measures which is boundedly complete without being complete is a slight modification of an example constructed for a different purpose by Girshick, Mosteller, and Savage (1946):

Example 3.1: Suppose W^x is the real line, ω is the open interval $0<\theta<1$, and P_θ^x assigns the measure $(1-\theta)^2\theta^x$ to the points $x = 0, 1, 2\ldots$, measure θ to the point $x = -1$, and measure zero to the complement of this set of points. The condition (3.2) then becomes

$$f(-1)\theta+\sum_{j=0}^{\infty} f(j)(1-\theta)^2\theta^j = 0.$$

We see that for $j = 0, 1, 2\ldots, f(j)$ is the coefficient of θ^j in the Taylor series about the origin for the function $-f(-1)\theta(1-\theta)^{-2}$. Since for $|\theta|<1$,

$$\theta(1-\theta)^{-2} = \sum_{j=0}^{\infty} j\,\theta^j,$$

it follows that (3.2) is satisfied if and only if

$$f(j) = -jf(-1), \qquad j = 0, 1, 2,\ldots. \qquad \ldots \quad (3.3)$$

Hence if $f(x)$ satisfies (3.2) and is bounded, $f(j) = 0$ for $j = -1, 0, 1, 2,\ldots$, that is, $f(x) = 0$ (*a.e.* $\mathfrak{P}^x$), and thus $\mathfrak{P}^x$ is boundedly complete. On the other hand, if $f(x)$ satisfies (3.3) and $f(-1) \neq 0$, then it satisfies (3.2) but not the condition $f(x) = 0$ (*a.e.* $\mathfrak{P}^x$), hence $\mathfrak{P}^x$ is not complete.

It is worth while to note that in general completeness or bounded completeness of a subset $\mathfrak{M}_1^x$ of $\mathfrak{M}^x$ does not imply the same for $\mathfrak{M}^x$, but that this implication does hold provided all null sets for $\mathfrak{M}_1^x$ are null sets for $\mathfrak{M}^x$.

We shall now give some simple examples of completeness. In each case the family $\mathfrak{P}^x$ whose completeness is discussed will be a family of probability measures that has been used extensively in statistics; it is of interest that in a number of these examples the question of completeness reduces to the problem of unicity for a transform that has been treated in the mathematical literature. In every case the space W^x will be the real line $-\infty<x<+\infty$, and the additive family $\mathfrak{F}^x$ on W^x may be taken as the class of Borel sets. The space ω of points θ may be taken as an appropriate set of real numbers. All the examples in which the family of measures is complete may be modified by replacing ω by certain proper subsets without destroying the completeness. We mention here that by means of some general theorems in Part II, one of which concerns completeness in product spaces, completeness may be proved for a large number of

[2] The property here called *bounded completeness* was called *completeness* by us in an earlier publication (1947).

families of measures in n-dimensional Euclidean spaces from these examples on the real line. Here and elsewhere in examples we use the "Nikodym derivative" notation

$$d\,P^x/d\mu^x = g(x)$$

to indicate that the measure P^x is absolutely continuous with respect to the measure μ^x and that

$$P^x(A) = \int_A g(x)d\mu^x$$

for all A in $\mathfrak{F}^x$.

Example 3.2: If $\mathfrak{P}^x$ is a family of normal probability distributions with mean θ and unit variance, then

$$d\,P_\theta^x/dx = (2\pi)^{-\frac{1}{2}}\exp\left[-\frac{1}{2}(x-\theta)^2\right].$$

If the condition (3.2) is satisfied by $f(x)$ we find

$$\int_{-\infty}^{+\infty} f(x)\exp\left[-\frac{1}{2}x^2+\theta x\right]dx = 0 \qquad \dots \quad (3.4)$$

tor $-\infty<\theta<\infty$. Now (3.4) is the bilateral Laplace transform of the function $f(x)$ $\exp(-\frac{1}{2}x^2)$, and from the unicity theorem of this transform it follows that $f(x)$ $\exp(-\frac{1}{2}x^2) = 0(a.e.\ L)$, where L denotes Lebesgue measure on $\mathfrak{F}^x$, and hence $f(x) = 0$ $(a.e.\ \mathfrak{P}^x)$. Thus $\mathfrak{P}^x$ is complete.

Example 3.3: For a family $\mathfrak{P}^x$ of normal distributions with zero mean and variance θ we have

$$d\,P_\theta^x/dx = (2\pi\theta)^{-\frac{1}{2}}\exp\left(-\frac{1}{2\theta}x^2\right) \quad \text{for } 0<\theta<\infty. \qquad \dots \quad (3.5)$$

This family is not complete or even boundedly complete because for every θ the density (3.5) is an even function of x, and hence (3.2) will be satisfied by any $f(x)$, which is an odd function of x and such that its product with (3.5) is integrable (L) for all $\theta>0$. If we transform to a new set $\mathfrak{P}^t$ of measures P_θ^t by the transformation $t = x^2$ we find

$$dP_\theta^t/dt = \begin{cases} (2\pi t\theta)^{-\frac{1}{2}}\exp(-\frac{1}{2\theta}t) \text{ for } t>0, \\ 0 \quad \text{for } t \leqslant 0. \end{cases}$$

Condition (3.2) written for $\mathfrak{P}^t$ instead of $\mathfrak{P}^x$ gives

$$\int_0^\infty f(t)t^{-\frac{1}{2}}\exp\left(-\frac{1}{2\theta}t\right)dt = 0 \qquad \dots \quad (3.6)$$

for $\theta>0$. Letting $\tau = (2\theta)^{-1}$, and applying the unicity theorem for the unilateral Laplace transform we find $f(t) = 0$ $(a.e.\ \mathfrak{P}^t)$, that is, $\mathfrak{P}^t$ is complete.

Example 3.4: For a family $\mathfrak{P}^x$ of Cauchy distributions with zero median we have

$$dP_\theta^x/dx = [\pi\theta^{\frac{1}{2}}(1+x^2/\theta]^{-1} \qquad \text{for } \theta>0.$$

For symmetry reasons $\mathfrak{P}^x$ may again be seen not to be boundedly complete. If, as before, we let $t = x^2$, the completeness of the resulting $\mathfrak{P}^t$ follows from the unicity property of the Stieltjes transform

$$\int_0^\infty \frac{f(t)dt}{t+\theta}$$

Example 3.5: For a family of gamma distributions (chi-square distributions if θ is a half integer),

$$dP_\theta^x/dx = \begin{cases} 2^{-\theta}x^{\theta-1}\exp(-\frac{1}{2}x)/\Gamma(\theta) & \text{if } x>0, \\ 0 \quad \text{if} \quad x \leqslant 0, \end{cases}$$

$\theta>0$, completeness follows from the unicity property of the Mellin transform

$$\int_0^\infty f(t)t^{\theta-1}\text{dt}.$$

Example 3.6: If P_θ^x is the uniform distribution on the interval $(0, \theta)$,

$$d\,P_\theta^x/dx = \begin{cases} 1/\theta \text{ if } 0<x<\theta, \\ 0 \text{ elsewhere}, \end{cases}$$

$\theta>0$, the completeness of $\mathfrak{P}^x$ follows from a theorem of Lebesgue : If

$$\int_0^\theta f(x)dx. = 0$$

for all θ in an interval, then $f(x) = 0$ (*a.e. L*) on the interval. The transform (3.1) of a function $f(x)$, which is involved in this case, is its indefinite integral.

Example 3.7: The family $\mathfrak{P}^x$ of uniform distributions on the intervals $(\theta, \theta+1)$,

$$dP_\theta{}^x/dx = \begin{cases} 1 \text{ if } \theta<x<\theta+1, \\ 0 \text{ elsewhere}, \end{cases}$$

$-\infty<\theta<\infty$, is not boundedly complete : If $f(x)$ is any periodic function with period 1 and

$$\int_0^1 f(x)dx = 0,$$

it is easily seen that

$$\int_{-\infty}^{+\infty} f(x) d\, P_\theta^x = 0$$

for all θ.

Example 3.8: Suppose $\mathfrak{P}^x$ is the family of Poisson distributions, so that P_θ^x assigns the measure $e^{-\theta}\theta^x/x!$ to the points $x = 0, 1, 2,\ldots$, and measure zero to the complement of this set. The condition (3.2) may be written

$$e^{-\theta} \sum_{j=0}^{\infty} f(j)\theta^j/j! = 0,$$

for $\theta > 0$. Since $f(j)$ is the coefficient of θ^j in the power series expansion of zero it follows that $f(x) = 0$ (*a.e.* $\mathfrak{P}^x$), and $\mathfrak{P}^x$ is complete.

Example 3.9: If $\mathfrak{P}^x$ is the family of binomial distributions corresponding to n independent trials with constant probability θ, $0<\theta<1$, P_θ^x assigns the probability

$$\binom{n}{x} \theta^x(1-\theta)^{n-x}$$

to the points $x = 0, 1,\ldots, n$ and zero probability elsewhere. Completeness of $\mathfrak{P}^x$ is implied by the theorem that if a polynomial of degree n vanishes for $n+1$ distinct values of the argument it is identically zero.

Example 3.10: Let $\mathfrak{P}^x$ be the family of hypergeometric distributions for fixed lot size N and fixed sample size n where θ is the number of defectives in the lot, $\theta = 0, 1,\ldots, N$; if P_θ^x is the probability of x defectives in the sample, P_θ^x assigns the discrete probabilites

$$\binom{\theta}{x} \binom{N-\theta}{n-x} \Big/ \binom{N}{n} \qquad \ldots \quad (3.7)$$

to the points $x = 0, 1,\ldots, n$. In (3.7) we understand

$$\binom{m}{r} = 0 \text{ if } r<0 \text{ or } r>m.$$

Condition (3.2) then becomes

$$\sum_{j=0}^{n} f(j)\binom{\theta}{x} \binom{N-\theta}{n-x} = 0, \qquad \ldots \quad (3.8)$$

and we find successively $f(0) = 0$, $f(1) = 0$, $\ldots$, $f(n) = 0$ by putting $\theta = 0, 1, \ldots, n$ in (3.8). Hence $\mathfrak{P}^x$ is complete.

Example 3.11: Let $X_1, X_2, \ldots$ be a sequence of independent random variables each being capable of taking on the values 1 and 0 with probability p and $1-p$ respectively. Consider a sequential sampling scheme on these variables where the decision on whether or not to take an $N+1^{st}$ observation, when N observations have already been taken is made according to the value of $\sum_{i=1}^{N} X_i$. Let n be the total number of observations taken in one experiment—n is a random variable — and let $X = \sum_{i=1}^{n} X_i$. The problem of bounded completeness for the family $\mathfrak{P}^x$ corresponding to a fixed stopping rule and the values $0<p<1$, was solved as follows in a series of papers by Girshick, Mosteller and Savage (1946), Wolfowitz (1946), and Savage (1947). For any N, consider $\left(\sum_{i=1}^{N} x_i,\ N-\sum_{i=1}^{N} x_i\right)$ as coordinates of a point in a plane, and define such a point to be accessible if it has positive probability under the sequential scheme being considered but is not a stopping point for this procedure so that, when this point is observed, another observation will be taken. As was shown in the above papers, a necessary and sufficient condition for bounded completeness is that, given any pair of accessible points P_1, P_2 corresponding to the same value of N (for any point P, N is the sum of the coordinates of P), all the points lying on the line segment connecting P_1 and P_2 and having integer valued coordinates are also accessible.

The concept of completeness developed in this section is related to that of minimality of a sufficient statistic, introduced at the end of secton 2, by

Theorem 3.1: If T *is a sufficient statistic for* $\mathfrak{P}^x$ *such that* $\mathfrak{P}^t$ *is boundedly complete, and if* U *is a minimal sufficient statistic for* $\mathfrak{P}^x$, *then* T *and* U *are equivalent* ($\mathfrak{P}^x$) *in the weak sense.*

The proof of this theorem can be based on the following

Lemma 3.1: *If* $T = t(X)$ *and* $T' = t'(X)$ *are two sufficient statistics for* $\mathfrak{P}^x$, *and if for every* A *in* $\mathfrak{F}^x$, $P(A|t)$ *and* $P(A|t')$, *considered as functions of* x *are equal* (*a.e.* $\mathfrak{P}^x$), *then* T *and* T' *are equivalent in the weak sense.*

To prove the lemma let $\mathfrak{F}^{x|t} = t^{-1}(\mathfrak{F}^t)$, and define $\mathfrak{F}^{x|t'}$ analogously. We shall prove that to any set A in $\mathfrak{F}^{x|t}$ there corresponds a set A' in $\mathfrak{F}^{x|t'}$ which differs from A by a null set for $\mathfrak{P}^x$. Let $\phi_A(x)$ be the characteristic function of the set A; then $\phi_A(x)$ depends on x through $t(x)$, say $\phi_A(x) = g(t(x))$. By (2.5) and (2.7),

$$P(A|t) = E(g(T)|t) = g(t) \qquad (\text{a.e. } \mathfrak{P}^t),$$

and thus

$$P(A|t(x)) = \phi_A(x) \qquad (\text{a.e. } \mathfrak{P}^x).$$

But by hypothesis

$$P(A|t(x)) = P(A|t'(x)) \qquad (\text{a.e. } \mathfrak{P}^x). \qquad \ldots \quad (3.9)$$

Define A' as the set in $\mathfrak{F}^{x|t'}$ where the right member of (3.9) equals 1. Now there exists a null set N for $\mathfrak{P}^x$ such that on $W^x - N$ the right member of (3.9) equals $\phi_A(x)$; hence the parts of A and A' in $W^x - N$ coincide, and the lemma is established.

Theorem 3.1 will thus be proved if we show that if there were an A_1 in $\mathfrak{F}^x$ for which $P(A_1/t(x))$ and $P(A_1|u(x))$, were not equal (a.e. $\mathfrak{P}^x$), then $\mathfrak{P}^t$ could not be boundedly complete. Consider the real-valued measurable ($\mathfrak{F}^x$) function $v(x)$ defined by

$$v(x) = P(A_1|t(x) - P(A_1|u(x)).$$

We note that $|v(x)| \leqslant 1$ (a.e. $\mathfrak{P}^x$), and that the set in $\mathfrak{F}^x$ where $v(x) \neq 0$ is not a null set for $\mathfrak{P}^x$. If $V = v(X)$, it follows from (2.2) that

$$E_\theta(V) = 0 \qquad \dots \quad (3.10)$$

for all θ in ω. Since U is a minimal sufficient statistic it is a function of T (a.e. $\mathfrak{P}^x$) and hence V is also a function of T (a.e. $\mathfrak{P}^t$). We can thus redefine $v(x)$ on a null set for $\mathfrak{P}^x$ so that the result is a measurable ($\mathfrak{F}^t$) function of t, say $f(t)$, with the properties that $|f(t)| \leqslant 1$, and the set in $\mathfrak{F}^t$ where $f(t) \neq 0$ is not a null set for $\mathfrak{P}^t$. This redefinition of $v(x)$ does not invalidate (3.10), and thus

$$\int_{W^t} f(t) dP_\theta^t = 0$$

for all θ in ω, and hence $\mathfrak{P}^t$ is not boundedly complete.

4. Similar Regions

A set A in $\mathfrak{F}^x$ is said to be a *similar region* of size α for the family $\mathfrak{P}^x$ of probability measures on $\mathfrak{F}^x$ if $P_\theta^x(A) = \alpha$ for all θ in ω. Neyman (1937) noted that if T is a sufficient statistic for $\mathfrak{P}^x$ and if the set A has the property

$$P(A|t) = \alpha \qquad \text{(a.e. } \mathfrak{P}^t), \qquad \dots \quad (4.1)$$

then A is a similar region of size α for $\mathfrak{P}^x$; this follows from (2.2). We whall say that a set A in $\mathfrak{F}^x$ has the *Neyman structure* with respect to the sufficient statistic T if it satisfies (4.1). Neyman did not investigate under what conditions, given a sufficient statistic T for $\mathfrak{P}^x$, all similar regions for $\mathfrak{P}^x$ have this structure with respect to T : this is of importance in the Neyman-Pearson theory of optimum tests, since there one wants to choose the "best" of *all* similar regions, and one therefore needs to know the totality of such regions. A partial answer is given by the following corollary to Theorem 4.1.

Corollary 4.1: *If T is a sufficient statistic for $\mathfrak{P}^x$, and if $\mathfrak{P}^t$ is boundedly complete, then a set A in $\mathfrak{F}^x$ is a similar region for $\mathfrak{P}^x$ if and only if it has the Neyman structure with respect to T.*

The situation in which the problem of similar regions arises is that sets A in $\mathfrak{F}^x$ are being considered as possible *critical regions* of statistical tests of a hypothesis : If the sample X falls in the critical region A the hypothesis is rejected by the test, while if it falls in $W^x - A$, the hypothesis is accepted. If A is a similar region for $\mathfrak{P}^x$, the probability of rejection is constant for all probability distribution in $\mathfrak{P}^x$. For many purposes it is convenient to employ a third category of points x such that when one of these points is observed one does not always reject or always accept, but rejects according to a chance method (say, with the help of a table of random mumbers), for which the probability of rejection is a predetermined number $\phi(x)$, $0<\phi(x)<1$. One may then extend the definition of the function $\phi(x)$ to the whole sample space W^x, by setting $\phi(x) = 1$ on the rejection set and $\phi(x) = 0$ on the acceptance set, and is thus led to the notion of critical function (Lehmann and Stein, 1948), which may be regarded as a special case of the randomized decision functions of Wald (1947): A *critical function* $\phi(x)$ is any measurable ($\mathfrak{F}^x$) function of x for which $0 \leqslant \phi(x) \leqslant 1$. Its use in testing hypotheses is that when $X=x$ one rejects the hypothesis with probability $\phi(x)$ according to a random process statistically independent of the random process governing X. The probability of rejection is then the expected value of $\Phi=\phi(X)$ calculated under P_θ^x, namely, $E_\theta(\Phi)$ as defined by (2.3). Critical regions are seen to correspond to the special case of critical functions which are characteristic functions of sets in W_x, that is, which take on only the values 0 and 1.

If $E_\theta(\Phi)$ is constant and equal to α for all θ in ω we shall say that $\phi(x)$ is a *similar critical function* of size α for $\mathfrak{P}^x$. Clearly, similar regions correspond to the special case of similar critical functions which are characteristic functions of sets: If A is a set in $\mathfrak{F}^x$ and ϕ_A is its characteristic function, then A is a similar region for $\mathfrak{P}^x$ if and only if ϕ_A is a similar critical function for $\mathfrak{P}^x$. We shall say that a critical function $\phi(x)$ has the *Neyman structure* with respect to a sufficient statistic T for $\mathfrak{P}^x$ if $E(\Phi|t) = \alpha$ (a.e. $\mathfrak{P}^t$). It is obvious from (2.5) that if a critical function ϕ is the characteristic function of a set A in $\mathfrak{F}^x$ then ϕ has the Neyman structure with respect to T if and only if A has the Neyman structure with respect to T. Therefore Corollary 4.1 follows from

Theorem 4.1: *If T is a sufficient statistic for $\mathfrak{P}^x$, a necessary and sufficient condition for all similar critical functions for $\mathfrak{P}^x$ to have the Neyman structure with respect to T is that $\mathfrak{P}^t$ be boundedly complete.*

We shall prove first sufficiency. Suppose $\phi(x)$ is a similar critical function of size α for $\mathfrak{P}^x$. Then from (2.6),

$$\int_{W^t} E(\Phi|t)d\,P_\theta^t = \alpha$$

for all θ in ω, so that

$$\int_{W^t} f(t)d\,P_\theta^t = 0 \qquad \dots \quad (4.2)$$

for all θ in ω, where

$$f(t) = E(\Phi|t)-\alpha. \qquad \dots \quad (4.3)$$

Since $0\leqslant\phi(x)\leqslant 1$, $0\leqslant E(\Phi)|t)\leqslant 1$ (a.e. $\mathfrak{P}^t$) by (2.8). Hence by taking a particular determination of $E\ (\Phi|t)$ in (4.3) and then redefining $f(t)$ on a null set for $\mathfrak{P}^t$, we may assume $f(t)$ to be bounded. Since $\mathfrak{P}^t$ is boundedly complete, (4.2) implies $f(t) = 0$ (a.e. $\mathfrak{P}^t$), that is, $\phi(x)$ has the Neyman structure with respect to T.

Next we shall prove necessity by showing that if $\mathfrak{P}^t$ were not boundedly complete there would exist similar critical functions (of every size α, $0<\alpha<1$) not having the Neyman structure with respect to T. If $\mathfrak{P}^t$ is not boundedly complete there exists a bounded measurable ($\mathfrak{F}^t$) function $f(t)$ such that

$$\int_{W^t} f(t)dP_\theta^t = 0$$

for all θ in ω, but

$$A_f = \{t|f(t) \neq 0\}$$

is not a null set for $\mathfrak{P}^t$. Suppose $|f(t)| < M$, and define

$$g(t) = cf(t)+\alpha,$$

where $0<c\leq M^{-1} \min \{\alpha, 1-\alpha\}$. Then $0\leqslant g(t)\leqslant 1$,

$$\int_{W^t} g(t)dP_\theta^t = \alpha \qquad \ldots \quad (4.4)$$

for all θ in ω, and

$$g(t) \neq \alpha \text{ on } A_f. \qquad \ldots \quad (4.5)$$

Now take as critical function $\phi(x) \equiv g(t(x))$. Then

$E(\Phi|t) = E(g(T)|t)$, and hence by (2.7)

$$E(\Phi|t) = g(t) \qquad (\text{a.e. } \mathfrak{P}^t). \qquad \ldots \quad (4.6)$$

We see from (2.6), (4.6) and (4.4) that $\phi(x)$ is a similar critical function of size α for $\mathfrak{P}^x$, and from (4.6) and (4.5) that it does not have the Neyman structure wsth respect to T.

It may be appropriate to give here a simple example of a family $\mathfrak{P}^x$ of distributions for which there exist similar regions which do not have the Neyman structure with respect to a sufficient statistic T.

Example 4.1: Let $X = (X_1, X_2, \ldots, X_n)$ be a random sample of size $n>1$ from the uniform distribution on the intervals $(\theta-\frac{1}{2}, \theta+\frac{1}{2})$. The sample space W^x may be taken as a Euclidean n-space, the additive family $\mathfrak{F}^x$ as the Borel sets in W^x, and ω as the real line $-\infty<\theta<+\infty$. The distribution P_θ^x of the sample is the uniform distribution on the n-dimensionsl cube $|x_i-\theta|<\frac{1}{2}$, $i = 1, 2, \ldots, n$. Let $t_1(x) = \min_i x_i$, $t_2(x) = \max_i x_i$. The transformation $t = t(x) = (t_1(x), t_2(x))$ maps W^x into the two dimensional Euclidean space W^t, and the statistic $T = t(X) = (T_1, T_2)$ has the probability density

$$dP_\theta^t/d\mu^t = \begin{cases} c(t_2-t_1)^{n-2} & \text{if } \theta-\frac{1}{2}<t_1\leqslant t_2<\theta+\frac{1}{2}, \\ 0 & \text{otherwise.} \end{cases} \qquad \ldots \quad (4.7)$$

where $d\mu^t = dt_1dt_2$. It is known (it also follows from Example 6.4) that T is a sufficient statistic for $\mathfrak{P}^x$; it will be shown in Example 5.3 that $\mathfrak{P}^t$ is not boundedly complete. Similar regions not having the Neyman structure with respect to T can be based on the range $R = T_2-T_1$. Since R has a continuous distribution independent of θ, there exists for any $\alpha(0<\alpha<1)$ a constant r_α such that for all θ the probability is α that $R<r_\alpha$. The set

$$A_\alpha = \{x|t_2(x)-t_1(x)<r_\alpha)\}$$

is thus a similar region. Since the characteristic function of the set A_α can be expressed as a function of t, it follows from (2.5) and (2.7) that $P(A_\alpha|t)$ takes only the values 0 and 1 (a.e. $\mathfrak{P}^t$), that is, A_α does not have the Neyman structure with respect to T.

If in this example we take $n = 1$, an even simpler illustration of similar regions not possessing the Neyman structure with respect to a sufficient statistic can be given. On the interval (0,1) in W^x take any Borel set A_1 of Lebesgue measure α and then define A so that its characteristic function is periodic with period 1 and coincides with that of A_1 in (0,1). Then $X = X$ is a sufficient statistic for $\mathfrak{P}^x$ and $P(A|x) = 0$ or 1 (a.e. $\mathfrak{P}^x$).

We conclude this section by noting the following "justification" of sufficient statistics: Let Ω be a set in the θ-space containing ω, let $\bar{\mathfrak{P}}^x=\{P_\theta^x|\theta\epsilon\Omega\}$, and suppose the statistic $T=t(X)$ is sufficient for $\bar{\mathfrak{P}}^x$. Then given any critical function $\phi(x)$ for testing $H_0:\theta\epsilon\omega$, there exists a critical function $\phi_1(x)$ depending on x only through $t(x)$, and having the same power function as $\phi(x)$. The proof is similar to that of Rao and Blackwell's theorem (Section 5): Let $\psi(t) = E(\Phi|t)$ and let $\phi_1(x) = \psi(t(x))$; ψ is independent of θ in Ω since T is sufficient for $\bar{\mathfrak{P}}^x$. The power (probability of rejecting H_0) of $\phi(x)$ is

$$E_\theta(\Phi_1) = E_\theta(\psi(T)) = E_\theta(E(\Phi|T)),$$

and by (2.6) this equals $E_\theta(\Phi)$.

It should be remarked that an even stronger justification of sufficient statistics for all problems of statistical inference was given by Halmos and Savage (1949) who pointed out that by means of a sufficient statistic and the use of random variables with known distribution it is possible to construct a statistic having the same distribution as the original sample. Their argument however presupposes that there exists a determination of the conditional probability $P(A|t)$ that is a probability measure (a.e. $\mathfrak{P}^t$); the validity of this supposition has been established by Doob (1948, p.399) in the case that W^x is a Euclidean space. The justification for sufficient statistics given by us above for the problem of testing has the advantage of being valid without any restriction on the sample space.

5. Unbiased Estimation

In this section $\mathfrak{P}^x = \{P_\theta^x | \theta\epsilon\omega\}$ will denote the family of all distributions to which attention is restricted a priori in a particular problem of statistical inference. In considering real-valued statistics as possible solutions of problems of unbiased estimation we shall restrict ourselves to those with finite second moments. Let $\mathfrak{V}$ be the class of all real-valued statistics $V = v(X)$ for which $v(x)$ is measurable ($\mathfrak{F}^x$) and $E_\theta(V^2)$ is finite for all θ in ω. For any V in $\mathfrak{V}$, $E_\theta(V)$ is also finite for all θ in ω. We shall say that a real-valued function $g(\theta)$ is *estimable* if there exists a V in $\mathfrak{V}$ such that $E_\theta(V) = g(\theta)$ for all θ in ω. An unbiased estimate of $g(\theta)$ which is of uniformly minimum variance will be called a minimum variance estimate of $g(\theta)$, that is, a statistic V in $\mathfrak{V}$ is a *minimum variance estimate* of $g(\theta)$ if

(*i*) $E_\theta(V) = g(\theta)$ for all θ in ω, and

(*ii*) $\text{Var}_\theta(V) \leqslant \text{Var}_\theta(V')$ for all θ in ω and all V' in $\mathfrak{V}$ statisfying (*i*).

By applying the notion of completeness to a result of Rao and Blackwell, one obtains easily the following

Theorem 5.1: *If there exists a sufficient statistic T for $\mathfrak{P}^x$ such that $\mathfrak{P}^t$ is complete then every estimable function has a minimum variance estimate, and a statisic V in $\mathfrak{V}$ is a minimum variance estimate of its expected value if and only if it is a function of T (a.e. $\mathfrak{P}^x$).*

The result of Rao (1945) and Blackwell (1947) of which this theorem is a consequence states that if T is a sufficient statistic for $\mathfrak{P}^x$ and V in $\mathfrak{V}$ an unbiased estimate of $g(\theta)$, then if ψ is defined by $\psi(t) = E(V|t)$, $\Psi = \psi(T)$ is also an unbiased estimate of $g(\theta)$ with $\text{Var}_\theta(\Psi) \leqslant \text{Var}_\theta(V)$. equality holding for all θ if and only if V is a function of T (a.e. $\mathfrak{P}^x$).

To prove Theorem 5.1 suppose $g(\theta)$ is an estimable function; then there exists a V in $\mathfrak{V}$ for which $E_\theta(V) = g(\theta)$. Let Ψ be defined as above. To see that Ψ is a minimum variance estimate of $g(\theta)$, suppose V' in $\mathfrak{V}$ is any other unbiased estimate of $g(\theta)$, and define $\psi'(t) = E(V'|t)$. Then by the Rao-Blackwell theorem $\Psi' = \psi'(T)$ is an unbiased estimate of $g(\theta)$ and $\text{Var}_\theta(\Psi') \leqslant \text{Var}_\theta(V')$ for all θ. Since $E_\theta(\Psi' - \Psi) = 0$,

$$\int_{W^t} [\psi'(t) - \psi(t)] dP_\theta^t = 0$$

for all θ in ω, and hence it follows from the completeness of $\mathfrak{P}^t$ that $\psi'(t) = \psi(t)$ (a.e. $\mathfrak{P}^t$). Thus $\text{Var}_\theta(\Psi) = \text{Var}_\theta(\Psi') \leqslant \text{Var}_\theta(V')$, and so Ψ is a minimum variance estimate of $g(\theta)$. The last part of the theorem is obtained from the condition for equality in the Rao-Blackwell theorem.

The application of Theorem 5.1 is illustrated by

Example 5.1: Suppose $\mathfrak{P}$ is the family of distributions of a random sample of size $n > 1$ from a normal population with mean θ_1 and variance θ_2. With $x = (x_1, x_2, \dots, x_n)$ a point in the Euclidean space W^x, $d\mu^x = dx_1\, dx_2 \dots dx_n$, and $\theta = (\theta_1, \theta_2)$ a point in the half plane $\omega = \{\theta \,|\, \theta_2 > 0\}$,

$$dP_\theta^x / d\mu^x = (2\pi\theta_2)^{-\frac{1}{2}n} \exp\left[-\frac{1}{2\theta_2} \sum_{i=1}^{n} (x_i - \theta_1)^2\right]. \quad \dots \quad (5.1)$$

It is well-known that with

$$t_1(x) = \sum_{i=1}^{n} x_i / n,$$

$$t_2(x) = \sum_{i=1}^{n} \left[x_i - t_1(x)\right]^2,$$

$$t(x) = (t_1(x), t_2(x)),$$

$T = (T_1, T_2) = t(X)$ is a sufficient statistic for $\mathfrak{P}^x$ (in Example 6.2 it will be derived as a minimal sufficient statistic), and that T_1 and T_2 are independently distributed with joint probability density

$$dP_\theta^t/d\mu^t = \begin{cases} \dfrac{t_2^{\frac{n-3}{2}} e^{-\frac{1}{2\theta_2}[n(t_1-\theta_1)^2+t_2]}}{(2\theta_2)^{\frac{n}{2}}(\pi/n)^{1/2}\Gamma\left(\frac{n-1}{2}\right)} & \text{if } t_2 > 0, \\ 0 \text{ if } t_2 \leqslant 0. \end{cases}$$

In Part II of this study it will be proved that $\mathfrak{P}^t$ is complete. From the completeness of $\mathfrak{P}^t$ it follows at once that the following functions of the sample mean T_1 and sample variance T_2/n are minimum variance estimates of the indicated functions of θ_1 and θ_2: T_1 of θ_1; $T_2/(n-1)$ and $c_n\sqrt{T_2}$ of θ_2 and $\sqrt{\theta_2}$, respectively, where

$$c_n = 2^{-\frac{1}{2}}\Gamma(\tfrac{1}{2}(n-1))\Gamma(\tfrac{1}{2}n);$$

$T_1+a_pc_n\sqrt{T_2}$ of the (lower) 100 p percent point of the population, where a_p is determined from normal probability tables so that

$$(2\pi)^{-\frac{1}{2}}\int_{-\infty}^{a_p} \exp(-\tfrac{1}{2}X^2)dX = p.$$

These are all special cases of the statement following from Theorem 5.1 that any real-valued function of (T_1, T_2) with finite variance is a minimum variance estimate of its expected value.

Theorem 5.1 : can be extended easily to the case of simultaneous estimation of several real valued-functions of θ if, following Cramér (1946), we express the concentration of an unbiased estimate about its mean, in terms of its ellipsoid of concentration. Let $g_1(\theta), \ldots, g_m(\theta)$ be m real-valued functions of θ, and let $g(\theta) = (g_1(\theta), \ldots, g_m(\theta))$. We shall say that $V = (V_1, \ldots V_m)$ is an unbiased estimate of $g(\theta)$ if V_i is an unbiased estimate of $g_i(\theta)$ for $i = 1, \ldots m$. The statistic V will be said to be an estimate of $g(\theta)$ with maximum concentration if it is an unbiased estimate of $g(\theta)$, if $E_\theta(V_i^2) < \infty$ for $i = 1, \ldots, m$ and all θ in ω, and if for any other unbiased estimate U, the concentration elliposid of V is contained in that of U.

Theorem 5.2 : *If there exists a sufficient statistic T for $\mathfrak{P}^x$ such that $\mathfrak{P}^t$ is complete, then every estimable function $g(\theta) = (g_1(\theta), \ldots, g_m(\theta))$ for which $g_1(\theta), \ldots, g_m(\theta)$, 1 are linearly independent, has an estimate with maximum concentration.*

The proof of this theorem follows easily from the following result[3] concerning quadratic forms: Let $\sum_{i,j=1}^{m} a_{ij}y_iy_j$, $\sum_{i,j=1}^{m} b_{ij}y_iy_j$ be two positive definite quadratic

[3] Pointed out to us by Professor E. W. Barankin.

forms, and let $\sum_{i,j=1}^{m} a^{ij}y_iy_j$, $\sum_{i,j=1}^{m} b^{ij}y_iy_j$ be the corresponding inverse forms. Then

$$\sum_{i,j=1}^{m} a_{ij}y_iy_j \leqslant \sum_{i,j=1}^{m} b_{ij}y_iy_j \text{ for all real } y\text{'s}$$

implies

$$\sum_{i,j=1}^{m} a^{ij}y_iy_j \geqslant \sum_{i,j=1}^{m} b^{ij}y_iy_j \text{ for all real } y\text{'s}.$$

Now let $V = (V_1,\ldots,V_m)$ be any unbiased estimate of $g(\theta)$, let $U_i = E(V_i|T)$, and let $U = (U_1,\ldots,U_m)$. Then for any real $y_1,\ldots y_m$, $\sum_{i=1}^{m} y_iV_i$ is an unbiased estimate of $\sum_{i=1}^{m} y_ig_i(\theta)$, and by Theorem 5.1, $\sum_{i=1}^{m} y_iU_i$ is a minimum variance estimate of $\sum_{i=1}^{m} y_ig_i(\theta)$. Comparing the variances of these two estimates we find

$$\sum_{i,j=1}^{m} \lambda^{U}_{ij}y_iy_j \leqslant \sum_{i,j=1}^{m} \lambda^{V}_{ij}y_iy_j$$

where λ^{U}_{ij} and λ^{V}_{ij} are the covariances of (U_i, U_j) and (V_i, V_j) respectively. Furthermore since $g_1(\theta),\ldots,g_m(\theta)$, 1 are linearly independent so are $V_1,\ldots,V_m,1$ and $U_1,\ldots,U_m,1$, and hence the above quadratic forms are positive definite. It follows that

$$\sum_{i,j=1}^{m} \lambda_U{}^{ij}y_iy_j \geqslant \sum_{i,j=1}^{m} \lambda_V{}^{ij}y_iy_j,$$

which proves the theorem since the ellipsoids of concentration of U and V are defined by the equations $\sum_{i,j=1}^{m} \lambda_U{}^{ij}y_iy_j = m+2$ and $\sum_{i,j=1}^{m} \lambda_V{}^{ij}y_iy_j = m+2$, respectively.

We shall now consider once more the problem of estimating a single real-valued function of θ, and we shall develop a formal theory of minimum variance estimates which answers certain questions about the class of estimable functions possessing minimum variance estimates, but which may be difficult to apply in specific problems. In addition to the class $\mathfrak{V}$ of statistics defined at the beginning of this section we define, for any sufficient statistic T, three classes $\mathfrak{V}_T$, $\mathfrak{V}_T{}^0$, and $\mathfrak{V}_T{}^1$. Here T will in general not be real-valued, and the application of the theory may be expected to be simplest if T is a minimal sufficient statistic for $\mathfrak{P}^x$. Let $\mathfrak{V}_T$ be the class of all statistics in $\mathfrak{V}$ which are functions of T (a.e. $\mathfrak{P}^x$). We define $\mathfrak{V}_T{}^0$ as the subclass of statistics V in $\mathfrak{V}_T$ for which $E_\theta(V) = 0$ for all θ in ω. If $\mathfrak{P}^t$ is complete, $\mathfrak{V}_T{}^0$ consists only of statistics $V = v(X)$ for which $v(x) = 0$ (a.e. $\mathfrak{P}^x$), and conversely. In statistical language we may say the $\mathfrak{V}_T{}^0$ consists of those functions of T which are estimates of zero (unbiased, with finite variance). Finally, $\mathfrak{V}_T{}^1$ is defined to be the subclass of $\mathfrak{V}_T$ whose members V satisfy the condition that VV^0 is in $\mathfrak{V}_T{}^0$ for every V^0 in $\mathfrak{V}_T{}^0$.

An equivalent condition is that $E_\theta(VV^0) = 0$, or $\mathrm{Cov}_\theta(VV^0) = 0$ for every θ in ω and every V^0 in $\mathfrak{V}_T^0$. The class $\mathfrak{V}_T^1$ always contains all real constants.[4]

Theorem 5.3: *A statistic V in $\mathfrak{V}$ is a minimum variance estimate of its expected value if and only if it is a member of $\mathfrak{V}_T^1$. The class $\mathfrak{G}$ of all estimable functions $g(\theta)$ possessing minimum variance estimates is thus obtained by applying the operator E_θ to the members of $\mathfrak{V}_T^1$. If we identify all V in $\mathfrak{V}$ which are equivalent ($\mathfrak{P}^x$) in the strict sense, then the correspondence between $\mathfrak{G}$ and $\mathfrak{V}_T^1$ is 1:1.*

To prove the theorem suppose first that $V = v(X)$ in $\mathfrak{V}$ is a minimum variance estimate of its expected value. Define $\psi(t) = E(V\,|\,t)$. Then $v(x) = \psi(t(x))$ (a.e. $\mathfrak{P}^x$) by Rao and Blackwell's theorem, and thus V is in $\mathfrak{V}_T$. Let V^0 be any element of $\mathfrak{V}_T^0$ and let $U = V + \lambda V^0$, where λ is a constant. Clearly U is also an unbiased estimate of $g(\theta) = E_\theta(V)$, and

$$\mathrm{Var}_\theta(U) - \mathrm{Var}_\theta(V) = \lambda^2 \mathrm{Var}_\theta(V^0) + 2\lambda \mathrm{Cov}_\theta(VV^0).$$

This quadratic function of λ cannot be negative since V is a minimum variance estimate, and consequently it is easily found in either of the cases $\mathrm{Var}_\theta(V^0) > 0$ or $= 0$, that $\mathrm{Cov}_\theta(VV^0) = 0$ for all θ in ω and V^0 in $\mathfrak{V}_T^0$. Hence V is in $\mathfrak{V}_T^1$.

Next suppose that V is in $\mathfrak{V}_T^1$ and that U is another unbiased estimate of $g(\theta) = E_\theta(V)$. Then if $w(t) = E(U|t)$, $W = w(T)$ is also unbiased, and $\mathrm{Var}_\theta(W) \leqslant \mathrm{Var}_\theta(U)$ by Rao and Blackwell's theorem. Also $W - V$ is in $\mathfrak{V}_T$, $E_\theta(W - V) = 0$ for all θ in ω, and so $V^0 = W - V$ is in $\mathfrak{V}_T^0$. We now have

$$\mathrm{Var}_\theta(U) \geqslant \mathrm{Var}_\theta(W) = \mathrm{Var}_\theta(V + V^0) = \mathrm{Var}_\theta(V) + \mathrm{Var}_\theta(V^0) \geqslant \mathrm{Var}_\theta(V).$$

Hence V is a minimum variance estimate.

The last statement in Theorem 5.3 follows from the following

Lemma 5.1: *If V and V^1 are minimum variance estimates of $g(\theta)$ then $V = V^1$ with probability one for all θ in ω.*

To prove this let $h(\theta) = \mathrm{Var}_\theta(V) = \mathrm{Var}_\theta(V^1)$. If for some θ, $h(\theta) = 0$, then with probability one for this θ, $V = g(\theta) = V^1$. If $h(\theta) > 0$, let $\rho(\theta)$ be the correlation coefficient of V and V^1, form the unbiased estimate $U = \frac{1}{2}(V + V^1)$, and note that

$$h(\theta) \leqslant \mathrm{Var}_\theta(U) = \mathrm{Var}_\theta(\tfrac{1}{2}V + \tfrac{1}{2}V^1) = \tfrac{1}{4}h(\theta) + \tfrac{1}{4}h(\theta) + \tfrac{1}{2}h(\theta)\rho(\theta).$$

Thus $\rho(\theta) \geqslant 1$; hence $\rho(\theta) = 1$, and there exist constants A_θ, B_θ such that the probability P_θ is one that

[4] It seems to us in the light of Theorem 5.3 that the term "efficient estimate" should be reserved for the members of $\mathfrak{V}_T^1$, and that (absolute, as opposed to relative) "efficiency" of an estimate V of $g(\theta)$ should be defined only when V is unbiased and $g(\theta)$ is a member of the class $\mathfrak{G}$ of the theorem ; its efficiency (a function of θ) may then be defined as $\mathrm{Var}_\theta(\Psi)/\mathrm{Var}_\theta(V)$ if $\mathrm{Var}_\theta(V) > 0$, 1 if $\mathrm{Var}_\theta(V) = 0$, where Ψ is the minimum variance estimate of $g(\theta)$. We feel that efficiency should not be defined in such a way that the efficiency of the minimum variance estimate Ψ may be less than 1, as is the case for example if efficiency is defined relative to the lower bound for the variance of unbiased estimates given by the Cramér-Rao inequality.

Hence

$$V^1 = A_\theta + B_\theta V.$$

$$E_\theta(V^1) = A_\theta + B_\theta E_\theta(V); \qquad \dots \quad (5.2)$$

also, $\mathrm{Var}_\theta(V^1) = B_\theta^2 \mathrm{Var}_\theta(V)$, so that $B_\theta^2 = 1$. If $B_\theta = -1$, $A_\theta = 2g(\theta)$ from (5.2), and with probability P_θ equal to one, $V^1 = 2g(\theta) - V$ and $U = g(\theta)$. Hence $\mathrm{Var}_\theta(U) = 0$; but $\mathrm{Var}_\theta(U) \geqslant h(\theta) > 0$. From this contradiction we see that $B_\theta = 1$, hence from (5.2), $A_\theta = 0$, and thus $V = V^1$ with probability P_θ equal to one.

We shall now give two examples of the application of Theorem 5.3 in the case where T is a sufficient statistic (actually minimal) and $\mathfrak{P}^t$ is not complete.

Example 5.2 : We take $t(x) = x$ in Example 3.1. In defining statistics $V = v(X)$, the definition of the function $v(x)$ matters only at the points $x = -1, 0, 1, 2, \dots$ $\mathfrak{V}_\Gamma = \mathfrak{V}$ is defined by the class of functions $v(x)$ for which the power series in θ

$$\sum_{j=0}^{\infty} [v(j)]^2 \theta^j$$

has a radius of convergence $\geqslant 1$. From the calculations of Example 3.1 we see that $\mathfrak{V}_T^0$ is defined by the class of functions $v(x)$ satisfying

$$v(j) = -jv(-1) \qquad (j = 0, 1, 2, \dots); \qquad \dots \quad (5.3)$$

for such functions the radius of convergence of the above series is equal to 1(∞, if $v(-1) = 0$). Finally $\mathfrak{V}_T^1$ is defined by the class of functions $v(x)$ satisfying the above series condition and such that $v(x)v^0(x)$ satisfies (5.3) for all $v^0(x)$ satisfying (5.3), that is,

$$v(j)v^0(j) = -jv(-1)\, v^0(-1) \qquad \dots \quad (5.4)$$

for all $v^0(x)$ such that

$$v^0(j) = -jv^0(-1). \qquad \dots \quad (5.5)$$

Combining (5.4) and (5.5), we find that $V = v(X)$ is in $\mathfrak{V}_T^1$ if and only if

$$v(j) = v(-1) \qquad (j = 1, 2, \dots),$$

with $v(0)$ arbitrary. Then for V in $\mathfrak{V}_T^1$,

$$E_\theta(V) = v(0)P_\theta\{X = 0\} + v(-1)P_\theta\{X \neq 0\}$$
$$= v(0)(1-\theta)^2 + v(-1)(2\theta - \theta^2).$$

The class $\mathfrak{G}$ of extimable functions $g(\theta)$ possessing minimum variance estimates is thus the two-parameter family of quadratic functions

$$g(\theta) = c_1 + c_2(1-\theta)^2,$$

and the minimum variance estimate of $g(\theta)$ is the statistic $V(X)$ taking on the value $c_1 + c_2$ if $X = 0$ and the value c_1 if $X \neq 0$.

Example 5.3: In this example we shall see that it is sometimes possible to prove that only constants have minimum variance estimates, without completely determining the class $\mathfrak{V}_T^0$ (although this would not be very difficult in the present example). As in Example 4.1, let $\mathfrak{P}^x$ be the family of distributions of a random sample of size $n>1$ from a rectangular population on the interval $(\theta-\frac{1}{2},\ \theta+\frac{1}{2})$. We define $T=(T_1,T_2)$ as in Example 4.1 and consider the class $\mathfrak{V}_T^0$. From (4.7) we see that a statistic of the type $V=f(T_1,T_2)$ (a.e. $\mathfrak{P}^x$) is in $\mathfrak{V}_T^0$ if and only if $E_\theta(V^2)<\infty$ and

$$\int_{\theta-\frac{1}{2}}^{\theta+\frac{1}{2}}\int_{\theta-\frac{1}{2}}^{t_2} f(t_1,t_2)(t_2-t_1)^{n-2}dt_1\,dt_2=0 \qquad \ldots \quad (5.6)$$

for all θ in $\omega=(-\infty,+\infty)$. We shall indicate how certain periodic solutions of (5.6) may be constructed; we remark in passing that there are also non-periodic solutions of a non-trivial kind.

Let

$$A_\theta=\{t|\theta-\tfrac{1}{2}<t_1\leqslant t_2\leqslant\theta+\tfrac{1}{2}\},$$

so that A_θ is a right triangle with hypotenuse on the 45^0 line, constituting the upper left half of a unit square S_θ centered at $(t_1,t_2)=(\theta,\theta)$, and with sides parallel to the axes. Hold $\theta=\theta_0$ fixed for the moment, and in A_{θ_0} define a Borel-measurable function $f=f(t_1,t_2)$ such that the condition (5.6) is satisfied for $\theta=\theta_0$, while

$$0<\int\int_{A_{\theta_0}} f^2(t_2-t_1)^{n-2}dt_1dt_2<\infty.$$

We next extend the definition of f to the square S_{θ_0} by defining f symmetrically in the lower right half,

$$f(t_2,t_1)=f(t_1,t_2),$$

and we then extend the definition of f to the strip $0\leqslant t_2-t_1<1$ by the periodicity condition

$$f(t_1+\nu,t_2+\mu)=f(t_1,t_2) \quad (\nu,\mu=0,\pm 1,\pm 2,\ldots).$$

With the aid of a figure showing the triangles A_{θ_0} and A_θ it is easy to see that (5.6) is satisfied for all θ, while

$$E_\theta(V^2)=E_{\theta_0}(V^2)<\infty.$$

We can now prove that the only estimable functions with minimum variance estimates are constants. Suppose $\psi(T_1,T_2)$ is in $\mathfrak{V}_T^1$ so that if $f(t_1,t_2)$ satisfies (5.6) and

$$E_\theta[f(T_1,T_2)]^2<\infty\ (\theta\epsilon\omega), \qquad \ldots \quad (5.7)$$

so does the product $\psi(t_1,t_2)f(t_1,t_2)$. It will suffice to show that in every A_θ, $\psi(t_1,t_2)$ is a constant (a.e. L), where L denotes Lebesgue measure; the constant must then be the same for all A_θ. Let θ_0 be any value of θ, henceforth held fixed, let A_+ be the part of A_{θ_0} where $\psi > 0$, and let A_- be the part of A_{θ_0} where $\psi < 0$. We shall show that it is impossible that $L(A_+)$ and $L(A_-)$ be both positive. If they were, we could define $f(t_1, t_2)$ in A_{θ_0} from

$$f(t_1,t_2)(t_2-t_1)^{n-2} = \begin{cases} 1/L(A_+) \text{ in } A_+, \\ -1/L(A_-) \text{ in } A_-, \\ 0 \text{ elsewhere in } A_{\theta 0}. \end{cases} \qquad \ldots \quad (5.8)$$

Since f satisfies (5.6) for $\theta = \theta_0$, it can be extended to an f statisfying (5.6) for all θ by the method described above. If A_+ or A_- have points within every ϵ-distance of the 45^0 line, the f defined in this way is not bounded and the condition (5.7) may not be satisfied. However, if $L(A_+)$ and $L(A_-)$ are positive, and if $A^\varepsilon{}_+$ and $A^\varepsilon{}_-$ are the parts of A_+ and A_- outside the strip $0 \leqslant t_2 - t_1 < \epsilon$, then for sufficiently small ϵ, $L(A_{\varepsilon+})$ and $L(A_{\varepsilon-})$ will also be positive. If the above definition of f is now modified by replacing A_+ and A_- in (5.8) by $A^\varepsilon{}_+$ and $A_{\varepsilon-}$, then f satisfies (5.6) and (5.7) for all θ. But then ψf cannot satisfy (5.6) for $\theta = \theta_0$ for in $A_{\varepsilon+}$ and $A^\varepsilon{}_-$, $\psi f > 0$, while in the rest of A_{θ_0}, $\psi f = 0$, and thus

$$\int\int_{A_{\theta 0}} \psi f (t_2 - t_1)^{n-2} dt_1 dt_2 > 0.$$

We have now shown that if $V = \psi((T_1, T_2)$ (a.e. $\mathfrak{P}^x$) is in $\mathfrak{V}_T{}^1$, then the sets in $A_{\theta 0}$ where $\psi(t_1, t_2) > 0$, and where $\psi(t_1, t_2) < 0$, cannot both have positive Lebesgue measure. But if V is in $\mathfrak{V}_T{}^1$ so is $V - c$ for every constant c, and so it follows that for every c the sets in A_{θ_0} where $\psi > c$ and where $\psi < c$ cannot both have positive measure. From this it can be shown that for some constant c, $\psi = c$(a.e. L) in A_{θ_0}, and this proves that if V is in $\mathfrak{V}_T{}^1$ then $V = c$ (a.e. $\mathfrak{P}^x$). Thus the only estimable functions with minimum variance estimates are the constants.

6. CONSTRUCTION AND EXISTENCE OF MINIMAL SUFFICIENT STATISTICS

So far it has not been necessary to impose any restrictions on the family $\mathfrak{P}^x$ of probability measures. However, in this section we shall assume that the measures in $\mathfrak{P}^x$ are all absolutely continuous with respect to some measure μ^x on $\mathfrak{F}^x$ which is independent of θ and has the property that W^x is a countable union of sets in $\mathfrak{F}^x$ of finite measure μ^x. This is equivalent to assuming the existence for all θ in ω of a function $p_\theta(x)$ integrable $(\mathfrak{F}^x, \mu^x)$ such that for all A in $\mathfrak{F}^x$

$$P_\theta{}^x(A) = \int_A p_\theta(x) d\mu^x.$$

We shall refer to this situation by saying there exists a *generalized probability density* $p_\theta(x)$ *with respect to* μ^x (it is the "Nikodym derivative" $dP_\theta{}^x/d\mu^x$ mentioned in section 3). This situation includes of course the two important cases with fixed sample size known as the continuous case, where μ^x is Lebesgue measure on a Euclidean space W^x, and the discrete[5] case, where for all θ the possible positions of the sample point X are included in a fixed countable set $\{x^i\}$ ($i = 1,2,3, \ldots$) independent of θ. In the latter case we may take for $\mu^x(A)$ the number of points x^i in A, and for $p_\theta(x^i)$, the probability that $X = x^i$ when the probability distribution of the sample is P_θ^x. The generalized density $p_\theta(x)$ is not uniquely determined by P_θ^x and μ^x; however, two determinations for the same θ are equal (a.e. μ^x and a.e. P_θ^x).

For most of the families $\mathfrak{P}^x$ ordinarily considered by statisticians there exists a generalized density $p_\theta(x)$ with respect to some μ^x, and if the "simplest" or "most natural" determination of $p_\theta(x)$ is used, a minimal sufficient statistic for $\mathfrak{P}^x$ can be found by applying to the family $\mathfrak{p} = \{p_\theta(x) | \theta \epsilon \omega\}$ an operation ϑ to be described below. Unfortunately the result of this operation does depend on which determination of the family $\mathfrak{p}$ is used. This introduces certain measure-theoretic difficulties, which can however be surmounted as we shall show later. We begin by defining the operation ϑ and by applying it to some examples of families of distributions of some statistical interest.

Let $\mathfrak{f}$ denote a family of real-valued functions $f(x)$ on W^x, and suppose these functions are indexed by a subscript θ taking on values in Λ, $\mathfrak{f} = \{f_\theta(x) | \theta \epsilon \Lambda\}$. The result of the operation ϑ on $\mathfrak{f}$ is a decomposition of W^x, to be denoted by $\vartheta(\mathfrak{f})$. For any point x^0 in W^x the element D of $\vartheta(\mathfrak{f})$ containing x^0, written $D(x^0)$, is defined as the set of all points x for which there exists a function $k(x, x^0) \neq 0$, not depending on θ, and such that $f_\theta(x) = k(x, x^0) f_\theta(x^0)$ for all θ in Λ. Roughly speaking, we may say that $D(x^0)$ consists of all x for which the ratio $f_\theta(x)/f_\theta(x^0)$ is independent of θ. We note that if x' is in $D(x^0)$, then x^0 is in $D(x')$; also, that

$$D^0 = \{x | f_\theta(x) = 0 \quad \text{for all } \theta \epsilon \Lambda\}$$

is an element of $\vartheta(\mathfrak{f})$. It may be shown (as in the proof of Theorem (6.1)) that if $\mathfrak{f}$ consists of a countable number of measurable ($\mathfrak{F}^x$) functions then $\vartheta(\mathfrak{f})$ is a measurable ($\mathfrak{F}^x$) decomposition.

We consider now five examples of applying the operation ϑ. In each case W^x is a Euclidean n-space, and $\mathfrak{F}^x$ may be taken as the family of Borel sets. In all but Example 6.1, μ^x will be taken as Lebsgue measure on $\mathfrak{F}^x$.

Example 6.1: Suppose $X = (X_1, X_2, \ldots, X_n)$ is a random sample from a binomial population with parameter θ, X_i taking on the values 1 and 0 with respective probabilities θ and $1-\theta$. Let μ^x assign measure 1 to each of the 2^n points in the set W_+^x consisting of the points $(x_1, x_2, \ldots, x_n)$ with $x_i = 0$ or 1, and measure zero to $W^x - W_+^x$. In this example we might take $\mathfrak{F}^x$ as the family of all subsets of W_+^x,

[5] In the general discrete case, while for each θ the possible values of X constitute a countable set A_θ, $A = \cup_\theta A_\theta$ can of course be non-denumerable. The case of non-denumerable A is not included in the present treatment.

or the family of all subsets of W^x, instead of the usual family of all Borel sets in W^x. For ω we take the open interval $0<\theta<1$, and for $p_\theta(x)$ the determination

$$p_\theta(x) = \begin{cases} \prod_{j=1}^{n} \theta^{x_i}(1-\theta)^{1-x_i} & \text{if } x \epsilon W_+^x, \\ 0 & \text{otherwise.} \end{cases} \quad \ldots \quad (6.1)$$

For this specification, $D^0 = W^x - W_+^x$. For x^0 not in D^0, x is in $D(x^0)$ if and only if there exists $k(x, x^0)$ such that $p_\theta(x) = k(x, x^0)p_\theta(x^0)$, and hence

$$[\theta/(1-\theta)]^{\Sigma_i x_i - \Sigma_i x_i^0} = k(x, x^0). \quad \ldots \quad (6.2)$$

The left member of (6.2) is independent of θ if and only if

$$\sum_{i=1}^{n} x_i = \sum_{i=1}^{n} x_i^0, \quad \ldots \quad (6.3)$$

and this conclusion would be valid even if (6.2) were required to hold only for two distinct values of θ instead of all values between 0 and 1. The decomposition resulting from the application of the operation ϑ to the family $\{p_\theta(x)\}$ of (6.1) may thus be described as follows: The decomposition is equivalent ($\mathfrak{P}^x$) in the strong sense to that associated with the statistic $\Sigma_i X_i$ or any 1:1 function of this. For any x^0 in W_+^x the element $D(x^0)$ consists of the $\binom{n}{\Sigma_i x_i^0}$ points in W_+^x satisfying (6.3). The decomposition contains $n+2$ elements D, namely D^0 and the $n+1$ sets D where $\sum_{i=1}^{n} x_i = \nu (\nu=0, 1, \ldots, n)$. The same decomposition is obtained if the operation ϑ is applied to any subset of $\{p_\theta x)$ consisting of two or more elements.

Example 6.2: Let X be a random sample from a normal population with mean θ_1 and variance θ_2 as in Example 5.1. With the determination (5.1) of the probability density $p_\theta(x)$ we find D^0 is the empty set. Since the denominator of the fraction $p_\theta(x)/p_\theta(x^0)$ cannot vanish, $D(x^0)$ is the set where this ratio is independent of θ. This is seen to be the same as the set where

$$-\frac{1}{2\theta_2}\left(\Sigma_i x_i^2 - \Sigma_i x_i^{0^2}\right) + \frac{\theta_1}{\theta_2}\left(\Sigma_i x_i - \Sigma_i x_i^0\right)$$

is independent of θ, namely, the set where

$$\Sigma_i x_i^2 = \Sigma_i x_i^{0^2} \text{ and } \Sigma_i x_i = \Sigma_i x_i^0.$$

The decomposition induced by the operation ϑ is thus that associated with the statistic $(\Sigma_i X_i, \Sigma_i X_i^2)$; this is the same as that associated with the statistic (T_1, T_2) defined in Example 5.1. It may be verified in this case that one obtains the same result if the operation ϑ is applied to a set of any three of the members of $\{p_\theta(x)\}$, say for $(\theta_1, \theta_2) = (\theta_{1i}, \theta_{2i})$, $i = 1,2,3$, providing the three points $(\theta_{1i}, \theta_{2i})$ are not collinear.

Example 6.3: Suppose X is a random sample of n from a Cauchy distribution with median θ, and $\omega=\{\theta \mid -\infty<\theta<+\infty\}$. We take for $p_\theta(x)$ the usual determination

$$p_\theta(x) = \pi^{-n} \prod_{j=1}^{n} [1+(x_j-\theta)^2]^{-1}.$$

Again D^0 is empty and $D(x^0)$ is the set where the ratio $p_\theta(x)/p_\theta(x_0)$ is independent of θ. This ratio may be written

$$\frac{\prod_{j=1}^{n} [(\theta-x_j^0-i)(\theta-x_j^0+i]}{\prod_{j=1}^{n} [(\theta-x_j-i)(\theta-x_j+i)]},$$

where $i = \sqrt{-1}$. If this quotient is independent of θ, the polynomials of degree $2n$ in θ in the numerator and denominator must have equal roots, from which we conclude that $(x_1, x_2, \ldots, x_n)$ must be a permutation of $(x^0_1, x_2^0, \ldots, x_n^0)$. The decomposition induced by ϑ may be regarded as that associated with the set of "order statistics" of the sample, namely $(Z_1, Z_2, \ldots, Z_n)$, where $Z_1 \leqslant Z_2 \leqslant \ldots \leqslant Z_n$ is a rearrangement of $X_1, X_2, \ldots, X_n$. The same result would be found for any subset of $\{p_\theta(x)\}$ corresponding to $2n+1$ distinct values of θ.

Example 6.4: As in Examples 4.1 and 5.3 let X be a random sample of n from the uniform distribution on $(\theta-\frac{1}{2}, \theta+\frac{1}{2})$, $-\infty<\theta<\infty$. We determine $p_\theta(x)$ as 1 if $|x_i-\theta|<\frac{1}{2}$ for all i, as 0 otherwise. With $t_1(x) = \min_i x_i$, $t_2(x) = \max_i x_i$, we see that $p_\theta(x) = 1$ if

$$\theta-\tfrac{1}{2}<t_1(x)\leqslant t_2(x)<\theta+\tfrac{1}{2}, \quad \ldots \quad (6.4)$$

$p_\theta(x) = 0$ otherwise. Therefore

$$D^0 = \{x | t_2(x)-t_1(x) \geqslant 1\}.$$

Now for any x^0 and x, x is in $D(x^0)$ if and only if $p_\theta(x)$ and $p_\theta(x^0)$ vanish for the same set of θ in ω. It follows that for x^0 not in D^0, x is in $D(x^0)$ if and only if (6.4) is true for all θ in ω which satisfy

$$\theta-\tfrac{1}{2}<t_1(x^0)\leqslant t_2(x^0)<\theta+\tfrac{1}{2}.$$

It follows that for x^0 not in D^0, x is in $D(x^0)$ if and only if $t_j(x) = t_j(x^0)$, $j = 1, 2$. The decomposition imposed by the operation ϑ is thus equivalent ($\mathfrak{P}^x$) in the strong sense to that associated with the statistic (T_1, T_2) of Examples 4.1 and 5.3. The same decomposition is obtained if the operation ϑ is applied to a denumerable subset of $\{p_\theta(x)\}$ corresponding to a subset of ω everywhere dense in ω.

Example 6.5: Denote by $(X_1, X_2, \ldots, X_m)$ and $(X_{m+1}, \ldots, X_n)$ independent random samples from normal populations with means θ_1 and $\theta_1+\delta$, and variances θ_2 and θ_3, respectively. The null hypothesis of the Behrens-Fisher problem then specifies $\delta = \delta_0$, where δ_0 is a known constant, so that the ω for the family $\mathfrak{P}^x$ of probability distributions under the null hypothesis is the subset of a three-dimensional Euclidean space satisfying $-\infty<\theta_1<+\infty, \theta_2>0, \theta_3>0$. If we determine the density $p_\theta(x)$ as

$$p_\theta(x) = (2\pi)^{\frac{1}{2}n}\theta_2^{\frac{1}{2}m}\theta_3^{\frac{1}{2}(n-m)}e^{-\frac{1}{2\theta_2}\sum_{i=1}^{m}(x_i-\theta_1)^2-\frac{1}{2\theta_3}\sum_{j=m+1}^{n}(x_j-\theta_1-\delta_0)^2}$$

it is easy to show that the result of applying the operation ϑ to this three-parameter family of density functions is the decomposition associated with the statistic (T_1, T_2, T_3, T_4) with four real components consisting of the two sample means and the two sample variances.

The measure-theoretic difficulty in applying the operation ϑ to construct a minimal sufficient statistic is the following: If a minimal sufficient statistic for $\mathfrak{P}^x$ exists then it is unique up to equivalence ($\mathfrak{P}^x$) in the strong sense. If the family $\mathfrak{P}^x$ of probability measures possesses a generalized density $p_\theta(x)$ with respect to the measure μ^x, then $p_\theta(x)$ is not unique; two determinations of $p_\theta(x)$ for fixed θ differ on a set A_θ which is in $\mathfrak{F}^x$ and for which $\mu^x(A_\theta) = 0$. Now if ω is non-denumerably infinite, the union of A_θ for θ in ω may not be a null set for μ^x; it need not even be measurable ($\mathfrak{F}^x$). The decomposition $\vartheta(\mathfrak{p})$ resulting from the application of the operation ϑ to the family of generalized densities $\mathfrak{p}=\{p_\theta(x)|\theta\epsilon\omega\}$ depends on which determination is chosen for $\mathfrak{p}$. As long as ω is countable it can be shown (Theorem 6.1) that the decomposition $\vartheta(\mathfrak{p})$ is measurable ($\mathfrak{F}^x$), and is associated with a minimal sufficient statistic for $\mathfrak{P}^x$, and it is easily seen that the two decompositions $\vartheta(\mathfrak{p})$ resulting from two different determinations of $\mathfrak{p}$ are equivalent (μ^x and $\mathfrak{P}^x$) in the strong sense. However, if ω is not countable, a pathological choice of the determination of $\mathfrak{p}$ could lead to a decomposition $\vartheta(\mathfrak{p})$ that is not measurable ($\mathfrak{F}^x$), and the equivalence of two $\vartheta(\mathfrak{p})$ resulting from two different determinations of $\mathfrak{p}$ could not be proved. This shows that the application of the operation ϑ need not necessarily lead to the minimal sufficient statistic.

This difficulty is resolved as follows: We restrict ourselves to families $\mathfrak{P}^x$ for which the family $\{p_\theta(x)\}$ of generalized densities is separable in a certain sense to be defined below; this separability *always* holds when W^x is a Euclidean space. There is then a countable subset of $\{p_\theta(x)\}$ dense in the whole set, and we apply the operation ϑ to this countable subset. (Recall that in the above examples we noted that in every case the same result was obtained by applying the operation θ to a suitable countable subset of the densities). On the basis of the theorems below it follows that the result of applying the operation ϑ to the countable subset of the densities is a measurable ($\mathfrak{F}^x$) decomposition which is associated with a minimal sufficient statistic for the whole set $\mathfrak{P}^x$ of distributions. The result needed for countable subsets is

Theorem 6.1: *Suppose $\mathfrak{P}^x$ is a countable set of probability measures on $\mathfrak{F}^x$, possessing a generalized probability density function $p_\theta(x)$ with respect to μ^x. If the operation ϑ is applied to a particular determination of the family $\{p_\theta(x)\}$, the resulting* *decomposition is measurable* ($\mathfrak{F}^x$) *and is associated with a minimal sufficient statistic for* $\mathfrak{P}^x$. *The decompositions resulting from applying the operation* ϑ *to two different determinations of the family* $\{p_\theta(x)\}$ *are equivalent* (μ^x *and* $\mathfrak{P}^x$) *in the strong sense.*

The proof of this theorem rests on the following result[6] of Halmos and Savage (1949) generalizing a theorem of Neyman (1935):

Theorem 6.2: *A necessary and sufficient condition for the statistic* $T=t(X)$ *to be a sufficient statistic for a set of probability measures* $\mathfrak{P}^x = \{P_\theta^x \mid \theta\epsilon\omega\}$ *possessing a generalized density with respect to* μ^x *is that there exist functions* $g_\theta(t)$ *and* $h(x)$ *such that*

$$p_\theta(x) = g_\theta(t(x))\,h(x) \qquad \ldots \quad (6.5)$$

for a suitable determination $p_\theta(x)$ *of this generalized density, where* $g_\theta(t(x))$ *is measurable* ($\mathfrak{F}^x$) *and* $h(x)$ *is integrable* ($\mathfrak{F}^x$, μ^x).

To prove Theorem 6.1 let us denote by θ_i the elements of the countable set ω, and by $f_i(x)$ some particular determination of the generalized density $p_{\theta_i}(x)$, $i = 1,2, \ldots$. Write $\mathfrak{f} = \{f_i(x)\}$, and $\vartheta(\mathfrak{f})$ for the decomposition generated by the operation ϑ on $\mathfrak{f}$.

We shall show first that $\vartheta(\mathfrak{f})$ is measurable ($\mathfrak{F}^x$). If $D_i^0 = \{x|f_i(x)=0\}$, then D_i^0 is measurable ($\mathfrak{F}^x$) since $f_i(x)$ is measurable ($\mathfrak{F}^x$); hence $D^0=\bigcap_i D_i^0$ is measurable ($\mathfrak{F}^x$). Suppose now that x^0 is not in D^0. Then there exists a smallest i for which $f_i(x^0)\neq 0$, say $I=I(x^0)$. Define $D_i(x^0)$ as the set where $f_I(x)\neq 0$ and

$$f_i(x) \mid f_I(x) = f_i(x^0) \mid f_I(x^0).$$

Since $f_i(x)$ and $f_I(x)$ are measurable ($\mathfrak{F}^x$) so is their quotient in $W^x-D_I^0$, the part of W^x where $f_I(x)\neq 0$, and hence the part $D_i(x^0)$ of $W^x-D_I^0$ where the quotient has a constant value is measurable ($\mathfrak{F}^x$). Therefore $\bar{D}(x^0)=\bigcap_i D_i(x^0)$ is measurable ($\mathfrak{F}^x$). If for any point x^0 in W^x, $D(x^0)$ denotes the element of $\vartheta(\mathfrak{f})$ containing x^0, it is easily verified that for x^0 not in $\bar{D}^0$, $D(x^0)=\bar{D}(x^0)$, and for x^0 in D^0, $D(x^0)=D^0$. This completes the proof that $\vartheta(\mathfrak{f})$ is measurable ($\mathfrak{F}^x$).

Let $\mathfrak{f}'=\{f'_i(x)\}$ be another determination of the generalized densities; then if N_i is the set where $f'_i(x)\neq f_i(x)$, N_i is in $\mathfrak{F}^x$ and $\mu^x(N_i)=0$. Hence $N=\mathrm{U}_i N_i$ is also a null set for μ^x. On W^x-N, $\mathfrak{f}$ and $\mathfrak{f}'$ are identical hence so are $\vartheta(\mathfrak{f})$ and $\vartheta(\mathfrak{f}')$. This proves the strong equivalence (μ^x and $\mathfrak{P}^x$) of the two decompositions.

Next we shall prove that, if $D(x)$ is the element of $\vartheta(\mathfrak{f})$ which contains x, and if $T=t(X)$ is a statistic associated with $v\,\vartheta(\mathfrak{f})$,—in particular we could take $t(x)=D(x)$ —then $f_i(x)$ may be factored in the form (6.5) of Theorem 6.2. To this end we choose in each element D of $\vartheta(\mathfrak{f})$ a single point $x^0=\chi(D)$. If the set of points x^0 thus chosen is denoted by W^0, this choice determines a function $x^0=\psi(x)$ from W^x into W^0,

[6] We are indebted to Professors Halmos and Savage for giving us a manuscript copy of their paper containing this result long before publication. The reader will not find the above Theorem 6.2 there in exactly the form stated here; the connection is discussed in the Appendix to the present paper.

since x in W^x determines $D=D(x)$ in ϑ ($\mathfrak{t}$), and D determines $x^0=\chi(D)=\chi(D(x))=\psi(x)$. Take any D in $\vartheta(\mathfrak{t})$ and hold it fixed. Then for all x in D and $x^0=\chi(D)$,

$$f_i(x)=k(x,x^0)f_i(x^0),$$

where $k(x,x^0)\neq 0$, and $i=1,2,\ldots$ This may be written

$$f_i(x)=k(x,\psi(x))f_i(\chi(D)).$$

For x in W^x-D^0 define $\bar{h}(x)=k(x, \psi(x))\neq 0$, and for x in D^0 define $\bar{h}(x)=0$. Define $\bar{g}_i(t)=f_i(\chi(D))$, where the elements D of ϑ ($\mathfrak{t}$) are the contours of $t(x)$. We now have

$$f_i(x)=\bar{g}_i(t(x))\bar{h}(x), \qquad \ldots (6.6)$$

and

$$\{x\,|\,\bar{h}(x)=0\}=D^0.$$

Now (6.6) is of the factored form (6.5) of Theorem 6.2 except that we do not know yet whether the factors can be determined so as to satisfy the measurability and integrability conditions of the theorem. For brevity in proving this let us denote by $\mathfrak{G}_t$ the family of sets in W^x each of which is a union of elements D of ϑ ($\mathfrak{t}$); the members of $\mathfrak{G}_t$ need not be in $\mathfrak{F}^x$. For $i=1,2,\ldots$, define

$$G_i=\{\,x\,|\,f_1(x)=f_2(x)=\ldots=f_{i-1}(x)=0,\ f_i(x)\neq 0\}.$$

It is clear that $\{G_i\}$ is a sequence of disjoint sets in $\mathfrak{F}^x$ with $\cup_i G_i = W^x-D^0$, and we shall now show that all G_i are in $\mathfrak{G}_t$: Let $H_i = \{x|\bar{g}_i(t(x)) = 0\}$, so H_i is in $\mathfrak{G}_t$. Let $K_i = \{x|f_i(x) = 0\}=\{x|\bar{g}_i(t(x))\bar{h}(x) = 0\} = H_i \cup D^0$, so K_i is also in $\mathfrak{G}_t$. Finally,

$$G_i=K_1\cap K_2\cap \ldots\cap K_{i-1}\cap (W^x-K_i)$$

is also in $\mathfrak{G}_t$. For x in $G_k(k = 1,2,\ldots)$ we have from (6.6)

$$f_k(x) = \bar{g}_k(t(x))\bar{h}(x)\neq 0. \qquad \ldots (6.7)$$

Combining this with (6.6), we get for x in G_k and $i=1, 2,\ldots$,

$$f_i(x) = [\bar{g}_i(t(x))\,|\,\bar{g}_k(t(x))]f_k(x). \qquad \ldots (6.8)$$

In G_k $(k = 1,2,\ldots)$ define

$$h(x) = 2^{-k}f_k(x), \qquad \ldots (6.9)$$

and in $t(G_k)$ define

$$g_i(t) = 2^k\bar{g}_i(t)\,|\,\bar{g}_k(t),$$

noting from (6.7) that $\bar{g}_k(t) \neq 0$ in $t(G_k)$. This defines $h(x)$ in $W^x - D^0$ and $g_i(t)$ in $t(W^x - D^0)$ for $i = 1, 2, \ldots$; in D^0 define

$$h(x) = 0, \quad \ldots \quad (6.10)$$

and in $t(D^0)$, $g_i(t) = 0$. To see that there is no inconsistency in our definition of $g_i(t)$ it is important to note that the sets $t(D^0)$ and $t(G_k)(k = 1,2,\ldots)$ are all disjoint: This is a consequence of D^0 and $G_1, G_2, \ldots$ being disjoint sets in $\mathfrak{G}_t$. From (6.8) we have

$$f_i(x) = g_i(t(x))h(x). \quad \ldots \quad (6.11)$$

From (6.9) and (6.10) we see that $h(x)$ is measurable ($\mathfrak{F}^x$) in each of the sets G_k and D^0, and since these form a countable disjoint covering of W^x by sets in $\mathfrak{F}^x$, $h(x)$ is measurable ($\mathfrak{F}^x$) in W^x. That it is also integrable ($\mathfrak{F}^x, \mu^x$) follows from

$$\int_{W^x} |h(x)| d\mu^x = \int_{W^x - D^0} h(x) d\mu^x = \sum_{k=1}^{\infty} \int_{G_k} f_k(x) d\mu^x$$

$$\leqslant \not\geqslant \sum_k 2^{-k} \int_{W^x} f_k(x) d\mu^x = \sum_k 2^{-k} < \infty .$$

Since $g_i(t(x)) = 0$ in D^0 while from (6.11) $g_i\,(t(x))$ is the quotient of measurable ($\mathfrak{F}^x$) functions with non-vanishing denominator in $W^x - D^0$, $g_i(t(x))$ is measurable ($\mathfrak{F}^x$) in W^x.

We have now shown all the hypotheses of Theorem 6.2 to be satisfied and hence $T = t(X)$ is a sufficient statistic for $\mathfrak{P}^x$; it remains only to prove that it is minimal. To do this we must show that if $T' = t'(X)$ is any other sufficient statistic for $\mathfrak{P}^x$, then it is possible to find a null set N'' for $\mathfrak{P}^x$ such that in $W^x - N''$ every contour of $t'(x)$ is contained in a contour of $t(x)$.

If T' is sufficient for $\mathfrak{P}^x$, then by Theorem 6.2 there exists a determination of the generalized density, say $f_i'(x)$, such that for $i = 1, 2, \ldots$,

$$f_i'(x) = g_i'(t'(x))h'(x),$$

where $h'(x)$ is measurable ($\mathfrak{F}^x$). The determination $\mathfrak{f}' = \{f'_i(x)\}$ may differ from the previous determination $\mathfrak{f} = \{f_i(x)\}$ to the extent that if $N_i' = \{x \mid f_i'(x) \neq f_i(x)\}$, then N_i' is a null set for μ^x. Hence $N' = \bigcup_i N_i'$ is a null set for μ^x and $\mathfrak{P}^x$, and $f_i' = f_i$ on $W^x - N'$. Let $N_0 = \{x | h'(x) = 0\}$. Then for all i, $f_i'(x) = 0$ on N_0,

$$P_{\theta_i}{}^x(N_0) = \int_{N_0} f_i'(x) d\mu^x = 0,$$

and so N_0 is null set for $\mathfrak{P}^x$. If $N'' = N' \cup N_0$, N'' is a null set for $\mathfrak{P}^x$, and on $W^x - N''$, $f_i'(x) = f_i(x)$ and $h'(x) \neq 0$.

We shall now show that in $W^x - N''$ every contour of $t'(x)$ is contained in a contour of $t(x)$. Let x^0 be any point in $W^x - N''$, and let $A(x^0)$ be the part in $W^x - N''$ of the contour of $t'(x)$ containing x^0,

$$A(x^0) = \{x | x \epsilon W^x - N'',\ t'(x) = t'(x^0)\}.$$

For all x in $A(x^0)$

$$f_i(x) = f_i'(x) = g_i'(t'(x^0))h'(x);$$

in particular,

$$f_i(x^0) = g_i'(t'(x^0))h'(x^0).$$

Since $h'(x^0) \neq 0$,

$$f_i(x) = k(x,x^0)f_i(x^0) \quad \ldots \quad (6.12)$$

for all x in $A(x^0)$, where

$$k(x,x^0) = h'(x)|h'(x^0) \neq 0.$$

It follows from (6.12) that in $W^x - N''$, $A(x^0)$ is contained in $D(x^0)$, the contour of $t(x)$ containing x^0. This completes the proof of Theorem 6.1.

In proving an existence theorem for minimal sufficient statistics for families $\mathfrak{P}^x$ which may be non-denumerably infinite, we shall employ the following two lemmas whose proof is obvious.

Lemma 6.1: *If the statistic T is sufficient for the family $\mathfrak{P}^x$ of probability measures it is sufficient for any subfamily of $\mathfrak{P}^x$.*

Lemma 6.2: *If T is a sufficient statistic for the family $\mathfrak{P}^x$, if it is a minimal sufficient statistic for the subfamily $\mathfrak{P}_1^x$ of $\mathfrak{P}^x$, and if every null set for $\mathfrak{P}_1^x$ is a null set for $\mathfrak{P}^x$, then T is a minimal sufficient statistic for $\mathfrak{P}^x$.*

Our extension of the existence theorem from the countable to the not necessarily countable case involves the notion of separability of a function space with respect to convergence in the mean of order one. If $f = f(x)$ and $g = g(x)$ are two real-valued functions on W^x, integrable $(\mathfrak{F}^x, \mu^x)$, we define their distance as

$$\delta_\mu (f, g) = \int_{W^x} |f(x -g(x)| d\mu^x.$$

We shall say that a family of real-valued functions, $\mathfrak{f} = \{f_\theta(x)|\theta\epsilon\omega\}$, each integrable $(\mathfrak{F}^x, \mu^x)$, is *separable* (μ^x) if there exists a fixed countable subset $\mathfrak{f}_1$ of $\mathfrak{f}$ such that for every f_θ in $\mathfrak{f}$ there is a sequence $\{g_i | i=1,2\ldots\}$ of functions g_i in $\mathfrak{f}_1$ for which $\delta_\mu (f,g_i) \to 0$ as $i \to \infty$. It will be convenient to say that the countable subset $\mathfrak{f}_1$ of this definition is *dense* (μ^x) in $\mathfrak{f}$. We remark that if W^x is a Euclidean space and $\mathfrak{F}^x$ is the family of Borel sets in W^x then any family $\{p_\theta(x)\}$ of generalized densities with respect to μ^x is separable: This is a consequence of the separability[7] (μ^x) of the family of all functions integrable $(\mathfrak{F}^x, \mu^x)$, and the separability of any subfamily of a separable family in a metric space. We can now conclude that the result of applying the operation ϑ in each of Examples 6.1 to 6.5 was to give a minimal sufficient statistic: This follows from (i) the remark just made about Euclidean W^x, (ii) the remarks in the examples, about the same decomposition being obtained if the operation ϑ is applied to certain countable subsets of the generalized densities, and (iii) the following theorem.

[7] This may be proved from the Approximation Theorem on page 4 of *Ergodentheorie* by Hopf (1948, Chelsea, N.Y.).

Theorem 6.3: *If the family* $\mathfrak{P}^x$ *of probability measures on* W^x *possesses a generalized density* $p_\theta(x)$ *with respect to* μ^x, *and if the family* $\mathfrak{p} = \{p_\theta(x) \mid \theta \epsilon \omega\}$ *of densities is separable* (μ^x), *then there exists a minimal sufficient statistic for* $\mathfrak{P}^x$, *and it may be constructed by applying the operation* ϑ *to any countable set* $\mathfrak{p}_1$ *dense* (μ^x) *in* $\mathfrak{p}$.

To prove the theorem let $\mathfrak{f}_1 = \{f_i(x)\}(i = 1,2,\ldots)$ be a particular determination of a countable subfamily $\mathfrak{p}_1$ of densities dense (μ^x) in $\mathfrak{p}$, and let $\mathfrak{P}_1{}^x$ be the countable subfamily of $\mathfrak{P}^x$ corresponding to $\mathfrak{p}_1$. In order to apply Lemma 6.2 we must show that if N is a null set for $\mathfrak{P}_1{}^x$ it is a null set for $\mathfrak{P}^x$. Let $P_{\theta_0}{}^x$ be any member of $\mathfrak{P}^x$, let $f_0(x) = dP_{\theta_0}{}^x/d\mu^x$, and let $\{g_i\}$ be a sequence in $\mathfrak{f}_1$ for which $\delta_\mu(f_0, g_i) \to 0$ as $i \to \infty$. We have

$$\int_N g_i(x) d\mu^x = 0$$

for all i. Now

$$\int_N f_0(x) d\mu^x = \left| \int_N f_0 d\mu^x \right| = \left| \int_N f_0 d\mu^x - \int_N g_i d\mu^x \right|$$

$$= \left| \int_N (f_0 - g_i) d\mu^x \right| \leqslant \int_N |f_0 - g_i| d\mu^x$$

$$\leqslant \int_{W^x} |f_0 - g_i| d\mu^x = \delta_\mu(f_0, g_i) \to 0.$$

Hence

$$\int_N f_0(x) d\mu^x = 0$$

and N is a null set for $\mathfrak{P}^x$.

Suppose $T = t(X)$ is a minimal sufficient statistic for $\mathfrak{P}_1^x$, obtained by applying the operation ϑ to $\mathfrak{f}_1$. To prove T is a minimal sufficient statistic for $\mathfrak{P}^x$ it remains only to prove that T is sufficient for $\mathfrak{P}^x$. That T is sufficient for $\mathfrak{P}_1{}^x$ means that there exists a real-valued function $P(A|t)$, measurable $(\mathfrak{F}^t)$ for fixed A, independent of θ, and such that

$$P^{\theta x}(A \cap t^{-1}(B)) = \int_B P(A|t) dP_\theta{}^t \quad \ldots \quad (6.13)$$

for all A in $\mathfrak{F}^x$, B in $\mathfrak{F}^t$, and θ in ω_1, where ω_1, is the subset of ω corresponding to $\mathfrak{P}_1{}^x$. To prove T sufficient for $\mathfrak{P}^x$ it thus suffices to show that if (6.13) is valid for all θ in ω_1, then it is valid for all θ in ω. Let us write B' as an abbreviation of $t^{-1}(B)$.

We may transform (6.13) to

$$\int_{A \cap B'} dP_\theta{}^x = \int_{B'} \pi(A, x) dP_\theta{}^x,$$

Where $\pi(A, x) = P(A|t(x))$, or

$$\int_{A \cap B'} p_\theta(x)\, d\mu^x = \int_{B'} \pi(A, x) p_\theta(x) d\mu^x. \quad \ldots \quad (6.14)$$

We note $0 \leqslant \pi(A,x) \leqslant 1$ (a.e. $\mathfrak{P}_1{}^x$) by (2.5) and (2.8), and hence (a.e. $\mathfrak{P}^x$). Let $\theta = \theta_0$

be any θ in ω, write $p_{\theta_0}(x) = g(x)$, and let $\{g_i\}$ be a sequence in $\mathfrak{f}_1$ for which $\delta_\mu(g, g_i) \to 0$ as $i \to \infty$. From

$$\int_{A \cap B'} g_i d\mu^x = \int_{B'} \pi g_i d\mu^x \qquad \ldots \quad (6.15)$$

we shall prove

$$\int_{A \cap B'} g d\mu^x = \int_{B'} \pi g d\mu^x \qquad \ldots \quad (6.16)$$

by letting $i \to \infty$ and showing that the limits of the right and left members of (6.15) are the corresponding members of (6.16).

$$\left| \int_{A \cap B'} g_i d\mu^x - \int_{A \cap B'} g d\mu^x \right| \leqslant \int_{A \cap B'} |g_i - g| d\mu^x$$

$$\leqslant \delta_\mu(g_i, g) \to 0.$$

$$\left| \int_{B'} \pi g_i d\mu^x - \int_{B'} \pi g d\mu^x \right| \leqslant \int_{B'} |\pi| \, |g_i - g| d\mu^x$$

$$\leqslant \int_{B'} |g_i - g| d\mu^x \leqslant \delta_\mu(g_i, g) \to 0.$$

This completes the proof of Theorem 6.3.

APPENDIX

With only notational changes, the result on sufficient statistics and the factorization of generalized probability densities stated by Halmos and Savage (1949), of which Theorem 6.2 above is a slightly different version, is their following

Corollary 1: *Let $\mathfrak{F}^x$ and $\mathfrak{G}^t$ be countably additive families of sets in the spaces W^x and W^t, respectively. Let $t = t(x)$ be a function from W^x onto W^t such that for every set B in $\mathfrak{G}^t$, $t^{-1}(B)$ is in $\mathfrak{F}^x$, and let $\mathfrak{G}^{x|t} = t^{-1}(\mathfrak{G}^t)$. Let $\mathfrak{M}^x = \{M_\theta^x | \theta \epsilon \omega\}$ be a family of finite measure on $\mathfrak{F}^x$ with all M_θ^x absolutely continuous with respect to the finite measure μ^x on $\mathfrak{F}^x$. Then a necessary and sufficient condition for $T = t(X)$ to be a sufficient statistic*[8] *for $\mathfrak{M}^x$ is that for every θ in ω, $f_\theta(x) = dM_\theta^x/d\mu^x$ be factorable in the form $f_\theta = g_\theta h$, where $0 \leqslant g_\theta$ is measurable ($\mathfrak{G}^{x|t}$), $0 \leqslant h$ and $0 \leqslant g_\theta h$ are integrable ($\mathfrak{F}^x$, μ^x), and $h(x) = 0$ (a.e. μ^x) on every null set for $\mathfrak{M}^x$.*

Some of the simplifications we have obtained arise from the restrictions that, first, we always take for $\mathfrak{G}^t$ the family $\mathfrak{F}^t$ of all sets in W^t whose pre-images are in

[8] If $\mathfrak{M}^x$ is not a family of probability measures, T is not a random variable. Halmos and Savage define statistic, conditional probability, and sufficient statistic for this case. We need not consider these definitions here except to remark that in the case that $\mathfrak{M}^x$ is a family of probability measures—the only case considered by us—the definitions agree with ours.

$\mathfrak{F}^x$, and that, secondly, we assume $\mathfrak{M}^x$ to be a family of probability measures. Theorem 6.2 follows from Corollary 1 by means of the following six steps:

1⁰. The assumption that $t(x)$ satisfies the condition that pre-images of sets in $\mathfrak{G}^t$ are in $\mathfrak{F}^x$ is automatically fulfilled for an arbitrary function $t(x)$ because we always take $\mathfrak{G}^t = \mathfrak{F}^t$ as just mentioned.

2⁰. It is pointed out by Halmos and Savage that in their work the assumption on μ^x that $\mu^x(W^x)$ is finite may be replaced by the assumption that W^x is a countable union of sets of finite measure (μ^x).

3⁰. The assumption that $g_\theta h$ is non-negative and integrable ($\mathfrak{F}^x$, μ^x) may be dropped because of our assumption that $\mathfrak{M}^x$ is a family of probability measures, so that $g_\theta h = f_\theta$ may be assumed non-negative and

$$\int_{W^x} g_\theta h d\mu^x = 1.$$

4⁰. The condition of non-negativeness on g_θ and h may be dropped, since we may assume f_θ non-negative and then redefine g_θ and h as $|g_\theta|$ and $|h|$ without affecting the product $g_\theta h$, the measurability of g_θ, or the integrability of h.

5⁰. The assumption that $g_\theta(x)$ is measurable ($\mathfrak{G}^{x/t}$) is equivalent to the assumption that it is of the form $\bar{g}_\theta(t(x))$, where $\bar{g}_\theta(t)$ is measurable ($\mathfrak{G}^t$) by Lemma 2 in the same paper by Halmos and Savage. But with our choice of $\mathfrak{G}^t = \mathfrak{F}^t$, $\bar{g}_\theta(t)$ is measurable ($\mathfrak{G}^t$) if and only if $\bar{g}_\theta(t(x))$ is measurable ($\mathfrak{F}^x$).

6⁰. The proof that the assumption $h(x)=0$ (a.e. μ^x) on every null set for $\mathfrak{M}^x$, may be dropped requires a little more effort. If h satisfies all the remaining conditions of Corollary 1 we shall show that $\bar{h}$ can be defined which also satisfies all the remaining conditions and furthermore $\bar{h}(x) = 0$ (a.e. μ^x) on every null set for $\mathfrak{M}^x$.

Let $f_\theta = g_\theta h$ denote a particular determination of $dM_\theta^x/d\mu^x$ for which $f_\theta \geqslant 0$. B Lemma 7 of Halmos and Savage, there exists a countable subset $\mathfrak{M}_1^x$ of $\mathfrak{M}^x$, say $\mathfrak{M}_1^x = \{M_\theta^x | \theta = \theta_1, \theta_2, \ldots\}$, such that all null sets for $\mathfrak{M}_1^x$ are null sets for $\mathfrak{M}^x$. Denote by A^0 the set in $\mathfrak{F}^x$ where $f_{\theta_i}(x) = 0$ for all i. Then A_0 is a null set for $\mathfrak{M}_1^x$ and hence for $\mathfrak{M}^x$. Define

$$\bar{h}(x) = \begin{cases} h(x) \text{ for } x \text{ in } W^x - A_0, \\ 0 \text{ for } x \text{ in } A_0. \end{cases}$$

For all A in $\mathfrak{F}^x$,

$$\int_A g_\theta h d\mu^x = M_\theta^x(A) = M_\theta^x(A - A_0) = \int_{A-A0} g_\theta h d\mu^x$$

$$= \int_{A-A0} g_\theta \bar{h} d\mu^x = \int_A g_\theta \bar{h} d\mu^x,$$

and hence $g_\theta h = g_\theta \bar{h}$ (a.e. μ^x). We may thus use $\bar{f}_\theta = g_\theta \bar{h} \geqslant 0$ instead of $f_\theta = g_\theta h$ as a determination of $dM_\theta^x/d\mu^x$, and $\bar{h}$ obviously satisfies the condition of integrability ($\mathfrak{F}^x$, μ^x) since h did.

It remains only to prove that if N is a null set for $\mathfrak{M}^x$ then $\bar{h}(x) = 0$ (a.e. μ^x) on N. Let N' be the part of N where $\bar{h} \neq 0$; we need to prove $\mu^x(N') = 0$. Now

$$0 = M_{\theta_i}{}^x(N') = \int_{N'} g_{\theta_i} \bar{h} d\mu^x.$$

Let N_i be the part of N' where $g_{\theta_i}\bar{h} > 0$, so $\mu^x(N_i) = 0$. On $N'-N_i$, $g_{\theta_i}\bar{h} = 0$, $\bar{h} \neq 0$, and thus $g_{\theta_i} = 0$. If $N'' = \mathrm{U}_i N_i$, then $\mu^x(N'') = 0$, and on $N' - N''$, $g_{\theta_i} = 0$ for all i. Hence $N'-N''$ is contained in A_0, and $\bar{h} = 0$ on $N'-N''$. But $\bar{h} \neq 0$ on N', and so $N'-N''$ is the empty set. Since N' contains N'', $N' = N''$, and therefore $\mu^x(N') = 0$.

References

Blackwell, David (1947): Conditional expectation and unbiased sequential estimation. *Annals of Math. Stat.*, **18**, 105-110.

Cramér, Harald (1946): *Mathematical Methods of Statistics*, Princeton.

Doob, J. (1948): Asymptotic properties of Markoff transition probabilities. *Trans. Amer. Math. Soc.*, **63**, 393-421.

Ghosh, M. N. (1948): On the problems of similar regions. *Sankhyā*, **8**, 329-338.

Girshick, M. A., Mosteller, F., and Savage, L.J. (1946): "Unbiased estimates for certain binomial sampling problems with applications," *Annals of Math. Stat.*, **17**, 13-23.

Halmos, Paul R. (1946): The theory of unbiased estimation. *Annals of Math. Stat.*, **17**, 34-43.

Halmos, P. R. and Savage, L. J. (1949): "Application of the Radon-Nikodym theorem to the theory of sufficient statistics," *Annals of Math. Stat.*, **20**, 225-241.

Hsu, P. L. (1941): "Analysis of variance from the power-function standpoint", *Biometrika*, **32**, 62-69.

Kolmogoroff, A. (1933): Grundbegriffe der Wahrscheinlichkeitsrechnung. *Ergebnisse der Mathematik und ihrer Grenzgebiete*, **2**, no. 3 Reprinted by Chelsea, New York (1946).

Koopman, B. O. (1936): On distributions admitting a sufficient statistic. *Trans. Amer. Math. Soc.*, **39**, 399-409.

Lehmann, E. L. (1947): On optimum tests of composite hypotheses with one constraint. *Annals of Math. Stat.*, **18**, 473-494.

Lehmann, E. L. and Scheffé, H. (1947): On the problem of similar regions. *Proc. Nat. Acad. Sci.*, **33**, 382-386.

Lehmann, E. L. and Stein, C. (1948): Most powerful tests of composite hypotheses, I. Normal Distributions. *Annals of Math. Stat.*, **19**, 495-516.

Neyman, J. (1937): Outline of a theory of statistical estimation based on the classical theory of probability. *Phil. Trans. Roy. Soc. London*, series A, **236**, 333-380.

Neyman, J. (1941): On a statistical problem arising in routine analyses and in sampling inspections of mass production. *Annals of Math. Stat.*, **12**, 46-76.

Neyman, J. and Pearson, E. S. (1933): On the problem of the most efficient tests of statistical hypotheses. *Phil. Trans. Roy. Soc. London*, series A, **231**, 289-337.

Rao, C. R. (1945): Information and accuracy attainable in the estimation of statistical parameters. *Bull. Calcutta Math. Soc.*, **37**, no. 3. 81-91.

Rao, C. R. (1947): Minimum variance and the estimation of several parameters. *Proc. Camb. Phil. Soc.*, **43**, 280-283.

RAO, C. R. (1949): Sufficient statistics and minimum variance estimates. *Proc. Camb. Phil. Soc.*, **45**, 213-218.

SAVAGE, L. J. (1947): A uniqueness theorem for unbiased sequential bionomial estimation. *Annals of Math. Stat.* **18**, 295-297.

SCHEFFÉ, HENRY (1942): On the theory of testing composite hypotheses with one constraint. *Annals of Math. Stat.*, **13**, 280-293.

SCHEFFÉ, HENRY (1943): On a measure problem arising in the theory of non-parametric tests. *Annals of Math Stat.*, **14**, 227-233.

SETH, G. R. (1949) abstract: Certain properties of the multiparameter unbiased estimates. *Annals of Math. Stat.*, **20**, 142.

WALD, ABRAHAM (1947): Foundations of a general theory of sequential decision functions. *Econometrica* **15**, 279-313.

WOLFOWITZ, J. (1946): On sequential binomial estimation," *Annals of Math. Stat.* **17**, 489-493.

SANKHYĀ

THE INDIAN JOURNAL OF STATISTICS

Edited by : P. C. MAHALANOBIS

ex VOL. 15 PART 3 1955

COMPLETENESS, SIMILAR REGIONS, AND UNBIASED ESTIMATION—PART II[1]

By E. L. LEHMANN and HENRY SCHEFFÉ

University of California, Berkeley

In Part I of this study (Lehmann and Scheffé, 1950), to which we shall refer as LSI, we defined and illustrated the concept of completeness of a family of measures and developed some of its statistical applications. In this part we shall state in section 7 some general results about complete families, showing how from given families one can construct others, and a specific result about the completeness of a certain parametric family. Uniformly most powerful unbiased tests for certain statistical hypotheses about this parametric family are considered in section 8. Some examples are given in section 9. As mentioned[2] in LSI, these results on testing are extensions and simplifications of earlier ones by Neyman (1941), Scheffé (1942), Lehmann (1947), Ghosh (1948), and Hoel (1948). Since then further papers have appeared on the subject by Nandi (1951) and Sverdrup (1953); however, the present version is in a number of ways more general than previous ones. The notation throughout will be that of LSI.

7. SOME FURTHER RESULTS ON COMPLETENESS

In this and the next section we shall restrict ourselves to families of measures $\mathfrak{M} = \{M_\theta \mid \theta \epsilon \omega\}$ for which the "label set" ω is a Borel set in Euclidean space and for which certain measurability assumptions with respect to θ are satisfied. Thus, if $\mathfrak{M}$ is a family $\mathfrak{M}^x = \{M_\theta^x \mid \theta \epsilon \omega\}$ of measures M_θ^x on the additive family $\mathfrak{F}^x$ of sets in the space W^x of points x, we shall denote by L^θ Lebesgue measure on the family $\mathfrak{L}^\theta$ of Borel subsets of the Borel set ω in a Euclidean space. We shall also assume that all the

[1] This paper was prepared with the partial support of the Office of Naval Research, U.S.A.

[2] We shall not give the derivation of the solutions of the differential equations of Neyman and Scheffé for the parametric family of probability densities, as announced in LSI, because this is lengthy, tedious, and now mainly of historical interest.

measures M^x for $\theta\epsilon\omega$ are absolutely continuous with respect to a common σ-finite measure μ^x on $\mathfrak{F}^x$. The measure μ^x is said to be σ-finite on $\mathfrak{F}^x$ if W^x is the union of a countable number of disjoint sets in $\mathfrak{F}^x$ on each of which μ^x is finite. Then there exists a function $g_\theta(x)$, measurable ($\mathfrak{F}^x$), which may be written $g_\theta(x) = dM^x_\theta/d\mu^x$, such that

$$M^x_\theta(A) = \int_A g_\theta(x) d\mu^x$$

for all $A\epsilon\mathfrak{F}^x$ and $\theta\epsilon\omega$; we shall assume further that $g_\theta(x)$ can be taken to be measurable ($\mathfrak{L}^\theta \times \mathfrak{F}^x$).

We shall define the family $\mathfrak{M}^x$ to be *strongly complete* if it satisfies all the assumptions of the preceding paragraph and

$$\int_{W^x} f(x) dM^x_\theta = 0 \qquad \text{(a.e. } L^\theta)$$

implies $f(x) = 0$ (a.e. $\mathfrak{M}^x$). The relation between the three kinds of completeness we have now defined is that strong completeness implies completeness implies bounded completeness. The reader may now wish to proceed directly to Theorem 7.3, which is the basis for the following sections of this paper. The developments up to that point are of some general interest but may be avoided.

The concept of strong completeness permits us to formulate the following theorem concerning families of product measures.

Theorem 7.1: *Let* $x = (x', x'')$, $W^x = W^{x'} \times W^{x''}$, $\mathfrak{F}^x = \mathfrak{F}^{x'} \times \mathfrak{F}^{x''}$, $\theta = (\theta', \theta'')$, $\omega = \omega' \times \omega''$, $\mathfrak{L}^\theta = \mathfrak{L}^{\theta'} \times \mathfrak{L}^{\theta''}$, $L^\theta = L^{\theta'} \times L^{\theta''}$. *Suppose* $\mathfrak{M}^x = \{M^x_\theta \mid \theta\epsilon\omega\}$, *where* $dM^x_\theta = g_\theta(x) d\mu^x$, μ^x *is σ-finite,*

$$g_\theta(x) = g'_{\theta'}(x') g''_{\theta''}(x''),$$

$g'_{\theta'}(x')$ is measurable ($\mathfrak{L}^{\theta'} \times \mathfrak{F}^{x'}$), and $g''_{\theta''}(x'')$ is measurable ($\mathfrak{L}^{\theta''} \times \mathfrak{F}^{x''}$). Suppose further there exists a measure $\mu^{x''}$ on $\mathfrak{F}^{x''}$ and a "conditional measure" $\mu^{x'|x''}$ on $\mathfrak{F}^{x'}$ depending on x'', such that whenever the integral

$$\int_{W^x} f(x', x'') g_\theta(x) d\mu^x \qquad \dots (7.1)$$

is defined[3] *it equals the iterated integral*

$$\int_{W^{x''}} F(\theta', x'') g''_{\theta''}(x'') d\mu^{x''}, \qquad \dots (7.2)$$

where

$$F(\theta', x'') = \int_{W^{x'}} f(x', x'') g'_{\theta'}(x') d\mu^{x'|x''} \qquad \dots (7.3)$$

[3] We allow the values $+\infty$ and $-\infty$.

is measurable $(\mathfrak{L}^{\theta'}\times\mathfrak{F}^{x''})$. *Then* $\mathfrak{M}^{x}$ *is strongly complete providing the same is true of* $\mathfrak{M}^{x''}=\{M^{x''}_{\theta''}|\theta''\epsilon w''\}$ *and (a.e.* $\mu^{x''}$*) of* $\mathfrak{M}^{x'|x''}=\{M^{x'|x''}_{\theta'}|\theta'\epsilon w'\}$, *where* $dM^{x''}_{\theta''}=g''_{\theta''}(x'')\,d\mu^{x''}$ *and* $dM^{x'|x''}_{\theta'}=g'_{\theta'}(x')d\mu^{x'|x''}$.

To prove the theorem we need to show that if the integral (7.1) vanishes (a.e. L^{θ}) then $f(x',x'')=0$ (a.e. $\mathfrak{M}^{x}$). Let N be the set in ω where the integral (7.1) does not vanish, so $N\epsilon\mathfrak{L}^{\theta}$ and $L^{\theta}(N)=0$. Then by the hypothesis of the theorem the iterated integral (7.2) exists and equals zero for $\theta\notin N$. Let S be the set of points (θ',x'') where $F(\theta',x'')$ defined in (7.3) does not vanish; $S\epsilon\mathfrak{L}^{\theta'}\times\mathfrak{F}^{x''}$. We show first that the product measure $L^{\theta'}\times M^{x''}_{\theta''}$ of S is zero for all $\theta''\epsilon\omega''$: Denote by $N_{\theta'}$ the section of N by $\theta'=$ constant, i.e., $N_{\theta'}=\{\theta''|(\theta',\theta'')\epsilon N\}$. Since $(L^{\theta'}\times L^{\theta''})(N)=0$, $L^{\theta''}(N_{\theta'})=0$ except for $\theta'\epsilon N'$, where $N'\epsilon\mathfrak{L}^{\theta'}$ and $L^{\theta'}(N')=0$. For $\theta'\notin N'$ we now have

$$\int_{W^{x'}} F(\theta',x'')dM^{x''}_{\theta''}=0 \qquad \text{(a.e. } L^{\theta''}),$$

and so by strong completeness of $\mathfrak{M}^{x''}$, $F(\theta',x'')=0$ (a.e. $\mathfrak{M}^{x''}$). Thus all sections of S by $\theta'=$ constant with $\theta'\notin N'$ have $M^{x''}_{\theta''}$ measure zero for all $\theta''\epsilon\omega''$. Hence $(L^{\theta'}\times M^{x''}_{\theta''})(S)=0$ for all $\theta''\epsilon\omega''$.

We may now conclude that if $S_{x''}$ denotes a section of S by $x''=$ constant, then the set N'' in $W^{x''}$ for which $L^{\theta'}(S_{x''})\neq0$ has $M^{x''}_{\theta''}$ measure zero for all $\theta''\epsilon\omega''$. In other words, except for $x''\epsilon N''$, a null set of $\mathfrak{M}^{x''}$,

$$F(\theta',x'')=\int_{W^{x'}} f(x',x'')dM^{x'|x''}_{\theta'}=0 \qquad \text{(a.e. } L^{\theta'}).$$

Hence by strong completeness of $\mathfrak{M}^{x'|x''}$, we have for $x''\notin N''$, $f(x',x'')=0$ (a.e. $\mathfrak{M}^{x'|x''}$), and from this we will now conclude that $f(x',x'')=0$ (a.e. $\mathfrak{M}^{x}$): We may calculate the measure M^{x}_{θ} of the set Q where $f(x',x'')\neq0$ as the integral of the characteristic function $\chi_Q(x',x'')$ of Q,

$$M^{x}_{\theta}(Q)=\int_{W^{x}}\chi_Q(x',x'')dM^{x}_{\theta},$$

and then express this as an iterated integral

$$M^{x}_{\theta}(Q)=\int_{W^{x''}}\left[\int_{W^{x'}}\chi_Q(x',x'')dM^{x'|x''}_{\theta'}\right]dM^{x''}_{\theta''},$$

as allowed by the hypotheses of the theorem. This gives

$$M^{x}_{\theta}(Q)=\int_{W^{x''}}[M^{x'|x''}_{\theta'}(Q_{x''})]dM^{x''}_{\theta''},$$

where $Q_{x''}$ is the section of Q by $x''=$ constant, and for $x''\notin N''$ the integrand vanishes.

If μ^x is itself a product of σ-finite measures $\mu^{x'}$ and $\mu^{x''}$, the conditional measure $\mu^{x'|x''}$ of Theorem 7.1 obviously exists and may be taken to be $\mu^{x'}$ by the Fubini theorem. This gives the following important corollary for product measures:

Corollary 7.1: *Suppose* $\mathfrak{M}^{x'} = \{M^{x'}_{\theta'} | \theta' \epsilon \omega'\}$ *and* $\mathfrak{M}^{x''} = \{M^{x''}_{\theta''} | \theta'' \epsilon \omega''\}$, *where* $dM^{x'}_{\theta'} = g'_{\theta'}(x')d\mu^{x'}$ *and* $dM^{x''}_{\theta''} = g''_{\theta''}(x'')d\mu^{x''}$, *are strongly complete families of measures. Let* $x = (x', x'')$, *etc. as in the first sentence of Theorem 7.1 and let* $\mathfrak{M}^x = \{M^x_\theta | \theta \epsilon \omega\}$, *where* $M^x_\theta = M^{x'}_{\theta'} \times M^{x''}_{\theta''}$. *Then the family* $\mathfrak{M}^x$ *of product measures is strongly complete.*

The following theorem is also useful.

Theorem 7.2: *If* $\mathfrak{M}^x = \{M^x_\theta | \theta \epsilon \omega\}$ *is strongly complete,*[4] *and if* $\mathfrak{K}^x = \{K^x_\theta | \theta \epsilon \omega\}$ *is defined by* $dK^x_\theta = u(\theta)v(x)dM^x_\theta$, *where* $u(\theta)$ *and* $v(x)$ *are non-negative,* $u(\theta)$ *is measurable* $(\mathfrak{L}^\theta)$, $u(\theta) \neq 0$ (a.e. L^θ), *and* $v(x)$ *is measurable* $(\mathfrak{F}^x)$, *then* $\mathfrak{K}^x$ *is strongly complete.*

The absolute continuity of the family $\mathfrak{K}^x$ of measures with respect to a σ-finite measure μ^x and the measurability $(\mathfrak{L}^\theta \times \mathfrak{F}^x)$ of $dK^x_\theta/d\mu^x$ follow from the corresponding properties of the family $\mathfrak{M}^x$ and the relation $dK^x_\theta/d\mu^x = u(\theta)v(x)dM^x_\theta/d\mu^x$. It will thus suffice to prove that if

$$\int_{W^x} f(x)dK^x_\theta = 0 \qquad \text{(a.e. } L^\theta), \qquad \dots (7.4)$$

then the set $H_f = \{x | f(x) \neq 0\}$ is a null set of $\mathfrak{K}^x$. We may write (7.4) in the form

$$\int_{W^x} f(x)v(x)dM^x_\theta = 0 \qquad \text{(a.e. } L^\theta). \qquad \dots (7.5)$$

Let $H_v = \{x | v(x) \neq 0\}$, $H_{fv} = \{x | f(x)v(x) \neq 0\}$, so $H_f \subset H_{fv} \cup (W^x - H_v)$. Since $\mathfrak{M}^x$ is complete we see from (7.5) that H_{fv} is a null set for $\mathfrak{M}^x$, and hence for $\mathfrak{K}^x$. Also, $W^x - H_v$ is a null set for $\mathfrak{K}^x$ since $v(x) = 0$ in $W^x - H_v$. Hence H_f is a null set for $\mathfrak{K}^x$.

A completeness result for a class of families of probability distributions of exponential form which has received much attention in statistical literature may be obtained by applying the above theorems to a result from the theory of Laplace transforms which we shall now state.

Lemma 7.1: *If* $\mathfrak{M}^x = \{M^x_\theta | \theta \epsilon \omega\}$ *is a family of measures on the real line* W^x *with a real parameter* θ *in a nondegenerate interval* ω, *and if the family has a density of the form*

$$dM^x_\theta / d\mu^x = C(\theta)e^{\theta x}h(x) \qquad \dots (7.6)$$

with respect to a σ*-finite measure* μ^x *on the class* $\mathfrak{F}^x$ *of Borel sets in* W^x, *then* $\mathfrak{M}^x$ *is strongly complete.*

[4] The proof will show that Theorem 7.2 remains true if "strongly complete" is replaced by "complete" in hypothesis and conclusion and the hypotheses $u(\theta) \geqslant 0$, $u(\theta) \neq 0$ (a.e. L^θ), $u(\theta)$ is measurable $(\mathfrak{L}^\theta)$ are replaced by the hypothesis $u(\theta) > 0$.

Since $C(\theta)e^{\theta x}h(x)$ is measurable $(\mathfrak{F}^x)$, so is $h(x)$. Since

$$C(\theta) = 1 \Big/ \int\limits_W e^{\theta x} h(x) d\mu^x,$$

it is measurable $(\mathfrak{L}^\theta)$. Strong completeness of $\mathfrak{M}^x$ will then follow from Theorem 7.2 if we can prove $\mathfrak{N}^x = \{N_\theta^x | \theta\epsilon\omega\}$ strongly complete, where $dN_\theta^x = e^{\theta x} d\mu^x$. But this is an immediate consequence of applying to the relation

$$\int\limits_{W^x} f(x) e^{\theta x} d\mu^x = 0 \qquad (\text{a.e. } L^\theta) \qquad \ldots \quad (7.7)$$

the uniqueness theorem for Laplace transforms, as given for example in Widder (1941, Ch. 6, Th. 6a). The theorem in Widder requires that $\alpha(t) = \int_0^t f(x) d\mu^x$ be of bounded variation in every finite interval. Let θ_0 be a value of θ for which

$$\int\limits_{W^x} f(x) e^{\theta_0 x} d\mu^x = 0. \qquad \ldots \quad (7.8)$$

Since the Lebesgue integral is an absolutely convergent integral, (7.8) implies

$$\int\limits_{W^x} |f(x) e^{\theta_0 x}| d\mu^x < \infty.$$

Let $I^t = \{t | a \leqslant t \leqslant b\}$ be any finite interval. The total variation of $\alpha(t)$ on I^t is

$$\int\limits_{I^x} |f(x)| d\mu^x \leqslant a^{-1} \int\limits_{I^x} |f(x) e^{\theta_0 x}| d\mu^x < \infty,$$

where $I^x = \{x | a \leqslant x \leqslant b\}$ and a is the minimum of $e^{\theta_0 x}$ on I^x. The theorem in Widder now tells us $\alpha(t) = 0$; thus $f(x)$ vanishes (a.e. μ^x), hence (a.e. $\mathfrak{M}^x$).

We remark that in the following theorems where a family of measures is a family of probability distributions (in which case we agreed in LSI to write $\mathfrak{P}^x$ instead of $\mathfrak{M}^x$), $\mathfrak{P}^x = \{P_\theta^x | \theta\epsilon\omega\}$ with $dP_\theta^x = p_\theta(x) d\mu^x$, it is not necessary to assume μ^x to be σ-finite if there exists a set W_+^x in $\mathfrak{F}^x$ such that for all $\theta\epsilon\omega$, $p_\theta(x) \neq 0$ (a.e. μ^x) in W_+^x and $p_\theta(x) = 0$ (a.e. μ^x) in $W^x - W_+^x$; for in this case μ^x is σ-finite in $W_+^x(\theta) = \{x | p_\theta(x) \neq 0\}$ (Halmos, 1950, sec. 25, Th. F), and hence in W_+^x, and we may assume $\mu^x(W^x - W_+^x) = 0$ without affecting the family $\mathfrak{P}^x$. It is also understood that the density $p_\theta(x)$ is measurable $(\mathfrak{F}^x)$ since it is integrable $(\mathfrak{F}^x, \mu^x)$.

Theorem 7.3: *If* $\mathfrak{P}^x = \{P_\theta^x | \theta\epsilon\omega\}$, $x = (x_1, \ldots, x_r)$, $\theta = (\theta_1, \ldots, \theta_r)$, *where* x_i *and* θ_i *are real* $(i = 1, \ldots, r)$, *if* ω *contains a nondegenerate r-dimensional interval, and if* P_θ^x *has a density of the form*

$$dP_\theta^x / d\mu^x = C(\theta_1, \ldots, \theta_r) h(x_1, \ldots, x_r) \exp\left(\sum_{i=1}^r \theta_i x_i\right) \qquad \ldots \quad (7.9)$$

with respect to a measure μ^x on the class $\mathfrak{F}^x$ of Borel sets in the r-dimensional Euclidean space W^x, then $\mathfrak{P}^x$ is strongly complete.

This theorem may be proved with the aid of Theorems 7.1 and 7.2 and Lemma 7.1. A shorter proof may be made along different lines by a method used by Sverdrup (1953, Th. 1) to obtain a similar result.

Theorem 7.3 contains as special cases the completeness results of Examples 3.2, 3.3, 3.5, 3.8, 3.9 of LSI, as well as many multidimensional examples that can be obtained from these. The example at the top of p. 322 of LSI, in which the sufficient statistic $T = (T_1, T_2)$ consists of the mean T_1 of a random sample from a normal population and the sum of squares T_2 of the deviations from T_1 does not fall directly under Theorem 7.3; however, if we let $U = (U_1, U_2)$, $U_1 = T_1$, $U_2 = T_2+nT_1^2$, then the completeness of $\mathfrak{P}^u$ is given immediately by Theorem 7.3, and it is obvious in general that if $t = t(u)$ and $\mathfrak{P}^u$ is complete, so is $\mathfrak{P}^t$.

8. UNIFORMLY MOST POWERFUL UNBIASED TESTS

Throughout this section we shall be concerned with the case where the sample point $X = (X_1, \ldots, X_n)$ in an n-dimensional Euclidean space W^x has a density

$$dP^x_{\vartheta,\theta}/d\mu^x = p_{\vartheta,\theta}(x) = C(\vartheta, \theta)h(x) \exp\left[\vartheta s(x)+\sum_{i=1}^{r}\theta_i t_i(x)\right] \quad \ldots (8.1)$$

with respect to a measure μ^x in the class $\mathfrak{F}^x$ of Borel sets in W^x. The statistical hypotheses considered will be statements about the real parameter ϑ of the form

H_1: $\vartheta \leqslant \vartheta_0$ against alternatives $\vartheta > \vartheta_0$,

H_2: $\vartheta = \vartheta_0$ against alternatives $\vartheta \neq \vartheta_0$,

H_3: $\vartheta_0 \leqslant \vartheta \leqslant \vartheta_1$ against alternatives $\vartheta < \vartheta_0$ or $\vartheta > \vartheta_1$,

H_4: $\vartheta \leqslant \vartheta_0$ or $\vartheta \geqslant \vartheta_1$ against alternatives $\vartheta_0 < \vartheta < \vartheta_1$.

The set of real parameters $(\theta_1, \ldots, \theta_r) = \theta$ are nuisance parameters $(r = 1, 2, \ldots)$. Denote by Ω the domain of (ϑ, θ). We shall assume the parameter space Ω is convex[5]. In the problem of testing H_i we shall assume there are points of Ω with $\vartheta > \vartheta_0$ for $i = 1, 2$; $\vartheta < \vartheta_0$ for $i = 2, 3$; $\vartheta > \vartheta_1$ for $i = 3$; $\vartheta \leqslant \vartheta_0$ and $\vartheta \geqslant \vartheta_0$ for $i = 4$. Let Ω_j denote the section of Ω by $\vartheta = \text{constant} = \vartheta_j (j = 0 \text{ or } 1)$, that is $\Omega_j = \{(\vartheta, \theta) \mid \vartheta = \vartheta_j, (\vartheta, \theta)\epsilon\Omega\}$, and let

$$\mathfrak{P}^x_j = \{P^x_{\vartheta,\theta} \mid (\vartheta, \theta)\epsilon\Omega_j\}. \quad \ldots (8.2)$$

It will be understood that statements about Ω_j, $\mathfrak{P}^x_j$, and $\mathfrak{P}^t_j$ (defined below) are made for $j = 0$ in the problem of testing hypotheses H_1 and H_2, for $j = 0$ and 1 in the case of H_3 and H_4. It is assumed that Ω_j contains a non-degenerate r-dimensional interval.

[5] The assumption that Ω is convex is made because it is brief to state and does not seem to exclude any interesting applications. The proof of Theorem 8.2 is valid with less restrictive assumptions on Ω.

The real-valued functions $h(x)$, $s(x)$, $t_i(x)$ are understood to be measurable ($\mathfrak{F}^x$). Let $t(x) = (t_1(x), \ldots, t_r(x))$; then clearly $T = t(X)$ is a sufficient statistic for $\mathfrak{P}^x_j$. We shall take for $\mathfrak{F}^t$ the family of Borel sets in the Euclidean space W^t, instead of the more inclusive additive family induced in W^t by $t(x)$. Let $P^t_{\vartheta,\theta}$ be the probability measure induced on $\mathfrak{F}^t$ by $P^x_{\vartheta,\theta}$ and $t(x)$, and write

$$\mathfrak{P}^t_j = \{P^t_{\vartheta,\theta} \mid (\vartheta, \theta)\epsilon\Omega_j\}. \qquad \ldots \quad (8.3)$$

In order to prove $\mathfrak{P}^t_j$ strongly complete we need

Lemma 8.1: *Let $t = t(x)$ be a measurable*[6] *transformation from* (W^x, $\mathfrak{F}^x$) *into* (W^t, $\mathfrak{F}^t$). *If the probability distribution of X is given by a density of the form*

$$dP^x_\theta/d\mu^x = q_\theta(t(x))h(x)$$

with respect to a measure μ^x on $\mathfrak{F}^x$, then there exists a measure ν^t on $\mathfrak{F}^t$ such that the distribution of $T = t(X)$ has the density

$$dP^t_\theta/d\nu^t = q_\theta(t)$$

with respect to ν^t.

This lemma may be proved by defining the measure ν^x on $\mathfrak{F}^x$ by $d\nu^x = h(x)d\mu^x$, and the measure ν^t on $\mathfrak{F}^t$ by $\nu^t(B) = \nu^x(t^{-1}(B))$ for all $B\epsilon\mathfrak{F}^t$. Then for all $B\epsilon\mathfrak{F}^t$ we have

$$P^t_\theta(B) = P^x_\theta(t^{-1}(B)) = \int_{t^{-1}(B)} q_\theta(t(x))h(x)d\mu^x = \int_{t^{-1}(B)} q_\theta(t(x))d\nu^x = \int_B q_\theta(t)d\nu^t,$$

which is the desired result.

From Lemma 8.1 and Theorem 7.3 we now have

Theorem 8.1: *If the sample point X has a probability density* (8.1) *and satisfies all the conditions stated in connection with* (8.1). *then, in the notation there defined, T is a sufficient statistic for $\mathfrak{P}^x_j$ with a probability density*

$$C(\vartheta_j, \theta) \exp\left(\sum_{i=1}^r \theta_i t_i\right)$$

with respect to a measure ν^t on the family $\mathfrak{F}^t$ of Borel sets in W^t, and the family $\mathfrak{P}^t_j$ is strongly complete.

In problems of testing statistical hypotheses about a parameter ϑ in the presence of nuisance parameters $(\theta_1, \ldots, \theta_r) = \theta$, we will denote the sample space by W^x, the parameter point by (ϑ, θ) the parameter space by Ω, the hypothesis by $H: \ \theta\epsilon\omega$,

[6] For any given additive classes $\mathfrak{F}^x$, $\mathfrak{F}^t$, $t = t(x)$ is said to be a *measurable transformation* if for every $A\epsilon\mathfrak{F}^t$, $t^{-1}(A)\epsilon\mathfrak{F}^x$. In the application of this lemma W^x, W^t will be Euclidean spaces and $\mathfrak{F}^x$, $\mathfrak{F}^t$ the classes of Borel sets in each.

where ω is a given subset of Ω. Any function $\varphi(x)$ measurable $(\mathfrak{F}^x)$, such that $0 \leqslant \varphi(x) \leqslant 1$, may be considered as a *critical function* for testing H: H is to be rejected with probability $\varphi(x)$ when $X = x$, so that the power of the test is $E_{\vartheta,\theta}(\Phi)$ with $\Phi = \varphi(X)$, as indicated in LSI, pp. 317-318, where the terminology of *similar* critical function, and *Neyman structure* is also defined. By the *size* of the test based on $\varphi(x)$ we mean $\sup E_{\vartheta,\theta}(\Phi)$ for $(\vartheta, \theta)\epsilon\omega$. The test is said to be *unbiased* if for every pair of parameter points $(\vartheta', \theta')\epsilon\omega$ and $(\vartheta'', \theta'')\epsilon\Omega-\omega$ we have $E_{\vartheta', \theta'}(\Phi) \leqslant E_{\vartheta'', \theta''}(\Phi)$. Since for any critical function $\varphi(x)$, the power $E_{\vartheta,\theta}(\Phi)$ is continuous (in fact, analytic) in ϑ for the family of distributions defined by (8.1) under the assumptions stated there, it follows that a critical function is similar for the family $\mathfrak{P}_0^x$ if it gives an unbiased test of H_1 or H_2, and similar for $\mathfrak{P}_0^x \cup \mathfrak{P}_1^x$ if it gives an unbiased test of H_3 or H_4. We can thus dispense with the classificatory terminology of "type B_1," "best one-sided similar," etc., and yet apply the results of LSI about similar critical functions to get the following theorem about uniformly most powerful unbiased tests.

Theorem 8.2: *If the sample point X has a probability density of the form* (8.1), *then for testing any of the hypotheses H_1, H_2, H_3, H_4 defined below* (8.1), *and for each $\alpha (0<\alpha<1)$, there exists a uniformly most powerful unbiased test of size α if the assumptions stated in connection with* (8.1) *are satisfied. The corresponding critical function $\varphi_i(x)$ for testing H_i may be constructed as follows: The function $\varphi_i(x)$ is defined by means of four functions of t, $v_i(t)$, $w_i(t)$ $\delta_i(t)$, $\gamma_i(t)$, satisfying the inequalities $v_i(t)<w_i(t)$, $0 \leqslant \delta_i(t) \leqslant 1$, $0 \leqslant \gamma_i(t) \leqslant 1$. Let $I_i(t)$ denote the interval from $v_i(t)$ to $w_i(t)$. For $i = 1, 2, 3$ we define $\varphi_i(x) = 0$ if $s(x)$ is inside the interval $I_i(t(x))$, $= 1$ if outside the interval; if $s(x)$ falls on an end point of $I_i(t(x))$ we define $\varphi_i(x) = \delta_i(t(x))$ or $\gamma_i(t(x))$ according as $s(x) = v_i(t(x))$ or $w_i(t(x))$. For $i = 4$ the definition of $\varphi_i(x)$ is similar except that we put $\varphi_i(x) = 1$ inside the interval, 0 outside. The four functions v_i, w_i, δ_i, γ_i are determined by the conditions $\mathfrak{C}_i$ below where Φ_i denotes $\varphi_i(X)$, S denotes $s(X)$, and $\mathfrak{P}_j^t$ is defined by* (8.3):

$$\mathfrak{C}_1: \quad v_1 = -\infty \text{ (hence, } \delta_1 \text{ need not be defined),}$$

$$E_{\vartheta_0\theta}(\Phi_1|t) = \alpha \qquad (a.e.\ \mathfrak{P}_0^t).$$

$$\mathfrak{C}_2: \quad E_{\vartheta_0,\theta}(\Phi_2|t) = \alpha \qquad (a.e.\ \mathfrak{P}_0^t),$$

$$E_{\vartheta_0,\theta}(S\Phi_2|t) = \alpha E_{\vartheta_0,\theta}(S|t) \qquad (a.e.\ \mathfrak{P}_0^t).$$

$$\mathfrak{C}_3,\ \mathfrak{C}_4: \quad E_{\vartheta_0,\theta}(\Phi_i|t) = \alpha \qquad (a.e.\ \mathfrak{P}_0^t).$$

$$E_{\vartheta_1,\theta}(\Phi_i|t) = \alpha \qquad (a.e.\ \mathfrak{P}_1^t).$$

The proof of the theorem will be facilitated by the following modified form of the Neyman-Pearson fundamental lemma.

Lemma 8.2: *Suppose $\mathfrak{P}^x = \{P_\eta^x | \eta\epsilon\Lambda\}$ is a family of probability measures on the additive family $\mathfrak{F}^x$ of sets in W^x with $dP_\eta^x = P_\eta(x)d\mu^x$, and $\mathfrak{P}_i^x = \{P_\eta^x | \eta\epsilon\Lambda_i\}$, where $\Lambda_i \subset \Lambda (i = 1, \ldots, k)$. Let $U_i = u_i(X)$ be a sufficient statistic for $\mathfrak{P}_i^x$. Suppose the following are given: $\eta^*\epsilon\Lambda$, $\eta_i\epsilon\Lambda_i$, constants α_i, and measurable $(\mathfrak{F}^x)$ functions $f_i(x) \geqslant 0$.*

Denote by $\mathfrak{P}_i^{u_i}$ *the family of measures on the additive family* $\mathfrak{F}_i^{u_i}$ *of sets in the range* W^{u_i} *of* $u_i(x)$, *where* $\mathfrak{F}_i^{u_i}$ *is such that* $u_i(x)$ *is a measurable transformation. If there exists a critical function* $\varphi^*(x)$ *and Borel-measurable functions* $g_i(u_i) \geqslant 0$ *such that*

$$E_{\eta_i}(f_i(X)\varphi^*(X)\,|\,u_i) = \alpha_i \quad (a.e.\ \mathfrak{P}_i^{u_i})(i = 1, \ldots, k) \qquad \ldots \quad (8.4)$$

and

$$\varphi^*(x) = \begin{cases} 1 \ \textit{where}\ p_{\eta^*}(x) > \sum\limits_{i=1}^{k} g_i(u_i(x))\, f_i(x)\, p_{\eta_i}(x), \\ 0 \ \textit{where}\ p_{\eta^*}(x) < \sum\limits_{i=1}^{k} g_i(u_i(x))\, f_i(x) p_{\eta_i}(x), \end{cases}$$

then, among all critical functions $\varphi(x)$ *satisfying* (8.4), $\varphi^*(x)$ *maximizes* $E_{\eta^*}(\varphi(X))$.

To prove the lemma suppose that $g_i(u_i)$ and $\varphi^*(x)$ satisfy the conditions of Lemma 8.2. Then β_i defined as

$$\beta_i = E_{\eta_i}(g_i(U_i)\, f_i(X)\varphi^*(X))$$

is finite since $\quad g_i(u_i(x))\, f_i(x)\varphi^*(x)p_{\eta_i}(x) \leqslant p_{\eta^*}(x).$

Hence $\quad \beta_i = E_{\eta_i}[g_i(U_i)E_{\eta_i}(\, f_i(X)\varphi(X)\,|\,u_i)] = \alpha_i E_{\eta_i}(g_i(U_i)).$

Now consider the wider class of critical functions $\varphi(x)$ satisfying

$$\int_{W^x} g_i(u_i(x))\, f_i\,(x)\varphi(x)p_{\eta_i}(x)d\mu^x = \beta_i \qquad (i = 1, \ldots, k). \qquad \ldots \quad (8.5)$$

It follows from the Neyman-Pearson lemma that in this wider class $\varphi^*(x)$ maximizes $E_{\eta^*}(\Phi(X))$.

We now prove Theorem 8.2 for testing hypothesis H_1. By the remark preceding the theorem, if $\varphi(x)$ gives an unbiased test then it is a similar critical function for the family $\mathfrak{P}_0^x$. By Theorem 8.1, the family $\mathfrak{P}_0^t$ generated by the sufficient statistic T is strongly complete. Hence by Theorem 4.1 of LSI, $\varphi(x)$ has the Neyman structure[6a] with respect to T, i.e.,

$$E_{\vartheta_0,\theta}(\varphi(X)\,|\,t) = \alpha \qquad (\text{a.e. } \mathfrak{P}_0^t). \qquad \ldots \quad (8.6)$$

Since T is sufficient for $\mathfrak{P}_0^x$, the left side is independent of θ for all $\varphi(x)$. Let (ϑ^*, θ^*) be any alternative with $\vartheta^* > \vartheta_0$, and let (ϑ_0, θ^0) be any point in Ω_0. We shall now apply Lemma 8.2 with $\Lambda = \Omega$, $k = 1$, $\Lambda_1 = \Omega_0$, $\eta^* = (\vartheta^*, \theta^*)$, $\eta_1 = (\vartheta_0, \theta^0)$, $\alpha_1 = \alpha$, $f_1 \equiv 1$, and $U_1 = T$. Then $\varphi(x)$ will maximize $E_{\vartheta^*, \theta^*}(\varphi(X))$ subject to (8.6) if it satisfies (8.6) and there exists a $g(t)$ such that $\varphi(x) = 1$ or 0 according as the ratio $p_{\vartheta^*,\theta^*}(x)/p_{\vartheta_0,\theta^0}(x)$ is $>$ or $< g(t(x))$. This condition reduces to $\varphi(x) = 1$ or 0 according as $s(x) >$ or $< w_1(t(x))$, where the function $w_1(t)$ may depend on ϑ^*, ϑ_0, θ^*, and θ^0.

[6a] We have only recently become aware of Bartlett's (1937) paper published at about the same time as Neyman's, in which he uses the same construction.

We must determine $w_1(t)$, and a suitable value $\gamma_1(t)$ for φ when $s = w_1(t)$, so that (8.6) is satisfied. This can be done so that $w_1(t)$ and $\gamma_1(t)$ do not depend on ϑ^*, θ^*, or θ^0, as follows: Write $F_t(s)$ for the conditional distribution function $P_{\vartheta_0,\theta}(S \leqslant s\,|\,t)$, which does not depend on θ, so that $F_t(s) = F_t(s+0)$ and $F_t(s) - F_t(s-0) = P_{\vartheta_0,\theta}(S = s\,|\,t)$. Then $w_1(t)$ and $\gamma_1(t)$ are to be determined so that

$$F_t(w_1 - 0) \leqslant 1 - \alpha \leqslant F_t(w_1), \qquad \ldots \quad (8.7)$$

$$F_t(w_1 - 0) + (1 - \gamma_1)[F_t(w_1) - F_t(w_1 - 0)] = 1 - \alpha. \qquad \ldots \quad (8.8)$$

This is clearly possible; in particular if $F_t(s)$ is continuous at w_1 defined by (8.7) then for such t the definition of $\gamma_1(t)$ is immaterial. Since the resulting most powerful test against (ϑ^*, θ^*) does not depend on $\vartheta^* > \vartheta_0$ or on θ^*, it is uniformly most powerful among those satisfying (8.6).

It remains to verify that the above test has the proper size, i.e.

$$E_{\vartheta,\theta}(\Phi) \leqslant \alpha \ \text{ for } \ \vartheta < \vartheta_0, \qquad \ldots \quad (8.9)$$

and is unbiased, i.e.

$$E_{\vartheta,\theta}(\Phi) \geqslant \alpha \ \text{ for } \ \vartheta > \vartheta_0. \qquad \ldots \quad (8.10)$$

The inequality (8.10) may be deduced from the fact that $\varphi(x) \equiv \alpha$ is in the class satisfying (8.6) of which the φ defined above is uniformly most powerful for $\vartheta > \vartheta_0$. The inequality (8.9) can be demonstrated by noting that the above φ minimizes $E_{\vartheta,\theta}(\Phi)$ for $\vartheta < \vartheta_0$ subject to (8.6) and again comparing it with $\varphi(x) \equiv \alpha$.

The proof that φ_1 thus constructed is Borel measurable, as are the other φ_i, is given in the appendix at the end of the paper.

We next prove Theorem 8.2 for hypothesis H_2. As before, unbiasedness implies (8.6). This time it also implies that

$$E_{\vartheta,\theta}(\Phi) = \int_{W^x} \varphi(x) p_{\vartheta,\theta}(x) d\mu^x \qquad \ldots \quad (8.11)$$

has a minimum at $\vartheta = \vartheta_0$. From well known properties of Laplace transforms, (8.11) has a derivative with respect to ϑ, which may be calculated by differentiation under the integral sign. Hence

$$\partial E_{\vartheta,\theta}(\Phi)/\partial\vartheta\,|_{\vartheta=\vartheta_0} = 0 \qquad \ldots \quad (8.12)$$

for all $(\vartheta_0, \theta) \epsilon \Omega_0$, and thus, differentiating (8.11) under the integral sign and using (8.1), we get from (8.12)

$$E_{\vartheta_0,\theta}(S\Phi) = -\alpha C^{-1}(\vartheta_0, \theta)\partial C(\vartheta, \theta)/\partial\vartheta\,|_{\vartheta=\vartheta_0}. \qquad \ldots \quad (8.13)$$

Since

$$C(\vartheta, \theta) = \left\{ \int_{W^x} h(x) \exp[\vartheta s(x) + \sum_{i=1}^{r} \theta_i t_i(x)] d\mu^x \right\}^{-1},$$

therefore,

$$\partial C(\vartheta, \theta)/\partial\vartheta = -C(\vartheta, \theta) E_{\vartheta,\theta}(S),$$

and so (8.13) may be written

$$E_{\vartheta_0,\theta}(S\Phi) = \alpha E_{\vartheta_0,\theta}(S),$$

or,

$$E_{\vartheta_0,\theta}[E_{\vartheta_0,\theta}(S\Phi - \alpha S\,|\,t)] = 0$$

for all θ such that $(\vartheta_0, \theta)\epsilon\Omega_0$. The last condition implies

$$E_{\vartheta_0,\theta}(S\Phi - \alpha S\,|\,t) = 0 \qquad (\text{a.e. } \mathfrak{P}_0^t) \qquad \ldots \quad (8.14)$$

since $\mathfrak{P}_0^t$ is strongly complete. Conditions (8.6) and (8.14) may be written

$$E_{\vartheta_0,\theta}(S^{i-1}\Phi\,|\,t) = \alpha_i \qquad (\text{a.e. } \mathfrak{P}_0^t) \quad (i = 1, 2), \qquad \ldots \quad (8.15)$$

where $\alpha_i = \alpha E_{\vartheta_0,0}(S^{i-1}\,|\,t)$, which does not depend on θ for $(\vartheta_0, \theta)\epsilon\Omega_0$ since T is sufficient for $\mathfrak{P}_0^t$.

Let (ϑ^*, θ^*) be an alternative in Ω with $\vartheta^* \neq \vartheta_0$, and let $(\vartheta_0, \theta^0)\epsilon\Omega_0$. We now apply Lemma 8.2 with $\Lambda = \Omega$, $k = 2$, $\Lambda_1 = \Lambda_2 = \Omega_0$, $\eta^* = (\vartheta^*, \theta^*)$, $\eta_1 = \eta_2 = (\vartheta_0, \theta^0)$, $f_i(x) = s^{i-1}(x)$, and $U_1 = U_2 = T$: $\varphi(x)$ will maximize $E_{\vartheta^*,\theta^*}(\varphi(X))$ subject to (8.14) if it satisfies (8.15) and there exist $g_1(t)$ and $g_2(t)$ such that $\varphi(x) = 1$ or 0 according as the ratio $p_{\vartheta^*,\theta^*}(x)/p_{\vartheta_0,\theta^0}(x)$ is $>$ or $<$ than $g_1(t(x)) + g_2(t(x))s(x)$. Since the ratio is a convex function of s and depends on x only through $s(x)$ and $t(x)$, the existence of such $g_1(t)$ and $g_2(t)$ is equivalent to that of $v_2(t)$ and $w_2(t)$ with $v_2(t) < w_2(t)$ such that $\varphi(x) = 1$ or 0 according as $s(x)$ lies outside or inside the interval from $v_2(t)$ to $w_2(t)$. At this stage of the argument $v_2(t)$ and $w_2(t)$ may depend on ϑ^*, ϑ_0, θ^*, and θ^0. It remains to show that the functions $v_2(t)$ and $w_2(t)$, and values $\delta_2(t)$ and $\gamma_2(t)$ for $\varphi(x)$ when $s(x) = v_2(t)$ or $w_2(t)$, can be determined so that they do not depend on ϑ^*, θ^*, or θ^0, and (8.15) is satisfied.

In terms of $v_2(t)$, $w_2(t)$, $\delta_2(t)$, $\gamma_2(t)$, and the conditional distribution function $F_t(s)$ introduced above (8.7), the conditions (8.15) become

$$F_t(v_2-0) + 1 - F_t(w_2) + \delta_2[F_t(v_2) - F_t(v_2-0)] + \gamma_2[F_t(w_2) - F_t(w_2-0)] = \alpha, \quad \ldots \quad (8.16)$$

$$\int_{-\infty}^{v_2-0} s\,dF_t + \int_{w_2+0}^{\infty} s\,dF_t + v_2\delta_2[F_t(v_2) - F_t(v_2-0)] + w_2\gamma_2[F_t(w_2) - F_t(w_2-0)] = \alpha_2, \quad \ldots \quad (8.17)$$

where (8.16) and (8.17) are to be satisfied (a.e. $\mathfrak{P}_0^t$). Since (8.16) and (8.17) do not depend on ϑ^*, θ^*, or θ^0 (F_t depends on ϑ_0), neither will solutions $v_2(t)$, $w_2(t)$, $\delta_2(t)$, and $\gamma_2(t)$, whose existence we shall now establish by a continuity argument: For each fixed t, let y be a variable with range $0 \leqslant y \leqslant \alpha$ and define v and δ as functions of y to satisfy

$$F_t(v-0) \leqslant y \leqslant F_t(v),$$

$$y = F_t(v-0) + \delta[F_t(v) - F_t(v-0)],$$

so that v is unique unless there is an interval where F_t is constant and equal to y, and δ is unique if $F_t(s)$ is discontinuous at $s = v$. Next determine w and γ similarly so that

$$F_t(w-0) \leqslant 1-\alpha+ y \leqslant F_t(w),$$

$$1-\alpha+y = F_t(w-0)+(1-\gamma)[F_t(w)-F_t(w-0)].$$

For any y $(0 \leqslant y \leqslant \alpha)$ the v, w, δ, γ thus defined will satisfy (8.16). Let $G_t(y)$ denote the value obtained when the v, w, δ, γ are substituted for the $v_2, w_2, \delta_2, \gamma_2$ in the left member of (8.17). Then $G_t(y)$ is a unique function of y even though v, w, δ, γ are in general not, and is continuous in y. If we can show (a.e. $\mathfrak{P}_0^t$) that $G_t(0) \geqslant \alpha_2$ and $G_t(\alpha) \leqslant \alpha_2$, then there must exist a y_t for which $G(y_t) = \alpha_2$. If we then define $v_2, w_2, \delta_2, \gamma_2$ as the v, w, δ, γ corresponding to $y = y_t$, both (8.16) and (8.17) will be satisfied.

To fill the above gap in the proof for H_2 we note that the test found above for for H_1 corresponds to $y = 0$, and this maximizes $E_{\vartheta,\theta}(\Phi)$ for every $\vartheta > \vartheta_0$ subject to (8.6), hence it also maximizes the derivative

$$\partial E_{\vartheta,\theta}(\Phi)/\partial\vartheta\Big|_{\vartheta=\vartheta_0} = E_{\vartheta_0,\theta}(S\Phi-\alpha S).$$

It must therefore maximize $E_{\vartheta_0,\theta}(S\Phi-\alpha S\,|\,t)$ (a.e. $\mathfrak{P}_0^t$). But this has the value 0 for $\Phi \equiv \alpha$. Consequently the Φ corresponding to $y = 0$ makes $E_{\vartheta_0,\theta}(S\Phi\,|\,t) \geqslant \alpha E_{\vartheta_0,\theta}(S\,|\,t)$, or $G_t(0) \geqslant \alpha_2$ (a.e. $\mathfrak{P}_0^t$). By similar reasoning about the test corresponding to $y = \alpha$ it can be shown $G_t(\alpha) \leqslant \alpha_2$ (a.e. $\mathfrak{P}_0^t$). Unbiasedness of the test follows as before from comparison with $\Phi \equiv \alpha$.

We shall treat together the cases of the theorem for hypotheses H_3 and H_4. By the remark preceding the theorem unbiasedness implies similarity for $\mathfrak{P}_0^x \cup \mathfrak{P}_1^x$, while sufficiency of T for $\mathfrak{P}_0^x$ and for $\mathfrak{P}_1^x$ and strong completeness of $\mathfrak{P}_0^t$ and of $\mathfrak{P}_1^t$ imply

$$E_{\vartheta_j,\theta}(\Phi\,|\,t) = \alpha \quad (\text{a.e. } \mathfrak{P}_j^t) \quad (j = 0, 1). \qquad \ldots \quad (8.18)$$

Let (ϑ^*,θ^*) be a point in Ω, $\vartheta^* \neq \vartheta_0$ or ϑ_1. We choose arbitrary θ^0 and θ^1 such that $(\vartheta_j, \theta^j) \epsilon \Omega_j$ $(j = 0, 1)$, and we again apply Lemma 8.2, this time with $k = 2$, $\Lambda_i = \Omega_{i-1}$, $\eta^* = (\vartheta^*, \theta^*)$, $\eta_i = (\vartheta_{i-1}, \theta^{i-1})$, $f_i \equiv 1$, $U_i = T$ $(i = 1, 2)$, to find that $\varphi(x)$ will maximize $E_{\vartheta^*,\theta^*}(\varphi(X))$ subject to (8.18) if it satisfies (8.18) and there exist $g_0(t), g_1(t)$ such that $\varphi(x) = 1$ or 0 according as

$$p_{\vartheta^*,\theta^*}(x) > \text{ or } < g_0(t(x))\, p_{\vartheta_0,\theta^0}(x)+g_1(t(x))\, p_{\vartheta_1,\theta^1}(x). \qquad \ldots \quad (8.19)$$

Substituting from (8.1) we find (8.19) equivalent to the existence of h_0 and h_1 such that $\varphi(x) = 1$ or 0 according as $H(s) <$ or > 1, where

$$H(s) = h_0 e^{\Delta_0 s}+h_1 e^{\Delta_1 s}, \qquad \ldots \quad (8.20)$$

$\Delta_i = \vartheta_i - \vartheta^*$, and where h_0 and h_1 may depend on $\vartheta^*, \vartheta_0, \vartheta_1, \theta^*, \theta^0, \theta^1$, as well as on t.

For any v, w with $-\infty \leqslant v < w \leqslant +\infty$ we shall define h_0, h_1 to be the unique solutions of the equations $H(v) = H(w) = 1$, namely

$$h_0 = (e^{\Delta_1 w} - e^{\Delta_1 v})/D, \quad h_1 = (e^{\Delta_0 v} - e^{\Delta_0 w})/D,$$

where
$$D = e^{\Delta_0 v + \Delta_1 w} - e^{\Delta_0 w + \Delta_1 v} > 0.$$

Suppose first $\vartheta_0 < \vartheta^* < \vartheta_1$, so $\Delta_0 < 0 < \Delta_1$, hence $h_0 > 0$, $h_1 > 0$. Then $H''(s) > 0$ for all s, and for this determination of h_0 and h_1, $H(s) < 1$ or > 1 according as s is inside or outside the interval (v, w). Suppose next $\vartheta_0 < \vartheta_1 < \vartheta^*$ so $\Delta_0 < \Delta_1 < 0$, hence $h_0 < 0$, $h_1 > 0$. Now $H(s) <$ or > 1 according as $\tilde{H}(s) >$ or < 1, where

$$\tilde{H}(s) = \tilde{h}_0 e^{\tilde{\Delta}_0 s} + \tilde{h}_1 e^{\tilde{\Delta}_1 s},$$

$$\tilde{h}_0 = -g_0/g_1 > 0, \; \tilde{h}_1 = 1/g_1 > 0, \; \tilde{\Delta}_0 = \Delta_0 - \Delta_1 < 0, \; \tilde{\Delta}_1 = -\Delta_1 > 0.$$

Applying the previous reasoning about $H(s)$ to $\hat{H}(s)$ we see that in the present case according as s is inside or outside the interval (v, w), $\tilde{H}(s)$ is < 1 or > 1, hence $H(s)$ is > 1 or < 1. The same result is found in the case $\vartheta^* < \vartheta_0 < \vartheta_1$. At this point we see that (8.18) is equivalent to $\varphi(x) = 1$ or 0 according as $s(x)$ is inside or outside an interval (v, w) if $\vartheta_0 < \vartheta^* < \vartheta_1$, the opposite if $\vartheta^* < \vartheta_0$ or $\vartheta^* > \vartheta_1$, where $v = v(t)$ and $w = w(t)$ may depend also on $\vartheta^*, \vartheta_0, \vartheta_1, \theta^*, \theta^0, \theta^1$.

Next we must show that $v(t), w(t), \delta(t), \gamma(t)$ may be chosen so that they do not depend on $\vartheta^*, \theta^*, \theta^0$, or θ^1 and (8.18) is satisfied, and that the resulting test has size α and is unbiased. Since this part of the proof is very similar for H_3 and H_4 we indicate it only for H_3. Let $F_{t,j}(s) = P_{\vartheta_j, \theta}(S \leqslant s \mid t)$, which does not depend on θ $(j = 0, 1)$. The conditions (8.18) are then equivalent to two equations similar to (8.16), namely,

$$F_{t,j}(v_3 - 0) + 1 - F_{t,j}(w_3) + \delta_3[F_{t,j}(v_3) - F_{t,j}(v_3 - 0)]$$
$$+ \gamma_3[F_{t,j}(w_3) - F_{t,j}(w_3 - 0)] = \alpha \; (j = 0, 1). \quad \ldots \quad (8.21_j)$$

First consider the solution (v, w, δ, γ) of (8.21_0) obtained by setting $v = -\infty$, which gives the uniformly most powerful unbiased test of H_1: $\vartheta \leqslant \vartheta_0$ against $\vartheta > \vartheta_0$ at size α. For this solution the left member of (8.21_1) is the conditional power of this test against $\vartheta = \vartheta_1$ and is therefore $\geqslant \alpha$ (a. e. $\mathfrak{P}_1^t$). Next consider the solution (v, w, δ, γ) of (8.21_0) obtained by setting $w = +\infty$, which gives the uniformly most powerful unbiased test of H_1': $\vartheta \geqslant \vartheta_0$ against $\vartheta < \vartheta_0$ at size α. For this (v, w) the left member of (8.21_1) is the conditional probability that this test will reject H_1' when $\vartheta = \vartheta_1$ and is therefore $\leqslant \alpha$ (a. e. $\mathfrak{P}_1^t$). If we now think of varying the solution (v, w, δ, γ) of (8.21_0) between these two extremes, it follows again from continuity considerations that there exists a solution of (8.21_0) for which (8.21_1) is also satisfied, and that the resulting $v = v_3(t)$, $w = w_3(t)$, $\delta = \delta_3(t)$, $\gamma = \gamma_3(t)$ do not depend on $\vartheta^*, \theta^*, \theta^0$, or θ^1. Verification of correct size and unbiasedness follow again from comparison with the test corresponding to $\Phi \equiv \alpha$.

9. Some examples

In applications of Theorem 8.2 the $\vartheta, \theta_1, \ldots, \theta_r$ of the density (8.1) are usually not the original parameters of the statistical problem but functions of these. Suppose that originally the parameters are $(\psi_1, \ldots, \psi_{r+1}) = \psi$ and that they appear in a density of the form

$$D(\psi)h(x) \exp \sum_{i=1}^{r+1} f_i(\psi)u_i(x).$$

If there exists a 1:1 transformation from $\psi_1, \ldots, \psi_{r+1}$ to $\vartheta, \theta_1, \ldots, \theta_r$ such that

$$\sum_{i=1}^{r+1} f_i(\psi)u_i(x) \equiv \vartheta s(x) + \sum_{i=1}^{r} \theta_i t_i(x) \qquad \ldots \quad (9.1)$$

identically in ψ and x, so that the density becomes of the form (8.1) and if the assumptions stated in connection with (8.1) are satisfied, then there exist uniformly most powerful unbiased tests of hypotheses of the form H_1, H_2, H_3, H_4 defined at the beginning of section 8, and Theorem 8.2 tells us how to construct them.

We are especially interested in determining the functions $s(x), t_1(x), \ldots, t_r(x)$, since Theorem 8.2 gives the test in terms of these functions. It is obvious that if we put $(\vartheta, \theta_1, \ldots, \theta_r)$ equal to a nonsingular linear transformation[7] of $(f_1(\psi), \ldots, f_{r+1}(\psi))$, then (9.1) will be satisfied. We may choose any constants $a_{11}, \ldots, a_{1,r+1}$ not all zero, and constants $\vartheta_0, \vartheta_1 (\vartheta_0 < \vartheta_1)$, and consider the hypotheses $H_i (i = 1, \ldots, 4)$ about

$$\vartheta \equiv \sum_{j=1}^{r+1} a_{1j} f_j(\psi). \qquad \ldots \quad (9.2)$$

Without affecting these hypotheses we can assume $\Sigma_{j=1}^{r+1} a_{ij}^2 = 1$, by redefinition of ϑ_0 and ϑ_1. We define the transformation from $(\psi_1, \ldots, \psi_{r+1})$ to $(\vartheta, \theta_1, \ldots, \theta_r)$ by the equations

$$a_{1i}\,\vartheta + \sum_{j=2}^{r+1} a_{ji}\theta_{j-1} = f_i(\psi) \quad (i = 1, \ldots, r+1), \qquad \ldots \quad (9.3)$$

where (a_{ij}) is any orthogonal matrix whose first row is $(a_{11}, \ldots, a_{1,r+1})$. Then (9.2) is satisfied. If we substitute (9.3) in (9.1) and collect coefficients of ϑ and of $\theta_1, \ldots, \theta_r$, we find that

$$s(x) = \sum_{j=1}^{r+1} a_{1j} u_j(x), \qquad \ldots \quad (9.4)$$

that is, $s(x)$ is the same linear form in the $u_j(x)$ as ϑ is in the $f_j(\psi)$, while $t_1(x), \ldots, t_r(x)$ are r linearly independent forms orthogonal to (9.4).

[7] It can be shown, under mild restrictions, that this is the most general transformation preserving (9.1).

Example 9.1: Suppose X is a random sample of n from a normal population with unknown mean μ and unknown variance σ^2. The density may be written $D \exp(f_1u_1+f_2u_2)$, where $f_1 = 1/\sigma^2$, $f_2 = \mu/\sigma^2$, $u_1 = -\frac{1}{2}\Sigma_1^n x_i^2$, $u_2 = \Sigma_1^n x_i$, and D is a function of the parameters. Set

$$\vartheta = A(1/\sigma^2)+B(\mu/\sigma^2),$$

where A and B are known constants. By specializing A, B, ϑ_0, ϑ_1 we get from Theorem 8.2 UMPU (uniformly most powerful unbiased) tests of any of the following hypotheses: $\mu = \mu_0$, $\mu \leqslant \mu_0$, $\sigma^2 = \sigma_0^2$, $\sigma^2 \geqslant \sigma_0^2$, σ^2 inside (or outside) an interval (σ_1^2, σ_0^2). We do not get a test of $\mu_0 \leqslant \mu \leqslant \mu_1$ by this method. It is curious that UMPU tests fall out for such apparently "unnatural" hypotheses as that the point (μ,σ) lies between two parabolas $\mu = \mu_0+c_1\sigma^2$ and $\mu = \mu_0+c_2\sigma^2$, tangent to each other at $(\mu_0, 0)$.

Example 9.2: For independent random samples from k normal populations with equal variance σ^2, the ith sample of size n_i from a population with mean μ_i, the density is of the form $D \exp[u_0(1/\sigma^2)+\Sigma_1^k u_i(\mu_i/\sigma^2)]$, and so we may take

$$\vartheta = A(1/\sigma^2)+\Sigma_1^k B_i(\mu_i/\sigma^2).$$

By specializing $A, B_1, \ldots, B_k, \vartheta_0, \vartheta_1$ we get UMPU tests of the hypotheses $\Sigma_1^k c_i\mu_i = c_0$, $\Sigma_1^k c_i\mu_i \leqslant c_0$, $\sigma^2 = \sigma_0^2$, $\sigma^2 \geqslant \sigma_0^2$, σ^2 inside (or outside) an interval (σ_1^2, σ_0^2).

Example 9.3: Suppose Example 9.2 is changed so that the variance of the ith population is σ_i^2 $(i = 1, \ldots, k)$. Then the density is $D \exp[\Sigma_1^k u_i'(1/\sigma_i^2) + \Sigma_1^k u_i(\mu_i/\sigma_i^2)]$, and we may take

$$\vartheta = \sum_1^k A_i(1/\sigma_i^2)+\sum_1^k B_i(\mu_i/\sigma_i^2),$$

yielding UMPU tests of the hypotheses that $\Sigma_1^k A_i(1/\sigma_i^2)$, a linear form in the precisions $1/\sigma_i^2$, is $= A$, $\leqslant A$, inside (or outside) an interval. For $k = 2$ this includes the hypotheses that σ_1^2/σ_2^2 is $= A$, $\leqslant A$, but not the hypothesis that σ_1^2/σ_2^2 is in a given interval.

Example 9.4: For independent random samples of sizes n_1 and n_2 from binomial populations with parameters p_1 and p_2 it is found that the density is of the form $Dh(x) \exp(f_1u_1+f_2u_2)$ with $f_i = \log[p_i/(1-p_i)]$, leading to

$$\vartheta = \log\left[\left(\frac{p_1}{1-p_1}\right)^{A_1}\left(\frac{p_2}{1-p_2}\right)^{A_2}\right].$$

Specializing $A_1 = -A_2 = 1$ yields UMPU tests of the hypotheses that the "odds ratio" $[p_1/(1-p_1)]/[p_2/(1-p_2)]$ is $= A$, $\leqslant A$, inside (outside) an interval. With $A = 1$ we get UMPU tests of the hypotheses $p_1 = p_2$, $p_1 \leqslant p_2$.

Example 9.5: For independent observations from two Poisson populations with parameters λ_1 and λ_2 the density is $Dh(x) \exp[x_1 \log \lambda_1 + x_2 \log \lambda_2]$, leading to

$$\vartheta = \log\left(\lambda_1^{A_1} \lambda_2^{A_2}\right).$$

With $A_1 = -A_2 = 1$, we get UMPU tests of the hypotheses that the ratio λ_1/λ_2 is $= A, \leqslant A$, inside (outside) an interval.

APPENDIX

MEASURABILITY OF PROPOSED CRITICAL FUNCTIONS

It is unfortunately necessary to show that our construction of the critical functions $\varphi_i(x)$ of Theorem 8.2 leads to B.m. (Borel measurable) functions. Since these are functions of the B.m. functions $s(x)$ and $t(x)$ it suffices to prove they are B.m. functions of (s, t). Following the above construction, the definitions of the φ_i are made as functions of s for each fixed t in a set which may differ from the t-space by a null set of $\mathfrak{P}^t$. In this connection, we note that the qualification "(a.e. $P^t_{\vartheta,\theta}$)" for some $(\vartheta, \theta) \in \Omega$ is equivalent to "(a.e. $\mathfrak{P}^t$)" since by (8.1) and the proof of Lemma 8.1, a null set of $P^t_{\vartheta,\theta}$ is of measure ν^t zero, where $d\nu^t = h(x)d\mu^x$, and is hence a null set of $\mathfrak{P}^t$.

Consider the critical function φ_1 for H_1. The conditional distribution function $F_t(s)$ introduced above (8.7) may be chosen to be B.m. in t (for each s) and non-decreasing and continuous from the right in s (for each t). It follows from the identity[8]

$$\{(s, t) | F_t(s) \geqslant c\} = \bigcap_n \bigcup_i \{(s, t) \,|\, 0 \leqslant r_i - s < 1/n, \; F_t(r_i) \geqslant c\},$$

where (r_i) denotes the rationals, that $F_t(s)$ is then B.m. in (s, t). Likewise, $F_t(s-0)$ is also B.m. in (s, t). Define $F_t^{-1}(y)$ as inf s for which $F_t(s) \geqslant y$; then $F_t^{-1}(y)$ and $F_t^{-1}(y-0)$ are B.m. in (y, t). The functions $w_1(t)$ and $\gamma_1(t)$ of (8.7) and (8.8) may be uniquely defined as $w_1(t) = F_t^{-1}(1-\alpha)$, and

$$\gamma_1(t) = 1 - \frac{F_t(w_1(t)-0)}{F_t(w_1(t)) - F_t(w_1(t)-0)},$$

where, here and below, if a denominator is zero the fraction is defined to be zero. Then $w_1(t)$ and hence $\gamma_1(t)$ are B.m. in t. Since $\varphi_1 = 1$ for $s > w_1(t)$, $\varphi_1 = 0$ for $s < w_1(t)$, and $\varphi_1 = \gamma_1(t)$ for $s = w_1(t)$, it follows that φ_1 is B.m. in (s, t).

In the proof for φ_2 we shall need that

$$g(x, t) = \int_{-\infty}^{x-0} s dF_t(s)$$

is B.m. in (x, t). This integral converges since $E_{\vartheta_0,\theta}$ is finite because of the exponential form of the density (8.1), hence the integral $g(\infty, t) = E_{\vartheta_0,\theta}(s|t)$ is absolutely

[8] This proof was pointed out to us by Professor L. M. LeCam.

convergent (a.e.$\mathfrak{P}^t$). Now suppose $x<+\infty$. We may express $g(x, t)$ as the limit of approximating sums $g_n(x, t)$ in which "subdivision of the ordinate" leads to

$$g_n(x, t) = \sum_{k=1}^{\infty}\left(x-\frac{k}{2^n}\right)\left[F_t\left(x-\frac{k-1}{2^n}-0\right)-F_t\left(x-\frac{k}{2^n}-0\right)\right]$$

which is B.m. in (x, t), hence $g(x, t)$ is also. That $g(\infty, t) = \alpha_2(t)/\alpha$ is B.m. in t follows from the fact that it is a conditional expectation.

For $0\leqslant y\leqslant\alpha$ define $v(y, t) = F_t^{-1}(y)$, $w(y, t) = F_t^{-1}(1-\alpha+y)$, and

$$\delta(y, t) = \frac{y-F_t(v(y, t)-0)}{F_t(v(y, t))-F_t(v(y, t)-0)},$$

$$\gamma(y, t) = 1-\frac{1-\alpha+y-F_t(w(y, t)-0)}{F_t(w(y, t))-F_t(w(y, t)-0)}.$$

These four functions are then B.m. in (y, t). Let $H(y, t)+\alpha_2(t)$ denote the left member of (8.17) with v_2, w_2, δ_2, γ_2 replaced by $v(y, t)$, $w(y, t)$, $\delta(y, t)$, $\gamma(y, t)$, respectively, i.e., $H(y, t)$ equals the $G_t(y)$ introduced above, diminished by $\alpha_2(t)$. Thus $H(0, t)\geqslant 0$, $H(\alpha, t)\leqslant 0$, and $H(y, t)$ is B.m. in (y, t) and continuous in y. Now define y_t as α on the set where $H(\alpha, t) = 0$ and otherwise as the inf of y for which $H(y, t)<0$. Because $H(y, t)$ is continuous in y, y_t satisfies the desired condition $H(y_t, t) = 0$, equivalent to (8.17). That y_t is B.m. in t may be seen as follows: $\{t| y_t<c, H(\alpha, t)<0\}$ is the set of t for which $H(\alpha, t)<0$ and $H(y,t)<0$ for some $y<c$, hence for some rational $y = r_n<c$, because of the continuity of $H(y, t)$ in y, and this is the union of the denumerable number of B.m. sets $\{t|H(r_n, t)<0, H(\alpha, t)<0\}$ for rational $r_n<c$. If $v_2(t)$, $w_2(t)$, $\delta_2(t)$, $\gamma_2(t)$ are defined as $v(y, t)$, $w(y, t)$, $\delta(y, t)$, $\gamma(y, t)$, which are B.m. in (y,t), with y replaced by y_t, which is B.m. in t, then it is clear these are then B.m. in t, and hence φ_2 is B.m. in (s, t).

The proof for φ_3 is rather similar to that for φ_2. We again define functions $v(y, t)$, $w(y, t)$, $\delta(y, t)$, $\gamma(y, t)$ by replacing the F_t in the previous definitions by $F_{t,1}$, where $F_{t,j}$ is defined above (8.21$_j$). We then define $H(y, t)$ as the left member of (8.21$_1$) diminished by α and with v_3, etc. replaced by $v(y, t)$, etc. The continuity of $H(y, t)$ in y now depends on the fact that for each t except a null set of $\mathfrak{P}^t$, the countable set of discontinuities in s of $F_{t,j}(s)$ is the same for $j = 0$ as for $j = 1$, and the set of intervals where $F_{t,j}$ is constant is the same for $j = 0$ and 1. To show this define A_j to be the set of points (s, t) where $F_{t,j}(s)-F_{t,j}(s-0) = 0$ and $F_{t,1-j}(s)-F_{t,1-j}(s-0) > 0$, so A_j is a B.m. set in (s, t). Let $f_{jk}(t)$ be the probability assigned by $F_{t,k}(s)$ to the cross-section of A_j on which $t =$ constant ($k = 0, 1$); so $f_{jj}(t) = 0$ and $f_{j,1-j}(t)>0$ for all non-empty cross-sections of A_j. We have

$$P^{s,t}_{\vartheta,\theta}(A_j) = \int_{W^t} f_{jk}(t)dP^t_{\vartheta,\theta}$$

for all $(\vartheta, \theta)\epsilon\Omega_k$. If B_j is the pre-image of A_j in W^x, so $P^{s,t}_{\vartheta,\theta}(A_j) = P^x_{\vartheta,\theta}(B_j)$, we see from (8.1) that if $P^x_{\vartheta,\theta}(B_j) = 0$ for some $(\vartheta, \theta)\epsilon\Omega$, then $h(x) = 0$ (a.e.μ^x) on B_j and hence $P^{\mathfrak{s}}_{\vartheta,\theta}(B_j) = 0$ for all $(\vartheta, \theta) \epsilon \Omega$. But $P^{s,j}_{\vartheta,\theta}(A_j) = 0$ for $(\vartheta, \theta) \epsilon \Omega_j$, therefore for $(\vartheta, \theta) \epsilon \Omega_{1-j}$, hence (a.e.$\mathfrak{P}^t$) $f_{j,1-j}(t) = 0$ and the cross-section of A_j for $t =$ constant is empty.

This proves the above assertion about the discontinuities of $F_{t,1}$ and $F_{t,2}$; to prove it for the intervals of constancy, denote by $F_{t,j}^{-1}(y)$ the inf of s for which $F_t(s) \geqslant y$, and by $\overline{F}_{t,j}^{-1}(y)$ the sup for $F_{t,j}(s) \leqslant y$. Write $K_j(s, t) = \overline{F}_{t,j}^{-1}(F_{t,j}(s) - F_{t,j}^{-1}(F_{t,j}(s))$, so $K_j(s, t)$ is B.m. in (s, t), and s_0 will be in an interval where $F_{t,j}(s)$ is constant if and only if $K_j(s_0, t) > 0$. Now let D_j be the set of points (s, t) where $K_j(s, t) > 0$ and $K_{1-j}(s, t) = 0$. Applying to D_j an argument similar to the above for A_j we can conclude the proof.

REFERENCES

BARTLETT, M. S. (1937): Properties of sufficiency and statistical tests. *Proc. Royal Soc.*, London, Series A, 160, 268-282.

GHOSH, M. N. (1948): On the problems of similar regions. *Sankhyā*, 8, 329-338.

HALMOS, PAUL R. (1950): *Measure Theory*, D. Van Nostrand Co., N.Y.C.

HOEL, PAUL G. (1948): On the uniqueness of similar regions. *Ann. Math. Stat.*, 19, 66-71.

LEHMANN, E. L. (1947): On optimum tests of hypotheses with one constraint. *Ann. Math. Stat.*, 18, 473-494.

LEHMANN, E. L. and SCHEFFÉ, HENRY (1950): Completeness, similar regions, and unbiased estimation. Part I. *Sankhyā*, 10, 305-340.

NANDI, H. K. (1951): On type B_1 and type B regions. *Sankhyā*, 11, 13-22.

NEYMAN, J. (1941): On a statistical problem arising in routine analyses and in sampling inspections of mass production. *Ann. Math. Stat.*, 12, 46-76.

SCHEFFÉ, HENRY (1942): On the theory of testing composite hypotheses with one constraint. *Ann. Math. Stat.*, 13, 280-293.

SVERDRUP, ERLING (1953): Similarity, unbiasedness, minimaxibility and admissibility of statistical test procedures. *Skandinavisk Aktuarietidskrift*, 36, 64-86.

WIDDER, DAVID VERNON (1941): *The Laplace Transform*, Princeton University Press.

Paper received : March, 1955.

ESTIMATES OF LOCATION BASED ON RANK TESTS

By J. L. Hodges, Jr.[1] and E. L. Lehmann[2]

University of California, Berkeley

1. Introduction and summary. A serious objection to many of the classical statistical methods based on linear models or normality assumptions is their vulnerability to gross errors. For certain testing problems this difficulty is successfully overcome by rank tests such as the two Wilcoxon tests or the Kruskal-Wallis H-test. Their power is more robust against gross errors than that of the t- and F-tests, and their efficiency loss is quite small even in the rare case in which the suspicion of the possibility of gross errors is unfounded.

For the corresponding problems of point estimation a beginning has been made to attack the difficulty by modifying the classical estimates either through removal or Winsorization of outlying observations; see for example Tukey (1960) and Anscombe (1960). It is the purpose of the present paper to explore a different approach to these problems of point estimation. In Sections 2–5 point estimates of location or shift parameter are defined in terms of rank test statistics such as the Wilcoxon or normal scores statistic, which are successful in providing robust power for the corresponding testing problems. In Sections 6 and 7, certain regularity and invariance properties of these estimates are proved. The distributions of the estimates are shown in Section 8 to be symmetric with respect to the parameter being estimated—and hence in particular to be unbiased—if the underlying distribution of the observations on which the estimate is based is symmetric. Without this assumption, the estimates are shown in Section 9 to be either exactly or approximately median unbiased for small samples and in Section 10 to be approximately normally distributed about the true parameter value for large samples. The variance of this asymptotic distribution depends of course on the underlying distribution of the observations, so that the estimates are not "distribution-free." In Section 9 there is also established a close relationship between the estimates and the corresponding upper and lower confidence bound for the parameter at confidence level $\frac{1}{2}$, with which the estimate coincide in many cases. Finally, in Section 11, it is proved that the asymptotic relative efficiency of the estimates to the classical linear estimates is the same as the Pitman efficiency of the rank tests on which they are based to the corresponding t-tests.

2. Point estimates based on test statistics. Let $X_1, \cdots, X_m; Y_1, \cdots, Y_n$ be independent random variables with distributions

$$P\{X_i \leqq u\} = F(u); \qquad P\{Y_j \leqq u\} = F(u - \Delta). \tag{2.1}$$

Received October 23, 1962.

[1] This paper was prepared with the partial support of the Office of Naval Research, Contract (Nonr-222-43). This paper in whole or in part may be reproduced for any purpose of the United States Government.

[2] Work done while author was Research Professor for the Adolph C. and Mary Sprague Miller Institute for Basic Research in Science, University of California, Berkeley.

Since the variables $X_1, \cdots, X_m$, and the variables $Y_1 - \Delta, \cdots, Y_n - \Delta$ obtained by shifting the Y-sample Δ to the left, are independently, identically distributed, it is natural to estimate Δ by the amount of shift needed to align a closely as possible the two sets $(X_1, \cdots, X_m)$ and $(Y_1 - \Delta, \cdots, Y_n - \Delta)$. A definition of alignment could for example be given with reference to the Wilcoxon statistic, by defining the two sets to be aligned if half of the non-zero differences $(Y_j - \Delta) - X_i$ are positive and half negative. There is either a unique such value of Δ, which would then serve as estimate, or an interval of such values; in the latter case, the midpoint of this interval provides a natural estimate. More generally, if a test of the hypothesis $\Delta = 0$ is based on a statistic whose distribution is symmetric about a point μ, the two sets could be defined to be in alignment when giving to the test statistic the value μ. To formalize these considerations let us either assume that F is known to belong to the class $\mathfrak{F}_0$ of all continuous distributions, or that it is known to belong to the class $\mathfrak{F}_1$ of all continuous distributions that are symmetric about zero. Consider a test statistic

$$h = h(X_1, \cdots, X_m; Y_1, \cdots, Y_n)$$

for the hypothesis $H: \Delta = 0$ against the alternatives $\Delta > 0$. We shall assume throughout that

(A) $h(x_1, \cdots, x_m, y_1 + a, \cdots, y_n + a)$ is a nondecreasing function of a for all x and y,

(B) when $\Delta = 0$, the distribution of $h(X_1, \cdots, X_m; Y_1, \cdots, Y_n)$ is symmetric about a fixed point μ (independent of F), (i) for all $F \varepsilon \mathfrak{F}_0$, or (ii) for all $F \varepsilon \mathfrak{F}_1$.

We shall use the notation $x = (x_1, \cdots, x_m)$ and $y = (y_1, \cdots, y_n)$ with the obvious conventions. Thus $x < x'$ means that the inequality holds for each coordinate; if a is a real number, then $x + a = (x_1 + a, \cdots, x_n + a)$; etc. The notation $P_0\{\cdot\}$ will be used to indicate that the probability in question is being computed for the case $\Delta = 0$. Let

$$\Delta^* = \sup\{\Delta: h(x, y - \Delta) > \mu\} \quad \text{and} \quad \Delta^{**} = \inf\{\Delta: h(x, y - \Delta) < \mu\} \tag{2.2}$$

and let

$$\hat{\Delta} = (\Delta^* + \Delta^{**})/2. \tag{2.3}$$

For suitable functions h we propose $\hat{\Delta}$ as estimate of the shift parameter Δ.

As a second problem suppose that $Z_1, \cdots, Z_N$ are independently distributed with common distribution

$$P\{Z_i \leqq u\} = F(u - \theta) \tag{2.4}$$

where F is continuous and symmetric about zero.

Considerations similar to those in the two-sample problem suggest basing an estimate on a test statistic $h = h(Z_1, \cdots, Z_N)$ for the hypothesis $\theta = 0$ against the alternatives $\theta > 0$. We shall assume throughout that

(C) $h(z_1 + a, \cdots, z_N + a)$ is a nondecreasing function of a for each z,

and

(D) for $\theta = 0$, the distribution of h is symmetric about a fixed point μ (independent of F) for all $F \varepsilon \mathfrak{F}_1$.

If μ is the median of $h(Z)$ when $\theta = 0$, let

$$(2.5)\quad \theta^* = \sup\{\theta : h(z - \theta) > \mu\} \quad \text{and} \quad \theta^{**} = \inf\{\theta : h(z - \theta) < \mu\}$$

and

$$(2.6)\quad \hat{\theta} = (\theta^* + \theta^{**})/2.$$

Then we propose $\hat{\theta}$ as the estimate of the location parameter θ.

3. A class of estimates for the two-sample problem. We shall be concerned primarily with estimates based on rank tests. An important class of rank statistics for the two-sample problem is given by

$$(3.1)\quad h(x, y) = \sum_{j=1}^{n} E_\Psi[V^{(s_j)}]$$

where $s_1, \cdots, s_n$ denote the ranks of $y_1, \cdots, y_n$ in the combined sample, and where $V^{(1)} < \cdots < V^{(m+n)}$ denote an ordered sample of size $m + n$ from a distribution Ψ.

The function h defined by (3.1) clearly satisfies requirement (A) of the preceding section. Conditions under which h satisfies (B) are given by the following lemma, in which h is not assumed to satisfy (3.1).

LEMMA 1. *The distribution of $h(X, Y)$ is symmetric about μ if any one of the following three conditions holds*:

(i) *h is a function only of the ranks and satisfies*

$$(3.2)\quad h(x, y) + h(-x, -y) = 2\mu \qquad \text{(a.e. } P_0)$$

(ii) *the sample sizes m and n are equal, and h satisfies*

$$(3.3)\quad h(x, y) + h(y, x) = 2\mu \qquad \text{(a.e. } P_0)$$

(iii) *the distribution F is symmetric about zero, and h satisfies* (3.2).

PROOF. (i) If $h(x, y) = g(s_1, \cdots, s_n)$, then $h(-x, -y) = g(m + n - s_n, \cdots, m + n - s_1)$ and $P_0\{h(X, Y) < \mu - a\} = P_0\{h(-X, -Y) > \mu + a\} = P_0\{g(m + n - S_n, \cdots, m + n - S_1) > \mu + a\} = P_0\{g(S_1, \cdots, S_n) > \mu + a\} = P_0\{h(X, Y) > \mu + a\}$. Here P_0 indicates that the probabilities are computed for $\Delta = 0$; the first equality follows from (3.2); the third equality follows from the fact that $(S_1, \cdots, S_n)$ and $(m + n - S_n, \cdots, m + n - S_1)$ have the same joint distribution when $\Delta = 0$.

(ii) When $m = n$ and $\Delta = 0$, the vectors (X, Y) and (Y, X) have the same joint distribution, and hence $P_0\{h(X, Y) < \mu - a\} = P_0\{h(Y, X) < \mu - a\} = P_0\{h(X, Y) > \mu + a\}$ where the second equality follows from (3.3).

(iii) This follows from the first equality in the proof of (i) and the fact that (X, Y) and $(-X, -Y)$ have the same distribution when F is symmetric about zero.

Conditions under which the function h defined by (3.1) satisfies (3.2) or (3.3) are given by the following lemma.

LEMMA 2. *Let h be defined by* (3.1). *Then*

(i) *if Ψ is symmetric about b, the function h satisfies* (3.2) *with $\mu = nb$*

(ii) *if $m = n$ and b denotes the expectation of Ψ the function h satisfies* (3.3) *with $\mu = \frac{1}{2}(m + n)b$.*

PROOF.

$$\text{(i) } h(x, y) + h(-x, -y) = \sum E_\Psi[V^{(s_j)}] + \sum E_\Psi[V^{(m+n-s_j)}]$$
$$= \sum E_\Psi[V^{(s_j)}] + \sum E_\Psi[2b - V^{(s_j)}] = 2nb.$$

(ii) $h(x, y) + h(y, x) = \sum E_\Psi[V^{(s_j)}] + \sum E_\Psi[V^{(r_i)}]$ where $r_1, \cdots, r_m$ de-denotes the ranks of the x's. The right hand side is equal to the sum of the expectations of all the V's and hence to $(m + n)b$.

It follows from Lemmas 1 and 2 that a function h given by (3.1) satisfies Condition B(i) of the preceding section if either Ψ is symmetric or the two sample sizes are equal.

Among the statistics given by (3.1) and satisfying B(i), we shall be particularly interested in the Wilcoxon statistic and the normal scores statistic obtained by taking for Ψ a rectangular or normal distribution respectively. The resulting estimates will be considered in more detail in the next section.

Suppose next that $m = n$. Let us denote by $\bar{}$ the average of the indicated variables and by $\tilde{}$ or med their median, so that for example

$$\bar{x} = (x_1 + \cdots + x_m)/m \tag{3.4}$$

and

$$\tilde{x} = \text{med } x = \begin{cases} x^{(k+1)} & \text{if } m = 2k + 1 \\ [x^{(k)} + x^{(k+1)}]/2 & \text{if } m = 2k \end{cases} \tag{3.5}$$

where $x^{(1)} < \cdots < x^{(m)}$ denote the ordered x's. Then both $h(x, y) = \bar{y} - \bar{x}$ and $h(x, y) = \tilde{y} - \tilde{x}$ satisfy (3.3) with $\mu = 0$ and $\Delta^* = \Delta^{**} = h$. The resulting estimates are therefore $\bar{Y} - \bar{X}$ and $\tilde{Y} - \tilde{X}$ respectively.

More generally suppose that in addition to (3.3), h satisfies the invariance relation

$$h(x, y + a) = h(x, y) + a \quad \text{for all real } a. \tag{3.6}$$

Assume without loss of generality that $\mu = 0$ since the function $h'(x, y) = h(x, y - \mu)$ satisfies (3.3) with $\mu = 0$. Condition (3.6) then implies that $\Delta^* = \Delta^{**} = h$ since for example

$$\Delta^{**}(x, y) = \inf\{\Delta : h(x, y - \Delta) < 0\} = \inf\{\Delta : h(x, y) < \Delta\} = h(x, y).$$

4. Estimates based on the Wilcoxon and normal scores statistics. Let $h(x, y)$ be the number of pairs (i, j) such that $x_i < x_j$ $(1 \leqq i \leqq m; 1 \leqq j \leqq n)$. The test based on this statistic is the Wilcoxon two-sample test in the Mann-Whitney form; it is equivalent to the test based on (3.1) with Ψ the rectangular distribution on (0, 1). The possible values of the function h, which satisfies assumption B(i), are the integers, 0, 1, $\cdots$, mn. To find an explicit expression for the estimate $\hat{\Delta}$ obtained from h through (2.2) and (2.3) denote by $W^{(1)} < \cdots < W^{(mn)}$ the ordered differences $Y_j - X_i$. It is convenient to distinguish two cases.

Suppose first that mn is odd, $mn = 2k + 1$, say. Then $\mu = k + \frac{1}{2}$, which is not a possible value of h, and

$$\begin{aligned} \Delta^{**} &= \inf\{\Delta : h(x, y - \Delta) < k + \tfrac{1}{2}\} \\ &= \inf\{\Delta : \text{Fewer than } k + \tfrac{1}{2} \text{ of the differences } Y_j - X_i \text{ exceed } \Delta\} \\ &= \inf\{\Delta : W^{(k+1)} \leqq \Delta\} = W^{(k+1)}. \end{aligned}$$

Similarly,

$$\begin{aligned} \Delta^{*} &= \sup\{\Delta : \text{More than } k + \tfrac{1}{2} \text{ of the differences } Y_j - X_i \text{ exceed } \Delta\} \\ &= \sup\{\Delta : W^{(k+1)} \leqq \Delta\} = W^{(k+1)}, \end{aligned}$$

so that $\hat{\Delta} = W^{(k+1)}$.

On the other hand, if mn is even, $mn = 2k$ say, then

$$\begin{aligned} \Delta^{**} &= \inf\{\Delta : W^{(k+1)} \leqq \Delta\} = W^{(k+1)} \\ \Delta^{*} &= \sup\{\Delta : W^{(k)} > \Delta\} = W^{(k)} \end{aligned}$$

and hence $\hat{\Delta} = [W^{(k)} + W^{(k+1)}]/2$.

Thus in both cases,

$$\hat{\Delta} = \text{med}\,(Y - X) \tag{4.1}$$

is the median of the set of mn differences $Y_j - X_i$.

Formula (4.1) will frequently not be the most convenient way of computing $\hat{\Delta}$ in practice. To illustrate how this can be done quite expeditiously consider the case $mn = 2k + 1$. Then k of the differences $(Y_j - \hat{\Delta}) - X_i$ are greater than, k less than, and one is equal to, zero. To obtain $\hat{\Delta}$, plot the two samples on two separate strips of paper. Sliding the Y-strip to the left decreases by one the number of positive differences $Y_j - X_i$ each time a Y moves over an X. By proceeding in this way, one rapidly finds the position in which k of the differences are positive, k negative, and one is zero. The difference of the origins of the two strips in this position is $\hat{\Delta}$. If $mn = 2k$, one finds in a similar way the shift Δ^* that produces k positive, $k - 1$ negative, and one zero difference, and the shift Δ^{**} that produces $k - 1$ positive, k negative, and one zero difference, and hence the estimate $\hat{\Delta} = (\Delta^* + \Delta^{**})/2$.

A somewhat similar procedure can be used to obtain the estimate based on the normal scores statistic $h = \sum E_\Phi[V^{(s_j)}]$. Suppose the y-strip is slid to the left so that $Y^{(j)}$ moves across an X. Then s_j is decreased by 1 so that h is decreased by $E_\Phi[V^{(s_j)}] - E_\Phi[V^{(s_j-1)}]$. Starting out with a position of the two strips which gives to h an approximately central value, the desired position can quickly be found using a table of expected normal order statistics.

5. Some estimates for the one-sample problem. Let $Z_1, \cdots, Z_N$ be identically distributed, their distribution given by (2.1) with F symmetric about zero. Let $s_1, \cdots, s_n$ denote the ranks of the positive Z's among the N absolute values $|Z_1|, \cdots, |Z_N|$. Here n is a random variable which for $\Delta = 0$ has the binomial distribution $Bi(N, \frac{1}{2})$. A class of rank tests is based on the test statistic

$$h(z) = \sum_{j=1}^{n} E_{\Psi}[V^{(s_j)}] \tag{5.1}$$

where $V^{(1)} < \cdots < V^{(N)}$ denote the ordered absolute values of a sample of size N from a distribution Ψ, so that they constitute an ordered sample from the distribution given by $\Psi^*(u) = \Psi(u) - \Psi(-u)$ for $u > 0$.

The function h defined by (5.1) clearly satisfied assumption (C) of Section 2. As in Lemma 1(iii) it is seen that a function h satisfies requirement (D) if

$$h(z) + h(-z) = 2\mu \qquad (\text{a.e. } P_0)\ . \tag{5.2}$$

It follows from (5.2) that any function h given by (3.1) satisfies requirement (D). We have in fact

LEMMA 3. *If h is given by* (5.1) *and $\theta = 0$, the distribution of h is symmetric about $\mu = \frac{1}{2}NE_{\Psi}|Z_1|$ for all $F \varepsilon \mathfrak{F}_1$.*

PROOF. Let $r_1, \cdots, r_m$ denote the ranks of the negative Z's among $|Z_1|, \cdots, |Z_N|$. Then $h(z) + h(-z) = \sum E_{\Psi}[V^{(s_j)}] + \sum E_{\Psi}[V^{(r_i)}] = NE_{\Psi}|Z_1|$.

An important special case is again that of the Wilcoxon test statistic corresponding to the choice of a rectangular distribution for Ψ. To obtain an explicit expression for the estimate θ in this case, it is convenient to use the equivalent form of the test statistic due to Tukey (1949), namely

$$h(z) = \text{Number of pairs } (i, j) \text{ with } 1 \leqq i \leqq j \leqq N \text{ such that } z_i + z_j > 0. \tag{5.3}$$

The possible values of h are the integers $0, 1, \cdots, N(N + 1)/2$. Let $W^{(1)} < \cdots < W^{(K)}$ be the $K = N(N + 1)/2$ averages $(z_i + z_j)/2$ with $i \leqq j$. Then it is seen quite similarly to the corresponding result in Section 4 that $\hat{\theta}$ is the median of the variables $W^{(1)}, \cdots, W^{(K)}$.

Another class of examples is obtained by taking for h a functon that satisfies (5.2) and is translation invariant in the sense that

$$h(z + a) = h(z) + a \quad \text{for all real } a. \tag{5.4}$$

As in the corresponding examples for the two-sample problem discussed at the end of Section 3, one can assume without loss of generality that $\mu = 0$ and then finds $\hat{\theta}(z) = h(z)$.

Examples of this are (i) $h(z) = \bar{z}$ and (ii) $h(z) = \tilde{z}$. It is interesting to see that $\tilde{z}$ is also obtained as the estimate corresponding to quite a different function h. For let $h(z)$ be the number of positive z's, so that the test defined by h is the sign test. Then h satisfies (5.2) with $\mu = N/2$ and a computation similar to that of the Wilcoxon case then shows that

$$\hat{\theta} = \tilde{Z}. \tag{5.5}$$

6. Regularity properties. In Section 2, estimates $\hat{\Delta}$ and $\hat{\theta}$ of a shift or location parameter were defined in terms of given test statistics, and these estimates were illustrated in Sections 3–5. The remainder of the paper is concerned with general properties of these estimates. In the present section, certain regularity properties are obtained; these are immediate consequences of the following theorem.

THEOREM 1. *Let h be a real-valued function on $(m + n)$-space such that $h(x_1, \cdots, x_m, y_1 + a, \cdots, y_n + a)$ is a non-decreasing function of a for all x and y, and let Δ^*, Δ^{**} be defined by (2.2). Suppose that $(X_1, \cdots, X_m, Y_1, \cdots, Y_n)$ is a random vector with joint distribution H. Then the distribution of Δ^* and Δ^{**} is (absolutely) continuous provided H is (absolutely) continuous.*

PROOF. For any fixed numbers $t_2, \cdots, t_n$, $h(x_1, \cdots, x_m, y_1, y_1 + t_2, \cdots, y_1 + t_n)$ is non-decreasing in y_1. Let $u(x_1, \cdots, x_m, t_2, \cdots, t_n)$ be such that $h(x_1, \cdots, x_m, y_1, y_1 + t_2, \cdots, y_1 + t_n) <$ or $\geqq \mu$ if y_1 is $<$ or $> u(x_1, \cdots, x_m; t_2, \ldots, t_n)$. Then

$$\begin{aligned}(6.1)\qquad &\Delta^{**}(x_1, \cdots, x_m, y_1, y_1 + t_2, \cdots, y_1 + t_n) \\ &\qquad = y_1 - u(x_1, \cdots, x_m; t_2, \cdots, t_n).\end{aligned}$$

Suppose now first that H is continuous. Since

$$\begin{aligned}(6.2)\qquad &\Delta^{**}(x_1, \cdots, x_m, y_1, \cdots, y_n) = c \Leftrightarrow \\ &\qquad y_1 = u(x_1, \cdots, x_m, y_2 - y_1, \cdots, y_n - y_1) + c,\end{aligned}$$

the set $S = \{(x_1, \cdots, x_m, y_1, \cdots, y_n): \Delta^{**}(x_1, \cdots, x_m, y_1, \cdots, y_n) = c\}$ intersects each line

$$\begin{aligned}(6.3)\qquad &L(x_1^0, \cdots, x_m^0, t_2^0, \cdots, t_n^0): x_1 = x_1^0, \cdots, x_m = x_m^0, \\ &\qquad y_2 = y_1 + t_2^0, \cdots, y_n = y_1 + t_n^0\end{aligned}$$

in the single point whose y_1-coordinate is $u(x_1^0, \cdots, x_m^0, t_2^0, \cdots, t_n^0) + c$. Consider now the probability of the set S. Each section of S with one of the lines (6.3) consists of a single point and by the assumed continuity of H therefore has probability zero. Since S is measurable, it follows from Fubini's theorem that $P(S) = 0$, as was to be proved.

Suppose now that H is absolutely continuous. Let A be any set on the real line with Lebesgue measure zero and let

$$S = \{(x_1, \cdots, x_m, y_1, \cdots, y_n): \Delta^{**}(x_1, \cdots, x_m, y_1, \cdots, y_n) \varepsilon A\}.$$

Since

$$\begin{aligned}&\Delta^{**}(x_1, \cdots, x_m, y_1, \cdots, y_n) \varepsilon A \\ &\qquad \Leftrightarrow y_1 - u(x_1, \cdots, x_m, y_2 - y_1, \cdots, y_n - y_1) \varepsilon A,\end{aligned}$$

the section of S with each line (6.3) has Lebesgue measure zero, and thus S itself has Lebesgue measure zero. It follows from the absolute continuity of H that $P(S) = 0$, as was to be proved.

In both cases, the proof for Δ^* is completely analogous. It follows therefore that the distribution of the estimate $\hat{\Delta}$ is (absolutely) continuous if F is (absolutely) continuous. The corresponding result for $\hat{\theta}$ follows by letting $m = 0$, $n = N$ in the lemma.

7. Invariance properties. A simple but useful property of the estimates $\hat{\Delta}$ and $\hat{\theta}$ is their translation invariance, namely

$$\hat{\Delta}(x, y + a) = \hat{\Delta}(x, y) + a \quad \text{for all real } a \tag{7.1}$$

and

$$\hat{\theta}(z + a) = \hat{\theta}(z) + a \quad \text{for all real } a. \tag{7.2}$$

These are immediate consequences of the definitions (2.2) and (2.5), which show that Δ^*, Δ^{**}, θ^* and θ^{**} are translation invariant.

It follows from (7.1) and (7.2) that

$$P_\Delta(\hat{\Delta} - \Delta \leqq u) = P_0(\hat{\Delta} \leqq u) \tag{7.3}$$

and

$$P_\theta(\hat{\theta} - \theta \leqq u) = P_0(\hat{\theta} \leqq u) \tag{7.4}$$

where the notation P_Δ and P_θ indicate, that the probabilities are computed assuming Δ and θ to be the true values of these parameters. Relations (7.3) and (7.4) show that when investigating distributional properties of the estimates, one may assume without loss of generality that $\Delta = 0$ or $\theta = 0$ respectively, since the distribution for the general case is obtained simply by translation.

Typically, the function h in the two-sample problem satisfies the invariance relation

$$h(x + a, y + a) = h(x, y) \quad \text{for all } a. \tag{7.5}$$

This relation holds in particular for any rank test. It is obvious from the definition that (7.5) implies the corresponding relationship for $\hat{\Delta}$.

8. Symmetry properties. If $\hat{\Delta}$ and $\hat{\theta}$ are to be reasonable estimates of Δ and θ, their distribution should in some sense be centered on the true parameter values. In the present section, we shall give conditions under which the distributions of $\hat{\Delta}$ and $\hat{\theta}$ are actually symmetric with respect to Δ and θ, so that in particular the estimates are unbiased. (These conditions are related to those given by van der Vaart (1950).

THEOREM 2. *The distribution of the estimate $\hat{\Delta}$ defined by* (2.2) *and* (2.3) *is symmetric about Δ if either one of the following conditions hold:*

(i) *The distribution F defined in* (2.1) *is symmetric and h satisfies* (3.2) *and* (7.5);

(ii) *the two sample sizes m and n are equal, and h satisfies* (3.3) *and* (7.5).

PROOF.

(i) By the results of Section 7, we may assume without loss of generality that $\Delta = 0$. Further, by (7.5), the distribution F may be assumed to be symmetric about zero. Since the random vectors (X, Y) and $(-X, -Y)$ then have the same distribution, $\hat{\Delta}(X, Y)$ and $\hat{\Delta}(-X, -Y)$ will also have the same distribution. To prove that $\hat{\Delta}(X, Y)$ and $-\hat{\Delta}(X, Y)$ have the same distribution (which is what the theorem claims), it is therefore only necessary to show that

$$\hat{\Delta}(-x, -y) = -\hat{\Delta}(x, y). \tag{8.1}$$

This equation follows from the relations

$$\Delta^{**}(-x, -y) = -\Delta^{*}(x, y) \quad \text{and} \quad \Delta^{*}(-x, -y) = -\Delta^{**}(x, y). \tag{8.2}$$

To see for example the first of these, note that

$$\Delta^{**}(x, y) = \inf\{\Delta : h(-x, y - \Delta) < \mu\} = \inf\{\Delta : h(x, y + \Delta) > \mu\}$$

while

$$-\Delta^{*}(x, y) = \inf\{-\Delta : h(x, y - \Delta) > \mu\} = \inf\{\Delta : h(x, y + \Delta) > \mu\}.$$

(ii) Assume again that $\Delta = 0$. Since then the vectors (X, Y) and (Y, X) have the same distribution, it is enough to show that

$$\hat{\Delta}(y, x) = -\hat{\Delta}(x, y). \tag{8.3}$$

This equation is an immediate consequence of the relations

$$\Delta^{*}(y, x) = -\Delta^{**}(x, y) \quad \text{and} \quad \Delta^{**}(y, x) = -\Delta^{*}(x, y)$$

which follow from (3.3) and (7.5) as the corresponding relations in the proof of (i) followed from (3.2).

COROLLARY. *If h is given by* (3.1), *then the distribution of* $\hat{\Delta}$ *is symmetric about* Δ *if either one of the following conditions holds:*

(i) *the distributions* F *and* Ψ *are symmetric*

(ii) *the sample sizes* m *and* n *are equal.*

PROOF. Since h depends only on the ranks, it satisfies (7.5). The result now follows from Lemma 2 and Theorem 2.

The requirement in part (i) of the Corollary is a restriction on the test, which is satisfied both for the Wilcoxon and the normal scores test. On the other hand, the symmetry condition for F concerns an unknown distribution, and it is therefore much less certain whether it is satisfied in any given situation. The assumption is however frequently not unreasonable if one might be willing to assume normality except for the possibility of symmetric gross errors.

If F is not symmetric, $\hat{\Delta}$ need no longer be either symmetric or unbiased. Consider for example the case that h is the Wilcoxon statistic and that $m = 1$. Then $\hat{\Delta} = \text{med}\,(Y - X) = \text{med}\,Y - X_1$. For large n, $\tilde{Y}$ is essentially equal to the median of F, and $\hat{\Delta}$ need clearly not be unbiased. We shall however show in the next section that Δ is typically at least approximately, and in many cases exactly, median unbiased.

Results analogous to Theorem 2 and its Corollary hold also for the one-sample problem.

THEOREM 3. *The distribution of the estimate $\hat{\theta}$ defined by (2.5) and (2.6) is symmetric about θ if*

(i) *F is symmetric about zero and h satisfies (5.2) and hence in particular if*

(ii) *h is given by (5.1).*

PROOF. The proof of (i) is exactly analogous to that of Theorem 2(i); part (ii) follows from (i) and Lemma 3.

As in the two-sample problem, the symmetry assumption for F is frequently not unreasonable. In the particular case of a paired comparison experiment, with $Z_i = Y_i - X_i$, it is of course guaranteed if the assignment of the two subjects within each pair to the two treatments is performed at random.

9. Median unbiasedness. In general, when the distribution of $\hat{\Delta}$ or $\hat{\theta}$ is not symmetric about the true value, the estimate will also not be unbiased, that is, its expectation will not coincide with the parameter being estimated. The estimate will, however, in many cases still be median unbiased in the sense that the median of the distribution will be equal to the true value of the parameter. This follows from the following two lemmas, the first of which will also be used in the succeeding sections.

LEMMA 4. *For any real number a, the estimates $\hat{\Delta}$ and $\hat{\theta}$ satisfy the inequalities*

$$P\{h(X, Y - a) < \mu\} \leqq P\{\hat{\Delta} < a\} \leqq P\{h(X, Y - a) \leqq \mu\} \tag{9.1}$$

and

$$P\{h(Z - a) < \mu\} \leqq P\{\hat{\theta} < a\} \leqq P\{h(Z - a) \leqq \mu\}. \tag{9.2}$$

PROOF. By the definition of Δ^* and Δ^{**}; $\Delta^{**} < a \Rightarrow h(x, y - a) < \mu \Rightarrow \Delta^{**} \leqq a$ and $\Delta^* > a \Rightarrow h(x, y - a) > \mu \Rightarrow \Delta^* \geqq a$. Since Δ^* and Δ^{**} have been shown to have continuous distributions, it follows that

$$P\{\Delta^{**} < a\} = P\{h(X, Y - a) < \mu\} \tag{9.3}$$

and

$$P\{\Delta^* < a\} = P\{h(X, Y - a) \leqq \mu\}, \tag{9.4}$$

and these imply (9.1).

The proof of (9.2) is exactly analogous.

LEMMA 5. *Let*

$$P_0\{h(X, Y) = \mu\} = \delta, \qquad P_0\{h(Z) = \mu\} = \epsilon. \tag{9.5}$$

Then

$$\tfrac{1}{2} - \delta/2 \leq P_\Delta\{\hat{\Delta} \leqq \Delta\} \leqq \tfrac{1}{2} + \delta/2 \tag{9.6}$$

and

$$\tfrac{1}{2} - \epsilon/2 \leqq P_\theta\{\hat{\theta} \leqq \theta\} \leqq \tfrac{1}{2} + \epsilon/2. \tag{9.7}$$

PROOF. The inequalities (9.6) follow directly from (9.3) and (9.4).

As an immediate consequence of Lemma 5, it is seen that $\hat{\Delta}$ and $\hat{\theta}$ are median unbiased if $P_0\{h(X, Y) = \mu\} = 0$ and $P_0\{h(Z) = \mu\} = 0$ respectively. Examples are the estimates based on the Wilcoxon statistics: in the two-sample problem if m and n are both odd, in the one-sample problem if N is odd, since in both these cases μ is an impossible value of h.

Even if $P_0(h = \mu)$ is not zero, it will typically be small, and it then follows from Lemma 5 that the probabilities of over- and under-estimation with the estimates $\hat{\Delta}$ and $\hat{\theta}$ will be close to $\frac{1}{2}$.

The fact that $\hat{\Delta}$ and $\hat{\theta}$ are either exactly or approximately median unbiased is related to their behaviour as upper or lower confidence bounds for Δ and θ. To discuss this relationship for the two-sample problem (the other case is completely analogous), consider the acceptance regions

$$A(\Delta_0) = \{(x, y): h(x, y - \Delta_0) < \mu\} \tag{9.8}$$

for testing $H(\Delta_0): \Delta = \Delta_0$ against $\Delta > \Delta_0$ on the basis of the test statistic h. The associated family of confidence sets is $\{\Delta: h(x, y - \Delta) < \mu\}$, so that Δ^{**} is a lower confidence bound for Δ at confidence level $P_0\{h(X, Y) < \mu\}$. Analogously Δ^* is an upper confidence bound for Δ at confidence level $P_0\{h(X, Y) < \mu\}$.

Suppose now that $\Delta^* = \Delta^{**}$, as is the case for example when h is the Wilcoxon statistic and $m \cdot n$ is odd. It then follows from (9.3) and (9.4) with $a = 0$ that $P_0\{h(X, Y) = \mu\} = 0$ and hence that $\hat{\Delta} = \Delta^* = \Delta^{**}$ is the lower and upper confidence bound for Δ with confidence coefficient $\frac{1}{2}$, based on the test statistic h.

10. Asymptotic normality. So far, we have discussed small-sample properties of the estimates $\hat{\Delta}$ and $\hat{\theta}$; in the remaining sections, we shall be concerned with their large-sample behaviour. For the two-sample problem, let $m(N)$ and $n(N)$ for $N = 1, 2, \cdots$, be a sequence of pairs of sample sizes tending to infinity in such a way that $m(N)/N \to \lambda$, say, and let Δ_N be a sequence of values of the parameter Δ. Also for the one-sample problem consider the sequence of sample sizes $N = 1, 2, \cdots$ and let θ_N be a sequence of values of θ. In both cases, we shall indicate the dependence of h and μ on N by writing h_N and μ_N.

THEOREM 4. *Let $a, c_1, c_2, \cdots$ be real constants, and let*

$$\Delta_N = -a/c_N \quad \text{or} \quad \theta_N = -a/c_N. \tag{10.1}$$

Let G be the continuous distribution function of a random variable with mean zero and unit variance, and suppose

$$\lim_{N\to\infty} P_N\{c_N(h_N - \mu_N) \leqq u\} = G((u + aB)/A) \tag{10.2}$$

where P_N indicates that the probability is computed for the parameter values Δ_N or θ_N and where h_N stands for $h_N(X_1, \cdots, X_{m(N)}; Y_1, \cdots, Y_{n(N)})$ or $h_N(Z_1, \cdots, Z_N)$. Then for any fixed Δ and θ

$$\lim_{N\to\infty} P_\Delta\{c_N(\hat{\Delta}_N - \Delta) \leqq a\} = G(aB/A) \tag{10.3}$$

or

$$\lim_{N\to\infty} P_\theta\{c_N(\hat{\theta}_N - \theta) \leqq a\} = G(aB/A). \tag{10.4}$$

PROOF. It is enough to give the proof for $\hat{\Delta}_N$, and we may in the proof let $\Delta = 0$. It follows from (9.1) that

$$\lim P_0\{c_N\hat{\Delta}_N \leqq a\} = \lim P_0\left\{h\left(X_1, \cdots, X_{m(N)}; Y_1 - \frac{a}{c_N}, \cdots, Y_{n(N)} - \frac{a}{c_N}\right) \leqq \mu_N\right\}$$

$$= \lim P_N\{h(X_1, \cdots, X_{m(N)}; Y_1, \cdots, X_{n(N)}) \leqq \mu_N\} = G\left(\frac{aB}{A}\right),$$

and this completes the proof.

Consider now the family of test statistics h given by (3.1). It then follows from the work of Chernoff and Savage (1958), see for example Theorem 7.1 of Puri (1962), that under suitable regularity conditions on Ψ

$$N^{\frac{1}{2}}[h_N(X, Y) - \mu_N] \tag{10.5}$$

satisfies the assumptions of Theorem 4, with G the standard normal distribution and with A and B given by

$$A^2 = \lambda(1 - \lambda)\left[\int_0^1 J^2(u)\, du - \left(\int_0^1 J(u)\, du\right)^2\right] \tag{10.6}$$

and

$$B = \lambda(1 - \lambda) \int \{dJ[F(x)]/dx\} dF(x) \tag{10.7}$$

where $J = \Psi^{-1}$.

Combining this result with Theorem 4, we have

THEOREM 5. *If h is given by* (3.1) *with Ψ satisfying the assumptions of Theorem* 7.1 *of Puri* (1963) *and if $m(N)/N \to \lambda$ as $N \to \infty$, then $N^{\frac{1}{2}}(\hat{\Delta}_N - \Delta)$ has a limiting normal distribution with mean zero and variance A^2/B^2 where A and B are given by* (10.6) *and* (10.7).

Consider now in particular the estimate (4.1) which is obtained from (3.1) by taking for Ψ the rectangular distribution on (0, 1). Since then $J(u) = u$ for $0 \leqq u \leqq 1$, we have $A^2 = \lambda(1 - \lambda)/12$ and $B = \lambda(1 - \lambda)\int f^2(x)\, dx$ where f is the density of F. The asymptotic variance of $N^{\frac{1}{2}}(\hat{\Delta} - \Delta)$ in this case is therefore

$$1\Big/\left[12\lambda(1 - \lambda)\left(\int f^2(x)dx\right)^2\right]. \tag{10.8}$$

11. Asymptotic efficiency. In basing the estimates $\hat{\Delta}$ and $\hat{\theta}$ on tests with desirable efficiency properties, it was the hope that these properties would be transferred to the estimates. That this hope is fulfilled under suitable regularity conditions follows from a result of Stuart (1954). The following theorem proves it under somewhat different assumptions.

THEOREM 6. *Let $\hat{\Delta}_N$ and $\hat{\Delta}_N'$ (or $\hat{\theta}_N$ and $\hat{\theta}_N'$) be estimates of Δ (or θ) based on sequences of test statistics h_N and h_N' satisfying the assumptions of Theorem* 4 *for the* *same limiting distribution G. Then the asymptotic relative efficiency of* $\hat{\Delta}'_N$ *relative to* $\hat{\Delta}_N$ *(or of* $\hat{\theta}'_N$ *relative to* $\hat{\theta}_N$*) in the sense of reciprocal ratio of asymptotic variances, is the same as the corresponding Pitman efficiency of the two sequences of tests based on* h'_N *and* h_N *provided the latter exists and* $c_N = c'_N = N^{\frac{1}{2}}$.

PROOF. Since G is assumed to have unit variance, it is seen from (10.3) that the asymptotic variance of $N^{\frac{1}{2}}\hat{\Delta}_N$ is A^2/B^2 and that of $N^{\frac{1}{2}}\hat{\Delta}'_N$ is A'^2/B'^2 so that the asymptotic relative efficiency of the second sequence of estimates with respect to the first is $\text{ARE}(\hat{\Delta}'_N ; \hat{\Delta}_N) = (A'^2/B'^2)/(A^2/B^2)$.

Consider on the other hand the power against the alternatives Δ_N given by (10.1) of the sequence of tests of $H:\Delta = 0$ against $\Delta < 0$ with rejection region $h_N < \mu_N$. By (10.2), this power tends to $G(aB/A)$. Similarly, the power of the corresponding sequence of tests based on h' and N' observations against the alternatives $\Delta'_{N'} = -a'/c'_{N'}$ tends to $G(a'B'/A')$. If we want the same limiting power against the same sequence of alternatives, we must have

$$a'/(N')^{\frac{1}{2}} = a/N^{\frac{1}{2}} \quad \text{and} \quad aB/A = a'B'/A'$$

and therefore $N'/N = (a'/a)^2 = (A'^2/B'^2)/(A^2/B^2)$. This completes the proof for $\hat{\Delta}_N$, and that for $\hat{\theta}_N$ is completely analogous.

It follows from this theorem (or more directly from (10.8)) that the asymptotic efficiency of the estimates (4.1) and (5.5) relative to the classical estimates $\bar{Y} - \bar{X}$ and $\bar{Z}$ is $12\sigma^2(\int f^2(x)\,dx)^2$, which in the case of normal F is $3/\pi \sim .955$. It is interesting to compare this value with the corresponding values for very small N. For $N = 1$ and 2, we have $\hat{\theta}_N = \bar{Z}$ so that the efficiency in these cases is 1. For $N = 3$, $\hat{\theta}_N$ is the median of the six quantities $Z_1, Z_2, Z_3, (Z_1 + Z_2)/2, (Z_1 + Z_3)/2, (Z_2 + Z_3)/2$. Let the ordered Z's be denoted by $Z^{(1)} < Z^{(2)} < Z^{(3)}$. Then

$$Z^{(1)} < (Z^{(1)} + Z^{(2)})/2 < Z^{(2)} < (Z^{(2)} + Z^{(3)})/2 < Z^{(3)}$$

and

$$Z^{(1)} < (Z^{(1)} + Z^{(2)})/2 < (Z^{(1)} + Z^{(3)})/2 < (Z^{(2)} + Z^{(3)})/2 < Z^{(3)}.$$

These inequalities show that $\hat{\theta}_3$ is the average of $Z^{(2)}$ and $(Z^{(1)} + Z^{(3)})/2$ so that

$$\hat{\theta}_3 = \tfrac{1}{4}(Z^{(1)} + 2Z^{(2)} + Z^{(3)}).$$

From a table of the covariances of normal order statistics, the efficiency of $\hat{\theta}_3$ is then seen to be .979.

REFERENCES

ANSCOMBE, F. J. (1960). Rejection of outliers. *Technometrics* **2** 123–147.

CHERNOFF, HERMAN and SAVAGE, I. RICHARD (1958). Asymptotic normality and efficiency of certain nonparametric tests. *Ann. Math. Statist.* **29** 972–994.

GOVINDARAJULU, Z. (1960). Central limit theorems and asymptotic efficiency for one-sample nonparametric procedures. Technical Report No. 11. Department of Statistics, Univ. of Minnesota.

LEHMANN, E. L. (1963). A class of selection procedures based on ranks. To appear in *Math. Ann.*

PURI, MADAN LAL (1962). Asymptotic efficiency of a class of c-sample tests. Submitted to *Ann. Math. Statist.*

STUART, ALAN (1954). Asymptotic relative efficiencies of distribution-free tests of randomness against normal alternatives. *J. Amer. Statist. Assoc.* **49** 147–157.

TUKEY, J. W. (1949). The simplest signed-rank tests. Mem. Report 17, Statistical Research Group, Princeton Univ.

TUKEY, J. W. (1960). A survey of sampling from contaminated distributions. *Contributions to Probability and Statistics*, Essays in honor of Harold Hotelling 448–485.

VAN DER VAART, H. R. (1950). Some remarks on the power function of Wilcoxon's test for the problem of two samples. I and II. *Indag. Math.* **12** 146–172.

The Annals of Mathematical Statistics
1969, Vol. 40, No. 5, 1523–1535

UNBIASED ESTIMATION IN CONVEX FAMILIES[1]

BY P. J. BICKEL AND E. L. LEHMANN

University of California, Berkeley

1. Introduction and summary. Suppose that we observe random variables $X_1, \cdots, X_n$ which are identically and independently distributed according to some distribution F where F ranges over a family $\mathfrak{F}$.

The following question was posed by J. Steffensen in this abstract context in [7].

(1) If a functional q is given on $\mathfrak{F}$, when does there exist a statistic $\delta(X_1, \cdots, X_n)$ such that,

$$E_F(\delta(X_1, \cdots, X_n)) = q(F), \tag{1.1}$$

for all $F \varepsilon \mathfrak{F}$? (As usual E_F denotes the expectation under the assumption that F is the common distribution of the X_i.) In other words, when does there exist an unbiased estimate of $q(F)$ based on n observations?

A question which naturally follows from question 1 was raised and considered by Halmos in [2].

(2) Let the *degree* of q be the smallest $n \leqq \infty$ such that (1.1) holds for all $F \varepsilon \mathfrak{F}$ and some δ. Characterize the degree of estimable functionals q for various families $\mathfrak{F}$. (By estimable we mean merely that (1.1) should hold for some finite n and δ.)

In his paper [2] Halmos dealt with question 2 as well as the more important issue of existence of best unbiased estimates in the context of families $\mathfrak{F}$ which are very large, and asked whether anything can be said about more interesting families $\mathfrak{F}$.

The next two sections of our paper deal exclusively with question 2 for families $\mathfrak{F}$ which are closed under finite mixtures (convex combinations). In Section 2, some examples lead to our main result, Theorem 2.1, which gives an elementary characterization of degree for functionals defined on families $\mathfrak{F}$ as above. The theorem is followed by further examples which illustrate its utility. (However, the required convexity property excludes the usual parametric families.) Theorem 2.2, an extension of a theorem of Halmos [2], characterizes the degree of a product of estimable functionals. It is applied to give the degree of the rth cumulant of a distribution when $\mathfrak{F}$ is the family of all distributions with moments of all orders.

The third section deals with the problem of degree when we observe samples from two or more populations with different distributions. Theorem 3.1 shows that the notion of degree can be successfully extended in this situation. The theorem is followed by several examples in which the degree of various functionals such as two-sample distance criteria is computed.

Received 16 July 1968; revised April 1969.

[1] The research of both authors was partially supported by Grant N00014-67-A 0114-0004 from the Office of Naval Research.

Finally, in Section 4 we address ourselves to question 1. Theorem 4.1 gives a simple characterization of estimable functionals when the family $\mathfrak{F}$ is $\mathfrak{F}(\mu)$ where,

$$\mathfrak{F}(\mu) = \{F:F \text{ is absolutely continuous with respect to } \mu\} \tag{1.2}$$

for a fixed σ finite measure μ. The characterization is suggested by Theorem 2.1. After some examples we state without proof a further theorem (4.2) along these lines and illustrate its applicability.

2. Degree.

We begin with some definitions. Throughout, $\mathfrak{F}$ denotes a family of distribution functions (usually on the real line).

DEFINITION 2.1. Suppose that q is a functional on $\mathfrak{F}$. We say q is *estimable with n observations* if there exists δ such that (1.1) is satisfied for all $F \varepsilon \mathfrak{F}$ and that q is *estimable* if it is estimable with n observations for some n. The *degree* of an estimable functional q is the smallest n for which q is estimable.

Do estimability and degree depend on the domain $\mathfrak{F}$ over which q is defined? Consider, for example, the class $\mathfrak{F}$ of all distributions with finite variance $\sigma^2(F)$ and let q be the standard deviation $\sigma(F)$. As we shall show shortly q is not estimable. It becomes estimable, however, if we restrict the domain to $\mathfrak{F}_0$ consisting of a single distribution $F_0 \varepsilon \mathfrak{F}$. Similarly $\sigma^2(F)$ itself will be seen to be of degree 2 in $\mathfrak{F}$; it is clearly of degree 0 in $\mathfrak{F}_0$ while it is of degree 1 in the class $\mathfrak{F}_1$ of all distributions $F \varepsilon \mathfrak{F}$ with $\int x\, dF(x) = a$ where a is any given constant.

DEFINITION 2.2. $\mathfrak{F}$ is *convex* if for every F, $G \varepsilon \mathfrak{F}$ and $0 \leqq \alpha \leqq 1$, $\alpha F + (1 - \alpha)G \varepsilon \mathfrak{F}$.

Given $\mathfrak{F}$ convex and a functional q on $\mathfrak{F}$ we define for fixed F and G in $\mathfrak{F}$ a function $Q_{F,G}$ on $[0, 1]$ by

$$Q_{F,G}(\alpha) = q[\alpha F + (1 - \alpha)G]. \tag{2.1}$$

If q is estimable then $Q_{F,G}$ is a polynomial for every F and G. For if (1.1) holds, then

$$\begin{aligned} q[\alpha F &+ (1 - \alpha)G] \\ &= \int_{-\infty}^{\infty} \cdots \int_{-\infty}^{\infty} \delta(x_1, \cdots, x_n)[\alpha\, dF(x_1) + (1 - \alpha)\, dG(x_1)] \\ &\qquad \cdots [\alpha\, dF(x_n) + (1 - \alpha)\, dG(x_n)]. \end{aligned} \tag{2.2}$$

It should now be clear that if $\mathfrak{F}$ is convex and the distributions $F \varepsilon \mathfrak{F}$ have finite variance $\sigma^2(F)$ then the standard deviation $\sigma(F)$ is not estimable for any n. The above remark and the further considerations of this paper do not require the distributions F to be univariate. Suppose, for example, that $\mathfrak{F}$ is the class of all bivariate distributions F, whose components have finite variances. Then the correlation coefficient of the components considered as a functional of F is not estimable. Similarly, if A and B denote fixed sets on the real line and (U, V) is a pair of random variables with joint distribution F and $\mathfrak{F}$ is the class of all F for which $P(V \varepsilon B) > 0$, then the conditional probability $P(U \varepsilon A \mid V \varepsilon B)$ is not estimable.

We shall return to the question of estimability in Section 4. In the present section we assume that q is estimable and try to determine its degree. In this connection (2.2) is again helpful. It shows that if there exists a pair F, G for which $Q_{F,G}$ is a polynomial of degree n, then the degree of q cannot be less than n. If then an estimate of q can be found, which is based on n observations, this establishes n as the degree of q.

EXAMPLE 2.1. As an example, consider the variance $\sigma^2(F)$ of a univariate distribution, with $\mathfrak{F}$ the class of all distributions with finite variance. Then Q is of degree 2 and hence q cannot be estimated with one observation. Since $\frac{1}{2}(X_2 - X_1)^2$ is an unbiased estimate based on two observations, this shows $\sigma^2(F)$ to be of degree 2.

EXAMPLE 2.2. For a second example let $q(F) = \int F\, dF$. The associated Q is,

$$(2.3)\quad Q_{F,G}(\alpha) = \alpha^2 \int (F - G)\, d(F - G) + \alpha\left[\int (F - G)\, dG + \int G\, d(F - G)\right] + \int G\, dG.$$

If $\mathfrak{F}$ is the class of all univariate continuous distributions $q(F)$ is identically equal to $\frac{1}{2}$ and hence of degree 0. On the other hand, if $\mathfrak{F}$ is the class of all distributions, Q is easily seen to be of degree 2 for suitable F, G. Since $\delta(X_1, X_2) = 1$ or 0 as $X_1 \leqq X_2$ or $X_1 > X_2$ is an estimate of $Q(F)$ based on two observations this shows that q is of degree 2. One can similarly show that if $\mathfrak{F}$ is the family of all continuous *bivariate* distributions then the degree of q is again 2.

The above examples suggest that, roughly speaking, a functional q is of degree n provided Q is a polynomial of degree n. This is made precise by the following theorem.

THEOREM 2.1. *Suppose $\mathfrak{F}$ is convex and q is an estimable functional on $\mathfrak{F}$. Then q is of degree n if and only if*

(i) *for every F and $G \varepsilon \mathfrak{F}$, $Q_{F,G}$ is a polynomial in α of degree $\leqq n$;*

and

(ii) *for every $F \varepsilon \mathfrak{F}$ there exists some G for which $Q_{G,F}(\alpha)$ is a polynomial of degree n in α.*

The essence of the proof is contained in the following lemma.

LEMMA 2.1. (i) *If q is of degree $\leqq n$, then $Q_{F,G}$ is a polynomial of degree $\leqq n$ for all F and G.*

(ii) *If there exists an F such that $Q_{F,G}$ is a polynomial of degree $<n$ for all G and if q is estimable, then q is of degree $<n$.*

PROOF. (i) By (2.2) if q is estimable with n observations, and $\delta(x_1, \cdots, x_n)$ is an estimate of q, then,

$$(2.4)\qquad Q_{F,G}(\alpha) = \textstyle\sum_{r=0}^{n} \alpha^r (1 - \alpha)^{n-r} C_r(F, G),$$

is a polynomial of degree $\leqq n$ and (i) is immediate.

(ii) We identify the coefficients C_r by,

$$(2.5)\quad C_r(F, G) = \textstyle\sum_{r=0}^{n} \sum_{\{i_1,\cdots,i_r\}} \int_{-\infty}^{\infty} \cdots \int_{-\infty}^{\infty} \delta(x_1, \cdots, x_n) \cdot \prod_{j=1}^{r} dF(x_{i_r}) \prod_{i \varepsilon B_n(i_1,\cdots,i_r)} dG(x_i),$$

where $B_n(i_1, \cdots, i_r) = \{1, \cdots, n\} - \{i_1, \cdots, i_r\}$ and $\{i_1, \cdots, i_r\}$ ranges over all distinct subsets of size r of $\{1, \cdots, n\}$.

The hypothesis of (ii) implies that for a fixed F

$$\sum_{r=0}^{n} (-1)^{n-r} C_r(F, G) = 0. \tag{2.6}$$

Now let us define,

$$\delta_{(i_1,\cdots,i_r)}(x_{j_1}, \cdots, x_{j_s}) = \int_{-\infty}^{\infty} \cdots \int_{-\infty}^{\infty} \delta(x_1, \cdots, x_n)\, dF(x_{i_1}) \cdots dF(x_{i_r}), \tag{2.7}$$

where $\{j_1, \cdots, j_s\} = B_n\{i_1, \cdots, i_r\}$. $\delta_{(i_1,\cdots,i_r)}$ is defined a.s. (F) by (2.7) and may be set equal to 0 when the right hand side of (2.7) is undefined. In terms of the $\delta_{(i_1,\cdots,i_r)}$, (2.6) may be rewritten as

$$\begin{aligned} &\int_{-\infty}^{\infty} \cdots \int_{-\infty}^{\infty} \delta(x_1, \cdots, x_n)\, dG(x_1) \cdots dG(x_n) \\ &= \int_{-\infty}^{\infty} \cdots \int_{-\infty}^{\infty} \Big[\sum_{r=1}^{n} (-1)^{n-r+1} \sum_{(i_1,\cdots,i_r)} \delta_{(i_1,\cdots,i_r)}(x_{j_1}, \cdots, x_{j_s})\Big] \\ &\quad \cdot dG(x_1) \cdots dG(x_n) \end{aligned} \tag{2.8}$$

which in turn may be read,

$$E_G(\delta(X_1, \cdots, X_n)) = E_G\Big[\sum_{r=1}^{n} (-1)^{n-r+1} \delta_{(i_1,\cdots,i_r)}(X_1, \cdots, X_s)\Big] \tag{2.9}$$

for all $G \in \mathfrak{F}$. Since $r \geqq 1$, s is less than n and we see that

$$\sum_{r=1}^{n} (-1)^{n-r+1} \delta_{(i_1,\cdots,i_r)}(X_1, \cdots, X_s)$$

is an estimate of q based on at most $n - 1$ observations. Assertion (ii) of the lemma follows.

Proof of Theorem 2.1. (a) Let q be of degree n. Then for every F there exists G such that Q is of degree n. If there did not, it would follow from part (i) of the lemma that there exists an F for which Q is of degree $<n$ for all G and hence by part (ii) of the lemma, q would be of degree $<n$.

(b) Suppose that conditions (i) and (ii) of the theorem hold and that q is estimable. Then it follows from part (i) of the lemma that q cannot be of degree $<n$ and from part (ii) that q cannot be of degree $>n$.

We have considered several examples in which condition (i) of Theorem 2.1 proved useful. Here are some further illustrations in which the full force of the theorem is used.

Example 2.3. Let $\mathfrak{F}$ be the family of all distributions F which are absolutely continuous with continuous densities f and suppose that we wish to estimate $q(F) = f(a)$ for some fixed known a. In this situation Q is evidently of degree $\leqq 1$ for all F, G. Yet q is not estimable. To see this, note that in view of Theorem 2.1, if q is estimable it must be estimable with 1 observation. Therefore, it suffices to exhibit a subfamily $\mathfrak{F}_0 \subset \mathfrak{F}$ for which q is not estimable with 1 observation. Let $\mathfrak{F}_0 = \{f: f(t) = \frac{1}{2}\lambda \exp(-\lambda|t|)$ for some $\lambda > 0$ and all $t\}$. If q were estimable with 1 observation in $\mathfrak{F}_0$ we would have,

$$\int_{-\infty}^{\infty} \delta(t)[\exp\{-\lambda(|t| - |a|)\}]\, dt = 1 \tag{2.10}$$

for some estimate δ and all $\lambda > 0$. Differentiating both sides we get,

$$\int_{-\infty}^{\infty} \delta(t)[|t| - |a|][\exp\{-\lambda(|t| - |a|)\}\, dt] = 0. \tag{2.11}$$

The uniqueness of the Laplace transform implies,

$$\delta(t)[|t| - |a|] = 0 \text{ a.e.} \tag{2.12}$$

and this is inconsistent with (2.10). (This result was proved in a different manner by Rosenblatt (1956) in [6]. The present proof is no simpler but has the slight advantage of not being restricted to nonnegative estimates.)

Example 2.4. Consider the following generalization of Tukey's [9] gross error model.

Let $\mathfrak{F} = \{F : F = (1 - \epsilon)\Phi + \epsilon H, H \varepsilon \mathfrak{H}, 0 \leqq \epsilon \leqq 1\}$ where Φ is the standard normal distribution, $\mathfrak{H}$ is a convex, complete (in the sense of [4]) family of distributions equivalent to Φ. Suppose, moreover, that $\mathfrak{F}$ is identifiable in the sense that,

$$(1 - \epsilon_1)\Phi + \epsilon_1 H_1 = (1 - \epsilon_2)\Phi + \epsilon_2 H_2 \tag{2.13}$$

implies $\epsilon_1 = \epsilon_2$ and $H_1 = H_2$. The identifiability and completeness conditions are satisfied by taking for $\mathfrak{H}$ all mixtures of rectangular distributions, all distributions with support bounded from below etc.

The functional to be estimated is $q((1 - \epsilon)\phi + \epsilon H) = \epsilon$. Clearly, $\mathfrak{F}$ is convex and Q is linear. Again, q is not estimable. For were it estimable by Theorem 2.1 it would be of degree 1. But,

$$(1 - \epsilon) \int_{-\infty}^{\infty} \delta(x)\, d\Phi(x) + \epsilon \int_{-\infty}^{\infty} \delta(x)\, dH(x) = \epsilon \tag{2.14}$$

implies

$$\int_{-\infty}^{\infty} \delta(x)\, d\Phi(x) = 0 \tag{2.15}$$

and

$$\int_{-\infty}^{\infty} \delta(x)\, dH(x) = 1. \tag{2.16}$$

It follows from the completeness of $\mathfrak{H}$ that these equations cannot be satisfied if $\mathfrak{H}$ is equivalent to Φ.

Example 2.5. For our final example let $\mathfrak{F}_0$ be the location parameter family of distributions generated by the standard Cauchy distribution F_0 whose density is given by,

$$f_0(t) = \pi^{-1}(1 + t^2)^{-1}. \tag{2.17}$$

Let $\mathfrak{F} = \mathfrak{C}(\mathfrak{F}_0)$ be the smallest convex family of distributions containing $\mathfrak{F}_0$. Suppose that for some $F \varepsilon \mathfrak{F}$

$$F(t) = \sum_{i=1}^{r} \alpha_i F_0(t - \theta_i) = \sum_{j=1}^{s} \beta_j F_0(t - \eta_j) \tag{2.18}$$

for all t, where the α's and β's are $\geqq 0$ and $\sum_{i=1}^{r} \alpha_i = \sum_{j=1}^{s} \beta_j = 1$. Then it follows easily from a consideration of the associated characteristic functions that

$r = s$ and the set of α's must coincide with the set of β's as must the set of θ's with the set of η's. In view of this we can define the "center of symmetry" $q(F)$ of a distribution F given by (2.18) by,

$$q(F) = \sum_{i=1}^{r} \alpha_i \theta_i . \tag{2.19}$$

The Q function corresponding to q is linear, but again q is not estimable since q is of degree >1 on $\mathfrak{F}_0$. This was shown in [1] by Ghosh and Singh using some results by Pollard.

Similar examples in which Q is of degree >1 for each pair F, G and q is not estimable, can easily be constructed.

Before proving our next result, we need an elementary lemma.

LEMMA 2.2. *Let $\mathfrak{F}$ be convex and q_1, q_2 defined on $\mathfrak{F}$ be estimable with n observations. Then either there exists $F \varepsilon \mathfrak{F}$ such that $q_1(F)q_2(F) \neq 0$ or one of q_1, q_2 is identically* 0.

PROOF. Either the second alternative of the lemma holds or we may assume there exist F^*, G^* such that $q_1(F^*) \neq 0$, $q_2(G^*) \neq 0$. Then,

$$q_1(\alpha F^* + (1 - \alpha)G^*)q_2(\alpha F^* + (1 - \alpha)G^*)$$

is a product of two polynomials of degree at most n neither of which vanishes identically. Hence there exists α_0 such that

$$q_1(\alpha_0 F^* + (1 - \alpha_0)G^*)q_2(\alpha_0 F^* + (1 - \alpha_0)G^*) \neq 0$$

and the lemma is proved.

We now establish a generalization of Theorem 2 of [2].

THEOREM 2.2. *Suppose $q_1, \cdots, q_r$ are estimable functions of degree $k_1, \cdots, k_r$ respectively, defined on a convex domain $\mathfrak{F}$. Then, $q = q_1 \cdots q_r$ is estimable of degree $k_1 + \cdots + k_r$.*

PROOF. It suffices to check the result for $r = 2$ since the general case then follows by induction.

Clearly, q has degree at most $k_1 + k_2$. To see this, note that if $\delta^{(1)}(X_1, \cdots, X_{k_1})$ and $\delta^{(2)}(X_1, \cdots, X_{k_2})$ are unbiased estmates of q_1 and q_2 based on k_1 and k_2 observations respectively, then $\delta(X_1, \cdots, X_{k_1+k_2}) = \delta^{(1)}(X_1, \cdots, X_{k_1}) \cdot \delta^{(2)}(X_{k_1+1}, \cdots, X_{k_1+k_2})$ is an unbiased estimate q based on $k_1 + k_2$ observations.

To show that the degree of q cannot be less than $k_1 + k_2$ let $K(F, G)$, $K_1(F, G)$, $K_2(F, G)$ denote the coefficients of $\alpha^{k_1+k_2}$, α^{k_1} and α^{k_2} in the $Q_{(F,G)}$ polynomials corresponding to q, q_1 and q_2 respectively. From (2.5) we see that for fixed F, $K_1(F, G)$ and $K_2(F, G)$ treated as functions of G are estimable with k_1 and k_2 observations respectively. Since q_1 and q_2 are of degree k_1 and k_2 respectively Theorem 2.1 implies that whatever be F, neither $K_1(F, G)$ nor $K_2(F, G)$ vanish identically. Hence, by Lemma 2.2 neither does $K(F, G) = K_1(F, G)K_2(F, G)$. Therefore, whatever be F, $q(\alpha F + (1 - \alpha)G)$ is of degree $(k_1 + k_2)$ for some G and the theorem follows from Theorem 2.1.

Along the same lines it is easy to see that if $\mathfrak{F}$ is convex, q_1 is of degree m on $\mathfrak{F}$ and q_2 is of degree n of $\mathfrak{F}$ with $m < n$ then $q_1 + q_2$ is of degree n on $\mathfrak{F}$.

EXAMPLE 2.6. As an application of this remark and Theorem 2.1, consider $\mathfrak{F} = \{F: \int_{-\infty}^{\infty} |x|^k \, dF(x) < \infty \text{ for all } k\}$. The rth *cumulant* of F, $K_r(F)$ is defined as the coefficient of $(it)^r/r!$ in the formal expansion of $\log \int_{-\infty}^{\infty} e^{itx} \, dF(x)$. Clearly,

$$K_r(F) = \sum_{\{(j_1, \cdots, j_r): \Sigma i j_i = r\}} c(j_1, \cdots, j_r) \alpha_1^{j_1}(F) \cdots \alpha_r^{j_r}(F) \tag{2.20}$$

where $\alpha_k(F) = \int_{-\infty}^{\infty} x^k \, dF(x)$. Since $c(r, 0, \cdots, 0) \neq 0$ and since $\alpha_k(F)$ is of degree 1, Theorem 2.2 and our remark show that $K_r(F)$ is of degree r.

3. Degree: Two-sample problem. Let us now extend the definition and characterization of degree to functionals of two distributions. The further extension to an arbitrary number of distributions is trivial. Suppose that we observe $X_1, \cdots, X_m$ independently and identically distributed according to $F_1 \in \mathfrak{F}_1$ and $Y_1, \cdots, Y_n$ independently and identically distributed according to $F_2 \in \mathfrak{F}_2$ where $\mathfrak{F}_1$ and $\mathfrak{F}_2$ are convex.

DEFINITION 3.1. A real functional q on $\mathfrak{F}_1 \times \mathfrak{F}_2$ is said to be estimable with (m, n) observations if there exists a statistic $\delta(X_1, \cdots, X_m; Y_1, \cdots, Y_n)$ such that

$$E_{(F_1, F_2)}[\delta(X_1, \cdots, X_m; Y_1, \cdots, Y_n)] = q(F_1, F_2) \tag{3.1}$$

for all $F_1 \in \mathfrak{F}_1$, $F_2 \in \mathfrak{F}_2$.

In generalization of (2.1) it is useful to introduce a function Q through

$$Q(\alpha_1, \alpha_2) = q[\alpha_1 F_1 + (1 - \alpha_1) G_1, \alpha_2 F_2 + (1 - \alpha_2) G_2] \tag{3.2}$$

where $0 \leqq \alpha_1, \alpha_2 \leqq 1$ and where we suppress the dependence of Q on F_i, G_i, $i = 1, 2$.

If q is estimable, it is seen as in Section 2 that $Q(\alpha_1, \alpha_2)$ is a polynomial for every F_1 and F_2. This remark is again enough to establish the non-estimability of many functions such as, for example, $q(F_1, F_2) = |F_1(x) - F_2(x)|$ for fixed x, or $q(F_1, F_2) = \int |F_1(x) - F_2(x)| \, d(F_1(x) + F_2(x))/2$.

Let us now consider how to define the degree of an estimable functional q. We might for example determine the smallest value n_0 of n for which there exists an m such that q is estimable with (m, n_0) observations and then the smallest value m_0 of m for which q is estimable with (m, n_0) observations, and define the degree of q on (m_0, n_0). If we interchange the role of m and n in this definition, it is not clear a priori that we shall arrive at the same answer. To show that this is so, and at the same time arrive at a more symmetric definition of degree, consider the partial ordering among pairs of integers (m, n) according to which $(m, n) < (m', n')$ provided $m \leqq m'$ and $n \leqq n'$ with at least one of the two inequalities being strict. If q is an estimable function, denote by S_q the set of all pairs (m, n) for which q is estimable. Then we shall show below that the set S_q has a least element with respect to the ordering $\leqq$; that is, there exists a pair of integers (m_0, n_0) in S_q such that $m \geqq m_0$, $n \geqq n_0$ for all $(m, n) \in S_q$ and that in fact S_q consists exactly of all pairs (m, n) with $m \geqq m_0$ and $n \geqq n_0$. This pair (m_0, n_0) will be called the *degree* of the functional q.

The existence of a least element easily follows from the following properties of S_q:

(i) $(m, n) \varepsilon S_q \Rightarrow (m + a, n + b) \varepsilon S_q$ for all $a, b \geqq 0$;

(ii) $(m, n + 1) \varepsilon S_q$ and $(m + 1, n) \varepsilon S_q \Rightarrow (m, n) \varepsilon S_q$.

PROOF OF (ii). If q is estimable with (a, b) observations, then Q defined by (3.2) is a polynomial in α_1, α_2 of degree at most (a, b). (We say that the degree of a polynomial in α_1, α_2 is (a, b) if the maximum power of α_1 that is present is a and the maximum power of α_2 that is present is b.) Thus for q to be estimable both with $(m, n + 1)$ and $(m + 1, n)$ observations, Q must be a polynomial of degree at most (m, n) whatever be F_i, G_i. Writing Q out explicitly, taking the α_1^{m+1} terms into account for the moment, we have

$$(3.3)\quad Q(\alpha_1, \alpha_2) = \sum_{k_1=0}^{m+1} \sum_{k_2=0}^{n} C_{k_1,k_2}(F_1, F_2, G_1, G_2)\alpha_1^{k_1}(1 - \alpha_1)^{m+1-k_1}\alpha_2^{k_2}(1 - \alpha_2)^{n-k_2}$$

where the C_{k_1,k_2} are defined by formulae analogous to (2.5) in terms of the estimate $\delta(X_1, \cdots, X_{m+1}, Y_1, \cdots, Y_n)$ based on $(m + 1, n)$ observations. Since Q is of degree at most (m, n) the coefficient of α_1^{m+1} must vanish so that,

$$(3.4)\quad \sum_{k_2=0}^{n} \alpha_2^{k_2}(1 - \alpha_2)^{n-k_2}\left(\sum_{k_1=0}^{m} (-1)^{m-k_1+1}C_{k_1,k_2}(F_1, F_2, G_1, G_2)\right) \equiv 0.$$

By the completeness of the binomial family of distributions this implies that,

$$(3.5)\quad \sum_{k_1=0}^{m+1} (-1)^{m+1-k_1}C_{k_1,k_2}(F_1, F_2, G_1, G_2) = 0$$

for all k_2 and in particular for $k_2 = 0$. By using the resulting equation, the result follows as did Lemma 2.1 from equation (2.6).

As in the univariate case, to determine the degree of a functional q, it is often enough to note that if Q is a polynomial of degree (m, n), the degree of q cannot be less than (m, n). If an estimate of q can then be found based on (m, n) observations, this establishes (m, n) as the degree of q. As an illustration, let $q = \int F_1\, dF_2$ with $\mathfrak{F}_1 = \mathfrak{F}_2$, the class of all continuous distributions. Here

$$Q(\alpha_1, \alpha_2) = \alpha_1\alpha_2 \int (F_1 - G_1)\, d(F_2 - G_2) + \alpha_1 \int (F_1 - G_1)\, dG_2 + \alpha_2 \int G_1 d(F_2 - G_2) + \int G_1\, dG_2$$

is clearly of degree (1, 1). Since

$$\delta(X_1, Y_1) = 1 \text{ or } 0 \text{ as } X_1 \leqq Y_1 \text{ as } X_1 > Y_1$$

is an unbiased estimate of q, the degree of q is (1, 1).

As another example, consider $q = F_1(c) - F_2(c)$ where c is any constant. Then

$$Q(\alpha_1, \alpha_2) = \alpha_1[F_1(c) - G_1(c)] - \alpha_2[F_2(c) - G_2(c)] + G_1(c) - G_2(c)$$

is of degree (1, 1). Also $I_{[X_1 \leqq c]} = I_{[Y_1 \leqq c]}$ where $I_{[E]}$ denotes the indicator function of the event E, is an unbiased estimate of q, and hence shows q to be of degree (1, 1). In the present case, it is noteworthy that the $\alpha_1\alpha_2$-term is missing in Q;

this makes it possible to estimate q effectively with one observation rather than two. For let X_1 or Y_1 be observed with probability $\frac{1}{2}$ each, and let $\delta' = I_{[X_1 \leqq c]}$ when the observed random variable is X_1 and $\delta' = I_{[Y_1 > c]}$ when the observed variable is Y_1. Then if $\delta = 2\delta' - 1$, we find

$$E(\delta') = \tfrac{1}{2}[F_1(c) + 1 - F_2(c)] \quad \text{and hence} \quad E(\delta) = q.$$

Of course, the variance of δ is larger, (in fact exactly twice as large), than that of the earlier estimate. (Such a reduction of the effective number of observations required to obtain an unbiased estimate of q is possible whenever q is the sum of a number of estimable functions q_i each of which has a total degree $m_i + n_i$ less than the total degree $m + n$ of q.)

As another example, suppose that X_1, X_2 and Y_1, Y_2 are independently distributed according to continuous distributions F_1 and F_2 and let q denote the probability that the two X's are either both smaller or both larger than the two Y's

$$\begin{aligned} q(F_1, F_2) &= P(X_1, X_2 < Y_1, Y_2 \quad \text{or} \quad Y_1, Y_2 < X_1, X_2) \\ &= \int (1 - F_2)^2\, dF_1^2 + \int (1 - F_1)^2\, dF_2 . \end{aligned}$$

Here it is easily seen that the coefficients of $\alpha_1^2\alpha_2$ and of $\alpha_1\alpha_2^2$ are in general different from zero while the coefficient of $\alpha_1^2\alpha_2^2$ is zero. The latter fact follows immediately from the well-known representation (Lehmann (1951))

$$q(F_1, F_2) = \tfrac{1}{3} + \tfrac{1}{2} \int (F_1 - F_2)^2\, d(F_1 + F_2).$$

It is clear in the present case that q can be estimated by observing with probability $\frac{1}{2}$ each, either two X's and one Y or two Y's and one X. This fact is related to an unbiased estimate of q proposed by Renyi (1953).

The above remarks extend in the obvious way to more than two samples, and the natural generalizations of Theorem 2.1 and 2.2 hold trivially also in the case of two or more samples.

4. Existence of unbiased estimates. In Theorem 2.1 we have shown that if q is estimable on a convex $\mathfrak{F}$ the Q functions must be polynomials. On the other hand, Examples 2.3–2.5 show that this condition is not sufficient for estimability. In this section we shall show that this requirement is essentially sufficient if $\mathfrak{F}$ is large enough. Our main result deals with the case $\mathfrak{F} = \mathfrak{F}(\mu)$ defined in the introduction.

THEOREM 4.1. *If q is a real functional on $\mathfrak{F}(\mu)$ then q is estimable with n observations if and only if,*

(i) *q is uniformly bounded on $\mathfrak{F}(\mu)$;*

(ii) *$Q_{(F,G)}$ is a polynomial of degree at most n in α for each F, $G \in \mathfrak{F}(\mu)$.*

We note that since $\mathfrak{F}(\mu)$ is convex it follows from Theorem 2.1 that q is of degree n if and only if (i) and (ii) hold and $Q_{F,G}(\alpha)$ is of degree n for every F and some G.

We need some preliminary definitions and results. Let $L_1(\mu) = \{f : \int |f|\, d\mu < \infty\}$

and $L_1^+(\mu) = \{f \varepsilon L_1(\mu): f \geqq 0\}$. Evidently $\mathfrak{F}(\mu) = \{f \varepsilon L_1^+(\mu): \int f\, d\mu = 1\}$ in the sense of the natural 1–1 correspondence between a distribution and its density.

DEFINITION 4.1. A functional B on $L_1^+(\mu)$ to R is said to be *linear* if $B(0) = 0$ and $B(a_1 f + a_2 g) = a_1 B(f) + a_2 B(g)$ for $f, g \varepsilon L_1^+(\mu)$, $a_1, a_2 \geqq 0$.

Similarly, a functional B on $L_1^+(\mu) \times \cdots \times L_1^+(\mu)$ is said to be *n-linear* if B is linear in each argument when the others are held fixed.

DEFINITION 4.2. An n linear functional on $L_1^+(\mu) \times \cdots \times L_1^+(\mu)$ is said to be *bounded* if,

$$(4.1)\quad \sup\{|B(f_1, \cdots, f_n)| : \|f_1\| = \cdots = \|f_n\| = 1, f_i \varepsilon L^+(\mu), i = 1, \cdots, n\} < \infty.$$

DEFINITION 4.3. A functional q on $L_1^+(\mu)$ is said to be a *polynomial* if for every, $f, g \varepsilon L_1^+(\mu)$, $q(f + ag)$ is a polynomial in a for $a \geqq 0$.

DEFINITION 4.4. A functional q on $L_1^+(\mu)$ is said to be *homogeneous of degree n* if $q(cf) = c^n q(f)$ for every $f \varepsilon L_1^+(\mu)$, $c \geqq 0$.

These definitions are all slight modifications of the standard definitions given for these terms when $L_1(\mu)$ rather than just $L_1^+(\mu)$ is the domain of interest (cf. Hille and Phillips [3], pages 40, 760). The following results established in [3], pages 760–764 for homogeneous polynomials on $L_1(\mu)$ and n linear functionals on $L_1(\mu) \times \cdots \times L_1(\mu)$ can easily be seen to continue to hold for the objects we have defined.

LEMMA 4.1. *If q is a polynomial, homogeneous of degree n on $L_1^+(\mu)$, there exists an n linear functional B on $L_1^+(\mu) \times \cdots \times L_1^+(\mu)$ such that $q(f) = B(f, f, \cdots, f)$.*

This is Theorem 26.2.2 of [3] specialized and modified.

LEMMA 4.2. *Suppose B is an n linear functional on $L_1^+(\mu) \times \cdots \times L_1^+(\mu)$ such that,*

$$(4.2)\qquad \sup\{B(f_1, \cdots, f_n) : \|f_i\| = 1\} < \infty,$$

whenever all co-ordinates except the ith *are fixed, for $i = 1, \cdots, n$. Then, B is bounded.*

REMARK. The proof requires an appropriate modification of the principle of uniform boundedness which is easily seen to be valid. This is essentially Theorem 26.2.5 of [3].

Finally, we need a deeper result whose statement and proof for the situation when the domain of interest is $L_1(\mu)$ may be found, for example, in Trèves [8] (Corollary 2, page 473).

LEMMA 4.3. *If B is a bounded n linear functional on $L_1^+(\mu) \times \cdots \times L_1^+(\mu)$ then there exists a measurable function δ on R^n such that,*

$$(4.3)\quad B(f_1, \cdots, f_n) = \int_{-\infty}^{\infty} \cdots \int_{-\infty}^{\infty} \delta(x_1, \cdots, x_n) f_1(x_1) \cdots f_n(x_n) \mu(dx_1) \cdots \mu(dx_n).$$

Furthermore,

$$(4.4)\qquad (\mu)\text{ess sup } |\delta| < \infty$$

where

(4.5) $(\mu)\text{ess sup } |\delta|$

$$= \inf_B \sup \{|\delta(x_1, \cdots, x_n)|: (x_1, \cdots, x_n) \in B, \mu^{(n)}(B^c) = 0\}$$

and $\mu^{(n)}$ is the product measure on R^n each of whose co-ordinate measures is μ and B^c is the complement of B.

Finally we define,

DEFINITION 4.5. The *n extension of q*, written $q^{(n)}$ is a functional on $L_1^+(\mu)$ given by,

$$q^{(n)}(f) = \|f\|^n q(f/\|f\|). \tag{4.6}$$

It is obvious that the n extension of any functional q on $\mathfrak{F}(\mu)$ is homogeneous of degree n.

We now proceed to the proof of Theorem 4.1.

PROOF.

Sufficiency. Construct $q^{(n)}$.

$$q^{(n)}(f + ag) = \|f + ag\|^n q(f + ag/\|f + ag\|). \tag{4.7}$$

Since f, g and a are $\geqq 0$, $\|f + ag\| = \|f\| + a\,\|g\|$.

Now if (ii) holds,

$$q\left(\frac{f + ag}{\|f + ag\|}\right) = q\left(\frac{f}{\|f\|}\left(\frac{\|f\|}{\|f + ag\|}\right) + \frac{g}{\|g\|}\left(\frac{a\|g\|}{\|f + ag\|}\right)\right) \tag{4.8}$$

is a polynomial of degree at most n in $a\,\|g\|/\|f + ag\|$. Expanding both factors in (4.7) suitably we conclude that $q^{(n)}$ is a polynomial. Since $q^{(n)}$ is homogeneous of degree n Lemma 4.1 yields an n linear functional B on $L_1^+(\mu) \times \cdots \times L_1^+(\mu)$ such that $q^{(n)}(f) = B(f, \cdots, f)$. From (i) we see that

$$\sup \{|B(f, \cdots, f)|: \|f\| = 1\} < \infty$$

and hence, by linearity, that B is bounded. Lemma 4.3 completes the proof of the sufficiency of our conditions.

Necessity. The necessity of (ii) is obvious. Given an estimate δ of q define an n linear functional B on $L_1^+(\mu) \times \cdots \times L_1^+(\mu)$ by,

(4.9) $B(f_1, \cdots, f_n)$

$$= \int_{-\infty}^{\infty} \cdots \int_{-\infty}^{\infty} \delta(x_1, \cdots, x_n) f_1(x_1) \cdots f_n(x_n)\mu(dx_1) \cdots \mu(dx_n).$$

B is well defined at least for $f_1 = \cdots = f_n$. Showing that B is finite on all $L_1^+(\mu) \times \cdots \times L_1^+(\mu)$ merely involves breaking δ up into its positive and negative parts and then noting that $B(c_1f_1 + \cdots + c_nf_n, \cdots, c_1f_1 + \cdots + c_nf_n)$ is finite for all $c_1, \cdots, c_n \geqq 0$. By Lemma 4.3 we see that our necessity assertion is equivalent to showing that B is bounded. By Lemma 4.2 and the Fubini Theorem we see that we need only prove the result for $n = 1$. But, in this case, define $B_m(f) = \int_{-\infty}^{\infty} \delta_m(x) f(x)\mu(dx)$

where

$$\delta_m = \delta \quad \text{if } |\delta| \leqq m,$$
$$= 0 \quad \text{otherwise.}$$

Then, $B_m(f)$ is a bounded linear functional and $B_m(f) \to B(f)$ for each f. The Banach-Steinhaus Theorem yields that B is bounded and the theorem follows.

Extension of the theorem to the k–sample case is easy.

We turn to some examples to which the theorem or its k–sample extension may be applied.

EXAMPLE 4.1. Let F_0 be a fixed distribution. Take μ to be Lebesgue measure and let $q(F)$ be the "goodness of fit" statistic.

$$q(F) = \int_{-\infty}^{\infty} (F - F_0)^2 h(F_0)\, dF_0 \tag{4.10}$$

where h is positive and bounded.

The requirements of the theorem are satisfied for $n = 2$ and q is estimable on $\mathfrak{F}(\mu)$ with two observations. Of course here, as is typically the case in a "natural' situation, direct construction of the estimate of q is easy. Theorem 2.1 guarantees that the degree is indeed 2 since the coefficient of α^2 in $Q_{F,G}(\alpha)$ is

$$\int_{-\infty}^{\infty} (F - G)^2 h(F_0)\, dF_0 .$$

EXAMPLE 4.2. On the other hand, if μ is Lebesgue measure,

$$\begin{aligned} q(F) &= \int_{-\infty}^{\infty} (F - F_0)^2\, dF \\ &= \int_{-\infty}^{\infty} F^2\, dF - 2\int_{-\infty}^{\infty} FF_0\, dF + \int_{-\infty}^{\infty} F_0^2\, dF \end{aligned} \tag{4.11}$$

is also of degree 2 since $\int_{-\infty}^{\infty} F^2\, dF = \frac{1}{3}$ and the coefficient of α^2 in $Q_{F,G}(\alpha)$ is $\int_{-\infty}^{\infty} (F - G)^2\, dF_0$.

The same considerations can be applied to two–sample distance criteria such as,

$$q(F_1, F_2) = \int_{-\infty}^{\infty} (F_1 - F_2)^2\, d(F_1 + F_2), \tag{4.12}$$

a functional we have already encountered in Section 3. Theorem 4.1 (or rather its extension) shows that q is estimable with (3,3) observations. Closer examination of Q in this case shows that the degree is in fact (2,2), and as was noted in Section 3, if an extraneous randomization is permitted only one observation from one population and two from the other are needed.

Finally, we give a theorem extending Theorem 4.1 to "unbounded" functionals defined on subdomains of $\mathfrak{F}(\mu)$.

Let $\{A_m\}$ be a nested sequence of measurable subsets of R^n, with $A_1 \subset A_2 \subset \cdots$ such that $\mu(R^n - \bigcup_m A_m) = 0$.

Let $\mathfrak{F}_m(\mu)$ be the set of all members of $\mathfrak{F}(\mu)$ which vanish off A_m.

THEOREM 4.2. *Let q be a functional defined on a domain $\mathfrak{F}$ such that,*

$$\mathfrak{F} \supset \bigcup_m \mathfrak{F}_m(\mu) \tag{4.13}$$

and

$$q \geqq 0. \tag{4.14}$$

Then q is estimable with n observations if and only if,

(i) *q is bounded on each $\mathfrak{F}_n(\mu)$.*

(ii) *$Q_{F,G}(\alpha)$ is a polynomial of degree at most n for each F, $G \varepsilon \mathfrak{F}$.*

(iii) *For any $f \varepsilon \mathfrak{F}$,*

$$q(f\psi_m/\|f\psi_m\|) \to q(f) \tag{4.15}$$

where ψ_m is the indicator of A_m.

The proof is straightforward. That (4.14) is needed in conjunction with the other conditions may be shown by example.

Using this result we can, for instance, drop the boundedness restriction on $h(F_0)$ in Example 4.1 and need only require that the functional q defined by (4.10) be defined on the set of all densities f such that the right hand side of (4.10) is finite.

Acknowledgment. We are indebted to L. LeCam for the simple proof of necessity in Theorem 4.1 we have given, and to G. Bredon and C. Moore for an earlier (unused) proof of the same part.

REFERENCES

[1] Ghosh, J. K. and Singh, R. (1966). Unbiased estimation of location and scale parameters. *Ann. Math. Statist.* **37** 1671–1675.

[2] Halmos, P. R. (1946). The theory of unbiased estimation. *Ann. Math. Statist.* **17** 34–43.

[3] Hille, E. and Phillips, R. S. (1957). Functional analysis and semi-groups. *Amer. Math. Soc. Coll.Publ.* **31**.

[4] Lehmann, E. L. and Scheffé, Henry (1950). Completeness, similar regions and unbiased estimation. *Sankhyā* **10** 305–340.

[5] Renyi, A. (1953). Neue Kriterien zum Vergleich Zweier Stichproben. *Magyar Tud. Akad. Mat. Fiz. Oszt. Közl.* **2** 243–265.

[6] Rosenblatt, Murray (1956). Remark on some nonparametric estimates of a density function. *Ann. Math. Statist.* **27** 832–837.

[7] Steffensen, J. (1930). *Some Recent Researches in the Theory of Statistics and Actuarial Science.* Cambridge Univ. Press.

[8] Trèves, J. F. (1967). *Topological Vector Spaces, Distributions, and Kernels.* Academic Press, New York.

[9] Tukey, John W. (1960). A survey of sampling from contaminated distributions. *Contributions to Probability and Statistics.* Stanford Univ. Press.

An Interpretation of Completeness and Basu's Theorem

E. L. LEHMANN*

In order to obtain a statistical interpretation of completeness of a sufficient statistic T, an attempt is made to characterize this property as equivalent to the condition that all ancillary statistics are independent of T. For technical reasons, this characterization is not correct, but two correct versions are obtained in Section 3 by modifying either the definition of ancillarity or of completeness. The light this throws on the meaning of completeness is discussed in Section 4. Finally, Section 5 reviews some of the difficulties encountered in models that lack completeness.

KEY WORDS: Ancillary statistic; Basu's theorem; Completeness; Conditioning; Unbiased estimation.

1. INTRODUCTION

Let X be a random observable distributed according to a distribution from the family $\mathcal{P} = \{P_\theta, \theta \in \Omega\}$. Recall the following definitions.

A statistic $V = V(X)$ is *ancillary* if its distribution does not depend on θ.

A statistic $T = T(X)$ is *sufficient* if the conditional distribution of X given T does not depend on θ.

These two concepts are complementary. If the rest of the data is discarded, a sufficient statistic by itself retains all the information about θ contained in the data; an ancillary statistic by itself contains no information about θ.

Since sufficiency reduces the data without loss of information, a sufficient statistic is of greatest interest when it is minimal, that is, when it provides the greatest reduction of the data. Formally, a sufficient statistic is *minimal* if it is a function of every other sufficient statistic. Although exceptions are possible (Landers and Rogge 1972; Pitcher 1975), minimal sufficient statistics typically exist and are easy to construct (Bahadur 1954; Lehmann and Scheffé 1950).

A sufficient statistic T is *complete* if

$$E_\theta f(T) = 0 \quad \text{for all } \theta \Rightarrow f(t) = 0 \text{ (a.e.}\mathcal{P}), \tag{1.1}$$

and is *boundedly complete* if (1.1) holds for all bounded f. A necessary condition for T to be complete or boundedly complete is that it is minimal.

The properties of minimality and completeness are of a rather different nature. Under mild assumptions a minimal sufficient statistic exists and whether or not T is minimal then depends on the choice of T. On the other hand, existence of a complete sufficient statistic is equivalent to the completeness of the minimal sufficient statistic, and hence is a property of the model $\mathcal{P}$. We shall say that a model is complete if its minimal sufficient statistic is complete.

It has been pointed out in the literature that definition (1.1) is purely technical and does not suggest a direct statistical interpretation. Since various aspects of statistical inference become particularly simple for complete models, one would expect this concept to carry statistical meaning. It is the principal aim of this article to attempt a statistical characterization of models which are complete or boundedly complete. In the last section, we shall briefly survey some of the difficulties presented by incomplete models.

2. AN EXAMPLE AND BASU'S THEOREM

In order to get some insight into the problem, consider the following example.

Example 1. Let $X_1, \ldots, X_n$ be iid with density $f(x_i - \theta)$, and contrast what happens when f is normal, exponential, or uniform on the one hand, and in most other cases, for example, when f is logistic, Cauchy, or double exponential on the other. In the first type of situation the data can be reduced without loss of information to a small number of sufficient statistics: one for the normal ($\bar{X}$) and exponential ($X_{(1)}$), and two for the uniform ($X_{(1)}, X_{(n)}$), where $X_{(1)} < \cdots < X_{(n)}$ denote the order statistics. In the second type of situation, the set of n order statistics is minimal sufficient. (Intermediate cases are of course possible, as is seen, for example, by considering $f(x - \theta) = Ce^{-(x-\theta)^{2k}}$.)

As will be seen later, this difference is related to the following property of the ancillary information. In any location problem, the $n - 1$ differences $Y_i = X_{(n)} - X_{(i)}$ ($i = 1, \ldots, n - 1$) constitute an $(n - 1)$-dimensional ancillary statistic. In the normal and exponential cases, the Y's are independent of the minimal sufficient statistic

* E. L. Lehmann is Professor, Department of Statistics, University of California, Berkeley, CA 94720. Research was supported in part by National Science Foundation Grant MCS79-03716 and Office of Naval Research Contract N00014-75-C-0444. The author wishes to thank the referee for his very careful reading of two versions of the manuscript, and in particular for his discovery of an error in the original version of Theorem 3.

© Journal of the American Statistical Association
June 1981, Volume 76, Number 374
Theory and Methods Section

and hence carry no information about θ either by themselves or in conjunction with the minimal sufficient statistic. On the other hand, when the n order statistics are minimal, the differences Y_i—although still ancillary—are functions of the minimal sufficient statistic. Thus, sufficiency has not been successful in squeezing out the ancillary material. By themselves, the Y's carry no information but in conjunction with the rest of the data they do.

The uniform case is intermediate. Here the Y's are not independent of the minimal sufficient statistic, but the ratios $Z_i = Y_i/Y_1$ $(i = 2, \ldots, n-1)$ are. If we represent the ordered sample by

$$(\tfrac{1}{2}[X_{(1)} \times X_{(n)}],\quad Y_1 = X_{(n)} - X_{(1)}, Z_2, \ldots, Z_{n-1})$$

we see that the minimal sufficient statistic $T = (X_{(1)}, X_{(n)})$ has pushed out the $n - 2$ Z's, which carry no information by themselves or in conjunction with the rest of the data, but not Y_1.

Note that the minimal sufficient statistic is complete in the normal and exponential cases, but not in the other four cases. This suggests that completeness of a minimal sufficient statistic T may be associated with the ability of T completely to rid itself of the ancillary part of the data by making it independent of T. As a further illustration of this idea, consider a sample from a uniform distribution on (θ_1, θ_2) where now both θ's are unknown. The minimal sufficient statistic continues to be $T = (X_{(1)}, X_{(n)})$. However, Y_1 is no longer ancillary, although the Z's continue to be so. In the present case, T is more successful in discarding ancillary material than it was in the one-parameter case, and T is now complete.

These comments are closely related to a theorem of Basu (1955, 1958), which states that when a sufficient statistic T is boundedly complete, then all ancillary statistics are independent of T. This would provide the desired characterization of complete models as those in which the minimal sufficient statistic is completely successful in discarding all ancillary material, if the converse were also true, that is, if the independence of all ancillary statistics from a sufficient statistic implied that T were boundedly complete. Unfortunately this is not correct (see Sec. 3, Ex. 2) but it contains a germ of truth.

3. SOME ADAPTATIONS OF BASU'S THEOREM

The converse of Basu's theorem fails because ancillarity is concerned with the whole distribution of a statistic, while completeness is a property dealing only with expectation. We shall now show that correct versions of the converse can be obtained by replacing either ancillarity with the corresponding first-order property or completeness with a condition reflecting the whole distribution.

Let us call a real-valued statistic $V = V(X)$ *first-order ancillary* if $E_\theta(V)$ is independent of θ. (A first-order ancillary is simply an unbiased estimator of zero plus a constant.) Then we have the following first-order version of Basu's theorem.

Theorem 1. A necessary and sufficient condition for a sufficient statistic T to be boundedly complete is that every bounded first-order ancillary V is uncorrelated (for all θ) with every bounded real-valued function of T.

The proof is based on the following characterization of all bounded, unbiased estimators of zero.

Lemma 1. If a sufficient statistic T is boundedly complete, then a bounded statistic $S = S(X)$ is an unbiased estimator of zero if and only if (iff)

$$\eta(t) = E[S \mid t] = 0 \text{ (a.e. } \mathcal{P}^T). \tag{3.1}$$

(Here (a.e. $\mathcal{P}^T$) means: except on a set N of values of T, which has probability zero for all θ.)

Proof of Lemma 1. If S is an unbiased estimator of zero, then $E(S) = E[\eta(T)] = 0$ for all θ, and (3.1) follows from completeness. Conversely, if (3.1) holds, taking the expectation of both sides shows S to be an unbiased estimator of zero.

Proof of Theorem 1. (a) *Necessity.* Suppose that T is boundedly complete and that V is a bounded first-order ancillary. Without loss of generality suppose that $E(V) = 0$. Then for any bounded $f(t)$ we have

$$\operatorname{cov}_\theta(f(T), V) = E_\theta[f(T)V] = E_\theta\psi(T)$$

where $\psi(t) = f(t)E[V \mid t]$. By the lemma, $E[V \mid t]$ and therefore $\psi(t)$ is zero (a.e. $\mathcal{P}^T$) and hence $\operatorname{cov}_\theta(f(T), V) = 0$ for all θ.

(b) *Sufficiency.* If T is not boundedly complete, there exists a bounded function f such that $E_\theta f(T) = 0$ for all θ but $f(T) \neq 0$ with positive probability for some θ_0. Put $V(X) = f(T)$. Then $\operatorname{cov}_{\theta_0}(V, f(T)) = E_{\theta_0}[f^2(T)] > 0$, and this proves the condition to be sufficient.

The analogous result holds for completeness instead of bounded completeness if attention is restricted to statistics with finite variance.

Instead of modifying the definition of ancillarity to obtain a converse of Basu's theorem, one can instead modify the definition of completeness. Quite generally, call a sufficient statistic $\mathcal{F}$-complete if $E_\theta f(T) = 0$ for all θ and $f \in \mathcal{F}$ implies $f(t) = 0$ (a.e. $\mathcal{P}^T$) where $\mathcal{F}$ is some given class of real-valued functions. Completeness and bounded completeness correspond respectively to the case that F is the class of all integrable or all bounded functions. The case that it is the class of all square integrable functions is relevant for the analog of Theorem 1 mentioned following the proof of that theorem.

If F_0 is the class of all two-valued functions, we have the following result, which provides a partial converse of Basu's theorem.

Theorem 2. If

$$\text{Every ancillary statistic is independent of } T, \tag{3.2}$$

then T is F_0-complete.

Proof. Suppose T is not F_0-complete. Then there exists a two-valued function f such that $E_\theta f(T) = 0$ for all θ, but f is not zero with probability one. Let $f(t) = -a$ when $t \in A$ and $f(t) = b$ otherwise. Then $P_\theta(T \in A) = b/(a + b)$ is independent of θ so that the indicator of A, $I_A(T)$, is an ancillary. This provides a contradiction.

Basu's theorem, together with Theorem 2, shows that

$$\text{Bounded completeness} \Rightarrow (3.2) \Rightarrow F_0\text{-completeness.} \tag{3.3}$$

That neither of these implications holds in the other direction is shown by the following example.

Example 2. Let X be distributed as

-5	-4	-3	-2	-1	1	2	3	4	5
$\alpha' p^2 q$	$\alpha' pq^2$	$\frac{1}{2}p^3$	$\frac{1}{2}q^3$	$\gamma' pq$	γpq	$\frac{1}{2}q^3$	$\frac{1}{2}p^3$	αpq^2	$\alpha p^2 q$

where the first row lists the values X takes on and the second row their probabilities and where α, γ, α', γ' are positive constants satisfying

$$\alpha + \gamma = \alpha' + \gamma' = \tfrac{3}{2}.$$

It is easy to check the following facts.

I If $0 < p < 1$, $q = 1 - p$, the probabilities add up to 1.

II $T = |X|$ is minimal sufficient.

III $P(X > 0) = \frac{1}{2}$ so that the indicator V of the set $\{X > 0\}$ is an ancillary statistic.

IV If $\alpha \neq \alpha'$, V is not independent of T, so that (3.2) does not hold.

V To see that T is F_0-complete, note that $Ef(T) = 0$ iff

$$f(1)[\gamma + \gamma']pq + f(2)q^3 + f(3)p^3 + (\alpha + \alpha')[f(4)pq^2 + f(5)p^2q] = 0.$$

This holds for all $0 < p < 1$ iff

$$f(2) = f(3) = 0, \quad f(1) = a, \qquad f(4) = f(5) = -\frac{\gamma + \gamma'}{\alpha + \alpha'}a \tag{3.4}$$

for some a. Since no two-valued f can satisfy (3.4), T is F_0-complete.

It now follows from (IV) that F_0-completeness does not imply (3.2). Also, since (3.4) shows T not to be boundedly complete, F_0-completeness does not imply bounded completeness.

That (3.2) does not imply bounded completeness can be seen from any example without bounded completeness in which no ancillary statistic exists and (3.2) therefore holds vacuously. A less trivial example is obtained by puting $\alpha = \alpha'$, $\gamma = \gamma'$ in Example 2. The totality of ancillary statistics is then obtained by selecting any of the 5 sign combinations ± 1, ± 2, ± 3, ± 4, ± 5, for example $+1$, -2, -3, $+4$, -5 and setting

$$V(+1) = V(-2) = V(-3) = V(4) = V(-5) = 1 \qquad V(-1) = V(+2) = V(+3) = V(-4) = V(5) = 0. \tag{3.5}$$

For any such V, $P(V = 1) = P(V = 0) = \frac{1}{2}$ and V is independent of T, so that (3.2) holds. On the other hand, T has been seen not to be boundedly complete.

That F_0-completeness is a very weak (and somewhat unnatural) condition is indicated by the fact that a sufficient statistic can be F_0-complete with being minimal. As a simple example, suppose that X and Y are independent Poisson variables with $E(X) = E(Y) = \lambda$. Then $T = X + Y$ is minimal sufficient and complete, but (X, Y) is F_0-complete. This follows from the facts that the conditional distribution of X given $X + Y = t$ is the binomial distribution $b(\frac{1}{2}, t)$ and that for no constant c does there exist a nontrivial two-valued function $f(x, y)$ such that $E[f(X, Y) \mid t] = c$ for all t.

Example 2, together with (3.3), shows that bounded completeness is a stronger condition than (3.2) while F_0-completeness is not strong enough. It is interesting that there exists an intermediate completeness condition which is exactly equivalent to (3.2). Although the condition is rather intractable and not likely to be of much use, we give it here for the sake of completeness.

Let F_1 be the class of all functions $f(t)$ which are the conditional expectation of a two-valued function of X given t. If $X = T$, then F_1 reduces to F_0.

Theorem 3. Condition (3.2) is necessary and sufficient for T to be F_1-complete in the model $\mathcal{P}$.

Proof. (a) Suppose T is F_1-complete with respect to $\mathcal{P}$ and that V is an ancillary statistic. Then for any set A, the probability $P(V \in A) = p$ is independent of θ, and $g(V) = I_A(V) - p$ is both a two-valued function of X and an unbiased estimator of zero. If $f(t) = E[g(V) \mid t]$, it follows as in the proof of Theorem 1(a) that $E_\theta f(T) = 0$ for all θ. Since $f \in F_1$, this implies

$$f(t) = P[V \in A \mid t] - p = 0 \text{ a.e.}$$

and hence the independence of V and T.

(b) Conversely, suppose that T is not F_1-complete, so that there exists $f \in F_1$ which is not 0 a.e. but such that $E_\theta f(T) = 0$ for all θ. Let g be a nontrivial two-valued function of X for which $E[g(X) \mid t] = f(t)$, say $g(x) = a$ if $x \in A$ and $g(x) = b$ otherwise. Then $E_\theta g(X) = 0$, the indicator V of A is an ancillary statistic as in the proof of Theorem 2, but by the assumption about f, V is not independent of T.

4. INTERPRETATION OF COMPLETENESS

Ancillary statistics by themselves have no value for estimating θ and the same is likely to be true of first-order ancillaries (since their expectation is independent of θ), although statistics with either property may be valuable when used in conjunction with other statistics. The results of the preceding section can be summarized roughly by saying that the various forms of completeness of a sufficient statistics T characterize the success of T in separating the informative part of the data from that part which by itself carries no (or little) information.

Basu's theorem explains how a complete sufficient statistic achieves this separation. It does this by making the ancillary part of the data independent of (or uncorrelated with) T. For the technical reasons mentioned in Section 2, this interpretation of completeness is not as clear-cut as one might like. Roughly speaking, however, a sufficient statistic T is complete if it contains no ancillary information or equivalently if all ancillary information is independent of (or at least uncorrelated with) T.

The idea that it is the lack of correlation between first-order ancillaries and functions of T that makes complete sufficient statistics so effective as a noise-free distillation of the data finds some support in a theorem of Lehmann and Scheffé (1950) and Rao (1952). Suppose T is not complete, and consider the functions $f(T)$ which are uncorrelated with all first-order ancillaries (all functions being assumed to be square integrable). One might then expect these f's (though not sufficient) to be particularly effective estimators. The theorem in question states in part that these functions are the only unbiased estimators with uniformly minimum variance (UMVU).

Unfortunately, it turns out that in many common problems there are no functions f for which the above property holds. For example, when $X_1, \ldots, X_n$ are iid according to the uniform distribution on $(\theta - \frac{1}{2}, \theta + \frac{1}{2})$, it was shown by Lehmann and Scheffé (1950, 1956) that no function of θ has a UMVU estimator. One might expect this property to hold fairly generally in location problems with incomplete minimal sufficient statistic. (For some partial results in this direction see Bondessen 1975.) The corresponding result was proved by Unni (1978) for the case that $X_1, \ldots, X_n$ is a sample from the normal distribution $N(r\sigma, \sigma^2)$, r = known, and for the Behrens-Fisher problem (i.e., the case of samples from two normal distributions $N(\xi, \sigma^2)$, $N(\xi, \tau^2)$ with unknown variances and common unknown mean.) Examples in which the minimal sufficient statistic is not complete but in which some parametric functions have UMVU estimators can of course be constructed (see for instance Ex. 5.2 of Lehmann and Scheffé 1950), but even then the dimension of the space of such functions is typically quite low.

5. ANCILLARY STATISTICS AND THE PRINCIPLE OF CONDITIONALITY

When a model is not complete, the minimal sufficient statistic will contain ancillary or at least first-order ancillary statistics, and as a result its dimension will typically be larger than that of the parameter space. The question then arises as to how to use the ancillary information to effect a further reduction of the data in order to facilitate the inference problem. Little is known about ways of utilizing first-order ancillaries (the theorems by Lehmann and Scheffé and by Rao mentioned in Section 4 can be viewed as a contribution in this direction) but there is considerable literature on the use of ancillary statistics. The ideas were first put forward by R.A. Fisher (1934, 1935) and an excellent discussion of the issues can be found in Cox and Hinkley (1974). Many additional results are given by Barndorff-Nielsen (1978). As a brief introduction to the subject, we shall here discuss a few examples.

Example 3. (a) In the standard introductory example, which is implicit in the writing of Fisher, the natural ancillary statistic is the sample size. To be specific, let N be a random variable taking on values 1, 2, ... with known probabilities $\pi_1, \pi_2, \ldots$. Having observed $N = n$, we perform n binomial trials with success probability p and let S_n be the number of successes in these trials. For the observation $X = (N, S_N)$, X is minimal sufficient and N is ancillary.

The natural estimator S_N/N of p is unbiased and has variance $pq\, E(1/N)$. It is, however, not clear that this is the appropriate measure of accuracy for the following reason. The most striking feature of the example is that N, while not providing any information about p, tells us how much information about p is contained in S_N. If the observed value n of N is large, S_N is highly informative; for small n it carries little information. For this reason it may be better to cite as measure of accuracy the conditional variance pq/n as appropriate in those cases in which n has been obtained.

If this argument is accepted, not only the variance but the whole problem should be viewed conditionally. This includes in particular the choice of estimate which in the present case would remain unchanged but in others might not. From this point of view, conditioning on N has the advantage of reducing the random part of the data from (N, S_N) to the simple one-dimensional statistic S_n. Since S_n is complete, the conditional model is complete, there exist UMVU estimators for p and pq/n, and so on.

(b) Example (a) may seem artificial (although very similar situations arise in sampling from finite populations) but some of its principal features are in fact quite typical. Consider for instance once more the location problem of Example 1. Here the differences Y_i provide the ancillary information which is analogous to N, and typically give an indication of the dispersion of the conditional distribution of an estimator of θ such as the mean or median. This is particuarly clear in the case of the uniform distribution. The minimal sufficient statistic can then be represented as (Z, Y) with $Z = \frac{1}{2}[X_{(1)} + X_{(n)}]$ and $Y = X_{(n)} - X_{(1)}$. The conditional distribution of Z, given $Y = y$, is the uniform distribution on $(\theta - \frac{1}{2}(1 - y), \theta + \frac{1}{2}(1 - y))$. Thus Y again measures the amount of information Z contains about θ and it is natural to restrict attention to the conditional distribution of Z given Y. In this case even the conditional model is not complete. (Additional ancillaries exist but are rather unsatisfactory. See Basu 1964, Sec. 4.)

(c) As a last illustration consider the ancillary statistic $V = I\{x > 0\}$ of Ex. 2. The probabilities of the conditional distribution of X given $V = 1$ are

$$\begin{matrix} 1 & 2 & 3 & 4 & 5 \\ (3-2\alpha)pq, & q^3, & p^3, & 2\alpha pq^2, & 2\alpha pq^2. \end{matrix} \qquad (5.1)$$

Given $V = 0$, they are the corresponding values with α replaced by α'. It is easy to see that an experiment with outcome probabilities (5.1) is more informative than the corresponding experiment with α' when $\alpha > \alpha'$. (The reason is that the events with probabilities $2\alpha pq^2$ and $2\alpha pq^2$ can be split at random into four events with probabilities $2\alpha' pq^2$, $2\alpha' p^2 q$, $2(\alpha - \alpha')pq^2$, $2(\alpha - \alpha')p^2 q$. The latter two have a combined probability of $2(\alpha - \alpha')pq$, which can be combined with $(3 - 2\alpha)pq$ to yield an event with probability $(3 - 2\alpha')pq$. Thus, from the α-experiment and a table of random numbers one can generate the α'-experiment but not vice versa.)

The above considerations suggest the conditionality principle, first proposed by Fisher (1935, 1936, 1956) and investigated as a fundamental principle by Birnbaum (1962): in the presence of an ancillary statistic V, statistical inference should be carried out not in the overall model but in the conditional model given V.

Apart from possible reservations which one may have about the principle itself, difficulties in implementing it arise from the fact that ancillary statistics are not unique. In Example 3(b), for instance, $\sum (X_i - \bar{X})^2$ is ancillary, as is $\sum | X_i - \bar{X} |$ or any other function of the differences. To avoid this difficulty it is tempting, in analogy with the definition of a minimal sufficient statistic, to call an ancillary statistic V maximal if every other ancillary statistic is a function of V. One would then condition on this maximal V, thereby carrying the conditioning process as far as possible.

Unfortunately, a maximal V in this strong sense often does not exist. Basu (1964) has therefore introduced a weaker concept and has defined V as maximal if there exists no ancillary statistic V' of which V is a function. Such maximal ancillaries, however, frequently are not unique. In searching for a way out of the difficulty, it is helpful to recall the motivation for conditioning which was given in the comments after Example 3, and which was stated by Fisher (1935, p. 48) as follows: "Ancillary statistics are only useful when different samples of the same size can supply different amounts of information, and serve to distinguish those which supply more from those which supply less."

Fisher's statement leads to the idea that the better an ancillary statistic is, in distinguishing the more informative samples from those that are less so, the more useful it is as a conditioning variable. An implementation of this criterion for choosing an ancillary has been provided by Cox (1971), who compares different ancillaries V in terms of the variance of the Fisher information of the conditional distribution of X given V. An ancillary V is preferred to V' if it has the larger such variance for all parameter values. Such a principle will resolve the choice of ancillary in many cases although it may of course turn out that it leads to different choices for different parameter values.

If $X = (Y, V)$, where V is ancillary, conditioning on V can be thought of as data reduction obtained by replacing the random quantity V by a known constant v, so that part of the randomness of the data is eliminated. It frequently happens that the conditional model for Y given v is complete, in which case conditioning is particularly rewarding. It is interesting to note that in this latter case the ancillary V is essentially maximal in the sense of Basu (1959). This result (to which we hope to return in a later paper) presents a considerable strengthening of Theorem 7 of Basu (1959).

In summary it is seen that models involving ancillaries contained in the minimal sufficient statistic are more difficult to analyze than complete models. Even in so relatively simple a situation as that of Example 1, no general agreement has been reached whether to assess the accuracy of the best estimator of θ (the Pitman estimator) conditionally or unconditionally. The situation becomes even more complicated when nuisance parameters are present, a possibility we have not considered here.

Commenting on his efforts to develop a satisfactory small-sample theory of estimation, Fisher wrote in 1956 (pp. 157 and 158): "The most important step which has been taken so far to complete the structure of the theory of estimation is the recognition of Ancillary statistics." Despite some important recent work on conditional inference by Efron and Hinkley (1978) and Hinkley (1977), we still seem to be far from a full understanding of this difficult topic and a satisfactory small-sample theory of models that lack completeness.

[*Received November 1979. Revised December 1980.*]

REFERENCES

BAHADUR, R.R. (1954), "Sufficiency and Statistical Decision Functions," *Annals of Mathematical Statistics*, 25, 423–462.

BARNARD, G.A., and SPROTT, D.A. (1971), "A Note on Basu's Examples of Anomalous Ancillary Statistics" (with discussion), in *Foundations of Statistical Inference*, eds. Godambe and Sprott, 163–176. Toronto: Holt, Rinehart and Winston.

BARNDORFF-NIELSEN, O. (1978), *Information and Exponential Families in Statistical Theory*, New York: John Wiley.

BASU, D. (1955, 1958), "On Statistics Independent of a Complete Sufficient Statistic," *Sankhyā*, 15, 377–380 and 20, 223–226.

——— (1959), "The Family of Ancillary Statistics," *Sankhyā*, 21, 247–256.

——— (1964), "Recovery of Ancillary Information," *Sankhyā*, Ser. A, 26, 3–16.

BIRNBAUM, A. (1962), "On the Foundations of Statistical Inference" (with discussion), *Journal of the American Statistical Association*, 57, 269–306.

BONDESSEN, L. (1975), "Uniformly Minimum Variance Estimation in Location Parameter Families," *Annals of Statistics*, 3, 637–660.

COX, D.R. (1971), "The Choice Between Alternative Ancillary Statistics," *Journal of the Royal Statistical Society*, Ser. B, 33, 251–255.

COX, D.R., and HINKLEY, D.V. (1974), *Theoretical Statistics*, London: Chapman and Hall.

EFRON, B., and HINKLEY, D.V. (1978), "Assessing the Accuracy of the Maximum Likelihood Estimator: Observed Versus Expected Fisher Information" (with comments), *Biometrika*, 65, 457–487.

FISHER, R.A. (1934), "Two New Properties of Mathematical Likelihood," *Proceedings of the Royal Statistical Society*, Ser. A, 144, 285–307.

——— (1935), "The Logic of Inductive Inference," *Journal of the Royal Statistical Society*, 98, 39–54.

——— (1936), "Uncertain Inference," *Proceedings of the American Academy of Arts and Science*, 71, No. 4, 245–258.

——— (1956), *Statistical Methods and Scientific Inference*, New York: Hafner.

HINKLEY, D.V. (1977), "Conditional Inference About a Normal Mean With Known Coefficient of Variation," *Biometrika*, 64, 105–108.

LANDERS, D., and ROGGE, L. (1972), "Minimal Sufficient σ-Fields and Minimal Sufficient Statistics. Two Counterexamples." *Annals of Mathematical Statistics*, 43, 2045–2049.

LEHMANN, E.L., and SCHEFFÉ, H. (1950, 195, 1956), "Completeness, Similar Regions, and Unbiased Estimation," *Sankhyā*, 10, 305–340; 15, 219–236; and 17, 250.

PITCHER, T.S. (1975), "Sets of Measures Not Admitting Necessary and Sufficient Statistics or Subfields," *Annals of Mathematical Statistics*, 26, 267–268.

RAO, C.R. (1952), "On Statistics With Uniformily Minimum Variance," *Science and Culture*, 17, 483–484.

UNNI, K. (1978), "The Theory of Estimation in Algebraic and Analytic Exponential Families With Applications to Variance Component Models," Ph.D. thesis, Indian Statistical Institute, Calcutta.

The Annals of Statistics
1981, Vol. 9, No. 4, 1119–1122

A MINIMAX PROPERTY OF THE SAMPLE MEAN IN FINITE POPULATIONS[1]

By P. J. Bickel and E. L. Lehmann

University of California, Berkeley

Consider the problem of estimating the mean of a finite population on the basis of a simple random sample. It was proved by Aggarwal (1954) that the sample mean minimizes the maximum expected squared error divided by the population variance τ^2. Aggarwal also stated, but did not successfully prove, that the sample mean minimizes the maximum expected squared error over the populations satisfying $\tau^2 \le M$ for any fixed positive M. It is the purpose of this paper to give a proof of this second result, and to indicate some generalizations.

1. Introduction. Consider a population $\{a_1, \cdots, a_N\}$ of real numbers. In order to estimate the population average, $\bar{a}$, a simple random sample of size n is drawn, say $Y_1, \cdots, Y_n$. We are concerned with the problem of determining an estimator $\delta = \delta(Y_1, \cdots, Y_n\}$ which minimizes the maximum expected squared error

$$\sup E(\delta - \bar{a})^2. \tag{1}$$

This problem is meaningless unless limitations are placed on the a's under which this minimax risk is finite. Two natural restrictions are

(i) $A \le a_i \le B$ where A, B are given

(ii) $\sum(a_i - \bar{a})^2 \le M$ where M is given.

A third possibility is

(iii) to minimize instead of (1)

$$\sup E\{(\delta - \bar{a})^2/\sum(a_i - \bar{a})^2\}, \tag{2}$$

in which case no restriction on the a's is required.

A minimax estimator for (i) was recently obtained by Hodges and Lehmann (1981). For problems (ii) and (iii), Aggarwal (1959) stated that the sample mean $\bar{Y}$ has the desired minimax properties. If

$$a_i = \theta + d_i, \qquad \theta = \bar{a}, \qquad d_i = a_i - \bar{a} \tag{3}$$

Aggarwal obtained the Bayes estimator δ_σ of θ when θ is $N(0, \sigma^2)$ and the d's have a suitable singular multivariate normal distribution, and showed that the risk of δ_σ tends to the risk of $\bar{Y}$ as $\sigma \to \infty$. His argument does prove property (iii) but does not establish (ii) since the prior distribution of the a's violates (ii). The purpose of the present paper is to show that $\bar{Y}$ is in fact minimax when the a's are subject to the restriction

$$\sum(a_i - \bar{a})^2 = 1. \tag{4}$$

It then immediately follows when 1 is replaced in (4) by an arbitrary M, and this establishes both (ii) and (iii).

Received July, 1980.

[1] Research partially supported by NSF Grant MCS79 03716 and Office of Naval Research Grant N00014 80 C 0163.

AMS 1970 *subject classifications.* Primary 62D05;secondary.62G05.

Key words and phrases. Finite population, simple random sampling, minimax estimator, means, labels, stratified sampling, sample design.

2. Reduction of the minimax problem. In order to minimize (1) subject to (4), consider the auxiliary problem in which the a's are given by (3) with θ unknown and the d's distributed according to a distribution Q over the intersection of the sphere $\sum d_i^2 = 1$ with the hyperplane $\sum d_i = 0$ to be specified below. We shall show that $\bar{Y}$ is a minimax estimator of θ for this auxiliary problem in which θ is the only unknown parameter, i.e., that $\bar{Y}$ minimizes

$$\sup_\theta \int E(\delta - \theta)^2 \, dQ(d_1, \cdots, d_N). \tag{5}$$

Since for $\delta = \bar{Y}$, (5) is equal to

$$\sup_\theta \sup_{\sum(d_i - \bar{d})^2 = 1} E(\delta - \theta)^2, \tag{6}$$

it will follow that $\bar{Y}$ is also minimax for the original problem under the restriction (4).

In the auxiliary problem if Q is invariant under permutation of the d_i, we can assume without loss of generality that

$$Y_i = \theta + D_i \qquad i = 1, \cdots, n, \tag{7}$$

where the joint distribution of $(D_1, \cdots, D_N)$ is Q. The problem of estimating θ on the basis of the variables (7) is that of estimating a location parameter θ when the joint density of the variables is known except for θ. It was shown by Girshick and Savage (1951) that the Pitman estimator is then minimax provided an estimator with finite risk exists. It will therefore be enough to show that $\bar{Y}$ is the Pitman estimator on the basis of (7).

Let us next specify the distribution of the D's. Let $X_1, \cdots X_N$ be i.i.d. $N(0, 1)$ and let

$$U_i = \frac{X_i - \bar{X}}{S_N}, \tag{8}$$

where $\bar{X} = (1/N) \sum_{i=1}^N X_i$, $S_N^2 = \sum_{i=1}^N (X_i - \bar{X})^2$. Then the U's satisfy $\sum U_i^2 = 1$ and $\sum U_i = 0$ and we shall take as distribution of $(D_1, \cdots, D_N)$ that of $(U_1, \cdots, U_N)$ so that the joint distribution of the variables $Y_1, \cdots, Y_n$ is the same as that of $U_1 + \theta, \cdots, U_n + \theta$.

The Pitman estimator of θ in this last representation is given by

$$\bar{Y} - E_0[\bar{Y} \mid Y_1 - \bar{Y}, \cdots, Y_n - \bar{Y}]. \tag{9}$$

Since the second term in (9) is equal to

$$E_0[\bar{U}_n \mid U_1 - \bar{U}_n, \cdots, U_n - \bar{U}_n] \tag{10}$$

where $\bar{U}_n = (U_1 + \cdots + U_n)/n$, it is enough to show that (10) is equal to zero.

3. A symmetry property. To show that (10) is equal to zero, we shall now prove the stronger property that

$$\mathscr{L}(\bar{U}_n \mid U_1 - \bar{U}_n, \cdots, U_n - \bar{U}_n) =' \mathscr{L}(-\bar{U}_n \mid U_1 - \bar{U}_n, \cdots, U_n - \bar{U}_n). \tag{11}$$

Condition (11) will follow if we show that

$$\mathscr{L}(\bar{X}_n - \bar{X} \mid X_1 - \bar{X}_n, \cdots, X_n - \bar{X}_n, S_N^2) \tag{12}$$

is symmetric about zero since symmetry about zero is preserved under mixing with respect to a scale parameter.

Let $(X_1, \cdots, X_n) \to (Z_1, \cdots, Z_n)$ be an orthogonal transformation such that $Z_1 = \sqrt{n}\, \bar{X}_n$ and let $Z_{n+i} = X_{n+i}$ $(i = 1, \cdots, N - n)$. Then

$$\bar{X}_n - \bar{X} = \frac{N - n}{n} \left(\frac{Z_1}{\sqrt{n}} - \bar{Z} \right)$$

where $\tilde{Z} = (Z_{n+1} + \cdots + Z_N)/(N - n)$, and a straight-forward calculation shows that

$$S_N^2 = \sum_{i=2}^n Z_i^2 + \sum_{i=n+1}^N (Z_i - \tilde{Z})^2 + \frac{n(N-n)}{N}\left(\tilde{Z} - \frac{Z_1}{\sqrt{n}}\right)^2.$$

Consider now the transformation

$$(13) \qquad (Z_1, \cdots, Z_N) \to (-Z_1, Z_2, \cdots, Z_n, -Z_{n+1}, \cdots, -Z_N)$$

and note that it has the following three properties:

(i) Under (12)

$$\frac{N-n}{N}\left(\frac{Z_1}{\sqrt{n}} - \tilde{Z}\right) \to -\frac{N-n}{N}\left(\frac{Z_1}{\sqrt{n}} - \tilde{Z}\right)$$

(ii) the transformation (13) leaves S_N^2 invariant.

(iii) The distribution of $(Z_1, \cdots, Z_N)$ is not changed by (13).

The distribution (12) in terms of the Z's becomes the conditional distribution

$$\mathscr{L}\left(\left(\frac{Z_1}{\sqrt{n}} - \tilde{Z}\right)\frac{N-n}{n} \mid Z_2, \cdots, Z_n, S_N^2\right)$$

which by (i) – (iii) is seen to be equal to

$$\mathscr{L}\left(-\left(\frac{Z_1}{\sqrt{n}} - \tilde{Z}\right)\frac{N-n}{n} \mid Z_2, \cdots Z_n, S_N^2\right).$$

Re-expressing this last distribution in terms of the X's completes the proof.

4. Some extensions.

(i) *Labels.* The minimax result proved in the preceding sections did not take into account the labels of the population elements, which potentially provide additional information. It is however easy to see that the result remains valid even when the labels of the sampled elements are considered as part of the data. The population then consists of the N pairs (i, a_i), $i = 1, \cdots, N$ and the data of the n sampled pairs. The problem of estimating $\bar{a}$ remains invariant under permutation of the labels. Since the estimand $\bar{a}$ remains invariant under these transformations, an estimand $\delta((i_1, a_{i_1}), \cdots, (i_n, a_{i_n}))$ (where $i_1, \cdots, i_n$ are the labels of the n sample elements) is invariant if it remains unchanged under all permutations of the first arguments, i.e., if it does not depend on the labels. Since the group of all permutations is finite there exists an invariant minimax estimator, and this completes the proof.

(ii) *Stratified sampling.* Suppose the population has been divided into s strata of sizes $N_1, \cdots, N_s$ from which simple random samples of sizes $n_1, \cdots, n_s$ are drawn. Let $Y_{i.}$ and $\theta_i = a_{i.}$ be the mean of the sample and of the population of the ith stratum respectively, and let $a_{..} = \sum N_i a_i / \sum N_i$ be the average of all $N = \sum N_i$ population elements a_{ij} ($j = 1, \cdots, N_i$; $i = 1, \cdots, s$). The natural estimator for estimating $a_{..}$ is $\delta = \sum (N_i/N) Y_{i}.$ and its variance is

$$\bar{\tau}^2 = \sum \frac{N_i^2}{N^2}\frac{N_i - n_i}{N_i - 1}\frac{1}{n_i}\tau_i^2,$$

where τ_i^2 is the population variance of the ith stratum. Then in generalization of the result for simple random sampling one can conclude that, subject to $\bar{\tau}^2 \leq M$, δ is minimax for estimating $a_{..}$ with squared error loss.

The result follows immediately from the following two facts, for both of which the loss is squared error.

(a) Let $(\mathbf{Y}_1, \theta_1), \cdots, (\mathbf{Y}_s, \theta_s)$ be s independent location families. Then the uniformly minimum risk translation equivariant estimator of $\theta = \sum c_i\theta_i$ is $\delta = \sum c_i\delta_i$ where δ_i is the Pitman estimator of θ_i based on $(\mathbf{Y}_i, \theta_i)$

(b) The Girshick-Savage argument extends to the situation described in (a), see for example, Kiefer (1957), and the uniformly minimum risk equivariant of θ is therefore also minimax.

(iii) *Design.* We can use an invariance argument as in Blackwell and Girshick (1954) to get a stronger minimax property.

In the situation of Section (1) let $\mathscr{P}$ be the class of all sampling designs in which a sample of size n is selected from a population of size N, i.e., members of $\mathscr{P}$ are probability measures on the finite set S whose members are the $\binom{N}{n}$ subsets of size n of the integers $\{1, \cdots, N\}$. A procedure for estimating $\bar{a}$ now consists of a pair $P \in \mathscr{P}$ and estimator δ and the risk is, with obvious notation,

$$R(a_1, \cdots, a_N, P, \delta) = E_p[\delta(Y_1, \cdots, Y_n) - \bar{a}]^2$$
$$= \sum P(s)[\delta(\{a_i\}:i \in s) - \bar{a}]^2.$$

It turns out that if P_0 is simple random sampling i.e., the uniform distribution on S, then $(P_0, \bar{Y})$ again minimizes $\max\{R(a_1, \cdots, a_N, P, \delta):\sum(a_i - \bar{a})^2 \leq M\}$ among all pairs (P, δ) as above. We need only invoke the results of Section 8.7 of Blackwell and Girshick to ensure that, for any fixed δ, (P_0, δ) has minimax risk within the class $\{(P, \delta):P \in \mathscr{P}\}$ and then apply our main theorem.

We have formulated this argument for populations without labels and for sampling plans where the order in which elements are drawn is disregarded, but it is clear that the minimax property continues to hold under these weaker conditions.

We can also generalize the result to stratified sampling in the obvious way obtaining minimaxity among all sampling plans taking n_i observations from stratum i, $i = 1, \cdots, s$. By a further invariance argument we can get an even more general result (but under more special assumptions). Suppose we are given a total sample size n and we are permitted to allocate $n_1, \cdots, n_s$ $(\sum n_i = n)$ observations to the various strata as we please and then choose a sampling plan and estimate $\bar{a}$. If the class of populations considered consists of all those with $\tau_i^2 \leq M$, $i = 1, \cdots, s$, and s divides n, then the optimal (minimax) plan and estimate is to sample $\frac{n}{s}$ observations from each stratum without replacement and then estimate $\bar{a}$ by the grand mean of the s samples.

REFERENCES

AGGARWAL, OM P. (1959). Bayes and minimax procedures in sampling from finite and infinite populations. *Ann. Math. Statist.* **30** 206–218.

BLACKWELL, DAVID and GIRSHICK, M. A. (1954). *Theory of Games and Statistical Decisions.* Wiley, New York.

GIRSHICK, M. A. and SAVAGE, L. J. (1951). Bayes and minimax estimates for quadratic loss functions. *Proc. Second Berkeley Symp Math. Statist. Prob.* 53–75.

HODGES, J. L., JR. and LEHMANN, E. L. (1981). Minimax estimation in simple random sampling (To appear in *Essays in Statistics and Probability* in honor of C. R. Rao (P. R. Krishnaiah, ed.). North-Holland.

KIEFER, J. (1957). Invariant minimax sequential estimation and continuous time processes. *Ann. Math Statist.* **28** 573–601.

DEPARTMENT OF STATISTICS
UNIVERSITY OF CALIFORNIA, BERKELEY
BERKELEY, CALIFORNIA 94720

Estimation With Inadequate Information

E. L. LEHMANN*

As a possible explanation for the existence of absurd UMVU estimators, a class of estimation problems is defined: estimating the probability of an "unobservable" event, for example, estimating the probability of an event of interest occurring over a long period of time from observations over a much shorter period. It is pointed out that in typical cases of this kind, no reasonable unbiased estimator can exist. Consideration of the maximum likelihood estimator suggests the possibility that in fact there may then often exist no reasonable estimators.

KEY WORDS: Unbiased estimation; Unobservable events; Inadequate information.

1. INTRODUCTION

The possibility of an absurd unbiased estimator with uniformly minimum variance (UMVU) is usually illustrated by one or another variant (see, e.g., Kendall and Stuart 1979, Ex.17.26) of the following situation.

Example 1. Let X have the Poisson distribution $P(\lambda)$ with $E(X) = \lambda$, and let the estimand be $g(\lambda) = e^{-b\lambda}$, $0 < b < \infty$. The equations

$$\sum \delta(x)\, \lambda^x \left(\frac{e^{-\lambda}}{x!}\right) = e^{-b\lambda} \qquad \text{for all } \lambda > 0$$

have the unique solution $\delta(x) = (1 - b)^x$. For $0 < b < 1$ this decreases strictly from 1 at $x = 0$ to 0 at $x = \infty$ and is a reasonable estimator of $g(\lambda)$, which decreases strictly from 1 at $\lambda = 0$ to 0 at $\lambda = \infty$. However, for $b > 1$, $\delta(x)$ oscillates between positive and negative values as x is even or odd and no longer appears to bear much relation to the function it estimates.

The purpose of this note is to suggest that this is not an isolated example but is rather a special case of a class of situations in which, in a certain sense, the available information is inadequate for the existence of reasonable unbiased (or possibly any) estimators.

2. PROBLEMS WITH INADEQUATE INFORMATION

Suppose that the observations are represented by X with a distribution P_θ, $\theta \in \Omega$, and that Y with distribution Q_θ is independent of X and unobserved. For the model (X, Y) with distribution $P_\theta \times Q_\theta$, $\theta \in \Omega$, we assume the existence of a complete sufficient statistic T. The estimand to be estimated on the basis of X is assumed to be of the form

$$g(\theta) = P_\theta(T \in A) \tag{1}$$

for some set A. Whether T falls into A cannot be observed on the basis of X. For short, the event $T \in A$ is therefore called *unobservable*.

To see how Example 1 fits into this scheme, suppose that $b > 1$ and that Y is distributed independently of X according to a Poisson distribution with $E(Y) = (b - 1)\lambda$. For (X, Y), the statistic $T = X + Y$ is sufficient and complete, and the estimand of Example 1 is $e^{-b\lambda} = P(T = 0)$. The situation may be described somewhat more concretely in terms of a Poisson process with parameter λ. The quantity $e^{-b\lambda}$ is the probability that no event occurs in a time interval of length b. This probability is estimated by observing the process for a time interval of length 1. No difficulty arises when $b < 1$ but X is no longer adequate when $b > 1$, and the inadequacy increases as b increases. Correspondingly, the oscillations of the estimator $(1 - b)^x$, and hence its unreasonableness, increase with $b > 1$. The case $b = 1$, in which $e^{-b\lambda} = P(X = 0)$ is marginal; the UMVU estimator is the indicator of the set $\{X = 0\}$. (Note: The term UMVU estimator may seem inappropriate here since there exists only one unbiased estimator. However, in this and the other situations considered here, X itself is typically a sufficient statistic for some observations $X_1, \ldots, X_n$; in the present case, the X_i might be the numbers of events occurring in n nonoverlapping intervals of length $1/n$.)

When estimating (1) on the basis of X, an unbiased estimator frequently will not exist, as is illustrated by the following two examples.

Example 2. Consider n binomial trials with success probability p, of which the first $m < n$ are observed but the remaining n-m are not. If X_i is 1 or 0 when the ith trial is, respectively, a success or a failure, let $X = X_1 + \cdots + X_m$, $Y = X_{m+1} + \cdots + X_n$, and $T = X + Y$ and consider the estimand $g(p) = P(T = n) = p^n$. No unbiased estimator of p^n exists since only polynomials of degree less than m have unbiased estimators based on X.

Example 3. Let $X_1, \ldots, X_m$ $(m > 1)$ be iid according to the normal distribution $N(\xi, \sigma^2)$ and consider the problem of estimating $\Phi((u - \xi)/\sigma) = P(X_i \leq u)$. This problem was treated by Kolmogorov (1950), Lieberman and Resnikoff (1955), and others. For the sake of simplicity sup-

* E.L. Lehmann is Professor of Statistics, Department of Statistics, University of California, Berkeley, CA 94720. This article was prepared under National Science Foundation Grant MCS-79-03716. The author would like to thank the referee for a very careful reading of the original version, which resulted in substantial improvements.

© Journal of the American Statistical Association
September 1983, Volume 78, Number 383
Theory and Methods Section

pose that σ^2 is known. Then $\bar{X}$ is a complete sufficient statistic, distributed as $N(\xi, \tau^2)$ with $\tau^2 = \sigma^2/m$. For $\tau = 1$, the problem reduces to that of estimating $\Phi((u - \xi)/b)$ on the basis of a single random variable X distributed as $N(\xi, 1)$, where $b = \sqrt{m} > 1$.

The corresponding problem with $b < 1$ is of the type considered here. This is seen by letting Y be distributed independently of X with normal distribution $N(\xi, \beta^2)$, β^2 known. Then $T = \alpha X + (1 - \alpha)Y$ with $\alpha = \beta^2/(1 + \beta^2)$ is a complete sufficient statistic distributed as $N(\xi, \alpha)$, and

$$g(\xi) = P(T < u) = \Phi\left(\frac{u - \xi}{\sqrt{\alpha}}\right), \qquad (2)$$

which reduces to the earlier problem for $\alpha = b^2$. We shall see below that no unbiased estimator of (2) exists for $\alpha < 1$.

There is, however, a class of problems, illustrated by Example 1, in which an unbiased estimator of (1) does exist. Later in this section, we show that in such cases $\delta(X)$ will always be unsatisfactory.

Theorem 1. $\delta(X)$ is an unbiased estimator of (1) if and only if

$$\begin{aligned} E[\delta(X) \mid t] &= 1 \quad \text{if} \quad t \in A \\ &= 0 \quad \text{if} \quad t \notin A \quad (\text{a.e. } \mathcal{P}^T). \end{aligned} \qquad (3)$$

Proof. Let $\eta(t) = E[\delta(X) \mid t]$. If $I_A(t)$ denotes the indicator of A and $\delta(X)$ is unbiased, we have $E_\theta[\eta(T)] = E_\theta[I_A(T)] = g(\theta)$, and hence by completeness of T, $\eta(t) = I_A(t)$ (a.e. $\mathcal{P}^T$). The converse is obvious.

Before discussing the general implications of (3), let us use this result to show that no unbiased estimator $\delta(X)$ of $g(\theta)$ exists in Example 3. An easy calculation shows that the conditional distribution of X, given $T = t$, is normal with mean t and constant variance, say γ^2. Equation (3) thus becomes

$$\begin{aligned} \frac{1}{\sqrt{2\pi}\,\gamma} \int_{-\infty}^{\infty} \delta(x) \exp\left(-\frac{1}{2\gamma^2}(x - t)^2\right) dx \\ = 1 \quad \text{if} \quad t \leqslant u \\ = 0 \quad \text{if} \quad t > u. \end{aligned}$$

This can have no solution since the left side is an analytic function of t. An analogous proof shows that no unbiased estimator $\delta(X)$ of $P_\sigma\,(T \leqslant r)$ exists when $T = X + Y$, where X and Y are independently distributed as $\sigma^2\chi_m^2$ and $\sigma^2\chi_n^2$, respectively.

As we have seen, unbiased estimators of (1) often do not exist. However, there are important classes of problems for which the mathematics forces the existence of such estimators, which are then liable to share some of the unfortunate features of the estimator $\delta(X)$ of Example 1.

Suppose that there exist $t_0 \in A$ and $t_1 \notin A$ such that the conditional distributions of X given t_0 and given t_1, respectively, are not mutually singular. Then if $0 < g(\theta) < 1$ for all θ, any unbiased estimator must take on values either both ≥ 1 and < 0 or both > 1 and ≤ 0. (Under weak additional assumptions it has to take on values both > 1 and < 0.)

That $\delta(X)$ takes on values outside the range of the estimand $g(\theta)$ is an unpleasant property shared by some other UMVU estimators (for example, by those of some variance components). What is more unusual is the oscillatory character of the estimator when one expects it to be monotone.

The following result shows that in many situations (3) precludes $\delta(X)$ from being monotone.

Theorem 2. Suppose X and T are real valued and that the conditional distribution of X given t is strictly stochastically increasing in t in the sense that $t < t'$ implies

$$P(X \leq x \mid t) > P(X \leq x \mid t') \qquad (4)$$

for all x not satisfying $P(X \leq x \mid t) = P(X \leq x \mid t') = 0$ or 1. Suppose in addition that for any $t < t'$ the conditional distributions of X, given t and t', are not mutually singular. Then if $\delta(X)$ is unbiased for estimating

$$g(\theta = P_\theta(T \leq r), \qquad (5)$$

it cannot be monotone.

Proof. By Theorem 1, δ satisfies

$$\begin{aligned} E[\delta(X) \mid t] &= 1 \quad \text{if} \quad t \leq r \\ &= 0 \quad \text{if} \quad t > r. \end{aligned} \qquad (6)$$

Suppose $\delta(X)$ is nonincreasing. Then $t < t'$ implies $E[\delta(X) \mid t] > E[\delta(X) \mid t']$ unless there exists a constant c such that

$$P[\delta(X) = c \mid t] = P[\delta(X) = c \mid t'] = 1.$$

It thus follows from (6) that $\delta(X) = 1$ with probability 1 under the conditional distributions of X, given t with $t \leq r$, and that $\delta(X) = 0$ with probability 1 under the conditional distributions X, given t with $t > r$. This violates the assumed nonsingularity of these conditional distributions.

3. SOME FURTHER EXAMPLES

In generalization of Example 1, consider the following problem.

Example 4. Let X be distributed as in Example 1, but let the estimand be

$$g_r(\lambda) = P(T = r) \quad \text{or} \quad h_r(\lambda) = P(T \leqslant r), \qquad (7)$$

where T is distributed as $P(b\lambda)$. Note that $h_r(\lambda)$ decreases from 1 to 0 as λ increases from 0 to infinity, while $g_r(\lambda)$ for $r > 0$ increases from 0 to a maximum at $\lambda = r/b$ and then decreases.

The UMVU estimators of $g_r(\lambda)$ and $h_r(\lambda)$ are, respectively,

$$\begin{aligned} \delta_r(x) &= \binom{x}{r} b^r(1 - b)^{x-r} && \text{if } x \geqslant r \\ &= 0 && \text{if } 0 \leqslant x < r \end{aligned} \qquad (8)$$

and

$$\eta_r(x) = \sum_{j=0}^{r} \delta_j(x). \tag{9}$$

It is easy to check that if $b > 1$, then in both cases the estimator will take on both positive and negative values for arbitrarily large values of x. In fact, for any given $b > 1$,

$$\frac{\delta_r(x+1)}{\delta_r(x)} \to -(b-1) \quad \text{as} \quad x \to \infty$$

so that for sufficiently large x, $\delta_r(x)$ oscillates between successive positive and negative values, and the magnitude of the oscillations increases with b. That $\eta_r(x)$ exhibits the same behavior follows from the fact that $\eta_r(x)/\delta_r(x) \to 1$ as $x \to \infty$.

Example 5. Let $X + m$ be the number of binomial trials required to obtain m successes, so that X has the negative binomial distribution

$$P(X = x) = \binom{m+x-1}{m-1} p^m q^x, \quad x = 0, 1, \ldots$$

and let

$$g(q) = (1-q)^N = P(T = 0),$$

where $T + N$ is the number of binomial trials required to obtain N successes (with the same success probability p). Then the UMVU estimator of $g(q)$ is given by

$$\sum \delta(x) \binom{m+x-1}{m-1} q^x \equiv (1-q)^{N-m}. \tag{10}$$

If $N < m$, say $m - N = k < m$, $\delta(x)$ reduces to

$$\delta(x) = \frac{k(k+1)\cdots(m-1)}{(x+k)(x+k+1)\cdots(x+m-1)},$$

which decreases from 1 at $x = 0$ to 0 as $x \to \infty$, as one would expect. On the other hand, consider the case $N > m$, $N - m = n > 0$, say. This corresponds to the situation in which $T = X + Y$, where $Y + n$ is the number of binomial trials required to obtain n successes. The assumptions of Theorem 1 thus hold, and it is easily checked that those of Theorem 2 do too. The estimator (10) now reduces to

$$\delta(x) = (-1)^x \frac{n(n-1)\cdots(n-x+1)}{m(m+1)\cdots(m+x-1)} \quad \text{for } x = 0, \ldots, n$$

and $\delta(x) = 0$ for $x > n$. As in the earlier examples, we find that the oscillations of $\delta(x)$ (for $x \leq n$) increase in magnitude with n, that is, as the unobserved information embodied in the estimand increases.

Example 6. Let $X_1, \ldots, X_n; X_{n+1}, \ldots, X_N$ be iid according to the uniform distribution on $(0, \theta)$, where it is assumed known that $\theta > 1$ but θ is otherwise unknown, and let

$$\begin{aligned} X &= 1 \quad \text{if } \max(X_1, \ldots, X_n) < 1 \\ &= \max(X_1, \ldots, X_n) \quad \text{otherwise} \\ Y &= 1 \quad \text{if } \max(X_{n+1}, \ldots, X_N) < 1 \\ &= \max(X_{n+1}, \ldots, X_N) \quad \text{otherwise} \end{aligned}$$

and $T = \max(X, Y)$. Then the factorization criterion shows T to be sufficient for θ, and it is easily checked that T is complete. Suppose only $X_1, \ldots, X_n$ are observed and that it is desired to estimate

$$g(\theta) = \frac{1}{\theta^N} = P_\theta(T < 1).$$

The estimator $\delta(X)$ given by

$$\delta(1) = 1; \; \delta(x) = \frac{n-N}{nx^N} \quad \text{for} \quad x > 1 \tag{11}$$

is UMVU. The example satisfies the conditions of Theorem 2 and although $g(\theta)$ is strictly decreasing, one can, in fact, easily see directly that $\delta(x)$ is not monotone since it is positive for $x = 1$ and negative but increasing for $x > 1$. The UMVU estimator is given by (11) also when $N < n$ but in that case is nonincreasing as one would expect.

4. CONCLUDING REMARKS

In the preceding sections we have indicated a class of estimation problems—estimating the probability of an unobservable event—that frequently lead to absurd UMVU estimators. The suggested explanation is that in these situations the data do not contain enough information to provide reasonable unbiased estimators. The question naturally arises whether in such cases any reasonable estimators exist. This requires another investigation.

As a first clue, one might consider the performance of the maximum likelihood estimator (MLE). It is the most widely used estimating procedure, cannot take on values outside the range of the estimand, and in all our examples is a monotone function of x when $g(\theta)$ is monotone in θ.

Example 1 (continued). In the context of Example 1, the MLE is $\delta(X) = e^{-bX}$, and

$$E\,\delta(X) = \exp[\lambda(e^{-b} - 1)] \equiv h(\lambda). \tag{12}$$

The bias of $\delta(X)$ is therefore

$$b(\lambda) = h(\lambda) - g(\lambda) = e^{-a\lambda} - e^{-b\lambda}, \tag{13}$$

where $a = 1 - e^{-b}$. One can easily see that $a < b$, so that on the average $\delta(X)$ overestimates $e^{-b\lambda}$ for all λ. The bias function $b(\lambda)$ tends to 0 as $\lambda \to \infty$ and has a unique maximum at

$$\lambda^* = \frac{\log b - \log a}{b - a}. \tag{14}$$

Table 1 shows the value of λ^* and the maximum bias

Table 1. Maximum Bias $b(\lambda^*)$ in Example 1

b	$\frac{1}{4}$	$\frac{1}{3}$	$\frac{1}{2}$	1	2	3	4	5
λ^*	4.250	3.250	2.249	1.247	.739	.561	.465	.403
$b(\lambda^*)$	.045	.060	.088	.167	.300	.401	.478	.537
$g(\lambda^*)$	.346	.339	.325	.287	.228	.186	.155	.133

Table 2. Maximum Bias $b(\lambda^*)$ in Example 6

N	1	2	4	6	8	10	12	15
				$n = 3$				
θ^*	1.732	1.500	1.333	1.260	1.217	1.188	1.167	—
$b(\theta^*)$	.192	.296	.422	.500	.555	.597	.630	—
$g(\theta^*)$	.577	.444	.316	.250	.208	.179	.157	—
				$n = 5$				
θ^*	1.357	1.291	1.250	1.200	1.170	1.149	1.133	1.116
$b(\theta^*)$	.217	.279	.328	.402	.457	.500	.535	.577
$g(\theta^*)$	.543	.465	.410	.335	.286	.250	.223	.192

$b(\lambda^*)$ for various values of b. It is striking to observe the deterioration of the situation for $b \geq 1$. Biases of .4 and .5 in estimating probabilities are hardly tolerable.

It is also noteworthy that for any fixed λ, the relative bias $b(\lambda)/e^{-\lambda} \sim e^{(b-1)\lambda}$ as $b \to \infty$ and hence tends to infinity exponentially as $b \to \infty$.

Example 6 (continued). The MLE of $1/\theta^N$ in the context of Example 6 is

$$\delta(X) = 1 \quad \text{if} \quad X < 1$$
$$= 1/X^N \quad \text{if} \quad X > 1$$

and its bias is

$$b(\theta) = \frac{N}{n - N}(\theta^{-N} - \theta^{-n}).$$

The bias is thus always positive. It is zero at $\theta = 1$ and tends to be zero as $\theta \to \infty$. It has a unique maximum at

$$\theta^* = \left(\frac{n}{N}\right)^{1/(n-N)}. \tag{15}$$

Table 2 shows for $n = 3$ and 5 the value of θ^* and the maximum bias for various values of N. As in the preceding example, the maximum bias increases steeply when the information becomes inadequate (i.e., for $N > n$).

The results of Sections 2 and 3 together with the rather limited numerical evidence just presented suggest that, in the situations considered here, reasonable estimators of the probabilities (1) may not exist. To prove that a reasonable estimator is incompatible with acceptable bias control, it would be necessary to determine the estimator that minimizes the maximum absolute bias subject to the appropriate monotonicity (and range) conditions. Beyond this, it would be important to determine the estimators that minimize the maximum risk without imposing such conditions.

Situations like those described here are not purely mathematical artifacts. In particular, the estimation of probabilities of minor or major catastrophes over long periods of time from observations over much shorter periods tend to be just of this type. However, when the probabilities being estimated are very small, the difficulty of the estimation problem may be overshadowed by the unreliability of the model in the extreme part of the distribution.

[*Received March 1981. Revised January 1983.*]

REFERENCES

KENDALL, M., and STUART, A. (1979), *The Advanced Theory of Statistics* (Vol. 2, 4th ed.), New York: Macmillan.

KOLMOGOROV, A.N. (1950), "Unbiased Estimates," *Izvestija Akademii Nauk Armjanskoĭ SSR, Serija Matematika*, 14, 303–326 (American Mathematical Society Translation 98).

LIEBERMAN, G.J., and RESNIKOFF, G.J. (1955), "Sampling Plans for Inspection by Variables," *Journal of the American Statistical Association*, 50, 457–516.

Part IV
Nonparametrics I

Erich Lehmann on rank methods

Willem R. van Zwet

Though statistical methods based on ranks, such as the sign test, have a history going back for centuries, it was the introduction of Wilcoxon's tests in Wilcoxon (1945) that started an avalanche of research on this subject. It is worth noting that Wilcoxon needed only a little over three pages introducing and discussing both his two-sample test and his one-sample test for symmetry, to set off a research effort that would fill thousands of journal pages. Almost from the beginning Erich Lehmann played a major role in the development of this subject. Eight of his papers will be discussed here, of which two are joint with J. L. Hodges, Jr. and one with C. Stein. The full content of these papers will not be described here. Instead, the focus will be on the main contributions of each paper to our present knowledge of rank tests. It all adds up to what is more or less a history of the subject. For ease of presentation attention will be focused on the two-sample problem. When needed, appropriate reference will be made to other models that are dealt with in a particular paper.

Let us introduce the necessary notation. The two samples are denoted by $X_1, X_2, \ldots, X_m$ and $Y_1, Y_2, \ldots, Y_n$ and for the combined sample of size $N = m + n$, we write $Z_1, \ldots, Z_N = X_1, \ldots, X_m, Y_1, \ldots, Y_n$. We assume that $Z_1, \ldots, Z_N$ are independent random variables, that $X_1, X_2, \ldots, X_m$ are identically distributed (i.d.) with common distribution function F, and that $Y_1, Y_2, \ldots, Y_n$ are i.d. with distribution function G. We suppose that F and G are continuous so that there are no equal values among $Z_1, \ldots, Z_N$. Let $Z_{(1)} < \cdots < Z_{(N)}$ denote the ordered sequence of $Z_1, \ldots, Z_N$, and R_j the rank of Y_j in this sequence, thus $Y_j = Z_{(R_j)}$. Wilcoxon's two sample test for $H_0 : F = G$ against $H_1 : F > G$ rejects H_0 for large values of the rank sum $W = \sum_{1 \le j \le n} R_j$. Here $F > G$ should be interpreted as $F(x) \ge G(x)$ for all x with strict inequality for at least one value of x, and hence on an interval. More generally, a rank test for this testing problem rejects H_0 for large values of $T = \sum_{1 \le j \le n} a(R_j)$, for an increasing sequence of scores $a(r)$. Obviously, under H_0,

W.R. van Zwet
Mathematical Institute, University of Leiden, P.O.Box 9512, 2300 RA Leiden, The Netherlands
e-mail: vanzwet@math.leidenuniv.nl

J. Rojo (ed.), *Selected Works of E. L. Lehmann*, Selected Works in Probability and Statistics, DOI 10.1007/978-1-4614-1412-4_30,

the vector $R = (R_1, \ldots, R_n)$ is distributed as a random sequence of length n drawn without replacement from the set $\{1, 2, \ldots, N\}$ and hence the distribution of W under H_0 is easy to find and independent of the underlying common distribution function F. In other words, rank tests are distribution-free and hence similar, in the sense that the probability of an error of the first kind is constant on the hypothesis H_0. For a given significance level α the critical value for the test based on W is therefore easy to determine.

Perhaps this is the place for a word about terminology. In the classical parametric model unknown distributions have a given parametric form with only a finite dimensional parameter left unspecified. The null hypothesis then sets restrictions on this finite dimensional parameter. In the model discussed above the unknown distributions F and G are left unspecified and under H_0 the common distribution $F = G$ is still unspecified. Neither the general model nor the null hypothesis can be described in terms of finite dimensional parameters, which is why such models and hypotheses are called *nonparametric*. This has led statisticians to call rank tests and other similar tests for such hypotheses *nonparametric tests*, which is clearly a misnomer. There is nothing parametric or nonparametric about a statistical test. The important property of such a test is simply that it deals with such large infinitely dimensional hypotheses by using test statistics that are distribution-free, in the sense that their distribution remains the same throughout the hypothesis. Hence *distribution-free tests* is clearly a more appropriate name for these tests.

Returning to the two-sample rank tests, the main problem is of course to get some idea of the power of these tests against simple alternatives (F, G). The case where F and G are normal distributions differing only in location is of course of special interest. This is the case where Student's test comes into its own and it would be of interest to compare the powers. Initially, the common view of rank tests was that it was a nice idea, but that restricting attention to ranks instead of the variables themselves would probably result in a very serious loss of power. Even Wilcoxon himself didn't sound too optimistic about this, even though he felt that his symmetry test would do better than the sign test.

1 On the theory of some non-parametric hypotheses

The first seven of Lehmann's papers discussed here were written between 1949 and 1963 and in these papers one can trace the progress of the research on this topic almost from its very beginning. These papers will therefore be discussed in chronological order. In 'On the theory of some non-parametric hypotheses' by Lehmann and Stein (1949) a first glimmer is shed on the problem, even though the paper deals primarily with nonparametric hypotheses rather than distribution-free tests. First of all the authors point out that instead of testing H_0, it is often more relevant to test the extended null hypothesis H_0^* that $Z_1, \ldots, Z_N$ are exchangeable, i.e. that their joint distribution is invariant under permutations of the variables. Of course for rank tests this makes no difference as the distribution of the ranks remains the same under the extended hypothesis H_0^*. For general tests, however, optimality properties of a test for H_0^* do not necessarily correspond to the same property of this test among the larger class of tests for the restricted null hypothesis H_0. However, the authors prove that this correspondence does exist if the tests for H_0 are restricted to distribution-free or similar tests. But this means that optimal tests for H_0^* can only be beaten by tests that are not even similar for the restricted hypothesis H_0. One may hope that this means that not much power is lost by using a test for H_0^* — such as a rank test — provided it is in some sense optimal for testing H_0^*. The paper further discusses the same phenomenon for more general pairs of null hypotheses and derives most powerful and most stringent tests for various testing problems.

2 Consistency and unbiasedness of certain nonparametric tests

In the second paper 'Consistency and unbiasedness of certain nonparametric tests' (Lehmann (1951)) the attack on the power of Wilcoxon's test is well under way. Mann and Whitney had proved that Wilcoxon's test is consistent, i.e. its power against a simple alternative with $F > G$ tends to 1 as $m, n \to \infty$. By noting that $F > G$ implies that there exists a function $h(x) > x$ such that $h(X_i)$ has distribution G, and that the test statistic is monotone, it is shown that the test is also unbiased against such alternatives, i.e. its power exceeds the level of significance α. The same proof shows that the power against alternatives $F < G$ is below α and hence the test may also be applied for testing the hypothesis $F \leq G$ against the alternative $F > G$. Unbiasedness and consistency are also proved for tests for the k-sample problem, tests for independence and the symmetry hypothesis.

Because W can be written as

$$W = \sum_{1 \leq i \leq m} \sum_{1 \leq j \leq n} I\{X_i < Y_j\} + \frac{n(n+1)}{2},$$

it is a two-sample U-statistic. Extending Hoeffding's results on one-sample U-statistics (cf. Hoeffding (1948)), asymptotic normality of W under alternatives is proved as m and n tend to infinity at the same rate. The news of this major advance traveled around the world fast. Already before Lehmann's paper appeared, my own teacher David van Dantzig reported to the Royal Netherlands Academy of Sciences that '... Mr. A.M. Mood kindly informed me that Mr. E. Lehmann has proved the important result that the Wilcoxon test criterion is asymptotically normally distributed even when the null hypothesis is not true.' (van Dantzig (1951))

3 The power of rank tests

The problem of power computation is attacked in a novel and imaginative way in the third paper 'The power of rank tests' (Lehmann (1953)). This paper is one of my all-time favorites and I have had the pleasure to speak about it in Berkeley on the occasion of Erich's 80th birthday in 1997. Power computations for alternatives such as shift — i.e. $G(x) = F(x-a)$ for $a > 0$ – look rather forbidding and the result will also depend on F as well as a. Lehmann argued that since one probably has only a vague idea what alternatives would be plausible in any particular case, one might as well choose a mathematically convenient alternative hypothesis. His suggestion is to choose a distribution function K on $(0,1)$ with density k, take $G(x) = K(F(x))$ and consider the alternative hypothesis $H_1^* : (F, K(F))$ for varying F. Since the ranks are invariant under the monotone transformation $x \mapsto F(x)$ of the Z's, this has the obvious advantage that that the distribution of the ranks under this alternative is the same as for the model (U, K) where the X_i have the uniform distribution function U on $(0,1)$ and the Y_j have distribution function K. Hence the power is constant on this alternative H_1^*. Application of a result of Hoeffding (1951) yields that under H_1^* and independent of F,

$$P(R_1 = r_1, \ldots, R_n = r_n \mid H_1^*) = \frac{m!n!}{(m+n)!}\mathbb{E}\left(\prod_{1\le j\le n} k(U_{r_j:N})\right), \tag{3.1}$$

where $U_{i:N}$ denotes the i-th order statistic of a sample of size N from the uniform distribution U on $(0,1)$. If $K(v) < v$ for $0 < v < 1$ we have $G < F$ and H_1^* is a proper subset of the full alternative $H_1 : F > G$. Obvious examples are $K(v) = v^a$ for $a > 1$. Especially for integer a, the right-hand side of (3.1) and consequently the exact power of Wilcoxon's test against H_1^* is easy to compute. One can now traverse the general alternative hypothesis $H_1 : F > G$ almost at will and compute the power of Wilcoxon's test for small samples. In the paper we find graphs of the alternatives that are dealt with and numerical values of the corresponding power. Also moments of W under H_1^* are easy to compute and together with the asymptotic normality proved in the previous paper, this yields large sample approximations for the distribution of W under alternatives, and hence for the power of W.

Finally (3.1) allows us to derive most powerful tests against simple alternatives. Under H_0, the probability of any set of ranks equals $m!n!/(m+n)!$ independent of F, and hence the Neyman-Pearson lemma tells us to reject H_0 for large values of the expected value on the right in (3.1). By taking $K(v) = K_p(v) = (1-p)v + pv^2$ and maximizing the derivative of the power against this alternative at $p = 0$, one finds the locally most powerful rank test which happens to be Wilcoxon's test. Other testing problems are also considered.

4 The efficiency of some nonparametric competitors of the t-test

The fourth paper 'The efficiency of some nonparametric competitors of the t-test' (Hodges and Lehmann (1956)) must be the one that caught the most attention, and the number 0.864 is probably engraved in many statisticians' minds. Consider the problem of testing $G = F$ against $G(x) = F(x - a)$, $a > 0$ for a given F with density f and variance σ^2. Pitman (1949) showed that for this problem the asymptotic relative efficiency (ARE) of Wilcoxon's test relative to the t-test equals $e_{W,t} = 12\sigma^2 \left(\int f^2(x)dx\right)^2$. It is easy to see that $e_{W,t}$ can be arbitrarily large for certain densities, indicating a superior performance of Wilcoxon's test. On the other hand the efficiency is less than 1 if $F = \Phi$, the standard normal distribution function, because asymptotically Student's test can not be beaten on its own turf. However, for the normal case $e_{W,t} = 3/\pi \approx 0.95$, which is still fairly close to 1. Lehmann asked what the worst performance of Wilcoxon's test relative to Student's test could be, i.e. what the minimum of $e_{W,t}$ would be as a function of f. He found that this equals $108/125 = 0.864$. This means that for large samples you never throw away more than 14 % of your data if you test for shift by using Wilcoxon's test instead of Student's, as was the general custom. For that relatively small price you buy a test that is valid regardless of the underlying distribution. This simple argument put a definite end to the idea that the cost of using ranks would be prohibitive.

5 Rank methods for combination of independent experiments in analysis of variance

In the paper 'Rank methods for combination of independent experiments in analysis of variance' by Hodges and Lehmann (1962) the authors turn to generalizations of Wilcoxon's two-sample test to randomized block designs. In this model the N available subjects are divided into k blocks, with N_i subjects in block i and hence $\sum_{1 \le i \le k} N_i = N$. The idea is that within each block subjects are believed to be more homogeneous than in the entire group. In block i, $m_i > 0$ and $n_i > 0$ subjects are randomly assigned to treatments A and B respectively, so $m_i + n_i = N_i$ and the two treatments correspond to two samples of sizes m_i and n_i. The problem is to test the hypothesis that there is no difference between the treatments, against shift alternatives where in each block the response to treatment B is stochastically larger than that to treatment A. One possible test statistic for this problem is a linear combination of Wilcoxon's two-sample statistics W_i for blocks $i = 1, \ldots, k$, with weights that are in some sense optimal. If the underlying distributions would be known to be normal with a common variance, an optimal statistic would be a linear combination of the differences of the sample means in blocks $1, \ldots, k$. Roughly speaking, for the normal distribution it turns out that if both the number of blocks k and the block sizes N_i tend to infinity, the ARE of this rank test with respect to the normal test is $3/\pi$. This equals the ARE of

Wilcoxon's two-sample test to Student's two-sample test. However, if the block sizes N_i equal 2 for all i and the number of blocks k tends to infinity, then this ARE equals $2/\pi$ which is the ARE of the sign test to Student's test. This is no surprise as $N_i = 2$ implies $m_i = n_i = 1$, and for two samples of size 1, Wilcoxon's test is equivalent to the sign test applied to the difference of the (single) responses to treatments B and A.

Another possibility is to align the blocks by subtracting the average response of all N_i subjects in block i from these N_i responses in that block. Having thus aligned the different blocks, one pools all N subjects and carries out Wilcoxon's two-sample test on the two samples of sizes $\sum_{1\leq i\leq k} m_i$ and $\sum_{1\leq i\leq k} n_i$ of subjects having received treatment A or B. The test is carried out as a conditional test given the division of the set of ranks $\{1, 2, \ldots, N\}$ of the N combined aligned responses over the k blocks. Critical values and powers are computed and asymptotic normality under the hypothesis is proved. If the block sizes N_i equal 2 for all i, then for the normal case the ARE of the aligned block test equals $3/\pi$ which is the ARE of Wilcoxon's test with respect to Student's test.

6 Robust estimation in analysis of variance

The sixth paper 'Robust estimation in analysis of variance' (Lehmann (1963)) is a follow-up on 'Estimates of location based on rank tests' (Hodges and Lehmann (1963)) which appeared in the immediately preceding issue of *Ann. Math. Statist.*. In this earlier paper the authors consider the two-sample shift model with $G(x) = F(x - \xi)$ and propose the median of the mn differences $(Y_j - X_i)$ as a robust estimator of ξ. The ARE of this estimator to the classical estimator which is the difference of the sample means is the same as the ARE of Wilcoxon's test to Student's test.

Now the set-up is generalized to the c-sample problem where for $i = 1, \ldots, c$, the i-th sample of size n_i consists of observations $X_{i,\alpha} = \xi_i + U_{i,\alpha}, \alpha = 1, \ldots, n_i$. The $U_{i,\alpha}$ are independent with median 0. The Hodges-Lehmann estimator of $(\xi_i - \xi_j)$ is given by $Y_{i,j} = \text{med}(X_{i,\alpha} - X_{j,\beta})$, the median of the $n_i n_j$ differences $(X_{i,\alpha} - X_{j,\beta})$ for $\alpha = 1, \ldots, n_i$, $\beta = 1, \ldots, n_j$. There is an amusing problem with this definition. It is not compatible in the sense that the estimator $Y_{i,k}$ of $(\xi_i - \xi_k)$ is generally not equal to the sum of the estimators $Y_{i,j}$ of $(\xi_i - \xi_j)$ and $Y_{j,k}$ of $(\xi_j - \xi_k)$. This becomes a nuisance when estimating general contrasts of the form $\sum a_i \xi_i$ with $\sum a_i = 0$ because there are many linear combinations of the $Y_{i,j}$ that can be used to estimate this contrast.

This problem is solved by estimating $(\xi_i - \xi_j)$ by minimizing

$$\sum_{i\neq j} \left(Y_{i,j} - (\xi_i - \xi_j)\right)^2,$$

which yields the difference of the two sample means $Z_{i,j} = Y_{i\cdot} - Y_{j\cdot}$ as an estimate. Now if a contrast is written in two different ways $\sum_{i,j} a_{i,j}(\xi_i - \xi_j) = \sum_{i,j} b_{i,j}(\xi_i - \xi_j)$ identically in $\xi_1, \ldots, \xi_c$, then obviously $\sum_{i,j} a_{i,j} Z_{i,j} = \sum_{i,j} a_{i,j}(Y_{i\cdot} - Y_{j\cdot}) =$

$\sum_{i,j} b_{i,j}(Y_{i\cdot} - Y_{j\cdot}) = \sum_{i,j} b_{i,j} Z_{i,j}$, so the estimator is unique. Joint asymptotic normality of appropriately normalized $Y_{i,j}$ is proved and with the same normalization, the $Z_{i,j}$ have the same limit distribution.

7 A class of selection procedures based on ranks

The seventh paper in this sequence 'A class of selection procedures based on ranks' (Lehmann (1963)) is devoted to the use of ranks in yet another area of statistics. Suppose we have samples of size n from each of c populations $\Pi_1, \ldots, \Pi_c$ which differ only in location. So for $i = 1, \ldots, c$ and $j = 1, \ldots, n$, the random variables $X_{i,j}$ are independent and $P(X_{i,j} \leq x) = F(x - \theta_i)$. One wishes to select the *best* population which is the one with the largest value of the parameter θ. The classical procedure is the *means procedure* that selects the population Π_i with largest mean $X_{i\cdot} = n^{-1} \sum_{1 \leq j \leq n} X_{i,j}$. If F is a normal distribution function this is the natural thing to do.

The performance of different selection procedures is compared as follows. Suppose that for $i = 1, 2, \ldots, c-1$, $\theta_i < \theta_c$ so that population Π_c is uniquely the best. A population Π_i is called *good* if $\theta_i \geq \theta_c - \Delta$ for some $\Delta > 0$. To ensure good behavior of a selection procedure one may require that

$$P(\text{selected population is good}) \geq \gamma, \tag{7.1}$$

for some $\gamma > 0$. To compare the asymptotic performance of two procedures A and B we consider the ratio $e = n_B/n_A$ of smallest sample sizes n_A and n_B needed to satisfy (7.1) for a given Δ and γ with these two procedures. The limit of e as $\Delta \to 0$ is the ARE of procedure A with respect to procedure B.

Let us now consider procedures based on ranks. Let $R_{i,j}$ denote the rank of $X_{i,j}$ among the combined samples $\{X_{i,j} : i = 1, \ldots, c, j = 1, \ldots, n\}$. Define $V_i = \sum_{1 \leq j \leq n} h(R_{i,j})$, for $i = 1, \ldots, c$ and an increasing function h, and select the population Π_i with the largest V_i. For $h(x) \equiv x$ we are dealing with a rank sum test which is similar to Wilcoxon's test. It is shown that in the case of normal F the ARE of this test to the means procedure equals $3/\pi$ which is the ARE of Wilcoxon's test to Student's test. Also the ARE in never smaller than 0.864 for any F and the entire theory for two-sample tests repeats itself is this very different situation.

8 Parametrics versus nonparametrics: two alternative methodologies

After this brief discussion of (a part of) Lehmann's work on rank tests it will be clear that he was one of the driving forces of this development. A masterful picture of what was achieved is given by Lehmann himself in the last paper of this sequence. 'Parametrics versus nonparametrics: two alternative methodologies' (Lehmann (2009)). Here we see Erich Lehmann in a role that we all know so well from his books, that of the superb expositor. In 9 pages he tells us all there is to know about this subject and it makes no sense to produce a further condensed version here. Just read this paper!

References

[1] van Dantzig, D. (1951). On the consistency and the power of Wilcoxon's two sample test. *Proc. Royal Netherlands Acad. Sci., Series A* **54** 1–8.

[2] Hodges, J. L. Jr. and Lehmann, E. L. (1956). The efficiency of some nonparametric competitors of the t-test. *Ann. Math. Statist.* **27** 324–335.

[3] Hodges, J. L. Jr. and Lehmann, E. L. (1962). Rank methods for combination of independent experiments in analysis of variance. *Ann. Math. Statist.* **33** 482–497.

[4] Hodges, J. L. Jr. and Lehmann, E. L. (1963). Estimates of location based on rank tests. *Ann. Math. Statist.* **34** 598–611.

[5] Hoeffding, W. (1948). A class of statistics with asymptotically normal distributions. *Ann. Math. Statist.* **19** 293–325.

[6] Hoeffding, W. (1951). Optimum nonparametric tests. *Proceedings of the Second Berkeley Symposium on Mathematical Statistics and Probability*, University of California Press, 83–92.

[7] Lehmann, E. L. and Stein, C. (1949). On the theory of some non-parametric hypotheses. *Ann. Math. Statist.* **20** 28–45.

[8] Lehmann, E. L. (1951). Consistency and unbiasedness of certain nonparametric tests. *Ann. Math. Statist.* **22** 165–179.

[9] Lehmann, E. L. (1953). The power of rank tests. *Ann. Math. Statist.* **24** 23–43.

[10] Lehmann, E. L. (1963). A class of selection procedures based on ranks. *Math. Annalen* **150** 268–275.

[11] Lehmann, E. L. (1963). Robust estimation in analysis of variance. *Ann. Math. Statist.* **34** 957–966.

[12] Lehmann, E. L. (2009). Parametrics versus nonparametrics: two alternative methodologies (with discussion). *J. Nonparametr. Stat.* **21** 397–405.

[13] Pitman, E. J. G. (1949). *Lecture notes on nonparametric statistical inference.* Columbia University.

[14] Wilcoxon, F. (1945). Individual comparisons by ranking methods. *Biometrics Bulletin* **1** 80–83.

ON THE THEORY OF SOME NON-PARAMETRIC HYPOTHESES

By E. L. Lehmann and C. Stein

University of California, Berkeley

Summary. For two types of non-parametric hypotheses optimum tests are derived against certain classes of alternatives. The two kinds of hypotheses are related and may be illustrated by the following example: (1) The joint distribution of the variables $X_1, \cdots, X_m, Y_1, \cdots, Y_n$ is invariant under all permutations of the variables; (2) the variables are independently and identically distributed. It is shown that the theory of optimum tests for hypotheses of the first kind is the same as that of optimum similar tests for hypotheses of the second kind. Most powerful tests are obtained against arbitrary simple alternatives, and in a number of important cases most stringent tests are derived against certain composite alternatives. For the example (1), if the distributions are restricted to probability densities, Pitman's test based on $\bar{y} - \bar{x}$ is most powerful against the alternatives that the X's and Y's are independently normally distributed with common variance, and that $E(X_i) = \xi$, $E(Y_i) = \eta$ where $\eta > \xi$. If $\eta - \xi$ may be positive or negative the test based on $|\bar{y} - \bar{x}|$ is most stringent. The definitions are sufficiently general that the theory applies to both continuous and discrete problems, and that tied observations present no difficulties. It is shown that continuous and discrete problems may be combined. Pitman's test for example, when applied to certain discrete problems, coincides with Fisher's exact test, and when $m = n$ the test based on $|\bar{y} - \bar{x}|$ is most stringent for hypothesis (1) against a broad class of alternatives which includes both discrete and absolutely continuous distributions.

1. Generalities. In the present paper we study the problem of determining optimum tests for certain non-parametric hypotheses. It is important in this connection to make some distinctions which are of lesser significance when the problem is approached from the intuitive point of view which has been customary in this field. Consider for example the hypothesis H that $Z_1, \cdots, Z_N$ are independently and identically distributed according to an unknown probability density function. All tests which have been suggested for testing H are valid also for testing the hypothesis H' that the unknown joint probability density function of the Z's is symmetric in its N arguments. On the other hand, tests which have optimum properties for testing H' against a certain class of alternatives will in general not possess the same properties when H' is replaced by H. From the present point of view the two hypotheses mentioned are essentially different. We shall be concerned in this paper primarily with generalizations of H', and we shall show that many of the tests suggested in the literature have optimum properties for testing hypotheses of this kind against certain classes of alternatives.

The corresponding general theory for hypotheses related to H is quite different.

J. Rojo (ed.), *Selected Works of E. L. Lehmann*, Selected Works in Probability and Statistics, DOI 10.1007/978-1-4614-1412-4_31, © Springer Science+Business Media, LLC 2012

However the two theories do coincide, provided tests of these latter hypotheses are restricted to similar regions. More specifically, all results on optimum tests of H' are equivalent to the corresponding results on optimum similar tests of H, and this equivalence holds also for many of the more general hypotheses considered in this paper.

It should be observed that in many experimental situations, the hypothesis H' that the joint distribution of the Z's is invariant under all permutations is more realistic than the hypothesis H that the Z's are independently and identically distributed. For example, suppose there is a block of land divided into $m + n$ plots, and the experimenter wants to test whether one of two fertilizers (used in fixed amounts) is more effective than the other in increasing the yield of a certain plant. Of the plots, m are chosen at random; fertilizer I is applied to these, and fertilizer II to the other n. If X_i denotes the yield from the ith plot to which fertilizer I has been applied and Y_j denotes the yield from the jth plot to which fertilizer II has been applied, where the plots are numbered at random, then the hypothesis that the two fertilizers are completely equivalent implies that the application of any permutation to $X_1, \cdots, X_m, Y_1, \cdots Y_n$ does not change their joint distribution. But it is not reasonable to suppose the X_i, Y_j are independently and identically distributed, since there may be intrinsic differences among the plots. For discussions of these and related points, see Fisher [1], Neyman [2], Pitman [3]. It may be that in many particular cases some hypothesis between the two is really appropriate but the hypothesis H is the only one that is evidently appropriate from a cursory inspection of the setup.

Many of the alternative hypotheses considered below, for example those involving normality, are dictated more by tradition and ease of treatment than by appropriateness in actual experiments. Thus this paper should not be considered as providing absolute justification for tests such as Pitman's but rather as suggesting a method of obtaining optimum non-parametric tests when the class of alternatives is fairly well specified.

Another possibility, first raised by Neyman [2], which has been ignored in this paper is the equality on the average of the two fertilizers but with fertilizer I having a larger dispersion than fertilizer II, or a distribution differing in some other characteristic. It would be reasonable to consider this as part of the hypothesis tested, but tests based on randomization may give a probability of rejection of the hypothesis of equivalence in this case which is much higher than the stated level of significance. We hope to return to problems of this type in later papers.

Let us make the following basic assumptions. $\mathfrak{Z}$ is a space of points z and $\mathfrak{A}$ is an additive class of subsets A of $\mathfrak{Z}$. Any member of $\mathfrak{A}$ will be said to be measurable. By a probability distribution we mean a measure F, defined over $\mathfrak{A}$ for which $F(\mathfrak{Z}) = 1$. We shall be concerned with two classes of probability distributions: One, the class of all distributions, and two, the class of distributions which are absolutely continuous with respect to a given measure μ, that is, the class of distributions F for which there exists a function f such that

$$F(A) = \int_A f(z)\,d\mu(z). \tag{1.1}$$

We shall call f a generalized probability density function with respect to μ. By Z we denote a random variable such that for any A in $\mathcal{A}$,

$$P\{Z \in A\} = F(A). \tag{1.2}$$

For most of the applications we shall take $\mathcal{Z}$ to be a Euclidean space, and $\mathcal{A}$ to be the class of all Borel sets. Then if μ is Lebesgue measure, (1.1) states that f is a probability density function in the usual sense. However, we shall have occasion to consider also some measures other than Lebesgue measure. By a hypothesis H we mean a class of probability distributions. Next we describe the hypotheses with which we shall be concerned. Let Π be a partition of $\mathcal{Z}$, that is, let Π be a class of mutually exclusive subsets S of $\mathcal{Z}$ such that every point z of $\mathcal{Z}$ lies in one of the sets S. If two points z_1 and z_2 lie in the same set S, we shall say that z_1 is equivalent to z_2 with respect to Π: $z_1 \sim z_2 \pmod{\Pi}$. The set of all points which are equivalent to z will be denoted by $T(z)$, the number of points of $T(z)$ by $n(z)$. Concerning Π we make the following assumptions:

(i) All sets in Π are finite, so that $n(z)$ is finite for all z.

(ii) If we define S_n as the union of all those sets S of Π which contain exactly n points, there exist mutually exclusive sets $S_n^{(1)}, \cdots, S_n^{(n)}$ which are measurable and such that every element S of Π containing exactly n points has one and only one point in common with each $S_n^{(i)}$.

We shall say that a measure μ is invariant under Π if the following condition holds: For all n and $i, j \leq n$, if S is any set contained in $S_n^{(i)}$ and if S' denotes the set of equivalent points in $S_n^{(j)}$, then $\mu(S) = \mu(S')$.

Given a partition Π satisfying (i) and (ii), we formulate the hypothesis H that the distribution F of Z is invariant under Π. We shall refer to H as the hypothesis of invariance under Π. We shall also consider the hypothesis of invariance under a partition for a class of generalized densities f. In this case we assume that the measure μ of (1.1) is given, and that Π, in addition to (i) and (ii) satisfies the condition:

(iii) The measure μ is invariant under Π. The hypothesis H in this case states that $z_1 \sim z_2 \pmod{\Pi}$ implies $f(z_1) = f(z_2)$.

By a test of a hypothesis H we mean (see [4]) a measurable function φ on $\mathcal{Z}$ to the interval [0, 1] which with every point $z \in \mathcal{Z}$ associates a probability $\varphi(z)$ of rejection. This definition, slightly more general than the usual one, is particularly useful in non-parametric work. Among other advantages it automatically takes care of the problem of tied observations. It also disposes of the difficulties encountered by Scheffé [5] in his treatment of the problem of similar regions, as will be shown in Lemma 1.

The size of a test φ is defined to be

$$\epsilon(\varphi) = \sup_{F \in H} \int \varphi(z)\,dF(z). \tag{1.3}$$

If in particular

$$\int \varphi \, dF = \epsilon(\varphi) \tag{1.4}$$

for all F in H, φ is said to be similar for testing H. Extending the terminology of Scheffé, we say that φ has structure $S(\epsilon)$ if for all z in S_n

$$\sum_{z' \epsilon T(z)} \varphi(z') = n\epsilon. \tag{1.5}$$

The following lemma extends a result of Scheffé.

LEMMA 1. *For testing a hypothesis of invariance, any test of structure $S(\epsilon)$ is similar and of size ϵ.*

PROOF. For any F in H and any φ

$$\int \varphi \, dF = \sum_{n=1}^{\infty} \sum_{i=1}^{n} \int_{S_n^{(i)}} \varphi \, dF = \sum_{n=1}^{\infty} \int_{S_n^{(1)}} [\sum_{z' \epsilon T(z)} \varphi(z')] \, dF(z). \tag{1.6}$$

But φ has structure $S(\epsilon)$ and hence (1.5) holds for all z. Therefore

$$\int \varphi \, dF = \sum_{n=1}^{\infty} n\epsilon \int_{S_n^{(1)}} dF = \epsilon. \tag{1.7}$$

We shall show next that for testing a hypothesis of invariance at level of significance ϵ, only tests of structure $S(\epsilon)$ need be considered. In order to make this result applicable both to hypotheses referring to the class of all distributions and to those referring to a class of generalized densities, we shall state it in an asymmetric form which when taken together with lemma 1 indicates the essential equivalence of the two types of hypotheses.

LEMMA 2. *If φ is any test of a hypothesis of invariance for the class of generalized densities with respect to a fixed measure μ, and if the size of φ is less than or equal to ϵ, then there exists a test φ_1 of structure $S(\epsilon)$ such that*

$$\int \varphi_1 \, dF \geq \int \varphi \, dF \tag{1.8}$$

for all probability distributions F.

PROOF. First we shall show that

$$\frac{1}{n(z)} \sum_{z' \epsilon T(z)} \varphi(z') \leq \epsilon \tag{1.9}$$

almost everywhere μ. For let A be the set of points z such that

$$\frac{1}{n(z)} \sum_{z' \epsilon T(z)} \varphi(z') > \epsilon \tag{1.10}$$

and suppose that $\mu(A)$ is positive. Let

$$f(z) = \begin{cases} \dfrac{1}{\mu(A)} & \text{if } z \epsilon A; \\ 0 & \text{elsewhere.} \end{cases} \tag{1.11}$$

Then f is in H since by definition of A, whenever z is in A, $T(z)$ is contained in A. But

$$\int \varphi f \, d\mu > \epsilon \int_A f \, d\mu = \epsilon, \tag{1.12}$$

in contradiction to the assumption that φ has size ϵ.

From (1.9) it follows easily that there exists a test φ_1 of structure $S(\epsilon)$ and such that for all z

$$\varphi_1(z) \geq \varphi(z). \tag{1.13}$$

Since condition (1.8) is then satisfied, this completes the proof.

Lemma 2 raises the question whether it is possible to reduce the problem of testing a hypothesis of invariance still further, or whether the tests of structure $S(\epsilon)$ form, what Wald [6] has called an essentially complete class of admissible tests. This question is answered by

THEOREM 1. *Let μ be a measure defined over $\mathfrak{A}$. Let Π_0 and Π_1 be two partitions of $\mathcal{Z}$ satisfying conditions* (i), (ii) *and* (iii), *and such that $z \sim z'$* (mod Π_1) *implies $z \sim z'$* (mod Π_0). *For the class of generalized densities with respect to μ denote by H_i ($i = 0, 1$) the hypothesis of invariance relative to Π_i. Then for testing H_0 against H_1 at level of significance ϵ, the totality of tests which* (a) *have structure $S(\epsilon)$, and for which* (b) *$z \sim z'$* (mod Π_1) *implies $\varphi(z) = \varphi(z')$, form an essentially complete class of admissible tests.*

PROOF. It is easily seen that we can restrict ourselves to that subclass of tests of structure $S(\epsilon)$ which possess property (b). For if φ is any test of structure $S(\epsilon)$ relative to Π_0, let

$$\varphi^*(z) = \frac{1}{n(z)} \sum_{z' \epsilon T(z)} \varphi(z). \tag{1.14}$$

Then clearly φ^* possesses property (b) and has structure $S(\epsilon)$. Furthermore if f is any probability density function of H_1, then

$$\int \varphi^* f \, d\mu = \int \varphi f \, d\mu, \tag{1.15}$$

so that φ and φ^* have identical power against H_1.

In order to complete the proof, we must show that if φ_1 and φ_2 are any two tests satisfying (a) and (b), and if φ_1 and φ_2 differ on a set of positive measure, there exists a probability density function f of H_1 for which

$$\int \varphi_1 f \, d\mu > \int \varphi_2 f \, d\mu. \tag{1.16}$$

Since both φ_1 and φ_2 have structure $S(\epsilon)$, the set A of points z for which

$$\varphi_1(z) > \varphi_2(z) \tag{1.17}$$

has positive measure. Also, because of (b), if two points are equivalent relative to Π_1, they are either both in A or both not in A. If $f(z)$ is defined as $1/\mu(A)$ for z in A and as zero elsewhere, then f is in H_1 and satisfies (1.16).

The theorem obtained from theorem 1 by letting the hypotheses H_0 and H_1 refer to the class of all probability distributions rather than to a particular class of generalized densities, is clearly also true, and cases between these two theorems could also be formulated.

Since the most powerful test φ for testing a hypothesis of invariance H_0 referring to a class of generalized densities against an alternative f from this class of densities has the correct size also for testing the wider hypothesis $\bar{H}_0$ referring to the class of all distributions, φ is also most powerful for testing $\bar{H}_0$ against f. The corresponding remark holds for most stringent tests. Therefore all optimum tests that will be derived in the sequel, through the use of theorems of this section, may be considered as tests of hypotheses referring to the class of all distributions: they are valid against these hypotheses, and no power is gained by restricting the hypothesis to the appropriate class of generalized densities.

2. Most powerful tests and most stringent tests. One of the main problems to be considered in this paper is the determination of a most powerful test of a hypothesis of invariance against a simple alternative. If we restrict our considerations to the class of generalized densities with respect to μ, a complete solution of this problem is given by the following

THEOREM 2. *Let H be the hypothesis of invariance under the partition* Π, *and let g be a probability density function not in H. For any z in S_n denote by $z^{(1)}, \cdots, z^{(n)}$ the n points of $T(z)$ arranged so that $g(z^{(1)}) \geq g(z^{(2)}) \geq \cdots \geq g(z^{(n)})$. For testing H against g a most powerful test of size ϵ is given by*

$$\varphi(z) = \begin{Bmatrix} 1 & \text{if} & g(z) > g(z^{(1+[\epsilon n])}) \\ a & \text{if} & g(z) = g(z^{(1+[\epsilon n])}) \\ 0 & \text{if} & g(z) < g(z^{(1+[\epsilon n])}) \end{Bmatrix} \quad \text{for } z \text{ in } S_n, \tag{2.1}$$

where $\sum_{i=1}^{n} \varphi(z^{(i)}) = n\epsilon$, $0 \leq a \leq 1$ and where a may depend on z through $T(z)$.

PROOF. First we observe that the number of $z^{(i)}$ for which $g(z^{(i)}) \geq g(z^{(1+[\epsilon n])})$ is greater than or equal to $1+[\epsilon n] \geq \epsilon n$ and that the number of $z^{(i)}$ for which $g(z^{(i)}) > g(z^{(1+[\epsilon n])})$ is less than or equal to $[\epsilon n] \leq \epsilon n$, so that there exists an a between 0 and 1 for which $\Sigma\varphi(z^{(i)}) = n\epsilon$. Since φ has structure $S(\epsilon)$, it follows from lemma 1 that it is similar and of size ϵ.

Let

$$g^*(z) = g(z^{(1+[\epsilon n])}) \quad \text{for} \quad z \in S_n. \tag{2.2}$$

To complete the proof consider first the special case that

$$\int g^*(z)\, d\mu(z) \tag{2.3}$$

vanishes. Then

$$\int \varphi g\, d\mu = \int g\, d\mu = 1 \tag{2.4}$$

that is, the test φ has power 1, and therefore is clearly most powerful. Assume next that the integral (2.3) is positive. Then g^* is proportional to a probability density function of H. For it is measurable and satisfies the symmetry condition required of a member of H, and the integral (2.3) is finite since

$$\begin{aligned}\sum_n \int_{S_n} g^*(z)\, d\mu\,(z) &\leq \sum_n \frac{1}{\epsilon n} \int_{S_n} \sum_{i=1}^{[\epsilon n]+1} g(z^{(i)})\, d\mu(z) \\ &\leq \frac{1}{\epsilon} \sum_n \int_{S_n} \frac{1}{n} \sum_{i=1}^{n} g(z^{(i)})\, d\mu(z) = \frac{1}{\epsilon}.\end{aligned} \tag{2.5}$$

The test φ therefore has the form of a probability ratio test. Since it is also similar, it follows from theorem 1 of [4] that φ is most powerful.

In practice one is usually interested in composite rather than simple alternatives. We shall therefore consider next the problem of deriving most stringent tests of hypotheses of invariance against certain classes of alternatives. This problem may be reduced to that of finding tests which maximize the minimum power over a class of alternatives by the following simple theorem of Hunt and Stein [7].

THEOREM 3. *Given a hypothesis H and a class of alternatives $\{g_\theta\}$, $\theta \in \Omega$, denote by $\beta^*(\theta)$ the envelope power function corresponding to the level of significance ϵ, that is, let*

$$\beta^*(\theta) = \sup_{\varphi} \beta(\varphi, \theta) \tag{2.6}$$

where $\beta(\varphi, \theta)$ stands for the power of the test φ against the alternative g_θ and where the least upper bound is taken over all tests φ of size ϵ. Let $\{\Omega_\delta\}$ be a class of mutually exclusive subsets of Ω such that $\bigcup \Omega_\delta = \Omega$ and such that $\beta^*(\theta)$ is constant on each Ω_δ. Denote by φ_δ a test which maximizes the minimum power over Ω_δ. If $\varphi_\delta = \varphi$ is independent of δ, then φ is most stringent[1] *for testing H against Ω at level of significance ϵ.*

For obtaining tests which maximize the minimum power over a class of alternatives to a hypothesis of invariance, we can state the following simple extension of theorem 2.

THEOREM 4. *Let H be a hypothesis of invariance, and let H_1 be the class of alternatives $\{g_\theta\}$, $\theta \in \Omega$. Suppose there exists a subset Ω' of Ω and a probability measure λ over Ω' such that for the test φ of size ϵ defined as in theorem 2 with*

$$g(z) = \int_{\Omega'} g_\theta(z)\, d\lambda(\theta), \tag{2.7}$$

the integral $\int \varphi g_\theta\, d\mu$ is constant for θ in Ω', and

$$\int \varphi g_\theta\, d\mu \geq \int \varphi g_{\theta'}\, d\mu \quad \text{for all} \quad \theta \in \Omega,\ \theta' \in \Omega'. \tag{2.8}$$

Then φ maximizes the minimum power over Ω at level of significance ϵ.

---

[1] A test is said to be most stringent [16] if it minimizes the maximum difference between envelope power and power, that is, if it minimizes Sup $[\beta^*(\theta) - \beta(\varphi, \theta)]$.

PROOF. By theorem 2, φ is a most powerful test for testing H against g, that is, for any φ' of size ϵ

$$\int \varphi'(z) \int_{\Omega'} g_\theta(z)\, d\lambda(\theta)\, d\mu(z) \leq \int \varphi(z) \int_{\Omega'} g_\theta(z)\, d\lambda(\theta)\, d\mu(z). \tag{2.9}$$

Consequently

$$\begin{aligned} \inf_{\theta\epsilon\Omega} \int \varphi'(z) g_\theta(z)\, d\mu(z) &\leq \int_{\Omega'} d\lambda(\theta) \int \varphi'(z) g_\theta(z)\, d\mu(z) \\ &= \int \varphi'(z)\, d\mu(z) \int_{\Omega'} g_\theta(z)\, d\lambda(\theta) \leq \int \varphi(z)\, d\mu(z) \int_{\Omega'} g_\theta(z)\, d\lambda(\theta) \\ &= \int_{\Omega'} d\lambda(\theta) \int \varphi(z) g_\theta(z)\, d\mu(z) = \inf_{\theta\epsilon\Omega} \int \varphi(z) g_\theta(z)\, d\mu(z). \end{aligned} \tag{2.10}$$

3. Normal alternatives. Let H be the hypothesis of invariance under Π, let $T(z)$ be the set of points equivalent to z (mod Π), and let f and g be two functions defined over $\mathfrak{Z}$. We shall write $f \sim g$ if there exists a function F such that

$$f(z) = F[g(z), T(z)], \tag{3.1}$$

where for any fixed $T(z)$, F is a strictly increasing function of g. We note that $f \sim g$ in the following two special cases:

(i) $f(z) = F[g(z)]$ where F is strictly increasing;

(ii) $f(z) = a(z)g(z) + b(z)$ where $a(z) > 0$ for all z, and where $z_1 \sim z_2$ (mod Π) implies $a(z_1) = a(z_2)$, $b(z_1) = b(z_2)$.

The usefulness of this notation stems from the following remark. Let g^* and φ be defined as in (2.2) and (2.1) respectively and let $f \sim g$. If the test ψ is obtained from φ by substituting f and f^* for g and g^* respectively, then $\psi = \varphi$.

The purpose of the present section is to obtain most powerful and most stringent tests of some hypotheses of invariance against certain classes of normal alternatives. In particular, problems will be exhibited for which various non-parametric tests suggested in the literature possess these optimum properties.

PROBLEM 1. Suppose that the random variables Z_{ij} ($j = 1, \cdots, s_i$; $i = 1, \cdots, m$) have a joint probability density function, and denote by H the hypothesis that this probability density is invariant under all permutations of the s_i arguments within the ith group for $i = 1, \cdots, m$. Consider the alternative H_1 that all variables are independently distributed with common variance σ^2, and that

$$E(Z_{ij}) = ax_{ij} + b_i, \tag{3.2}$$

where a, the b's and the x's are assumed known and where, without essential loss of generality, we assume $a > 0$. Assume further that

$$\sum_{j=1}^{s_i} x_{ij} = 0. \tag{3.3}$$

In order to obtain the most powerful test of H against H_1, we apply theorem 2 with

$$g(z) = c \exp\left[-\frac{1}{2\sigma^2}\Sigma\Sigma(z_{ij} - ax_{ij} - b_i)^2\right] \tag{3.4}$$
$$\sim \Sigma\Sigma(ax_{ij} + b_i)z_{ij} \sim \Sigma\Sigma x_{ij}z_{ij}.$$

The most powerful test is therefore given by (2.1), if we replace $g(z)$ by $\Sigma\Sigma x_{ij}z_{ij}$. This test being independent of σ^2, the b's and $a > 0$, it is uniformly most powerful against the class of alternatives obtained from H_1 by not specifying the values of these parameters but restricting a to be positive.

If we drop the restriction $a > 0$, a uniformly most powerful test no longer exists; we shall instead obtain the most stringent test against this extended class of alternatives, using theorems 3 and 4. Clearly the envelope power function is constant on the surfaces $|a|/\sigma^2$ = constant. Take as the Ω of theorem 4, the set consisting of the two points $(a, b_1, \cdots b_m, \sigma)$ and $(-a, b_1, \cdots, b_m, \sigma)$. Let λ assign the probability $\frac{1}{2}$ to each of the two points. Then the function g of (2.7) becomes

$$\frac{1}{2}\left(\frac{1}{\sqrt{2\pi}\sigma}\right)^{\Sigma s_i} \exp\left\{\frac{1}{2\sigma^2}\Sigma\Sigma(z_{ij} - ax_{ij} - b_i)^2\right\} + \frac{1}{2}\left(\frac{1}{\sqrt{2\pi}\sigma}\right)^{\Sigma s_i}$$
$$\exp\left\{-\frac{1}{2\sigma^2}\Sigma\Sigma(z_{ij} + ax_{ij} - b_i)^2\right\} \tag{3.5}$$
$$\sim \exp\{\Sigma\Sigma z_{ij}(ax_{ij} + b_i)\} + \exp\{\Sigma\Sigma z_{ij}(-ax_{ij} + b_i)\}$$
$$\sim \exp\{\Sigma\Sigma a x_{ij}z_{ij}\} + \exp\{-\Sigma\Sigma a x_{ij}z_{ij}\} \sim |\Sigma\Sigma x_{ij}z_{ij}|$$

The power of the test φ obtained by substituting this expression for g in (2.1) is the same at both points of Ω. For this test is most powerful for testing H against the simple alternatives H' that the density of the Z's is given by the first member of (3.5). But under the transformation $Z'_{ij} = -Z_{ij} + 2b_i$, H and H' and therefore the test φ are left invariant, while the two points of Ω are permuted.

Condition (2.8) of theorem 4 is therefore satisfied, and hence φ maximizes the minimum power over Ω. Since furthermore φ is independent of the particular set Ω chosen, it follows from theorem 3 that φ is most stringent for the problem under consideration. In case condition (3.3) is not satisfied, let $x'_{ij} = x_{ij} - x_{i\cdot}$. Then $\Sigma\Sigma x'_{ij} = 0$ and $E(Z_{ij}) = ax'_{ij} + b'_i$.

Therefore the test criterion (3.5) becomes

$$|\Sigma\Sigma z_{ij}(x_{ij} - x_{i\cdot})| = |\Sigma\Sigma(z_{ij} - z_{i\cdot})(x_{ij} - x_{i\cdot})|. \tag{3.6}$$

Some special cases of problem 1 are of particular interest.

a) Suppose that the variables of the ith group fall into two subgroups, and write for Z_{ij} : U_{ij} when $j = 1, \cdots, k_i$; V_{ij-k_i} when $j = k_i + 1, \cdots, k_i + l_i (k_i + l_i = s_i)$. Let

$$x_{ij} = \begin{cases} 0 & \text{for} \quad j = 1, \cdots, k_i; \\ 1 & \text{for} \quad j = k_i + 1, \cdots, k_i + l_i. \end{cases} \tag{3.7}$$

Then the alternatives ascribe to the variables normal distributions with common variance and such that

$$E(U_{ij}) = b_i\,; \qquad E(V_{ij}) = b_i + a. \tag{3.8}$$

The criterion becomes

$$\Sigma\Sigma z_{ij}(x_{ij} - x_i) = \sum_{i=1}^{m}\left(\frac{k_i}{k_i + l_i}\Sigma v_{ij} - \frac{l_i}{k_i + l_i}\Sigma u_{ij}\right) = \Sigma \frac{1}{\frac{1}{k_i} + \frac{1}{l_i}}(v_i - u_i) \tag{3.9}$$

or

$$\left| \Sigma \frac{1}{\frac{1}{k_i} + \frac{1}{l_i}}(v_i - u_i) \right| \tag{3.10}$$

according as a is restricted to positive values or not.

b) If we specialize still further and let $m = 1$, we are dealing with a problem which would coincide with the two sample problem if we added independence to the assumptions of the hypothesis. (3.10) becomes $| \bar{v} - \bar{u} |$, the criterion suggested by Pitman [3].

c) If instead of m we set $k_i = l_i = 1$ for $i = 1, \cdots, m$ we are testing interchangeability within each pair (u_i, v_i) against normal alternatives under which the means of U_i and V_i are different, the difference being independent of i. The criterion $| \Sigma (v_i - u_i) |$ to which (3.10) reduces was first suggested by R. A. Fisher [1].

d) As a last example set $m = 1$ in the original problem. Under the hypothesis the joint density of $Z_1, \cdots, Z_s$ is symmetric in its s arguments, while under the alternatives the Z's are normally distributed with common variance and mean $ax_i + b$. The criterion reduces to $| \Sigma (z_i - \bar{z})(x_i - \bar{x}) |$ which was proposed by Pitman [3].

We therefore see that several non-parametric tests which have been discussed in the literature are most powerful one-sided, or most stringent for testing a hypothesis of invariance against certain classes of normal alternatives. In a later section we shall indicate to what extent these results remain valid if to these hypotheses we add the assumption of independence.

The remaining problems will be considered somewhat more briefly since the proofs follow the same pattern as in problem 1.

PROBLEM 2. The conditions of problem 1*d*) are satisfied in particular if $x_1, \cdots, x_s$ are values taken on by random variables $X_1, \cdots, X_s$ and if under the alternatives the pairs (X_i, Z_i) have a common bivariate normal distribution with $\sigma_X^2 = \sigma_Z^2$. We are then concerned with a problem related to that of testing for absence of interclass correlation. For the corresponding intraclass problem, we consider random variables $X_1, \cdots, X_s, Z_1, \cdots, Z_s$, and test the hypothesis that the joint density of the $2s$ variables is symmetric in all its arguments, against the alternatives that the pairs (X_i, Z_i) have a common bivariate normal distribution, the means and variances of the X's and Z's being the same. We shall only consider the case of positive correlation. Clearly, the criterion will be $\Sigma\, x_i z_i$ as in the one sided case of problem (d). However the tests differ, in that this expression must now be compared not only with the $s!$ expressions obtained by permuting the z's among themselves, but instead with the $(2s)!/2^s s!$ expressions obtained by considering all possible ways in which s pairs can be formed from the complete set of $2s$ observations.

PROBLEM 3. Consider once more the hypothesis that the joint density of $Z_1, \cdots, Z_n$ is symmetric in its n arguments, and consider the alternatives that the Z's are normally distributed with positive circular serial correlation. Then

$$(3.11) \quad g(z) = C \exp\left\{-\frac{1}{2\sigma^2} \sum_{i=1}^{n} [(x_i - \xi) - \delta(x_{i+1} - \xi)]^2\right\} \sim \sum z_i z_{i+1}$$

where $z_{n+1} = z_1$. The test based on this criterion, which was proposed by Wald and Wolfowitz [8], is therefore most powerful against the above class of alternatives.

PROBLEM 4. As a last problem, we shall test the hypothesis H that the joint density of $Z_1, \cdots, Z_n$ is symmetric in its n arguments and symmetric about each coordinate hyperplane, that is, invariant under the transformation $x_i' = -x_i$, $x_j' = x_j$ for all $j \neq i$, for $i = 1, \cdots, n$. This will be tested against the alternatives that the Z's are independently, identically distributed according to a normal distribution with non-zero mean. If we restrict this mean to positive values, we get

$$(3.12) \qquad g(z) = \frac{1}{(\sqrt{2\pi\sigma})^n} \exp\left\{-\frac{1}{2\sigma^2} \Sigma(z_i - \xi)^2\right\} \sim \Sigma\, z_i .$$

If on the other hand both positive and negative values are allowed for the mean, the most stringent test is based on the statistic $|\Sigma\, z_i|$.

This test may be appropriate for some situations in which it is customary to use the sign test.

4. Binomial and other non-normal alternatives.

In the present section we shall be concerned mainly with generalisations of problems 1b) and 1c) of section 3. As described there, the hypotheses referred to the class of all probability densities in the usual sense. However, as was pointed out at the end of section 2, the same tests may be considered as referring to much wider hypotheses. If they are interpreted in this way, it is possible to greatly widen the class of alternatives without destroying the optimum properties of the tests.

Let $Z = (X_1, \cdots, X_n, Y_1, \cdots, Y_n)$ and denote by Π the partition under which two points z and z' are equivalent if they are obtainable from each other by a permutation of coordinates. Let H_0 be the hypothesis of invariance under Π. This is a generalization of the hypothesis of complete symmetry referring to a class of probability densities. Consider as alternative the class of distributions defined by

$$(4.1) \quad P\{Z \in A\} = \int_A C \exp\{\theta_1 \Sigma x_i + \theta_2 \Sigma y_i + \Sigma r(x_i) + \Sigma r(y_i)\}\, d\mu(z).$$

where the θ's are any real numbers, where μ is the $2n$th power of any one dimensional measure ν (and therefore invariant under Π), and where r is any ν-measurable function, subject only to the condition that the integral (4.1) converges when taken over the whole space.

We first consider the one-sided case $\theta_2 > \theta_1$. Using theorem 2 for a particular θ_1, θ_2, r and μ, we then have

$$
\begin{aligned}
g(z) &= C \exp \{\theta_1 \Sigma x_i + \theta_2 \Sigma y_i + \Sigma r(x_i) + \Sigma r(y_i)\} \\
&\sim \theta_1 \Sigma x_i + \theta_2 \Sigma y_i \sim \theta_1 \Sigma x_i + \theta_2 \Sigma y_i - \tfrac{1}{2}(\theta_1 + \theta_2)\Sigma[x_i + y_i] \\
&= \tfrac{1}{2}(\theta_1 - \theta_2)[\Sigma x_i - \Sigma y_i] \sim \Sigma y_i - \Sigma x_i .
\end{aligned} \tag{4.2}
$$

Since this test does not depend on θ_1, θ_2, r or μ, it is uniformly most powerful against the one-sided class of alternatives $\theta_2 > \theta_1$.

Dropping the restriction $\theta_2 > \theta_1$, we apply theorem 4 with Ω the set consisting of the two points $\theta_1, \theta_2, r, \mu$ and $\theta_2, \theta_1, r, \mu$. At these two points the envelope power function obviously takes on the same value. If for λ we select the distribution, which assigns equal probabilities to both points, then

$$
\begin{aligned}
g(z) &\sim \exp \{\theta_1 \Sigma x_i + \theta_2 \Sigma y_i\} + \exp \{\theta_2 \Sigma x_i + \theta_1 \Sigma y_i\} \\
&\sim \exp \{\tfrac{1}{2}(\theta_1 - \theta_2)[\Sigma x_i - \Sigma y_i]\} + \exp \{\tfrac{1}{2}(\theta_2 - \theta_1)[\Sigma x_i - \Sigma y_i]\} \\
&\sim | \Sigma x_i - \Sigma y_i | \sim | \bar{y} - \bar{x} | .
\end{aligned}
$$

The power of this test clearly is the same against both points of Ω. Since furthermore the test does not depend on the θ's, r, or μ, it is most stringent against H_1.

A univariate distribution such that

$$
P\{X \epsilon A\} = \int_A C \exp \{\theta x + r(x)\}\, d\nu(x) \tag{4.4}
$$

has been called Laplacian by Tweedie [9], who has studied these distributions in a different connection. Among others, the normal and χ^2, the binomial and Poisson distributions are Laplacian. To obtain, for example, the distribution of a characteristic variable, take for ν the measure ν^* which assigns to a set D the values 0. 1 or 2 according as D contains none, one or both of the points $x = 0$ and $x = 1$, and take as density the function

$$
p^x(1 - p)^{1-x} = (1 - p)e^{x \log(p/1-p)} \tag{4.5}
$$

For comparison with tests which have been considered in the literature, one can specialize the problem just considered, so that the hypothesis $\bar{H}_0$ and the class of alternatives $\bar{H}_1$ consist only of those members of H_0 and H_1 which are generalized densities with respect to a fixed measure μ. One can specialize even further and take as alternative any subset of $\bar{H}_1$ provided with any point θ_1, θ_2, r, it also contains the point θ_2, θ_1, r. The test clearly will not change with these specializations, and the test based on (4.3) will therefore possess the same optimum properties with respect to these special problems as with respect to the problem for which it was originally derived.

If in particular one selects for ν the measure ν^* mentioned above, one obtains the problem for which R. A. Fisher proposed the test based on (4.3). It follows that this test, Fisher's exact test, is most stringent in connection with the following problem: The random variables $X_1, \cdots, X_n, Y_1, \cdots, Y_n$ are characteristic variables, that is, they can take on only the values 0 and 1. If we let (4.6) $P\{X_1 = x_1, \cdots, Y_n = y_n\} = P(x_1, \cdots, y_n)$, the hypothesis states that the function P is invariant under all permutations of its arguments. An equivalent formulation is that the probability (4.6) depends only on $\Sigma x_i + \Sigma y_i$, the total number of "successes". Fisher's exact test is most stringent against the alternative that the X's and Y's are samples from two distinct populations of characteristic variables, that is, two populations corresponding to distinct probabilities of success.

Problem 1c) of section 3 can be extended quite analogously. Put again $Z = (X_1, \cdots, X_n, Y_1, \cdots, Y_n)$, and denote by Π the partition under which two points z and z' are equivalent provided they can be obtained from each other by a permutation of coordinates in which only the coordinates within pairs (X_i, Y_i) are interchanged. Consider the hypothesis of invariance under Π with reference to the class of all distributions and as alternative the class of distributions given by

$$(4.7) \qquad P\{Z \epsilon A\} = \int_A C \exp\left\{\sum_{i=1}^{n} [\theta_1 x_i + \theta_2 y_i + r(x_i, y_i)]\right\} d\mu(z).$$

The θ's here are any real numbers, μ is the $2n$th power of any one-dimensional measure ν, and r is any ν-measurable function such that (a) the integral (4.7) converges when A is the whole space, and such that (b) $r(x, y) = r(y, x)$.

Clearly in the one-sided case $\theta_2 > \theta_1$ we will again find $g(z) \sim \Sigma y_i - \Sigma x_i \sim \bar{y} - \bar{x}$, so that the associated test is uniformly most powerful against this one-sided class of alternatives, while the test based on $|\bar{y} - \bar{x}|$ is again most stringent against the full alternative H_1.

The class of distributions (4.7) contains the distributions (4.1) as a special case. If (X_i, Y_i) $i = 1, \cdots, n$ is a sample from a bivariate normal distribution with $\sigma_X^2 = \sigma_Y^2$, we get another case of (4.7).

As a last somewhat more special problem we mention a discrete analogue of problem 4 of section 3. Let $Z = (Z_1, \cdots, Z_n)$ and consider the class of generalized densities given by

$$(4.8) \qquad P\{Z \epsilon A\} = \int_A P(z_1, \cdots, z_n)\, d\mu(z)$$

where μ is the nth power of ν^*. Let H_0 be the hypothesis that P is invariant under permutations of the coordinates and under the group generated by the transformations $z_i' = 1 - z_i$, $z_j' = z_j$ $j \neq i$ for $i = 1, \cdots, n$. This is an extension of the hypothesis that the probability of success in a binomial distribution equals $\frac{1}{2}$. The test of H_0 against the alternatives that $Z_1, \cdots Z_n$ is a sample of a characteristic variable is based on Σz_i or $| \Sigma z_i |$ as $P\{Z_i = 1\}$ is restricted to be greater than $\frac{1}{2}$ or is not so restricted. In the first case the test is most powerful, in the second most stringent.

5. Hypotheses of invariance for independent variables. To the results obtained so far, a different interpretation can be given, which throws some light on certain related problems. Theorem 2 gave sufficient conditions for a test to be most powerful against a simple alternative H_1 for the hypothesis H_0 of invariance under a partition Π. However, if taken in conjunction with section 1, the theorem can be intepreted as giving sufficient conditions for a test to be the most powerful test of structure $S(\epsilon)$ with respect to Π against H_1. That is, the theorem is really independent of the hypothesis, and depends solely on the alternative and on the class of tests admitted into competition, in our case the class of all tests having structure $S(\epsilon)$ with respect to Π. The same remark obviously also applies to most stringent tests.

Let us now consider a special class of partitions. Let Z stand for the m groups of random variables $(Z_{i1}, \cdots, Z_{is_i})$ $(i = 1, \cdots, m)$ and let Π denote the partition under which two points z and z' are equivalent provided they can be obtained from each other by a permutation of coordinates which however permutes only the coordinates within the m groups. Let μ be the power of a one-dimensional measure ν, and assume that the probability distribution of Z is absolutely continuous with respect to μ and that the Z's are independently distributed, so that

$$P\{Z \in A\} = \int_A \prod_{i,j} [f_{ij}(z_{ij})\, d\nu(z_{ij})]. \tag{5.1}$$

Under these assumptions consider the hypothesis H that f_{ij} is independent of j, that is, that the Z's are identically distributed within each group. It easily can be shown that not all admissible tests of H that have size ϵ, have structure $S(\epsilon)$. However a generalization of a result of Feller [10] and Scheffé [5] for the case $m = 1$ and $\mu =$ Lebesgue measure, states that the only tests which are of size ϵ and similar for H, are the tests of structure $S(\epsilon)$ with respect to Π [11]. It follows that any test which is most powerful or most stringent for testing the hypothesis H' of invariance under Π for the class of generalised densities with respect to μ, has the same property relative to the class of all tests which are similar for testing H.

As an example, take problem 1b) of section 3. Here μ is Lebesgue measure, m is 1, and we put

$$Z_{ij} = \begin{cases} U_j & \text{for } j = 1, \cdots, k \\ V_{j-k} & \text{for } j = k+1, \cdots, k+l = s. \end{cases} \tag{5.2}$$

It was shown in section 3 that the test based on $| \bar{u} - \bar{v} |$, Pitman's test, is most stringent for testing the hypothesis that the joint density of the U's and V's is symmetric in its $k + l$ arguments against the alternative that the variables are independently normally distributed with common variance and such that $E(U_i) = \xi$, $E(V_i) = \eta$ where ξ and η are any distinct real numbers. It follows now that the same test is most stringent similar for testing against the same class of alternatives the hypothesis that $U_1, \cdots, U_k, V_1, \cdots, V_l$ are independently distributed, all with the same probability density. This is the hypothesis for which Pitman proposed his test, and the result just stated is a partial solution of the problem recently raised by Wilks [12], to determine the class of alternatives for which Pitman's test is satisfactory.

If we modify the example by taking for μ instead of Lebesgue measure the $k + l$th power of the measure ν^* of section 4, we are dealing with characteristic variables $U_1, \cdots, U_k, V_1, \cdots, V_l$. We have shown earlier that if $k = l$ the test based on $|\bar{u} - \bar{v}|$ is most stringent for testing the hypothesis of complete permutability against the alternative that the U's and V's are samples from two distinct populations of characteristic variables. If we add to this hypothesis the assumption of independence of all variables, we obtain a parametric problem, namely essentially the problem of testing equality of probability of success in two binomial populations corresponding to the same number of trials. It now follows that the test based on $|\bar{u} - \bar{v}|$ is most stringent for this problem. As is well known, it is also the uniformly most powerful, unbiased similar test.

These two examples suffice to illustrate the type of result that can be obtained. It should perhaps be mentioned that the equivalence discussed at the beginning of this section, can be utilized also in the opposite direction. The fact, for example, that the test based on $|\bar{u} - \bar{v}|$ is known to be uniformly most powerful unbiased similar for testing equality of probability of success in two populations of characteristic variables from which the U's and V's are samples, proves that this test is uniformly most powerful unbiased for testing the hypothesis of complete symmetry for the joint generalized density of the U's and V's.

6. Extension to infinite equivalence classes. The definition of a hypothesis of invariance given in section 1—in spite of the restriction to finite equivalence classes—was sufficiently general to cover the non-parametric problems that we wanted to study. It is possible however to extend the definition so as to allow infinite equivalence classes. In this concluding section we shall briefly outline a theory based on such a broader definition. This will enable us to point out a relationship between the approach of the present paper and the standard parametric theory.

Let $\mathcal{Z}$ be a space of points z and $\mathfrak{A}$ an additive class of subsets of $\mathcal{Z}$. We define a partition of $\mathcal{Z}$ into subsets $\{S_t\}$ as follows: Let $\mathcal{T}$ be some space, and for each $t \in \mathcal{T}$ let S_t be a measurable subset of $\mathcal{Z}$ (i.e. an element of $\mathfrak{A}$) such that the S_t are mutually exclusive and exhaustive. Let $\mathcal{C}_0$ be the class of all $C_0 \in \mathfrak{A}$ which can be expressed in the form

$$C_0 = \bigcup_{t \in D_0} S_t \tag{6.1}$$

and let $\mathfrak{D}_0$ be the class of all D_0 occurring in such relationships. For each $t \in \mathcal{T}$ let G_t be a specified probability measure over $\mathfrak{A}_t$, where $\mathfrak{A}_t$ is the class of A_t such that $A_t \in S_t$, $A_t \in \mathfrak{A}$. Let Z be a random variable distributed over $\mathcal{Z}$ according to an unknown probability measure F. Let $\psi(z)$ be that $t \in \mathcal{T}$ for which $z \in S_t$, and let $T = \psi(Z)$. Let H_0 be the hypothesis that for each $t \in \mathcal{T}$ the conditional distribution of Z given $Z \in S_t$ is G_t, i.e. that there exists a probability measure Q_0 over $\mathfrak{C}_0$ such that for all $A \in \mathfrak{A}$

$$F(A) = \int G_t(A \cap S_t)\, dQ_0(t). \tag{6.2}$$

It is seen that we have essentially the situation described in section 1, except that there we assumed further that each S_t was finite and for all t, G_t assigned equal probabilities to all points of S_t.

We say that a test φ of H_0 has structure $S(\epsilon)$ if the conditional expectation $E_t[\varphi(Z)]$ of $\varphi(Z)$ given $Z \in S_t$ satisfies

$$E_t[\varphi\,(z)] = \int_{S_t} \varphi\, dG_t = \epsilon \quad \text{for all} \quad t. \tag{6.3}$$

The lemmas and theorems stated below are straight-forward generalizations of those in section 1 so that no proof will be given.

Lemma 1′. *Any test φ of structure $S(\epsilon)$ with respect to H_0 is similar and of size ϵ for H_0.*

Lemma 2′. *If φ is any test of H_0 of size $\leq \epsilon$, there exists a test φ, of H_0 having structure $S(\epsilon)$ and such that*

$$\int \varphi_1\, dF \geq \int \varphi\, dF \tag{6.4}$$

for all probability measures F, for which the conditional distribution of Z given $Z \in S_t$ is absolutely continuous with respect to G_t for all t.

Suppose next there is defined another partition of $\mathcal{Z}$ into sets $\{S'_u\}$ by means of a space $\mathcal{U}$, and let $\mathfrak{C}_1$, $\mathfrak{D}_1$ and $\mathfrak{A}_u$ refer to this second partition. We shall assume that for every $t \in \mathcal{T}$, $u \in \mathcal{U}$ either $S'_u \subset S_t$ or $S'_u \cap S_t$ is empty. Let G'_u be a specified probability measure over $\mathfrak{A}_u$ and suppose that for each $t \in \mathcal{T}$ here exists a probability measure Q_t such that for all $A_t \in \mathfrak{A}$,

$$G_t(A_t) = \int G'_u(A_t \cap S'_u) dQ_t(u). \tag{6.5}$$

If H_1 denotes the hypothesis that for each $u \in \mathcal{U}$ the conditional distribution of Z given $Z \in S'_u$ is G'_u, we can state

Theorem 1′. *For testing H_0 against H_1 at level of significance ϵ, the totality of tests φ which have structure $S(\epsilon)$ and for which $z, z' \in S'_u$ implies $\varphi(z) = \varphi(z')$ form an essentially complete class of admissible tests.*

Let F_1 be a distribution not in H_0, and for each $t \in \mathcal{T}$ let G_{1t} be the conditional distribution of Z given $Z \in S_t$. We suppose that for each $t \in \mathcal{T}$, G_{1t} is chosen to be a true probability measure, which is possible in most cases of practical interest (see Doob [13] for a discussion of this point). Then we have the equivalent of theorem 2:

THEOREM 2′. *Let*

$$G_{1t}(A_t) = \int_{A_t - H_t} g_t \, dG_t + G_{1t}(A_t \cap H_t) \tag{6.6}$$

for all $A_t \subset S_t$*, where in accordance with the Radon-Nikodym Theorem* [14], g_t *is a non-negative function integrable over* S_t*, and* $H_t \subset S_t$ *has* G_t *measure* 0 *and does not depend on* A_t. *For testing* H_0 *against* H_1*, a most powerful test of size* ϵ *is given by* $\varphi(z) = \varphi_t(z)$ *for* $z \in S_t$ *where*

$$\varphi_t(z) = \begin{cases} 1 & \text{if } z \in H_t \\ 1 & \text{if } g_t(z) > c_t \\ a_t & \text{if } g_t(z) = c_t \\ 0 & \text{if } g_t(z) < c_t \end{cases} \tag{6.7}$$

where c_t *and* a_t *are so chosen that* φ *has structure* $S(\epsilon)$.

Theorems 3 and 4 require no modification.

As in the case of finite equivalence classes the results just outlined can be interpreted differently. Again the theorems are really independent of the hypotheses, but depend only on the alternatives and on the class of tests admitted into competition. This class of tests φ is in the present case defined by condition (6.4), that the conditiona expectation of φ given $Z \in S_t$ equals ϵ. But this is just the condition which in the standard approach to the problem of testing a composite parametric hypothesis for which T is a sufficient statistic, by means of similar regions is frequently found to be the necessary and sufficient condition for φ to be similar. (See for example [15]). For these cases therefore the hypotheses of the present section represent non-parametric analogues to which the same tests apply with the same optimum properties but without the a priori restriction to similar regions.

As a simple illustration of this remark, let $Z = (Z_1, \cdots, Z_n)$, and let $T = \sum_{i=1}^{n} Z_i^2$. For the conditional distribution of Z given $T = t$ take the uniform distribution over the sphere $T = t$, and for μ take Lebesgue measure. Then the hypothesis H states merely that the joint probability density of the Z's is a function only of $\sum_{i=1}^{n} Z_i^2$. If we add to this the assumption of independence of the Z's, we obtain the new hypothesis H' that the Z's are a sample from a normal distribution with zero mean. The tests φ for which the conditional expectation over each sphere is ϵ, constitute the only admissible tests of H and the only admissible similar tests of H'. If as alternatives we consider that the Z's are a sample from a normal distribution with mean $\xi > 0$, the test

$$\frac{\bar{x}}{\sqrt{\Sigma(x_i - \bar{x})^2}} \geqslant C \tag{6.8}$$

is uniformly most powerful for H and uniformly most powerful similar for H'. If we do not restrict ξ to positive values, the test

$$\left| \frac{\bar{x}}{\sqrt{\Sigma(x_i - \bar{x})^2}} \right| \geq C', \tag{6.9}$$

Student's test, is uniformly most powerful unbiased and most stringent for testing H, uniformly most powerful unbiased similar and most stringent similar for testing H'.

REFERENCES

[1] R. A. Fisher, *Design of Experiments*, Oliver and Boyd, Edinburgh, 1935.

[2] J. Neyman, K. Iwaskiewicz and St. KolodziECZYK, "Statistical problems in agricultural experimentation," *Roy. Stat. Soc. Jour. Suppl.*, Vol. 2 (1935), p. 107.

[3] E. J. G. Pitman, "Significance tests which may be applied to samples from any proportion, *Roy. Stat. Soc. Jour. Suppl.*, Vol. 4 (1937), p. 119; II. The correlation coefficient test, *Roy. Stat. Soc. Jour. Suppl.*, Vol. 4 (1937), p. 225; III. The analysis of variance test, *Biometrika*, Vol. 29 (1938), p. 322.

[4] E. L. Lehmann and C. Stein, "Most powerful tests of composite hypotheses. I. Normal distributions," *Annals of Math. Stat.*, Vol. 19 (1948).

[5] H. Scheffé, "On a measure problem arising in the theory of non-parametric tests," *Annals of Math. Stat.*, Vol. 14 (1943), p. 227.

[6] A. Wald, "An essentially complete class of admissible decision functions," *Annals of Math. Stat.*, Vol. 18 (1947), p. 549.

[7] G. Hunt and C. Stein, "Most stringent tests of statistical hypotheses," *unpublished.*

[8] A. Wald and J. Wolfowitz, "An exact test for randomness in the non-parametric case, based on serial correlation," *Annals of Math. Stat.*, Vol. 14 (1943), p. 378.

[9] M. C. K. Tweedie, "Functions of a statistical variate with given means, with special reference to Laplacian distributions," *Cam. Phil. Soc. Proc.*, Vol. 43 (1947), p. 41.

[10] W. Feller, "Note on regions similar to the sample space," *Stat. Res. Memoirs*, Vol. 2 (1938), p. 117.

[11] E. L. Lehmann and H. Scheffé, "Completeness, similar regions and unbiased estimation," *unpublished.*

[12] S. S. Wilks, "Order Statistics," *Am. Math. Soc. Bull.*, Vol. 54 (1948), p. 6.

[13] J. L. Doob, "Asymptotic properties of Markoff transition probabilities," *Trans. Amer. Math. Soc.*, Vol. 63 (1948), footnote p. 399.

[14] S. Saks, *Theory of the Integral*, Stechert, 1937.

[15] J. Neyman and E. S. Pearson, "On the problem of the most efficient tests of statistical hypotheses," *Roy. Soc. London Phil. Trans.*, Ser. A., Vol. 231 (1933), p. 289.

[16] A. Wald, *On the Principles of Statistical Inference*, Notre Dame Mathematical Lectures, Number 1.

CONSISTENCY AND UNBIASEDNESS OF CERTAIN NONPARAMETRIC TESTS

BY E. L. LEHMANN

University of California, Berkeley

1. Summary. It is shown that there exist strictly unbiased and consistent tests for the univariate and multivariate two- and k-sample problem, for the hypothesis of independence, and for the hypothesis of symmetry with respect to a given point. Certain new tests for the univariate two-sample problem are discussed. The large sample power of these tests and of the Mann-Whitney test are obtained by means of a theorem of Hoeffding. There is a discussion of the problem of tied observations.

2. Introduction. The purpose of the present paper is to investigate the existence and various properties of strictly unbiased and of consistent tests for testing certain nonparametric hypotheses. The problems that will be considered are the two-sample and k-sample problem, the hypothesis of independence and the hypothesis of symmetry with respect to a given point.

A sequence of tests is said to be consistent against a certain class of alternatives if for each alternative the power of the test tends to one as the sample sizes tend to infinity. A test will be said to be strictly unbiased if the power for each alternative exceeds the level of significance.

Consistency being a rather weak property, which one would expect most sequences of tests to satisfy for the class of alternatives for which they are designed, it is important to obtain some more detailed information concerning the power of the various tests under consideration. Because of the tremendous variety of the alternatives it seems fairly hopeless to get a comprehensive view of the achievements of most tests when the samples are small. This in spite of the fact that it is occasionally possible to write down the power explicitly (for example in the simplest cases of the tests discussed by Mathisen [1]). On the other hand, the large sample distribution of a number of test statistics may be found by means of the asymptotic theorems of Hoeffding [2]. Asymptotically, the power then usually involves only a few parameters and a large sample comparison of various different tests becomes possible.

3. Two-sample problem: specific classes of alternatives. We shall discuss in detail only one of the problems mentioned, the two-sample problem, and indicate only briefly certain extensions to the other problems. In the two-sample problem one is given independent samples $X_1, \cdots, X_m$ and $Y_1, \cdots, Y_n$ from populations with unknown cumulative distribution functions F and G respectively, and it is desired to test the hypothesis $F = G$. In this connection various classes of alternatives are possible.

It may, for example, be known that unless $F = G$, the Y's tend to be larger than the X's. For this problem it has been proposed as a test to consider the

J. Rojo (ed.), *Selected Works of E. L. Lehmann*, Selected Works in Probability and Statistics, DOI 10.1007/978-1-4614-1412-4_32, © Springer Science+Business Media, LLC 2012

number of pairs X_i, Y_j for which $X_i < Y_j$, and to reject the hypothesis if this number is too large. This test was proved consistent by Mann and Whitney [3] against the alternatives that

$$F(t) > G(t) \quad \text{for all} \quad t. \tag{3.1}$$

Actually their proof shows that the test is consistent[1] against all alternatives for which $P(Y_j > X_i) > \frac{1}{2}$.

We shall now prove that this test is also unbiased against the alternatives satisfying (3.1).[2] This is true not only for this test but also for those proposed by Thompson [4] and for tests based on randomisation of such statistics as $\bar{y} - \bar{x}$. In fact we have

THEOREM 3.1. *Let w be any similar region for testing $H: F = G$ on the basis of $X_1, \cdots, X_m; Y_1, \cdots, Y_n$. Suppose w is such that $(x_1, \cdots, x_m; z_1, \cdots, z_n) \varepsilon w$ and $y_i \geqq z_i$ for $i = 1, \cdots, n$ implies $(x_1, \cdots, x_m; y_1, \cdots, y_n) \varepsilon w$. Then the test is unbiased against all continuous alternatives F, G satisfying* (3.1).

PROOF. Suppose that $X_1, \cdots, X_m, Y_1, \cdots, Y_n$ are independent and all have the same c.d.f. F and that G is such that (3.1) holds. Then we shall construct $Y_i = f(Z_i)$ such that $Y_i > Z_i$ for $i = 1, \cdots, n$ and such that the Y's have c.d.f. G. Thus the probability of $(X_1, \cdots, X_m; Z_1, \cdots, Z_n) \varepsilon w$ equals the level of significance, α say, while the probability of $(X_1, \cdots, X_m; Y_1, \cdots, Y_n) \varepsilon w$ equals the power of the test against the alternative (F, G). But since $Y_i > Z_i$ for all i, the test rejects for the X's and Y's whenever it rejects for the X's and Z's, and hence the power is $\geqq \alpha$.

The function f is easily defined by the equation

$$G(f(z)) = F(z).$$

(When this does not define $f(z)$ uniquely, it does not matter which of the possible definitions is used.) That $y = f(z) > z$ follows from assumption (3.1).

The theorem as stated refers only to tests in which no randomisation is allowed, but the extension to randomised tests is immediate. Also, as we shall show later, the assumption of continuity of F and G may be omitted.

Theorem 3.1 may be used also to widen the applicability of the tests to which it refers. So far, we have taken the hypothesis to state that X and Y have the same distribution. This formulation may arise, for example, when one is faced with the question whether a treatment, known to be harmless, has a beneficial effect: Either it has no effect so that $F = G$, or it has a good effect. If, on the other hand, the comparison is between two different treatments one may wish to test hypothesis H' that Y tends to be smaller than X, against the alternatives that it tends to be larger. The hypothesis would then be

$$H': F(t) \leqq G(t) \quad \text{for all} \quad t.$$

---

[1] This was also noticed by van Dantzig who points it out in a paper "On the consistency and the power function of Wilcoxon's two sample test," to be published in *Proc. Roy. Inst. Acad. Sci.*, 1951.

[2] For alternatives (F, G) differing only in location this was proved by Van der Vaart [26].

There is of course no nontrivial similar region for this problem, however any region w satisfying the condition of Theorem 3.1 and such that $P(w) = \alpha$ whenever $F \neq G$ clearly will be of size α for testing H' i.e. $P(w)$ will be $\leqq \alpha$ whenever $F(t) \leqq G(t)$ for all t.

Returning to the Mann-Whitney test, let us define V by

$$mnV = \text{number of pairs } X_i, Y_j \text{ with } X_i < Y_j. \tag{3.2}$$

It was shown by Mann and Whitney that V is asymptotically normally distributed when $F = G$ and $m, n \to \infty$ in an arbitrary manner. From a result of Hoeffding (Theorem 7.3 of [2]) it follows that asymptotic normality holds also when $F \neq G$ provided m/n remains constant as $m, n \to \infty$.

We shall apply Hoeffding's theorem to prove asymptotic normality not only of V, but of a large class of statistics connected with the two-sample problem. We begin by stating Hoeffding's theorem, somewhat specialised and with slight changes of notation:

Let $Z_1, \cdots, Z_n$ be independently, identically distributed chance vectors with real components, let $s \leqq n$ and let $\phi(Z_1, \cdots, Z_s)$ be a real valued symmetric function of its s arguments such that $E[\phi(Z_1, \cdots, Z_s)]^2 < \infty$, and let us write

$$E\phi(Z_1, \cdots, Z_s) = \theta.$$

Let

$$U_n = \binom{n}{s}^{-1} \Sigma\phi(Z_{\alpha_1}, \cdots, Z_{\alpha_s}),$$

where the summation extends over all subscripts $1 \leqq \alpha_1 < \cdots < \alpha_s \leqq n$, and let

$$U_n' = U_n + R_n,$$

where R_n is a random variable for which

$$E(nR_n^2) \to 0 \quad \text{as} \quad n \to \infty.$$

Then $\sqrt{n}(U_n' - \theta)$ is asymptotically normally distributed. Further, if we put

$$\psi(z_1) = E[\phi(z_1, Z_2, \cdots, Z_s) - \theta],$$

the limiting distribution of $\sqrt{n}(U_n' - \theta)$ is nondegenerate provided $E[\psi(Z_1)]^2 > 0$.

We can now state

THEOREM 3.2. *Let* $X_1, \cdots, X_m$; $Y_1, \cdots, Y_n$ *be independently distributed with* c.d.f.'s F, G *respectively. Let* $t(x_1, \cdots, x_r, y_1, \cdots, y_r)$ *be symmetric in the* x's *alone and in the* y's *alone. Suppose that*

$$Et(X_1, \cdots, X_r, Y_1, \cdots, Y_r) = \theta(F, G) = \theta,$$

$$E[t(X_1, \cdots, X_r, Y_1, \cdots, Y_r)]^2 = M < \infty.$$

Let $m/n = c$, *and let* n *be sufficiently large so that* $r \leqq m, n$. *Define*

$$U_n' = \binom{m}{r}^{-1}\binom{n}{r}^{-1} \Sigma t(X_{\alpha_1}, \cdots, X_{\alpha_r}, Y_{\beta_1}, \cdots, Y_{\beta_r}),$$

where the summation is extended over all subscripts $1 \leqq \alpha_1 < \cdots < \alpha_r \leqq m$; $1 \leqq \beta_1 < \cdots < \beta_r \leqq n$. *Then, as* $n \to \infty$, $\sqrt{n}(U'_n - \theta)$ *is asymptotically normally distributed.*

PROOF. For the sake of simplicity we shall give the proof only in the case $m = n$. Let $Z_i = (X_i, Y_i)$ and define

$$\phi(Z_1, \cdots, Z_{2r}) = \binom{2r}{r}^{-1} \Sigma t(X_{i_1}, \cdots, X_{i_r}, Y_{j_1}, \cdots, Y_{j_r})$$

summed over all sets of indices for which $(i_1 < \cdots < i_r, j_1 < \cdots < j_r)$ is a permutation of $(1, \cdots, 2r)$. Further let

$$U_n = \binom{n}{2r}^{-1} \Sigma\phi(Z_{\gamma_1}, \cdots, Z_{\gamma_{2r}})$$

summed over all γ's such that $1 \leqq \gamma_1 < \cdots < \gamma_{2r} \leqq n$.

Clearly $\binom{n}{r}^2 U'_n$ is the sum of all possible t-terms, while $\binom{2r}{r}\binom{n}{2r} U_n$ is the sum of only those t-terms in which the X's and Y's have no common subscript. Hence, since $\binom{n}{2r}\binom{2r}{r} = \binom{n}{r}\binom{n-r}{r}$, we have

$$U'_n = \binom{n-r}{r}\binom{n}{r}^{-1} U_n + \binom{n}{r}^{-2} W_n,$$

where W_n is a sum of $\left[\binom{n}{r}\binom{n}{r} - \binom{n}{r}\binom{n-r}{r}\right]$ t-terms, and we can write

$$U'_n = U_n + D_n,$$

where

$$D_n = \left[\binom{n-r}{r} - \binom{n}{r}\right]\binom{n}{r}^{-1} U_n + \binom{n}{r}^{-2} W_n.$$

Since for any real numbers $t_1, \cdots, t_k$ we have $(t_1 + \cdots + t_k)^2 \leqq k(t_1^2 + \cdots + t_k^2)$ we see that

$$E(D_n^2) \leqq \left[\binom{n}{r} - \binom{n-r}{r}\right]^2 \binom{n}{r}^{-2} M + \left[\binom{n}{r} - \binom{n-r}{r}\right]\binom{n}{r}^{-3} M.$$

But, as $n \to \infty$, $\sqrt{n}\left[\binom{n}{r} - \binom{n-r}{r}\right]\binom{n}{r}^{-1} \to 0$. Hence $E(nD_n^2) \to 0$ and the result follows.

Let us now consider the application of this theorem to the Mann-Whitney statistic. We define

$$t(x, y) = \begin{cases} 1 & \text{if} \quad y > x, \\ 0 & \text{if} \quad y \leqq x. \end{cases}$$

Then $U'_n = V_{m,n}$ and asymptotic normality follows since $E\ell^2(X, Y) \leq 1$. It remains to check under what conditions $E\psi^2(Z_1) > 0$. Since we have $s = 2r = 2$,

$$2\psi(z_1) = P(Y_2 > x_1) + P(y_1 > X_2) - 2P(Y > X).$$

Hence $E\psi^2(Z) = 0$ is equivalent to $F(Y) - G(X) =$ constant with probability 1, or $P(Y > x) + P(y > X) =$ constant except on a set of points (x, y) that has probability zero. It is easy to see that this is satisfied if and only if $P(Y > X)$ is 1 or 0.

So far we have considered the hypothesis $H: F = G$ against the alternatives that the Y's tend to be larger than the X's. As a second example we shall consider testing H, or even the wider hypothesis H' that F and G differ only in location (i.e., that $F(x) = G(x + d)$ for some d), against the alternative that the Y's are more spread out than the X's (in a sense to be defined below). In analogy with the Mann-Whitney test let $W_{m,n}$ be the proportion of quadruples X_i, X_j, Y_k, Y_l for which $| Y_l - Y_k | > | X_j - X_i |$. We reject H if $W_{m,n}$ is too large. This test is unbiased against all alternatives (F, G) for which $F(x_1) = G(y_1)$, $F(x_2) = G(y_2)$ implies $| x_1 - x_2 | < | y_1 - y_2 |$. The test is consistent against the wider class of alternatives for which $P(| Y' - Y | > | X' - X |) > \frac{1}{2}$ where X, X', Y, Y' are independently distributed with c.d.f. F, G, respectively. The proof of unbiasedness is quite analogous to the one given previously, and we shall therefore omit it.

We shall however indicate the proof of consistency, and refer in this connection to the closely related remarks by Hoeffding [5] on the construction of consistent sequences of tests.

We first state for reference the following trivial

LEMMA 3.1. *Let $\theta = f(F, G)$ be a real valued function such that $f(F, F) = \theta_0$ for all (F, F) in a class $\mathcal{C}_0$. Let $T_{m,n} = t_{m,n}(X_1, \cdots, X_m, Y_1, \cdots, Y_n)$ be a sequence of real valued statistics such that $T_{m,n}$ tends to θ in probability as* min $(m, n) \to \infty$. *Suppose that $f(F, G) > \theta_0 (\neq \theta_0)$ for all (F, G) in a class $\mathcal{C}_1$. Then the sequence of tests which reject when $T_{m,n} - \theta_0 > C_{m,n}$ (when $| T_{m,n} - \theta_0 | > C'_{m,n}$) is consistent for testing H: $\mathcal{C}_0$ at every fixed level of significance against the alternatives $\mathcal{C}_1$.*

For proof one need only to notice that a fixed level of significance $\neq 0$ implies that $C_{m,n} \to 0$ $(C'_{m,n} \to 0)$ as $m, n \to \infty$.

In the applications we have in mind, $E(T_{m,n})$ is usually independent of m and n, and is easy to find. On the other hand some work is required to determine $\sigma^2(T_{m,n})$. It is therefore of interest to notice that the evaluation of $\sigma^2(T_{m,n})$ is frequently not necessary to prove consistency. To this end we shall state the following lemma, which is a generalisation of a theorem of Halmos [6], and which follows easily from Theorem 5.1 of [7]. A simple proof will be given in [8].

LEMMA 3.2 (Lehmann-Scheffé). *Let $f(F, G)$ be a real valued function defined for all continuous* c.d.f.'s *F and G. There exists at most one function $t_{m,n}$ such that $t_{m,n}(X_1, \cdots, X_m, Y_1, \cdots, Y_n)$ is symmetric in the first m and in the last n arguments and is an unbiased estimate of $f(F, G)$ for all continuous (or even ab-*

solutely continuous) c.d.f.'s F, G. *If such a function* $t_{m,n}$ *exists, (and has finite variance), it has among all unbiased estimates of* $f(F, G)$ *uniformly smallest variance.*

For the application to be made here we need the slightly stronger statement that the conclusion of the Lemma remains valid if $t_{m,n}(X_1, \cdots, X_m; Y_1, \cdots, Y_n)$ is an unbiased estimate of $f(F, G)$ for all continuous c.d.f.'s F, G for which

$$P(|Y' - Y| > |X' - X|) > \tfrac{1}{2}.$$

This generalization follows immediately from the proof of the Lemma given in [8].

The proof of consistency of the proposed test is now immediate. For let $W'_{m,n}$ be the proportion of quadruples for which the Y's are further apart than the X's among the independent quadruples $X_1, X_2, Y_1, Y_2; X_3, X_4, Y_3, Y_4; \cdots$. Then

$$E(W'_{m,n}) = E(W_{m,n}) = P(|Y' - Y| > |X' - X|),$$

and hence by Lemma 3.2,

$$\sigma^2(W_{m,n}) \leqq \sigma^2(W'_{m,n}). \tag{3.3}$$

But $\sigma^2(W'_{m,n})$ obviously tends to zero as $m, n \to \infty$.

We remark finally that the large sample distribution of $W_{m,n}$, by Theorem 3.1, is again normal. Degeneracy occurs only if either F or G are one-point distributions.

As a last problem in this section we shall consider the hypothesis $F = G$ against the combined class of alternatives that the Y's are larger than the X's or more spread out. In such a problem it seems important not only to decide whether F and G are equal but, in case the hypothesis is rejected, for which of the two possible reasons it is rejected or whether it is rejected for both of them. (See in this connection the discussion by Berkson [9]). Thus one is really dealing with a multidecision problem. One must decide between

d_0: Accepting the hypothesis $H: F = G$,

d_1: Rejecting H for the reason that the Y's are larger than the X's,

d_2: Rejecting H for the reason that the Y's are more spread out than the X's,

d_3: Rejecting H for both reasons.

It is desired to find a decision procedure under which the probability of taking decision d_0 is $1 - \alpha$ when $F = G$ while the probability of taking the appropriate of the decisions d_1, d_2, d_3 when the hypothesis is false tends to 1 as the sample sizes tend to infinity. Let us recall the statistics $V_{m,n}$ and $W_{m,n}$ introduced in connection with the previous problems and let us denote $E(V_{m,n})$ and $E(W_{m,n})$ by θ and η respectively. One may then accept H when $V_{m,n} \leqq a_{m,n}$, $W_{m,n} \leqq b_{m,n}$, or take one of the remaining three decisions according as to which one of the three complementary inequalities holds. The constants $a_{m,n}$ and $b_{m,n}$ are not completely determined by the equation

$$P(V_{m,n} \leqq a_{m,n}, W_{m,n} \leqq b_{m,n} \mid F = G) = \alpha.$$

One may specify some additional restriction, such as

$$P(V_{m,n} \leqq a_{m,n} \mid F = G) = P(W_{m,n} \leqq b_{m,n} \mid F = G).$$

It is easy to prove that the above procedure has the consistency property asked for. This follows from Lemmas 3.1 and 3.2 generalised to the case that the function $f(F, G)$ of these lemmas is vector valued instead of real valued. The function $t_{m,n}$ of Lemma 3.2 is then also vector valued and instead of its variance one may consider its ellipsoid of concentration (see [10] and Theorem 5.2 of [7]). In the present case we notice that $(V_{m,n}, W_{m,n})$ is a symmetric estimate of (θ, η) and hence has a uniformly smallest ellipsoid of concentration. But one can easily construct unbiased estimates of θ and η based on independent samples, whose concentration ellipsoid has both axes tending to zero as the sample sizes increase indefinitely and so consistency can be proved by the device used after Lemma 3.2.

4. Two-sample problem: general class of alternatives. In the present section we shall consider the problem of testing the hypothesis $F = G$ against the class of all continuous alternatives $F \neq G$. One might argue that this should not be treated as a hypothesis-testing problem. For Berkson's argument seems to apply: The question is not only whether or not the hypothesis is true. If it is false, it is necessary to decide what alternative hypothesis is correct. While in some situations, this criticism seems to be valid, there are others in which it does not seem to apply.

The two-sample problem may arise in the following two quite different settings.

A: Two production processes, treatments or populations are available, and it is desired to decide whether one is better than the other. In this case the populations F and G are in competition, and the main problem is that of ranking them. Here the notion of such a ranking automatically suggests some specific class or classes of alternatives to the hypothesis that the populations do not differ.

B: The two populations coexist. There is no question of which is preferable, but we wish to know whether the two can be treated as one. One may, for example, want to know whether the output of two different machines can be treated as a uniform product, or whether data obtained under two different experimental setups or by two different investigators may be pooled. These problems really are two-decision problems: The data can or can not be pooled. An explanation of why they can not be pooled is not necessarily of interest.

In connection with the present problem Wald and Wolfowitz [11] proposed as test statistic the total number of runs of the ordered x's and y's, the hypothesis to be rejected if the number of runs is too small. The authors proved their test consistent, under the assumption of constant ratio of sample sizes m/n, against alternatives of all shapes restricted only by mild assumptions, concerning existence and positiveness of the probability densities. It was also proved in their paper that the test statistic has an asymptotically normal distribution when the hypothesis is true. More recently Wolfowitz [12] proved that the limiting distribution is normal even when $F \neq G$, and obtained the asymptotic variance for this case. It follows from his results that the test is in general not consistent if $m/n \to 0$ or ∞. This is actually what one would expect since when m/n is sufficiently extreme the maximum number of runs will in general occur with near-certainty whether the hypothesis is true or false.

Another test suitable for this problem is that of Smirnov [13] based on the maximum difference between the two sample cumulative distribution functions. For the given samples $X_1, \cdots, X_m; Y_1, \cdots, Y_n$ let

$$\phi_m(t) = \phi(X_1, \cdots, X_m; t) = \frac{1}{m} \text{ (number of } X\text{'s} \leqq t),$$

$$\psi_n(t) = \psi(Y_1, \cdots, Y_n; t) = \frac{1}{n} \text{ (number of } Y\text{'s} \leqq t),$$

be the two sample cumulative distribution functions. It follows from a theorem of Glivenko-Cantelli [14], that $\sup_t |\phi_m(t) - F(t)|$ and $\sup|\psi_n(t) - G(t)|$ tend to zero in probability as $\min(m, n) \to \infty$. From this it is easily seen that $\sup|\phi_m(t) - \psi_n(t)|$ is a consistent estimate of $\sup|F(t) - G(t)|$, and hence that Smirnov's test is consistent against all alternatives $F \neq G$ as $\min(m, n) \to \infty$. A different proof of this fact was given recently by Massey [25].

The large sample distribution of $\sup|\phi_m(t) - \psi_n(t)|$ was obtained by Smirnov, for the case that $F = G$, a simpler proof having recently been given by Feller [15] (see also Doob [16] and Smirnov [17]). Although the large sample distribution is not known when $F \neq G$, Massey [25] obtained a lower bound for the power of Smirnov's test, which may permit comparing this test with others.

While these two generally consistent tests are known for the two-sample problem, very little work has been done on the existence of unbiased tests for this or other nonparametric problems. Mann [18] proved unbiasedness of a test for randomness against a certain class of trends. Hoeffding [5] proved the non-existence for the hypothesis of independence of unbiased critical regions based on ranks, corresponding to certain very small levels of significance.

As far as the two-sample problem is concerned, Smirnov's test is easily shown to be biased on the basis of an example given by Massey for the problem of goodness of fit. On the other hand, it seems very possible that the Wald-Wolfowitz run test is unbiased whenever the two samples are of equal size. We have not proved this but shall now construct a test for the two sample problem that is strictly unbiased.

Lemma 4.1 *Let $X, X'; Y, Y'$ be independently drawn from populations with continuous cumulatives F, G respectively, and let us denote for any random variables $U, U'; V, V'$ the event* $\max(U, U') < \min(V, V')$ *by $U, U' < V, V'$. Then*

$$p = P((X, X' < Y, Y') + (Y, Y' < X, X')) = \tfrac{1}{3} + 2 \int (F - G)^2 \, d\left(\frac{F + G}{2}\right),$$

and hence p attains its minimum value $\frac{1}{3}$ if and only if $F = G$.

PROOF. Since F and G are continuous,

$$p = \int (1 - F)^2 \, dG^2 + \int (1 - G)^2 \, dF^2 = 2 + \int (F^2 \, dG^2 + G^2 \, dF^2)$$

$$- 4 \int FG \, d(F + G) = 2 + \int d(F^2 G^2) - 4 \int FG \, d(F + G)$$

$$= 3 - 2 \int [(F+G)^2 - (F - G)^2] \, d\left(\frac{F + G}{2}\right)$$

$$= \tfrac{1}{3} + 2 \int (F - G)^2 \, d\left(\frac{F + G}{2}\right).$$

To prove the second part of the lemma, we must show that $\Delta = \int (F - G)^2 d(F + G) = 0$ implies $F = G$. Now $\Delta = 0$ implies $F(x) = G(x)$ except possibly on a set N such that $\int_N dF = \int_N dG = 1$. Suppose that $F(x_1) \neq G(x_1)$, $G(x_1) - F(x_1) = \eta > 0$ say. Then by continuity there exists $x_0 < x_1$ such that $G(x_0) = F(x_0) + \eta/2$ and $F(x) < G(x)$ for $x_0 \leqq x \leqq x_1$. Since $G(x_1) - G(x_0) > 0$, it follows that $\Delta > 0$.

It is now clear that there exists a strictly unbiased test of $H:F = G$ if $m, n \geqq 2$. For we can consider the number of quadruples $X_{2i-1}, X_{2i}; Y_{2i-1}, Y_{2i}$ for which either the two X's fall below the two Y's or vice versa. These may be regarded as the successes in independent trials with probability $p = \frac{1}{3} + 2\Delta$ of success, and the problem reduces to that of testing $H:p = \frac{1}{3}$ against alternatives $p > \frac{1}{3}$.

The unbiased test just described has the pleasant property that its power is a strictly increasing function of $\Delta = \int (F - G)^2 \, d\,\dfrac{F + G}{2}$, which seems a reasonable measure of the degree of difference of F and G. On the other hand one would not expect this test to be very efficient. More reasonable use of the data seems to be made if one modifies the test in the direction of the Mann-Whitney test described earlier. One would then compare each pair of X's with each pair of Y's, and reject H if among the $\binom{m}{2}\binom{n}{2}$ possible quadruples $X_i, X_j; Y_k, Y_l$ it happened too frequently that both X's lie on the same side of both Y's.

This test is no longer unbiased, but it is still consistent as follows from the argument given in the previous section. Further, the test retains the property that the statistic on which it is based provides an unbiased estimate, in fact the minimum variance unbiased estimate, of the quantity $\int (F - G)^2 \, d\,\dfrac{F + G}{2}$. Finally, it is again easily seen that the distribution of the test statistic is approximately normal, degeneracy occurring only if $P(Y > X)$ equals 1 or 0.

The test can be expressed in a form more convenient for computation in terms of the ranks of one of the sets of variables. Let $r_1 < r_2 < \cdots < r_n$ be the ranks of the n Y's among the totality of $m + n$ observations, and denote by $Q_{m,n}$ the number of quadruples X_i, X_j, Y_k, Y_l for which both X's lie on the same side of both Y's. Then it is easily seen that

$$Q_{m,n} = \sum_{k=1}^{n} \left[(n - k) \binom{r_k - k}{2} + (k - 1) \binom{m - r_k + k}{2} \right].$$

From this it follows by easy computation that

$$2Q_{m,n} = (n - 1) \sum_{k=1}^{n} r_k^2 - 2(n + m - 2)\Sigma k r_k - (n - 2m + 1)\Sigma r_k$$

$$+ (n + 2m - 3) \frac{n(n + 1)(2n + 1)}{6} + (n + m^2 - 3m + 1) \frac{n(n + 1)}{2}.$$

It may perhaps be worth noting that the first of the two tests described in this section can also be used as the basis of a sequential test of the two sample problem. This is clear since the problem is simply reduced to that of testing a simple binomial situation against a one-sided class of alternatives. The sequential probability ratio test to which one is led in this manner of course is again unbiased and has a power function that is strictly increasing in $\int (F - G)^2 \, d \frac{F + G}{2}$.

The measure of discrepancy

$$\int (F - G)^2 \, d \frac{F + G}{2}$$

utilised in the tests of the present section, suggests using $\int (\phi_m - \psi_n) \, d \left(\frac{\phi_m + \psi_n}{2} \right)$ as test statistic. It should be pointed out that tests of this kind have been studied in connection with the closely related problem of goodness of fit by Cramér [19] and von Mises [20]. In the present case, let us denote the x's and y's in order of magnitude by $x^{(1)} < x^{(2)} < \cdots < x^{(m)}$; $y^{(1)} < y^{(2)} < \cdots < y^{(n)}$, let m_1 be the number of x's $< y^{(1)}$, m_2 the number of x's between $y^{(1)}$ and $y^{(2)}$ etc., and define $n_1, n_2, \cdots$ analogously. Then it is easily seen that

$$\int (\phi_m - \psi_n)^2 \, d(\phi_m + \psi_n)$$

$$= \frac{1}{n} \left[\left(\frac{m_1}{m} - \frac{1}{n} \right)^2 + \left(\frac{m_1 + m_2}{m} - \frac{2}{n} \right)^2 + \cdots + \left(\frac{m_1 + \cdots + m_n}{m} - 1 \right)^2 \right]$$

$$+ \frac{1}{m} \left[\left(\frac{n_1}{n} - \frac{1}{m} \right)^2 + \cdots \right].$$

Tests of this type have been proposed by Dixon and by Mood [21], but have not been studied thoroughly.

Finally it should be mentioned that one might also try the method of randomisation, which has been considered by Pitman [22] and others in connection with specific classes of alternatives, for the present problem. One statistic which may be suitable for this purpose if $m = n$ is $\sum_{i=1}^{n} (Y^{(i)} - X^{(i)})^2$.

5. Discontinuous distributions. So far, we have assumed F and G to be continuous. This assumption is obviously not satisfied in practice, and we must therefore consider the difficulties introduced by discontinuities. (These difficulties were investigated in connection with various estimation problems by Scheffé and Tukey [23]).

Let us restrict our attention to rank tests and introduce the convention that tied observations are ordered at random. Thus if $X_{i_1} = \cdots = X_{i_s} = Y_{j_1} = \cdots = Y_{j_t}, s + t = r$, we perform an experiment with $r!$ possible and equally probable outcomes. We then establish a 1:1 correspondence between the $r!$ possible orderings of r objects and these $r!$ outcomes, and treat the X's and Y's as if they had occurred in the order indicated by this experiment. If the X's and Y's have the same distribution it is then clear that the distribution of any rank statistic of the X's and Y's is what it would be if this common distribution were continuous since in both cases each possible ordering of the X's and Y's is again equally probable.

In order to see that various unbiasedness results of the preceding and following sections remain valid, we state the following

LEMMA 5.1. *Let $\mathcal{E} = \mathcal{E}(X_1, \cdots, X_m, Y_1, \cdots, Y_n)$ be a random event depending only on the ranking of the X's and Y's. Suppose that F and G may have discontinuities and that in case of ties the event $\mathcal{E}$ is defined by ordering the tied observations at random. Then there exist continuous* c.d.f.'s $F^* = F^*(F, G)$ *and* $G^* = G^*(F, G)$ *such that*

$$P_{F,G}(\mathcal{E}) = P_{F^*,G^*}(\mathcal{E}).$$

and that $F^* = G^*$ if and only if $F = G$.

PROOF. We shall only give the construction of F^*, G^*; the remainder of the proof then follows easily.

Consider the (denumerable) totality of points that are points of discontinuity of either F or G, and suppose these points have been numbered: $x_1, x_2, \cdots$. Consider first x_1 and define two new c.d.f.'s F_1, G_1 as follows:

$$\begin{aligned} F_1(x) &= F(x + \tfrac{1}{4}) \quad \text{if} \quad x < x_1 - \tfrac{1}{4} \\ &= F(x_1^-) + \frac{x - (x_1 - \frac{1}{4})}{\frac{1}{2}} [F(x_1) - F(x_1^-)] \quad \text{if} \quad |x_1 - x| \leqq \tfrac{1}{4} \\ &= F(x - \tfrac{1}{4}) \quad \text{if} \quad x > x_1 + \tfrac{1}{4}. \end{aligned}$$

G_1 is defined analogously in terms of G. What this construction does is to push F and G apart at x_1 symmetrically by a total amount of $\frac{1}{2}$, and to distribute the probability at x_1 uniformly over the gap thus created.

In the same way we now push F_1 and G_1 apart at the second discontinuity (in its new position) by a total amount of $1/2^2$ and distribute the amount of jump uniformly over the gap, thus obtaining F_2 and G_2. Then the sequence F_1, $F_2, \cdots$ will converge to a continuous distribution F^* and analogously for the G's and F^*, G^* will have the desired properties.

It follows from this lemma that the unbiased test of the hypothesis $F = G$ discussed in Section 4 remains strictly unbiased when the assumption of continuity of F and G is dropped. On the other hand, the power is no longer such a simple function of F and G. In fact let X, X'; Y, Y' denote as before independent random variables with distributions F and G respectively and denote by $X, X' < Y, Y'$ that this ordering occurred after randomisation of ties. Then it is not difficult to show that

$$P((X, X' < Y, Y') + (Y, Y' < X, X')) = \tfrac{1}{3} + 2\Delta',$$

where

$$3\Delta' = \int [(F - G)^2 + (F^- - G^-)^2 + (F - G)(F^- - G^-)] \, d\, \frac{F + G}{2}.$$

Here $F^-(x) = F(x^-)$, $G^-(x) = G(x^-)$.

6. Existence of unbiased tests for the hypothesis of independence and some other nonparametric problems. In this last section we shall briefly consider some more complicated nonparametric problems. Our aim is to prove for all these problems the existence of strictly unbiased and consistent tests. The problem is treated purely theoretically in that no effort is made to construct tests that make good use of the data and that are convenient to apply, but that instead the sole purpose is to exhibit tests possessing the properties asked for.

For the hypothesis of independence Hoeffding proposed a test that he proved consistent against all alternatives with continuous joint and marginal probability densities. In this connection he also considered the problem of unbiasedness and proved the nonexistence of unbiased critical regions based on rank for certain small levels of significance. This negative result seems to contradict those of the present section. This is however not so. Hoeffding restricted his attention to critical regions while we are here admitting also randomised tests. It should be pointed out in this connection that, while randomisation was used in previous sections only in a trivial manner, namely so as to get the exact level of significance, we shall here make very heavy use of this device. This could be avoided in part, however the tests would then become more complicated. Further if the problem is reduced, as is done here, to that of testing equality of two binomial p's, randomisation is needed to get an exactly similar test.

The hypothesis of independence states that the joint c.d.f. equals the product of the two marginal c.d.f.'s. Thus if $(X_i^{(1)}, X_i^{(2)})$, $i = 1, 2, \cdots$, are independently drawn from a bivariate distribution F, it is equivalent to the hypothesis that the pair $(X_1^{(1)}, X_1^{(2)})$ comes from the same bivariate population as the pair $(X_2^{(1)}, X_3^{(2)})$. It is therefore clear that if we can prove the existence of strictly unbiased and consistent tests for the bivariate two-sample problem, this will imply the existence of tests with these properties for the hypothesis of independence. The same remark clearly applies to hypothesis of independence (both complete independence and independence of sets of variates) in more than two variables.

Consider now samples $X_i = (X_i^{(1)}, \cdots, X_i^{(k)})$ $i = 1, 2, \cdots$ and $Y_j = (Y_j^{(1)}, \cdots, Y_j^{(k)})$ $j = 1, 2, \cdots$ from two k-variate distributions F and G. The work of section 4 suggests utilising the expression

$$\int (F - G)^2 \, d\left(\frac{F + G}{2}\right) = \int (F^2 + G^2) \, d\left(\frac{F + G}{2}\right) - 2\int FG \, d\left(\frac{F + G}{2}\right).$$

All that is necessary is to construct events A and B such that

$$p_1 = P(A) = \int \frac{F^2 + G^2}{2} \, d\left(\frac{F + G}{2}\right),$$

$$p_2 = P(B) = \int FG \, d\left(\frac{F + G}{2}\right).$$

The hypothesis $H: F = G$ will then be reduced to $H': p_1 = p_2$ to be tested against alternatives $p_1 > p_2$. The events A and B may be defined as follows:

A: With probability $\frac{1}{2}$ observe either X_1, X_2 or Y_1, Y_2 and with probability $\frac{1}{2}$ observe either X_3 or Y_3. Denote the three variables that are observed by Z_1, Z_2, Z_3, and define A as the event

$$Z_1^{(i)}, Z_2^{(i)} \leqq Z_3^{(i)} \qquad \text{for} \quad i = 1, \cdots, k.$$

B: Observe X_4, Y_4 and with probability $\frac{1}{2}$ either X_5 or Y_5. If the last of these variables is denoted by Z_5, define B as the event

$$X_4^{(i)}, Y_4^{(i)} \leqq Z_5^{(i)} \qquad \text{for} \quad i = 1, \cdots, k.$$

It should be mentioned that instead of observing five random vectors some of which may be either X's or Y's, we could have obtained a test with the desired property based on ten observations, five X's and five Y's.

To complete the proof we must show that the hypotheses H and H' are really equivalent, that is, that $p_1 = p_2$ if and only if $F = G$. For the case that F and G are continuous this follows immediately by an argument similar to the one given in the univariate case, and it is easy to show it even without this restriction.

It is clear that one can generalise further and instead of two samples consider s samples. For this purpose one may replace $\int (F - G)^2 \, d\left(\frac{F + G}{2}\right)$ for example by $\sum_{i=1}^{s} (F_i - \bar{F})^2 \, d\bar{F}$ where $\bar{F}$ is the average of the s c.d.f.'s. Alternatively, one may utilise the expression $\sum_{i<j} \int (F_i - F_j)^2 \, d\left(\frac{F_i + F_j}{2}\right)$.

As a last problem let us consider a sample $X_1, \cdots, X_n$ from an unknown univariate c.d.f. F, assumed to be continuous. It is desired to test the hypothesis H of symmetry with respect to the origin, i.e., that $F(x) = 1 - F(-x)$ for all x. Smirnov [24] recently proposed $\max_x \{ | N^+(x) - N^-(x) | \}$ as a test statistic where $N^+(x)$, $N^-(x)$ denote the number of x's contained in the intervals $(0, x)$, $(-x, 0)$ respectively.

The work of Section 4 suggests considering 4 X's (X_i, X_j, X_k, X_l) and defining the following two events.

A: Exactly two of the four X's are positive.

B: If A is satisfied, and $X_i, X_j < 0 < X_k, X_l$, say, the event B is said to occur if neither

$$|X_i|, |X_j| < X_k, X_l \qquad \text{nor} \qquad X_k, X_l < |X_i|, |X_j|.$$

Then if $F(0) = p_0$, $P(A) = 6p_0^2q_0^2$ takes on its maximum value 3/8 if and only if $p_0 = 1/2$. Further, $P(B \mid A)$ takes on its maximum value 2/3 if and only if the conditional distribution F^* of $-X$ given $X < 0$ is the same as that, G^*, of X given $X > 0$. Thus

$$P(AB) = 6p_0^2q_0^2\left\{\frac{2}{3} - 2\int(F^* - G^*)^2\, d\left(\frac{F^* + G^*}{2}\right)\right\}$$

takes on its maximum value 1/4 if and only if the hypothesis of symmetry holds.

If we apply this method to independent quadruples, we obtain a test that is strictly unbiased and consistent. If we apply it to all possible quadruples the test remains consistent and may be a reasonable test for the hypothesis in question. Hoeffding's theory can again be applied to the asymptotic distribution problem.

REFERENCES

[1] H. C. Mathisen, "A method of testing the hypothesis that two samples are from the same population," *Annals of Math. Stat.*, Vol. 14 (1943), pp. 188–194.

[2] W. Hoeffding, "A class of statistics with asymptotically normal distributions," *Annals of Math. Stat.*, Vol. 19 (1948), pp. 293–325.

[3] H. B. Mann and D. R. Whitney, "On a test of whether one of two random variables is stochastically larger than the other," *Annals of Math. Stat.*, Vol. 18 (1947), pp. 50–60.

[4] W. R. Thompson, "Biological applications of normal range and associated significance tests in ignorance of original distribution forms," *Annals of Math. Stat.*, Vol. 9 (1938), pp. 281–287.

[5] W. Hoeffding, "A non-parametric test of independence," *Annals of Math. Stat.*, Vol. 19 (1948), pp. 546–557.

[6] P. R. Halmos, "The theory of unbiased estimation," *Annals of Math. Stat.*, Vol. 17 (1946), pp. 34–43.

[7] E. L. Lehmann and H. Scheffé, "Completeness, similar regions, and unbiased estimation—Part I," *Sankhyā*, Vol. 10 (1950), pp. 305–340.

[8] E. L. Lehmann and H. Scheffé, "Completeness, similar regions, and unbiased estimation, Part II," unpublished.

[9] J. Berkson, "Comments on Dr. Madow's 'Note on tests of departure from normality' with some remarks concerning tests of significance," *Jour. Am. Stat. Assn.*, Vol. 36 (1941), p. 539.

[10] H. Cramér, "Contributions to the theory of statistical estimation," *Skandinavisk Aktuarietidskrift*, Vol. 29 (1946), p. 85.

[11] A. Wald and J. Wolfowitz, "On a test whether two samples are from the same population," *Annals of Math. Stat.*, Vol. 11 (1940), pp. 147–162.

[12] J. Wolfowitz, "Non-parametric statistical inference," *Proceedings of the Berkeley Symposium on Mathematical Statistics and Probability*, University of California Press, Berkeley, 1949.

[13] N. V. Smirnov, "On the estimation of the discrepancy between empirical curves of distribution for two independent samples," *Bull. Math. Univ. Moscow, Serie Int.*, Vol. 2 (1939).

[14] M. Fréchet, *Recherches Théoriques Modernes sur la Théorie des Probabilités, Vol. 1*, p. 260, Gauthier-Villars, Paris, 1937.

[15] W. Feller, "On the Kolmogorov-Smirnov limit theorems for empirical distributions," *Annals of Math. Stat.*, Vol. 19 (1948), pp. 177–189.

[16] J. L. Doob, "Heuristic approach to the Kolmogorov-Smirnov theorems," *Annals of Math. Stat.*, Vol. 20 (1949), pp. 393–403.

[17] N. V. Smirnov, "Approximate laws of distribution of random variables from empirical data," *Uspehi Matem. Nauk*, Vol. 10 (1944), p. 179.

[18] H. B. Mann, "Non-parametric tests against trend," *Econometrica*, Vol. 13 (1945), pp. 245–259.

[19] H. Cramér, "On the composition of elementary errors," *Skandinavisk Aktuarietidskrift*, Vol. 11 (1928), p. 13 and p. 141.

[20] R. von Mises, *Wahrscheinlichkeitsrechnung und ihre Anwendung in der Statistik und theoretischen Physik*, Franz Deuticke, Leipzig, 1931.

[21] W. Dixon, "A criterion for testing the hypothesis that two samples are from the same population," *Annals of Math Stat.*, Vol. 11 (1940), pp. 199–204.

[22] E. J. G. Pitman, "Significance tests which may be applied to samples from any populations," *Jour. Roy. Stat. Soc., Suppl.*, Vol. 4 (1937), pp. 225–232.

[23] H. Scheffé and J. W. Tukey, "Non-parametric estimation. I. Validation of order statistics," *Annals of Math. Stat.*, Vol. 16 (1945), pp. 187–192.

[24] N. V. Smirnov, "Sur un critère de symetrie de la loi de distribution d'une variable aléatoire," *C. R. Acad. Sci. URSS*, Vol. 56 (1947), p. 11.

[25] F. J. Massey, Jr., "A note on the power of a non-parametric test," *Annals of Math. Stat.*, Vol. 21 (1950), pp. 440–443.

[26] H. R. Van der Vaart, "Some remarks on the power function of Wilcoxon's test for the problem of two samples. I," *Indagationes Math.*, Vol. 12 (1950), pp. 146–158.

THE POWER OF RANK TESTS[1]

By E. L. Lehmann

Stanford University and University of California, Berkeley

1. Summary. Simple nonparametric classes of alternatives are defined for various nonparametric hypotheses. The power of a number of such tests against these alternatives is obtained and illustrated with some numerical results. Optimum rank tests against certain types of alternatives are derived, and optimum properties of Wilcoxon's one- and two-sample tests and of the rank correlation test for independence are proved.

2. Introduction. The most pressing need in the theory and practice of nonparametric tests at this time seems to be for results concerning the power of such tests, particularly those based on ranks. This would provide a basis for comparing the many different tests proposed as well as for determining the sample sizes necessary to distinguish significant departures from a hypothesis with a reasonable degree of certainty.

The chief problem one is faced with when investigating the power of a nonparametric test is the choice of suitable alternatives. Even in the simplest problems the variety of alternatives is so great that it is clearly impossible to consider all of them. In the past, investigators have concentrated on alternatives postulating normal distributions for the random variables in question. These alternatives, which unfortunately are rather difficult to handle mathematically, must, of course, be studied if one wishes to find out how nonparametric methods compare with procedures based on normal theory. On the other hand, when comparing different rank tests, one is no longer tied to normal alternatives, but it would on the contrary seem rather desirable to make the comparisons in terms of nonparametric classes of alternatives.

As a specific example, consider the one-sided two-sample problem, and suppose that on the basis of samples $X_1, \cdots, X_m; Y_1, \cdots, Y_n$ from cumulative distribution functions F and G respectively we wish to test the hypothesis $H: F = G$ against the alternatives that $G(x) \leqq F(x)$ for all x. If among these alternatives we look for some simple subclasses, parametric theory suggests

$$G(x) = F(x - a) \qquad \text{for some } a > 0. \tag{2.1}$$

But under such alternatives, the distribution of the ranks will depend not only on a, but also on F, nor, in general, would a be a suitable measure of the difference

Received 8/7/52.

[1] This work was sponsored in part by the Office of Naval Research. Part of the computations were carried out in connection with a project on the evaluation of rain making undertaken by the Statistical Laboratory, University of California for the State Department of Public Works.

of F and G. The situation is similar to the corresponding one for normal distributions with different means and, say, common but unknown variance.

We shall in the present paper discuss mathematically "natural" nonparametric alternatives against which the distribution of the ranks is constant. Once these have been defined, it is relatively simple, on the basis of a theorem of Hoeffding, to obtain the power of any rank test and also to derive tests possessing various optimum properties.

The classes of alternatives with which we shall be dealing involve arbitrary functions for which one must make a definite choice in order to get specific power-results. This choice is here made solely on the grounds of simplicity for the resulting calculations. We do not, of course, claim that these are the alternatives that actually prevail when the hypothesis is not true. Rather, it seems that where nonparametric methods are appropriate, one usually does not have very precise knowledge of the alternatives. What is then required are alternatives representative of the principal types of deviation from the hypothesis, in terms of which one can study, at least in outline, the ability of various tests to detect such deviations. Such an approach is here presented, and the computations are carried through for a few examples. However, in order to get a valid comparison of such tests as the Wald-Wolfowitz run test and the Smirnov two-sample test, for example, much more systematic computation is required. Such computations seem entirely feasible and would seem to be a worthwhile undertaking.

I should like to express my gratitude to Miss E. L. Scott for her help in setting up and supervising the computations for Table 1 and to Mrs. M. Vasilewskis who carried out these computations, as well as to Mr. H. Wagner and Mr. J. Rosenbaum on whose computations Fig. 2 and 3 are based.

3. The hypothesis of randomness. While we shall be concerned mainly with the two-sample problem, it is convenient to present some preliminary considerations in the more general notation of the hypothesis of randomness. We shall here make the assumption, to hold throughout the paper, that all distribution functions that we consider are to be continuous.

Let f_i $(i = 1, \cdots, N)$ be continuous, nondecreasing functions defined over the interval $[0, 1]$ such that $f_i(0) = 0, f_i(1) = 1$. Let $Z_1, Z_2, \cdots, Z_N$ be independent random variables distributed according to cumulative distribution functions $F_1, \cdots, F_N$. We shall denote by $\mathfrak{F}(f_1, \cdots, f_N)$ the family of all $(F_1, \cdots, F_N)$ such that $F_i = f_i(F)$ where F runs through all continuous cdf's. The classes $\mathfrak{F}(f_1, \cdots, f_N)$ for different choices of the functions $f_1, \cdots, f_N$ then define a partition of the family of all N-tuples $(F_1, \cdots, F_N)$ of the kind described. It should perhaps be pointed out that different N-tuples f_i do not necessarily generate different families of F's. If f_1 is strictly increasing on $[0, 1]$, a natural normalization would be to take $f_1(x) = x, 0 \leq x \leq 1$. If $(F_1, \cdots, F_N)$ belongs to the class $\mathfrak{F}(f_1, \cdots, f_N)$ we shall write

$$F_1:F_2: \cdots :F_N = f_1:f_2: \cdots :f_N. \tag{3.1}$$

We shall now show that the distribution of the ranks of the Z's is constant within each family $\mathfrak{F}(f_1, \cdots, f_N)$.

LEMMA 3.1. *If F is a continuous cdf and if the cdf of Z is given by $P(Z \leqq z) = f(F(z))$ where f is nondecreasing on $[0, 1]$ with $f(0) = 0, f(1) = 1$, then the cdf of $F(Z)$ is f.*

PROOF. When $f(u) = u, 0 \leqq u \leqq 1$ this result is well known and implies in our case that $f(F(Z))$ is uniformly distributed over $[0, 1]$. Therefore

$$P(f(F(Z)) < f(u)) \leqq P(F(Z) \leqq u) \leqq P(f(F(Z)) \leqq f(u))$$

and the first and third member equal $f(u)$.

Let us denote the ranks of the N variables $Z_1, \cdots, Z_N$ by $T_1, \cdots, T_N$. Then we have

LEMMA 3.2. *If $Z_1, \cdots, Z_N$ are independent, the distribution of $T_1, \cdots, T_N$ is constant within each family $\mathfrak{F}(f_1, \cdots, f_N)$.*

PROOF. Clearly

$$P(F(Z_{i_1}) < \cdots < F(Z_{i_N})) \leqq P(T_{i_1} = 1, \cdots, T_{i_N} = N) \leqq P(F(Z_{i_1}) \leqq \cdots \leqq F(Z_{i_N})).$$

But the first and third members of this inequality are independent of F and equal since by Lemma 3.1 the distribution of the $F(Z_i)$ is independent of F and continuous.

As an immediate consequence of this lemma we have

THEOREM 3.1. *Given any functions $f_1^0, \cdots, f_N^0$ and any rank test of the hypothesis $H:(F_1, \cdots, F_N) \in \mathfrak{F}(f_1^0, \cdots, f_N^0)$, the power of this test depends only on $F_1: \cdots :F_N$. That is, if $F_1: \cdots :F_N = F_1': \cdots :F_N'$ so that $(F_1, \cdots, F_N)$ and $(F_1', \cdots, F_N')$ belong to the same class $\mathfrak{F}(f_1, \cdots, f_N)$ the test has the same power against these two alternatives. Furthermore, given any class of alternatives $K:(F_1, \cdots, F_N) \in \mathfrak{F}(f_1', \cdots, f_N')$ there exists a uniformly most powerful rank test for testing H against K.*

PROOF. The first statement is just a specialization of Lemma 3.2. Since the distribution of the ranks is simple both under H and K, the second statement as well as a method of constructing the most powerful rank test follow from the Neyman-Pearson fundamental Lemma.

In order to apply this theorem we require the distribution of $(T_1, \cdots, T_N)$ for the $(f_1, \cdots, f_N)$ of our choice. The relevant result was obtained by Hoeffding ([1], p. 88). Instead of stating it here we shall in the next section give its specialization for the two-sample problem.

4. The two-sample problem. Let $X_1, \cdots, X_m$ and $Y_1, \cdots, Y_n$ be independently distributed with cdf's F and G respectively. We wish to test the hypothesis $H:F = G$. The classes $\mathfrak{F}(f_1, \cdots, f_N)$, in the present case, involve only two functions f and g and may be written as $\mathfrak{F}(f, g)$. To simply our notation, we shall assume that f is strictly increasing. Then $\mathfrak{F}(f, g)$ may be represented by a single function g and is given by $\mathfrak{F}(g) = \{(F, g(F))\}$ where the domain of F is as before the totality of continuous cdf's, and where g is a continuous nondecreasing function with $g(0) = 0$, $g(1) = 1$.

Let us denote the ordered X's and Y's by $X^{(1)} < X^{(2)} < \cdots < X^{(m)}$ and $Y^{(1)} < \cdots < Y^{(n)}$ and the ranks of the X's and Y's in the combined sample by $R_1 < \cdots < R_m$ and $S_1 < \cdots < S_n$ respectively. The complete set of the ranks is, of course, determined by the ranks of the Y's alone. We shall assume here that the function g is differentiable on [0, 1] with derivative g'. Specializing

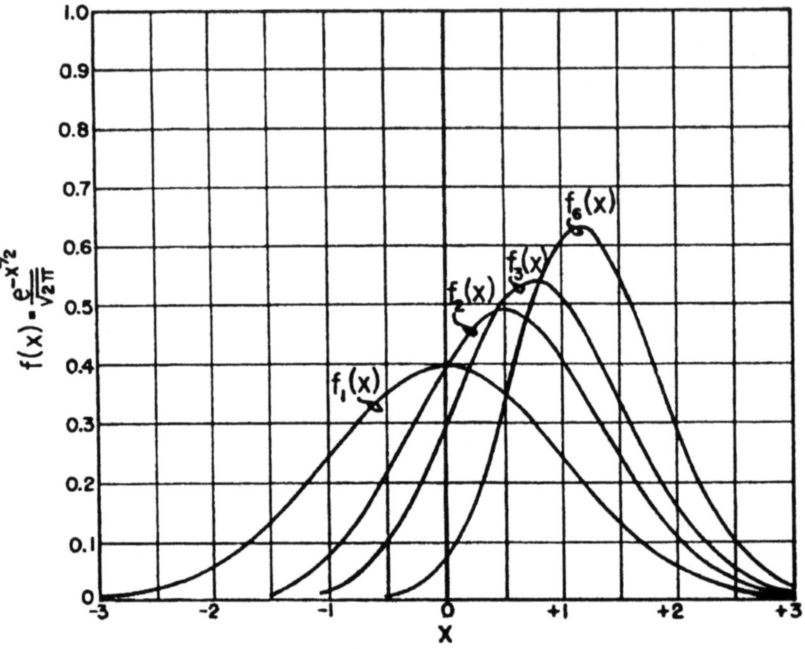

FIG. 1-A

a theorem of Hoeffding ([1], p. 88) to the present case we find that when F and $G = g(F)$ are the cdf's of the X's and Y's, then

$$P(S_1 = s_1, \cdots, S_n = s_n) = \frac{1}{\binom{m+n}{m}} E_F[g'(F(Y^{(s_1)})) \cdot \cdots \cdot g'(F(Y^{(s_n)}))]$$

where the expectation is computed under the assumption that F is the true distribution of both the X's and Y's. Since in this case $F(Y)$ is uniformly distributed over [0, 1], we get

$$P(S_1 = s_1, \cdots, S_n = s_n) = \frac{1}{\binom{m+n}{m}} E[g'(U^{(s_1)}) \cdot \cdots \cdot g'(U^{(s_n)})] \tag{4.1}$$

where $U^{(s_1)}, \cdots, U^{(s_n)}$ are the s_1^{st} to s_n^{th} order statistics in a sample of $m + n$ variables distributed uniformly over [0, 1].

Since the probability distribution of the ranks can be expressed so simply in terms of g it is seen that the difficulty in obtaining power results for a specific alternative is directly related to the complexity of the function g involved. This explains why the investigation for normal alternatives has proved so difficult. When F and G are two distinct normal cdf's, the function $g = G(F^{-1})$ is not particularly easy to handle.

Consider now the one-sided alternatives $G(x) \leqq F(x)$. To this corresponds a function g such that $g(x) \leqq x$, $0 \leqq x \leqq 1$. The simplest choice in view of (4.1) seems to be $g(x) = x^k$; $k > 1$. The associated problem is that of testing $H:G = F$ against the alternatives $K:G = F^k$. In addition to mathematical simplicity, this choice has the advantage of admitting a simple interpretation of the alternatives. Suppose that k is an integer. Then F^k is the distribution of the

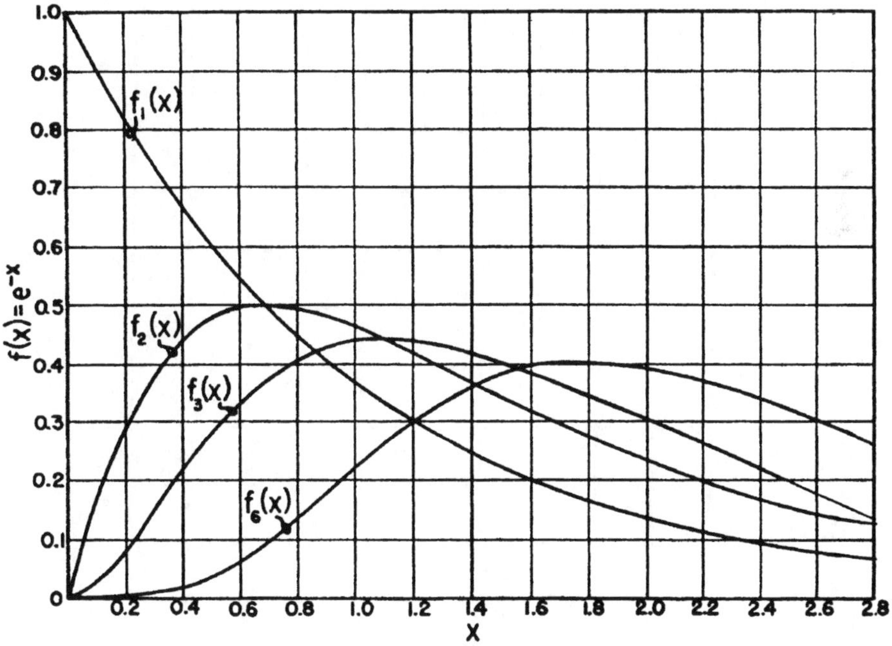

FIG. 1-B

maximum of k independent variables having distribution F. Thus under the alternative, the X's have distribution F while the distribution of the Y's is the same as that of the maximum of k X's.

In order to give an idea of how much larger the Y's are than the X's, note that if $G = F^k$, $P(X < Y) = \int F\,dG = k/k + 1$. In Fig. 1, we have assumed that the distribution of X is given by the densities

$$f_1(x) = \frac{1}{\sqrt{2\pi}} e^{-x^2/2}; \qquad f_1(x) = e^{-x}, \qquad 0 \leqq x; \qquad f_1(x) = 1, \qquad 0 \leqq x \leqq 1$$

respectively and show the density of f_k of Y when $G = F^k$ for $k = 2, 3$ and 6. In terms of the present frame of reference the distance of the density f_k from f_1 is the same in all three cases, since in each of them f_k is the density of the maximum of k observations from f_1 and since further in all three cases, every rank test has the same probability of detecting the hypothesis to be false when f_1 is the density of the X's and f_k that of the Y's.

It is clear that a similar interpretation of the alternatives F^k can be given when k instead of being an integer is any rational number. Altogether, we may think of the class of alternatives $G = F^k$, as a one parameter family of nonparametric classes of alternatives. The distribution of the ranks under these alternatives is now easily determined from (4.1).

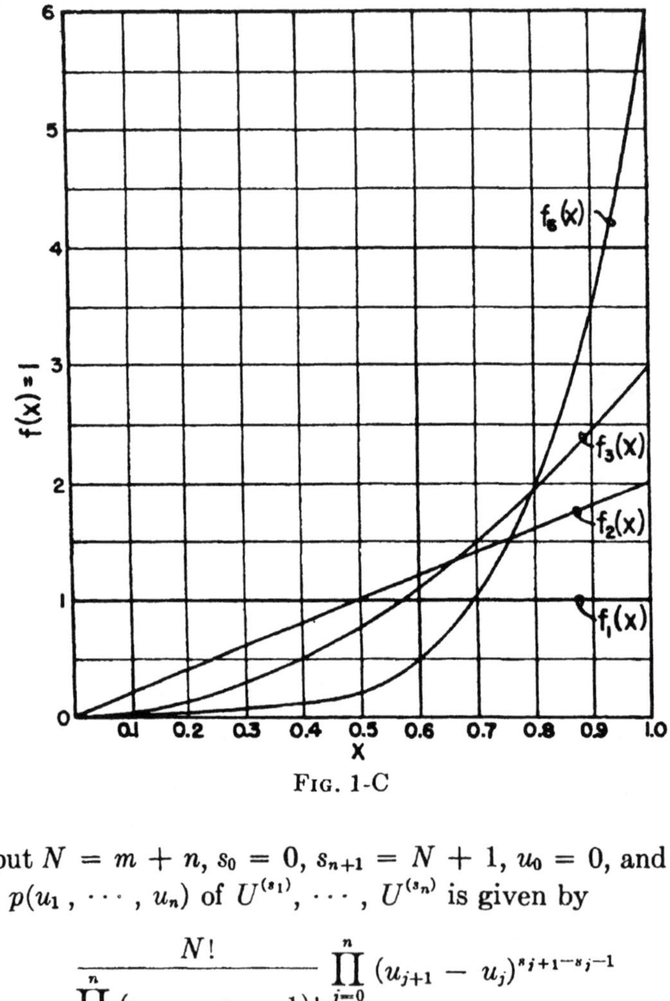

FIG. 1-C

For if we put $N = m + n$, $s_0 = 0$, $s_{n+1} = N + 1$, $u_0 = 0$, and $u_{n+1} = 1$, the joint density $p(u_1, \cdots, u_n)$ of $U^{(s_1)}, \cdots, U^{(s_n)}$ is given by

$$\frac{N!}{\prod_{i=0}^{n} (s_{i+1} - s_i - 1)!} \prod_{j=0}^{n} (u_{j+1} - u_j)^{s_{j+1} - s_j - 1} \tag{4.2}$$

over the region $0 = u_0 \leqq u_1 \leqq \cdots \leqq u_{n+1} = 1$. If here we make a transformation to new variables $V_1, \cdots, V_n$ defined by

$$u_i = v_i v_{i+1} \cdots v_n \qquad (i = 1, \cdots, n) \tag{4.3}$$

and put $v_0 = 0$, $v_{n+1} = 1$, it is seen that the joint density of the V's is

$$\frac{N!}{\prod_{i=0}^{n}(s_{i+1} - s_i - 1)!}\prod_{j=1}^{n} v_j^{s_j-1}(1 - v_j)^{s_{j+1}-s_j-1} \tag{4.4}$$

over the region $0 \leqq v_j \leqq 1$, $j = 1, \cdots, n$ so that the V's are independently distributed according to Beta-distributions, that is, as are single order statistics from a uniform distribution.

TABLE 1

Test	$m = n = 4$		$m = n = 6$	
	$\beta(F^2)$	$\beta(F^3)$	$\beta(F^2)$	$\beta(F^3)$
T_1	.23	.33	.29	.45
T_2	.31	.47	.38	.59
T_3	.32	.49	.41	.64
T_4	.14	.20	.17	.29
T_5	.15	.22	.21	.36
T_6	.19	.32	.25	.44

Since $U^{(s_1)} \cdot \cdots \cdot U^{(s_n)} = V_1 \cdot V_2^2 \cdot \cdots \cdot V_n^n$, we have when $G = F^k$ and hence $g'(u) = ku^{k-1}$,

$$P(S_1 = s_1, \cdots, S_n = s_n) = \frac{k^n}{\binom{m+n}{m}} E[(U^{(s_1)} \cdot \cdots \cdot U^{(s_n)})^{k-1}]$$

$$= \frac{k^n}{\binom{m+n}{m}} \prod_{j=1}^{n} E(V_j^{jk-j}) = \frac{k^n}{\binom{m+n}{m}} \prod_{j=1}^{n} \frac{\Gamma(s_j + jk - j)}{\Gamma(s_j)} \cdot \frac{\Gamma(s_{j+1})}{\Gamma(s_{j+1} + jk - j)}. \tag{4.5}$$

In particular, when $k = 2$ so that $G = F^2$,

$$P(S_1 = s_1, \cdots, S_n = s_n) = \frac{2^n}{\binom{m+n}{m}} \cdot \frac{s_1(s_2 + 1) \cdots (s_n + n - 1)}{(m + n + 1)(m + n + 2) \cdots (m + 2n)}. \tag{4.6}$$

Using (4.6) (or more generally (4.5) or (4.1)) one can now compute the power of various rank tests against the alternatives in question. One must list the sets $(s_1, \cdots, s_n)$ making up the critical region and then sum the right-hand side of (4.6) over these values of the ranks. In this manner Table 1 was com-

puted, which gives the power of six different rank tests $T_1 - T_6$ against the alternatives $G = F^2$ and $G = F^3$ at level of significance $\alpha = .1$. Since the computation of the exact power rapidly increases in difficulty with the sample size, these computations have been carried through only for the cases $m = n = 4$ and $m = n = 6$.
The above tests are defined as follows.

T_1: One-sided median test. Rejects H when too many Y's exceed the median of the combined sample. See Mood ([2], p. 394) and Westenberg [18].

T_2: One-sided Wilcoxon test [3], [4]. Rejects when $S_1 + \cdots + S_n$ is too large, or equivalently when there are too many pairs X_i, Y_j ($i = 1, \cdots, m$; $j = 1, \cdots, n$) with $X_i < Y_j$.

T_3: This is the most powerful rank test for testing $G = F$ against $G = F^2$ (see Section 6). It rejects when $S_1(S_2 + 1) \cdot \cdots \cdot (S_n + n - 1)$ is too large.

T_4: Wald-Wolfowitz run test [5]. Rejects when the total number of runs of X's and Y's is too small.

T_5: Two-sided median test ([2], p. 394). Rejects when either too many X's or Y's exceed the median of the combined sample.

T_6: Two-sided Wilcoxon test [3], [4]. Rejects when

$$\left| S_1 + \cdots + S_n - \binom{m + n + 1}{2} \right|$$

is too large.

Although it is not shown in Table 1, we mention for later reference also

T_7: Smirnov two-sample test [6]. Rejects when $\sup_t$

$$| F_{X_1, \cdots, X_m}(t) - G_{Y_1, \cdots, Y_n}(t) |$$

is too large where $F_{X_1, \cdots, X_m}(t)$ and $G_{Y_1, \cdots, Y_n}(t)$ are the sample cumulative distribution functions of the X's and Y's respectively.

Of course, not all of these tests are directly comparable. While the first three are aimed only at alternatives under which the Y's tend to be larger than the X's, the fifth and sixth test are designed against two-sided alternatives, and the fourth and seventh against arbitrary deviations from the hypothesis.

5. Large sample power. For alternatives of the type $G = F^k$ it also becomes a relatively easy task to compute the approximate power of certain rank tests using large sample theory. Some results obtained in this way are shown in Fig. 2 and 3.

Fig. 2 gives the power of various tests against the alternative $g(F) = F^2$ for different sample sizes n. The lowest of the four curves (labeled β_4) corresponds to the run test (the subscripts refer to the numbering of the tests in the previous section) and is based on theory not yet completely verified. The next curve,

β_7, gives a lower bound to the power of the Smirnov test, while the two upper curves show the power of the two-sided and one-sided Wilcoxon test. In Fig. 3 are shown the corresponding curves for $g(F) = F^3$, except that the run test has been omitted.

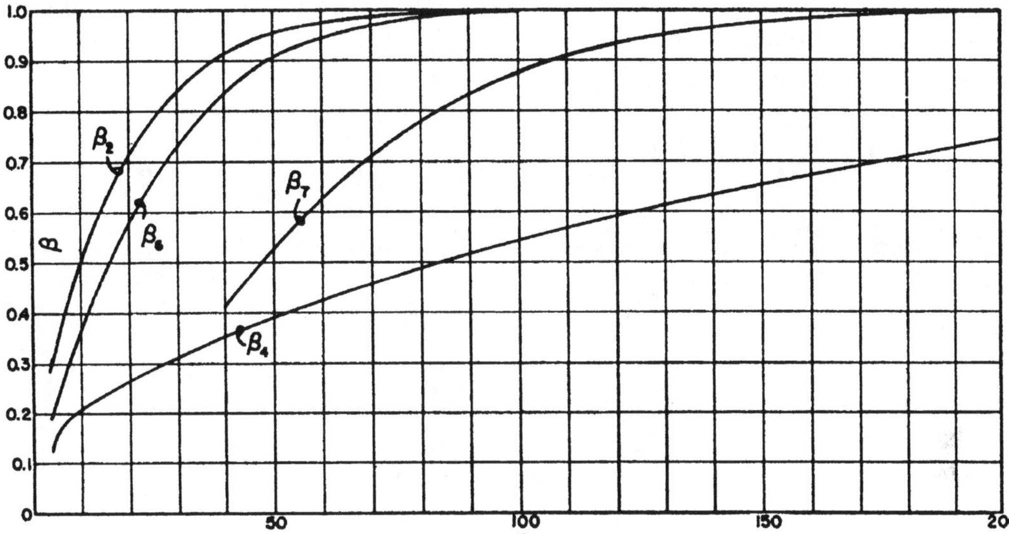

FIG. 2. $\alpha = 0.1 \qquad g(F) = F^2$

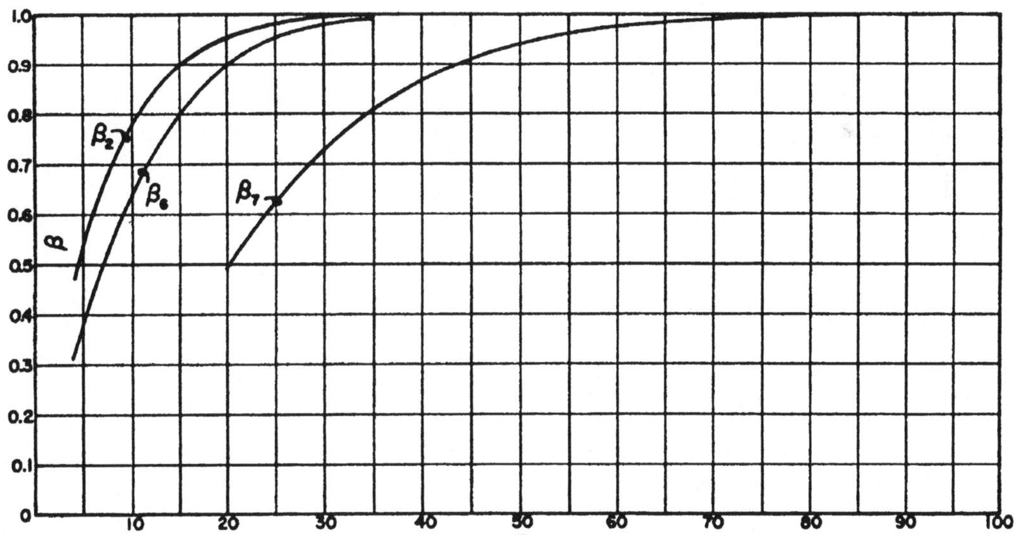

FIG. 3. $\alpha = 0.1 \qquad g(F) = F^3$

The remainder of this section is devoted to a discussion of the formulae from which Fig. 2 and 3 were computed.

For the Wilcoxon statistic U which counts the number of pairs X_i, Y_j with $X_i < Y_j$ it was proved by Mann and Whitney [4] that for large samples

$$\frac{\dfrac{U}{mn} - E\left(\dfrac{U}{mn}\right)}{\sigma\left(\dfrac{U}{mn}\right)}$$

is approximately normally distributed when $F = G$, and the proof of the corresponding fact when $F \neq G$ was given in [7]. Mann and Whitney also gave the first two moments of U as

$$E\left(\frac{U}{mn}\right) = \int F\,dG$$

$$(5.1)\qquad mn\sigma^2\left(\frac{U}{mn}\right) = \left[\frac{m+n+1}{12} + (m-1)(\lambda - \epsilon_1) + (n-1)(\lambda - \epsilon_2) - \lambda^2(m+n-1)\right]$$

where

$$\lambda = \tfrac{1}{2} - \int F\,dG$$

$$\epsilon_1 = \tfrac{1}{3} - \int F^2\,dG, \qquad \epsilon_2 = \tfrac{1}{3} - \int (1-G)^2\,dF.$$

(Note that the notation used here differs from that in [4].)

If $G = F^k$ we have

$$\int F\,dG = \frac{k}{k+1}, \qquad \int F^2\,dG = \frac{k}{k+2}, \qquad \int G^2\,dF = \frac{1}{2k+1},$$

and hence on substituting in (5.1)

$$(5.4)\qquad \begin{aligned} E\left(\frac{U}{mn}\right) &= \frac{k}{k+1} \\ mn\sigma^2\left(\frac{U}{mn}\right) &= \frac{k}{(k+1)^2}\left[\frac{m-1}{k+2} + \frac{k(n-1)}{2k+1} + 1\right]. \end{aligned}$$

The theory of the run test was developed by Wald and Wolfowitz [5] and certain extensions were given in [8]. If W denotes the total number of runs of X's and Y's and if $m/n = \gamma$ it was shown in [5] that when $F = G$, $(W/m - E(W/m))/\sigma(W/m)$ is asymptotically normally distributed. It was also proved that when $G = g(F)$ where the derivative g' of g is continuous and positive on $0 < x < 1$, then

$$(5.3)\qquad E\left(\frac{W}{m}\right) \to 2\int_0^1 \frac{g'(x)}{\gamma + g'(x)}\,dx \qquad \text{as } m \to \infty.$$

In [8] Wolfowitz stated that the distribution of W is approximately normal even when $g(x) \not\equiv x$, and he derived the asymptotic formula

$$\tag{5.4} \begin{aligned} \tfrac{1}{4}\sigma^2\left(\frac{W}{\sqrt{m}}\right) \to \int_0^1 \frac{\gamma g'^2}{(\gamma + g')^3}\,dx + \int_0^1 \frac{g'(\gamma^3 + g'^3)}{(\gamma + g')^4}\,dx \\ - \left[\int_0^1 \frac{g'^2}{(\gamma + g')^2}\,dx\right]^2 - \gamma^3\left[\int_0^1 \frac{g'}{(\gamma + g')^2}\,dx\right]^2. \end{aligned}$$

When $G = F^k$, then $g(x) = x^k$ and $g'(x) = kx^{k-1}$, and the integrals on the righthand side of (5.3) and (5.4) can be evaluated without much difficulty. The power of the run test against $g(F) = F^2$ shown in Fig. 2 was computed in this manner. Since then it has been pointed out to me by R. Savage that when the limit result for

$$\left(\frac{W}{m} - E\left(\frac{W}{m}\right)\right) \Big/ \sigma\left(\frac{W}{m}\right)$$

we replace

$$E(W/m)$$

by

$$2\int_0^1 g'(x)/(\gamma + g'(x))\,dx,$$

the error is of the order

$$\sqrt{m}\left[E(W/m) - 2\int_0^1 g'(x)/(\gamma + g'(x))\,dx\right],$$

as is seen from (5.4). Thus (5.3) is not enough to guarantee the validity of this substitution. However, the numerical results obtained seemed sufficiently interesting to leave them in, in the hope that a proof of their validity will soon be forthcoming.

The large sample distribution of the Smirnov statistic has not yet been investigated when $F \neq G$. However, it was pointed out by Massey [9] that a lower bound to the power can be obtained simply by the inequality

$$\tag{5.5} \begin{aligned} P(\sup_t \mid F_{X_1,\cdots,X_m}(t) = G_{Y_1,\cdots,Y_n}(t) \mid \geq C) \\ \geq P(\mid F_{X_1,\cdots,X_m}(t_0) - G_{Y_1,\cdots,Y_n}(t_0) \mid \geq C) \end{aligned}$$

where t_0 is any particular value of t. If $F(x) = x$ $(0 \leqq x \leqq 1)$, $G(x) = F^k(x) = x^k$ and we take for t_0 the point of maximum difference between F and G, we get $t_0 = 1/k - 1\sqrt{k}$. Now $F_{X_1,\cdots,X_m}(t_0)$ and $G_{Y_1,\cdots,Y_n}(t_0)$ are the proportion of successes in m and n binomial trials with probability of success equal to $F(t_0) = t_0$ and $G(t_0) = t_0^k$ respectively. Thus for moderately large sample sizes

$$F_{X_1,\cdots,X_m}(t_0) - F_{Y_1,\cdots,Y_n}(t_0)$$

is approximately normally distributed with mean $t_0 - t_0^k$ and variance

$$(t_0(1 - t_0)/m) + (t_0^k(1 - t_0^k)/n).$$

The constant C of (5.5) can be obtained from [10] or [11].

6. Optimum rank tests for the two-sample problem. We next consider the problem of determining the optimum rank test of $H:G = F$ against $K:G = g(F)$. Under the hypothesis all $\binom{m+n}{m}$ possible combinations of $s_1, \cdots, s_n$ are equally likely while their probabilities under K are given by (4.1). Thus the problem, in terms of ranks, reduces to that of testing a simple hypothesis against a simple alternative, and its solution is given by the fundamental lemma of Neyman and Pearson. The most powerful test rejects when the ratios of the probabilities is too large. Since the denominator of this ratio is constant (independent of $s_1, \cdots, s_n$), this is equivalent to rejecting when (4.1) is too large.

If we take, for example, $g(F) = F^2$ the most powerful rank test rejects when

$$s_1(s_2 + 1) \cdots (s_n + n - 1) > C.$$

The power of this test is shown against $G = F^2$ and $G = F^3$ in Table 1.

Since usually one does not have any precise alternatives in mind, it is perhaps more interesting to turn the problem around and to investigate what optimum properties (if any) are possessed by some of the standard rank tests. This gives an indication of the type of deviation from the hypothesis that the test under consideration is particularly suited to detect and therefore of the circumstances in which the application of the test is appropriate. As one such example, we shall discuss here the Wilcoxon test.

Consider to this end the one-parameter family of nonparametric alternatives given by

$$g_p(F) = qF + pF^2, \qquad 0 \leqq p \leqq 1, \qquad p + q = 1. \tag{6.1}$$

If $\beta(p)$ denotes the power of a test against $g_p(F)$ we shall show that the one-sided Wilcoxon test among all rank test maximizes $\beta'(0)$, the slope of the power function at the hypothetical point. It is thus "locally most powerful" just against the type of alternative we have been considering.

To prove this result we must consider $\beta(p)$. Since $d/du\ (g_p(u)) = q + 2pu$ it follows from (4.1) that

$$P(S_1 = s_1, \cdots, S_n = s_n \mid p) = \frac{1}{\binom{m+n}{m}} E[(q + 2pU^{(s_1)}) \cdot \cdots \cdot (q + 2pU^{(s_n)})]$$

and hence that

$$\frac{d}{dp} P(S_1 = s_1, \cdots, S_n = s_n \mid p)\Bigg|_{p=0} = \frac{1}{\binom{m+n}{m}} E\left[\sum_{i=1}^{n} (2U^{(s_i)} - 1)\right]$$

$$= \frac{1}{\binom{m+n}{m}} \left[\frac{2}{m+n+1} \sum_{i=1}^{n} s_i - n\right].$$

Therefore

$$\beta'(0) = \sum \frac{1}{\binom{m+n}{m}} \left[\frac{2}{m+n+1} \sum_{i=1}^{n} s_1 - n\right], \tag{6.2}$$

where the summation extends over the sets $(s_1, \cdots, s_n)$ that form the critical region. It follows as before from the fundamental lemma of Neyman and Pearson that we maximize $\beta'(0)$ at a fixed level of significance by rejecting H when the right-hand side of (6.2), and hence $\sum_{i=1}^{n} s_i$, is too large. This is the desired result.

The above property of the Wilcoxon two-sample test can be generalized in various directions, which we shall sketch only briefly. First, an analogous property holds for any test whose region of rejection is of the form

$$h(s_1) + h(s_2) + \cdots + h(s_n) \geqq C. \tag{6.3}$$

In any such case one can find a function h^* for which the test given by (6.3) maximizes the slope of the power function $\beta(p)$ against the alternatives $g_p(F) = qF + ph^*(F)$ at $p = 0$. A particular example of (6.3) is Mood's median test T_1, which rejects when the number of s_i exceeding a given constant is too large. However, in most cases, and this seems to include the one under consideration, the function h^* is too complicated to be very enlightening and to warrant the tedious computations necessary to obtain it. The existence of h^* follows from the fact that the s_i can only take on a finite number of values so that without loss of generality, the function h in (6.3) may be taken to be a polynomial. Furthermore, as in the case of the Wilcoxon statistic the test maximizing $\beta'(0)$ against the alternatives $qF + ph^*(F)$ is given by the rejection region

$$E[h^{*\prime}(U^{(s_1)}) + \cdots + h^{*\prime}(U^{(s_n)})] > C.$$

To complete the proof it is enough to show that there exists a polynomial h^* and constants $a > 0$, b such that $h^*(0) = 0$, $h^*(1) = 1$, $h^{*\prime}(u) \geqq 0$ for $0 \leqq u \leqq 1$ and $E[h^{*\prime}(U^{(s)})] = a[h(s) + b]$. Now from the fact (see (4.5)) that

$$E[(U^{(s)})^k] = \frac{s(s+1)\cdots(s+k-1)}{(m+n+1)\cdots(m+n+k)}$$

it is seen that there exists a polynomial P for which $E[P'(U^{(s)})] = h(s)$. Putting $h^*(s) = a[P(s) + bs] + c$ we need to show only that given any polynomial P

there exist constants $a > 0, b, c$ such that $a\,P(0) + c = 0$, $a[P(1) + b] + c = 1$ and $a[P'(s) + b] \geqq 0$ for $0 \leqq s \leqq 1$, and this is easily verified.

Another extension of the above result concerns a problem different from but closely related to the two-sample problem. (In this connection see Hemelrijk [19]). Let $Z_1, \cdots, Z_N$ be identically and independently distributed with cdf M. The hypothesis to be tested is that M is symmetric about the origin, that is, that for all z, $M(z) + M(-z) = 1$. If we assume M to be continuous, put $1 - M(0) = \rho$ and denote by F and G the conditional distributions of Z given that $Z > 0$ and of $-Z$ given that $Z < 0$, the hypothesis is equivalent to the two statements $\rho = \frac{1}{2}$, $F = G$. Let m and n be the number of positive and negative Z's respectively, and denote by $X_1, \cdots, X_m$ and $Y_1, \cdots, Y_n$ the positive Z's and the absolute values of the negative Z's respectively, in their original order of subscripts. Consider now the probability of any particular set of ranks of the Y's under some alternative. Given m and n, this is independent of ρ and is given by (4.1) when $G = g(F)$. In addition, n is a binomial variable with probability ρ of success. Thus we get

$$P(\text{The number of } Z\text{'s} > 0 \text{ is } n \text{ and } S_1 = s_1, \cdots, S_n = s_n) = \rho^n(1-\rho)^{N-n}E[g'(U^{(s_1)}) \cdot \cdots \cdot g'(U^{(s_n)})]. \tag{6.4}$$

If in particular, one considers alternatives with $\rho = \frac{1}{2}$, the right-hand side of (6.4) becomes $2^{-N}\, E[g'(U^{(s_1)}) \cdot \cdots \cdot g'(U^{(s_n)})]$, which formally differs from (4.1) only by a multiplicative constant. Thus any optimum test of the two-sample problem derived on the basis of (4.1) gives rise to a dual one for the hypothesis of symmetry. As an example, the Wilcoxon two-sample test which rejects when $s_1 + \cdots + s_n > C$, under translation in the above manner becomes a test of the hypothesis of symmetry also proposed by Wilcoxon [3] and recently shown by Tukey [12] to be equivalent to a test proposed independently by Walsh [13]. This test is now seen to maximize $\beta'(0)$ against the alternatives according to which $M(0) = \frac{1}{2}$, the conditional distribution of Z given $Z > 0$ is F and that of $-Z$ given $Z > 0$ is $qF + pF^2$.

As another application of this approach, let us once more consider the two-sample problem, but this time with a two-sided class of alternatives. For simplicity, we take $m = n$, and we assume that either the X's are distributed according to F and the Y's according to $qF + pF^2$ or vice versa. Let $\beta(p)$ denote the power of a rank test against the first of these alternatives and $\beta^*(p)$ that against the second. We shall then maximize the average power $\frac{1}{2}[\beta(p) + \beta^*(p)]$ at $p = 0$. Since it turns out that $\beta'(0) + \beta^{*\prime}(0) = 0$ this is equivalent to maximizing $\beta''(0) + \beta^{*\prime\prime}(0)$.

From (4.1) we see that the sum of the probabilities of $R_1 = r_1, \cdots, R_n = r_n$, $S_1 = s_1, \cdots, S_n = s_n$ under the two alternatives is

$$\begin{aligned} E[(q + 2pV^{(s_1)}) \cdots (q + 2pV^{(s_n)})] &+ E[(q + 2pU^{(r_1)}) \cdots (q + 2pU^{(r_n)})] \\ &= 2 + pE\left[\sum_{i=1}^{n} (2U^{(s_i)} - 1) + \sum_{i=1}^{n} (2U^{(r_i)} - 1)\right] \\ + p^2\, E[\sum_{i<j} (2V^{(s_i)} - 1)(2V^{(s_j)} - 1) &+ 2 \sum_{i<j} (U^{(r_i)} - 1)(U^{(r_j)} - 1)] + o(p^2). \end{aligned}$$

Now

$$\sum_{i=1}^{n} [E(U^{(s_i)}) + E(V^{(r_i)})] = \frac{\sum r_i + \sum s_i}{2n+1} = n$$

so that the coefficient of p is zero. The coefficient of p^2 is, except for a constant, given by

$$4 \sum_{i<j} [E(V^{(s_i)} V^{(s_j)}) + E(U^{(r_i)} U^{(r_j)})] - 2 \sum_{i<j} E[V^{(s_i)} + V^{(s_j)} + U^{(r_i)} + U^{(r_j)}]$$

$$= 4 \sum_{i<j} \frac{s_i(s_j+1) + r_i(r_j+1)}{(2n+1)(2n+2)} - 4 \sum_{i \neq j} \frac{s_i + s_j + r_i + r_j}{(2n+1)}.$$

Using the fact that $\sum_{i<j} s_i = \sum (n-i)s_i$ and that $\sum (r_i + s_i)$ as well as $\sum [r_i^2 + s_i^2)$ are constants, the coefficient of p^2 is, except for a constant,

$$\text{(6.5)} \qquad \frac{2}{(2n+1)(2n+2)} [(\textstyle\sum s_i)^2 + (\sum r_i)^2 + \sum (s_i - i)^2 + \sum (r_i - i)^2].$$

Thus we maximize the average power at $p = 0$ by maximizing (6.5) or equivalently

$$\text{(6.6)} \qquad \left[\sum s_i - \frac{n(2n+1)}{2}\right]^2 + \left[\sum r_i - \frac{n(2n+1)}{2}\right]^2 + \sum (s_i - i)^2 + \sum (r_i - i)^2.$$

Rather surprisingly this is not the two-sided Wilcoxon test which is given just by the first two terms of (6.6).

This result can be given a slightly different form. Let us write the alternative $qF + pF^2$ in the form

$$\text{(6.7)} \qquad g_\theta(F) = \frac{1}{1+\theta} (F + \theta F^2)$$

where we shall be interested only in values of θ close to zero, and where we may then consider also negative values of θ. If $\beta(\theta)$ denotes the power of some rank test against θ we may in analogy to the type A tests of Neyman and Pearson [14] maximize $\beta''(0)$ subject to $\beta(0) = \alpha$, $\beta'(0) = 0$. This will clearly again lead to (6.6).

Such a parametric approach can be carried further. Consider, for example, samples from k populations $F_1, \cdots, F_k$ and the hypothesis $H: F_1 = \cdots = F_k$. If we then consider alternatives of the form $F_i = (1/1 + \theta_i)(F + \theta_i F^2)$ with $\sum \theta_i = 0$, we can, for example, maximize the average power over the sphere $\sum \theta_i^2 = \delta^2$ for small δ. This is analogous to a formulation given by Wald [15] for the normal case, and leads to an extension of (6.6).

7. The hypothesis of independence. To illustrate the general approach of this paper with another example, consider a sample $(X_1, Y_1), \cdots, (X_n, Y_n)$ from a bivariate distribution. The hypothesis to be tested is that X and Y are independent. Nonparametric alternatives to this will be defined by means of a function h of two variables such that h is a continuous cdf over the unit square $0 \leqq x, y \leqq 1$ (so that $h(0, 0) = 0$, $h(1, 1) = 1$). A nonparametric class $\mathcal{H}(h)$ of bivariate distributions is then formed by the totality of distributions $h(F, G)$ where F, G are arbitrary continuous univariate cdf's. Suppose now that the X's are ordered and that in this ordering the rank of X_i is R_i $(i = 1, \cdots, n)$. Similarly, we shall denote by S_i the rank of Y_i among the Y's. We then have, analogously to the corresponding result in Section 3,

THEOREM 7.1. *For the distributions of a class $\mathcal{H}(h)$ the distribution of the R's and S's is constant, that is, independent of F and G.*

This follows from

LEMMA 7.1. *If h, F and G are continuous and if* $P(X \leqq x, Y \leqq y) = h(F(x), G(y))$ *then*

$$P(F(X) \leqq u, G(Y) \leqq v) = h(u, v). \tag{7.1}$$

PROOF. Let x, y be such that $F(x) = u$, $G(y) = v$. Then

$$P(X < x, Y < y) \leqq P(F(X) \leqq u, G(Y) \leqq v) \leqq P(X \leqq x, Y \leqq y).$$

But

$$P(X < x, Y < y) = P(X \leqq x, Y \leqq y) = h(F(x), G(y)) = h(u, v).$$

Again we can write down the distribution of the R's and S's using Hoeffding's theorem. In fact if $h'(u, v) = (\partial^2/\partial u \partial v)h(u, v)$ we have

$$\begin{aligned} &P(R_1 = r_1, \cdots, R_n = r_n; S_1 = s_1, \cdots, S_n = s_n) \\ &\qquad = E[h'(U^{(r_1)}, V^{(s_1)}) \cdot \cdots \cdot h'(U^{(r_n)}, V^{(s_n)})] \end{aligned} \tag{7.2}$$

where $U_1, \cdots, U_n$; $V_1, \cdots, V_n$ are two independent samples from a uniform distribution on [0, 1] and $U^{(r_i)}$, $V^{(s_j)}$ are the associated order statistics. It should be noted that in (7.2) it is not assumed that either the r_i or the s_j are arranged in natural order. Alternatively we may take $r_i = i$ and define s_i as the rank of the Y that is associated with the ith smallest X. In this notation only the S_i remain random and instead of (7.2) we get

$$P(S_1 = s_1, \cdots, S_n = s_n) = \frac{1}{n!} E[h'(U^{(1)}, V^{(s_1)})) \cdot \cdots \cdot h'(U^{(n)}, V^{(s_n)})]. \tag{7.3}$$

Perhaps the simplest choice for h would seem to be

$$h'(u, v) = u + v; \qquad h(u, v) = \tfrac{1}{2}(uv^2 + u^2v). \tag{7.4}$$

This corresponds to the family of cdf's $H(x, y) = \frac{1}{2}[F(x)G^2(y) + F^2(x)G(y)]$ and can be interpreted similarly to the alternatives discussed in previous sections. The observations X, Y are drawn with probability $\frac{1}{2}$ each from two bivariate populations with independent components. According to one of these X is an observation from F while Y is the larger of two observations from G; according to the other the situation is just reversed. A general family of alternatives could be obtained in this way, given by

$$H(x, y) = \sum_{i=1}^{l} p_i F^{a_i}(x) G^{b_i}(y), \qquad p_i \geqq 0, \qquad \sum p_i = 1.$$

The distribution of the ranks under such alternatives can be written down on the basis of (4.5) and (7.3). However, even in the simplest cases such as (7.4) the resulting expressions are quite complex except for very small values of n.

It seems of interest that one of the best known tests for independence, that based on the rank correlation coefficient, possesses an optimum property similar to the ones derived in Section 6 for the Wilcoxon and related tests. For let

$$h_p(u, v) = quv + pu^2v^2. \tag{7.4}$$

Using (7.2) we see that the probability of $R_i = r_i$, $S_j = s_j$ $(i, j = 1, \cdots, n)$ is

$$E[(q + 4pU^{(r_1)}V^{(s_1)}) \cdot \cdots \cdot (q + 4pU^{(r_n)}V^{(s_n)})].$$

Differentiating this with respect to p and setting $p = 0$ we get

$$E\left[\sum_{i=1}^{n} (-1 + 4U^{(r_i)}V^{(s_i)})\right] = -n + \frac{4}{(n+1)^2} \sum_{i=1}^{n} r_i s_i .$$

Thus the test that maximizes the slope of the power function against the alternatives (7.4) at $p = 0$ rejects when $\sum r_i s_i$ is too large, and hence when the rank correlation coefficient is too large. More generally, if $h_p(u, v) = quv + pg_1(u)g_2(v)$, the test that maximizes the slope of the power function rejects when $g_1'(U^{(r_i)})g_2'(V^{(s_i)})$ is too large. In this manner we obtain a generalization of the result connected with (6.3).

REFERENCES

[1] W. Hoeffding, "Optimum nonparametric tests," *Proceedings of the Second Berkeley Symposium on Mathematical Statistics and Probability*, University of California Press, (1951), pp. 83–92.

[2] A. M. Mood, *Introduction to the theory of statistics*, McGraw-Hill Book Co., 1950.

[3] F. Wilcoxon, "Individual comparison by ranking methods," *Biometrics*, Vol. 1 (1945), pp. 80–83.

[4] H. B. Mann and D. R. Whitney, "On a test of whether one of two random variables is stochastically larger than the other," *Ann. Math. Stat.*, Vol. 18 (1947), pp. 50–60.

[5] A. Wald and J. Wolfowitz, "On a test whether two samples are from the same population," *Ann. Math. Stat.*, Vol. 11 (1940), pp. 147–162.

[6] N. V. Smirnov, "On the estimation of the discrepancy between empirical curves of distribution for two independent samples," *Bull. Math. Univ. Moscow, Serie Int.*, Vol. 2 (1939), pp. 3–16.

[7] E. L. Lehmann, "Consistency and unbiasedness of certain nonparametric tests," *Ann. Math. Stat.*, Vol. 22 (1951), pp. 165–179.

[8] J. Wolfowitz, "Nonparametric statistical inference," *Proceedings of the Berkeley Symposium on Mathematical Statistics and Probability*, University of California Press, (1949), pp. 93–113.

[9] F. MASSEY, "A note on the power of a nonparametric test," *Ann. Math. Stat.*, Vol. 21 (1950), pp. 440–443.

[10] N. SMIRNOV, "Table for estimating the goodness of fit of empirical distributions," *Ann. Math. Stat.*, Vol. 19 (1948), pp. 279–281.

[11] F. MASSEY, "The distribution of the maximum deviation between two sample cumulative step functions," *Ann. Math. Stat.*, Vol. 22 (1951), pp. 125–130.

[12] J. TUKEY, "The simplest signed-rank tests," Stat. Res. Group, Princeton University, *Memorandum Report 17*, 1949, unpublished.

[13] J. WALSH, "Some significance tests for the median which are valid under very general conditions," *Ann. Math. Stat.*, Vol. 20 (1949), pp. 64–81.

[14] J. NEYMAN AND E. S. PEARSON, "Contributions to the theory of testing statistical hypotheses. I. Unbiased critical regions of type A and type A_1," *Stat. Res. Memoirs*, Vol. 1 (1936), pp. 1–37.

[15] A. WALD, "On the power function of the analysis of variance test," *Ann. Math. Stat.*, Vol. 13 (1942), pp. 434–439.

[16] E. L. LEHMANN, "Some principles of the theory of testing hypotheses," *Ann. Math. Stat.*, Vol. 21 (1950), pp. 1–26.

[17] P. R. HALMOS, *Measure Theory*, D. Van Nostrand, 1950.

[18] J. WESTENBERG, "Significance test for median and interquartile range in samples from continuous populations of any form," *Koninklijke Nederlandse Akademie van Wetenschappen Proc.*, Vol. 51 (1948), pp. 252–261.

[19] J. HEMELRIJK, "A family of parameter-free tests for symmetry with respect to a given point," *Nederl. Akad. Wetensch., Proc.*, Vol. 12 (1950), Part I, pp. 340–350, Part II, pp. 419–431.

THE EFFICIENCY OF SOME NONPARAMETRIC COMPETITORS OF THE t-TEST

By J. L. Hodges, Jr., and E. L. Lehmann[1]

University of California, Berkeley

0. Summary. Consider samples from continuous distributions $F(x)$ and $F(x - \theta)$. We may test the hypothesis $\theta = 0$ by using the two-sample Wilcoxon test. We show in Section 1 that its asymptotic Pitman efficiency, relative to the t-test, never falls below 0.864. This result also holds for the Kruskal-Wallis test compared with the F-test, and for testing the location parameter of a single symmetric distribution.

A number of alternative notions of asymptotic efficiency are compared in Section 2. In this connection, certain difficulties arise because power is not necessarily a convex function of sample size. As an alternative to the Pitman notion of asymptotic efficiency, we consider in Section 3 one based on the speed with which power at a fixed alternative tends to 1. In particular we obtain, for the sign test relative to the t in normal populations, the limit as $n \to \infty$ of the sequence of power efficiency functions. It is noted that certain interchanges of limit passages are not always possible.

1. Minimum Pitman efficiency of the Wilcoxon and sign tests. For comparing the large sample power of two sequences of tests, the concept of asymptotic relative efficiency was developed by Pitman [1]. An exposition of his work, together with some extensions, was recently given in [2] and [3]. Applications to a number of specific problems are made in [4] and [5].

Let $\beta_N(\theta)$ and $\beta_N^*(\theta)$ denote the power functions of two tests, say A and A^*, based on the same set of N observations, against a parametric family of alternatives labeled by θ, and let θ_0 be the value of θ specified by the hypothesis. We shall assume that all tests are at level of significance α. Let β be a specified power with $\alpha < \beta < 1$. Consider a sequence of alternatives θ_N such that

$$\beta_N(\theta_N) \to \beta, \qquad \text{as } N \to \infty, \tag{1.1}$$

and a sequence $N^* = h(N)$ such that

$$\beta_{N^*}^*(\theta_N) \to \beta, \qquad \text{as } N \to \infty. \tag{1.2}$$

Then if

$$e_{A^*,A} = \lim_{N \to \infty} \frac{N}{N^*} \tag{1.3}$$

Received February 25, 1955.

[1] Prepared with partial support of the Office of Naval Research. This paper may be reproduced in whole or in part for any purpose of the United States Government.

exists, and is independent of α, β and the particular sequences $\{\theta_N\}$ and $\{h(N)\}$ chosen, then $e_{A^*,A}$ is defined to be the *relative asymptotic efficiency* of the test A^* with respect to the test A. Under weak assumptions (1.1) implies that $\theta_N \to \theta_0$, and in the most common cases it turns out that θ_N tends to θ_0 at the rate $N^{-1/2}$. Usually the N observations constitute a sample, or are divided into two samples of sizes m and n with $m + n = N$. In the latter case we assume that m/n tends to some limit ρ, $0 < \rho < \infty$, as N tends to ∞. In many problems, including those we study, $e_{A^*,A}$ is independent of ρ.

Pitman gave a method for obtaining the limit (1.3), and evaluated it for a number of problems. Consider in particular the case of samples $X_1, \cdots, X_m$ and $Y_1, \cdots, Y_n$ from continuous distributions F and G and the hypothesis $H{:}F = G$. We shall be concerned with the narrower alternatives that G differs from F only by a shift, so that $G(u) = F(u - \theta)$ for all u. The discussion applies to both the one-sided case $\theta > \theta_0 = 0$ and the two-sided case $\theta \neq \theta_0 = 0$. If F is a normal distribution, the appropriate test is Student's t-test. A nonparametric test proposed by Wilcoxon is based on the rank sum of the Y's among the set of N-ordered observations. Pitman computed the relative asymptotic efficiency of the Wilcoxon test relative to the t-test as

$$e_{w,t} = 12\sigma^2\left[\int f^2(x)\,dx\right]^2, \tag{1.4}$$

where f is the probability density of the distribution F, and σ^2 is the common variance of the X's and Y's. Some particular values given by Pitman are $e_{w,t} = 3/\pi \sim .95$ when f is a normal density, $e_{w,t} = 1$ for the case of a uniform distribution, and $e_{w,t} = 81/64$ when $f(x) = x^2e^{-x}/\Gamma(3)$ for $x \geqq 0$. All of these values are surprisingly high and raise the question as to how low e can actually drop. We shall prove, below, the following theorem.

THEOREM 1. *Let N^* satisfy (1.2) where the tests A and A^* are the (two-sample) t-test and Wilcoxon test, respectively, for testing against shift of a continuous distribution F. Then* (a)

$$\liminf_{N\to\infty} N/N^* \geqq 108/125 = 0.864 \tag{1.5}$$

whatever F may be.

Furthermore, (b) *the lower limit is attained for the distribution with density* (1.12, 1.13). *for which $e = .864$.*

PROOF. It was shown by Andrews [5] that if F is continuous, and

$$\lim_{\theta\to\infty} \int \frac{1}{\theta}\,[F(x + \theta) - F(x)]\,dF(x) = c, \tag{1.6}$$

then the efficiency, given by (1.3), exists and is $12c^2\sigma^2$. This proof also shows that quite generally

$$\liminf_{N\to\infty} \frac{N}{N^*} \geqq 12\sigma^2\left[\liminf_{\theta\to 0} \frac{1}{\theta}\int [F(x + \theta) - F(x)]\,dF(x)\right]^2. \tag{1.7}$$

By Fatou's lemma, the right-hand side is greater than or equal to

$$12\sigma^2\left\{\int\left[\liminf_{\theta\to 0}\frac{F(x+\theta)-F(x)}{\theta}\right]dF(x)\right\}^2. \tag{1.8}$$

It follows further, from the Decomposition Theorem of De La Vallée Poussin (see [6], p. 127), that when F has a singular component,

$$\lim\,[F(x+\theta)-F(x)]/\theta = \infty$$

on a set of positive F-measure, so that (1.8), and hence e, is infinite. We may therefore assume that except on a set of F-measure zero, the density $F(x) = F'(x)$ exists, in which case (1.8) becomes

$$12\sigma^2\left[\int f^2(x)\,dx\right]^2. \tag{1.9}$$

If $\sigma^2 = \infty$, then it follows from (1.6) that $e = \infty$, so that we may assume σ^2 to be finite. Since (1.9) is invariant under a change of location or sale, we may take $\sigma^2 = 1$, and the problem of minimizing (1.9) then reduces to that of minimizing

$$\int f^2(x)\,dx \tag{1.10}$$

subject to the conditions

$$\int xf(x)\,dx = 0; \qquad \int f(x)\,dx = \int x^2f(x)\,dx = 1; \qquad f(x) \geqq 0 \text{ for all } x. \tag{1.11}$$

According to the method of undetermined multipliers, it is sufficient to minimize

$$\int [f^2(x) + 2b(x^2 - a^2)f(x)]\,dx.$$

For nonnegative f, this is achieved by setting

$$f(x) = b(a^2 - x^2), \qquad \text{if } x^2 \leqq a^2, \tag{1.12}$$

and $f(x) = 0$ otherwise. The constants a and b are determined from (1.11) to be

$$s = \sqrt{5}, \qquad b = \tfrac{3}{20}\sqrt{5}, \tag{1.13}$$

and with these values, (1.9) becomes equal to 108/125, which is therefore a lower bound to (1.7). Since for the density (1.12) the limit of (1.6) may be taken under the integral sign, it is seen that the efficiency exists in this case, and equals the lower bound, which therefore cannot be improved.

To the extent that the above concept of efficiency adequately represents what happens for the sample sizes and alternatives arising in practice, this result shows that use of the Wilcoxon test instead of the Student's t-test can never entail a serious loss of efficiency for testing against shift. (On the other hand, it is obvious from (1.4) that the Wilcoxon test may be infinitely more efficient than the t-test.)

It should be mentioned that there are rank tests: that of Fisher and Yates, which has been discussed by Hoeffding [7], Terry [8], and Dwass [9]; and that of van der Waerden [10], for which the asymptotic efficiénty relative to the t-test is 1 when F is normal, and is conjectured to be >1 when F is not normal. Should this be correct, then for these tests the lower bound .864 in (1.5) would be replaced by the even better value 1.

The conclusion of Theorem 1 also applies to the H-test of Kruskal and Wallis [11] for testing equality of k distributions $F_1, \cdots, F_k$, which are assumed to differ only in location. This follows from the fact that Andrews' work, quoted above, was carried out for this more general problem, and that in particular formulae (1.6) and (1.7) hold for all values of k.

Another application is to the case of a single sample $X_1, \cdots, X_N$ from a distribution $F(x - \theta)$, where F is symmetric about 0. The hypothesis to be tested is $H:\theta = 0$, and if F is known to be normal, the one-sample t-test is appropriate. The Wilcoxon test for this problem is based on the rank sum of the positive X's among the ordered absolute values $|X_1|, \cdots, |X_N|$. Pitman showed that (1.4) also applies in this case, and the considerations of Andrews can be used to generalize this again to (1.6) and (1.7).

A particularly simple test of the hypothesis $H:\theta = 0$ in the one-sample problem is the sign test, based on the number of positive observations. For asymptotic efficiency of the sign test, relative to the t-test, Pitman obtained the result

$$e_{s,t} = 4\sigma^2 f^2(0), \tag{1.14}$$

which is valid whenever the derivative $F'_{(0)} = f(0)$ of F at the origin exists. A particular value given by Pitman is $e = 2/\pi$ in case of a normal distribution. In the present case there is, of course, no positive lower bound, since $e = 0$ when $f(0) = 0$. If the distribution F is assumed to possess a unimodal density (in the weak sense that $0 \leqq |x| < |x'|$ implies $f(x') \leqq f(x)$), then it is easily seen that $e \geqq \frac{1}{3}$, the value $\frac{1}{3}$ being attained for the case of a rectangular distribution. For let $f(0) = 1$ without loss of generality, since (1.14) is invariant under a change of scale. Then we must minimize

$$\int (x^2 - a^2)f(x)\, dx$$

subject to $0 \leqq f(x) \leqq 1$, and this is achieved by putting $f(x) = 1$ when $|x| \leqq a$ and $f(x) = 0$ otherwise.

It may be questioned whether the high efficiency of Wilcoxon relative to t established by Theorem 1 is the result of the particular alternatives considered. It is therefore of interest to make the comparison for other than shift alternatives. We shall now consider what may be called mixture or contamination alternatives. In the two-sample problem this takes the form

$$X_1, \cdots, X_m : F,$$
$$Y_1, \cdots, Y_n : (1 - \theta)F + \theta G.$$

In the one-sample problem, the form is

$$Z_1, \cdots, Z_N: (1-\theta)F + \theta G, \qquad F(z) + F(-z) = 1.$$

In both cases we take $G \leqq F$ and test the hypothesis $\theta = 0$.

Mixture alternatives may be reasonable in many situations. For example, a treatment may be effective in only a proportion θ of the population of subjects. Thus, cancer operations are effective only if metastasis has not occurred; vitamin therapy is useful only if there is a vitamin deficiency.

If we let (m and) n tend to infinity (at the same rate), while θ tends to 0, with F and G fixed, we can compute the limiting efficiency of the Wilcoxon test relative to the t-test (or, equivalently, to the test based on $\bar{Y} - \bar{X}$) from Pitman's formula

$$e = \lim \left[\frac{\sigma_N(\theta_0)}{\sigma_N^*(\theta_0)} \frac{\mu_N^{*\prime}(\theta_0)}{\mu_N'(\theta_0)}\right]^2,$$

where T_N and T_N^* are the statistics on which the tests are based and it is assumed that

$$\frac{T_N - \mu_N(\theta)}{\sigma_N(\theta)}, \qquad \frac{T_N^* - \mu_N^*(\theta)}{\sigma_N^*(\theta)}$$

have the limiting distribution $N(0, 1)$. If $T_N = \bar{Y} - \bar{X}$ and mnT_N^* is the Mann-Whitney form of the Wilcoxon statistic, one obtains

$$\begin{aligned}
\sigma_N^{*2}(0) &= (1/12)(1/m + 1/n),\\
\sigma_N^2(0) &= \sigma^2(1/m + 1/n),\\
\mu_N^*(\theta) &= P\{X \leqq Y\} = (1-\theta)\int F\,dF + \theta \int F\,dG,\\
\mu_N(\theta) &= E(Y) - E(X) = \theta\left[\int x\,dG - x\,dF\right].
\end{aligned}$$

It follows that

$$e = 12\sigma^2\left(\frac{\int (F-G)\,dF}{\int x d\,(G-F)}\right)^2 = 12\sigma^2\left(\frac{\int (F-G)\,dF}{\int (F-G)\,dx}\right)^2, \tag{1.15}$$

where, as before, σ^2 is the variance of an observation from F. The equality of the denominators in (1.15) follows by viewing each as an expression for the (signed) area between F and G. The above computation can be made rigorous by the methods of [5].

It is clear that $\int(F - G)\,dF \leqq \frac{1}{2}$, while $\int(F - G)\,dx$ may be arbitrarily large if G is far to the right of F. Thus (1.15) has no positive lower bound, which corresponds to our intuition that the Wilcoxon test, like any rank test, is insensitive to the size of large deviations.

If we particularize to $G(x) = F(x - \Delta)$, Δ fixed, we obtain

$$12\sigma^2\left[\int\left(\frac{F(x) - F(x - \Delta)}{\Delta}\right) dF(x)\right]^2. \tag{1.16}$$

As $\Delta \to 0$, (1.16) agrees (under suitable regularity conditions) with (1.4), so there is no finite upper bound to (1.15). We observe that

$$\int [F(x) - F(x - \Delta)]\, dF(x) = \Pr\{|X_1 - X_2| \leqq \Delta\},$$

and that $\Pr\{|X_1 - X_2| \leqq \Delta\}/\Delta$ will be a decreasing function of Δ whenever $X_1 - X_2$ is unimodal. Thus in particular, if X itself is unimodal, the efficiency decreases as Δ increases [12], so that the performance of Wilcoxon relative to t is often less good against contamination with a shift than against shift itself. For example, if F is normal and $\Delta = \sigma$, (1.16) has the value 0.812; for $\Delta = 2\sigma$ it is 0.533; and it tends to 0 as $\Delta \to \infty$.

We remark that the greater sensitivity of the t-test to contamination is not an unmixed blessing, as the contamination may, in some cases, represent gross errors of observation rather than the true effect of the treatment. In fact, insensitivity to large deviations is one of the advantages of nonparametric tests.

2. Alternative notions of asymptotic efficiency. The result obtained above suggests that if Pitman efficiency is taken as a guide, one may prefer the Wilcoxon test to the t-test in almost all problems of testing against shift. But how reliable is Pitman efficiency? Dixon [13], [14] has emphasized that a comprehensive efficiency comparison of two tests cannot be made with a single number. Suppose that a test A of level α and using N observations has power $\beta_A(N, \alpha, \theta)$ against alternative θ. If test A^*, also of level α, requires N^* observations to produce the same power at the same alternative, we define the efficiency of A^* relative to A in these circumstances to be the ratio N/N^*, and denote it by $e_{A^*,A}(N, \alpha, \theta)$. The complete comparison of A^* with A would require the evaluation of this "power efficiency function" for all values of its three arguments.

We note that the definition of N^* just given is not quite complete. There usually will not exist an integer N^* such that $\beta_{A^*}(N^*, \alpha, \theta) = \beta_A(N, \alpha, \theta)$, but rather an N_0 such that $\beta_{A^*}(N_0, \alpha, \theta) < \beta_A(N, \alpha, \theta) < \beta_{A^*}(N_0 + 1, \alpha, \theta)$. Dixon suggests that N^* be defined by inverse interpolation of $\beta_{A^*}(N^*, \alpha, \theta)$ as a function of N^*; specifically, he proposes polynomial interpolation of N^* against $\Phi^{-1}(\beta_{A^*}(N^*, \alpha, \theta))$ in [13], [14]. We feel that this method, while yielding "smooth" results, lacks any operational or functional meaning. Instead, we prefer to define N^* to be $N_0 + p$, where the test A^* has power $\beta_A(N, \alpha, \theta)$ if its number of observations is randomly chosen with probability p of being $N_0 + 1$ and probability $1 - p$ of being N_0. Thus, our N^* is the expected number of observations required with test A^* to match the power of test A, when randomizing between consecutive integers. (Our definition implies linear interpolation.)

We note in this connection a curious fact. For some tests, and specifically for the t-test against normal shift, $\beta(N, \alpha, \theta)$ is not always convex as a function of N.

Thus, if we wish to attain a stated power for stated α and δ with smallest expected number of observations, we would *not* randomize between consecutive integers! However, as our main objective is to define an N^* which gives the desired power, and the randomization is introduced only out of necessity, we shall use the definition given.

It might be felt that the question of the definition of N^* is too trivial to require so much discussion, and indeed if N is large this is so. But efficiency comparisons are often made for small N and here (especially with β large) the precise definition of N^* becomes important. To illustrate the point we present, below, the efficiency figures given by Dixon [14] for Wilcoxon against t for normal shift of amount δ, equal samples of 5, $\alpha = 4/126$, and the corresponding values as computed by our definition. (We are not able to obtain a worthwhile figure for $\delta = 4$, since the value of β_w is not given by Dixon to enough decimal places.) It is seen that Dixon's conclusion that the "power efficiency decreases slightly for more distant alternatives" is dependent on his method of interpolation for N^*. With our definition, the efficiency rises as

δ	.5	1	1.5	2	2.5	3	3.5	4	4.5
β_w	.072	.210	.431	.674	.858	.953	.988	.998	.9996
e (Dixon's paper)	.97	.97	.96	.95	.94	.94	.93	.92	.91
e (this paper)	.968	.978	.961	.956	.960	.960	.964	.976 ± .01	—

δ is increased beyond about 3, and appears never to fall below about 0.96, while the efficiency as computed by Dixon reaches .91 at $\delta = 4.5$ and seems still to be dropping. Similar results hold for the sign test as discussed in Section 3.

Depending as it does on three arguments, the function $e_{A^*,A}$ is difficult of complete evaluation, and interest has centered on finding simpler quantities which will serve to represent its general behavior. It is obvious that the Pitman efficiency denoted above by $e_{A^*,A}$ is $\lim_{N\to\infty} e_{A^*,A}(N, \alpha, \theta_A)$, where θ_N satisfies (1.1).

A second kind of efficiency limit is considered by Dixon [13], who evaluates for the sign test compared with the t-test the limit $e_{s,t}(N, \alpha, \infty)$ (which he denotes by E_∞). This limit would be of interest if we were concerned with small N, moderate α, but β very near to 1. (He also obtains $e_{s,t}(N, \alpha, 0)$ and finds that $\lim_{N\to\infty} e_{s,t}(N, \alpha, 0)$ is, for his problem, equal to the Pitman efficiency.)

It is clear that a wide choice of limiting values of $e_{A^*,A}(N, \alpha, \theta)$ might be defined, many of them pertinent in one situation or another. We wish next to call attention to one possibility which is in a sense intermediate between those of Pitman and Dixon and which seems to help to round out some comparisons. Instead of letting $\theta \to 0$ as does Pitman, or $\theta \to \infty$ as does Dixon, we hold θ as well as α fixed and let $N \to \infty$. This limit we denote by $e_{A^*A}(\infty, \alpha, \theta)$. It is closely related to the "index" of Chernoff [15], differing mainly in that Chernoff requires that $\alpha \to 0$, so that α and $1 - \beta$ remain of the same order. Our limit is presumably pertinent when one is interested in large samples and the region of high power, but its main interest seems to reside in the fact that it can, in some cases, be computed and serves to give the limit as $N \to \infty$ of sequences of efficiency curves of the form computed by Dixon for small N, permitting interpolation for moderate N.

3. Limiting efficiencies for the sign test. All of the tests we shall now consider (sign, normal, t) arise in both one-sided and two-sided versions. However, it is true for all of them that as the power tends to 1, the probability of type II error for the one-sided test of level α is asymptotically equivalent to that for the corresponding two-sided test of level 2α. The reason for this is simply that the two tests have identical critical values, and that one of the two tails in the two-sided test is dominant. This consideration simplifies the efficiency comparisons made below.

We are interested in the limiting behavior as $N \to \infty$ of the probability of second-kind error of the sign test. Suppose X is binomial for N trials with success probability p. We may test $H:p = r$ against the alternatives $p < r$. The test accepts if $X \geqq a_N$, where $a_N = rN - c\sqrt{N} + d_N$. That d_N is bounded follows easily from the fact that the error of the normal approximation to the binomial is of order $1/\sqrt{N}$ (see, for example, [16], p. 129). (Using this critical value, the level of significance tends to $\Phi(-c)$.) The probability of second-kind error is then

$$P_{\text{II}} = \sum_{x \geqq a_N} \pi(x),$$

$$\text{where } \pi(x) = \binom{N}{x} p^x (1-p)^{N-x}.$$

We can study the behavior of P_{II} by separately considering the initial term $\pi(a_N)$, and the ratio of the sum to this initial term.

LEMMA 3.1. *If $N \to \infty$ and $a/N \to r > p$, then*

$$\sum_{x \geqq a} \pi(x)/\pi(a) \to r(1-p)/(r-p).$$

PROOF. Since $R(x) = \pi(x+1)/\pi(x)$ is strictly decreasing, $[R(a)]^c > \pi(a+c)/\pi(a) > [R(a+b-1)]^c$, where $0 < c \leqq b$. Summing for $0 \leqq c \leqq b$ we get

$$\frac{1-[R(a)]^{b+1}}{1-R(a)} > \frac{\sum_a^{a+b} \pi(x)}{\pi(a)} > \frac{1-[R(a+b-1)]^{b+1}}{1-R(a+b-1)}. \tag{3.1}$$

As $N \to \infty$, $b \to \infty$, and $a/N \to r$, we have $R(a) \to (1-r)/r \cdot p/(1-p) < 1$, so that the upper bound in (3.1) tends to $r(1-p)/(r-p)$. If, in addition, we require $b/N \to \infty$, the lower bound has the same limit. Since $R(x)$ is decreasing for $x > a$, we have $\sum_{x > a+b} \pi(x)/\sum_a^{a+b} \pi(x) \to 0$, from which the result follows.

LEMMA 3.2. *If $a_N = rN - c\sqrt{N} + d_N$ with d_N bounded, then*

$$\pi(a_N) \sim \frac{\exp[-c^2/2r(1-r)]}{\sqrt{N}\sqrt{2\pi}\sqrt{r(1-r)}} \left[\left(\frac{p}{r}\right)^r \left(\frac{1-p}{1-r}\right)^{1-r}\right]^N \left[\frac{r(1-p)}{p(1-r)}\right]^{c\sqrt{N}-d_N}$$

as $N \to \infty$.

The proof consists in using Stirling's formula and simplifying.

Combining Lemmas 3.1 and 3.2, we see that

$$\sqrt[N]{P_{\text{II}}} \to \left(\frac{p}{r}\right)^r \left(\frac{1-p}{1-r}\right)^{1-r} \qquad \text{as } N \to \infty. \tag{3.2}$$

Note that the limit depends on the hypothesis r and alternative p, but not on α. Since it happens in each of the three problems dealt with in this section that $\sqrt[N]{P_{\text{II}}}$ tends to a positive limit as $N \to \infty$, we shall give to this limit a name, referring to it as the *base* of P_{II}. The base is essentially the quantity ρ discussed by Chernoff [15]. In fact, the limit in (3.2) is the value obtained in [15] for ρ in the binomial case. However, Chernoff's ρ involves $\alpha \to 0$, whereas our α is fixed, and as he considers a much more general problem his results are less sharp. A similar remark applies to the normal test, below.

We shall also need the bases for the normal and t-tests. Consider the problem of testing that the mean of a normal population of unit variance is zero, against the alternative that the mean is $\delta > 0$. From a sample $X_1, \cdots, X_N$ we compute $N\bar{X} = \sum X_i$ and reject if $\sqrt{N}\bar{X} > K$, where $\Phi(K) = 1 - \alpha$. The power is $\beta(N, \alpha, \delta) = 1 - \Phi(K - \sqrt{N}\,\delta)$. If we fix α and δ and let $N \to \infty$, $1 - \beta = P_{\text{II}}$ is equivalent (in the sense of ratio) to

$$(1/\sqrt{N}\delta)\cdot(1/\sqrt{2\pi})\exp[-(\tfrac{1}{2})(\sqrt{N}\delta - K)^2].$$

The limit of $\sqrt[N]{P_{\text{II}}}$ is thus $\exp[-(\frac{1}{2})\delta^2]$, as given by Chernoff. This is our base, say $b_N(\alpha, \delta)$, which turns out to depend on δ but not on α.

Now suppose that the variance is unknown. We estimate it by $s^2/(N-1)$, where $s^2 = \sum(X_i - \bar{X}_N)^2 : \chi^2_{N-1}$, form $t_N = \sqrt{N}\bar{X}_N/(s/\sqrt{N-1})$, and reject if $t_N > K_N$, where $K_N \to K$. The power is

$$\beta_N^*(\delta) = P\left\{\sqrt{N}\bar{X}_N > \frac{sK_N}{\sqrt{N-1}}\,\middle|\,\delta\right\}$$

$$= P\left\{\sqrt{N}(\bar{X}_N - \delta) > -\delta\sqrt{N} + \frac{sK_N}{\sqrt{N-1}}\,\middle|\,\delta\right\}.$$

Thus

$$1 - \beta_N^*(\delta) = \int_0^\infty \Phi\left(-\delta\sqrt{N} + \frac{sK_N}{\sqrt{N-1}}\right) p_{\chi_{N-1}}(s)\,ds.$$

We first consider an upper bound. Break the integration at $(N-1)^{2/3}$ to get $1 - \beta_N^*(\delta) < \Phi(-\delta\sqrt{N} + K_N(N-1)^{1/6}) + P(\chi_{N-1} > (N-1)^{2/3}$. The first term has base $\exp[-(\frac{1}{2})\delta^2]$ as before; the second has base 0, since (writing $N - 1 = m$)

$$\int_{m^{2/3}}^\infty c_m u^{m-2}\exp(-\tfrac{1}{2}u^2)\,du < \exp(-\tfrac{1}{4}m^{3/4}(\sqrt{2})^{m-1}$$

$$\int_{m^{2/3}}^\infty c_m\left(\frac{u}{\sqrt{2}}\right)^{m-2}\exp\left[-\tfrac{1}{2}\left(\frac{u}{\sqrt{2}}\right)^2\right]d\left(\frac{u}{\sqrt{2}}\right) < 2^{(m-1)/2}\exp(-\tfrac{1}{4}m^{3/4}).$$

Take the $1/(m+1)$ power and pass to limit to get 0.

A straightforward calculus argument permits us to verify the

LEMMA. *If base* $\{a_N\} \leq$ *base* $\{b_N\}$, *then base* $\{a_N + b_N\}$ = *base* $\{b_N\}$.

With the aid of this lemma we see that our upper bound has the same base $\exp[-(\frac{1}{2})\delta^2]$ as does P_{II} for the normal test. But since the normal test is more powerful than the t, it follows that the base of the t-test is also P_{II}.

We now apply these results to make limiting efficiency statements for fixed δ with $N \to \infty$. Suppose, for a standard test, that $\sqrt[N]{1-\beta(\delta)} \to A(\delta)$ as $N \to \infty$; while for a second test, suppose that $\sqrt[N]{1-\beta^*(\delta)} \to A^*(\delta)$. If we define $N^*(N)$ as in Section 2, it is easy to see that $N^*(N)/N \to (\log A^*(\delta))/(\log A(\delta))$. Thus, for the t-test compared to the normal, $e_{t,g}(\infty, \alpha, \delta) = 1$ for all α, δ. It follows that the comparison of sign to t will be the same as that of sign to normal; and as the latter is simpler, we shall examine it.

Let $X_1, \cdots, X_N$ be a sample from a normal population of unit variance. We may test the hypothesis that $E(X_i) = 0$ against the alternative that $E(X_i) = \delta > 0$ by using $\bar{X}_N$, in which case $\sqrt[N]{1-\beta_N(\delta)} \to \exp[-(\frac{1}{2})\delta^2]$ as seen above. We could also employ the sign test, rejecting the hypothesis if too many of the X_i are positive. The number of positive signs is binomial, with $p = \frac{1}{2}$ under the hypothesis, $p = \Phi(\delta)$ under the alternative. Therefore, for the sign test, we have from (3.2) the base $2\sqrt{\Phi(\delta)\,[1-\Phi(\delta)]}$. Thus for the sign test relative to the normal (and hence to the t),

$$e(\infty, \alpha, \delta) = \frac{2\log 2 + \log \Phi(\delta) + \log[1-\Phi(\delta)]}{-\delta^2}. \tag{3.3}$$

This quantity is seen to be independent of α but dependent on δ. As $\delta \to 0$, $e(\infty, \alpha, \delta) \to 2/\pi$, which agrees with the Pitman efficiency. As $\delta \to \infty$, $e(\infty, \alpha, \delta) \to \frac{1}{2}$. A few values of (3.3) are shown in the table. It is notable that (3.3) is very flat for δ in the range of interest, thus giving results in good agreement with those obtained from the simpler Pitman limit.

TABLE

δ	$1-\Phi(\delta)$	$e_{s,t}(\infty, \alpha, \delta)$
0	.50	.637
.253	.40	.636
.524	.30	.634
1.645	.05	.614
3.090	.001	.578
3.719	.0001	.566
∞	0	.500

The curve (3.3) may be regarded as the limit as $N \to \infty$ of the power efficiency function, values of which for the sign test relative to the t-test have been given by Dixon [13]. It appears from Dixon's charts that for fixed α, the actual power efficiency curve decreases smoothly toward its limit (3.3), making it possible to interpolate for intermediate N and thus to obtain rough values of the power of the t-test from binomial tables.

As a curiosity we finally examine the limit of $e(N, \alpha, \delta)$ as $\delta \to \infty$ with N, α fixed—a limit which Dixon denotes by E_∞. Our tool is an analog of basis: we recall that for the normal test, $1 - \beta_N(\delta) = \exp[-(\frac{1}{2})N\delta^2]\cdot f(N, \alpha, \delta)$, where $\log f(N, \alpha, \delta) = 0(N\delta^2)$. Restricting ourselves to even N, we can use formula (10) of [17] to show that for the t-test

$$1 - \beta_N(\delta) = \exp[-(1 - x)N\delta^2/4]\cdot g(N, \alpha, \delta),$$

where again $\log g(N, \alpha, \delta) = 0(N\delta^2)$ and x is the α-point on the beta distribution.

Lemma 3.3. *Suppose that two tests A and B are available for each sample size N, and that*

$$1 - \beta_A(N, \alpha, \delta) = \exp(-a_N\delta^2)f(N, \alpha, \delta),$$
$$1 - \beta_B(N, \alpha, \delta) = \exp(-bN\delta^2)g(N, \alpha, \delta),$$

where $\log f(N, \alpha, \delta)$ *and* $\log g(N, \alpha, \delta)$ *are* $o(\delta^2)$.

Suppose that $\beta_B(N, \alpha, \delta)$ *is strictly monotonely increasing in* N, $\beta_B(1, \alpha, \delta) = \alpha$, $\beta_B(N, \alpha, \delta) \to 1$ *as* $N \to \infty$. *Then* $e_{A,B}(N, \alpha, \infty) = (N_0 + 1)/N$, *where* N_0 *is the greatest integer less than* a_N/b.

Proof. We shall first assume that a_N/b is not an integer, so that there exists an integer N_0 with $N_0 < a_N/b < N_0 + 1$. Examining the ratio $[1 - \beta_A(N)]/[1 - \beta_B(m)] = [f(N)/g(m)]\exp[(bm - a_N)\delta^2]$, we see that for all sufficiently large δ,

$$\beta_B(N_0, \alpha, \delta) < \beta_A(N, \alpha, \delta) < \beta_B(N_0 + 1, \alpha, \delta).$$

Recalling our definition of efficiency, we see that if $p(\delta)$ is defined by

$$p(\delta)\beta_B(N_0, \alpha, \delta) + [1 - p(\delta)]\beta_B(N_0 + 1, \alpha, \delta) = \beta_A(N, \alpha, \delta), \tag{3.4}$$

then

$$e_{A,B}(N, \alpha, \delta) = \frac{N_0 + 1 - p(\delta)}{N}.$$

If we solve (3.4) for $p(\delta)$ and let $\delta \to \infty$, we find that $p(\delta) \to 0$. Therefore $e_{A,B}(N, \alpha, \infty) = (N_0 + 1)/N$.

In the remaining case, in which a_N/b is an integer, let $N_0 + 1 = a_N/b$. A similar analysis then produces the same limiting formula.

It is convenient to introduce the convention that $[u]$ denotes the greatest integer less than u. Then we see that

$$e_{t,g}(N, \alpha, \infty) = \frac{\left[\frac{1 - x}{2}N\right] + 1}{N}.$$

It is notable that this limiting efficiency is a discontinuous function of α. Given any N, there exists an $\alpha_0(N)$ such that for $\alpha < \alpha_0(N)$, $e_{t,g}(N, \alpha, \infty) = 1$. But if we fix α and let $N \to \infty$, $e_{t,g}(N, \alpha, \infty)$ tends to a limit less than 1. Thus,

$$\lim_{N\to\infty}\lim_{\delta\to\infty} e_{t,g}(N, \alpha, \delta) \neq \lim_{\delta\to\infty}\lim_{N\to\infty} e_{t,g}(N, \alpha, \delta).$$

Now consider the behavior of $\beta_s(N, \alpha, \delta)$ as $\delta \to \infty$. For an individual observation, the probability p of a positive sign is $\Phi(\delta) \sim 1 - (1/\delta)\varphi(\delta)$ for large δ. Examination of Lemmas 3.1 and 3.2 shows that the assumptions of Lemma 3.3 are met by the sign test with a_m the critical value for the number of positive signs. Therefore

$$e_{s,t}(N, \alpha, \infty) = \frac{\left[\frac{2a_N}{1 - x}\right] + 1}{N}.$$

This formula is not comparable to the E_∞ of Dixon, since our definition of N^* is not the same as his.

REFERENCES

[1] E. J. G. Pitman, Lecture notes on nonparametric statistical inference, Columbia University, 1949.

[2] G. E. Noether, "On a theorem of Pitman," *Ann. Math. Stat.*, Vol. 26 (1955), pp. 64–68.

[3] W. Hoeffding and J. Rosenblatt, "The efficiency of tests," *Ann. Math. Stat.*, Vol. 26 (1955), pp. 52–63.

[4] A. Mood, "On the asymptotic efficiency of certain non-parametric two-sample tests," *Ann. Math. Stat.*, Vol. 25 (1954), pp. 514–522.

[5] F. C. Andrews, "Asymptotic behavior of some rank tests for analysis of variance," *Ann. Math. Stat.*, Vol. 25 (1954), pp. 724–735.

[6] S. Saks, *Theory of the Integral*, Warsaw-Lvov, 1937.

[7] W. Hoeffding, "'Optimum' nonparametric tests," *Proceedings of the Second Berkeley Symposium on Mathematical Statistics and Probability*, University of California Press, 1951.

[8] M. E. Terry, "Some rank order tests which are most powerful against specific parametric alternatives," *Ann. Math. Stat.*, Vol. 23 (1944), pp. 346–366.

[9] M. Dwass, "The large sample power of rank order tests in the two-sample problem," *Ann. Math. Stat.*, this issue.

[10] B. L. van der Waerden, "Ein neuer Test für das Problem der zwei Stich proben," *Math. Ann.* Vol. 126 (1953), pp. 93–107.

[11] W. H. Kruskal and W. A. Wallis, "Use of ranks in one-criterion analysis of variance," *Jour. Am. Stat. Assoc.*, Vol. 47 (1952), pp. 583–621.

[12] J. L. Hodges, Jr. and E. L. Lehmann, "Matching in paired comparisons," *Ann. Math. Stat.*, Vol. 25 (1954), pp. 787–791.

[13] W. J. Dixon, "Power functions of the sign test and power efficiency for normal alternatives," *Ann. Math. Stat.*, Vol. 24 (1953), pp. 467–473.

[14] W. J. Dixon, "Power under normality of several nonparametric tests," *Ann. Math. Stat.*, Vol. 25, (1954), pp. 610–614.

[15] H. Chernoff, "A measure of asymptotic efficiency for tests of a hypothesis based on the sum of observations," *Ann. Math. Stat.*, Vol. 23 (1952), pp. 493–507.

[16] J. V. Uspensky, *Introduction to Mathematical Probability*, McGraw-Hill Book Company, Inc., 1937, New York.

[17] W. L. Nicholson, "A computing formula for the power of the analysis of variance test," *Ann. Math. Stat.*, Vol. 25 (1954), pp. 607–610.

RANK METHODS FOR COMBINATION OF INDEPENDENT EXPERIMENTS IN ANALYSIS OF VARIANCE[1]

By J. L. Hodges, Jr. and E. L. Lehmann

University of California, Berkeley

1. Introduction and summary. It is now coming to be generally agreed that in testing for shift in the two-sample problem, certain tests based on ranks have considerable advantage over the classical t-test. From the beginning, rank tests were recognized to have one important advantage: their significance levels are exact under the sole assumption that the samples are randomly drawn (or that the assignment of treatments to subjects is performed at random), whereas the t-test in effect is exact only when we are dealing with random samples from normal distributions. On the other hand, it was felt that this advantage had to be balanced against the various optimum properties possessed by the t-test under the assumption of normality. It is now being recognized that these optimum properties are somewhat illusory and that, under realistic assumptions about extreme observations or gross errors, the t-test in practice may well be less efficient than such rank tests as the Wilcoxon or normal scores test [6], [7].

Rank tests were naturally developed first for the simple two-sample problem,[2] but in practice experiments for the evaluation of population or treatment differences are seldom of this form. To secure the advantages of increased homogeneity and the resultant increased precision, it is more customary to stratify the populations or to divide the experimental subjects into blocks, using a (generalized) randomized block design. In other experiments where additivity of certain effects is assumed, a design of the Latin square type may be appropriate.

It is the purpose of the present paper to provide a method for constructing rank tests for such designs. The basic idea of the method is described in Section 3. The main body of the paper is concerned with the application of the method to the comparison of two treatments using a Wilcoxon-type test statistic. The exact distribution of this statistic is discussed in Section 4, and its asymptotic distribution in Sections 5 and 6. Some remarks concerning the efficiency of the test are given in Section 7. Finally, Section 8 illustrates the application of the method to the comparsion of more than two treatments, for such designs as incomplete blocks and Latin squares.

2. Tests based on independent rankings. A simple method for dealing with at least some cases of the problem described in the introduction is based on separate rankings of the observations in each block.

Received January 30, 1960.

[1] This paper was prepared with the partial support of the Office of Naval Research (nonr-222-43). This paper in whole or in part may be reproduced for any purpose of the United States Government.

[2] And for the case of paired comparisons, which also turns out to lend itself particularly well to a rank analysis.

J. Rojo (ed.), *Selected Works of E. L. Lehmann*, Selected Works in Probability and Statistics, DOI 10.1007/978-1-4614-1412-4_35, © Springer Science+Business Media, LLC 2012

To fix ideas, let us consider the generalized randomized block design for comparing two treatments. Suppose that N subjects are available for the experiment. They are divided into n blocks, each block consisting of subjects thought to be homogeneous with regard to the response to be observed. The N_i subjects in the ith block are then randomly divided into groups of sizes s_i and t_i ($s_i + t_i = N_i$), and the groups are given treatments A and B respectively. We want a test of the hypothesis that the treatments do not differ, which will be sensitive to the shift alternatives that the response to treatment B tends consistently to be higher than the response to A.

A rank test for this problem was proposed by Wilcoxon [10]. The observations in each block are ranked separately, a Wilcoxon statistic (for example, the sum of the B-ranks) $W^{(i)}$ is formed for the ith block, and the test statistic is the sum of these W's. In a recent investigation [3], van Elteren has considered the more general statistic $\sum_{i=1}^{n} c_i W^{(i)}$ and has shown that in a certain sense the optimum constants are

$$c_i = 1/(s_i + t_i + 1). \tag{2.1}$$

What is the efficiency of this test relative to the appropriate t- or normal test, which is based on the statistic

$$\sum \frac{s_i\, t_i}{s_i + t_i} (Y_{i\cdot} - X_{i\cdot}) \Big/ \sum \frac{s_i\, t_i}{s_i + t_i} \tag{2.2}$$

where $X_{i\cdot}$ and $Y_{i\cdot}$ denote the average of the A- and B-responses in the ith block? To answer this, suppose that the Wilcoxon experiment consists of r' replications of some given set of blocks (s_i, t_i), and that the t-experiment consists of $r'' = g(r')$ replications of the same set where r'' will be determined below. Suppose that the X's and Y's in the ith block are normally distributed with means ξ_i and $\xi_i + \Delta$ respectively and with common variance 1. For fixed values of α and β, let $\Delta = \Delta(r')$ be determined so that the power of the Wilcoxon-van Elteren level α test tends to β as $r' \to \infty$. Let $r'' = g(r')$ be determined, so that the power of the level α t-test (based on r'' replications) against the same sequence of alternatives $\Delta(r')$ also tends to β. We then say that the relative asymptotic efficiency of the Wilcoxon-van Elteren test to the t-test is e if

$$\lim_{r' \to \infty} (r'/r'') = e,$$

independent of α and β.

THEOREM 1. *For block sizes* (s_i, t_i) *the relative asymptotic efficiency of the Wilcoxon-van Elteren to the t-test based on the statistic* (2.2) *is*

$$e = \frac{3}{\pi} \sum \frac{s_i\, t_i}{s_i + t_i + 1} \Big/ \sum \frac{s_i\, t_i}{s_i + t_i}. \tag{2.3}$$

PROOF. Since the method of proof is well known (see for example [7]), it will be enough here to sketch the proof. The statistic (2.2) (and therefore asymptotically as $\sum s_i$ and $\sum t_i \to \infty$ also the corresponding t-statistic) is normally distributed with mean Δ and variance $[\sum s_i t_i/(s_i + t_i)]^{-1}$.

Analogously, the Wilcoxon statistic in the ith block, for small Δ has approximately mean $s_i t_i(\frac{1}{2} + \frac{1}{2}\Delta/\sqrt{\pi})$ and null variance $s_i t_i(s_i + t_i + 1)/12$ and the linear combination of these statistics with weights (2.1) therefore has approximately mean $(\frac{1}{2} + \frac{1}{2}\Delta/\sqrt{\pi}) \sum s_i t_i/(s_i + t_i + 1)$ and null variance

$$\sum s_i t_i(s_i + t_i + 1)/12.$$

If the Wilcoxon statistics are based on (s_i', t_i') and the t-test on (s_i'', t_i'') observations, and if $\sum s_i'$, $\sum t_i'$ are sufficiently large for the normal approximation to be close, the power of the two tests for small Δ will be approximately the same if

$$\Delta^2 \sum \frac{s_i'' t_i''}{s_i'' + t_i''} = \frac{\Delta^2}{4\pi} \frac{\left(\sum \frac{s_i' t_i'}{s_i' + t_i' + 1}\right)^2}{\frac{1}{12} \sum \frac{s_i' t_i'}{s_i' + t_i' + 1}},$$

that is, if

$$\sum \frac{s_i'' t_i''}{s_i'' + t_i''} = \frac{3}{\pi} \sum \frac{s_i' t_i'}{s_i' + t_i' + 1}. \tag{2.4}$$

Suppose now that the Wilcoxon experiment consists of r' replications of some given set of blocks (s_i, t_i); and that the t-experiment consists of r'' replications of the given set. Then (2.4) becomes

$$r'' \sum \frac{s_i t_i}{s_i + t_i} = r' \frac{3}{\pi} \sum \frac{s_i t_i}{s_i + t_i + 1}$$

and this completes the proof.

We note two interesting special cases of (2.3). If the block size $N_i = s_i + t_i$ is constant, $N_i = k$ say, then (2.3) reduces to

$$e_k = \frac{3}{\pi} \frac{k}{k+1} \tag{2.5}$$

regardless of how the numbers s_i, t_i are chosen in each block. Table 2.1 gives this efficiency for a number of different values of k.

Another case of interest is that of a single block. The test then is the ordinary Wilcoxon test, and as $s + t \to \infty$, the efficiency $(3/\pi)(s + t)/(s + t + 1)$ tends to the value $3/\pi$, ordinarily given as the asymptotic relative efficiency of the Wilcoxon to the t-test in this case.

For the comparison of more than two, say $c + 1$, treatments in a randomized block design ($c + 1$ subjects in each block to which the $c + 1$ treatments are assigned at random), an analogous rank test was proposed by Friedman [5],[3] and his test was generalized to balanced incomplete randomized blocks by Durbin [2] and to general blocks by Benard and van Elteren [1]. In these tests,

[3] For other rank procedures for this problem see Walsh [9].

TABLE 2.1

k	2	3	4	5	10	15	∞
e_k	$\frac{2}{\pi} = .637$	.716	.764	.796	.868	.895	$\frac{3}{\pi} = .955$

the ranking is done separately within each block, and the test statistic is a suitable quadratic form in the sum of the ranks for the various treatments. When the block size is k, it was shown by van Elteren and Noether [4] that the asymptotic relative efficiencies of Friedman's and Durbin's tests relative to the appropriate F-tests is also given by (2.5).

For the case of two treatments and blocks of size $k = 2$, the efficiency $2/\pi$ is not surprising since the Wilcoxon sum and Friedman's test then reduce to the one- or two-sided sign test for matched pairs. It is seen from the table of e_k that the efficiency remains unpleasantly low as long as the blocks are small. This is unfortunate since it is often desirable to use rather small blocks either because the natural blocks are small (for example, litters) or because small blocks are required to achieve within-block homogeneity. In such cases, tests based on independent rankings leave much to be desired.

There are of course situations in which separate rankings are all that can be obtained. An example is the case in which the blocks correspond to different observers, each of whom makes a comparison of the different treatments assigned at random to different sets of subjects. If this comparison can be made only in the form of a ranking, the basic data are just the separate rankings and the over-all evaluation must be based on these data. However, if instead each observer assigns scores to the different treatments, even if the method of scoring is not the same for each observer, the method to be outlined in the next section can be applied.

3. Ranking after alignment. While in the case of a matched pairs experiment ($s_i = t_i = 1$), the efficiency of the Wilcoxon-sum test is only $2/\pi$, there does of course in this case exist a rank test, Wilcoxon's one-sample test, whose efficiency is $3/\pi$ in normal populations. In this test the absolute differences in response are ranked for the n pairs, and with each rank is associated the sign of the corresponding response difference. The test statistic is the sum of the ranks corresponding to the positive (or negative) differences.

It seems natural to ask why this test has efficiency $3/\pi$ as compared with the efficiency $2/\pi$ for the Wilcoxon-sum test. The reason appears to be essentially that the former test pays attention to certain interblock comparisons which are entirely ignored by the latter test. The main objective of the present paper is to find rank procedures for more general designs that will preserve the interblock comparisons in the hope that this will lead to higher efficiency.

We can best introduce the method by continued consideration of the com-

TABLE 3.1

Block	Treatment A	Treatment B	Mean
1	98, 169	28, 113	102
2	259	168, 128	185
3	81, 120	24, 102, 8	67

parison of two treatments. Suppose that treatments A and B are compared by random assignment to the subjects in three blocks with $s_1 = t_1 = 2$; $s_2 = 1$, $t_2 = 2$; $s_3 = 2$, $t_3 = 3$, and that the results are observed as given in Table 3.1. From each observation we subtract the block mean, arrange the residuals in order of increasing size, and rank them:

Residual	−74	−59	−57	−43	−17	−4	11	14	35	53	67	74
Rank	1	2	3	4	5	$\underline{6}$	7	$\underline{8}$	9	$\underline{10}$	$\underline{11}$	$\underline{12}$

The ranks corresponding to treatment A are underlined; their sum

$$6 + 8 + 10 + 11 + 12 = 47$$

is the value of a test statistic which will be denoted by W.

Let us first observe that W provides an exact test of the null hypothesis that the treatments do not differ. Under that hypothesis, the labels A and B are meaningless and could as well be assigned at the end of the experiment, after the data are recorded. Each block has certain ranks, as tabulated below.

Block	Ranks
1	1, 6, 7, 11
2	3, 5, 12
3	2, 4, 8, 9, 10

The random assignment of two A-labels to the four ranks of Block 1 produces two A-ranks, whose sum will be denoted by W_1. W_1 is equally likely to have any of the values $1 + 6 = 7$, $1 + 7 = 8$, $1 + 11 = 12$, $6 + 7 = 13$, $6 + 11 = 17$ or $7 + 11 = 18$. Similarly W_2, the rank labeled A in Block 2, takes on the values 3, 5, 12 with probability $\frac{1}{3}$ each, while W_3 is equally likely to have any of the 10 values $2 + 4 = 6$, 10, 11, 12, 12, 13, 14, 17, 18, 19. The statistic $W = W_1 + W_2 + W_3$ is the sum of three independent integer-valued random variables. The maximum value of W is 49, and one sees by inspection that the only value of W_1, W_2, W_3 for which $W \geqq 47$ are $W_1 = 17$ or 18, $W_2 = 12$, $W_3 = 17$, 18 or 19. Thus

$$P(W \geqq 47) = \frac{2 \cdot 3}{\binom{4}{2}\binom{3}{1}\binom{5}{2}} = \frac{6}{180} = \frac{1}{30}.$$

We can now state the idea of the procedure. The first step is to bring the observations in the various blocks into alignment with one another. In the example above this was done by subtracting from each observation the mean observation in its block, but in some cases other methods for alignment might be better, such as subtracting a trimmed or Winsorized mean.[4] The important point is that the treatments must be ignored when the alignment is made.

Once the observations are aligned they are pooled and ranked without regard to their blocks. Then the ranks are labeled according to the treatment given to the corresponding observation. Under the null hypothesis of no treatment effect, the assignment of labels to the ranks pertaining to each block is done at random, and may be thought of as having been done after the ranks for each block are determined. The partition of the ranks in each block into label-groups is independent, and has in each block a known distribution that depends only on the design employed at the beginning of the experiment in assigning treatments to subjects.

Finally, we may compute from the labeled ranks any rank statistic appropriate for the alternatives against which it is desired to test the null hypothesis. In the example we used the Wilcoxon statistic and computed its one-sided significance probability, but one may in other cases want a two-sided test, or use the normal scores test statistic, or indeed whatever statistic seems appropriate. In any case, the statistic will have an exact null distribution whose computation depends on the known distribution of the labels among the block ranks.

The procedure is thus very flexible. We have already mentioned that the method of alignment and the choice of test statistic may be adjusted to the problem. Still another possibility is that of first transforming the data, where a different transformation may be used in each block. None of these devices affects the exactness of the null distribution. They do however influence the power of the test, and the choices should be made with the view of providing large power against the alternatives of interest.

The null distributions of all the tests suggested above are based on the independent random assignment of treatments in each block. One can of course also apply the tests to a different experimental situation, in which the subjects in the different blocks are randomly sampled from different populations, but in this case the null distributions discussed above are conditional distributions, given the responses of the sampled subjects.

4. Rank sum analysis of two treatments in a block design. The computation of significance probability in the example considered above was carried out by inspection. The various equally likely samples were examined to determine the number of them at least as significant as the one observed. While such a procedure can be employed quite generally with small designs or with highly significant results, it is not feasible for hand computation in other cases. Fortunately the method readily adapts to automatic machine computation, and it would be easy to write programs for the routine calculation of exact significance probabilities.

---

[4] For Trimming and Winsorizing see Sect. 14 of "The future of data analysis" by John W. Tukey (1962). *Ann. Math. Statist.* **33** 1–67.

In the special case of the comparison of two treatments, enumeration by hand can be organized so that it is applicable in designs considerably larger than that considered in Section 3. Suppose that the subjects are divided into several blocks, that there are two treatments A and B applied with unrestricted randomization within blocks, and that the test statistic W is the sum of the ranks after alignment associated with, say, treatment A. (This is the situation illustrated by the example in Section 3.) Let n denote the number of blocks, let there be N_i subjects in the ith block of which s_i are allocated to the first treatment, and let W_i be the sum of the ranks after alignment of these s_i subjects. Then $W = W_1 + \cdots + W_n$.

Under the null hypothesis the W_i are independent random variables. The possible values of W_i are integers, and its null distribution can easily be written down by considering the $\binom{N_i}{s_i}$ equally likely choices of ranks of the first treatment. The calculation of the null distribution of W now requires the convolution of the k independent integer-valued random variables. There are $\prod \binom{N_i}{s_i}$ equally likely cases, and we have only to count those giving $W = w$ to find $P(W = w)$. A convenient layout for this count essentially like that required in the exact use of the Wilcoxon sum statistic W_s, may be presented through an example.

Suppose there are $k = 5$ blocks whose sizes, allocations, and ranks after alignment are as given in Table 4.1. The ranks of subjects receiving treatment A are underlined, and the observed value of W is 79, which is 22 greater than its smallest possible value 57. To compute $P(W \leqq 79)$ it will only be necessary to consider the lower tail for a range of 22 units at each step.

We first write down by inspection the distribution of $W_i - \min W_i = U_i$ for each block, multiplying each probability by $\binom{N_i}{s_i}$ so that only integers need be written. The results are shown in the top section of Table 4.2. For example, when two ranks are chosen from those in Block 1, the six equally likely values

TABLE 4.1

i	N_i	s_i	$\binom{N_i}{s_i}$	Ranks after alignment					W_i	$\min W_i$
1	4	2	6	$\underline{4}$	5	$\underline{17}$	20		21	9
2	5	3	10	$\underline{6}$	$\underline{7}$	$\underline{12}$	13	15	25	25
3	3	1	3	9	$\underline{10}$	16			10	9
4	4	2	6	2	$\underline{8}$	$\underline{11}$	21		19	10
5	5	2	10	$\underline{1}$	$\underline{3}$	14	18	19	4	4
			$\prod \binom{N_i}{s_i} = 10800$						$W = 79$	$\min W = 57$

TABLE 4.2

	u = 0					5					10					15					20			Check Column
i = 1	1												1	1		1	1							
2	1	1		1			1	1	1	2	1					1								
3	1	1						1																
4	1			1						1				1						1			1	
5	1											1		1		1	1	1	1					
	1												1	1		1	1							5
		1												1	1		1	1						5
				1												1	1		1	1				5
							1												1	1		1	1	5
								1												1	1		1	4
									1												1	1		3
										2												2	2	6
											1												1	2
																1								1
	1	1		1			1	1	1	2	1		1	2	1	3	3	1	2	3	2	4	5	36
		1	1		1			1	1	1	2	1		1	2	1	3	3	1	2	3	2	4	31
								1	1		1			1	1	1	2	1		1	2	1	3	16
	1	2	1	1	1		1	3	3	3	4	1	1	4	4	5	8	5	3	6	7	7	12	83
				1	2	1	1	1		1	3	3	3	4	1	1	4	4	5	8	5	3	6	57
										1	2	1	1	1		1	3	3	3	4	1	1	4	36
														1	2	1	1	1		1	3	3	3	16
																				1	2	1	1	5
																							1	1
	1	2	1	2	3	1	2	4	3	5	9	5	5	10	7	8	16	13	11	20	18	15	27	188
												1	2	1	2	3	1	2	4	3	5	9	5	38
														1	2	1	2	3	1	2	4	3	5	24
																1	2	1	2	3	1	2	4	16
																	1	2	1	2	3	1	2	12
																		1	2	1	2	3	1	10
																			1	2	1	2	3	9
	1	2	1	2	3	1	2	4	3	5	9	6	7	12	11	13	22	22	22	33	34	35	47	297
	1	3	4	6	9	10	12	16	19	24	33	39	46	58	69	82	104	126	148	181	215	250	297	

for W_1 are $4 + 5 = 9, 4 + 17 = 21, 5 + 17 = 22, 4 + 20 = 24, 5 + 20 = 25$ and $17 + 20 = 37$. Therefore the values of U_1 are $9 - 9 = 0$, $21 - 9 = 12$, $22 - 9 = 13$, $24 - 9 = 15$, $25 - 9 = 16$ and $37 - 9 = 28$. The last of these is outside the range of interest; the other five are represented in the row of Table 4.2 corresponding to $i = 1$.

The convolution of the five random variables U_i is displayed in the remainder of Table 4.2. The second section shows the convolution of U_1 and U_2. The rows correspond to values of U_2, the columns to values of $U_1 + U_2$. Thus, the first row of the second section gives for each possible value of $u_1 + u_2$ the number of cases for which $u_1 = 0$; the second row those for which $u_1 = 1$ (these are obtained from the first row by shifting it to the right by one unit); etc. By adding the number in each column, we obtain the distribution of $U_1 + U_2$ $\left(\text{with each probability multiplied by } \binom{N_1}{s_1}\binom{N_2}{s_2}\right)$, which is given in the top line of the third section. Similarly, the third section represents the convolution of $U_1 + U_2$ with U_3, etc. The final results are in the next to last line, so that $P(\sum U_i = 22) = P(W = 79) = 47/10800$. Cumulating, the last line gives $P(W \leqq 79) = 297/10800 = .0275$.

The computation requires only additions and proceeds quite rapidly. If desired a check is provided by the row sums as shown on the right. The method will be effective so long as the observed value w of W is not too much larger than its minimum value (or too much smaller than its maximum, if $P(W \geqq w)$ is to be computed).

In principle, the exact joint distribution of the rank sums could be similarly handled if there are more than two treatments, but the labor is much greater. For example, if there are treatments A, B, C, with rank sums W, X, Y, we may regard (W, X) as the sum of n independent vectors (W_i, X_i) each of which is integer-valued and has a distribution whose terms when multiplied by the multinomial coefficient $\binom{N_i}{s_i, t_i}$ are integers. However, each step in the convolution requires a double summation, so that the layout is less simple than that shown above.

5. Normal approximation to the null distribution of W. With designs much larger than that used for illustration above, exact hand computation becomes excessively lengthy. Fortunately, in such cases a simple normal approximation may serve. Recall that $W = W_1 + \cdots + W_n$ is the sum of n independent random variables W_i. Each W_i, as the sum of a sample drawn from a fixed set of integers, has easily calculated moments, so the low order moments of W may be readily obtained. The W_i will usually have variances of about the same magnitude; thus, if n is reasonably large one may expect W to be approximately normal. This hope is reinforced by a theorem proved in the next section.

Let us illustrate the method on the example of the preceding section. If the N_i ranks in Block i are denoted by r_{ij}, $j = 1, \cdots, N_i$, then $EW_i = s_i \sum r_{ij}/N_i$ while

TABLE 5.1

i	N_i	s_i	Ranks					EW_i	var W_i
1	4	2	4	5	17	20		23	67
2	5	3	6	7	12	13	15	31.8	18.36
3	3	1	9	10	16			11.67	9.56
4	4	2	2	8	11	21		21	63
5	5	2	1	3	14	18	19	22	85.8
								$E(W) = 109.47$	$\text{var}(W) = 243.72$
									$\sigma(W) = 15.61$

TABLE 5.2

w	$P(W \leqq w)$	Normal approximation	Error
79	.0275	.0274	.0001
78	.0231	.0236	.0005
77	.0199	.0203	.0004
76	.0168	.0173	.0005
75	.0137	.0148	.0011
74	.0117	.0125	.0008

$$\text{var}\,(W_i) = \frac{s_i\,t_i}{N_i - 1}\,[(\textstyle\sum r_{ij}^2)/N_i - (\sum r_{ij}/N_i)^2],$$

where in each case the summation is for j from 1 to N_i. Finally $EW = \sum EW_i$ and $\text{var}(W) = \sum \text{var}(W_i)$, as shown in Table 5.1. As W is integer-valued, it is natural to use a continuity correction. We approximate $P(W \leqq 79)$ by

$$\Phi\left(\frac{79.5 - 109.47}{15.61}\right) = \Phi(-1.9199) = .0274.$$

The excellent agreement of this value with the correct .0275 is something of an accident, as the following table shows. However, even with only 5 blocks the approximation would be accurate enough for most purposes (see Table 5.2).

We remark that the third and fourth moments of W are also easily available if it is thought desirable to use an approximation based on more than two moments.

6. Asymptotic null distribution of the blocked Wilcoxon statistic. In the preceding section we have discussed the exact null distribution of the blocked Wilcoxon statistic, as well as the normal approximation. In the present section we shall show that this distribution when normalized in the usual way tends to the standard normal distribution. In the present section, we shall consider the limiting behavior of this distribution as the number of blocks becomes large. As before, we shall consider the responses in each block (without regard as to which belongs to treatment and which to control) as fixed, and we shall refer to the totality of these responses as the *configuration*. The only randomness is that resulting from the independent random assignments of treatments in each block. We shall prove below under certain assumptions that the null distribution of the blocked Wilcoxon statistic when normalized in the usual way tends to the standard normal distribution and that the convergence is *uniform* in the configurations.

This is proved by means of the Berry-Esseen theorem (cf., Theorem B on p. 288 of [8]) which may be stated as follows:

Let $W_1, W_2, \cdots$ be independently distributed, with means $\mu_1, \mu_2, \cdots$, and let

$$E(W_i - \mu_i)^2 = \sigma_i^2, \qquad E|W_i - \mu_i|^3 = \beta_i, \qquad S_n^2 = \sum_{i=1}^{n} \sigma_i^2.$$

Let F_n denote the c.d.f. of $\sum_{i=1}^{n} (W_i - \mu_i)/S_n$, and Φ the standard normal c.d.f. There exists a constant $c < \infty$ such that for all x:

$$|F_n(x) - \Phi(x)| \leqq \frac{c}{S_n^3} \sum_{i=1}^{n} \beta_i. \tag{6.1}$$

We are concerned with the comparison of a treatment with a control in n blocks. As before, suppose that the ith block contains N_i experimental subjects, of which t_i are selected at random to receive the treatment, with the remaining s_i serving as controls $(s_i + t_i = N_i,\ i = 1, \cdots, n)$. We shall assume that

$$N_i \leqq k \quad \text{for all} \quad i. \tag{6.2}$$

LEMMA 1. *Let W_i denote the sum of the ranks (after alignment) of the control responses in the ith block and let β_i denote the absolute* 3rd *moment of W_i. Then under assumption* (6.2) *there exists a constant $0 < b < \infty$ independent of the configuration such that*

$$\sum_{i=1}^{n} \beta_i \leqq bn^4. \tag{6.3}$$

PROOF. Note that

$$E|W_i - \mu_i|^3 \leqq s_i \max |r_{ij} - \mu_i|^3 \leqq k \cdot N^3,$$

where $r_{i1}, \cdots, r_{iN_i}$ denote the ranks in the ith block and where $N = \sum N_i$.

Since $N \leqq kn$, it follows that

$$\sum_{i=1}^{n} \beta_i \leqq n \cdot k^4 n^3 = k^4 n^4$$

as was to be proved.

We note that (6.3) is valid without any assumptions regarding the method of alignment, and that it does not require complete randomization within each block but would be equally valid under any method of restricted randomization. On the other hand, to obtain a lower bound for S_n^2 we make the assumption:

(A) After alignment, each block contains at least one observation above and one below the origin.

This assumption is satisfied for all reasonable methods of alignment such as alignment on the mean, on a censored or Winsorized mean, or on the median.

In addition, we shall suppose that:

(B) Within each block, complete randomization is employed.

LEMMA 2. *Let* S_n^2 *be the sum of the variances of the variables* W_i *of Lemma* 1. *Under assumption* (6.2), (A) *and* (B), *there exists a constant* $0 < a < \infty$ *independent of the configuration such that*

$$S_n^2 \geqq an^3. \tag{6.4}$$

PROOF. Under assumption (B), it is well known that

$$\sigma_i^2 = \frac{N_i - 1}{N_i - s_i} s_i \tau_i^2 \qquad \text{where} \qquad \tau_i^2 = \sum_{j=1}^{N_i} (r_{ij} - \bar{r}_i)^2/N_i .$$

Since this is a minimum when $s_i = 1$, in which case $\sigma_i^2 = \tau_i^2$, it is enough to prove that $\sum_{i=1}^n \tau_i^2 \geqq an^3$. Now

$$\tau_i^2 \geqq (1/N_i) \min_j (r_{ij} - \bar{r})^2 \geqq (1/k) \min_j (r_{ij} - \bar{r})^2$$

since of the r_{ij} at least one lies on either side of $\bar{r}$ and hence at least one must be closer to $\bar{r}$ than to $\bar{r}_i$. Since the N ranks r_{ij} are distinct integers, it follows that

$$\sum_i \min_j (r_{ij} - \bar{r})^2 \geqq 2\left(1^2 + 2^2 + \cdots + \left[\frac{n-2}{2}\right]^2\right)$$

and since the right-hand side is of the order n^3, this completes the proof.

THEOREM 2. *Under assumptions* (6.2), (A) *and* (B), *the distribution* F_n *of* $\sum_{i=1}^n (W_i - \mu_i)/S_n$ *tends to the standard normal distribution as* $n \to \infty$, *and the convergence is uniform in the configuration.*

PROOF. Combining (6.3) and (6.4) with (6.1), we see that

$$|F_n(x) - \Phi(x)| \leqq (bc/a^{\frac{3}{2}})(1/n^{\frac{1}{2}}),$$

which proves the desired result.

In order to apply (6.1) it is of course not necessary that (6.3) and (6.4) be satisfied but only that

$$\sum_{i=1}^n \beta_i/S_n^3 \to 0 \quad \text{as} \quad n \to \infty. \tag{6.5}$$

If for example, the ranks in the first block are $1, \cdots, N_1$; those in the second block $N_1 + 1, \cdots, N_1 + N_2$; etc., both S_n^2 and $\sum_{i=1}^n \beta_i$ are of order n, and hence (6.5) holds without (6.4) being satisfied.

On the other hand, although assumptions (A) and (B) are certainly not necessary for the validity of Theorem 2, some conditions on the method of alignment and randomization are required as the following two examples will show.

EXAMPLE 6.1. Suppose $N = 2n$ and the ith block consists of the ranks $(2i, 2i + 1)$ for $i = 1, \cdots, n - 1$ while the nth block consists of the ranks 1 and $2n$. Within each block, let one of the ranks be assigned at random to treatment, the other to control.

EXAMPLE 6.2. Suppose $N = 4n$ and the ranks in block 1 are $1, 2, 4n - 1, 4n$; those in block 2 are $3, 4, 4n - 3, 4n - 2$; etc., and suppose that in each block except the first one, the probability is $\frac{1}{2}$ that the two outside ranks belong to treatment and $\frac{1}{2}$ that the inner two ranks belong to treatment.

In the first of these examples, condition (B) is satisfied but not (A); in the second example, condition (A) is satisfied but not (B). For both examples, it is clear that $\sum W_i$ does not tend to be normally distributed as $n \to \infty$.

7. Efficiency. Some indications regarding the efficiency of the W-test discussed in the preceding sections can be obtained by considering the special case of paired comparisons $(s_i = t_i = 1)$.[5] Let W be the rank sum oe the treated subjects, after each block (of two) has been aligned on the midpoint between the two observations X (control) and Y (treated) in this block. To determine the efficiency of the W-test, we shall obtain a relation between W and the Wilcoxon one-sample statistic W'.

Suppose without loss of generality that the n blocks are numbered in order of increasing absolute difference $Y - X$ between treated and control response. Let

$$r_i = i, \qquad s_i = 0 \quad \text{if in the } i\text{th block} \quad X_i < Y_i,$$

$$r_i = 0, \qquad s_i = i \quad \text{otherwise.}$$

Further, let r_i^* denote the rank of the Y-observation in the ith block after alignment. Then if we have $X_i < Y_i$, clearly

$$r_i^* = n + i = n + r_i,$$

since after alignment Y_i exceeds all n negative observations and is the ith smallest of the positive observations.

Similarly, if $Y_i < X_i$, it is seen that

$$r_i^* = n + 1 - i = n + 1 - s_i.$$

The Wilcoxon statistic after alignment is then

$$W = \sum_{i=1}^{n} r_i^* = \sum{}^{(1)} (n + r_i) + \sum{}^{(2)} (n + 1 - s_i)$$

where $\sum^{(1)}$ and $\sum^{(2)}$ extend over all blocks with $Y > X$ and $Y < X$ respectively. If S is the number of blocks in which $Y > X$, that is, the sign statistic,

[5] In this case the efficiency computation is very much simpler than it is in general since the conditional null distribution of the test statistic given the configuration is independent of the configuration.

we have

$$W = \sum_{i=1}^{n} r_i + nS + (n + 1)(n - S) - \sum_{i=1}^{n} s_i$$

$$= \sum_{i=1}^{n} r_i - \sum_{i=1}^{n} s_i - S + n(n + 1).$$

Finally, since $\sum_{i=1}^{n} r_i + \sum_{i=1}^{n} s_i = 1 + 2 \cdots + n = \frac{1}{2}n(n + 1)$, we get the relationship

$$W = 2W' - S + \tfrac{1}{2}[n(n + 1)]$$

where $W' = \sum_{i=1}^{n} r_i$ is the one-sample Wilcoxon statistic.

Since the variance of W' is of order n^3 and the variance of S only of order n, it is seen that W and W' are asymptotically equivalent, and that the two associated tests have the same Pitman efficiency. In particular, the asymptotic efficiency of W relative to the corresponding t-test is $3/\pi$.

Suppose now that the block sizes are even but larger than 2 and that $s_i = t_i$ for all i. We can then obtain a test of asymptotic efficiency $3/\pi$ relative to the t-test by pairing control and treatment observations within each block at random and applying the W'-test to the resulting pairs. It seems plausible that an efficient method of alignment of the block as a whole followed by an application of the W'-test should be more efficient than this rather arbitrary procedure,[6] and preliminary work for the case of normal distributions with alignment on the block mean suggests that this is indeed the case.

This slight gain in efficiency appears however to decrease and tend to zero if instead of a large number of small blocks we are dealing with a small number of large blocks. As the block size tends to infinity, the additional information gained from intrablock comparisons above that provided by interblock comparisons, seems to tend to zero, with the efficiency of the W'-method tending to the efficiency $3/\pi$ found in Section 2 as the limiting efficiency for the Wilcoxon-van Elteren method based on independent rankings.

We hope to amplify these remarks, which are based partly on heuristic reasoning and partly on preliminary computations, in a later paper.

8. Several treatments. We conclude with two examples which indicate how the technique of aligned ranks may be applied when more than two treatments are to be compared.

EXAMPLE 8.1. *Incomplete block design.* Suppose three treatments are compared on three matched pairs, with these results

---

[6] Another interesting possibility would be to replace random pairing by a pairing of the smallest control with the smallest treatment observation, of the second smallest control with the second smallest treatment observation, etc., and then to compute the one-sample Wilcoxon statistic for these pairs.

	Treatment		
Block	A	B	C
1	131	115	—
2	—	151	141
3	131	—	105

If we align on block mean, the ranks are as follows

	Treatment		
Block	A	B	C
1	5	2	—
2	—	4	3
3	6	—	1

Let us use the Kruskal-Wallis statistic, which tests against unspecified differences in the treatments. The treatments A, B, C have rank sums 11, 6, 4 respectively, and the sum of squared differences from the mean of 7 is

$$(11 - 7)^2 + (6 - 7)^2 + (4 - 7)^2 = 26.$$

There are $2^3 = 8$ equally likely ways to label the ranks in the three blocks, and the actual labeling is seen by inspection to give the largest value of the statistic. Thus the significance probability is $\frac{1}{8}$.

EXAMPLE 8.2. *Latin squares.* To illustrate the application of our method to Latin squares, suppose that three treatments A, B, C are compared in two 3×3 squares, yielding these observations

Square 1			Square 2		
B 4.461	C 2.798	A 7.402	C 5.424	B 9.670	A 9.669
A 3.412	B 2.405	C 5.227	B 5.062	A 9.368	C 5.710
C 3.454	A 2.169	B 6.717	A 6.605	C 7.786	B 7.427

When the row and column effects have been removed from each square in the usual way, these residuals are obtained:

Square 1

B .025 (10)	C −.319 (5)	A .293 (13)
A .182 (12)	B .494 (15)	C −.676 (4)
C −.208 (6)	A −.174 (7)	B .382 (14)

Square 2

C −1.114 (2)	B −.112 (8)	A 1.226 (18)
B .065 (11)	A 1.127 (17)	C −1.192 (1)
A 1.049 (16)	C −1.014 (3)	B −.034 (9)

After each residual is written its rank in the residual pool. The rank sums of treatments A, B, C are respectively

$$U = 32 + 51 = 83 \qquad V = 39 + 28 = 67 \qquad W = 15 + 6 = 21$$

According to the null hypothesis of no treatment effect, the 18 ranks are fixed numbers to which the labels A, B, C were attached in a pattern of restricted randomization. In a 3×3 square there are only two basic patterns, but in addition the labels may be arranged in $3! = 6$ orders, giving $2 \cdot 6 = 12$ cases in all. As each square has 12 cases, there are $12^2 = 144$ cases, which are equally likely in the usual design. Whatever test statistic may be used, its null distribution may be obtained by calculating its value for each of these cases.

For example, suppose we again use the Kruskal-Wallis statistic,

$$T = (U - 57)^2 + (V - 57)^2 + (W - 57)^2.$$

With the actual data, $T = 2072$. To determine the significance of this value, notice that in Square 1 the three rank sums must be either 15, 32, 39 or else 21, 31, 34, depending on which basic pattern is used. Similarly, in Square 2 we should have either 25, 28, 32 or 6, 28, 51. Now T is symmetric in U, V, W, which means that we may attach the labels A, B, C arbitrarily in one square. There are 24 equally likely values of T, according to the four choices of pattern and the six orderings in Square 2. Inspection shows that only one of these arrangements,

$$U = 32 + 28 = 60 \qquad V = 39 + 51 = 90 \qquad W = 15 + 6 = 21$$

yields a larger value of T than that actually observed. Thus the experiment has significance probability $2/24 = 0.083$.

If there are many blocks or squares, direct enumeration would have to be done on an automatic computing machine. Fortunately there is a relatively simple normal approximation. In Block i or Square i, the sums of ranks of treatments A and B gives a vector U_i, V_i whose early moments are easy to compute. From these the moments of $U = \sum U_i$, $V = \sum U_i$ may be obtained. If one uses as test statistic an appropriately chosen quadratic function of U and V, the null distribution would be approximately χ_2^2. This method readily extends to blocks or squares with more than three treatments.

REFERENCES

[1] BENARD, A. and ELTEREN, PH. VAN (1953). A generalization of the method of m-rankings. *Indag. Math.* **15** 358–369.

[2] DURBIN, J. (1951). Incomplete blocks in ranking experiments. *British J. Psych.* **4** 85–90.

[3] ELTEREN, PH. VAN (1960). On the combination of independent two sample tests of Wilcoxon. *Bull. Inst. Internat. Statist.* **37** 351–361.

[4] ELTEREN, PH. VAN and NOETHER, G. E. (1959). The asymptotic efficiency of the χ_n^2-test for a balanced incomplete block design. *Biometrika* **46** 475–477.

[5] FRIEDMAN, M. (1937). The use of ranks to avoid the assumption of normality implicit in the analysis of variance. *J. Amer. Statist. Assoc.* **32** 675–698.

[6] HODGES, J. L., JR. and LEHMANN, E. L. (1956). The efficiency of some non-parametric competitors of the t-test. *Ann. Math. Statist.* **27** 324–335.

[7] HODGES, J. L., JR. and LEHMANN, E. L. (1961). Comparison of the normal scores and Wilcoxon tests. *Proc. Fourth Berkeley Symp. Math. Statist. Prob.* **1** 307–317. Univ. of California Press.

[8] LOÈVE, M. (1960). *Probability Theory*, 2nd ed. Van Nostrand, Princeton.

[9] WALSH, JOHN E. (1959). Exact nonparametric tests for randomized blocks. *Ann. Math. Statist.* **30** 1034–1040.

[10] WILCOXON, F. (1946). Individual comparisons of grouped data by ranking methods. *J. of Entomology* **39** 269–270.

ROBUST ESTIMATION IN ANALYSIS OF VARIANCE[1]

By E. L. Lehmann

University of California, Berkeley

1. Summary. In linear models with several observations per cell, estimates of all contrasts are given whose small and large sample behaviour is analogous to that of the estimate of a shift parameter proposed in [2]. In particular, the asymptotic efficiency of these estimates relative to the standard least squares estimates, as the number of observations in each cell gets large, is shown to be the same as the Pitman efficiency of the Wilcoxon test relative to the t-test.

2. A compatible set of difference estimates. Consider the c-sample model, in which the observations are

$$X_{i\alpha} = \xi_i + U_{i\alpha} \, (\alpha = 1, \cdots, n_i \, ; i = 1, \cdots, c) \tag{2.1}$$

where the variables $U_{i\alpha}$ are independently distributed with distributions F_i having a common median. Let

$$Y_{ij} = \text{med}\,(X_{i\alpha} - X_{j\beta}) \tag{2.2}$$

be the median of the $n_i n_j$ differences $X_{i\alpha} - X_{j\beta}$ $(\alpha = 1, \cdots, n_i \, ; \beta = 1, \cdots, n_j)$. It was shown in [2] that the estimate Y_{ij} of $\xi_i - \xi_j$ has more robust efficiency than, and hence may be preferable to, the standard estimate $T_{ij} = X_{i\cdot} - X_{j\cdot}$, where $X_{i\cdot} = \sum X_{i\alpha}/n_i$.

Unfortunately, unlike the estimates T_{ij}, the estimates Y_{ij} are incompatible in the sense (see [4]) that they do not satisfy the linear relations satisfied by the differences they estimate. This leads to ambiguities. To estimate, for example, $\xi_3 - \xi_1$, we will get a different answer if we estimate this difference directly by Y_{31} or if we take the sum of the estimates Y_{32} and Y_{21} of $\xi_3 - \xi_2$ and $\xi_2 - \xi_1$. This is particularly awkward in the estimation of more general contrasts, where there is no longer a single "natural" estimate. Thus, for estimating $(\xi_3 + \xi_4) - (\xi_1 + \xi_2)$ one will get the estimate $Y_{31} + Y_{42}$ when writing the contrast as $(\xi_3 - \xi_1) + (\xi_4 - \xi_2)$, but $Y_{41} + Y_{32}$ when writing it as $(\xi_4 - \xi_1) + (\xi_3 - \xi_2)$.

To avoid this difficulty, we shall replace the *raw* estimates Y_{ij} by *adjusted* estimates Z_{ij} which have the form $\hat{\xi}_i - \hat{\xi}_j$ or equivalently which satisfy the set of restrictions $Z_{ij} + Z_{jk} + Z_{ki} = 0$ for all i, j, k.

If we suppose that any serious effect of gross errors has been eliminated by replacing the estimates T_{ij} by Y_{ij}, it is natural to adjust these latter estimates by minimizing the sum of squares

$$\sum_{i \neq j} [Y_{ij} - (\xi_i - \xi_j)]^2. \tag{2.3}$$

Received December 20, 1962.

[1] This research was done while the author was a Professor of the Aldolph C. and Mary Sprague Miller Institute for Basic Research in Science, University of California, Berkeley.

If Y_{ii} is defined to be zero for all i and $\xi_i - \xi_j$ denote the minimizing values of (2.3), differentiation of (2.3) gives $\sum_{j=1}^{c} (Y_{ij} - \xi_i + \xi_j) = 0$ and hence $\xi_i - (1/c) \sum_{j=1}^{c} \xi_j = Y_{i\cdot}$, so that

$$Z_{ij} = \xi_i - \xi_j = Y_{i\cdot} - Y_{j\cdot}\,. \tag{2.4}$$

We shall in the next sections show that the estimates Z_{ij} have essentially the same small- and large-sample properties that had been proved for the estimates Y_{ij} in [2]. They are affected however by a peculiarity which may be considered a disadvantage: The estimate of $\xi_i - \xi_j$ depends not only on the observations from the ith and jth sample but also on those of the other samples, which are quite unrelated to ξ_i and ξ_j.

Suppose for example that $c = 3$ and that $x_{11} = -4, x_{12} = 0, x_{13} = 6; x_{21} = -10, x_{22} = 3, x_{23} = 7; x_{31} = -7, x_{32} = 0, x_{33} = 8$. Then the raw estimates of $\xi_3 - \xi_1$, $\xi_3 - \xi_2$ and $\xi_2 - \xi_1$ are 0, 1 and 1 respectively. Since the data suggest that ξ_3 is larger than ξ_2 and ξ_2 is larger than ξ_1, this is taken into account in arriving at the adjusted estimate for $\xi_3 - \xi_1$. Instead of estimating this difference to be zero, the adjusted estimate is $\frac{1}{3}[2\cdot 0 + 1 + 1] = \frac{2}{3}$. As will be seen later, the size of the influence of the other samples becomes small as the sample sizes increase, tending in fact to zero in probability at a rate $o_p(N^{-\frac{1}{2}})$ where N is the total sample size.

While the adjustment of the estimates does not affect their asymptotic efficiency, we unfortunately have no results concerning the effect on small-sample efficiency. It should be pointed out however that the adjustment may result in a serious loss of efficiency when the sample sizes are widely disparate. This may be illustrated on the case $c = 3$. The variance of the raw estimate Y_{12} of $\xi_1 - \xi_2$ tends to zero as n_1 and n_2 tend to infinity. This is however not true of the variance of $Z_{12} = \frac{1}{3}(2Y_{12} + Y_{13} - Y_{23})$ unless n_3 also tends to infinity.

3. The estimation of contrasts. A contrast is a linear function of the ξ's, say $\sum c_i\xi_i$, with $\sum c_i = 0$. Since this is a function only of the differences of the ξ's, it can be written in the form

$$\theta = \sum \sum d_{ij}(\xi_i - \xi_j). \tag{3.1}$$

As estimate of the contrast (3.1) we propose

$$\hat{\theta} = \sum \sum d_{ij}Z_{ij} = \sum \sum d_{ij}(Y_{i\cdot} - Y_{j\cdot}). \tag{3.2}$$

The representation (3.1) is not unique; however the estimate (3.2) is independent of the representation. For if $\sum \sum d_{ij}(\xi_i - \xi_j) = \sum \sum e_{ij}(\xi_i - \xi_j)$ for all vectors $(\xi_1, \cdots, \xi_c)$, then obviously

$$\sum \sum d_{ij}(Y_{i\cdot} - Y_{j\cdot}) = \sum \sum e_{ij}(Y_{i\cdot} - Y_{j\cdot})$$

for all vectors $(Y_{1\cdot}, \cdots, Y_{c\cdot})$.

As an example, consider a two-way layout with several observations per cell. Changing the notation of (2.1), let the observable random variables be $X_{ij\alpha} =$

$\xi_{ij} + U_{ij\alpha}$ where the U's are independent and identically distributed. In the usual decomposition into main effects and interactions, let $\xi_{ij} = \mu + \alpha_i + \beta_j + \gamma_{ij}$ where

$$\mu = \xi_{..}, \qquad \alpha_i = \xi_{i.} - \xi_{..}, \qquad \beta_j = \xi_{.j} - \xi_{..}, \tag{3.3}$$
$$\gamma_{ij} = \xi_{ij} - \xi_{i.} - \xi_{.j} + \xi_{..}\,.$$

If we denote the raw estimate of $\xi_{ij} - \xi_{kl}$ by V_{ijkl}, so that V_{ijkl} is the median of all differences $X_{ij\alpha} - X_{kl\beta}$ between observations in the ijth and the klth cell, then the adjusted estimate of $\xi_{ij} - \xi_{kl}$ is by (2.4), $W_{ijkl} = \hat{\xi}_{ij} - \hat{\xi}_{kl} = V_{ij..} - V_{kl..}$. This leads to the following estimates for the contrasts α_i, β_j and γ_{ij}:

$$\hat{\alpha}_i = V_{i...} - V_{....}, \qquad \beta_j = V_{.j..} - V_{....}, \tag{3.4}$$
$$\hat{\gamma}_{ij} = V_{ij..} - V_{i...} - V_{.j.} + V_{....},$$

where actually $V_{....} = 0$.

As an illustration consider the data given by Brownlee (*Statistical Theory and Methodology in Science and Engineering*, Wiley, New York, 1960, p. 379) on the per cent reduction in blood sugar a certain time after injection of insulin into rabbits, which are shown in the table below. There are two factors involved: the dose at three levels, and the preparation of insulin at two levels, A and B. The number of observations (rabbits) per cell is four.

	Preparation A			*Preparation B*		
Dose	2.29	3.63	5.75	2.29	3.63	5.75
	17	64	62	33	41	56
	21	49	72	37	64	62
	49	34	61	40	34	57
	54	63	91	16	64	72

The following table shows three estimates for the difference of any two cell means $\xi_{ij} - \xi_{kl}$: The raw estimate $V_{ijkl} = \text{med}\,(X_{ij\alpha} - X_{kl\beta})$, the adjusted estimate $W_{ijkl} = V_{ij..} - V_{kl..}$, and the classical estimate $X_{ij.} - X_{kl.}$.

	$\xi_{11} - \xi_{12}$	$\xi_{11} - \xi_{13}$	$\xi_{11} - \xi_{21}$	$\xi_{11} - \xi_{22}$	$\xi_{11} - \xi_{23}$	$\xi_{12} - \xi_{13}$	$\xi_{12} - \xi_{21}$
V	−14.5	−40.5	7	−15	−29	−18	23.5
W	−17.33	−36.16	4.42	−15.91	−27.00	−18.83	21.75
Classical	−17.25	−36.25	3.75	−15.5	−26.5	−19	21

	$\xi_{12} - \xi_{22}$	$\xi_{12} - \xi_{23}$	$\xi_{13} - \xi_{21}$	$\xi_{13} - \xi_{22}$	$\xi_{13} - \xi_{23}$	$\xi_{21} - \xi_{22}$	$\xi_{21} - \xi_{23}$	$\xi_{22} - \xi_{23}$
V	0	−8	37	24	5.5	−24	−27	−11.5
W	1.42	−9.67	40.58	20.25	9.16	−20.33	−31.42	−11.09
Classical	1.75	−0.25	40	20.75	9.75	−19.25	−30.25	−11

At least in this example, adjustment has moved the estimates closer to the classical ones. This is perhaps not surprising since the adjustment puts the estimates onto the same linear surface on which the classical estimates are lying.

From this table we obtain the estimates (3.4) of the α's, β's and γ's, which are shown below together with the corresponding classical estimates

	$\alpha_1 = -\alpha_2$	β_1	β_2	β_3	$\gamma_{11} = -\gamma_{21}$	$\gamma_{21} = -\gamma_{22}$	$\gamma_{13} = -\gamma_{23}$
Estimates (3.4)	2.50	−17.54	1.29	16.25	−.29	−1.79	2.08
Classical Estimates	2.54	−17.17	1.08	16.08	−.67	−1.67	2.33

Returning now to the general case, the following theorem gives two conditions under which the estimate $\hat{\theta}$ of any contrast θ is symmetrically distributed about the true value of the contrast, and hence in particular is unbiased.

THEOREM 1. *For any contrast θ given by* (3.1), *the distribution of the estimate $\hat{\theta}$ given by* (3.2) *is symmetric about θ if either*

(i) *the distributions F_i defined in relation to* (2.1) *are symmetric or*

(ii) *all sample sizes $n_1, \cdots, n_c$ are equal.*

PROOF.

(i) Let X, Y, Z denote the matrices $(X_{i\alpha})$, (Y_{ij}), (Z_{ij}) respectively and let $Y = f(X)$ and $Z = g(Y)$. Then it follows from 2(i) of [2] that $f(-X) = -f(X)$ and from Equation (2.4) of the present paper that $g(-Y) = -g(Y)$. Hence we have $g[f(-X)] = -g[f(X)]$ and therefore

$$\hat{\theta}(-X) = -\hat{\theta}(X). \tag{3.5}$$

The result now follows from (3.5) as did Corollary (i) to Theorem 2 in [2] from the Equation (8.1) of that paper, with Ψ here being the rectangular distribution.

(ii) The proof of (ii) is exactly analogous, with Equation (8.3) of [2] in place of (8.1).

4. Asymptotic distribution and efficiency. In the remainder of the paper we shall assume that the variables $U_{i\alpha}$ have a common distribution F. To obtain the asymptotic distribution of the adjusted estimates Z_{ij}, we begin by determining the asymptotic distribution of the raw estimates Y_{ij}. This is given by the following theorem, where the sample sizes n_i are assumed to tend to infinity in such a way that $n_i = \rho_i N$, $N \to \infty$.

THEOREM 2.

(i) *The joint distribution of $(V_1, \cdots, V_{c-1})$ where*

$$V_i = N^{\frac{1}{2}}[Y_{ic} - (\xi_i - \xi_c)] \tag{4.1}$$

is asymptotically normal with zero mean and covariance matrix

$$\begin{aligned} \text{Var}(V_i) &= (1/12)(1/\rho_i + 1/\rho_c)/(\textstyle\int f^2(x)\,dx)^2 \\ \text{Cov}(V_i, V_j) &= (1/12\rho_c)/(\textstyle\int f^2(x)\,dx)^2. \end{aligned} \tag{4.2}$$

Here the density f of F is assumed to satisfy the regularity conditions of Lemma 3(a) *of* [1].

(ii) *For any i and j*

$$N^{\frac{1}{2}} Y_{ij} \sim N^{\frac{1}{2}}(Y_{ic} - Y_{jc}) \tag{4.3}$$

where $\sim$ indicates that the difference of the two sides tends to zero in probability.

The proof of (i) rests on the following lemma.

LEMMA 1. *Suppose the variables $X_{i\alpha}$ have the distribution specified in connection with (2.1) with fixed F but a sequence of means $(\xi_1, \cdots, \xi_c) = (\xi_1^{(N)}, \cdots, \xi_c^{(N)})$ satisfying $\xi_i^{(N)} - \xi_c^{(N)} = -a_i/N^{\frac{1}{2}}$. Let h_{ij} be defined by*

$$n_i n_j h_{ij} = \text{Number of pairs } (\alpha, \beta) \text{ such that } X_{i\alpha} < X_{j\beta} \tag{4.4}$$

and let $\mu_{ij} = E(h_{ij}) = P(X_{i\alpha} < X_{j\beta})$ where actually both h_{ij} and μ_{ij} depend on N. Then the variables $(W_1, \cdots, W_{c-1})$ given by

$$W_i = N^{\frac{1}{2}}(h_{ic} - \mu_{ic}) \qquad i = 1, \cdots, c-1 \tag{4.5}$$

have a joint asymptotic normal distribution as $N \to \infty$, with zero mean and covariance matrix

$$\operatorname{Var}(W_i) = [1/\rho_i + 1/\rho_c]/12; \qquad \operatorname{Cov}(W_i, W_j) = 1/12\rho_c. \tag{4.6}$$

The proof of this lemma is given in the appendix.

PROOF OF THEOREM 2(i). By (9.1) of [2],

$$\lim P\{N^{\frac{1}{2}}[Y_{ic} - (\xi_i - \xi_c)] \leqq a_i \text{ for all } i\} = \lim P_N\{N^{\frac{1}{2}}(h_{ic} - \tfrac{1}{2}) \leqq 0 \text{ for all } i\}$$

where P_N indicates that the probability is computed for a sequence of means satisfying (4.4). Since under the assumed regularity conditions

$$N^{\frac{1}{2}}(\mu_{ic} - \tfrac{1}{2}) \to -a_i \textstyle\int f^2(x)\, dx \qquad \text{as} \qquad N \to \infty$$

(see for example Lemma 3(a) of [1]), it follows that

$$\lim P\{N^{\frac{1}{2}}[Y_{ic} - (\xi_i - \xi_c)] \leqq a_i \text{ for all } i\}$$
$$= \lim P_N\{N^{\frac{1}{2}}(h_{ic} - \mu_{ic}) \leqq a_i \textstyle\int f^2(x)\, dx \text{ for all } i\}.$$

By Lemma 1, this is equal to $Q(a_1, \cdots, a_{c-1})$ where Q is the $(c-1)$-dimensional multivariate normal distribution with zero mean and covariance matrix (4.2), as was to be proved.

Part (ii) of Theorem 2 can also be proved through U-statistics. Alternatively it is a simple consequence of the following lemma which was communicated to me by Professor L. LeCam.

LEMMA 2. (LeCam). *Let $(A_1^{(N)}, A_2^{(N)}, A_3^{(N)})$ be a sequence of random vectors such that for any $1 \leqq i < j \leqq 3$, the pair $(A_i^{(N)}, A_j^{(N)})$ converges in law to a bivariate distribution with mean (η_i, η_j) and covariance matrix $\begin{pmatrix} \sigma_{ii} & \sigma_{ij} \\ \sigma_{ij} & \sigma_{jj} \end{pmatrix}$. If the σ's satisfy*

$$\sum_i \sigma_{ii} + 2 \sum_{i<j} \sigma_{ij} = 0, \tag{4.7}$$

then $\sum_{i=1}^3 A_i^{(N)} \to \sum_{i=1}^3 \eta_i$ in probability.

PROOF. Since each sequence $\{A_i^{(N)}\}$ has a non-degenerate limit distribution, there exists a sequence $\{N_k\}$ such that the distribution of $(A_1^{(N_k)}, A_2^{(N_k)}, A_3^{(N_k)})$ converges completely, say that $(A_1^{(N_k)}, A_2^{(N_k)}, A_3^{(N_k)})$ tends in law to (A_1, A_2, A_3). The covariance matrix of this limiting distribution by assumption satisfies (4.7) and hence the variance of $\sum A_i$ is

$$\text{Var}(\sum A_i) = \sum_i \sigma_{ii} + 2\sum\sum_{i<j} \sigma_{ij} = 0,$$

so that $\sum A_i = \sum \eta_i$ with probability one. Since this argument applies to every converging subsequence, it follows that $\sum A_i^{(N)} \to 0$ in probability, as was to be proved.

PROOF OF THEOREM 2(ii). Let

$$A_1^{(N)} = N^{\frac{1}{2}}[Y_{ij} - (\xi_i - \xi_j)]; \qquad A_2^{(N)} = N^{\frac{1}{2}}[Y_{jc} - (\xi_j - \xi_c)];$$
$$A_3^{(N)} = N^{\frac{1}{2}}[Y_{ci} - (\xi_c - \xi_i)].$$

Since each pair of the Y's have one subscript in common, and using the fact that $Y_{ji} = -Y_{ij}$, it follows from part (i) of Theorem 2 that the assumptions of Lemma 2 are satisfied with

$$\sigma_{11} = \frac{1}{12}\left(\frac{1}{\rho_i} + \frac{1}{\rho_j}\right) \Big/ \left(\int f^2(x)\,dx\right)^2, \qquad \sigma_{22} = \frac{1}{12}\left(\frac{1}{\rho_j} + \frac{1}{\rho_c}\right) \Big/ \left(\int f^2(x)\,dx\right)^2,$$

$$\sigma_{33} = \frac{1}{12}\left(\frac{1}{\rho_i} + \frac{1}{\rho_c}\right) \Big/ \left(\int f^2(x)\,dx\right)^2, \qquad \sigma_{12} = -\frac{1}{12\rho_j} \Big/ \left(\int f^2(x)\,dx\right)^2,$$

$$\sigma_{13} = -\frac{1}{12\rho_i} \Big/ \left(\int f^2(x)\,dx\right)^2, \qquad \sigma_{23} = -\frac{1}{12\rho_c} \Big/ \left(\int f^2(x)\,dx\right)^2,$$

so that (4.6) holds. Hence $N^{\frac{1}{2}}(Y_{ij} + Y_{jc} + Y_{ci})$ tends to zero in probability, which is equivalent to (4.3).

The following corollary shows the asymptotic distribution of the adjusted estimate Z_{ij} to be the same as that of the raw estimate Y_{ij} given by Theorem 2; the corollary also establishes the statement at the end of Section 2, that the adjustments tend to become small as $N \to \infty$.

COROLLARY. *The difference $N^{\frac{1}{2}}(Z_{ij} - Y_{ij})$ tends to zero in probability for all i, j.*

PROOF. By definition, $Z_{ij} = \sum_\alpha (Y_{i\alpha} - Y_{j\alpha})/c$. Using the fact that by Theorem 2(ii), $N^{\frac{1}{2}}Y_{j\alpha} \sim N^{\frac{1}{2}}(Y_{jc} - Y_{\alpha c})$ and $N^{\frac{1}{2}}Y_{i\alpha} \sim N^{\frac{1}{2}}(Y_{ic} - Y_{\alpha c})$, we see that $N^{\frac{1}{2}}Z_{ij} \sim N^{\frac{1}{2}}(Y_{ic} - Y_{jc}) \sim N^{\frac{1}{2}}Y_{ij}$.

We are now in a position to prove that the asymptotic efficiency of the estimate $\hat{\theta}$ of a contrast θ given by (3.2) and (3.1), relative to the classical least squares estimate of θ is the same as the Pitman efficiency of the Wilcoxon test relative to the t-test.

THEOREM 3. *The asymptotic efficiency of the estimate $\hat{\theta} = \sum\sum d_{ij}Z_{ij}$ of $\theta = \sum\sum d_{ij}(\xi_i - \xi_j)$ relative to the estimate $\sum\sum d_{ij}(X_{i\cdot} - X_{j\cdot})$ is*

$$e = 12\sigma^2\left[\int f^2(x)\,dx\right]^2 \tag{4.8}$$

where $\sigma^2 = \text{Var}(X_{i\alpha})$.

PROOF. Let $T_{ij} = X_{i\cdot} - X_{j\cdot}$ and let $V'_i = N^{\frac{1}{2}}[T_{ic} - (\xi_i - \xi_c)]$. Then the variables $(V'_1, \cdots, V'_{c-1})$ have an asymptotic normal distribution with zero mean and covariance matrix

$$\text{Var}(V'_i) = \sigma^2(1/\rho_i + 1/\rho_c), \qquad \text{Cov}(V_i, V_j) = \sigma^2/\rho_c .$$

Furthermore, $T_{ij} = T_{ic} - T_{jc}$. Comparing this with the asymptotic distribution of the Z_{ij}, which is given by Theorem 2 with Y_{ij} replaced by Z_{ij} throughout, the result follows.

So far, the means $\xi_1, \cdots, \xi_c$ were assumed to be completely unrestricted. However, in more general linear models, they may be required to satisfy certain linear restrictions. These may then affect the classical estimate. However, whenever this estimate is a linear function of the differences $X_{i\cdot} - X_{j\cdot}$, then the estimate obtained from it by replacing $X_{i\cdot} - X_{j\cdot}$ by $Y_{i\cdot} - Y_{j\cdot}$ has asymptotic relative efficiency (4.1). This is an immediate consequence of Theorem 2.

As an example suppose that $E(X_{i\alpha}) = \xi_i$ $(i = 1, \cdots, 4)$ where $\xi_1 = \mu + \gamma$, $\xi_2 = \mu + \delta, \xi_3 = \mu + \gamma + \delta, \xi_4 = \mu - \gamma - \delta$. Then the classical estimate of γ is

$$\tilde{\gamma} = (10X_{1\cdot} - 8X_{2\cdot} + 3X_{3\cdot} - 5X_{4\cdot})/18,$$

which is a linear function of the differences $X_{i\cdot} - X_{j\cdot}$ since the sum of the coefficients is zero. Replacing $X_{i\cdot} - X_{j\cdot}$ by $Y_{i\cdot} - Y_{j\cdot}$, we obtain the estimate

$$\begin{aligned}\hat{\gamma} &= (10Y_{1\cdot} - 8Y_{2\cdot} + 3Y_{3\cdot} - 5Y_{4\cdot})/18 \\ &= (18Y_{12} + 7Y_{13} + 15Y_{14} - 11Y_{23} - 3Y_{24} + 8Y_{34})/72\end{aligned}$$

and the asymptotic efficiency of $\hat{\gamma}$ relative to $\tilde{\gamma}$ is given by (4.8).

5. An alternative approach. There exists another class of estimates of the parameters in a linear model, not restricted to contrasts and not suffering from the peculiarity discussed at the end of Section 2. However, these estimates appear to have the efficiency of those treated in the main body of this paper only when the underlying distribution F is symmetric.

Consider once more the c-sample model (2.1). In [2] it is shown that the estimate

$$\xi_i^* = \text{med}\,[(X_{i\alpha} + X_{i\beta})/2] \tag{5.1}$$

has asymptotic efficiency (4.8) relative to the standard estimate $X_{i\cdot}$ as $n_i \to \infty$ provided the distribution F of the X's is symmetric. Hence the estimate $\sum c_i\xi_i^*$ of $\sum c_i\xi_i$ has asymptotic efficiency (4.8) relative to the standard estimate $\sum c_iX_{i\cdot}$ when F is symmetric.

As an example in which the method of Sections 2–4 is not applicable consider a weighing design, in which the sum and the difference of two objects A and B is each weighed n times, so that $E(X_{1\alpha}) = \xi + \eta$, $E(X_{2\alpha}) = \eta - \xi$ for $\alpha = 1, \cdots, n$, and suppose that the distribution F of the X's is symmetric. Then med $[(X_{1\alpha} + X_{1\beta})/2]$ and med $[(X_{2\alpha} + X_{2\beta})/2]$ are independent estimates of $\xi + \eta$ and $\eta - \xi$ with asymptotic efficiency (4.8) relative to $X_{1\cdot}$ and $X_{2\cdot}$. Hence

$$\tfrac{1}{2}\{\text{med}\,[(X_{1\alpha} + X_{1\beta})/2] + \text{med}\,[(X_{2\alpha} + X_{2\beta})/2]\}$$

and

$$\tfrac{1}{2}\{\text{med}\,[(X_{1\alpha} + X_{1\beta})/2] - \text{med}\,[(X_{2\alpha} + X_{2\beta})/2]\}$$

are estimates of η and ξ having asymptotic efficiency (4.8) relative to the standard estimates $\frac{1}{2}(X_{1\cdot} + X_{2\cdot})$ and $\frac{1}{2}(X_{1\cdot} - X_{2\cdot})$.

6. Appendix: generalized U-statistics. The following theorem which is a slight extension of Hoeffding's theorem on U-statistics [3], is known to many workers in nonparametric inference. It is stated here solely for the purpose of convenient reference; the proof exactly parallels Hoeffding's argument. (Part (i) for $s = 2$ is stated in Fraser: *Nonparametric Methods in Statistics*, p. 229.)

(i) *Let* $\varphi(x_1, \cdots, x_{m_1}; y_1, \cdots, y_{m_2}; z_1, \cdots, z_{m_3}; \cdots)$ *be a function symmetric in the* x's, *in the* y's, *in the* z's, $\cdots$. *Let* $X_1, \cdots, X_{n_1}; Y_1, \cdots, Y_{n_2}; Z_1, \cdots, Z_{n_3}; \cdots$ *be samples of sizes* $n_i = \rho_i N$ *from distributions* $F, G, H, \cdots$ *and consider the generalized* U*-statistic*

$$U = \binom{n_1}{m_1}^{-1}\binom{n_2}{m_2}^{-1}\binom{n_3}{m_3}^{-1} \sum \varphi(X_{\alpha_1}, \cdots, X_{\alpha_{m_1}}; Y_{\beta_1}, \cdots, Y_{\beta_{m_2}}; Z_{\gamma_1}, \cdots, Z_{\gamma_{m_3}}; \cdots)$$

where the summation extends over all

$$\binom{n_1}{m_1}\binom{n_2}{m_2}\binom{n_3}{m_3}\cdots$$

sets of subscripts satisfying $1 \leqq \alpha_1 < \alpha_2 < \cdots < \alpha_{m_1} \leqq n_1$; $1 \leqq \beta_1 < \beta_2 < \cdots < \beta_{m_2} \leqq n_2$; $\cdots$. *Let* $E(U) = \theta(F, G, H, \cdots)$. *Then as* $N \to \infty$, $Z = N^{\frac{1}{2}}(U - \theta)$ tends in law to $N(0, \sigma^2)$ *with*

$$\sigma^2 = (m_1^2/\rho_1)\zeta_{10\cdots0} + (m_2^2/\rho_2)\zeta_{010\cdots0} + \cdots + (m_s^2/\rho_s)\zeta_{00\cdots1}.$$

Here $\zeta_{c_1,\cdots,c_s}$ is the covariance of

$$\varphi(X_1, \cdots, X_{c_1}, X_{c_1+1}, \cdots, X_{m_1}; Y_1, \cdots, Y_{c_2}, Y_{c_2+1}, \cdots, Y_{m_2}; \cdots)$$

with $\varphi(X_1, \cdots, X_{c_1}, X'_{c_1+1}, \cdots, X'_{m_1}; Y_1, \cdots, Y_{c_2}, Y'_{c_2+1}, \cdots, Y'_{m_2}; \cdots)$ where $X_1, \cdots, X_{m_1}, X'_{c_1+1}, \cdots, X'_{m_1}$ are independently distributed according to F; $Y_1, \cdots, Y_{m_2}, Y'_{c_2+1}, \cdots, Y'_{m_2}$ according to G; etc.

(ii) *Let* $U^{(i)}$ $(i = 1, \cdots, r)$ *be* r *generalized* U*-statistics defined in terms of functions* $\varphi^{(i)}$, *where the numbers* $m_1, m_2, \cdots$ *may depend on* i *but not the sample sizes* $n_1, n_2, \cdots$. *The joint distribution of the variables* $Z^{(1)} = N^{\frac{1}{2}}(U^{(1)} - \theta^{(1)})$, $\cdots$, $Z^{(r)} = N^{\frac{1}{2}}(U^{(r)} - \theta^{(r)})$ *is asymptotically normal, with mean zero and covariance matrix* $\Sigma = (\sigma_{ij})$ *where*

$$\sigma_{ij} = (m_1^{(i)}m_1^{(j)}/\rho_1)\zeta_{10\cdots0}^{(i,j)} + \cdots + (m_s^{(i)}m_s^{(j)}/\rho_s)\zeta_{00\cdots1}^{(i,j)}.$$

Here $\zeta_{c_1,\cdots,c_s}^{(i,j)}$ is the covariance of

$$\varphi^{(i)}(X_1, \cdots, X_{c_1}, X_{c_1+1}, \cdots, X_{m_1}; Y_1, \cdots, Y_{c_2}, Y_{c_2+1}, \cdots, Y_{m_2}; \cdots)$$

with

$$\varphi^{(j)}(X_1, \cdots, X_{c_1}, X'_{c_1+1}, \cdots, X'_{m_1}; Y_1, \cdots, Y_{c_2}, Y'_{c_2+1}, \cdots, Y'_{m_2}; \cdots).$$

(iii) *Suppose that the distributions $F, G, H, \cdots$ depend on N. Let $\theta_N^{(i)} = E_N[U^{(i)}]$ and let Σ_N denote the covariance matrix Σ computed for $(F_N, G_N, H_N, \cdots)$. Then if Σ_N tends to a non-singular covariance matrix Σ as $N \to \infty$, it remains true that*

$$N^{\frac{1}{2}}(U^{(1)} - \theta_N^{(1)}), \cdots, N^{\frac{1}{2}}(U^{(r)} - \theta_N^{(r)})$$

tend in law to a normal distribution with zero mean and covariance matrix Σ.

As an application, we now prove Lemma 1. It will be convenient to use the present notation rather than that of Section 4.

For simplicity of notation consider the case of $c = 3$ samples $X_1, \cdots, X_{n_1}$; $Y_1, \cdots, Y_{n_2}$; $Z_1, \cdots, Z_{n_3}$. Let

$$\varphi^{(1)}(x_1, y_1, z_1) = \begin{matrix} 1 \text{ if } x_1 < z_1 \\ 0 \text{ otherwise} \end{matrix}$$

$$\varphi^{(2)}(x_1, y_1, z_1) = \begin{matrix} 1 \text{ if } y_1 < z_1 \\ 0 \text{ otherwise} \end{matrix}$$

and consider a sequence of distributions

$$F_N(x) = F(x - a_1N^{-\frac{1}{2}}), \qquad G_N(y) = F(y - a_2N^{-\frac{1}{2}}), \qquad H_N(z) = F(z)$$

for the variables X_α, Y_β and Z_γ. Then $n_1n_3U^{(1)}$ = Number of pairs (α, γ) such that $X_\alpha < Z_\gamma$ and $n_2n_3U^{(2)}$ = Number of pairs (β, γ) such that $Y_\beta < Z_\gamma$.

To apply (iii), it is only necessary to compute the limiting covariance matrix Σ.

Now

$$\zeta_{100}^{(1)} = E\varphi^{(1)}(X_1, Y_1, Z_1)\varphi^{(1)}(X_1, Y_1', Z_1') - [P(X_1 < Z_1)]^2$$
$$= P\{X_1 < Z_1, Z_1'\} - [P(X_1 < Z_1)]^2 \to \tfrac{1}{3} - \tfrac{1}{4} = \tfrac{1}{12}.$$

Similarly $\zeta_{010}^{(1)} \to \frac{1}{12}$ while $\zeta_{001}^{(1)} = 0$. Thus $\sigma_{11} = (1/\rho_1 + 1/\rho_3)/12$. Exactly similarly, $\sigma_{22} = (1/\rho_2 + 1/\rho_3)/12$. Finally,

$$\zeta_{100}^{(1,2)} = P\{X_1 < Z_1 \text{ and } Y_1' < Z_1'\} - P(X_1 < Z_1)P(Y_1 < Z_1) = 0 = \zeta_{010}^{(1,2)}$$

while

$$\zeta_{001}^{(1,2)} = P\{X_1 < Z_1 \text{ and } Y_1' < Z_1\} - P(X_1 < Z_1)P(Y_1 < Z_1) \to \tfrac{1}{12}$$

so that $\sigma_{12} = 1/12\rho_3$. This verifies the covariance matrix (4.6) and completes the proof of Lemma 1.

REFERENCES

[1] Hodges, J. L., Jr. and Lehmann, E. L. (1961). Comparison of the normal scores and Wilcoxon tests. *Proc. Fourth Berkeley Symp. Math. Statist. Prob.* **1** 307–318.

[2] Hodges, J. L., Jr. and Lehmann, E. L. (1963). Estimates of location based on rank tests. *Ann. Math. Statist.* **34** 598–611.

[3] Hoeffding, Wassily (1948). A class of statistics with asymptotically normal distribution. *Ann. Math. Statist.* **19** 293–325.

[4] Lehmann, E. L. (1957). A theory of some multiple decision problems, I. *Ann. Math. Statist.* **28** 1–25.

LEHMANN, E. L.
Math. Annalen 150, 268—275 (1963)

A Class of Selection Procedures Based on Ranks*

For B. L. VAN DER WAERDEN on the occasion of his 60th birthday

By

E. L. LEHMANN in Berkeley/Calif.

Let X_{ij} ($j = 1, \ldots, n$; $i = 1, \ldots, s$) be independent samples from populations with cumulative distribution functions $F(x - \Theta_i)$. For selecting the population with the highest Θ-value, procedures based on the ranks of the observations are compared with the means procedure which selects the population with the largest mean $\overline{X}_i$. The asymptotic relative efficiency of two selection procedures is defined as the ratio of the sample sizes required to achieve the same minimum probability of selecting a "good" population. It is shown that the asymptotic relative efficiency of the procedures based on ranks relative to the means procedure is the same as that of the associated tests in the two — or c-sample problem. If the ratio of the sample sizes is equal to this efficiency, the two procedures being compared are shown to have the same asymptotic performance characteristic.

1. *The means procedure.* Let X_{ij} ($j = 1, \ldots, n$; $i = 1, \ldots, s$) be independent samples from populations Π_i, with distributions

$$P(X_{ij} \leq x) = F(x - \Theta_i)\,, \tag{1}$$

and consider the problem of selecting the population with the largest Θ-value. When F is normal, the natural procedure is the *means procedure* M which selects Π_i if

$$\overline{X}_i = \max_k \overline{X}_k \tag{2}$$

where $\overline{X}_k = \sum_{j=1}^{n} X_{kj}/n$. Strong optimum properties of this procedure were proved by BAHADUR [1] and BAHADUR and GOODMAN [2].

The feature of the means procedure (2) on which attention has been focused in the literature (see for example BECHOFER [3] and HALL [5]), is the sample size n required to guarantee some desirable property, for example that

$$P\,(\text{selected distribution is good}) \geq \gamma\,, \tag{3}$$

where the ith population is considered good if Θ_i is sufficiently close to the

* This paper was begun with the partial support of the Office of Naval Research (Nonr-222-43) and was completed while the author was Research Professor for the Adolph C. and Mary Sprague Miller Institute for Basic Research in Science, University of California, Berkeley.

largest Θ-value, say if

$$\Theta_i \geqq \Theta_{\max} - \Delta^* \tag{4}$$

where Δ^* is a given constant.

If we do not wish to rely on the assumption of normality, we can find a large-sample solution, which depends only on the variance σ^2 of F. To this end, consider a sequence of situations for increasing n, and define the ith population as good if

$$\Theta_i \geqq \Theta_{\max} - \Delta^{(n)}, \tag{5}$$

where $\Delta^{(n)}$ will be defined below. This definition seems rather opportunistic since it appears that our idea of what Θ-values are acceptable is changing with the sample size. This is of course not the case: the sequence is only a mathematical device for approximating the actual situation. If in a concrete case, the definition (4) applies with a given value of Δ^*, then $\Delta^{(n)}$ will be identified with Δ^*.

Suppose now without loss of generality that $\Theta_s = \Theta_{\max}$. Then for all F the left hand side of (3) takes on its maximum value when

$$\Theta_1 = \cdots = \Theta_{s-1} = \Theta_s - \Delta^{(n)}, \tag{6}$$

and the sample size is therefore determined by the condition

$$P(\overline{X}_s = \overline{X}_{\max}) = \gamma \text{ when } (\Theta_1, \ldots, \Theta_s) \text{ satisfies (6)}. \tag{7}$$

The large-sample solution of the sample size problem follows from the following lemma.

Lemma 1. *For fixed* γ, *and with „goodness" of a population defined by* (5), *let* n *be determined so that* (6) *and* (7) *hold. Then as* $n \to \infty$,

$$\Delta^{(n)} = \frac{\Delta\sigma}{\sqrt{n}} + o\left(\frac{1}{\sqrt{n}}\right). \tag{8}$$

Here σ^2 is the variance of F and Δ is determined by the condition

$$Q\left(\frac{\Delta}{\sqrt{2}}, \ldots, \frac{\Delta}{\sqrt{2}}\right) = \gamma \tag{9}$$

where Q is the cumulative distribution function of a normally distributed vector $(U_1, \ldots, U_{s-1})$ with

$$E(U_i) = 0,\quad \operatorname{Var}(U_i) = 1,\quad \operatorname{Cov}(U_i, U_j) = \frac{1}{2} \quad \text{for all } i, j. \tag{10}$$

Proof. Let $Y_i = (\overline{X}_i - \overline{X}_s)\sqrt{\frac{n}{2\sigma^2}}$ and let $(\Theta_1^{(n)}, \ldots, \Theta_s^{(n)})$ be a sequence of parameter points satisfying (6). Then equation (7) is equivalent to

$$P(Y_i \leqq 0 \quad \text{for all } i) = \gamma$$

and hence by the central limit theorem (and the fact that the convergence is uniform in the arguments of the cumulative distribution function) to

$$\lim_{n\to\infty} P\left(U_i \leqq (\Theta_s^{(n)} - \Theta_i^{(n)})\sqrt{\frac{n}{2\sigma^2}} \quad \text{for all } i\right)$$
$$= \lim_{n\to\infty} P\left(U_i \leqq \Delta^{(n)}\sqrt{\frac{n}{2\sigma^2}} \quad \text{for all } i\right) = \gamma.$$

If Δ is defined by (9), this equation will be satisfied if and only if (8) holds.

18*

Suppose now that we are given a value Δ^* and wish to find the smallest sample size n for which (3) holds with "goodness" defined by (4). It follows from the lemma that a large sample solution is obtained by putting $\Delta\sigma/\sqrt{n} = \Delta^*$ or

$$n = \left(\frac{\Delta\sigma}{\Delta^*}\right)^2. \tag{11}$$

Since Δ is determined by γ, this defines n as a function of Δ^*, γ and σ.

2. *Procedures based on scores.* Suppose the observations X_{ij} are ranked and the rank of X_{ij} is denoted by R_{ij}. Let us consider the scores procedures which are obtained by replacing X_{ij} in (2) by a score $h(R_{ij})$. More particularly we shall assume that h is defined by

$$h(r) = E_{F_0}(Z^{(r)}) \tag{12}$$

where $Z^{(1)} < \cdots < Z^{(N)}$ is an ordered sample from a given distribution F_0. The resulting F_0*-scores procedure* $S(F_0)$ then selects the ith population if

$$V_i = \max_k V_k \tag{13}$$

where

$$V_i = \sum_j h(R_{ij}). \tag{14}$$

Standard terminology suggests calling these procedures nonparametric or distribution-free. However, there is no significance level or similar quantity to be computed, which for procedure (2) would depend on the underlying distribution F but would be independent of F for (13). The terminology is therefore inappropriate in the present case, and so is the associated justification of (13). The relative merits of (13) and (2) must instead be decided on the basis of the sample sizes required to achieve (3), (or some similar criterion).

Let us therefore consider the sample size $m = g(n)$ required by the scores procedure (13) if it is to satisfy (3) with the definition of "good" still being given by (5) and (8). Since h is nondecreasing, it is easily seen that also for this procedure the left hand side of (3) takes on its minimum value when $(\Theta_1, \ldots, \Theta_s)$ satisfies (6), and hence that for large samples the sample size m is determined by the condition

$$\lim_{n\to\infty} P(V_s = V_{\max}) = \gamma \quad \text{when} \quad (\Theta_1, \ldots, \Theta_s) \quad \text{satisfies (6)}. \tag{15}$$

The following lemma is the analogue of Lemma 1 for the present case.

Lemma 2. *For fixed* γ, *let* m *be determined so that* (15) *holds, and suppose that* F *and* $J = F_0^{-1}$ *satisfy the regularity conditions of Theorem 6.2 and Lemma 7.2 of* PURI [8]. *Then as* $m \to \infty$

$$\Delta^{(m)} = \frac{\Delta}{\sqrt{m}} \frac{A}{\int \frac{d}{dx}\{J[F(x)]\}\, dF(x)} + o\left(\frac{1}{\sqrt{m}}\right) \tag{16}$$

where

$$A^2 = \int J^2(u)\, du - \left(\int J(u)\right)^2. \tag{17}$$

The proof of this lemma will be given at the end of section 4.

If m and n are determined by (3) and if the same definition of good is applied in both cases, the quantity $\Delta^{(m)}$ defined by (16) must agree with $\Delta^{(n)}$ defined by (8), and hence

$$\lim_{n\to\infty} \frac{n}{m} = \frac{\sigma^2}{A^2}\left(\int \frac{d}{dx}\{J\left[F(x)\right]\}\, dF(x)\right)^2. \tag{18}$$

This relative efficiency $e_{s(F_0),M}$ of the F_0-scores procedure $s(F_0)$ to the means procedure M is the same as that found by CHERNOFF and SAVAGE [4] for the corresponding tests in the two-sample problem and shown by PURI [8] to be valid also for the c-sample problem.

Let us now consider two special cases:

(i) If F_0 is a rectangular distribution, (13) reduces to the rank-sum procedure R, which selects the ith population when $\sum_j R_{kj}$ takes on its maximum value for $j = i$. The efficiency (18), then is known to satisfy $e_{R,M}(F) \geqq .864$ for all F; $e_{R,M}(F) = 3/\pi \sim .955$ when F is normal; and $e_{R,M}(F) > 1$ for many non-normal distributions (HODGES and LEHMANN [6]).

(ii) If F_0 is a normal distribution, the efficiency $e_{S(\Phi),M}$ of the resulting normal scores procedure is known to satisfy $e_{S(\Phi),M}(F) \geqq 1$ for all F, and $e_{S(\Phi),M}(F) = 1$ if and only if F is normal.

Thus from the efficiency point of view both the rank procedure and the normal scores procedure (or the asymptotically equivalent procedure based on VAN DER WAERDENs X-test [9]) appear to be advantageous compared with the means procedure, unless one can be reasonably sure of the absence of gross errors and other departures of normality. (For an efficiency comparison of the rank sum to the normal scores procedure see HODGES and LEHMANN [7].)

Let us next consider the problem of sample size determination with a scores procedure. Suppose we are given a value Δ^* and wish to find the smallest sample size m for which (3) holds when the definition of "good" is given by (4). This is approximately achieved by putting

$$m = \left(\frac{\Delta}{\Delta^*}\right)^2 \frac{A^2}{\left(\int \frac{d}{dx}\{F_0^{-1}\left[F(x)\right]\}\, dF(x)\right)^2}.$$

To be specific, let us consider the rank-sum procedure for which this equation reduces to

$$m = \left(\frac{\Delta}{\Delta^*}\right)^2 \frac{1}{12\left(\int f^2(x)\, dx\right)^2}$$

where f denotes the probability density of F.

Suppose we expect F to be approximately normal with variance τ^2. For this distribution the sample size m becomes

$$m = \left(\frac{\Delta}{\Delta^*}\right)^2 \cdot \frac{\pi\tau^2}{3}.$$

This sample size determination is considerably more robust against gross errors or other deviations in the tail behaviour of F than that for the means procedure since $\int f^2(x)\, dx$ is much less sensitive to such deviations than is the variance of the X's.

3. *Comparison of performance characteristics.* If $m = g(n)$ is determined so that (18) holds, then for large n the procedures $S(F_0)$ and M have approximately the same minimum probability γ of selecting a good population. This, however, still leaves open the relation of the performance of the two procedures (as characterized by the probability of selecting a good population) for parameter points not satisfying (6).

To discuss this relationship, consider any sequence of parameter points satisfying

$$\Theta_s^{(n)} - \Theta_i^{(n)} = \Delta_i^{(n)} = \frac{\Delta_i \sigma}{\sqrt{n}} + o\left(\frac{1}{\sqrt{n}}\right). \tag{19}$$

We shall further restrict attention to points for which $\Delta_i \neq \Delta$ for all i to avoid ambiguities as to whether or not a population is good. Without loss of generality suppose that

$$\Theta_1^{(n)}, \ldots, \Theta_r^{(n)} < \Theta_s^{(n)} - \Delta^{(n)} < \Theta_{r+1}^{(n)}, \ldots, \Theta_{s-1}^{(n)} \leqq \Theta_s^{(n)} \tag{20}$$

so that for the means procedure M

$$\lim_{n \to \infty} P \text{ (selected population is good)} = \sum_{i=r+1}^{s} P_i(\Delta_1, \ldots, \Delta_{s-1}) \tag{21}$$

where

$$P_i(\Delta_1, \ldots, \Delta_{s-1}) = \lim_{n \to \infty} P(\overline{X}_i = \overline{X}_{\max}) . \tag{22}$$

Since the joint limiting distribution of the variables

$$Y_i = (\overline{X}_i - \overline{X}_s) \sqrt{\frac{n}{2\sigma^2}}$$

is the same as that of an $(s-1)$-dimensional normal vector $(U_1, \ldots, U_{s-1})$ with

$$E(U_i) = -\frac{\Delta_i}{\sqrt{2}}, \quad \operatorname{Var}(U_i) = 1, \operatorname{Cov}(U_i, U_j) = \frac{1}{2} \quad \text{for all} \quad i, j \tag{23}$$

it follows that

$$P_s(\Delta_1, \ldots, \Delta_{s-1}) = P(U_j \leqq 0 \quad \text{for all} \quad j) = Q\left(\frac{\Delta_1}{\sqrt{2}}, \ldots, \frac{\Delta_{s-1}}{\sqrt{2}}\right) \tag{24}$$

and

$$P_i(\Delta_1, \ldots, \Delta_{s-1}) = P(U_i = U_{\max} \text{ and } U_i > 0) \quad \text{for all} \quad i = 1, \ldots, s-1 . \tag{25}$$

Consider now the F_0-scores procedure based on samples of size $m = g(n)$ where m again satisfies (18). Then the limiting behaviour of the procedure can be seen from the following theorem, which will be proved in the next section.

Theorem 1. *For* $n = 1, 2, \ldots$ *let* X_{ij} $(j = 1, \ldots, m = g(n);\ i = 1, \ldots, s)$ *be independently distributed according to* $F(x - \Theta_i^{(n)})$. *Suppose that the sequence of parameter points* $\Theta^{(n)} = (\Theta_1^{(n)}, \ldots, \Theta_s^{(n)})$ *satisfies the assumptions of Theorem 6.2 and Lemma 7.2 of* [8]. *Let* $m = g(n)$ *be determined so that* (18) *holds. Then if*

$$T_i^{(n)} = (V_i - V_s)/\sqrt{2mA^2} \tag{26}$$

where V_i is defined by (14) *and* (12), *the joint limiting distribution of the random variables* $(T_1^{(n)}, \ldots, T_{s-1}^{(n)})$ *is the distribution of an* $(s-1)$*-dimensional normal vector* $(U_1, \ldots, U_{s-1})$ *satisfying* (23).

It follows from this theorem that the F_0-scores procedure based on samples of size m, with $m = g(n)$ determined in accordance with (18), satisfies (21) where the P_i are given by (24) and (25). For sequences $\Theta^{(n)}$ satisfying (19), this procedure therefore has the same asymptotic performance characteristic as the means procedure based on samples of size n. This statement holds not only for sequences $(\Theta_1^{(n)}, \ldots, \Theta_s^{(n)})$ for which all differences $\Theta_s^{(n)} - \Theta_i^{(n)}$ tend to zero at the rate $1/\sqrt{n}$ as assumed in (19). This follows from the fact that the behaviour for differences tending to zero faster than this rate is obtained by putting $\Delta_i = 0$ in (19), and sequences tending to zero more slowly than $1/\sqrt{n}$ (or tending to a positive limit) by putting $\Delta_i = \infty$.

4. *Proof of Theorem 1 and Lemma 2.* Under the assumptions of Theorem 1, let

$$F_j(x) = F\left((x + \frac{\Delta_j}{\sqrt{m}}\right) \tag{27}$$

and

$$\mu_j = \int J[H(x)]\, dF_j(x) \tag{28}$$

where $H(x) = \sum_{j=1}^{s} F_j(x)/s$. Then PURI [8] has shown that the variables

$$[V_i - m\,\mu_i(\Theta^{(n)})]/\sqrt{m} \tag{29}$$

asymptotically have a joint normal distribution with zero means and covariance matrix

$$\sigma_{ij} = \left(\delta_{ij} - \frac{1}{s}\right) A^2, \tag{30}$$

where the δ_{ij} are the Kronecker deltas and where A^2 is given by (17).

We require the following lemma, the proof of which is immediate, but which is the key for the simple comparability of the limit distributions of the variables $(T_1^{(n)}, \ldots, T_{s-1}^{(n)})$ and $(Y_1, \ldots, Y_{s-1})$.

Lemma 3. *Let the covariance matrix of the random vector* $(W_1, \ldots, W_s)$ *be given by*

$$\sigma_{ij} = \begin{matrix} \tau^2 & \text{if} & i = j \\ \varrho\tau^2 & \text{if} & i \neq j \end{matrix} \tag{31}$$

and let $Z_i = W_i - W_s$ $(i = 1, \ldots, s)$. *Then*

$$\operatorname{Cov}(Z_i, Z_j) = \begin{matrix} 2\tau^2(1-\varrho) & \text{if} & i = j \\ \tau^2(1-\varrho) & \text{if} & i \neq j \end{matrix}$$

and hence the correlation coefficient of Z_i *and* Z_j *is* 1/2 *regardless of the values of* ϱ *and* τ^2.

It follows from this lemma that two different random vectors having the structure of the Z's assumed in the lemma and having normal limit distributions can, as in the univariate case, be compared by comparing their means and variances.

To prove the theorem, consider the random variables $T_i^{(n)}$ defined by (26). By Lemma 3, the variance of the limiting distribution of $T_i^{(n)}$ is $2\tau^2(1-\varrho)$ where by (30),

$$\tau^2 = (s-1)/2s \quad \text{and} \quad \varrho = -1/(s-1)\,. \tag{32}$$

The variance of the limiting distribution of $T_i^{(n)}$ is therefore seen to be equal to 1.

It only remains to show that the mean of $T_i^{(n)}$ in the limiting distribution is equal to $-\Delta_i/\sqrt{2}$, i.e. that for all $i = 1, \ldots, s-1$,

$$\lim_{n\to\infty} \sqrt{m}\,[\mu_i(\Theta^{(n)}) - \mu_s(\Theta^{(n)})]/\sqrt{2A^2} = -\Delta_i/\sqrt{2}\,. \tag{33}$$

Let $\Delta_s = 0$ and $\bar{\Delta} = (\Delta_1 + \cdots + \Delta_s)/s$. Then we shall show that for all $i = 1, \ldots, s$

$$\lim_{n\to\infty} \sqrt{n}\,[\mu_i(\Theta^{(n)}) - \mu_i(0)] = \sigma(\bar{\Delta} - \Delta_i) \int \frac{d}{dx}\{J[F(x)]\}\,dF(x)\,. \tag{34}$$

Equation (33) is then an immediate consequence of (34) and (18).

To see (34), note that its left hand side can be written as

$$\sqrt{n} \int \left\{ J\left[\frac{1}{s}\sum_{j=1}^{s} F(x + \Theta_i^{(n)} - \Theta_j^{(n)})\right] - J[F(x)]\right\} dF(x)$$

$$= \int A_n(x)\, B_n(x)\, dF(x)\,,$$

where

$$A_n(x) = \frac{J\left[\frac{1}{s}\sum_{j=1}^{s} F\left(x + \frac{(\Delta_j - \Delta_i)\sigma + R_n}{\sqrt{n}}\right)\right] - J[F(x)]}{\frac{1}{s}\sum_{j=1}^{s} F\left(x + \frac{(\Delta_j - \Delta_i)\sigma + R_n}{\sqrt{n}}\right) - F(x)}$$

and

$$B_n(x) = \sqrt{n}\left[\frac{1}{s}\sum_{j=1}^{s} F\left(x + \frac{(\Delta_j - \Delta_i)\sigma R_n}{\sqrt{n}}\right) - F(x)\right].$$

Here R_n is independent of x and tends to zero as $n \to \infty$. Under the assumed regularity conditions,

$$\lim_{n\to\infty} A_n(x) = \frac{d}{du} J(u)|_{u=F(x)}, \quad \lim_{n\to\infty} B_n(x) = \sigma(\bar{\Delta} - \Delta_i)\, F'(x)$$

and, since differentiation under the integral is permitted, this completes the proof of (34) and hence of Theorem 1.

To prove Lemma 2, consider a sequence $\Theta^{(m)}$ of parameter points satisfying

$$\Theta_1^{(m)} = \cdots = \Theta_{s-1}^{(m)} = \Theta_s^{(m)} - \Delta^{(m)}.$$

It follows as before from Puri's theorem that the random variables $[(V_i - V_s) - m(\mu_i(\Theta^{(m)}) - \mu_s(\Theta^{(m)}))]/\sqrt{2mA^2}$ have the limiting distribution of an $(s-1)$-dimensional normal vector $(U_1, \ldots, U_{s-1})$ satisfying (10). Condition (15) with n in (6) replaced by m is equivalent to

$$\lim_{m\to\infty} P\left(U_i \leq (\mu_s(\Theta^m) - \mu_i(\Theta^{(m)}))\sqrt{m/2A^2} \quad \text{for all} \quad i\right) = \gamma\,.$$

Since μ_i is independent of i, this reduces to

$$\lim_{m \to \infty} [\mu_s(\Theta^{(m)}) - \mu_i(\Theta^{(m)})] \sqrt{m/2A^2} = \Delta/\sqrt{2} \quad \text{for all} \quad i = 1, \ldots, s-1 .$$

By the argument used to prove (34), this condition is seen to be equivalent to (16), which completes the proof.

References

[1] Bahadur, Raghu Raj: On a problem in the theory of k populations. Ann. Math. Stat. **21**, 362—375 (1950).

[2] —, and L. A. Goodman: Impartial decision rules and sufficient statistics. Ann. Math. Stat. **23**, 553—562 (1952).

[3] Bechofer, R. E.: A single sample multiple decision procedure for ranking means of normal populations with known variances. Ann. Math. Stat. **25**, 273—289 (1954).

[4] Chernoff, H., and I. R. Savage: Asymptotic normality and efficiency of certain nonparametric test statistics. Ann. Math. Stat. **29**, 972—994 (1958).

[5] Hall, W. J.: The most-economical character of some Bechofer and Sobel decision rules. Ann. Math. Stat. **30**, 964—969 (1959).

[6] Hodges, J. L., jr, and E. L. Lehmann: The efficiency of some nonparametric competitors of the t-test. Ann. Math. Stat. **27**, 324—335 (1956).

[7] — — Comparison of the normal scores and Wilcoxon tests. Proc. 4th Berk. Symposium on Math. Stat. and Prob. **1**, 307—317 (1961).

[8] Puri, M. L.: Asymptotic efficiency of a class of c-sample tests. Submitted to Ann. Math. Stat.

[9] van der Waerden, B. L.: Ein neuer Test für das Problem der zwei Stichproben. Math. Ann. **126**, 93—107 (1953).

(Received October 15, 1962)

Journal of Nonparametric Statistics
Vol. 21, No. 4, May 2009, 397–405

Taylor & Francis
Taylor & Francis Group

Parametric versus nonparametrics: two alternative methodologies

Erich L. Lehmann*

Department of Statistics, University of California, Berkeley, CA 94720, USA

(Received 17 July 2008; final version received 8 October 2008)

This article compares parametric and nonparametric approaches to statistical inference. It considers their advantages and disadvantages, and their areas of applicability. Although there is no clear comprehensive conclusion, the article finds that in simple problems in which Wilcoxon type tests and estimators apply, they may be recommended as the methods of choice.

Keywords: Wilcoxon t and F tests; least squares; robustness; adaptation

1. Introduction

The comparison of parametric and nonparametric approaches is a huge subject, and here, the author considers only how it applies to the unpaired comparison of two treatments. However, very similar results apply to paired comparisons, some analysis of variance problems, regression and independence.

The standard model for the comparison of two treatments assumes that $N = m + n$ subjects are drawn at random from some population of subjects and that, of these, m are assigned to treatment A (for example, control) and n to treatment B (for example, a new treatment).

The control responses $X_1, \ldots, X_m$ are assumed to be i.i.d. according to a distribution F, and the treatment responses $Y_1, \ldots, Y_n$ according to G. The hypothesis to be tested is

$$H : G = F, \text{ no treatment effect.}$$

A general class of alternatives specifies that the Ys are stochastically larger than the Xs. A special case is the shift model

$$G(y) = F(y - \Delta), \quad \Delta > 0$$

This model seems rather special, but is much more general than the usual normal model and will be assumed throughout this paper.

*Email: shaffer@stat.berkely.edu

ISSN 1048-5252 print/ISSN 1029-0311 online

DOI: 10.1080/10485250902842727
http://www.informaworld.com

2. Four tests

The two approaches, parametric and nonparametric, will be compared in terms of the following four tests:

(i) *The classical t-test*: This test rejects when $t > t_\alpha$, where t denotes the two-sample t-statistic and t_α is the upper α-critical value of the t-distribution with $m + n - 2$ degrees of freedom.
(ii) *The Permutation t-test* (t^*) : Given the $m + n$ responses

$$(z_1, \ldots, z_N) = (x_1, \ldots, x_m; y_1, \ldots, y_n),$$

the statistic t can take on $\binom{N}{n}$ equally likely values corresponding to the possible assignments of the zs to xs and ys. If $\alpha = K/\binom{N}{n}$, the (t^*)-test rejects H for the K largest of these values.

The other two tests are rank tests. The N observations are ranked from 1 (the smallest) to N (the largest). Let $S_1, \ldots, S_n$ denote the ranks of the Ys.

(i) *The Wilcoxon test* (W) [1]: This test rejects when

$$S_1 + \ldots + S_n > C.$$

(ii) *The Normal Scores test* (NS) [2]: This test rejects when

$$h(S_1) + \ldots + h(S_n) > C',$$

where

$$h(S) = E(Z^{(S)})$$

is the expected value of the Sth smallest of N independent $N(0, 1)$ variables.

3. Validity of the tests

(i) The t-test is exactly valid only under the additional assumption that the common null distribution F of the Xs and Ys is normal [1]. It was this dependence on the assumption of normality that first aroused interest in permutation and rank tests.

(ii)–(iv) The other three tests, and in fact all permutation and rank tests, are valid if $X_1, \ldots, X_m; Y_1, \ldots, Y_n$ are i.i.d. according to any continuous distribution F.

From this point of view, the t-test is at a disadvantage, while there is no difference between the other three. The Wilcoxon test became popular (particularly in the social sciences) because of its independence from the normality assumption and its simplicity. However, there was a concern that the test would be very inefficient because it replaced the original values by their ranks, thereby discarding a lot of information.

4. Asymptotic relative efficiency

To assess the extent of this loss of efficiency, Pitman in 1947 and 1948, in courses at Chapel Hill and Columbia, introduced the concept of asymptotic relative efficiency (ARE). His approach was first published by Noether in 1950, who also extended it in 1955 [3,4].

Definition of ARE *Let n_1, n_2 be the numbers of observations required by two tests to achieve the same results (i.e., same level and power). Then*

$$\text{ARE} = e_{1,2} = \lim \left(\frac{n_2}{n_1}\right) \quad \textit{as } n_1, n_2 \to \infty.$$

In our applications, the test statistics are asymptotically normal, and the ARE is inversely proportional to the ratio of the asymptotic variances.

5. ARE of t^* to t

Assuming that one is dealing with two independent samples, the asymptotic normality of t is obvious; that of t^* was first proved by Wald and Wolfowitz [5], and under more general conditions by Noether [6], Hoeffding [7,8] and Hajek [9,10]. It follows from that work that

$$e_{t^*,t}(F) = 1 \quad \text{for essentially all } F.$$

In fact, the two tests reject when respectively

$$t > t_\alpha \quad \text{and } t > t^*_\alpha,$$

where t^*_α is random, and

$$t^*_\alpha - t_\alpha \xrightarrow{P} 0 \quad \text{as } n_1, n_2 \longrightarrow \infty.$$

Thus, except for small n_1, n_2, there is a very little difference between the tests.

6. ARE of Wilcoxon and normal scores to t

Pitman obtained a general formula for $e_{W,t}(F)$ in the shift model and calculated it for the normal case. He found that

$$e_{W,t}(N) = \frac{3}{\pi} = 0.955.$$

This was very surprising since it showed that W is nearly as powerful as t in the situations for which t was designed and in which it is optimal.

Pitman's formula was

$$e_{W,t}(F) = 12\sigma^2\left(\int f^2\right)^2,$$

where σ^2 is the variance of F and f is density. From this, the ARE can be calculated for special cases, for example

F	Logistic	Double exponential	Uniform	Exponential
$e_{W,t}(F)$	1.097	1.5	1.0	3.0

Furthermore, it is easily seen that $e_{W,t}(F)$ can be ∞. On the other hand, it was shown by Hodges and Lehmann [11] that for all F

$$e_{W,t}(F) \geq 0.864.$$

Thus Wilcoxon can be infinitely more efficient than t but never much less efficient.

Hodges and Lehmann conjectured that the situation was even more favorable for the NS relative to the t-test, namely that

$$e_{NS,t}(F) \geq 1 \quad \text{for all } F,$$

with equality holding only when F is normal. This was proved by Chernoff and Savage [12], and by Gastwirth and Wolff [13].

7. ARE of Wilcoxon to normal scores

With Wilcoxon and NS, both being serious competitors of t, it is of interest to me to compare these two tests directly. Such a comparison was provided by Hodges and Lehmann [14]. The following are some values of $e_{\mathrm{W,NS}}(F)$:

Uniform	Normal	Logistic	Double exponential	Cauchy
0	0.955	1.05	1.18	1.41

A sharp upper bound is $6/\pi = 1.91$.

The general conclusion is NS tends to be better for short-tailed distributions, and W for long-tailed ones. Some theoretical support for this conclusion is provided by Van Zwet [15].

8. Estimation

Comparisons have so far been made in terms of the power of the tests and have proved surprisingly favorable to the rank tests. However, the t-test also has an advantage. It is closely related to the associated point and interval estimates; Wilcoxon and NS are just tests. However, a similar development is possible for rank-based methods in the shift model $G(y) = F(y - \Delta)$.

By the Neyman theory, tests, for Δ can be converted into confidence sets for Δ. These are simply the set of parameter values of Δ_0 for which $H : \Delta = \Delta_0$ is accepted. This construction was worked out by Lincoln Moses (in Walker and Lev [16], and some theory developed by Lehmann [17]).

A method for deriving point estimates of Δ from rank tests was given by Hodges and Lehmann [17]. Both the resulting point and interval estimates share the AREs (relative to the normal theory methods) of the associated rank tests.

In the Wilcoxon case, all three (tests (in the Mann–Whitney form), point estimates and confidence intervals) are based on the quantiles of the $m + n$ differences $Y_j - X_i$ and form a cohesive whole like the classical normal methods.

9. Comparison of the four tests

On the basis of the AREs one might make the following recommendations.

(i) If firmly convinced of normality (without gross errors), the t-test is best.
(ii) If fairly confident of normality, but want an insurance policy, t^* seems a good solution.
(iii) If the tails of F are believed to be as short as or shorter than those of the normal, the NS test offers the greatest advantage.

(iv) If the tails of F are believed to be as heavy as or heavier than those of the normal, the Wilcoxon test would be the first choice.

Since (i) and (ii) rarely hold, the NS and Wilcoxon tests and associated estimates provide an alternative to the classical normal methods that often should be preferred.

The following are some weak points in the earlier-mentioned argument.

10. Unrealistic distributions

The distributions considered in Sections 6 and 7 (such as uniform, double exponential and Cauchy) are not the most likely alternatives to the normal distribution to be met in practice. A more plausible family is the so-called Tukey model

$$F = (1 - \epsilon)N(0, 1) + \epsilon N(0, \tau^2),$$

in which the normal distribution is contaminated by gross errors. Here are some AREs for this case.

$\tau = 3$				
ϵ	0	0.05	0.1	0.2
W/t	0.955	1.2	1.37	1.58
W/NS	0.955	1.04	1.09	1.48

$\tau = 4$				
ϵ	0	0.05	0.1	0.2
W/t	0.955	1.46	1.82	2.20
W/NS	0.955	1.06	1.12	1.20

The tables show a clear superiority of Wilcoxon over t in the presence of 5% or more of gross errors, which increases with the size of the errors. They also show a superiority of the NS test over t, but the superiority is less strong.

11. Asymptotic versus small-sample results

All the AREs are asymptotic and therefore could be quite misleading. What is particularly suspicious is that the actual efficiency depends on the level α and the power β to be achieved while the ARE is a single number.

This problem was studied for F = normal by Dixon [19] and Hodges and Lehmann [11] with the following results for $m = n = 5$, $\alpha = 0.032$

β	0.07	0.21	0.43	0.67	0.86	0.95	0.99	∞ (ARE)
$e_{W,t}$	0.97	0.98	0.96	0.96	0.96	0.96	0.96	0.955

The efficiency is surprisingly stable over the whole range. The table suggests that at least in the neighborhood of the normal distribution it is fairly reliable.

12. Sampling versus randomisation

The ARE results were recently attacked as irrelevant to most experimental work by Ludbrook and Dudley [20] since they assume that the observations are samples from a population. These authors point out that most medical research is based on randomisation rather than on sampling. This process is represented by a *randomisation model* in which the N subjects are considered as fixed. Randomness then enters only through the random assignment of subjects, m to control and n to treatment.

For rank tests the null distribution is the same for the sampling and randomisation models. For permutation tests, the randomisation distribution coincides with the conditional distribution (given the subjects) of the sampling situation. The null distribution is therefore the same in both cases. The difference comes in when considering power, but the issue is deeper. It arises from the fact that in the randomisation case any inference applies only to the subjects at hand, and it cannot be generalised to other subjects.

Ludbrook and Dudley [21] admit this drawback but state:

> 'This needs not deter experimeters from inferring that their results are applicable to similar patients, ... though their argument must be verbal rather than statistical.'

Unfortunately, verbal arguments are not enough for power comparisons. However, the reference to similar patients suggests that (and, in fact, is meaningful only if) these patients form a population from which the subjects can be considered to be (at least approximately) a random sample. Only with such an assumption can the inference be extended to this population.

Such an assumption is clearly less reliable than randomisation. The theory therefore involves two stages.

1. Rely on randomisation to control the level.
2. Use the sampling model for choice of test.

For the second stage, it is not necessary to know the population F in detail, but only roughly (*e.g.*, heavy tails).

Note that in any case, a choice has to be made, and for making this it is good to receive some guidance from statistical theory.

13. Using more than one test

The problem so far has been which of the many tests to use. However, the availability of several tests also opens up other possibilities, both for misuse and for improvement.

(i) *Misuse*: When considering two tests, say t and Wilcoxon, it is tempting to carry out both at level α (*e.g.* 5%) but report only the more significant. This is, of course, not legitimate since the level of the resulting test is no longer α. The true level α^* that depends on F has been studied by Jiang [22]. The following is a numerical example for large sample sizes, when the tests used are W and t and $\alpha = 0.05$.

F	Normal	Double exponential	t_3	Sharp upper bound
α^*	0.059	0.066	0.074	0.097

Thus the increase in α is not negligible. Examples in which two tests are used are given in Gastwirth [23, Chapter 11].

(ii) *Improvement through adaptation*: If we decide to use the Wilcoxon or NS test because we believe F to be heavy or light tailed respectively, we would have made a poor choice if our belief was mistaken. This difficulty can be avoided by letting the data guide us in our choice. Such an approach was suggested by Hajek [24,25] and further elaborated by Hogg [26,36].

To be specific, let S be a selection criteria (*e.g.* the coefficient of kurtosis) with high or low values, indicating heavy or short tails. We shall use W when S is large, NS when S is small (both at level α). If S is a symmetric function of the N variables $(X_1, \ldots, X_m; Y_1, \ldots, Y_n)$, it is independent of the ranks and the combined test has level α.

Simulation studies indicate the excellent performance of such a procedure over a broad spectrum of distributions.

14. Full adaptation

Full adaptation is in the same spirit as the partially adaptive procedures in that it uses the data to adapt the procedure to the unknown F. But it is much more ambitious because it attempts to determine the best test not among a finite number of tests (*e.g.* Wilcoxon and NS) but among all tests. That this may be possible was first proposed by Stein [27].

Let us assume a shift model as before. Stein's procedure consists of two steps:

(i) Estimate the unknown density f by a suitable density estimate $\hat{f}$.
(ii) Perform the test that would be optimal if $\hat{f}$ were the true density.

Then it is hoped that the resulting test will be asymptotically efficient for all f, *i.e.* have ARE $= 1$ relative to the best test when f is true, whatever f may be.

A rigorous proof of such a result was given by Hajek [10]. However, he added that it is 'of little use in practice because of slow convergence' (*i.e.* because it requires very large sample sizes before the asymptotic results apply).

Stein-type adaptation has been much more fully explored for estimation than for testing. Stone [28] surprisingly found excellent results even for $n = 40$. A general theory of adaptive estimation was developed by Bickel [29] and by Bickel *et al.* [30], whose conditions are given, under which full adaptation is possible. The corresponding problems for testing constitute an open area for research.

15. Extension

As mentioned in Section 1, the results discussed here apply to other simple situations, *e.g.* s-sample problems, simple linear regression, the measurement and testing of dependence. They do, however, not apply to more complex situations as linear models with interactions and multivariate problems. This obviously limits their usefulness.

16. The advantage of simplicity

In the simple situations in which they do apply, Wilcoxon-type (but not NS) methods have the advantage of great simplicity and ease of interpretation. This is of particular importance for the many nonstatisticians who in their work apply such methods and for the introductory courses in which they learn about them. This fact was stressed by Noether [31] who compared the difficulty of interpreting the sample correlation coefficient r with the simple interpretations of Kendall's rank correlation in terms of concordant and discordant pairs.

Noether implemented his ideas in an elementary text [32] in which for every problem he discusses first a Wilcoxon-type test with associated point and interval estimation, and follows it with the corresponding normal theory.

17. Conclusions

The issue of parametric versus nonparametric inference is too complex to permit neat, comprehenive conclusions, but here are a few concluding comments.

Least squares estimation (Legendre, Gauss) and the associated normal theory tests of Fisher have for a long time been, and still are, the standard. They are convenient and we have gotten used to them. However, we also know that they suffer from a defect: They are too strongly influenced by outlying observations. In testing, this results in low power.

Under the headings of robust and nonparametric inference, many modifications and alternatives have been developed to eliminate or to diminish this effect [33]. In general there is no clear winner.

For simple problems to which Wilcoxon-type tests and estimators apply, I agree with Noether that they have much to recommend them as the methods of choice.

Acknowledgements

The author thanks Ralph D'Agostino and his committee for inviting him to give the 2000 Noether Lecture on which this work is based, and three referees for many helpful comments.

Note

1. The t-test is however asymptotically distribution-free. [34,35].

References

[1] F. Wilcoxon, *Individual comparisons by ranking methods*, Biometrics 1 (1945), pp. 80–83.
[2] R.A. Fisher and F. Yates, *Statistical Tables for Biological, Agricultural, and Medical Research (Table XX)*, Oliver and Boyd, London, 1938.
[3] G.E. Noether, *Asymptotic properties of the Wald–Wolfowitz test of randomness*, Amer. Math. Statist. 21 (1950), pp. 231–246.
[4] G.E. Noether, *On a theorem of Pitman*, Ann. Math. Statist. 26 (1955), pp. 64–68.
[5] A. Wald, and J. Wolfowitz, *Statistical tests based on permutations of the observations*, Ann. Math. Statistics 15 (1944), pp. 358–372.
[6] G.E. Noether, *On a theorem by Wald and Wolfowitz*, Ann. Math. Statist. 20 (1949), pp. 455–458.
[7] W. Hoeffding, *A combinatorial central limit theorem*, Ann. Math. Statist. 22 (1951), pp. 558–566.
[8] W. Hoeffding, *The large-sample power of tests based on permutations*, Ann. Math. Statist. 23 (1952), pp. 169–192.
[9] J. Hajek, *Some extensions of the Wald–Wolfowitz–Noether theorem*, Ann. Math. Statist. 32 (1961), pp. 506–523.
[10] J. Hajek, *Asymptotically most powerful rank order tests*, Ann. Math. Statist. 33 (1962), pp. 1124–1147.
[11] J.L. Hodges, Jr. and E.L. Lehmann, *The efficiency of some nonparametric competitors of the t-test*, Ann. Math. Statist. 27 (1956), pp. 274–335.
[12] H. Chernoff and I.R. Savage, *Asymptotic normality and efficiency of certain nonparametric test statistics*, Ann. Math. Statist. 29 (1958), pp. 972–994.
[13] J.L. Gastwirth and S. Wolff, *An elementary method for obtaining lower bounds on the asymptotic power of rank tests*, Ann. Math. Statist. 39 (1968), pp. 2128–2130.
[14] J.L. Hodges, Jr. and E.L. Lehmann, *Comparison of the Normal Scores and Wilcoxon Tests*, Proceedings of the 4th Berkeley Symposium on Mathematical Statistics Vol. I, 1961, pp. 307–317.

[15] W.R. Van Zwet, *Convex transformations: a new approach to skewness and kurtosis*, Statist. Neerlandica 18 (1964), pp. 433–444.
[16] H.M. Walker and J. Lev, *Statistical Inference,* Holt, New York, 1953.
[17] E.L. Lehmann, *Nonparametric confidence intervals for a shift parameter*, Ann. Statist. 33 (1963), pp. 1138–1154.
[18] J.L. Hodges, Jr. and E.L. Lehmann, *Estimates of location based on rank tests*, Ann. Math. Statist. 34 (1963), pp. 598–611.
[19] W.J. Dixon, *Power under normality of several nonparametric tests*, Ann. Math. Statist. 25 (1954), pp. 610–614.
[20] J. Ludbrook and H. Dudley, *Response (to letters concerning Ludbrook and Dudley (1998))*, Amer. Statist. 54 (2000), p. 87.
[21] J. Ludbrook and H. Dudley, *Why permutation tests are superior to t and F tests in biomedical research*, Amer. Statist. 52 (1998), pp. 127–132.
[22] J. Jiang, *Sharp upper and lower bounds for asymptotic levels of some statistical tests*, Statist. Probab. Letters 35 (1997), pp. 395–400.
[23] J.L. Gastwirth, *Statistical Reasoning in Law and Public Policy*, Academic Press, Boston, 1988.
[24] J. Hajek, *Nonparametric Statistics,* Holden-Day, San Francisco, 1969.
[25] J. Hajek, *Miscellaneous problems of rank tests*, in *Nonparametric Techniques in Statistical Inference,* Puri, ed. Cambridge University Press, 1970.
[26] R.V. Hogg and R.V. Lenth, *A review of some adaptive statistical techniques*, Comm. Statist Theory Methods. 13 (1984), pp. 1551–1579.
[27] C. Stein, *Efficient Nonparametric Testing and Estimation*, Proceedings of the Third Berkeley Symposium Mathematical Statistics Vol. 1, pp. 187–195.
[28] C.J. Stone, *Adaptive maximum likelihood estimators of a location parameter*, Ann. Statist. 3 (1975), pp. 267–284.
[29] P.J. Bickel, *On adaptive estimation*, Ann. Statist. 10 (1982), pp. 647–671.
[30] P.J. Bickel, C.A.J. Klaassen, Y. Ritov, and J. Wellner, *Efficient and Adaptive Estimation for Semiparametric Models,* Johns Hopkins, University Press, Baltimore, 1993.
[31] G.E. Noether, *The role of nonparametrics in introductory statistics courses*, Ann. Math. Statist. 34 (1980), pp. 22–23.
[32] G.E. Noether, *Introduction to Statistics: The Nonparametric Way,* Springer Verlag, New York, 1976, 1991.
[33] D. Birkes and Y. Dodge, *Alternative Methods of Regression,* Wiley, New York, 1993.
[34] E.L. Lehmann, *Nonparametrics: Statistical Methods Based on Ranks*. Springer, New York, 2006.
[35] E.L. Lehmann and J.P. Romano, *Testing Statistical Hypotheses*, 3rd ed., Springer, New York, 2005.
[36] R.V. Hogg, *Adaptive robust procedures*, J. Amer. Statist. Assoc. 69 (1974), pp. 909–923.

Journal of Nonparametric Statistics
Vol. 21, No. 4, May 2009, 425–426

Rejoinder

Erich L. Lehmann*

Department of Statistics, University of California, Berkeley, CA 94720, USA

(Received 21 February 2009; final version received 21 February 2009)

I am grateful to the four discussants for their interesting and helpful comments. Their contributions have a common structure. They take cognisance of the narrow focus of my survey, and discuss various aspects and issues that I did not cover. Let me be more specific.

1. Testing for goodness (or lack) of fit

Professor Eubank points to the distinction between nonparametric and distribution-free, and to the fact that the first of these covers, for example, nonparametric function estimation. So as not to go too far afield, he focuses on a situation at the intersection of the two concepts, testing for goodness-of-fit. He considers various ways to compare different such tests. These comparisons constitute a difficult problem on which Professor Eubank states that 'the final word has yet to be spoken'. (A discussion of some of the relevant issues can be found in Chapter 14 of Lehmann and Romano [1].)

The same testing problem is also considered in Professor Hart's contribution where, in particular, he reviews the order selection test proposed by Eubank and Hart.

2. Full adaptation

This is another topic considered by Professor Hart. It was discussed in my survey for the case of a two-sample shift model, where the fully adaptive test does not seem very practicable. Hart treats, instead, a different setting where an adaptive rank test seems useful. It is a multiple testing situation that is of considerable practical importance.

3. Multiple comparisons

This is an extension considered by three of the discussants. I have already mentioned Professor Hart's multiple comparisons example in the context of full adaptation.

*Email: shaffer@stat.berkeley.edu

ISSN 1048-5252 print/ISSN 1029-0311 online

DOI: 10.1080/10485250902846926
http://www.informaworld.com

Multiple comparisons is the central topic of the contribution by Professors Lin and Dicker. The authors discuss protein data for 1450 proteins from four obese and four lean mice, and rise the question whether Wilcoxon rank-sum statistics could be used instead of t-statistics. They enumerate some of the difficulties with such an approach. To begin with, the Bonferroni correction is useless since no p-value could ever be significant. Even the FDR is too conservative. Other difficulties are different variances and lack of independence. Although they do not mention it, a way of dealing with these issues through resampling has been proved by Westfall and Young [2]. However, they do not consider rank-based methods.

Finally, Professor Romano considers the efficiency of Wilcoxon to t in a simple multiple comparisons setting. In this situation, he finds the efficiency to be the same as it is for individual tests.

4. Infinite variance

An additional topic treated by Professor Romano is the efficiency of Wilcoxon to t when the underlying distributions have infinite variances. He shows that under fairly general conditions this efficiency is infinite.

Through their extensions, the discussants have greatly broadened the coverage of my survey and increased its utility. They have also raised many questions, which I hope others will take up.

I am most grateful to them and to the Editor, Professor Rojo, for their help.

References

[1] E.L. Lehmann and J.P. Romano, *Testing Statistical Hypotheses*, 3rd ed., Springer, New York, 2005.
[2] P.H. Westfall and S.S. Young, *Resampling-Based Multiple Testing*, Wiley, New York, 1993.

Part V
Nonparametrics II

Descriptive statistics for nonparametric models. The impact of some Erich Lehmann's papers

Hannu Oja

1 Introduction

In this review we discuss the six papers: Lehmann (1963) and Bickel and Lehmann (1974, 1975, 1976a, 1976b, 1976c).

The first paper deals with confidence intervals based on nonparametric tests, and the other papers discuss descriptive statistics for nonparametric models.

In an estimation problem, sample statistics $\hat{\theta}$ may often be seen as values of a functional $\theta(F)$ at the sample cumulative distribution function F_n. For example, for the sample mean $\bar{X} = \frac{1}{n}\sum_{i=1}^{n} X_i$ and sample variance $S^2 = \frac{1}{n}\sum_{i=1}^{n}(X_i - \bar{X})^2$ the functionals are the mean functional $\mu(F) = \int x dF(x)$ and variance functional $\sigma^2(F) = \int (x - \mu(F))^2 dF(x)$, respectively. Under general assumptions on F, two functionals $\theta_1(F)$ and $\theta_2(F)$ may give the same value (e.g., mean functional and median functional under the symmetry assumption) but the statistical properties (e.g. efficiency and robustness) of the sample statistics $\hat{\theta}_1 = \theta_1(F_n)$ and $\hat{\theta}_2 = \theta_2(F_n)$ may be totally different. How should one then choose between the estimates/functionals $\theta_1(F)$ and $\theta_2(F)$? When can one be convinced that, in wide nonparametric models, the functional provides a measure of a natural aspect of the population values? These questions are posed and discussed in the series of the last five papers.

2 Nonparametric confidence intervals for a shift parameter

In this paper, Lehmann considers the two-sample location problem and gives exact expressions as well as large-sample approximations for the nonparametric confidence

H. Oja
Tampere School of Health Sciences, 33014 University of Tampere, Finland
e-mail: Hannu.Oja@uta.fi

J. Rojo (ed.), *Selected Works of E. L. Lehmann*, Selected Works in Probability and Statistics, DOI 10.1007/978-1-4614-1412-4_40, © Springer Science+Business Media, LLC 2012

intervals for a shift parameter Δ. The confidence interval is obtained by "inverting" the two-sample Wilcoxon-Mann-Whitney test which is strictly distribution-free test for testing the null hypothesis that all the observations are i.i.d.

The paper proposes that different confidence intervals (with the same covering probabilities) should be compared using the lengths of the intervals. It is proven that, in regular cases, the length of the confidence interval standardized in an appropriate way converges in probability to a constant, and these constants can be used for asymptotic efficiency comparisons. The asymptotical efficiencies of the intervals based on Wilcoxon test and on the t test are then compared and shown to be the same as in the testing case. As a consequence of the results in the paper, a consistent estimate for the important constant $[\int f^2(x)dx]^{-1}$ is found. This estimation technique is still used in practice; see e.g. the recent book by Hettmansperger (1988).

3 Descriptive statistics for nonparametric models

3.1 The literature before Bickel and Lehmann

We first briefly discuss the literature just before the publication of these papers. von Mises (1947) may be the first one to develop a general theory of statistical functionals. Tukey (1962) discussed the properties of several location functionals in case of "spotty data". Huber (1964) gave a new approach for robust estimation, introduced the M-estimates for location and found the asymptotic theory for the estimates.

He also found the "most robust" estimate in a contaminated normal distribution. (The estimate minimizes the maximum asymptotic variance in a contaminated normal neighborhood model.) Instead of considering asymptotic variances in neighborhood models, robustness is often seen as a continuity requirement of the functional θ. Small changes in F should result in small changes in $\theta(F)$. For first alternative formulations of this requirement and their analysis, see Hampel (1971). In his review, Huber (1972) discussed an early history of robust estimation and considered three classes of location estimates in semiparamateric models, the maximum likelihood type estimates (M-estimates), linear combinations of order statistics (L-estimates), and the estimates derived from rank tests (R-estimates). In almost all studies discussed so far, the assumed model was semiparametric, with natural location parameter θ, and the location functionals $T(F)$ were used in the estimation of unknown θ.

Bickel and Lehmann then introduced the idea that one should, in wide nonparametric models, define measures (functionals) of different characteristics of the population and then use the corresponding sample statistics as estimators. The series of papers on descriptive statistics in nonparametric models together with van Zwet's (1964) work on skewness was a main source of inspiration for future research in the area.

3.2 Statistical models

When Bickel and Lehmann wrote the series of papers on descriptive statistics in nonparametric models, the statisticians and the practical users of statistical testing and estimation procedures did not trust any more in the dogma of normally distributed errors and often did not even believe in any parametrical statistical model.

Bickel and Lehmann discussed the model selection problem and distinguished a number of possibilities: Parametric models (e.g., normal model), neighborhood models (e.g., contaminated normal model), semiparametric models with some natural parameters (e.g. symmetry assumption), or totally nonparametric models. In the totally nonparametric model one can describe different interesting properties of the distribution F just by defining and using a corresponding measure or functional $\theta(F)$. Classical moment based measures of location, scale, skewness and kurtosis, μ, σ, β_1 and β_2 serve as examples here. Bickel and Lehmann find it convenient, however, first to try to define when a distribution G possesses the interesting attribute more strongly than a distribution F. Also, the measures should satisfy certain equivariance conditions.

3.3 Partial orderings of distributions

The idea to order the distributions with respect to the considered property is not very old. Mann and Whitney (1947) introduced the notion of stochastic ordering. Birnbaum (1948) proposed a dispersion ordering. The skewness and kurtosis orderings by van Zwet (1964) have proved useful and now accepted widely. The first step in the general strategy by Bickel and Lehmann is to try to define when a distribution G possesses the interesting attribute more strongly than a distribution F. This is then denoted by $F \prec G$ and it is required that a measure of this property must preserve this ordering, that is,

$$\theta(F) \leq \theta(G) \text{ whenever } F \prec G.$$

The authors then introduce and discuss natural orderings for location, dispersion (symmetric models) and spread (asymmetric models). Skewness is not discussed, and measures of kurtosis are seen as ratios of two dispersion measures.

Inspired by this work Oja (1981) considers the concepts of location, scale, skewness and kurtosis of univariate distributions. For two random variables with strictly increasing cumulative distribution functions F and G, these comparisons were made using functions

$$R(x) = G^{-1}(F(x)) \text{ and } \Delta(x) = R(x) - x.$$

Doksum (1975) called $\Delta(x)$ the shift function as $X \sim F$ when shifted by $\Delta(X)$ has the distribution G, that is,

$$X \sim F \;\Rightarrow\; R(X) = X + \Delta(X) \sim G.$$

See also Doksum and Sievers (1976). It is then interesting to note that the two distributions can be naturally ordered using the function $\Delta(x)$: The orderings for location ("stochastic ordering"), spread (Bickel and Lehmann), and skewness (van Zwet) are given by

$$\Delta(x) \geq 0, \quad \text{for all } x, \qquad \Delta'(x) \geq 0, \quad \text{for all } x, \qquad \text{and } \Delta''(x) \geq 0, \quad \text{for all } x.$$

The ordering given by $\Delta^{(3)}(x) \geq 0$ may be seen as an ordering for kurtosis, and so on. Orderings of skewness are used widely also in reliability theory, see Barlow and Proschan (1975). The approach based on partial orderings have inspired statisticians for example to define and consider, also in the multivariate case, dispersion orderings (see e.g. Oja (1983), Zuo and Serfling (2000), and Romanazzi (2009), as well as kurtosis orderings (see Wang and Serfling (2005) and Wang (2009) and references therein). See also Serfling (2004).

3.4 Equivariance of the measures under linear transformations

In the following we write F_X for the cdf of X. Bickel and Lehmann required that a *measure of location* $\mu(F)$ should satisfy

1. $\mu(F) \leq \mu(G)$ whenever $F \prec_{loc} G$ and
2. $\mu(F_{aX+b}) = a\mu(F_X) + b$ for all a and b.

Similarly it was required that a *measure of spread* $\sigma(F)$ satisfies

1. $\sigma(F) \leq \sigma(G)$ whenever $F \prec_{spread} G$ and
2. $\mu(F_{aX+b}) = |a|\mu(F_X)$ for all a and b.

Bickel and Lehmann consider only location and spread (scale) measures. This is sufficient, however, as secondary properties, like skewness and kurtosis, can often be seen trough the comparisons of the behavior of different location and scale measures. Let μ_1 and μ_2 be two different location measures, and σ, σ_1 and σ_2 different scale measures. Then

$$\beta_1(F) = \frac{\mu_1(F) - \mu_2(F)}{\sigma(F)} \quad \text{and} \quad \beta_2(F) = \frac{\sigma_1(F)}{\sigma_2(F)}$$

can be used to measure skewness and kurtosis, respectively. The measures of skewness and kurtosis then naturally satisfy invariance properties

$$\beta_1(F_{aX+b}) = sgn(a)\beta_1(F_X) \quad \text{and} \quad \beta_2(F_{aX+b}) = \beta_2(F_X),$$

respectively.

One can easily find in the literature a lot of studies on location, scale, skewness and kurtosis including robust measures of skewness and kurtosis Brys *et al.* (2004) and Brys *et al.* (2006). Also equivariance properties for multivariate location vectors, scatter matrices and regression coefficient matrices are commonly required in the literature. See e.g. Maronna *et al.* (2006). Natural multivariate skewness and kurtosis measures were proposed by Mardia (1970). Nordhausen *et al.* (2011) considered multivariate skewness and kurtosis measures based on two different location vectors and two different scatter matrices.

3.5 Robustness and efficiency comparisons

How can one then compare different measures of location, dispersion and spread? Is the comparison always possible? The comparison of two measures is definitely reasonable if their values coincide under our model assumptions. Under the symmetry assumption for example, the sample mean and the sample median estimate the same population quantity and their comparison as estimates is reasonable. But what about measures of spread or measures of location for asymmetric distributions?

Bickel and Lehmann wish to find measures which are robust and at the same time could be estimated efficiently. In the location case efficiency means a guaranteed high efficiency relative to the mean (using the asymptotic variance). For dispersion and spread measures, the comparison was to the standard deviation (and the criterion was the standardized asymptotic variance). According to their general strategy, the authors wished to have a partial ordering for robustness as well: A measure μ_2 is called more robust than μ_1 if and only if μ_2 is continuous with respect to the topology induced by μ_1. Three classes of location estimates, L-estimates, R-estimates and M-estimates were compared. Trimmed standard deviations and pth power deviations were compared to the standard deviation in the dispersion case.

In the community of robust statisticians, the measures of location and spread (vectors and matrices in the multivariate case) are usually chosen to satisfy some equivariance properties, but relevant partial orderings are not discussed. The measures of spread (or scatter matrices in the multivariate case) are often rescaled so that they estimate the standard deviation (or covariance matrix) in the (multivariate) normal case. Then of course the efficiency comparisons of different estimates become reasonable in a neighborhood of a normal model. The most popular tools for robustness used today are the breakdown point, the influence function, gross-error sensitivity, asymptotic bias, etc.

4 Final comments

When Bickel and Lehmann wrote the series of papers on descriptive statistics in nonparametric models, the statisticians and the users of statistical testing and estimation procedures in practice did not trust any more in the dogma of normally distributed observations. As shown by my short review, the papers considered here had a very strong impact on future ideas and work on developing new measures and estimates for the description of population and data under general nonparametric models. Personally speaking, these papers together with van Zwet's (1964) work on skewness were the main source of inspiration in my own thesis work, Oja (1981), on the concepts of location, scale, skewness and kurtosis of the univariate distributions and also in my later work on analysis methods for multivariate data.

References

[1] Barlow, R. E. and Proschan, F. (1975). *Statistical Theory of Reliability and Life Testing*. Holt, Rinehart and Winston, New York.

[2] Bickel, P. J. and Lehmann, E. L. (1974). Measures of location and scale. *Proc. Prague Symposium on Asymptotic Statist*, J. Hajek, Ed., Charles University, Prague, **1**, 23–36.

[3] Bickel, P. J. and Lehmann, E. L. (1975). Descriptive statistics for nonparametric models. I. Introduction. *Ann. Statist.*, **3**, 1038–1044.

[4] Bickel, P. J. and Lehmann, E. L. (1976a). Descriptive statistics for nonparametric models II. Location. *Ann. Statist.*, **4**, 1045–1069.

[5] Bickel, P. J. and Lehmann, E. L. (1976b). Descriptive statistics for nonparametric models. III. Dispersion. *Ann. Statist.*, **4**, 1139–1158.

[6] Bickel, P. J. and Lehmann, E. L. (1976c). Descriptive statistics for nonparametric models, IV. Spread. In *Contributions to Statistics* , Jaroslav Hajek Memorial Volume, Prague.

[7] Birnbaum, Z. W. (1943). On random variables with comparable peakedness. *Ann. Math. Stat.* **19**, 76–81.

[8] Brys, G., Hubert, M. and Stryuf, A. (2004). A robust measure of skeness. *Journal of Computational and Graphical Statistics* **13**, 996–1017.

[9] Brys, G., Hubert, M. and Stryuf, A. (2006). Robust measures of tailweight. *Computational Statistics & Data Analysis* **50**, 733–759.

[10] Doksum, K. A. (1975). Measures of location and asymmetry. *Scand. J. Statist.* **2**, 11-22.

[11] Doksum, K. A. and Sievers, G. L. (1975). Plotting with confidence: Graphical comparisons of two populations. *Biometrika*, **63**, 421–434.

[12] Hampel, F. R. (1971). A general qualitative definition of robustness. *Ann. Math. Stat.*, **42**, 1887–1896.

[13] Hettmansperger, T. P. and McKean, J. W. (1998). *Robust Nonparametric Statistical Methods*. Arnold, London.

[14] Huber, P. J. (1964). Robust estimation of a location parameter. *Ann. Math. Stat.*, **35**, 73–101.

[15] Huber, P. J. (1972). Robust statistics: A review. *Ann. Math. Stat.*, **43**, 1041–1067.

[16] Huber, P. J. (1981). *Robust Statistics*. Wiley, New York.

[17] Lehmann, E. L. (1963). Nonparametric confidence intervals for a shift parameter. *Ann. Math. Stat.*, **34**, 1507–1512.

[18] Mann, H. B. and Whitney, D. R. (1947). On tests of whether one of two random variables is stochastically larger than the other. *Ann. Math. Stat.*, **18**, 50–60.

[19] Mardia, K. V. (1970). Measures of multivariate skewness and kurtosis with applications. *Biometrika*, **57**, 519–530.

[20] Maronna, R. A., Martin, R. D. and Yohai, V. Y. (2006). *Robust Statistics. Theory and Methods*. Wiley, New York.

[21] Nordhausen, K., Oja, H. and Ollila, E. (2011). Multivariate Models and the First Four Moments. In Hunter, D. R., Richards, D. S. R. and Rosenberger, J. L. (editors) "Nonparametric Statistics and Mixture Models: A Festschrift in Honor of Thomas P. Hettmansperger", 267–287, World Scientific, Singapore.

[22] Oja, H. (1981). On location, scale, skewness and kurtosis of univariate distributions. *Scandinavian Journal of Statistics*, **8**, 154–168.

[23] Oja, H. (1983). Descriptive statistics for multivariate distributions. *Stat. Prob. Letters* **1**, 327–332.

[24] Romanazzi, M. (2009). Data depth, random simplices and multivariate dispersion. *Stat. Prob. Letters* **79**, 1473–1479.

[25] Serfling, R. (2004). Nonparametric multivariate descriptive measures based on spatial quantiles. *Journal of Statistical Planning and Inference*, bf 123, 259-278.

[26] Tukey, J. W. (1962). The future of data analysis. *Ann. Math. Stat.*, **33**, 1–67.

[27] van Zwet, W. R. (1964). *Convex transformations of random variables.* Math Centrum, Amsterdam.

[28] von Mises, R. (1947). On the asymptotic distribution of differentiable statistical functions. *Ann. Math. Stat.*, **18**, 309–348.

[29] Wang, J. (2009). A family of kurtosis orderings for multivariate distributions. *Journal of Multivariate Analysis*, **100**, 509–517.

[30] Wang, J. and Serfling, R. (2005). Nonparametric multivariate kurtosis and tailweight measures. *Journal of Nonparametric Statistics*, **17**, 441–456.

[31] Zuo, Y. and Serfling, R. (2000). Nonparametric notions of multivariate scatter measure and more scattered based on statistical depth functions. *Journal of Multivariate Analysis*, **75**, 62–78.

NONPARAMETRIC CONFIDENCE INTERVALS FOR A SHIFT PARAMETER[1]

By E. L. Lehmann

University of California, Berkeley

Summary. Exact expressions and large-sample approximations are given for the nonparametric confidence intervals for a shift parameter Δ, which are obtained from the two-sample Wilcoxon test. These intervals are shown to have the same asymptotic efficiency relative to the standard confidence intervals for Δ as the Wilcoxon test has relative to Student's t-test. As a consequence of this result, a constant multiple of the length of the nonparametric intervals is shown to be a consistent estimator of the quantity $1/\int f^2(x)\,dx$.

Let $X_1, \cdots, X_m$ and $Y_1, \cdots, Y_n$ be independent observations from distributions $P[X_i \leqq x] = F(x)$ and $P[Y_j \leqq y] = F(y - \Delta)$ respectively, where F is assumed to be continuous but otherwise unknown. Exact confidence intervals for Δ can be based on nonparametric tests such as the Wilcoxon test. Let $U(\Delta)$ be the number of differences $Y_j - X_i$ which exceed Δ, and suppose that the two-sided symmetric level α test of the hypothesis $H(\Delta_0) : \Delta = \Delta_0$ accepts $H(\Delta_0)$ when

$$C_\alpha \leqq U(\Delta_0) \leqq mn - C_\alpha . \tag{1}$$

If the ordered set of mn differences $Y_j - X_i$ is denoted by

$$D^{(1)} < \cdots < D^{(mn)},$$

the confidence intervals obtained from (1) by solving for Δ_0 and replacing Δ_0 by Δ are seen to be

$$\Delta_L < \Delta < \Delta_U \tag{2}$$

where

$$\Delta_L = D^{(C_\alpha)} \quad \text{and} \quad \Delta_U = D^{(mn+1-C_\alpha)}. \tag{3}$$

For given sample sizes m and n, a constant C_α for which (1) has exactly probability $1 - \alpha$ may not exist. For the large-sample problems to be considered below it is enough that the constants $C_\alpha = C_\alpha(m, n)$ are chosen in such a way that the probability $1 - \alpha(m, n)$ of (1) tends to the specified value $1 - \alpha$ as m and n tend to infinity.

For small and moderate values of m and n, the critical value C_α can be read from tables of the null distribution of the Wilcoxon statistic. For large m and n, one has the following approximation.

Received February 26, 1963.

[1] This research was done while the author was a Professor of the Adolph C. and Mary Sprague Miller Institute for Basic Research in Science, University of California, Berkeley.

LEMMA 1. *The critical value C_α is for large m and n equal to*

$$C_\alpha = mn/2 - K_{\alpha/2}[mn(m+n)/12]^{\frac{1}{2}} + o[mn(m+n)]^{\frac{1}{2}} \tag{4}$$

where $K_{\alpha/2}$ is the $100a/2$ upper percentage point of the standard normal distribution.

PROOF. Since $\Delta_L = \sup\{\Delta : U(\Delta) > mn - C_\alpha\}$ we have $\alpha/2 = \lim P_\Delta[\Delta < \Delta_L] = \lim P_\Delta[U(\Delta) > mn - C_\alpha]$ and the result now follows from the fact that the Wilcoxon statistic $U(0)$ for $\Delta = 0$ has a normal limiting distribution with mean $mn/2$ and variance $mn(m + n + 1)/12$.

What is the efficiency of the intervals (2) relative to the standard intervals

$$\bar{Y} - \bar{X} - K_{\alpha/2}(1/m + 1/n)^{\frac{1}{2}}S < \Delta < \bar{Y} - \bar{X} + K_{\alpha/2}(1/m + 1/n)^{\frac{1}{2}}S \tag{5}$$

where S^2 is the usual estimate of the variance σ^2 of F? Suppose first that efficiency is measured in terms of the probability of covering false values, say in terms of the probability that the intervals cover the value $\Delta + rN^{-\frac{1}{2}}$, where $N = m + n$. Then it follows from the correspondence between confidence intervals and the tests on which they are based, and the known efficiency properties of the Wilcoxon test, that if the intervals defined by (2) and (3) are based on $m = \rho N$ and $n = (1 - \rho)N$ observations, and the intervals (5) on $m' = \rho N'$ and $n' = (1 - \rho)N'$ observations, they will have the same limiting probability of covering the values $\Delta + rN^{-\frac{1}{2}}$, provided

$$N'/N \to 12\sigma^2 \left(\int f^2(x)\, dx\right)^2 \tag{6}$$

as $N \to \infty$, where f is the density of the distribution F. In this sense, (6) is the asymptotic relative efficiency of the two sets of intervals.

Alternatively, efficiency might be measured in terms of the lengths of the intervals. To carry out this comparison, we shall now prove the following result.

THEOREM 1. *If $L = \Delta_U - \Delta_L$ is the length of the intervals defined by (2) and (3) and based on $m = \rho N$ and $n = (1 - \rho)N$ observations, then*

$$N^{\frac{1}{2}}L \to K_{\alpha/2}/[3\rho(1-\rho)]^{\frac{1}{2}} \int f^2(x)\, dx \tag{7}$$

in probability as $N \to \infty$.

To prove this result, we shall require the following lemmas.

LEMMA 2. *Let $T_{1N}, \cdots, T_{NN}$ $(N = 1, 2, \cdots)$ be a sequence of sets of random variables (not necessarily identically or independently distributed). Suppose for any fixed a and b the number of T's between a/N^β and b/N^β equals $[k(b - a) + R_N]N^\gamma$ where R_N tends to zero in probability uniformly for all a, b satisfying $-A < a < b < A$ for any finite A. Let U_N be a sequence of random variables (possibly depending on $T_{N1}, \cdots, T_{NN}$) which are bounded in probability, and let Z_N be a value such that exactly $cN^\gamma + o(N^\gamma)$ of the T's lie between U_N/N^β and Z_N/N^β. Then $Z_N - U_N \to c/k$ in probability.*

PROOF. Suppose to the contrary that with probability exceeding ϵ, the inequality $Z_N - U_N > (c + \delta)/k$ holds for some positive δ and ϵ for arbitrarily large N. (The proof for the case $Z_N - U_N < (c - \delta)/k$ is completely analogous). Then with probability $> \epsilon$ the number of T's between U_N/N^β and $[U_N + (c + \delta)/k]/N^\beta$ is at most $cN^\gamma + o(N^\gamma)$.

On the other hand, there exists A so large that with probability exceeding $1 - \epsilon/4$ we have $-A < U_N < U_N + (c + \delta)/k < A$ for all sufficiently large N. Given that this inequality holds, the probability is $> 1 - \epsilon/4$ that the number of T's between U_N/N^β and $[U_N + (c + \delta)/k]/N^\beta$ exceeds $[c + \delta/2]N^\gamma$ for all N sufficiently large. Hence the over-all probability that this number exceeds $(c + \delta/2)N^\gamma$ is greater than $(1 - \epsilon/4)^2 > 1 - \epsilon/2$, which gives the desired contradiction.

LEMMA 3. *Let W_N be the number of differences $Y_j - X_i$ satisfying $aN^{-\frac{1}{2}} < Y_j - X_i < bN^{-\frac{1}{2}}$, and let the distribution G of $Y_j - X_i$ have density g. Then*

$$N^{-\frac{3}{2}}W_N = \rho(1 - \rho)g(0)(b - a) + R_N \tag{8}$$

where R_N tends to zero uniformly for $-A < a < b < A$.

PROOF. Let I_{ij} be equal to 1 if $aN^{-\frac{1}{2}} < Y_j - X_i < bN^{-\frac{1}{2}}$ and equal to zero otherwise. Then

$$E[N^{-\frac{3}{2}}W_N] = \rho(1 - \rho)(b - a)[G(bN^{-\frac{1}{2}}) - G(aN^{-\frac{1}{2}})]/(b - a)N^{-\frac{1}{2}}$$
$$\rightarrow \rho(1 - \rho)(b - a)g(0),$$

and the convergence is uniform in $-A < a < b < A$. To prove the desired result, it is enough to show that $\text{Var}[N^{-\frac{3}{2}}W_N] = mnN^{-3}[\text{Var}(I_{12}) + (n - 1)\text{Cov}(I_{12}, I_{13}) + (m - 1)\text{Cov}(I_{12}, I_{32})]$ tends to zero uniformly for a, b in any finite interval. Since $\text{Var}(I_{12})$ is uniformly bounded, it only remains to show that the two covariances tend to zero uniformly. This follows, for example, from

$$|\text{Cov}(I_{12}, I_{13})| \leqq P[|Y_2 - X_1| < AN^{-\frac{1}{2}} \text{ and } |Y_3 - X_1| < AN^{-\frac{1}{2}}]$$
$$+ P^2[|Y_1 - X_1| < AN^{-\frac{1}{2}}] \rightarrow 0$$

for the first covariance, and analogously for the other.

LEMMA 4. *As $N \rightarrow \infty$, the random variable $N^{\frac{1}{2}}(\Delta_L - \Delta)$ has a normal limiting distribution with mean $-K_{\alpha/2}/[12\rho(1 - \rho)]^{\frac{1}{2}}\int f^2(x)\,dx$ and variance $1/12\rho(1 - \rho)[\int f^2(x)\,dx]^2$.*

PROOF. Assuming without loss of generality that $\Delta = 0$, it is seen as in the proof of Lemma 1 that for any constant v, $P[N^{\frac{1}{2}}\Delta_L > v] = P[U(vN^{-\frac{1}{2}}) > mn - C_\alpha]$. The variables $N^{-\frac{3}{2}}[U(vN^{-\frac{1}{2}}) - \frac{1}{2}mn]$ have a normal limit distribution with mean

$$\lim_{N\rightarrow\infty} N^{-\frac{3}{2}}mn\{P[Y_1 - X_1 > vN^{-\frac{1}{2}}] - \tfrac{1}{2}\}$$
$$= \lim \rho(1 - \rho)N^{\frac{1}{2}} \int [F(y - vN^{-\frac{1}{2}}) - F(y)]\,dF(y) = -\rho(1 - \rho)v \int f^2(y)\,dy$$

and variance

$$\lim N^{-3}mn(m + n + 1)/12 = \rho(1 - \rho)/12.$$

Using Lemma 1, we therefore have

$$\lim P[N^{\frac{1}{2}}\Delta_L > v] = \lim P[U(vN^{-\frac{1}{2}}) - \tfrac{1}{2}mn > K_{\alpha/2}[mn(m+n)/12]^{\frac{1}{2}} + o(N^{\frac{3}{2}})$$

$$= 1 - \Phi\left(K_{\alpha/2} + v[12\rho(1-\rho)]^{\frac{1}{2}} \int f^2(x)\, dx\right)$$

where Φ is the standard normal cumulative distribution function. This implies the desired result.

PROOF OF THEOREM 1. Take for the set $T_{1N}, \cdots, T_{NN}$ of Lemma 2 the differences $Y_j - X_i$, and for U_N and Z_N of Lemma 2 the statistics $N^{\frac{1}{2}}\Delta_L$ and $N^{\frac{1}{2}}\Delta_U$. Then

(i) the number of T's between $U_N/N^{\frac{1}{2}}$ and $Z_N/N^{\frac{1}{2}}$ is $2K_{\alpha/2}[\rho(1-\rho)/12]^{\frac{1}{2}}N^{\frac{3}{2}} + o(N^{\frac{3}{2}})$, by Lemma 1;

(ii) the number of T's between $aN^{-\frac{1}{2}}$ and $bN^{-\frac{1}{2}}$ is $[\rho(1-\rho)g(0)(b-a) + R_N] \cdot N^{\frac{3}{2}}$, with R_N tending to zero in probability uniformly for a, b in any finite interval by Lemma 3;

(iii) U_N is bounded in probability by Lemma 4.

Hence Lemma 2 is applicable with $\beta = \frac{1}{2}$, $\gamma = \frac{3}{2}$, $c = 2K_{\alpha/2}[\rho(1-\rho)/12]^{\frac{1}{2}}$ and $k = \rho(1-\rho)g(0)$. Since $g(0) = \int f^2(x)\, dx$, this establishes the desired result.

The length L' of the standard intervals (4) is given by $L' = 2K_{\alpha/2}S/[\rho(1-\rho)N]^{\frac{1}{2}}$, and it therefore follows from Theorem 1 that the ratio of the squares of the lengths L'^2/L^2 tends in probability to the right hand side of (6). Alternatively, if the intervals (2) are based on N and the intervals (5) on N' observations, the ratio L'/L will tend in probability to one, provided (6) holds. Thus the right hand side of (6) is a reasonable measure of efficiency also when the comparison is made in terms of the lengths of the intervals.

It is interesting to compare the intervals (Δ_L, Δ_U) with the asymptotically distribution-free intervals, say $\Delta_L^* < \Delta < \Delta_U^*$, proposed in [1]. These were centered at the median med$(Y_j - X_i)$ of the mn differences $Y_j - X_i$ and had length $K_{\alpha/2}/[12NT_N\rho(1-\rho)]^{\frac{1}{2}}$ where $(T_N)^{\frac{1}{2}}$ is any consistent estimator of $\int f^2(x)\, dx$. It is seen from the results of Section 5 of [1] and from Lemma 4 and Theorem 1 of the present paper that the joint limiting distribution of $(N^{\frac{1}{2}}(\Delta_L - \Delta), N^{\frac{1}{2}}(\Delta_U - \Delta))$ is the same as that of $(N^{\frac{1}{2}}(\Delta_L^* - \Delta), N^{\frac{1}{2}}(\Delta_U^* - \Delta))$, so that the two intervals have the same asymptotic behaviour. It is a consequence of this fact and the results of [1] that the right hand side of (6) also measures the efficiency of the present intervals in terms of the measure of accuracy

$$W(\Delta; \Delta_L, \Delta_U) = a(\Delta - \Delta_L)^2 + b(\Delta_U - \Delta)^2$$

in a sense made precise in Section 5 of [1].

We note finally that Theorem 1 proves

$$N^{\frac{1}{2}}(\Delta_U - \Delta_L)[3\rho(1-\rho)]^{\frac{1}{2}}/K_{\alpha/2} \tag{9}$$

to be a consistent estimator of $1/\int f^2(x)\, dx$. Thus (9) may be taken as the statistic $(T_N)^{\frac{1}{2}}$ required for (Δ_L^*, Δ_U^*).

The above results are easily extended to the one-sample problem. Let $Z_1, \cdots, Z_N$ be independent observations from a symmetric distribution F with median θ. Let $U'(\theta)$ be the number of averages $(Z_i + Z_j)/2$ with $i < j$ which exceed θ, and suppose that the two-sided symmetric level α test of the hypothesis $H'(\theta_0) : \theta = \theta_0$ accepts $H'(\theta_0)$ when

$$C'_\alpha \leqq U'(\theta_0) \leqq \binom{N}{2} - C'_\alpha . \tag{10}$$

(This test is asymptotically equivalent to the Wilcoxon one-sample test, which could be used equally well throughout the argument.) If the ordered set of $\binom{N}{2}$ averages $(Z_i + Z_j)/2$ $(i < j)$ is denoted by $E^{(1)} < \cdots < E^{(M)}$ where $M = \binom{N}{2}$, then the confidence intervals obtained from (10) are seen to be

$$\theta_L < \theta < \theta_U \tag{11}$$

where

$$\theta_L = E^{(C'_\alpha)} \quad \text{and} \quad \theta_U = E^{(M+1-C'_\alpha)}. \tag{12}$$

For large N, one obtains in generalization of Lemma 1 the following approximation for C'_α.

LEMMA 1′. *The critical value C'_α is for large N equal to*

$$C'_\alpha = N(N-1)/4 - K_{\alpha/2}(N-1)(N/12)^{\frac{1}{2}} + o(N^{\frac{3}{2}}) \tag{13}$$

where $K_{\alpha/2}$ has the same meaning as in Lemma 1.

The proofs of Lemma 1′, and of Lemmas 3′ and 4′ below will be omitted since they are exactly parallel to those of the corresponding unprimed lemmas. The efficiency results for the two-sample case also extend to the present problem, with the efficiency again being given by formula (6). The efficiency comparison in terms of length is based on the following.

THEOREM 1′. *If $L' = \theta_U - \theta_L$ is the length of the intervals defined by* (12) *and* (13), *then $N^{\frac{1}{2}}L' \to K_{\alpha/2}/3^{\frac{1}{2}}\int f^2(x)\,dx$ in probability as $N \to \infty$.*

The proof is completely analogous to that of Theorem 1, and is based on Lemma 2 and the following two lemmas.

LEMMA 3′. *Let W'_N be the number of pairs $(Z_i + Z_j)/2$ satisfying $a/N^{\frac{1}{2}} < (Z_i + Z_j)/2 < b/N^{\frac{1}{2}}$ and let the distribution G of $(Z_i + Z_j)/2$ have density g. Then*

$$N^{-\frac{3}{2}}W'_N = g(0)(b-a)/2 + R_N \tag{14}$$

where R_N tends to zero uniformly for $-A < a < b < A$.

LEMMA 4'. *As $N \to \infty$, the random variable $N^{\frac{1}{2}}(\theta_L - \theta)$ has a normal limiting distribution with mean $-K_{\alpha/2}/(12)^{\frac{1}{2}} \int f^2(x)\,dx$ and variance $1/12(\int f^2(x)\,dx)^2$.*

As a consequence of Theorem 1', we note finally that for symmetric distributions F

$$(3N)^{\frac{1}{2}}(\theta_U - \theta_L)/K_{\alpha/2} \tag{15}$$

is a consistent estimator of $1/\int f^2(x)\,dx$.

REFERENCE

[1] LEHMANN, E. L. (1963). Asymptotically nonparametric inference: An alternative approach to linear models. *Ann. Math. Statist.* **34** 1494–1506.

The Annals of Statistics
1975, Vol. 3, No. 5, 1038–1044

DESCRIPTIVE STATISTICS FOR NONPARAMETRIC MODELS I. INTRODUCTION

BY P. J. BICKEL[1] AND E. L. LEHMANN[2]

University of California, Berkeley

An overview is given of an approach to the definition of descriptive characteristics of populations and to their estimation. The emphasis is on the robustness and efficiency of the estimators. Detailed summaries will be found in successive papers of the series dealing with the problems of location, scale and kurtosis.

Introduction. Descriptive statistics deals with measures of different aspects of a population (or a distribution of population values). The population may be finite, as is the case for example when it consists of a set of data, or it may be infinite. Typical examples of descriptive measures are the mean and median as measures of location, the standard deviation or interquartile range as measures of scale, and the classical measures of skewness, kurtosis and correlation.

When defining such measures, one usually has in mind not a single population to which the measure is to be applied but a family of such populations. In particular, in statistics one is typically concerned with a family to which the given population is assumed to belong. The choice of suitable measures then depends strongly on the nature of this family. We shall distinguish a number of possibilities.

1. Parametric models. Here the distribution in question is characterized by a small number of (natural) parameters. Examples are the family of normal distributions, characterized by expectation and variance; the family of Pearson curves, characterized by the first four moments, and hence by the classical measures of location, scale, skewness and kurtosis; or the family of stationary, two-state Markov chains, characterized by initial and transition probabilities.

2. Nonparametric neighbourhood models. It has long been recognized that the validity of parametric models can at best be approximate, and that the discrepancies due to gross errors and other impurities can have a serious effect on the descriptive parameters and their estimates. This has led to the suggestion of models in which the main body of the observations come from a distribution belonging to a given parametric family (for example, the normal family) but a

Received April 1973; revised September 1974.

[1] This research was partially supported by the Office of Naval Research, Contract, NONR N00014-69-A-0200-1038.

[2] Work done while the author was Research Professor for the Adolph C. and Mary Sprague Miller Institute for Basic Research in Science, University of California, Berkeley.

AMS 1970 *subject classifications.* Primary 62G99; Secondary 62G05, 62G20, 62G35.

Key words and phrases. Descriptive statistics, nonparametric models, robustness, estimation, asymptotic efficiency, standardized variance, location, scale, kurtosis.

small proportion represent impurities about whose distribution no parametric assumptions are made. The result is a parametric model contaminated by a small nonparametric admixture, a model which one could describe as a nonparametric neighbourhood of a parametric family.

The parameters of interest in such a model are still the parameters of the parametric part of the model. However, in estimating or testing these parameters one must now guard against the nonparametric disturbances. An important and satisfactory body of inference in such models has been constructed by Huber (1964). This work is still in progress; a survey of the present state is given in Huber's Wald lecture (1972).

3. Nonparametric models with natural parameters. The above approach seems very well suited to the many situations in which extensive experience with the type of data under consideration makes an approximately parametric model reliable either for the measurements themselves or for suitable (known) transforms of these measurements. However, assumptions such as approximate normality will often be unwarranted, for example, in sociological or psychological investigations, where the observations by the nature of the situations are less well controlled. In such cases, one may prefer a totally nonparametric model, in which the observations may, for example, be assumed to be independently, identically distributed according to a distribution F, of which nothing is assumed except possibly certain smoothness or symmetry conditions.

However, even in such apparently totally nonparametric models there may exist natural parameters describing important aspects of the model, and the estimation or testing of such parameters has in fact been a very active field of study in recent years.

The case which has received the greatest attention is that in which F is a univariate distribution which is assumed to be symmetric. The natural location parameter for F is then its center of symmetry. A survey of the work concerning the estimation of this center, is given by Andrews *et al.* (1972). Another example is the two-sample shift problem where two distributions are assumed to differ only in location and where the difference in location then constitutes a natural shift parameter (see, for example, Høyland (1965)). More generally, there is a natural measure of contrasts in linear models, even when the (common) shape of the symmetric error distribution is unknown (see, for example, Lehmann (1963) and Spjøtvoll (1968)).

4. Neighbourhoods of nonparametric models with natural parameters. It seems natural to consider the extension of model 3. which parallels the extension 2. of model 1. A typical example would be to assume that we are dealing with a sample from an unknown symmetric distribution F which has a slight amount of asymmetric contamination; the parameter of interest would be the center of symmetry of F. Problems of this kind do not seem to have been considered in the literature.

The four types of models above share the important feature that there exist natural measures of the aspect of the model under consideration. There are many possible estimators of these measures, and a central problem is to choose among these.

Let us now turn to situations in which such natural measures do not exist. (The model may still possess natural parameters, but these do not correspond to the feature of interest. This is the case, for example, when we are concerned with the scale of a symmetric distribution.) It is here that the descriptive aspect of the problem becomes nontrivial. This problem has been pointed out and was discussed from a somewhat similar point of view by Takeuchi (1967).

How do we measure the location of a nonsymmetric distribution? How do we measure or describe its spread, its skewness or kurtosis? What do we even mean by kurtosis: does it describe the tail-behavior of a distribution, its peakedness, or, as has recently been argued (Darlington (1970), Chissom (1970)), its tendency to biomodality?

It is this kind of question, of course, with which we must begin the description of a particular aspect of a distribution, and we shall find it convenient to answer it by defining in each case when a distribution G (or a random variable Y with distribution G) possesses the attribute under consideration more strongly than a distribution F (or random variable X).

The literature contains many examples of such partial orderings. We may, for example, say that Y is "to the right" of X if Y is stochastically larger than X (a concept introduced by Mann and Whitney (1947) and studied further by Lehmann (1955)). A stronger definition is obtained by requiring monotone likelihood ratio. Partial orderings corresponding to the comparison of the skewness of two distributions, or of their kurtosis were introduced by van Zwet (1964) and by Barlow and Proschan (1966). Comparisons for dispersion were discussed by Z. W. Birnbaum (1948) and for the degree of positive association between the two components of a bivariate distribution by Lehmann (1966), Esary, Proschan and Walkup (1967), and Esary and Proschan (1972).

Suppose now that there has been defined a partial ordering, with $F \prec G$ meaning that G possesses the attribute under consideration more strongly than F. Then the first condition required of a measure θ of this attribute is that

$$\theta(F) \leqq \theta(G) \quad \text{whenever} \quad F \prec G\,. \tag{1}$$

A second condition characterizes the behaviour of $\theta(F)$ (which we shall also denote by $\theta(X)$ when X is a random variable with distribution F) under linear transformations. Thus, a measure of location should satisfy

$$\theta(aX + b) = a\theta(X) + b \qquad \text{for all} \quad a\,, \quad b\,, \tag{2}$$

and a measure of scale

$$\theta(aX + b) = |a|\theta(X) \qquad \text{for all} \quad a \neq 0 \quad \text{and all} \quad b\,; \tag{3}$$

A referee has pointed out that (3) is evidently desirable in the scale case if

$b = 0$ and $a > 0$. However he feels, and we agree, that if, for example, X is known to be positive, the constraints imposed by (3) for $b \neq 0$ and/or $a < 0$ are much less appealing. We here assume that no such restriction on X is given and that X must be considered totally unknown. In this case we feel that (3) is acceptable as stated.

To illustrate a slightly more complicated situation, suppose that the quantity in question is a standardized measure of location, of which $E(X)/[\text{Var}(X)]^{\frac{1}{2}}$ is a typical example. We should characterize this as a ratio $\theta = \theta_1/\theta_2$ of a location parameter θ_1 satisfying (2), divided by a scale parameter θ_2 satisfying (3).

Another example of this kind is provided by a measure of kurtosis, which in a latter part of this paper we shall define as the ratio $\theta = \theta_1/\theta_2$ of two scale parameters which satisfies (1) when $F \prec G$ is the partial order introduced by Barlow and Proschan (1966) or the stronger ordering introduced by van Zwet (1964).

Once the conditions have been laid down which a measure θ is to satisfy, there typically will exist an infinity of measures satisfying these conditions. In the location case, for example, the functionals

$$\theta(F) = \int_0^1 F^{-1}(t)\, dK(t) \tag{4}$$

where K is any distribution function on (0, 1) which is symmetric with respect to $\frac{1}{2}$ defines a large (but by no means exhaustive) class of such measures.

How should one choose among this great variety of possibilities? An additional condition one might like to impose is that of robustness; that is, small changes in F should result in small changes of θ. An early discussion of the need for such a condition was given by Bahadur and Savage (1956). For a rigorous formulation of the condition, which is essentially a continuity requirement, and a detailed analysis of the problem see Hampel (1968, 1971). Such a condition would rule out, for example, expectation as a measure of location and standard deviation in the scale case. Even with this restriction, there will still be available an infinity of measures. The location measures (4), for example, will be robust provided K is continuous and assigns probability 1 to some interval (u, v) with $0 < u < v < 1$.

To make a choice among these measures let us now recall that it will be necessary to estimate the chosen $\theta(F)$ from the data, and let us compare different θ's in terms of the accuracy with which they can be estimated. The most natural estimator of $\theta(F)$ is $\theta(\hat{F})$ where $\hat{F}$ denotes the empirical cdf; that is, the values of θ for the whole population is estimated by its value for the subpopulation made up of the sample values. One might conjecture that in a very strong sense $\theta(\hat{F})$ is also the best estimator. This conjecture is invalidated by the phenomenon of superefficiency which we exemplify in BL II, Section 4. Nevertheless, the estimators $\theta(\hat{F})$ are the ones with which we shall be concerned.

At this point, a difficulty arises, which we have only been able to overcome in each case separately through an *ad hoc* argument adjusted to the case. This difficulty (to which reference is made also in Daniell (1920) and Tukey (1960)) concerns the role played by the functionals $h(\theta)$, where h is any strictly increasing function. In a sense, these functionals are all equivalent since each is determined by any other. However, they differ widely in the ease with which they can be estimated. Suppose now that we wish to compare two functionals θ_1 and θ_2 which are not equivalent in this sense, and suppose we use variance (or asymptotic variance) as our measure of accuracy. Then it will clearly make a rather essential difference whether we compare θ_1 with θ_2, or with $\theta_2/100$. In fact, we can make any given θ as accurate as we please by dividing it by a sufficiently large constant.

This problem is greatly simplified when θ is required to satisfy equivariance conditions such as (2) or (3). For if θ satisfies these conditions, then the only functions for which the condition can hold are linear functions. Let us now briefly indicate how the difficulty remaining after this reduction can be resolved in the three cases which will be studied in the later parts of this paper.

(i) *Location.* If θ is a functional satisfying (2), then the only functions h for which $h(\theta)$ satisfies (2) are

$$h(\theta) = \theta + b \tag{5}$$

where b is an arbitrary constant. Since our estimator of (5) is $\hat{\theta} + b$ and since addition of a constant does not affect the variance of the estimator, the choice of a member of (5) is immaterial.

(ii) *Scale.* If θ is a functional satisfying (3), the only functions h for which $h(\theta)$ satisfies (3) are

$$h(\theta) = a\theta\,, \quad a > 0\,. \tag{6}$$

For a given θ satisfying (3), we could thus choose instead $a\theta$, which would be estimated by $a\hat{\theta}$, and thereby multiply the variance by a^2. These choices will become equivalent if we measure the accuracy of $\hat{\theta}$ not by its variance (or asymptotic variance) but by its *standardized variance*

$$\operatorname{Var}(\hat{\theta})/\theta^2 \tag{7}$$

which is invariant under the transformations (6). With this definition of accuracy, it is now possible to compare different measures θ_1 and θ_2.

There is an alternative way of arriving at (7). If the distribution of X is symmetric about μ, attention can be restricted to the variables $|X - \mu|$. By taking logarithms, the scale problem is then reduced to the location problem (i). If θ is any measure of scale and $\log \theta$ the corresponding location measure, the latter is estimated by $\log \hat{\theta}$; the asymptotic variance of $\log \hat{\theta}$ under suitable regularity conditions is just the asymptotic variance of $\hat{\theta}$ divided by θ^2.

(iii) *Kurtosis.* As mentioned above, we shall define a measure of kurtosis as a suitable ratio of the form $\theta(F) = \tau_2(F)/\tau_1(F)$ where τ_1 and τ_2 are measures of scale and hence satisfy (3). If τ_1 and τ_2 are two such functionals, the only functions $h_1(\tau_1)$ and $h_2(\tau_2)$ with which they could be replaced are $h_i(\tau_i) = a_i \tau_i$; hence the only functions of θ with which we need to be concerned are those satisfying (6). The same argument as in (ii) shows that these choices become equivalent provided we use the standardized variance or asymptotic variance as our measure of accuracy.

At this point, we find ourselves in the following position. We have available a large class of measures of the characteristic in question. From this class we should like to choose a member which is robust and the estimator of which has good accuracy globally, i.e. for a large (nonparametric) family of distributions F. In the symmetric location case this turned out to be fairly easy. For estimating the center of symmetry of a symmetric distribution there exist a variety of estimators which are nearly as efficient as the mean for all F, and more efficient for many F. This is a piece of good luck, which one cannot expect in general. In the symmetric location case, a theoretically even more appealing possibility exists if one is willing to forego robustness. It was shown by Stein (1956), van Eeden (1970) and Takeuchi (1971) that in this case it is possible to find a fully efficient estimator, essentially by estimating the shape of F and adapting the estimator of the center to this estimated shape. The efficiency of this procedure depends heavily on the symmetry of F, and it seems unlikely that a similar approach would work in the cases considered here.

How should a measure be chosen if it cannot be simultaneously robust and estimated with high efficiency (either absolute or relative to a standard estimator) for all distributions satisfying the assumptions of the model? Two possibilities are the following.

(i) We can relax the condition of robustness. Instead of requiring the measure (and its estimator) to be totally robust, we may be satisfied if it is significantly more robust than the standard measure. (For details of such a comparative concept of robustness, see Section 4 of the following second part of this paper).

(ii) Alternatively, we can put additional restrictions on the model, which one could expect to be satisfied in practice, (for example, restrictions on the heaviness of the tails of the distributions). Such restrictions may make it easier for the estimators to be either robust or to have high efficiency for all distributions in the restricted model, or both.

REFERENCES

ANDREWS, D. F., BICKEL, P. J., HAMPEL, F. R., HUBER, P. J., ROGERS, W. H. and TUKEY, J. W. (1972). *Robust Estimation of Location.* Princeton Univ. Press.

BAHADUR, RAGHU RAJ and SAVAGE, LEONARD J. (1956). The nonexistence of certain statistical procedures in nonparametric problems. *Ann. Math. Statist.* **27** 1115–1122.

BARLOW, RICHARD E. and PROSCHAN, FRANK (1966). Tolerance and confidence limits for classes of distributions based on failure rates. *Ann. Math. Statist.* **37** 1593–1601.

BIRNBAUM, Z. W. (1943). On random variables with comparable peakedness. *Ann. Math. Statist.* **19** 76–81.

CHISSOM, BRAD S. (1970). Interpretation of the kurtosis statistic. *Amer. Statist.* **24** (4) 19–23.

DANIELL, P. J. (1920). Observations weighted according to order. *Amer. J. Math.* **42** 222–336.

DARLINGTON, R. B. (1970). Is kurtosis really peakedness? *Amer. Statist.* **24** (2) 19–20.

ESARY, J. D. and PROSCHAN, F. (1972). Relationships among some concepts of bivariate dependence. *Ann. Math. Statist.* **43** 651–655.

ESARY, J. D., PROSCHAN, F. and WALKUP, D. W. (1967). Association of random variables with applications. *Ann. Math. Statist.* **38** 1466–1474.

HAMPEL, F. R. (1968). Contributions to the theory of robust estimation. Ph. D. Dissertation, Univ. of California, Berkeley.

HAMPEL, F. R. (1971). A general qualitative definition of robustness. *Ann. Math. Statist.* **42** 1887–1896.

HØYLAND, ARNLJOT (1965). Robustness of the Hodges-Lehmann estimator for shift. *Ann. Math. Statist.* **36** 174–197.

HUBER, P. J. (1964). Robust estimation of a location parameter. *Ann. Math. Statist.* **35** 73–101.

HUBER, P. J. (1972). Robust statistics: A review. *Ann. Math. Statist.* **43** 1041–1067.

LEHMANN, E. L. (1955). Ordered families of distributions. *Ann. Math. Statist.* **26** 399–419.

LEHMANN, E. L. (1963). Robust estimation in analysis of variance. *Ann. Math. Statist.* **34** 957–966.

LEHMANN, E. L. (1966). Some concepts of dependence. *Ann. Math. Statist.* **37** 1137–1153.

MANN, HENRY B. and WHITNEY, D. R. (1947). On tests of whether one of two random variables is stochastically larger than the other. *Ann. Math. Statist.* **18** 50–60.

SPJØTVOLL, EMIL (1968). A note on robust estimation in analysis of variance. *Ann. Math. Statist.* **39** 1486–1492.

STEIN, CHARLES (1956). Efficient nonparametric testing and estimation. *Proc. Third Berkeley Symp. on Math. Stat. and Prob.* 187–195, Univ. of California Press.

TAKEUCHI, K. (1967). Robust estimation and robust parameter. (Unpublished).

TAKEUCHI, K. (1971). A uniformly asymptotically efficient estimator of a location parameter. *J. Amer. Statist. Assoc.* **66** 292–301.

TUKEY, J. W. (1960). A survery of sampling from contaminated distributions. *Contributions to Probability and Statistics*. Ed. I. Olkin, Stanford Univ. Press.

TUKEY, J. W. (1962). The future of data analysis. *Ann. Math. Statist.* **33** 1–67.

VAN EEDEN, CONSTANCE (1970). Efficiency—robust estimation of location. *Ann. Math. Statist.* **41** 172–181.

VAN ZWET, W. R. (1964). *Convex transformations of random variables.* Math. Centrum, Amsterdam.

DEPARTMENT OF STATISTICS
UNIVERSITY OF CALIFORNIA
BERKELEY, CALIFORNIA 94720

The Annals of Statistics
1975, Vol. 3, No. 5, 1045-1069

DESCRIPTIVE STATISTICS FOR NONPARAMETRIC MODELS II. LOCATION

By P. J. Bickel[1] and E. L. Lehmann[2]

University of California, Berkeley

Measures of location (without assumption of symmetry) are defined as functionals satisfying certain equivariance and order conditions. Three classes of such measures are discussed whose estimators are respectively linear functions of order statistics, R-estimators and M-estimators. It is argued that such measures can be compared in terms of the (asymptotic) efficiencies of their estimators. Of the three classes considered, it is found that trimmed expectations (and certain other weighted quantiles) are the only ones which are both robust and whose estimators have guaranteed high efficiency relative to the mean $\bar{X}$ for all underlying distributions.

1. Conditions for a location parameter. Since the publication of two fundamental papers by Tukey (1960 and 1962), much work has been done on the problem of robust estimation culminating in the Princeton report "Robust Estimation of Location" and Huber's Wald lecture (1972) "Robust Statistics: A review". Most of this work has been concerned with the estimation of location parameters, and in keeping with the program suggested by Tukey (1962, Section 13), the effort to date has been concentrated nearly entirely on the case in which the error distributions are assumed to be symmetric, or at least nearly so. One of the chief advantages of this restriction, as Tukey pointed out in recommending it, lies in the fact that it is then clear what is being estimated: for a symmetric distribution, the only natural location parameter is its center of symmetry.

It is interesting to note that this point was actually made as early as 1920 in a paper by Daniell, recently brought to the attention of present-day statisticians by Stephen Stigler. In this paper, in which Daniell develops a surprisingly modern and comprehensive theory of linear functions of order statistics for estimating location and scale parameters, the author writes: "...however, we shall only consider cases in which the theoretical distribution is symmetrical, and this for logical reasons. It is useless to compare the relative merits of the various kinds of average, for example, the mean and the median, unless they all tend to coincide when n increases indefinitely."

Received April 1973; revised September 1974.

[1] This research was partially supported by the Office of Naval Research, Contract NONR N00014-69-A-0200-1038.

[2] Work done while the author was Research Professor for the Adolph C. and Mary Sprague Miller Institute for Basic Research in Science, University of California, Berkeley.

AMS 1970 *subject classifications.* Primary 62G99; Secondary 62G05, 62G20, 62G35.

Key words and phrases. Measures of location, robustness, efficiency, linear combinations of order statistics, trimmed means, estimators derived from rank tests, Huber's maximum likelihood estimators, heavy-tailed distributions, Tukey model.

As was indicated in BL I[3] and will be discussed in more detail in Section 5, we in fact disagree with this last statement and believe that meaningful comparisons of this kind can, and should, be made. However, we must first lay the groundwork by defining the parameters we wish to estimate.

By a measure of location (or *location parameter*) we shall mean a functional $\mu(F)$ defined over a suitably large class of distributions (the precise definition of the class will depend on the context), which satisfies the three conditions below (these were already indicated briefly in BL I). In stating these conditions and elsewhere, we shall sometimes find it convenient to write $\mu(X)$ for $\mu(F)$ where X is a random variable with distribution F.

(i) If the parameter is to indicate location, it should take on larger values for random variables which typically are larger. Formally we shall require that $\mu(X) \leqq \mu(Y)$ whenever Y is stochastically larger than X.

(ii) More specific behaviour is required of μ under change of location or scale, namely

$$\mu(aX + b) = a\mu(X) + b \qquad \text{if} \quad a > 0\,. \tag{1.1}$$

(iii) Finally, it seems natural to require a measure of location to change sign under reflection w.r.t. the origin, i.e., to satisfy

$$\mu(-X) = -\mu(X)\,. \tag{1.2}$$

While condition (i) seems to us very intuitive, some referees have objected to it on the grounds that location is sometimes considered as referring only to the central part of a distribution. The following example shows that no reasonable version of (i) is possible if that position is adopted since truncation of two stochastically ordered distributions on a common point may reverse the ordering.

Let $G(t) = t$ for $0 < t < 1$ and let

$$\begin{aligned} F(t) &= t \qquad \text{for} \quad 0 < t < t_0\,(< 1) \\ &= t_0 + \frac{B - t_0}{A - t_0}(t - t_0) \qquad \text{for} \quad t_0 < t \leqq t_1 = t_0 + \frac{(A - t_0)(1 - t_0)}{B - t_0} \\ &= 1 \qquad \text{for} \quad t > t_1\,, \end{aligned}$$

where $t_0 < A < B < 1$. It is then easily checked that $G(t) < F(t)$ for $t_0 < t < 1$, but that $F^*(t) < G^*(t)$ for $0 < t < A$, where F^*, G^* denote the conditional distributions given that the random variable is $< A$.

Related to this remark is the fact that certain important classes of location measures that have been discussed in the literature do not satisfy condition (i). This is true in particular for the measures corresponding to Huber's M-estimators when scale is estimated simultaneously with location.

THEOREM 1. *Conditions* (i)—(iii) *imply the following four additional desirable requirements.*

[3] BL I refers to part I of the present paper.

1. *If F is symmetric* w.r.t. θ, *then* $\mu(F) = \theta$;
2. *In particular, if* $X = c$ *with probability* 1, $\mu(X) = c$.
3. *If* $a \leqq X \leqq b$ *with probability* 1, *then* $a \leqq \mu(X) \leqq b$;
4. *If X is stochastically positive, then* $\mu(X) \geqq 0$.

PROOF. 1. By (i), it is enough to prove 1. for $\theta = 0$. But if X and $-X$ have the same distribution, it follows from (iii) that $\mu(X) = -\mu(X) = 0$, as was to be proved.

3. If $a \leqq X \leqq b$ with probability 1, then X is stochastically larger than the constant variable a and smaller than b. The assertion now follows from (i) and 2.

4. By definition, X is stochastically positive if there exists a random variable U, symmetric about 0 and such that X is stochastically larger than U. By the first part of the theorem, $\mu(U) = 0$ and the result now follows from (i).

The following three examples show that conditions (i)—(iii) are independent.

(a) The functional $\mu(F) = F^{-1}(a)$ with $0 < a < 1$, $a \neq \frac{1}{2}$, satisfies (i) and (ii) but not (iii);

(b) the functional $\mu(F) = 2E(X) - F^{-1}(\frac{1}{2})$ satisfies (ii) and (iii) but not (i);

(c) finally, $\mu(X) = [E(X)]^3$ satisfies (i) and (iii) but not (ii).

Let us now mention a few standard measures of location, which satisfy all three conditions:

(a) The expectation of F

$$\mu_1(F) = \int x\, dF(x)\,; \tag{1.3}$$

(b) the median of F

$$\mu_2(F) = F^{-1}(\tfrac{1}{2}) \tag{1.4}$$

and the average of the 1st and 3rd quartile

$$\mu_3(F) = \tfrac{1}{2}[F^{-1}(\tfrac{1}{4}) + F^{-1}(\tfrac{3}{4})]\,. \tag{1.5}$$

An important class of location measures are the symmetrically trimmed expectations

$$E_\alpha(X) = \frac{1}{1-2\alpha} \int_{u_\alpha}^{u_{1-\alpha}} x\, dF(x) \tag{1.6}$$

where $F(u_\alpha) = \alpha$, $F(u_{1-\alpha}) = 1 - \alpha$. The mean and median are the limiting cases corresponding respectively to $\alpha = 0$ and $\alpha = \frac{1}{2}$.

(c) As a last example, we mention the pseudo-median $\mu_4(F)$ defined by Høyland (1965) as the median of the distribution of $\frac{1}{2}(X_1 + X_2)$ where X_1, X_2 are independently distributed according to F. If F is continuous, $\mu_4(F)$ is the solution of the equation

$$\int F(2\theta - x)\, dF(x) = \tfrac{1}{2}\,. \tag{1.7}$$

In the next section, we shall consider two classes of location measures, which contain the above examples as special cases.

Two additional properties, which seem attractive although not as compelling as (i)—(iii) are

(iv) If $\mu(F) \leqq \mu(G)$ and $H = \alpha F + (1 - \alpha)G$, then $\mu(F) \leqq \mu(H) \leqq \mu(G)$.
(v) If $\mu(X) \leqq \mu(Y)$ and h is a nondecreasing function, then $\mu[h(X)] \leqq \mu[h(Y)]$.

If $G(x) \leqq F(x)$ for all x, then both (iv) and (v) are consequences of (i). However, in general both of these conditions are very restrictive. Of the examples (1.3) to (1.7) it is not difficult to see that μ_2 satisfies both (iv) and (v), that μ_1 satisfies (iv) but not (v) and that none of the other measures satisfy either (iv) or (v). For most location parameters neither (iv) or (v) will hold; for instance, the median is the only location parameter of the type defined by (3.2) that satisfies (v). We shall not consider these conditions any further here.

There is one last condition, with which we shall be concerned. We shall be interested in measures which are robust in the sense that small changes in F will not result in large changes in μ. Mathematically, this means that μ should be continuous with respect to a suitable metric such as that of Lévy, and then a fortiori also that of Kolmogorov. For a detailed discussion of this and related concepts, see Hampel (1968, 1971); the idea is fore-shadowed in Bahadur and Savage (1956). It is well-known and not difficult to see, (and will follow from the results of Section 4) that of the examples above, $\mu_1(F)$ is not robust, but $\mu_2(F)$ to $\mu_4(F)$ are under mild restrictions on F, and so is $F_\alpha(X)$ for any $0 < \alpha < \frac{1}{2}$.

2. Some examples. Let us now examine somewhat more closely the location parameters which were given as examples in the preceding section. Consider first the three parameters μ_1, μ_2 and μ_4, i.e. the expectation, median and pseudo median. What can be said about their relative positions? Clearly, $\mu_2(F)$ can be either larger or smaller than $\mu_4(F)$ since $\mu_2(F) < \mu_4(F)$ implies that $\mu_4(-F) < \mu_2(-F)$. Furthermore, regardless of their relative positions, $\mu_1(F)$ can be to the left of both, to the right of both, or between them. This follows from the fact that the position of $\mu_1(F)$ can be moved from $-\infty$ to $+\infty$ by moving an arbitrarily small probability mass, which would result in only very small changes of $\mu_2(F)$ and $\mu_4(F)$.

Nevertheless, $\mu_4(F)$ can be bounded between certain quantiles. In fact, if F is continuous

$$F^{-1}\left(1 - \frac{1}{2^{\frac{1}{2}}}\right) \leqq \mu_4(F) \leqq F^{-1}\left(\frac{1}{2^{\frac{1}{2}}}\right). \tag{2.1}$$

To see this let $\mu_4(F) = \theta$ and let $p = P(X < \theta)$, $q = P(X > \theta)$. Then

$$P\left[\frac{X + X'}{2} < \theta\right] \geqq p^2, \qquad P\left(\frac{X + X'}{2} > \theta\right) \geqq q^2$$

and the result follows from the definition of $\mu_4(F)$.

To get an idea of the values of the parameters μ_1, μ_2 and μ_4 in particular situations, consider as a first example the case that F is a χ^2-distribution with ν degrees

TABLE 2.1

Location parameters for χ_ν^2

$\mu_1(F) = \nu$	1	2	3	4	5
$\mu_4(F)$	.69	1.68	2.67	3.67	4.67
$\mu_2(F)$	.45	1.39	2.37	3.36	4.35

TABLE 2.2

Differences of location parameters for χ_ν^2

ν	1	2	3	4	5	∞
$\mu_1(F) - \mu_4(F)$	.31	.32	.33	.33	.33	$\frac{1}{3}$
$\mu_2(F) - \mu_4(F)$	.24	.29	.30	.31	.32	$\frac{1}{3}$

of freedom. Then the table above gives the values of $\mu_1(F) = \nu$, $\mu_4(F)$ and $\mu_2(F)$ for $\nu = 1, 2, 3, 4$ and 5. In spite of the rather different shapes of the distributions (which however are all skewed to the right), the differences $\mu_1(F) - \mu_4(F)$ and $\mu_4(F) - \mu_2(F)$ are remarkably constant, as is shown in Table 2.2 above. The limiting value as $\nu \to \infty$, shown in the last column comes from the well-known Cornish–Fisher expansion for the solution of the equation $P(\chi_\nu^2 \leq x^2_{\nu,\varepsilon}) = \varepsilon$:

$$x^2_{\nu,\varepsilon} = \nu\left[1 + 3w_\varepsilon\left(\frac{2}{9\nu}\right)^{\frac{1}{2}} - \frac{2}{3\nu} + \frac{2w_\varepsilon^2}{3\nu} + o\left(\frac{1}{\nu}\right)\right]$$

where $w_\varepsilon = \Phi^{-1}(\varepsilon)$ with Φ denoting the standard normal distribution. Putting $\varepsilon = \frac{1}{2}$, we see that

$$\mu_2(F) = x^2_{\nu,\frac{1}{2}} = \nu - \tfrac{2}{3} + o(1)\,.$$

Now $\mu_4(F)$ is the median of $(X + X')/2$ where X and X' are independently distributed as $\chi^2_{2\nu}$. The median of $X + X'$ thus satisfies

$$x^2_{2\nu,\frac{1}{2}} = 2\nu - \tfrac{2}{3} + o(1)$$

and

$$\mu_4(F) = \nu - \tfrac{1}{3} + o(1)\,.$$

Comparison with $\mu_2(F)$ and $\mu_1(F) = \nu$ gives the desired result.

It is interesting to note that in the above examples and those computed below, μ_4 is always between μ_1 and μ_2. This is not too surprising (although all orders are possible) since $\mu_1(X) = \mu_1[\frac{1}{2}(X_1 + X_2)]$ and $\mu_4(X) = \mu_2[\frac{1}{2}(X_1 + X_2)]$. By the central limit theorem one might expect $\frac{1}{2}(X_1 + X_2)$ to be closer to symmetry than is X, and μ_2 to be closer to μ_1 for $\frac{1}{2}(X_1 + X_2)$ than for X.

As a second example, consider the family of distributions

$$F_k = \gamma N(1, 1) + (1 - \gamma)N(k, k^2) \tag{2.2}$$

where $N(a, b^2)$ denotes a normal distribution with mean a and standard deviation b. The following table shows the values of μ_1, μ_2 and μ_4 for $\gamma = \frac{3}{4}$ and $k =$ 2, 3, 5 and 10. The last column gives the limiting value as $k \to \infty$. To see how

TABLE 2.3
Location parameter for normal mixtures

k	2	3	5	10	∞
μ_1	1.25	1.50	2	3.25	∞
μ_2	1.14	1.19	1.26	1.27	1.29
μ_4	1.18	1.27	1.35	1.45	1.53

this is obtained for μ_2 and μ_4, consider for example μ_2 and let u_k be the solution of the equation

$$\gamma\Phi(u-1) + (1-\gamma)\Phi\left(\frac{u-k}{k}\right) = \tfrac{1}{2} \tag{2.3}$$

with $\gamma = \frac{3}{4}$. As $u \to \infty$, the first term on the left hand side tends to $\frac{3}{4}$ which is too large. Thus u_k must remain bounded as $k \to \infty$. This means that the second term tends to $\frac{1}{4}\Phi(-1) = .0397$ and u_∞ is the solution of the equation

$$\tfrac{3}{4}\Phi(u-1) = .4603\,.$$

The argument for μ_4 is quite analogous.

Consider next the family (2.2) for a value of γ less than $\frac{1}{2}$. This time, as $k \to \infty$ any finite value of u is too small so that $u_k \to \infty$. The first term then tends to γ, and u_k must satisfy the equation

$$\Phi\left(\frac{u_k}{k} - 1\right) \to \frac{\frac{1}{2}-\gamma}{1-\gamma}\,.$$

Proceeding in the same way for $\mu_4(F)$, we find that as $k \to \infty$, we have for example for $\gamma = \frac{1}{4}$

$$\frac{\mu_1(F_k)}{k} \to .75\,, \qquad \frac{\mu_2(F_k)}{k} \to .069\,, \qquad \frac{\mu_4(F_k)}{k} \to .705\,.$$

As a last example, consider the Gumbel distribution with density

$$f(x) = e^{-x}e^{-e^{-x}}\,, \qquad -\infty < x < \infty\,.$$

Here the values of μ_1, μ_4 and μ_2 are respectively

$$\mu_1 = .577\,, \qquad \mu_4 = .464\,, \qquad \mu_2 = .366\,.$$

Of those, μ_1 and μ_2 can be found for example in Johnson and Kotz (1970). To calculate μ_4 we note that,

$$\begin{aligned} P\left[\frac{X_1+X_2}{2} \leq y\right] &= \int_{-\infty}^{\infty} e^{-e^{-(2y-x)}-x-e^{-x}}\,dx \\ &= \int_0^\infty \exp -\left\{\frac{e^{-2y}}{w} + w\right\} dw \\ &= e^{-2y}\int_0^\infty e^{-e^{-2y}z}e^{-1/z}\,dz \\ &= 2e^{-y}K_1(2e^{-y}) \end{aligned}$$

TABLE 2.4
Trimmed expectations for exponential distributions

α	0	.01	.07	.03	.05	.1	.5
$E_\alpha(X)$	1	.963	.939	.919	.888	.831	.693

where K_1 is the Bessel function of the second kind with imaginary argument (cf. Churchill (1958)). The value of μ_4 is now obtainable from the British Association Tables **6** (1958).

So far, we have compared the parameters μ_1, μ_2 and μ_4 for a number of different distributions. Let us next illustrate the trimmed expectations $E_\alpha(X)$ defined by (1.6). As a first illustration, let F be the exponential distribution with density e^{-x}, $x > 0$. Then $E_\alpha(X)$ is shown above for a number of values of α. The values corresponding to $\alpha = 0$ and $\alpha = .5$ are of course the expectation and median respectively. The pseudo median in this case is .84.

As a second illustration, consider the mixture $F(x) = \frac{3}{4}\Phi(x) + \frac{1}{4}\Phi(x - 1)$. Some values of $E_\alpha(X)$ in this case are

TABLE 2.5
Trimmed expectations for a normal mixture

α	0	.05	.1	.15	.5
$E_\alpha(X)$	.25	.248	.239	.238	.23

The pseudo median is .243.

3. Three classes of location parameters. Among the estimators of the center of symmetry of a symmetric distribution F, three classes have been found to be of particular interest (see for example Jaeckel (1971), Andrews et al. (1972), and Huber (1972)): linear combinations of order statistics, estimators derived from rank tests, and the maximum likelihood type estimators introduced by Huber. We shall now consider the quantities estimated by these estimators when F is no longer assumed to be symmetric, under restrictions which assure that they are measures of location in the sense of Section 1.

(A) If $\mu(X)$ satisfies conditions (i) and (ii) of Section 1 but not necessarily condition (iii), then it is easily seen that

$$\frac{1}{2}[\mu(X) - \mu(-X)] \tag{3.1}$$

satisfies all three conditions, i.e. is a location parameter. Furthermore, if $\{\mu_i(X)\}$ is a countable collection of functionals which satisfy (i)—(iii) then so does $\sum \alpha_i \mu_i(X)$ for any nonnegative α's which add up to 1. This latter remark clearly extends also to noncountable collections.

As an application of these remarks consider the functional $\mu(X) = F^{-1}(a)$ for some fixed a between 0 and 1. This clearly satisfies conditions (i) and (ii). Since $\mu(-X) = -F^{-1}(1 - a)$, it follows that $\frac{1}{2}[F^{-1}(a) + F^{-1}(1 - a)]$ is a location parameter for any $0 < a < 1$. Forming convex combinations of these parameters, we then see that the weighted quantile average

$$\mu(F) = \int_0^1 F^{-1}(t)\, dK(t) \tag{3.2}$$

is a location parameter, where K is any distribution function on (0, 1) which is symmetric with respect to $\frac{1}{2}$.

Important special cases of (3.2) are the symmetrically trimmed expectations (1.6), as is seen by rewriting (1.6) as

$$E_\alpha(X) = \frac{1}{1-2\alpha} \int_\alpha^{1-\alpha} F^{-1}(t)\, dt\,, \tag{3.3}$$

and their limiting cases, the expectation and the median.

(B) To obtain the second class of location parameters, consider the probability

$$P(|X-\theta| \leqq x) = F(x+\theta) - F(-x+\theta)\,. \tag{3.4}$$

It seems natural to define as center of F the number θ for which the probabilities (3.4) are, in some average sense, as large as possible. Let L be an increasing convex function on [0, 1] which is bounded and such that $L(0) = 0$. Define $\mu(F)$ as the number θ that maximizes

$$\int_0^\infty \{L[F(x+\theta) - F(-x+\theta)] - L[F(x) - F(-x)]\}\, dx\,. \tag{3.5}$$

Here the subtraction of $L[F(x) - F(-x)]$ under the integral sign is intended to aid convergence. It may be shown that if L' is bounded in absolute value, the functional (3.5) is always defined. (See Knüsel (1969) for a closely related remark.)

It is sometimes more convenient to extend L so that it is an even function on $[-1, 1]$, and it is then equivalent to maximize

$$\begin{aligned}\int_{-\infty}^\infty &\{L[F(x+\theta) - F(-x+\theta)] - L[F(x) - F(-x)]\}\, dx \\ &= \int_{-\infty}^\infty \{L[F(x) - F(2\theta - x)] - L[F(x) - F(-x)]\}\, dx\,.\end{aligned} \tag{3.6}$$

Suppose that F has a density f, L is continuously differentiable, that $M = L'$, and that (3.6) can be differentiated under the integral sign. It can then be shown exactly as was done by Knüsel (1969) for a related but somewhat different class of functionals that the equation

$$\int_{-\infty}^\infty L'[F(x) - F(2\theta - x)] f(2\theta - x)\, dx = 0\,,$$

or, equivalently

$$\int_{-\infty}^\infty L'[F(x) - F(2\theta - x)] f(x)\, dx = 0 \tag{3.7}$$

has at least one solution. If L' is strictly increasing, this solution is unique and hence is the desired maximizing value. (These measures are the quantities estimated by estimators derived from rank tests; see for example Andrews et al. (1972)).

In particular if $L(y) = y^2$, it is seen that (3.7) reduces to

$$\int F(2\theta - x) f(x)\, dx = \tfrac{1}{2} \tag{3.8}$$

so that θ is the pseudo median $\mu_4(F)$ of F. As another example suppose that in (3.5) we take $L(y) = y$. Then differentiation shows θ to be the median $\mu_2(F)$.

To see that the measure $\mu(F)$ uniquely defined through (3.7) is a location parameter for F continuous, rewrite (3.7) as

$$\int L'[t - F(2\theta - F^{-1}(t))]\, dt = 0 \,. \tag{3.9}$$

To check condition (i), suppose that $G(x) \leqq F(x)$ for all x. Then for any fixed value of θ, we have $F[2\theta - F^{-1}(t)] \geqq G[2\theta - G^{-1}(t)]$, and since L' is an increasing function, it follows that

$$\int L'[t - F(2\theta - F^{-1}(t))]\, dt \leqq \int L'[t - G(2\theta - G^{-1}(t))]\, dt \,.$$

In this last inequality, let $\theta = \mu(G)$ so that the right hand side is zero. Since the left hand side is a decreasing function of θ, the value $\mu(F)$ of θ which will make the left hand side equal to zero will be less than or equal to $\mu(G)$, as was to be proved. That conditions (ii) and (iii) are also satisfied, is easily seen by replacing $F(x)$ by $F((x - b)/a)$ in (3.7).

(C) The third class is obtained from the quantities estimated by Huber's M-estimators. These are the quantities $\theta = \mu(F)$ which minimize

$$\int \rho(x - \theta)\, dF(x) \tag{3.10}$$

where we shall assume ρ to be positive, even, convex and twice differentiable with derivative $\rho' = \psi$. (It is possible to enlarge the domain of definition of $\theta(F)$ by minimizing $\int_{-\infty}^{\infty} [\rho(x - \theta) - \rho(x)]\, dF(x)$ instead, as was done by Knüsel (1969).) Essentially we shall now show that only a one-parameter subclass of (C) satisfies the conditions of Section 1.

THEOREM 2. *Suppose that $\mu(F)$ is defined as minimizing* (3.10) *on a set $\mathscr{F}$ which is convex, contains all point masses, is closed under changes of scale, and contains a distribution F° symmetric about* 0 *such that*

$$\int \psi'(x)\, dF^\circ(x + t) < \infty \quad \text{for all } t \quad \text{and}$$

$$\frac{d}{dt} \int \psi(x - t)\, dF^\circ(x) = -\int \psi'(x - t)\, dF^\circ(x) \,.$$

For any given distribution F, denote by F_σ the distribution defined by $F_\sigma(x) = F(x/\sigma)$. Suppose that

$$\mu(F_\sigma) = \sigma \mu(F) \qquad \text{for all } F \in \mathscr{F}, \quad \sigma > 0 \,. \tag{3.11}$$

(This is part of the assumption (1.1)). *Then*

$$\psi(x) = c|x|^\alpha \operatorname{sgn} x \qquad \text{for some } \alpha > 0 \,, \quad c > 0 \,. \tag{3.12}$$

PROOF. The measure $\mu(F)$ is the solution of

$$\int \psi(x - \theta)\, dF(x) = 0 \tag{3.13}$$

Using our assumptions, we calculate

$$\lim_{\varepsilon \to 0} \frac{\mu[(1 - \varepsilon)F^\circ + \varepsilon\delta_x] - \mu(F^\circ)}{\varepsilon} = \frac{\psi(x)}{\int_{-\infty}^{\infty} \psi'(x)\, dF^\circ(x)}$$

and similarly

$$\lim_{\varepsilon\to 0} \frac{\mu([(1-\varepsilon)F^\circ + \varepsilon\delta_x]_\sigma) - \mu(F^\circ)}{\varepsilon} = \frac{\psi(\sigma x)}{\int_{-\infty}^{\infty} \psi'(\sigma y)\, dF^\circ(y)} .$$

From (3.11)

$$\frac{\psi(\sigma x)}{\int_{-\infty}^{\infty} \psi'(\sigma y)\, dF^\circ(y)} = \frac{\sigma\psi(x)}{\int_{-\infty}^{\infty} \psi'(y)\, dF^\circ(y)} .$$

Differentiating both sides with respect to x for $x > 0$ leads to

$$\frac{\psi'(\sigma x)}{\int \psi'(\sigma y)\, dF^\circ(y)} = \frac{\psi'(x)}{\int \psi'(y)\, dF^\circ(y)}$$

and hence

$$\frac{\psi'(\sigma x)}{\psi'(x)} = \frac{\psi(\sigma x)}{\sigma\psi(x)} \quad \text{for} \quad x > 0 .$$

Put $x = 1$ to get

$$\frac{\psi'(\sigma)}{\psi(\sigma)} = \frac{\psi'(1)}{\psi(1)} \frac{1}{\sigma} \qquad \text{or} \qquad \log \psi = \frac{\psi'(1)}{\psi(1)} \log \sigma + c , \quad \sigma > 0$$

and this completes the proof.

REMARKS.

(1) The conclusion of this theorem was stated by Takeuchi (1967) without proof or regularity conditions.

(2) One of the referees has pointed out that the theorem also holds under a weaker set of conditions not including differentiability.

4. Robustness. Following Hampel (1968, 1971) and earlier writers, we shall define a functional μ to be robust at a point F_0 if μ is continuous at F_0 with respect to Lévy (or equivalently Prohorov) distance. However, this concept does not completely correspond to our intuitive notion of robustness since even the median is not continuous at all F. An easy example is provided by distributions for which $F(x) = \frac{1}{2}$ on a nondegenerate interval. On the other hand, the median is robust at every point of the set of all distributions having a unique median.

(A) Consider now the functionals μ defined by (3.2) and suppose that K is continuous. Then a necessary and sufficient condition for μ to be continuous at all points of $\mathscr{F}$ is that K assigns probability zero to the intervals $(0, \alpha)$ and $(1 - \alpha, 1)$ for some $\alpha > 0$.

The necessity of this condition (which does not require the continuity of K) was pointed out by Huber (1972). Sufficiency follows from the fact that convergence in law of F_n implies bounded convergence, except on a countable set, for F_n^{-1} on the intervals $(\alpha, 1 - \alpha)$.

(B) For measures μ defined through (3.7), suppose that L is bounded and L' is nondecreasing. Huber (1972) has shown that μ is continuous at any distribution F for which (3.6) can be differentiated under the integral sign and such that the resulting equation (3.7) has a unique solution. An alternative argument employs the following representation which may have some independent interest.

Define a distribution Q on the unit square by the cdf.

$$Q(u, v) = \frac{1}{2L(1)} \int_0^u [L'(v - t) - L'(-t)]\, dt \,. \tag{4.1}$$

Then $\mu(F)$ is the median of the distribution of $\frac{1}{2}[F^{-1}(U) + F^{-1}(V)]$ where (U, V) has distribution Q. Since Q is continuous, convergence in law of the cdf F_n to F implies convergence in law of $\frac{1}{2}[F_n^{-1}(U) + F_n^{-1}(V)]$ to $\frac{1}{2}[F^{-1}(U) + F^{-1}(V)]$. It follows that μ is continuous at any F for which it is uniquely defined.

(C) Of the measures defined by (3.12) and (3.13), it is easy to see that none are robust. However, it is clear that in a suitable sense if $\alpha < \beta$, μ_α is more robust than μ_β.

It seems reasonable that if ν is to be considered more robust than μ, then for any sequence F_n tending in law to F for which $\mu(F_n) \to \mu(F)$, we should also have $\nu(F_n) \to \nu(F)$. Unfortunately this definition seems to lead to noncomparability for the nonrobust functionals we have considered. All we need do is to consider sequences F_n converging in law to F such that $\mu(F_n) \to \mu(F)$ "fortuitously". For example let F_n assign mass $1 - [1/n]^{1/\alpha} - [1/n]^{1+1/\alpha}$ to 0 and mass $[1/n]^{1/\alpha}$ to n and $[1/n]^{1+1/\alpha}$ to $-n^2$. Then F_n tends to point mass at 0 and the expectations under F_n converge to that under F. However, $\mu_\alpha(F_n)$ converges to 1. To remedy this difficulty we formulate the following notion of comparative robustness, which gives us some needed control over the tails of the F_n. For any cdf F and any constant $M > 0$ let us define,

$$\begin{aligned}
\bar{F}(x, M) &= 0\,, & x &< 0\,, \\
&= F(M)\,, & 0 \leqq x &\leqq M \\
&= F(x)\,, & x &\geqq M \\
\underline{F}(x, M) &= F(x)\,, & x &\leqq -M\,, \\
&= F(-M)\,, & -M \leqq x &< 0\,, \\
&= 1\,, & x &\geqq 0\,.
\end{aligned}$$

Given a functional T we define a mode of convergence and hence a topology on its set of definition $\mathscr{F}$ (which as usual is assumed to be large enough) as follows.

$$F_n \to_T F \Leftrightarrow$$

(i) $F_n \to F$ in law;

(ii) $T(F_n) \to T(F)$;

(iii) $\lim_{M \uparrow \infty} \limsup_n \{T(\bar{F}_n(\cdot, M)) + |T(\underline{F}_n(\cdot, M))|\} = 0$.

(Note that by our assumptions if T is a measure of location, $T(\bar{F}) \geqq 0$, $T(\underline{F}) \leqq 0$).

DEFINITION. We shall call T_2 more robust than T_1 if and only if T_2 is continuous with respect to the topology induced by T_1.

This permits us to make comparisons between different measures of type (A) and (C) which are not robust in the absolute sense considered previously. For convenience in what follows we suppose all functionals to be defined on the space of distributions with compact support.

THEOREM 3. *Let T_1 and T_2 be two (L) measures defined by* (3.2), *say,*

$$T_i(F) = \int_0^1 F^{-1}(t)\, dK_i(t) \quad i = 1, 2\,, \tag{4.2}$$

and suppose that K_1, K_2 are continuous probability measures. Then T_2 is more robust than T_1 if there exists $0 < \varepsilon \leqq \frac{1}{2}$ such that

$$\begin{aligned} K_1(t) &\geqq K_2(t) \quad \text{for all} \quad 0 \leqq t \leqq \varepsilon\,, \qquad \text{and} \\ K_1(t) &\leqq K_2(t) \quad \text{for all} \quad 1-\varepsilon \leqq t \leqq 1\,. \end{aligned} \tag{4.3}$$

PROOF. Note that

$$\begin{aligned} T_i[\bar{F}(\cdot, M)] &= \int_{F(M)}^1 F^{-1}(t)\, dK_i(t)\,, \qquad \text{and} \\ T_i[\underline{F}(\cdot, M)] &= \int_0^{F^{-1}(M)} F^{-1}(t)\, dK_i(t)\,. \end{aligned} \tag{4.4}$$

Therefore, for functionals defined by (3.2), (with T in place of μ), if (i) holds it follows that (ii) and (iii) are equivalent to

$$\lim_{\varepsilon\to\infty} \limsup_n \{|\textstyle\int_\varepsilon^1 F_n^{-1}(t)\, dK(t)| + |\int_0^\varepsilon F_n^{-1}(t)\, dK(t)|\} = 0\,. \tag{4.5}$$

Now suppose that $F_n \to_{T_1} F$. To show that $T_2(F_n) \to T_2(F)$ note that

$$\begin{aligned} |T_2(F_n) - T_2(F)| \leqq{} & [\textstyle\int_\varepsilon^{1-\varepsilon} |F_n^{-1}(t) - F^{-1}(t)|\, dK_2(t)] \\ & + |\textstyle\int_0^\varepsilon F_n^{-1}(t)\, dK_2(t)| + |\int_0^\varepsilon F^{-1}(t)\, dK_2(t)| \\ & + |\textstyle\int_\varepsilon^1 F_n^{-1}(t)\, dK_2(t)| + |\int_\varepsilon^1 F^{-1}(t)\, dK_2(t)| \end{aligned} \tag{4.6}$$

and that the first term on the right hand side of (4.6) converges to 0 for each fixed ε by dominated convergence.

Therefore, we need only establish that

$$\lim_{\varepsilon\to 0} \limsup_n \{|\textstyle\int_\varepsilon^1 F_n^{-1}(t)\, dK_2(t)| + |\int_0^\varepsilon F_n^{-1}(t)\, dK_2(t)|\} = 0\,. \tag{4.7}$$

To prove (4.7) begin by supposing that $0 < F(0-) \leqq F(0) < 1$. Then, for ε sufficiently small, $F_n^{-1}(t) \geqq 0$ for all $t \geqq 1-\varepsilon$ and $\leqq 0$ for all $t \leqq \varepsilon$ and all n. Hence,

$$\begin{aligned} |\textstyle\int_0^\varepsilon F_n^{-1}(t)\, dK_2(t)| &= -\textstyle\int_0^\varepsilon F_n^{-1}(t)\, dK_2(t) \leqq -\int_0^\varepsilon F_n^{-1}(t)\, dK_1(t) \\ &= |\textstyle\int_0^\varepsilon F_n^{-1}(t)\, dK_1(t)| \end{aligned}$$

by our hypothesis on the K_i. Similar inequalities hold for $\int_{1-\varepsilon}^1$ and (4.7) follows from the hypothesis $F_n \to_{T_1} F$ in view of (4.5).

The two cases when $F(0) = 1$ or $F(0-) = 0$ are argued similarly. □

Consider next the estimators of type (C) satisfying (3.12).

THEOREM 4. *Suppose that $0 < \alpha_1 < \alpha_2$ and T_α corresponds to ψ_α. Then T_{α_1} is more robust than T_{α_2}.*

This result is an easy consequence of the following lemma.

LEMMA. *If* $\alpha > 0$, $F_n \to_{T_\alpha} F \Leftrightarrow$

(4.8 a) $$F_n \to F \quad \textit{in law}$$

and

(4.8 b) $$\lim_{M\uparrow\infty} \limsup_n \int_{[|x|\geq M]} |x|^\alpha \, dF_n(x) = 0 .$$

PROOF. Suppose that $F_n \to F$ in law and that (4.8 a) and (4.8 b) hold. Then

$$\int \psi_\alpha(x-\theta)\, d\bar{F}_n(x, M) \to \int \psi_\alpha(x-\theta)\, d\bar{F}(x, M)$$

for all θ, M and hence

(4.9) $$T_\alpha(\bar{F}_n(\cdot, M)) \to T_\alpha(\bar{F}(\cdot, M))$$

Convergence similarly holds for $T_\alpha(\underline{F}_n(\cdot, M))$ and $T_\alpha(F_n)$. Since

$$\lim_{M\uparrow\infty} \{|T_\alpha(\bar{F}(\cdot, M))| + |T_\alpha(\underline{F}(\cdot, M))|\} = 0$$

and $T_\alpha(\bar{F}_n(\cdot, M))$ is decreasing in M for fixed n it can easily be argued that (4.9) implies $\lim_M \limsup_n |T_\alpha(\bar{F}_n(\cdot, M))| = 0$. Similarly,

$$\lim_M \limsup_n |T_\alpha(\underline{F}_n(\cdot, M))| = 0$$

and the "only if" part of the lemma follows. Conversely, suppose that $F_n \to_{T_\alpha} F$. Then, for M sufficiently large and all n, by the definition of T_α, for any $\varepsilon > 0$, the solutions $T_\alpha(\bar{F}_n(\cdot, M))$ and $T_\alpha(\underline{F}_n(\cdot, M))$, of

$$\int_M^\infty (x-\theta)^\alpha \, dF_n(x) = \theta^\alpha F(M)$$

and

$$\int_{-\infty}^{-M} (\theta - x)^\alpha \, dF_n(x) = |\theta|^\alpha (1 - F(-M))$$

lie in the interval $(-\varepsilon, \varepsilon)$.
Then

$$\int_M^\infty |x|^\alpha \, dF_n(x) \leqq 2^\alpha \varepsilon^\alpha$$

and

$$\int_{-\infty}^{-M} |x|^\alpha \, dF_n(x) \leqq 2^\alpha \varepsilon^\alpha .$$

The lemma follows.

5. Estimation. The measures defined by (3.2) and (3.7) all satisfy conditions (i)—(iii) of Section 1, and are robust under mild conditions. There is therefore little to choose among them on this basis. We have, however, so far neglected an important aspect of these measures: how to estimate them.

The situation here is rather different from that in the symmetric location case. There are many reasonable estimators of the center of symmetry of a symmetric distribution, and much work has been devoted to comparing the accuracies of different such estimators. Without the assumption of symmetry, there is typically only one natural estimator of a location parameter $\mu(F)$, namely μ evaluated at the empirical distribution function.

The problem of choosing among different estimators of the same location parameter therefore does not arise. Instead, we are faced with the problem of comparing location parameters in terms of the ease with which they can be estimated. It may at first seem strange to compare estimators of different parameters. However, such comparisons appear justified, philosophically because primarily we are not comparing the estimators but just these parameters, and mathematically because a shift in origin leaves the problem of estimating location essentially unchanged. This latter point becomes clearer through a comparison with the scale problem. Here the corresponding fact is not true since the formal expression for the accuracy with which a scale parameter can be estimated depends on the unit of measurement. (See also the related discussion in BL I.)

NOTE. We stated above that typically there is only one natural estimator of a location parameter $\mu(F)$, namely $\mu(\hat{F})$ where $\hat{F}$ is the empirical cdf. However, unless additional restrictions are imposed on the estimators, the phenomenon of superefficiency can occur. Consider, for example, the problem of estimating the median of a completely unknown continuous strictly increasing distribution function. If T_n is an estimator based on n observations from F, one might conjecture that, asymptotically, the sample median is uniformly best among all estimators which are consistent estimators of the population median, and which satisfy the equivariance condition

$$T_n(aX_1 + b, \cdots, aX_n + b) = aT_n(X_1, \cdots, X_n) + b .$$

Consider however the estimator T_n^* defined by

$$\begin{aligned} T_n^*(X_1, \cdots, X_n) &= \hat{\mu}_1 \qquad \text{if} \quad \inf_{\mu,\sigma} \sup_x \left|\hat{F}_n(x) - \Phi\left(\frac{x-\mu}{\sigma}\right)\right| \leqq n^{-\frac{1}{4}} \\ &= \hat{\mu}_2 \qquad \text{otherwise} \end{aligned}$$

where $\hat{\mu}_1$ and $\hat{\mu}_2$ denote the sample mean and sample median respectively. Then it is easy to see that T_n^* is equivariant, consistent for μ_2 and superefficient when F is normal. In view of the results of Le Cam, Huber, Hájek (see Hájek (1970)), one might conjecture that such phenomena can be excluded by requiring continuity (in a suitable sense) of the asymptotic variance of the estimator.

We shall now consider from the above point of view the three classes of estimators obtained from the classes (A), (B) and (C) of location measures described in Section 3 by replacing the true distribution function F in $\mu(F)$ by the empirical distribution function $\hat{F}_n$.

5 A. Linear functions of order statistics. The estimators corresponding to the measures (3.2) are linear functions of order statistics (see for example Andrews et al. (1972)). Let us compare the estimators of two such measures

$$\mu_i(F) = \textstyle\int_0^1 F^{-1}(t)\, dK_i(t) \qquad i = 1, 2 \tag{5.1}$$

where the K_i are two distributions on (0, 1) with densities $K_i'(t)$.

THEOREM 5. *Suppose that the densities K_i' satisfy*

$$0 \leqq K_2'(t)/K_1'(t) \leqq A \qquad \textit{where} \quad 1 < A . \tag{5.2}$$

Then if T_i is the estimator of $\mu_i(F)$ obtained by replacing F by the empirical cdf, *the asymptotic variances of T_i satisfy*

$$(5.3)\qquad \sigma_2^2(F) \leqq A^2\sigma_1^2(F) \qquad \textit{for all} \quad F\,,$$

under the regularity conditions permitting the representation (5.5) *below. Under these conditions $1/A^2$ is then a lower bound for the efficiency of T_2 relative to T_1,* i.e.

$$(5.4)\qquad e_{T_2,T_1}(F) \geqq \frac{1}{A^2} \qquad \textit{for all} \quad F\,,$$

and the lower bound is sharp in the sense that if A is the sharp bound on K_2'/K_1' there exists a sequence F_n for which the efficiency tends to $1/A^2$.

PROOF. The asymptotic variance of $n^{\frac{1}{2}}T_i$ (cf. Huber (1972)) is

$$(5.5)\qquad \sigma_i^2(F) = \operatorname{Var}[U_i(T)]$$

where T is uniform over (0, 1) and

$$(5.6)\qquad U_i(t) = \int_{\frac{1}{2}}^{t} \frac{K_i'(t)}{f[F^{-1}(t)]}\,dt\,.$$

Let T_1, T_2 be independent, uniform over (0, 1). Then

$$\begin{aligned}2\sigma_i^2(F) &= \operatorname{Var}[U_i(T_2) - U_i(T_1)] = E[U_i(T_2) - U_i(T_1)]^2 \\ &= E\left(\int_{T_1}^{T_2} \frac{K_i'(s)}{f(F^{-1}(s))}\,ds\right)^2.\end{aligned}$$

Since $K_2'(t) \leqq AK_1'(t)$ it follows that

$$\sigma_2^2(F) \leqq A^2\sigma_1^2(F)\,.$$

Let us next show that this upper bound for σ_2^2 is sharp. For this purpose suppose that

$$K_2'(t) = (1 - \varepsilon)AK_1'(t) \qquad \text{for} \quad a \leqq t \leqq b\,,$$

and let us write

$$2\sigma_i^2(F) = E\left(\int_{F^{-1}(T_1)}^{F^{-1}(T_2)} K_i'[F(w)]\,dw\right)^2 = 2\iint_{0<t_1<t_2<1} \left(\int_{F^{-1}(t_1)}^{F^{-1}(t_2)} K_i'[F(w)]\,dw\right)^2 dt_1\,dt_2\,.$$

Then

$$\sigma_2^2(F) \geqq \iint_{0<t_1<t_2<1} A^2(1-\varepsilon)^2\left(\int_{\max(F^{-1}(t_1),F^{-1}(a))}^{\min(F^{-1}(t_2),F^{-1}(b))} K_1'[F(w)]\,dw\right)^2 dt_1\,dt_2\,.$$

Suppose we choose F so that $F^{-1}(a) = u$, $F^{-1}(b) = v$ and we put all mass to the left of u and the right of v very close to u and v respectively. Then $t_1 < a$ implies that $F^{-1}(T_1)$ is close to $F^{-1}(a)$ with high probability and $t_2 > b$ implies that $F^{-1}(T_2)$ is close to $F^{-1}(b)$ with high probability so that for a suitable sequence F_n

$$\iint_{0<t_1<t_2<1} \left(\int_{\max}^{\min} K_1'[F(w)]\,dw\right)^2 dt_1\,dt_2 = \iint \left(\int_{F^{-1}(t_1)}^{F^{-1}(t_2)} K_1'[F(w)]\,dw\right)^2 dt_1\,dt_2$$

and this completes the proof.

In the above, we have not used the symmetry of K. If K is symmetric about $\frac{1}{2}$, it is easily seen that we can choose the sequence F_n to be symmetric, so that A^2 is the sharp lower bound even if F is restricted to be symmetric. Thus, provided (5.2) holds, the lower bound is the same when F is permitted to be asymmetric as under the assumption of symmetry.

As an example, consider the weight functions K_i with densities

$$K_1'(t) \equiv 1\,, \qquad K_2'(t) = 6t(1-t) \qquad \text{for} \quad 0 \leqq t \leqq 1\,. \tag{5.7}$$

The estimator correspoding to K_2 was proposed by Bickel (1973), and it follows from Theorem 5 that its asymptotic efficiency relative to $\bar{X}$ is always $\geqq (\frac{4}{6})^2 = \frac{4}{9}$.

As an important class of estimators whose comparison with $\bar{X}$ is covered by Theorem 5, let us now consider the trimmed means $\bar{X}_\alpha$. These estimate the measures $E_\alpha(X)$ defined by (1.6) from which they are obtained by replacing F by $\hat{F}_n$. For this comparison, (5.2) holds with $A = 1 - 2\alpha$. The efficiency $e_\alpha(F)$ of $\bar{X}_\alpha$ to $\bar{X}$ thus satisfies

$$e_\alpha(F) \geqq (1-2\alpha)^2 \qquad \text{for all} \quad F\,. \tag{5.8}$$

For symmetric F, this was shown earlier by Bickel (1965).

If more is known about F, it may of course be possible to sharpen the bound (5.8). Typically, one would expect $e_\alpha(F)$ to increase as the tail of the distribution becomes heavier. An ordering of distributions which reflect this property was introduced by Barlow and Proschan (1966) (see also Doksum (1969)), and Lawrence (1966), who suggest that if G has heavier tails than F one would expect that

$$G^{-1}(u)/F^{-1}(u) \quad \text{is increasing}\,. \tag{5.9}$$

THEOREM 6. *If F and G are symmetric distributions satisfying* (5.9), *then*

$$e_\alpha(F) \leqq e_\alpha(G) \qquad \textit{for any} \quad 0 < \alpha \leqq \tfrac{1}{2}\,. \tag{5.10}$$

The proof is based on the following lemmas which will be useful also for proving some later results.

LEMMA 1. *Let F, G be two distributions defined on* $(0, \infty)$ *such that the* cdf *defined by*

$$F^*(t) = \int_0^t a(x)\,dF(x) \Big/ \int_0^\infty a(x)\,dF(x) \tag{5.11}$$

is stochastically smaller than the cdf *G^* obtained when F is replaced by G, for all nonnegative functions a which are integrable with respect to F and G. Then*

$$\frac{\int b(x)\,dF(x)}{\int a(x)\,dF(x)} \leqq \frac{\int b(x)\,dG(x)}{\int a(x)\,dG(x)} \tag{5.12}$$

for all nonnegative functions a, b for which $b(x)/a(x)$ is nondecreasing.

PROOF. The left hand side can be written as

$$\int r(x)\,dF^*(x)$$

where $r(x) = b(x)/a(x)$, and the result follows from a well known property of stochastically ordered distributions.

LEMMA 2. *The conclusion of Lemma* 1 *will hold if* $b(x)/a(x)$ *is nondecreasing and if the functions* a, F *and* G *satisfy one of the following three conditions*:

(a) $dG(x)/dF(x)$ *is nondecreasing*;[4]

(b) *The function* a *is nondecreasing and* F *is obtained from* G *by truncation at some point* u: i.e.,

$$F(x) = G(x) \quad \text{for} \quad x < u$$
$$F(x) = 1 \quad \text{for} \quad x \geqq u\,.$$

(c) F *and* G *are obtained from a distribution* H *by truncation at points* $u < v$, *respectively*.

PROOF. (a) Since $dG^*(x)/dF^*(x)$ is proportional to $dG(x)/dF(x)$, it is nondecreasing, and hence G^* is stochastically larger than F^*.

(b) For any $t < u$, since $F(x) = G(x)$ for $x < u$,

$$\frac{\int_0^t a(x)\,dG(x)}{\int_0^\infty a(x)\,dG(x)} = \frac{\int_0^t a(x)\,dF(x)}{\int_0^\infty a(x)\,dG(x)}\,.$$

Since $a(x)$ is nondecreasing, $\int_0^\infty a(x)\,dG(x) \geqq \int_0^\infty a(x)\,dF(x)$, and hence $G^*(t) \leqq F^*(t)$ for $t < u$.

For any $t \geqq u$, we have

$$\frac{\int_0^t a(x)\,dG(x)}{\int_0^\infty a(x)\,dG(x)} \leqq 1 = \frac{\int_0^t a(x)\,dF(x)}{\int_0^\infty a(x)\,dF(x)}$$

and this completes the proof.

(c) This follows from (b) since F can be obtained from G through truncation.

PROOF OF THEOREM 6. Without loss of generality suppose that F and G are symmetric about 0, and let

$$\tilde{F}(x) = 2F(x) - 1\,, \qquad \tilde{G}(x) = 2G(x) - 1\,, \quad 0 < x < \infty\,. \tag{5.13}$$

If in (5.5) and (5.6) we substitute $dK_1 = dt$ and

$$dK_2 = dt/(1-2\alpha)\,, \quad \alpha \leqq t \leqq 1-\alpha$$
$$= 0 \quad \text{otherwise,}$$

we obtain for the efficiency $e_\alpha(F)$

$$e_\alpha(F) = (1-2\alpha)^2 \int_0^1 [\tilde{F}^{-1}(t)]^2\,d\Lambda_2(t) / \int_0^1 [\tilde{F}^{-1}(t)]^2\,d\Lambda_1(t)\,, \tag{5.14}$$

where Λ_2 and Λ_1 are respectively the uniform distribution on (0, 1) and the uniform distribution truncated at $1 - 2\alpha$. The result now follows from Lemma 2(b) with

$$\Lambda_1 \text{ in place of } F\,; \qquad \Lambda_2 \text{ in place of } G\,;$$
$$b(t) = [\tilde{G}^{-1}(t)]^2\,, \qquad a(t) = [\tilde{F}^{-1}(t)]^2\,.$$

[4] This part of the lemma is given in Goldstein (1973).

To apply the lemma, we need only check that

$$\frac{b(t)}{a(t)} = \left[\frac{\tilde{G}^{-1}(t)}{\tilde{F}^{-1}(t)}\right]^2 \quad \text{is nondecreasing for} \quad t > 0\,,$$

which is assured by the assumptions of the theorem; and that $a(t) = [\tilde{F}^{-1}(t)]^2$ is nondecreasing for $t > 0$, which is obvious. Finally, Λ_1 is obtained from Λ_2 through truncation, and this completes the proof for $\alpha < \frac{1}{2}$. For the case $\alpha = \frac{1}{2}$, the result follows by letting $\alpha \to \frac{1}{2}$ in the inequality (5.10).

A number of interesting examples of this theorem are obtained by considering the families $\mathscr{G}(F)$ of distribution G obtained as scale mixtures from a given distribution F, which is symmetric about 0,

$$(5.15) \qquad G(x) = \int_0^\infty F\left(\frac{x}{\sigma}\right) d\nu(\sigma)$$

where ν is an arbitrary probability distribution on $(0, \infty)$. It will be proved in a later paper of this series that F and any $G \in \mathscr{G}(F)$ satisfy (5.9), and hence that

$$(5.16) \qquad e_\alpha(G) \geqq e_\alpha(F) \qquad \text{for all} \quad G \in \mathscr{G}(F)\,,$$

provided the one-parameter family of distributions $F(\sigma x)$, $0 < \sigma < \infty$, have densities with monotone likelihood ratio. This condition holds, for example, when F is normal, a mixture of two normals with the same mean, or when it is double exponential. For the first of these cases, the family $\mathscr{G}(F)$ has been shown in Andrews et al. (1972) to be a very wide class of heavy-tailed distributions. It was recently pointed out by Stigler (1973) that this family was in fact proposed as representing distributions with tails heavier than the normal by Newcomb (1882 and particularly 1886).

The following table shows the lower bound $e_\alpha(F)$ for $\alpha = .05, .1, .15, 25, .5$ and the following four distributions:

(1) The standard normal distribution $N(0, 1)$;

(2a) and (2b) The contaminated normal distribution (Tukey model)

$$(1 - \varepsilon)N(0, 1) + \varepsilon N(0, \sigma^2) \qquad \text{for} \quad \varepsilon = .05 \quad \text{and} \quad .25 \quad \text{and} \quad \sigma = 3\,;$$

(3) The double exponential distribution.

These efficiencies are obtained from the corresponding asymptotic variances in Dachs (1972). The last column, for the sake of comparison, gives the universal lower bound $(1 - 2\alpha)^2$.

TABLE 5.1

Efficiency lower bound $e_\alpha(F)$

α	(1)	(2a)	(2b)	(3)	$(1-2\alpha)^2$
.05	.971	1.186	1.402	1.212	.81
.1	.943	1.197	1.622	1.342	.64
.15	.909	1.197	1.786	1.449	.49
.25	.833	1.085	1.667	1.626	.25
.5	.637	.833	1.327	2	0

From Theorem 6 it also follows easily that if F is uniform, inequality (5.10) holds whenever G has a symmetric unimodal density. This result was proved differently by Bickel (1965).

We mention finally that the bound (5.8) and Theorem 6 have obvious extensions to the comparison of two trimmed means, say $\bar{X}_\alpha$ and $\bar{X}_\beta$. If $\alpha > \beta$ and $e_{\alpha,\beta}(F)$ denotes the efficiency of $\bar{X}_\alpha$ w.r.t. $\bar{X}_\beta$ we have from Theorem 5,

$$e_{\alpha,\beta}(F) \geqq \left(\frac{1-2\alpha}{1-2\beta}\right)^2 \tag{5.17}$$

and from Lemma 2b that

$$e_{\alpha,\beta}(F) \leqq e_{\alpha,\beta}(G) \tag{5.18}$$

whenever F and G satisfy (5.9).

Another interesting class of examples is provided by t-distributions with low degrees of freedom. Takeuchi (1973) in his review of Andrews et al. (1972) expresses the view that in the choice of estimator "at least four cases should be distinguished": (i) from the normal to the t-distribution with four to five degrees of freedom; (ii) to the t with two degrees of freedom and grossly contaminated normals; (iii) to the Cauchy; and (iv) to beyond the Cauchy. For the Cauchy and t with 2 degrees of freedom the efficiency $e_\alpha(F)$ are of course infinite. The following table gives the values of $e_\alpha(F)$ for $\alpha = .05$ when F is the t-distribution with $\nu = 3$, 4 or 5 degrees of freedom:

TABLE 5.2
$e_\alpha(F)$ *for t-distribution with* ν d.f.

ν	3	4	5	∞
$e_\alpha(F)$	1.70	1.32	1.20	.97

It follows from our earlier results and the work of Van Zwet (1964) that if F and G are t-distributions with $\nu > \nu'$ degrees of freedom respectively, then (5.16) will hold. Takeuchi's remarks and Tables 5.1 and 5.2 suggest that a 5% trimmed mean seems to be a highly satisfying competitor to the mean.

5B. *R*-estimators. Consider next the estimators obtained from (3.9) by replacing the true distribution function F in $\mu(F)$ by the empirical distribution function $\hat{F}_n$. Such an estimator may not be defined since the equation (3.9) may not have a solution. However, we can always define it as the median of $\frac{1}{2}[F_n^{-1}(U) + F_n^{-1}(V)]$ where (U, V) has the cdf (4.1). The resulting estimators form the class known as R-estimators, which can be derived from rank tests and are described, for example, in Section 4.3 of Huber's survey paper (1972). The functions L' of (3.7) and J of Huber's paper are related through the equation $L'(u) = J[\frac{1}{2}(u+1)]$. In particular, if $L(y) = |y|$, the estimator reduces to $\hat{\mu}_2$, for $L(y) = y^2$, it reduces to $\hat{\mu}_4 = \text{med}_{i \leq j}\,[(X_i + X_j)/2]$, while for $L(y) = \Phi^{-1}[\frac{1}{2}(y+1)]$ it reduces to the estimator based on Normal Scores.

The attraction of these latter two estimators for estimating the center of symmetry of a symmetric distribution, stems from the fact that their asymptotic efficiency relative to $\bar{X}$ is respectively $\geqq .864$ and $\geqq 1$ for all symmetric F. As was pointed out by Høyland (1965), the first of these inequalities breaks down completely when F is not restricted to be symmetric. In that case, there exist distributions for which the asymptotic efficiency of $\hat{\mu}_4$ relative to $\bar{X}$ is zero.

We shall now prove

THEOREM 7. *For any R-estimator there exist distributions F such that the asymptotic efficiency of the R-estimator relative to $\bar{X}$ is zero.*

The discrepancy of this result with those in the symmetric case is perhaps not too surprising since the signed-rank tests from which these estimators derive are tests of symmetry with respect to a given point.

PROOF. Recall (for example from Huber (1972)) that the variance of the R-estimator corresponding to L is, under suitable regularity conditions,

$$\sigma^2(\textstyle\int U'(x)f(x)\,dx)^2/\mathrm{Var}\,[U(X)] \tag{5.19}$$

where X is distributed according to F and where

$$U(x) = \int^x L''[F(y) - F(2\theta - y)]f(2\theta - y)\,dy\,, \tag{5.20}$$

θ being the solution of (3.9). We shall determine F in such a way that

$$\int L'[F(x) - F(-x)]f(x)\,dx = 0 \tag{5.21}$$

and

$$\int U'(x)f(x)\,dx = 0\,, \tag{5.22}$$

while $\mathrm{Var}\,[U(X)] > 0$.

If (5.21) holds, (5.22) becomes

$$\int L''[F(x) - F(-x)]f(x)f(-x)\,dx = 0\,, \tag{5.23}$$

and this is satisfied whenever $f(x)f(-x) = 0$ for all x. For example, we can take (as Høyland did)

$$\begin{aligned} f(x) &= p \quad \text{if} \quad -1 \leqq x \leqq 0\,, \\ &= q \quad \text{if} \quad 1 \leqq x \leqq 2\,; \qquad p + q = 1\,, \end{aligned} \tag{5.24}$$

and (5.23) will be satisfied.

To show that there exist p for which (5.21) holds, note that (5.21) states that the solution θ of (3.7) is zero. Now for $p = 1$, this solution must lie between -1 and 0, and for $p = 0$ it must lie between 1 and 2 (by Theorem 1, part 2). The existence of the required p then follows by continuity, and this completes the proof.

When F is not restricted to be symmetric, the asymptotic efficiency of, for example, $\hat{\mu}_2$ or $\hat{\mu}_4$ relative to $\bar{X}$, can take on any value between 0 and ∞. It may be argued that this result is not very meaningful since distributions such as (5.24) are not likely to obtain in practice. What can be said if F is restricted to some more reasonable class, say unimodal distribution? We shall now show that for $\hat{\mu}_2$ the lower bound is then positive; and this is also true for $\hat{\mu}_4$ if F is restricted to be strongly unimodal.

The asymptotic efficiency of $\hat{\mu}_2$ relative to $\hat{\mu}_1 = \bar{X}$ is

$$e_{2,1} = 4\sigma^2 f^2[F^{-1}(\tfrac{1}{2})] \,. \tag{5.25}$$

Without loss of generality suppose that $F^{-1}(\frac{1}{2}) = 0$ and that the mode of F is $\leqq 0$ so that $f(x)$ is nonincreasing for all $x \geqq 0$. Then

$$e_{2,1} \geqq 4f^2(0) \textstyle\int_0^\infty (x - \mu)^2 f(x)\, dx \,,$$

where μ is the expectation of F. Let g be the probability density defined by

$$\begin{aligned} g(x) &= 2f(x) \qquad &&\text{for} \quad x \geqq 0 \\ &= 0 &&\text{for} \quad x < 0 \,. \end{aligned}$$

Denoting the mean of g by ν, we have

$$e_{2,1} \geqq \tfrac{1}{2} g^2(0) \textstyle\int_0^\infty (x - \nu)^2 g(x)\, dx \,. \tag{5.26}$$

We shall now prove that the right hand side of (5.26) is minimized, among all densities on $(0, \infty)$ which are nonincreasing, when g is the density of a uniform distribution whose left hand end point is zero. This is a consequence of the following lemma.

LEMMA 3. *Subject to* $\int_0^\infty x g(x)\, dx = \nu$, *if* $g(x) = 0$ *for* $x < 0$ *and is nonincreasing for* $x > 0$, *the integral* $\int_0^\infty x^2 g(x)\, dx$ (*and hence the variance of* g) *is minimized when* $g(x) = 1/2\nu$ *for* $0 < x < 2\nu$.

The proof of this lemma is completely analogous to the proof of the second part of Theorem 8 below, and we shall therefore omit it here.

To prove the desired result note that for fixed ν, the uniform distribution also minimizes $g(0)$. The right hand side of (5.26) is of course scale invariant, and hence is minimized for the uniform distribution over any interval $(0, a)$. The associated minimum value of (5.26) is $\frac{1}{24}$.

The lower bound $\frac{1}{24}$ is presumably not sharp. To see what values can be obtained, consider the density

$$\begin{aligned} f(x) &= \tfrac{1}{2}a \qquad &&\text{if} \quad -a \leqq x < 0 \\ &= 1 &&\text{if} \quad 0 \leqq x < \tfrac{1}{2} \,. \end{aligned} \tag{5.27}$$

Formula (5.25) is not applicable because of the discontinuity of f at zero. However, f can be approximated by a sequence of densities, continuous in $(-a, \frac{1}{2})$ and with $f(0) = 1$. Letting $a \to 0$, it is then seen that $e_{2,1} \to \frac{5}{48}$, so that the sharp lower bound lies between $\frac{2}{48} = .057$ and $\frac{5}{48} = .104$.

Much higher values for the lower bound are obtained if F is restricted to the subfamilies (5.15) considered in Section 5A. That (5.16) holds for $\hat{\mu}_2$ when F and G satisfy (5.9) and hence for the four examples considered in Table 5.1 is in fact the case $\alpha = \frac{1}{2}$ of Theorem 6. The values of the efficiency $e_{2,1}$ of $\hat{\mu}_2$ relative to $\bar{X}$ for these four distributions F (and hence the lower bounds for the associated (F)) are shown in the last row of Table 5.1.

Consider next $\hat{\mu}_4$. The asymptotic efficiency $e_{4,1}$ of this estimator relative to $\bar{X}$ is given by (see again Andrews et al. (1972))

$$e_{4,1} = \sigma^2(\textstyle\int f(-x)f(x)\,dx)^2/\int [\frac{1}{2} - F(-x)]^2 f(x)\,dx \tag{5.28}$$

provided $\mu_4(F) = 0$. Now $\int f(-x)f(x)\,dx$ is the density of $X + X'$ at 0, where X, X' are independent identically distributed with density f. If the density of $X + X'$ is unimodal, since the denominator of (5.28) is bounded above by $\frac{1}{4}$, we can apply the preceding result to get the crude lower bound .021. This bound holds whenever the density of $X + X'$ is unimodal and hence by a well-known result of Ibragimov (1956) whenever f is strongly unimodal.

Although crude, the bound is of the right order of magnitude. To see this, consider the following example. Let

$$\begin{aligned} f(x) &= q/a \quad \text{if} \quad -a < x < 0 \\ &= p/b \quad \text{if} \quad 0 < x < b\,, \qquad p + q = 1\,. \end{aligned} \tag{5.29}$$

It is easily computed that the pseudo median of F is zero when $q \to \frac{1}{2}2^{\frac{1}{2}}$ as $a \to 0$, and that then $e_{4,1} \to .071$.

We shall finally show that (5.16) holds when e_α is replaced by $e_{4,1}$, the efficiency of $\hat{\mu}_4$ relative of $\bar{X}$ when F is normal. If F is a distribution symmetric with respect to 0 and if $F * F$ denotes the convolution of F with itself, the efficiency $e_{4,1}$ can be written as

$$e_{4,1}(F) = 6[(F * F)'(0)]^2 \textstyle\int x^2 d(F * F)(x)\,. \tag{5.30}$$

Consider now the class $\mathscr{G}(\Phi)$ defined by (5.15) with F as the standard normal distribution Φ. Then when $F \in \mathscr{G}(\Phi)$, also $F * F \in \mathscr{G}(\Phi)$ and hence

$$\inf_{F \in \mathscr{G}(\Phi)} [e_{4,1}(F)] \geq \inf_{F \in \mathscr{G}(\Phi)} 6[f(0)]^2 \textstyle\int x^2\,dF(x)\,. \tag{5.31}$$

However, except for the constant the expression being minimized on the right hand side of (5.31) is just the minimum efficiency of $e_{2,1}(F)$ over $\mathscr{G}(\Phi)$, and this minimum was shown earlier to be attained at $F = \Phi$. It thus follows that

$$\inf_{F \in \mathscr{G}(\Phi)} [e_{4,1}(F)] = 6[\varphi(0)]^2 = \frac{3}{\pi} = .955\,.$$

5C. *M*-estimators. Consider finally the class of "(M)" location parameters characterized by Theorem 2. Let $\mu_\alpha(F)$ minimize $\int_{-\infty}^{\infty} |t - \mu|^{\alpha+1}\,dF(x)$ where $0 < \alpha \leq 1$. (Note that μ_1 here is the mean). Let T_α be the estimator obtained by replacing F by the empirical cdf in μ_α. The asymptotic variance of $n^{\frac{1}{2}}(T_\alpha - \mu_\alpha(F))$ is then under some regularity conditions given by (cf. Huber (1964))

$$\sigma_\alpha^2(F) = \frac{\int_{-\infty}^{\infty} \psi_\alpha^2(t - \mu_\alpha(F))\,dF(t)}{(\int_{-\infty}^{\infty} \psi_\alpha'(t - \mu_\alpha(F))\,dF(t))^2} \qquad \text{for} \quad \alpha > 0\,, \tag{5.32}$$

where ψ_α is given in (3.12).

The following result is the analogue of the lower bound for the efficiency of trimmed means.

THEOREM 8. *If* $0 < \alpha \leqq 1$,

$$\sigma_\alpha^2(F) \leqq \sigma_1^2(F)/\alpha^2 ; \tag{5.33}$$

that is the efficiency of T_α *to* $\bar{X}$ *is at least* α^2 *whatever be the underlying distribution for which the representation* (5.32) *of the asymptotic variance is valid.*

If F is known to be unimodal and symmetric about some point and (5.32) *is valid,*

$$\sigma_\alpha^2(F) \leqq \frac{3}{2\alpha + 1} \sigma_1^2(F) ; \tag{5.34}$$

that is, the efficiency of T_α *to* $\bar{X}$ *for symmetric unimodal distributions is at least* $(2\alpha + 1)/3$. *Both bounds are sharp. They are achieved for* (5.33) *when F assigns mass* $\frac{1}{2}$ *to each of* ± 1 *and for* (5.34) *when F is the uniform distribution on an interval.*

PROOF. To prove (5.33) we note that if we fix $\sigma_2^2(F) = 1$ since $2\alpha - 1 \leqq \alpha \leqq 1$

$$\begin{aligned} &\int_{-\infty}^{\infty} \psi_\alpha^2(t - \mu_\alpha(F))\, dF(t) \\ &\quad \leqq \left[\int_{-\infty}^{\infty} |t - \mu_\alpha(F)|^\alpha\, dF(t)\right]^2 \leqq \left[\int_{-\infty}^{\infty} |t - \mu_1(F)|^\alpha\, dF(t)\right]^2 \\ &\quad \leqq \left[\int_{-\infty}^{\infty} (t - \mu_1(F))^2\, dF(t)\right]^\alpha = 1 . \end{aligned}$$

Similarly,

$$\begin{aligned} \int_{-\infty}^{\infty} \psi_\alpha'(t - \mu_\alpha(F))\, dF(t) &= \alpha \int_{-\infty}^{\infty} \frac{1}{|t - \mu_\alpha(F)|^{1-\alpha}}\, dF(t) \\ &\geqq \alpha \Big/ \int_{-\infty}^{\infty} |t - \mu_\alpha(F)|^{1-\alpha}\, dF(t) \\ &\geqq \alpha \Big/ \left[\int_{-\infty}^{\infty} (t - \mu_1(F))^2\, dF(t)\right]^{(1-\alpha)/2} = \alpha , \end{aligned}$$

and (5.33) follows. That the bound is sharp is clear. A proof of (5.34) is given by Loynes (1970) who also proved (5.33) under the restriction of symmetry.

We can generalize Loynes' result somewhat along the lines of Theorem 6. Suppose that F is symmetric about 0 and that $\mathscr{G}(F)$ is defined as in the remarks following that theorem. Then for $G \in \mathscr{G}(F)$ we have $\mu_\alpha(G) = 0$ and if G is given by (5.15)

$$\sigma_\alpha^2(G) = \frac{\int_0^\infty \psi_\alpha^2(t)\, d\nu(t) \int_{-\infty}^{\infty} |t|^{2\alpha}\, dF(t)}{\left[\int_0^\infty \psi_\alpha'(t)\, d\nu(t)\right]^2 \left[\int_{-\infty}^{\infty} |t|^{\alpha-1}\, dF(t)\right]^2} .$$

Hence,

$$\sigma_\alpha^2(G) \leqq \frac{\sigma_1^2(G)}{\alpha^2} \cdot \frac{\int_{-\infty}^{\infty} |t|^{2\alpha}\, dF(t)}{\left[\int_{-\infty}^{\infty} |t|^{\alpha-1}\, dF(t)\right]^2 \int_{-\infty}^{\infty} t^2\, dF(t)} .$$

More generally, if $0 < \alpha \leqq \beta \leqq 1$, by Lemma 1 of Loynes

$$\sigma_\alpha^2(G) \leqq \sigma_\beta^2(G) \cdot \frac{\beta^2 \int_{-\infty}^{\infty} |t|^{2\alpha}\, dF(t) \left[\int_{-\infty}^{\infty} |t|^{\beta-1}\, dF(t)\right]^2}{\alpha^2 \left[\int_{-\infty}^{\infty} |t|^{\alpha-1}\, dF(t)\right]^2 \int_{-\infty}^{\infty} |t|^{2\beta}\, dF(t)}$$

with equality if and only if there exists σ such that $G(x) = F(x/\sigma)$ for all x.

If F is the uniform distribution on $(-1, 1)$, $\mathscr{G}(F)$ is the family of symmetric unimodal distributions and we obtain Loynes' Theorem 2:

$$\sigma_\alpha^2(G) \leqq \frac{2\alpha + 1}{2\beta + 1} \sigma_\beta^2(G)$$

with equality if and only if G is uniform.

If F is a normal distribution with mean zero, we find

$$\sigma_\alpha^2(G) \leqq \frac{\Gamma(\alpha + \frac{1}{2})}{\Gamma(\beta + \frac{1}{2})} \left[\frac{\beta\Gamma(\beta/2)}{\alpha\Gamma(\alpha/2)}\right]^2 \sigma_\beta^2(G)$$
$$= \frac{\Gamma(\alpha + \frac{1}{2})}{\Gamma(\beta + \frac{1}{2})} \left[\frac{\Gamma(\beta/2 + 1)}{\Gamma(\alpha/2 + 1)}\right]^2 \sigma_\beta^2(G)$$

with equality if and only if G is normal. Passage to the limit as $\alpha \to 0$ is easily seen to be justified and the results then agree with those for the efficiency of the median with respect to the mean stated after Theorem 6.

6. Conclusion. We set out in this paper to find measures of location in the asymmetric case, which would be robust and at the same time could be estimated efficiently. If by efficient we mean a guaranteed high efficiency relative to the mean, we have shown that measures estimated by R-estimators are robust but do not satisfy this efficiency criterion, while the situation is just the reverse for the rather narrow class of location measures estimated by M-estimators. However, within the class of weighted quantiles, we have found measures satisfying both criteria. In particular, this is the case for the trimmed means, which appear to constitute a simple and satisfactory solution to our problem.

REFERENCES

ANDREWS, D. F., BICKEL, P. J., HAMPEL, F. R., HUBER, P. J., ROGERS, W. H. and TUKEY, J. W. (1972). *Robust estimates of location*. Princeton Univ. Press.

BAHADUR, RAGHU RAJ and SAVAGE, LEONARD J. (1956). The nonexistence of certain statistical procedures in nonparametric problems. *Ann. Math. Statist.* **27** 1115–1122.

BARLOW, RICHARD E. and PROSCHAN, FRANK (1966). Tolerance and confidence limits for classes of distributions based on failure rate. *Ann. Math. Statist.* **37** 1593–1601.

BICKEL, P. J. (1965). On some robust estimates of location. *Ann. Math. Statist.* **43** 847–858.

BICKEL, P. J. (1973). On some analogues of linear combinations of order statistics in the linear model. *Ann. Statist.* **1** 597–616.

BICKEL, P. J. and LEHMANN, E. L. (1975). Descriptive statistics for nonparametric models. I. Introduction. *Ann. Statist.* **3** 1038–1044.

British Association for the Advancement of Science (1958). *Mathematical Tables* **6** Cambridge Univ. Press.

CHURCHILL, RUEL V. (1958). *Operational mathematics*. 2nd ed. McGraw-Hill, New York.

DACHS, J. N. W. (1972). Asymptotic expansions for M-estimators. Unpublished Ph.D. thesis. Univ. of California, Berkeley.

DANIELL, P. J. (1920). Observations weighted according to order. *Amer. J. Math.* **42** 222–236.

DOKSUM, KJELL (1969). Starshaped transformations and the power of rank tests. *Ann. Math. Statist.* **40** 1167–1176.

HAMPEL, F. R. (1968). Contributions to the theory of robust estimation. Ph.D. dissertation, Univ. of California, Berkeley.

HAMPEL, F. R. (1971). A general qualitative definition of robustness. *Ann. Math. Statist.* **42** 1887–1896.

HØYLAND, ARNLJOT (1965). Robustness of the Hodges-Lehmann estimator for shift. *Ann. Math. Statist.* **36** 174–197.

HUBER, P. J. (1964). Robust estimation of a location parameter. *Ann. Math. Statist.* **35** 73–101.

HUBER, P. J. (1972). Robust statistics: A review. *Ann. Math. Statist.* **43** 1041–1067.

HUBER, P. J., *Robust Statistics*. A book in preparation.

IBRAGIMOV, J. A. (1956). On the composition of unimodal distributions (Russian). *Teor. Verojatnost. i Primenen.* **1** 283–288.

JAECKEL, L. A. (1971). Robust estimates of location: symmetry and asymmetric contamination. *Ann. Math. Statist.* **42** 1020–1034.

JOHNSON, NORMAN L. and KOTZ, SAMMUEL (1970). *Continuous univariate distributions 1*. Houghton and Mifflin, Boston.

GOLDSTEIN, MATTHEW (1973). An application of ordered expectations. *Amer. Statist.* **27** (1) 30.

KNÜSEL, L. F. (1969). Uber minimum-distance Schätzungen. Thesis, ETH, Zürich.

LAWRENCE, MICHAEL (1966). Inequalities and tolerance limits for *s*-ordered distributions. Unpublished report.

LOYNES, R. M. (1970). On the asymptotic relative efficiencies of certain location parameter estimates. *J. Roy. Statist. Soc. Ser. B.* 134–136.

NEWCOMB, S. (1882). Discussion and results of observations on transits of Mercury from 1677 to 1881. Astronomical papers **1** 363–487.

NEWCOMB, S. (1886). A generalized theory of the combination of observations so as to obtain the best result. *Amer. J. Math.* **8** 343–366.

STIGLER, STEPHEN M. (1972). Simon Newcomb, Percy Daniell, and the history of robust estimation 1885–1920. Unpublished report of the Univ. of Wisconsin, Dept. of Statistics.

TAKEUCHI, K. (1967). Robust estimation and robust parameter. Unpublished.

TAKEUCHI, K. (1973). Review of robust estimates of location. Andrews et al. *Science* **179** 469.

TUKEY, J. W. (1960). A survey of sampling from contaminated distributions. *Contributions to Probability and Statistics*. (I. Olkin, ed.) Stanford Univ. Press.

TUKEY, J. W. (1962). The future of data analysis. *Ann. Math. Statist.* **33** 1–67.

VAN ZWET, W. R. (1964). *Convex Transformations of Random Variables*. Math. Centrum, Amsterdam.

DEPARTMENT OF STATISTICS
UNIVERSITY OF CALIFORNIA
BERKELEY, CALIFORNIA 94720

The Annals of Statistics
1976, Vol. 4, No. 6, 1139–1158

DESCRIPTIVE STATISTICS FOR NONPARAMETRIC MODELS. III. DISPERSION

BY P. J. BICKEL AND E. L. LEHMANN[1]

University of California, Berkeley

Measures of dispersion are defined as functionals satisfying certain equivariance and order conditions. In the main part of the paper attention is restricted to symmetric distributions. Different measures are compared in terms of asymptotic relative efficiency, i.e., the inverse ratio of their standardized variances. The efficiency of a trimmed to the untrimmed standard deviation turns out not to have a positive lower bound even over the family of Tukey models. Positive lower bounds for the efficiency (over the family of all symmetric distributions for which the measures are defined) exist if the trimmed standard deviations are replaced by pth power deviations. However, these latter measures are no longer robust, although for $p < 2$ they are more robust than the standard deviation. The results of the paper suggest that a positive bound to the efficiency may be incompatible with robustness but that trimmed standard deviations and pth power deviations for $p = 1$ or 1.5 are quite satisfactory in practice.

1. Measures of dispersion. In analogy with the definition of a measure of location, we shall define a measure of dispersion to be a functional (defined over a sufficiently large family of distributions) which satisfies certain invariance conditions and which in addition has the property of assigning a larger value to G than to F if G is more dispersed than F. In the present paper we shall consider the problem for *symmetric* distributions (see BL IV for the asymmetric case) and assume that X is a random variable whose distribution F is symmetric about μ. It then seems natural to interpret dispersion in terms of the distance of X from μ, that is, in terms of the magnitude of $|X - \mu|$, and to consider Y as *more dispersed* about ν than X about μ if

$$|Y - \nu| \text{ is stochastically larger than } |X - \mu| . \tag{1.1}$$

(This is essentially the "peakedness"-ordering introduced by Z. W. Birnbaum (1948).)

Note that

(a) any symmetric random variable is more dispersed than a constant;
(b) aX is more dispersed than X if $a > 1$.

If F and G are symmetric about 0 with densities f and g, a simple sufficient

Received February 1975; revised April 1976.

[1] This paper was prepared with support of U.S. Office of Naval Research Contract No. N00014-69-A-0200-1038/NR042-036 and National Science Foundation Grant No. GP-38485.

AMS 1970 *subject classifications.* 62G05, 62G20, 62G30, 62G35.

Key words and phrases. Dispersion, estimation, standardized asymptotic variance, asymptotic efficiency, Tukey model, standard deviation, trimmed standard deviation, pth power deviation, substitution of an estimate of location, scaling a positive distribution.

condition for (1.1) with $\mu = \nu = 0$ is

$$g(x)/f(x) \quad \text{is increasing for} \quad x > 0\,. \tag{1.2}$$

If F and G are symmetric about zero, and G is more dispersed than F, and if

$$H_\theta(x) = \theta G(x) + (1 - \theta)F(x)\,, \tag{1.3}$$

then H_θ is more dispersed than F for any $0 < \theta < 1$. As an illustration, note that a standard normal distribution contaminated with another normal distribution with zero mean and variance > 1 (Tukey model) is more dispersed than the uncontaminated standard normal distribution.

An important class of examples is provided by the following result, which is a generalization of a lemma of Birnbaum (1948).

THEOREM 1. *Let X_i, Y_i $(i = 1, 2)$ be independent with distributions F_i, G_i $(i = 1, 2)$ which are symmetric about zero, and suppose that*

(i) *Y_i is more dispersed than X_i for $i = 1, 2$ and*
(ii) *F_1 and G_2 have unimodal densities and possibly some probability mass at zero.*

Then $Y_1 + Y_2$ is more dispersed than $X_1 + X_2$.

PROOF. Consider the probability

$$P(|X_1 + X_2| < c) = 2 \int_0^\infty [F_1(x + c) - F_1(x - c)]\, dF_2(x)\,.$$

The unimodality of F_1 implies that the integrand on the right-hand side is a decreasing function of x. From the fact that (F_2, G_2) satisfies (1.1), it then follows that this last integral is decreased when F_2 is replaced by G_2. Thus,

$$\begin{aligned} P(|X_1 + X_2| \leqq c) &\geqq 2 \int_0^\infty [F_1(x + c) - F_1(x - c)]\, dG_2(x) \\ &= 2 \int_0^\infty [G_2(x + c) - G_2(x - c)]\, dF_1(x)\,. \end{aligned}$$

Repeating the argument (this time using the unimodality of G_2), we arrive at the desired result. Birnbaum has shown that Theorem 1 no longer holds when assumption (ii) is dropped.

Consider now a functional $\tau(F)$ [also denoted by $\tau(X)$ when X is a random variable with distribution F] defined over a sufficiently large class of symmetric distributions which is closed under changes of location and scale. We shall require τ to be nonnegative and to satisfy

$$\tau(aX) = |a|\tau(X) \qquad \text{for} \quad a > 0 \tag{1.4}$$

and

$$\tau(X + b) = \tau(X) \qquad \text{for all} \quad b\,. \tag{1.5}$$

It follows from (1.5) and the symmetry of F that

$$\tau(-X) = \tau(X) \tag{1.6}$$

so that (1.4) holds for all $a \neq 0$.

From (1.4) and (1.5) it is easily seen that

$$\tau(c) = 0 \quad \text{for any constant} \quad c\,. \tag{1.7}$$

For by (1.4), we have $\tau(0) = \tau(2 \times 0) = 2\tau(0)$ and hence $\tau(0) = 0$, and by (1.5), $\tau(c) = \tau(0)$. The converse, that $\tau(X) = 0$ requires X to be a constant with probability 1, will in general not hold. An example is provided by the trimmed standard deviation defined in Section 3 below.

A nonnegative functional τ satisfying (1.4) and (1.5) will be called a *measure of dispersion* if it satisfies in addition

$$\tau(F) \leqq \tau(G) \quad \text{whenever} \quad G \quad \text{is more dispersed than} \quad F\,. \tag{1.9}$$

Note that if $\tau(F)$ is a measure of dispersion, so is $k\tau(F)$ for any $k > 0$.

A large and important class of dispersion measures is provided by the functionals

$$\tau(F) = \{\textstyle\int_0^1 [F_*^{-1}(t)]^\gamma \, d\Lambda(t)\}^{1/\gamma} \tag{1.10}$$

where F is assumed to be symmetric about μ, F_* denotes the distribution of $|X - \mu|$, Λ is any probability distribution on (0, 1) and γ any positive number.

That (1.10) satisfies (1.4) and (1.5) is easily checked; that it satisfies (1.9) follows from the fact that $F_*^{-1}(t) \leqq G_*^{-1}(t)$ for all t when G_* is stochastically larger than F_*.

A special case of (1.10) is the standard deviation (SD) of F defined as

$$\mathrm{SD}(F) = [\textstyle\int (x - \mu)^2 \, dF(x)]^{\frac{1}{2}}\,. \tag{1.11}$$

This is easily seen to be given by (1.10) with $\gamma = 2$ and Λ the uniform distribution on (0, 1). The following three important classes of measures, all special cases of (1.10) provide the alternatives to the standard deviation with which we shall be concerned.

(i) A generalization $\tau(F; p)$ of the standard deviation is the pth *power deviation* obtained by replacing γ by p in (1.10) and letting Λ be the uniform distribution on (0, 1).

(ii) The doubly trimmed standard deviation $\tau(F; \alpha, \beta)$ is given by (1.10) with $\gamma = 2$ and Λ the uniform distribution on $(\alpha, 1 - \beta)$. The most important example of this is the case $\alpha = 0$.

(iii) The αth quantile is obtained from (1.10) by letting Λ assign probability 1 to the point α. The resulting measure is independent of γ.

The standard deviation is of course a member of both (i) and (ii). The αth quantile is the limit of the doubly trimmed standard deviation as $\beta \to \alpha$.

2. Estimation. A most important aspect in comparing two measures of scale $\tau_i(F)$ and $\tau_j(F)$ is the accuracy with which they can be estimated. Unfortunately, it is no longer possible to compare these accuracies directly in terms of the asymptotic variances of the estimators. This is clearly seen by considering the case $\tau_2 = c\tau_1$ where c is any positive constant. If τ_1 is a possible measure of scale and its estimator is δ_1, one would be equally happy to use τ_2 and estimate it by $\delta_2 = c\delta_1$; of course the asymptotic variance does not remain the same but gets multiplied by c^2. For this reason, a natural measure of accuracy of an estimator of $\tau(F)$ with asymptotic variance $n^{-1}v^2(F)$ is not $n^{-1}v^2(F)$ itself but the scale invariant *standardized asymptotic variance* (already proposed by Daniell (1920))

$$v^2(F)/\tau^2(F)\,. \tag{2.1}$$

The asymptotic efficiency $e_{2,1}$ of δ_2 (estimating τ_2) to δ_1 (estimating τ_1) will then be defined as

$$e_{2,1}(F) = \frac{v_1^2(F)}{\tau_1^2(F)}\Big/\frac{v_2^2(F)}{\tau_2^2(F)}\,. \tag{2.2}$$

If $(n_i)^{\frac{1}{2}}(\delta_i - \tau_i)$ is asymptotically normally distributed for $i = 1, 2$ as the number n_i of observations tends to infinity, the usual argument shows that the asymptotic efficiency (2.2) is the limiting ratio of the numbers of observations required by the two estimators to achieve the same standardized variance.

That the above definitions are reasonable can be seen from another point of view. The logarithm of δ is an estimator of the location parameter $\log\tau(F)$. Suppose that the distribution of $n^{\frac{1}{2}}(\delta - \tau)$ tends to the normal distribution with zero mean and variance v^2. Then the distribution of $n^{\frac{1}{2}}(\log\delta - \log\tau)$ tends to the normal distribution with mean zero and variance $v^2/\tau^2(F)$; that is, $v^2/\tau^2(F)$ is the asymptotic variance of the location estimate $\log\delta$.

When studying the estimators of functionals such as those defined in Section 1, it is convenient first to consider F restricted to distributions which are symmetric about a *known* point μ. On the other hand, in order even to define the estimators of τ we wish to study, it is necessary to extend τ to asymmetric distributions. In all the examples to be considered here, there is a natural functional form of τ which applies also to asymmetric distributions. Given this extension, we define as estimator of $\tau(F)$ the functional τ evaluated at the empirical distribution function $\hat{F}$ of $X_1, \cdots, X_n$. In what follows, we shall assume without loss of generality that the known value μ of the center of symmetry of F is $\mu = 0$.

The point of view in the present paper will be that the standard measure of scale is the standard deviation $\tau(F; 0, 0)$ which (since $\mu = 0$) is estimated by

$$\delta_1 = (\textstyle\sum_{i=1}^n X_i^2/n)^{\frac{1}{2}}\,. \tag{2.3}$$

This estimator is well known to be very unsatisfactory because of its extreme sensitivity to outlying observations. We shall therefore look at the other functionals under consideration as competitors of τ_1 and hence shall be interested principally in comparing their behavior with that of τ_1. Unfortunately, we are only able to make these comparisons asymptotically. For $\sum X_i^2/n$, it is of course obvious from the central limit theorem that $n^{\frac{1}{2}}[\sum X_i^2/n - \sigma^2]$ is asymptotically normal with zero mean and variance Var (X^2) provided the later variance is finite. Here $\sigma^2 = E(X^2)$ denotes the variance of X.

It follows that $n^{\frac{1}{2}}(\delta_1 - \sigma)$ is asymptotically normal with mean zero and variance Var $(X^2)/4\sigma^2$. Thus the standardized variance is

$$\frac{v_1^2(F)}{\tau_1^2(F)} = \frac{\text{Var}\,(X^2)}{4\sigma^4} = \frac{1}{4}\left[\frac{E(X^4)}{\sigma^4} - 1\right]. \tag{2.4}$$

In the next sections we shall obtain the corresponding expansion for the estimators of some other functionals, and then study the efficiencies (2.2).

3. The doubly trimmed standard deviation. To replace the standard deviation as a measure of scale, we shall seek among the measures discussed in Section 1 one which would be more robust but which still can be estimated fairly efficiently. Encouraged by the results of BL II, we shall begin this search by studying the trimmed standard deviations. In this and the succeeding section we shall carry out the necessary efficiency investigation. The robustness properties follow by arguments completely analogous to those of BL II.

As before, let the random variables X_i be independently distributed according to a distribution F, which is symmetric with respect to the origin. Let $Y = X^2$ and denote the distribution of Y by G. We can then write $\tau^2(F; \alpha, \beta)$ (with α, $\beta \leq \frac{1}{2}$) as

$$\tau^2(F; \alpha, \beta) = \frac{1}{1 - \alpha - \beta} \int_\alpha^{1-\beta} G^{-1}(t)\, dt = \frac{1}{1 - \alpha - \beta} \int_{u_\alpha}^{u_{1-\beta}} y\, dG(y) \tag{3.1}$$

where u_α is the αth percentile of G.

Consider now the estimator

$$\tau^2(\hat{F}, \alpha, \beta) \equiv \hat{\tau}^2(\alpha, \beta)\,. \tag{3.2}$$

This is obtained by trimming off the 100 α% observations with the smallest absolute values and the 100 β% observations with the largest absolute values and then computing the standard deviation of the remaining observations. In terms of the Y's, $\hat{\tau}^2$ is a doubly trimmed mean. Its expectation is given by(3.1) and the asymptotic variance of $n^{\frac{1}{2}}\hat{\tau}^2$ is (see, for example, BL II)

$$\begin{aligned} w^2_{\alpha,\beta} = \frac{1}{(1 - \alpha - \beta)^2}\{\textstyle\int_{u_\alpha}^{u_{1-\beta}} [y - c(\alpha, \beta)]^2\, dG(y) \\ + \alpha[u_\alpha - c(\alpha, \beta)]^2 + \beta[u_{1-\beta} - c(\alpha, \beta)]^2\}\,, \end{aligned} \tag{3.3}$$

where

$$c(\alpha, \beta) = \int_{u_\alpha}^{u_{1-\beta}} y\, dG(y) + \alpha u_\alpha + \beta u_{1-\beta}\,, \tag{3.4}$$

and

$$u_t = G^{-1}(t) = \left[F^{-1}\left(\frac{t+1}{2}\right)\right]^2. \tag{3.5}$$

(This formula holds if u_α, $u_{1-\beta}$ are uniquely defined and G is continuous at u_α, $u_{1-\beta}$ (see Stigler (1973)).)

The standardized asymptotic variance of $\hat{\tau}$ is thus w^2/τ^4 where τ^2 is given by (3.1).

To get an idea of the behavior of such procedures with respect to the untrimmed standard deviation we consider some representative cases, namely the (singly) trimmed standard deviations with $\beta = .1, .2, \alpha = 0$. The tables below computed by Winston Chow and W. Carmichael show the efficiencies (2.2) of these estimators with respect to the standard deviation for the following three classes of distributions.

(a) t-distributions with 5, 6, 7, 8, 9, 10, 25, 50 degrees of freedom.

(b) Symmetric beta distributions with densities

$$f(x) = [B(r, r)]^{-1}(\tfrac{1}{4} - t^2)^{r-1}, \qquad |t| \leqq \tfrac{1}{2} \tag{3.6}$$

for $r = \frac{1}{2}$, 2, 4, 6, 8, 10, 20, 30.

(c) Tukey normal gross error distributions, i.e.,

$$F(x) = (1 - \varepsilon)\Phi(x) + \varepsilon\Phi\left(\frac{x}{\lambda}\right), \tag{3.7}$$

for $\varepsilon = .025, .075, .10, .20, .40, .50$ and $\lambda = 2, 4, 6$.

These models were selected as representing a range of long and short tailed distributions and for ease of computation.

The figures suggest that, on the whole, for heavy-tailed distributions both trimmed SD's are better than the untrimmed SD and that for the ranges considered $\beta = .1$ is preferable to $\beta = .2$. For light-tailed distributions such as the Beta-distribution, the untrimmed SD does best, $\beta = .1$ does better than $\beta = .2$, and $\beta = .1$ performs reasonably well.

A surprising feature of Table 3.3 are the extremely high efficiency values for small $\varepsilon > 0$ and large λ. These seem to arise in cases where there is enough trimming to insure with very high probability that only a small proportion of

TABLE 3.1

Asymptotic efficiency of β-trimmed with respect to untrimmed SD: *t-distribution with f degrees of freedom*

f	5	6	7	8	9	10	25	50	∞
$\beta = .1$	2.35	1.56	1.29	1.16	1.08	1.03	.85	.81	.78
$\beta = .2$	2.11	1.36	1.11	.99	.92	.87	.69	.66	.63

TABLE 3.2

Asymptotic efficiency of β-trimmed with respect to untrimmed SD: *Beta distribution with density* (3.6)

r	.5	1.0	2.0	4.0	6.0	8.0	10.0	0.0	30.0	∞
$\beta = .1$	1.30	.86	.72	.70	.71	.72	.73	.75	.76	.78
$\beta = .2$	.83	.57	.52	.54	.55	.57	.57	.60	.61	.63

TABLE 3.3

Asymptotic efficiency of β-trimmed with respect to untrimmed SD: *Tukey model* (3.7)

ε	λ					
	$\beta = .1$			$\beta = .2$		
	2	4	6	2	4	6
0	.78	.78	.78	.63	.63	.63
.025	.98	3.97	10.04	.80	3.27	8.36
.075	1.19	4.01	6.48	.99	3.57	6.03
.10	1.23	3.48	4.79	1.03	3.27	4.93
.20	1.21	1.46	.76	1.05	2.04	2.26
.40	.96	.63	.50	.87	.68	.38
.50	.87	.63	.55	.79	.52	.36

the gross errors is retained in the trimmed sample. The standardized variance of the untrimmed SD then rises very sharply with λ while that of the trimmed SD is affected only little as λ gets large. The curious "dip" which occurs in the neighborhood of $\varepsilon = \beta$ is predicted by the asymptotic theory of the next section.

4. Nonexistence of a lower bound. The numerical results of the preceding section are encouraging and raise the hope that, as in the case of the trimmed means as measures of location, the efficiencies of the trimmed to the untrimmed SD have a positive lower bound. Unfortunately, this turns out not to be the case even if attention is restricted to unimodal distributions. In fact, even within the class of Tukey models (3.7), the efficiencies can take on arbitrarily small values.

THEOREM 2. *Let $e(\beta;\varepsilon,\lambda)$ denote the asymptotic efficiency of the trimmed standard deviation $\hat{\tau}(0,\beta)$ relative to the untrimmed standard deviation, in the Tukey model* (3.7). *Then*

$$\lim_{\lambda\to\infty} e(\beta;\beta,\lambda) = 0\,. \tag{4.1}$$

The proof of this result is based on the following two lemmas.

LEMMA 1. *For the Tukey model* (3.7), *the standardized asymptotic variance of the* SD

$$\frac{1}{4}\left[\frac{E(X^4)}{E^2(X^2)} - 1\right] \tag{4.2}$$

is bounded above uniformly in λ for any fixed ε.

PROOF. For the model (3.7), the standardized variance (4.2) becomes

$$\frac{1}{4}\left(\frac{3[(1-\varepsilon)+\varepsilon\lambda^4]}{[(1-\varepsilon)+\varepsilon\lambda^2]^2} - 1\right).$$

Since this is continuous in λ we need only consider its behavior as $\lambda \to \infty$ when it is clearly bounded.

LEMMA 2. *If* $\varepsilon = \beta$ *in* (3.7) *and if* $u_{1-\beta} = u_{1-\beta}^{(\lambda)}$ *is given by* (3.5), *then as a function of* λ,

$$u_{1-\beta}^{(\lambda)} \sim 2 \log \lambda \qquad \text{as} \quad \lambda \to \infty .$$

PROOF. Let

$$x(\lambda) = (u_{1-\beta}^{(\lambda)})^{\frac{1}{2}} .$$

Then $x(\lambda)$ satisfies

$$(1 - \beta)\Phi(x) + \beta\Phi\left(\frac{x}{\lambda}\right) = 1 - \frac{\beta}{2} . \tag{4.3}$$

First note that as $\lambda \to \infty$ we must have

$$x(\lambda) \to \infty , \tag{4.4}$$

and

$$\frac{x(\lambda)}{\lambda} \to 0 . \tag{4.5}$$

For if (4.4) does not hold there exists a sequence λ_n such that

$$x(\lambda_n) \to c < \infty .$$

Then,

$$(1 - \beta)\Phi(x(\lambda_n)) + \beta\Phi\left(\frac{x(\lambda_n)}{\lambda_n}\right) \to (1 - \beta)\Phi(c) + \frac{\beta}{2} < 1 - \frac{\beta}{2} ,$$

and this contradicts (4.3). Similarly if (4.4) holds but (4.5) is violated we could find a sequence $\{\lambda_n'\}$ such that

$$\lim_n (1 - \beta)\Phi(x(\lambda_n')) + \beta\Phi\left(\frac{x(\lambda_n')}{\lambda_n'}\right) > 1 - \beta ,$$

and we would again have a contradiction.

Now, by (4.4) and (4.5),

$$1 - \Phi[x(\lambda)] = \frac{\varphi[x(\lambda)]}{x(\lambda)}[1 + o(1)] , \tag{4.6}$$

and

$$\Phi\left(\frac{x(\lambda)}{x}\right) - \frac{1}{2} = \frac{\varphi(0)x(\lambda)}{\lambda}[1 + o(1)] . \tag{4.7}$$

Substituting (4.6) and (4.7) in (4.3), cancelling, and then taking logs we obtain

$$\frac{-x^2(\lambda)}{2} - \log x(\lambda) = \log x(\lambda) - \log \lambda + \log \frac{\beta}{1 - \beta} + o(1) . \tag{4.8}$$

By (4.4)

$$\log x(\lambda) = o(x^2(\lambda)) ,$$

and we conclude that

$$x^2(\lambda)(1 + o(1)) = 2 \log \lambda \tag{4.8a}$$

which was to be proved. □

PROOF OF THEOREM 2. In view of Lemma 1 we only need to show that

$$w^2_{0,\beta} \to \infty \tag{4.9}$$

while

$$\tau^2(F; 0, \beta) = O(1) \quad \text{as} \quad \lambda \to \infty . \tag{4.10}$$

To show (4.10) we calculate,

$$\begin{aligned}(1-\beta)\tau^2(F; 0, \beta) &= (1-\beta) \textstyle\int_{-x(\lambda)}^{x(\lambda)} x^2\varphi(x)\,dx + \frac{\beta}{\lambda} \textstyle\int_{-x(\lambda)}^{x(\lambda)} x^2\varphi\left(\frac{\lambda}{x}\right) dx \\ &\leqq (1-\beta) + 2\beta\lambda^2 \textstyle\int_0^{x(\lambda)/\lambda} y^2\varphi(y)\,dy \\ &\leqq (1-\beta) + \frac{2\beta\varphi(0)}{3}\frac{x^3(\lambda)}{\lambda},\end{aligned}$$

and the result follows. This argument and Lemma 2 show that

$$c(0, \beta) = (1-\beta)\tau^2(F; 0, \beta) + \beta u_{1-\beta} \sim \beta u_{1-\beta}, \tag{4.11}$$

where $c(\alpha, \beta)$ is given by (3.4).

Finally, from (4.11)

$$(1-\beta)^2 w^2_{0,\beta} \geqq \beta(u_{1-\beta} - c(0, \beta))^2 \sim \beta(1-\beta)^2 u^2_{1-\beta}$$

which tends to ∞, and (4.9) and the theorem follow. □

REMARK 1. The same arguments show that the result of Theorem 2 continues to hold if $\tau^2(\hat{F}; 0, \beta)$ is replaced by $\tau^2(\hat{F}; \alpha, \beta)$ for any $\alpha < 1 - \beta$, and if as before $\varepsilon = \beta$.

REMARK 2. In addition to predicting that the efficiency of the β-trimmed SD to the SD tends to 0 as $\tau \to \infty$ for $\varepsilon = \beta$ the asymptotic theory also gives a positive limit for $\varepsilon \neq \beta$. Roughly speaking, for $\varepsilon > \beta$ the behavior is governed by the contaminant while for $\varepsilon < \beta$ it is governed by the main portion. Clearly Table 3.3 reflects this only very crudely. For the values of τ we have given, the efficiency at $\varepsilon = \beta$ remains quite large. However, further numerical calculations for extremely large values of τ (of the order of 50 and above) confirm the asymptotic theory although convergence is slow. It is seen from the tables that small efficiency values do appear for smaller values of τ and $\varepsilon > \beta$. These presumably reflect small positive limits which are approached more rapidly than the asymptotic value for $\varepsilon = \beta$.

REMARK 3. An interesting limiting case of the doubly trimmed standard deviations arises as both α and β tend to $\frac{1}{2}$. The limiting functional is naturally taken to be

$$\tau(F; \tfrac{1}{2}, \tfrac{1}{2}) = (G^{-1}(\tfrac{1}{2}))^{\frac{1}{2}}, \tag{4.12}$$

the median of the distribution of $|X - \mu|$ which, since X is symmetric about μ, coincides with

$$\tfrac{1}{2}[F^{-1}(\tfrac{3}{4}) - F^{-1}(\tfrac{1}{4})],$$

the *interquartile range of F*. The estimator $\tau(\hat{F}; \frac{1}{2}, \frac{1}{2})$ has asymptotic standardized variance

$$\frac{1}{16x^2 f^2(x)}, \tag{4.13}$$

where $F(x + \mu) = \frac{3}{4}$.

Again for the Tukey model with $\varepsilon = .5$ the asymptotic efficiency of this estimator relative to the SD tends to zero as $\lambda \to \infty$. Numerical values of the efficiency similar to those shown in Table 3.3 are given in Table 4.1 computed by Winston Chow. They are clearly much less satisfactory than the corresponding values for $\beta = .1$ in Table 3.3. In particular, the low values at the uncontaminated normal make this measure unsuitable.

TABLE 4.1

Asymptotic efficiency of $\tau(\hat{F}; \frac{1}{2}, \frac{1}{2})$ *with respect to the* SD *for the Tukey model* (3.7)

	λ		
ε	2	4	6
0	.37	.37	.37
.1	.62	2.05	3.16
.2	.65	1.43	1.78
.3	.61	1.01	1.11
.4	.56	.73	.70
.5	.51	.52	.43

The result of Theorem 2 is rather disappointing, particularly since it is in such contrast to the general (asymmetric) location case. One may ask whether a positive lower bound can be obtained if the trimmed SD is replaced by a measure of the form (1.10) with $\gamma = 2$ but some other Λ. An essentially negative answer (if attention is restricted to robust measures) is provided by the following generalization of Theorem 2.

THEOREM 3. *Suppose that $\tau(F)$ is given by* (1.10) *where $\gamma = 2$ and the weight function Λ has a density Λ' which vanishes outside $[\alpha, 1-\beta]$ with $0 < \alpha < 1 - \beta < 1$, is bounded, and is bounded away from 0 inside. Then the relative efficiency of $\tau(\hat{F})$ relative to the standard deviation tends to 0 as $\lambda \to \infty$ in the Tukey model* (3.7) *with $\varepsilon = \beta$.*

For the proof we require the following lemma.

LEMMA 3. *If τ_1^2 and τ_2^2 are two functionals of the form* (1.10) *with $\gamma = 2$ and weight functions Λ_1, Λ_2, and if the Λ_i have densities Λ_i' satisfying*

$$0 \leqq a \leqq \Lambda_2'(t)/\Lambda_1'(t) \leqq A < \infty, \tag{4.14}$$

then the efficiency of the estimator of τ_2 to that of τ_1 satisfies

$$e_{2,1}(F) \geqq a^2/A^2. \tag{4.15}$$

PROOF. Condition (4.14) clearly implies that $\tau_2(F) \geqq a\tau_1(F)$ for all F. On the other hand, it follows from Theorem 5 of BL II that $v_2^2(F) \leqq A^2 v_1^2(F)$, and the conclusion follows.

The theorem now follows on taking τ_1 and τ_2 as $\tau(\hat{F}; \alpha, \beta)$ and the SD respectively and applying Remark 1 following Theorem 2.

5. The *p*th power deviations. Although we have not found a measure of dispersion which is robust in the sense of BL (I, II) and whose efficiency relative to the SD has a positive lower bound, it is possible to find measures which are more robust (in the sense of BL II) than the SD and which achieve such a lower bound. In fact, restrict attention to distributions with finite pth moment and consider the pth absolute power deviation

$$\tau^2_{(p)} = (E|X - \mu|^p)^{1/p} \tag{5.1}$$

for $1 \leqq p \leqq 2$, and the associated estimator

$$\delta_{(p)} = \left(\frac{1}{n} \sum |X_i - \mu|^p\right)^{1/p}, \tag{5.2}$$

where from now on we shall assume without loss of generality that $\mu = 0$.

Below we shall obtain lower bounds for the efficiency of (5.2) relative to the SD for the family $\mathscr{F}_0$ of all symmetric distributions. This bound can be improved if we restrict attention to the family $\mathscr{F}_1$ of symmetric unimodal distributions and improved still further for the family $\mathscr{F}_2$ of scale mixtures of normal distributions with a common mean. As in Sections 3 and 4 the robustness properties of these measures follow as in Theorem 3 of BL II, and we shall not discuss them here.

THEOREM 4. *Let $e(F; p, q)$ denote the efficiency of the* p*th relative to the* q*th absolute power deviation where $1 \leqq p < q$. Then*

$$\begin{aligned} \inf_{\mathscr{F}_i} e(F; p, q) &= p^2/q^2 \qquad \textit{for} \quad i = 0 \\ &= \frac{p^2}{q^2}\frac{2p+1}{2q+1}\left(\frac{q+1}{p+1}\right)^2 \qquad \textit{for} \quad i = 1 \\ &= \frac{p^2}{q^2}\frac{\Gamma(q+\frac{1}{2})}{\Gamma^2((q+1)/2)}\frac{\Gamma^2((p+1)/2)}{\Gamma(p+\frac{1}{2})} \qquad \textit{for} \quad i = 2\,. \end{aligned} \tag{5.3}$$

Before proving these results, let us note that for $p = 1$, $q = 2$ we obtain for the efficiency of the absolute deviation to the standard deviation the values

$$\begin{aligned} \tfrac{1}{4} &= .250 \qquad \text{for} \quad i = 0 \\ \tfrac{27}{80} &= .337 \qquad \text{for} \quad i = 1 \\ \frac{1.5}{\pi} &= .477 \qquad \text{for} \quad i = 2\,. \end{aligned}$$

These bounds are rather low and can be improved (at the cost of some robustness) by taking a larger value of p. For example, if $p = 1.5$, the values are

$$\begin{aligned} .563 \quad & \text{for} \quad i = 0 \\ .648 \quad & \text{for} \quad i = 1 \\ .851 \quad & \text{for} \quad i = 2\,. \end{aligned}$$

The proof of all three parts of the theorem hinge on the following lemma, which is equivalent to the case $i = 0$.

LEMMA 4. *Let V be any nonnegative random variable and*

$$\mu_\beta = E(V^\beta) < \infty\,. \tag{5.4}$$

Then, if $1 \leqq \alpha \leqq \beta$,

$$\frac{\mu_{2\alpha}}{\mu_\alpha^{\ 2}} \leqq \frac{\mu_{2\beta}}{\mu_\beta^{\ 2}} \tag{5.5}$$

with equality if and only if V is a positive constant with probability 1.

PROOF. To see this, note that $\log \mu_\alpha$ is a convex function of α (see for example Loève (1955), page 156). From this it follows easily that

$$\frac{\log \mu_{2\beta} - \log \mu_{2\alpha}}{2(\beta - \alpha)} \geqq \frac{\log \mu_\beta - \log \mu_\alpha}{\beta - \alpha}$$

and hence that $\mu_{2\beta}/\mu_\beta^{\ 2} \geqq \mu_{2\alpha}/\mu_\alpha^{\ 2}$ as was to be proved.

PROOF OF THEOREM 4(i). By the central limit theorem we find easily that if the required moments exist, we have

$$e(F; p, q) = \frac{p^2}{q^2}\left\{\left(\frac{E|X|^{2q}}{E^2|X|^q} - 1\right)\Big/\left(\frac{E|X|^{2p}}{E^2|X|^p} - 1\right)\right\}. \tag{5.6}$$

The result then follows immediately from the lemma.

PROOF OF THEOREM 4 (ii, iii). Note first that both of these families are of the following type:

$$\mathscr{F} = \{F : X \sim F,\ X = ZW,\ Z \text{ is fixed and } W \text{ is independent of } Z \text{ and varies freely}\}. \tag{5.7}$$

In case (ii) this is achieved by taking Z to be the uniform distribution on $(-1, 1)$ (see Feller (1966), page 155). In case (iii) we just take Z to be a standard normal variable. To complete the proof we need the following lemma.

LEMMA 5. *Let $\mathscr{F}$ be as in* (5.7) *with Z symmetric about* 0. *Let*

$$\nu_\alpha = E|Z|^\alpha\,. \tag{5.8}$$

Then for $1 \leqq p < q$ and $\nu_{2q} < \infty$,

$$\inf_{\mathscr{F}} e(F; p, q) = \frac{p^2}{q^2}\frac{\nu_{2q}}{\nu_q^{\ 2}}\frac{\nu_p^{\ 2}}{\nu_{2p}}\,. \tag{5.9}$$

PROOF. Let

$$w_\alpha = \frac{E|W|^{2\alpha}}{E^2|W|^\alpha}.$$

Then from (5.6) and (5.7)

$$e(F; p, q) = \frac{p^2}{q^2}\frac{(\nu_{2q}/\nu_q^{\,2})w_q - 1}{(\nu_{2p}/\nu_p^{\,2})w_p - 1} = \frac{p^2}{q^2}\frac{(\nu_{2q}/\nu_q^{\,2})\rho - (1/w_p)}{(\nu_{2p}/\nu_p^{\,2}) - (1/w_p)},$$

where $\rho = w_q/w_p$ and where $X = ZW$ has distribution F. Keeping w_p fixed, this is minimized by taking for ρ its minimum value which is 1 by Lemma 4. Since by Lemma 4, $\nu_{2q}/\nu_q^{\,2} \geqq \nu_{2p}/\nu_p^{\,2}$, we now obtain a lower bound for the efficiency by letting $w_p \to \infty$. To show that this lower bound is sharp we need to exhibit a sequence of distributions for W such that $\rho \to 1$ and $w_p \to \infty$. Take

$$(5.10)\qquad \begin{aligned} W &= 1 \quad \text{with probability} \quad \pi \\ &= a \quad \text{with probability} \quad 1 - \pi\,. \end{aligned}$$

Then,

$$w_p = \frac{\pi + (1 - \pi)a^{2p}}{(\pi + (1 - \pi)a^p)^2}.$$

As $a \to \infty$,

$$w_p \to \frac{1}{1 - \pi} \qquad \text{and} \quad \rho \to 1\,.$$

Now letting $\pi \to 1$ we can extract the requisite sequence of distributions. The lemma follows. □

The theorem now follows from the standard formulas for the absolute moments of uniform and standard normal variables. □

REMARK. Note that for case (iii) of formula (5.3) for the trimmed standard deviations, an extremal sequence is provided by an appropriate sequence of Tukey models. In fact, the distributions defined by (5.10) are of this type.

As we have noted, the lower bound to the efficiency of the mean deviation is rather low even for $\mathscr{F}_2$. However, as we found for trimmed standard deviations, in reasonable situations the bound is very conservative. Here are some numerical results for the pth power deviation for $p = 1$, 1.5 and a selection of the distributions given in Tables 3.1–3.3.

Qualitatively the behavior of these measures closely parallels that of the corresponding Tables 3.1–3.3. For reasonable distributions the mean deviation

TABLE 5.1

Asymptotic efficiency of p*th power deviations with respect to standard deviation t-distribution with f degrees of freedom*

f	5	10	25	50	∞
$p = 1$	2.35	1.12	.94	.91	.88
$p = 1.5$	1.88	1.12	1.01	.99	.97

TABLE 5.2

Asymptotic efficiency of pth *power deviation with respect to* SD:
Beta distribution with density (3.6)

r	.5	1.0	2.0	4	6	8	10	20	30	∞
$p = 1$	.53	.60	.68	.75	.78	.80	.81	.84	.85	.88
$p = 1.5$	.76	.80	.85	.89	.92	.93	.94	.96	.97	.97

TABLE 5.3

Asymptotic efficiency of pth *power deviation witn respect to standard deviation. Tukey model* (3.7)

	λ					
ε	$p = 1$			$p = 1.5$		
	2	4	6	2	4	6
0	.88	.88	.88	.97	.97	.97
.025	1.06	3.08	5.18	1.10	1.99	2.33
.075	1.22	2.53	2.66	1.18	1.61	1.51
.10	1.25	2.21	2.14	1.19	1.48	1.35
.20	1.24	1.50	1.31	1.17	1.20	1.09
.40	1.09	1.04	.92	1.08	1.02	.96
.50	1.03	.95	.85	1.05	.99	.94

particularly seems to do even better than the trimmed deviations. Of course, for sufficiently heavy tailed distributions, for example t-distributions with sufficiently low degrees of freedom, it can break down badly.

6. Measuring the scale of positive random variables. The concepts and results developed so far also apply to a somewhat different problem. Consider a random variable X with distribution F, which is known to be positive. Then one may be interested in scaling this distribution by defining a suitable measure σ of its distance from zero. Of such a measure we shall require (in analogy with the earlier axioms for τ)

$$\sigma(aX) = a\sigma(X) \quad \text{for} \quad a > 0 \tag{6.1}$$

and

$$\sigma(Y) \geqq \sigma(X) \quad \text{if} \quad Y \quad \text{is stochastically larger than} \quad X\,. \tag{6.2}$$

There is a simple correspondence between such measures and the earlier measures τ, defined by

$$\tau(X) = \sigma(|X - \mu|) \tag{6.3}$$

where $\mu = \mu_F$ as before denotes the center of symmetry of X. Given a measure σ defined over positive random variables and satisfying (6.1) and (6.2), let X be a symmetric variable whose center of symmetry is denoted by μ. Then (6.3) defines a measure satisfying (1.4), (1.5) and (1.9). Conversely, let τ be a measure of dispersion defined over symmetric random variables, and let Z be any positive random variable. Extend Z to negative values so that it is symmetric about 0 and denote the resulting random variable by X. Then $\mu = 0$ and $|X - \mu| = |X| = Z$. The measure $\sigma(Z)$ defined through (6.3) satisfies (6.1) and (6.2).

Examples of scale measures σ are provided by the median of a positive random variable X, by the first moment $E(X)$, by the square root of the second moment $(E(X^2))^{\frac{1}{2}}$, or by trimmed versions of these latter measures. Using (6.3), it is a trivial matter to adapt the results of Sections 3 and 4 to comparisons of the estimator of $(E(X^2))^{\frac{1}{2}}$ with its trimmed versions. This is, however, not quite appropriate since the scale measures of greatest interest for positive random variables are those corresponding to $E(X)$ and its trimmed versions.

We believe that the results of numerical and theoretical comparisons of the sample mean and its trimmed competitors qualitatively will be quite similar to those obtained for the sample SD and its trimmed competitors, but we have not carried out this program. In addition to this, Theorem 4 of Section 5 reveals that the sample mean, which is the estimator of $E(X)$, has for $p > 1$ efficiency bounded from below with respect to estimators of $[E(X^p)]^{1/p}$ for the families

$\mathscr{F}_1 = \{$All distributions on the positive axis with finite second moment$\}$

$\mathscr{F}_2 = \{$All members of $\mathscr{F}_1$ with monotone nonincreasing densities$\}$

$\mathscr{F}_3 = \{$All members of $\mathscr{F}_2$ which are scale mixtures of half normal distributions$\}$.

7. Unknown center of symmetry. In estimating dispersion we have so far restricted attention to symmetric distributions and have assumed the center of symmetry to be known. If the center μ of symmetry is unknown, it is tempting to estimate μ by a suitable estimate of location and to substitute this for μ in the estimator of τ. The question naturally arises what effect this has on the asymptotic distribution and hence on the efficiency of the estimator. For the measures of dispersion considered in this paper it turns out that under suitable regularity conditions the asymptotic distribution is unchanged by this substitution. We have not investigated the small sample behavior of these procedures.

The theorem below gives a simple sufficient condition for substitution of an estimate of location to work in this sense. We subsequently check that this condition is satisfied by trimmed standard deviations and pth power deviations among others.

To formalize the process of substitution we appeal to the discussion of Section 6 in which we indicated that there is a 1–1 correspondence between measures of dispersion τ for symmetric distributions and measures of scale σ for positive random variables via (6.3). Let us start then with such a σ defined for nonnegative variables. For reasons similar to those given at the beginning of Section 2, we need to consider extensions of σ to larger families of distributions, namely to the family of all distributions F for which $\sigma(|X|)$ is defined. To avoid proliferation of notation we also call the extension σ and define it by

$$\sigma(X) = \sigma(|X|) .$$

For example, if $\sigma^2(X) = \int_0^\infty x^2\, dF$ for nonnegative variables, the extension of σ^2 is the second moment. If $\sigma(X)$ is the median of the distribution on $(0, \infty)$, the extension is the median of the distribution of $|X|$.

Let $\mu(F)$ be a measure of location satisfying (1.1) and (1.2) of BL II. Suppose also that $X_1, \cdots, X_n$ are independently distributed each with distribution F and let $\hat{F}$ be the empirical cdf. Define

$$\hat{\mu} = \mu(\hat{F})$$

as the usual estimator of $\mu(F)$ and let

$$F_\mu(x) = F(x + \mu)$$

denote the cdf of $X - \mu$. Note that the empirical cdf of $X_1 - \hat{\mu}, \cdots, X_n - \hat{\mu}$ is then $\hat{F}_{\hat{\mu}}$.

If F is symmetric about μ the measure of dispersion corresponding to σ by (6.3) is

$$\tau(F) = \sigma(F_\mu) .$$

The estimator of τ we have used for known μ is $\sigma(\hat{F}_\mu)$. Failing this knowledge we use $\hat{\tau}$ defined by

$$\hat{\tau} = \sigma(\hat{F}_{\hat{\mu}}) .$$

For example if

$$\mu(F) = \int_{-\infty}^\infty x\, dF(x) ,$$
$$\sigma^2(F) = \int_0^\infty x^2\, dG(x) ,$$

where G is the cdf of $|X|$, then

$$\sigma^2(\hat{F}_\mu) = \frac{1}{n} \sum (X_i - \mu)^2$$

and $\hat{\tau}$ is the sample SD.

If

$$\mu(F) = F^{-1}(\tfrac{1}{2}) ,$$
$$\sigma(F) = G^{-1}(\tfrac{1}{2}) ,$$

then $\hat{\tau}$ is the median of the absolute deviations from the median of the observations.

THEOREM 5. *Suppose that the underlying distribution F is symmetric about μ and suppose without loss of generality that $\mu = 0$. Suppose further that*:

(7.1) $$\sigma(F_\mu) \text{ is differentiable in } \mu \text{ at } \mu = 0 ,$$

(7.2) $$\lim_{M\to\infty} \limsup_n P[n^{\frac{1}{2}}|\hat{\mu}| \geqq M] = 0 ,$$

(7.3) $$\lim_{\delta\downarrow 0} \limsup_n P[\sup\{n^{\frac{1}{2}}|\sigma(\hat{F}_\mu) - \sigma(F_\mu) - \sigma(\hat{F}) + \sigma(F)| : |\mu| \leqq \delta\} \geqq \varepsilon] = 0 \quad \text{for all } \varepsilon > 0 .$$

Then,

$$n^{\frac{1}{2}}(\hat{\tau} - \sigma(\hat{F})) \to_P 0 . \tag{7.4}$$

If we assume that $\tau(F)$ is positive and

$$\sigma(\hat{F}) \to_P \tau(F) , \tag{7.5}$$

then we can arrive at conclusion (7.4) *even if throughout* (7.3) *we replace σ by σ^p where $p > 0$.*

NOTE. $\hat{\mu}$ will satisfy (7.2) provided that $n^{\frac{1}{2}}\hat{\mu}$ has a limiting normal distribution.

By Slutsky's theorem, (7.4) implies that $n^{\frac{1}{2}}(\hat{\tau} - \sigma(F))$ and $n^{\frac{1}{2}}(\sigma(\hat{F}) - \sigma(F))$ have the same limiting distribution. But since F is symmetric about 0,

$$\sigma(F) = \tau(F) ,$$

and

$$\sigma(\hat{F}) = \tau(\hat{F}) ,$$

the estimate we would use if the center of symmetry of F were known. Thus, (7.4) implies that $n^{\frac{1}{2}}(\hat{\tau} - \tau(F))$ and $n^{\frac{1}{2}}(\tau(\hat{F}) - \tau(F))$ have the same limiting distribution. That is, if (7.4) holds substitution works.

PROOF. From (7.2) and (7.3),

$$n^{\frac{1}{2}}[\hat{\tau} - \sigma(F_{\hat{\mu}}) - \sigma(\hat{F}) + \sigma(F)] \to_P 0 . \tag{7.6}$$

Since $\hat{\mu} \to_P 0$ by (7.2) we can apply (7.1) to conclude that

$$n^{\frac{1}{2}}\left[(\sigma(F_{\hat{\mu}}) - \sigma(F)) - \hat{\mu}\left\{\frac{\partial\sigma}{\partial\mu}(F_\mu)\Big|_{\mu=0}\right\}\right] \to_P 0 . \tag{7.7}$$

If F is symmetric about 0, if follows from (1.4) and (1.6) that $\sigma(F_\mu)$ is an even function of μ, and hence that $(\partial\sigma/\partial\mu)(F_\mu)|_{\mu=0} = 0$. Substituting in (7.7) and (7.6) completes the proof of (7.4).

If now σ is replaced by σ^p in (7.3) we can imitate the proof of (7.4) exactly to conclude that

$$n^{\frac{1}{2}}(\sigma^p(\hat{F}_{\hat{\mu}}) - \sigma^p(\hat{F})) \to_P 0 . \tag{7.8}$$

But,

$$n^{\frac{1}{2}}(\hat{\tau} - \sigma(\hat{F})) = \frac{1}{p}\,\tilde{\sigma}^{p-1} n^{\frac{1}{2}}(\sigma^p(\hat{F}_{\hat{\mu}}) - \sigma^p(\hat{F})) ,$$

where $\tilde{\sigma}$ lies between $\sigma(\hat{F}_{\hat{\mu}})$ and $\sigma(\hat{F})$. By (7.5) and (7.8)

$$\tilde{\sigma} \to_P \sigma(F) > 0$$

and (7.4) follows. □

EXAMPLES.

(i) *The* pth *power deviations.* Define $\tau(H) = (\int_{-\infty}^{\infty} |x|^p \, dH(x))^{1/p}$ for all H such that $\int |x|^p \, dH(x) < \infty$. Then $\tau(F_\mu)$ is the pth power deviation. We shall check that (7.3) holds for τ^p under the condition

$$\int |x|^{2p}\, dF(x) < \infty . \tag{7.9}$$

For $\mu < \nu$, we find

$$E(n^{\frac{1}{2}}(\tau^p(\hat{F}_\nu) - \tau^p(\hat{F}_\mu) - \tau^p(F_\nu) + \tau^p(F_\mu)))^2 = \text{Var}\,(|X_1 - \nu|^p - |X_1 - \mu|^p) .$$

If $p > 1$ this variance is bounded by

$$\begin{aligned} E(|X_1 - \nu|^p - |X_1 - \mu|^p)^2 &\leqq p^2(\nu - \mu)^2 \sup \{E|X_1 - \lambda|^{2p-2} : \lambda \in [\mu, \nu]\} \\ &\leqq C(\nu - \mu)^2 \quad \text{if } |\nu|, |\mu| \leqq M , \end{aligned}$$

where C is a constant depending on M. If $p = 1$,

$$\begin{aligned} \text{Var}\,(|X_1 - \nu| - |X_1 - \mu|) &\leqq E[(\nu - \mu)(2I_{[X_1<\mu]} - 1) + 2(\nu - X_1)I_{[\mu<X_1<\nu]}]^2 \\ &\leqq C(\nu - \mu)^2 \quad \text{if } |\nu|, |\mu| \leqq M . \end{aligned}$$

In any event

$$E(n^{\frac{1}{2}}(\tau^p(\hat{F}_\nu) - \tau^p(\hat{F}_\mu) - \tau^p(F_\nu) + \tau^p(F_\mu)))^2 \leqq C(\nu - \mu)^2$$

and an argument using Theorem 12.3, page 95 in [3] completes the proof. Since (7.9) implies (7.5) as well as the asymptotic normality of $\tau(\hat{F})$ we see that any measure $\hat{\mu}$ satisfying (7.2), for instance the mean, can be substituted without affecting the asymptotic behavior of the estimators.

(ii) (L) *estimators*. Suppose τ is given by

$$\tau(H) = [\textstyle\int_0^1 |H^{-1}(t)|^r \, d\tilde{\Lambda}(t)]^{1/r} \tag{7.10}$$

where $\tilde{\Lambda}$ is symmetric about $\frac{1}{2}$.

If F is symmetric about μ, and if we define $\Lambda(t) = 2\hat{\Lambda}((1 + t)/2) - 1$, then $\tau(F_\mu)$ is the measure defined in (1.10).

Suppose that

(7.11) Λ places no mass outside an interval $(\alpha/2, 1 - \alpha/2)$, $0 < \alpha < \frac{1}{2}$

(7.12) F is differentiable with derivative f which is positive and continuous.

We shall now show that (7.3) holds for τ^r under these assumptions, so that we may safely substitute an estimate of location in the singly and doubly trimmed standard deviations.

Here is a sketch proof.

Let

$$G(x, \mu) = P[|X_1 - \mu|^r \leqq x] ,$$

and let $\hat{G}(x, \mu)$ be the empirical cdf of $|X_1 - \mu|^r, \cdots, |X_n - \mu|^r$. Define

$$G(x) = G(x, 0) , \qquad \hat{G}(x) = \hat{G}(x, 0) .$$

Finally let $G^{-1}(t, \mu)$, $\hat{G}^{-1}(t, \mu)$ be the corresponding inverses (in x).

Note that

$$\tau(F_\mu) = 2 \int_0^{1-\alpha} G^{-1}(t, \mu)\, d\tilde{\Lambda}\left(\frac{1+t}{2}\right) = \int_0^{1-\alpha} G^{-1}(t, \mu)\, d\Lambda(t)\,,$$

and similar assertions hold for $\hat{F}$, $\hat{F}_\mu$, etc.

Then,

$$\begin{aligned} &|\tau^r(\hat{F}_\mu) - \tau^r(\hat{F}) - \tau^r(F_\mu) + \tau^r(F)| \\ &\quad = |\textstyle\int_0^{1-\alpha} (\hat{G}^{-1}(t, \mu) - \hat{G}^{-1}(t) - G^{-1}(t, \mu) + G^{-1}(t))\, d\Lambda(t) \\ &\quad \leq \sup\{|\hat{G}^{-1}(t, \mu) - \hat{G}^{-1}(t) - G^{-1}(t, \mu) + G^{-1}(t)| : 0 \leq t \leq 1 - \alpha\}\,. \end{aligned} \tag{7.13}$$

Therefore, we need only establish that

$$\begin{aligned} \lim\nolimits_{\delta \downarrow 0} \limsup\nolimits_n P[\sup\{n^{\frac{1}{2}}|\hat{G}^{-1}(t, \mu) - \hat{G}^{-1}(t) - G^{-1}(t, \mu) \\ + G^{-1}(t)| : 0 \leq t \leq 1 - \alpha, |\mu| \leq \delta\} \geq \varepsilon] = 0\,. \end{aligned} \tag{7.14}$$

By a standard but tedious argument (see [1] and [8]), we can show that (7.14) follows if, for every M

$$\sup\{|\hat{G}^{-1}(t, \mu) - G^{-1}(t, \mu)| : 0 \leq t \leq 1 - \alpha, |\mu| \leq M\} \to_P 0 \tag{7.15}$$

and

$$j(t, \mu) = \frac{\partial G^{-1}}{\partial t}(t, \mu) \quad \text{is continuous on} \tag{7.16}$$

$$\left\{(t, \mu) : 0 \leq t \leq 1 - \frac{\alpha}{2},\ |\mu| \leq M\right\}$$

and if in addition

$$\begin{aligned} \lim\nolimits_{\delta \downarrow 0} \limsup\nolimits_n P\Bigg[\sup\Bigg\{n^{\frac{1}{2}}\Bigg|\frac{\hat{G}(x, \mu) - G(x, \mu)}{j(G(x, \mu), \mu)} \\ - \frac{\hat{G}(x) - G(x)}{j(G(x), 0)}\Bigg| : |x| \leq G^{-1}(1 - \alpha), |\mu| \leq \delta\Bigg\} \geq \varepsilon\Bigg] = 0\,. \end{aligned} \tag{7.17}$$

Now, (7.15) follows from the fact that for every M and M'

$$\sup\{|\hat{G}(x, \mu) - G(x, \mu)| : x \leq M', |\mu| \leq M\} \to_P 0\,. \tag{7.18}$$

This assertion is a consequence of the Glivenko–Cantelli theorem and the definition of $\hat{G}$, G.

Since

$$j(t, \mu) = \frac{1}{g(G^{-1}(t, \mu), \mu)} = [f(G^{-1}(t, \mu) + \mu) + f(-G^{-1}(t, \mu) + \mu]^{-1}\,,$$

assertion (7.16) is immediate from assumption. Finally, (7.17) follows from (7.16), the tightness of the empirical process

$$\begin{aligned} \lim\nolimits_{\delta \downarrow 0} \limsup\nolimits_n P[\sup\{n^{\frac{1}{2}}|\hat{F}(x + \mu) - F(x + \mu) - \hat{F}(x) \\ + F(x)| : |\mu| \leq \delta, |x| \leq M\} \geq \varepsilon] = 0\,, \end{aligned}$$

and the boundedness (in probability) of $\sup_x n^{\frac{1}{2}}|\hat{G}(x) - G(x)|$. Therefore, (7.13) holds for τ defined by (7.10). As before (7.1) and (7.5) are usually obvious under our assumptions and we conclude that substitution of location estimates satisfying (7.2) such as the median is legitimate.

REFERENCES

[1] BICKEL, P. J. (1973). On some analogues to linear combinations of order statistics in the linear model. *Ann. Statist.* **1** 597–616.

[2] BICKEL, P. J. and LEHMANN, E. L. (1975). Descriptive statistics for nonparametric models. I. Introduction. *Ann. Statist.* **3** 1038–1044; II. Location. *Ann. Statist.* **3** 1045–1069; IV. Spread. To appear in Hájek volume. These papers are referred to as BL I, II, IV.

[3] BILLINGSLEY, P. (1968). *Weak Convergence of Probability Measures.* Wiley, New York.

[4] BIRNBAUM, Z. W. (1948). On random variables with comparable peakedness. *Ann. Math. Statist.* **37** 1593–1601.

[5] DANIELL, P. J. (1920). Observations weighted according to order. *Amer. J. Math.* **42** 222–236.

[6] FELLER, WILLIAM (1966). *An Introduction to Probability Theory and Its Applications* **2**. Wiley, New York.

[7] LOÈVE, MICHEL (1955). *Probability Theory.* Van Nostrand, New York.

[8] PYKE, R. and SHORACK, G. (1968). Weak convergence of a two-sample empirical process and a new approach to Chernoff–Savage theorems. *Ann. Math. Statist.* **39** 755–771.

[9] STIGLER, STEPHEN M. (1973). The asymptotic distribution of the trimmed mean. *Ann. Statist.* **1** 474–477.

DEPARTMENT OF STATISTICS
UNIVERSITY OF CALIFORNIA
BERKELEY, CALIFORNIA 94720

DESCRIPTIVE STATISTICS FOR NONPARAMETRIC MODELS IV. SPREAD

by

P. J. BICKEL (1)*, AND E. L. LEHMANN (2)**

1. Ordering by spread

In the preceding paper of this series [1] (to which we refer as BL III) we studied the dispersion of a symmetric distribution about its center of symmetry. In the present paper we study a related aspect of dispersion, which does not require the assumption of symmetry but which even in the symmetric case does not coincide with the concept considered in BL III. Roughly speaking, instead of looking at dispersion relative to a fixed point, we now consider the spread of a random variable throughout its distribution. The difference is perhaps best explained in terms of an example.

Example 1. Let,

$$f_p(x) = \tfrac{1}{2}p \quad \text{if} \quad |x| \leqq 1, \tag{1.1}$$
$$= \tfrac{1}{2}(1 - p) \quad \text{if} \quad 1 < |x| \leqq 2,$$

for $0 < p < 1$. Let X have distribution F with density f_p, Y have distribution G with density f_{1-p} for $p > \frac{1}{2}$. Then it follows from (1.2) of BL III that $|Y|$ is stochastically larger than $|X|$ and hence Y is more dispersed about 0 than X according to the definition of BL III. This corresponds to our intuitive feeling that in a global sense G is more dispersed than F since it can be obtained by pushing some of the central mass of F into the tails. Yet, locally for $1 < |X| < 2$, G is more concentrated than F since it has a higher uniform density there.

The basic definition of BL III for calling a symmetric distribution G more dispersed about its center of symmetry than a distribution F about its center of symmetry is equivalent to

$$G^{-1}(v) - G^{-1}(\tfrac{1}{2}) \gtreqless F^{-1}(v) - F^{-1}(\tfrac{1}{2}) \quad \text{as} \quad v \gtrless \tfrac{1}{2}. \tag{1.2}$$

* This paper was prepared with support of U.S. Office of Naval Research Contract No. N00014-69-A-0200-1038/NR042-036.

** This research was prepared with support of National Science Foundation Grant No. GP-38485.

In the present paper, we shall call an arbitrary (i.e. not necessarily symmetric) distribution G *more spread out* than a distribution F if

$$G^{-1}(v) - G^{-1}(u) \geqq F^{-1}(v) - F^{-1}(u) \quad \text{for all} \quad 0 < u < v < 1\,, \tag{1.3}$$

where F^{-1} is defined by $F^{-1}(u) = \sup\{x : F(x) \leqq u\}$, that is, if two percentage points of G are at least as far apart as the corresponding percentage points of F. A concept which contains the essence of this definition was introduced by Brown and Tukey (1946).

Note that the definition (1.3) reflects the greater concentration of F throughout, and it of course does not require symmetry, thus fulfilling the two desiderata mentioned above. Furthermore, a comparison of (1.3) with (1.2) shows that for symmetric distributions (1.2) is satisfied whenever (1.3) holds; that (1.3) is in fact more stringent follows from Example 1.

Two properties of (1.2) noted for symmetric distributions in BL III are seen to be implied for arbitrary distributions by (1.3), namely,

(a) Any random variable is more spread out than a constant;

(b) aX is more spread out than X if $a > 1$.

Note however, that the ordering is not invariant under monotone transformations: $G^{-1}(u) - F^{-1}(u)\uparrow$ does not imply $h[G^{-1}(u)] - h[F^{-1}(u)]\uparrow$.

In BL III it was noted that if F and G are symmetric about 0 with densities f and g satisfying: $g(x)/f(x)$ is increasing for $x > 0$, then (1.2) holds. Example 1 shows that these conditions are not enough to insure (1.3).

Example 2. Let X take on the values $a < b$ with probabilities p and q $(0 < p < 1)$. Then

$$F^{-1}(v) - F^{-1}(u) = \begin{cases} 0 & \text{if} \quad p \leqq u \quad \text{or} \quad v < p\,, \\ b - a & \text{if} \quad u < p \leqq v\,. \end{cases} \tag{1.4}$$

From (1.4) it is easily seen that no continuous strictly increasing distribution can be either more or less spread out than F in the sense of (1.3). More generally, continuous distributions and discrete distributions are not comparable. This illustrates how strong a requirement (1.3) is. We shall show below that it is nevertheless satisfied in many cases. However, it is convenient first to give some alternative expressions for condition (1.3).

Suppose that F^{-1} and G^{-1} are differentiable. Dividing both sides of (1.3) by $v - u$, it is then clear that (1.3) implies

$$\frac{\mathrm{d}}{du}[G^{-1}(u)] \geqq \frac{\mathrm{d}}{du}[F^{-1}(u)] \quad \text{for all} \quad u \tag{1.5}$$

and conversely (1.5) implies (1.3). Thus, G is more spread out than F if G^{-1} is nowhere steeper than F^{-1}.

Evaluating the derivatives in (1.5), we see that another form of the condition is

$$g[G^{-1}(u)] \leqq f[F^{-1}(u)] \quad \text{for all} \quad u\,. \tag{1.6}$$

This is a so-called tail ordering condition introduced by Doksum (1969).

We want to stress that in our opinion this ordering corresponds to spread as property (b) above indicates. Tailweight should be specified by a scale-free ordering such as van Zwet's [5] or Lawrence's [4] and measured by scale-free functionals such as the kurtosis.

Lemma 2.2 of Doksum [3] states that, for distributions symmetric about 0, G ordered with respect to F in Lawrence's sense and $g(0) \leqq f(0)$ implies that G is more dispersed than F. This is consistent with our point of view. To see this note that $1/f(0)$ is a measure of scale and the condition $g(0) \leqq f(0)$ added to the scale-free ordering of Lawrence creates a new ordering of spread which possesses properties (a) and (b).

We mention finally a form which for the sake of simplicity we shall state under the additional assumption that F and G are strictly increasing.

Theorem 1. *If F and G are strictly increasing then G is more spread out than F if and only if there exists a strictly increasing function h such that*

(i) *$x < x'$ implies $h(x') - h(x) \geqq x' - x$*

and

(ii) *if X has distribution F, then $h(X)$ has distribution G.*

Proof. Suppose first that such a function exists. Then (1.3) follows from the relation $G^{-1}(u) = h[F^{-1}(u)]$. Conversely, if (1.3) holds, it is easily seen that the function $h(x) = G^{-1}[F(x)]$ has the desired property.

Theorem 1 shows that G being more spread out than F means that one can get from F to G by spreading all pairs of points further apart. Note also that Theorem 1 shows that our dispersion ordering depends on $G^{-1}F$ only.

Example 3. Let F be the uniform distribution on (0, 1). Since the right-hand side of (1.6) is then 1 for all $0 < u < 1$, the condition for G with density g to be more spread out than F is

$$g(x) \leqq 1 \quad \text{for all} \quad x\,. \tag{1.7}$$

Similarly, G is less spread out than F if and only if its support is an interval (a, b) of length <1 and if in this interval $g(x) \geqq 1$ for all x.

Example 4. Let $F = \Phi$ be the standard normal distribution. Then G is more spread out than Φ if and only if the normal probability plot of G, $y = \Phi^{-1}\, G(x)$ has slope $\geqq 1$ at all points.

Example 5. Let F be the double exponential distribution with density $\frac{1}{2}\exp(-|x|)$. Then

$$(1.8) \qquad f[F^{-1}(u)] = \begin{cases} u & \text{if } 0 < u < \frac{1}{2}, \\ 1-u & \text{if } \frac{1}{2} < u < 1 \end{cases}$$

and G is more spread out than F if

$$(1.9) \qquad g(x) \leqq \begin{cases} G(x) & \text{for all } x < 0, \\ 1 - G(x) & x > 0. \end{cases}$$

Alternatively the failure rate of G must be at most 1 when x is interpreted as time running in either direction from the origin.

Example 6. Let F be the logistic distribution with density $e^{-x}(1 + e^{-x})^{-2}$. Then, G is more spread out than F if and only if $g(x) \leqq G(x)(1 - G(x))$ for all x and less spread out if and only if the reverse inequality holds for all x.

From (1.9) it follows easily, for example, that a normal distribution with sufficiently small variance is less spread out than F, but that a normal distribution can never be more spread out than F no matter how large its variance. The situation is just the reverse in the case of a Cauchy distribution, which is more spread out than F if its scale is sufficiently large, but which can never be less spread out than F.

Finally, the logistic distribution provides an example of a distribution which is more spread out than F for sufficiently large scale and less spread out than F for sufficiently small scale.

An interesting connection between the dispersion ordering of BL III and the present ordering by spread is given by the following result.

Theorem 2. *If Y is more spread out than X and if Y', Y'' and X', X'' are independent copies of Y and X respectively, then $Y'' - Y'$ is more dispersed than $X'' - X'$.*

Proof. Let h be the function guaranteed by Theorem 1. Then $|Y'' - Y'| = |h(X'') - h(X')| \geqq |X'' - X'|$ and hence $|Y'' - Y'|$ is stochastically larger than $|X'' - X'|$.

That the converse of Theorem 2 does not hold is shown by the following generalization of Example 1.

Example 7. Let $f = f_p$, $g = f_{p'}$ be defined by (1.1). Then if $p > p' \geqq \frac{1}{2}$, it is easily seen that $X'' - X'$ is less dispersed than $Y'' - Y'$; on the other hand, G is not more spread out than F.

2. Measures of spread

The axioms for a measure of spread coincide with those for a measure of dispersion given in BL III except that in (1.9) of that paper dispersion ordering is replaced by the present spread ordering and that minor differences result from the dropping of the assumption of symmetry. For the sake of completeness, we shall now restate the full set of axioms.

A *measure of spread* is a functional $\Delta(F)$ (also denoted by $\Delta(X)$ where X is a random variable with distribution F) defined over a sufficiently large class of distributions which is closed under changes of location and scale. We shall require Δ to be nonnegative and to satisfy

$$\Delta(aX) = |a|\,\Delta(X) \quad \text{for} \quad a > 0\,, \tag{2.1}$$

$$\Delta(X + b) = \Delta(X) \quad \text{for all} \quad b\,, \tag{2.2}$$

and

$$\Delta(-X) = \Delta(X)\,. \tag{2.3}$$

As before, these conditions imply that

$$\Delta(c) = 0 \quad \text{for any constant} \quad c\,. \tag{2.4}$$

The converse: $\Delta(X) = 0$ implies $X = c$, is false for measures of spread as well as for measures of dispersion. Let Δ be the measure given in (2.12) below and F assign mass $\frac{1}{4}$ each to the points ± 1 and $\frac{1}{2}$ to 0. Then $\Delta(F) = 0$. A nonnegative functional satisfying (2.1), (2.2) and (2.3) will be called a measure of spread if it satisfies in addition

$$\Delta(F) \leqq \Delta(G) \quad \text{whenever } G \text{ is more spread out than} \quad F\,. \tag{2.5}$$

Note that if $\Delta(F)$ is a measure of spread, so is $\varkappa\,\Delta(F)$ for any $\varkappa > 0$.

A large and interesting class of measures of spread is obtained when the following theorem is applied to some of the results of BL III.

Theorem 3. *Let $\tau(X)$ be a measure of dispersion in the sense of* BL III *and let X', X'' be two independent copies of X. Then*

$$\Delta(F) = \tau(X'' - X') \tag{2.6}$$

is a measure of spread.

Proof. That Δ satisfies (2.1)–(2.3) is obvious from the fact that τ satisfies the corresponding conditions. To prove (2.5), suppose that G is more spread out than F. Then $X'' - X'$ and $Y'' - Y'$ are symmetric about 0 and it follows from Theorem 2 that $Y'' - Y'$ is more dispersed than $X'' - X'$ and hence from the fact that τ is a measure of dispersion that $\Delta(F) = \tau(X'' - X') \leqq \tau(Y'' - Y') = \Delta(G)$ as was to be proved.

Example 8. As a first example let $\tau(F)$ be the standard deviation of X, which was seen in BL III to be a measure of dispersion. Then

$$\Delta(F) = \{E(X'' - X')^2\}^{1/2} = \sqrt{(2)}\,[\tau(F)]\,.$$

It follows that $\Delta(F)$ is a measure of spread, and hence also that

$$\Delta(F) = SD(F) \tag{2.7}$$

is a measure of spread although no longer restricted to symmetric F.

An obvious generalization, obtained by starting with the pth power deviations considered in BL III are the measures

$$\Delta_p(F) = \{E\,|\,X'' - X'\,|\,^p\}^{1/p}\,. \tag{2.8}$$

Example 9. Similarly, by starting with $\tau(F) = \text{med}\,|X|$, we find that

$$\Delta(F) = \text{med}\,|X'' - X'| \tag{2.9}$$

is a measure of spread.

Example 10. A class of examples not having the above structure is given by

$$\Delta(F) = F^{-1}(t) - F^{-1}(1-t) \quad \text{for any} \quad t > \tfrac{1}{2}\,, \tag{2.10}$$

which obviously satisfies (2.1)–(2.3) and (2.5). The same is true of the more general class

$$\Delta(F) = \left[\int_{1/2}^{1} [F^{-1}(t) - F^{-1}(1-t)]^{\gamma}\, d\Lambda(t)\right]^{1/\gamma} \tag{2.11}$$

where Λ is any finite measure on $(\tfrac{1}{2}, 1)$.

A case of particular interest is $t = \tfrac{3}{4}$, and hence

$$\Delta(F) = F^{-1}(\tfrac{3}{4}) - F^{-1}(\tfrac{1}{4}) \tag{2.12}$$

the interquartile range of F. In BL III, this was seen to be a measure of dispersion for symmetric distribution; it now follows that it is a measure of spread for arbitrary distributions. More generally, the class of measures (2.11) when restricted to symmetric distributions for suitable Λ coincides with the class of dispersion measures (1.10) of BL III. Note, however, that even for $\gamma = 2$ and Λ = Lebesgue measure, the measure (2.11) is not a multiple of the SD when F is asymmetric.

That not every measure of spread, when restricted to symmetric distributions, reduces to a measure of dispersion is shown by the following example.

Example 11. Let X take on the values $-1, 0, 1$ with probabilities $p/2$, $1-p$, $p/2$ respectively. Then

$$E\{|X'' - X'|^r\} = 2p\;1 - p) + 2^{r-1}p^2$$

and it is easily seen that for $r < 1$, the left-hand side is not an increasing function of p. Since $|X|$ is stochastically increasing with p, this means that $[E\{|X'' - X'|^r\}]^{1/r}$ is not a measure of dispersion for symmetric distributions although it is a measure of spread.

3. Choice of measure

The choice of a measure of spread, as was the case for measures of location and of dispersion will be based largely on the accuracy with which it can be estimated, and the appropriate measure of accuracy is the same of that discussed in BL III for dispersion, the *standardized asymptotic variance*, i.e. the asymptotic variance of the estimator of $\Delta(F)$ divided by $\Delta^2(F)$. As in the earlier paper one would again try to find measures which behave satisfactorily (a) in relation to robustness, either by being robust or at least more robust than $\Delta_1(F)$; and (b) possessing good efficiency relative to $\Delta_1(F)$, ideally for all F, but if this cannot be achieved at least for the type of F likely to occur in practice.

We have not carried out this program in the present case. The work of BL III suggests that in terms of the indicated properties no completely satisfactory measure is likely to exist. It suggests further that the following two types of measures may be reasonable compromise solutions.

(i) *The pth power measures* (2.8). While not robust, for $p < 2$ these measures are presumably more robust than $\Delta(F)$, which corresponds to the case $p = 2$. The natural estimator of $\Delta(F)$ is $\hat{\Delta}(F)$ or the asymptotically equivalent statistic

$$\hat{\Delta}_p = \left\{\frac{1}{\binom{n}{2}} \sum_{i<j}\sum |X_j - X_i|^p\right\}^{1/p}. \tag{3.1}$$

Since this is a U-statistic, it follows from the work of Hoeffding that $\sqrt{(n)}\,.\,[\hat{\Delta} - \Delta(F)]$ is asymptotically normal. It seems plausible to conjecture that for values of $p \geqq 1.5$ the asymptotic efficiency of $\hat{\Delta}_p$ relative to the standard deviation $\hat{\Delta}_2$ will be reasonably high at least for typical distributions.

(ii) *Trimmed standard deviations*. There are two possible versions of trimmed standard deviations in this context.

(a) We can take $\gamma = 2$ and Λ the uniform distribution on $(\frac{1}{2}, 1 - \beta)$ in (2.11).

(b) We can consider $\tau(X - X', \alpha, \beta)$ where $\tau(X, \alpha, \beta)$ is given by (3.1) of BL III.

(As usual we let X represent its distribution.)

It is easy to see that if $0 < \alpha < 1 - \beta < 1$ both of these measures are robust. Their estimates are given by (essentially),

(a) $$[n(\tfrac{1}{2} - \beta)]^{-1/2}\Big[\sum_{k=n/2}^{n(1-\beta)} [X_{(k)} - X_{(n-k+1)}]^2\Big]^{1/2}$$

(b) $$\left[\binom{n}{2}(1 - \beta - \alpha)\right]^{-1/2}\Big[\sum_{k=\binom{n}{2}}^{\binom{n}{2}(1-\beta)} (X - X')^2_{(k)}\Big]^{1/2}$$

where $X_{(1)} \leqq \ldots \leqq X_{(n)}$ are the order statistics of $X_1, \ldots, X_n$ and $(X - X')_{(1)} \leqq \ldots \ldots \leqq (X - X')_{\binom{n}{2}}$ are the order statistics of the pseudo sample $X_i - X_j$, $i < j$. Asymptotic normality with variance of the order $1/n$ of these estimates is evident only for case (a) when $\gamma = 1$. It seems plausible that this property holds in general and that the resulting efficiencies with respect to the S.D. have the same general numerical features as we found for their analogues in studying dispersion. However, we do not pursue this. We do not know a fortiori which of the measures (a) or (b) is preferable and leave these interesting questions open.

References

[1] Bickel, P. J. - Lehman, E. L. (1976). Descriptive statistics for nonparametric models III. Dispersion. *Ann. Statist.*, **4**, 1139—1158.

[2] Brown, G. - Tukey, J. W. (1946). Some distributions of sample means. *Ann. Math. Statist.* **7**, 1—12.

[3] Doksum, K. (1969). Starshaped transformations and the power of rank tests. *Ann. Math. Statist.*, **40**, 1167—1176.

[4] Lawrence, M. J. (1975). Inequalities of *s*-ordered distributions. *Ann. Statist.*, **3**, 413—428.

[5] Zwet, van, W. R. (1964). Convex transformations of random variables. *Math. Centrum*, Amsterdam.

(1), (2) UNIVERSITY OF CALIFORNIA, BERKELEY, CALIFORNIA, U.S.A.

Received October 1975

Part VI
Asymptotics

Erich Lehmann's Work on Asymptotics

David Draper

Erich Lehmann's contributions to mathematical statistics have been both broad and deep, and his influence on applied statistics has been more widespread than perhaps even he knew. As an example, I used `Google scholar` to look at the citation patterns of Chernoff and Lehmann (1954), one of the articles I discuss below. As of this writing (November 2011), this paper has been cited

- quite steadily from 1954 to the present, in fact a total of 327 times, by workers in archaeology, astronomy, biology, biostatistics, business, computer engineering, computer science, ecology, econometrics, education, electrical engineering, environmental sciences, finance, genetics, gerontology, machine learning, mathematical statistics, mechanical engineering, oncology, operations research, physics, probability theory, psychology, reliability theory, remote sensing, sports sciences, traffic safety, and urban planning;
- by researchers writing in English, French, German, Italian, Korean, Romanian, Russian, and Spanish;
- by investigators working in Australia, Belgium, Canada, China, Denmark, France, Germany, Greece, India, Italy, Japan, Korea, the Netherlands, New Zealand, Romania, Russia, Spain, Taiwan, the U.K., and the U.S.; and
- by both frequentists and Bayesians.

A small (non-random) sample of the titles of the works that cite this paper, chosen to illustrate the breadth of application areas, includes the following (alphabetically by the first letter of the title):

- "Antecedents and implications of search engine use as pre-purchase information tools";

D. Draper
Department of Applied Mathematics and Statistics, Baskin School of Engineering,
University of California, 1156 High Street, Santa Cruz CA 95064 USA
e-mail: draper@ams.ucsc.edu

J. Rojo (ed.), *Selected Works of E. L. Lehmann*, Selected Works in Probability and Statistics, DOI 10.1007/978-1-4614-1412-4_46, © Springer Science+Business Media, LLC 2012

- "A downscaling approach for air quality at a mid-latitude site using circulation patterns and surface meteorology";
- "Bayesian item fit analysis for unidimensional item response theory models";
- "Communication of emotion in mediated and technology-mediated contexts: face-to-face, telephone, and instant messaging";
- "Gender equity and `foxsports.com`: a coverage analysis of the 2007 NCAA Division I basketball tournament";
- "Guiding architectural SRAM models";
- "Is the universe expanding?";
- "Labor market segmentation: the case of Ukraine and Russia";
- "Mechanical state estimation for overhead-transmission power lines with level spans";
- "Nonlinear multisystem physiological dysregulation associated with frailty in older women";
- "Opinion retrieval from blogs";
- "Principle of detailed balance and convergence assessment of Markov Chain Monte Carlo methods and simulated annealing" (this topic is absolutely central to contemporary Bayesian computing);
- "Probabilistic approach for durability design of reinforced concrete in marine environments";
- "Production of antibody fragments in *Arabidopsis* seeds";
- "Statistical problems in ancient numismatics";
- "Supermarket customers segments stability";
- "Transcriptomic and metabolite analyses of Cabernet Sauvignon grape berry development";
- "Universal Donsker classes and metric entropy";
- "Vergleich verschiedener Postremissionsstrategien bei dur akuten myeloischen Leukämie mit normalem Karyotyp"; and
- "Whom we laugh with affects how we laugh."

In what follows I examine a portion of Lehmann's contributions to asymptotic methods in statistics, by revisiting and commenting upon six of his papers in this field.

Chernoff and Lehmann (1954) *The use of maximum likelihood estimates in* χ^2 *tests for goodness of fit*. The problem considered by the authors of this paper is the use of the χ^2 machinery to test for goodness of fit to a specified distribution such as the Poisson or normal, a method that at the time they wrote the paper had been in use for more than 50 years (Pearson (1900)). Suppose (to illustrate the issues) that you have data values $(y_1, \ldots, y_n)$ on the real line that you think are like a random sample from a normal distribution with unknown mean μ and standard deviation σ, and you want to make a repeated-sampling calculation that will assess the plausibility of this model. If you knew μ and σ, Pearson would tell you to (a) partition the real line into k intervals (although he was not so clear on suggesting what k should be and what cut-points to use); (b) calculate

$$D = \sum_{j=1}^{k} \frac{(n_j - np_j)^2}{np_j}, \tag{1}$$

where n_j is the number of data values in interval j and p_j is the probability that the $N(\mu, \sigma^2)$ distribution assigns to that same interval; and (c) refer D to its asymptotic χ^2 distribution with $(k-1)$ degrees of freedom to judge whether the assumed model is plausible. If you don't know μ and σ, it's natural (in this frequentist setting) to estimate them from the sample $(y_1, \ldots, y_n)$, using a method such as maximum likelihood, and then pretend that

$$\hat{D} = \sum_{j=1}^{k} \frac{(n_j - n\hat{p}_j)^2}{n\hat{p}_j} \tag{2}$$

also has a large-sample χ^2 distribution, this time with $(k-s-1)$ degrees of freedom, where $\hat{p}_j$ is the MLE of p_j and where (in this case) $s = 2$ (the number of parameters estimated). Chernoff and Lehmann note that "this is in fact the procedure recommended in many textbooks," and it is the purpose of their paper to show that this approach may in practice be flawed, because "the test statistic $[\hat{D}]$ ... is stochastically larger than would be expected under the χ^2 theory." They work out the actual asymptotic distribution of $\hat{D}$ and provide some numerical examples of how far wrong one can go using the χ^2_{k-s-1} distribution; for instance, with the normal setup above and the $k = 4$ intervals obtained by cutting the real line at $\{-1, 0, +1\}$ they obtain a lower bound of 0.12 for the actual significance level of a nominal 0.05 test, showing that "in the normal case the use of maximum likelihood estimates in χ^2 may lead to [a serious] underestimate of the probability of type I error." The reverberations of this elegant paper, only eight journal pages long, are still being felt today.

Fix, Hodges and Lehmann (1959): *The restricted χ^2 test*. This is another paper about χ^2, but in this case the authors are interested in examining variants on the usual χ^2 method that have greater power against a class of specified alternatives than the unrestricted χ^2 test (while of course at the same time sacrificing power against other alternatives). Since the basic idea of hypothesis tests is to construct a measure of the distance between {how the data came out} and {how the data should have come out if the null hypothesis were true}, and since the usual χ^2 measure is clearly constructed to be a distance between observed and expected counts, it's natural for Fix, Hodges and Lehmann (FHL) to adopt a geometric perspective in examining their restricted χ^2 tests, and they do so in a way that again brings to mind the word "elegant." Letting $X = (X_1, \ldots, X_k)$ be multinomial with parameters n and $p = (p_1, \ldots, p_k)$, FHL note that the usual χ^2 measure when testing the simple hypothesis that $p = \pi$ can be written as

$$n \sum_{i=1}^{k} \frac{(R_i - \pi_i)^2}{\pi_i}, \tag{3}$$

where $R = (R_1, \ldots, R_k) = \frac{X}{n}$. This is a special case of the general expression

$$D_c(a, b) = n \sum_{i=1}^{k} \frac{(a_i - b_i)^2}{c_i} \tag{4}$$

for vectors a, b and c of values between 0 and 1 and summing to 1; in particular (3) is $D_\pi(R, \pi)$. When the null hypothesis is composite, say $p \in T$ where T is a t-dimensional surface on the hyperplane $\sum_{i=1}^k p_i = 1$, it's again natural to use the data X to find the best estimate π^* of the location of p if the null were true, and to base the test on $D_{\pi^*}(R, \pi^*)$. FHL note that Pearson (1900) got the asymptotic distribution of this test statistic wrong and that the error was not corrected until Fisher did so about 25 years later (in fact this was one of the strongest battles in Fisher's relentless campaign to show that he was the rising young star in British statistics and Pearson was yesterday's man).

Fix, Hodges, and Lehmann (1959) assume that the alternative hypotheses can be represented by a surface S of dimension s that contains the null surface T. Neyman (1949) had shown earlier that if P is a good estimate of p under restriction T and Q is a good estimate of p under restriction S (where "good" includes possibilities such as maximum likelihood and minimum χ^2), then $U(R) = D_R(R, P) - D_R(R, Q)$ is an asymptotically valid statistic for testing $H_0 : p \in T$ against $H_A : p \in S$ with limiting null distribution χ^2_{s-t}. FHL investigate the behavior of this test against a sequence of alternative values of p approaching T at a $\sqrt{n}$ rate and prove a theorem, identifying the non-central χ^2 distribution obeyed by $U(R)$ asymptotically, that permits an investigation of the power of the test. In numerical work that was not easy for the late 1950s, they provide new tables of the non-central χ^2 distribution and a plot of the power of the test against the degrees of freedom $f = (s - t)$ for a variety of values of the non-centrality parameter λ. They conclude the paper by illustrating the use of Neyman's test in a setting in which the null hypothesis is that events of interest occur uniformly in time against the alternative that their occurrence is cyclical, noting that at significance level 0.01 for a particular alternative considered the restricted test has power 0.90 when the usual χ^2 test would only have power 0.62.

There are many points of interest in this paper even with 40 years of hindsight: it features a non-Bayesian use of prior information (in the precise formulation of the alternative hypothesis); the proofs, which involve a nice degree of geometric insight, are discussed in a clear, intuitive way rather than enclosing them in a straitjacket of formalism; FHL display an appealing command of the history of statistics in their narrative (for example, they note at one point in a proof that the idea they're about to mention dates back to Gauss); in an interesting demonstration of flexibility on the setting of type *I* and type *II* error targets, the numerical tables give entries for significance levels from 0.001 all the way up to 0.5; and again the question of what to choose for k comes up (and this time FHL have useful advice to offer the reader).

Lehmann (1963a) and **Lehmann (1964):** *Asymptotically nonparametric inference: an alternative approach to linear models* and *Asymptotically nonparametric inference in some linear models with one observation per cell.* I've known these papers rather well for a long time; they were the basis for part of my dissertation work with Lehmann in the late 1970s and early 1980s (Draper (1988)). The context of these articles is as follows.

It's hard to argue with the viewpoint that parametric frequentist inferential methods were greatly strengthened in the 1920s and 1930s by the work of Fisher on the one hand (e.g., Fisher (1922)) and Neyman and Pearson on the other hand (e.g., Neyman and Pearson (1933)). Given a particular parametric model

$$(y_i|\theta, f) \overset{\text{IID}}{\sim} f(\cdot|\theta), \tag{5}$$

where θ is a vector of real numbers and the functional form of f is assumed known, these investigators sought methods of estimation and hypothesis testing that were in some sense "best" in repeated sampling from the presumed model (5). As people started to use these methods, a new class of questions arose: what should you do if you don't know the "right" parametric model? One idea would naturally be to continue to use one of the methods that Fisher and Neyman-Pearson had developed for standard choices of f, and then to investigate how badly these methods perform when f was not the actual data-generating mechanism; this approach (studying the *robustness* of the standard methods) was pursued vigorously by many people, including Pearson (1931) himself. A second idea would be to look for new methods that worked rather well "in the neighborhood of f" and never did much worse than the "best" methods (if somehow you knew what f really was). This gave rise to *nonparametric* (or *distribution-free*) techniques, such as the Wilcoxon (1945) rank tests in the one- and two-sample problems. These tests were shown by Pitman (1949) and others to have an *efficiency* (the asymptotic ratio of sample sizes needed to achieve the same power against the same alternative at the same significance level) (a) of approximately 96% when f is normal and (b) never lower than about 86% no matter what f is, but they had the limitation that, as tests, their final products (P-values) live on the probability scale, rather than on the scale of the data (where most good scientific inference actually resides (or at least *should* reside) in practice). Lehmann wondered: how can point and interval estimates (on the data scale) be developed that in some sense arise naturally from the nonparametric tests?

It's a common idea in (frequentist) statistics that if you have a good method for estimating a parameter then you can derive a good test from that estimate; it's equally true (but less often used) that if you have a good test you can work backwards from it (*invert* it) to get a good (point or interval) estimate. In the early 1960s, during a highly productive period, Lehmann — sometimes working alone (e.g., Lehmann (1963b)), sometimes with Joe Hodges (e.g., Hodges and Lehmann (1963)) — used this idea, of inverting the Wilcoxon procedures, to produce rank-based point and interval estimates in a large variety of interesting inferential problems. The two papers I discuss here are examples of this: Lehmann (1964) and Lehmann (1963c) cover analysis of variance models with one and several observations per cell, respectively.

Following Draper (1988), the model that Lehmann (1963c) considers for ANOVA with several observations per cell can be written

$$Y_{ij} = \mu_i + e_{ij} \, , \left\{ \begin{array}{l} i = 1, \ldots I \\ j = 1, \ldots, n_i \end{array} \sum_{i=1}^{I} n_i = N \right\} , \tag{6}$$

in which the e_{ij} are IID continuous real-valued random variables with density g satisfying

$$\theta = \int_{-\infty}^{\infty} g^2(y)\, dy < \infty \tag{7}$$

and $\sigma^2 = V(e_{ij}) < \infty$; here μ_i is a measure of center for the ith of I total cells, Y_{ij} is the jth of the n_i observations in cell i and N is the total number of observations. This looks like a one-way ANOVA model, but more complicated layouts can be handled just by numbering the cells from 1 to I.

Lehmann noted that most inference in ANOVA is based on contrasts among the cell centers, and any contrast

$$\phi = \sum_{i=1}^{I} c_i\, \mu_i, \quad \sum_{i=1}^{I} c_i = 0, \tag{8}$$

can be expressed in terms of the cell centers:

$$\sum_{i=1}^{I} c_i\, \mu_i = \sum_{i=1}^{I-1} \sum_{j=i+1}^{I} b_{ij} (\mu_i - \mu_j) \tag{9}$$

(note that the b_{ij} are not unique). To estimate $(\mu_i - \mu_j)$, Hodges and Lehmann (1963) showed that inverting the Wilcoxon rank-sum test yields the robust estimate

$$T_{ij} = \text{med}\{Y_{ik} - Y_{jl} \colon k = 1, \ldots, n_i; l = 1, \ldots, n_j\}, \tag{10}$$

which inherits the efficiency properties of the Wilcoxon test noted above. This turns out to be unsatisfactory as a basis for inference in ANOVA, however, because the T_{ij} don't satisfy the linearity constraints obeyed by the quantities they're trying to estimate:

$$(\mu_i - \mu_j) + (\mu_j - \mu_k) = (\mu_i - \mu_k), \quad \text{but} \quad T_{ij} + T_{jk} \neq T_{ik}, \tag{11}$$

since the operations of subtraction and taking a median don't commute. Lehmann's suggested fix for this problem, improved slightly a few years later by Spjøtvoll (1968), was to express $(\mu_i - \mu_j)$ as $[(\mu_i - \bar{\mu}) - (\mu_j - \bar{\mu})]$, where $\bar{\mu} = \frac{1}{N} \sum_{i=1}^{I} n_i \mu_i$ is the grand mean, and estimate $(\mu_i - \bar{\mu})$ by $\bar{T}_i = \frac{1}{N} \sum_{k=1}^{I} n_k T_{ik}$; this linearizes the estimates by making all comparisons relative to $\bar{\mu}$. With this approach, $(\mu_i - \mu_j)$ can now be estimated by $W_{ij} = (\bar{T}_i - \bar{T}_j)$, leading to the contrast estimate

$$\hat{\phi} = \sum_{i=1}^{I-1} \sum_{j=i+1}^{I} b_{ij} W_{ij}; \tag{12}$$

the above linearization ensures that even though the b_{ij} are not unique, all choices of b_{ij} lead to the same value of $\hat{\phi}$.

It's now straightforward to "robustify" virtually any standard normal-theory-based ANOVA procedure, simply by (a) replacing the usual quantity $(\bar{Y}_i - \bar{Y})$ — in which $\bar{Y}_i = \frac{1}{n_i}\sum_{j=1}^{n_i} Y_{ij}$ and $\bar{Y} = \frac{1}{N}\sum_{i=1}^{I} n_i \bar{Y}_i$ — by its rank-based analogue $\bar{T}_i$, and (b) replacing the normal-theory estimated error variance $\hat{\sigma}^2$ by its analogue $\hat{\sigma}_R^2$. Lehmann showed that the asymptotic analogue of σ^2 in this procedure is

$$\sigma_R^2 = \frac{1}{12\,\theta^2} = \frac{1}{12(\int g^2)^2}. \tag{13}$$

In my dissertation work with Lehmann I developed two methods for estimating $\int g^2$ — one (suggested by Lehmann (1963b)) derived from the lengths of confidence intervals based on the same differences $(Y_{ik} - Y_{jl})$ whose median is T_{ij} (see equation (10)), and another involving density estimation — and conducted a large simulation study to see how this machinery works in small samples. The conclusions were that interval estimates from this approach (a) have about the same coverage as the standard methods with normal and non-normal data, (b) are about 2% wider when the data really are normal, but (c) can be noticeably narrower with samples from non-normal distributions (e.g., normal-theory intervals are about 40% wider than the rank-based intervals with data from t distributions with small degrees of freedom).

In parallel with this paper, Lehmann (1964) is principally interested in ANOVA models with one observation per cell:

$$X_{i\alpha} = \nu + \xi_i + \mu_\alpha + U_{i\alpha}, \quad \sum_{i=1}^{c} \xi_i = \sum_{\alpha=1}^{N} \mu_\alpha = 0; \tag{14}$$

here the factor of interest, represented by ξ, is at c levels, μ represents a blocking factor at N levels, and the $U_{i\alpha}$ are IID with CDF F. Lehmann again focuses on contrasts $\theta = \sum_{i=1}^{c} c_i \xi_i$ (with $\sum_{i=1}^{c} c_i = 0$); he defines $V_\alpha = \sum_{i=1}^{c} c_i X_{i\alpha}$ and notes that these random variables are also IID, with CDF (say) G_c. If F (and therefore G_c) are symmetric, then a natural place to start in estimating θ from a rank-based point of view is with the *Walsh averages* med$\left[\frac{V_\alpha + V_\beta}{2}\right]$, where the median is taken over all $\alpha \leq \beta$. As in the paper discussed above, these quantities are linearly incompatible, so Lehmann again focuses instead on $(\xi_i - \xi_j)$ and estimates these differences with

$$Y_{ij} = \text{med}\left[\frac{(X_{i\alpha} - X_{j\alpha}) - (X_{i\beta} - X_{j\beta})}{2}\right]. \tag{15}$$

Even these quantities are not compatible, so he linearizes them with

$$Z_{ij} = (Y_{i\cdot} - Y_{j\cdot}), \tag{16}$$

where the dot signifies averaging with respect to the indicated subscript, and bases his inferences about contrasts on the Z_{ij}. The asymptotic distribution of the Z_{ij} again involves $\int g^2$, where this time g is the density of $(X_{i\alpha} - X_{j\beta})$, and efficiency gains relative to the standard normal-theory methods are similar to those found in Lehmann (1963a). Taken together, these two papers — and the other related articles published around the same time — provide the foundation for *R-estimation*, an approach to robust inference that's competitive with the other two leading methods developed subsequently (*L-estimation* (e.g., Bickel (1973)), based on linear combinations of order statistics, and *M-estimation* (e.g., Huber (1973)), which generalizes maximum likelihood in a robust fashion).

Hodges and Lehmann (1970): *Deficiency*. In this paper Hodges and Lehmann (HL) are interested in comparing the number of observations required by two different inferential methods to achieve the same accuracy, as a measure of how much one method is better than the other. As they note, if methods A and B require n and $k_n > n$ observations to arrive at the same level of performance, one natural way to summarize this is through the ratio $\frac{k_n}{n}$, whose limit e as $n \to \infty$ will generally be stable (this is the basis of the familiar concept of *asymptotic relative efficiency* (ARE)). If $e > 1$, its value can often give reliable guidance as to how much method A will be better than B with finite (and even rather small) n; but what if (as frequently occurs) $e = 1$? It may still be that A is better than B in small samples, and a natural way to quantify this is through the difference $(k_n - n)$, which HL call the *deficiency* of B relative to A; if $d = \lim_{n\to\infty}(k_n - n)$ exists, they call this the *asymptotic deficiency*.

The paper is devoted to an exploration of this idea in a series of examples, using expected squared error of point estimators as the performance measure. They show, for instance, that if you and I are both estimating the population variance σ^2 based on a sample $(X_1, \ldots, X_n)$, and I pretend that I know the population mean ξ and use $M_n = \frac{1}{n}\sum_{i=1}^n (X_i - \xi)^2$ while you make no such assumption and use $M_n' = \frac{1}{n-1}\sum_{i=1}^n (X_i - \bar{X})^2$ (where $\bar{X} = \frac{1}{n}\sum_{i=1}^n X_i$), both of us will be using estimators that are unbiased (in repeated sampling) with the same large-sample variance $\frac{\gamma\sigma^4}{n}$, where $(\gamma + 1) = \frac{\mu_4}{\sigma^4}$ is the standardized fourth central moment of the population distribution F. Thus in this case $e = 1$, but they show that the asymptotic deficiency of M_n' relative to M_n is $\frac{2}{\gamma}$. With normal data $\gamma = 2$, so (as they put it) in that case "it costs only one observation to protect against an erroneous value of ξ," whereas the deficiency can be arbitrarily large for non-normal populations since γ can be as close to 0 as you might want to make it.

Other examples analyzed include comparisons of (a) biased and unbiased estimates when drawing inferences about the population variance, (b) medians versus quasi-medians in estimating the center of symmetry of a symmetric distribution, (c) several methods for creating confidence sets for normal means, (d) the one-sample t test versus the analogous z procedure obtained by pretending the population variance is known, and (e) Bayesian versus unbiased estimation of a normal mean. I'm a bit uncomfortable with their only other example not on this list, however: with X as the (binomial) number of successes in n IID Bernoulli trials with unknown success probability p, they compare the usual unbiased estimator $M_n = \frac{X}{n}$ with the minimax estimator

$$M_n' = \frac{\sqrt{n}}{1+\sqrt{n}} M_n + \frac{1}{2(1+\sqrt{n})}; \tag{17}$$

Table 1 *Expected squared errors for the unbiased (V_n) and minimax (V_n') estimators of a Bernoulli/binomial p, for various values of n and p.*

	$p = 0.25$		$p = 0.4999$		$p = 0.5$	
n	V_n	V_n'	V_n	V_n'	V_n	V_n'
1	$1.88 \cdot 10^{-1}$	$6.25 \cdot 10^{-2}$	$2.50 \cdot 10^{-1}$	$6.25 \cdot 10^{-2}$	$2.50 \cdot 10^{-1}$	$6.25 \cdot 10^{-2}$
10	$1.88 \cdot 10^{-2}$	$1.44 \cdot 10^{-2}$	$2.50 \cdot 10^{-2}$	$1.44 \cdot 10^{-2}$	$2.50 \cdot 10^{-2}$	$1.44 \cdot 10^{-2}$
100	$1.88 \cdot 10^{-3}$	$2.07 \cdot 10^{-3}$	$2.50 \cdot 10^{-3}$	$2.07 \cdot 10^{-3}$	$2.50 \cdot 10^{-3}$	$2.07 \cdot 10^{-3}$
1,000	$1.88 \cdot 10^{-4}$	$2.35 \cdot 10^{-4}$	$2.50 \cdot 10^{-4}$	$2.35 \cdot 10^{-4}$	$2.50 \cdot 10^{-4}$	$2.35 \cdot 10^{-4}$
10,000	$1.88 \cdot 10^{-5}$	$2.45 \cdot 10^{-5}$	$2.50 \cdot 10^{-5}$	$2.45 \cdot 10^{-5}$	$2.50 \cdot 10^{-5}$	$2.45 \cdot 10^{-5}$

these estimators have expected squared errors

$$V_n = \frac{p(1-p)}{n} \quad \text{and} \quad V_n' = \frac{1}{4(1+\sqrt{n})^2}, \tag{18}$$

respectively. You can see that if $p \neq \frac{1}{2}$ the asymptotic relative efficiency of M_n' to M_n is $\frac{p(1-p)}{4} < 1$, so that for large n the unbiased estimator is better (with expected squared error as the figure of merit); but with $p = \frac{1}{2}$ the ARE is 1 and $V_n' < V_n$, and HL show that in this case the asymptotic deficiency of M_n relative to M_n' is infinite. This seems misleading to me from a practical point of view, as illustrated in Table 1. HL note that "the deficiency computation thus shows that the [unbiased] estimator requires a much larger sample size than the minimax [estimator] if they are to have the same expected squared error at $p = \frac{1}{2}$, in spite of the fact that the corresponding asymptotic efficiency is 1." While this may strictly speaking be true, to me a better summary of what's going on (with reference to Table 1) would be just to say that (a) for p near 0.5, M_n' is better for all realistic sample sizes n, but the amount that it's better goes to 0, and (b) for p not near 0.5, M_n' is better for small n and then from a certain point on (as n grows), M_n is better.

Lehmann and Loh (1990): *Pointwise versus uniform robustness of some large-sample tests and confidence intervals*. This paper is about the sensitivity of hypothesis tests to departures from the distributional assumptions made to derive them. Lehmann and Loh (LL) note, for instance, that the one-sample t test of the null hypothesis that the population mean is 0 against positive alternatives is designed to have exact type-I error rate α under normality, and it's natural to wonder how it behaves when the data are not normal. If the data $(X_1, \ldots, X_n)$ are IID from CDF F, then as long as F has finite variance the first-order asymptotic answer is pleasant: with $\alpha_n(F)$ as the probability of rejecting $H_0 : \mu(F) = 0$ (where $\mu(F)$ is the mean of F), it's easy to show that $\alpha_n(F) \to \alpha$ as $n \to \infty$, which people typically describe by saying that the the level of the t test is *robust* against non-normality. But **Q:** how big does n need to be for $\alpha_n(F)$ to be close to α?

Looking at the problem a bit more generally, LL consider testing

$$H_0'\text{: the mean } \mu_F \text{ of the unknown } F \text{ is } 0 \tag{19}$$

against the alternatives (indexed by F) that μ_F is positive. When testing H_0' the level of the t test with n observations is $\alpha_n = \sup_{\mathcal{F}} \alpha_n(F)$, with $\mathcal{F}$ as the set of all CDFs with mean 0 and finite variance. It would be nice if $\lim_{n\to\infty} \alpha_n = \alpha$; if this were true we could say that the t test is *uniformly robust* over $\mathcal{F}$. But it's been known for a long time (Bahadur and Savage (1956)) that $\sup_{\mathcal{F}} \alpha_n(F) = 1$ for all $\alpha > 0$, so it's necessary to seek the answer to question **Q** above a bit differently: the Bahadur-Savage result says that, given a small $\epsilon > 0$, you cannot find a sample size n_0 such that

$$|\alpha_n(F) - \alpha| < \epsilon \text{ for all } n > n_0 \tag{20}$$

when F is allowed to vary over $\mathcal{F}$, but can you find a *subset* $\mathcal{F}_0$ of $\mathcal{F}$ (that's big enough to be relevant to actual statistical practice) for which (20) is true? LL devote much of this paper to showing, unhappily, that across natural and obvious choices of $\mathcal{F}_0$ — for example, all distributions in a fixed Kolmogorov-Smirnov neighborhood of normal densities with mean 0, and all continuous distributions with support contained in a bounded interval — the t test fails to achieve uniform robustness; in fact, the only class $\mathcal{F}_0$ that LL can find that works is the set of absolutely continuous $F \in \mathcal{F}$ with the rth standardized absolute moment about the mean uniformly bounded for some $r > 2$. They also obtain a negative result for inference about the success parameter p in Bernoulli/binomial sampling. Overall the paper serves to show how hard it is to achieve uniform robustness as a performance criterion for repeated-sampling hypothesis tests.

An overview. Together (as of this writing) these papers have been cited more than 560 times, by researchers working in a wide variety of fields; the resulting citation rate per paper (about 93) is above average for an author who is himself highly-cited (the citation software `Publish or Perish` finds that Lehmann has been cited almost 22,000 times over a 66-year period, with an average number of citations per paper of about 62). All of these papers are well-written, which (besides high impact) is another general feature of Lehmann's work. Two threads run through these articles, and serve to draw my comments to a close.

- They're all written from the repeated-sampling (frequentist) point of view, although Bayesian methods sometimes get a mention. A Bayesian would approach the problems addressed in these articles differently — for example, (1) today I would solve the linear-model problems in Lehmann (1963b), Lehmann (1964) not by seeking robust estimates but by direct Bayesian parametric or nonparametric modeling of the (non-normal) error distributions, which would yield inferential procedures with repeated-sampling robustness properties similar to those exhibited by Lehmann's methods, and (2) in my view many uses of hypothesis-testing in practice should really be reformulated as Bayesian decision-theory problems with utility structures (a) that are sensitive to the real-world context of the problem and (b) that will often be found to differ from the loss function inherent in the usual type *I* and *II* error story — but everyone would agree that the problems posed here (Does this model fit the data? Does this procedure still work well, even though some of the assumptions on which it's based are violated? How much better is method A than B in extracting signal from the data?) are important.

- They all feature an interest in how often statistical procedures get the right answer. This is a fundamental scientific question to which, in my view, all statisticians — whether they work in the Bayesian or frequentist paradigms — need to pay constant attention (I learned the importance of this topic both from Lehmann and from his mentor Jerzy Neyman). I began my research career in the frequentist paradigm (which was the only story readily available at Berkeley, at least when I was there in the late 1970s and early 1980s), and came to appreciate the value of Bayesian methodology and thinking later (in the mid 1980s). The standard position taken by many statisticians in the 20th century was that you had to (a) choose between the Bayesian and frequentist paradigms and then (b) defend your chosen approach against attacks from people who had chosen the other paradigm, but this is a flawed formulation of the problem: when you look closely you find that each paradigm has both strengths and weaknesses, so — in my view — in the 21st century it's the job of all working statisticians to try to construct a fusion of the two paradigms that emphasizes the strengths and de-emphasizes the weaknesses. My own personal fusion, combining elements of my Berkeley training and post-Berkeley study, is (i) to reason in a Bayesian way when formulating my inferences, predictions and decisions (because the Bayesian approach seems to me to be the most successful method so far invented for capturing and quantifying all relevant sources of uncertainty, whether arising from an inherently repeatable process or not) and (ii) to reason in a frequentist way when evaluating the procedures in (i) (because we all need to pay attention to how often we get the right answer, and this is an inherently frequentist question). I'm not sure what Lehmann's position was on this issue toward the end of his long career, but I *am* sure that, if he were still here to give us his thoughts on the subject, they would be well worth listening to.

References

[1] Bahadur, R. R. and Savage, L. J. (1956). The Nonexistence of Certain Statistical Procedures in Nonparametric Problems. *The Annals of Mathematical Statistics*, **27**, 1115–1122.

[2] Bickel, P. J. (1973). On Some Analogues to Linear Combinations of Order Statistics in the Linear Model. *Ann. Statist.* **1**, 597–616.

[3] Chernoff, H., and Lehmann E. L. (1954). The use of maximum likelihood estimates in χ^2 tests for goodness of fit. *Ann. Math. Statist.*, **2**, 5579–586.

[4] Draper, D. (1988). Rank-Based Robust Analysis of Linear Models. I. Exposition and Review. *Statistical Science*, **3**, 239–257.

[5] Fisher, R. A. (1922) On the Mathematical Foundations of Theoretical Statistics. *Phil. Trans. R. Soc. Lond. A*, 309–368

[6] Fix, E., Hodges, J. L., and Lehmann, E. L. (1959). The restricted chi-square test, Probability and Statistics. *The Harald Cramer Volume*, 92–107.

[7] Hodges, J. L., and Lehmann, E. L. (1963). Estimates of location based on rank tests. *Ann. Math. Stat.*, **34** (1963), 593–611.

[8] Hodges, J. L., and Lehmann, E. L. (1970). Deficiency. *The Annals of Mathematical Statistics*, **41**, 783–801.

[9] Huber, P. J. (1973). Robust Regression: Asymptotics, Conjectures and Monte Carlo. *The Annals of Statistics*, **1**, 799–821.

[10] Lehmann, E. L. (1963a). Asymptotically nonparametric inference: an alternative approach to linear models. *Ann. Math. Stat.*, **34**, 1494–1506.

[11] Lehmann, E. L. (1963b). Nonparametric confidence intervals for a shift parameter. *Ann. Math. Stat.*, **34**, 1507–1512.

[12] Lehmann, E. L. (1963c). Robust estimation in analysis of variance. *Ann. Math. Stat.*, **34** 957–966.

[13] Lehmann, E. L. (1964). Asymptotically nonparametric inference in some linear models with one observation per cell. *Ann. Math. Stat.*, **35**, 726–734.

[14] Lehmann, E. L. and Loh, W-Y. (1990). Pointwise vs. uniform robustness of some large-sample tests and confidence intervals. *Scand. J. Statist.*, **17**, 177–187.

[15] Neyman, J. and Pearson, E. S. (1933). On the Problem of the Most Efficient Tests of Statistical Hypotheses. *Phil. Trans. R. Soc. Lond. A*, 289–337.

[16] Neyman, J., (1949). Contribution to the theory of χ^2 test. In: *J. Neyman, ed., Proceeding of the First Berkeley Symposium on Mathematical Statistics and Probability*, 239-273.

[17] Pearson, K. (1900). On the criterion that a given system of deviations from the probable in the case of a correlated system of variables is such that it can be reasonably supposed to have arisen from random sampling. *Philosophy Magazine, 5th Series*, **50**, 157–172.

[18] Pearson, E. S. (1931). The Analysis of Variance in Cases of Non-normal Variation. *Biometrika*, 114–133.

[19] Pitman, E. J. G. (1949). Lecture notes on nonparametric statistical inference. Unpublished.

[20] Spjøtvoll, A. (1968). A Note on Robust Estimation in Analysis of Variance. *Ann. Math. Statist.*, **39**, 1486–1492.

[21] Wilcoxon, F. (1945). Individual Comparisons by Ranking Methods. *Biometrics Bulletin* **1**, 80–83.

THE USE OF MAXIMUM LIKELIHOOD ESTIMATES IN χ^2 TESTS FOR GOODNESS OF FIT[1]

By Herman Chernoff and E. L. Lehmann

Stanford University and University of California

Summary. The usual test that a sample comes from a distribution of given form is performed by counting the number of observations falling into specified cells and applying the χ^2 test to these frequencies. In estimating the parameters for this test, one may use the maximum likelihood (or equivalent) estimate based (1) on the cell frequencies, or (2) on the original observations. This paper shows that in (2), unlike the well known result for (1), the test statistic does not have a limiting χ^2-distribution, but that it is stochastically larger than would be expected under the χ^2 theory. The limiting distribution is obtained and some examples are computed. These indicate that the error is not serious in the case of fitting a Poisson distribution, but may be so for the fitting of a normal.

1. Introduction. When using χ^2 for testing that a sample comes from a distribution of specified functional form such as a Poisson or normal distribution, the problem arises as to what estimates of the population parameters to use. If only the numbers m_i of observations falling into the ith of the k cells are available, there is no difficulty. Let p_i ($i = 1, \cdots, k$) denote the probability of an observation falling into the ith cell, and let $\tilde{p}_i$ be any best asymptotically normal (b.a.n.) estimate of p_i such as the minimum χ^2 or maximum likelihood estimate. Then it is known [1], [2] that under suitable regularity conditions the asymptotic distribution of

$$\tilde{R} = \sum (m_i - m\tilde{p}_i)^2/n\tilde{p}_i \tag{1}$$

is that of χ^2 with $k - s - 1$ degrees of freedom, where s is the number of (independent) population parameters being estimated.

If, however, the original observations $x_1, \cdots, x_n$ are available, one is tempted to use more efficient estimates, such as the maximum likelihood estimates $\hat{p}_i$ based on all the data. One may reasonably expect this procedure to provide more powerful tests than those based only on the m_i; at the same time the estimates usually are simpler and easier to obtain. This is in fact the procedure recommended in many textbooks, particularly for the fitting of Poisson distributions, either as an approximation to the one with known theory described above, or more often without comment.

It is the purpose of the present paper to obtain the distribution of

$$\hat{R} = \sum (m_i - n\hat{p}_i)^2/n\hat{p}_i, \tag{2}$$

Received 7/24/53.

[1] This work was prepared with the partial support of the Office of Naval Research.

which differs from that of $\tilde{R}$. If we let

$$R = \sum (m_i - np_i)^2/np_i, \tag{3}$$

which has a limiting χ^2-distribution with $k - 1$ degrees of freedom, we shall show that the limiting distribution of $\hat{R}$ lies between those of $\tilde{R}$ and of R. More specifically, we shall show in Section 3 that under suitable regularity conditions we have

THEOREM 1. *The asymptotic distribution of $\hat{R}$ is that of*

$$\sum_{i=1}^{k-s-1} y_i^2 + \sum_{i=k-s}^{k-1} \lambda_i y_i^2 \tag{4}$$

where the y_i are independently normally distributed with mean zero and unit variance, and the λ_i are between 0 and 1 and may depend on the s parameters $\theta_1, \cdots, \theta_s$.

This result indicates that the recommended procedure of rejecting the hypothesis of goodness of fit when $\hat{R} > C$, where C is obtained from the χ^2-distribution with $k - s - 1$ degrees of freedom, will lead to a probability of rejection which, when the hypotheses is true, is greater than the desired level of significance. However, a numerical investigation of a few special cases indicates that, at least in the Poisson problem, this excess of probability of type I error will be so small as not to be serious. The situation appears to be not quite so favorable in the normal case.

Throughout this paper, the notation and background material given in Section 2 of the preceding paper [3] will be used.

2. Example. Before proceeding to the main result, let us treat the special example where the observations are independently and normally distributed with unknown mean and variance 1, and where the cells are $(-\infty, 0)$ and $(0, \infty)$. In this case it is obvious that $\tilde{R} = 0$. However,

$$\hat{R} = \sum_{i=1}^{2} \frac{(m_i - n\hat{p}_i)^2}{n\hat{p}_i} = \frac{(m_1 - n\hat{p}_1)^2}{n\hat{p}_1\hat{p}_2}$$

where

$$p_1(\mu) = \int_{-\infty}^{0} \frac{1}{\sqrt{2\pi}} e^{-(y-\mu)^2/2}\, dy = \int_{-\infty}^{-\mu} \frac{1}{\sqrt{2\pi}} e^{-y^2/2}\, dy$$

$$p_2(\mu) = 1 - p_1(\mu), \qquad \hat{p}_i = p_i(\bar{x}).$$

We have

$$\hat{R} = \frac{1}{p_1p_2 + o_p(1)}\left(\frac{m_1 - np_1}{\sqrt{n}} - \frac{n(\hat{p}_1 - p_1)}{\sqrt{n}}\right)^2 = \frac{(\epsilon - \nu)^2}{p_1p_2}[1 + o_p(1)]$$

where $\epsilon = (m_1 - np_1)/\sqrt{n}$ and $\nu = \sqrt{n}\,(\hat{p}_1 - p_1)$. Using the first order Taylor expansion of $p_1(\bar{x})$ about $p_1(\mu)$, we have

$$\nu = -\sqrt{n}\,(\bar{x} - \mu)\, e^{-\mu^2/2}/\sqrt{2\pi} + o_p(1) = \nu' + o_p(1)$$

where $\nu' = -\sqrt{n}(\bar{x} - \mu)\, e^{-\mu^2/2}/\sqrt{2\pi}$. Let $g(x) = 1$ for $x < 0$, and $= 0$ otherwise. The central limit theorem tells us that, since

$$(m_1, n\bar{x}) = \sum_{\alpha=1}^{n} [g(x_\alpha), x_\alpha], \qquad d\infty(\epsilon, \nu') = N(0, \Sigma)$$

where $N(0, \Sigma)$ denotes the normal distribution with mean 0 and covariance matrix Σ given by

$$\Sigma = \begin{pmatrix} p_1 p_2 & e^{-\mu^2}/2\pi \\ e^{-\mu^2}/2\pi & e^{-\mu^2}/2\pi \end{pmatrix}.$$

Hence

$$d\infty(\epsilon, \nu) = N(0, \Sigma), \qquad d\infty(\epsilon - \nu) = N(0, p_1p_2 - e^{-\mu^2}/2\pi)$$

and in particular $\epsilon - \nu = O_p(1)$. It follows that $d\infty(\hat{R}) = d(\lambda y^2)$, where $d(y) = N(0, 1)$ and $\lambda = 1 - e^{-\mu^2}/2\pi p_1 p_2 < 1$. The fact that $\lambda \geqq 0$ follows from the fact that Σ is nonnegative definite.

NOTE. A general proof of Theorem 1 cannot be based solely on the fact that $\hat{p}_1$ is a better estimate of p_1 than $\tilde{p}_1$ is. Suppose, in fact, that we use $p_1^* = p_1(2 - \bar{x})$ as our estimate of p_1. In the event that $\mu = 1$, p_1^* has the same distribution as $\hat{p}_1$. The above argument repeated for R^* shows that ν would be replaced by $-\nu$ and λ by $\lambda^* = 1 + 3e^{-\mu^2}/2\pi p_1 p_2$.

3. The general case. We shall now prove Theorem 1 under the following regularity conditions:

(i) The $p_i(\theta)$ satisfy the condition on pages 426–427 of Cramér's *Mathematical Methods of Statistics.*

(ii) Let $z = (z_1, \cdots, z_k)$ where $z_i = 1$ if the observation falls in the ith cell and 0 otherwise. Let $f(z, \theta) = \prod p_i^{z_i}$, and let us assume that the value w of our chance variable x determines z, and that the density of x is given by

$$f^*(w, \theta) = \prod p_i^{z_i}\, g(w|z, \theta)$$

where g is the conditional density of x given z. Then we assume that f^* satisfies the condition $\mathcal{R}$ of the preceding paper [3].

Let

$$m_i - np_i = \sqrt{np_i}\,\epsilon_i, \tag{5}$$

$$n(\tilde{p}_i - p_i) = \sqrt{np_i}\,\tilde{\nu}_i, \tag{6}$$

$$n(\hat{p}_i - p_i) = \sqrt{np_i}\,\hat{\nu}_i. \tag{7}$$

Then

$$R = \sum \epsilon_i^2 = \epsilon'\epsilon, \tag{8}$$

$$\tilde{R} = \sum (\epsilon_i - \tilde{\nu}_i)^2\, [1 + o_p(1)], \tag{9}$$

$$\hat{R} = \sum (\epsilon_i - \hat{\nu}_i)^2\, [1 + o_p(1)]. \tag{10}$$

We shall first compute $\tilde{\nu}_i$ to show that $\tilde{R}$ is asymptotically a sum of squares of the components of a normally distributed chance variable, and then do the same for $\hat{R}$. We have

$$\frac{\partial \log f(z, \theta)}{\partial \theta_j} = \sum_{i=1}^{k} \frac{z_i}{p_i} \frac{\partial p_i}{\partial \theta_j}.$$

The information matrix referred to in Section 2 of [3] is given by

$$\tilde{J} = \left\| \sum_{r=1}^{k} \frac{1}{p_r} \frac{\partial p_r}{\partial \theta_i} \frac{\partial p_r}{\partial \theta_j} \right\| = D'D \tag{11}$$

where

$$D = \left\| \frac{1}{\sqrt{p_i}} \frac{\partial p_i}{\partial \theta_j} \right\|. \tag{12}$$

The corresponding A vector $\tilde{A}$ has elements

$$\tilde{A}_i = \frac{1}{n} \sum_{r=1}^{k} \frac{m_r}{p_r} \frac{\partial p_r}{\partial \theta_i} = \sum_{r=1}^{k} \frac{m_r - np_r}{np_r} \frac{\partial p_r}{\partial \theta_i}.$$

Therefore

$$\tilde{A} = (1/\sqrt{n}) D'\epsilon \tag{13}$$

and

$$\tilde{\nu}_i = \frac{\sqrt{n}(\tilde{p}_i - p_i)}{\sqrt{p_i}} = \sum_{j=1}^{s} \sqrt{n}(\tilde{\theta}_j - \theta_j) \frac{1}{\sqrt{p_i}} \frac{\partial p_i}{\partial \theta_j} + o_p(1),$$

$$\tilde{\nu} = D\sqrt{n}(\tilde{\theta} - \theta) + o_p(1) = D\tilde{J}^{-1}D'\epsilon - o_p(1).$$

Finally

$$\tilde{R} = (\tilde{F}\epsilon)'(\tilde{F}\epsilon) + o_p(1) \tag{14}$$

where

$$\tilde{F} = I - D\tilde{J}^{-1}D'. \tag{15}$$

Now

$$\frac{\partial \log f^*(w, \theta)}{\partial \theta_j} = \sum_{i=1}^{k} \frac{z_i}{p_i} \frac{\partial p_i}{\partial \theta_j} + \frac{\partial \log g(w \mid z, \theta)}{\partial \theta_j}. \tag{16}$$

Since the conditional expectation, given z, of

$$\left[\sum_{i=1}^{k} \frac{z_i}{p_i} \frac{\partial p_i}{\partial \theta_j} \right] \cdot \frac{\partial \log g(w \mid z, \theta)}{\partial \theta_l}$$

is zero, we have

$$\hat{J} = \tilde{J} + J^* \tag{17}$$

$$\hat{A} = \tilde{A} + A^* \tag{18}$$

where

$$J^* = \left\| E\left[\frac{\partial \log g(x \mid z, \theta)}{\partial \theta_i} \cdot \frac{\partial \log g(x \mid z, \theta)}{\partial \theta_j}\right]\right\|, \tag{19}$$

$$A_i^* = \frac{1}{n} \sum_{\alpha=1}^{n} \frac{\partial \log g(x \mid z^{(\alpha)}, \theta)}{\partial \theta_i} \tag{20}$$

and $z^{(\alpha)}$ is the αth observation on z. Now

$$\hat{\nu}_i = \sqrt{n}\,\frac{\hat{p}_i - p_i}{\sqrt{p_i}} = \sum \sqrt{n}(\hat{\theta}_j - \theta_j)\,\frac{1}{\sqrt{p_i}}\frac{\partial p_i}{\partial \theta_j} + o_p(1),$$

$$\hat{\nu} = D\sqrt{n}(\hat{\theta} - \theta) + o_p(1). \tag{21}$$

$$\hat{\nu} = D(\bar{J} + J^*)^{-1}(D'\epsilon + \sqrt{n}A^*) + o_p(1).$$

Hence

$$\hat{R} = (\hat{F}\epsilon + \hat{G}\eta)'(\hat{F}\epsilon + \hat{G}\eta) + o_p(1) \tag{22}$$

where $\eta = \sqrt{n}A^*$, while $\hat{F} = I - D(\bar{J} + J^*)^{-1} D'$ and $\hat{G} = D(\bar{J} + J^*)^{-1}$. The asymptotic distributions of R, $\bar{R}$, and $\hat{R}$ are those of $\epsilon'\epsilon$, $(\bar{F}\epsilon)'(\bar{F}\epsilon)$, and $(\hat{F}\epsilon + \hat{G}\eta)'(\hat{F}\epsilon + \hat{G}\eta)$, respectively. To find these distributions we must know the asymptotic distribution of (ϵ, η). Applying the central limit theorem to

$$\left[z_1, z_2, \cdots, z_k, \frac{\partial \log g(w \mid z, \theta)}{\partial \theta_1}, \cdots, \frac{\partial \log g(w \mid z, \theta)}{\partial \theta_s}\right]$$

we see that

$$d\infty(\epsilon, \eta) = N\left[0, \begin{pmatrix} I - qq' & 0 \\ 0 & J^* \end{pmatrix}\right] \tag{23}$$

where q is the vector whose ith component is $\sqrt{p_i}$. (Note that $D'q = 0$.)

From one of the Mann-Wald results it follows that the asymptotic distributions we desire are those obtained by assuming that (ϵ, η) actually have the above joint normal distribution. That is we assume that

$$d(\epsilon) = N(0, \Sigma), \qquad d(\bar{F}\epsilon) = N(0, \bar{\Sigma}), \qquad d(\hat{F}\epsilon + \hat{G}\eta) = N(0, \hat{\Sigma})$$

where

$$\Sigma = I - qq' \tag{24}$$

$$\bar{\Sigma} = I - qq' - D\bar{J}^{-1} D' \tag{25}$$

$$\hat{\Sigma} = I - qq' - D(\bar{J} + J^*)^{-1} D'. \tag{26}$$

If for symmetric matrices we write $K \geqq L$ whenever $K - L$ is nonnegative definite, then

$$\Sigma \geqq \hat{\Sigma} \geqq \bar{\Sigma}. \tag{27}$$

We digress to present

LEMMA 1. *If* $d(y) = N(0, U)$ *where the characteristic roots of* U *are* $\lambda_1, \lambda_2, \cdots, \lambda_k$, *then*

$$d(y'y) = d(\Sigma \lambda_i z_i^2)$$

where $d(z) = N(0, I)$.

PROOF. Expressing U in canonical form, we have $U = P\Lambda P'$, where P is orthogonal and Λ is the diagonal matrix whose diagonal elements are the λ_i. Since U is nonnegative definite, the λ_i are nonnegative and we may define $\Lambda^{\frac{1}{2}}$ in the obvious way. Let $d(z) = N(0, I)$ and $y^* = P\Lambda^{\frac{1}{2}}z$. Then $d(y^*) = N(0, U)$ and $d(y^{*\prime}y^*) = d(y'y)$. But

$$y^{*\prime}y^* = z'\Lambda^{\frac{1}{2}}P'P\Lambda^{\frac{1}{2}}z = z'\Lambda z$$

and the lemma follows.

As a consequence of this lemma, it follows that the distributions of R $\tilde{R}$, and $\hat{R}$ are those of $z'\Lambda z$, $z'\tilde{\Lambda}z$, and $z'\hat{\Lambda}z$ where Λ, $\tilde{\Lambda}$, $\hat{\Lambda}$ are the diagonal matrices of characteristic values corresponding to R, $\tilde{R}$, and $\hat{R}$, respectively. From the known results on R and $\tilde{R}$ it follows that Σ has for characteristic roots $k - 1$ ones and 1 zero, while $\tilde{\Sigma}$ has for characteristic roots $k - s - 1$ ones and $s + 1$ zeros. Since $\Sigma \geqq \hat{\Sigma} \geqq \tilde{\Sigma}$, it follows that $\hat{\Sigma}$ has for characteristic roots: $k - s - 1$ ones, 1 zero, and s roots $\lambda_1, \lambda_2, \cdots, \lambda_s$ between zero and one. Our Theorem 1 follows.

REMARK. A direct proof of the above-mentioned properties of the characteristic roots of Σ and $\tilde{\Sigma}$ may be given by showing that qq' and $D\tilde{J}^{-1}D'$ are projection operators on orthogonal manifolds of dimensions 1 and s respectively, that is,

$$(qq')(qq') = q(\sum p_i)q' = qq'$$

$$(D\tilde{J}^{-1}D')(D\tilde{J}^{-1}D') = D\tilde{J}^{-1}\tilde{J}\tilde{J}^{-1}D' = D\tilde{J}^{-1}D'$$

$$(D\tilde{J}^{-1}D')(qq') = 0.$$

The roots $\lambda_1, \cdots, \lambda_s$ which determine the distribution of the test criterion $\hat{R}$ can be obtained from

THEOREM 2. *If* $\mu_i = 1 - \lambda_i$, *then the* μ_i *are the characteristic roots of the determinantal equation* $|\tilde{J} - \mu\hat{J}| = 0$.

PROOF. We shall use the fact that if the vectors $t_1, \cdots, t_k$ form an orthonormal basis of k-dimensional space, then the matrix $\sum \tau_i t_i t_i'$ has the characteristic roots $\tau_1, \cdots, \tau_k$. This implies, in particular, that $\sum t_i t_i'$ is the identity matrix.

Given $\hat{J}$ and $\tilde{J}$, there exists a nonsingular $(s \times s)$ matrix S and a diagonal matrix M such that

$$\tilde{J}^{-1} = SS', \qquad \hat{J}^{-1} = SMS'$$

where the diagonal elements of M are the roots of $|\hat{\mathcal{J}}^{-1} - \mu\tilde{\mathcal{J}}^{-1}| = 0$ and hence of $|\tilde{\mathcal{J}} - \mu\hat{\mathcal{J}}| = 0$, and are all between 0 and 1, since $\hat{\mathcal{J}} \leqq \tilde{\mathcal{J}}$.

If $u_1, \cdots, u_s$ are the columns of DS,

$$D\tilde{\mathcal{J}}^{-1}D' = (DS)(DS)' = \sum u_i u_i'.$$

Since $(DS)'(DS) = S'\tilde{\mathcal{J}}S = I$ and $D'q = 0$, it follows that $q, u_1, \cdots, u_s$ are mutually orthogonal unit vectors. If we let $v_1, \cdots, v_{k-s-1}$ be a complementary set of orthogonal unit vectors we have

$$\hat{\Sigma} = I - qq' - DSMS'D' = I - qq' - \sum_{i=1}^{s} \mu_i u_i u_i'$$

$$= \sum_{j=1}^{k-s-1} v_j v_j' + \sum_{i=1}^{s} (1 - \mu_i) u_i u_i'.$$

It follows that the characteristic roots of $\hat{\Sigma}$ consist of $k - s - 1$ ones, one zero, and $\lambda_i = 1 - \mu_i$ for $i = 1, \cdots, s$.

4. Some Numerical Examples. By using the maximum likelihood estimates based on the full sample, one is operating at a higher significance level than the one stated. One can, however, on the basis of the above results, make an adjustment which asymptotically provides the correct value.

Given θ, let $C(\theta)$ be such that

$$P\left\{\sum_{i=1}^{k-s-1} y_i^2 + \sum_{i=k-s}^{k-1} \lambda_i(\theta) y_i^2 \neq C(\theta)\right\} = \alpha.$$

Clearly $C(\theta)$ is a continuous function of θ and hence $C(\hat{\theta}) \to C(\theta)$ in probability as $n \to \infty$. It follows that the probability of

$$\sum_{i=1}^{k-s-1} y_i^2 + \sum_{i=k-s}^{k-1} \lambda_i(\theta) y_i^2 \geqq C(\hat{\theta})$$

tends to α as $n \to \infty$. Here $C(\hat{\theta})$ can be computed, at least in theory, to an arbitrary degree of accuracy using the results of Pitman and Robbins [4].

Theoretically, the error committed by using the maximum likelihood estimates based on the full sample without an adjustment can be quite serious in the case of a small number of cells. For example, if $s = 1$ and $\lambda(\theta)$ is close to 1, we have essentially one extra degree of freedom, and when the number k of cells is small so that $k - 2 = 1$, 2 or 3, the actual probability α^* of type I error would vary from 15 per cent to 10 per cent when the level of significance is supposed to be $\alpha = 5$ per cent.

In practice, however, at least for fitting a Poisson distribution, the error does not appear to be so serious. Some values of $\lambda(\theta)$ and the true probability of rejection $\alpha^*(\theta)$ in the Poisson case are given below for groupings $x = 0, 1, \geqq 2$ and $x = 0, 1, 2, \geqq 3$, and level of significance supposed to be $\alpha = .05$.

x	$0,1, \geqq 2$		$0,1,2, \geqq 3$	
θ	1	2	2	3
$\lambda(\theta)$	.12	.35	.14	.32
$\alpha^*(\theta)$	.054	.067	.055	.065

As a second example, consider the fitting of a normal distribution with mean ζ and variance σ^2, both unknown. For the case of the four cells $(-\infty, -1)$, $(-1, 0)$, $(0, 1)$, $(1, \infty)$ and two combinations of ζ and σ we obtain the following values for the two roots λ_1 and λ_2 :

$$\zeta = 0, \quad \sigma = 2.5; \quad \lambda_1 = .80, \quad \lambda_2 = .20$$

$$\zeta = .5, \quad \sigma = 2.0; \quad \lambda_1 = .74, \quad \lambda_2 = .15.$$

The probability α^* is then given by

$$\alpha^* = P\{U + \lambda_1 V + \lambda_2 W \geqq C_\alpha\}$$

where U, V, W are χ^2 variables with 1 degree of freedom and C_α is such that $P\{U \geqq C_\alpha\} = \alpha$. As a lower bound of α^* in the first case we have computed $P\{U + .8V \geqq C_{.05}\} = .12$. This indicates that in the normal case the use of maximum likelihood estimates in χ^2 may lead to a more serious underestimate of the probability of type I error.

REFERENCES

[1] J. NEYMAN, "Contribution to the theory of the χ^2-test," *Proceedings of the Berkeley Symposium on Mathematical Statistics and Probability*, University of California Press, 1949.

[2] H. CRAMÉR, *Mathematical Methods of Statistics*, Princeton University Press, 1946.

[3] H. CHERNOFF, "On the distribution of the likelihood ratio," *Ann. Math. Stat.*, Vol. 25 (1954), pp. 573–578.

[4] E. J. G. PITMAN AND H. ROBBINS, "Application of the method of mixtures to quadratic forms in normally correlated variables," *Ann. Math. Stat.*, Vol. 20 (1949), pp. 552–560.

(b) It was desirable to explore the sensitivity of his z-test to departures from normality in the population, i.e. the question which was later to be termed by G. E. P. Box that of *robustness*.

Thus, with respect to Pearson, F. N. David's assessment seems essentially correct: the main ideas leading to Pearson's research were indeed provided by Student.

5.2 Gosset and Fisher

A corresponding conclusion does not apply to Gosset's influence on Fisher. It is of course true that the central idea of Gosset's 1908 papers—the possibility of determining the exact distribution of various statistics by assuming a known (e.g., normal) underlying distribution—provided the inspiration, and their mathematical incompleteness the opportunity, for Fisher's basic paper (1915). And in addition Gosset provided the impetus for Fisher's later distributional work by urging him to determine the distribution of regression and correlation coefficients.

However, Fisher also made many highly original and influential contributions to small-sample theory that owed nothing to Student, such as variance-stabilizing and normalizing transformation, permutation tests, the design of experiments (including randomization), the concepts of sufficiency and of likelihood.

5.3 Gosset and Neyman

Although no conversations or correspondence between Neyman and Student have been reported, Student of course affected Neyman indirectly through his influence on Pearson. However, Neyman's and Pearson's recollections of the origin of the Neyman–Pearson theory are at variance. While Pearson attributed the basic idea leading to their work to Student's letter of 1926, Neyman cites

> ...remarks of Borel that served as an inspiration to Egon S. Pearson and myself in our effort to build a frequentist theory of testing hypothesis. [Neyman (1977)]

Neyman's first reference to Borel, who incidentally does not mention alternatives explicitly, occurs in Neyman (1929). The discrepancy between these two different views has been discussed in some detail in Lehmann (1993). Pearson's recollections are quite specific and because he was clearly the leader in the work up to 1928 (this was to change later), it seems fair to consider Student's contribution as the decisive one.

5.4 Concluding Remark

This paper has a limited aim: to assess the contributions made by Student to the three stages of small-sample theory listed at the end of Section 1. Thus, in particular, it is not an account of Student's work as a whole and does not cover his remaining papers, the wealth of ideas and suggestions contained in his correspondence and his work for Guinness both as a statistician and a brewer. Such a more comprehensive account would, I believe, support the statement made by Fisher (1939) in his obituary of Student in which he describes Gosset as "one of the most original minds in contemporary science."

ACKNOWLEDGMENT OF SOURCES

The primary sources for Gosset are not only his papers but also his correspondence, of which Pearson (1939) says, "My real understanding of Gosset as a statistician began, as no doubt for many others, when I joined that wide circle of his scientific correspondents." Gosset's biographer, McMullen (1939), notes, "'Student' had many correspondents, mostly agricultural and other experimenters, in different parts of the world. He took immense pains with these...[and they] contain some of his clearest writing... ."

A discussion of the letters exchanged between Gosset and Pearson (including some important extracts) can be found in Pearson (1939) and (1990).

Some early correspondence between Gosset and Fisher is discussed in Pearson (1968), and the bulk of the surviving letters from 1915 (when Fisher was 25) to shortly before Gosset's death in 1937 were privately published by Gosset's lifelong employer, the firm of Arthur Guinness Sons and Co. (Gosset, 1970). Unfortunately most of Fisher's letters are lost, although there are some important exceptions. On the other hand, the volume includes valuable summaries of the letters (sometimes with added comments) that Fisher made at a later date. Some of the most important passages from the letters are reprinted, with comments, in Pearson (1990).

In addition to the papers and letters, there exists a wealth of secondary literature. A long obituary, "W. S. Gosset, 1876–1937," was published in the 1939 volume of *Biometrika* in two parts: "'Student' as a man" by Gosset's brewery colleague, L. McMullen, and "'Student' as a statistician" by E. S. Pearson.

There are other obituaries, book chapters, and so on, but we mention here only the most important of these: the book *'Student'–A Statistical Biography of William Sealy Gosset* (Pearson, 1990), which was "edited and augmented" from an incomplete

ASYMPTOTICALLY NONPARAMETRIC INFERENCE: AN ALTERNATIVE APPROACH TO LINEAR MODELS[1]

BY E. L. LEHMANN[2]

University of California, Berkeley

1. Summary. As an alternative to the standard analysis of variance, procedures are developed for linear models with several observations per cell which are asymptotically distribution-free and whose asymptotic efficiency relative to the standard procedures is the same as that of the Wilcoxon test relative to Student's t-test. Specific procedures discussed are (i) tests of linear hypotheses, (ii) confidence intervals for any contrast, (iii) simultaneous intervals for all contrasts.

2. Introduction. One of the basic problems of statistics is the two-sample location problem, which arises in the comparison of two populations. Let $X_1, \cdots, X_m$ and $Y_1, \cdots, Y_n$ be two samples and suppose that

$$P\{X_i \leqq x\} = F(x) \quad \text{and} \quad P\{Y_j \leqq y\} = F(y - \Delta). \tag{2.1}$$

The classical estimate of Δ is of course $\bar{Y} - \bar{X}$, while the hypothesis $H : \Delta = 0$ was tested during the 19th century by referring the test statistic $t = (\bar{Y} - \bar{X})/(1/m + 1/n)^{\frac{1}{2}}S$ to the normal distribution to get an approximate significance level which, for large samples and F with finite second moment, is justified by the central limit theorem.

A new approach to testing—small sample theory—was initiated by Student (1908) who conjectured the exact form of the distribution of t for the case that F is normal. His conjecture was proved by R. A. Fisher (1923) who went on to develop the exact distributions of a number of other test statistics under the assumption of normality. (An earlier such derivation had been that of Helmert (1876) of the χ^2-distribution of S^2).

Attention was thus focused on the possibility of *exact tests*, i.e., tests whose significance level could be exactly computed under suitable assumptions, for example the assumption of normality. This aspect was further emphasized through the development in the 1930's by Neyman and Pearson [18] of a theory of optimum tests. Tests were sought which are most powerful in a clearly defined parametric model within the class of all similar tests, i.e., tests whose significance level is exactly equal to the specified value independent of any nuisance parameters of the model.

The emphasis on exactness raised the question what to do when the form of

Received January 29, 1963.

[1] This research was done while the author was a Professor of the Adolph C. and Mary Sprague Miller Institute for Basic Research in Science, University of California.

[2] I am grateful to Professor John W. Tukey for a critical reading of the manuscript and several helpful suggestions.

the distribution could not be assumed to be known, and hence no particular parametric model could be postulated. In response to this need, nonparametric tests were developed, such as the tests proposed by Wilcoxon (1945) (see also [13]), whose levels do not depend on the underlying distribution and are therefore exact for a broadly nonparametric model, for example model (2.1) where F is only required to be continuous.

Such nonparametric tests were regarded rather skeptically at first. Since they use only the order relationship between the observations, it seemed obvious that these tests discard so much of the information contained in the data that there must result a considerable loss of power. However, in this case intuition proved unreliable. Work by Pitman (1949) and others showed for example that, asymptotically, the Wilcoxon test is about 95.5% efficient in the case of a normal distribution[3]; that it is more efficient than the t-test if the normal distribution is seriously disturbed by gross errors; and that its asymptotic efficiency relative to the t-test is never (i.e., for no F) less than 86.4% but can be arbitrarily high and even infinite.

The Wilcoxon test is thus seen to be superior to the t-test in two respects.

(i) Its significance level is more robust in that it is exactly equal to the stated value for all continuous distributions F. (It has this value also in much more general models, which require no assumption of independence of the variables or identity of their distribution.)

(ii) Its power is more robust, particularly against distributions with heavy tails.

Of these two advantages, the second is perhaps the more important.

If it is only a question of testing the hypothesis $\Delta = 0$ in model (2.1), a nonparametric test such as the Wilcoxon or Normal Scores test would nearly always seem preferable to the t-test. The objection is however frequently raised that occasions for such isolated tests are rare; that the nonparametric tests are not adaptable to point estimation, multiple decision problems, etc.; and that they are not suitable for a flexible use in linear models such as is provided by the standard analysis of variance.

It is the purpose of the present paper to describe procedures possessing this flexibility and enjoying the above property (ii), of nonparametric methods although unfortunately not property (i). The approach is here developed only for linear models in which there are several observations per cell. The significance level, confidence level, or other probabilities to be controlled, have their nominal values asymptotically as the numbers of observations in all cells become large; the procedures are therefore asymptotically distribution-free. They are in this sense comparable to the corresponding classical procedures. While the latter are "exact" under the assumption of normality, this assumption is of course never exactly fulfilled and in particular is likely to be disturbed by gross errors. In

[3] The exact values obtained by Klotz (1963) for the one-sample problem in the normal case for sample sizes 5 to 10 are even slightly higher.

practice, the classical procedures therefore also have the correct levels only asymptotically.

For some special linear hypotheses, nonparametric tests have been proposed by Friedman (1937), Durbin (1951), Kruskal and Wallis (1952), Benard and Van Elteren (1953), Hodges and Lehmann (1962), and others. This type of test seems however not applicable to many of the standard problems such as, for example, that of testing for the absence of interactions in a two-way layout. A general class of nonparametric procedures, applicable to a variety of linear hypotheses, has been proposed by Mood (1950) and Brown and Mood (1951). In the particular case of the c-sample problem, the asymptotic efficiency of the Brown-Mood median test relative to the classical F-test has been shown by Andrews (1954) to be the same as that of the sign test relative to Student's t-test in the one-sample problem. (For other efficiency results concerning the Brown-Mood tests see Bhapkar (1963)). In contrast, the procedures proposed here have the same asymptotic efficiency relative to the classical procedures as the Wilcoxon test has relative to the t-test.

3. Linear models with several observations per cell. Let the observable random variables be $X_{i\alpha}$, and suppose they are of the form

$$X_{i\alpha} = \xi_i + U_{i\alpha} \qquad (\alpha = 1, \cdots, n_i\,; i = 1, \cdots, c) \tag{3.1}$$

where the U's are independently distributed with common distribution F having density f, and the ξ's are unknown constants. Let

$$Y_{ij} = \operatorname{med}(X_{i\alpha} - X_{j\beta}) \tag{3.2}$$

be the median of the $n_i n_j$ differences $X_{i\alpha} - X_{j\beta}$ $(\alpha = 1, \cdots, n_i\,; \beta = 1, \cdots, n_j)$, and let

$$Z_{ij} = Y_{i\cdot} - Y_{j\cdot}, \tag{3.3}$$

where $Y_{i\cdot} = \sum Y_{ij}/c$. It was shown in [15] that for $n_i = \rho_i N$ and N tending to infinity, the variables $N^{\frac{1}{2}}[Z_{ic} - (\xi_i - \xi_c)]$ have a limiting normal distribution with zero means and covariance matrix $\Sigma = (\sigma_{ij})$ where

$$\begin{aligned} \sigma_{ii} &= \frac{1}{12}\left(\frac{1}{\rho_i} + \frac{1}{\rho_c}\right) \Big/ \left[\int f^2(x)\,dx\right]^2 \\ \sigma_{ij} &= \frac{1}{12\rho_c} \Big/ \left(\int f^2(x)\,dx\right)^2 \end{aligned} \qquad \text{for all } i, j < c. \tag{3.4}$$

Also for any i and j

$$N^{\frac{1}{2}} Z_{ij} = N^{\frac{1}{2}}(Z_{ic} - Z_{jc}). \tag{3.5}$$

Consider now instead the statistics $T_{ij} = X_{i\cdot} - X_{j\cdot}$. The variables $N^{\frac{1}{2}}[T_{ic} - (\xi_i - \xi_c)]$ are asymptotically normally distributed with zero means and covariance matrix Σ', which is related to Σ through

$$\Sigma' = k^2\Sigma \tag{3.6}$$

with $k^2 = 12\sigma^2[\int f^2(x)\,dx]^2$. Also of course, $N^{\frac{1}{2}}T_{ij} = N^{\frac{1}{2}}(T_{ic} - T_{jc})$.

These facts may be summarized in

THEOREM 1. *The matrices with elements* $kN^{\frac{1}{2}}[Z_{ij} - (\xi_i - \xi_j)]$ *and* $N^{\frac{1}{2}}[T_{ij} - (\xi_i - \xi_j)]$ *respectively, have the same limiting distribution.*

The parallel notation in the definitions of the statistics Z_{ij} and T_{ij}, frequently makes it easy to write down the estimates, tests, etc. based on the Z's simply by substituting in the corresponding classical procedures. However, in one respect the parallel notation is deceptive. The averaging indicated by a dot in $X_{i\cdot}$ is over the n_i observations in the ith cell; the corresponding average in $Y_{i\cdot}$, on the other hand, is taken over the c cells. This difference sometimes leads to certain dissimilarities in the notation as for instance in the example of Section 4 below.

A principal purpose in introducing the estimates Y_{ij} is to provide protection against the possibility of gross errors. It should be pointed out in this connection that for $n_i = n_j = 2$, Y_{ij} coincides with the usual estimate $T_{ij} = X_{i\cdot} - X_{j\cdot}$ and hence does not provide such protection. More generally, little protection is afforded by Y_{ij} unless at least both $n_i \geqq 3$ and $n_j \geqq 3$.

Let us now see how the estimates Z_{ij} of $(\xi_i - \xi_j)$ can be used to obtain large sample tests of linear hypotheses concerning contrasts in the model (3.1). Consider first the classical test. If F were normal with known σ, attention could be restricted to the statistics $X_{i\cdot} = \xi_i + U_{i\cdot}$. A linear hypothesis H specifies that the vector $(\xi_1, \cdots, \xi_c)$ of means of $(X_{1\cdot}, \cdots, X_{c\cdot})$ lies in a $(c - r)$-dimensional linear subspace Π_ω of c-space. The test is based on the statistic

$$\sum n_i(X_{i\cdot} - \hat{\xi}_i)^2/\sigma^2 \tag{3.7}$$

where $(\hat{\xi}_1, \cdots, \hat{\xi}_c)$ is the projection of $(X_{1\cdot}, \cdots, X_{c\cdot})$ on Π_ω.

Under H, the statistic (3.7) has a χ^2-distribution with r degrees of freedom if F is normal, and has this as limiting distribution for any F with finite variance. Let

$$X_{i\cdot} - \hat{\xi}_i = \sum_{j=1}^{c-1} a_{ij}(X_{j\cdot} - X_{c\cdot}), \tag{3.8}$$

so that (3.7) may be written as

$$\sum n_i[\sum a_{ij}(X_{j\cdot} - X_{c\cdot})]^2/\sigma^2. \tag{3.9}$$

From (3.8) it follows that

$$\sum a_{ij}(\xi_c - \xi_j) = 0 \quad \text{for all} \quad (\xi_1, \cdots, \xi_c) \in \Pi_\omega, \tag{3.10}$$

and hence from Theorem 1 that the statistics $N^{\frac{1}{2}}\sum a_{ij}T_{jc}$ $(i = 1, \cdots, c - 1)$ under H have the same joint limiting distribution as the statistics $kN^{\frac{1}{2}}\sum a_{ij}Z_{jc}$. Since (3.9) has a χ^2-distribution with r degrees of freedom under H, we have therefore shown that the limiting distribution under H of

$$k^2 \sum n_i(\sum a_{ij} Z_{ij})^2/\sigma^2 = 12\left(\int f^2(x)\, dx\right)^2 \sum n_i(\sum a_{ij} Z_{ij})^2 \tag{3.11}$$

is χ^2 with r degrees of freedom. The statistic $\sum n_i(\sum a_{ij}Z_{ij})^2$ can be obtained very simply from the classical statistic $\sum n_i(X_{i\cdot} - \xi_i)^2$. The latter is a quadratic form in the variables $(X_{1\cdot}, \cdots, X_{c\cdot})$ and it is only necessary to replace these by the variables $(Y_{1\cdot}, \cdots, Y_{c\cdot})$, as is illustrated in Section 4 below.

It has been tacitly assumed in the above argument that $\sigma^2 < \infty$ and

$$\int f^2(x)\, dx < \infty. \tag{3.12}$$

Actually, the latter assumption alone is enough to insure the result; for it follows from (3.4) and (3.5) that the joint limiting distribution of the variables $kN^{\frac{1}{2}} \cdot [Z_{ij} - (\xi_i - \xi_j)]$ is the same as the distribution of $N^{\frac{1}{2}}[(X'_{i\cdot} - X'_{j\cdot}) - (\xi_i - \xi_j)]$ where the variables $X'_{i\alpha}$ are independently normally distributed with mean ξ_i and unit variance. Assumption (3.12) clearly holds whenever the density f is bounded.

The quantity (3.11) is not yet a usable test statistic since it depends on the unknown quantity $\int f^2(x)\, dx$. The following are two ways out of this difficulty.

(i) Let T_N be any consistent estimate of $1/\int f^2(x)\, dx$ as the sample sizes n_i tend to infinity. Then the test statistic

$$W = 12\sum n_i(\sum a_{ij}Z_{ij})^2/T_N^2 \tag{3.13}$$

under H has a limiting χ^2-distribution with r degrees of freedom. The rejection region $W > C$, where C is the 100α upper percentage point of the χ^2-distribution with r degrees of freedom, therefore provides a large-sample level α test of H.

Two such consistent estimates are provided in the paper following the present one. If $T^{(i)}$ denotes the estimate (15) of that paper based on the observations from the ith cell, one may take for T_N the average $\sum T^{(i)}/c$. Alternatively if $T^{(i,j)}$ is the estimate (9) of the succeeding paper based on the observations from the ith and jth cell, one may take for T_N the average of the $T^{(i,j)}$'s.

(ii) In many cases, it is possible to obtain a quadratic form, say Q', which is asymptotically independent of $Q = \sum n_i(\sum a_{ij}Z_{ij})^2$, and which after multiplication with $12(\int f^2(x)\, dx)^2$ is asymptotically distributed as χ^2 with, say, r' degrees of freedom, regardless of whether the hypothesis being tested is true or not. The statistic

$$W' = Q/r \div Q'/r' \tag{3.14}$$

under H then has a limiting F distribution with r and r' degrees of freedom and hence provides a large-sample test of H.

Such quadratic forms will be available whenever all the contrasts in a subspace of the full contrast space may be assumed to be zero. If, for example, in a two-way layout it is assumed that the interactions are zero, the quadratic form that would have been appropriate for testing the hypothesis that the interactions are zero could be used for Q'.

Other such quadratic forms can be obtained by dividing the observations in each cell into two groups (say, at random, with the sizes of the subgroups equal or differing by one). For the ith cell one may then set up the spurious hypothesis H_i that there is no difference in location between the distributions of the observations in the two subgroups. Let the variables in the two subgroups of the ith cell be denoted by X'_{ij} ($j = 1, \cdots, n'_i$) and X''_{ik} ($k = 1, \cdots, n''_i$) with $n'_i + n''_i = n_i$. Then the test statistic appropriate for H_i is

$$Q_i = n_i Y_i'^2/[(1/n'_i) + (1/n''_i)], \tag{3.15}$$

where $Y'_i = \operatorname{med}(Y'_{ij} - Y''_{ik})$. On multiplication with $12(\int f^2(x)\,dx)^2$, Q_i has a limiting χ^2-distribution with one degree of freedom. Furthermore, $Q_1, \cdots, Q_c$ are asymptotically independent of each other, and of Q and of Q'. If the sample sizes n_i are sufficiently large for this method to be applicable ($\geqq 6$ to make the half-group size $\geqq 3$, and preferable $\geqq 8$ so that the χ^2-approximation is based on half-group sizes $\geqq 4$), then $Q_1 + \cdots + Q_c$ provides a quadratic form with c degrees of freedom, which may be added to the form Q' suggested above or be used in its place if $\xi_1, \cdots, \xi_c$ are completely unknown.

The above suggestion of dividing the observations in each cell into two groups was based on the assumption, corresponding to typical practice, that the number of observations per cell is not very large. Of course, if many observations are available for a cell, they can be divided into more than two subgroups, with a corresponding gain in degrees of freedom for the resulting quadrating form. In fact, suppose that the observations in the ith cell are divided into $k_i + 1$ subgroups, and let Q_i be the quadratic form appropriate for testing the hypothesis H_i that there is no difference in location between the distributions of the observations in these subgroups. Then the same asymptotic results (as the numbers of observations in all subgroups become large) hold as in the case of two subgroups with the only exception that the limiting χ^2-distribution of Q_i now has k_i degrees of freedom, and that of $\sum Q_i$ therefore $\sum k_i$ degrees of freedom.

The approach of this section also leads to large-sample confidence sets for any contrast or set of contrasts. Consider for example a single contrast

$$\theta = \sum c_i \xi_i = \sum \sum d_{ij}(\xi_i - \xi_j), \qquad (\textstyle\sum c_i = 0). \tag{3.16}$$

The standard acceptance region of the hypothesis $\theta = \theta_0$ (for known σ) is $|\sum \sum d_{ij}(X_{i\cdot} - X_{j\cdot}) - \theta_0|/\sigma(\sum c_i^2/n_i)^{\frac{1}{2}} \leqq C$ and the associated confidence statement

$$|\textstyle\sum \sum d_{ij}[(X_{i\cdot} - X_{j\cdot}) - (\xi_i - \xi_j)]| \leqq (C\sigma/N^{\frac{1}{2}})(\sum c_i^2/\rho_i)^{\frac{1}{2}} \tag{3.17}$$

has limiting probability $1 - \alpha$. By Theorem 1, the statement

$$(12N)^{\frac{1}{2}} \int f^2(x)\,dx |\textstyle\sum\sum d_{ij}[(Y_{i\cdot} - Y_{j\cdot}) - (\xi_i - \xi_j)]| \leqq C(\sum c_i^2/\rho_i)^{\frac{1}{2}}$$

also has limiting probability $1 - \alpha$. Substituting the consistent estimate T_N of $1/\int f^2(x)\,dx$, we therefore get the asymptotic confidence intervals at level $1 - \alpha$

$$\sum\sum d_{ij}(Y_{i\cdot} - Y_{j\cdot}) - [C/(12N/T_N^2)^{\frac{1}{2}}][\sum (c_i^2/\rho_i)]^{\frac{1}{2}} \leqq \theta \leqq \sum\sum d_{ij}(Y_{i\cdot} - Y_{j\cdot}) + [C/(12N/T_N^2)^{\frac{1}{2}}][\sum (c_i^2/\rho_i)]^{\frac{1}{2}}. \tag{3.18}$$

Alternatively, let Q' be a quadratic form which is asymptotically independent of $\sum\sum d_{ij}(Y_{i\cdot} - Y_{j\cdot})$ and which on multiplication by $12(\int f^2(x)\,dx)^2$ is asymptotically distributed as χ^2 with r' degrees of freedom. Then

$$N^{\frac{1}{2}}[\sum\sum d_{ij}(Y_{i\cdot} - Y_{j\cdot}) - \theta_0]/(\sum c_i^2/\rho_i)^{\frac{1}{2}} \div (Q'/r')^{\frac{1}{2}}$$

is asymptotically distributed as Student's t with r' degrees of freedom. Hence if C is the critical value for a two-sided t-test at level α,

$$\sum\sum d_{ij}(Y_{i\cdot} - Y_{j\cdot}) \pm C(\sum c_i^2/\rho_i)^{\frac{1}{2}}(Q')^{\frac{1}{2}}/r'N^{\frac{1}{2}} \tag{3.19}$$

will provide another set of asymptotic confidence intervals for θ at level $1 - \alpha$.

In exactly the same manner, one can obtain simultaneous approximate confidence statements based on the Y's for all contrasts, analogous to Scheffé's S-method. For let C' be determined so that the limiting probability is $1 - \alpha$ that (3.18) holds for all matrices (d_{ij}) when C is replaced by C'. Then it follows as before that the limiting probability is $1 - \alpha$, that (3.18) will hold for all matrices (d_{ij}), and hence for all contrasts θ, when C is replaced by C'. The same remark applies of course to the intervals (3.19).

4. An example. As an illustration, consider a two-way layout with n observations per cell. With a slight change of notation, let the observable variables be denoted by

$$X_{ij\alpha} = \xi_{ij} + U_{ij\alpha} \qquad (i = 1, \cdots, a; j = 1, \cdots, b; \alpha = 1, \cdots, n), \tag{4.1}$$

where the U's are independently and identically distributed. In the usual decomposition into main effects and interactions, let

$$\xi_{ij} = \mu + \alpha_i + \beta_j + \gamma_{ij} \quad (\sum_i \alpha_i = \sum_j \beta_j = \sum_i \gamma_{ij} = \sum_j \gamma_{ij} = 0) \tag{4.2}$$

and let $Y_{ijkl} = \text{med}(X_{kl\beta} - X_{ij\alpha})$ be the median of all differences between observations in the klth and ijth cell. Then the adjusted estimates of the α's, β's and γ's are

$$\hat{\alpha}_i = Y_{i\cdots} - Y_{\cdots\cdot}, \qquad \hat{\beta}_j = Y_{\cdot j\cdot\cdot} - Y_{\cdots\cdot}$$
$$\hat{\gamma}_{ij} = Y_{ij\cdot\cdot} - Y_{i\cdots} - Y_{\cdot j\cdot\cdot} + Y_{\cdots\cdot}, \tag{4.3}$$

where actually $Y_{\cdots\cdot} = 0$. Let T_N be the estimate of $1/\int f^2(x)\,dx$ which is discussed in the preceding section. Using method (i) above, we can then base a test of the hypothesis $H: \alpha_1 = \cdots = \alpha_a = 0$ on the statistic $W = 12bn\sum (Y_{i\cdots} - Y_{\cdots\cdot})^2/T_N^2$ which under H has asymptotically a χ^2-distribution with $a - 1$ degrees of freedom. If C is the upper 100 α-percentage point of this distribution, the rejection region $W \geqq C$ will provide an asymptotically distribution-free test of H.

Corresponding confidence sets of the α's can be based on the acceptance region $12bn \sum (Y_{i\ldots} - Y_{\ldots.} - \alpha_i^0)^2/T_N^2 \leqq C$ of the hypothesis $\alpha_1 = \alpha_1^0, \cdots, \alpha_a = \alpha_a^0$. The resulting confidence sets for $(\alpha_1, \cdots, \alpha_a)$ are the intersections of the hyperspheres

$$\sum [\alpha_i - (Y_{i\ldots} - Y_{\ldots.})]^2 \leqq CT_N^2/12bn \tag{4.4}$$

with the hyperplanes

$$\sum \alpha_i = 0. \tag{4.5}$$

In an exactly analogous manner we obtain tests of hypotheses specifying the β's or γ's, or confidence sets for these parameters. The sums of squares on which the test statistics are based, are the terms of a decomposition, exactly analogous to the usual one in the analysis of variance of a two-way layout, namely

$$\begin{aligned} \sum \sum (Y_{ij..} - Y_{\ldots.})^2 = \sum \sum (Y_{ij..} - Y_{i\ldots} - Y_{.j..} + Y_{\ldots.})^2 \\ + b\sum (Y_{i\ldots} - Y_{\ldots.})^2 + a\sum (Y_{.j..} - Y_{\ldots.})^2. \end{aligned} \tag{4.6}$$

Note that this is the decomposition corresponding to that of analysis of variance when the variance is known, since we take the $Y_{ij..}$ as our basic variables, the averaging over the cells being required for robustness and forming the basis of the large-sample theory.

Instead of (i), we may also use method (ii) above, where however the only available quadratic form for the denominator of (3.14) is $Q_1 + \cdots + Q_c$ with Q_i defined by (3.15). The changes which this modification requires in the tests and confidence procedures are obvious.

The situation is somewhat different if one is willing to assume the γ's to be zero. The estimates $\hat{\alpha}_i$ and $\hat{\beta}_j$ are then still given by (4.3). However, for the quadratic form Q' one may now take

$$Q' = n\sum \sum \hat{\gamma}_{ij}^2 = n\sum \sum (Y_{ij..} - Y_{i\ldots} - Y_{.j..} + Y_{\ldots.})^2 \tag{4.7}$$

with $(a - 1)(b - 1)$ degrees of freedom. A test of the hypothesis $H: \alpha_1 = \cdots = \alpha_a = 0$ is then given by the rejection region

$$Q/(a - 1) \div Q'/(a - 1)(b - 1) > C \tag{4.8}$$

where $Q = bn\sum (Y_{i\ldots} - Y_{\ldots.})^2$ and Q' is given by (4.7), and where C is the $100\alpha\%$ upper percentage point of the F-distribution with $(a - 1)$ and $(a - 1)(b - 1)$ degrees of freedom.

The analogy between the test proposed here and the corresponding classical test becomes even closer if instead of with Model I we are concerned with a Model II analysis. Suppose that ξ_{ij}, instead of by (4.2), is given by

$$\xi_{ij} = \mu + A_i + B_j \tag{4.9}$$

where the A's and B's are jointly independent random variables, independent of the U's, the A's having common distribution G_A and the B's common distribution G_B, say. The test (4.8) is then clearly still (asymptotically) valid for testing the hypothesis that G_A is concentrated on a single point, and it is exactly analogous to the corresponding classical test.

As an illustration of the above procedures, consider the two-way layout from Brownlee (1960) p. 379 with $a = 2$, $b = 3$ and $n = 4$, used as an example in [15]. Suppose that the interactions can be assumed to be zero and that we wish to test the hypothesis $\alpha_1 = \alpha_2 = 0$ in the model given by (4.1) and (4.2), or the hypothesis that G_A is concentrated on a single point in model (4.9). In [15] the following estimates were obtained for the α's, β's and γ's.

	$\alpha_1 = -\alpha_2$	β_1	β_2	β_3	$\gamma_{11} = -\gamma_{21}$	$\gamma_{12} = -\gamma_{22}$	$\gamma_{13} = -\gamma_{23}$
Estimates (4.3)	2.50	−17.54	1.29	16.25	−.29	−1.79	2.08
Classical estimates	2.54	−17.17	1.08	16.08	−.67	−1.67	2.33

We thus get the following decomposition of $\sum\sum (Y_{ij\cdot\cdot} - Y_{\cdot\cdot\cdot\cdot})^2$ and $\sum\sum (X_{ij\cdot} - X_{\cdot\cdot\cdot})^2$.

	$bn \sum \hat{\alpha}_i^2/(a-1)$	$an \sum \hat{\beta}_j^2/(b-1)$	$n \sum\sum \hat{\gamma}_{ij}^2/(a-1)(b-1)$
Decomposition (4.5)	150.00	2293.51	30.46
Classical Analysis of Variance	154.84	2218.17	34.67

In this example, the number of observations per cell is too small for use of the quadratic forms (3.15). Using therefore the test (4.8), we find the significance probability to be .16. For the corresponding classical test in model (4.9), the significance probability is .17; on the other hand, for the classical test in model (4.1), where in the denominator one can use the residual sum of squares, we find the significance probability to be .39.

5. Asymptotic efficiency. Let us now determine the asymptotic efficiency of the test $W > C$ with W given by (3.13). Suppose that the classical statistic (3.7) is based on $n_i' = \rho_i N'$ ($i = 1, \cdots, c$) observations $X'_{i\alpha}$, while (3.13) and (3.14) continue to be based on $n_i = \rho_i N$ observations $X_{i\alpha}$, both the $X'_{i\alpha}$ and $X_{i\alpha}$ being given by (3.1). Then if $T'_{ij} = X'_{i\cdot} - X'_{j\cdot}$, it follows from Theorem 1 that the matrices with elements

$$(5.1) \qquad kN^{\frac{1}{2}}[Z_{ij} - (\xi_i - \xi_j)] \quad \text{and} \quad (N')^{\frac{1}{2}}[T'_{ij} - (\xi_i - \xi_j)]$$

respectively, have the same limiting distribution.

Consider now the linear hypothesis H of Section 3. By (3.10) the null distribution of the test criteria (3.13) and (3.7) may be computed for $\xi_1 = \cdots =$

$\xi_c = 0$, so that under H the matrices with elements $kN^{\frac{1}{2}}Z_{ij}$ and $(N')^{\frac{1}{2}}T'_{ij}$ have the same limiting distribution. As usual, we shall compare the power of the two tests against a sequence of alternatives K_N whose difference from H is of the order $1/N^{\frac{1}{2}}$. Because of (3.10), we may assume without loss of generality that the vectors $(\xi_1, \cdots, \xi_c) = (\xi_1^{(N)}, \cdots, \xi_c^{(N)})$ of K_N satisfy

$$\xi_i^{(N)} - \xi_j^{(N)} = (b_i - b_j)/N^{\frac{1}{2}} + o(1/N^{\frac{1}{2}}). \tag{5.2}$$

If $Z_{ij}^* = Z_{ij} - (\xi_i - \xi_j)$ and $T_{ij}^* = T'_{ij} - (\xi_i - \xi_j)$, we then have $kN^{\frac{1}{2}}Z_{ij} = kN^{\frac{1}{2}}Z_{ij}^* + k(b_i - b_j) + o(1)$ and $(N')^{\frac{1}{2}}T'_{ij} = (N')^{\frac{1}{2}}T_{ij}^* + (N'/N)^{\frac{1}{2}}(b_i - b_j) + (N')^{\frac{1}{2}}o(1/N^{\frac{1}{2}})$. It follows from the asymptotic equivalence of the matrices (5.1), that the matrices with elements

$$kN^{\frac{1}{2}}Z_{ij} \quad \text{and} \quad (N')^{\frac{1}{2}}T'_{ij} \tag{5.3}$$

have the same limit distribution under K_N provided $N' = g(N)$ is determined in such a way that

$$N'/N \to k^2 \quad \text{as} \quad N \to \infty. \tag{5.4}$$

If (5.4) holds, the two test statistics therefore have the same limit distribution under K_N and hence have the same limiting power against these alternatives. This shows that k^2 is the Pitman efficiency of the test based on (3.13) relative to the classical test based on (3.7), or to the corresponding test with σ replaced by S. This is of course also the asymptotic efficiency of the test based on (3.14), provided each cell is divided into sufficiently many subgroups so that the number of degrees of freedom r' of Q' tends to infinity with the sample sizes.

To discuss the efficiency of the confidence intervals (3.18) we may use the relationship between the power of a test and the probability of false parameter values being covered by the associated confidence sets. From this relationship it follows that k^2 is also the asymptotic efficiency of the confidence intervals (3.18) relative to the intervals (3.17) in the following sense. The intervals (3.18) based on N' observations have the same limiting probability of covering values $\theta^{(N)}$ of the contrast differing from the true value by a term of order $1/N^{\frac{1}{2}}$, i.e., of the form $\theta^{(N)} = \theta + \Delta/N^{\frac{1}{2}} + o(1/N^{\frac{1}{2}})$, provided (5.4) holds. Again, the result is unchanged if in (3.17) the standard deviation σ is replaced by a consistent estimate S.

Some authors prefer to measure the accuracy of confidence intervals by their length rather than by the probability of covering false values. (For a discussion of this issue see Pratt [20], [21] and Madansky [16].) The length of the intervals (3.18) based on N observations is

$$L = 2C(\sum c_i^2/\rho_i)^{\frac{1}{2}}/(12NT_N^2)^{\frac{1}{2}} \tag{5.5}$$

while that of the intervals (3.17), based on N' observations and with σ replaced by S, is

$$L' = 2C(\sum c_i^2/\rho_i)^{\frac{1}{2}}S/(N')^{\frac{1}{2}}. \tag{5.6}$$

The ratio L'/L tends in probability to $k(N/N')^{\frac{1}{2}}$. Thus, if $N' = N$, k^2 is the efficiency of the intervals (3.18) to the intervals (3.17) in the sense that it is the limit of the ratio of the squares of the lengths of the intervals. Alternatively, it is the efficiency in the sense that if the intervals are based on $k\sum \rho_i N$ an $\sum \rho_i N$ observations respectively, then the limiting ratio of the length is 1.

A third approach to the comparison of two sets of confidence intervals can be based on a definition of accuracy suggested by Wolfowitz (1950). If (θ_L, θ_U) are confidence intervals for a parameter θ, Wolfowitz proposed as a measure of accuracy the quantity

$$W(\theta; \theta_L, \theta_U) = a(\theta_L - \theta)^2 + b(\theta_U - \theta)^2. \tag{5.7}$$

Let us denote by $(\theta_{LX}, \theta_{UX})$ and $(\theta_{LY}, \theta_{UY})$ the confidence intervals for θ defined by (3.17) and (3.18) respectively. Then $N^{\frac{1}{2}}(\theta_{LX} - \theta)$ has a limiting normal distribution with mean and variance $\gamma = C\sigma(\sum c_i^2/\rho_i)^{\frac{1}{2}}$ and $\tau^2 = \sigma^2 \sum c_i^2/\rho_i$ and the difference $D = N^{\frac{1}{2}}[(\theta_{UX} - \theta) - (\theta_{LX} - \theta)]$ tends to $\lambda = 2C\sigma(\sum c_i^2/\rho_i)^{\frac{1}{2}}$. (If in (3.18) σ is replaced by S, the difference D tends to λ in probability.) By Theorem 1, $N^{\frac{1}{2}}(\theta_{LY} - \theta)$ then has a normal limiting distribution with mean γ/k and variance τ^2/k^2, and the difference $N^{\frac{1}{2}}[(\theta_{UY} - \theta) - (\theta_{LY} - \theta)]$ tends in probability to λ/k. Suppose now that the intervals $(\theta_{LX}, \theta_{UX})$ are based on N' observations where $N'/N \to k^2$ as $N \to \infty$. Then it is easily seen that the quantities $NW(\theta; \theta_{LX}, \theta_{UX})$ and $NW(\theta; \theta_{LY}, \theta_{UY})$ have the same limit distribution, so that again k^2 appears as a reasonable measure of asymptotic efficiency. The same result clearly holds for more general measures of accuracy.

In a similar sense, the comparison of the two methods for simultaneous estimation of several contrasts also leads to efficiency k^2. As an illustration consider the confidence spheres for the effects $(\alpha_1, \cdots, \alpha_c)$ discussed in Section 4. If accuracy is measured by the probability of covering false values, it follows of course from the corresponding result for tests that the asymptotic efficiency of the spheres given by (4.4) and (4.5) relative to the classical spheres is k^2.

An alternative comparison is in terms of the volumes of the spheres. The spheres defined by (4.4) and (4.5) are centered at the point $(Y_{1\cdots} - Y_{\cdots\cdot}, \cdots, Y_{c\cdots} - Y_{\cdots\cdot})$ and the square of their radius is $R^2 = C/12bnT_N^2$. On the other hand, the standard confidence spheres are centered on the point $(X_{1\cdot\cdot} - X_{\cdots}, \cdots, X_{c\cdot\cdot} - X_{\cdots})$ and the square of their radius is $R'^2 = CS^2/bn'$ if the number of observations per cell is n'. It follows that if n' is determined so that $n'/n \to k^2$ as $n \to \infty$, the ratio of the volumes of the two spheres tends to 1.

In analogy to (5.7), one might measure the accuracy of the confidence spheres by

$$\begin{aligned} W = {} & a[\text{min squared distance of } (\alpha_1, \cdots, \alpha_c) \text{ from the sphere}] \\ & + b[\text{max squared distance of } (\alpha_1, \cdots, \alpha_c) \text{ from the sphere}]. \end{aligned} \tag{5.8}$$

Then NW is a function of $N^{\frac{1}{2}}R$ and $N^{\frac{1}{2}}(\hat{\alpha} - \alpha)$ where R is the radius and $\hat{\alpha} = (\hat{\alpha}_1, \cdots, \hat{\alpha}_c)$ the center of the sphere. From this it is easily seen as before that

NW has the same limit distribution for the spheres defined by (4.5) and (4.6) and based on N observations, as for the standard spheres based on N' observations when $N'/N \to k^2$.

Finally, for Scheffé's method of judging all contrasts the comparison may be made, as suggested in [12], in terms of the lengths of the corresponding intervals. Since these are given for all contrasts by (5.5) and (5.6), the remarks made in this connection for a single contrast also apply to the present case.

Throughout the present paper, the procedures have been based on the two-sample estimates of contrasts discussed in Sections 2 to 4 of [15]. Instead, they can of course be based on the one-sample estimates of Section 5 of [15]. All results will then hold with the obvious modifications, subject only to the restriction that the distribution F of the variables $U_{i\alpha}$ must be symmetric.

REFERENCES

[1] Andrews, Fred C. (1954). Asymptotic behavior of some rank tests for analysis of variance. *Ann. Math. Statist.* **25** 724–736.

[2] Benard, A. and Van Elteren, Ph. (1953). A generalization of the method of m rankings. *Indag. Math.* **15** 358–369.

[3] Bhapkar, V. P. (1963). The asymptotic power and efficiency of Mood's test for two-way classification. *J. Indian Statist. Assoc.* **1** 24–31.

[4] Brown, G. W. and Mood, A. M. (1951). On median tests for linear hypotheses. *Proc. Second Berkeley Symp. Math. Statist.and Prob.* **II** 159–166.

[5] Brownlee, K. A. (1960). *Statistical Theory and Methodology in Science and Engineering.* Wiley, New York.

[6] Durbin, J. (1951). Incomplete blocks in ranking experiments. *British J. Psych.* **4** 85–90.

[7] Fisher, R. A. (1923). Note on Dr. Burnsides' recent paper on error of observation. *Proc. Cambridge Philos. Soc.* **21** 655–658.

[8] Friedman, M. (1937). The use of ranks to avoid the assumption of normality implicit in the analysis of variance. *J. Amer. Statist. Assoc.* **32** 675–699.

[9] Helmert, F. R. (1876). Die Genauigkeit der Formel von Peters zur Berechnung des wahrscheinlichen Beobachtungsfehlers direkter Beobachtungen gleicher Genauigkeit. *Astr. Nachr.* **88** 113–132.

[10] Hodges, J. L., Jr. and Lehmann, E. L. (1956). The efficiency of some nonparametric competitors of the t-test. *Ann. Math. Statist.* **27** 324–335.

[11] Hodges, J. L., Jr. and Lehmann, E. L. (1962). Rank methods for combination of independent experiments in the analysis of variance. *Ann. Math. Statist.* **33** 482–497.

[12] Klotz, Jerome (1963). Small sample power and efficiency for the one sample Wilcoxon and normal scores test. *Ann. Math. Statist.* **34** 624–632.

[13] Kruskal, W. H. (1957). Historical notes on the Wilcoxon unpaired two-sample test. *J. Amer. Statist. Assoc.* **52** 356–360.

[14] Kruskal, William H., and Wallis, Allen W. (1952). Use of ranks on one-criterion variance analysis. *J. Amer. Statist. Assoc.* **47** 583–621.

[15] Lehmann, E. L. (1963). Robust estimation in analysis of variance. *Ann. Math. Statist.* **34** 957–966.

[16] Madansky, Albert (1962). More on length of confidence intervals. *J. Amer. Statist. Assoc.* **57** 586–589.

[17] Mood, Alexander McFarlane (1950). *Introduction to the Theory of Statistics*. McGraw-Hill, New York.

[18] NEYMAN, J. and PEARSON, E. S. (1933). On the problem of the most efficient tests of statistical hypotheses. *Philos. Trans. Roy. Soc. Ser. A* **231** 289–337.

[19] PITMAN, E. J. G. (1949). Lecture notes on nonparametric statistical inference. (Unpublished.)

[20] PRATT, JOHN W. (1961). Length of confidence intervals. *J. Amer. Statist. Assoc.* **56** 549–567.

[21] PRATT, JOHN W. (1963). Shorter confidence intervals for the mean of a normal distribution with known variance. *Ann. Math. Statist.* **34** 574–586.

[22] SCHEFFÉ, HENRY (1953). A method for judging all contrasts in the analysis of variance. *Biometrika* **40** 87–104.

[23] STUDENT (1908). On the probable error of a mean. *Biometrika* **6** 1–25.

[24] WILCOXON, FRANK (1945). Individual comparisons by ranking methods. *Biometrics* **1** 80–83.

[25] WOLFOWITZ, J. (1950). Minimax estimates of the mean of a normal distribution with known variance. *Ann. Math. Statist.* **21** 218–230.

ASYMPTOTICALLY NONPARAMETRIC INFERENCE IN SOME LINEAR MODELS WITH ONE OBSERVATION PER CELL[1]

By E. L. Lehmann

University of California, Berkeley

1. Introduction and summary. In [5] and [6] the author proposed point estimates, tests and confidence procedures for the parameters in a linear model, which have the same asymptotic efficiency (relative to the corresponding classical methods) as the Wilcoxon test has relative to the t-test. Here, "asymptotic" refers to the case that the numbers of observations per cell tend to infinity; in practice, they should presumably be at least equal to four.

In the present paper, we shall consider experiments with only one observation per cell. It then of course becomes necessary to impose some restrictions on the parameters of the model. We shall suppose that an experiment, concerned with various factors at several levels, is replicated at different levels of a nuisance factor (i.e. in different "blocks") and that this nuisance factor does not interact with the factors of interest. Nonparametric procedures for this situation were considered earlier by Friedman (1937) and in [3], where, however, only tests were proposed for the hypothesis of no effect of the factors of interest. In the present paper we shall be concerned primarily with estimating arbitrary contrasts in the factors of interest, and also with the problem of testing such contrasts.

Let us assume for the observations $X_{i\alpha}(i = 1, \cdots, c;\ \alpha = 1, \cdots, N)$ the model

$$X_{i\alpha} = \nu + \xi_i + \mu_\alpha + U_{i\alpha} \left(\sum \xi_i = \sum \mu_\alpha = 0\right) \tag{1.1}$$

where the ξ's are the parameters of interest, the μ's are the effects of the nuisance factor (or factors), and the U's are independently distributed according to a common continuous distribution F. Except for the assumption of equal sample sizes, this agrees with the model of [5] and [6] when the μ's are assumed to be zero. We shall be interested in inference methods concerning the ξ's, and their asymptotic properties as $N \to \infty$.

2. Estimation of a single contrast. Let $\theta = \sum c_i\xi_i(\sum c_i = 0)$ be any contrast in the ξ's, and let $V_\alpha = \sum c_i X_{i\alpha}$. Then $V_1, \cdots, V_N$ are independently, identically distributed; let their common distribution function be G_c. A possible

---

Received 23 August 1963.

[1] This work was begun while the author was a Professor of the Adolph C. and Mary Sprague Miller Institute for Basic Research in Science, University of California, Berkeley, and was completed with the partial support of the Office of Naval Research, Contract Nonr-222-(43). This paper in whole or in part may be reproduced for any purpose of the United States Government.

estimate of θ is that proposed in [4], namely

$$\text{med}\,[\tfrac{1}{2}(V_\alpha + V_\beta)] \tag{2.1}$$

where the median is taken over all $\alpha \leqq \beta$. Suppose that G_c is symmetric (as will of course in particular be the case when F is symmetric). It then follows from the results of [4] that this estimate is symmetric about θ, and that its asymptotic efficiency relative to the classical estimate $\sum c_i X_{i\cdot}$ (with $X_{i\cdot} = \sum X_{i\alpha}/N$) is $12\tau_c^2(\int g_c^2(x)\,dx)^2$ where g_c is the density and τ_c^2 the variance of G_c. This efficiency is therefore always $\geqq .864$ and is $3/\pi$ when F (and hence G_c) is normal.

As in [5], the estimates (2.1) for different contrasts are incompatible, and it may be desirable to replace them by a mutually compatible system. Such a system can, as in [5], be based on the estimates of the differences $\xi_i - \xi_j$ which by specialization of (2.1) are given by

$$Y_{ij} = \text{med}\,[\tfrac{1}{2}(X_{i\alpha} - X_{j\alpha} + X_{i\beta} - X_{j\beta})]. \tag{2.2}$$

We shall denote the common distribution of the difference $X_{i\alpha} - X_{j\beta}$ of any two X's by G, and note that G is symmetric without any assumption about F. As basis for a discussion of a compatible set of estimates based on the Y's we require the joint asymptotic distribution of the Y's, which is given by the following theorem.

THEOREM 1. *If the density g of G satisfies the regularity conditions of Lemma 3(a) of [2], the joint limiting distribution of the random variables Y_{ij} is the $\binom{c}{2}$-variate normal distribution with zero mean and covariance matrix $\Sigma^* = (\sigma^*_{ij,kl})$ where the variances are given by*

$$\sigma_{ij,ij} = 1/12(\textstyle\int g^2(x)\,dx)^2 \qquad \textit{for all} \quad i, j \tag{2.3}$$

and the covariances by

$$\begin{aligned} \sigma_{ij,kl} &= 0, && \textit{if} \quad i, j, k, l \quad \textit{are distinct},\\ &= [\lambda(F) - \tfrac{1}{4}]/(\textstyle\int g^2(x)\,dx)^2, && \textit{if} \quad i = k \quad \textit{or} \quad j = l,\\ &= [\tfrac{1}{4} - \lambda(F)]/(\textstyle\int g^2(x)\,dx)^2, && \textit{if} \quad i = l \quad \textit{or} \quad j = k, \end{aligned} \tag{2.4}$$

and where

$$\lambda(F) = P(X_1 < X_2 + X_3 - X_4 \quad \text{and} \quad X_1 < X_5 + X_6 - X_7), \tag{2.5}$$

$X_1, \cdots, X_7$ being independent random variables with distribution F.

PROOF. Let $\binom{N}{2} U^{(ij)}$ be equal to the number of pairs (α, β) with $\alpha < \beta$ and such that

$$X_{j\alpha} - X_{i\alpha} + X_{j\beta} - X_{i\beta} > 2a_{ij}/N^{\frac{1}{2}},$$

and let Q^* denote the $\binom{c}{2}$-variate normal distribution with zero mean and covariance matrix Σ^*. Then the argument leading to (10.3) of [4] and used in proving Theorem 4 of [5], shows that

$$(2.6)\quad \begin{aligned}&\lim P\{N^{\frac{1}{2}}[Y_{ij} - (\xi_i - \xi_j)] \leqq a_{ij} \quad \text{for all} \quad 1 \leqq i < j \leqq c\}\\ &= \lim P_0\{N^{\frac{1}{2}}[U^{(ij)} - \tfrac{1}{2}] \leqq 0 \quad \text{for all} \quad 1 \leqq i < j \leqq c\},\end{aligned}$$

where P_0 indicates that the probability is being computed under the assumption $\xi_1 = \cdots = \xi_c$.

To evaluate the right hand side of (2.6) we shall use the fact that the theorem on U-statistics stated in the appendix of [5] holds if in Part (iii) the distributions $F, G, H, \cdots$ are independent of N but instead the functions $\phi^{(i)}$ depend on N, provided only these functions are uniformly bounded.

Let $X_\alpha = (X_{1\alpha}, \cdots, X_{c\alpha})$ and define

$$\begin{aligned}\phi_N^{(ij)}(x_\alpha, x_\beta) &= 1 \qquad \text{if} \quad x_{j\alpha} - x_{i\alpha} + x_{j\beta} - x_{i\beta} > 2a_{ij}/N^{\frac{1}{2}}\\ &= 0 \qquad \text{otherwise.}\end{aligned}$$

Then the quantities ζ defined in the appendix of [5] are given by

$$\zeta_N^{(ij)} = \text{Cov}\,[\phi_N^{(ij)}(X_1, X_2), \phi_N^{(ij)}(X_1, X_3)]$$

which tends to $\zeta^{(ij)} = \frac{1}{12}$, while $\zeta_N^{(ij,kl)}$ tends to

$$\begin{aligned}\zeta^{(ij,kl)} &= 0 && \text{if} \quad i, j, k, l \quad \text{are distinct}\\ &= \lambda(F) - \tfrac{1}{4} && \text{if} \quad i = k \quad \text{or} \quad j = l\\ &= \tfrac{1}{4} - \lambda(F) && \text{if} \quad i = l \quad \text{or} \quad j = k.\end{aligned}$$

Hence, if we put $\theta_N^{(ij)} = EU^{(ij)}$, it follows that the variables $N^{\frac{1}{2}}(U^{(ij)} - \theta_N^{(ij)})$ have a joint normal limiting distribution with zero mean and covariance matrix $(\sigma_{ij,kl}) = 4(\zeta^{(ij,kl)})$. Now

$$\theta_N^{(ij)} = P\{X_{j1} - X_{i1} + X_{j2} - X_{i2} > 2a_{ij}/N^{\frac{1}{2}}\}$$

and a standard argument shows that

$$N^{\frac{1}{2}}(\tfrac{1}{2} - \theta_N^{(ij)}) \to 2a_{ij} \int g^2(x)\, dx.$$

This proves that the right hand side or (2.6) tends to

$$Q(2a_{12} \int g^2(x)\, dx, 2a_{13} \int g^2(x)\, dx, \cdots, 2a_{c-1c} \int g^2(x)\, dx)$$

where Q is the $\binom{c}{2}$-variate normal distribution with zero mean and covariance matrix $(\sigma_{ij,kl})$, and the assertion of the theorem follows.

In the above argument it was tacitly assumed that the covariance matrix Σ^* is non-singular as is typically the case. Should there exist cases in which Σ^* is singular, the argument could be modified along the lines of the proof of Theorem 2 of [5].

For all that follows it is essential to have some knowledge regarding the quantity $\lambda(F)$ defined by (2.5). We therefore prove next the following inequalities.

THEOREM 2. *For all distributions F we have*

$$(2.7)\qquad \tfrac{1}{4} \leqq \lambda(F) \leqq \tfrac{7}{24}.$$

PROOF.

(i) The left inequality is a consequence of the following more general one. Let X_1, X_2, Y_1, Y_2 be independently distributed, the X's with distribution F_X and the Y's with distribution F_Y, then

$$P(X_1 < Y_1 \text{ and } X_1 < Y_2) \geqq P(X_1 < Y_1 \text{ and } X_2 < Y_2). \tag{2.8}$$

This follows immediately from the fact that the inequality (2.8) is equivalent to

$$\int [1 - F_Y(x)]^2 \, dF_X(x) \geqq \{\int [1 - F_Y(x)] \, dF_X(x)\}^2$$

which is a consequence of the Schwarz' inequality. Identifying the left hand side of (2.8) with that of (2.5) with $Y_1 = X_2 + X_3 - X_4$ and $Y_2 = X_5 + X_6 - X_7$, we find that the left hand side of (2.5) is $\geqq [P(X_1 + X_4 < X_2 + X_3)]^2 = \frac{1}{4}$, as was to be proved.

(ii) To prove the right inequality, consider the covariance matrix defined by (2.3) and (2.4) for the random variables Y_{12}, Y_{13}, Y_{23}. Putting $\gamma = 3[4\lambda(F) - 1]$, the determinant of this matrix is seen to be proportional to $(1 + \gamma)^2(1 - 2\gamma)$. This can be nonnegative only if either $\gamma = -1$ and hence $\lambda = \frac{1}{6}$, which is ruled out by (i), or if $\gamma \leqq \frac{1}{2}$ and hence $\lambda \leqq \frac{7}{24}$, as was to be proved.

Three values of $\lambda(F)$ are

(2.9)

F	Normal	Rectangular	Cauchy
$\lambda(F)$	.2902	.2909	.2879

These values are all only slightly below the upper bound $\frac{7}{24} = .2917$.

3. A compatible set of estimates for all contrasts. We have seen above (in the proof of (ii) of Theorem 2) that for any distribution F with $\lambda(F) \neq \frac{7}{24}$, the joint limiting distribution of the three random variables

$$N^{\frac{1}{2}}[Y_{ij} - (\xi_i - \xi_j)], \qquad N^{\frac{1}{2}}[Y_{jk} - (\xi_j - \xi_k)], \qquad N^{\frac{1}{2}}[Y_{ki} - (\xi_k - \xi_i)] \tag{3.1}$$

is nondegenerate. In particular, the variables therefore do not satisfy the condition $N^{\frac{1}{2}}(Y_{ij} + Y_{jk} + Y_{ki}) \to 0$ in probability. They are thus not only incompatible but even asymptotically incompatible in the sense of [5]. As in [5], we may wish to replace the incompatible estimates Y_{ij} of $\xi_i - \xi_j$ by the compatible estimates

$$Z_{ij} = Y_{i\cdot} - Y_{j\cdot} \tag{3.2}$$

where the dot as usual denotes averaging with respect to the indicated subscript. It follows from the asymptotic incompatibility of the Y's that in contrast to the corresponding result of [5], the Z's and the Y's are not asymptotically equivalent. In the present section, we shall investigate the asymptotic behaviour of the Z's and in particular shall prove that asymptotically the Z's are at least as efficient as the Y's for all distributions F.

LEMMA 1. *If the covariance matrix of the variables $N^{\frac{1}{2}}Y_{ij}$ is given by* (2.3) *and* (2.4), *and if we put*

$$(3.3) \qquad \operatorname{Var}(N^{\frac{1}{2}}Y_{i\cdot}) = \sigma_0^2, \qquad \operatorname{Cov}(N^{\frac{1}{2}}Y_{i\cdot}, N^{\frac{1}{2}}Y_{j\cdot}) = \rho\sigma_0^2,$$

then

$$(3.4) \qquad \sigma_0^2 = [(c-1)/c^2]\{\tfrac{1}{12} + (c-2)[\lambda(F) - \tfrac{1}{4}]\}/[\textstyle\int g^2(x)\,dx]^2$$

and

$$(3.5) \qquad \rho = -1/(c-1).$$

PROOF. An easy calculation shows that

$$\operatorname{Var}(N^{\frac{1}{2}}Y_{i\cdot}) = (c-1)c^{-2}[\operatorname{Var}(N^{\frac{1}{2}}Y_{ij}) + (c-2)\operatorname{Cov}(N^{\frac{1}{2}}Y_{ij}, N^{\frac{1}{2}}Y_{ik})]$$

and (3.4) follows by substituting from (2.3) and (2.4). Equation (3.5) is a consequence of the fact that $\sum Y_{i\cdot} = 0$ and the joint distribution of $(Y_{1\cdot}, \cdots, Y_{c\cdot})$ is symmetric.

THEOREM 3. *Suppose that the density g of G satisfies the regularity conditions of Lemma* 3a *of* [2]. *Then the set of random variables*

$$(3.6) \qquad N^{\frac{1}{2}}[Z_{ic} - (\xi_i - \xi_c)], \qquad i = 1, \cdots, c-1$$

is asymptotically (as $N \to \infty$) normally distributed with mean zero and covariance matrix $\Sigma = (\sigma_{ij})$,

$$(3.7) \qquad \sigma_{ii} = 2\sigma_0^2(1-\rho), \qquad \sigma_{ij} = \sigma_0^2(1-\rho)$$

where σ_0^2 and ρ are given by (3.4) *and* (3.5).

PROOF. Since the Z_{ic} are linear functions of the Y's, joint asymptotic normality follows from Theorem 1. That the covariance of the limiting distribution is given by (3.7), is an immediate consequence of Lemma 1.

We are now in position to compute the asymptotic efficiency e^* of Z_{ij} relative to Y_{ij} (in the sense of reciprocal of the ratio of the asymptotic variances).

THEOREM 4. *The efficiency e^* is given by*

$$(3.8) \qquad e_\lambda^*(c) = c/24\{\tfrac{1}{12} + (c-2)[\lambda(F) - \tfrac{1}{4}]\}$$

and is $\geqq 1$ (i.e. Z_{ij} is at least as efficient as Y_{ij}) for all F.

PROOF. Formula (3.8) follows directly from (2.3) and (3.7). For $c = 2$, the quantity (3.8) is equal to 1, as it should be since in this case $Y_{ij} = Z_{ij}$. For $c > 2$, (3.8) is seen to be $\geqq 1$ since $\lambda(F) \leqq \frac{7}{24}$.

For any fixed $\lambda < \frac{7}{24}$, the efficiency $e_\lambda^*(c)$ is an increasing function of c, and for $\lambda < \lambda'$, we have $e_\lambda^*(c) > e_{\lambda'}^*(c)$ for all $c > 2$. For the values of λ found in (2.9), the variation in e^* with changing c is seen to be slight. In the normal case, for example, e^* increases from $e^*(3) = 1.012$ to $e^*(\infty) = 1.036$. For the asymptotic efficiency e' of Z_{ij} relative to the classical estimate $X_{i\cdot} - X_{j\cdot}$, given by

$$e' = 12\tau^2\left(\textstyle\int g^2(x)\,dx\right)^2 c/24\{\tfrac{1}{12} + (c-2)[\lambda(F) - \tfrac{1}{4}]\},$$

we have of course a corresponding increase. In particular, in the normal case, we find for e' the values

c	3	5	10	∞
e'	.9664	.9755	.9826	.9893 .

The efficiency result stated in Theorem 4 holds not only for the estimation of the difference $\xi_i - \xi_j$ but immediately generalizes to the estimation of any contrast $\theta = \sum c_i\xi_i = \sum \sum d_{ij}(\xi_i - \xi_j)$.

COROLLARY 1. *Let $\hat{\theta}_1$ be the estimate of θ defined by* (2.1) *and $\hat{\theta}_2$ the estimate $\sum \sum d_{ij}(Y_{i\cdot} - Y_{j\cdot})$. Then the asymptotic efficiency e^* of $\hat{\theta}_2$ relative to $\hat{\theta}_1$ is given by* (3.8)

PROOF. Let e' and e denote the asymptotic efficiency of $\hat{\theta}_2$ relative to the classical estimate $\hat{\theta}_0 = \sum \sum d_{ij}(X_{i\cdot} - X_{j\cdot})$ and of $\hat{\theta}_1$ relative to $\hat{\theta}_0$ respectively. Then it was seen in Section 2 that $e(F) = 12\tau^2 (\int g^2(x)\, dx)^2$, independent of the coefficient vector $\mathbf{c} = (c_1, c_2, \cdots)$. On the other hand, since the covariance matrix of the asymptotic distribution of the Z's is proportional to that of the classical estimates $X_{i\cdot} - X_{j\cdot}$, e' is independent of $\mathbf{c}$. Thus $e^* = e'/e$ is also independent of $\mathbf{c}$ as was to be proved.

4. Tests and confidence procedures. So far, we have discussed only the problem of estimating contrasts of the ξ's. The corresponding testing and confidence problems can now be treated quite similarly to their treatment in [6]. Analogously to (3.9) of [6], the classical test of $H: (\xi_1, \cdots, \xi_c) \in \pi_w$ concerning contrasts of the ξ's, is based on the statistic $N \sum (X_{i\cdot} - \tilde{\xi}_i)^2/\sigma^2$ where π_w is a $(c - r)$-dimensional subspace of c-space and where $(\tilde{\xi}_1, \cdots, \tilde{\xi}_c)$ is the projection of $(X_{1\cdot}, \cdots, X_{c\cdot})$ on π_w. This statistic can be written in the form $N \sum [\sum a_{ij}(X_{j\cdot} - X_{c\cdot})]^2/\sigma^2$, and under H has a limiting χ^2 distribution with r degrees of freedom. It then follows as in [6] that under H

$$
\begin{aligned}
&N \sum (\sum a_{ij}Z_{ij})^2/\sigma_0^2 \\
&\quad = 12c^2 (\textstyle\int g^2(x)\, dx)^2 N \sum (\sum a_{ij}Z_{ij})^2/ \\
&\qquad\qquad (c-1)\{1 + 12(c-2)[\lambda(F) - \tfrac{1}{4}]\}
\end{aligned} \tag{4.1}
$$

also has a limiting χ^2-distribution with r degrees of freedom.

We can now apply either of the approaches (i) or (ii) of Section 3 of [6].

(i) Since the variables $(X_{i\alpha} - X_{j\alpha}) - (\xi_i - \xi_j)$ constitute a sample from the symmetric distribution G, the estimate (15) of [7] is a consistent estimate of $1/\int g^2(x)\, dx$. If we denote this estimate by T_{ij}, we can replace the unknown quantity $\int g^2(x)\, dx$ in (4.1) by the average of the T_{ij}'s without changing the asymptotic distribution of the statistic. Clearly, it is necessary also to replace the unknown $\lambda(F)$ by a consistent estimate. Such an estimate, which incidentally is unbiased, is obtained by considering for each sextuple $(i, j, j'; \alpha, \beta, \beta')$ whether or not the simultaneous inequalities

$$X_{i\alpha} < X_{j\alpha} + X_{i\beta} - X_{j\beta}, \quad X_{i\alpha} < X_{j'\alpha} + X_{i\beta'} - X_{j'\beta'}$$

are satisfied. The proportion of cases in which these inequalities hold is the desired estimate. The total number of sextuples will typically be very large, and one may be satisfied with basing this estimate on a subset.

(ii) Suppose that all contrasts in a subspace of the ξ's—of dimension r', say—are zero. Let Q'' be the quadratic form $N \sum [\sum a'_{ij}(X_{j\cdot} - X_{c\cdot})]^2$ appropriate in the classical analysis of variance for testing the hypothesis that these contrasts are indeed zero, and let Q' be obtained from Q'' by substituting Z_{ij} for $X_{i\cdot} - X_{j\cdot}$. Then

$$N \sum \left(\sum a_{ij} Z_{ij}\right)^2 / r \div Q'/r' \tag{4.2}$$

has, under H, a limiting F-distribution with r and r' degrees of freedom.

As in Section 5 of [6], it is easily seen that the asymptotic efficiency of the test discussed above under (i) relative to the classical F-test is the efficiency e^* given by (3.8) and that the corresponding efficiency of the test based on (4.2) tends to e^* as r' tends to infinity.

The method suggested in [6] for getting additional degrees of freedom for the quadratic form Q' by introducing spurious further parameters and then testing that they are zero, leads to difficulties in the present situation since it changes the dimensionality c of the space of parameters of interest and hence the factor $1 + 12(c - 2)(\lambda - \frac{1}{4})$ in (4.1).

Let us now illustrate the test based on (4.2). Suppose that in each of N blocks, we have a two-way layout, with one observation per cell; that both factors of the layout are of interest but not the block effect; and that we may assume all three factors to be additive. Changing slightly the notation of (1.2), we may then write

$$X_{ij\alpha} = \gamma + \xi_i + \zeta_j + \mu_\alpha \qquad i = 1, \cdots, a; j = 1, \cdots, b; \ \alpha = 1, \cdots, N \tag{4.3}$$

where $\sum \xi_i = \sum \zeta_j = \sum \mu_\alpha = 0$. From (2.2) we get

$$Y_{ij,kl} = \text{med}\left[\tfrac{1}{2}(X_{ij\alpha} - X_{kl\alpha} + X_{ij\beta} - X_{kl\beta})\right] \tag{4.4}$$

and $Z_{ij,kl} = Y_{ij\cdot\cdot} - Y_{kl\cdot\cdot}$. Substituting the Y's for the X's in the formulae $\hat{\xi}_i = X_{i\cdot\cdot} - X_{\cdot\cdot\cdot}$, $\hat{\zeta}_j = X_{\cdot j\cdot} - X_{\cdot\cdot\cdot}$ giving the classical estimates, we find

$$\hat{\xi}_i = Y_{i\cdot\cdot\cdot} - Y_{\cdot\cdot\cdot\cdot}, \qquad \hat{\zeta}_j = Y_{\cdot j\cdot\cdot} - Y_{\cdot\cdot\cdot\cdot} \tag{4.5}$$

where actually $Y_{\cdot\cdot\cdot\cdot} = 0$. If the interactions γ_{ij} had not been assumed to be zero, their estimates would be $\hat{\gamma}_{ij} = Y_{ij\cdot\cdot} - Y_{i\cdot\cdot\cdot} - Y_{\cdot j\cdot\cdot} + Y_{\cdot\cdot\cdot\cdot}$.

For testing the hypothesis $H: \xi_1 = \cdots = \xi_a = 0$, we can then use the test statistic

$$bN \sum \hat{\alpha}_i^2/(a - 1) \div N \sum \sum \hat{\gamma}_{ij}^2/(a - 1)(b - 1)$$

which under H has a limiting F-distribution with $a - 1$ and $(a - 1)(b - 1)$ degrees of freedom.

The classical test of $H: (\xi_1, \cdots, \xi_c) \in \Pi_\omega$ is the same whether the block effects μ_α in (1.1) are known to be zero as was assumed in [6] or whether they are unknown. It follows that the quadratic form $N \sum \left(\sum a_{ij} Z_{ij}\right)^2 = N \sum \sum a_{ij}(Y_{i\cdot} - Y_{j\cdot})^2$ is also formally the same in both cases. Actually, the two statistics are of course different since Y_{ij} is defined by (2.2) in the present case but as med $(X_{i\alpha} - X_{j\beta})$ or as $\text{med}_{\alpha \leq \beta}\left[\frac{1}{2}(X_{i\alpha} + X_{i\beta})\right] - \text{med}_{\gamma \leq \delta}\left[\frac{1}{2}(X_{j\gamma} + X_{j\delta})\right]$ in [6].

Since the asymptotic efficiency $e^* = e^*(F)$ of the present procedures has been shown to be uniformly higher than the efficiency $e = e(F)$ of the procedures proposed in [6], it is of course tempting to apply the procedures of the present paper also in the case where the μ_α's are zero. The only disadvantages of this approach appear to be (a) the restriction to equal sample sizes, (b) the difficulty in using possible additional degrees of freedom for Q'.

5. Several observations per cell. If instead of just one observation we had many observations in each cell, the ξ's and μ's could be estimated or tested by the methods of [5] and [6]. We shall here consider the case where there are only few, perhaps two or three, observations in each cell and where again we are concerned only with the ξ's. Let the observations be denoted by

$$X_{i\alpha r} = \nu + \xi_i + \mu_\alpha + U_{i\alpha r} \tag{5.1}$$

where $i = 1, \cdots, a$; $\alpha = 1, \cdots, N$; $r = 1, \cdots, m_\alpha$. The numbers of observations in different cells of the same block must therefore be equal but they may differ from block to block.

With the $X_{i\alpha r}$ defined by (5.1), the $N' = \sum m_\alpha$ differences $(X_{i\alpha r} - X_{j\alpha r}) - (\xi_i - \xi_j)$ constitute a sample from the distribution G. If we denote the N' differences $X_{i\alpha r} - X_{j\alpha r}$ by $D_\beta(\beta = 1, \cdots; N')$ and define $Y_{ij} = \text{med}_{\beta \leqq \beta'} [\frac{1}{2}(D_\beta + D_{\beta'})]$ and $Z_{ij} = Y_{i\cdot} - Y_{j\cdot}$, the asymptotic theory established in Sections 2 and 3 applies with N' in place of N.

This method has an unattractive feature, the random pairing of the observations in the $i\alpha$th and $j\alpha$th cell. Such pairing seems not to make the best possible use of the data since it does not utilize all possible differences between observations in the two cells. We shall however now show that the asymptotic efficiency of the proposed procedure relative to the classical procedure is as before given by e^*. This is in fact immediately seen by noting that both the present estimate Y_{ij} and the classical estimate $X_{i\cdots} - X_{j\cdots}$ of $\xi_i - \xi_j$ (and their distributions) are unchanged if the quantity μ_α in (5.1) is replaced by $\mu_{\alpha r}$. The earlier efficiency results therefore apply with N' instead of N, and since they do not involve the sample size, this establishes the desired result.

One might expect that the efficiency loss due to random pairing would become serious as the number of observations per cell increases. This seems however not to be the case. For consider the case of a single block with a large number of observations in each cell. If the observations in the ith cell are denoted by $X_{ir}(r = 1, \cdots, m)$, pairing leads to the differences $X_{ir} - X_{jr}$, and the estimate $Y_{ij} = \text{med}\, [\frac{1}{2}(X_{ir} - X_{jr} + X_{is} - X_{js})]$ of $\xi_i - \xi_j$. The asymptotic efficiency of this as $m \to \infty$ is $e = 12\tau^2 (\int g^2(x)\, dx)^2$.

It was assumed above that the numbers of observations in different cells of the same block must be equal. What should be done if, for example, in a given block it was intended to have two observations per cell, but the second observation in one of the cells is missing? A simple possibility, which presumably would not seriously affect the performance, consists in simply duplicating the single observation and proceeding as before.

REFERENCES

[1] FRIEDMAN, M. (1937). The use of ranks to avoid the assumption of normality implicit in the analysis of variance. *J. Amer. Statist. Assoc.* **32** 675–701.

[2] HODGES, J. L., JR. and LEHMANN, E. L. (1961). Comparison of the normal scores and Wilcoxon tests. *Proc. Fourth Berkeley Symp. Math. Statist. Prob.* **1** 307–317.

[3] HODGES, J. L., JR. and LEHMANN, E. L. (1962). Rank methods for combination of independent experiments in analysis of variance. *Ann. Math. Statist.* **33** 482–497.

[4] HODGES, J. L., JR. and LEHMANN, E. L. (1963). Estimates of location based on rank tests. *Ann. Math. Statist.* **34** 598–611.

[5] LEHMANN, E. L. (1963). Robust estimation in analysis of variance. *Ann. Math. Statist.* **34** 957–966.

[6] LEHMANN, E. L. (1963). Asymptotically nonparametric inference: an alternative approach to linear models. *Ann. Math. Statist.* **34** 1494–1506.

[7] LEHMANN, E. L. (1963). Nonparametric confidence intervals for a shift parameter. *Ann. Math. Statist.* **34** 1507–1512.

The Annals of Mathematical Statistics
1970, Vol. 41, No. 3, 783–801

DEFICIENCY

BY J. L. HODGES, JR.[1] AND E. L. LEHMANN[2]

University of California, Berkeley

1. Introduction and summary. Consider a statistical procedure (Method A) which is based on n observations, and a less effective procedure (Method B) which requires a larger number k_n of observations to give equally good performance. Comparison of the two methods involves the comparison of k_n with n, and this can be carried out in various ways. Perhaps the most natural quantity to examine is the difference $k_n - n$, the number of additional observations required by the less effective method. Such difference comparisons have been performed from time to time. (See, for example, Fisher (1925), Walsh (1949) and Pearson (1950).) Historically, however, comparisons have been based mainly on the ratio k_n/n. Thus, Fisher (1920), in comparing the mean absolute deviation with the mean squared deviation as estimates of a normal scale, found this ratio to be 1/1.14. Similarly in 1925 he found a large-sample ratio of $2/\pi$ for median compared with mean for estimating normal location, and the same value was found by Cochran (1937) for the sign test relative to the t-test in the normal case.

The reason for using the ratio rather than the difference in these cases is of course that the ratio is stable in large samples, so that a single limit value, say $e = \lim_{n\to\infty} n/k_n$, known as the *asymptotic relative efficiency* or ARE of B with respect to A, conveys a great deal of useful information in a compact form. When $e < 1$, the difference $k_n - n$, is not a useful measure in large samples because it tends to infinity with n.

The situation is however different, and in a sense even the reverse, in the many important statistical problems in which $e = 1$. It is then possible also for the difference to be stable, and the main purpose of this paper is to point out a number of problems in which this is the case. For the additional number $k_n - n$ of observations needed by Method B we suggest the term *deficiency*. If it exists, the limit value $d = \lim_{n\to\infty}(k_n - n)$, will be called the *asymptotic deficiency*. The number d summarizes the comparison much more revealingly in these cases than does the fact that e equals 1. (If it is not known a priori, the latter does not even tell us which of the two procedures is better.) Of course, d is less easy to compute than e since it in effect requires computing an additional term in the asymptotic expansions.

2. General properties of deficiency. We shall now introduce several conceptual and theoretical aspects of deficiency, using as the motivating topic the comparison of two methods of point estimation in terms of expected squared error. Suppose

Received February 17, 1969.

[1] Part of this research was done while this author was a Professor of the Adolph C. and Mary Sprague Miller Institute for Basic Research in Science, University of California, Berkeley.

[2] The research of this author was partially supported by the Office of Naval Research, Contract NONR N00014-67-A-0114-0004.

that A and B are methods of point estimation and let their expected squared errors be denoted by V_n for Method A and by V_n' for Method B, when each estimator is based on n observations. Then for each sample size n for A consider the equivalent sample size $k = k_n$ for B, such that V_k' is equal to V_n or as nearly equal as possible. In the problems with which we shall be concerned, V_n and V_n' will typically be of the form

$$V_n = \frac{c}{n^r} + \frac{a}{n^{r+1}} + o\left(\frac{1}{n^{r+1}}\right) \quad \text{and} \tag{2.1}$$

$$V_n' = \frac{c}{n^r} + \frac{b}{n^{r+1}} + o\left(\frac{1}{n^{r+1}}\right) \tag{2.2}$$

with $r > 0$. Here the positive coefficients of $1/n^r$ are assumed to be the same since otherwise the asymptotic efficiency of one procedure relative to the other would be different from 1. It occasionally happens that the second terms are of order $1/n^{r+s}$ with $s \neq 1$, and we shall see below how to modify the results for this case.

Let us assume, as will typically be the case, that both $\{V_n\}$ and $\{V_n'\}$ are strictly decreasing sequences. Since V_n and V_n' are inherently nonnegative, it then follows that V_n and V_n' are positive for all n.

In spite of the integral nature of sample size, let us for a moment continualize the problem by defining V_k' for nonintegral k in such a way that it is a continuous, strictly-decreasing function of k. (There are clearly many such functions; we shall discuss a particular choice below.) Given such a function, let k_n be the (unique) solution of the equation $V'_{k_n} = V_n$. As $n \to \infty$, it follows from (2.1) and (2.2) that $V_n \to 0$, hence $V'_{k_n} \to 0$, and thus $k_n \to \infty$. Equations (2.1) and (2.2) show further that

$$\frac{1}{n^r}\left[1 + \frac{a+o(1)}{cn}\right] = \frac{1}{k_n^r}\left[1 + \frac{b+o(1)}{ck_n}\right]$$

and hence that

$$k_n/n \to 1. \tag{2.3}$$

Defining d_n by

$$k_n = n + d_n, \tag{2.4}$$

the equation preceding (2.3) may be rewritten as

$$1 + \frac{d_n}{n} = \left[1 + \frac{b+o(1)}{cn}\right]^{1/r}\left[1 + \frac{a+o(1)}{ck_n}\right]^{-1/r} = 1 + \frac{b}{rcn} - \frac{a}{rck_n} + o\left(\frac{1}{n}\right).$$

By (2.3), this shows that

$$d_n \to \frac{b-a}{cr} \quad \text{as} \quad n \to \infty. \tag{2.5}$$

The same argument also shows that when the second term of (2.1) and (2.2) is respectively a/n^{r+s} and b/n^{r+s}, then $d_n \to \infty$ or $d \to 0$ as $s < 1$ or $s > 1$.

Formula (2.5), which is independent of the particular continuous function which is interpolated between V_k for integral k, suggests that the difference between the numbers of observations required by the two procedures tends to a limit as the sample sizes tend to infinity. However, it unjustifiably treats k as a continuous variable and consequently is difficult to interpret. The difficulty can be avoided by the method of stochastic interpolation introduced in [5], according to which sample size k or $k+1$ is chosen with probability $1-\pi$ and π respectively. This yields a continuous expected sample size $k+\pi$. Let us suppose that the performance of the resulting procedure is measured by $V'_{k+\pi} = (1-\pi)V'_k + \pi V'_{k+1}$, as will be the case when V' is a probability, an expectation or a variance. This defines V' as a continuous strictly decreasing function of its subscript as required, and it follows that

$$d_n = k_n + \pi - n \tag{2.6}$$

tends to

$$d = (b-a)/rc, \tag{2.7}$$

which we shall call the *asymptotic expected deficiency* (AED) of procedure B relative to A, or simply the deficiency of B relative to A when no confusion is likely.

Formula (2.7) shows that if V_n and V_n' are given by (2.1) and (2.2), then deficiency has the following two properties, which are analogous to the corresponding properties for efficiency.

(i) *Reflexivity.* If d is the AED of B with respect to A, then the AED of A with respect to B is $-d$.

(ii) *Transitivity.* If d_1 is the AED of B with respect to A, and d_2 the AED of C with respect to B (where the accuracy of C is of the same form as those of A and B), then the AED of C with respect to A is d_1+d_2.

If interpolation is not acceptable, and only integral sample sizes are permitted, the equivalent sample size k_n may be defined as the integer(s) k for which V_k' is closest to V_n. For fixed n and $d = k-n$ we find

$$\begin{aligned} V'_{n+d} - V_n &= c\left[\frac{1}{(n+d)^r} - \frac{1}{n^r}\right] - \frac{a}{(n+d)^{r+1}} - \frac{b}{n^{r+1}} + o\left(\frac{1}{n^{r+1}}\right) \\ &= [crd - ab + o(1)]\frac{1}{n^r} = \left[d - \frac{ab}{cr} + o(1)\right]\frac{cr}{n^r}. \end{aligned}$$

For all sufficiently large n, V'_{n+d} and V_n are therefore made most nearly equal by giving d the integral value closest to ab/cr (or one of the two nearest integers if $ab/cr = 1/2, 3/2, 5/2, \cdots$). We shall call this value of d the *asymptotic integral deficiency* of B with respect to A. While for simplicity we have presented the discussion of deficiency in the context of comparing the squared errors of point estimates, much of it applies to other measures of performance. Examples will be given in Sections 4 and 5.

3. Further examples of point estimation. In the present section, we shall illustrate the use of deficiency on some additional problems of point estimation. As measure of accuracy of an estimator, we shall continue to take the expected squared error. Of course in the case of unbiased estimators, this is the variance.

EXAMPLE 3.1. *Estimation of variance; price of not knowing the mean.* Observations $X_1, X_2, \cdots$ are drawn from a distribution F with expectation ξ and variance σ^2. If ξ is given, the customary estimator for σ^2 based on n observations is

$$M_n = \frac{1}{n}\sum_{i=1}^{n}(X_i-\xi)^2. \tag{3.1}$$

This estimator is unbiased and consistent, and its variance is

$$V_n = \gamma\sigma^4/n \tag{3.2}$$

where $\gamma+1 = \mu_4/\sigma^4$ is the standardized fourth central moment of F.

The estimator M_n is of course valid only if the value given for ξ happens to be correct. For this reason, even when a value for ξ is available from theory or past experience, one may often prefer an estimator which is robust against errors in ξ. The conventional robust estimator is

$$M_n' = \frac{1}{n-1}\sum_{i=1}^{n}(X_i-\bar{X})^2 \quad \text{where} \quad \bar{X} = \frac{1}{n}\sum X_i. \tag{3.3}$$

This estimator is also unbiased and its variance (see Cramér (1946) formula (27.4.2)) may be written as

$$V_n' = \sigma^4 \frac{\gamma(n-1)+2}{n(n-1)}. \tag{3.4}$$

In making a choice between these estimators a highly relevant factor is the price one must pay for the robustness. The moments of any distribution F satisfy $\mu_4 \geqq \sigma^4$ and hence $\gamma \geqq 0$. In fact, $\gamma > 0$ except for one special case which we shall discuss separately. Assuming γ to be positive, the variances V_n and V_n' satisfy (2.1) and (2.2) with $r = 1$, $c = \gamma$, $a = 0$ and $b = 2$. The asymptotic relative efficiency of M' relative to M is therefore 1, and the asymptotic expected deficiency is $d = 2/\gamma$.

When F is normal, $\gamma = 2$ and hence $d = 1$. In this case the deficiency of M' with respect to M is equal to 1 in a stronger sense than that we have been discussing, since for normal samples M_n and M'_{n+1} have exactly the same distribution for all sample sizes rather than merely matching the variances in large samples.

While in the normal case it costs only one observation to protect against an erroneous value given for ξ, the deficiency of M' relative to M can be arbitrarily large for certain nonnormal populations since γ can be arbitrarily near zero. It is easy to see that the limiting value $\gamma = 0$ arises only when F puts probability 1/2 at each of the points $\xi-\sigma$ and $\xi+\sigma$. In this case $\operatorname{Var}(M_n) = 0$ for each value of n, corresponding to the fact that $M_n = \sigma^2$ with probability 1. On the other hand, $\operatorname{Var}(M_k') = 2\sigma^2/k(k-1)$ is always positive, approaching 0 as $k \to \infty$. One might thus say that $k_n = \infty$ for every n, so that the ARE is 0 while the AED is ∞. This agrees in a limiting sense with the conclusion that the deficiency is $2/\gamma$.

In statistical practice, special interest attaches to populations that are generally similar to the normal but depart from it by having somewhat heavier tails—corresponding to the presence of an occasional gross error. In such populations one would expect to find γ greater than its value 2 in the normal case, and hence the integral deficiency would be 0 or 1. It is reassuring that the large-sample cost of the protection afforded by use of the robust estimator M' will not exceed one observation in these frequently-encountered heavy-tailed populations.

EXAMPLE 3.2. *Estimation of variance; price of unbiasedness.* In the preceding example, suppose that ξ is unknown but that instead of (3.3) we are willing to consider any estimator of the form

$$M_c = \sum(X_i - \bar{X})^2/(n+c). \tag{3.5}$$

If $c \neq -1$, this will not be unbiased but may have a smaller expected squared error than the unbiased estimator. The deficiency of M_{-1} relative to M_c will then indicate how many observations one loses by insisting on unbiasedness, and thereby provides a basis for deciding whether or not the price is too high.

From (3.4), one easily finds

$$E(M_c - \sigma^2)^2 = \frac{\sigma^4}{n(n+c)^2}\{(n-1)[\gamma(n-1)+2]+n(c+1)^2\} \tag{3.6}$$

and hence

$$E(M_c - \sigma^2)^2 = \sigma^4\left[\frac{\gamma}{n}+\frac{(c+1)^2-2\gamma+2-2c\gamma}{n^2}+o\left(\frac{1}{n^2}\right)\right]. \tag{3.7}$$

As an example, consider the classical estimator M_0. The deficiency of M_{-1} relative to M_0 is given by (2.7) with $r = 1$, $c = \gamma$, $a = 3-2\gamma$ and $b = 2$, and hence is equal to

$$d = (2\gamma - 1)/\gamma. \tag{3.8}$$

The classical M_0 is thus better than M_{-1} when $\gamma > \frac{1}{2}$, with the situation reversed when $\gamma < \frac{1}{2}$. When F is normal, in particular, $\gamma = 2$ and $d = -3/2$. One can therefore save an expected 1.5 observations by using the biased estimator M_0. The best value of c in the normal case is $c = 1$ for which $d = 2$ and which therefore provides an additional saving of .5 observations.

EXAMPLE 3.3. *Median versus quasi-median.* Consider the problem of estimating the center θ of a symmetric distribution. Let $X_1, X_2, \cdots$ be a sample from the distribution, and denote the ordered sample by $Y_1 \leqq Y_2 \leqq \cdots$. Suppose that the sample size is odd, and consider as two unbiased estimators of θ: the median $M' = Y_{h+1}$ based on $k = 2h+1$ observations, and the quasi-median $M = (Y_m + Y_{m+2})/2$ based on $n = 2m+1$ observations. It was pointed out in [7] on the basis of a heuristic argument that up to terms of order n^{-2}, $\operatorname{Var} M = \operatorname{Var} M'$ provided $k = n+2$, for all F satisfying suitable regularity conditions. For these F, the AED of M' with respect to M is therefore equal to two.

EXAMPLE 3.4. *Comparison of the minimax and unbiased estimator of binomial p.* Let X denote the number of successes in n binomial trials with success probability p. Then the unbiased estimator of p with uniformly minimum variance is $M_n = X/n$ while the minimax estimator is

$$M_n' = \frac{\sqrt{n}}{1+\sqrt{n}} \frac{X}{n} + \tfrac{1}{2} \cdot \frac{1}{1+\sqrt{n}},$$

and these estimators have expected squared errors

$$V_n = \frac{pq}{n} \quad \text{and} \quad V_n' = \frac{1}{4(1+\sqrt{n})^2}$$

respectively. If $p \neq \frac{1}{2}$, the ARE of V_n' to V_n is $pq/4 < 1$ so that for large n, M_n is superior to M_n'. However, for $p = \frac{1}{2}$, the situation is reversed. The ARE is 1, and $V_n' < V_n$. To determine the deficiency in this case, we must compare $V_n = 1/4n$ with

$$V_n' = \frac{1}{4n}\left[1 - \frac{2}{\sqrt{n}} + o\left(\frac{1}{\sqrt{n}}\right)\right].$$

Here the remark applies which was made following (2.5). We have $r = 1$ and $s = \frac{1}{2}$ and the asymptotic deficiency of V_n relative to V_n' is infinite. In fact, if we equate V_{kn} with V_n', we find $k_n = n + 2n^{\frac{1}{2}} + 1$ so that $k_n - n = 2n^{\frac{1}{2}} + 1$. The deficiency computation thus shows that the classical estimator requires a much larger sample size than the minimax estimates if they are to have the same expected squared error at $p = \frac{1}{2}$, in spite of the fact that the corresponding asymptotic efficiency is 1.

4. Confidence sets for normal means. Let $X_{ij}, j = 1, \cdots, n; i = 1, \cdots, p$, be independent observations from p normal distributions with expectations $\xi_1, \cdots, \xi_p$ and common variance σ^2, and suppose we wish to obtain a confidence set for the vector $(\xi_1, \cdots, \xi_p)$. It is sometimes possible to treat σ as known, the value being taken from past experience or theoretical considerations. The best confidence sets then are spheres centered at $(X_{1\cdot}, \cdots, X_{p\cdot})$ where $X_{i\cdot} = \sum X_{ij}/n$ and with a fixed radius computed from the χ^2-distribution. The confidence coefficient of this procedure could of course be seriously invalid if the given value of σ should happen to be incorrect. For this reason, it is often preferred to ignore the given information and instead estimate σ from the data. The resulting confidence sets, which are based on the F-distribution, will again be spheres centered on $(X_{1\cdot}, \cdots, X_{p\cdot})$ but the radius will now be a random variable. Estimation of σ involves some loss of effectiveness; the resulting confidence sets will tend to be larger. The choice between the two procedures will depend mainly on two considerations: the degree of reliance placed on the given σ, and the cost of ignoring it. As a contribution to the second aspect we shall in the present section determine the deficiency of the F- with respect to the χ^2-confidence sets, where we shall take as our measure of performance the volume or expected volume of the confidence spheres.

For known σ, the confidence sets at confidence level $1-\alpha$ are given by

$$n\sum(\xi_i - X_{i\cdot})^2/\sigma^2 \leqq v_p. \tag{4.1}$$

Here v_p is determined by the equation

$$\Psi(v_p) = 1-\alpha \tag{4.2}$$

where Ψ denotes the cumulative distribution function of the χ^2-distribution with p degrees of freedom. For simplicity of notation, let us choose the scale so that $\sigma = 1$.

If σ is unknown, the confidence sets are

$$n\sum(\xi_i - X_{i\cdot})^2 \leqq CT \tag{4.3}$$

where $T = \sum\sum(X_{ij} - X_{i\cdot})^2/p(n-1)$. If we put $t = p(n-1)$, then tT has the χ^2-distribution with t degrees of freedom. The constant C (which of course depends on t) is determined by the equation

$$E\Psi(CT) = 1-\alpha. \tag{4.4}$$

As $t \to \infty$, $T \to \sigma^2 = 1$ in probability and hence $C \to v_p$. For a comparison of the two families of confidence sets, it is necessary to find the difference between C and v_p to terms of order $1/t$. Since T is close to 1, this can be accomplished by expanding Ψ about C, and for this purpose we note that the density ψ of Ψ is of the form

$$\psi(y) = Ky^{\frac{1}{2}p-1}e^{-\frac{1}{2}y} \tag{4.5}$$

and hence that its derivative is

$$\psi'(y) = [(\tfrac{1}{2}p-1)y^{-1} - \tfrac{1}{2}]\psi(y). \tag{4.6}$$

Thus

$$\begin{aligned}\Psi(CT) &= \Psi[C + C(T-1)] \\ &= \Psi(C) + C(T-1)\psi(C) + \tfrac{1}{2}C^2(T-1)^2\psi'(C) + \tfrac{1}{6}\psi''(U)C^3(T-1)^3\end{aligned}$$

where U depends on T.

Since $E(T-1) = 0$, $E(T-1)^2 = 2/t$ and $E|T-1|^3 = O(t^{-2})$, it can be shown that

$$\begin{aligned}1-\alpha &= \Psi(C) + \tfrac{1}{t}C^2[(\tfrac{1}{2}p-1)C^{-1} - \tfrac{1}{2}]\psi(C) + o(t^{-1}) \\ &= \Psi[C + Ct^{-1}\{(\tfrac{1}{2}p-1) - \tfrac{1}{2}C\} + o(t^{-1})].\end{aligned}$$

This result is obvious for values of p for which $\psi''(y)$ is bounded i.e. for $p \geqq 6$, and easily follows by truncation for the remaining values of p. Comparing this equation with (4.2), we find $v_p = C + Ct^{-1}[(\frac{1}{2}p-1) - \frac{1}{2}C] + o(t^{-1})$ and hence

$$C = v_p - v_p t^{-1}[(\tfrac{1}{2}p-1) - \tfrac{1}{2}v] + o(t^{-1}). \tag{4.7}$$

Since the radius of the confidence spheres (4.1) is $r = (v_p/n)^{\frac{1}{2}}$ when $\sigma = 1$, their volume is given by

$$V_n = K_p r^p = K_p(pv_p/N)^{\frac{1}{2}p} \tag{4.8}$$

where $N = pn$ is the total sample size and K_p is a constant. Similarly, the radius of the confidence spheres (4.3) is $r' = (CT/n)^{\frac{1}{2}}$, so that their expected volume is

$$E(V_n') = K_p(C/n)^{\frac{1}{2}p}E(T^{\frac{1}{2}p}). \tag{4.9}$$

Now

$$E(T^{\frac{1}{2}p}) = 1 + A_p t^{-1} + o(t^{-1}) \tag{4.10}$$

where we shall determine A_p below. Hence, using the fact that $t = p(n-1)$, we have

$$E(V_n') = K_p\left(\frac{pv_p}{N}\right)^{p/2}\left\{1 - \frac{p}{2N}\left[\left(\frac{p}{2}-1\right) - \frac{v_p}{2}\right] + \frac{A_p}{N} + o\left(\frac{1}{N}\right)\right\}. \tag{4.11}$$

It follows that V_n and $E(V_n')$ satisfy (2.1) and (2.2) with $r = \frac{1}{2}p$, $c = (pv_p)^{\frac{1}{2}p}K_p$, $a = 0$ and $b = (pv_p)^{\frac{1}{2}p}K_p\{A_p - \frac{1}{2}p[(\frac{1}{2}p-1) - \frac{1}{2}v_p]\}$. By (2.7), the AED is therefore

$$d_p = 2p^{-1}A_p - (\tfrac{1}{2}p - 1) + \tfrac{1}{2}v_p\,. \tag{4.12}$$

To complete the evaluation of d, it remains to find A_p. Now the moments of T are given by (see for example [6])

$$E(T^{\frac{1}{2}p}) = \frac{(t+2a-2)(t+2a-4)\cdots t}{t^a} = 1 + \frac{a(a-1)}{t} + o\left(\frac{1}{t}\right) \qquad \text{if} \quad p = 2a$$

and

$$E(T^{\frac{1}{2}p}) = \frac{(t+2a-1)(t+2a-3)\cdots(t+1)}{t^a}\left[1 - \frac{1}{4t}, + o\left(\frac{1}{t}\right)\right]$$

$$= 1 + (a^2 - \tfrac{1}{4})\frac{1}{t} + o\left(\frac{1}{t}\right) \qquad \text{if} \quad p = 2a+1.$$

Substituting the values $A_{2a} = a(a-1)$ and $A_{2a+1} = a^2 - \frac{1}{4}$ into (4.12) shows that for all p the deficiency of the F- to the χ^2-confidence sets is

$$d_p = v_p/2. \tag{4.13}$$

The deficiency thus increases with $1-\alpha$ and with p. The values of d corresponding to a number of values of α and p are shown in the table below. It is interesting to note how large some of these values are.

p	α				
	.1	.05	.025	.01	.005
1	1.353	1.921	2.512	3.317	3.940
2	2.303	2.996	3.689	4.605	5.298
5	4.618	5.535	6.416	7.543	8.375
10	7.994	9.154	10.242	11.605	12.594

The result extends without any difficulty to the case of unequal sample sizes. If the number of observations in the ith sample is n_i, the total number of observations $N = \sum n_i$ and $\rho_i = n_i/N$, the confidence sets for known σ are

$$N \sum \rho_i(\xi_i - X_{i\cdot})^2/\sigma^2 \leqq v_p. \tag{4.14}$$

These are ellipsoids whose volume is $K_p(v_p/N)^p/\rho_1 \cdots \rho_p$ instead of (4.8). Similarly, the expected volume of the ellipsoids for unknown σ is $K_p(C/N)^p E(T^p)/\rho_1 \cdots \rho_p$ and (4.13) now follows exactly as before.

5. Student's t-test versus the $\bar{X}$-test. In the preceding sections we were concerned with the comparison of two procedures the performance of which was measured by a single real number: A variance, or volume of a confidence set, which happened to be independent of the unknown parameters, or which we considered only at specified values of the parameters. This is however atypical; more usually, performance is measured by a function of the parameters and an equivalent sample size will have to be defined in terms of an overall matching of two such functions.

In the present section we shall consider a testing problem, where the performance is described by the power functions of the tests in question but where through special circumstances the difficulty alluded to need be confronted only partially. The problem is the normal one-sample location problem. The variables $X_1, X_2, \cdots$ are assumed to be independently normally distributed with expectation ξ and variance σ^2, and the hypothesis $H : \xi = 0$ is to be tested against the alternatives $\xi > 0$. As in the preceding section, we shall compare the test that is appropriate when a given value of σ can be relied upon, with the t-test, which places no reliance on such an assumed value of σ, in terms of deficiency. The determination of this deficiency enables us to throw some fresh light on the interesting solution to the cost problem given by John Walsh (1949). (We are indebted to Prof. J. Hemelrijk for pointing out to us the connection of our results with the work of Walsh.)

When σ is known, the hypothesis H is accepted at level α on the basis of n observations $X_1, \cdots, X_n$ when

$$n^{\frac{1}{2}}\bar{X}_n/\sigma \leqq u_\alpha \tag{5.1}$$

where

$$u_\alpha = \Phi^{-1}(1-\alpha). \tag{5.2}$$

For simplicity of notation, we shall suppress the subscript α of u in the following discussion and suppose furthermore that $\sigma = 1$. The acceptance probability of the test (5.1) is then

$$\beta(\xi, n) = \Phi(u - n^{\frac{1}{2}}\xi) = \Phi(u - \theta) \tag{5.3}$$

where $\theta = n^{\frac{1}{2}}\xi$.

Consider now the t-test based on k observations, with acceptance region

$$k^{\frac{1}{2}}\bar{X}_k/S_k \leqq c_k, \tag{5.4}$$

where $S_k^2 = \sum(X_i - \bar{X}_k)^2/(k-1)$. The acceptance probability of this second test can be written as

$$\gamma(\xi, k) = E\Phi(c_k S_k - k^{\frac{1}{2}}\xi). \tag{5.5}$$

Let us put $\eta = k^{\frac{1}{2}}\xi$, add and subtract c_k within the parenthesis, and expand the right-hand side about $c_k - \theta$. Suppressing the subscript k in c and S, formula (5.5) then becomes

$$\gamma(\xi, k) = E\{\Phi(c-\eta)+\varphi(c-\eta)c(S-1)+\tfrac{1}{2}\varphi'(c-\eta)c^2(S-1)^2+\tfrac{1}{6}\varphi''(T)c^3(S-1)^3\},$$

where T depends on S and η. From the facts that

$$E(S_k) = 1+\frac{a}{k}+O\left(\frac{1}{k^2}\right), \qquad \operatorname{Var}(S_k) = \frac{1}{2k}+O\left(\frac{1}{k^2}\right)$$

and $E|S_k - E(S_k)|^3 \leqq M/k^2$, it then follows that

$$\gamma(\xi, k) = \Phi(c-\eta)+\frac{ac}{k}\varphi(c-\eta)+\frac{c^2}{4k}\varphi'(c-\eta)+O\left(\frac{1}{k^2}\right).$$

Here the boundedness of φ'' insures that the error term is of order $1/k^2$ uniformly in η. Using the fact that $\varphi'(x) = -x\varphi(x)$ and once more the expansion of Φ, γ can be now written as

$$\gamma(\xi, k) = \Phi\left[\left(c+\frac{ac}{k}-\frac{c^3}{4k}\right)-\eta\left(1-\frac{c^2}{4k}\right)\right]+O\left(\frac{1}{k^2}\right).$$

Putting $\eta = 0$ in this equation yields

$$1-\alpha = \Phi(u) = \Phi\left(c+\frac{ac}{k}-\frac{c^3}{4k}\right)+O\left(\frac{1}{k^2}\right)$$

and hence

$$u = c+\frac{ac}{k}-\frac{c^3}{4k}+O\left(\frac{1}{k^2}\right),$$

and thus in particular also $u = c+O(1/k)$. Substitution in the last expression for γ and the equation $\eta = (k/n)^{\frac{1}{2}}\theta$ finally gives the approximation

$$\gamma(\xi, k) = \Phi\left[u-\left(\frac{k}{n}\right)^{\frac{1}{2}}\theta\left(1-\frac{u^2}{4k}\right)\right]+O\left(\frac{1}{k^2}\right), \tag{5.6}$$

where the error term is of order $1/k^2$ uniformly in θ.

If we now equate expressions (5.3) for $\beta(\xi, n)$ and (5.6) for $\gamma(\xi, k)$ to find the value $k = k_n$ for which the two tests give the same power (up to terms of order $1/n$) for a fixed value of θ, we find in analogy to (2.5) that

$$d_n = k_n - n \to u^2/2. \tag{5.7}$$

This result is independent of θ and hence of ξ, and we see in fact that the maximum difference $\sup|\beta(\xi, n)-\gamma(\xi, k_n)|$ is of order $1/n^2$ if and only if k_n satisfies (5.7).

As did (2.5), relation (5.7) suffers from the difficulty that k is not a continuous variable. This can be avoided as before by stochastic interpolation. Instead of a fixed sample size k, we take either k or $k+1$ observations with probability $1-\pi$ and π respectively, the acceptance probability of the resulting test being

$$(1-\pi)\gamma(\xi, k)+\pi\gamma(\xi, k+1).$$

If this is equated with $\beta(\xi, n)$ we find as in Section 2 that the difference between the expected sample size $k_n+\pi$ for the t-test and the sample size n for the normal test tends to

$$d = \tfrac{1}{2}u_\alpha{}^2, \tag{5.8}$$

which is therefore the AED in the present case. The following are the values of d corresponding to a number of values of α.

α	.10	.05	.025	.01	.005
d	.821	1.353	1.921	2.706	3.3175

In large samples, those values may be interpreted as the price, measured by the number of additional observations, required to protect oneself against possible error in the given value of σ. In our judgment, the values of d are small enough so that it would rarely be reasonable to rely on the given σ.

Since the value of d is asymptotic, the question arises how good the approximation is for a given finite value of n or k. Some indications regarding the accuracy of the approximation for small samples will be given at the end of the section.

In the foregoing, we were concerned with testing the hypothesis $\xi = 0$ against the alternatives $\xi > 0$. By symmetry, (5.8) is clearly also the deficiency when the problem is that of testing $\xi = 0$ against the alternatives $\xi < 0$. Consider now the problem of testing $\xi = 0$ against the two-sided alternatives $\xi \neq 0$ at significance level α. The power function of the symmetric two-sided normal or t-test is just the sum of the power functions of the corresponding two one-sided tests at level $\alpha/2$. Since an expected addition to n of $d = \frac{1}{2}u_{\alpha/2}^2$ observations will bring both of the one-sided t-power functions into coincidence with the corresponding normal power functions, it is seen that the deficiency in the two-sided case is just

$$d = \tfrac{1}{2}u_{\alpha/2}^2. \tag{5.9}$$

This two-sided testing problem is closely related to the problem of obtaining confidence intervals for ξ, which in turn is the special case $p = 1$ of the problem treated in Section 4. The deficiency found there was $\frac{1}{2}v$, with v defined by $\Psi(v) = 1-\alpha$ where Ψ is the cumulative distribution function of a χ^2 variable with 1 degree of freedom. Clearly $v^{\frac{1}{2}} = u_{\alpha/2}$, and hence the deficiency (5.9) agrees with the deficiency (4.13) for $p = 1$.

Procedures intermediate between the one-sided and symmetric two-sided tests are the asymmetric two-sided tests. The corresponding comparisons in these cases are quite different from those treated above, because it is not possible to match the power curves to the same order as in the one-sided case.

Let us return now to the one-sided case and, as in Section 2, consider the alternative approach of integral deficiency, in which the equivalent sample size is defined as the integer k_n for which $\gamma(\xi, k)$ is closest to $\beta(\xi, n)$. To make this precise, it is necessary to specify what is meant by the functions $\gamma(\xi, k)$ and $\beta(\xi, n)$ being as close as possible. Since (5.3) and (5.6) express β and γ conveniently in terms of θ, we could specify a value of θ and ask that the acceptance probabilities be as close as possible for this particular value of θ. However, such a choice would typically have to be rather arbitrary (although in the present case the large-sample result would be independent of θ) and we shall instead define the equivalent sample size to be the integer k_n which minimizes

$$\sup_\xi \left|\gamma(\xi, k)-\beta(\xi, n)\right|. \tag{5.10}$$

Here the supremum is taken over $\xi > 0$ since the tests are one-sided.

To examine the behavior of $\gamma(\xi, k)$ for k near n, let us write $k = n+d$ where d is now an integer. Then

$$\left(\frac{k}{n}\right)^{\frac{1}{2}} = \left(\frac{n+d}{n}\right)^{\frac{1}{2}} = 1+\frac{d}{2n}+O\left(\frac{1}{n^2}\right)$$

so that by (5.6)

$$\begin{aligned}\gamma(\xi, n+d) &= \Phi\left[u-\theta\left(1+\frac{d}{2n}-\frac{u^2}{4n}\right)\right]+O\left(\frac{1}{n^2}\right) \qquad (5.11)\\ &= \Phi(u-\theta)+\frac{1}{n}(\tfrac{1}{2}u^2-d)\cdot\tfrac{1}{2}\theta\varphi(u-\theta)+O\left(\frac{1}{n^2}\right),\end{aligned}$$

and hence

$$\gamma(\xi, n+d)-\beta(\xi, n) = (\tfrac{1}{2}u^2-d)\tfrac{1}{2}\theta\varphi(u-\theta)\cdot\frac{1}{n}+O\left(\frac{1}{n^2}\right).$$

Let us now find the value of d for which $\gamma(\xi, n+d)$ fits most closely to $\beta(\xi, n)$ in the sense of minimizing (5.10).

It is necessary to distinguish three cases, and we first suppose that for some integer d_0,

$$d_0 < \tfrac{1}{2}u^2 < d_0+\tfrac{1}{2}. \tag{5.12}$$

The other two cases: $d_0-\frac{1}{2} < \frac{1}{2}u^2 < d_0$ and $d_0+\frac{1}{2} = \frac{1}{2}u^2$ can be handled similarly. Furthermore, we shall assume that $\alpha < \frac{1}{2}$ so that $u > 0$; again, the other case is completely analogous. The function $\frac{1}{2}\theta\varphi(u-\theta)$ then achieves a unique maximum, say h, at a point $\theta = \theta_0 > 0$. The value of h is also the maximum of $\left|\frac{1}{2}\theta\varphi(u-\theta)\right|$. If for the moment we ignore the uniform error terms of order $1/n^2$, we find that the acceptance curves of the three t-tests $\gamma(\xi, n+d_0-1)$, $\gamma(\xi, n+d_0)$ and $\gamma(\xi, n+d_0+1)$ all attain their maximum distance from $\beta(\xi, n)$ at θ_0, where they are respectively: low by $(d_0+1-\frac{1}{2}u^2)h/n$; high by $(\frac{1}{2}u^2-d_0)h/n$; and high by $(\frac{1}{2}u^2-d_0+1)h/n$. Of these three curves, $\gamma(\xi, n+d_0)$ is thus closest to $\beta(\xi, n)$ and it is closer than its nearer competitor by $[2(d_0-\frac{1}{2}u^2)+1]h/n = A$. For values of k other than these three, the superiority of fit of $\gamma(\xi, n+d_0)$ is still greater: from (5.11), it is seen that for $\xi > 0$, $\gamma(\xi, n+d)$ falls monotonely as d increases, so that $\gamma(\xi, n+d_0+2)$, $\gamma(\xi, n+d_0+3), \cdots$ are still further below $\beta(\xi, n)$ than is $\gamma(\xi, n+d_0)$, while $\gamma(\xi, n+d_0-2)$, $\gamma(\xi, n+d_0-3), \cdots$ are still further above $\beta(\xi, n)$ than is $\gamma(\xi, n+d_0-1)$.

To summarize: if we ignore the error terms, the t-acceptance-curve which best fits $\beta(\xi, n)$ is $\gamma(\xi, n+d_0)$, and its superiority of fit over its nearest rival is proportional to $1/n$. However, we know that the error terms tend to zero uniformly at rate $1/n^2$. For n sufficiently large and with k_n/n near 1, the true acceptance curves will all differ from their approximations by less than $\frac{1}{2}A$. It follows that for sufficiently large n, the equivalent sample size is $k_n = n+d_0$. Dropping assumption (5.12), the equivalent sample size is then finally $k_n = n+d$ where d is the integer(s) closest to the AED namely $\frac{1}{2}u_\alpha{}^2$.

The problem of determining the cost (in terms of the number of lost observations) was treated as early as 1949 by John E. Walsh [10]. His solution was based on a formula of Johnson and Welch (1939). By a heuristic argument, in which a chi-variable is treated as if it were normal, these authors were led to

$$\Phi\left\{k^{\frac{1}{2}}\xi\left[1-\frac{u^2}{2(k-1)}\right]^{\frac{1}{2}}-u\right\} \tag{5.13}$$

as an approximation for the acceptance probability of the t-test. Walsh identified this expression with the acceptance probability (5.3) of the $\bar{X}$-test, treating n as if it were a continuous variable. This leads at once to the expression

$$k-n = \tfrac{1}{2}u^2k/(k-1) \tag{5.14}$$

which Walsh refers to as approximately the number of "sample values 'wasted' by using a t-test." The quantity $k-n$ is an approximation to what we have called deficiency.

In order to compare (5.13) with our analysis, note that it can be written as

$$\tfrac{1}{2}u^2\frac{k}{k-1} = \tfrac{1}{2}u^2+\frac{1}{2k}u^2+O\left(\frac{1}{k^2}\right). \tag{5.15}$$

Walsh notes that the first term on the right-hand side can be used to approximate the number of "wasted" observations if k is not too small. It is of course just our AED (5.8). From this point of view we have provided a more rigorous justification of Walsh's formula, which is perhaps also more easily interpreted: With stochastic interpolation (or some other equivalent interpolation method) the dominant term of Walsh's formula for the number of wasted observations is the limit value of the difference of equivalent sample sizes.

We are however not able to give a similar justification to the second term of (5.15). In fact, the correct term of order $1/k$ for the deficiency depends on the $1/k^2$-term in the power of t. Unfortunately, the normal approximation (5.13) has an error of this order. From the results of Section 6 of [6] it is easy to see that, for fixed α and fixed noncentrality parameter δ, the acceptance probability of the t-test with f degrees of freedom is given by

$$\Phi\left[u-\delta+\frac{\delta u^2}{4f}+\frac{6u^2\delta-5u^4\delta+4u^3\delta^2}{96f^2}\right]+O\left(\frac{1}{f^3}\right). \tag{5.16}$$

On the other hand, the approximation (5.13), on expansion, can be written as

$$\Phi\left[u-\delta+\frac{\delta u^2}{4f}+\frac{3u^4\delta}{96f^2}\right]+O\left(\frac{1}{f^3}\right).$$

This differs from the correct value by a term of order $1/f^2$.

One could of course use (5.16) to derive an expression for the deficiency accurate to terms of order $1/n$. The deficiency to this order depends on the precise manner in which the acceptance curves are matched. In particular, if they are matched at a fixed power π (as well as at the null hypothesis) the $1/n$ term of the expansion for k_n-n would depend on the value of π.

We conclude with a numerical illustration of the comparison of t with $\bar{X}$ in the one-sided case. The table below shows power values of the $\bar{X}$-test at level $\alpha = .05$, for $n = 4$ and $n = 8$ observations. Also shown in each case are the corresponding power values of the stochastically-interpolated t-test, as well as the power for the bracketing values, the best fitting (with minimax matching) interpolated sample sizes being $k_4 = 5.6377$ and $k_8 = 9.5149$ respectively. The small-sample expected deficiencies (1.6377 at $n = 4$ and 1.5149 at $n = 8$) are close enough to the asymptotic value (1.3528), so that the asymptotic theory provides reasonable guidance even at these very small sample sizes. The approach to the limit is also consistent with the fact that the difference between the small-sample value and the asymptotic value is of order $1/n$. Note that the matching is satisfactorily close (the maximum difference is .00670 at $n = 4$ and .00176 at $n = 8$), and is decreasing consistently with the theoretical rate $1/n^2$.

6. Bayes versus unbiased estimation of a normal mean. In a deficiency investigation in which the performance of the procedures in question depends on a parameter, this will typically also be true of the deficiency. For an overall comparison of the procedures it is then necessary to adjust the sample sizes so as to provide an optimum fit for the two performance curves, for example, by minimizing their maximum difference, This approach which was illustrated in the preceding section, arises also in the following example.

Let $X_1, \cdots, X_n$ be independently normally distributed with mean θ and common variance which without essential loss of generality we shall take to be equal to 1, and consider the problem of estimating θ, with squared error as loss. The standard estimator is $\delta = \bar{X} = \sum X_i/n$ with risk equal to

$$R_n(\theta) = n^{-1}. \tag{6.1}$$

TABLE 1

Power of $\bar{X}$-test with n observations, matched by t-test with k observations, $\alpha = .05$

$n = 4$

	t-power:			$\bar{X}$-power:	
$4^{\frac{1}{2}}\Delta$	$k = 5$	$k = 6$	$k = 5.6377$	$n = 4$	difference
−.8	.00797	.00606	.00675	.00725	−.00050
0	.05	.05	.05	.05	0
.8	.18549	.21366	.20345	.19910	.00435
1.2	.29964	.35457	.33467	.32821	.00646
1.4	.36606	.43494	.40998	.40328	.00670
1.6	.43638	.51779	.48830	.48211	.00619
2.0	.57976	.67695	.64174	.63876	.00298
2.4	.71191	.80806	.77322	.77491	−.00169
2.8	.81942	.89964	.87058	.87598	−.00540
3.0	.86193	.93094	.90594	.91231	−.00637
3.2	.89690	.95406	.93335	.94004	−.00669
3.4	.92485	.97046	.95394	.96038	−.00644
4.0	.97484	.99361	.98681	.99074	−.00393
4.8	.99585	.99949	.99817	.99920	−.00103

$n = 8$

	t-power:			$\bar{X}$-power:	
$8^{\frac{1}{2}}\Delta$	$k = 9$	$k = 10$	$k = 9.5149$	$n = 8$	difference
−.8	.00762	.00663	.00711	.00725	−.00014
0	.05	.05	.05	.05	0
.8	.19303	.20704	.20024	.19910	.00114
1.2	.31588	.34312	.32991	.32821	.00170
1.4	.38750	.42156	.40504	.40328	.00176
1.6	.46308	.50316	.48372	.48211	.00161
2.0	.61522	.66238	.63950	.63876	.00074
2.4	.75098	.79649	.77441	.77491	−.00050
2.8	.85572	.89223	.87452	.87598	−.00146
3.0	.89485	.92547	.91062	.91231	−.00169
3.2	.92563	.95022	.93829	.94004	−.00175
3.4	.94898	.96791	.95873	.96038	−.00165
4.0	.98634	.99308	.98981	.99074	−.00093
4.8	.99849	.99947	.99899	.99920	−.00021

Most Bayesian analyses presuppose that θ has a prior normal distribution, say with mean μ and variance τ^2. The Bayes estimator resulting from this assumption is

$$\delta' = (\mu + n\tau^2\bar{X})/(n\tau^2+1) \tag{6.2}$$

and its risk function is

$$R_n'(\theta) = [n\tau^4 + (\theta-\mu)^2]/(n\tau^2+1)^2. \tag{6.3}$$

The estimator δ' arises also in a different context. Suppose there are available m earlier observations $U_1, \cdots, U_m$ from a normal distribution with unit variance and hopefully the same mean θ. Then instead of δ one might wish to use the estimator $(m\bar{U}+n\bar{X})/(m+n)$. For given $\bar{U}$, this coincides with δ' if we put $\tau^2 = m^{-1}$ and $\mu = \bar{U}$.

In trying to compare the two estimators δ and δ', we note first that for every value of θ, the asymptotic efficiency of δ' relative to δ is 1. Efficiency thus suggests that there is little to choose between the estimators; as we shall see, a deficiency analysis shows instead that the differences between the two estimators may be quite striking and that their comparison presents a rather complex problem.

In view of the equivalence, established in Section 2, of stochastic interpolation with the formally simpler treatment of sample size as a continuous variable, let us suppose that δ is based on $n+d$ observations without restricting d to integral values. Then

$$R_{n+d}(\theta) = \frac{1}{n} - \frac{d}{n^2} + o\left(\frac{1}{n^2}\right) \tag{6.4}$$

while

$$R_n'(\theta) = \frac{1}{n} + \frac{1}{n^2\tau^4}[(\theta-\mu)^2 - 2\tau^2] + o\left(\frac{1}{n^2}\right). \tag{6.5}$$

Thus, putting $D(\theta) = n^2\tau^4[R_n'(\theta) - R_{n+d}(\theta)]$, we have

$$D(\theta) = (\theta-\mu)^2 - 2\tau^2 + d\tau^4. \tag{6.6}$$

Suppose now that we wish to determine the deficiency $d = d(\theta)$ at a fixed given value of θ. This is obtained by setting $D(\theta) = 0$, and yields

$$d = \frac{2}{\tau^2} - \frac{(\theta-\mu)^2}{\tau^4}. \tag{6.7}$$

As θ tends to $+\infty$ or $-\infty$, d tends to $-\infty$, so that for large values of $|\theta|$, δ is greatly superior to δ'. The maximum value of d occurs for $\theta = \mu$ and is $2/\tau^2$; for $\theta = \mu$ and small values of τ, δ' is therefore the much better estimator. (Note however that d takes on large negative values for any value of $\theta \neq \mu$ if τ is small.)

In view of this very strong dependence of d on θ, one cannot expect to find a compromise which is generally satisfactory. Nonetheless, it is interesting to see what results are obtained by different criteria.

As a first possibility, let us suppose that θ is really a random variable with the postulated prior distribution. Then the average number of observations lost as a result of using δ instead of δ' is obtained by determining d so that $ER_{n+d}(\theta) = ER_n'(\theta)$, where the expectation is taken with respect to the assumed normal distribution of θ. Then

$$ER_n'(\theta) = \frac{1}{n} - \frac{1}{\tau^2 n^2} \quad \text{while} \quad ER_{n+d}(\theta) = \frac{1}{n} - \frac{d}{n^2}$$

and the Bayes deficiency is therefore $d = \tau^{-2}$. It is interesting to note that in the non-Bayesian model with m prior observations, $\tau^{-2} = m$ is just the number of observations lost if δ is used instead of δ'.

At the other extreme, in a sense, is the minimax criterion according to which d is determined so as to minimize $\sup|R_{n+d}(\theta) - R_n'(\theta)|$. Neglecting terms of order $o(1/n^2)$, this means minimizing $\sup|D(\theta)|$, where $D(\theta)$ is given by (6.6). Unfortunately, it is seen that this maximum discrepancy is infinite for all values of d, so that the minimax criterion breaks down.

A possible approach to the problem is presented by a compromise between the Bayes and minimax principles. Prior information rarely prescribes a specific distribution. Rather it may indicate that θ lies in an interval, say $I = (\mu - A, \mu + A)$ and that values closer to μ are more likely than more distant values. If symmetry with respect to μ seems reasonable, mathematical convenience suggests a normal distribution which is centered at μ and assigns high probability to the interval I. In comparing δ with δ' we may then be willing to restrict attention to θ in I and minimize $\sup|D(\theta)|$ for $\theta \in I$.

Since both $D(\mu)$ and $D(\mu - A)$ are increasing functions of d, it is easily seen from (6.6) that the unique minimax solution is obtained by putting $D(\mu) = -D(\mu - A)$. The resulting deficiency is

$$d = \frac{2}{\tau^2} - \frac{A^2}{2\tau^4} \tag{6.8}$$

and the corresponding minimax value is

$$\sup_{\theta \in I} |D(\theta)| = A^2/2. \tag{6.9}$$

It follows from (6.5) that $d > 0$ (i.e. the Bayes estimator is "better" in the sense being considered), $d = 0$ or $d < 0$ as

$$\tau < A/2, \quad = A/2, \quad \text{or} \quad > A/2. \tag{6.10}$$

This suggests that, when fitting a prior distribution to a given A, one should choose for τ a value less than $A/2$. Condition (6.10) can be given an alternative formulation. Suppose in deciding on a prior distribution we determine τ, for a given value of A, so that $P\{|\theta - \mu| \leqq A\} = \gamma$ where γ is some fixed probability level typically close to 1. Then $A/\tau = u$ where $\Phi(u) = (1+\gamma)/2$. Thus

$$d = \frac{2}{\tau^2}\left(1 - \frac{u^2}{4}\right). \tag{6.11}$$

The most interesting case is that of small τ, since in this case there is a substantial difference between the two estimators δ and δ' being compared. As $\tau \to 0$, $d \to +\infty$ or $-\infty$ as u is < 2 or > 2, and remains $= 0$ when $u = 2$. The condition $u = 2$ means that (approximately) $\gamma = .95$. Thus, if we fit a normal prior to the given interval I, it should be such as to assign to I a probability of at least .95.

7. Further possibilities. There are many examples similar to the ones treated in the preceding sections, which could be handled in a completely analogous manner. The approach of Examples 1 and 2 of Section 3 to the estimation of variance extends, for instance, to the estimation of covariance and of higher moments. Similarly, the result of Section 4 concerning confidence sets for the simultaneous estimation of several means generalizes to confidence sets for the mean of a multivariate normal distribution. The method for obtaining the deficiency of Student's one-sample t-test applies also to the other t-tests, for example, the two-sample t-test or the Bartlett-Scheffé test for the Behrens–Fisher problem. More generally, it applies also to substitute t-tests in which the usual denominator is replaced by some other estimator of the standard deviation, provided only this estimator is independent of the numerator. (We plan to discuss such substitute t-tests in this manner in a separate paper.) Finally, the comparison in Section 6 of the standard estimator of a normal mean with a Bayes estimator extends to certain other situations involving exponential families, for example, the estimation of binomial p.

The extensions mentioned above require no new techniques and there are other applications for which this is true, for example, to comparisons in certain design-problems. More interesting perhaps are a number of problems in which the deficiency concept appears to be useful but where its application presents certain technical difficulties stemming from the fact that the computation of deficiency requires higher-order asymptotic terms than we encounter in the usual efficiency analyses. We shall mention only three problems of this type, all of a nonparametric nature: (i) What is the deficiency of the normal scores test or of Van der Waerden's X-test with respect to the t-test? (ii) What is the deficiency of the test based on Kendall's rank correlation coefficient with respect to that based on Spearman's coefficient? (iii) If $X_1, X_2, \cdots$ is a sample from a symmetric distribution centered on θ, what is the deficiency of the Wilcoxon test of $H: \theta = \theta_0$ and of the associated estimator of θ based on all averages $(X_i + X_j)/2$ with $i \leqq j$ relative to that based only on the averages with $i < j$?

In addition to providing useful information for specific comparisons, the deficiency concept raises a number of more general questions, of which we shall again mention three. (i) In parametric situations such as that of Example (i) of the preceding paragraph, there often exist rank tests of efficiency 1. Do there also exist rank tests of deficiency zero? (ii) The asymptotic relative efficiency of two tests is typically the same as that of two estimators based on them. Does this result in any sense carry over to deficiency? (iii) In a situation involving several parameters, consider the likelihood ratio test for the hypothesis specifying the value of one of the parameters when the other parameters are either known or unknown. Under suitable regularity conditions, is it true that there is then a finite deficiency, and can one give a simple formula for it? (A simple counter example is provided by the problem of estimating the range a of the rectangular distribution $R(\theta, \theta+a)$. Here the asymptotic efficiency of the best unbiased estimator with unknown θ to the corresponding estimator with known θ is $2^{-\frac{1}{2}}$. However, in this case the estimators of both a and θ are superefficient.)

REFERENCES

[1] Cochran, W. G. (1937). The efficiencies of the binomial series tests of significance of a mean and of a correlation coefficient. *J. Roy. Statist. Soc. Ser. A* **100** 69–73.

[2] Cramér, Harold (1946). *Mathematical Methods of Statistics*. Princeton Univ. Press.

[3] Fisher, R. A. (1920). A mathematical examination of the methods of determining the accuracy of an observation by the mean error and by the mean square error. *Monthly Notices Roy. Astron. Soc.* **80** 758–770.

[4] Fisher, R. A. (1925). Theory of statistical estimation. *Proc. Cambridge Philos. Soc.* **22** 700–725.

[5] Hodges, J. L., Jr. and Lehmann, E. L. (1956). The efficiency of some nonparametric competitors of the t-test. *Ann. Math. Statist.* **27** 324–335.

[6] Hodges, J. L., Jr. and Lehmann, E. L. (1967a). Moments of chi and power of t. *Proc. Fifth Berkeley Symp. Math. Statist. Prob.* **1** 187–201.

[7] Hodges, J. L., Jr. and Lehmann, E. L. (1967b). On medians and quasi-medians. *J. Amer. Statist. Assoc.* **62** 926–931.

[8] Johnson, N. L. and Welch, B. L. (1939). Applications of the non-central t-distribution. *Biometrika* **31** 362–389.

[9] Pearson, E. S. (1950). Some notes on the use of range. *Biometrika* **37** 88–92.

[10] Walsh, John E. (1949). On the "information" lost by using a t-test when the population variance is known. *J. Amer. Statist. Assoc.* **44** 122–125.

Scand J Statist 17: 177–187, 1990

Pointwise Versus Uniform Robustness of Some Large-sample Tests and Confidence Intervals

E. L. LEHMANN

University of California, Berkeley

WEI-YIN LOH

University of Wisconsin, Madison

ABSTRACT. The true level $\alpha_n(F)$ of a test at nominal level α and sample size n is said to be pointwise robust if $\alpha_n(F)\to\alpha$ as $n\to\infty$ for each distribution F. If F is unknown and only assumed to belong to the class $\mathcal{F}$ of distributions, one may wish to consider the level to be defined by $\alpha_n=\sup_{\mathcal{F}}\alpha_n(F)$. The test is said to be uniformly robust over $\mathcal{F}$ if $\alpha_n\to\alpha$ as $n\to\infty$. Corresponding definitions apply to the coverage probabilities of confidence intervals.

The purpose of this paper is to illustrate the fact that it is typical for many large-sample tests to be pointwise robust but not uniformly robust, even for seemingly narrow classes $\mathcal{F}$. Two cases, the one-sample t-test and confidence intervals for the binomial proportion p, are treated in detail. It is shown that the t-test is not uniformly robust if $\mathcal{F}$ consists of distributions within a Kolmogorov neighbourhood of an arbitrary distribution or those with densities supported in a finite interval. The test is uniformly robust if $\mathcal{F}$ is a set of distributions with uniformly bounded standardized moments. In the binomial case it is shown that the standard confidence intervals constructed using the delta method are not uniformly robust if $\mathcal{F}$ consists of all binomial distributions with $0<p<1$.

Key words: Kolmogorov distance, significance level

1. Introduction

Let $\mathbf{X}_n=(X_1,\ldots,X_n)$ be i.i.d. according to $N(\theta,\sigma^2)$, the normal distribution with mean θ and variance σ^2, and let $\Phi(\cdot)$ be the standard normal c.d.f. The level-α t-test of $\mathcal{H}$: $\theta=0$ against $\theta>0$ rejects when

$$t_n=(n)^{1/2}s_n^{-1}\bar{X}\geq c_n \tag{1}$$

where $\bar{X}=n^{-1}\Sigma_{i=1}^n X_i$ and $s_n^2=(n-1)^{-1}\Sigma_{i=1}^n(X_i-\bar{X})^2$, and where

$$c_n\to z_\alpha \quad \text{as} \quad n\to\infty$$

with z_α defined by $1-\Phi(z_\alpha)=\alpha$.

Suppose now that the assumption of normality is not justified (as it never is), and that instead $\mathbf{X}_n$ is a sample from some other distribution F with mean $\mu_F=0$. Let $\alpha_n(F)$ denote the probability of rejecting the hypothesis that μ_F is 0. Then a standard result (see, e.g. Lehmann, 1986) states that

$$\alpha_n(F)\to\alpha \quad \text{as} \quad n\to\infty \tag{2}$$

for any F with finite variance. This is often described by saying that the level of the t-test is *robust* against non-normality. For any such F, the rejection probability under $\mathcal{H}$ will then be close to α if n is sufficiently large. How large n needs to be for the actual level $\alpha_n(F)$ to be close to the nominal level α has been much studied in the literature. Some recent references are Cressie (1980), Benjamini (1983), Herrendoerfer, Rasch & Feige (1983), and Herrendoerfer & Feige (1985). A universal answer could be given if the convergence of $\alpha_n(F)$ to α is uniform in F. Otherwise, the answer must depend on F.

There is another way of looking at the robustness of the t-test. If F is unknown, one may wish to consider the t-test as a test of the hypothesis

$$\mathcal{H}': \text{the mean } \mu_F \text{ of the unknown distribution } F \text{ is } 0 \tag{3}$$

against the alternatives $\mathcal{H}'$ that μ_F is positive. As a test of $\mathcal{H}'$ the level of the t-test is then

$$\alpha_n = \sup_{\mathcal{F}} \alpha_n(F),$$

where $\mathcal{F}$ is the class of all distributions F with mean zero and finite variance. Robustness of the t-test in this setting would require

$$\alpha_n \to \alpha \quad \text{as} \quad n \to \infty. \tag{4}$$

If (4) holds, we shall say that the test is *uniformly robust* over $\mathcal{F}$. Clearly (4) implies that the convergence of $\alpha_n(F)$ to α is uniform over $\mathcal{F}$.

Unfortunately, as was shown in Bahadur & Savage (1956) in a more general context, (4) does not hold. On the contrary, for the t-test (1), one has by Bahadur & Savage (1956) that

$$\sup_{\mathcal{F}} \alpha_n(F) = 1 \tag{5}$$

for every value of α. (The same result then follows trivially for the two-sided t-test. See Gleser & Hwang, 1987, for a similar result in an errors-in-variables model.)

In the light of (5) the question about the sample size required for the level $\alpha_n(F)$ of the t-test to be reasonably close to α, must be approached somewhat differently. Given a small $\varepsilon>0$, no sample size n_0 can insure that $|\alpha_n(F)-\alpha|<\varepsilon$ for all $n>n_0$. However, one may hope that such an n_0 will exist if F is restricted to a suitable subset $\mathcal{F}_0$ of $\mathcal{F}$ that includes all distributions likely to obtain in the given situation. We shall examine this possibility for several definitions of $\mathcal{F}_0$ in the next section.

Let us briefly mention two other approaches to the problem.

1. Although the t-test does not satisfy (4), there might exist another test that does. Suppose φ is any test that in the original normal model has power $\beta>\alpha$ against some normal alternative $N(\theta, \sigma^2)$ with $\theta>0$. Then given any $\varepsilon>0$ it follows from Bahadur & Savage (1956) that there exists a distribution $F \in \mathcal{F}$ with mean zero and $\alpha_n(F)>\beta-\varepsilon$ and hence such a test will satisfy $\alpha_n \geq \beta$. In particular, if its power tends to 1 as $\theta \to \infty$, then the test will satisfy (5).
2. Without much consideration, we generalized the original hypothesis $\mathcal{H}$ that the mean of the assumed normal distribution is zero to the hypothesis $\mathcal{H}'$ defined by (3). However, under the assumption of normality, θ is not only the mean of the X's; it is also the median, the symmetrically trimmed mean, and so on, since all these measures of location coincide for a symmetric distribution.

   If such an alternative generalization is adopted, the results can be different. For example, let ν be the median of F, so that $F(\nu)=1/2$. Then the sign test can be used to test $\mathcal{H}''$: $\nu=0$ against the alternatives that $\nu>0$. If S_n denotes the number of positive X's, then under $\mathcal{H}''$, S_n has the binomial distribution corresponding to n trials and success probability $p=1/2$, and its level is therefore identically equal to α, independent of F, and hence trivially satisfies $\alpha_n=\alpha$.

   Whether the mean, median, or some other measure of location is the most appropriate extension of θ (in the normal case) when arbitrary distributions F are considered possible depends on the context of the particular problem. For a discussion, see for example Lehmann (1975, pages 84–85).

The discussion of robustness of the significance level of the one- and two-sided t-test translates in the obvious way to the confidence coefficients of the corresponding confidence bounds and intervals. Furthermore, the discussion extends easily to other tests concerning normal means, for example, the linear models F-test.

The situation is different for the corresponding tests or confidence sets for a normal variance, the ratio of two variances, or the correlation coefficient. These procedures are not even robust, i.e. don't even satisfy (2) (cf. Miller, 1986). They can be made robust, essentially through studentization. The resulting robust procedures behave similarly to the t-test and are not uniformly robust.

2. Uniform and non-uniform robustness of the t-test

In this section, we shall consider some families $\mathscr{F}_0$ of "reasonable" distributions and for each ask whether it entails

$$\sup_{\mathscr{F}_0} \alpha_n(F) \to \alpha \quad \text{as} \quad n \to \infty. \tag{6}$$

2.1. Kolmogorov-Smirnov neighbourhoods

It would seem natural to hope for the validity of (6) when $\mathscr{F}_0$ is the family of all distributions not differing much from the family of normal distributions with mean zero. One well-known measure of closeness between two distributions is the Kolmogorov–Smirnov distance. It turns out that not only the t-test, but any non-trivial unbiased test of the hypothesis that $\mu_F=0$ is not uniformly robust over Kolmogorov–Smirnov neighbourhoods.

To see this, let $\Delta>0$ and let F_0 be a fixed distribution with mean zero and bounded density function $f_0(x)$, and let $\mathscr{F}_{01}$ be the family of all distributions for which

$$|F(x)-F_0(x/\sigma)|<\Delta \quad \text{for all } x \text{ and some } \sigma>0. \tag{7}$$

Let $U(a, h)$ denote the uniform distribution on the interval $(a-h, a+h)$ and let $\omega=\sup_x f_0(x)$. Consider the mixture c.d.f

$$F(x)=(1-\lambda)F_0(x-\xi_2)+\lambda G_{\xi_1,\lambda}(x), \tag{8}$$

where

$$0<\lambda<1, \quad 0<\xi_2<\omega^{-1}\Delta,$$

$$\xi_1=-\lambda^{-1}(1-\lambda)\xi_2,$$

and $G_{\xi_1,\lambda}(x)$ is the cdf of the $U(\xi_1, \lambda)$ distribution.

Because for this F,

$$|F(x)-F_0(x)|\leq|(1-\lambda)F_0(x)-F_0(x)|+(1-\lambda)|F_0(x)-F_0(x-\xi_2)|+\lambda G_{\xi_1,\lambda}(x)$$
$$\leq 2\lambda+(1-\lambda)\omega\xi_2$$

it follows from the above conditions on ξ_1 and ξ_2 that (8) belongs to $\mathscr{F}_{01}$ for all λ sufficiently small.

Let φ be any test of the hypothesis $\mathscr{H}'$: $\mu_F=0$ and let $\beta_n(\omega^{-1}\Delta; \varphi)$ be its power against the alternative $F_0(x-\omega^{-1}\Delta)$. Suppose $\mathscr{S}$ denotes the rejection region of the φ-test and let $\alpha_n^{(\varphi)}(F)$ be the probability that φ rejects $\mathscr{H}'$ based on a sample of size n from F. Then

$$\alpha_n^{(\varphi)}(F)\geq(1-\lambda)^n \Pr_{\xi_2}\{(X_1, \ldots, X_n)\in\mathscr{S}\}$$

where $\Pr_{\xi_2}$ indicates that the probability is calculated under $F_0(x-\xi_2)$. It follows by letting $\lambda\to 0$ and $\xi_2\to\omega^{-1}\Delta$ that

$$\sup_{\mathscr{F}_{01}} \alpha_n^{(\varphi)}(F)\geq\beta_n(\omega^{-1}\Delta;\varphi) \quad \forall\ F_0, \varphi, n, \text{ and } \quad \Delta>0. \tag{9}$$

We therefore have proved the following result.

Theorem 1

Let F_0 be any distribution with mean zero and bounded density function $f_0(x)$ and let $\omega=\sup_x f_0(x)$. Given $\Delta>0$, let $\mathscr{F}_{01}$ be the Kolmogorov–Smirnov neighbourhood (7) around the scale-family induced by F_0. Then (9) obtains for any test φ of the hypothesis that the mean is zero.

Specializing φ to be the t-test and F_0 to be the standard normal distribution yields the following corollary.

Corollary 1

Let F_0 be the standard normal distribution. Then

$$\sup_{\mathscr{F}_{01}} \alpha_n(F)\to 1 \quad \text{as} \quad n\to\infty, \quad \forall\ 0<\alpha<1.$$

2.2. *Uniformly bounded support*

Since $\xi_1\to-\infty$ as $\lambda\to 0$ in the proof of the theorem, it might be thought that the cause of the difficulty with the family defined in (7) is that it is possible to cause a perturbation arbitrarily far out in the tail of the distribution. This may lead to the conjecture that the difficulty will disappear if $\mathscr{F}_0$ is restricted to contain only distributions with uniformly bounded support.

Theorem 2

Let $\mathscr{F}_{02}$ be the class of all absolutely continuous distributions in $\mathscr{F}$ whose support is contained in a bounded interval. Then for every $0<\alpha\leq 1$,

$$\sup_{\mathscr{F}_{02}} \alpha_n(F)=1 \text{ for all sufficiently large } n. \tag{10}$$

Proof. Without loss of generality, assume that the support of each $F\in\mathscr{F}_{02}$ is contained in the interval $(-1, 1)$. Consider the distribution

$$F=(1-\lambda)U(-\lambda, \lambda/2)+\lambda U(1-\lambda, \lambda/2).$$

It is easily verified that $F\in\mathscr{F}_{02}$ for $0<\lambda<2/3$. Let $\mathscr{A}$ denote the event that the entire sample $X_1, \ldots, X_n$ comes from the first component $U(-\lambda, \lambda/2)$. When $\mathscr{A}$ occurs, the minimum value of $|\bar{X}|$ is $\lambda/2$ and the maximum value of s_n is $\lambda/2$ if n is odd, and $(\lambda/2)\sqrt{n(n-1)^{-1}}$ if n is even. This yields $|t_n|\geq\sqrt{n-1}$. Therefore for all large n (independent of λ)

$$\alpha_n(F)\geq\Pr(\mathscr{A})=(1-\lambda)^n. \tag{11}$$

Letting $\lambda\to 0$ in (11) yields (10). □

2.3. *Uniformly bounded standardized moments*

Let σ_F denote the standard deviation of F and let $\varkappa_r(F)=E_F|X-\mu_F|^r/\sigma_F^r$ denote the rth standardized absolute moment about the mean of F. Let $\mathscr{F}_{03}$ be a class of absolutely continuous $F\in\mathscr{F}$ with $\varkappa_s(F)$ uniformly bounded for some $s=2+\delta$, $\delta>0$. It turns out that

$$\sup_{\mathscr{F}_{03}} \alpha_n(F)\to\alpha \quad \text{as} \quad n\to\infty. \tag{12}$$

Result (12) is easily proved using a Berry–Esséen theorem for the t-statistic in Chung (1946) if we assume that $\varkappa_8(F)$ is uniformly bounded. A proof of the more general result was communicated to us by T. Severini. We give a sketch of it via the following lemmas.

Lemma 1

Let $Y_1, Y_2, \ldots,$ be a sequence of i.i.d. random variables with means 0 and distribution $F\in\mathscr{C}$, where $\mathscr{C}$ satisfies

$$\sup_{\mathscr{C}} E|Y_1|^{1+\delta}<\infty$$

for some $\delta>0$. Then for every $\varepsilon>0$,

$$\lim_{n\to\infty}\sup_{\mathscr{C}} \Pr\left(\left|n^{-1}\sum_{i=1}^{n} Y_i\right|>\varepsilon\right)=0.$$

Proof. The result is a consequence of Chebychev's inequality if $\delta=1$. For the general case, define $Z_j=Y_j$ if $|Y_j|\leqslant n$, and 0 otherwise, for $j=1, \ldots, n$. Then $\operatorname{Var}(Z_j)\leqslant n^{-1\delta}M$ and $E|Y_j-Z_j|\leqslant n^{-\delta}M$, where $M=\sup_{\mathscr{C}}E|Y_1|^{1+\delta}$. Hence for any $\varepsilon>0$,

$$\Pr\left\{\left|n^{-1}\sum_{j=1}^{n}(Y_j-Z_j)\right|>\varepsilon\right\}\leqslant n^{-\delta}\varepsilon^{-1}M$$

and

$$\Pr\left(\left|n^{-1}\sum_{j=1}^{n} Z_j\right|>\varepsilon\right)\leqslant n^{-\delta}\varepsilon^{-2}M$$

with both inequalities holding uniformly over $\mathscr{C}$. This implies that

$$\sup_{\mathscr{C}}\Pr\left(\left|n^{-1}\sum_{j=1}^{n} Y_j\right|>2\varepsilon\right)\leqslant n^{-\delta}M\varepsilon^{-2}(1+\varepsilon)$$

for all n and every $\varepsilon>0$, and the lemma follows. □

Lemma 2

Let $Y_1, Y_2, \ldots$ be a sequence of i.i.d. random variables with mean η, variance τ^2, and distribution $F\in\mathscr{D}$, where $\mathscr{D}$ satisfies

$$\sup_{\mathscr{D}} E|\tau^{-1}(Y_1-\eta)|^{2+\delta}<\infty$$

for some $\delta>0$. Let $\bar{Y}_n=n^{-1}\Sigma_{j=1}^{n}Y_j$ and $\hat{\tau}_n^2=n^{-1}\Sigma_{j=1}^{n}(Y_j-\bar{Y}_n)^2$. Then for every $\varepsilon>0$,

$$\sup_{\mathscr{D}}\Pr(\tau^{-1}|\hat{\tau}_n-\tau|>\varepsilon)\to 0 \quad \textit{as} \quad n\to\infty.$$

Proof. Let $Z_j=\tau^{-1}(Y_j-\eta)$. Then $E(Z_j^2-1)=0$ and

$$\sup_{\mathscr{D}} E|Z_j^2-1|^{1+\delta/2}\leqslant 2^{\delta/2}\left(\sup_{\mathscr{D}} E|Z_j|^{2+\delta}+1\right)<\infty.$$

It follows from lemma 1 that, for every $\varepsilon>0$,

$$\sup_{\mathscr{D}} \Pr\left(\left|n^{-1}\sum_{=1}^{n} Z_j^2-1\right|>\varepsilon\right)\to 0$$

as $n\to\infty$. Similarly, because $EZ_j=0$ and $\sup_{\mathscr{D}} E|Z_j|^{1+\delta}<\infty$, lemma 1 implies that

$$\sup_{\mathscr{D}} \Pr\left(\left|n^{-1}\sum_{j=1}^{n} Z_j\right|>\varepsilon\right)\to 0$$

as $n\to\infty$ for every $\varepsilon>0$. Finally, because

$$\tau^{-2}\hat{\tau}_n^2=n^{-1}\sum_{j=1}^{n} Z_j^2-\left(n^{-1}\sum_{j=1}^{n} Z_j\right)^2,$$

we have

$$\sup_{\mathscr{D}} \Pr\left(\left|\tau^{-2}\hat{\tau}_n^2-1\right|>\varepsilon\right)\to 0$$

for every $\varepsilon>0$. □

Proof of (12). Let $W_n=n^{-1/2}(\Sigma_{i=1}^n X_i-n\mu_F)$, $G_{n,F}(x)=\Pr_F\{\sigma_F^{-1}W_n\leq x\}$, and $\hat{G}_{n,F}(x)=\Pr_F\{s_n^{-1}W_n\leq x\}$. If $\varkappa_{2+\delta}(F)$ is uniformly bounded for some $\delta>0$, the Berry–Esséen theorem for sums of independent random variables (see, e.g. Prakasa Rao, 1987, page 25) implies that

$$\sup_{\mathscr{F}_{03}} \sup_{x} |G_{n,F}(x)-\Phi(x)|\to 0 \quad \text{as} \quad n\to\infty.$$

It therefore follows that, for every $\varepsilon>0$,

$$\begin{aligned}
|\hat{G}_{n,F}(x)-\Phi(x)| &\leq |G_{n,F}(x)-\Phi(x)|+|\hat{G}_{n,F}(x)-G_{n,F}(x)| \\
&\leq |G_{n,F}(x)-\Phi(x)|+\Pr(\sigma_F^{-1}|s_n-\sigma_F|>\varepsilon) \\
&\quad +|\Pr(\{s_n^{-1}W_n\leq x\}\cap\{\sigma_F^{-1}|s_n-\sigma_F|\leq\varepsilon\})-G_{n,F}(x)| \\
&\leq |G_{n,F}\{x(1+\varepsilon)\}-\Phi\{x(1+\varepsilon)\}|+2|G_{n,F}(x)-\Phi(x)| \\
&\quad +|G_{n,F}\{x(1-\varepsilon)\}-\Phi\{x(1-\varepsilon)\}|+2\Pr(\sigma_F^{-1}|s_n-\sigma_F|>\varepsilon) \\
&\quad +|\Phi\{x(1+\varepsilon)\}-\Phi(x)|+|\Phi\{x(1-\varepsilon)\}-\Phi(x)| \\
&\to 0
\end{aligned}$$

uniformly over x and $\mathscr{F}_{03}$ as $n\to\infty$ and $\varepsilon\to 0$. □

Following are some examples of $\mathscr{F}_{03}$ that are determined by the shapes of the distributions. Throughout, S_F will denote the support of F.

2.3.1. Symmetric distributions defined by shape

Let F_0 be a fixed, symmetric, absolutely continuous distribution with mean zero and finite $(2+\delta)$th moment, for some $\delta>0$. Let $\mathscr{F}_{04}$ be the family of absolutely continuous symmetric distributions with zero mean that precede F_0 according to the s-ordering $<_s$ of van Zwet (1964):

$F<_sG\Leftrightarrow G^{-1}F(x)$ is concave–convex about the point of symmetry of F.

This is a partial ordering on the set of symmetric distributions according to shape. Intuitively, $F<_s G$ if F is less peaked as well as less heavily-tailed than G. Van Zwet (1964, theorem 2.3.2) showed that $\varkappa_r(F)$ is isotonic with respect to the s-ordering for all $r\geq 2$, i.e.

$$F<_s G \Rightarrow \varkappa_r(F)\leq\varkappa_r(G)\ \forall r\geq 2.$$

Hence

$$\sup_{\mathscr{F}_{04}} \varkappa_{2+\delta}(F)\leq\varkappa_{2+\delta}(F_0),$$

and (6) holds for $\mathscr{F}_0=\mathscr{F}_{04}$.

Some examples are:

1. $\mathscr{F}_{04}=\{F: f(x)$ is symmetric and U-shaped$\}$, because every symmetric distribution with a U-shaped density s-precedes the uniform distribution.
2. Define $\bar{F}(x)=2F(x+\mu_F)-1$. We may let $\mathscr{F}_{04}=\{F$: F is symmetric and $\bar{F}$ is an increasing failure rate distribution (IFR)$\}$, because if $\bar{F}$ is IFR, then $F<_s$ double exponential.
3. $\mathscr{F}_{04}=\{F$: F is a symmetric and strongly unimodal distribution$\}$, because the family of symmetric strongly unimodal distributions is a subfamily of the family of F such that $\bar{F}$ is IFR (see Loh, 1984). Recall that a non-degenerate strongly unimodal distribution is one whose density is absolutely continuous and log-concave on its support (Ibragimov, 1956). This family includes the uniform, normal, logistic, and double exponential distributions.

2.3.2. Neighbourhoods around normal defined by moment conditions

Given δ, $\Delta>0$, let $\mathscr{F}_{05}$ be the family of distributions in $\mathscr{F}$ which are close to the normal scale-family in the sense that

$$|\varkappa_{2+\delta}(F)-\varkappa_{2+\delta}(\Phi)|<\Delta. \tag{13}$$

Then $\varkappa_{2+\delta}(F)$ is uniformly bounded and (6) holds for $\mathscr{F}_0=\mathscr{F}_{05}$.

2.3.3. Neighbourhoods around non-normal F

It is clear that the result of section 2.3.2 remains valid if in (13) the normal c.d.f. Φ is replaced by any non-degenerate distribution F_1 with finite $(2+\delta)$th moment for some $\delta>0$. This fact may be described by saying that the t-test is *locally uniformly robust* for the neighbourhoods defined by

$$|\varkappa_{2+\delta}(F)-\varkappa_{2+\delta}(F_1)|<\Delta$$

for some $\delta>0$.

3. Intervals for binomial p

Let Y have a binomial distribution $B(p, n)$ with parameters p and n. Exact confidence intervals for p exist and can be found in published tables. We discuss in this section some problems concerning large-sample intervals for p from the point of view of sections 1 and 2. We will say that a nominal $100(1-\alpha)\%$ confidence interval I_n for p constructed from n observations is *uniformly robust* over a class of distributions $\mathscr{F}$ if

$$\inf_{\mathscr{F}} \Pr_p(p\in I_n)\rightarrow 1-\alpha \quad \text{as} \quad n\rightarrow\infty. \tag{14}$$

3.1. Normal-approximation interval

Consider the standard $100(1-\alpha)\%$ "normal-approximation" interval I_n with endpoints

$$\hat{p}\pm z_{\alpha/2}n^{-1/2}\sqrt{\hat{p}(1-\hat{p})}, \tag{15}$$

where $\hat{p}=Y/n$. This interval reduces to the singleton $\{0\}$ or the empty set (depending on whether I_n is taken to be closed or open) whenever $Y=0$. Suppose that I_n is the open interval and let $\mathscr{G}_0$ be the set of all distributions $B(p, n)$ with $0<p<1$. The interval is clearly pointwise consistent for each p in $(0, 1)$. Because $\Pr_p(\hat{p}=0)\to 1$ as $p\to 0$, however (see e.g. Blyth & Still, 1983, p. 114),

$$\inf_{p\in(0,1)} \Pr_p(p\in I_n)=0 \text{ for every } n.$$

Therefore (14) is not satisfied, i.e. I_n is not uniformly robust over $\mathscr{G}_0$.

Let $0<a<b<1$. It turns out that

$$\lim_{n\to\infty}\ \inf_{p\in(a,b)} \Pr_p(p\in I_n)=1-\alpha \tag{16}$$

and

$$\lim_{n\to\infty}\ \sup_{p\in(a,b)} \Pr_p(p\in I_n)=1-\alpha. \tag{17}$$

That is, I_n is *locally robust.*

Proof of (16) *and* (17). Let $q=1-p$, $c=z_{\alpha/2}$ and $\gamma_n=\Pr_p(p\in I_n)$. Then for large enough n,

$$\gamma_n=\Pr_p\left\{\hat{p}-p\in\frac{(1-2p)c^2n^{-1}\pm cn^{-1/2}\sqrt{4p(1-p)+c^2n^{-1}}}{2(1+c^2n^{-1})}\right\}$$

$$\geqslant\Pr_p\left\{\frac{(Y-np)}{\sqrt{npq}}\in\frac{(1-2p)c^2(npq)^{-1/2}\pm c[2-c(npq)^{-1}]}{2(1+c^2n^{-1})}\right\}$$

$$>\Pr_p\left\{Z\in\frac{(1-2p)c^2(npq)^{-1/2}\pm c[2-c(npq)^{-1/2}]}{2(1+c^2n^{-1})}\right\}-\frac{C\varrho_n}{\sigma_n^3n^{1/2}}.$$

Here σ_n^2 and ϱ_n denote the variance and third absolute moment about the mean of Y respectively, Z denotes the $N(0, 1)$ random variable, and C is a universal constant. The second inequality follows from the Berry–Esséen theorem (see e.g. Feller, 1966, p. 515). Because p is bounded away from 0 and 1, the two quantities $(npq)^{-1/2}$ and $\varrho_n/\sigma_n^3n^{1/2}$ tend to zero as $n\to\infty$. Thus

$$\liminf_{n\to\infty}\ \inf_{p\in(a,b)} \Pr_p(p\in I_n)\geqslant 1-\alpha. \tag{18}$$

Similarly,

$$\gamma_n<\Pr_p\left\{Z\in\frac{(1-2p)c^2(npq)^{-1/2}\pm c[2+c(npq)^{-1/2}]}{2(1+c^2n^{-1})}\right\}+\frac{C\varrho_n}{\sigma_n^3n^{1/2}},$$

which implies that

$$\limsup_{n\to\infty}\ \sup_{p\in(a,b)} \Pr_p(p\in I_n)\leqslant 1-\alpha. \tag{19}$$

The desired results now follow from (18) and (19). □

The proof shows that it is not necessary for a and b to be held fixed as n increases. The same results hold if we allow the neighbourhood to be (a_n, b_n), where $a_n \to 0$ and $b_n \to 1$ in such a way that

$$\sup_{p \in (a_n, b_n)} (npq)^{-1} = o(1), \qquad \sup_{p \in (a_n, b_n)} \varrho_n \sigma_n^{-3} = o(n^{1/2}).$$

Suppose now that the family $\mathscr{G}_0$ is enlarged to include the degenerate binomial distributions corresponding to $p=0$ and $p=1$. Let $\mathscr{G}_1$ denote the enlarged family. The interval I_n is still pointwise consistent over $\mathscr{G}_1$. However, it is no longer locally uniformly robust at $B(0, n)$ and $B(1, n)$. Thus I_n is locally uniformly robust at each distribution in $\mathscr{G}_0$ but not $\mathscr{G}_1$. This shows that local uniform robustness can be rather a fragile property which may be destroyed by slight perturbations of the problem.

3.2. *Arcsine transformation*

It may be thought that the failure of the interval (15) to be uniformly robust over $\mathscr{G}_0$ and locally uniformly robust over $\mathscr{G}_1$ is because the width of the interval shrinks to zero when $Y=0$ or n. The *arcsine* transformation is sometimes recommended because $\arcsin(\sqrt{\hat{p}})$ is more nearly normally distributed than is $\hat{p}$ and also because the variance of the former is approximately a constant equal to $(4n)^{-1}$ (see e.g. Johnson & Kotz, 1969, p. 65). Applying the normal approximation to $\arcsin(\sqrt{\hat{p}})$, we get the following approximate $100(1-\alpha)\%$ confidence interval, J_n say, for p:

$$\sin^2\{\arcsin(\sqrt{\hat{p}}) - z_{\alpha/2}(2n^{1/2})^{-1}\}^+ < p < \sin^2\{\arcsin(\sqrt{\hat{p}}) + z_{\alpha/2}(2n^{1/2})^{-1}\}.$$

Note that the width of J_n does not vanish when $\hat{p}=0$ or 1. Unfortunately, J_n is also not uniformly robust over $\mathscr{G}_0$ when $0<\alpha<1/2$ because

$$\limsup_{n\to\infty} \inf_{p \in (0,1)} \Pr_p(p \in J_n) < 1-\alpha, \quad 0<\alpha<1/2. \tag{20}$$

Proof of (20). When $Y=0$, J_n reduces to the interval $(0, u_n)$, where $u_n = \sin^2\{z_{\alpha/2}(2n^{1/2})^{-1}\}$. Let $p=u_n$ and $v_n = 1-u_n$. Then the probability that J_n does not contain p is at least v_n^n, which for large n is approximately $\{1 - z_{\alpha/2}^2(4n)^{-1}\}^n$. The last expression converges as $n\to\infty$ to $\exp(-4^{-1} z_{\alpha/2}^2)$. It follows that

$$\limsup_{n\to\infty} \inf_{p \in (1,1)} \Pr_p(p \in J_n) \leq 1 - \exp(-4^{-1} z_{\alpha/2}^2) \tag{21}$$

Using the inequality (see Abramowitz & Stegun, 1965, p. 933, formula 26.2.23)

$$z_\alpha < (-2 \log \alpha)^{1/2} + 4.5 \times 10^{-4} \quad \text{for } 0<\alpha<1/2$$

it can be verified that the right side of (21) is less than $1-\alpha$ for $0<\alpha<1/2$. □

Table 1 gives values of the upper bound in (21) for some common values of α.

As in section 3.1, it can be shown that J_n is locally uniformly robust over $\mathscr{G}_0$ but not $\mathscr{G}_1$.

4. Conclusion

Although the main emphasis of this paper is on the asymptotic behaviour of the t-test and binomial confidence intervals, the problems they illustrate are typical of many statistical

Table 1. *Some values of the bound in (21)*

$1-\alpha$	$1-\exp(-4^{-1}z^2_{\alpha/2})$
0.70	0.24
0.80	0.34
0.90	0.49
0.95	0.62
0.98	0.74
0.99	0.81

procedures derived through delta method arguments. That is, such procedures are usually pointwise robust but not uniformly robust.

To illustrate this somewhat more broadly, suppose that $\mathbf{X}_i=(X_{i1}, \ldots, X_{ip})$, $i=1, \ldots, n$ are n i.i.d. vectors and let $T_i=h(X_i)$ with $\eta=E(T_i)$ and $\tau^2=\text{Var}(T_i)<\infty$, both independent of i. Then tests or confidence intervals for η can be based on $n^{1/2}(\bar{T}-\eta)\hat{\tau}^{-1}$ where $\hat{\tau}$ is a consistent estimator of τ, and it follows from the results for the t-test that the procedures are pointwise but not uniformly robust. The following are two applications.

1. Tests or confidence intervals for the variance of i.i.d. variables X_i, based on

$$n^{-1/2}\hat{\tau}^{-1}\sum(X_i-\bar{X})^2=n^{1/2}\hat{\tau}^{-1}\left(\sum X_i^2/n\right)-n^{-1/2}\hat{\tau}^{-1}(n^{1/2}\bar{X})^2.$$

Since the left side is independent of $\xi=E(X_i)$, suppose without loss of generality that $\xi=0$. Then the second term on the right side tends to 0 in probability and the robustness results of the preceding paragraph apply with $p=1$ and $T_i=X_i^2$.

2. Tests or confidence intervals for the correlation coefficient of a bivariate distribution based on

$$\frac{\Sigma(X_{i1}-X_{\cdot 1})(X_{i2}-X_{\cdot 2})}{n^{1/2}\hat{\tau}}=\frac{n^{1/2}(\Sigma X_{i1}X_{i2}/n}{\hat{\tau}}-\frac{(n^{1/2}X_{\cdot 1})(n^{1/2}X_{\cdot 2})}{n^{1/2}\hat{\tau}}$$

with $p=2$, $T_i=X_{i1}X_{i2}$ and $X_{\cdot j}=n^{-1}\Sigma_i X_{ij}$.

The corresponding binomial result also holds much more generally. Let $X_1, \ldots, X_n$ be n i.i.d. random variables whose possible values are the integers $\geqslant a$ and whose distribution is a one-parameter exponential family

$$\Pr_\theta(X=x)=\exp\{\eta(\theta)x-A(\theta)\}h(x);\quad x=a, a+1, \ldots; \theta\in(\theta_0, \theta_1).$$

Suppose that

$$\Pr_\theta(X=a)\to 1 \quad \text{as} \quad \theta\to\theta_0 \tag{22}$$

and let $\xi=E_\theta(X)$. Then $\xi\to a$ as $\theta\to\theta_0$. Let $\tau^2=\text{Var}_\theta(\bar{X})=n^{-1}\text{Var}_\theta(X)=n^{-1}g(\xi)$, say, where g is a continuous function. Equation (22) implies that $g(a)=0$. Let $\hat{\tau}^2=n^{-1}g(\bar{X})$. The confidence interval for ξ

$$I_n=\bar{X}\pm z_\alpha\sqrt{g(\bar{X})/n}$$

is pointwise but not uniformly robust because $\Pr_\theta(\xi\in I_n)\to 0$ as $\theta\to\theta_0$. The Poisson, truncated Poisson, and negative binomial are included in this situation.

Distribution-free tests such as the Wilcoxon and other rank tests are somewhat safer in this respect because of their (trivially achieved) uniform robustness. However, these distribution-free tests are not without problems of their own. For example, the size of the two-sample Wilcoxon test is not even pointwise robust if the population distributions are not identical, whereas the two-sample t-test is, as long as the population means are the same.

Extensions of the present results to the k-sample F test are treated in Chen & Loh (1990) and a more detailed analysis of the binomial confidence interval problem is given in Chen (1990).

Acknowledgements

We thank T. Severini for help in the proof of equation (12) and an associate editor for many helpful suggestions.

The first author's research was supported by National Science Foundation grant DMS 84-01388. The second author's research was supported by National Science Foundation grants DMS 85-02303 and DMS 88-03271, Air Force Office of Scientific Research grant ISSA 86-0068, and an IBM Junior Faculty Research Fellowship at the T. J. Watson Research Centre.

References

Abramowitz, M. & Stegun, I. A. (1965). *Handbook of mathematical functions*. Dover, New York.

Bahadur, R. R. & Savage, L. J. (1956). The nonexistence of certain statistical procedures in non-parametric problems. *Ann. Math. Statist.* **27,** 1115–1122.

Benjamini, Y. (1983). Is the t-test really conservative when the parent distribution is long-tailed? *J. Amer. Statist. Assoc.* **78,** 645–654.

Blyth, C. R. & Still, H. A. (1983). Binomial confidence intervals. *J. Amer. Statist. Assoc.* **78,** 108–116.

Chen, H. (1990). The accuracy of approximate intervals for a binomial parameter. *J. Amer. Statist. Assoc.* **85,** 514–518.

Chen, H. & Loh, W.-Y. (1990). Uniform robustness against nonnormality of the t and F tests. To appear.

Chung, K. L. (1946). The approximate distribution of Student's statistic. *Ann. Math. Statist.* **17,** 447–465.

Cressie, N. (1980). Relaxing assumptions in the one-sample t-test. *Austr. J. Statist.* **22,** 143–153.

Feller, W. (1966). *An introduction to probability theory and its applications*. Volume II. Wiley, New York.

Gleser, L. J. & Hwang, J. T. (1987). The nonexistence of $100(1-\alpha)\%$ confidence sets of finite expected diameter in errors-in-variables and related models. *Ann. Statist.* **15,** 1351–1362.

Herrendoerfer, G. & Feige, K. D. (1985). A combinatorial method in robustness research and two applications. In *Robustness of statistical methods and nonparametric statistics* (eds. D. Rasch & M. L. Tiku), 53–57, Reidel Publ. Co, Holland.

Herrendoerfer, G., Rasch, D. & Feige, K. D. (1983). Robustness of statistical methods II. Methods for the one-sample problem. *Biometrical J.* **4,** 327–343.

Ibragimov, I. A. (1956). On the composition of unimodal distributions. *Theor. Probab. Appl.* **1,** 255–260.

Johnson, N. L. & Kotz, S. (1969). *Distributions in statistics—discrete distributions*. Wiley, New York.

Lehmann, E. L. (1975). *Nonparametrics: statistical methods based on ranks*. Holden Day, San Francisco.

Lehmann, E. L. (1986). *Testing statistical hypotheses*. Second edition. Wiley, New York.

Loh, W.-Y. (1984). Bounds on AREs for restrictred classes of distributions defined via tail-orderings. *Ann. Statist.* **12,** 685–701.

Miller, R. G., Jr. (1986). *Beyond ANOVA, basics of applied statistics*. Wiley, New York.

Prakasa Rao, B. L. S. (1987). *Asymptotic theory of statistical inference*. Wiley, New York.

van Zwet, W. R. (1964). *Convex transformations of random variables*. Math. Centrum, Amsterdam.

Received July 1988, in final form October 1989

E. L. Lehmann, Department of Statistics, University of California, Berkeley, CA 94720.

Part VII
Multiple Decisions

Erich Lehmann's contributions to multiple decision making

Juliet Popper Shaffer

1 Introduction

Much of statistical theory and methodology deals with testing a single hypothesis or specifying a single confidence set. But many, probably most studies deal with a number of questions, requiring more than one test or confidence set. In fact, within the last 20 years, advances in scientific and computer technology has made it possible to collect and analyze massive amounts of data, resulting in decision-making for thousands, even millions, of items.

Should each hypothesis or confidence set be treated individually, without taking the number of such inferences into account? Under what conditions would that be satisfactory, and when would it be inadvisable? If inadvisable, what alternative approach should be taken? These considerations constitute the subject of multiple inference, to which Erich Lehmann has made many contributions.

There was sporadic work in this area earlier, but concentrated efforts to deal with multiplicity issues in a comprehensive way began aound 1950, with papers by Tukey, Duncan, Scheffé, and others. See, e.g. Hochberg and Tamhane (1987) for descriptions. After a brief discussion in his 1952 paper "Testing multiparameter hypotheses," Erich's detailed work on this problem began with his 1957 papers.

In these comments, a consistent notation will be used, rather than changing notation as it appears in each paper.

J.P. Shaffer
Department of Statistics, University of California, Berkeley, CA 94720-3860
e-mail: shaffer@stat.berkeley.edu

J. Rojo (ed.), *Selected Works of E. L. Lehmann*, Selected Works in Probability and Statistics, DOI 10.1007/978-1-4614-1412-4_52, © Springer Science+Business Media, LLC 2012

2 The 1957 papers

In the first of two papers, "A theory of multiple decision problems" I and II in the Annals of Statistics (Lehmann, 1957a), Erich begins with an example of two independent, normally-distributed random variables X_1 and X_2 with unit variance, where $E(X_i) = \theta_i, i = 1, 2$. If we were interested in testing $H : \theta_1 = 0$ and $\theta_2 = 0$ versus $(\bar{H})$: at least one $\theta_i \neq 0$, it would seem sensible to use the spherically symmetric rejection region $x_1^2 + x_2^2 > c$. However, if upon rejection more detailed statements were desired, we would have a multiple decision problem. The parameter space can be partitioned into the nine regions corresponding to all combinations of the decisions for each θ_i: that it equals 0, is less than 0, or is greater than 0. Then a multiple decision procedure should permit partitioning the regions of the parameter space to allow the possibility of each of these decisions. These can be generated by testing, for each i, the hypotheses $H_{i1} : \theta_i \leq 0$ and $H_{i2} : \theta_i \geq 0$, with acceptance of both H_{i1} and H_{i2} equivalent to the decision $\theta_i = 0$.

The proposed loss structure is as follows:

Loss = 0 for a correct decision, a for rejecting a true hypothesis, b for accepting a false hypothesis. The loss table for the set of decisions is obtained by adding the losses for the individual decisions. If each hypothesis H_{i1} and H_{i2} is tested at level $b/(a+b)$ using the standard t-test, the natural resulting multiple test procedure is shown to be unbiased and to have uniformly minimum risk.

The procedure is described in much more general terms, given any composite hypothesis that, if rejected, can be partitioned into a set of detailed hypotheses in this way. The only requirements are that each separate hypothesis be tested by a method with uniformly minimum risk, and that the set of separate decisions is *compatible*, *i.e.*, that there are no sets in the decision space with positive probability corresponding to sets in the parameter space with zero probability, and vice versa.

The formulation is shown to apply to a large number of problems, including point estimation, goodness of fit, and equivalence (rather than equality) testing. Thus, a good deal of literature in general statistics is encompassed.

In the second of these papers (Lehmann, 1957b), Erich reformulates the problem, noting that the earlier formulation does not generalize to a number of areas including the very important area of comparison of pairs of parameters, in which the compatibility assumption is not satisfied.

For example, suppose we have random samples from three normal populations and want to test the three pairs of hypotheses

(i) $\theta_1 - \theta_2 \leq (\geq) 0$,
(ii) $\theta_1 - \theta_3 \leq (\geq), 0$,
(iii) $\theta_2 - \theta_3 \leq (\geq) 0$.

If the unformly-minimum-risk tests of each of the six hypotheses are used, it is possible to arrive at the decision

$$\theta_1 < \theta_3, \theta_1 = \theta_2, \theta_2 = \theta_3, \tag{2.1}$$

while there is no set in the parameter space corresponding to this decision. Thus, this set of decisions is not compatible.

In the second 1957 paper, the decision to accept a hypothesis is interpreted, not as an assertion that the hypothesis is true, but as providing no information about the value of the parameter involved. Instead of the conclusion (1) in this example, a decision with positive probability although there is no corresponding state of nature, the conclusion would be

$$-\infty < \theta_1 - \theta_2 < \infty, -\infty < \theta_2 - \theta_3 < \infty, \quad \theta_1 < \theta_3. \tag{2.2}$$

The reformulation includes the possibility of such decisions, treated as partial joint decisions, and the earlier optimality properties still hold.

Erich acknowledges early formulations by Duncan that are similar, but not identical, to this approach. (Duncan's loss functions consider the magnitude as well as the direction of departure from the null hypothesis.) In return, Duncan, in a series of papers beginning in the 1960s (see Duncan, 1961), developed a Bayesian procedure for comparing means of normal distributions based on the assumption that the population means are realizations of independent normally-distributed random variables. He used Erich's theorem to show that his procedure has minimum Bayes risk. Shaffer (1999) studied Duncan's work in an empirical Bayes and frequentist context. Pennello (2007) has provided an overview of Duncan's k-ratio Bayes rule approach and further developments arising out of it.

3 Optimum significance levels for multistage comparison procedures

Much of the work in multiple testing has used not the additive risk formulation of the 1957 papers, but rather has been based on the idea first proposed by Tukey of controlling the familywise error rate ($FWER$), *i.e.*, the probability of one or more errors in the whole set of comparisons, at some small value α, typically 0.05. Therefore the level at which any hypothesis is tested may depend on the number of and relationships among tested hypotheses, rather than only the individual hypothesis loss structure. In the so-called stepwise methods, the level for testing any hypothesis also depends on the outcomes of tests of other hypotheses. This paper (Lehmann and Shaffer, 1979) deals with multiple range procedures, one type of stepwise method applicable to comparisons of the effects of treatments. Some optimality conditions are proposed and a procedure derived that turns out to be almost identical to a widely-used procedure proposed by Tukey, who did not note any optimality properties. The multiple range methods begin by testing the equality of the full set of treatments, comparing the treatments with the largest and smallest sample means using the upper-α significance level of the range or studentized range, proceeding to successively smaller ranges only if the including ranges have been declared significant. The paper finds the required significance levels for each successive range test, with the criterion of maximizing the levels for the smallest ranges. An alternative optimality criterion, maximizing the successive levels for the largest ranges, was investigated by Finner (1990).

Many later developments in stepwise testing of pairwise comparisons with $FWER$ control have been discussed in textbooks (Hochberg and Tamhane, 1987, Westfall and Young, 1993, Hsu, 1996) and in articles (Shaffer, 1995, Tamhane, 1996).

4 On optimality of stepdown and stepup multiple test procedures

4.1 Previous related work

In the 1952 paper mentioned in Section 1, "Testing multiparameter hypotheses", instead of considering hypotheses $H_i : \theta_i \leq \theta_{i_0}, i = 1, ..., s$, with the desire to find all false H_i, Erich considers the hypothesis H_1 or H_2 or $\cdots H_s$ are true, with the alternative that all H_i are false. If $\theta_{io} = 0$ for all i, the alternative is equivalent to stating that $min(\theta_i) > 0$; *i.e.*, that all θ_i are positive. This kind of test also has important applications. For example, in medicine it is vital to know whether a drug is helpful not only on the average but also for subgroups of a population, divided by age, gender, ethnicity, etc., *i.e.*, whether the effects are positive for all designated groups.

For simplicity, assume two groups, and assume X_1 and X_2 are independently distributed $N(\theta_i, 1)$, $i = 1, 2$, respectively. In considering this type of hypothesis, Erich defines a *monotone* procedure as one for which, if x_1 and x_2 are in the rejection region Ω and $x_1' \geq x_1, x_2' \geq x_2$, then x_1' and x_2' are in Ω. This seems desirable if large values of the test statistics are more likely with large values of the parameters, as in this example. Then Erich proves that the test procedure: Reject H if $min(x_i) > u_\alpha$, where u_α is the upper-α critical value of the normal distribution, is the uniformly most powerful monotone test with $FWER = \alpha$ for the hypothesis.

Note that the optimal monotone test is extremely conservative for small values of the parameters: Power has a lower limit of α^2 if the test statistics are independent, as θ_1 and θ_2 approach 0 from above, and a lower limit of α^s for s such hypotheses. He considers the monotonicity requirement to be sensible, but points out that more powerful procedures are possible when this requirement is dropped, and describes one such procedure (p. 542-43), which was adopted by Roger Berger (1989) and generalized in a series of papers. The idea is to reject when *both* test statistics are within certain designated small regions. There is related work by Gutmann (1987) and Zelterman (1990). Perlman and Wu (1999) has additional references and alleges that such procedures, although uniformly more powerful than monotone procedures, defy common sense and should not be used.

4.2 Application to this paper

Holm (1979) introduced a general stepdown procedure with $FWER$ control that can be used with any set of hypotheses, not only pairwise equality comparisons. It is a modification of the well-known Bonferroni method, sometimes called the simultaneous or singlestep method, in which each of s hypotheses is rejected at level α if its associated p-value is $\leq \alpha/s$. Assume the hypotheses are numbered in the order of their p-values, *i.e.*, H_i is the hypothesis indexed by p_i, $p_1 \leq p_2 \leq \ldots \leq p_s$. In Holm's stepdown procedure at level α, H_i' is rejected if and only if $p_i < \alpha/(s-i+1), \forall i \leq i'$. The procedure can be carried out by beginning with H_1 and successively rejecting the ordered hypotheses as long as each successive hypothesis is rejected.

With this procedure, $FWER$ is $\leq \alpha$ with no restrictive conditions. Later, a stepup procedure was introduced for normally-distributed statistics (Dunnett and Tamhane, 1992), and under more general conditions by Hochberg (1988), the latter based on work by Simes (1986). There is a great deal of literature dealing with these methods, as well as a further extension to generalized step-up-down methods, introduced by Tamhane, Liu, and Dunnett (1998). In contrast with the Holm stepdown method, the stepup and generalized step-up-down methods require some restrictions on the joint distributions of the test statistics for $FWER$ control. See Sarkar (2007, 2008) for a review of these methods and required conditions.

Lehmann, Romano, and Shaffer (2005) described and proved the first optimality results for $FWER$-controlling stepup and stepdown methods based on individual p-values. Under monotonicity conditions including that noted above in the discussion of the 1952 paper, and considering one-sided hypotheses, it is shown that the procedures are optimal in the sense that stepdown procedures maximize the minimum probability of rejecting at least one false hypothesis, and stepup procedures maximize the minimum probability of rejecting all false hypotheses. As of this writing, a similar minimax result has not been extended to the step-up-down procedures, but it seems a reasonable possibility that it could be.

5 Generalizations of the Family-wise Error Rate

Although tests of large numbers of hypotheses had been common in areas such as survey sample analyses and epidemiological studies, the somewhat complex structure of these studies had led to either ignoring the multiplicity issues or treating them in an ad hoc way, dividing up the tests into families and using $FWER$ control within those families. The development within the last 20 years of technology for mass observations in fields like genomics, neuroscience, and astronomy has led to the desire for joint tests of thousands, even millions, of hypotheses as a single family. The problem of error buildup in these cases, both rejection of true hypotheses and acceptance of false ones, was obvious and couldn't be ignored. Furthermore, for most of these applications, the consequences of rejecting true hypotheses were likely to be less serious than in previous multiple hypothesis tests, since apparent discoveries would result in followup studies before any action would be taken. Familywise error control, by assuring very low probability of even a single error of rejecting a true hypothesis, results in a correspondingly low probability of rejecting false hypotheses, *i.e.*, of power.

For these new applications, the $FWER$-controlling methods seemed too conservative, but the need for some multiple error control was clear. As a result, interest in multiple testing blossomed, and a series of alternative criteria have been developed and explored.

Although there were earlier explorations of alternative criteria (Victor, 1982, Hommel and Hoffman, 1988, Seeger, 1968, Eklund and Seeger, 1965), the first paper to receive widespread attention was that of Benjamini and Hochberg (1995), who proposed use of the false discovery rate (*FDR*) (earlier treated less formally in Seeger, 1968, and Eklund and Seeger, 1965)), an alternative, more liberal, criterion than the *FWER*, and presented a method, based on an equality of Simes (1986), for controlling it.

Letting R = the number of rejections, V = the number of false rejections (*i.e.*, rejections of true hypotheses), the FDR is defined as $E(V/R|R > 0) * Pr(R > 0)$. This equals the *FWER* if all hypotheses are true, but allows some false rejections as long as their expected proportion is no greater than a specified, usually small, value. Many related measures have since been explored in detail, including: control of the positive *FDR* (Storey), control of $Pr(V \geq k)$, called the $k - FWER$, where k is a specified integer (note: $k - FWER = FWER$ when $k = 1$), $Pr(V/R > \gamma)$, where V/R is called the false discovery proportion (*FDP*), and γ is a specified proportion, various measures related to the probabilities of nonrejection of true hypotheses and balances between nonrejection of true and rejection of false hypotheses, Bayes and empirical Bayes approaches, and methods justified by good asymptotic properties in relation to many of these. See Farcomeni (2008), a review article with many references.

Lehmann and Romano (2005) proposed methods of controlling the number of false rejections - some of their results for this criterion were obtained earlier by Hommel and Hoffman (1988)-, and associated methods of controlling the *FDP*. The methods can be contrasted with those proposed by van der Laan, Dudoit, and Pollard (2004). Romano and Wolf (2007) compares the two types of methods, generalizations, and alternatives theoretically and empirically. See also Sarkar (2007, 2009).

Let $p_i, i = 1, \ldots, s$, be the ordered p-values of the s hypotheses being tested, $p_1 \leq p_2 \leq \ldots, p_s$, with the correspondingly ordered hypotheses $H_i, \ldots H_s$. Let $a_i, i = 1, \ldots, s$ be constants associated with the respective p-values.

k-FWER control: In Method 1, $a_i = k\alpha/s,\ \forall i$, and all hypotheses for which $p_i < a_i$ are rejected. When $k = 1$, this is the well-known Bonferroni method of controlling the FWER, and is often called the simultaneous or single-step method.

In Method 2, $a_i = k\alpha/s, i \leq k$ and $a_i = k\alpha/(s + k - i), i > k$. Reject all hypotheses $H_1, \ldots, H_j$, where j is the largest integer such that $p_i \leq a_i\ \forall i = 1, \ldots, j$. When $k = 1$, it corresponds to the Holm method for $FWER$ control, and thus is a generalization of the stepdown method to control of the $k - FWER$.

One somewhat unpleasant consequence of sole reliance on $k - FWER$ control is the fact that one can reject $k - 1$ hypotheses without any consideration of the associated

p-values and still meet the criterion. Even if monotonicity in the p-values is required, their specific values can be ignored. The paper discusses this possibility but notes that k or more hypotheses can be rejected if and only if $p_k \leq k\alpha/s$. At any rate, it seems desirable for some additional criteria besides $k - FWER$ control to be imposed.

As is typical in papers authored or co-authored by Erich, an optimality result is proved: No a_i can be increased without violating $k - FWER$ control for some joint distribution of the test statistics.

Romano has since explored more powerful modifications (Romano and Shaikh, 2006a, 2006b), Romano and Wolf (2007). As noted above, the latter article includes comparisons of a number of $k - FWER$ and FDP-controlling methods.

6 Summary

Erich worked early in the area of multiple decision making. As shown in the discussions of the individual papers, his work has been influential in many facets of this topic. His papers have given results for types of problems of practical importance that are the subject of ongoing research. His work in this area always involves optimality considerations. One expression that can never be used in relation to Erich's work is: *ad hoc*.

References

[1] Benjamini, Y. and Hochberg, Y. (1995). "Controlling the false discovery rate." *Journal of the Royal Statistical Society Series B*, 57, 289-300.

[2] Berger, R. L. (1989), "Uniformly more powerful tests for hypotheses concerning linear inequalities and normal means." *Journal of the American Statistical Association*, 84, 192-199.

[3] Duncan, D. B. (1961). "Bayes rules for a common multiple comparison problem and related Student-t problems." *Annals of Mathematical Statistics*, 32, 1013-1033.

[4] Dunnett, C. W. and Tamhane, A. C. (1992). "A step-up multiple testi procedure." *Journal of the American Statistical Association*, 87, 162-170.

[5] Eklund, G. and Seeger, P. (1965). Massignifikansanalys. *Statistisk Tidskrift, 3rd series-4*, 355-365.

[6] Farcomeni, A. (2008). "A review of modern multiple hypothesis testing, with particular attention to the false discovery proportion." *Statistical Methods in Medical Research*, 17, 347-388.

[7] Finner, H. (1990). "Some new inequalities for the range distribution, with appolication ot he determination of optimum significance levels of multiple range tests." *J. Amer. Statist. Assoc.*, 85, 191-194.

[8] Gutmann, S. (1987). "Tests uniformly more powerful than uniformly most powerful monotone tests." *Journal of Statistical Planning and Inference*, 17, 279-292.

[9] Hochberg, Y. (1988). "A sharper Bonferroni procedure for multiple tests of significance." *Biometrika*, 75, 800-802.

[10] Hochberg, Y. and Tamhane, A. (1987). "Multiple Comparison Procedures." Wiley, New York.

[11] Holm, S. (1979). "A simple sequentially rejective multiple test procedure". *Scandinavian Journal of Statistics*, 6, 65-70.

[12] Hommel, G. and Hoffman, T. (1988). "Controlled uncertainty." In *Multiple Hypotheses Testing (P. Bauer, G.Hommel, and E. Sonneman, Eds.)*, 154-161. Springer, Heidelberg.

[13] Hsu, J. C. (1996). Multiple Comparisons: Theory and Methods. Chapman & Hall, New York.

[14] Lehmann, E. L. (1952). "Testing multiparameter hypotheses". *Annals of Mathematical Statistics*, 23, 541-552

[15] Lehmann, E. L. (1957a). "A theory of some multiple decision problems I". *Annals of Mathematical Statistics*, 28, 1-25.

[16] Lehmann, E. L. (1957b). "A theory of some multiple decisions problems II". *Annals of Mathematical Statistics*, 28, 547-572.

[17] Lehmann, E. L. and Shaffer, J. P. (1979). "Optimum signficance levels for multistage comparison procedures". *Annals of Statistics*, 7, 27-45.

[18] Lehmann, E. L. and Romano, J. P. (2005). "Generalizations of the familywise error rate". *Annals of Statistics*, 33, 1138-1154.

[19] Lehmann, E. L., Romano, J. P., and Shaffer, J. P. (2005). "On optimality of stepdown and stepup multiple test procedures". *Annals of Statistics* 33, 1138-1154.

[20] Pennello, G. A. (2007). "Duncan's k-ratio Bayes rule approach to multiple comparisons: An overview." *Biometrical Journal*, 49, 78-93.

[21] Perlman, M. D. and Wu, L. (1999). "The emperor's new tests" (with discussion). *Statistical Science*, 14, 355-381.

[22] Romano, J. P. and Shaikh, A. M. (2006a). "On stepdown control of the false discovery proportion." In *Optimality: The Second Erich L. Lehmann Symposium*, (J. Rojo, Ed), 33-50.

[23] Romano, J. P. and Shaikh, A. M. (2006b). "Stepup procedures for control of generalizations of the family error rate." *Annals of Statistics*, 34, 1850-1873.

[24] Romano, J. P. and Wolf, M. (2007). "Control of generalized error rates in multiple testing." *Annals of Statistics*, 35, 1378-1408.

[25] Sarkar, S. K. (2007). "Stepup procedures controlling generalized FWER and generalized FDR." *Annals of Statistics*, 2007, 35, 2405-2420.

[26] Sarkar, S. K. (2008). "On the Simes inequality and its genralization." In *Beyond Parametrics in Interdisciplinary Research: Festschrift in Honor of Professor Pranab K. Sen*, IMS Collections Vol. 1, 231-242.

[27] Sarkar, S. K. (2009). "On Methods Controlling the False Discovery Rate (with discussion)." *Sankhya, Ser A*. In Press.

[28] Seeger, P. (1968). "A note on a method for the analysis of significances en masse." *Technometrics*, 10, 586-593.

[29] Shaffer, J. P. (1995). "Multiple hypothesis testing." In *Annual Review of Psychology*, 46, 561-584.

[30] Shaffer, J. P. (1999). "A semi-Bayesian study of Duncan's Bayesian multiple comparison procedure." *Journal of Statistical Planning and Inference*, 82, 197-213.

[31] Simes, R. (1986). "An improved Bonferroni procedure for multiple tests of significance." *Biometrika*, 73, 751-754.

[32] Tamhane, A. C. (1996). "Multiple Comparisons." In Handbook of Statistics, vol. 13 (Eds. S. Ghosh and C. R. Rao), 587-630.

[33] Tamhane, A. C., Liu, W. and Dunnett, C. W. (1998). "A generalized step-up-down multiple test procedure". *Canadian Journal of Statistics*, 26, 353-363.

[34] van der Laan, M., Dudoit, S., and Pollard, K. (2004). "Multiple testing, Part III, Procedures for control of the generalized familywise error rate and proportion of false positives.

[35] Victor, N. (1982). "Exploratory data analysis and clinical research." *Methods of Information in Medicine*, 21, 53-54.

[36] Westfall, P. H. and Young, S. S. (1993). *Resampling-based Multiple Testing: Examples and Methods for p-value Adjustment.* Wiley, New York.

[37] Zelterman, D. (1990). "On tests for qualitative interactions." *Statistics and Probability Letters*, 10, 59-63.

A THEORY OF SOME MULTIPLE DECISION PROBLEMS, I[1]

By E. L. Lehmann

University of California, Berkeley

Summary. A class of multiple decision procedures is described and its members are shown to possess uniformly minimum risk among all procedures that are unbiased with respect to a certain loss function. This provides a justification for a number of procedures considered by Tukey, Duncan, and others, for certain classes of point estimates, and for some nonparametric decision procedures based on sample cumulative distribution functions and related to tests of the Kolmogoroff-Smirnoff type.

1. Introduction. As has frequently been pointed out, many statistical situations, which it is customary to treat by means of hypothesis testing, really involve a choice between more than two decisions. In such problems, when the hypothesis is rejected, one wants to know in which of a number of possible ways the actual situation differs from the one postulated by the hypothesis. By formulating the problem as one involving only two decisions one not only neglects to differentiate between certain alternative decisions, which may differ considerably in their consequences, but one may also be led to an inappropriate acceptance region for the hypothesis.

As an example suppose that X and Y are independently normally distributed with unit variance and means ξ and η. While there are situations in which one only wishes to determine whether the hypothesis H: $\xi = \eta = 0$ is true or not, it is perhaps more common that in case of rejection one will want to know whether it is ξ or η that is different from 0, or both, and of the nonzero means whether they are positive or negative. Here the first formulation implies complete spherical symmetry between the alternatives, and the appropriate acceptance region is

$$x^2 + y^2 \leqq c.$$

On the other hand, when the choice lies between the nine indicated decisions, it seems most natural to accept H when

$$\max(|x|, |y|) \leqq k,$$

and in case of rejection to divide the rejection region into the eight subregions shown in Fig. 1 corresponding to the eight possible alternative decisions. This procedure actually will be justified later in terms of a specific loss function.

Received February 8, 1956.

[1] Work done while the author was a Fellow of the John Simon Guggenheim Memorial Foundation.

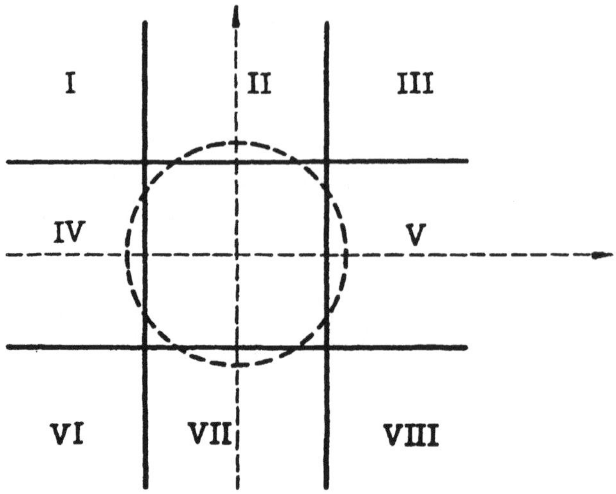

FIG. 1

One of the attractions of formulating statistical problems in terms of hypothesis testing is the resulting structural simplicity. However, at the same time this reduction to a choice between only two decisions frequently causes complications by creating a class of alternatives which combines too many different elements. In many such cases, if one is willing to forego structural simplicity and to divide the class of alternatives into its natural components, one obtains a multiple decision problem, which admits a simpler and more natural solution than the apparently less complex testing problem.

As an example consider the comparison of k variances $\sigma_1^2, \cdots, \sigma_k^2$ on the basis of samples $X_{ij}(j = 1, \cdots, n; i = 1, \cdots, k)$ from normal populations $N(\xi_i, \sigma_i^2)$. There does not seem to exist any really convincing solution to the small sample problem of testing the hypothesis $H: \sigma_1 = \cdots = \sigma_k$. On the other hand, there exists a natural multiple decision problem based on the comparisons of the different pairs (σ_i, σ_j). The associated acceptance region for H is the one discussed by Hartley [6].

In the present paper, a general class of multiple decision problems is described together with procedures that seem appropriate for these problems. The method of constructing the procedures is not new, and is in fact the one used in most cases of multiple comparisons treated in the literature. It was mentioned explicitly in 1950 by Howard Levene in a seminar lecture at Columbia University, and was recently stated, with only a minor difference, by Duncan in [3]. It is also closely related to a principle of test construction proposed by Roy [12], and utilized further by Bose and Roy [2]. As will be shown, the method is applicable not only to the typical multiple comparison problems, but also to problems of point estimation and various nonparametric problems.

The main purpose of the present paper is to prove an optimum property of the above procedures, namely that they are unbiased (in a sense introduced by the author in [9]), and that among all unbiased procedures they uniformly minimize the risk. It should be mentioned that in the application to specific cases a number of distributional problems arise, concerning error control and other properties of the operating characteristic of the procedures. These problems, which require separate treatment for each example, are not considered here but instead the paper is concerned only with general aspects of the method. In particular, the main result involves assumptions concerning the distribution of the observable random variables only indirectly in that the existence of tests of related hypotheses H is assumed, which possess certain optimum properties. Apart from this, the conditions concern only the structure of the multiple decision problems in terms of the hypotheses H.

2. A class of multiple decision procedures. Let X be a random observable, the distribution of which depends on the parameter θ, and suppose that a number of different hypotheses concerning θ are of interest, say $H_\gamma : \theta \varepsilon \omega_\gamma$, $\gamma \varepsilon \Gamma$. The class of alternatives K_γ to H_γ is that θ lies in the complement of ω_γ, which will be denoted by ω_γ^{-1}. When considering these hypotheses simultaneously one will wish to determine whether all of them are true, all of them false, or in the intermediate cases, exactly which of the hypotheses hold and which do not.

One is therefore faced with a multiple decision problem in which the different possible decisions correspond to the statements that a certain set of the hypotheses H_γ is correct while the remaining ones are false, or equivalently, that θ lies in a certain set, say Ω_i, $i \epsilon I$, determined by these conditions. The sets Ω_i, which are the atoms of the field of sets generated by the sets ω_γ, are formally given by

$$\Omega_i = \bigcap_{\gamma \varepsilon \Gamma} \omega_\gamma^{x_{i\gamma}}, \tag{2.1}$$

where the x's indicate which of the hypotheses are true ($x_{i\gamma} = 1$) and which are false ($x_{i\gamma} = -1$) for the given Ω_i. If, as is frequently the case, some of the intersections formally defined by (2.1) are empty, we shall restrict the Ω's to denote the nonempty ones, and shall require that none of the possible decisions should correspond to the empty intersections. Authors dealing with specific multiple comparison problems have frequently not insisted on this restriction, and this is also the point in which Duncan's definition, referred to above, differs from the one given here.

To specify a loss function, suppose that the losses for the individual testing problems are a_γ and b_γ for falsely rejecting and accepting the hypothesis H_γ, and that in the simultaneous consideration of these problems the losses are additive. If then $\theta \varepsilon \Omega_i$ and the decision d_k is taken that $\theta \varepsilon \Omega_k$, the resulting loss is

$$w_{ik} = \sum_{\gamma \epsilon \Gamma} (\epsilon_{ik\gamma}\, a_\gamma + \epsilon_{ki\gamma} b_\gamma), \tag{2.2}$$

where

$$\epsilon_{ik\gamma} = \begin{cases} 1 \text{ if } x_{i\gamma} = 1, \quad x_{k\gamma} = -1, \\ 0 \text{ otherwise.} \end{cases} \tag{2.3}$$

Formula (2.2) expresses the fact that w_{ik} is the sum of all those a_γ for which H_γ is true ($x_{i\gamma} = 1$) but rejected ($x_{k\gamma} = -1$), plus the sum of the b_γ for which H_γ is falsely ($x_{i\gamma} = -1$) accepted ($x_{k\gamma} = 1$). The risk is thus simply a weighted sum of the probabilities of error. Slightly more generally (taking the a's and b's to be given) one may wish to put, in the case that Γ is finite,

$$w_{ik} = \sum_{\gamma\varepsilon\Gamma} \nu_\gamma(\epsilon_{ik\gamma} a_\gamma + \epsilon_{ki\gamma} b_\gamma),$$

where the ν's are any positive weights, and in the general case

$$w_{ik} = \int_\Gamma (\epsilon_{ik\gamma} a_\gamma + \epsilon_{ki\gamma} b_\gamma)\, d\mu(\gamma). \tag{2.4}$$

Suppose now that attention is restricted to nonrandomized procedures. This involves no essential loss of generality since it can be achieved for all decision problems with which we shall be concerned, by adjoining to the original random variables a continuous variable which is independent of them. Then a decision procedure for the given multiple decision problem is a partition of the sample space into sets D_i such that, when the observation falls into D_i, the decision $d_i : \theta\varepsilon\Omega_i$ is taken. It is natural to try to relate these decision procedures to the tests of the hypotheses H_γ. Let $\mathfrak{A}$ be a family of such tests with acceptance regions A_γ for H_γ and rejection regions A_γ^{-1}, and consider the decision procedure that $\mathfrak{A}$ induces through the relation

$$D_i = \bigcap_{\gamma\varepsilon\Gamma} A_\gamma^{x_{i\gamma}}. \tag{2.5}$$

Here it may happen that

$$P(\bigcup_{i\varepsilon I} D_i) < 1.$$

This is the case when the union of those intersections $\bigcap_\gamma A_\gamma^{x_{i\gamma}}$, for which the corresponding intersection (2.1) is empty, has positive probability. The partition (2.5) is then not a procedure for the given decision problem. When the induced procedure satisfies

$$P_\theta(\bigcup_{i\varepsilon I} D_i) = 1 \qquad \text{for all } \theta, \tag{2.6}$$

the family $\mathfrak{A}$ is said to be *compatible*. Since the acceptances and rejections making up the intersection (2.5) are consistent with each other if the corresponding intersection (2.1) is nonempty, and are otherwise mutually contradictory, compatibility is equivalent to the condition that the simultaneous application of the tests A_γ not lead to any inconsistencies.

In the case that Γ is noncountable, a further complication is the possibility that a set of measurable A_γ's through (2.5) may give rise to a nonmeasurable D_i. However, barring such measurability difficulties, which it is usually easy to eliminate in specific problems, one has the following result.

THEOREM 1. *Relation* (2.5) *defines a* 1:1 *correspondence between the decision* *procedures for the problem induced by the hypotheses H_γ, $\gamma\varepsilon\Gamma$ and the compatible families $\mathfrak{A}$ of tests of these hypotheses.*

PROOF. Clearly, if $\mathfrak{A}$ is compatible, (2.5) does define a valid decision procedure. Conversely there always exists one, and essentially only one, compatible $\mathfrak{A}$ leading to a given decision procedure. To see this, suppose first that $\cup D_i$ is the whole sample space X. The desired result is then a consequence of the fact that (2.5) implies and is implied by

$$A_\gamma = \bigcup_{\{i:x_{i\gamma}=1\}} D_i, \tag{2.7}$$

which we shall now prove.

Assume first that (2.5) holds, and let $x\varepsilon A_\gamma$. Since x is in some D_i, it must be in some D_i with $x_{i\gamma} = 1$, and we have $A_\gamma \subseteq \cup_{\{i:x_{i\gamma}=1\}} D_i$. On the other hand,

$$\bigcup_{\{i:x_{i\gamma}=1\}} D_i = \bigcup_{\{i:x_{i\gamma}=1\}} \bigcap_{\delta\varepsilon\Gamma} A_\delta^{x_{i\delta}} \subseteq \bigcup_{\{i:x_{i\gamma}=1\}} A_\gamma^{x_{i\gamma}} = A_\gamma .$$

Thus (2.5) implies (2.6).

Suppose conversely that (2.6) holds. Then $A_\gamma^{-1} = \cup_{\{i:x_{i\gamma}=-1\}} D_i$, and hence

$$\bigcap_{\gamma\varepsilon\Gamma} A_\gamma^{x_{k\gamma}} = \bigcap_{\gamma\varepsilon\Gamma} \bigcup_{\{i:x_{i\gamma}=x_{k\gamma}\}} D_i .$$

Now a point belongs to the set on the right-hand side of this equation if and only if it lies in a set D_i for which $x_{i\gamma} = x_{k\gamma}$ for all γ. But this holds if and only if $i = k$, so that

$$\bigcap_{\gamma\varepsilon\Gamma} \bigcup_{\{i:x_{i\gamma}=x_{k\gamma}\}} D_i = D_k ,$$

and (2.6) implies (2.5), as was to be proved.

In case

$$\cup D_i = X - N,$$

where N is a null set for all distributions P_θ, relation (2.5) clearly holds also if each A_γ is replaced by $A_\gamma \cap (X - N)$. Applying the correspondence just proved to the space $X - N$, we see that (2.7) is replaced by

$$A_\gamma \cap (X - N) = \cup D_i .$$

Thus the D's determine the A's on the set $X - N$ and it is seen that, except on a null set, (2.5) and (2.7) establish the desired 1:1 correspondence quite generally.

For later reference we also give the relationship of the multiple decision procedure to the tests of the individual hypotheses in the case that they may be randomized. Suppose, for this purpose only, that Γ and I are countable. The tests are then described by means of critical functions φ_γ, where $\varphi_\gamma(x)$ denotes the probability with which H_γ is rejected when x is observed. Similarly, a procedure for the given multiple decision problem is a function ψ of the two arguments i and x, the value $\psi_i(x)$ of which is the probability with which decision D_i

is taken when x is observed. If $\Phi = \{\varphi_\gamma, \gamma\varepsilon\Gamma\}$ is a family of tests of the hypotheses H_γ, equations (2.5) and (2.7) become

$$\psi_i = \prod_{\gamma\varepsilon\Gamma} \varphi_\gamma^{-x_{i\gamma}},$$

where $\varphi_\gamma^{-1} = 1 - \varphi_\gamma$, and

$$\varphi_\gamma = \sum_{i\varepsilon I} \psi_i^{(1-x_{i\gamma})/2}$$

3. Some classification problems. We shall now illustrate the concepts of the previous section with some classes of examples. It will be shown later that for many distributions of interest (normal, binomial, Poisson, etc.) the indicated procedures possess the optimum property of uniformly minimizing the risk among all unbiased procedures, provided the levels α_γ of the tests φ_γ of H_γ are related to the losses a_γ and b_γ through the equation

$$\alpha_\gamma = b_\gamma / (a_\gamma + b_\gamma). \tag{3.1}$$

(i) *Three-decision problems.* Perhaps the simplest problem involving more than two decisions, and one which has been previously treated in somewhat similar terms by the author [8] and by Duncan [3], is that of deciding whether a real-valued parameter θ is less than, equal to, or greater than a specified value θ_0. This may be generated by the hypotheses $H_1: \theta \geqq \theta_0$, $H_2: \theta \leqq \theta_0$, which according to (2.1) lead to the choice between the three parameter sets $\Omega_0 = \omega_1\omega_2: \theta = \theta_0$; $\Omega_1 = \omega_1\omega_2^{-1}: \theta > \theta_0$; $\Omega_2 = \omega_1^{-1}\omega_2: \theta < \theta_0$. If the losses a and b of false rejection and acceptance are taken to be the same in the two component problems, the loss function is given by

	d_2	d_0	d_1
$\theta < \theta_0$	0	b	$a + b$
$\theta = \theta_0$	a	0	a
$\theta > \theta_0$	$a + b$	b	0

Here the greatest weight is attached to the losses resulting from taking decision d_1 when d_2 is correct and vice versa, which are neglected in the usual formulation of testing $H: \theta = \theta_0$.

In several cases of interest, for example those of a binomial or Poisson population, or of a normal population with θ either the mean or the variance where the other parameter may be unknown, the tests φ_1 and φ_2 of H_1 and H_2 depend on a common statistic T, and have rejection regions $T \leqq C_1$ for H_1 and $T \geqq C_2$ for H_2, where the constants are determined by

$$P_{\theta_0}\{T \leqq C_1\} = P_{\theta_0}\{T \geqq C_2\} = \alpha. \tag{3.2}$$

Of the intersections $\omega_1^{x_1}\omega_2^{x_2}$, only $\omega_1^{-1}\omega_2^{-1}$ is empty. The corresponding action, which would consist in the simultaneous rejection of both hypotheses, is impossible when $\alpha < \frac{1}{2}$ since then $C_1 < C_2$. Subject to this restriction, the pair of tests (φ_1, φ_2) is therefore compatible. The induced decision procedure consists simply in applying the "equal-tails" test of the hypothesis $H:\theta = \theta_0$ at level 2α and drawing the indicated conclusions. It is of interest to note that this two-sided test usually does not coincide with the standard unbiased test to which one is led by combining the two decisions d_1 and d_2. If the losses are not the same for the two hypotheses, but are a_i, b_i for H_i, equation (3.2) is replaced by

$$P_{\theta_0}\{T \leqq C_1\} = \alpha_1, \qquad P_{\theta_0}\{T \geqq C_2\} = \alpha_2,$$

and the condition for compatibility becomes $\alpha_1 + \alpha_2 < 1$.

Frequently one wishes to determine not whether θ is exactly equal to θ_0 but only whether this equality holds approximately. The corresponding three-decision problem of deciding whether $\theta < \theta_1$, $\theta_1 \leqq \theta \leqq \theta_2$ or $\theta > \theta_2$ (where $\theta_1 < \theta_0 < \theta_2$), can be generated by the hypotheses $H_2: \theta \leqq \theta_2$ and $H_1: \theta \geqq \theta_1$. The tests φ_1 and φ_2 then have the same form as before with the constants C_1, C_2 determined by

$$P_{\theta_1}\{T \leqq C_1\} = \alpha_1, \qquad P_{\theta_2}\{T \geqq C_2\} = \alpha_2.$$

They are compatible with the given decision problem if $\alpha_1 + \alpha_2 < 1$, and if the cumulative distribution function $F_\theta(c) = P_\theta\{T \leqq c\}$ is for each c a decreasing function of θ. For we then have

$$C_1 = F_{\theta_1}^{-1}(\alpha_1) < F_{\theta_2}^{-1}(\alpha_1) \leqq F_{\theta_2}^{-1}(1 - \alpha_2) = C_2,$$

and hence $C_1 < C_2$, which was seen above to be the condition for compatibility.

In the above two examples the assumptions concerning the form of the tests φ_1 and φ_2 was unnecessarily restrictive. As an illustration of a somewhat more general situation consider the case of two independent binomial variables X_1 and X_2, and the problem of deciding whether $p_2 < p_1$, $p_2 = p_1$ or $p_2 > p_1$. The best unbiased tests of the one-sided hypotheses $p_2 \leqq p_1$ and $p_2 \geqq p_1$ at level α reject the hypotheses when X_2 is too large, respectively too small, on each line segment $X_1 + X_2 = \text{const.}$, where the cutoff points are determined so that the conditional probability in each tail is equal to α. It is seen as before that this pair of tests is compatible provided $\alpha < \frac{1}{2}$. The induced three-decision procedure consists in performing the two-sided test conditionally on each line segment as an "equal tails" test at level 2α, and then making the indicated statement.

(ii) *Classification of two independent parameters.* Let ξ, η be two real-valued parameters, and consider the problem of classifying ξ and η as being $\leqq$ or $>$ than ξ_0 and $\leqq$ or $>$ than η_0 respectively, so that the choice lies between the four parameter sets $\Omega_1: \xi \leqq \xi_0, \eta \leqq \eta_0$; $\Omega_2: \xi \leqq \xi_0, \eta > \eta_0$; $\Omega_3: \xi > \xi_0, \eta \leqq \eta_0$; $\Omega_4: \xi > \xi_0, \eta > \eta_0$. This clearly can be generated by the hypotheses $H_1: \xi \leqq \xi_0$, $H_2: \eta \leqq \eta_0$. The problem of compatibility does not arise here since none of the intersections (2.1) is empty. The procedure consists in carrying out the two tests separately, and combining the results in the obvious fashion.[2] Examples in which this would be appropriate are that of a normal population $N(\xi, \sigma^2)$ where it is desired to compare ξ and σ^2 with ξ_0 and σ_0^2 respectively, or that of a bivariate normal population where both means are to be compared with certain standards.

The method applies of course equally well when more than two parameters are to be classified. An example is the preference ordering of m objects by n judges. If we assume that the judges constitute a sample from a population with probability p_{ij} of preferring object i to object j, we can test each of the hypotheses $H_{ij} : p_{ij} \leqq \frac{1}{2}$ by means of a sign test. The result of applying the set of all of these tests is a judgment concerning each pair (i, j) that either i is preferred to j, or j to i, or that neither is preferred to the other. In general one will of course not obtain a simple ordering but a complex comparison, which may be represented by a preference polygon as shown for example in Fig. 1 of [7].

(iii) *Comparing several populations.* Let samples be given from s populations with distributions depending on the parameters $\theta_1, \cdots, \theta_s$ and possibly certain nuisance parameters, which may or may not be common to the different populations, and consider the hypotheses $H_{ij} : \theta_i \leqq \theta_j$. To be specific, let the distributions be normal with means θ_i and common variance σ^2, and let the rejection region for H_{ij} be, in the usual notation, $\bar{X}_i - \bar{X}_j \leqq C_{ij}S$ with

$$C_{ij} = C[(1/n_i) + (1/n_j)]^{1/2}.$$

We shall again assume that the level of the tests is less than $\frac{1}{2}$, so that the constants C_{ij} are positive. The procedure, which is essentially the one proposed by Tukey in [14], leads to the decision $\theta_i = \theta_j$ when $| \bar{X}_i - \bar{X}_j | \leqq C_{ij}S$, and it is seen that the system is not compatible since with positive probability

$$| \bar{X}_i - \bar{X}_j | \leqq C_{ij}S, \qquad | \bar{X}_j - \bar{X}_k | \leqq C_{jk}S$$

but $| \bar{X}_i - \bar{X}_k | > C_{ik}S$ while the associated parameter sets $\theta_i = \theta_j$, $\theta_j = \theta_k$ and $\theta_i \neq \theta_k$ have an empty intersection. A justification of the resulting inconsistencies may be obtained if one interprets the acceptance of a hypothesis sufficiently loosely. For such inconsistencies occur only if at least one of the hypotheses is accepted. If, for example, H_{ij} and H_{jk} are both rejected, we have

$$\bar{X}_i - \bar{X}_k > (C_{ij} + C_{jk})S > C_{ik}S.$$

Therefore, H_{ik} is then also rejected corresponding to the fact that $\theta_i > \theta_j$, $\theta_j > \theta_k$ implies $\theta_i > \theta_k$.

A perhaps more satisfactory solution is obtained if one replaces the hypotheses H_{ij} by $H'_{ij} : \theta_i \leqq \theta_j + \Delta$, $(\Delta > 0)$, with rejection regions $\bar{X}_i - \bar{X}_j > C'_{ij}S$, where the constants are determined so that the probability of rejection is α when $\theta_i = \theta_j + \Delta$. It is easily checked that this system is compatible since H'_{ik} may be false when both H'_{ij} and H'_{jk} are true. Each difference $\theta_i - \theta_j$ is

[2] A different justification of the resulting acceptance region for the combined hypothesis was given by the author in [10].

now classified as being either $< -\Delta$, between $-\Delta$ and Δ, or $> \Delta$. Since the significance statement obtained for the differences $\theta_i - \theta_j$ in this manner are self-consistent, they lead to a classification of the s populations of the kind described by Duncan in Section 3 of [3].

As another example consider samples of equal size from $N(\xi_i, \sigma_i^2)$, and the problem of classifying the populations according to their variances. This can be generated by the hypotheses $H_{ij}: \sigma_i \leqq \sigma_j$, for which the rejection regions are $S_i/S_j > C$. In particular the decision that all of the σ's are equal is taken when $\max(S_i/S_j) \leqq C$, so that the present procedure constitutes a refinement of the test for equality of variances discussed by Hartley [6]. For the same reason as in the preceding example the system is incompatible. But for $\alpha < \frac{1}{2}$ inconsistencies can again occur only if at least one acceptance is involved since the rejection of both H_{ij} and H_{jk} implies $S_i/S_k > C^2 > C$, and hence the rejection of H_{ik}. As before the inconsistencies may be avoided altogether by replacing the H_{ij} by the hypotheses $H'_{ij}: \sigma_i \leqq \delta\sigma_j$ $(\delta > 1)$, and in this way one obtains, as in the case of the means, a satisfactory classification procedure for the variances.

4. Estimation. (i) *Point estimation.* Let θ be a continuous real-valued parameter, and consider the decision problem generated by the set of hypotheses $H(\theta_0): \theta \leqq \theta_0$. If θ^* denotes the true value of the parameter, the hypotheses $H(\theta_0)$ with $\theta^* \leqq \theta_0$ are true while those with $\theta^* > \theta_0$ are false. The associated intersection (2.1) is therefore

$$\bigcap_{\theta_0 \geqq \theta^*} \omega(\theta_0) \cap \bigcap_{\theta_0 < \theta^*} \omega^{-1}(\theta_0). \tag{4.1}$$

Since $\omega(\theta_0)$ is the interval $\theta \leqq \theta_0$, the set (4.1) consists of the single point θ^*, or in the case that nuisance parameters are present, of the totality of points for which $\theta = \theta^*$. In the induced multiple decision problem the possible decisions therefore correspond exactly to the possible true values of θ, that is, the problem is one of point estimation.

Suppose now, as in Section 2, that the tests of $H(\theta_0)$ are nonrandomized. If then $A(\theta_0)$ is the acceptance region for $H(\theta_0)$, a necessary and sufficient condition for compatibility is that, except on a null set,

$$A(\theta_0) \subseteq A(\theta_1), \quad \text{whenever} \quad \theta_0 \leqq \theta_1, \tag{4.2}$$

and

$$\bigcap_{\theta > \theta_0} A(\theta) = A(\theta_0). \tag{4.3}$$

That this condition is necessary is obvious since the corresponding relationships do hold for the $\omega(\theta)$'s. To prove sufficiency one must, by the criterion given at the end of Section 2, show that each sample point lies in one of the intersections (4.1) with $\omega(\theta)$ replaced by $A(\theta)$. Consider now the set of θ's for which the sample point is in $A(\theta)$, and let $\hat{\theta}$ be its greatest lower bound. Then by (4.2) and (4.3), $H(\theta_0)$ is accepted for $\theta_0 \geqq \hat{\theta}$ and rejected for $\theta_0 < \hat{\theta}$, and hence the sample point lies in the intersection

$$\bigcap_{\theta_0 \geqq \hat{\theta}} A(\theta_0) \cap \bigcap_{\theta_0 < \hat{\theta}} A^{-1}(\theta_0),$$

as was to be proved. The decision taken in this case is that θ lies in the set

$$\bigcap_{\theta_0 \geqq \hat{\theta}} \omega(\theta_0) \cap \bigcap_{\theta_0 < \hat{\theta}} \omega^{-1}(\theta_0) = \{\hat{\theta}\},$$

that is, that θ equals $\hat{\theta}$, which therefore is a point estimate of θ. The relationship

$$\hat{\theta} \leqq \theta_0 \Leftrightarrow X \varepsilon A(\theta_0)$$

shows furthermore that measurability of the sets $A(\theta_0)$ implies that of the function $\hat{\theta}$, and conversely. One also sees from it that $\hat{\theta}$ is a lower confidence limit for θ with confidence coefficient $1 - \alpha$, if all the tests are carried out at level α.

If $\theta^* < \hat{\theta}$, the hypothesis $H(\theta_0)$ is incorrectly rejected for $\theta^* \leqq \theta_0 < \hat{\theta}$ and never falsely accepted. Thus in accordance with (2.4) the loss may be taken as $a(\hat{\theta} - \theta^*)$, where the loss for false rejection is assumed to be a for all of the hypotheses $H(\theta)$. Similarly, when $\hat{\theta} < \theta^*$, the loss is $b(\theta^* - \hat{\theta})$. The loss is therefore the absolute error, multiplied by a or b as θ is an over- or underestimate. If these two kinds of error are considered of equal importance, the loss is simply proportional to the absolute error.

If the losses a and b are the same for the different hypotheses $H(\theta)$, and if (3.1) is assumed to hold, then the different tests must be carried out at a constant level of significance α. Under this assumption, the optimum one-sided tests satisfy (4.2) and (4.3) in many of the standard problems, in particular when one is dealing with an exponential family of distributions. On the other hand, these conditions may also hold in cases in which the losses a and b and hence also the level α at which $H(\theta)$ is tested, vary with θ. As an example suppose that the tests have acceptance regions of the form $T \leqq C(\theta)$, where the c.d.f. $F_\theta(c) = P_\theta\{T \leqq c\}$ is for each c a continuous and decreasing function of θ. Then $C(\theta) = F_\theta^{-1}[\alpha(\theta)]$, and conditions (4.2) and (4.3) are satisfied provided $C(\theta)$ is an increasing function of θ, which is continuous on the right, or equivalently if $\alpha(\theta)$ is decreasing and continuous on the right.

Slightly more generally, one can take the losses for over- and underestimation to be

$$a \int_{\theta^*}^{\theta} d\mu(\theta), \qquad b \int_{\theta^*}^{\hat{\theta}} d\mu(\theta), \tag{4.4}$$

with μ not necessarily Lebesgue measure. In case of a scale parameter, for example, an appropriate loss function may be given by

$$a \int_{\theta^*}^{\theta} \frac{1}{\theta}\, d\theta = a \log (\hat{\theta}/\theta^*), \qquad b \int_{\theta}^{\theta^*} \frac{1}{\theta}\, d\theta = b \log (\theta^*/\hat{\theta}).$$

(ii) *Point estimation after a preliminary test of significance.* It is frequently of interest to obtain a point estimate of a parameter θ after one has tested, and rejected, some hypothesis concerning it. If for example a new treatment is being compared with a standard one, the hypothesis may be tested that the new treatment does not represent an improvement, that is, that $\theta = \eta - \xi \leqq 0$, where η and ξ denote the means of the new and old treatments. In case the hypothesis is rejected one requires an estimate of $\eta - \xi$.

A procedure for testing the hypothesis $H:\theta \leqq \theta_0$, and estimating θ in case of rejection, can be generated by the set of hypotheses $H(\theta_1):\theta \leqq \theta_1$, with $\theta_1 \geqq \theta_0$. If the tests are carried out at a constant level α, the procedure consists in performing the usual test of $H(\theta_0)$ and in case of rejection estimating θ by the estimate $\hat{\theta}$ of (i), that is, by the lower confidence limit corresponding to confidence coefficient $1 - \alpha$. A drawback of this method is the limitation it imposes on the levels $\alpha(\theta)$. The conservative attitude reflected by the customary choice of a small level α for testing H suggests that also in the estimation part of the problem an overestimate should be considered more serious than the corresponding underestimate. However, one would usually still wish to test H at a lower level than is desirable for the construction of the estimate. Unfortunately such a combination of levels leads to an incompatible procedure.

In some cases a procedure with the desired properties can be obtained by a slight modification of the construction given above. To illustrate the method consider a single normal variable X with mean θ and unit variance. The hypothesis $H:\theta \leqq 0$ is to be tested at level $\alpha < \frac{1}{2}$ with the acceptance region $X \leqq C$ and in case of rejection θ is to be estimated by X which corresponds to the level $\frac{1}{2}$. This may be generated by the family of hypotheses $H(\theta_1):\theta \leqq \theta_1$, with $\theta_1 = 0$ and $\theta_1 \geqq C$, at the levels $\alpha(0) = \alpha$, $\alpha(\theta_1) = \frac{1}{2}$ for $\theta_1 \geqq C$. In a similar manner one can generate a joint testing and estimation procedure, in which the level for the estimation part of the problem is higher than that of the test, in the case of a binomial or Poisson variable. Another example in which this is possible is that of the ratio τ^2/σ^2 of two variances (for example in components of variance problems), where one wishes to test $H:\tau^2/\sigma^2 \leqq k$, and in case of rejection requires a point estimate of the variance ratio. The method, however, does not appear to be applicable without further modification to the case of a sample $X_1, \cdots, X_n$ from $N(\xi, \sigma^2)$ on the basis of which one wishes to test $H:\xi \leqq 0$ and in case of rejection to use, say $\bar{X}$, as an estimate of ξ. While this problem may be generated by the class of hypotheses $N(\xi_1):\xi \leqq \xi_1$ with $\xi_1 = 0$ and $\xi_1 \geqq CS$, this class depends on the random variable S, and can therefore be determined only after the observations have been taken.

The indicated difficulty usually does not exist if the estimation problem arises when H is accepted rather than when it is rejected. An example of this occurs when one wishes to test the hypothesis that a drug has a significant toxic effect ($H:\theta \geqq \theta_0$), and in case H is accepted wants to estimate the size of this effect.

5. Some nonparametric problems. (i) *Testing for goodness of fit.* Let

$$X_1, \cdots, X_n$$

be independently distributed with cumulative distribution function F, and consider the problem of deciding whether $F = F_0$, or, if this is judged not to be the case, of determining the sets of points u for which $F(u)$ is $<$, $=$, and $>F_0(u)$. This problem may be generated by the set of hypotheses $H_+(u):F(u) \geqq F_0(u)$ and $H_-(u):F(u) \leqq F_0(u)$. If in order to be specific we assume F_0 to be continuous and strictly increasing, the set of u for which the true $F(u)$ exceeds $F_0(u)$ is a union of intervals each of which is open on the right. It is necessary for compatibility that the corresponding condition hold for the set of u at which $F(u)$ is judged to exceed $F_0(u)$, that is, for the set of u for which the sample point is in $A_+(u) \cap A_-^{-1}(u)$. A similar condition must be satisfied by the sets $A_+^{-1}(u) \cap A_-(u)$. These conditions are clearly also sufficient for compatibility since given any two such unions of intervals, there exists a cumulative distribution function F which is in the desired relationship to F_0.

The best unbiased tests of the hypotheses $H_+(u)$ and $H_-(u)$ are the appropriate one-sided sign tests, which reject the hypotheses if the number $X(u)$ of observations $\leqq u$ satisfies

$$X(u) < a(u) \quad \text{and} \quad X(u) > b(u), \tag{5.1}$$

respectively. In order to achieve desired levels of significance $\alpha_+(u)$ and $\alpha_-(u)$ it may be necessary to introduce an auxiliary random variable Z, distributed uniformly on (0, 1), and to reject $H_+(u)$ and $H_-(u)$ as

$$X(u) + Z < a(u) \quad \text{and} \quad X(u) + Z > b(u). \tag{5.2}$$

Unfortunately the usual choice of levels, $\alpha_+(u) = \alpha_-(u) = \alpha$, is not satisfactory for the present problem. In fact, with this choice the tests (5.2) will always lead to rejection for u sufficiently large and sufficiently small respectively. The difficulty stems from the circumstance that for sufficiently extreme u, $X(u)$ tends to the sure variable n or 0 and hence contains no information, so that the decisions in the extreme tails depend solely on the value of Z. Since one is usually not even particularly interested in the behavior of F in the extreme tails, it is natural to avoid this difficulty by choosing $\alpha_+(u)$ and $\alpha_-(u)$ in such a way that they tend to 0 as u tends to $\pm\infty$. This will be the case for example if in (5.1) one sets

$$a(u) = F_0(u) - \Delta \quad \text{and} \quad b(u) = F_0(u) + \Delta,$$

so that the acceptance of all of the hypotheses simultaneously reduces to that of Kolmogoroff's test of the hypothesis $F = F_0$.

One obtains a completely analogous problem, only without the complications caused by the behavior in the tails, if the observations are grouped. The procedure will then decide for each interval whether the hypothesis $p_i = p_{i0}$ is to be accepted or whether the observed frequency in the ith interval indicates that p_i exceeds or falls short of its hypothetical value. This is a special case of the classification problems considered in (ii) of Section 3.

(ii) *The two-sample problem.* The problem of deciding whether two unknown cumulative distribution functions F and G are equal, or in the contrary case of determining for each u whether $G(u)$ is $<$, $=$, or $>$ $F(u)$, may be generated by the hypotheses $H_+(u){:}G(u) \geqq F(u)$ and $H_-(u){:}G(u) \leqq F(u)$. Let $X_1, \cdots, X_m$, and $Y_1, \cdots, Y_n$ be samples from F and G, and denote by $X(u)$ and $Y(u)$ the number of observations in these samples that are $\leqq u$. The appropriate tests of $H_+(u)$ and $H_-(u)$ are the standard one-sided tests for equality of two binomial distributions, and with $\alpha_+(u) = \alpha_-(u) = \alpha < \frac{1}{2}$ they are clearly compatible. This choice of levels, as in the previous case, puts the weight of the decision in the tails on an irrelevant random experiment. But this is less serious in the present problem since the very small value of α that is required for satisfactory error control implies that one will only rarely reject the hypotheses in the tails, where $X(u) = Y(u) = 0$ or $X(u) = m$, $Y(u) = n$.

(iii) ***Estimating a cumulative distribution function.*** Let $X_1, \cdots, X_n$ be independently distributed with cumulative distribution function F, and consider the hypotheses $H(u_0, p_0){:}F(u_0) \leqq p_0$. As in Section 4(i), if F^* denotes the true c.d.f., the hypotheses $H(u_0, p_0)$ for which $F^*(u_0) \leqq p_0$ are true, and those with $F^*(u_0) > p_0$ are false. The associated intersection (2.1) is therefore

$$\bigcap_{\{(u_0,p_0):p_0 \geqq F^*(u_0)\}} \omega(u_0, p_0) \bigcap_{\{(u_0,p_0):p_0 < F^*(u_0)\}} \omega^{-1}(u_0, p_0). \tag{5.3}$$

Since $\omega(u_0, p_0)$ is the set of all F for which $F(u_0) \leqq p_0$, the first and second member of (5.3) are the sets of all F satisfying $F(u) \leqq F^*(u)$ for all u and $F(u) \geqq F^*(u)$ for all u, respectively. The set (5.3) therefore contains as its only element the c.d.f. F^*.

It is seen that for each fixed u_0, if we set $\theta = F(u_0)$, we are dealing with the problem of Section 4(i), so that in particular the family of sign tests of the hypotheses $H(u_0, p)$ based on the binomial variable $X(u_0)$ leads to the estimate $\hat{\theta} = \hat{F}(u_0)$ derived there. However, for compatibility one must now add the requirement that as u varies the $\hat{F}(u)$ should constitute a c.d.f. This condition is violated in a rather trivial way if one puts $\alpha(u, p) \equiv \alpha$. For when $X(u)$ is 0 or n, the estimate $\hat{F}(u)$ is not 0 and 1 but only close to these values. One can achieve compatibility by putting $\alpha(u, p) = \alpha(u)$, and letting $\alpha(u)$ tend to 0 as u tends to ∞, and 1 as u tends to $-\infty$. Since it is enough to make this change in the extreme tails, it need not affect the result in practice.

6. Restricted products of decision problems. The method, described in Section 2, of generating a multiple decision problem from a set of hypotheses is a special case of the following process. Consider the definition of a general decision problem in terms of a family of distributions $\mathcal{P} = \{P_\theta, \theta \varepsilon \Omega\}$, a space of possible decisions $\mathfrak{D} = \{d\}$ and a loss function $W(\theta, d)$. Suppose that $\mathcal{P}$ is fixed but that two different decision spaces $\mathfrak{D}'$, $\mathfrak{D}''$ with the loss functions W', W'' are of interest. From the two associated decision problems one can form a new problem, which consists in the simultaneous consideration of the two given ones, and may be termed their *product*. Its decision space is the Cartesian product $\mathfrak{D}' \times \mathfrak{D}''$ and the loss resulting from the decision $d = (d', d'')$ is

$$W(\theta, d) = W'(\theta, d') + W''(\theta, d''), \tag{6.1}$$

or slightly more generally $W(\theta, d) = \rho W'(\theta, d') + (1 - \rho)W''(\theta, d'')$. A typical illustration is the first example of Section 1, where the component problems are concerned with the classification of ξ, respectively η, as negative, zero, or positive, and where the product problem is that of simultaneously classifying both parameters.

This concept is however not general enough to cover most of the other problems considered in the previous sections. Consider for example the two hypotheses $H':\theta \leqq \theta_0$ and $H'':\theta \geqq \theta_0$. If we denote the decisions to accept H' and H'' by d_0' and d_0'' and the decisions to reject by d_1' and d_1'', the product problem offers the choice of the four decisions (d_0', d_0''), (d_0', d_1''), (d_1', d_0''), (d_1', d_1''). Of these the first corresponds to the parameter point $\theta = \theta_0$, the second and third to the sets $\theta < \theta_0$ and $\theta > \theta_0$, while the last one combines two inconsistent decisions and hence corresponds to an empty set in the parameter space. In order to obtain the problem of choosing only between the first three of these possibilities, one must eliminate the point (d', d'') from the decision space $\mathfrak{D}$. In general, we shall speak of a *restricted product* if in a product problem some of the decision pairs (d', d'') are omitted from $\mathfrak{D}' \times \mathfrak{D}''$, so that $\mathfrak{D}$ is a subset of $\mathfrak{D}' \times \mathfrak{D}''$. Given any procedures δ', δ'' for the problems with decision spaces $\mathfrak{D}'$ and $\mathfrak{D}''$, let $\delta = (\delta', \delta'')$ be the procedure that takes decision (d', d'') when $\delta' = d'$ and $\delta'' = d''$. In conformance with our earlier terminology we shall say that the pair (δ', δ'') is *compatible* with the given set of restrictions if

$$P_\theta\{(\delta'(X), \delta''(X)) \in \mathfrak{D}\} = 1 \qquad \text{for all } \theta,$$

that is, if the probability is zero of the procedure (δ', δ'') leading to one of the forbidden elements of $\mathfrak{D}' \times \mathfrak{D}''$. Under suitable measurability conditions there is then again a 1:1 correspondence between compatible pairs of decision procedures for the component problems and decision procedures for the restricted product problem. The proof is exactly as that of Theorem 1.

$$\left[\frac{f(x, \theta_1)}{f(x, \theta_0)}\right]^\rho \left[\frac{f(x, \theta_{-1})}{f(x, \theta_0)}\right]^{1-\rho} = \left[\frac{r(\theta_1)}{r(\theta_0)}\right]^\rho \left[\frac{r(\theta_{-1})}{r(\theta_0)}\right]^{1-\rho} = h. \tag{4.12}$$

Hence, for each m and all $x^{(m)}$,

$$\min\left[\frac{p_{1m}(x^{(m)})}{p_{0m}(x^{(m)})}, \quad \frac{p_{-1m}(x^{(m)})}{p_{0m}(x^{(m)})}\right] \leqq h^{m\epsilon} \tag{4.13}$$

where $\epsilon = \min\,[1/2\rho, 1/2(1 - \rho)]$. Equation (4.10) follows at once from equation (4.13). This completes the proof of Lemma 4.2.

We now prove some consequences of Assumptions A, B, and C. In what follows $\xi^{(0)}$ is a fixed a priori probability measure all of whose components are positive. The reader should recall the italicized statement a few lines above Assumption A. Write $D^{(0)} = \xi_{-1}^{(0)}/\xi_1^{(0)}$, where $\xi^{(0)} = (\xi_{-1}^{(0)}, \xi_0^{(0)}, \xi_1^{(0)})$.

CORRECTION

Page 14, formula (4.12) through page 17, line 23 should be exchanged with page 70, line 8 through page 72, next-to-last line.

LEMMA 4.3. *Under Assumption* A *there exist positive constants* b_m *and* a_m $(m = 0, 1, 2, \cdots)$ *with* $b_m \leqq a_m$ *and such that*

$$\xi^{(m)} \varepsilon C_1 \quad \text{if and only if} \quad p_{1m}(x^{(m)}) / p_{-1m}(x^{(m)}) \geqq a_m, \tag{4.14}$$

$$\xi^{(m)} \varepsilon C_{-1} \quad \text{if and only if} \quad p_{1m}(x^{(m)}) / p_{-1m}(x^{(m)}) \leqq b_m, \tag{4.15}$$

with strict inequality holding if and only if $\xi^{(m)}$ *is an interior point of the appropriate* C_i.

(Of course, the values a_m, b_m depend on $\xi^{(0)}$.)

PROOF: The method of proof is similar to that of Theorem 3. Let $x^{(m)}$ and $y^{(m)}$ be such that (4.3) holds, and let $\xi^{(m)}(x^{(m)})$, $\xi^{(m)}(y^{(m)})$ be the a posteriori probability measures corresponding to observed values $x^{(m)}$, $y^{(m)}$. Equation (4.15) will follow if we can show that $\xi^{(m)}(x^{(m)}) \varepsilon C_{-1}$ implies $\xi^{(m)}(y^{(m)}) \varepsilon C_{-1}$. (The reader will be aided in what follows if he draws a picture.) Now, (4.3) says that the line $V_0\xi^{(m)}(y^{(m)})$ lies toward V_{-1} from (or on) the line $V_0\xi^{(m)}(x^{(m)})$. Moreover, (4.3) implies (4.4) and (4.5), which say that the line $V_1\xi^{(m)}(y^{(m)})$ lies toward V_{-1} from (or on) the line $V_1\xi^{(m)}(x^{(m)})$ and that the line $V_{-1}\xi^{(m)}(y^{(m)})$ lies toward V_0 from (or on) the line $V_{-1}\xi^{(m)}(x^{(m)})$. Hence, $\xi^{(m)}(y^{(m)})$ lies inside or on the triangle T whose vertices are V_{-1}, $\xi^{(m)}(x^{(m)})$, and the intercept of $V_1\xi^{(m)}(x^{(m)})$ with $V_{-1}V_0$. Since T is contained in the triangle $V_0\xi^{(m)}(x^{(m)})V_{-1}$ which (by convexity) is contained in C_{-1}, (4.15) is proved. Moreover, since the last part of Assumption A implies that $\xi^{(m)}(y^{(m)})$ could lie on the line $V_{-1}\xi^{(m)}(x^{(m)})$ only if $\xi^{(m)}(y^{(m)})$ is V_{-1} or $\xi^{(m)}(x^{(m)})$, it is clear that $\xi^{(m)}(y^{(m)})$ is a boundary point of C_{-1} if and only if either $\xi^{(m)}(y^{(m)}) = V_{-1}$ (see the italicized remark a few lines above Assumption A in this case) or else $\xi^{(m)}(x^{(m)})$ is a boundary point of C_{-1} and $\xi^{(m)}(y^{(m)}) = \xi^{(m)}(x^{(m)})$; the latter implies equality in (4.3). Thus (modulo the italicized remark), defining b_m^* to be (for fixed $\xi^{(0)}$) the supremum (over $x^{(m)}$) of those values $p_{1m}(x^{(m)}) / p_{-1m}(x^{(m)})$ for which $\xi^{(m)}(x^{(m)}) \varepsilon C_{-1}$, and taking $b_m = b_m^*$ if $\xi^{(m)}(x^{(m)}) \neq V_{-1}$ is on the boundary of C_{-1} for some $x^{(m)}$ and $b_m^* < b_m <$ infimum of those $p_{1m}(x^{(m)}) / p_{-1m}(x^{(m)})$ for which $\xi^{(m)}(x^{(m)}) \not\varepsilon C_{-1}$ otherwise, we see from the previous sentence and the fact that $p_{1m}(y^{(m)}) / p_{-1m}(y^{(m)}) < p_{1m}(x^{(m)}) / p_{-1m}(x^{(m)})$ if $\xi^{(m)}(x^{(m)}) \neq \xi^{(m)}(y^{(m)}) \varepsilon T$, that the last part of the lemma as it applies to (4.15) is proved. Equation (4.14) (and the corresponding last part) is proved similarly.

LEMMA 4.4. *Under Assumptions* A *and* B, *there exist constants* $b_m \leqq D^{(0)} \leqq a_m$ *of Lemma* 4.3 *satisfying* $b_m \leqq b_{m+1}$, $a_m \geqq a_{m+1}$, *for* $m = 0, 1, 2, \cdots$.

PROOF: We shall prove the assertion regarding the b_m, a similar proof applying for the a_m. Keeping $\xi^{(0)}$ fixed as before, in order to prove $b_m \leqq b_{m+1}$ it clearly suffices to prove that $\xi^{(m)} \varepsilon C_{-1}$ and $p_{1,m+1}(x^{(m+1)}) / p_{-1,m+1}(x^{(m+1)}) \leqq p_{1m}(x^{(m)}) / p_{-1,m}(x^{(m)})$ imply that $\xi^{(m+1)} \varepsilon C_{-1}$ (the case where either ratio is 0/0 or where both are 0 is easily disposed of); i.e., that $\xi^{(m)} \varepsilon C_{-1}$ and $f_1(x_{m+1}) \leqq f_{-1}(x_{m+1})$ imply $\xi^{(m+1)} \varepsilon C_{-1}$. The last inequality says that the line $V_0\xi^{(m+1)}$ lies toward V_{-1} from (or on) the line $V_0\xi^{(m)}$; by (4.8) and (4.9), it implies that the line $V_{-1}\xi^{(m+1)}$ lies toward V^0 from (or on) the line $V_{-1}\xi^{(m)}$. Thus, $\xi^{(m+1)}$ lies in the triangle

$V_0\xi^{(m)}V_{-1}$ and hence, by convexity, $\xi^{(m+1)} \varepsilon C_{-1}$. The remaining part of the lemma follows at once from the fact that $\xi_1^{(m)} \leqq \xi_{-1}^{(m)}$ is equivalent to $p_{1m}/p_{-1m} \leqq D^{(0)}$.

If $f_1(x)/f_{-1}(x)$ cannot take on a suitably dense set of values, the a_m and b_m might (for fixed $\xi^{(0)}$) not be unique and might correspond to $\xi^{(m)}$ in the interior of the C_i or the complement of $C_{-1} \cup C_1$. If this is not the case, the previous paragraph and the fact that $V_0PV_{-1} \varepsilon C_{-1}$ show that we can strengthen the weak inequality of Lemma 4.4. One possible formulation of this result is the following:

LEMMA 4.5. *If $b_m < D^{(0)}$ (resp., $a_m > D^{(0)}$) and if for every open interval J containing b_m (resp., a_m) the ratio $p_{1m}(x^{(m)})/p_{-1,m}(x^{(m)})$ takes on values in $J - \{b_m\}$ (resp., $J - \{a_m\}$) with positive probability under H_1 and H_{-1} (so that b_m, a_m are unique), and if for every open interval J' containing $D^{(0)}$ as a left (resp., right) end-point $f_1(x)/f_{-1}(x)$ takes on values in J' with positive probability under H_1 and H_{-1}, then $b_m < b_{m+1}$(resp., $a_m > a_{m+1}$). In particular, in case* (3) *of Lemma* 4.2, *if μ is equivalent to Lebesgue measure (or if Lebesgue measure is absolutely continuous with respect to μ) on the real line, the a_m and b_m are unique and this last result holds.*

In fact, it remains only to prove the last assertion of the lemma, which follows at once from the fact that $e^{(\theta_1-\theta_{-1})x}$ takes on values in any interval of positive numbers with positive probability under H_1 and H_{-1}, if Lebesgue measure is absolutely continuous with respect to μ.

LEMMA 4.6. *Under Assumption* C, *for any $\xi^{(0)}$ all of whose components are (or in fact, for which $\xi_0^{(0)}$ is) positive, there is an integer $N = N(\xi^{(0)})$ such that every Bayes solution with respect to $\xi^{(0)}$ requires fewer than N observations with probability one under f_1, f_{-1}, and f_0.*

PROOF: Fix $\xi^{(0)}$. Since P is a positive distance from V_0, there is clearly a positive number c such that every Bayes solution must stop with probability one whenever either $p_{1m}(x^{(m)})/p_{0m}(x^{(m)}) < c$ or else $p_{-1m}(x^{(m)})/p_{0m}(x^{(m)}) < c$. The desired result now follows at once from (4.10). (Note again the remark made in italics just before Assumption A).

We may now summarize our results:

THEOREM 4. *Under Assumptions* A, B, *and* C (*in particular, under* (1), (2), *or* (3) *of Lemma* 4.2), *any procedure which minimizes $A_0(\delta)$ subject to* (4.1) *is a* GSPRT *of H_1 against H_{-1} with $b_m \leqq b_{m+1} \leqq D^{(0)} \leqq a_{m+1} \leqq a_m$ for $m = 0, 1, 2, \cdots, N$ and some $D^{(0)}$, which stops with probability one under $f_i (i = 0, \pm 1)$ after N or fewer observations. Under additional conditions specified in Lemma* 4.5 *the values a_m, $b_m (m \geqq 1)$ will be unique and $a_m > a_{m+1}$ or $b_m < b_{m+1}$ unless $a_m = D^{(0)}$ or $b_m = D^{(0)}$, where $D^{(0)}$ corresponds to the a priori distribution with respect to which the optimum procedure is Bayes.*

REMARKS, GENERALIZATIONS, ETC.

1. Of course, a GSPRT of Theorem 4 involves a randomization rule for all $m \leqq N$, including a possibly randomized starting rule ($m = 0$) if a_0 or $b_0 = D^{(0)}$. If μ is nonatomic, there will clearly exist an optimum GSPRT involving no randomization, except possibly in the starting rule. The lack of uniqueness of

the a_m and b_m in cases not covered by Lemma 4.5 is of course inessential, reflecting only that certain intervals of values of p_{1m}/p_{-1m} have probability zero under all f_i.

2. In all of the above, the X_i are random elements whose range is immaterial (not necessarily real) as long as the appropriate assumptions are satisfied. To conserve space we have not included statements about the obvious sets of measure zero where various conditions may be permitted to fail.

3. As an example of what can happen when our assumptions are not satisfied, we mention briefly the following example: Suppose $f_j(x) = 1 / \pi[1 + (x - j)^2]$, $j = 0, \pm 1$, μ = Lebesgue measure. In this Cauchy case it is easy to see that, for $\xi^{(0)}$ with all components positive, the set of possible $\xi^{(1)}$ values is a simple closed curve minus the point $\xi^{(0)}$, and lies entirely in the interior of the triangle $V_0V_1V_{-1}$. Assumption A is not satisfied, and there is no reason why the result of Lemma 4.3. should be valid. Also, since (e.g.) $f_j(\sqrt{3/2})f_j(-\sqrt{3/2})$ does not depend on j, there is no reason why the result of Lemma 4.6 should hold here.

4. Remarks analogous to those of Section 2 can be made here: for concave (resp., convex) nondecreasing $c(n)$, the minimization of a linear combination such as (4.2) with $A_0(\delta)$ replaced by $E_0c(n)$ under δ may be compared in an obvious fashion to the minimization when $c(n)$ is replaced by the linear homogeneous cost function $c_m^*(n)$ passing through $(1, c(m + 1) - c(m))$ (i.e., to the solution of the problem we have considered): the stopping region will now change with m, being contained in (resp., containing) that fixed region for the problem concerning $c_m^*(n)$.

To prove the procedures unbiased, we note that unbiasedness of a two-decision procedure φ with losses a and b, by (6.3) is equivalent to

$$E_\theta\varphi(X) \begin{matrix} \leqq \\ \geqq \end{matrix} b/(a + b) \quad \text{for} \quad \begin{matrix} \theta \in \omega \\ \theta \in \omega^{-1} \end{matrix}, \tag{7.4}$$

that is, to the Neyman-Pearson condition of unbiasedness at the level

$$\alpha = b/(a + b). \tag{7.5}$$

The result now follows from (iv) of Section 6 since in all of the examples the procedures were obtained as products of unbiased tests.

Unfortunately, as has already been pointed out, it is in general not true that unbiasedness of a product implies the same property for the component problems. Suppose however that for every test of $H: \theta \in \omega$, the power function $E_\theta\varphi(X)$ is a continuous function of θ. Then unbiasedness of φ entails the somewhat weaker condition of *similarity on the boundary*, namely

$$E_\theta\varphi(X) = \alpha \quad \text{for} \quad \theta \in \Lambda, \tag{7.6}$$

where Λ is the common boundary of ω and ω^{-1}. For an important class of testing problems, there exists not only among all unbiased tests but also among the larger class of tests satisfying (7.6), one that uniformly maximizes $E_\theta\varphi(X)$ for $\theta \in \omega^{-1}$ and uniformly minimizes it for $\theta \in \omega$. This test therefore, among all those that are similar on the boundary, uniformly minimizes the risk (7.1). As was shown in

CORRECTION

Page 14, formula (4.12) through page 17, line 23 should be exchanged with page 70, line 8 through page 72, next-to-last line.

[11], this is the case in particular when $\theta = (\theta_1, \cdots, \theta_s)$, the distributions of Ω form an exponential family and ω is of the form $\theta_1 \leqq \theta_1^0$ or $\theta_1 \geqq \theta_1^0$.

The desired optimum property of the various procedures discussed in the earlier sections, for finite Γ is an easy consequence of the above remarks and the following theorem.

THEOREM 2. *Let $\{H_\gamma : \theta \varepsilon \omega_\gamma, \gamma \varepsilon \Gamma\}$ be a finite family of hypotheses, and suppose that for each γ the test φ_γ^0 uniformly minimizes the risk among all tests that are similar on the boundary at level $\alpha_\gamma = b_\gamma/(a_\gamma + b_\gamma)$, and that the family $\{\varphi_\gamma^0, \gamma \varepsilon \Gamma\}$ is compatible. Suppose further that the following structural assumption is satisfied.*

(*) *For every $\gamma_0 \varepsilon \Gamma$ and each common boundary point θ of ω_{γ_0} and $\omega_{\gamma_0}^{-1}$ there exist intersection sets Ω_i and Ω_j of the form (2.1) such that θ is also a common boundary point of Ω_i and Ω_j, and such that $x_{i\gamma} = x_{j\gamma}$ for all $\gamma \neq \gamma_0$ but that $x_{i\gamma_0} \neq x_{j\gamma_0}$.*

Then if $E_\theta \varphi_\gamma(X)$ is continuous in θ for each γ, the product procedure ψ^0 given by

$$\psi_k^0 = \prod_{\gamma \varepsilon \Gamma} (\varphi_\gamma^0)^{x_{k\gamma}}$$

is unbiased, and uniformly minimizes the risk among all unbiased decision procedures of the restricted product problem, the components of which are the problems of testing H_γ with losses a_γ and b_γ.

PROOF. Let $\mathcal{C}$ be the class of all decision procedures the component tests of which are similar on the boundary at level α_γ. It follows from the assumptions made and from (iii) of Section 6 that the procedure ψ^0 uniformly minimizes the risk within $\mathcal{C}$. Since the tests φ_γ^0 are unbiased—as is seen by comparison with the tests $\varphi_\gamma(x) \equiv \alpha_\gamma$—the same is true of ψ^0. Let $\mathcal{C}_0$ denote the class of all unbiased procedures of the product problem. We shall now show that $\mathcal{C}_0 \subseteq \mathcal{C}$, which will complete the proof.

Let ψ be any procedure belonging to $\mathcal{C}_0$, let γ_0 be any element of Γ and θ_0 any boundary point of ω_{γ_0} and $\omega_{\gamma_0}^{-1}$. Let Ω_i and Ω_j be the sets, the existence of which is guaranteed by (*), and assume without loss of generality that $x_{i\gamma_0} = 1$. Then unbiasedness of ψ implies that for any θ in Ω_i,

$$E_\theta \sum\nolimits_\gamma \tfrac{1}{2}[(x_{i\gamma} + 1)a_\gamma \varphi_\gamma(X) - (x_{i\gamma} - 1)b_\gamma \varphi_\gamma^{-1}(X)]$$
$$\leqq E_\theta \sum\nolimits_\gamma \tfrac{1}{2}[(x_{j\gamma} + 1)a_\gamma \varphi_\gamma(X) - (x_{j\gamma} - 1)b_\gamma \varphi_\gamma^{-1}(X)],$$

where the φ_γ are the component tests of ψ. Since $x_{i\gamma} = x_{j\gamma}$ for $\gamma \neq \gamma_0$, $x_{i\gamma_0} = 1$, $x_{j\gamma_0} = -1$, this reduces to

$$a_{\gamma_0} E_\theta \varphi_{\gamma_0}(X) \leqq b_{\gamma_0} E_\theta[1 - \varphi_{\gamma_0}(X)].$$

Analogously the opposite inequality is seen to hold for any θ in Ω_j. Because of the continuity of $E_\theta \varphi_{\gamma_0}(X)$ it follows that equality must hold on the boundary of Ω_i and Ω_j, and hence in particular for $\theta = \theta_0$. Thus

$$E_\theta \varphi_{\gamma_0}(X) = \frac{b_{\gamma_0}}{a_{\gamma_0} + b_{\gamma_0}} = \alpha_{\gamma_0} \tag{7.7}$$

for every boundary point of ω_{γ_0} and $\omega_{\gamma_0}^{-1}$, as was to be proved.

The assumptions of this theorem may be weakened slightly at one point, which is important for applications. If Λ_γ is the common boundary of ω_γ and ω_γ^{-1}, it is not necessary for (*) to be satisfied at every point of Λ_γ but enough if it holds for the points of a dense subset Λ_γ^0. The proof then shows that $E_\theta\varphi_\gamma(X) = \alpha_\gamma$ for all $\theta \in \Lambda_\gamma^0$, and by continuity the equality holds as before for all points of Λ_γ.

The remaining conditions being automatically satisfied for any problem that is generated by one-sided hypotheses concerning one of the parameters in an exponential family, it is necessary only to verify (*) in order to prove the desired optimum property for the various examples of Section 3. This requires no further reference to the possible distributions of the observable random variables, since (*) concerns only the structure of the multiple decision problem, that is, of the sets Ω_i, not the distributions that are represented by the points $\theta \in \Omega$.

In the first example of Section 3, the only boundary point of ω_i and ω_i^{-1} ($i = 1, 2$) is θ_0. But this is also the boundary point of $\Omega_0 = \omega_i\omega_j$ and $\Omega_i = \omega_i^{-1}\omega_j$ ($i \neq j$), and hence (*) is satisfied.The result here is slightly stronger than the one given by the author in [8] since the condition of unbiasedness can be seen to be less stringent than the restriction imposed on the procedure in [8]. The checking of (*) is exactly analogous in the second version of this example, in which θ_0 is replaced by the interval $\theta_1 \leqq \theta \leqq \theta_2$.

In example (ii), the common boundary points of ω_1 and ω_1^{-1} for example, are the points with $\xi = \xi_0$. Let (ξ_0, η) be any such point, and suppose without loss of generality that $\eta \leqq 0$. Then (ξ_0, η) is also a boundary point of $\Omega_1 = \omega_1\omega_2$ and $\Omega_3 = \omega_1^{-1}\omega_2$, as was to be proved. The other cases are verified analogously.

In example (iii) where ω_{ij} is the parameter set $\theta_i \leqq \theta_j + \Delta$, consider the common boundary of, say, ω_{21} and ω_{21}^{-1}. By the remark following Theorem 2, we may restrict attention to points of this boundary satisfying

$$\theta_{i_1} < \cdots < \theta_{i_r} < \theta_2 = \theta_1 + \Delta < \theta_{j_1} < \cdots < \theta_{j_{s-r-2}}$$

and we may assume further that all of the differences $\theta_2 - \theta_{i_t}$ ($t = 1, \cdots, r$) are $\neq \Delta$. Let

$$\theta^0 = (\theta_1^0, \cdots, \theta_s^0) \in \omega_{12} \cap \omega_{21} \cap \bigcap \omega_{ij}^{x_{ij}}$$

be any specific such point, and consider the points $(\theta_1^0, \theta_2, \cdots, \theta_s^0)$ with $0 < \theta_2^0 - \theta_2 < \epsilon$. If ϵ is sufficiently small the relationship between all pairs of coordinates will be the same as before, except that $\theta_2 <$ instead of $= \theta_1 + \Delta$. These points are therefore in the intersection.

$$\omega_{12}^{-1} \cap \omega_{21} \cap \bigcap \omega_{ij}^{x_{ij}},$$

and since θ^0 is a boundary point of this set as well as of the intersection differing from the present one only in the factor ω_{12}, (*) is verified.

8. Optimality of the procedures of Sections 4 and 5. A basic assumption of Theorem 2 is the finiteness of the set Γ, and the theorem is therefore not applicable to any of the problems of Sections 4 and 5. Since the assumption was used however only in the proof of the relationship

$$\mathfrak{C}_0 \subseteq \mathfrak{C}, \tag{8.1}$$

it will be enough in the following to prove (8.1) in each case. This means showing for these problems that unbiasedness of a decision procedure implies that all of its component tests are similar on the boundary.

(i) *Estimation.* For the problem of estimating a real-valued parameter θ, let the risk function of an estimate Y, in accordance with Section 4, be given by

$$a \int_\theta^\infty (y - \theta)\, dP_\theta(y) + b \int_{-\infty}^\theta (\theta - y)\, dP_\theta(y),$$

where P_θ denotes the probability distribution of Y. The condition of unbiasedness then becomes

$$\begin{aligned} &a \int_\theta^\infty (y - \theta)\, dP_\theta(y) + b \int_{-\infty}^\theta (\theta - y)\, dP_\theta(y) \\ &\qquad \leqq a \int_{\theta'}^\infty (y - \theta')\, dP_\theta(y) + b \int_{-\infty}^{\theta'} (\theta' - y)\, dP_\theta(y) \end{aligned} \tag{8.2}$$

for all θ, θ'. In the case that $a = b$, this states that the estimate, on the average is closer to the true value θ than to any other value θ'.

In the following we shall restrict attention to estimates Y with finite risk. If $\theta < \theta'$, (8.2) then reduces to

$$\begin{aligned} &a\left[\int_\theta^{\theta'} (y - \theta)\, dP_\theta(y) + (\theta' - \theta) \int_{\theta'}^\infty dP_\theta(y)\right] \\ &\qquad \leqq b\left[\int_\theta^{\theta'} (\theta' - y)\, dP_\theta(y) + (\theta' - \theta) \int_{-\infty}^\theta dP_\theta(y)\right]. \end{aligned}$$

Dividing both sides by $\theta' - \theta$, and letting θ' tend to θ, we see that

$$0 \leqq \frac{1}{\theta' - \theta} \int_\theta^{\theta'} (y - \theta)\, dP_\theta(y) \leqq P_\theta\{\theta < Y \leqq \theta'\} \to 0,$$

and that similarly also the first term on the right-hand side tends to zero. In the limit we therefore get

$$aP_\theta\{Y > \theta\} \leqq bP_\theta\{Y \leqq \theta\}$$

or

$$P_\theta\{Y > \theta\} \leqq \frac{b}{a + b} = \alpha.$$

By letting θ' tend to θ from below, we find analogously that unbiasedness of the estimate Y implies

$$P_\theta\{Y \geqq \theta\} \geqq \alpha$$

Suppose now that the distributions of Ω constitute an exponential family, and that they possess densities with respect to Lebesgue measure. This is the case not only for families of univariate and multivariate normal distributions, of gamma distributions, etc., but also when one is dealing with binomial or Poisson variables to which, as a randomization device, one adds a variable that is uniformly distributed over (0, 1). If then $P_\theta\{Y = c\}$ is positive for some θ, it is positive for all θ. Hence

$$P_\theta\{Y > \theta\} = \alpha, \tag{8.3}$$

except possibly for a countable set of parameter values, and it follows from a theorem of Scheffé [13] that (8.3) must hold for all θ. Since it was shown in Section 4 that $Y > \theta_0$ is the rejection region of the hypothesis $H(\theta_0):\theta \leqq \theta_0$, and since θ_0 is the only common boundary point of $\omega(\theta_0)$ and $\omega^{-1}(\theta_0)$, this completes the proof of (8.1).

(ii) *Estimation after a preliminary test of significance.* For the problem generated by the hypotheses $H(\theta_1):\theta \leqq \theta_1(\theta_0 \leqq \theta_1)$, and with constant losses a and b, the risk is

$$a \int_{\theta+}^{\infty} (y - \theta)\, dP_\theta(y) + b \int_{\theta_0+}^{\theta} (\theta - y)\, dP_\theta(y) + b(\theta - \theta_0) \int_{-\infty}^{\theta_0} dP_\theta(y) \quad \text{if } \theta_0 < \theta,$$

and

$$a \int_{\theta_0+}^{\infty} (y - \theta_0)\, dP_\theta(y) \qquad \text{if } \theta \leqq \theta_0.$$

It is seen exactly as before that unbiasedness implies (8.3) for all $\theta > \theta_0$, and that this holds also for $\theta = \theta_0$ follows again from Scheffé's theorem.

Consider instead the modified problem of Section 4 (ii) generated by the hypotheses $H(\theta_1):\theta \leqq \theta_1$ with $\theta_1 = 0$ or $\theta_1 \geqq C$, and with losses a_0, b_0 for $H(0)$ and a, b for the remaining hypotheses. A compatible procedure accepts $H(0)$ when a statistic $Y \leqq C$, and otherwise takes Y as an estimate of θ. The associated risk is

$$\int_{C+}^{\infty} [a(y - C) + \rho a_0]\, dP_\theta(y) \qquad \text{if } \theta \leqq 0,$$

$$b_0 \rho \int_{-\infty}^{C} dP_\theta(y) + a \int_{C+}^{\infty} (y - C)\, dP_\theta(y) \qquad \text{if } 0 < \theta \leqq C,$$

$$a \int_{\theta+}^{\infty} (y - \theta)\, dP_\theta(y) + b \int_{C+}^{\theta} (\theta - y)\, dP_\theta(y) + \int_{-\infty}^{C} [b(\theta - C) + b_0 \rho]\, dP_\theta(y) \qquad \text{if } C < \theta.$$

Here we have taken the measure μ of (2.4) to be Lebesgue measure for the hypotheses $H(\theta_1)$ with $\theta_1 \geqq C$ and to assign measure ρ to the hypothesis $H(0)$.

For $\theta \geqq C$ we see as before that unbiasedness implies (8.3), while for $\theta = 0$ we find

$$a_0 P_0\{Y > C\} = b_0 P_0\{Y \leqq C\}$$

and hence

$$P_0\{Y > C\} = \alpha_0$$

by considering the condition of unbiasedness for $\theta = 0$ and $\theta' \downarrow 0$ and vice versa.

(iii) *Testing for goodness of fit.* The problem, as described in Section 5, is generated by the hypotheses $H_-(u):F(u) \leqq F_0(u)$ and $H_+(u):F(u) \geqq F_0(u)$. Let $d_-(u)$, $d_0(u)$, $d_+(u)$ denote the decisions that the true $F(u)$ is $<$, $=$, $>$ than $F_0(u)$ and let $\psi_-(x, u)$, $\psi_0(x, u)$, $\psi_+(x, u)$ be the probabilities with which these decisions are taken when $x = (x_1, \cdots, x_n)$ is observed. Then the corresponding over-all probabilities of these decisions are $P_-(u) = P_-(u, F) = E_F\psi_-(X, u)$, $P_0(u)$ and $P_+(u)$. If the losses resulting from false rejection and acceptance are $a(u)$ and $b(u)$ for both $H_-(u)$ and $H_+(u)$, and if we put

$$\begin{aligned} R_-(u, F) &= R_-(u) = [a(u) + b(u)]P_+(u) + b(u)P_0(u), \\ R_0(u, F) &= R_0(u) = a(u)[P_-(u) + P_+(u)], \\ R_+(u, F) &= R_+(u) = [a(u) + b(u)]P_-(u) + b(u)P_0(u), \end{aligned}$$

the risk function is

$$R(F) = \int_{S_-} R_-(u)\, d\mu(u) + \int_{S_0} R_0(u)\, d\mu(u) + \int_{S_+} R_+(u)\, d\mu(u),$$

where S_-, S_0 and S_+ are the sets on which the true $F(u)$ is $<$, $=$, and $>$ than $F_0(u)$. In particular

$$R(F_0) = \int_{-\infty}^{\infty} R_0(u, F_0)\, d\mu(u).$$

We shall assume in the following that $R(F_0)$ is finite, that $a(u)$ and $b(u)$ are bounded in every finite interval, and for the sake of convenience also that F_0 possesses a probability density.

The condition of unbiasedness becomes in the present case

$$\int_{S_-} R_-(u, F)\, d\mu(u) + \int_{S_0} R_0(u, F)\, d\mu(u) + \int_{S_+} R_+(u, F) d\mu(u)$$

$$\leqq \int_{S'_-} R_-(u, F)\, d\mu(u) + \int_{S'_0} R_0(u, F)\, d\mu(u) + \int_{S'_+} R_+(u, F)\, d\mu(u),$$

where S'_-, S'_0 and S'_+ are the sets on which some alternative c.d.f. F' is $<$, $=$, and $>$ than F_0. Consider this condition now for some F and F', both of which agree with F_0 except on a finite interval I on which $F < F_0$ and $F' > F_0$. It then reduces to

$$\int_I R_-(u, F)\, d\mu(u) \leqq \int_I R_+(u, F)\, d\mu(u).$$

If, holding I fixed, one considers a sequence of such distributions F, which possess probability densities that tend to the density of F_0, it follows from Scheffé's theorem that

$$\int_I R_-(u, F_0)\, d\mu(u) \leqq \int_I R_+(u, F_0)\, d\mu(u),$$

and since I was an arbitrary interval, that

$$R_-(u, F_0) \leqq R_+(u, F_0) \qquad \text{a.e.}$$

Analogously one sees that the reverse inequality must hold, so that

$$R_-(u, F_0) = R_+(u, F_0) \qquad \text{a.e.}$$

If in the above argument F is replaced by F_0, one finds

$$\int_I R_0(u, F_0)\, d\mu(u) \leqq \int_I R_+(u, F_0)\, d\mu(u)$$

for all I, and hence

$$R_0(u, F_0) \leqq R_+(u, F_0) \qquad \text{a.e.}$$

Similarly, on replacing F' by F_0 one gets

$$\int_I R_+(u, F)\, d\mu(u) \leqq \int_I R_0(u, F)\, d\mu(u)$$

for all I, which by the same argument as before leads to

$$R_+(u, F_0) \leqq R_0(u, F_0).$$

Thus, for almost all u,

$$R_-(u, F_0) = R_0(u, F_0) = R_+(u, F_0),$$

which implies

$$P_-(u, F_0) = P_+(u, F_0) = \frac{b(u)}{a(u) + b(u)} = \alpha(u),$$

as was to be proved.

A very similar proof applies in the two-sample problem discussed in Section 5(ii), and we shall therefore not give the details.

(iv) *Estimating a cumulative distribution function.* The problem of estimating a c.d.f. was treated from a minimax point of view by Agarwal [1] for several loss functions all of which differ from the one below. Following our earlier definitions, we take the risk function here to be

$$\int_{-\infty}^{\infty}\left\{\int_{F(u)}^{\infty} a(u)[y - F(u)]\, dP_u(y) + \int_{-\infty}^{F(u)} b(u)[F(u) - y]\, dP_u(y)\right\} d\mu(u),$$

where $P_u = P_{u,F}$ is the distribution of the estimate $Y(u)$ of $F(u)$. As in the problem of estimating a single parameter, we shall restrict attention to estimates with finite risk, and the condition of unbiasedness then becomes

$$\int_{-\infty}^{\infty}\left\{a(u)\int_{F(u)}^{F'(u)} [y - F(u)]\, dP_u(y) + [F'(u) - F(u)]\int_{F'(u)}^{\infty} dP_u(y)\right\} d\mu(u)$$

$$\leqq \int_{-\infty}^{\infty}\left\{b(u)\int_{F(u)}^{F'(u)} [F'(u) - y]\, dP_u(y) + [F'(u) - F(u)]\int_{-\infty}^{F(u)} dP_u(y)\right\} d\mu(u),$$

where the probabilities are computed with respect to F, and where F' is any alternative c.d.f.

We shall consider first the case that F is the uniform distribution on (0, 1), that $I = (u_0, u_1)$ is any subinterval of (0, 1), and that F' is any continuous c.d.f. such that $F'(u) = F(u) + \Delta$ for $u_0 < u < u_1$ and $F'(u) = F(u)$ for $u < u_0 - \Delta$ and $u > u_1 + \Delta$. On dividing by Δ, and letting Δ tend to zero, (8.4) is then seen to reduce to

$$\int_I a(u)P_u\{Y(u) > F(u)\}\, d\mu(u) \leqq \int_I b(u)P_u\{Y(u) \leqq F(u)\}\, d\mu(u),$$

and since this holds for all I, to

$$P_u\{Y(u) > F(u)\} \leqq \frac{b(u)}{a(u) + b(u)} = \alpha(u) \qquad \text{a.e.}$$

On letting F' tend to F from below one finds similarly

$$P_u\{Y(u) \geqq F(u)\} \geqq \alpha(u) \qquad \text{a.e.}$$

In exactly the same manner these two inequalities are seen to hold also for any F belonging to the family $\mathfrak{F}$ of mixtures of uniform distribution over nonoverlapping intervals. By considering a countable dense subset of $\mathfrak{F}$, for example mixtures with rational weights of uniform distributions over intervals with rational endpoints, it is seen that there exists a null set N such that for any $u \not\in N$ the two inequalities hold for all $F \in \mathfrak{F}$, and it follows by an argument similar to that given in (i) of this section that for all $u \not\in N$, and all $F \in \mathfrak{F}$

$$P_{u,F}\{Y(u) > F(u)\} = \alpha(u).$$

For $F \in \mathfrak{F}$ and any fixed $u \not\in N$, the common boundary points of $\omega(u_0, p_0)$: $F(u_0) \leqq p_0$ and $\omega^{-1}(u_0, p_0)$ are exactly the distributions of $\mathfrak{F}$ for which $F(u_0) = p_0$. For these we then have

$$P_{u,F}\{Y(u_0) > p_0\} = \alpha(u_0),$$

and since the left-hand side is the probability of rejecting the hypothesis $H(u_0, p_0)$, this completes the proof of (8.1). The desired optimum property of the procedure now follows from the fact, proved by Fraser [4], [5], that for the family of distributions $\mathfrak{F}$, the one-sided sign test uniformly minimizes the probabilities of error among all unbiased tests of the hypothesis $H(u_0, p_0)$.

REFERENCES

[1] Om P. Aggarwal, "Some minimax invariant procedures for estimating a cumulative distribution function," *Ann. Math. Stat.*, Vol. 26 (1955), pp. 450–463.

[2] R. C. Bose and S. N. Roy, "Simultaneous confidence interval estimation," *Ann. Math. Stat.*, Vol. 24 (1953), pp. 513–536.

[3] David B. Duncan, "Multiple range and multiple F-tests," *Biometrics*, Vol. 11 (1955), pp. 1–42.

[4] D. A. S. Fraser, "Completeness of order statistics," *Canadian J. Math.*, Vol. 6 (1953), pp. 42–45.

[5] D. A. S. Fraser, "Non-parametric theory: Scale and location parameters," *Canadian J. Math.*, Vol. 6 (1953), pp. 46–68.

[6] H. O. Hartley, "Maximum F ratio as a short-cut test for heterogeneity of variance," *Biometrika*, Vol. 37 (1950), pp. 308–312.

[7] M. G. Kendall, "Further contributions to the theory of paired comparisons," *Biometrics*, Vol. 11 (1955), pp. 43–62.

[8] E. L. Lehmann, "Some principles of the theory of testing hypothesis," *Ann. Math. Stat.*, Vol. 21 (1950), pp. 1–26.

[9] E. L. Lehmann, "A general concept of unbiasedness," *Ann. Math. Stat.*, Vol. 22 (1951), pp. 587–592.

[10] E. L. Lehmann, "Testing multiparameter hypotheses," *Ann. Math. Stat.*, Vol. 23 (1952), pp. 541–552.

[11] E. L. Lehmann and Henry Scheffé, "Completeness, similar regions, and unbiased estimation, Part II," *Sankhyā*, Vol. 15 (1955), pp. 219–236.

[12] S. N. Roy, "On a heuristic method of test construction and its use in multivariate analysis," *Ann. Math. Stat.*, Vol. 24 (1953), pp. 220–238.

[13] Henry Scheffé, "A useful convergence theorem for probability distributions," *Ann. Math. Stat.*, Vol. 18 (1947), pp. 434–438.

[14] J. W. Tukey, "Quick and dirty methods in statistics, Part II, Simple analyses for standard designs," *Proc. Fifth Annual Convention, Am. Soc. for Quality Control* (1951), pp. 189–197.

A THEORY OF SOME MULTIPLE DECISION PROBLEMS*. II

By E. L. Lehmann[1]

University of California, Berkeley

Summary. The theory of Part I is extended to problems in which it is permitted not to come to a definite conclusion regarding one or more of the questions under consideration. Some problems are also investigated in which, from a single set of observations, one wishes to answer a number of questions in sequence. Here the nature of the question at a later stage will depend on the answers obtained at the earlier stages.

9. Decision procedures permitting partial conclusions. It is a common feature of all the problems treated in Part I, that a fixed partition of the parameter space Ω into sets Ω_t is given, and that one wishes to determine which of these sets contains the true parameter point. There are however many statistical problems, such as the estimation by confidence sets, in which the possible decisions do not correspond to the sets of a fixed partition. In particular, this is the case in the field of statistical inference, when the statistician is free to decide how sharp a statement he can reliably make on the basis of the observations. We shall show in the present section how such problems may be generated by the simultaneous consideration of a number of two-decision problems as in Part I, if one suitably modifies the interpretation of the decisions involved.

Previously we were concerned with testing a set of hypotheses $H_\gamma : \theta \, \varepsilon \, \omega_\gamma$, so that in each component problem the choice lay between the two decisions $\theta \, \varepsilon \, \omega_\gamma$ (acceptance of H_γ) and $\theta \, \varepsilon \, \omega_\gamma^{-1}$ (rejection of H_γ). Suppose now instead that the statistician is asked only whether the data reject the hypothesis, and that in case they do not, no alternative positive statement is required. The choice may then be said to lie between the statements $\theta \, \varepsilon \, \omega_\gamma^{-1}$ and $\theta \, \varepsilon \, \omega_\gamma^{0} = \Omega$.

This actually appears to be the point of view taken by Duncan ("Multiple range and multiple F tests", *Biometrics*, Vol. 11 (1955), pp. 1–42) in his formulation of this class of multiple decision problems to which reference is made in section 1 of the present paper.

If one considers simultaneously a number of such problems, one is faced with a multiple decision problem in which the different possible decisions correspond to the statements that a certain number of the hypotheses H_γ are false, but where nothing is said regarding the remaining hypotheses. This is equivalent to the statement that the parameter point θ lies in the set

$$\Omega_i = \bigcap\nolimits_\gamma \omega_\gamma^{y_i} \tag{9.1}$$

Received October 12, 1956.

* To rectify a printing error in Part I, in the March, 1957 issue of these *Annals* replace page 14, formula (4.12) through page 17, line 23 by page 70, line 8 through page 72, next-to-last line. Ed.

[1] Work done while the author was a Fellow of the John Simon Guggenheim Memorial Foundation.

J. Rojo (ed.), *Selected Works of E. L. Lehmann*, Selected Works in Probability and Statistics, DOI 10.1007/978-1-4614-1412-4_54, © Springer Science+Business Media, LLC 2012

where the y's are -1 for each rejected H_γ and 0 for the others. In the particular case that all of the y's are zero, we have $\Omega_i = \Omega$, and thus make no statement whatever about the position of θ. As before, it may of course happen that some of the formal intersections (9.1) are empty, and we shall then restrict the Ω's to denote the nonempty ones and require that none of the possible decisions should correspond to these empty intersections. If we assume that in the simultaneous consideration of a number of problems the losses are additive so that the total loss is the sum (or a weighted sum) of the losses of the component problems, and if we suppose the losses to be a_γ for rejecting H_γ when $\theta \,\varepsilon\, \omega_\gamma$, b_γ for not rejecting it when $\theta \,\varepsilon\, \overset{-1}{\omega}_\gamma$, and zero in the other two cases, the total loss is again given by (2.2) and (2.3) with

$$x_{i\gamma} = 2y_{i\gamma} + 1. \tag{9.2}$$

Since in particular the loss function of the basic two-decision problem is unchanged, the associated optimum unbiased procedure of Part I will retain its optimum property in spite of the reinterpretation of one of the decisions. It now leads to the statements $\theta \,\varepsilon\, \overset{-1}{\omega}_\gamma$ and $\theta \,\varepsilon\, \overset{0}{\omega}_\gamma$ as $X \,\varepsilon\, \overset{-1}{A}_\gamma$ or $X \,\varepsilon\, A_\gamma$ where A_γ is the acceptance region of the best unbiased test of H_γ. The simultaneous carrying out of a number of these tests then still leads to a procedure in which the decision $d_i\colon \theta \,\varepsilon\, \Omega_i$ is taken when X falls in the set

$$D_i = \textstyle\bigcap_\gamma A^{x_i} \tag{9.3}$$

but where Ω_i is now defined by (9.1) and (9.2) instead of (2.1).

As has already been pointed out the sets Ω_i, which define the possible decisions, no longer constitute a partition of the parameter space. Instead, they are generated through intersections from the class $\{\overset{-1}{\omega}_\gamma, \gamma \,\varepsilon\, \Gamma\}$, that is, they constitute the smallest class that is closed under intersections and contains the sets $\overset{-1}{\omega}_\gamma$. It may happen that two of these Ω's are equal, $\Omega_i = \Omega_j$, say, and one would then wish to identify the associated decisions. On the other hand, viewing the problem as a product one must consider all of the formal intersections (9.1) as distinct. Otherwise the definition of the loss function, for example, would become ambiguous since the losses resulting from decisions d_i and d_j when θ is in some Ω_h would usually not be the same even though $\Omega_i = \Omega_j$. Fortunately, the difficulty arises in the applications we wish to make only in cases in which it can be overcome by a natural further restriction of the decision space. Suppose namely that H_γ and H_δ are two hypotheses with

$$\overset{-1}{\omega}_\gamma \subset \overset{-1}{\omega}_\delta$$

so that the two intersections $\overset{-1}{\omega}_\gamma \cap \overset{-1}{\omega}_\delta$ and $\overset{-1}{\omega}_\gamma \cap \overset{0}{\omega}_\delta$ are identical. It then seems reasonable that whenever the data lead to the rejection of H_γ one would also wish to reject the more restrictive hypothesis H_δ. (In part I this was actually part of the compatibility requirement.) With such tests the decisions $\theta \,\varepsilon\, \overset{-1}{\omega}_\gamma \cap \overset{0}{\omega}_\delta$ would never be reached, and the conflict would thus be avoided. For this reason we shall eliminate from the list of permissible decisions not only the formal intersections

(9.1) that are empty, but also those for which the intersection Ω_i equals some other Ω_j satisfying $y_{j\gamma} \leqq y_{i\gamma}$ for all γ. It follows from the general discussion of Sec. 7 that (9.3) defines a 1:1 correspondence between families of tests which are compatible with this set of restrictions, and decision procedures for the restricted product problem. It is useful to note further that for these restrictions the formation of restricted products of decision problems also retains the property of being commutative and associative, which it obviously had in the problems of Part I.

Let $\mathcal{P}_+(\theta_0)$ denote the two-decision problem discussed above, in which the choice lies between the statements $\theta > \theta_0$ and $-\infty < \theta < \infty$, and let $\mathcal{P}_-(\theta_0)$ denote the dual problem in which the first of these possibilities is replaced by the statement $\theta < \theta_0$. By considering these two problems simultaneously one obtains a three-decision problem with loss function

(9.4)

	d_2 $\theta < \theta_0$	d_0 $-\infty < \theta < \infty$	d_1 $\theta > \theta_0$
$\theta < \theta_0$	0	b	$a + b$
$\theta = \theta_0$	a	0	a
$\theta > \theta_0$	$a + b$	b	0

As an example suppose that θ measures the difference in quality of two products which are being compared by an impartial research organization. The decisions d_1 and d_2 claim superiority for one or the other of the products, while d_0 states that the data are inconclusive and that neither of the two products can be ascertained to be better than the other. It is an advantage of such a formulation over the more conventional one in which d_0 is replaced by the statement $\theta = \theta_0$, that it enables the statistician to control the probability of error. In the standard situation with D_2 and D_1 given by

$$D_2: T \leqq C_1 \quad \text{and} \quad D_1: T \geqq C_1' \tag{9.5}$$

where $P_\theta\{T \leqq C_1\}$ and $P_\theta\{T \geqq C_2'\}$ are $< \alpha$ for $\theta > \theta_0$ and $\theta < \theta_0$ respectively, and where $P_{\theta_0}\{T \leqq C_1\} = P_{\theta_0}\{T \geqq C_2'\} = \alpha$, the maximum probability of error occurs when $\theta = \theta_0$, and is 2α. (A very similar formulation was discussed by Bahadur, "A property of the t-statistic," *Sankhya*, Vol. 12 (1952), pp. 79–88.)

The loss function (9.4) is not appropriate in situations in which a definite decision is preferred to d_0 even when $\theta = \theta_0$. A formulation which is more suitable for this case is obtained if in the one-sided problems $\mathcal{P}_-(\theta_0)$ and $\mathcal{P}_+(\theta_0)$ one replaces the decisions $\theta < \theta_0$ and $\theta > \theta_0$ by $\theta \leqq \theta_0$ and $\theta \geqq \theta_0$ respectively, so that these two component problems are given by

(9.6a)

	$-\infty < \theta < \infty$	$\theta \leqq \theta_0$
$\theta > \theta_0$	0	a
$\theta \leqq \theta_0$	b	0

and $(b < a)$

(9.6b)

	$-\infty < \theta < \infty$	$\theta \geqq \theta_0$
$\theta < \theta_0$	0	a
$\theta \geqq \theta_0$	b	0

The simultaneous consideration of these two problems leads to a four-decision problem with loss table

(9.7)

	d_2 $\theta \leqq \theta_0$	d_0 $-\infty < \theta < \infty$	d_1 $\theta \geqq \theta_0$	d_3 $\theta = \theta_0$
$\theta < \theta_0$	0	b	$a + b$	a
$\theta = \theta_0$	b	$2b$	b	0
$\theta > \theta_0$	$a + b$	b	0	a

It turns out that this formulation leads to essentially the same solution as the previous one, with D_2 and D_1 given by (9.5), decision d_0 being taken when $C_1 < T < C_1'$, and decision d_3 not occurring at all.

We mention finally that still another problem leads to the same solution, namely that given by

(9.8)

	d_2 $\theta \leqq \theta_0$	d_0 $-\infty < \theta < \infty$	d_1 $\theta \geqq \theta_0$
$\theta < \theta_0$	0	b	a'
$\theta = \theta_0$	0	b'	0
$\theta > \theta_0$	a'	b	0

$$\begin{pmatrix} b < a' \\ b < b' \end{pmatrix}.$$

The level α of (9.5) is in this case given by

$$\alpha = \frac{b' - b}{a' + 2(b' - b)}.$$

10. Partial classification of one or more parameters. (i) Let θ be a real parameter and suppose that we wish to determine, as far as possible, its position relative to two given values $\theta_1 < \theta_2$. A procedure may be generated by considering simultaneously the four problems $\mathcal{P}_\pm(\theta_i)$, $i = 1, 2$. The resulting problem offers the choice between the decisions

(10.1)

$$\begin{aligned}
&d_1: \{\theta < \theta_1\} = \{\theta < \theta_1\} \cap \{\theta < \theta_2\} \\
&d_2: \{\theta < \theta_2\} = \{-\infty < \theta < \infty\} \cap \{\theta < \theta_2\} \\
&d_3: \{\theta_1 < \theta < \theta_2\} = \{\theta_1 < \theta\} \cap \{\theta < \theta_2\} \\
&d_4: \{\theta > \theta_1\} = \{\theta > \theta_1\} \cap \{-\infty < \theta < \infty\} \\
&d_5: \{\theta > \theta_2\} = \{\theta > \theta_2\} \cap \{\theta > \theta_1\} \\
&d_6: \{-\infty < \theta < \infty\} = \{-\infty < \theta < \infty\} \cap \{-\infty < -\theta < \infty\}.
\end{aligned}$$

Here the sets $\{\theta < \theta_1\}$ and $\{\theta > \theta_2\}$ may also be represented as the intersections $\{\theta < \theta_1\} \cap \{-\infty < \theta < \infty\}$ and $\{\theta > \theta_2\} \cap \{-\infty < \theta < \infty\}$. However these decisions are ruled out by the convention of the previous section.

Suppose now that the tests of the four generating hypotheses have rejection regions

$$T \leqq C_1, \qquad T \leqq C_1', \qquad T \geqq C_2, \qquad T \geqq C_2'$$

for $H:\theta \leqq \theta_1$, $\theta \geqq \theta_1$, $\theta \leqq \theta_2$, $\theta \geqq \theta_2$ respectively, where the constants are determined by

$$P_{\theta_1}\{T \leqq C_1\} = P_{\theta_1}\{T \geqq C_1'\} = P_{\theta_2}\{T \leqq C_2\} = P_{\theta_2}\{T \geqq C_2'\} = \alpha.$$

Compatibility requires that the intersections

$$\{T \leqq C_i\} \cap \{T \geqq C_i'\}, \qquad \{T \leqq C_1\} \cap \{C_2 < T < C_2'\}$$
$$\text{and } \{C_1 < T < C_i'\} \cap \{T \geqq C_2'\}$$

should be empty, and hence that

$$C_1 < C_1', \qquad C_2 < C_2', \qquad C_1 < C_2 \quad \text{and} \quad C_1' < C_2'.$$

These conditions are satisfied if $\alpha < \frac{1}{2}$, and if for each fixed C

$$P_\theta\{T \leqq C\} > P_{\theta'}\{T \leqq C\} \quad \text{when} \quad \theta < \theta'. \tag{10.2}$$

According to (9.3) the resulting procedure is given by

$$\begin{array}{ll} D_1: T \leqq C_1 & D_4: \max(C_1', C_2) \lesssim T < C_2' \\ D_2: C_1 \lesssim T < \min(C_1', C_2) & D_5: T \geqq C_2' \\ D_3: C_1' \leqq T \leqq C_2 & D_6: C_2 < T < C_1' \end{array} \tag{10.3}$$

where $\lesssim$ is $<$ when $C_1' < C_2$ and $\leqq$ when $C_1' \geqq C_2$. Depending on the sign of the difference $C_2 - C_1'$ and hence on the distance between θ_1 and θ_2, only one of the decisions d_3 and d_6 will occur. For intermediate values of T the positive statement $\theta_1 < \theta < \theta_2$ will be made only if θ_1 and θ_2 are not too close. Otherwise such T-values will leave the position of θ in doubt.

If (10.3) holds, the probability of the procedure leading to a false statement never exceeds 2α. For the probability of error is equal to

$$P_\theta\{T \geqq C_1'\} \leqq P_{\theta_1}\{T \geqq C_1'\} = \alpha \qquad \text{for} \quad \theta < \theta_1$$

$$P_{\theta_1}\{T \leqq C_1\} + P_{\theta_1}\{T \geqq C_1'\} = 2\alpha \qquad \text{for} \quad \theta = \theta_1$$

$$P_\theta\{T \leqq C_1\} + P_\theta\{T \geqq C_2\} \leqq P_{\theta_1}\{T \leqq C_1\} + P_{\theta_2}\{T \geqq C_2\} = 2\alpha$$
$$\text{for} \quad \theta_1 < \theta < \theta_2,$$

and similarly for $\theta \geqq \theta_2$.

In the usual applications T is a function of a sample, and as the sample size increases T tends to θ in probability. The procedure is then consistent in the sense that

$$P(D_i) \to 1 \quad \text{for} \quad \theta \in \Omega_i \qquad (i = 1, 2, 3).$$

Except when θ is exactly equal to θ_1 or θ_2, the probability is therefore 1 that one of the three sharp statements d_1, d_3, or d_5 will be made, and that this statement will be correct. Since

$$P_{\theta_1}(D_2) = P_{\theta_2}(D_4) = 1 - 2\alpha,$$

it also follows that

$$P_{\theta_1}(D_2) = P_{\theta_2}(D_4) \to 1$$

if one lets α tend to zero as n tends to infinity.

A slightly different procedure results for problem (10.1) when one is concerned with a sample $X_1, \cdots, X_n$ from $N(\theta, \sigma^2)$. Here the tests of the four generating hypotheses have rejection regions

$$(\bar{X} - \theta_1)/S \leqq -C, \qquad (\bar{X} - \theta_1)/S \geqq C,$$
$$(\bar{X} - \theta_2)/S \leqq -C, \qquad (\bar{X} - \theta_2)/S \geqq C,$$

and the induced multiple decision procedure is given by (10.3) with

$$T = \bar{X}/S, \qquad C_i = -C + \frac{\theta_i}{S}, \qquad C_i' = C + \frac{\theta_i}{S}. \tag{10.4}$$

Both of the decisions d_3 and d_6 occur in this procedure with positive probability, d_3 only in cases in which $S \leqq (\theta_2 - \theta_1)/C$ and d_6 only when the opposite inequality holds. The remarks concerning error control and consistency require no change.

As an application consider the comparison of two normal populations $N(\xi, \sigma^2)$ and $N(\eta, \sigma^2)$ on the basis of samples $X_1, \cdots, X_m$ and $Y_1, \cdots, Y_n$. With $\theta = \eta - \xi$, $\theta_1 = -\Delta$, $\theta_2 = \Delta$, the possible decisions become

$$d_1: \eta - \xi < -\Delta, \qquad d_3: -\Delta < \eta - \xi < \Delta, \qquad d_5: \eta - \xi > \Delta,$$
$$d_2: \eta - \xi < \Delta, \qquad d_4: \eta - \xi > -\Delta, \qquad d_6: -\infty < \eta - \xi < \infty.$$

Here d_1 states that ξ is significantly larger than η, d_2 that η is not significantly larger than ξ, d_3 that the two means do not differ significantly, etc. The procedure is given by (10.3) with $T = (\bar{Y} - \bar{X})/S$ and

$$C_1 = -C - \frac{\Delta}{S}, \qquad C_2 = -C + \frac{\Delta}{S}, \qquad C_1' = C - \frac{\Delta}{S}, \qquad C_2' = C + \frac{\Delta}{S}.$$

Problem (10.1) leads to still another type of procedure in the case of two independent Poisson variables, say X and Y, where one wishes to classify the ratio $\rho = \lambda/\mu$ of the parameters of the two distributions with respect to two values $\rho_1 < \rho_2$. Here the tests of the four generating hypotheses $\rho \leqq \rho_i$, $\rho \geqq \rho_i (i = 1, 2)$ are carried out conditionally, given the value of $X + Y$. The conditional distribution of Y given $X + Y = m$ is the binomial distribution $b(p, m)$ corresponding to m trials and probability $p = \mu/(\lambda + \mu)$ of success. The conditional situation is therefore of the kind discussed at the beginning of this section, and leads to the procedure (10.3) with $T = Y$ and $\theta = p$.

Whether d_3 or d_6 occurs depends here on m, d_3 being associated with large values of m. To see this, let $F_{p,m}(t)$ denote the conditional cumulative distribution function, given $X + Y = m$, of $Y + U$ where U is independent of Y and is uniformly distributed on (0, 1). Then $C_1' = C'(m)$ and $C_2 = C_2(m)$ are determined by

$$F_{p_2,m}(C_2) = 1 - F_{p_1,m}(C_1') = \alpha. \tag{10.5}$$

We shall show that $C_1'(m_1) < C_2(m_1)$ implies that also $C_1'(m_2) < C_2(m_2)$ for all $m_2 > m_1$. Since $F_{p_2,m}(C_2) = \alpha$, $F_{p_1,m}(C_2)$ is the power of the most powerful level α test for testing p_2 against p_1 in a binomial distribution $b(p, m)$. If $C_1'(m_1) < C_2(m_1)$ it follows from (10.9) that in the case of m trials this power is greater than $1 - \alpha$; but then it must also exceed $1 - \alpha$ for $m_2 > m_1$ trials so that

$$F_{p_1,m_2}(C_1') = 1 - \alpha < F_{p_1,m_2}(C_2),$$

and hence $C_1'(m_2) < C_2(m_2)$.

It is interesting to note that in all of these problems the choice between decision d_3 and d_6 depends on the distance between θ_1 and θ_2 relative to the amount of information the data contain for the problem.

While the classification of θ with respect to a single value θ_0, or two values $\theta_1 < \theta_2$, are the most interesting cases, let us consider briefly also the problem of classifying θ with respect to a countable set of values $\cdots < \theta_{-2} < \theta_{-1} < \theta_0 < \theta_1 < \cdots$. This is generated by the problems $\mathcal{P}_{\pm}(\theta_i)$, $i = 0, \pm 1, \pm 2, \cdots$. Supposing that $\theta_i \to \pm\infty$ as $i \to \pm\infty$ and letting $\theta_{-\infty} = -\infty$, $\theta_{\infty} = \infty$, the possible decisions consist of the totality of statements $\theta_i < \theta < \theta_j$. The decision $\theta_i < \theta < \theta_j$ corresponds to the individual decisions that $\theta < \theta_k$ for $k \geqq j$ and $\theta > \theta_k$ for $k \leqq i$, and that the position of θ is left in doubt with respect to the points θ_k with $i < k < j$.

The limiting case of this problem, in which one wishes to classify θ with respect to all possible values of θ_0, is obtained by considering simultaneously the problems $\mathcal{P}_{\pm}(\theta_0)$ for all θ_0. The possible decisions then consist of the totality of statements $\underline{\theta} < \theta < \bar{\theta}$, and if $a_\gamma = a$, $b_\gamma = b$ for all γ, (9.3) yields precisely the standard confidence intervals for θ with confidence coefficient $1 - 2\alpha$. The loss function resulting from the additivity assumption is in the simplest case[2]

$$\begin{array}{lll} (a + b)(\underline{\theta} - \theta) + b(\bar{\theta} - \underline{\theta}) & \text{if} & \theta < \underline{\theta} \\ b(\bar{\theta} - \underline{\theta}) & \text{if} & \underline{\theta} < \theta < \bar{\theta} \\ (a + b)(\theta - \bar{\theta}) + b(\bar{\theta} - \underline{\theta}) & \text{if} & \theta > \bar{\theta}. \end{array} \tag{10.6}$$

[2] A loss function of this type was suggested by Wolfowitz [4].

More generally one may replace $\bar{\theta} - \underline{\theta}$ by $\int_{\underline{\theta}}^{\bar{\theta}} d\mu(\theta)$, and similarly for the lengths of the other intervals. Unfortunately the condition of unbiasedness becomes very difficult to interpret in the present problem, and it is doubtful that for the standard distributions the procedure is unbiased with uniformly minimum risk, as is the case with the other problems of this section.

It should be pointed out that all these procedures concerning a single real-valued parameter θ can be obtained from the standard confidence intervals concerning θ. When classifying θ with respect to θ_1 and θ_2 for example, one can state $\theta < \theta_1$ if $\bar{\theta} < \theta_1$, and $\theta_1 < \theta < \theta_2$ if $\theta_1 < \underline{\theta} < \bar{\theta} < \theta_2$. On the other hand, if $\underline{\theta} < \theta_1 < \bar{\theta} < \theta_2$, only the conclusion $\theta < \theta_2$ is possible, the relation of θ to θ_1 being left in doubt. This approach, however, does not yield any optimum properties for these procedures, and does in fact not carry over to problems involving more than one parameter.

In Example (iii) of Sec. 3 we considered the comparison of a number of normal populations $N(\theta_i, \sigma^2)$. The consistency difficulties that occurred in combining the decisions $\theta_i \leqq \theta_j$ for different pairs (i, j) disappear if one treats the problem from the present point of view, which is exactly that from which the problem was treated by Duncan, "Multiple range and multiple F-tests," *Biometrics*, Vol. 11 (1955), pp. 1–42. For each pair (θ_i, θ_j) the possible decisions are now $\theta_i < \theta_j$, $\theta_i > \theta_j$, and $-\infty < \theta_j - \theta_i < \infty$ instead of the earlier $\theta_i = \theta_j$. Since there is no loss in omitting the vacuous statements, the total decisions consist in the ordering of some but not necessarily all of the pairs (θ_i, θ_j). In the case of three populations, for example, the following decision types will occur:

(a) $\theta_i < \theta_j < \theta_k$, (c) $\theta_i < \theta_j$, $\theta_i < \theta_k$

(b) $\theta_i < \theta_k$, $\theta_j < \theta_k$, (d) $\theta_i < \theta_j$

(e) no statement.

We shall now show that for this procedure the probability of error can be controlled through the choice of α, and that its maximum is in fact attained when all of the θ's are equal and is then given by

$$P\left\{\frac{|\bar{X}_j - \bar{X}_i|}{S} > C_{ij} \quad \text{for some} \quad i, j\right\}. \tag{10.7}$$

In the particular case of equal sample sizes this becomes

$$P\left\{\frac{\max |\bar{X}_j - \bar{X}_i|}{S} > C\right\} \tag{10.8}$$

where C is the cut-off point of the one-sided t-test at level α.[3] The probability of error is the probability measure of the set

$$\cup\left\{\frac{\bar{X}_j - \bar{X}_i}{S} > C_{ij}\right\} + \cup\left\{\frac{\bar{X}_\ell - \bar{X}_k}{S} < C_{k\ell}\right\} \tag{10.9}$$

[3] For a table of the values C for which this probability is 1 per cent or 5 per cent see [2], where also a number of related tables are discussed.

where the first union is extended over all pairs (i, j) for which $\theta_j \leqq \theta_i$ and the second union over all pairs (k, ℓ) for which $\theta_\ell \geqq \theta_k$. Let $X_i^* = X_i - \theta_i$, so that the X_i^* are distributed as $N(0, \sigma^2)$. Then $\theta_j \geqq \theta_i$ and $\bar{X}_j - \bar{X}_i > C_{ij}S$ imply $X_j^* - X_i^* > C_{ij}S$, and $\theta_\ell \leqq \theta_k$ and $\bar{X}_\ell - \bar{X}_k < C_{k\ell}S$ imply $X_\ell^* - X_k^* < C_{k\ell}S$. The probability of the set (10.9) is therefore not decreased if one replaces the X's by X^*'s, nor if one extends the union over all pairs (i, j) and (k, ℓ). But this is equivalent to evaluating the probability of (10.9) under the assumption that all of the θ's are equal, which completes the proof.

11. Decision problems with simple loss functions. As a tool for proving the procedures of Secs. 9 and 10 to be unbiased with uniformly minimum risk, we shall now give an extension of Theorem 2 (Sec. 7), which is valid for a rather general class of decision problems. We shall say that a decision problem $\mathcal{P}$ is *simple* if it satisfies the following two conditions.

(a) Its loss function $W(\theta, d)$, considered for fixed d as a function of θ, has sets of constancy independent of d, that is, there exists a partition Π of the parameter space Ω into sets Θ_i, $i \varepsilon I$, such that $W(\theta, d)$ is independent of θ on each Θ_i. We may then write

$$W(\theta, d) = V_i(d) \quad \text{for} \quad \theta \varepsilon \Theta_i . \tag{11.1}$$

(b) With respect to some convergence notion in Ω, $\theta_n \to \theta_0$ implies $E_{\theta_n}\psi(X) \to E_{\theta_0}\psi(X)$ for each integrable ψ or, if all of the functions V_i are bounded, for each bounded ψ.

We shall require the following properties of simple decision problems.

(i) *For any procedure δ the risk function $R_\delta(\theta)$ is continuous on each set of the partition* Π. The risk function is given by

$$R_\delta(\theta) = E_\theta V_i[\delta(X)] \quad \text{for} \quad \theta \varepsilon \Theta_i . \tag{11.2}$$

Hence θ_n, $\theta_0 \varepsilon \Theta_i$ and $\theta_n \to \theta_0$ imply

$$R_\delta(\theta_n) = E_{\theta_n} V_i[\delta(X)] \to E_{\theta_0} V_i[\delta(X)] = R_\delta(\theta_0),$$

as was to be proved.

(ii) *Unbiasedness of a procedure δ implies the continuity of its risk function.* By (i) it is enough to prove this for boundary points of Π. Let θ_0 be such a boundary point, and suppose that $\theta_0 \varepsilon \Theta_i$ and $\theta_0 = \lim_{n\to\infty} \theta_n$ with $\theta_n \varepsilon \Theta_j$. Unbiasedness implies

$$E_{\theta_0} V_i[\delta(X)] \leqq E_{\theta_0} V_j[\delta(X)],$$

$$E_{\theta_n} V_i[\delta(X)] \geqq E_{\theta_n} V_j[\delta(X)] \qquad \text{for } n = 1, 2, \cdots$$

and hence also

$$E_{\theta_0} V_i[\delta(X)] \geqq E_{\theta_0} V_j[\delta(X)].$$

It follows that

$$R_\delta(\theta_0) = E_{\theta_0} V_i[\delta(X)] = E_{\theta_0} V_j[\delta(X)] = \lim_{n\to\infty} R_\delta(\theta_n).$$

(iii) *Any restricted product of simple decision procedures is again simple.*

We can now prove the main result of this section, which provides a sufficient condition for unbiasedness of a product procedure to imply continuity of the risk functions of all of the component procedures.

(iv) Let $\mathcal{P}$ be a restricted product of a finite number of simple decision problems $\mathcal{P}_\gamma$, and suppose that the partitions Π_γ of the component problems into sets $\Theta_{i\gamma}$, $i \varepsilon I_\gamma$ satisfy the following condition.

> (*) Let $\Theta_{i\gamma_0}$, $\Theta_{j\gamma_0}$ be any two sets of Π_{γ_0} with common boundary points, let θ_0 be any such point, and assume without loss of generality that $\theta_0 \varepsilon \Theta_{i\gamma_0}$. Then there exists a sequence of points $\theta_n \varepsilon \Theta_{j\gamma_0}$ such that $\theta_n \to \theta_0$ and
>
> $$\theta_n \sim \theta_0(\Pi_\gamma) \quad \text{for all} \quad \gamma \neq \gamma_0,$$
>
> where $\theta \sim \theta'(\Pi)$ indicates that the two points lie in the same set of the partition.

Under these assumptions, if the risk function $R_\delta(\theta)$ of a product procedure δ is continuous, so are the risk functions $R_{\delta_\gamma}(\theta)$ of the components δ_γ of δ.

To see this, let γ_0 be any fixed value of γ, θ_0 any boundary point of Π_{γ_0}, and $\{\theta_n\}$ the sequence guaranteed by (*). Since $R_\delta(\theta) = \sum_\gamma R_{\delta_\gamma}(\theta)$ is continuous, we have

$$\sum\nolimits_\gamma R_{\delta_\gamma}(\theta_n) \to \sum\nolimits_\gamma R_{\delta_\gamma}(\theta_0).$$

Also, for each $\gamma \neq \gamma_0$ all of the points θ_n lie in the same set of the partition Π_γ with θ_0, so that by (i)

$$R_{\delta_\gamma}(\theta_n) \to R_{\delta_\gamma}(\theta_0) \quad \text{for all} \quad \gamma \neq \gamma_0 .$$

It follows that $R(\delta_{\gamma_0},\theta_n) \to R(\delta_{\gamma_0},\theta_0)$, as was to be proved, where we have written $R(\delta, \theta)$ for $R_\delta(\theta)$.

It is convenient that (*) depends only on the partitions Π_γ, not on the values the loss functions W_γ take on over these partitions. For applications it is further important to note that (*) may be weakened slightly. Let $\Lambda_{i\gamma}$ be the set of common boundary points of $\Theta_{i\gamma}$ and $\Theta_{j\gamma}$ that belongs to $\Theta_{i\gamma}$. Then in order to ensure (iv) it is sufficient if (*) holds on a dense subset of each $\Lambda_{ij\gamma}$. This is an immediate consequence of (i).

Consider now any problem $\mathcal{P}$, which is a restricted product of a finite number of problems $\mathcal{P}_\pm(\theta_\gamma)$, and which satisfies (*). For the component problems continuity of the risk function is equivalent to similarity on the boundary at level $\alpha_\gamma = b_\gamma/(a_\gamma + b_\gamma)$. Hence a product procedure δ uniformly minimizes the risk among all unbiased procedures of $\mathcal{P}$ provided each component procedure δ_γ uniformly minimizes the risk among all procedures of $\mathcal{P}_\pm(\theta_\gamma)$ that are similar on the boundary at level α_γ. Since $\mathcal{P}_\pm(\theta_\gamma)$ is formally equivalent to the problem of testing $\theta \leqq \theta_\gamma$ or $\theta \geqq \theta_\gamma$, this is the case in particular if the possible distributions of the observable random variables constitute an exponential family, and the procedures δ_γ are the best unbiased tests of the hypotheses in question. Under these conditions it is then only necessary to verify (*) in order to establish the desired optimum property for the resulting δ.

As an example consider problem (9.4), which is generated by $\mathcal{P}_{\pm}(\theta_0)$. Here $\mathcal{P}_{-}(\theta_0)$ induces the partition $\theta \leqq \theta_0$, $\theta > \theta_0$, with θ_0 as its only boundary point. Let θ_n be any sequence of points greater than and tending to θ_0. Then all the points θ_n and θ_0 lie in the set $\theta \geqq \theta_0$, and hence $\theta_n \sim \theta_0$ with respect to the partition induced by $\mathcal{P}_{+}(\theta_0)$, which completes the verification of (*). The argument is exactly the same for problem (9.6) and (10.1).

In the example of Sec. 10 leading to procedure (10.7), a typical partition is $\theta_2 \leqq \theta_1$, $\theta_2 > \theta_1$. Attention may be restricted to boundary points $\theta^{(0)}$ with coordinates

$$\theta_{i_1}^{0} < \cdots < \theta_{i_r}^{(0)} < \theta_1^{(0)} = \theta_2^{(0)} < \theta_{j_1}^{(0)} < \cdots < \theta_{j_{s-r-2}}^{(0)}.$$

For these, (*) is satisfied by the sequence of points $\theta^{(n)}$ with coordinates $\theta_i^{(n)} = \theta_i^{(0)}$ for $i \neq 2$, and $\theta_2^{(n)}$ between $\theta_1^{(0)}$ and $\theta_{j_1}^{(0)}$ and tending to $\theta_1^{(0)}$.

12. Consecutive decisions in a single experiment. The multiple decision problems treated in the previous sections were generated by the simultaneous consideration of a number of simpler component problems. We shall now suppose that these separate problems arise not in parallel but in sequence. A single sample is available for investigating a number of questions that are potentially of interest and are taken up one by one. Whether a given question is relevant, or which of a number of possible alternative formulations is appropriate at a certain stage, depends on the decisions reached up to that point.

(i) As an example suppose that independent variables $X_1, \cdots, X_n$ from a normal distribution $N(\xi, \sigma^2)$ are measurements on an experimental batch of a new product of quality ξ. The product is of no interest unless $\xi > \xi_0$, so that one will wish to test first of all the hypothesis $H_1: \xi \leqq \xi_0$. If the quality is found satisfactory, that is, if H_1 is rejected, it becomes necessary to investigate the variability of the product. One will then test $H_2: \sigma \geqq \sigma_0$, and in case this hypothesis is accepted one will try to reduce σ, for example by using less variable materials. The problem of testing H_2 arises here only in case H_1 is rejected.

(ii) Suppose that two treatments are being compared on a number of different categories of patients. Let the observed effect of treatment $i (i = 1, 2)$ on the kth patient of the jth category be distributed as $N(\xi_{ij}, \sigma^2)$ where

$$\xi_{ij} = \eta + \lambda_i + \mu_j + \nu_{ij} \; (\textstyle\sum_i \lambda_i = \sum_j \beta_j = \sum_i \nu_{ij} = \sum_j \nu_{ij} = 0).$$

Here λ_i is the main effect of treatment i and ν_{ij} the interaction between the ith treatment and the jth category. One may believe in the possibility of the interactions being negligible and hence wish first to test the hypothesis $H_1: \nu_{ij} = 0$ for all i, j. In case H_1 is accepted the λ's are the objects of primary interest, and the problem becomes that of deciding whether $\lambda_2 - \lambda_1$ is $<$, $=$, or > 0, or to estimate this difference either by confidence intervals or by a point estimate. On the other hand, if H_1 is rejected one will be concerned less with the over-all effects of the treatments which is measured by the λ's than with the treatment differences $\xi_{2j} - \xi_{1j}$ for each category.

More generally, let there be given a first problem $\mathcal{P}'$, in which the possible decisions are $d'_1, \cdots, d'_m$ and the loss function is $W'(\theta, d)$. If decision d_i is taken, a second problem $\mathcal{P}''_i$ arises with possible decisions $d''_{ij} (j = 1, \cdots, n_i)$ and loss function $W''_i(\theta, d)$. The combination of these two problems leads to a two-stage problem with decisions $d_{ij} = (d'_i, d''_{ij})$. One may of course continue in this manner and suppose that decision (d'_i, d''_{ij}) gives rise to a further problem $\mathcal{P}'''_{ij}$ with decisions d'''_{ijk}. However, it is enough to treat the case of two levels since the discussion then extends immediately to the more general situation by induction.

In specifying a loss function we shall assume that even if a wrong decision is taken at the first step, so that the second problem is not the most appropriate one or perhaps need not have been considered at all, it is still desirable to do as well with respect to it as is possible. Thus in example (ii) above, if one has incorrectly decided that the interactions are negligible, one will in the estimation of $\theta_2 - \theta_1$ still wish to obtain as good an estimate as possible, and analogously if H_1 has wrongly been rejected. Whether the assumption holds in Example (i), that is, whether one would wish to control the variability of the new product after having mistakenly judged its quality to be satisfactory, appears to depend on the circumstances of the problem.

With this assumption, a natural loss function for the compound problem is

$$W(\theta, d_{ij}) = W'(\theta, d'_i) + W''_i(\theta, d''_{ij}). \tag{12.1}$$

The possibility of not considering a second problem in case a certain decision d'_i is taken at the first step, is included in this formulation. One need then only take as problem $\mathcal{P}''_i$ the vacuous decision problem, that is, set $n_i = 1$ and $W''_i(\theta, d''_{i1}) = 0$. Suppose in particular, as was the case in Example (i), that a second problem occurs only for one of the decisions of the first stage, say d'_1. The possible decisions of the compound problem are then

$$d_{ij} = (d'_1, d''_{1j}), \qquad j = 1, \cdots, n$$

and

$$d_2 = d'_2, \cdots, d_m = d'_m,$$

and the loss function is given by

$$\begin{aligned} W(\theta, d_i) &= W'(\theta, d'_i), && i = 2, \cdots, m; \\ W(\theta, d_{1j}) &= W'(\theta, d'_1) + W''(\theta, d''_{1j}), && j = 1, \cdots, n. \end{aligned} \tag{12.2}$$

Returning now to the general case, suppose that there exist a satisfactory procedure δ' for $\mathcal{P}'$, which takes decision d'_i when $X \varepsilon D'_i$, and that the problems $\mathcal{P}''_1, \cdots, \mathcal{P}''_m$ are all different. It then seems natural to retain δ' as first step of the compound procedure, and to consider the problems at the second level relative to the circumstances in which they occur, namely conditionally given that $X \varepsilon D'_1, \cdots, X \varepsilon D'_m$ respectively. Suppose further that for each $i = 1, \cdots, m$ there exists a satisfactory procedure δ''_i for $\mathcal{P}''_i$ when the distribution of X is the conditional distribution given $X \varepsilon D'_i$. Such a procedure consists of a partition of the new sample space D'_i into regions D''_{ij} in which the decisions

d''_{ij} are taken. Together, the sets $D''_{ij}(j = 1, \cdots, n_i; i = 1, \cdots, m)$ form a partition of the original sample space and a solution of the compound problem, with decision $d_{ij} = (d'_i, d''_{ij})$ being taken when $X \varepsilon D''_{ij}$. One can of course again include in the formulation the possibility of ruling out some of the decisions d_{ij}, and the resulting compatibility questions can be treated exactly as before. However, this possibility seems to be less important in the present context and in order not to complicate the discussion unnecessarily we shall assume that no restrictions are imposed on the compound decision problem.

As an application consider the case that m and the n_i are equal to two, so that each of the component problems is one of hypothesis testing. Suppose that the hypotheses in question concern the parameters θ_i in an exponential family

$$dP_\theta(x) = C(\theta_1, \cdots, \theta_r)e^{\Sigma\theta_i T_i(x)}\, d\nu(x).$$

There then exist uniformly most powerful unbiased tests of the hypotheses $\theta_i \leqq \theta_i^0$ and more generally of $\sum c_i\theta_i \leqq c_0$. (See, for example, [2].) Since after truncation on a fixed set D the family of distributions P_θ retains its property of forming an exponential family, such optimum unbiased tests will in particular also exist at the second stage after a preliminary test of significance has been performed as a first step. This will however not coincide with the standard optimum test for the corresponding problem without truncation.

We have assumed so far that the problems occurring at the second stage are all distinct. Suppose now instead that $\mathcal{P}''_1$ is appropriate when decisions $d'_1, \cdots, d'_{r_1}$ are taken in the first problem, that $\mathcal{P}''_2$ corresponds to decisions $d'_{r_1+1}, \cdots, d'_{r_1+r_2}$, etc. One would then consider the problems $\mathcal{P}''_1, \mathcal{P}''_2, \cdots$ conditionally given that $X \varepsilon D'_1 + \cdots + D'_{r_1}$, $X \varepsilon D'_{r_1+1} + \cdots + D'_{r_1+r_2}, \cdots$, and otherwise proceed as before. If one is dealing in particular with the product of two decision problems so that all of the problems at the second level are the same, one considers this common problem $\mathcal{P}''$ given that $X \varepsilon D'_1 + \cdots + D'_m$, that is, unconditionally. The procedure therefore reduces to the product of the procedures for $\mathcal{P}'$ and $\mathcal{P}''$, so that the present theory agrees with that given earlier for products of decision problems.

Unfortunately the properties of the conditional procedures considered in the present section are not as satisfactory as of those discussed in the earlier parts of this paper. To be specific, let $\mathcal{P}'$ and $\mathcal{P}''_1, \cdots, P''_m$ define a two-stage problem, in which the components $\mathcal{P}''_i$ of the second stage are distinct. The risk function of a procedure δ with components $\delta', \delta''_1, \cdots, \delta''_m$ is then

$$R_\delta(\theta) = R_{\delta'}(\theta) + \textstyle\sum_{i=1}^m P_\theta(D'_i)R_{\delta''_i|\delta'}(\theta) \tag{12.3}$$

where the notation $R_{\delta''_i|\delta'}$ is used to indicate that this risk component is computed conditionally given $X \varepsilon D'_i$ and hence depends on δ' as well as on δ''_i.

It is clear from (12.3) that unbiasedness of δ' and $\delta''_1, \cdots, \delta''_m$ implies that of δ. Also it is again true for most problems of interest that unbiasedness of δ implies either unbiasedness or at least similarity on the boundary for $\delta', \delta''_1, \cdots, \delta''_m$. However, the basic comparison of two procedures in terms of their components is no longer simple. In particular,

$$R_{\delta'}(\theta) \leqq R_{\psi'}(\theta), \qquad R_{\delta_i''|\delta'}(\theta) \leqq R_{\psi_i''|\psi'}(\theta)$$

does not in general imply $R_\delta(\theta) \leqq R_\psi(\theta)$. If one wishes to minimize $R_\delta(\theta)$ then, given δ', one must select δ_i'' to minimize $R_{\delta_i''|\delta'}(\theta)$. But the best choice of δ' is not necessarily that which minimizes $R_{\delta'}(\theta)$ since the choice of δ' influences not only $R_{\delta'}(\theta)$ but also the second components of (12.3) and in particular the conditional risks $R_{\delta_i|\delta'}$. As a result of these complications it turns out that for the problem under consideration there usually does not exist among the unbiased procedures one that uniformly minimizes the risk. Of the procedures, the components of which have this optimum property we can only say that they are unbiased, and within the class of all unbiased procedures admissible.

13. Some examples of conditional procedures. Although we have found no satisfactory justification for the procedures discussed in the preceding section, they are rather natural from the Neyman-Pearson point of view, and we shall briefly illustrate them here with a few examples leaving a more detailed discussion and comparison with alternative procedures for a later paper.

(i) The problem mentioned at the beginning of Sec. 12 is concerned with testing, on the basis of a normal sample, the two hypotheses $H_1: \xi \leqq \xi_0$ and $H_2: \sigma \geqq \sigma_0$, where H_2 is assumed to be of interest only in case H_1 is rejected. If without loss of generality we put $\xi_0 = 0$, the best unbiased procedure for testing H_1 is Student's t-test with rejection region

$$\bar{X}/S \geqq C \tag{13.1}$$

and size $\alpha_1 = b_1/(a_1 + b_1)$. With this as first step, the condition of unbiasedness implies in the usual way that the rejection region R_2 of H_2 must satisfy

$$P_{\sigma_0}(R_2 \mid S \leqq \bar{x}/C \mid \bar{x}) \overset{(\bar{x})}{=} \alpha_2 . \tag{13.2}$$

Applying the fundamental lemma of Neyman and Pearson one sees that the uniformly most powerful unbiased conditional test of H_2 has a rejection region of the form

$$S^2 < f(\bar{X}). \tag{13.3}$$

Here the function f is defined by

$$\int_0^{f(u)} p_{\sigma_0}(z)\, dz = \alpha_2 \int_0^{u^2/C^2} p_{\sigma_0}(z)\, dz, \qquad u > 0, \tag{13.4}$$

where p_{σ_0} is the probability density of S^2 when $\sigma = \sigma_0$.

The resulting compound procedure then decides between the three possible conclusions

$d_1: \xi \leqq \xi_0$—the new product is not of satisfactory quality,

$d_2: \xi > \xi_0,\ \sigma \geqq \sigma_0$—the quality of the new product is satisfactory but its variability must be reduced,

$d_3: \xi > \xi_0,\ \sigma < \sigma_0$—the new product is satisfactory with regard to both quality and variability, as the sample falls into the corresponding one of the regions

$$D_1: \bar{X} \mid S \leqq C,$$
$$D_2: \bar{X} \mid S > C, \qquad S^2 \geqq f(\bar{X})$$
$$D_3: \bar{X} \mid S > C, \qquad S^2 < f(\bar{X}).$$

These decision regions are illustrated for the case $n = 10$, $\alpha_1 = \alpha_2 = .05$ in the figure.

(ii) As a somewhat more complex example consider two treatments which are being compared on a number of different categories of patients. Let the observed effect Y_{ijk} of treatment i $(i = 1, 2)$ on the kth patient $(k = 1, \cdots, n)$ in the jth category be distributed as $N(\xi_{ij}, \sigma^2)$, and let

$$\xi_{ij} = \eta + \lambda_i + \mu_j + \nu_{ij} \left(\sum_i \lambda_i = \sum_j \mu_j = \sum_i \nu_{ij} = \sum_j \nu_{ij} = 0\right)$$

where λ_i is the main effect of treatment i and ν_{ij} the interaction between the ith treatment and jth category. One may here wish to test first the hypothesis of no interaction

$$H_1: \nu_{ij} = 0 \qquad \text{for all } i, j.$$

If H_1 is accepted one will be interested in the difference of the λ's and wish either to test it or alternatively to estimate it by confidence intervals or point estimate. We shall here suppose that we then want to test

$$H_2: \lambda_2 - \lambda_1 \leqq 0.$$

On the other hand, if H_1 is rejected one will usually be concerned less with comparing the over-all effects of the two treatments, which is measured by the λ's, than with a comparison of the treatment effects ξ_{2j} and ξ_{1j} separately for each category j. In particular one may be interested to test the set of hypotheses

$$H_{3j}: \xi_{2j} - \xi_{1j} = (\lambda_2 + \nu_{2j}) - (\lambda_1 + \nu_{1j}) \leqq 0.$$

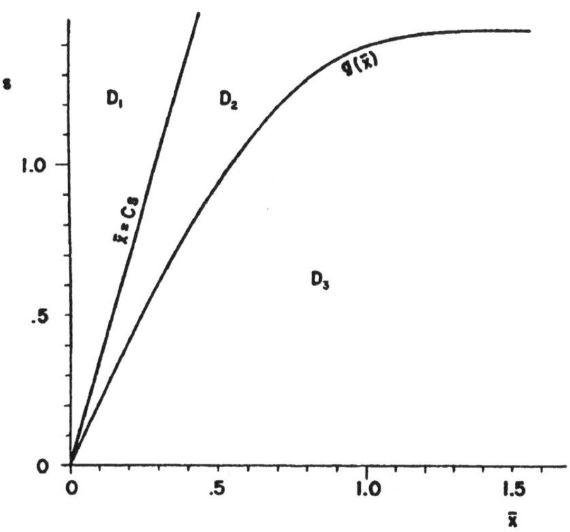

FIG. 1

Although there exists no uniformly most powerful unbiased test of H_1 it seems natural to start out with the standard test of this hypothesis which is uniformly most powerful invariant, and is given by the acceptance region

$$S_1^2/S_0^2 \leqq C \tag{13.5}$$

where

$$S_0^2 = \sum\sum\sum (Y_{ijk} - Y_{ij.})^2$$
$$S_1^2 = \sum\sum (Y_{ij.} - Y_{i..} - Y_{.j.} + Y_{...})^2.$$

The hypothesis H_2 should then be considered conditionally, given $S_1^2/S_0^2 \leqq C$. A routine application of the theory of unbiased tests and similar regions[4] shows that a necessary condition for the rejection region R_2 to be unbiased is

$$P_{\lambda_2=\lambda_1}\{R_2 \mid s_1^2, s_0^2 + s_2^2, S_1^2/S_0^2 \leqq C\} = \alpha_2$$

for all values of s_1^2 and $s_0^2 + s_2^2$ where

$$S_2^2 = \sum_{i=1}^2 (Y_{i..} - Y_{...})^2 = \tfrac{1}{2}(Y_{2..} - Y_{1..})^2.$$

If one puts

$$U = (Y_{2..} - Y_{1..})/S_0, \qquad V = S_0^2 + S_2^2, \qquad W = (S_0^2 + S_2^2)/S_1^2,$$

it follows from the fundamental lemma that the uniformly most powerful (conditional) test of H_2 has the rejection region

$$U \geqq k(V, W)$$

where k is determined by

$$P_{\lambda_1=\lambda_2}\{U \geqq k(V, W) \mid V = v, W = w \text{ and } U^2 \leqq Cw - 1\} \overset{(v,w)}{=} \alpha_2 .$$

Now when $\lambda_1 = \lambda_2$, the variable U is independent of V so that k depends only on w, $k(v, w) = f(w)$ say. The rejection region then becomes

$$U \geqq f(w), \tag{13.6}$$

where f is determined by

$$P_{\lambda_1=\lambda_2}\{U < f(w) \mid U^2 \leqq Cw - 1\} \overset{(w)}{=} (1 - \alpha_2).$$

Since acceptance of H_1 is equivalent to $u^2 \leqq Cw - 1$ it implies in particular that $0 \leqq Cw - 1$. The defining condition for f may therefore be written as

$$\int_{-(Cw-1)^{1/2}}^{f(w)} p_U(u)\, du = (1 - \alpha_2) \int_{-(Cw-1)^{1/2}}^{(Cw-1)^{1/2}} p_U(u)\, du \tag{13.7}$$

where $p_U(u)$ is the probability density of U when $\lambda_1 = \lambda_2$, that is, essentially, the density of a t-distribution with $2m(n - 1)$ degrees of freedom.

---

[4] The proof requires the easily shown fact that the family of noncentral χ^2-distributions with a fixed number of degrees of freedom is boundedly complete.

Let us now consider in a similar manner the hypothesis

$$H_{31}:(\lambda_2 - \lambda_1) + (\nu_{21} - \nu_{11}) \leqq 0,$$

conditionally given that $S_1^2/S_0^2 > C$. Let $\hat{\lambda}_i$ and $\hat{\nu}_{ij}$ be the least squares estimates of the corresponding parameters so that

$$S_1^2 = \sum_{i=1}^{2}\sum_{j=1}^{m}\hat{\nu}_{ij}^2$$

and let

$$Z_1 = \tfrac{1}{2}(\hat{\lambda}_2 - \hat{\lambda}_1 - \hat{\nu}_{21} + \hat{\nu}_{11})$$

$$Z_2 = \tfrac{1}{2}(\hat{\lambda}_2 - \hat{\lambda}_1 + \hat{\nu}_{21} - \hat{\nu}_{11})$$

$$S_1'^2 = S_1^2 - (\hat{\nu}_{21}^2 - \hat{\nu}_{11})^2 = S_1^2 - (Z_2 - Z_1)^2.$$

If $\zeta_i = E(Z_i)$, the hypothesis becomes $\zeta_2 \leqq 0$, and unbiasedness of the conditional test of H_{31} with rejection region R_{31} implies

$$P_{\zeta_2=0}\{R_{31} \mid z_1, s_0^2 + z_2^2, s_1'^2 \text{ and } S_1^2/S_0^2 > C\} \equiv \alpha_{31}.$$

The condition $S_1^2 > CS_0^2$ is equivalent to $(Z_2 - Z_1)^2 > CS_0^2 - S_1'^2$, which may be rewritten as

$$\left(Z_2 - \frac{Z_1}{1+C}\right)^2 + \frac{CZ_1^2}{(1+C)^2} > \frac{C}{1+C}T^2 - S_1'^2,$$

where $T^2 = S_0^2 + Z_2^2$. This is satisfied for all values of Z_2 if

$$\frac{C}{1+C}T^2 - S_1'^2 - \frac{CZ_1^2}{(1+C)^2} \leqq 0 \tag{13.8}$$

and otherwise for the values of Z_2, for which either

$$\frac{Z_2}{T} > \frac{Z_1}{(1+C)T} + \sqrt{\frac{C}{1+C} - \frac{S_1'^2}{T^2} - \frac{C}{(1+C)^2}\frac{Z_1^2}{T^2}} = K_2\left(\frac{Z_1}{T}, \frac{S_1'}{T}\right)$$

or

$$\frac{Z_2}{T} < \frac{Z_1}{(1+C)T} - \sqrt{\frac{C}{1+C} - \frac{S_1'^2}{T^2} - \frac{C}{(1+C)^2}\frac{Z_1^2}{T^2}} = K_1\left(\frac{Z_1}{T}, \frac{S_1'}{T}\right).$$

Since Z_2/T is independent of Z_1/T and S_1'/T when $\zeta_2 = 0$, the uniformly most powerful unbiased test of H_{31} given $S_1^2/S_0^2 > C$ is then given by the rejection region

$$\frac{Z_2}{T} \geqq K\left(\frac{Z_1}{T}, \frac{S_1'}{T}\right)$$

with the function K defined as follows. When $(Z_1/t,\ S_1'/t)$ satisfies (13.8), we have

$$K\left(\frac{z_1}{t}, \frac{s_1'}{t}\right) = K$$

where

$$\int_K^1 p_R(r)\, dr = \alpha_{31},$$

$p_R(r)$ being the probability density of

$$R = \frac{Z_2}{T} = \frac{\frac{1}{2}(\hat{\lambda}_2 - \hat{\lambda}_1 - \hat{\nu}_{21} + \hat{\nu}_{11})}{\sqrt{S_0^2 + \frac{1}{4}(\hat{\lambda}_2 - \hat{\lambda}_1 + \hat{\nu}_{21} - \hat{\nu}_{11})^2}}$$

when $\zeta_2 = 0$. For all other values of $(z_1/t, s_1'/t)$, $K(z_1/t, s_1'/t)$ is given by

$$\int_{K((z_1/t),(s_1'/t))}^1 p_R(r)\, dr = \alpha_{31}\left[\int_{-1}^{K_1((z_1/t),(s_1'/t))} p_R(r)\, dr + \int_{K_2((z_2/t),(s_1'/t))}^1 p_R(r)\, dr\right].$$

(iii) As an example involving more than two stages let us consider the determination of the degree of a regression polynomial. Let $Y_1, \cdots, Y_n$ be independently normally distributed with constant variance σ^2 and means

$$\eta_i = E(Y_i) = e_s + e_{s-1}x_i + \cdots + e_0 x_i^s \qquad (i = 1, \cdots, n).$$

We shall assume that a polynomial of degree s will in any case be adequate for our purposes, and wish to determine the smallest degree $r \leqq s$ that would also be satisfactory. It is convenient for this purpose to express the regression polynomial in terms of the orthogonal polynomials P_i defined by

$$\sum_{k=1}^n P_i(x_k)P_j(x_k) = \begin{cases} 1 & \text{for } i = j \\ 0 & \text{for } i \neq j. \end{cases}$$

Writing

$$\eta_i = c_s + c_{s-1}P_1(x_i) + \cdots + c_0 P_s(x_i),$$

we test successively the hypotheses $H_0\!: c_0 = 0$, $H_1\!: c_1 = 0, \cdots$, continuing as long as no rejection occurs.

The problem can be stated in the following canonical form: $X_1, \cdots, X_s$ and S_0^2 are independent variables, with X_i being distributed as $N(\xi_i, \sigma^2)$ and with S_0^2/σ^2 having a χ^2-distribution with $n_0 = n - s$ degrees of freedom. One wishes to test consecutively the hypotheses $H_i\!:\xi_i = 0$, $(i = 1, \cdots, s)$ continuing as long as no rejection occurs. Slightly more generally one might have variables $X_{ij}\!:N(\xi_{ij}, \sigma^2)$, $(j = 1, \cdots, n_i; i = 1, \cdots, s)$ and S_0^2 and consider consecutively the hypotheses $H_i\!:\xi_{ij} = 0$ for $j = 1, \cdots, n_i$. Invariance reduces the problem to the statistics S_0^2 and $S_i^2 = \sum_{j=1}^{n_i} X_{ij}^2$, which are independent and where S_i^2/σ^2 has a χ^2 distribution with n_i degrees of freedom when H_i is true.

Let

$$U_i = \frac{S_i^2}{S_0^2}, \qquad V_i = \frac{S_i^2}{S_{i-1}^2}, \qquad W_i = V_i\left(1 + \frac{1}{U_i}\right) = \frac{S_i^2 + S_0^2}{S_{i-1}^2}.$$

We shall now show inductively that the best unbiased conditional test of H_i, given the acceptance of $H_1, \cdots, H_{i-1}$, has an acceptance region of the form

$$U_i \leqq f_i(W_i, V_{i-1}, \cdots, V_2). \tag{13.9}$$

The functions f_i are defined by

$$\begin{cases} f_1(w_1) = C, \\ \displaystyle\int_0^{f_i(w_i, v_{i-1}, \cdots, v_2)} p_i(r)\, dr = (1 - \alpha_i) \int_0^{w_i h_{i-1}(v_{i-1}, \cdots, v_2) - 1} p_i(r)\, dr \quad (i \geqq 2) \end{cases} \tag{13.10}$$

where p_i is the probability density of S_i^2/S_0^2 when H_i is true, h_i is defined by

$$\begin{cases} h_1 = C, \\ h_i(v_i, \cdots, v_2) = \min[v_i h_{i-1}(v_{i-1}, \cdots, v_2), g_i(v_i, \cdots, v_2)] \quad (i \geqq 2) \end{cases} \tag{13.11}$$

and g_i is a function taking on values in the space of u_i and defined by

$$u = g_i(v_i, \cdots, v_2) \leftrightarrow v_i = \frac{f_i^{-1}(u, v_{i-1}, \cdots, v_2)}{1 + \dfrac{1}{u}} \quad (i \geqq 2) \tag{13.12}$$

where f_i^{-1} denotes the inverse of f_i considered for fixed values of $v_{i-1}, \cdots, v_2$ as a function of u. It is seen that (13.10), (13.11) and (13.12) define f_i in terms of f_{i-1}.

The best unbiased test of H_1 has the acceptance region $u_1 \leqq C$ and hence satisfies (13.9). Suppose now that the acceptance regions $A_1, \cdots, A_{i-1}$ have been shown to be given by (13.9). To prove (13.9) for A_i we need the fact that

$$A_1 \cap \cdots \cap A_{i-1} = \{(u, v): u_{i-1} \leqq h_{i-1}(v_{i-1}, \cdots, v_2)\} \tag{13.13}$$

and since this is true for $i = 2$, we may accept (13.13) as part of our induction hypothesis. The condition of unbiasedness implies that A_i should satisfy

$$P_{H_i}\{A_i \mid s_0^2 + s_i^2, s_1^2, \cdots, s_{i-1}^2 \text{ and } U_{i-1} \leqq h_{i-1}(V_{i-1}, \cdots, V_2)\} = 1 - \alpha_i.$$

Now $U_{i-1} \leqq h_{i-1}$ is equivalent to

$$U_i \leqq W_i h_{i-1}(V_{i-1}, \cdots, V_2) - 1$$

so that the best unbiased acceptance region for H_i is

$$U_i \leqq K_i(W_i, S_1^2, \cdots, S_{i-1}^2)$$

where

$$P_{H_i}\{U_i \leqq K_i \mid s_0^2 + s_i^2, s_1^2, \cdots, s_{i-1}^2 \text{ and } U_i \leqq W_i h_{i-1}(V_{i-1}, \cdots, V_2) - 1\} = 1 - \alpha_i.$$

Since U_i is independent of $W_i, V_{i-1}, \cdots, V_2$ when H_i is true, it is seen that K_i depends only on $w_i, v_{i-1}, \cdots, v_2$ and that A_i is given by (13.9).

To complete the proof, it only remains to verify (13.13) for the set

$A_1 \cap \cdots \cap A_i$, which is the intersection of

$$A_1 \cap \cdots \cap A_{i-1}: u_{i-1} \leqq h_{i-1}(v_{i-1}, \cdots, v_2)$$

and

$$A_i: u_i \leqq f_i(w_i, v_{i-1}, \cdots, v_2).$$

The inequality describing A_i is equivalent to

$$f_i^{-1}(u_i, v_{i-1}, \cdots, v_2) \leqq w_i = v_i\left(1 + \frac{1}{u_i}\right)$$

and hence to

$$A_i: u_i \leqq g_i(v_i, \cdots, v_2)$$

by (13.12). Since the inequality describing $A_1 \cap \cdots \cap A_{i-1}$ is equivalent to

$$A_1 \cap \cdots \cap A_{i-1}: u_i \leqq v_i h_{i-1}(v_{i-1}, \cdots, v_2),$$

the intersection $A_1 \cap \cdots \cap A_i$ is given by (13.13), as was to be proved.

A closely related problem arises in the study of components of variance when one is dealing with a hierarchical classification. Here the problem reduces to independent statistics $S_1^2, S_2^2, \cdots$ where S_i^2/σ_i^2 has a χ^2-distribution with n_i degrees of freedom. It follows from the underlying model that $\sigma_1 \leqq \sigma_2 \leqq \cdots$, and one is interested in testing successively the hypotheses $H_1: \sigma_2 = \sigma_1$, $H_2: \sigma_3 = \sigma_2, \cdots$, continuing only if all the previous hypotheses were accepted. If $H_1, \cdots, H_i$ are true, the distribution of $S_1^2, \cdots, S_i^2$ is the same as in the preceding problem, and it is easily seen that exactly the same procedure is applicable to the present situation.

14. Minimizing the maximum risk. For problems of the kind discussed in the preceding section unbiasedness is closely related to the minimax property. We begin by considering the two-decision problem of testing a hypothesis $H: \theta \in \omega$. Let d_0 and d_1 denote the decisions of accepting and rejecting H, and let the losses of false rejection and acceptance be a and b respectively. Then we have

(i) A necessary condition for a procedure δ to be minimax is that it be unbiased.

(ii) This condition is also sufficient if the probability $P_\theta(A)$ of any set A is continuous in θ and if the common boundary of ω and ω^{-1} is nonempty.

To see (i) note that unbiasedness of a procedure δ_0 implies

$$P_\theta(d_1) \leqq \frac{b}{a+b} \text{ for } \theta \in \omega, \qquad P_\theta(d_0) \leqq \frac{a}{a+b} \text{ for } \theta \in \omega^{-1}. \tag{14.1}$$

Since the risk function of any procedure δ is given by

$$R_\delta(\theta) = \begin{cases} aP_\theta(d_1) & \text{for } \theta \in \omega \\ bP_\theta(d_0) & \text{for } \theta \in \omega^{-1}, \end{cases} \tag{14.2}$$

δ_0 satisfies

$$\sup R_{\delta_0}(\theta) \leqq \frac{ab}{a+b}. \tag{14.3}$$

Any minimax procedure δ_1 must then also satisfy (14.3) and hence by (14.2) must also be unbiased. Suppose next that the assumptions of (ii) hold, let θ_0 be a common boundary point of ω and ω^{-1}, and let δ_0 be unbiased. Then (14.1) and $P_\theta(d_0) = 1 - P_\theta(d_1)$ show that

$$P_{\theta_0}(d_1) = \frac{b}{a+b} \text{ and hence } R_{\delta_0}(\theta_0) = ab/(a+b).$$

Thus for any unbiased procedure δ_0 we have

$$\sup R_{\delta_0}(\theta) = \frac{ab}{a+b}. \tag{14.4}$$

By (i), this relation holds in particular for any minimax procedure, so that $ab/(a+b)$ is the minimax risk and (ii) follows from (14.4).

We shall now extend (ii) to the case of a number of hypotheses $H_i:\xi_i(\theta) \in \omega_i$ $(i = 1, \cdots, s)$ to be tested in sequence, where each hypothsis H_i is tested only if a particular prescribed chain of decisions has been reached on $H_1, \cdots, H_{i-1}$. For these problems we have

THEOREM 3. *Every unbiased procedure minimizes the maximum risk provided* (i) $P_\theta(A)$ *is a continuous function of* θ *for each* A, (ii) *the* 2^s *intersections* $\bigcap_{i=1}^s \omega_i^{x_i}$, *(each* $x_i = \pm 1$*), are all nonempty,* (iii) *there exists at least one parameter point* θ_0 *which lies on the common boundary of* ω_i *and* ω_i^{-1} *for all* i, (iv) *the losses* a_i *and* b_i *for falsely rejecting and accepting* H_i *satisfy condition* (14.10) *below.*

PROOF. Let us write a_i^{-1} for b_i, let d_i and d_i^{-1} denote the decisions of rejecting and accepting H_i, and let $y_i = +1$ if H_{i+1} is considered in case H_i is rejected, and $y_i = -1$ in the contrary case. If $\theta \in \bigcap_{i=1}^s \omega_i^{x_i}$, the risk of a procedure δ is then given by

$$\begin{aligned} R_\delta(\theta) &= a_1^{x_1}P(d_1^{x_1}) + P(d_1^{y_1})\{a_2^{x_2}P(d_2^{x_2} \mid d_1^{y_1}) + P(d_2^{y_2} \mid d_1^{y_1}) \\ &\quad [a_3^{x_3}P(d_3^{x_3} \mid d_1^{y_1}d_2^{y_2} + \cdots]\} \\ &= a_1^{x_1}P(d_1^{x_1}) + a_2^{x_2}P(d_1^{y_1})P(d_2^{x_2} \mid d_1^{y_1}) \\ &\quad + a_3^{x_3}P(d_1^{y_1})P(d_2^{y_2} \mid d_1^{y_1})P(d_3^{x_3} \mid d_1^{y_1}d_2^{y_2}) + \cdots . \end{aligned}$$

By comparing each of these expressions for which $x_i = 1$ with the corresponding one for which $x_i = -1$ but all the other x's are the same, it is seen that unbiasedness implies and hence is equivalent to the conditions

$$P_\theta(d_i \mid d_1^{y_1} \cdots d_{i-1}^{y_{i-1}}) \leqq \frac{b_i}{a_i + b_i} \text{ for } \theta \in \omega_i$$

$$P_\theta(d_i^{-1} \mid d_1^{y_1} \cdots d_{i-1}^{y_{i-1}}) \leqq \frac{a_i}{a_i + b_i} \text{ for } \theta \in \omega_i^{-1}$$

for $i = 1, \cdots, s$. Putting $r_i = a_i b_i / (a_i + b_i)$ we see that the risk function of any unbiased procedure satisfies

$$\sup R_\delta(\theta) \leqq r_1 + \frac{a_1^{y_1}}{a_1 + b_1} r_2 + \frac{a_1^{y_1}}{a_1 + b_1} \frac{a_2^{y_2}}{a_2 + b_2} r_3 + \cdots + \frac{a_1^{y_1}}{a_1 + b_1} \cdots \frac{a_{s-1}^{y_{s-1}}}{a_{s-1} + b_{s-1}} r_s = r^*.$$

It now remains to show that r^* is the minimax risk.

Consider any procedure δ, and let

$$\alpha_i = P_{\theta_0}(d_i \mid d_1^{y_1} \cdots d_{i-1}^{y_{i-1}}).$$

Then it follows from assumptions (i), (ii) and (iii) that

$$\text{(14.6)} \quad \sup R_\delta(\theta) \geqq \max_{x_1, \cdots, x_s} [(a_1 \alpha_1)^{x_1} + \alpha_1^{y_1}(a_2 \alpha_2)^{x_2} + \cdots + \alpha_1^{y_1} \cdots \alpha_{s-1}^{y_{s-1}}(a_s \alpha_s)^{x_s}]$$

since the 2^s expressions in brackets are the possible values of $\lim R_\delta(\theta)$ as θ tends to θ_0. Let us now minimize the right-hand side of (14.6). Given any $\alpha_1, \cdots, \alpha_{s-1}$, $x_1, \cdots, x_{s-1}$, the maximum of the pair of expressions

$$(a_1\alpha_1)^{x_1} + \cdots + \alpha_1^{y_1} \cdots \alpha_{s-2}^{y_{s-2}}(a_{s-1}\alpha_{s-1})^{x_{s-1}} + \alpha_1^{y_1} \cdots \alpha_{s-1}^{y_{s-1}}(a_s\alpha_s),$$

$$(a_1\alpha_1)^{x_1} + \cdots + \alpha_1^{y_1} \cdots \alpha_{s-2}^{y_{s-2}}(a_{s-1}\alpha_{s-1})^{x_{s-1}} + \alpha_1^{y_1} \cdots \alpha_{s-1}^{y_{s-1}}(a_s\alpha_s)^{-1},$$

is clearly minimized by minimizing

$$\max[a_s\alpha_s, a_s^{-1}(1 - \alpha_s)].$$

This gives $\alpha_s = \alpha_s^*$ where we put

$$\text{(14.7)} \qquad \alpha_i^* = \frac{a_i^{-1}}{a_i + a_i^{-1}} = \frac{b_i}{a_i + b_i},$$

so that

$$\text{(14.8)} \qquad a_i\alpha_i^* = (a_i\alpha_i^*)^{-1}.$$

We can now proceed inductively. Suppose that it has already been shown for any fixed $\alpha_1, \cdots, \alpha_i$ that the right-hand side of (14.6) is minimized by $\alpha_j = \alpha_j^*$ for $j = i + 1, \cdots, s$. Consider now the minimization of the maximum of the quantities

$$(a_1\alpha_1)^{x_1} + \cdots + \alpha_1^{y_1} \cdots \alpha_{i-1}^{y_{i-1}}(a_i\alpha_i) + \alpha_1^{y_1} \cdots \alpha_i^{y_i}[(a_{i+1}\alpha_{i+1}^*)^{x_{i+1}} + (\alpha_{i+1}^*)^{y_{i+1}}(a_{i+1}\alpha_{i+2}^*)^{x_{i+2}} + \cdots],$$

$$(a\alpha_1)^{x_1} + \cdots + a_1^{y_1} \cdots \alpha_{i-1}^{y_{i-1}}(a_i\alpha_i)^{-1} + \alpha_1^{y_1} \cdots \alpha_i^{y_i} [(a_{i+1}\alpha_{i+1}^*)^{x_{i+1}} + (\alpha_{i+1}^*)^{y_{i+1}}(a_{i+2}\alpha_{i+2}^*)^{y_{i+2}} + \cdots],$$

where the expression in brackets by (14.8) is independent of the x's and hence equal to

$$(14.9)\quad k_i = a_{i+1}\alpha^*_{i+1} + (\alpha^*_{i+1})^{y_{i+1}} a_{i+2}\alpha^*_{i+2} + \cdots + (\alpha^*_{i+1})^{y_{i+1}} \cdots (\alpha^*_{s-1})^{y_{s-1}} a_s \alpha^*_s .$$

Eliminating the common terms at the beginning, and the common factor $\alpha_1^{y_1} \cdots \alpha_{i-1}^{y_{i-1}}$ of the terms involving α_i, we see that it is enough to minimize

$$\max[a_i\alpha_i + k_i\alpha_i^{y_i}, b_i\alpha_i^{-1} + k_i\alpha_i^{y_i}].$$

This is achieved by equating the two quantities, which gives $\alpha_i = \alpha_i^*$, provided the coefficient of α_i in $b_i + (k_i - b_i)\alpha_i$ is <0 in case $y_i = 1$, or that the coefficient of α_i in $a_i\alpha_i + k_i(1 - \alpha_i)$ is >0 in case $y_i = -1$. Thus the right-hand side of (14.6) is minimized by putting $\alpha_i = \alpha_i^*$ for all i, provided

$$(14.10)\quad a_{i+1}\alpha^*_{i+1} + (\alpha^*_{i+1})^{y_{i+1}} a_{i+2}\alpha^*_{i+2} + \cdots + (\alpha^*_{i+1})^{y_{i+1}} \cdots (\alpha^*_{s-1})^{y_{s-1}} a_s\alpha^*_s < a_i^{-y_i} .$$

Putting $\alpha_i = \alpha_i^*$ in the right-hand side of (14.6) one then sees that the 2^s quantities in brackets become equal, and that their common value is r^*. Thus, for any procedure δ, we have

$$\sup R_\delta(\theta) \geqq r^*$$

and the desired result now follows from comparison with (14.5).

COROLLARY. *The conclusion of Theorem* 3 *holds if* (14.10) *is replaced by either*

$$(14.11)\qquad y_i = 1 \text{ for all } i \quad\text{and}\quad b_1 \geqq b_2 \geqq \cdots \geqq b_s$$

or

$$(14.12)\qquad y_i = -1 \text{ for all } i \quad\text{and}\quad a_1 \geqq a_2 \geqq \cdots \geqq a_s .$$

PROOF. It is necessary only to show that each of the conditions (14.11) and (14.12) implies (14.10). If all the y's are $+1$ the left-hand side of (14.10) may be rewritten as

$$b_{i+1}(\alpha^*_{i+1})^{-1} + b_{i+2}(\alpha^*_{i+2})^{-1}\alpha^*_{i+1} + \cdots + b_s(\alpha^*_s)^{-1}\alpha^*_{i+1} \cdots \alpha^*_{s-1} .$$

If (14.11) holds, this is

$$\leqq b_{i+1}[(\alpha^*_{i+1})^{-1} + \alpha^*_{i+1}(\alpha^*_{i+2})^{-1} + \cdots + \alpha^*_{i+1} \cdots \alpha^*_{s-1}(\alpha^*_s)^{-1}]$$

$$= b_{i+1}[1 - \alpha^*_{i+1}\alpha^*_{i+2} \cdots \alpha^*_s] < b_{i+1} \leqq b_i .$$

Similarly, if (14.12) holds, the left-hand side of (14.10) is

$$\leqq a_{i+1}[\alpha^*_{i+1} + (\alpha^*_{i+1})^{-1}\alpha^*_{i+2} + \cdots + (\alpha^*_{i+1} \cdots \alpha^*_{s-1})^{-1}\alpha^*_s]$$

$$= a_{i+1}[1 - (\alpha^*_{i+1} \cdots \alpha^*_s)^{-1}] < a_i .$$

If the assumptions of Theorem 3 are not satisfied it is sometimes possible to prove a slightly weaker result. We shall illustrate this with the simplest case of Example (ii) of Sec. 12. Here one is concerned with four means ξ_{ij} $(i, j = 1, 2)$ given by

$$\xi_{11} = \lambda + \mu + \nu, \qquad \xi_{12} = \lambda - \mu - \nu$$

$$\xi_{21} = -\lambda + \mu - \nu, \qquad \xi_{22} = -\lambda - \mu - \nu.$$

The first hypothesis tested is $H_0: \nu = 0$. In case of acceptance this is followed by $H_1: \lambda \leqq 0$, while in case of rejection one is interested in the two hypotheses $H_2: \xi_{11} \leqq \xi_{21}$ which is equivalent to $\lambda + \nu \leqq 0$ or $H_3: \xi_{12} \leqq \xi_{22}$ which is equivalent to $\lambda - \nu \leqq 0$. The feature which complicates this problem is that when H_0 is true the three remaining hypotheses must either all be true or all be false. Thus the condition corresponding to (ii) of Theorem 3 is not satisfied.

Let us denote the decision of rejecting or accepting H_i by d_i and $\bar{d}_i^{-1}$ as before, and consider the class $\mathfrak{C}_0$ of procedures satisfying

$$(14.13)\quad \begin{cases} P(d_0) \leqq \dfrac{b_0}{a_0 + b_0} \text{ if } \nu = 0, \quad P(\bar{d}_0^{-1}) \leqq \dfrac{a_0}{a_0 + b_0} \text{ if } \nu \neq 0 \\ P(d_1 \mid \bar{d}_0^{-1}) \leqq \dfrac{b_1}{a_1 + b_1} \text{ if } \lambda + \nu \leqq 0, \\ \qquad P(\bar{d}_1^{-1} \mid \bar{d}_0^{-1}) \gtreqless \dfrac{a_1}{a_1 + b_1} \text{ if } \lambda + \nu > 0 \\ P(d_i \mid d_0) \leqq \dfrac{b_i}{a_i + b_i} \text{ if } H_i \text{ is true,} \\ \qquad P(\bar{d}_i^{-1} \mid d_0) \leqq \dfrac{a_i}{a_i + b_i} \text{ if } H_i \text{ is false } (i = 2, 3), \end{cases}$$

We shall then prove that any element of $\mathfrak{C}_0$, which is a subclass of the class of unbiased procedures, is minimax, under the additional assumption that

$$a_i = a \quad \text{and} \quad b_i = b \text{ for all } i.$$

For any procedure δ, consider the error probabilities $\alpha_0 = P(d_0)$, $\alpha_1 = P(d_1 \mid \bar{d}_0^{-1})$ $\alpha_i = P(d_i \mid d_0)$ $(i = 2, 3)$, evaluated for $\lambda = \nu = 0$ and some fixed values μ_0 and σ_0 of μ and σ. Then the possible value of $\lim R_\delta(\theta)$ as $\theta = (\lambda, \mu, \nu, \sigma)$ tends to $\theta_0 = (0, \mu_0, 0, \sigma_0)$ are

$$a_1(1 - \alpha_0)\alpha_1 + \alpha_0[a_0 + a_2\alpha_2 + a_3\alpha_3] \qquad \text{for } \theta \in \omega_0\omega_1\omega_2\omega_3$$

$$b_1(1 - \alpha_0)(1 - \alpha_1) + \alpha_0[a_0 + b_2(1 - \alpha_2) + b_3(1 - \alpha_3)] \qquad \text{for } \theta \in \omega_0\bar{\omega}_1^{-1}\omega_2\bar{\omega}_3^{-1}$$

$$(1 - \alpha_0)[b_0 + a_1\alpha_1] + \alpha_0[a_2\alpha_2 + a_3\alpha_3] \qquad \text{for } \theta \in \bar{\omega}_0^{-1}\omega_1\omega_2\omega_3$$

$$(1 - \alpha_0)[b_0 + a_1\alpha_1] + \alpha_0[a_2\alpha_2 + b_3(1 - \alpha_3)] \qquad \text{for } \theta \in \bar{\omega}_0^{-1}\omega_1\omega_2\bar{\omega}_3^{-1}$$

$$(1 - \alpha_0)[b_0 + a_1\alpha_1] + \alpha_0[b_2(1 - \alpha_2) + a_3\alpha_3] \qquad \text{for } \theta \in \bar{\omega}_0^{-1}\omega_1\bar{\omega}_2^{-1}\omega_3$$

$$(14.14)\quad (1 - \alpha_0)[b_0 + a_1\alpha_1] + \alpha_0[b_2(1 - \alpha_2) + b_3(1 - \alpha_3)] \qquad \text{for } \theta \in \bar{\omega}_0^{-1}\omega_1\bar{\omega}_2^{-1}\bar{\omega}_3^{-1}$$

$$(1 - \alpha_0)[b_0 + b_1(1 - \alpha_1)] + \alpha_0[a_2\alpha_2 + a_3\alpha_3] \qquad \text{for } \theta \in \bar{\omega}_0^{-1}\bar{\omega}_1^{-1}\omega_2\omega_3$$

$$(1 - \alpha_0)[b_0 + b_1(1 - \alpha_1)] + \alpha_0[a_2\alpha_2 + b_3(1 - \alpha_3)]$$

$$\text{for } \theta \, \varepsilon \, \omega_0^{-1}\omega_1^{-1}\omega_2\omega_3^{-1}$$

$$(1 - \alpha_0)[b_0 + b_1(1 - \alpha_1)] + \alpha_0[b_2(1 - \alpha_2) + a_3\alpha_3]$$

$$\text{for } \theta \, \varepsilon \, \omega_0^{-1}\omega_1^{-1}\omega_2^{-1}\omega_3$$

$$(1 - \alpha_0)[b_0 + b_1(1 - \alpha_1)] + \alpha_0[b_2(1 - \alpha_2) + b_3(1 - \alpha_3)]$$

$$\text{for } \theta \, \varepsilon \, \omega_0^{-1}\omega_1^{-1}\omega_2^{-1}\omega_3^{-1}.$$

With $u = \alpha_0$, $v = (1 - \alpha_0)\alpha_1$, $w = \alpha_0\alpha_2$, $z = \alpha_0\alpha_3$ and $a_i = a$, $b_i = b$ these quantities become

$$
\begin{array}{l}
a[u + v + w + z] \\
(a + b)u - bv - bw - bz + b \\
-bu + av + aw + az + b \\
av + aw - bz + b \\
av - bw + az + b \\
bu + av - bw - bz + b \\
-2bu - bv + aw + az + 2b \\
-bu - bv + aw - bz + 2b \\
-bu - bv - bw + az + 2b \\
-2bu - bv - bw - bz + 2b.
\end{array}
\tag{14.15}
$$

From the first form of these 10 quantities it is seen that they all become equal for $\alpha_i = \alpha_i^* = b_i / (a_i + b_i)$. Let the corresponding values of (u, v, w, z) be (u_0, v_0, w_0, z_0). We shall now show that if we change from (u_0, v_0, w_0, z_0) to some other point (u_1, v_1, w_1, z_1) at least one of the 10 quantities will be increased so that (u_0, v_0, w_0, z_0) and hence $\alpha_i = \alpha_i^*$ minimizes their maximum. If in (14.15) all possible sign combinations were occurring, this would be obvious. For then in at least one row all of the four increments $\pm(u_1 - u_0)$, $\pm(v_1 - v_0)$, $\pm(w_1 - w_0)$, $\pm(z_1 - z_0)$ would be positive. But of the 16 possible combinations only 12 occur (with rows 4 and 5 each counting for two combinations, the missing combinations being $- + - -$, $+ - + +$, $+ - + -$, $+ - - +$.

As an example let us consider the first possibility. Suppose that $u_1 = u_0 - \xi$, $v_1 = v_0 + \eta$, $w_1 = w_0 - \Delta$, $z_1 = z_0 - \zeta$, with ξ, η, Δ, ζ all nonnegative. The total change in the first and tenth rows of (14.15) will be

$$a(-\xi + \eta - \Delta - \zeta) \quad \text{and} \quad b(2\xi - \eta + \Delta + \zeta).$$

Since both of these are to be $\leqq 0$, we have

$$2\xi + \Delta + \zeta \leqq \eta \leqq \xi + \Delta + \zeta$$

and hence $\xi = 0$. But then the change in the sixth row will be positive unless also $\eta = \Delta = \zeta = 0$. The other three possibilities can be ruled out in a similar manner so that $\alpha_i = \alpha_i^*$ minimizes the maximum of the 10 quantities in question. If r^* denotes the common value $\alpha_0^*[a_0 + a_1\alpha_1^* + a_2\alpha_2^* + a_3\alpha_3^*]$ of these quantities, it only remains to show that (14.13) implies $R_\delta(\theta) \leqq r^*$ for all θ. This is clearly the case since for each i the coefficient of α_i in (14.14) is $\geqq 0$ in the lines corresponding to $\theta \; \varepsilon \; \omega_i$ and $\leqq 0$ otherwise.

The result that the natural combination of a number of best unbiased or minimax tests leads to a minimax procedure for the compound problem, does unfortunately not hold even in the simplest cases in which the different hypotheses concern the same parameter. Consider for example problem (i) of section 3, in which the simultaneous consideration of $H_1{:}\,\theta \geqq \theta_0$ and $H_2{:}\,\theta \leqq \theta_0$ leads to the choice between the three decisions $d_2{:}\,\theta < \theta_0$, $d_0{:}\,\theta = \theta_0$ and $d_1{:}\,\theta > \theta_0$. If the losses of false rejection and acceptance are a and b for both hypotheses the risk function is given by

$$R_\delta(\theta) = \begin{cases} b\,P_\theta(d_0) + (a + b)P_\theta(d_1) & \text{for} \quad \theta < \theta_0 \\ a[P_\theta(d_1) + P_\theta(d_2)] & \text{for} \quad \theta = \theta_0 \\ bP_\theta(d_0) + (a + b)P_\theta(d_2) & \text{for} \quad \theta > \theta_0 . \end{cases}$$

If $\alpha_i = P_{\theta_0}(d_i)$ we have

$$\sup R_\delta(\theta) \geqq \max[a\alpha_1 - b\alpha_2 + b,\ a(\alpha_1 + \alpha_2),\ a\alpha_2 - b\alpha_1 + b]. \tag{14.16}$$

For a given value of $\alpha_1 + \alpha_2$, the maximum of the first and third terms on the right-hand side is minimized by equating them which gives $\alpha_1 = \alpha_2$ and reduces the right-hand side of (14.16) to

$$\max[(a - b)\alpha + b,\ 2a\alpha].$$

In the usual case that $a > b$, this is minimized for $\alpha = 0$, that is, for $\alpha_0 = 1$, $\alpha_1 = \alpha_2 = 0$. In the standard case that the family P_θ is homogeneous this implies $P_\theta(d_0) = 1$ for all θ, the risk function of which is

$$R_\delta(\theta) = \begin{cases} b & \text{if} \quad \theta \neq \theta_0 \\ 0 & \text{if} \quad \theta = \theta_0 \end{cases}$$

which is then clearly minimax.

REFERENCES

[1] E. L. Lehmann, "A theory of some multiple decision problems. I.," *Am. Math. Stat.*, Vol. 28 (1957), pp. 1–25.

[2] E. L. Lehmann and H. Scheffé, "Completeness, similar regions and unbiased estimation," *Sankhyā*, Vol. 10 (1950), pp. 305–339.

[3] Joyce M. May, "Extended and corrected tables of the upper percentage points of the Studentized range," *Biometrika*, Vol. 39 (1952), pp. 192–193.

[4] J. Wolfowitz, "Minimax estimates of the mean of a normal distribution with known variance," *Ann. Math. Stat.*, Vol. 21 (1950), pp. 218–230.

The Annals of Statistics
1979, Vol. 7, No. 1, 27–45

OPTIMUM SIGNIFICANCE LEVELS FOR MULTISTAGE COMPARISON PROCEDURES[1]

BY E. L. LEHMANN AND JULIET POPPER SHAFFER
University of California, Berkeley

The framework for multistage comparison procedures in the present paper is roughly that introduced by Duncan and treated more fully by Tukey. In the present paper we consider the problem of finding the optimum allocation of nominal significance levels for successive stages. The optimum procedure we obtain when the number s of treatments is odd, and the compromise procedure we propose for even s, essentially agree with a procedure suggested by Tukey. The agreement is exact when s is even and close when s is odd. The results of the present paper apply among others to the problem of distinguishing normal distributions with known variances, multinomial distributions, Poisson distributions, and distributions in certain nonparametric settings. However, they do not apply exactly to the comparison of normal distributions with a common unknown variance. When the variances are completely unknown, the method applies in principle but faces the difficulty that no exact test is then available for testing the homogeneity of a set of means.

1. Introduction. The framework for multistage comparison procedures in the present paper is roughly that introduced by Duncan [1, 2] and treated more fully by Tukey [9, Part F]. It is also used in the recent paper by Einot and Gabriel [3] which is concerned with comparing the power of several such procedures corresponding to different allocation of nominal levels. In the present paper we shall consider the problem of finding an optimum such allocation. Our formulation differs slightly from that of Einot and Gabriel in the standardization used to insure comparability of the different procedures. For this purpose, Einot and Gabriel fix the probability of at least one false significance statement when all the treatments are equal; instead, we prefer to use the maximum probability of at least one false significance statement. These coincide in many but not in all cases (for example, not in the Newman-Keuls allotment). Einot and Gabriel also compare the power of Studentized range and F-statistics. Such a comparison is outside the scope of the present paper, which is concerned only with the allocation of nominal significance levels among the different stages.

The optimum procedure we obtain when the number s of treatments is odd, and the compromise procedure we propose for even s, essentially agree with a procedure suggested by Tukey. The agreement is exact when s is even and close when s is odd. Tukey's purpose, to obtain "constant standard significance level," although

Received January 1977; revised August 1977.

[1]This paper was prepared with support of U.S. Office of Naval Research contract number N00014-75-C-0444/NR 042-036. Also this research was prepared with the support of National Science Foundation grant number MCS76-10238.

AMS 1970 *subject classifications.* 62F99 62G99 62J15.

Key words and phrases. Multistage comparison procedures, optimal significance levels, multiple comparisons, maximum error probability, power.

having a different motivation, thus turns out to be close to that of maximizing power.

The results of the present paper apply among others to the problem of distinguishing normal distributions with known variances, multinomial distributions, Poisson distributions, and distributions in certain nonparametric settings. However, they do not apply exactly to the comparison of normal distributions with a common unknown variance (or to other situations in which estimation of a common nuisance parameter destroys the independence of test statistics from nonoverlapping sets of samples). When the variances are completely unknown, the method applies in principle but faces the difficulty that no exact test is then available for testing the homogeneity of a set of means. These complications are discussed in part (i) of Section 7.

The general formulation of multistage comparison procedures adopted by Duncan [2], Tukey [9], Einot and Gabriel [3], and in the present paper runs into a difficulty when the test statistics used are the (Studentized) sample ranges. The resulting procedure may then not agree with the multiple range procedures as commonly defined and used, unless some additional conditions are satisfied. This difficulty is considered in part (iv) of Section 7.

2. Multistage procedures. Consider s independent random quantities X_i, $i = 1, \cdots, s$ (they may, for example, be vector-valued) with distributions P_i ranging over sets $\mathcal{P}_i$. We shall be concerned with the problem of distinguishing among functions g_i defined over $\mathcal{P}_i$ by means of a multistage procedure based on the X_i.

A multistage procedure is defined through a series of rejection regions. The first stage consists of a test of the hypothesis

$$H_s : g_1(P_1) = \cdots = g_s(P_s) \tag{2.1}$$

by means of a rejection region R_s in the sample space of $(X_1, \cdots, X_s)$. If $(X_1, \cdots, X_s) \in R_s$, the hypothesis H_s is rejected or, as we shall say, the homogeneity of the set $\{g_1, \cdots, g_s\}$ is rejected and the procedure goes on to stage 2 in order to get more detail about the reasons for rejection. If $(X_1, \cdots, X_s)$ falls into the complement $\bar{R}_s$ of R_s, we shall employ the traditional terminology of hypothesis testing and say that H_s or the homogeneity of the set $(g_1, \cdots, g_s)$ is accepted. This, of course, is not meant to imply that the data have convinced us of the validity of (2.1) but only that they have not enabled us to reject (2.1). In that case no further tests will be made.

The second stage consists of tests of the s hypotheses

$$H_{s-1;i} : g_1 = \cdots = g_{i-1} = g_{i+1} = \cdots = g_s \tag{2.2}$$

by means of rejection regions $R_{s-1;i}$. If the $(s-1)$-tuple $(X_1, \cdots, X_{i-1}, X_{i+1}, \cdots, X_s)$ falls into the complement of $R_{s-1;i}$, homogeneity of the set $\{g_1, \cdots, g_{i-1}, g_{i+1}, \cdots, g_s\}$ is accepted and no subset will be tested. If on the other hand $(X_1, \cdots, X_{i-1}, X_{i+1}, \cdots, X_s)$ falls into $R_{s-1;i}$, homogeneity of the set $\{g_1, \cdots, g_{i-1}, g_{i+1}, \cdots, g_s\}$ is rejected and one proceeds to the third stage.

This procedure is continued until nothing is left to be tested. As a shorthand notation we shall sometimes write H_k and R_k instead of the more complete $H_{k;\, i_1, i_2, \cdots}$ and $R_{k;\, i_1, i_2, \cdots}$ when the additional subscripts are clear from the context.

As an illustration we give in Figures 1(a) and 1(b) examples of possible outcomes of such a procedure for the case $s = 4$ where homogeneity has been accepted for the underlined sets and rejected for those not underlined.

$$1234$$

$$123 \qquad 124 \qquad 134 \qquad 234$$

$$\underline{12} \quad 13 \qquad 14 \qquad \underline{23} \qquad 24 \quad 34$$

FIG. 1(a)

$$1234$$

$$\underline{123} \qquad \underline{124} \qquad 134 \qquad \underline{234}$$

$$\underline{12} \quad \underline{13} \qquad \underline{14} \qquad \underline{23} \qquad \underline{24} \quad \underline{34}$$

FIG. 1(b)

Typically, the rejection regions are defined in terms of statistics T such as ranges or χ^2-statistics in the case of normal means, Kruskal-Wallis and Smirnov statistics in the nonparametric case, etc. These are computed for the variables in question and provide measures of inhomogeneity. Then R_s is the set for which $T_s(X_1, \cdots, X_s) \geqslant C_s$; $R_{s-1;\, i}$ is the set for which $T_{s-1;\, i}(X_1, \cdots, X_{i-1}, X_{i+1}, \cdots, X_s) \geqslant C_{s-1;\, i}$; and so on. For the T's and C's we shall frequently delete the subscripts behind the semi-colon as we did for the H's and R's. In the most commonly used procedures, the statistics T are obtained from a single functional (for example, the range) evaluated at the respective empirical cumulative distribution functions, but this need not be the case and will not be assumed here.

In the present paper we shall suppose that the statistics $T_s, T_{s-1}, \cdots$ are given and we shall be concerned with the best choice of the critical values $C_s, C_{s-1}, \cdots$. After some preparation in the next sections, a solution of this problem will be obtained in Section 4.

It follows from the definition of stagewise procedures given above that they have the following properties.

(i) Rejection of homogeneity for any set of g's implies rejection of homogeneity for all sets containing it.

(ii) Acceptance of homogeneity for any set of g's implies acceptance of homogeneity for any set contained in it.

(iii) Homogeneity of the set $\{g_{i_1}, \cdots, g_{i_k}\}$ can be rejected only when $X_{i_1}, \cdots, X_{i_k}$ falls into R_k.

A procedure satisfying properties (i) and (ii) has been termed coherent by Gabriel.

3. Separability. We require another property of multistage procedures which is not implied by the definition. (For the normal location case this essentially reduces to condition (X) proposed by Tukey in Chapter 30 of [9]).

Suppose the g's fall into t distinct groups of sizes $v_1, v_2, \cdots, v_t$ respectively ($\Sigma v_i = s$), say

$$g_{i_1} = \cdots = g_{i_{v_1}};\ g_{i_{v_1+1}} = \cdots = g_{i_{v_1+v_2}};\ \cdots \tag{3.1}$$

where $(i_1, i_2, \cdots, i_s)$ is a permutation of $(1, \cdots, s)$. We shall say that the configuration (3.1) is *separable* by a given multistage procedure if there exists a sequence of distributions $(P_1^{(m)}, \cdots, P_s^{(m)})$ with $P_i^{(m)} \in \mathcal{P}_i$ and satisfying (3.1) such that any hypothesis

$$g_{j_1} = \cdots = g_{j_r} \tag{3.2}$$

with subscripts from at least two of the t distinct subsets of (3.1) is rejected with probability tending to 1 as $m \to \infty$. If every configuration (3.1) is separable by a given multistage procedure, we call the procedure *separating*.

EXAMPLE 1. Let $X_i = (X_{i1}, \cdots, X_{in})$ with the X_{ij} independently normally distributed with mean θ_i and known common variance σ^2. Let $g_i(P_i) = \theta_i$ and let $X_{i\cdot}$ denote the mean of the ith sample. If each of the statistics $T_s, T_{s-1}, \cdots$ is for example the range of the $X_{i\cdot}$'s in question, or the sum of the squared deviations of the $X_{i\cdot}$'s from their mean, the procedure is clearly separating since such a T tends in probability to infinity if the difference between at least two of the involved θ's tends to infinity.

EXAMPLE 2. Let $X_i = (X_{i1}, \cdots, X_{in})$ with the X_{ij} independently distributed according to unknown continuous distributions F_i. Let $g_i(P_i) = F_i$ and suppose that for testing the equality of, say, r of the F_i, the associated rn variables X_{ij} are ranked and the test is based on the Kruskal-Wallis statistic. For testing H: $F_1 = \cdots = F_r$, this is given by

$$K = \frac{12n}{rn(rn+1)} \Sigma\left(R_{i\cdot} - \frac{rn+1}{2}\right)^2 = \frac{12}{rn^2(rn+1)} \Sigma R_i^2 - 3(rn+1), \tag{3.3}$$

where R_i is the rank-sum of the ith sample in the ranking of the rn variables, and $R_{i\cdot} = R_i/n$.

We shall now show that for any given critical values $C_2, \cdots, C_s$ this procedure is separating for sufficiently large n, provided there exist distributions $F_i^0 \in \mathcal{F}_i$ (the class of F's corresponding to $\mathcal{P}_i$) such that

$$P\left[X_{i\alpha} < X_{j\beta}\right] = 1 \quad \text{whenever} \quad i < j, \tag{3.4}$$

where the $X_{i\alpha}$ and $X_{j\beta}$ are distributed according to F_i^0 and F_j^0 respectively. It will be an obvious consequence that the conclusion continues to hold if there exist sequences $F_i^{(m)} \in \mathcal{F}_i$ for which (3.4) holds in the limit. In particular it therefore holds for any location family given by $F_i(x) = F(x - \theta_i)$ with unrestricted θ's.

To prove this result consider a set of g's containing at least one element each from two of the subsets defined by (3.1). Suppose the lowest (according to the ordering (3.4)) and next lowest subsets from which the set contains elements are the ith and the jth, respectively, and that the set consists of $\mathbf{a}$ elements from the ith, $\mathbf{b}$ from the jth and $\mathbf{c}$ from the other (larger) subsets. Then we shall have $\mathbf{a}$ samples which are smaller than all other observations, $\mathbf{c}$ which are larger than all other observations, and $\mathbf{b}$ in the middle.

The ranks are: in the bottom group $1, \cdots, an$; in the middle group $an + 1, \cdots, (a+b)n$; in the top group $(a+b)n+1, \cdots, (a+b+c)n$. Since the sum of the ranks in each of the three groups is fixed, ΣR_i^2 is minimized when the $\mathbf{a}$ rank-sums in the bottom group are equal; the $\mathbf{b}$ in the middle group are equal; and the $\mathbf{c}$ in the top group are equal. An easy calculation shows that the resulting minimum value of K is

$$(3.5)\quad K_0(a, b, c) = 3n^2(a+b)(a+c)(b+c)/(a+b+c)\left[(a+b+c)n+1\right],$$

so that

$$(3.6)\qquad K \geqslant K_0(a, b, c)$$

for all samples that are possible under the given configuration. Since K has a limiting χ^2-distribution with $a+b+c-1$ degrees of freedom under H as $n \to \infty$, it follows that K exceeds the given critical value with probability 1 for sufficiently large n.

To complete the proof, we must show that it is possible to choose n large enough so that $K_0(a, b, c)$ exceeds the critical value for all choices of a, b and c that correspond to possible subsets of the given configuration which contain at least two distinct elements. Since the number of such choices is finite, the proof is complete.

EXAMPLE 3. In the situation of the preceding example, suppose that the test statistic for testing (3.2) instead of K is Kiefer's generalization [5] of the Smirnov statistic

$$(3.7)\quad K' = n \sup_x \Sigma\left[F_i^*(x) - \bar{F}^*(x)\right]^2 = n \sup_x \left[\Sigma F_i^{*2}(x) - r\bar{F}^{*2}(x)\right]$$

where F_i^* denotes the empirical cdf of the ith sample and where $\bar{F}^* = \Sigma_{i=1}^r F_i^*/r$ is the empirical cdf of all rn observations.

Again the problem is to find a lower bound for K' in the case that $\mathbf{a}$ samples fall below and $\mathbf{c}$ samples above all other observations, with the remaining $\mathbf{b}$ forming a middle group. Let z be a point such that

$$P(X_{i\alpha} \leqslant z) = P(X_{j\beta} > z) = 1 \qquad \text{for all} \quad i = 1, \cdots, a$$
$$j = a+1, \cdots, a+b+c.$$

Then $F_i^*(z) = 1$ and $F_j^*(z) = 0$ for all $i \leqslant a$ and $a < j \leqslant a+b+c$ and $\bar{F}^*(z) = a/(a+b+c)$. It follows that

$$(3.8)\qquad K' \geqslant K_0'(a, b, c) = na(b+c)/(a+b+c).$$

Since K' has a nondegenerate limit distribution, it follows as in Example 3 that K' is separating for sufficiently large n, provided there exist distributions $F_i^0 \in \mathfrak{F}_i$ satisfying (3.4), or sequences $F_i^{(m)} \in \mathfrak{F}_i$ for which (3.4) holds in the limit.

The separating property of the procedure based on K' holds also for (sequences of) distributions other than those defined by (3.4). An obvious example is provided by symmetric distributions, symmetric about a common point μ, with different scales.

EXAMPLE 4. Let $(X_{i1}, \cdots, X_{id})$, $i = 1, \cdots, s$ be s independent multinomial vectors with distributions

(3.9)

$$P(X_{i1} = x_{i1}, \cdots, X_{id} = x_{id}) = \frac{n_i!}{x_{i1}! \cdots x_{id}!} p_{i1}^{x_{i1}} \cdots p_{id}^{x_{id}}, \qquad \left(\Sigma_{\alpha=1}^{d} p_{i\alpha} = 1\right).$$

The problem is that of distinguishing the vectors $p_i = (p_{i1}, \cdots, p_{id})$.

Consider first the binomial case $d = 2$. It is clear that two binomial distributions can be distinguished with probability 1 if and only if one of the two binomial p's is zero and the other is one. It follows that when $d = 2$, $s \geqslant 3$, no procedure can be separating.

It might appear at first sight that the arc sin transformation provides an asymptotic way out, since it approximately transforms the binomial distributions into normal distributions with known variance and hence into the situation of Example 1. Notice, however, that Example 1 requires the normal means to tend to $-\infty$ or $+\infty$; this corresponds to letting the p's tend to 0 and 1, which is just where the approximation breaks down.

An extension of the above argument shows that no procedure can be separating in the more general multinomial situation (3.9) when $s > d$. On the other hand, any reasonable procedure will be found to be separating when $s \leqslant d$ by letting $p_{ii} = 1$ for $i = 1, \cdots, d$ and $p_{i\alpha} = 0$ otherwise.

EXAMPLE 5. Let X_i be s independent Poisson variables with parameters $E(X_i) = \lambda_i$ and consider the problem of distinguishing among the λ's. Suppose the procedure is based on exact tests of the hypotheses

$$\lambda_{i_1} = \cdots = \lambda_{i_r}$$

which are carried out conditionally given

$$X_{i_1} + \cdots + X_{i_r} = w.$$

To prove separability of a configuration of s λ's, let the distinct values $\lambda_i^{(m)}$ tend to infinity in such a way that the ratios $\lambda_i^{(m)}/\lambda_j^{(m)}$ all tend to limits $\neq 1$. Consider now testing the homogeneity of any subset of the given configuration which contains at least two distinct values. As $m \to \infty$, the sum w' of the associated Poisson variables tends to ∞ in probability. The conditional distribution of these X's given w' is a multinomial distribution corresponding to w' trials and with probabilities of which at least two tend to different limits. Then any standard test of homogeneity will reject with probability tending to 1 as $w' \to \infty$ and this proves that such a procedure is separating.

4. Significance levels. The choice of the critical values C_k is typically made in terms of the probability of rejecting homogeneity for the subset $\{g_{i_1}, \cdots, g_{i_k}\}$ when in fact

$$g_{i_1} = \cdots = g_{i_k}. \tag{4.1}$$

The value of this probability depends also on the remaining g's or P's since an earlier hypothesis may be accepted thereby depriving the hypothesis (4.1) of the chance of being tested. A quantity of interest in this connection is

$$\alpha_k = P_{g_{i_1} = \cdots = g_{i_k}}\left[(X_{i_1}, \cdots, X_{i_k}) \in R_k\right], \tag{4.2}$$

where R_k may of course also depend on $i_1, \cdots, i_k$ but where we shall assume that the probability on the right hand side is the same for all $(P_{i_1}, \cdots, P_{i_k})$ satisfying (4.1).

It follows from property (iii) of Section 2 that

$$P\left[\text{rejecting homogeneity of } \{g_{i_1}, \cdots, g_{i_k}\}\right] \leqq \alpha_k \tag{4.3}$$

for all $(g_1, \cdots, g_s)$ satisfying (4.1). Consider now the configuration (3.1) with $v_1 = k$ and $v_2 = v_3 = \cdots = 1$, and suppose this configuration is separable by the given procedure. Then there exists a sequence of distributions $(P_1^{(m)}, \cdots, P_s^{(m)})$ satisfying (4.1) and for which the left hand side of (4.3) will tend to α_k. It follows that

$$\sup P\left[\text{rejecting homogeneity of } \{g_{i_1}, \cdots, g_{i_k}\}\right] = \alpha_k \tag{4.4}$$

where the sup is taken over all $(P_1, \cdots, P_s)$ satisfying (4.1). The relation (4.4) thus holds in particular whenever the given procedure is separating. For the normal case this result was already noted by Tukey in Chapter 30 of [6]. The levels α_k are called apparent significance levels by Tukey [9], k-mean significance levels by Duncan [2], and nominal levels by Einot and Gabriel [3].

An aspect of a multiple comparison procedure which is of great interest is the maximum probability of rejecting the homogeneity of at least one set of g's which are in fact equal, say

$$\alpha_0 = \sup \alpha(P_1, \cdots, P_s) \tag{4.5}$$

where for a given s-tuple $(P_1, \cdots, P_s)$

$$\alpha(P_1, \cdots, P_s) = P\left[\text{at least one false rejection}\right] \tag{4.6}$$

and where the sup in (4.5) is taken over all possible distributions $P_i \in \mathcal{P}_i$.

The relation of α_0 to the significance levels $\alpha_2, \cdots, \alpha_s$ can be seen from the following result which underlies all that follows, and which for the normal case was first stated informally by Tukey in Chapter 30 of [9].

THEOREM 1. *If the configuration of* $(g_1, \cdots, g_s)$ *is given by* (3.1) *and is separable by the given procedure,*

$$\sup \alpha(P_1, \cdots, P_s) = 1 - \Pi_{i=1}^t(1 - \alpha_{v_i}) \tag{4.7}$$

where $\alpha_1 = 0$ *and where the supremum is taken over all s-tuples* $(g_1, \cdots, g_s)$ *which satisfy* (3.1).

The proof of this theorem is essentially the same as that given for Theorem 1 in [7] and will therefore be omitted here. It follows from (4.7) that

$$\alpha_0 = \sup_{v_1, \cdots, v_t}\left[1 - \Pi_{i=1}^t(1 - \alpha_{v_i})\right] \tag{4.8}$$

holds for all configurations provided the given procedure is separating. It may be of interest to note that separability of the configuration is not only sufficient for Theorem 1 but also necessary.

The remarkable feature of these results is the fact that the suprema in question are independent of the statistics T defining the procedure, and depend only on the values $\alpha_2, \cdots, \alpha_s$. Our primary concern in this paper is the choice of these α's.

In analogy with standard practice in hypothesis testing, we shall impose a bound on one of the rejection probabilities, namely α_0, thereby insuring comparability of different procedures. We shall thus fix a value, say α_0^* and restrict attention to procedures satisfying

$$\alpha_0 \leqslant \alpha_0^*. \tag{4.9}$$

Subject to this condition, we should like to maximize the "power" of the procedure, that is, the probability of detecting existing inhomogeneities. Since we are assuming the statistics T to be given and only the critical values C to be at our disposal, it is clear that power is maximized by maximizing the α's.

From (4.8) we see that the problem is that of maximizing $\alpha_2, \cdots, \alpha_s$ subject to

$$\Pi_{i=1}^t(1 - \alpha_{v_i}) \geqslant 1 - \alpha_0^* \text{ for all } v_1, \cdots, v_t \text{ satisfying } \Sigma_{i=1}^t v_i = s. \tag{4.10}$$

Let us begin with α_s. If the left hand side of (4.10) is to involve α_s, we must have $t = 1$, $v_1 = s$ and the only restriction on α_s is $\alpha_s \leqslant \alpha_0^*$, so that we shall maximize α_s by putting

$$\alpha_s = \alpha_0^*. \tag{4.11}$$

Analogously, we see that we can also put

$$\alpha_{s-1} = \alpha_0^* \tag{4.12}$$

without imposing any additional restrictions on $\alpha_2, \cdots, \alpha_{s-2}$. This shows that for $s = 3$, there exists a uniformly best choice of the α's, namely $\alpha_2 = \alpha_3 = \alpha_0^*$.

Consider next the case $s = 4$. The possible configurations are $t = 1$, $v_1 = 4$; $t = 2$, $\{v_1, v_2\} = \{1, 3\}$ or $\{2, 2\}$; $t = 3$, $\{v_1, v_2, v_3\} = \{1, 1, 2\}$ and $t = 4$, $v_1 = v_2 = v_3 = v_4 = 1$ and it is seen that a uniformly best choice of the α's is given by $\alpha_3 = \alpha_4 = \alpha_0^*$, $\alpha_2 = 1 - (1 - \alpha_0^*)^{\frac{1}{2}}$.

This happy state of affairs however does not extend to higher values of s. For $s = 5$, for example, it is no longer possible simultaneously to maximize α_2 and α_3. The maximum value of α_3 is α_0^* but this requires $\alpha_2 = 0$; alternatively the maximum value of α_2 is $1 - (1 - \alpha_0^*)^{\frac{1}{2}}$ in which case α_3 cannot exceed $1 - (1 - \alpha_0^*)^{\frac{1}{2}}$.

When a uniformly best choice of the α's does not exist, it is convenient to introduce the following definition.

DEFINITION. A set $(\alpha_2, \cdots, \alpha_s)$ for which α_0 satisfies (4.9) will be called inadmissible if there exists another set $(\alpha_2', \cdots, \alpha_s')$ for which α_0' also satisfies (4.9) and such that

(4.13) $\quad \alpha_i \leqslant \alpha_i'$ for all i, with strict inequality for at least some i.

This use of the term is analogous to that which is customary in the theory of hypothesis testing.

THEOREM 2. *Any admissible procedure which is separating satisfies*

$$\alpha_2 \leqslant \alpha_3 \leqslant \cdots \leqslant \alpha_s. \tag{4.14}$$

PROOF. If (4.14) does not hold, there exists k such that $\alpha_{k+1} < \alpha_k$. For any such k, consider the procedure in which $\alpha'_i = \alpha_i$ for $i \neq k + 1$ and $\alpha_{k+1}' = \alpha_k$. Then clearly $\alpha_0' \geqslant \alpha_0$. To show that $\alpha_0' \leqslant \alpha_0^*$, we need only show that $\Pi_{i=1}^t(1 - \alpha_{v_i}') \geqslant 1 - \alpha^*_0$ for all $(v_1, \cdots, v_t)$.

If none of the v's is equal to $k + 1$, $\alpha_{v_i}' = \alpha_{v_i}$ for all $i = 1, \cdots, t$ and the result holds. Otherwise, replace each v that is equal to $k + 1$ by two v's: one equal to k and one equal to 1, and denote the resulting set by $v_1', \cdots, v_{t'}'$. Then $\Pi_{i=1}^t(1 - \alpha_{v_i}') = \Pi_{i=1}^{t'}(1 - \alpha_{v_i'}) \geqslant 1 - \alpha_0^*$, and this completes the proof.

Having fixed $\alpha_0 = \alpha_s = \alpha_{s-1}$ at α_0^*, how should we choose the remaining α's? In order to have a reasonable chance of detecting existing inhomogeneities irrespective of their pattern we should like to have none of the α's too small. In fact, at the end of the section we shall propose a set of α-values which for $s \leqslant 21$ will keep α_0 less than .1 without letting any of $\alpha_2, \cdots, \alpha_s$ fall below .01. These considerations suggest a minimax approach: namely subject to $\alpha_0 \leqslant \alpha_0^*$ to maximize $\min(\alpha_2, \cdots, \alpha_s)$. In view of Theorem 2, this is achieved by maximizing α_2.

THEOREM 3. *The maximum value of α_2 for a multistage procedure which is separating and satisfies* (4.9) *is given by*

$$\alpha_2 = 1 - (1 - \alpha_0^*)^{[s/2]^{-1}} \tag{4.15}$$

where $[A]$ as usual denotes the largest integer $\leqslant A$.

PROOF. Instead of fixing α_0 at α_0^* and maximizing α_2, it is more convenient to invert the problem and fix α_2 at, say, α and minimize α_0. The theorem will be proved by showing that the resulting minimum value of α_0 is

$$\alpha_0^* = 1 - (1 - \alpha)^{[s/2]}. \tag{4.16}$$

Suppose first that s is even. Since α_2 is fixed at α, it then follows from the fact that the procedure has been assumed to be separating that the right hand side of (4.7) can be made arbitrarily close to α_0^*. This is seen by using the separability of the configuration $v_1 = \cdots = v_{[s/2]} = 2$.

When s is odd, we put an additional v equal to 1 and use separability of the resulting configuration.

In either case it follows that

$$\sup_{v_1, \cdots, v_t}\left[1 - \Pi_{i=1}^{t}(1 - \alpha_{v_i})\right] = \sup \alpha(P_1, \cdots, P_s) \geqslant \alpha_0^*. \tag{4.17}$$

To complete the proof, we must exhibit values of $\alpha_2, \cdots, \alpha_s$ for which the left hand side of (4.17) equals α_0^*. To this end consider the procedure defined by

$$\alpha_3 = \cdots = \alpha_s = \alpha. \tag{4.18}$$

The right hand side of (4.7) then becomes $1 - (1 - \alpha)^{t'}$ where

$$t' = \text{number of } v\text{'s} > 1. \tag{4.19}$$

This quantity takes on its maximum value for $t' = [s/2]$ and this completes the proof.

Theorems 2 and 3 provide some guidance for the choice of the levels $\alpha_2, \cdots, \alpha_s$ and α_0. If α_2 is fixed at a conventional level α independent of s, then as s increases, so does α_0 (and hence α_{s-1} and α_s) to values which soon become intolerably high. Conversely, if α_0 is fixed at a conventional level independent of s, the level α_2 soon becomes so small that there is virtually no chance of detecting moderate differences between pairs when they exist. It turns out however that for $s \leqslant 21$, one can let α_2 vary from .01 to .05 while at the same time keeping α_0 between .05 and .10. The remaining α's, by Theorem 2, will then also lie between these limits.

While 10% is larger than is conventionally recommended for a single significance test, it must be remembered that α_0 is the maximum probability of at least one false significance statement when many comparisons are being made. We believe that in these circumstances such a higher level often can be tolerated, and therefore propose the sets of levels of α_2 and α_0 shown in Table 1.

Having decided on α_0 and α_2, how should one choose the remaining α's? We shall show in the next section that when s is odd there is a uniformly best choice

TABLE 1

	$\alpha_2 = .05$		$\alpha_2 = .03$		$\alpha_2 = .02$					
s	2	3	4	5	6	7	8	9	10	11
α_0	.05	.05	.059	.059	.059	.059	.078	.078	.096	.096

	$\alpha_2 = .01$									
s	12	13	14	15	16	17	18	19	20	21
α_0	.059	.059	.068	.068	.077	.077	.086	.086	.096	.096

for $\alpha_3, \cdots, \alpha_s$. The result is not quite as clear-cut when s is even. However, we shall see that even in that case, admissibility determines rather narrow limits within which $\alpha_2, \cdots, \alpha_s$ must lie.

5. Most powerful test when s is odd. We shall now prove the result indicated at the end of the preceding section: that if s is odd and α_2 and α_0 have been fixed at α and α_0^* (given by (4.16)) respectively it is possible to maximize all the remaining α's simultaneously. Subject to the two restrictions, the resulting choice thus provides a uniformly most powerful set of α's.

To simplify the notation, instead of the probability $\alpha(P_1, \cdots, P_s)$ of at least one error, let us work with its complement

(5.1) $$\gamma(P_1, \cdots, P_s) = P(\text{no error}).$$

If we then put

(5.2) $$\gamma_k = 1 - \alpha_k$$

so that in particular

(5.3) $$\gamma_2 = \gamma = 1 - \alpha,$$

it follows from (4.10) and (4.16) that

(5.4) $$\Pi_{i=1}^t \gamma_{v_i} \geqslant \gamma^{[s/2]} = \gamma^{(s-1)/2}.$$

THEOREM 4. *Subject to* (5.3) *and* (5.4), *the minimum value of* γ_i $(i = 3, \cdots, s)$ *is, for any odd* s,

(5.5) $$\gamma_i^* = \gamma^{[i/2]},$$

and this value can be attained simultaneously for all i.

PROOF. (i) Consider any procedure satisfying (5.3) and (5.4). First let i be odd and consider the configuration in which $v_1 = i$ and all the remaining v's are equal to 2. Then by (5.4)

$$\gamma_i \gamma^{(s-i)/2} \geqslant \gamma^{(s-1)/2}$$

and hence

(5.6) $$\gamma_i \geqslant \gamma_i^*.$$

A completely analogous argument shows that (5.6) also holds when i is even.

(ii) Consider now the procedure defined by

(5.7) $$\gamma_i = \gamma_i^*.$$

This clearly satisfies (5.3) and it only remains to check that it also satisfies (5.4). Since

$$\Pi_{i=1}^t \gamma_{v_i} = \gamma^{\Sigma_{i=1}^t [v_i/2]},$$

we only need to show that

$$\Sigma_{i=1}^t [v_i/2] \leqslant (s-1)/2.$$

Now

$$
\begin{aligned}
[v_i/2] &= \tfrac{1}{2}v_i && \text{when} \quad v_i \quad \text{is even} \\
&= \tfrac{1}{2}(v_i - 1) && \text{when} \quad v_i \quad \text{is odd}
\end{aligned}
$$

so that $\Sigma[v_i/2] = (s - b)/2$, where b is the number of odd v's (including ones). Since s is odd, $b \geqslant 1$ and this completes the proof.

The levels defined by (5.7) are close to the levels

$$
\begin{aligned}
\gamma_i &= \gamma^{i/2}, && 1 < i < s - 1 \\
&= \gamma^{s/2}, && i = s - 1, s
\end{aligned} \tag{5.8}
$$

proposed by Tukey in Chapter 31 of [9]; see also Ryan [8]. These levels are admissible but, of course, do not satisfy (4.15). Other widely discussed sets of levels have been proposed: $\gamma_i = \gamma^{i-1}$ by Duncan, and $\gamma_i = \gamma$ by Newman and Keuls. (See Einot and Gabriel [3] for references.) Neither of these sets is admissible according to our definition. (Actually, the procedures of these authors are for unknown σ while our discussion assumes σ to be known. See Section 6.)

6. Choice of $\alpha_3, \cdots, \alpha_{s-1}$ when s is even. When s is odd, there exists a uniformly best choice of $\gamma_3, \cdots, \gamma_{s-1}$ satisfying

$$
\Pi_{i=1}^{t}\gamma_{v_i} \geqslant \gamma^{[s/2]} \tag{6.1}
$$

where $\gamma_2 = \gamma$. The following theorem shows that this is not the case when s is even.

THEOREM 5. *Let s be even and let the γ's satisfy* (5.3) *and* (6.1).

(i) *The values of $\gamma_3, \cdots, \gamma_{s-1}$ are bounded below by*

$$
\begin{aligned}
\gamma_i &\geqslant \gamma^{i/2} && \text{when } i = \text{even and when} \quad i = \text{odd and} \quad 3 \leqslant i \leqslant s/2 \\
&\geqslant \gamma^{(i+1)/2} && \text{when } i = \text{odd and} \quad s/2 < i \leqslant s - 1.
\end{aligned} \tag{6.2}
$$

(ii) *For any given i, there exists a set of γ's for which γ_i attains the lower bound* (6.2).

(iii) *If $s > 6$, the lower bounds* (6.2) *cannot be attained simultaneously for all i.*

PROOF. (i) follows directly from (6.1) with the following values of the v's:

i = even: $\quad v_1 = i$, the other v's = 2;

i = odd, $\quad 1 < i \leqslant s/2$; $\quad v_1 = v_2 = i$, the other v's = 2;

i = odd, $\quad s/2 < i \leqslant s - 1$: $\quad v_1 = 1$, $v_2 = i$, the other v's = 2.

(ii) For the cases with $\gamma_i = \gamma^{i/2}$, this is seen by putting $\gamma_j = \gamma^{j/2}$ for all j, since then

$$
\Pi_{i=1}^{t}\gamma_{v_i} = \gamma^{\frac{1}{2}\Sigma_{i=1}^{t}v_i} = \gamma^{s/2}
$$

which verifies (6.1).

For the cases with $\gamma_i = \gamma^{(i+1)/2}$, it is seen by putting

$$\begin{aligned}
\gamma_i &= \gamma^{(i+1)/2} & & \\
\gamma_j &= \gamma^{(j-1)/2} & &\text{for all odd } j \neq i \\
\gamma_j &= \gamma^{j/2} & &\text{for all even } j.
\end{aligned}$$

To verify (6.1) one need only note that at most one γ can be equal to $\gamma^{(i+1)/2}$ and that the subscript of at least one other γ must be odd.

(iii) If $s > 6$, $t = 2$, $v_1 = 3$, $v_2 = s - 3$, then v_2 is odd and $> s/2$, and we have

$$\Pi_{i=1}^{t} \gamma_{v_i} < \gamma^{s/2}$$

if the γ_{v_i} are set equal to their lower limits for all i, so that (6.1) is not satisfied.

It is seen from the proof of (iii) that one cannot lower any of the γ_i with i = odd and $s/2 < i < s - 1$ below $\gamma^{i/2}$ without increasing at least one of the γ's with i = odd and $< s/2$ by the same (or a larger) factor above its minimum level. Under these circumstances, a reasonable compromise seems to be to put

$$\begin{aligned}
\gamma_i = \gamma_i^* &= \gamma^{i/2} & &\text{for all } 1 < i < s - 1 \\
&= \gamma^{s/2} & &\text{for } i = s - 1, s.
\end{aligned} \tag{6.3}$$

This rule, which was proposed by Tukey [9], is clearly admissible. As in the case of odd s, the sets of levels proposed by Duncan and by Newman and Keuls are not admissible.

The values (6.3) do not constitute the only admissible set of γ's, and the question arises as to how far an admissible set can deviate from the compromise solution (6.3). A partial answer is obtained by supplementing the lower bounds (6.2) by a corresponding set of upper bounds for all admissible sets of γ's. Together, the two sets of bounds can be summarized in the following theorem.

THEOREM 6. *Let s be even and let the γ's satisfy* (5.3) *and* (6.1). *Then any admissible set of γ's satisfies, for $i = 3, \cdots, s$*

$$\begin{aligned}
&\gamma_i = \gamma_i^* \text{ when } i = \text{even and } > \frac{s}{2} - 4 \\
&\text{or } i = \text{odd and } = \frac{s}{2} \text{ or } = s - 1
\end{aligned} \tag{6.4}$$

$$\begin{aligned}
&\gamma_i^* \leqslant \gamma_i \leqslant \gamma_i^* \gamma^{-\frac{1}{2}} = \gamma_{i-1}^* \text{ when } i = \text{even and } \leqslant \frac{s}{2} - 4 \\
&\text{or } i = \text{odd and } < \frac{s}{2}
\end{aligned} \tag{6.5}$$

and

$$\gamma_{i+1}^* = \gamma_i^* \gamma^{\frac{1}{2}} \leqslant \gamma_i \leqslant \gamma_i^* \quad \text{when } i = \text{odd and } \frac{s}{2} < i < s - 1. \tag{6.6}$$

Before turning to the proof of this result, let us consider its implications. The bounds show that an admissible γ_i can differ from γ_i^* by at most $\gamma^{i/2} - \gamma^{(i+1)/2}$ or $\gamma^{(i-1)/2} - \gamma^{i/2}$. For $i \geqslant 3$, the largest possible value of $|\gamma_i - \gamma_i^*| = |\alpha_i - \alpha_i^*|$ is thus

$\gamma(1-\gamma^{\frac{1}{2}})$ which for all $0<\gamma<1$ is $\leqslant \frac{1}{2}(1-\gamma) = \frac{1}{2}\alpha$. For the values of α proposed in Section 3 an admissible γ_i can therefore differ from γ_i^* by at most .01 for $s = 8$ and 10, and by at most .005 for any even s between 12 and 20.

The other point worth noting is that Theorem 2 is a consequence of Theorem 4 (for s = odd) and Theorem 6 (for s = even). In the former case this is obvious; in the latter it follows from the fact that if the lower and upper bounds for γ_i are denoted by $\underline{\gamma}_i$ and $\bar{\gamma}_i$, then

$$\bar{\gamma}_{i+1} \leqslant \underline{\gamma}_i \leqslant \bar{\gamma}_i \leqslant \gamma_{i-1}.$$

SKETCH OF PROOF OF THEOREM 6. To establish the upper bounds $\bar{\gamma}_i$ stated in Theorem 6, a single method of proof suffices for all cases. Consider any set $(\gamma_3, \cdots, \gamma_i, \cdots, \gamma_s)$ with $\gamma_i > \bar{\gamma}_i$ and satisfying (6.1)· then the set in which γ_i is replaced by $\bar{\gamma}_i$ also satisfies (6.1) and hence proves the first set inadmissible. We shall not carry out this check but should perhaps explain the somewhat surprising quantity $(s/2) - 4$ which appears in (6.4) and (6.5) when i is even. This stems from the fact that if $\gamma_i > \gamma_i^*$ for an even i, it follows from Theorem 5 that admissibility requires a compensating odd j with $\gamma_j < \gamma_j^*$. Since the v's must add up to s which is even, at least one other v must be odd. If the only other odd v is equal to 1, the left hand side of (6.1) is $\geqslant \gamma^{s/2}$ even when $\gamma_i = \gamma_i^*$. It thus requires an odd factor with $v \geqslant 3$ to justify letting $\gamma_i > \gamma_i^*$. Hence $i + j + 3 \leqslant s$ where $j > s/2$. Thus $\gamma_i > \gamma_i^*$ is possible only if $i < s - \frac{s}{2} - 3 = \frac{s}{2} - 3$, or $i \leqslant \frac{s}{2} - 4$ as was to be proved.

7. Applications and further extensions.

(i) *The normal case*. The results of the preceding sections apply to the normal location problem with known variances σ_i^2. Let us now consider the more important situation in which the variances are unknown.

(a) Suppose first that the unknown variances are assumed to be equal (or have known ratios). If the usual procedure for estimating the unknown σ^2 is followed, the relations between the α's obtained in the preceding sections will no longer hold exactly. In particular, if we define $\alpha_3, \cdots, \alpha_s$ as before in terms of $\alpha = \alpha_2$, the quantity α_0^* given by (4.16) will continue to be a sharp upper bound for α_0. However, given the same value of $\alpha_s = \alpha_{s-1}$ we can now raise $\alpha_2, \ldots, \alpha_{s-2}$ without raising $\alpha_0^* = \alpha_s$. This follows from the fact that the s Studentized sample means will now be positively dependent instead of independent as were the means without Studentization (see [7]). The dependence, and hence the permissible increase in $\alpha_2, \cdots, \alpha_{s-1}$ is, however, likely to be so small when σ^2 is estimated with at least moderate degrees of freedom, as to be of no practical importance. The conclusions may therefore be expected to hold to a close approximation unless the number of degrees of freedom is quite small.

(b) If a common variance is not assumed (as it often should not be), the situation is somewhat different. No exact test of homogeneity is then available; however a test proposed by Welch [10] seems to provide a highly satisfactory approximate test. (See Kohr and Games [6] for a comparison with several other tests.) The values of $\alpha_2, \cdots, \alpha_s$ defined in Sections 4, 5 and 6 are then no longer exact and in fact depend slightly on the true values of the variance ratios. However if none of the sample sizes is too small, the approximation can be expected to be close and to that degree the results of the paper continue to apply.

(c) Another problem of interest in the normal case with unknown variances is the comparison of the variances instead of the means. The results of the paper apply to this problem with any of the standard tests of homogeneity of variances.

(ii) *Lowering* α_s. We saw earlier that subject to (4.10) the maximum value of α_s is given by (4.11) and that setting α_s at that value puts no new restrictions on the remaining α's. However, some practitioners attach special importance to α_s, the probability of falsely rejecting the overall null hypothesis $g_1 = \cdots = g_s$, and would prefer a smaller value for this probability. What are the consequences of lowering α_s from the value α_0 of Table 1 to a lower value such as .05 or even .01? Theorem 1 shows that such a lowering, with values of $\alpha_2, \cdots, \alpha_{s-1}$ remaining unchanged, will not affect any of the maximum probabilities (4.7), except when $g_1 = \cdots = g_s$, although it will lower the probabilities for all actual configurations (3.1). The resulting procedure will of course no longer be admissible according to our definition of this concept. In spite of this, one may wish to impose the additional requirement that the probability of falsely rejecting the overall null hypothesis not exceed the new lower value assigned to α_s. In this case, if the original procedure is admissible under the requirements (5.3) and (6.1), the procedure obtained by lowering α_s to this new value will clearly be admissible among the procedures satisfying (5.3), (6.1) and the additional bound on α_s.

(iii) *Deleting stages; Fisher's LSD test*. The following method, not covered by the multistage procedures described in Section 2, although it is concerned with the comparison of a set of normal means, is often referred to as Fisher's least significant difference test. It consists of first carrying out an F-test say, at level α, of the hypothesis H_s: $g_1 = \cdots = g_s$; if H_s is accepted, the procedure stops and homogeneity of the set $\{g_1, \cdots, g_s\}$ is accepted; in case of rejection, all $\binom{s}{2}$ pairs $\{g_i, g_j\}$ are compared at level α by means of t-tests. (If σ^2 is known, the F- and t-tests are replaced by the corresponding χ^2 and normal tests.)

Let us now consider the more general set-up of Section 2 but restrict ourselves to the following two stages. First, the overall test of H_s is performed at level, say, α'. If H_s is accepted, the procedure stops and homogeniety of the full set $\{g_1, \cdots, g_s\}$ is accepted. In case of rejection, all pairs are tested by means of the rejection regions R_2, each at level α. The probability of at least one false rejection is then given by the formula

$$(7.1) \quad \sup P[\text{at least one false rejection}]$$

$$\begin{aligned} &= \alpha' &&\text{if } g_1 = \cdots = g_s \\ &= 1 - \Pi_{i=1}^{t}(1 - \alpha_{v_i}) &&\text{if (3.1) holds with } t > 1, \end{aligned}$$

where the sup extends over all $(g_1, \ldots, g_s)$ satisfying the condition indicated on the right hand side of (7.1) and where α_k is given by

$$(7.2) \qquad \alpha_k = P\left[(X_i, X_j) \in R_2 \text{ for at least one pair } 1 \leqslant i < j \leqslant k\right].$$

Consider a fixed value of

$$(7.3) \qquad \alpha_0 = \max\left[\alpha';\ 1 - \Pi_{i=1}^{t}(1 - \alpha_{v_i}), \text{ all configurations}\right].$$

The most powerful choices of α and α' are then clearly given by

$$(7.4) \qquad \alpha' = \alpha_0$$

and

$$(7.5) \qquad \max\left[1 - \Pi_{i=1}^{t}(1 - \alpha_{v_i})\right] = \alpha_0$$

where the maximum is taken with respect to all configurations $(v_1, \cdots, v_t)$ with $\Sigma_{i=1}^{t} v_i = s$ and $t > 1$.

(iv) *A subset option.* The generalization (iii), required to cover the least significant difference and related procedures, consisted in allowing the deletion of some of the stages in the multistage procedure of Section 2.

The most commonly used multistage comparison procedures, the multiple range tests as customarily described (discussed for example in [7]), also do not fall within the framework of our analysis but for a different reason. To cover procedures of this kind, let us consider a multistage procedure with the following added *subset option*. When homogeneity of a set $\{P_{i_1}, \cdots, P_{i_k}\}$ is rejected we permit the additional conclusion that a subset of these P's, selected on the basis of the data from a specified family of subsets, is also rejected. If such a subset is rejected, all parameter sets containing this subset are also rejected and will not be tested any more. Actually, in what follows we shall apply this option only to the case that the subsets are pairs. The procedures we shall discuss are somewhat less powerful than those of Sections 5 and 6; however, they have the advantage that whenever homogeneity of a set is rejected, at least one pair in this set will also be declared to be inhomogeneous.

Let us assume that $X_i = (X_{i1}, \cdots, X_{in})$ and that the rejection regions R_k of Section 2 are defined in terms of statistics $T_s, T_{s-1}, \cdots$ and critical values $C_s, C_{s-1}, \cdots$ as described there but that now the T's and C's depend only on the size of the subset being tested, so that they do not require any additional subscripts.

If the procedure has not already stopped at an earlier stage, homogeneity of a pair (X_i, X_j) is then rejected whenever

$$(7.6) \qquad T_2(X_i, X_j) \geqslant C_2.$$

We shall restrict attention to procedures in which the remaining T's are defined by

$$(7.7) \qquad T_k(X_1, \cdots, X_k) = \max\left[T_2(X_i, X_j);\ 1 \leqslant i < j \leqslant k\right],$$

and then adopt the following rule: whenever homogeneity of a set $\{g_{i_1}, \cdots, g_{i_k}\}$ is rejected because $T_k(X_{i_1}, \cdots, X_{i_k})$ is too large, then homogeneity of the pair of g's corresponding to the two X's for which $T_2(X_{i_\mu}, X_{i_\nu})$ is maximum, is also rejected.

A well-known example of a procedure of this kind in the normal case is that based on the range: that is, in which

$$T_2(X_i, X_j) = |X_{j\cdot} - X_{i\cdot}| \tag{7.8}$$

where $X_{i\cdot}$ and $X_{j\cdot}$ denote the means of the ith and jth sample.

Another illustration is provided if in the notation of the Smirnov example (Example 3) of Section 4 we take

$$T_2(X_i, X_j) = \sup_x|F_i^*(x) - F_j^*(x)|. \tag{7.9}$$

As a third example, suppose that for each i and j the $2n$ observations constituting the ith and jth samples are ranked from 1 to $2n$ and that $R_{i\cdot}$ and $R_{j\cdot}$ denote the average rank of the ith and jth sample in this ranking. Then we could take

$$T_2(X_i, X_j) = |R_{j\cdot} - R_{i\cdot}|. \tag{7.10}$$

Consider finally the multinomial example of Section 3 with $n_i = n$ for all i. Here we can take, for instance for $T_2(X_i, X_j)$ the usual χ^2-statistic for comparing two multinomial distributions. If n is sufficiently large, C_2 can be obtained approximately from the χ^2-distribution with $d - 1$ degrees of freedom and the resulting procedure satisfies the assumptions made above.

With this type of procedure, properties (i) and (ii) of Section 2 continue to hold; however, property (iii) need no longer be valid, since it may now happen that homogeneity of a set $\{g_{i_1}, \cdots, g_{i_k}\}$ is rejected at an early stage in spite of the fact that $(X_{i_1}, \cdots, X_{i_k})$ falls into $\bar{R}_k$. As shown in the case of multiple range tests in [7], Theorem 1 need then no longer be true, and the analysis of the preceding sections is no longer applicable.

Whether or not (iii), and hence Theorem 1, holds in the present situation typically will depend on the critical values C. However, a sufficient condition for them to hold is

$$C_2 \leqslant \cdots \leqslant C_s. \tag{7.11}$$

Under mild additional assumptions which are satisfied in all our examples, (7.11) is also necessary. These statements follow from the argument given for the case of range-based procedures in [7].

The values of the C's of course depend on the α's, and unfortunately it turns out that for the α's given by (5.3), (5.7) and (6.3) the C's will not always satisfy (7.11). There is then no guarantee that α_0 will have the prescribed value α_0^*. To avoid this difficulty we can proceed as follows. We can replace C_j by

$$C_j' = \max(C_2, \cdots, C_j) \tag{7.12}$$

so that the resulting C_j' will satisfy (7.11). We shall now consider the properties of this approach. Note that for this discussion, we need not concern ourselves with the subset option. We are only concerned with the relationships of the α_k and α_k'

defined by (7.12) and

(7.13)

$$\alpha_k = P[T_k(X_1, \cdots, X_k) \geqslant C_k], \qquad \alpha'_k = P[T_k(X_1, \cdots, X_k) \geqslant C'_k]$$

where the probabilities are computed assuming $g_1 = \cdots = g_k$.

We shall begin by proving the following lemma.

LEMMA. *If $\alpha_2, \cdots, \alpha_s$ are any levels satisfying $\alpha_2 = \alpha$, $\alpha_k \leqslant \alpha_{k+1}$ for all k and if α_0 is equal to α_0^* given by* (4.16), *then the α_k and α'_k satisfy the following relationships*:

$$\alpha'_2 = \alpha_2, \qquad \alpha'_k \leqslant \alpha_k \qquad \text{for} \quad k = 3, \cdots, s \tag{7.14}$$

and

$$\alpha'_2 \leqslant \alpha'_3 \leqslant \cdots \leqslant \alpha'_s \leqslant \alpha'_0 = \alpha_0. \tag{7.15}$$

PROOF. The relations (7.14) follow from the facts that $C'_2 = C_2$ and $C'_k \geqslant C_k$ for $k > 2$. To prove (7.15) note that $C'_{k+1} = \max(C'_k, C_{k+1})$.

If $C'_k \leqslant C_{k+1}$ we have $\alpha'_{k+1} = \alpha_{k+1} \geqslant \alpha_k \geqslant \alpha'_k$ as desired. If $C_{k+1} < C'_k$, we have

$$\alpha'_{k+1} = P[T_{k+1}(X_1, \cdots, X_{k+1}) \geqslant C'_k] > P[T_k(X_1, \cdots, X_k) \geqslant C'_k] = \alpha'_k$$

and this shows that $\alpha'_k \leqslant \alpha'_{k+1}$ for all $k = 2, \cdots, s$. That $\alpha'_s \leqslant \alpha'_0$ is obvious. Finally, it follows from (7.14) that $\alpha'_0 \leqslant \alpha_0$ and hence from the proof of Theorem 3 that $\alpha'_0 = \alpha_0$.

Since the levels α'_k satisfy the requirements

$$\alpha = \alpha'_2 \leqslant \alpha'_k \leqslant \alpha'_0 = \alpha_0^* \tag{7.16}$$

for all k, the procedure defined by the C'_k is a possible procedure satisfying the conditions imposed earlier. Is it admissible within the class satisfying the additional restrictions (7.11)?

We shall now show that this is the case when s is odd. The proof breaks down for even s, and we shall deal with this case in a more pragmatic manner.

THEOREM 7. *Let s be odd. Then the procedure Π' obtained by applying* (7.12) *to the procedure defined by* (5.7) *is admissible, and is in fact the unique admissible procedure among all those satisfying* (7.16) *and* (7.11).

PROOF. Suppose that Π' is not the unique admissible procedure. Then we shall show that there exists a procedure Π'' with levels α'' satisfying (7.16) and (7.11) and such that

$$\alpha''_k > \alpha_k \qquad \text{for some} \quad k. \tag{7.17}$$

This contradicts the consequence of Theorem 4 that there exists a unique admissible procedure satisfying (7.16), and hence proves that the α'-procedure is admissible.

To prove the existence of the required α''_j, note that if Π' is not the unique admissible procedure satisfying (7.11) and (7.16), there exists a procedure Π'' satisfying these conditions and such that $\alpha'_k < \alpha''_k$ for some k. If $\alpha'_k = \alpha_k$, then

$\alpha_k < \alpha_k''$ as was to be proved. If $\alpha_k' < \alpha_k$, then by (7.12) there exists $j < k$ such that $C_j = C_k'$. Since $C_k'' < C_k'$ we have by (7.11)

$$C_j'' < C_k'' < C_k' = C_j$$

and hence $\alpha_j'' > \alpha_j$, as was to be proved.

The above argument makes essential use of the uniqueness of the admissible solution in the case that s is odd and no longer applies when s is even and $\geqslant 8$. As an example of what may occur when s is even, we will consider the normal location problem with known variance σ^2. Easy calculations (using Harter's tables [4]) show that for the even values $s = 8, 10, \cdots, 20$, reversals of the C's occur only in the cases $s = 8$, $C_7^* < C_6^*$ and $s = 12$, $C_{11}^* < C_{10}^*$ and possibly in the cases $s = 10$, (C_9^*, C_8^*); $s = 14$, (C_{13}^*, C_{12}^*) and $s = 16$, (C_{15}^*, C_{14}^*). In these latter three cases, the C^*'s are equal to the two decimal places recorded in the tables. In each of these cases (7.12) would require increasing C_{s-1} to the value of C_{s-2}^*, and no other C-value could be decreased to compensate for this increase without violating (6.1). Therefore, in this normal location problem, the C'-procedures are admissible, at least for all $s \leqslant 21$.

Acknowledgment. The authors wish to thank Emil Spjøtvoll for pointing out a possible generalization of their original results which greatly improved the paper. They would also like to thank the referee for a most careful reading of the manuscript and a number of very helpful comments.

REFERENCES

[1] DUNCAN, C. B. (1952). Testing the homogeneity of treatment means in an analysis of variance of engineering data. Paper presented at annual meetings, Amer. Statist. Assoc., Chicago, December 1952.

[2] DUNCAN, D. B. (1955). Multiple range and multiple F-tests. *Biometrics* **11** 1–42.

[3] EINOT, ISRAEL and GABRIEL, K. R. (1975). A study of the powers of several methods of multiple comparisons. *J. Amer. Statist. Assoc.* **70** 574–583.

[4] HARTER, H. L. (1969). *Order Statistics and their use in Testing and Estimation.* Vol. I. Aerospace Research Laboratories, Office of Aerospace Research, U.S. Air Force.

[5] KIEFER, J. (1959). K-sample analogues of the Kolmogorov-Smirnov and Cramer-von Mises tests. *Ann. Math. Statist.* **30** 420–447.

[6] KOHR, RICHARD L. and GAMES, PAUL A. (1974). Robustness of the analysis of variance, the Welch procedure and a Box procedure to heterogeneous variances. *J. Experimental Education* **43** 61–69.

[7] LEHMANN, E. L. and SHAFFER, JULIET POPPER (1977). On a fundamental theorem in multiple comparisons. *J. Amer. Statist. Assoc.* **72** 576–578.

[8] RYAN, THOMAS A. (1960). Significance tests for multiple comparison of proportions, variances, and other statistics. *Psychological Bulletin* **57** 318–328.

[9] TUKEY, J. W. (1953). The problem of multiple comparisons. Unpublished manuscript, Princeton Univ.

[10] WELCH, B. L. (1951). On the comparison of several mean values: an alternative approach. *Biometrika* **38** 330–336.

DEPARTMENT OF STATISTICS
UNIVERSITY OF CALIFORNIA
BERKELEY, CALIFORNIA 94720

The Annals of Statistics
2005, Vol. 33, No. 3, 1084–1108
DOI 10.1214/009053605000000066

ON OPTIMALITY OF STEPDOWN AND STEPUP MULTIPLE TEST PROCEDURES

BY E. L. LEHMANN, JOSEPH P. ROMANO AND JULIET POPPER SHAFFER

University of California, Berkeley, Stanford University and University of California, Berkeley

Consider the multiple testing problem of testing k null hypotheses, where the unknown family of distributions is assumed to satisfy a certain monotonicity assumption. Attention is restricted to procedures that control the familywise error rate in the strong sense and which satisfy a monotonicity condition. Under these assumptions, we prove certain maximin optimality results for some well-known stepdown and stepup procedures.

1. Introduction. For classical single-stage multiple comparison procedures, a number of optimality results are available. (See, e.g., [6], Chapter 11, and [11], Chapter 7 particularly Sections 7.9 and 7.10.) However, no such literature exists for the more recent stepdown and stepup methods. It is the purpose of the present paper to establish optimality properties for procedures of the latter kind.

Our setup and conditions are those of Lehmann [9], who discusses such an optimality result for the testing of two hypotheses. For the general problem of testing k null hypotheses $H_1, \dots, H_k$, consider k random variables $X_1, \dots, X_k$; typically, these are test statistics for the individual hypotheses $H_1, \dots, H_k$. We assume that $X = (X_1, \dots, X_k)$ has some k-dimensional joint cumulative distribution function $F_\theta(\cdot)$ indexed by $\theta = (\theta_1, \dots, \theta_k)$ in $\mathbf{R}^k$. The null hypothesis H_i states $\theta_i \le 0$, which is being tested against the alternatives $\theta_i > 0$.

Stepdown procedures were initiated by Holm [7], while the stepup approach can be found in [2, 5, 8, 14, 16]. Background material on stepwise procedures is provided by Hochberg and Tamhane [6] and Westfall and Young [18]. Roughly speaking, stepdown procedures start by rejecting the most significant hypothesis (corresponding to the largest X_i) and then they sequentially consider the most significant of the remaining hypotheses. Alternatively, stepup procedures start with the least significant hypothesis (corresponding to the smallest X_i).

Our optimality results require crucial monotonicity assumptions and restrictions. We say that a region $\mathcal{M}$ of x values is a monotone increasing region if $x = (x_1, \dots, x_k) \in \mathcal{M}$ and $x_i \le y_i$ implies that $y = (y_1, \dots, y_k)$ is also in $\mathcal{M}$. We

Received May 2003; revised August 2004.
AMS 2000 subject classifications. Primary 62J15; secondary 62F03.
Key words and phrases. Familywise error rate, maximin, monotonocity, multiple testing, stepdown and stepup procedures.

assume of our model that increased values of θ lead to higher values of X, specifically, that if $\theta_i \le \gamma_i$, then

$$\int_{\mathcal{M}} dF_\theta(x_1,\dots,x_k) \le \int_{\mathcal{M}} dF_\gamma(x_1,\dots,x_k) \tag{1}$$

for every monotone increasing region $\mathcal{M}$. This assumption holds, in particular, if the distributions F_θ have densities p_θ with (increasing) monotone likelihood ratio; that is, if $x = (x_1,\dots,x_k)$, $y = (y_1,\dots,y_k)$, $\theta = (\theta_1,\dots,\theta_k)$ and $\theta' = (\theta'_1,\dots,\theta'_k)$, then

$$\frac{p_{\theta'}(x)}{p_\theta(x)} \le \frac{p_{\theta'}(y)}{p_\theta(y)}$$

whenever $x_i \le y_i$ for all i and $\theta_j \le \theta'_j$ for all j. This notion of monotonicity was studied in [10]; other notions of stochastic ordering are discussed in [12].

In addition to condition (1), we will assume an analogous monotonicity property for the distribution of $(\delta_1 X_1,\dots,\delta_k X_k)$, for any $\delta_i \in \{-1, 1\}$. Specifically, for every monotone increasing region $\mathcal{M}$ and $\delta_i\theta_i \le \delta_i\gamma_i$,

$$\int_{\mathcal{M}} dF_\theta(\delta_1 x_1,\dots,\delta_k x_k) \le \int_{\mathcal{M}} dF_\gamma(\delta_1 x_1,\dots,\delta_k x_k). \tag{2}$$

For example, the condition for $(-X_1,\dots,-X_k)$ means that for any monotone decreasing region $\mathcal{M}'$ (the complement of a monotone increasing region), the inequality (1) is reversed; that is, the probability of the event $\{X \in \mathcal{M}'\}$ increases as θ decreases (in each component).

Under these assumptions, we shall restrict attention to decision rules satisfying the following monotonicity condition. A decision rule D based on X states for each possible value x of X the subset $I = I_x$ of $\{1,\dots,k\}$ of values i for which the hypothesis H_i is rejected. A decision rule D is said to be monotone if

$$x_i \le y_i \quad \text{for } i \in I_x \quad \text{but } y_i < x_i \quad \text{for } i \notin I_x$$

implies that $I_x = I_y$. Thus, the subset of x values that results in rejecting all hypotheses is a monotone increasing region. More generally, fix $I \subset \{1,\dots,k\}$ and, based on a monotone decision rule, let $\mathcal{M}_I$ denote the set of x values such that $I_x = I$. If $\delta_i = 1$ for $i \in I$ and $\delta_i = -1$ otherwise, then

$$\{(\delta_1 x_1,\dots,\delta_k x_k) : (x_1,\dots,x_k) \in \mathcal{M}_I\}$$

is a monotone increasing set. By assumption (2), the probability of this set is increasing in $\delta_i\theta_i$.

Among all monotone decision rules that provide strong control of the family-wise error rate (FWER), that is, of the probability of committing a Type 1 error by wrongly rejecting one or more true hypotheses, under any configuration of true and false null hypotheses we shall show how to maximize certain aspects of the power of the procedures (i.e., of the probability of correctly rejecting false hypotheses).

However, we note that we are not restricting attention to any kind of stepwise procedure; rather, the resulting optimal procedures take the form of well-known stepwise procedures, which will be fully described later.

Here the restriction to monotone procedures is not just for convenience—the results are not true without this restriction. It is, in fact, possible to improve the rejection probability without violating the error control by adding small implausible pieces to the rejection regions, resulting in decision rules that are very counterintuitive. That this is possible is due to the fact that the bound for the error control is not attained but only approached in the limit as some parameter values tend to ∞ or $-\infty$. For a discussion of the pros and cons of such counterintuitive decision rules with references to the literature, see [13].

To conclude this introduction, we mention some situations in which the present approach does and some in which it does not apply. As a first example, consider a paired comparison experiment with pairs of observations (Y_i, Z_i). Let $E(Y_i) = \mu_i$ and $E(Z_i) = \nu_i$, and consider testing the hypotheses $\theta_i = \nu_i - \mu_i = 0$ against the alternatives $\theta_i > 0$. If we put $X_i = Z_i - Y_i$ and base our inferences on the X's, this reduces to the situation considered here. This example can be extended to the comparison of two treatments with m_i and n_i observations ($i = 1, \dots, k$), respectively, on k subjects. Another application is the comparison of k treatments with a control. Here $\theta_i = \mu_i - \mu_0$, where the μ_i ($i = 1, \dots, k$) and μ_0 are the means for the k treatments and the control, respectively.

On the other hand, the approach does not apply to the comparison of k treatments, that is, the hypothesis $H : \mu_1 = \cdots = \mu_k$, where in the case of rejection one wishes to determine the pairs $i < j$ for which $\mu_i < \mu_j$. As in the preceding examples, the hypothesis can be reduced to $H : \theta_2 = \cdots = \theta_k = 0$ with, for example, $\theta_i = \mu_i - \mu_1$. However, with the resulting procedure, one can only determine the significant differences $\mu_j - \mu_i$ with $i = 1$ and not those with $1 < i < j$.

In Section 2 we treat the case $k = 2$ separately. In Section 3 we consider general k for stepdown procedures, but make a further exchangeability assumption. The corresponding results for stepup procedures are then provided in Section 4, though a further assumption of monotonicity of critical values is invoked. Section 5 is a brief conclusion and all proofs are deferred to Section 6.

Distributional assumptions. We suppose $(X_1, \dots, X_k)$ has a joint cumulative distribution function $F_\theta(\cdot)$, indexed by $\theta = (\theta_1, \dots, \theta_k)$ in $\mathbf{R}^k$. The parameter space is a finite or infinite open rectangle with

$$\theta_i^L < \theta_i < \theta_i^U, \qquad i = 1, \dots, k.$$

Similarly, the sample space is assumed to be a finite or infinite open rectangle with

$$x_i^L < X_i < x_i^U,$$

independent of θ. We further assume the distribution of any subcollection $\{X_i, i \in I\}$ depends only on those θ_i with $i \in I$, and that X_i tends in probability to x_i^L as $\theta_i \to \theta_i^L$ and X_i tends in probability to x_i^U as $\theta_i \to \theta_i^U$.

To ease the notation, we assume here and in the remainder of the paper that θ_i varies in all of $\mathbf{R}$, so that $\theta_i^L = -\infty$ and $\theta_i^U = \infty$. We also simplify the notation by taking $x_i^L = -\infty$ and $x_i^U = \infty$. In addition, we assume that the joint distribution of X has a density with respect to Lebesgue measure; this is used only so that the critical constants of the optimal procedures can be obtained for control at a given level α to be achieved exactly, but this hypothesis can certainly be weakened. In order for the critical constants to be uniquely defined, we further assume that the joint density is positive on its (assumed rectangular) region of support, but this can be weakened as well.

2. The case $k = 2$. We are testing hypotheses H_1 and H_2 with H_i corresponding to $\theta_i \leq 0$. Let $\omega_{0,0}$ denote the part of the parameter space where both H_1 and H_2 are true; let $\omega_{0,1}$ correspond to the part where H_1 is true and H_2 is not true; similarly for $\omega_{1,0}$ and $\omega_{1,1}$.

A decision rule D analogously divides up the sample space into regions $d_{0,0}$, $d_{0,1}$, $d_{1,0}$ and $d_{1,1}$. For example, $d_{0,1}$ corresponds to the region in the sample space where H_1 is declared true and H_2 is declared false. Also, let d_i be the region where H_i is rejected, so $d_1 = d_{1,0} \cup d_{1,1}$ and $d_2 = d_{0,1} \cup d_{1,1}$.

We will restrict attention to rules D that are

$$\text{monotone} \tag{3}$$

and such that the

$$\text{FWER} \leq \alpha. \tag{4}$$

For $\varepsilon = (\varepsilon_1, \varepsilon_2)$ with $\varepsilon_i > 0$, consider subsets of elements (θ_1, θ_2) defined by

$$A_1(\varepsilon) = \{\theta_1 > \varepsilon_1\} \cup \{\theta_2 > \varepsilon_2\} \tag{5}$$

and

$$A_2(\varepsilon) = \{\theta_1 > \varepsilon_1\} \cap \{\theta_2 > \varepsilon_2\}. \tag{6}$$

A decision rule is deemed good if the quantities

$$\inf_{\theta \in A_1(\varepsilon)} P_\theta\{d_{0,0}^c\} \tag{7}$$

and

$$\inf_{\theta \in A_2(\varepsilon)} P_\theta\{d_{1,1}\} \tag{8}$$

are large. As we will see, it is *not* possible to find a rule D satisfying (3) and (4) that maximizes (7) and (8) simultaneously.

In order to appreciate the criteria (7) and (8), first suppose $\theta \in A_1(\varepsilon)$. Then at least one θ_i is positive and so we would not want to conclude that both θ_i are ≤ 0; rather, we wish to conclude $d_{0,0}^c$. Thus, maximizing (7) maximizes the minimum probability that we do not conclude $d_{0,0}$ as θ varies in $A_1(\varepsilon)$. Similarly, if $\theta \in A_2(\varepsilon)$, then both θ_i are positive, and so we wish to maximize the (minimum) chance that we make the decision $d_{1,1}$.

In addition, we also consider the following notion of optimality. Again suppose $\theta \in A_1(\varepsilon)$, so that at least one θ_i is positive. Then, as above, we do not want to make the decision $d_{0,0}$. However, we also do not wish to make the decision $d_{0,1}$ if, in fact, H_1 is false and H_2 is true; we would rather make the correct decision $d_{1,0}$. So, we also consider the probability of maximizing

$$\inf_{\theta \in A_1(\varepsilon)} P_\theta\{\text{reject at least one false hypothesis}\}. \tag{9}$$

In other words, the criterion (7) maximizes the minimum probability of rejecting exactly least one hypothesis (regardless of which are true and false), while criterion (9) maximizes the minimum probability of rejecting at least one *false* hypothesis. The latter criterion seems more compelling, though the former criterion might be justified in a situation where it is important to know that the joint null hypothesis (i.e., the global hypothesis that both hypotheses are true) is not true. In any case, we shall see that the same optimal procedure D arises from both criteria.

THEOREM 2.1. *Consider the case $k = 2$ under the assumptions given in Section* 1.

(i) *A rule D satisfying* (3) *and* (4) *maximizes* (7) *if*

$$d_{0,0}^c = \{X_1 > a_1 \text{ or } X_2 > a_2\} \tag{10}$$

and $\{X_i > a_i\} \subset d_i$, where a_1 and a_2 are determined so that

$$P_{0,0}\{X_1 > a_1 \text{ or } X_2 > a_2\} = \alpha \tag{11}$$

and

$$P_{\varepsilon_1}\{X_1 > a_1\} = P_{\varepsilon_2}\{X_2 > a_2\}. \tag{12}$$

Its minimum (rejection) probability over $A_1(\varepsilon)$ is given by $P_{\varepsilon_1}\{X_1 > a_1\}$.

(ii) *The (stepdown) rule D satisfying* (3), (4) *and* (10) *that maximizes* (8) *is given by*

$$d_{0,1} = \{X_1 < b_1, X_2 \geq a_2\}, \tag{13}$$

$$d_{1,0} = \{X_1 \geq a_1, X_2 < b_2\}, \tag{14}$$

$$d_{1,1} = \{X_1 \geq b_1, X_2 \geq b_2\} \cap d_{0,0}^c, \tag{15}$$

where b_i satisfies

$$P_0\{X_i \geq b_i\} = \alpha \tag{16}$$

(and so $b_i < a_i$). The minimum probability of $d_{1,1}$ over $A_2(\varepsilon)$ is

$$P_{\varepsilon_1,\varepsilon_2}\{X_1 > a_1, X_2 > b_2 \cup X_1 > b_1, X_2 > a_2\}.$$

(iii) *The result* (i) *holds for D if criterion* (7) *is replaced by* (9), *and* (12) *is also the maximum value of criterion* (9).

Note that once $d_{0,0}$ and $d_{1,1}$ are determined, so are $d_{0,1}$ and $d_{1,0}$ (by monotonicity).

The procedure D of Theorem 2.1 is an example of a stepdown procedure. It starts by rejecting the most significant hypothesis (corresponding to the largest X_i) and it then sequentially considers the most significant of the remaining hypotheses. Alternatively, stepup procedures start with the least significant hypothesis (corresponding to the smallest X_i), and an optimality result is now given for such a procedure.

REMARK 2.1. The proof shows that the optimal procedure D in (i) and (ii) is the unique rule satisfying (3) and (4) which maximizes (7), in the sense that if E is any other such rule, then $e_{0,0} \triangle d_{0,0}$ has Lebesgue measure 0, where $A \triangle B$ denotes the symmetric difference between sets A and B. Similarly, a rule E satisfying (3), (4) and (10) maximizing (8) must satisfy $e_{1,1} \triangle d_{1,1}$ has Lebesgue measure 0.

Also, notice that the optimal procedure D does not depend on ε. It follows that D is admissible in the following sense. Suppose there exists another monotone rule E that controls the FWER, and such that

$$P_\theta\{d_{0,0}^c\} \leq P_\theta\{e_{0,0}^c\} \qquad \text{for all } \theta \in \omega_{0,0}^c, \tag{17}$$

with strict inequality for some $\theta \in \omega_{0,0}^c$. Taking the infimum of both sides over $\theta \in A_1(0)$, it follows that E must also be optimal in the sense of Theorem 2.1(i). But, by uniqueness, $e_{0,0} \triangle d_{0,0}$ has Lebesgue measure 0, which implies the $\leq$ in (17) is an equality. A similar admissibility result for the region $d_{1,1}$ can be stated as well.

Analogous uniqueness and admissibility results hold for all the optimal procedures presented later on. For a discussion of admissibility in multiple testing problems, see [3].

THEOREM 2.2. *Consider the case $k = 2$ under the assumptions given in Section* 1.

(i) *A rule D satisfying* (3) *and* (4) *maximizes* (8) *if $d_{1,1}$ is given by*

$$d_{1,1} = \{X_1 > b_1, X_2 > b_2\}, \tag{18}$$

and $d_i \subset \{X_i > b_i\}$, *where* b_i *satisfies* (16) (*so it is the same constant as in Theorem* 2.1). *Its minimum probability over* $A_2(\varepsilon)$ *is given by* $P_{\varepsilon_1,\varepsilon_2}\{X_1 > b_1, X_2 > b_2\}$.

(ii) *The* (*stepup*) *rule D satisfying* (3), (4) *and* (18) *that maximizes* (7) *is given by*

$$d_{0,1} = \{X_1 < b_1, X_2 \geq \tilde{a}_1\}, \tag{19}$$

$$d_{1,0} = \{X_1 \geq \tilde{a}_1, X_2 < b_2\}, \tag{20}$$

$$d_{0,0} = \{X_1 \leq \tilde{a}_1, X_2 \leq \tilde{a}_2\} \cap d_{1,1}^c, \tag{21}$$

where $\tilde{a}_i$ *is determined so that*

$$P_{0,0}\{d_{0,0}^c\} = \alpha \tag{22}$$

and

$$P_{\varepsilon_1}\{X_1 \geq \tilde{a}_1\} = P_{\varepsilon_2}\{X_2 \geq \tilde{a}_2\}. \tag{23}$$

The value of (23) *is the minimum probability of D over* $A_1(\varepsilon)$.

(iii) *The result* (ii) *holds for D if criterion* (7) *is replaced by* (9), *and* (23) *is also the maximum value of criterion* (9).

REMARK 2.2. Note that $b_i < a_i < \tilde{a}_i$. Also, the best minimum probability over $A_1(\varepsilon)$ in the case of Theorem 2.1 exceeds the best in the case of Theorem 2.2, but it reverses for Theorem 2.2.

REMARK 2.3. A remark analogous to Remark 2.1 applies to the optimal procedure in Theorem 2.2.

REMARK 2.4. It is now clear that, subject to (3) and (4), we cannot find a rule to maximize both (7) and (8). By Theorem 2.1(i) and Theorem 2.2(ii), such a rule D would have to satisfy

$$d_{0,0} = \{X_1 \leq a_1 \text{ and } X_2 \leq a_2\}$$

and

$$d_{1,1} = \{X_1 > b_1, X_2 > b_2\}$$

simultaneously, which is impossible because these two sets have a nontrivial intersection as $b_i < a_i$.

REMARK 2.5. The results of this paper do not hold without the monotonicity assumption. For example, consider part (i) of Theorem 2.2. Suppose further that X_1 and X_2 are independent with X_i normally distributed with mean θ_i and variance 1. Then $b_i = b = z_{1-\alpha}$, the $1 - \alpha$ quantile of the standard normal distribution. The probability of $d_{1,1}$ under (θ_1, θ_2) with both $\theta_i > 0$ is always less than α and approaches α as either $\theta_i \to \infty$. Therefore, by adding to $d_{1,1}$ a small enough region in the southwest quadrant, one can increase the rejection probability without violating the level constraint; see Section 4 of [13]. Such a procedure is not monotone. Similarly, regarding the problem addressed in (i) of Theorem 2.1, [9], Section 3, shows that the maximin test is not monotone.

3. General *k* stepdown. Consider testing k null hypotheses $H_1, \dots, H_k$ with H_i corresponding to $\theta_i \le 0$. In this section and the next, we add a symmetry condition for the joint distribution of $(X_1, \dots, X_k)$. Specifically, we assume that the joint distribution of $(X_1, \dots, X_k)$ under $\theta_i = \theta$ (some value independent of i) is exchangeable. This is not a crucial assumption (and actually only needs to hold at $\theta = 0$ or $\theta = \varepsilon$, where ε is given in the statement of the theorems), but it reduces the number of critical values from order 2^k to k. The results should generalize, but at the expense of more complicated notation.

Let

$$X_{r_1} \ge X_{r_2} \ge \cdots \ge X_{r_k}$$

denote the ordered X-values, and let $H_{r_1}, \dots, H_{r_k}$ denote their corresponding null hypotheses.

For any (monotone) decision rule E, let $E_{k,j}$ denote the event that E rejects at least j of the null hypotheses. For $\varepsilon > 0$, let

$$A_j(\varepsilon) = \{(\theta_1, \dots, \theta_k) : \text{at least } j \ \theta_i \text{ satisfy } \theta_i > \varepsilon\}.$$

Consider the monotone stepdown decision rule D that rejects $H_{r_1}, \dots, H_{r_j}$ and accepts the remaining null hypotheses if $X_{r_i} \ge c_{k,i}$ for $1 \le i \le j$, but $X_{r_{j+1}} < c_{k,j+1}$, where the $c_{k,j} = c_{k,j}(\alpha)$ are determined by

$$P_{\underbrace{0,\dots,0}_{k-j+1 \text{ times}}} \{X_i > c_{k,j} \text{ for some } i, 1 \le i \le k-j+1\} = \alpha. \tag{24}$$

Then

$$D_{k,j} = \{X_{r_i} \ge c_{k,i}, 1 \le i \le j\}.$$

Note that (24) implies the important relationship

$$c_{k,j} = c_{k-1,j-1} \tag{25}$$

if $k \ge j \ge 2$. Also note that, for fixed k, $c_{k,j}$ is nonincreasing in j.

Since the constants $c_{k,j}$ depend only on $k - j$, we may more succinctly define

$$f_{k-i+1} \equiv c_{k,i}, \tag{26}$$

where the f_j are determined by

$$P_{\underbrace{0,\dots,0}_{j \text{ times}}}\{\max(X_1, \dots, X_j) > f_j\} = \alpha. \tag{27}$$

The procedure D then rejects $H_{r_1}, \dots, H_{r_j}$ if and only if $X_{r_i} \ge f_{k-i+1}$ for $1 \le i \le j$.

LEMMA 3.1. *Suppose the assumptions of Section* 1 *and the symmetry condition described at the beginning of this section hold.*

(i) *The above decision rule D controls the FWER at level* α.

(ii) *Define*

$$\beta_{k,j}(\alpha, \varepsilon) = \inf_{\theta \in A_j(\varepsilon)} P_\theta\{D_{k,j}\};$$

that is, $\beta_{k,j}(\alpha, \varepsilon)$ *is the minimum probability of* $D_{k,j}$ *over* $A_j(\varepsilon)$. *Then*

$$\beta_{k,j}(\alpha, \varepsilon) = P_{\underbrace{\varepsilon,\dots,\varepsilon}_{j \text{ times}}}\{S_{k,j}\}, \tag{28}$$

where

$$S_{k,j} = \{X_{\pi_j(1)} > f_k, \dots, X_{\pi_j(j)} > f_{k-j+1} \text{ for some permutation } \pi_j \text{ of } \{1, \dots, j\}\}. \tag{29}$$

So (28) *is the minimum probability over* $A_j(\varepsilon)$ *not only of rejecting at least j hypotheses, but also of rejecting at least j false hypotheses.*

THEOREM 3.1. *Suppose the assumptions of Section* 1 *and the symmetry condition described at the beginning of this section hold.*

(i) *Among monotone decision rules E that control the FWER, D maximizes*

$$\inf_{\theta \in A_1(\varepsilon)} P_\theta\{E_{k,1}\}. \tag{30}$$

Also, D maximizes

$$\inf_{\theta \in A_2(\varepsilon)} P_\theta\{E_{k,2}\}$$

among such rules E that also satisfy $E_{k,2} \subset D_{k,1}$. *In general, for* $j = 2, \dots, k$, *D maximizes*

$$\inf_{\theta \in A_j(\varepsilon)} P_\theta\{E_{k,j}\} \tag{31}$$

among monotone rules E that control the FWER and satisfy

$$E_{k,j} \subset D_{k,j-1}. \tag{32}$$

Therefore, for any other rule E, *we must have*

$$\inf_{\theta \in A_j(\varepsilon)} P_\theta(E_{k,j}) < \beta_{k,j}(\alpha, \varepsilon)$$

for at least one j.

(ii) *D also is optimal in the sense that it maximizes*

$$\inf_{\theta \in A_j(\varepsilon)} P_\theta\{\textit{reject at least } j \textit{ false hypotheses}\}$$

subject to (32).

REMARK 3.1. The procedure D is essentially unique (up to sets of Lebesgue measure 0), as described in Remark 2.1, and an admissibility result analogous to that described in Remark 2.1 holds as well.

REMARK 3.2. For fixed k, the optimal constants with $c_{k,j} = f_{k-j+1}$ are given by the values

$$c_{k,1}, c_{k,2}, \ldots, c_{k,k}. \tag{33}$$

But, since $c_{k,2} = c_{k-1,1}$, $c_{k,3} = c_{k-2,1}$, and so on, the sequence (33) is equivalent to

$$c_{k,1}, c_{k-1,1}, c_{k-2,1}, \ldots, c_{1,1}.$$

This is just a sequentially rejective procedure of the kind proposed by Holm [7]: after the first step using the critical value $c_{k,1}$, reduce the number of hypotheses from k to $k-1$ and repeat the first step but now with $c_{k-1,1}$, and so on. In the case where the X_i have a uniform (0, 1) marginal distribution under the null hypothesis so that we translate everything into p-values and reject for *small* values, Holm [7] used $c_{k,1} = \alpha/k$ since he assumed only the marginal distributions to be known (and strong error control follows by Bonferroni). Our $c_{k,1}$ would then be determined by

$$P_{\underbrace{0,\ldots,0}_{k \text{ times}}}\{X_i \le c_{k,1} \text{ for one or more values of } i : 1 \le i \le k\} = \alpha$$

or, equivalently,

$$P_{\underbrace{0,\ldots,0}_{k \text{ times}}}\{\min(X_1, \ldots, X_k) \le c_{k,1}\} = \alpha.$$

If we further assume independence of the p-values, then the critical constants $c_{k,j}$ satisfy

$$1 - (1 - c_{k,j})^{k-j+1} = \alpha.$$

Thus, the Holm principle remains in effect, except that instead of using $c_{k,1} = \alpha/k$, the independence assumption implies the exact critical values $c_{k,1} = 1 - (1-\alpha)^{1/k}$.

4. General k stepup. Assume the conditions imposed in the previous section. We are testing null hypotheses $H_1, \dots, H_k$ with H_i corresponding to $\theta_i \le 0$. Let

$$X_{(1)} \le X_{(2)} \le \cdots \le X_{(k)}$$

denote the ordered X-values; in the notation of the previous section, $X_{(j)} = X_{r_{k-j+1}}$.

Consider the following monotone stepup decision rule D for appropriately chosen constants $d_1, \dots, d_k$ (to be specified shortly, but assumed nondecreasing). If $X_{(1)} > d_1$, then reject all null hypotheses. Otherwise, if $X_{(1)} \le d_1$ but $X_{(2)} > d_2$, reject the $k-1$ hypotheses corresponding to the $k-1$ largest X's. In general, for the smallest j such that $X_{(j)} > d_j$, reject the $k-j+1$ hypotheses corresponding to the $k-j+1$ largest X's and accept the remaining. (Note that the constants d_j should perhaps be written as $d_{k,j}$ to show the dependence on k; however, we will see that $d_{k,j}$ will be chosen to be independent of k and so we just abbreviate to d_j.)

The above rule rejects at least j null hypotheses for the set $D_{k,j}$ defined by

$$D_{k,j} = \{X_{(1)} > d_1\} \cup \cdots \cup \{X_{(k-j+1)} > d_{k-j+1}\}.$$

Equivalently, at least $k-j+1$ hypotheses are accepted if $D_{k,j}^c$ occurs, where

$$D_{k,j}^c = \{X_{(1)} \le d_1\} \cap \cdots \cap \{X_{(k-j+1)} \le d_{k-j+1}\}.$$

Evidently,

$$D_{k,j+1} \subset D_{k,j}.$$

The constants d_j are determined so that

$$P_{\underbrace{0,\dots,0}_{j \text{ times}}}\{L_j\} = 1 - \alpha, \tag{34}$$

where

$$L_j = \{X_{\pi(1)} \le d_1, \dots, X_{\pi(j)} < d_j \text{ for some permutation of } \{1, \dots, j\}\}. \tag{35}$$

Note that the constant d_j does not depend on k as reflected in the notation. Also, $d_1 = c_{k,k} = c_{1,1} = f_1$, where $c_{1,1}$ and f_1 are the constants (24) and (26) of the previous section. However, as pointed out by an anonymous referee, in the case $k > 2$, it need not be the case that $d_{k,j}$ is nondecreasing in j. A counterexample is provided in [4]; some further references on the monotonicity of critical values are [1] and [15]. In order to prove our results, we need to assume the monotonicity holds.

LEMMA 4.1. *Assume the conditions of Lemma* 3.1. *Also assume that the constants d_j used in the procedure D are nondecreasing in j.*

(i) *The above decision rule D controls the FWER at level α.*

(ii) *Define*

$$\tilde{\beta}_{k,j} = \inf_{\theta \in A_j(\varepsilon)} P_\theta\{D_{k,j}\};$$

that is, $\tilde{\beta}_{k,j}(\alpha, \varepsilon)$ *is the minimum probability of* $D_{k,j}$ *over* $A_j(\varepsilon)$. *Then*

$$\tilde{\beta}_{k,j}(\alpha, \varepsilon) = P_{\underbrace{\varepsilon,\dots,\varepsilon}_{j \text{ times}}}\{\min(X_1, \dots, X_j) > d_{k-j+1}\}. \tag{36}$$

The minimum probability over $A_j(\varepsilon)$ *of rejecting at least* j *false hypotheses is also given by* (36).

THEOREM 4.1. *Assume the conditions of Theorem* 3.1. *Also assume that the constants* d_j *used in the procedure* D *are nondecreasing in* j.

(i) *Among monotone decision rules* E *that control the FWER at level* α, D *maximizes*

$$\inf_{\theta \in A_k(\varepsilon)} P_\theta\{E_{k,k}\}. \tag{37}$$

Also, D *maximizes*

$$\inf_{\theta \in A_{k-1}(\varepsilon)} P_\theta\{E_{k,k-1}\}$$

among rules that satisfy $D_{k,k} \subset E_{k,k-1}$. *In general, for* $j = k-1, \dots, 1$, D *maximizes*

$$\inf_{\theta \in A_j(\varepsilon)} P_\theta\{E_{k,j}\} \tag{38}$$

among monotone rules E *that control the FWER and satisfy*

$$D_{k,j+1} \subset E_{k,j}. \tag{39}$$

Therefore, for any other rule E, *we must have*

$$\inf_{\theta \in A_j(\varepsilon)} P_\theta(E_{k,j}) < \tilde{\beta}_{k,j}(\alpha, \varepsilon)$$

for at least one j.

(ii) D *also is optimal in the sense that it maximizes*

$$\inf_{\theta \in A_j(\varepsilon)} P_\theta\{\textit{reject at least } j \textit{ false hypotheses}\}$$

subject to (39).

REMARK 4.1. Again, the procedure D is unique up to sets of Lebesgue measure 0, and it is admissible; see Remark 2.1.

REMARK 4.2. Letting

$$X_{j:1} \le \cdots \le X_{j:j}$$

denote the ordered values of just the first j X's, the constants d_j are determined by

$$\underbrace{P_{0,\dots,0}}_{j \text{ times}}\{X_{j:1} \le d_1, \dots, X_{j:j} \le d_j\} = 1 - \alpha.$$

If we compare this with (27), we see that $f_j < d_j$, except when $j = 1$, in which case $f_1 = d_1$.

5. Conclusions. Stepdown and stepup methods were proposed as intuitively appealing by Holm, Hochberg, Dunnett and Tamhane, and others. The present paper, treating the case of one-sided alternatives only, used optimality criteria that seemed reasonable and were not selected to justify predetermined solutions. It is gratifying that the results confirm the intuition of the originators of these methods. Even though our assumptions are strong, some stepwise methods can now be viewed as asymptotically optimal, such as the stepup method of Dunnett and Tamhane [2]. Outside the strong assumptions imposed in this paper, Westfall and Young [18] give general resampling methods to approximate the critical values of stepdown procedures, while Troendle [17] addresses the corresponding problem for stepup procedures.

6. Proofs and auxiliary results.

PROOF OF THEOREM 2.1. First, observe that for the procedure D given in (i), for $\theta \in \omega_{0,0}$,

$$P_\theta\{d_{0,0}^c\} \le P_{0,0}\{d_{0,0}^c\} = P_{0,0}\{X_1 > a_1 \text{ or } X_2 > a_2\} = \alpha$$

by choice of a_i. For this D, by monotonicity, the inf over $\theta \in A_1(\varepsilon)$ in (7) occurs at $(\theta_1, \theta_2) = (\varepsilon_1, -\infty)$ or $(-\infty, \varepsilon_2)$; this is a shorthand notation so that

$$P_{\varepsilon_1,-\infty}\{d_{0,0}^c\} = \lim_{\theta_2 \to -\infty}\{d_{0,0}^c\}.$$

But then

$$P_{\varepsilon_1,-\infty}\{d_{0,0}^c\} = P_{\varepsilon_1}\{X_1 > a_1\}$$

and

$$P_{-\infty,\varepsilon_2}\{d_{0,0}^c\} = P_{\varepsilon_2}\{X_2 > a_2\}.$$

So, the value of criterion (7) for the procedure D is indeed given by (12). Similarly, the value of criterion (9) for D is also (12). Indeed, as $\theta_1 \to -\infty$, the chance that H_1 is incorrectly rejected tends to 0.

To prove (i), suppose E is another decision rule satisfying (3) and (4). Assume there exists $(x_1, x_2) \notin d_{0,0}$, but $(x_1, x_2) \in e_{0,0}$. Then there exists at least one component with $x_i > a_i$, say $x_1 > a_1$. Hence,

$$P_{\varepsilon_1,-\infty}\{e_{0,0}) \geq P_{\varepsilon_1,-\infty}\{X_1 \leq x_1, X_2 \leq x_2\} = P_{\varepsilon_1}\{X_1 \leq x_1\} > P_{\varepsilon_1}\{X_1 \leq a_1\}.$$

Therefore,

$$P_{\varepsilon_1,-\infty}\{e_{0,0}^c\} < P_{\varepsilon_1,-\infty}\{X_1 \leq a_1\} = P_{\varepsilon_1}\{X_1 \leq a_1\},$$

so that E has a smaller value of criterion (7) than does a claimed optimal D. So it must be the case that $e_{0,0} \subset d_{0,0}$. But, if $e_{0,0}$ is strictly contained in $d_{0,0}$ such that the set difference $e_{0,0} \triangle d_{0,0}$ has positive Lebesgue measure, then its region for rejecting $\omega_{0,0}$, namely, $e_{0,0}^c$, is bigger than $d_{0,0}^c$, implying

$$P_{0,0}\{e_{0,0}^c\} > P_{0,0}\{d_{0,0}^c\} = \alpha.$$

The conclusion is that an optimal region D must have the stated region (10) $d_{0,0}^c$.

To prove (ii), let us first check that the claimed solution controls the FWER. For $\theta \in \omega_{0,0}$,

$$P_\theta\{d_{0,0}^c\} \leq \alpha$$

as previously argued. For $\theta \in \omega_{0,1}$,

$$P_\theta\{\text{Type 1 error}\} = P_\theta\{d_{1,1} \cup d_{1,0}\} \leq P_\theta\{X_1 \geq b_1\} \leq P_0\{X_1 \geq b_1\} = \alpha$$

similarly for $\omega_{1,0}$.

The goal now is to find D satisfying (3), (4) and $d_{0,0}^c$ given by (10) to maximize (8). Consider another rule E satisfying (3), (4) and $e_{0,0} = d_{0,0}$. Suppose there exists $(x_1, x_2) \in e_{1,1}$ such that $x_i < b_i$ for some i, say $i = 1$. Then

$$P_{0,\infty}\{e_{1,1}\} \geq P_{0,\infty}\{X_1 \geq x_1, X_2 \geq x_2\} = P_0\{X_1 \geq x_1\} > P_0(X_1 \geq b_1) = \alpha,$$

which would contradict strong control. So $e_{1,1} \subset d_{1,1}$. But you cannot take away points from $d_{1,1}$ without lowering the minimum power at $(\theta_1, \theta_2) = (\varepsilon, \varepsilon)$.

To prove (iii), simply observe, for any θ,

$$P_\theta\{\text{rejecting at least one false } H_i\} \leq P_\theta\{\text{rejecting at least one } H_i\},$$

and so

$$\inf_{\theta \in A_1(\varepsilon)} P_\theta\{\text{rejecting at least one false } H_i\} \leq \inf_{\theta \in A_1(\varepsilon)} P_\theta\{\text{rejecting at least one } H_i\}.$$

But the right-hand side is $P_{\varepsilon_1}\{X_1 > a_1\}$, and so it suffices to show that D satisfies

$$\inf_{\theta \in A_1(\varepsilon)} P_\theta\{D \text{ rejects at least one false } H_i\} = P_{\varepsilon_1}\{X_1 > a_1\}.$$

But the earlier argument for (12) showed this to be the case. □

PROOF OF THEOREM 2.2. To prove (i), suppose E is another rule satisfying (3) and (4) which rejects both hypotheses if $(X_1, X_2) \in e_{1,1}$. Suppose there exists $(x_1, x_2) \in e_{1,1}$ such that $x_i < b_i$ for some i, say $i = 1$. Then

$$P_{0,\infty}\{e_{1,1}\} \geq P_{0,\infty}\{X_1 \geq x_1, X_2 \geq x_2\} = P_0\{X_1 \geq x_1\} > P_0\{X_1 \geq b_1\} = \alpha,$$

which would contradict E control of the FWER. So, $e_{1,1} \subset d_{1,1}$. But you cannot take away any point from $d_{1,1}$ without lowering the minimum power at $(\varepsilon_1, \varepsilon_2)$.

To prove (ii), note that, for the claimed solution the value of (7) is given by

$$\inf_{\theta:\theta\in A_1(\varepsilon)} P_\theta(d_{0,0}^c\} = P_{\varepsilon_1,-\infty}\{d_{0,0}^c\} = P_{\varepsilon_1}\{X_1 > \tilde{a}_1\}.$$

We now seek to determine $d_{0,0}$ [like Theorem 2.1(i) with the added constraint that $d_{0,0} \subset d_{1,1}^c$]. To prove optimality of the claimed solution, suppose E is another rule satisfying (3), (4) and $e_{1,1} = d_{1,1}$, with $d_{1,1}$ given by (18). Suppose $(x_1, x_2) \notin d_{0,0}$, but $(x_1, x_2) \in e_{0,0}$, so that $x_i > \tilde{a}_i$ for some i, say $i = 1$. Then

$$\begin{aligned} P_{\varepsilon_1,-\infty}\{e_{0,0}\} &\geq P_{\varepsilon_1,-\infty}\{X_1 \leq x_1, X_2 \leq x_2\} \\ &= P_{\varepsilon_1}\{X_1 \leq x_1\} > P_{\varepsilon_1}\{X_1 > \tilde{a}_1\}. \end{aligned}$$

Therefore,

$$P_{\varepsilon,-\infty}\{e_{0,0}^c\} < P_\varepsilon\{X_1 > \tilde{a}_1\},$$

so that E cannot be optimal. So it must be the case that $e_{0,0} \subset d_{0,0}$. But if $e_{0,0}$ is strictly contained in $d_{0,0}$, its region for rejecting $\omega_{0,0}$, namely, $e_{0,0}^c$, is bigger than $d_{0,0}^c$, in which case

$$P_{0,0}\{e_{0,0}^c\} > P_{0,0}\{d_{0,0}^c\} = \alpha,$$

a contradiction of strong control.

Finally, we check that D itself exhibits control of the FWER. For $\theta \in \omega_{0,0}$, the probability of a Type 1 error is $\leq \alpha$ because of (22). For $\theta = (\theta_1, \theta_2) \in \omega_{0,1}$,

$$\begin{aligned} P_\theta\{\text{Type 1 error}\} &\leq P_{0,\infty}\{X_1 > b_1, X_2 > b_2 \cup X_1 \geq \tilde{a}_1, X_2 < b_2\} \\ &= P_0\{X_1 > b_1\} = \alpha, \end{aligned}$$

as required.

The proof of (iii) is completely analogous to the proof of Theorem 2.1(iii). □

PROOF OF LEMMA 3.1. To prove (i), suppose $H_1, \ldots, H_p$ are true and $H_{p+1}, \ldots, H_k$ are false. A Type 1 error occurs if any of $H_1, \ldots, H_p$ are rejected. For the rule D, the set where a rejection of any of $H_1, \ldots, H_p$ occurs is a monotone increasing set, and so the probability of this event is largest under this configuration of true and false hypotheses when

$$(\theta_1, \ldots, \theta_k) = (\underbrace{0, \ldots, 0}_{p \text{ times}}, \underbrace{\infty, \ldots, \infty}_{k-p \text{ times}}),$$

and this probability is equal to

$$P_{\underbrace{0,\dots,0}_{p \text{ times}}}\{X_i > f_p \text{ for some } i = 1, \dots, p\} = \alpha$$

by (27) with $j = p$.

To prove (ii), note that the minimum power occurs when θ is one of the $\binom{k}{j}$ points with j values of ε and $k - j$ values of $-\infty$, such as

$$w_{k,j} = w_{k,j}(\varepsilon) = (\underbrace{\varepsilon, \dots, \varepsilon}_{j \text{ times}}, \underbrace{-\infty, \dots, -\infty}_{k-j \text{ times}}). \tag{40}$$

Then, $P_{w_{k,j}}(D_{k,j})$ reduces to $\beta_{k,j}(\alpha, \varepsilon)$ as claimed. Also, for such a configuration $w_{k,j}$, only the j hypotheses $H_1, \dots, H_j$ can be rejected, and so the minimum probability of rejecting at least j hypotheses is the same as the minimum probability of rejecting exactly j hypotheses (and it is also equal to the probability of rejecting exactly j false hypotheses). □

Before the proof of Theorem 3.1, we need two lemmas. We will make use of the following notation. If R is any region in $\mathbf{R}^k$, let

$$R^z = \{(x_1, \dots, x_{k-1}) : (x_1, \dots, x_{k-1}, z) \in R\}.$$

LEMMA 6.1. *Let R be any monotone rejection region in $\mathbf{R}^k$ [so $x = (x_1, \dots, x_k) \in R$ implies $y \in R$ if $y_i \geq x_i$ for all i].*

(i) *If $z_1 < z_2$, then $R^{z_1} \subset R^{z_2}$.*
(ii) *R^z, $\bigcup_z R^z$ and $\bigcap_z R^z$ are all monotone rejection regions in $\mathbf{R}^{k-1}$.*

PROOF. If $z_1 < z_2$ and $(x_1, \dots, x_{k-1}) \in R^{z_1}$, then $(x_1, \dots, x_{k-1}, z_1) \in R$. By monotonicity, $(x_1, \dots, x_{k-1}, z_2) \in R$, and so $(x_1, \dots, x_{k-1}) \in R^{z_2}$. The proof of (ii) is just as easy. □

LEMMA 6.2. *Assume the distributional assumptions given at the end of Section 1. Let R be any monotone rejection region in $\mathbf{R}^k$. Then for any values of the parameters $\theta_1, \dots, \theta_{k-1}$,*

$$P_{\theta_1,\dots,\theta_{k-1},\infty}(R) = P_{\theta_1,\dots,\theta_{k-1}}\left\{\bigcup_z R^z\right\} \tag{41}$$

and

$$P_{\theta_1,\dots,\theta_{k-1},-\infty}\{R\} = P_{\theta_1,\dots,\theta_{k-1}}\left\{\bigcap_z R^z\right\}. \tag{42}$$

PROOF. To prove (41),

$$P_{\theta_1,\dots,\theta_{k-1},\infty}\{R\} = \lim_{\theta_k\to\infty} P_{\theta_1,\dots,\theta_k}\{(X_1,\dots,X_{k-1}) \in R^{X_k}\}$$

$$= \lim_{\theta_k\to\infty} P_{\theta_1,\dots,\theta_k}\{(X_1,\dots,X_{k-1}) \in R^{X_k}, X_k \geq z\}$$

$$\leq P_{\theta_1,\dots,\theta_{k-1}}\Big\{(X_1,\dots,X_{k-1}) \in \bigcup R^z\Big\}.$$

Also, for every z,

$$P_{\theta_1,\dots,\theta_{k-1},\infty}\{R\} = \lim_{\theta_k\to\infty} P_{\theta_1,\dots,\theta_k}\{(X_1,\dots,X_{k-1}) \in R^{X_k}, X_k \geq z\}$$

$$\geq P_{\theta_1,\dots,\theta_{k-1}}\{(X_1,\dots,X_{k-1}) \in R^z\}$$

and so

$$P_{\theta_1,\dots,\theta_{k-1},\infty}\{R\} \geq P_{\theta_1,\dots,\theta_{k-1}}\Big\{(X_1,\dots,X_{k-1}) \in \bigcup R^z\Big\},$$

and (41) follows.

To prove (42),

$$P_{\theta_1,\dots,\theta_{k-1},-\infty}(R)$$

$$= \lim_{\theta_k\to-\infty} P_{\theta_1,\dots,\theta_k}\{(X_1,\dots,X_{k-1}) \in R^{X_k}\}$$

$$= \lim_{\theta_k\to-\infty} P_{\theta_1,\dots,\theta_k}\{(X_1,\dots,X_{k-1}) \in R^{X_k} X_k, \leq z\}$$

$$\leq P_{\theta_1,\dots,\theta_{k-1}}\{(X_1,\dots,X_{k-1}) \in R^z\}$$

for every z. Let $z \to -\infty$, so that R^z decreases to $\bigcap R^z$. Then we can conclude

$$P_{\theta_1,\dots,\theta_{k-1},-\infty}(R) \leq P_{\theta_1,\dots,\theta_{k-1}}\Big\{(X_1,\dots,X_{k-1}) \in \bigcap R^z\Big\}.$$

Also,

$$P_{\theta_1,\dots,\theta_{k-1},-\infty}\{R\} = \lim_{\theta_k\to\infty} P_{\theta_1,\dots,\theta_k}\{(X_1,\dots,X_{k-1}) \in R^{X_k}\}$$

$$\geq P_{\theta_1,\dots,\theta_{k-1}}\Big\{(X_1,\dots,X_{k-1}) \in \bigcap R^z\Big\},$$

and the result follows. □

Next, given a monotone rejection region R, define

$$U^1(R) = \bigcup_z R^z,$$

$$U^2(R) = U^1(U^1(R))$$

and

$$U^j(R) = U^1(U^{j-1}(R)).$$

Similarly, let

$$I^1(R) = \bigcap_z R^z$$

and

$$I^j(R) = I^1(I^{j-1}(R)).$$

By applying Lemma 6.2 repeatedly, we also obtain

$$P_{\theta_1,\dots,\theta_{k-j},\underbrace{\infty,\dots,\infty}_{j \text{ times}}}\{R\} = P_{\theta_1,\dots,\theta_{k-j}}\{U^j(R)\} \tag{43}$$

and

$$P_{\theta_1,\dots,\theta_{k-j},\underbrace{-\infty,\dots,-\infty}_{j \text{ times}}}\{R\} = P_{\theta_1,\dots,\theta_{k-j}}\{I^j(R)\}. \tag{44}$$

PROOF OF THEOREM 3.1. (i) Note, for any monotone rule E, the smallest probability of $E_{k,j}$ over $A_j(\varepsilon)$ occurs when $\theta = w_{k,1}$ defined in (40), as well as when θ is any permutation of $w_{k,1}$. Furthermore, for any monotone rule E that controls the FWER, we must have

$$P_\theta\{E_{k,j}\} \le \alpha$$

when θ is

$$v_{k,j} = (\underbrace{\infty,\dots,\infty}_{j-1 \text{ times}},\ \underbrace{0,\dots,0}_{k-j+1 \text{ times}}), \tag{45}$$

or permutations of $v_{k,j}$.

To prove the optimality result (30), consider another rule E, with $E^c_{k,1}$ the subset of $\mathbf{R}^k$ that accepts all null hypotheses. Suppose there exists $x = (x_1,\dots,x_k) \notin D^c_{k,1}$, but $x \in E^c_{k,1}$. Then there exists at least one component of x, say x_1, with $x_1 > c_{k,1}$. By monotonicity, the set

$$L(x) = \{y \in \mathbf{R}^k : y_i \le x_i\} \tag{46}$$

is also in $E^c_{k,1}$. Then

$$P_{w_{k,1}}\{E^c_{k,1}\} \ge P_{w_{k,1}}\{L(x)\} = P_\varepsilon\{X_1 \le x_1\} > P_\varepsilon\{X_1 \le c_{k,1}\} = 1 - \beta_{k,1}(\alpha,\varepsilon),$$

and so the smallest power of E over $A_1(\varepsilon)$ satisfies

$$P_{w_{k,1}}\{E_{k,1}\} < \beta_{k,1}(\alpha,\varepsilon).$$

Therefore, in order for E to be optimal we must have

$$E_{k,1}^c \subset D_{k,1}^c.$$

But if $D_{k,1}$ is a proper subset of $E_{k,1}$ (except for a set with 0 Lebesgue measure), then

$$P_{\underbrace{0,\dots,0}_{k \text{ times}}}\{E_{k,1}\} > P_{\underbrace{0,\dots,0}_{k \text{ times}}}\{D_{k,1}\} = \alpha,$$

a contradiction if E controls the FWER. Therefore, (30) is proved.

To prove the result (31) with $j = k$, let E be any other monotone decision rule which has strong control and satisfies the constraint

$$E_{k,k} \subset D_{k,k-1} = \{X_{r_1} \geq c_{k,1}, \dots, X_{r_{k-1}} \geq c_{k,k-1}\}.$$

Suppose $E_{k,k}$ includes a point $y = (y_1, \dots, y_k)$, where $y_i \geq c_{k,i}$ for $i = 1, \dots, k-1$ and $y_k < c_{k,k}$. Then

$$\begin{aligned} P_{\underbrace{\infty,\dots,\infty}_{k-1 \text{ times}},0}\{E_{k,k}\} &\geq P_{\underbrace{\infty,\dots,\infty}_{k-1 \text{ times}},0}\{X_{r_1} \geq y_1, \dots, X_{r_{k-1}} \geq y_{k-1}, X_k \geq y_k\} \\ &= P_0\{X_k \geq y_k\} > P_0\{X_k > c_{k,k}\} = \alpha, \end{aligned}$$

a contradiction of strong control. So such a point y cannot be in $E_{k,k}$, nor can any permutation of the coordinates of y (by a similar argument). Therefore, $E_{k,k}$ can at most include $D_{k,k}$. But taking away any points from $D_{k,k}$ could only lower the minimum power at $(\varepsilon, \dots, \varepsilon)$, and so $D_{k,k}$ is optimal.

To prove the result (31) with $1 < j < k$, let E be any other monotone decision rule which has strong control and satisfies the constraint (32). Let

$$X_{j:1} \geq X_{j:2} \geq \cdots \geq X_{j:j} \tag{47}$$

denote the ordered values of $X_1, \dots, X_j$. Since E has strong control, it follows by (43) that

$$P_{\underbrace{0,\dots,0}_{k-j+1 \text{ times}}}\{U^{j-1}(E_{k,j})\} = \alpha.$$

Hence, $U^{j-1}(E_{k,j})$ can be viewed as a rejection region in $\mathbf{R}^{k-j+1}$ for the case with k and j replaced by $k' = k - j + 1$ and $j' = 1$. [Note that if $E_{k,j}$ satisfies the constraint $E_{k,j} \subset D_{k,j-1}$, then

$$U^{j-1}(E_{k,j}) \subset U^{j-1}(D_{k,j-1}) = \mathbf{R}^{k-j+1},$$

so the constraint is vacuous.] It follows that

$$P_{0,\underbrace{-\infty,\dots,-\infty}_{k-j \text{ times}}}\{U^{j-1}(E_{k,j})\} \leq \beta_{k-j+1,1}(\alpha, 0)$$

or

$$P_{0,\underbrace{-\infty,\ldots,-\infty}_{k-j\text{ times}},\underbrace{\infty,\ldots,\infty}_{j-1\text{ times}}}\{E_{k,j}\} \le \beta_{k-j+1,1}(\alpha,0).$$

By the same reasoning applied to any permutation of

$$\theta = (0, \underbrace{-\infty,\ldots,-\infty}_{k-j\text{ times}}, \underbrace{\infty,\ldots,\infty}_{j-1\text{ times}}),$$

$$P_{0,\underbrace{\infty,\ldots,\infty}_{j-1\text{ times}},\underbrace{-\infty,\ldots,-\infty}_{k-j\text{ times}}}\{E_{k,j}\}$$

$$= P_{0,\underbrace{\infty,\ldots,\infty}_{j-1\text{ times}}}\{I^{k-j}(E_{k,j})\} \le \beta_{k-j+1,1}(\alpha,0).$$

So $I^{k-j}(E_{k,j})$ is a rejection region in $\mathbf{R}^j$ that controls the Type 1 error at the point

$$(0, \underbrace{\infty,\ldots,\infty}_{j-1\text{ times}})$$

(as well as at permutations of its coordinates), not at level α, but at level $\beta_{k-j+1,1}(\alpha,0)$. [In words, if you use the rule E which is originally designed to test k hypotheses, but you ignore the last $k-j$ hypotheses, the overall probability of a Type 1 error for testing the j hypotheses is reduced to $\beta_{k-j+1,1}(\alpha,0)$.] Also, note that the constraint $E_{k,j} \subset D_{k,j-1}$ implies

$$I^{k-j}(E_{k,j}) \subset I^{k-j}(D_{k,j-1}) = \{X_{j:1} \ge c_{k,1},\ldots,X_{j:j-1} \ge c_{k,j-1}\}.$$

(Note that $c_{k,j}$ always refers to the critical values based on the given value of α, so its dependence on α is suppressed.) Then, by the case with k and j replaced by j and j (already proved above) and α replaced by $\beta_{k-j+1,1}(\alpha,0)$, it follows that

$$P_{\underbrace{\varepsilon,\ldots,\varepsilon}_{j\text{ times}}}\{I^{k-j}(E_{k,j})\} \le \beta_{j,j}(\beta_{k-j+1,1}(\alpha,0),\varepsilon)$$

or

$$P_{\underbrace{\varepsilon,\ldots,\varepsilon}_{j\text{ times}},\underbrace{-\infty,\ldots,-\infty}_{k-j\text{ times}}}\{E_{k,j}\} \le \beta_{j,j}(\beta_{k-j+1,1}(\alpha,0),\varepsilon). \tag{48}$$

We must argue that the right-hand side of (48) is $\beta_{k,j}(\alpha,\varepsilon)$. But notice that if we apply the above reasoning to $E = D$, the inequalities are all equalities. Indeed,

$$U^{j-1}(D_{k,j}) = \{\text{at least one of } X_1,\ldots,X_{k-j+1} \ge c_{k,j}\}$$

and the optimal minimum power (with $\varepsilon = 0$) for the subproblem with $k' = k-j+1$, $j' = 1$ and $\alpha' = \alpha$ is $\beta_{k-j+1,1}(\alpha,0)$. Also,

$$I^{k-j}(D_{k,j}) = \{X_{j:1} \ge c_{k,1},\ldots,X_{j:j} \ge c_{k,j}\}$$

is optimal for the case $k'' = j'' = j$ at the level $\alpha'' = \beta_{k-j+1,1}(\alpha, 0)$. Indeed, checking the level condition,

$$P_{0,\underbrace{\infty,\ldots,\infty}_{j-1\text{ times}}}\{I^{k-j}(D_{k,j})\} = P_0\{X_1 \geq c_{k,j}\}$$

$$= P_0\{X_1 \geq c_{k-j+1,1}\} = \beta_{k-j+1,1}(\alpha, 0).$$

So, by the case $k'' = j'' = j$,

$$P_{\underbrace{\varepsilon,\ldots,\varepsilon}_{j\text{ times}}}\{I^{k-j}(E_{k,j})\} \leq P_{\underbrace{\varepsilon,\ldots,\varepsilon}_{j\text{ times}}}\{I^{k-j}(D_{k,j})\}$$

$$= P_{\underbrace{\varepsilon,\ldots,\varepsilon}_{j\text{ times}}}\{X_{j:1} \geq c_{k,1}, \ldots, X_{j:j} \geq c_{k,j}\} = \beta_{k,j}(\alpha, \varepsilon).$$

The proof of (ii) is completely analogous to the proof of Theorem 2.1(iii), with the help of Lemma 3.1(ii). □

PROOF OF LEMMA 4.1. To prove (i), suppose $H_1, \ldots, H_p$ are true and $H_{p+1}, \ldots, H_k$ are false. A Type 1 error occurs if any of $H_1, \ldots, H_p$ are rejected. For the rule D, the set where any of $H_1, \ldots, H_p$ is rejected is a monotone increasing set (invoking the monotonicity of critical values). Hence the probability of this event is largest under this configuration of true and false hypotheses when

$$(\theta_1, \ldots, \theta_k) = (\underbrace{0, \ldots, 0}_{p\text{ times}}, \underbrace{\infty, \ldots, \infty}_{k-p\text{ times}}),$$

and this probability is equal to

$$P_{\underbrace{0,\ldots,0}_{p\text{ times}},\underbrace{\infty,\ldots,\infty}_{k-p\text{ times}}}\{\text{reject any of } H_1, \ldots, H_p\}$$

$$= P_{\underbrace{0,\ldots,0}_{p\text{ times}},\underbrace{\infty,\ldots,\infty}_{k-p\text{ times}}}\{\text{reject any of } H_1, \ldots, H_p \quad \cap \text{ reject all of } H_{p+1}, \ldots, H_k\}, \tag{49}$$

because as $(\theta_1, \ldots, \theta_k) \to (\underbrace{0, \ldots, 0}_{p\text{ times}})$, $X_{(p+1)} > d_{p+1}$ with probability tending to one, and so the hypotheses $H_{p+1}, \ldots, H_k$ are rejected with probability tending to one. Then (49) is bounded above by

$$P_{\underbrace{0,\ldots,0}_{p\text{ times}},\underbrace{\infty,\ldots,\infty}_{k-p\text{ times}}}\{\text{at least } k - p + 1 \text{ rejections}\}$$

$$= P_{\underbrace{0,\ldots,0}_{p\text{ times}},\underbrace{\infty,\ldots,\infty}_{k-p\text{ times}}}\{D_{k,k-p+1}\}$$

$$= 1 - P_{\underbrace{0,\dots,0}_{p \text{ times}},\underbrace{\infty,\dots,\infty}_{k-p \text{ times}}}\{X_{(1)} \le d_1, \dots, X_{(p)} \le d_p\}$$

$$= 1 - P_{\underbrace{0,\dots,0}_{p \text{ times}}}\{L_j\} = \alpha,$$

by (34) and (35).

To prove (ii), note that the minimum power occurs when θ is one of the $\binom{k}{j}$ points with j values of ε and $k - j$ values of $-\infty$, such as $w_{k,j}$ given by (40). Then

$$P_{w_{k,j}}(D_{k,j}) = 1 - P_{w_{k,j}}\{X_{(1)} \le d_1, \dots, X_{(k-j+1)} \le d_{k-j+1}\}$$
$$= 1 - P_{\underbrace{\varepsilon,\dots,\varepsilon}_{j \text{ times}}}\{\{X_1 \le d_{k-j+1}\} \cup \cdots \cup \{X_j \le d_{k-j+1}\}\},$$

which reduces to $\tilde{\beta}_{k,j}(\alpha, \varepsilon)$ as claimed. □

PROOF OF THEOREM 4.1. To prove (37) (the case $j = k$), first observe that

$$D_{k,k} = \{X_{(1)} > d_1\}.$$

Consider another monotone rule E, and suppose there exists some point $x = (x_1, \dots, x_k)$ with $x \in E_{k,k}$ but $x \notin D_{k,k}$. Then there exists at least one component of x, say x_1, with $x_1 < d_1$. By monotonicity the set

$$M(x) = \{y \in \mathbf{R}^k : y_i \ge x_i\}$$

is also in $E_{k,k}$. Then

$$P_{0,\underbrace{\infty,\dots,\infty}_{k-1 \text{ times}}}\{E_{k,k}\} \ge P_{0,\underbrace{\infty,\dots,\infty}_{k-1 \text{ times}}}\{M(x)\}$$
$$= P_0\{X_1 \ge x_1\} > P_0\{X_1 \ge d_1\} = \alpha,$$

which would contradict strong control. So we must have $E_{k,k} \subset D_{k,k}$. But then

$$P_{\underbrace{\varepsilon,\dots,\varepsilon}_{k \text{ times}}}\{E_{k,k}\} \le P_{\underbrace{\varepsilon,\dots,\varepsilon}_{k \text{ times}}}\{D_{k,k}\},$$

and so (37) is proved.

To prove the result (38) in the case $j = 1$, the constraint is that $E_{k,1}$ must contain $D_{k,2}$, or, equivalently,

$$E_{k,1}^c \subset D_{k,2}^c = \bigcap_{i=1}^{k-1}\{X_{(i)} \le d_i\}.$$

Suppose $x = (x_1, \dots, x_k) \in E_{k,1}^c$ but $x \notin D_{k,1}^c$. For the sake of argument, assume the x_i are nondecreasing in i with $x_i \le d_i$ for $i = 1, \dots, k-1$ (so the constraint is satisfied), but $x_k > d_k$. Then $x \in E^c_{k,1}$ implies $L(x) \in E^c_{k,1}$, where $L(x)$ is defined in (46). So

$$P_{\underbrace{-\infty,\dots,-\infty}_{k-1 \text{ times}},\varepsilon}\{E^c_{k,1}\} \geq P_{\underbrace{-\infty,\dots,-\infty}_{k-1 \text{ times}},\varepsilon}\{L(x)\}$$

$$= P_\varepsilon\{X_k \leq x_k\} > P_\varepsilon\{X_k \leq d_k\}.$$

Therefore

$$P_{\underbrace{-\infty,\dots,-\infty}_{k-1 \text{ times}},\varepsilon}\{E_{k,1}\} < P_\varepsilon\{X_k > d_k\} = \tilde{\beta}_{k,1}(\alpha, \varepsilon),$$

and so $E_{k,1}$ is less powerful than $D_{k,1}$. Therefore such a point x cannot exist in order for $E_{k,1}$ to be optimal. (A similar argument applies to any permutation of the coordinates of x.) Then $x \in D_{k,1}$ implies $x \in E_{k,1}$. But adding any points x to $D_{k,1}$ would increase the probability of rejection when $\theta = (0, \dots, 0)$, and this would contradict the level constraint. So the case $j = 1$ is proved.

To prove (38) for $1 < j < k$, let E be any other monotone decision rule which has strong control and satisfies the constraint (39). Since the set $E_{k,j}$ cannot have probability greater than α when $\theta = v_{k,j}$, where $v_{k,j}$ is given by (45), we must have

$$\alpha \geq P_{v_{k,j}}\{E_{k,j}\} = P_{\underbrace{0,\dots,0}_{k-j+1 \text{ times}}}\{U^{j-1}(E_{k,j})\}$$

by (43). Therefore $U^{j-1}(E_{k,j})$ is a region in $\mathbf{R}^{k-j+1}$ which has rejection probability α when $\theta = (0, \dots, 0)$. Note that the constraint $E_{k,j} \supset D_{k,j+1}$ implies the region $U^{j-1}(E_{k,j})$ must contain

$$U^{j-1}(D_{k,j+1}) = \{X_{k-j+1:k-j+1} > d_1\} \cup \cdots \cup \{X_{k-j+1:2} > d_{k-j}\}.$$

Therefore, by the case considered above with $k' = k - j + 1$ and $j' = 1$, the optimal region in $\mathbf{R}^{k-j+1}$ is

$$\{X_{k-j+1:k-j+1} > d_1\} \cup \cdots \cup \{X_{k-j+1:1} > d_{k-j+1}\},$$

which, in fact, is equal to $U^{j-1}(D_{k,j})$. So

$$P_{0,\underbrace{-\infty,\dots,-\infty}_{k-j \text{ times}}}\{U^{j-1}(E_{k,j})\} \leq P_{0,\underbrace{-\infty,\dots,-\infty}_{k-j \text{ times}}}\{U^{j-1}(D_{k,j})\}$$

$$= P_0\{X_1 > d_{k-j+1}\} = \tilde{\beta}_{k-j+1,1}(\alpha, 0).$$

Using (43) and applying the argument to any permutation of $v_{k,j}$, we have

$$P_{0,\underbrace{\infty,\dots,\infty}_{j-1 \text{ times}},\underbrace{-\infty,\dots,-\infty}_{k-j \text{ times}}}\{E_{k,j}\} \leq \tilde{\beta}_{k-j+1,1}(\alpha, 0),$$

or by (44),

$$P_{\underbrace{0,\infty,\ldots,\infty}_{j-1}}\{I^{k-j}(E_{k,j})\} \le \tilde{\beta}_{k-j+1,1}(\alpha,0).$$

So $I^{k-j}(E_{k,j})$ is a rejection region in $\mathbf{R}^j$ that controls the Type 1 error at

$$(0,\underbrace{\infty,\ldots,\infty}_{j-1 \text{ times}})$$

(as well as permutations of its coordinates), not at level α, but at level $\tilde{\beta}_{k-j+1,1}(\alpha,0)$. [Also note that the constraint $E_{k,j} \supset D_{k,j+1}$ implies

$$I^{k-j}(E_{k,j}) \supset I^{k-j}(D_{k,j+1}) = \varnothing,$$

which is always satisfied.] By the case with $k'' = j'' = j$ and $\alpha'' = \tilde{\beta}_{k-j+1,1}(\alpha,0)$ considered above,

$$I^{k-j}(D_{k,j}) = \{\min(X_1,\ldots,X_j) > d_{k-j+1}\}$$

is optimal for this case and so

$$P_{\underbrace{\varepsilon,\ldots,\varepsilon}_{j \text{ times}}}\{I^{k-j}(E_{k,j})\} \le P_{\underbrace{\varepsilon,\ldots,\varepsilon}_{j \text{ times}}}\{I^{k-j}(D_{k,j})\} = \tilde{\beta}_{k,j}(\alpha,\varepsilon)$$

by Lemma 4.1(ii). Therefore

$$P_{\underbrace{\varepsilon,\ldots,\varepsilon}_{j \text{ times}},\underbrace{-\infty,\ldots,-\infty}_{k-j \text{ times}}}\{E_{k,j}\} \le \tilde{\beta}_{k,j}(\alpha,\varepsilon),$$

as was to be proved.

The proof of (ii) is analogous to the proof of Theorem 3.1(ii). □

Acknowledgment. Special thanks to an anonymous referee for valuable comments and references.

REFERENCES

[1] DALAL, S. and MALLOWS, C. (1992). Buying with exact confidence. *Ann. Appl. Probab.* **2** 752–765.

[2] DUNNETT, C. and TAMHANE, A. (1992). A step-up multiple test procedure. *J. Amer. Statist. Assoc.* **87** 162–170.

[3] FINNER, H. (1994). Testing multiple hypotheses: General theory, specific problems, and relationship to other multiple decision procedures. Habilitationsschrift, Fachbereich IV-Mathematik, Univ. Trier.

[4] FINNER, H. and ROTERS, M. (1998). Asymptotic comparison of step-down and step-up multiple test procedures based on exchangeable test statistics. *Ann. Statist.* **26** 505–524.

[5] HOCHBERG, Y. (1988). A sharper Bonferroni procedure for multiple tests of significance. *Biometrika* **75** 800–802.

[6] HOCHBERG, Y. and TAMHANE, A. (1987). *Multiple Comparison Procedures*. Wiley, New York.

[7] HOLM, S. (1979). A simple sequentially rejective multiple test procedure. *Scand. J. Statist.* **6** 65–70.

[8] HOMMEL, G. (1989). A comparison of two modified Bonferroni procedures. *Biometrika* **76** 624–625.

[9] LEHMANN, E. (1952). Testing multiparameter hypotheses. *Ann. Math. Statist.* **23** 541–552.

[10] LEHMANN, E. (1955). Ordered families of distributions. *Ann. Math. Statist.* **26** 399–419.

[11] LEHMANN, E. (1986). *Testing Statistical Hypotheses*, 2nd ed. Springer, New York.

[12] MARSHALL, A. (1991). Multivariate stochastic orderings and generating cones of functions. In *Stochastic Orders and Decision Under Risk* (K. Mosler and M. Scarsini, eds.) 231–247. IMS, Hayward, CA.

[13] PERLMAN, M. and WU, L. (1999). The emperor's new tests (with discussion). *Statist. Sci.* **14** 355–381.

[14] ROM, D. (1990). A sequentially rejective test procedure based on a modified Bonferroni inequality. *Biometrika* **77** 663–665.

[15] SARKAR, S. (2000). A note on the monotonicity of the critical values of a step-up test. *J. Statist. Plann. Inference* **87** 241–249.

[16] SIMES, R. (1986). An improved Bonferroni procedure for multiple tests of significance. *Biometrika* **73** 751–754.

[17] TROENDLE, J. (1996). A permutational step-up method of testing multiple outcomes. *Biometrics* **52** 846–859.

[18] WESTFALL, P. and YOUNG, S. (1993). *Resampling-Based Multiple Testing.* Wiley, New York.

E. L. LEHMANN
J. P. SHAFFER
DEPARTMENT OF STATISTICS
UNIVERSITY OF CALIFORNIA, BERKELEY
BERKELEY, CALIFORNIA 94720
USA
E-MAIL: shaffer@stat.berkeley.edu

J. P. ROMANO
DEPARTMENT OF STATISTICS
STANFORD UNIVERSITY
STANFORD, CALIFORNIA 94305-4065
USA
E-MAIL: romano@stat.stanford.edu

The Annals of Statistics
2005, Vol. 33, No. 3, 1138–1154
DOI 10.1214/009053605000000084

GENERALIZATIONS OF THE FAMILYWISE ERROR RATE

BY E. L. LEHMANN AND JOSEPH P. ROMANO

University of California, Berkeley and Stanford University

Consider the problem of simultaneously testing null hypotheses $H_1, \dots, H_s$. The usual approach to dealing with the multiplicity problem is to restrict attention to procedures that control the familywise error rate (FWER), the probability of even one false rejection. In many applications, particularly if s is large, one might be willing to tolerate more than one false rejection provided the number of such cases is controlled, thereby increasing the ability of the procedure to detect false null hypotheses. This suggests replacing control of the FWER by controlling the probability of k or more false rejections, which we call the k-FWER. We derive both single-step and stepdown procedures that control the k-FWER, without making *any* assumptions concerning the dependence structure of the p-values of the individual tests. In particular, we derive a stepdown procedure that is quite simple to apply, and prove that it cannot be improved without violation of control of the k-FWER. We also consider the false discovery proportion (FDP) defined by the number of false rejections divided by the total number of rejections (defined to be 0 if there are no rejections). The false discovery rate proposed by Benjamini and Hochberg [*J. Roy. Statist. Soc. Ser. B* **57** (1995) 289–300] controls $E(FDP)$. Here, we construct methods such that, for any γ and α, $P\{FDP > \gamma\} \le \alpha$. Two stepdown methods are proposed. The first holds under mild conditions on the dependence structure of p-values, while the second is more conservative but holds without any dependence assumptions.

1. Introduction. In this paper, we will consider the general problem of simultaneously testing a finite number of null hypotheses H_i, $i = 1, \dots, s$. We shall assume that tests for the individual hypotheses are available and the problem is how to combine them into a simultaneous test procedure. The easiest approach is to disregard the multiplicity and simply test each hypothesis at level α. However, with such a procedure the probability of one or more false rejections increases with s. When the number of true hypotheses is large, we shall be nearly certain to reject some of them.

A classical approach to dealing with this problem is to restrict attention to procedures that control the probability of one or more false rejections. This probability is called the familywise error rate (FWER). Here the term "family" refers to the collection of hypotheses $H_1, \dots, H_s$ that is being considered for joint testing. Which tests are to be treated jointly as a family depends on the situation.

Received December 2003; revised June 2004.
AMS 2000 subject classifications. Primary 62J15; secondary 62G10.
Key words and phrases. Familywise error rate, multiple testing, p-value, stepdown procedure.

Once the family has been defined, control of the FWER (at joint level α) requires that

$$\text{FWER} \le \alpha \tag{1}$$

for all possible constellations of true and false hypotheses. A quite broad treatment of methods that control the FWER is presented in [4].

Safeguards against false rejections are of course not the only concern of multiple testing procedures. Corresponding to the power of a single test, one must also consider the ability of a procedure to detect departures from the hypothesis when they do occur. When the number of tests is in the tens or hundreds of thousands, control of the FWER at conventional levels becomes so stringent that individual departures from the hypothesis have little chance of being detected. For this reason, we shall consider an alternative to the FWER that controls false rejections less severely and consequently provides better power.

Specifically, we shall consider the k-FWER, the probability of rejecting at least k true null hypotheses. Such an error rate with $k > 1$ is appropriate when one is willing to tolerate one or more false rejections, provided the number of false rejections is controlled.

More formally, suppose data X is available from some model $P \in \Omega$. A general hypothesis H can be viewed as a subset ω of Ω. For testing $H_i : P \in \omega_i$, $i = 1, \dots, s$, let $I(P)$ denote the set of true null hypotheses when P is the true probability distribution; that is, $i \in I(P)$ if and only if $P \in \omega_i$. Then, the k-FWER, which depends on P, is defined to be

$$k\text{-FWER} = P\{\text{reject at least } k \text{ hypotheses } H_i \text{ with } i \in I(P)\}. \tag{2}$$

Control of the k-FWER requires that k-FWER $\le \alpha$ for all P, that is,

$$P\{\text{reject at least } k \text{ hypotheses } H_i \text{ with } i \in I(P)\} \le \alpha \qquad \text{for all } P. \tag{3}$$

Evidently, the case $k = 1$ reduces to control of the usual FWER.

We will also consider control of the *false discovery proportion* (FDP), defined as the total number of false rejections divided by the total number of rejections (and equal to 0 if there are no rejections). Given a user specified value $\gamma \in (0, 1)$, the measure of error control we wish to control is $P\{FDP > \gamma\}$ and we derive methods where this is bounded by α.

Recently, there has been a flurry of activity in finding methods that control error rates that are less stringent than the FWER, which is no doubt inspired by the FDR controlling method of Benjamini and Hochberg [1] and applications such as genomic studies where s is so large that control of the FWER is too stringent. For example, Genovese and Wasserman [3] study asymptotic procedures that control the FDP (and the FDR) in the framework of a random effects mixture model. These ideas are extended in [9], where in the context of random fields the number of null hypotheses is uncountable. Korn, Troendle, McShane and Simon [8] provide methods that control both the k-FWER and FDP; they provide some justification for their methods, but they are limited to a multivariate permutation model. Alternative methods of control of the k-FWER and FDP are given in van der Laan, Dudoit and Pollard [13]; they include both finite sample and asymptotic results. Surprisingly, the methods presented here are distinct from the above techniques. Our methods are not asymptotic and hold under either mild or no assumptions, as long as p-values are available for testing each individual hypothesis.

Before describing methods that provide control of the k-FWER and FDP, we first recall the notion of a p-value, since multiple testing methods are often described by the p-values of the individual tests. Consider a single null hypothesis $H : P \in \omega$. Assume a family of tests of H, indexed by α, with level α rejection regions S_α satisfying

$$P\{X \in S_\alpha\} \le \alpha \qquad \text{for all } 0 < \alpha < 1,\ P \in \omega, \tag{4}$$

and

$$S_\alpha \subset S_{\alpha'} \qquad \text{whenever } \alpha < \alpha'. \tag{5}$$

Then the p-value is defined by

$$\hat{p} = \hat{p}(X) = \inf\{\alpha : X \in S_\alpha\}. \tag{6}$$

The important property of a p-value that will be used later is the following.

LEMMA 1.1. *Assume $\hat{p}$ is defined as above.*

(i) *If $P \in \omega$, then*

$$P\{\hat{p} \le u\} \le u. \tag{7}$$

(ii) *Furthermore,*

$$P\{\hat{p} \le u\} \ge P\{X \in S_u\}. \tag{8}$$

Therefore, if the S_α are such that equality holds in (4), *then $\hat{p}$ is uniformly distributed on* (0, 1) *when $P \in \omega$.*

PROOF. Assume $P \in \omega$. To prove (i), note that the event $\{\hat{p} \le u\}$ implies $\{X \in S_{u+\varepsilon}\}$ for any small $\varepsilon > 0$. Therefore,

$$P\{\hat{p} \le u\} \le P\{X \in S_{u+\varepsilon}\} \le u + \varepsilon$$

by assumption (4). Now let $\varepsilon \to 0$. To prove (ii), the event $\{X \in S_u\}$ implies $\{\hat{p} \le u\}$, and so (8) follows. □

Two classic procedures that control the FWER are the Bonferroni procedure and the Holm procedure. The Bonferroni procedure rejects H_i if its corresponding p-value satisfies $\hat{p}_i \le \alpha/s$. Assuming $\hat{p}_i$ satisfies

$$P\{\hat{p}_i \le u\} \le u \qquad \text{for any } u \in (0, 1) \text{ and any } P \in \omega_i, \tag{9}$$

the Bonferroni procedure provides strong control of the FWER. Unfortunately, the ability of the Bonferroni procedure to detect cases in which H_i is false will typically be very low since H_i is tested at level α/s which—particularly if s is large—is orders smaller than the conventional α levels.

For this reason procedures are prized for which the levels of the individual tests are increased over α/s without an increase in the FWER. It turns out that such a procedure due to Holm [5] is available under the present minimal assumptions.

The Holm procedure can conveniently be stated in terms of the p-values $\hat{p}_1, \dots, \hat{p}_s$ of the s individual tests. Let the ordered p-values be denoted by $\hat{p}_{(1)} \leq \cdots \leq \hat{p}_{(s)}$, and the associated hypotheses by $H_{(1)}, \dots, H_{(s)}$. Then the Holm procedure is defined stepwise as follows:

Step 0. Let $k = 0$.

Step 1. If $\hat{p}_{(k+1)} > \alpha/(s-k)$, go to step 2. Otherwise set $k = k+1$ and repeat step 1.

Step 2. Reject $H_{(j)}$ for $j \leq k$ and accept $H_{(j)}$ for $j > k$.

The Bonferroni method is an example of a *single-step* procedure, meaning any null hypothesis is rejected if its corresponding p-value is less than or equal to a common cutoff value (which in the Bonferroni case is α/s). The Holm procedure is a special case of a class of *stepdown* procedures, which we now briefly describe. Let

$$\alpha_1 \leq \alpha_2 \leq \cdots \leq \alpha_s \tag{10}$$

be constants. If $\hat{p}_{(1)} > \alpha_1$, reject no null hypotheses. Otherwise, if

$$\hat{p}_{(1)} \leq \alpha_1, \dots, \hat{p}_{(r)} \leq \alpha_r, \tag{11}$$

reject hypotheses $H_{(1)}, \dots, H_{(r)}$ where the largest r satisfying (11) is used. That is, a stepdown procedure starts with the most significant p-value and continues rejecting hypotheses as long as their corresponding p-values are small. The Holm procedure uses $\alpha_i = \alpha/(s-i+1)$.

2. Control of the *k*-FWER. The usual Bonferroni procedure compares each p-value $\hat{p}_i$ with α/s. Control of the k-FWER allows one to increase α/s to $k\alpha/s$, and thereby greatly increase the ability to detect false hypotheses. That such a simple modification results in control of the k-FWER is seen in the following result.

THEOREM 2.1. *For testing $H_i : P \in \omega_i$, $i = 1, \dots, s$, suppose $\hat{p}_i$ satisfies* (9). *Consider the procedure that rejects any H_i for which $\hat{p}_i \leq k\alpha/s$.*

(i) *This procedure controls the k-FWER, so that* (3) *holds. Equivalently, if each of the hypotheses is tested at level $k\alpha/s$, then the k-FWER is controlled.*

(ii) *For this procedure, the inequality* (3) *is sharp in the sense that there exists a joint distribution for $(\hat{p}_1, \dots, \hat{p}_s)$ for which equality is attained in* (3).

PROOF. (i) Fix any P and suppose H_i with $i \in I = I(P)$ are true and the remainder false, with $|I|$ denoting the cardinality of I. Let N be the number of false rejections. Then, by Markov's inequality,

$$P\{N \ge k\} \le \frac{E(N)}{k} = \frac{E[\sum_{i\in I(P)} I\{\hat{p}_i \le k\alpha/s\}]}{k} = \sum_{i\in I(P)} \frac{P\{\hat{p}_i \le k\alpha/s\}}{k}$$

$$\le \sum_{i\in I(P)} \frac{k\alpha/s}{k} = |I(P)|\frac{\alpha}{s} \le \alpha.$$

To prove (ii), consider the following construction. Pick k indices at random without replacement from $\{1, \dots, s\}$. Call them J. Given $i \in J$, let $\hat{p}_i = U_1$, where U_1 is uniform on $(0, k/s)$, that is, $U_1 \sim U(0, k/s)$. Given $i \notin J$, let $\hat{p}_i = U_2$, where U_2 is independent of U_1 and $U_2 \sim U(k/s, 1)$. Then, unconditionally,

$$\hat{p}_i \sim \frac{k}{s} U\left(0, \frac{k}{s}\right) + \left(1 - \frac{k}{s}\right) U\left(\frac{k}{s}, 1\right) \sim U(0, 1).$$

Indeed, if $u \le k/s$,

$$P\{\hat{p}_i \le u\} = P\{i \in J\} \cdot P\{U_1 \le u\} = \frac{k}{s} \cdot \frac{u}{k/s} = u$$

and if $u \ge k/s$,

$$P\{\hat{p}_i \le u\} = P\{i \in J\} \cdot 1 + P\{i \notin J\} \cdot P\{U_2 \le u\} = \frac{k}{s} + \left(1 - \frac{k}{s}\right) \cdot \frac{u - k/s}{1 - k/s} = u.$$

Now exactly k of the $\hat{p}_i$ are less than or equal to k/s by construction. The probability that these are all less than or equal to $\alpha k/s$ is

$$P\left\{U_1 \le \frac{\alpha k}{s}\right\} = \frac{\alpha k/s}{k/s} = \alpha. \qquad \square$$

As is the case for the Bonferroni method, the above single-stage procedure can be strengthened by a Holm type of improvement. Consider the stepdown procedure described in (11), where now we specifically consider

$$\alpha_i = \begin{cases} \dfrac{k\alpha}{s}, & i \le k, \\[2ex] \dfrac{k\alpha}{s + k - i}, & i > k. \end{cases} \tag{12}$$

Of course, the α_i depend on s and k, but we suppress this dependence in the notation.

THEOREM 2.2. *For testing $H_i : P \in \omega_i, i = 1, \ldots, s$, suppose $\hat{p}_i$ satisfies* (9). *The stepdown procedure described in* (11) *with α_i given by* (12) *controls the k-FWER, that is,* (3) *holds.*

PROOF. Fix any P and let $I(P)$ be the indices of the true null hypotheses. Assume $|I(P)| \geq k$ or there is nothing to prove. Order the p-values corresponding to the $|I(P)|$ true null hypotheses; call them

$$\hat{q}_{(1)} \leq \cdots \leq \hat{q}_{|I(P)|}.$$

Let j be the smallest (random) index satisfying $\hat{p}_{(j)} = \hat{q}_{(k)}$, so

$$k \leq j \leq s - |I(P)| + k \tag{13}$$

because the largest possible index j occurs when all the smallest p-values correspond to the $s - |I(P)|$ false null hypotheses and the next $|I(P)|$ p-values correspond to the true null hypotheses. So $\hat{p}_{(j)} = \hat{q}_{(k)}$. Then our generalized Holm procedure commits at least k false rejections if and only if

$$\hat{p}_{(1)} \leq \alpha_1, \qquad \hat{p}_{(2)} \leq \alpha_2, \qquad \ldots, \qquad \hat{p}_{(j)} \leq \alpha_j,$$

which certainly implies that

$$\hat{q}_{(k)} = \hat{p}_{(j)} \leq \alpha_j = \frac{k\alpha}{s + k - j}.$$

But by (13),

$$\frac{k\alpha}{s + k - j} \leq \frac{k\alpha}{|I(P)|}.$$

So the probability of at least k false rejections is bounded above by

$$P\left\{\hat{q}_{(k)} \leq \frac{k\alpha}{|I(P)|}\right\}.$$

By Theorem 2.1(i) the chance that the kth largest among $I(P)$ p-values is less than or equal to $k\alpha/|I(P)|$ is less than or equal to α. □

REMARK 2.1. Evidently, one can always reject the hypotheses corresponding to the smallest $k - 1$ p-values without violating control of the k-FWER. However, it seems counterintuitive to consider a stepdown procedure whose corresponding α_i are not monotone nondecreasing. In addition, automatic rejection of $k - 1$ hypotheses, regardless of the data, appears at the very least a little too optimistic. To ensure monotonicity, our stepdown procedure uses $\alpha_i = k\alpha/s$. Even if we were to adopt the more optimistic strategy of always rejecting the hypotheses corresponding to the first $k - 1$ hypotheses, we could still only reject k or more hypotheses if $\hat{p}_{(k)} \leq k\alpha/s$, which is also true for the specific procedure of Theorem 2.2.

REMARK 2.2. If the p-values have discrete distributions, it is possible that there may be ties among them. However, the proof remains valid regardless of how tied p-values are ordered because monotonicity of the α_i ensures that all hypotheses with a common tied p-value will be rejected if any of them are rejected.

The question naturally arises whether it is possible to improve the procedure further by increasing the critical values $\alpha_1, \alpha_2, \ldots$ without violating control of the k-FWER (3). By the previous remark we can always increase α_i to 1 for $i < k$. A more interesting question is whether we can increase α_i for $i \geq k$. We will show that this is not possible by exhibiting for each $i \geq k$ a joint distribution of the p-values for which

$$P\{\hat{p}_{(1)} \leq \alpha_1, \hat{p}_{(2)} \leq \alpha_2, \ldots, \hat{p}_{(i-1)} \leq \alpha_{i-1}, \hat{p}_{(i)} \leq \alpha_i\} = \alpha. \tag{14}$$

Moreover, changing α_i to $\beta_i > \alpha_i$ results in the right-hand side being greater than α. Thus, with $i \geq k$, one cannot increase α_i without violating the k-FWER. Then, having picked $\alpha_1, \ldots, \alpha_k, \ldots, \alpha_{i-1}$, the largest possible choice for α_i is as stated in the algorithm.

THEOREM 2.3. (i) *Let the α_i be given in* (12). *For any $i \geq k$ there exists a joint distribution for $\hat{p}_1, \ldots, \hat{p}_s$ such that $s + k - i$ of the $\hat{p}_i$ are uniformly distributed on* $(0, 1)$ *and* (14) *holds.*

(ii) *For testing $H_i : P \in \omega_i$, $i = 1, \ldots, s$, suppose $\hat{p}_i$ satisfies* (9). *For the stepdown procedure* (11) *with α_i given in* (12), *one cannot increase even one of the constants α_i (for $i \geq k$) without violating the k-FWER.*

Before proving the theorem, we make use of the following lemma.

LEMMA 2.1. *Fix k, u and constants $0 < \beta_1 \leq \beta_2 \leq \cdots \leq \beta_k \leq u$. Assume for every $j = 2, \ldots, k$,*

$$\frac{j(\beta_j - \beta_{j-1})}{\beta_j} \leq 1. \tag{15}$$

Then there exists a joint distribution for $(\hat{q}_1, \ldots, \hat{q}_k)$ satisfying the $\hat{q}_i$ are marginally uniform on $(0, u)$ such that the ordered values $\hat{q}_{(1)} \leq \cdots \leq \hat{q}_{(k)}$ satisfy

$$P\{\hat{q}_{(1)} \leq \beta_1, \ldots, \hat{q}_{(k)} \leq \beta_k\} = \beta_k / u. \tag{16}$$

PROOF. The proof is by induction on k. The result clearly holds for $k = 1$. With probability β_k/u we will construct $(\hat{q}_1, \ldots, \hat{q}_k)$ equal to $(\tilde{q}_1, \ldots, \tilde{q}_k)$, where $\tilde{q}_i \sim U(0, \beta_k)$ for $i = 1, \ldots, k$ and such that their ordered values $\tilde{q}_{(1)} \leq \cdots \leq \tilde{q}_{(k)}$ satisfy

$$P\{\tilde{q}_{(1)} \leq \beta_1, \ldots, \tilde{q}_{(k)} \leq \beta_k\} = 1. \tag{17}$$

But, with probability $1 - \beta_k/u$, construct the $\tilde{q}_j$ to be conditionally distributed as $U(\beta_k, u)$. Then unconditionally the $\hat{q}_j$ satisfy (16) and are marginally distributed as $U(0, u)$. So it suffices to construct the $\tilde{q}_j$ satisfying $\tilde{q}_j \sim U(0, \beta_k)$ and (17).

Let $\beta_0 = 0$ and for $i = 1, \ldots, k$ let $E_i = \{(\beta_{i-1}, \beta_i]\}$ and $p_i = \beta_i - \beta_{i-1}$. First construct $Y_1, \ldots, Y_{k-1}$, each taking values in $(0, \beta_{k-1}]$ such that their ordered values $Y_{(1)} \leq \cdots \leq Y_{(k-1)}$ satisfy

$$P\{Y_{(1)} \leq \beta_1, \ldots, Y_{(k-1)} \leq \beta_{k-1}\} = 1 \tag{18}$$

and Y_i is uniform on $(0, \beta_{k-1}]$. This is possible by the inductive hypothesis, since we can assume the result holds for $k - 1$ as long as $\beta_1, \ldots, \beta_k$ and u satisfy the stated conditions; in particular, we apply the result with $u = \beta_{k-1}$. Next, let Y_k be uniform on E_i with probability θp_i for $i = 1, \ldots, k-1$ and let it be uniform on E_k with probability $1 - \theta\beta_{k-1}$, where θ satisfies

$$\theta = \frac{1}{\beta_{k-1}}\left[1 - \frac{k(\beta_k - \beta_{k-1})}{\beta_k}\right]. \tag{19}$$

Finally, let $\tilde{q}_1, \ldots, \tilde{q}_k$ be a random permutation of $Y_1, \ldots, Y_k$. Because of (18) and the fact that $Y_k \leq \beta_k$, the ordered values of $Y_1, \ldots, Y_k$ and hence the ordered values of $\tilde{q}_1, \ldots, \tilde{q}_k$ satisfy (17). Furthermore, it is easy to check that $\tilde{q}_i$ falls in E_j with probability p_j and so $\tilde{q}_i$ is $U(0, \beta_k)$. Indeed, if $j < k$, the probability that $\tilde{q}_i$ falls in E_j, conditional on $\tilde{q}_i$ not being equal to Y_k, is p_i/β_{k-1} and is θp_i in the latter case, which unconditionally is

$$\frac{k-1}{k} \cdot \frac{p_i}{\beta_{k-1}} + \frac{1}{k}\theta p_i = p_i,$$

and similarly for the probability that $\hat{q}_i$ falls in E_k. The only detail that remains is to note that this construction with θ defined in (19) is possible only if θp_i and $1 - \theta\beta_{k-1}$ are all values in $(0, 1)$. But

$$1 - \theta\beta_{k-1} = \frac{k(\beta_k - \beta_{k-1})}{\beta_k},$$

which is certainly ≥ 0 since $\beta_k \geq \beta_{k-1}$. It is also ≤ 1 by the assumption (15). Also,

$$\theta p_i = \frac{p_i}{\beta_{k-1}} \cdot \left[1 - \frac{k(\beta_k - \beta_{k-1})}{\beta_k}\right].$$

But the first factor p_i/β_{k-1} is in $(0, 1)$ as is the latter by the above, and so the product is in $(0, 1)$. □

PROOF OF THEOREM 2.3. The case $i = k$ follows from the construction in the proof of Theorem 2.1. Let the first $i - k$ of the $\hat{p}_j$ be identically equal to 0. (Actually, rather than point mass at 0, any distribution supported on $[0, \alpha_1)$ will do.) For the remaining $s' = s + k - i$ p-values $\hat{p}_j$, $j = i - k + 1, \ldots, s$, randomly choose k indices from $i - k + 1, \ldots, s$. The k that are chosen will be marginally $U(0, k/s')$ and have a joint distribution which will be specified below; the remaining $s - i$ can be taken to be distributed as $U(k/s', 1)$.

Let $\hat{q}_1, \ldots, \hat{q}_k$ denote the k observations that are marginally $U(0, k/s')$. We need to specify the joint distribution of $\hat{q}_1, \ldots, \hat{q}_k$ so that their ordered values $\hat{q}_{(1)} \leq \cdots \leq \hat{q}_{(k)}$ satisfy

$$P\{\hat{q}_{(1)} \leq \alpha_{i-k+1}, \hat{q}_{(2)} \leq \alpha_{i-k+2}, \ldots, \hat{q}_{(k)} \leq \alpha_i\} = \alpha \tag{20}$$

(because $\hat{q}_{(j)} = \hat{p}_{(j+i-k)}$ for $j = 1, \ldots, k$). So the problem reduces to constructing a joint distribution for $(\hat{q}_1, \ldots, \hat{q}_k)$ satisfying (20) subject to the constraint that $\hat{q}_j$ is marginally distributed as $U(0, k/s')$. To do this, apply Lemma 2.1 with $u = k/s'$ and $\beta_j = \alpha_{i-k+j}$. We need to verify the conditions of the lemma, which reduces to showing

$$\frac{j(\alpha_{i-k+j} - \alpha_{i-k+j-1})}{\alpha_{i-k+j}} \leq 1 \tag{21}$$

for $i \geq k$ (and s and k fixed). But, if $i - k + j - 1 \leq k$, then the left-hand side of (21) is 0; otherwise it is easily seen to simplify to

$$\frac{j}{s + 2k - i - j} \leq \frac{j}{s + k - j} \leq k/s, \tag{22}$$

where the first inequality holds because $i \geq k$ and the second because $j \leq k$. But $k/s \leq 1$ and so the conditions of the lemma are satisfied. Therefore, we can conclude that the left-hand side of (20) is given by

$$\frac{\beta_k}{u} = \frac{\alpha_i}{k/s'} = \alpha,$$

and (i) is proved.

To prove (ii), the construction used in (i) can be used even if α_i is replaced by $\bar{\alpha}_i > \alpha_i$, as long as such a switch still allows one to appeal to the lemma. However, the same argument works as long as $\bar{\alpha}_i$ does not get bigger than $s/k \cdot \alpha_i$, so that the argument leading to (22) being less than or equal to 1 still applies. For such an $\bar{\alpha}_i$, the argument for (i) then shows that, if the left-hand side of (14) has α_i replaced by $c\alpha_i$ for some $1 < c < s/k$, then the right-hand side of (14) will be $c\alpha > \alpha$, which would violate control of the k-FWER. □

3. Control of the false discovery proportion. The number k of false rejections that one is willing to tolerate will often increase with the number of hypotheses rejected. So it might be of interest to control not the number of false rejections (sometimes called false discoveries) but the proportion of false discoveries. Specifically, let the *false discovery proportion* (FDP) be defined by

$$FDP = \begin{cases} \dfrac{\text{Number of false rejections}}{\text{Total number of rejections}}, & \text{if the denominator is greater than 0,} \\ 0, & \text{if there are no rejections.} \end{cases} \tag{23}$$

Thus FDP is the proportion of rejected hypotheses that are rejected erroneously. When none of the hypotheses is rejected, both numerator and denominator of that proportion are 0; since in particular there are no false rejections, the FDP is then defined to be 0.

Benjamini and Hochberg [1] proposed to replace control of the FWER by control of the *false discovery rate* (FDR), defined as

$$FDR = E(FDP). \tag{24}$$

The FDR has gained wide acceptance in both theory and practice, largely because Benjamini and Hochberg proposed a simple stepup procedure to control the FDR. Unlike control of the k-FWER, however, their procedure is not valid without assumptions on the dependence structure of the p-values. Their original paper assumed the very strong assumption of independence of p-values, but this has been weakened to include certain types of dependence; see [2]. In any case, control of the FDR does not prohibit the FDP from varying, even if its average value is bounded. Instead, we consider an alternative measure of control that guarantees the FDP is bounded, at least with prescribed probability. That is, for a given γ and α in $(0, 1)$, we require

$$P\{FDP > \gamma\} \le \alpha. \tag{25}$$

To develop a stepdown procedure satisfying (25), let F denote the number of false rejections. At step i, having rejected $i - 1$ hypotheses, we want to guarantee $F/i \le \gamma$, that is, $F \le \lfloor \gamma i \rfloor$, where $\lfloor x \rfloor$ is the greatest integer less than or equal to x. So, if $k = \lfloor \gamma i \rfloor + 1$, then $F \ge k$ should have probability no greater than α; that is, we must control the number of false rejections to be less than or equal to k. Therefore, we use the stepdown constant α_i with this choice of k (which now depends on i); that is,

$$\alpha_i = \frac{(\lfloor \gamma i \rfloor + 1)\alpha}{s + \lfloor \gamma i \rfloor + 1 - i}. \tag{26}$$

We give two results that show the stepdown procedure with this choice of α_i satisfies (25). Unfortunately, like FDR control, some assumptions on the dependence of p-values are required, at least by our method of proof. Later, we will modify the method so we can dispense with the dependence assumptions. As before, $\hat{p}_1, \dots, \hat{p}_s$ denotes the p-values of the individual tests. Also, let $\hat{q}_1, \dots, \hat{q}_{|I|}$ denote the p-values corresponding to the $|I| = |I(P)|$ true null hypotheses. So $q_i = p_{j_i}$, where $j_1, \dots, j_{|I|}$ correspond to the indices of the true null hypotheses. Also, let $\hat{r}_1, \dots, \hat{r}_{s-|I|}$ denote the p-values of the false null hypotheses. Consider the following condition: for any $i = 1, \dots, |I|$,

$$P\{\hat{q}_i \le u | \hat{r}_1, \dots, \hat{r}_{s-|I|}\} \le u; \tag{27}$$

that is, conditional on the observed p-values of the false null hypotheses, a p-value corresponding to a true null hypothesis is (conditionally) dominated by the uniform distribution, as it is unconditionally in the sense of (7). No assumption is made regarding the unconditional (or conditional) dependence structure of the true p-values, nor is there made any explicit assumption regarding the joint structure of the p-values corresponding to false hypotheses, other than the basic assumption (27). So, for example, if the p-values corresponding to true null hypotheses are independent of the false ones, but have arbitrary joint dependence within the group of true null hypotheses, the above assumption holds.

THEOREM 3.1. *Assume condition* (27). *Then the stepdown procedure with* α_i *given by* (26) *controls the FDP in the sense of* (25).

PROOF. Assume the number of true null hypotheses is $|I(P)| > 0$ (or there is nothing to prove) and the number of false null hypotheses is $f = s - |I(P)|$. The argument is conditional on the $\{\hat{r}_i\}$. Let

$$\hat{r}_{(1)} \le \hat{r}_{(2)} \le \cdots \le \hat{r}_{(f)}$$

denote the ordered values of the $\hat{r}_i$ and similarly for the $\hat{q}_i$. Let $\alpha_0 = 0$ and define R_i to be the number of $\hat{r}_i$ in the interval $(\alpha_{i-1}, \alpha_i]$. (Actually, assume R_1 includes the value 0 as well.) Given the values of $\hat{r}_1, \ldots, \hat{r}_f$, it may be impossible to have $FDP > \gamma$, that is,

$$P\{FDP > \gamma | \hat{r}_1, \ldots, \hat{r}_f\} = 0.$$

Otherwise, let $j = j(\hat{r}_1, \ldots, \hat{r}_f)$ be defined as

$$j = \min\left\{m : m - \sum_{i=1}^{m} R_i > m\gamma\right\}. \tag{28}$$

To interpret this, given the p-values of the false hypotheses, j is the smallest critical index (depending only on the $\hat{r}_i$) where it is possible to have $FDP > \gamma$, except whenever there are several p-values within an interval (α_{i-1}, α_i) we consider the index of the largest one. The point of the construction is that if the stepdown procedure stops at an index $m < j$, then $m - \sum_i R_i/m \le \gamma$ and so $FDP \le \gamma$. On the other hand, if the event $FDP > \gamma$ occurs, then there must be a rejection of a true null hypothesis at step j.

For example, if $s = 100$, $f = 5$ and $\gamma = 0.1$, then if all five of the $\hat{r}_i$ are less than α_1, then we define $j = 6$ even though the smallest true p-value could be the smallest among the 100. So the FDP could be greater than 0.1 after the first step of the algorithm if $\hat{q}_{(1)} < \hat{r}_{(1)}$, but even if this is the case, we then know we will reject at least six total hypotheses. So the important point here is that, given such a configuration of $\{\hat{r}_i\}$, in order for FDP to be greater than 0.1, it must be the case that we reject a true null hypothesis at step 6.

Note that, with j so defined, $R_j = 0$. For if $\sum_{i=1}^{j} R_i = j - k$ with $k/j > \gamma$ and $R_j > 0$, then

$$\sum_{i=1}^{j-1} R_i = j - k - R_j \leq j - 1 - k$$

and $k/(j-1) > \gamma$, so that $m = j - 1$ satisfies the criterion. Furthermore, we also have $\sum_{i=1}^{j} R_i = j - k$ (so not $< j - k$), where $k/j > \gamma$, because if $\sum_{i=1}^{j} R_i < j - k \leq j - 1 - k$ say, then $k/(j-1) > \gamma$ if $k/j > \gamma$ and so j can again be reduced to $j - 1$.

In addition, at the index j it must be the case that

$$k = k(j) = j - \sum_{i=1}^{j} R_i = 1 + \lfloor \gamma j \rfloor.$$

But $k > \gamma j$ implies $k \geq \lfloor \gamma j \rfloor + 1$. But if $k > \lfloor \gamma j \rfloor + 1$, then $k - 1 \geq \lfloor \gamma j \rfloor + 1$ and so

$$\frac{k-1}{j-1} \geq \frac{\lfloor \gamma j \rfloor + 1}{j-1} > \gamma,$$

the last equality trivially following from $1 + \lfloor \gamma j \rfloor \geq \gamma j > \gamma(j-1)$.

We can now complete the argument. At the index j we must have $k = j - \sum_{i=1}^{j} R_i = 1 + \lfloor \gamma j \rfloor$ of the $\hat{q}_i$ being $\leq \alpha_j$. But from Theorem 2.1 (applied conditional on the $\hat{r}_i$),

$$\begin{aligned} &P\{\text{at least } k(j) \text{ of the } \hat{q}_i \leq \alpha_j | \hat{r}_1, \ldots, \hat{r}_f\} \\ &\leq \frac{|I|\alpha_j}{k(j)} \\ &= \frac{|I|(\lfloor \gamma j \rfloor + 1)\alpha}{k(j)(s + \lfloor \gamma j \rfloor + 1 - j)} = \frac{|I|\alpha}{s + \lfloor \gamma j \rfloor + 1 - j}. \end{aligned}$$

But $|I| \leq s - \sum_{i=1}^{j} R_i = s - j + k$, so the above probability is less than or equal to

$$\frac{s - j + k}{s + \lfloor \gamma j \rfloor + 1 - j} \cdot \alpha = \alpha.$$

Therefore,

$$P\{FDP > \gamma | \hat{r}_1, \ldots, \hat{r}_f\} \leq \alpha,$$

which of course implies $P\{FDP > \gamma\} \leq \alpha$. □

Next, we prove the same stepdown procedure controls the FDP in the sense of (25) under an alternative assumption. Here, the assumption only involves the dependence of the p-values corresponding to true null hypotheses.

THEOREM 3.2. *Consider testing s null hypotheses, with $|I|$ of them true. Let $\hat{q}_{(1)} \le \cdots \le \hat{q}_{(|I|)}$ denote their corresponding ordered p-values. Set $M = \min(\lfloor \gamma s \rfloor + 1, |I|)$.*

(i) *For the stepdown procedure with α_i given by* (26),

$$P\{FDP > \gamma\} \le P\left\{\bigcup_{i=1}^{M}\left\{\hat{q}_{(i)} \le \frac{i\alpha}{|I|}\right\}\right\}. \tag{29}$$

(ii) *Therefore, if the joint distribution of the p-values of the true null hypotheses satisfies Simes inequality, that is,*

$$P\left\{\left\{\hat{q}_{(1)} \le \frac{\alpha}{|I|}\right\} \cup \left\{\hat{q}_{(2)} \le \frac{2\alpha}{|I|}\right\} \cup \cdots \cup \{\hat{q}_{(|I|)} \le \alpha\}\right\} \le \alpha,$$

then $P\{FDP > \gamma\} \le \alpha$.

PROOF. Let j be the smallest (random) index where the FDP exceeds γ for the first time at step j; that is, the number of false rejections corresponding to the first $j-1$ rejections divided by j exceeds γ for the first time at j. If j is such that $\gamma j < 1$, then $FDP > \gamma$ at step j implies $\hat{p}_{(j)} \le \alpha_j$. But this implies

$$\hat{q}_{(1)} \le \alpha_j = \frac{\alpha}{s+1-j} \le \frac{\alpha}{|I|},$$

because the number of true null hypotheses $|I|$ necessarily satisfies $|I| \le s - (j-1)$ for such a j.

Similarly, if j is such that $1 \le \gamma j < 2$, then we must have $\hat{p}_{(i)} \le \alpha_i$ and $\hat{p}_{(j)} \le \alpha_j$ for some $i < j$, where i, j correspond to true null hypotheses. But for such a j, $\alpha_j = 2\alpha/(s+2-j)$, and so we must have $\hat{q}_{(2)} \le 2\alpha/(s-j+2)$. But, by definition of j, we must have $|I| \le s-(j-2)$ and so $\hat{q}_{(2)} \le 2\alpha/|I|$.

Continuing in this way, if $m-1 \le \gamma j < m$, the event $FDP > \gamma$ at step j implies $\hat{q}_{(m)} \le m\alpha/|I|$. The largest value of j is of course s and so the largest possible m is $\lfloor \gamma s \rfloor + 1$. Also, we cannot have $m > |I|$. So, with M as in the statement of the theorem,

$$\begin{aligned} P\{FDP > \gamma\} &\le \sum_{m=1}^{M} P\left\{\hat{q}_{(m)} \le \frac{m\alpha}{|I|}, m-1 \le \gamma j < m\right\} \\ &\le \sum_{m=1}^{M} P\left\{\bigcup_{i=1}^{M}\left\{\hat{q}_{(i)} \le \frac{i\alpha}{|I|}\right\}, m-1 \le \gamma j < m\right\} \\ &\le P\left\{\bigcup_{i=1}^{M}\left\{\hat{q}_{(i)} \le \frac{i\alpha}{|I|}\right\}\right\}. \end{aligned}$$

Part (ii) follows trivially. □

In fact, there are many joint distributions of positively dependent variables for which Simes inequality is known to hold. In particular, Sarkar and Chang [11] and Sarkar [10] have shown that the Simes inequality holds for the family of distributions which is characterized by the multivariate positive of order 2 condition, as well as some other important distributions.

Theorem 3.2 points toward a method that controls the FDP without any dependence assumptions. One simply needs to bound the right-hand side of (29). In fact, Hommel [6] has shown that

$$P\left\{\bigcup_{i=1}^{|I|}\left\{\hat{q}_{(i)} \leq \frac{i\alpha}{|I|}\right\}\right\} \leq \alpha \sum_{i=1}^{|I|} \frac{1}{i}.$$

This suggests we replace α by $\alpha(\sum_{i=1}^{|I|}(1/i))^{-1}$. But of course $|I|$ is unknown. So one possibility is to bound $|I|$ by s, which then results in replacing α by α/C_s, where

$$C_j = \sum_{i=1}^{j}(1/i). \tag{30}$$

As is well known, $C_s \approx \log(s + 0.5) + \zeta_E$, with $\zeta_E \approx 0.5772156649$ known as Euler's constant. Clearly, changing α in this way is much too conservative and results in a much less powerful method. However, notice in (29) that we really only need to bound the union over $M \leq \lfloor \gamma s + 1 \rfloor$ events. Therefore, we need to slightly generalize the inequality by Hommel [6], which is done in the following lemma.

LEMMA 3.1. *Suppose* $\hat{p}_1, \dots, \hat{p}_t$ *are p-values in the sense that* $P\{\hat{p}_i \leq u\} \leq u$ *for all* i *and* u *in* $(0, 1)$. *Let their ordered values be* $\hat{p}_{(1)} \leq \cdots \leq \hat{p}_{(t)}$. *Let* $0 = \beta_0 \leq \beta_1 \leq \beta_2 \leq \cdots \leq \beta_m \leq 1$ *for some* $m \leq t$.

(i) *Then*

$$P\{\{\hat{p}_{(1)} \leq \beta_1\} \cup \{\hat{p}_{(2)} \leq \beta_2\} \cup \cdots \cup \{\hat{p}_{(m)} \leq \beta_m\}\} \leq t \sum_{i=1}^{m} (\beta_i - \beta_{i-1})/i. \tag{31}$$

(ii) *As long as the right-hand side of* (31) *is less than or equal to* 1, *the bound is sharp in the sense that there exists a joint distribution for the p-values for which the inequality is an equality.*

PROOF. Let J be the smallest (random) index j among $1 \leq j \leq m$ for which $\hat{p}_{(j)} \leq \beta_j$; define J to be $t + 1$ if $\hat{p}_{(j)} > \beta_j$ for all $1 \leq j \leq m$. Let $\theta_k = P\{J = k\}$. Then the left-hand side of (31) is equal to

$$P\left\{\bigcup_{k=1}^{m}\{J = k\}\right\} = \sum_{k=1}^{m} \theta_k,$$

since the events $\{J = k\}$ are disjoint. We wish to bound $\sum_k \theta_k$. For any $1 \le j \le m$,

$$\sum_{k=1}^{j} JI\{J = k\} = JI\{J \le j\} \le S_j,$$

where S_j is the number of p-values $\le \beta_j$. Taking expectations yields

$$\sum_{k=1}^{j} k\theta_k \le t\beta_j, \qquad j = 1, \dots, m. \tag{32}$$

For $j = 1, \dots, m-1$, multiply both sides of (32) by $1/[j(j+1)]$, and for $j = m$, multiply both sides by $1/m$; then sum over j to yield

$$\sum_{j=1}^{m-1} \frac{1}{j(j+1)} \sum_{k=1}^{j} k\theta_k + \frac{1}{m} \sum_{k=1}^{m} k\theta_k \le \sum_{j=1}^{m-1} \frac{t\beta_j}{j(j+1)} + \frac{t\beta_m}{m}. \tag{33}$$

By changing the order of summation, the left-hand side of (33) becomes

$$\sum_{k=1}^{m-1} k\theta_k \left(\frac{1}{k} - \frac{1}{m}\right) + \frac{1}{m} \sum_{k=1}^{m} k\theta_k = \sum_{k=1}^{m} \theta_k.$$

The right-hand side of (33) is easily seen to be the right-hand side of (31) and (i) follows.

To prove (ii), we construct $\hat{p}_1, \dots, \hat{p}_t$ as follows. Let U_i be uniform in I_i and let U_{m+1} be uniform in $(\beta_m, 1)$. Let p be equal to the right-hand side of (31), assumed less than or equal to 1. Let $\pi_1, \dots, \pi_m$ be probabilities summing to 1, with $\pi_i \propto (\beta_i - \beta_{i-1})/i$. Then, with probability $\pi_i p$, randomly pick i indices and let those p-values be equal to U_i, and the remaining $t - i$ p-values equal to U_{m+1}. With the remaining probability $1 - p$, let all p-values be equal to U_{m+1}. With this construction it is easily checked that $\hat{p}_i$ is uniform on (0, 1) and the left-hand side of (31) is equal to the right-hand side of (31). □

Theorem 3.2 and Lemma 3.1 now lead to the following result.

THEOREM 3.3. *For testing $H_i : P \in \omega_i$, $i = 1, \dots, s$, suppose $\hat{p}_i$ satisfies* (9). *Consider the stepdown procedure with constants $\alpha_i' = \alpha_i / C_{(\lfloor \gamma s \rfloor + 1)}$, where α_i is given by* (26) *and C_j is defined by* (30). *Then $P\{FDP > \gamma\} \le \alpha$.*

PROOF. By Theorem 3.2(i), $P\{FDP > \gamma\}$ is bounded by the right-hand side of (29) with α replaced by $\alpha / C_{\lfloor \gamma s \rfloor + 1}$, which is further bounded by the same expression with M replaced by $\lfloor \gamma s \rfloor + 1$. Then apply Lemma 3.1 with $t = |I|$ and $\beta_i = i\alpha / (C_{\lfloor \gamma s \rfloor + 1} |I|)$. □

It is of interest to compare control of the FDP with control of the FDR. Some obvious connections between methods that control the FDP in the sense of (25) and methods that control its expected value, the FDR, can be made. Indeed, for any random variable X on $[0, 1]$, we have

$$E(X) = E(X|X \leq \gamma)P\{X \leq \gamma\} + E(X|X > \gamma)P\{X > \gamma\}$$
$$\leq \gamma P\{X \leq \gamma\} + P\{X > \gamma\},$$

which leads to

$$\frac{E(X) - \gamma}{1 - \gamma} \leq P\{X > \gamma\} \leq \frac{E(X)}{\gamma}, \tag{34}$$

with the last inequality just Markov's inequality. Applying this to $X = FDP$, we see that, if a method controls the FDR at level q, then it controls the FDP in the sense $P\{FDP > \gamma\} \leq q/\gamma$. Obviously, this is very crude because if q and γ are both small, the ratio can be quite large. The first inequality in (34) says that if the FDP is controlled in the sense of (25), then the FDR is controlled at level $\alpha(1 - \gamma) + \gamma$, which is greater than or equal to α but typically only slightly. These crude arguments suggest that control of the FDP is perhaps more stringent than control of the FDR.

The comparison of actual methods, however, is complicated by the fact that the FDR controlling procedure of Benjamini and Hochberg [1] is a stepup procedure, but we have only considered stepdown procedures. It is interesting to note that, in order to make our procedure work without any dependence assumptions, we needed to change α to $\alpha/C_{\lfloor \gamma s \rfloor + 1}$. Benjamini and Yekutieli [2] show that the Benjamini–Hochberg procedure that controls the FDR at level q can also work without dependence assumptions, if you replace q by q/C_s. Clearly, this is a more drastic change since C_s is typically much larger than $C_{\lfloor \gamma s \rfloor + 1}$. Such connections need to be explored more fully.

4. Conclusions. We have seen that a very simple stepdown procedure is available to control the k-FWER under absolutely no assumptions on the dependence structure of the p-values. Furthermore, control of the k-FWER provides a measure of control for the *actual* number of false rejections, while the number of false rejections in the case of the FDR can vary widely. We have also considered two stepdown methods that control the FDP in the sense of (25). The first method provides control under very reasonable types of dependence assumptions, while the second holds in general.

Acknowledgments. We thank Juliet Shaffer and Michael Wolf for some helpful discussion and references. We also thank the referees and an Associate Editor for many helpful suggestions that greatly improved the clarity of the paper. Thanks to Wenge Guo for pointing out an error in an earlier version.

After the revision and acceptance of this paper, we became aware of the work by Hommel and Hoffman [7] which has much overlap with the results in Section 2, and we'd like to thank Helmut Finner for pointing out this oversight. In particular, Hommel and Hoffman [7] provide Theorem 2.1(i) with proof, Theorem 2.2 (stated but no proof) and a weaker version of Theorem 2.3(ii) (stated but no proof). They attribute the idea of controlling the number of false hypotheses to Victor [14], who also suggested control of the FDP. However, Hommel and Hoffman did not further discuss control of the FDP as they "could not find suitable procedures satisfying this criterion." As far as we know, the three theorems in Section 3 which address control of the FDP are new.

REFERENCES

[1] BENJAMINI, Y. and HOCHBERG, Y. (1995). Controlling the false discovery rate: A practical and powerful approach to multiple testing. *J. Roy. Statist. Soc. Ser. B* **57** 289–300.

[2] BENJAMINI, Y. and YEKUTIELI, D. (2001). The control of the false discovery rate in multiple testing under dependency. *Ann. Statist.* **29** 1165–1188.

[3] GENOVESE, C. and WASSERMAN, L. (2004). A stochastic process approach to false discovery control. *Ann. Statist.* **32** 1035–1061.

[4] HOCHBERG, Y. and TAMHANE, A. (1987). *Multiple Comparison Procedures.* Wiley, New York.

[5] HOLM, S. (1979). A simple sequentially rejective multiple test procedure. *Scand. J. Statist.* **6** 65–70.

[6] HOMMEL, G. (1983). Tests of the overall hypothesis for arbitrary dependence structures. *Biometrical J.* **25** 423–430.

[7] HOMMEL, G. and HOFFMAN, T. (1988). Controlled uncertainty. In *Multiple Hypotheses Testing* (P. Bauer, G. Hommel and E. Sonnemann, eds.) 154–161. Springer, Heidelberg.

[8] KORN, E., TROENDLE, J., MCSHANE, L. and SIMON, R. (2004). Controlling the number of false discoveries: Application to high-dimensional genomic data. *J. Statist. Plann. Inference* **124** 379–398.

[9] PERONE PACIFICO, M., GENOVESE, C., VERDINELLI, I. and WASSERMAN, L. (2004). False discovery control for random fields. *J. Amer. Statist. Assoc.* **99** 1002–1014.

[10] SARKAR, S. (1998). Some probability inequalities for ordered MTP_2 random variables: A proof of the Simes conjecture. *Ann. Statist.* **26** 494–504.

[11] SARKAR, S. and CHANG, C. (1997). The Simes method for multiple hypothesis testing with positively dependent test statistics. *J. Amer. Statist. Assoc.* **92** 1601–1608.

[12] SIMES, R. (1986). An improved Bonferroni procedure for multiple tests of significance. *Biometrika* **73** 751–754.

[13] VAN DER LAAN, M., DUDOIT, S. and POLLARD, K. (2004). Multiple testing. Part III. Procedures for control of the generalized family-wise error rate and proportion of false positives. Working Paper Series, Paper 141, Div. Biostatistics, Univ. California, Berkeley.

[14] VICTOR, N. (1982). Exploratory data analysis and clinical research. *Methods of Information in Medicine* **21** 53–54.

DEPARTMENT OF STATISTICS
UNIVERSITY OF CALIFORNIA
BERKELEY, CALIFORNIA 94720
USA

DEPARTMENT OF STATISTICS
STANFORD UNIVERSITY
STANFORD, CALIFORNIA 94305-4065
USA
E-MAIL: romano@stat.stanford.edu

Part VIII
Orderings of Probability Distributions

Erich Lehmann's Contributions to Orderings of Probability Distributions

Marco Scarsini

The initial contributions to the theory of comparison of experiments appeared in the late forties/early fifities (see, e.g., Bohnenblust, Shapley, and Sherman,, 1949; Stein, 1951; Blackwell, 1951, 1953).

At around the same time, motivated by some problems in hypothesis testing, Erich Lehmann introduced some basic concepts in the theory of comparison of distributions. As later developments show, the two theories are strictly connected and they often examine conditions on two probability measures P and Q under which $\int \phi \, \mathrm{d}P \leqslant \int \phi \, \mathrm{d}Q$ for all functions ϕ in some class.

The success of the concept has been enormous both in probability theory and in applications in statistics, decision theory, economics, operations research, and reliability theory. We refer to Müller and Stoyan (2002); Shaked and Shanthikumar (2007) for an extensive analysis of stochastic orders and their applications.

Testing multiparameter hypotheses

Lehmann (1952) is the first paper where a monotonicity condition is considered, in the framework of hypothesis testing. The hypothesis studied in this paper concerns different parameters. It could be, for instance, of the type $H : \theta_1 \leqslant \theta_1^*, \theta_2 \leqslant \theta_2^*$. For this type of hypotheses an unbiased test does not exist in general.

Therefore a restriction of monotonicity is imposed, that is, the rejection region for H is assumed to be an increasing set. This makes sense when for $\theta_1 \leqslant \theta_1'$ and $\theta_2 \leqslant \theta_2'$ we have that $F_{\theta_1\theta_2}$ is stochastically dominated by $F_{\theta_1'\theta_2'}$. The uniformly most powerful test among all monotonic regression regions is found and it has a simple natural form.

M. Scarsini
Dipartimento di Economia e Finanza, LUISS, Viale Romania 12, I–00197 Roma, Italy
e-mail: marco.scarsini@luiss.it

J. Rojo (ed.), *Selected Works of E. L. Lehmann*, Selected Works in Probability and Statistics, DOI 10.1007/978-1-4614-1412-4_58, © Springer Science+Business Media, LLC 2012

Ordered families of distributions

Lehmann (1955) introduces what is now known as the *usual stochastic order* and the *monotone likelihood ratio order*.

Using the same notation as in Lehmann's paper, two distributions P_θ and $P_{\theta'}$ can be compared for instance according to the following four criteria:

(i) There exists a random vector Z and two functions g, g' such that $g(Z)$ has distribution P_θ, $g'(Z)$ has distribution $P_{\theta'}$ and $g(z) \leqslant g'(z)$ for all z.

(ii) For every increasing function ϕ,

$$\mathbb{E}_\theta \phi(X) \leqslant \mathbb{E}_{\theta'} \phi(X).$$

(iii) For every increasing set S,

$$P_\theta(S) \leqslant P_{\theta'}(S).$$

(iv) Let P_θ and $P_{\theta'}$ have densities p_θ and $p_{\theta'}$ with respect to some common measure μ. Then the ratio $p_{\theta'}(x)/p_\theta(x)$ is increasing.

The three conditions (i), (ii), (iii) are equivalent and this has been proved later in great generality by Kamae, Krengel, and O'Brien (1977) using results by Strassen (1965). Lehmann (1955) shows the equivalence of (ii) and (iii) and provides two counterexamples (Examples 2.2 and 2.3 in the paper) to the equivalence of (i) and (ii). Unfortunately there is a mistake, since the two distributions used in the counterexample are actually not ordered.

Condition (iv) was studied extensively by Karlin and Rubin (1956a,b). For univariate distributions this condition is stronger than (i), but for multivariate distributions this is not the case, as Example 2.4 rightly shows. The sufficient multivariate condition that implies (i) is the following

$$p_{\theta'}(x)p_\theta(y) \leqslant p_\theta(x \wedge y)p_{\theta'}(x \vee y),$$

where $\wedge$ and $\vee$ indicate the componentwise minimum and maximum, respectively. This condition was studied by Karlin and Rinott (1980) and Whitt (1982).

Comparing location experiments

Lehmann (1988) starts with a review of the classical theory of comparison of experiments developed in the late forties by several scholars and culminated with Blackwell (1951, 1953). We refer to the book by Torgersen (1991) and to the nice survey by Le Cam (1996).

An experiment $\mathscr{E}$ is defined as $\{X, \mathscr{P}\}$, where X is a random quantity and $\mathscr{P} = \{P_\theta,$
$\theta \in \Omega\}$ is a parametric family of possible distributions of X.

Call $\mathscr{F} = \{Y, \mathscr{Q} = \{Q_\theta, \theta \in \Omega\}\}$ another experiment where P_θ and Q_θ are indexed by the same state of nature. The experiment $\mathscr{F}$ is more informative than the experiment $\mathscr{E}$ if there exists a random quantity Z with known distribution and a function h such that for all $\theta \in \Omega$

$$Y \text{ is distributed as } Q_\theta \text{ implies that } h(Y, Z) \text{ is distributed as } P_\theta.$$

Blackwell proved that $\mathscr{F}$ is more informative than $\mathscr{E}$ if an only if for any decision procedure based on X and any loss function there exists a (possibly randomized) decision procedure based on Y that guarantees a weakly smaller risk.

Some examples show that the above criterion is too demanding and does not allow to compare experiments that should be intuitively comparable. Lehmann (1988) therefore suggests a weaker criterion that involves considering only a sub-class of decision problems, in particular the class $\mathscr{M}$ of monotone procedures introduced by Karlin and Rubin (1956b). In these procedures the correct action is a nondecreasing function $\alpha(\theta)$ and the loss function $L(\theta, d)$ is nondecreasing in $d - \alpha(\theta)$ for $d - \alpha(\theta) > 0$ and nonincreasing in $d - \alpha(\theta)$ for $d - \alpha(\theta) < 0$, with $L(\theta, \alpha(\theta)) = 0$.

Define $F_\theta(t) = P_\theta(X \leqslant t)$ and $G_\theta(t) = Q_\theta(Y \leqslant t)$, and let F_θ and G_θ have densities with monotone likelihood ratio. The main theorem states that Y is more effective than X with respect to the class $\mathscr{M}$ if and only if $h_\theta(t) = G_\theta^{-1}(F_\theta(t))$ is nondecreasing in θ for each t. This condition corresponds to the comparison of the two P-P plots $(F_{\theta_1}, F_{\theta_2})$ and $(G_{\theta_1}, G_{\theta_2})$, for $\theta_1 < \theta_2$, and, unlike Blackwell's condition, is easy to verify.

For location families with strongly unimodal densities the condition of the theorem reduces to domination in term of the "more spread out" order studied by Bickel and Lehmann (1979).

This idea has found extensive application in the economic literature (see, for instance, Persico, 2000; Levin, 2001; Athey and Levin, 2001; Bergemann and Välimäki, 2002; Quah and Strulovici, 2009, and references therein). In particular Jewitt (2007) examines conditions under which the Lehmann criterion is applicable to various economic problems.

Invariant directional orderings

Several stochastic orderings have been studied and used since Lehmann (1955). Some are stronger than others. For instance, the likelihood ratio order is stronger than the hazard order, which in turn is stronger than the usual stochastic order. These three orders are defined in totally different ways and the above chain of implications is not immediate. Lehmann and Rojo (1992) provide a unified way to define these three orders and many others.

The paper considers a class $\mathscr{F}$ of continuous strictly increasing distribution functions and a preorder $S \subset \mathscr{F} \times \mathscr{F}$.

A preorder S is invariant under monotone transformations if $(F, G) \in S$ implies $(F\psi^{-1}, G\psi^{-1}) \in S$ for all $\psi \in \Psi$, the class of strictly increasing continuous functions. If S is invariant under monotone transformations and $(F, G) \in S$, then $\mathscr{O}(F, G) \subset S$, where

$$\mathscr{O}(F, G) = \{(F\psi^{-1}, G\psi^{-1}) : \psi \in \Psi\}$$

is the orbit of (F, G) under the group of transformations Ψ.

Each orbit $\mathscr{O}(F, G)$ can be labeled by the function $k : [0, 1] \to [0, 1]$ defined as $k(u) = GF^{-1}(u)$. Different orderings can be characterized through properties of the function k. So a preorder S can be defined as

$$S = \{(F, G) : k(F, G) \in \mathscr{K}\}$$

for some class $\mathscr{K}$ that contain the identity function and is closed under composition. Different choices of $\mathscr{K}$ produce the likelihood ratio order, the hazard rate order, and the usual stochastic order.

This approach allows even to order ordered pairs. Let $(F_1, G_1), (F_2, G_2) \in S$. Then G_2 is said to be further to the right of F_2 than G_1 is of F_1 if $(k_1, k_2) \in S$, where $k_i = G_i F_i^{-1}$.

The paper also defines metrics that are coherent with the various orders, that is, metrics d_S that are invariant with respect to monotone transformations and such that $d_S(F_1, G_1) \leqslant d_S(F_2, G_2)$ whenever $(k_1, k_2) \in S$.

The technique discovered in this paper has been used by Bartoszewicz and Benduch (2009) to define orders based on a generalization of the total time on test transform.

References

ATHEY, S. and LEVIN, J. (2001) The Value of Information in Monotone Decision Problems. Working Paper 98-24, MIT.

BARTOSZEWICZ, J. and BENDUCH, M. (2009) Some properties of the generalized TTT transform. *J. Statist. Plann. Inference* **139**, 2208–2217.

BERGEMANN, D. and VÄLIMÄKI, J. (2002) Information acquisition and efficient mechanism design. *Econometrica* **70**, 1007–1033.

BICKEL, P. J. and LEHMANN, E. L. (1979) Descriptive statistics for nonparametric models. IV. Spread. In *Contributions to statistics*, 33–40. Reidel, Dordrecht.

BLACKWELL, D. (1951) Comparison of experiments. In *Proceedings of the Second Berkeley Symposium on Mathematical Statistics and Probability, 1950*, 93–102. University of California Press, Berkeley and Los Angeles.

BLACKWELL, D. (1953) Equivalent comparisons of experiments. *Ann. Math. Statistics* **24**, 265–272.

BOHNENBLUST, H. F., SHAPLEY, L. S., and SHERMAN, S. (1949) Reconnaissance in game theory. Technical Report RM 208, RAND Corporation, Santa Monica, CA. URL http://www.rand.org/pubs/research_memoranda/RM208.

JEWITT, I. (2007) Information order in decision and agency problems. Mimeo, Nuffield College.

KAMAE, T., KRENGEL, U., and O'BRIEN, G. L. (1977) Stochastic inequalities on partially ordered spaces. *Ann. Probability* **5**, 899–912.

KARLIN, S. and RINOTT, Y. (1980) Classes of orderings of measures and related correlation inequalities. I. Multivariate totally positive distributions. *J. Multivariate Anal.* **10**, 467–498.

KARLIN, S. and RUBIN, H. (1956a) Distributions possessing a monotone likelihood ratio. *J. Amer. Statist. Assoc.* **51**, 637–643.

KARLIN, S. and RUBIN, H. (1956b) The theory of decision procedures for distributions with monotone likelihood ratio. *Ann. Math. Statist.* **27**, 272–299.

LE CAM, L. (1996) Comparison of experiments—a short review. In *Statistics, probability and game theory*, volume 30 of *IMS Lecture Notes Monogr. Ser.*, 127–138. Inst. Math. Statist., Hayward, CA. URL http://dx.doi.org/10.1214/lnms/1215453569.

LEHMANN, E. L. (1952) Testing multiparameter hypotheses. *Ann. Math. Statistics* **23**, 541–552.

LEHMANN, E. L. (1955) Ordered families of distributions. *Ann. Math. Statist.* **26**, 399–419.

LEHMANN, E. L. (1988) Comparing location experiments. *Ann. Statist.* **16**, 521–533.

LEHMANN, E. L. and ROJO, J. (1992) Invariant directional orderings. *Ann. Statist.* **20**, 2100–2110.

LEVIN, J. (2001) Information and the market for lemons. *Rand J. Econom.* **32**, 657–666.

MÜLLER, A. and STOYAN, D. (2002) *Comparison methods for stochastic models and risks.* Wiley Series in Probability and Statistics. John Wiley & Sons Ltd., Chichester.

PERSICO, N. (2000) Information acquisition in auctions. *Econometrica* **68**, 135–148.

QUAH, J. K.-H. and STRULOVICI, B. (2009) Comparative statics, informativeness, and the interval dominance order. *Econometrica* **77**, 1949–1992.

SHAKED, M. and SHANTHIKUMAR, J. G. (2007) *Stochastic orders.* Springer Series in Statistics. Springer, New York. URL http://dx.doi.org/10.1007/978-0-387-34675-5.

STEIN, C. (1951) Notes on a Seminar on Theoretical Statistics. I. Comparison of experiments. University of Chicago.

STRASSEN, V. (1965) The existence of probability measures with given marginals. *Ann. Math. Statist.* **36**, 423–439.

TORGERSEN, E. (1991) *Comparison of statistical experiments*, volume 36 of *Encyclopedia of Mathematics and its Applications.* Cambridge University Press, Cambridge.

WHITT, W. (1982) Multivariate monotone likelihood ratio and uniform conditional stochastic order. *J. Appl. Probab.* **19**, 695–701.

TESTING MULTIPARAMETER HYPOTHESES[1]

By E. L. Lehmann

Stanford University and University of California, Berkeley

1. Summary. Let the distribution of some random variables depend on real parameters $\theta_1, \cdots, \theta_s$ and consider the hypothesis $H: \theta_i \leqq \theta_i^*, i = 1, \cdots, s$. It is shown under certain regularity assumptions that unbiased tests of H do not exist. Tests of minimum bias and other types of minimax tests are derived under suitable monotonicity conditions. Certain related multidecision problems are discussed and two-sided hypotheses are considered very briefly.

2. Introduction. The extensive literature on optimum tests has been concerned mainly with hypotheses specifying a set of values for a single real valued parameter. Important exceptions are some cases that can be reduced to the one-parameter situation by the principle of invariance, such as the linear (univariate) hypothesis and Hotelling's T^2-problem. These have been used to illustrate a number of different principles, the successful application of which however seems to rest on the symmetry whose full exploitation makes the problems uniparametric. Another exception is the theory of tests with local optimum properties, initiated by Neyman and Pearson [1] and recently developed further by Isaacson [2].

We shall here concern ourselves mainly with hypotheses which, rather than specifying the values of the parameter in question, state that these parameters do not exceed certain bounds. The following examples illustrate the way in which such problems arise.

Example 2.1. Let p and p' denote the number of major and minor defects in a lot. Then the lot will be considered acceptable provided $p \leqq p_0$ and $p' \leqq p_0'$, where $p_0 < p_0'$.

Example 2.2. It may be desired to compare some new treatments with a standard one. Here the hypothesis would specify that none of the new treatments is better by more than a given amount than the standard.

Example 2.3. Let $x_1, \cdots, x_n$ be a sample from a normal distribution with mean ξ and variance σ^2. The population in question may be considered adequate if $\xi \leqq \xi_0$ and $\sigma \leqq \sigma_0$.

In some of the above examples we are dealing with *bona fide* testing problems while in others we are faced with a choice among more than two decisions. Which of these is the case cannot always be seen from the mathematical formulation alone. Thus in Example 2.1 it clearly depends on the disposition that is made of a rejected lot. If there is complete screening, the reason for rejection is immaterial. If on the other hand a lot rejected for major defects is treated differently from one rejected only for minor defects the decision problem becomes more complicated.

[1] Work done under the sponsorship of the Office of Naval Research.

J. Rojo (ed.), *Selected Works of E. L. Lehmann*, Selected Works in Probability and Statistics, DOI 10.1007/978-1-4614-1412-4_59, © Springer Science+Business Media, LLC 2012

We shall in the following concern ourselves mainly with the one-sided case of the straightforward testing problem. The two-sided situation and the multi-decision problem will be discussed only rather briefly. For simplicity we shall take the number of parameters to be two. The extension to the higher dimensional cases is immediate.

3. Unbiasedness and the minimax principle. The success of the concept of unbiasedness in the one-parameter case suggests the use of this approach also for the present problems. Unfortunately it turns out that in general unbiased tests of the hypotheses in question do not exist. Let us consider the case of two parameters θ_1, θ_2 and the hypothesis $H: \theta_1 \leqq \theta_1^*, \theta_2 \leqq \theta_2^*$. We shall assume that the power function $\beta(\theta_1, \theta_2)$ of any test is analytic in θ_1 and θ_2 in the sense that it can be expanded in an absolutely convergent double power series. Then we shall show that for any unbiased test we have $\beta(\theta_1, \theta_2) \equiv \alpha$, so that any unbiased test is equivalent to the trivial one that rejects with probability α regardless of the observations. This incidentally proves this trivial and most unsatisfactory test to be admissible for the problem under consideration.

Without loss of generality assume that $\theta_1^* = \theta_2^* = 0$. Then unbiasedness states that $\beta(\theta_1, \theta_2) \leqq \alpha$ in the third quadrant of the θ_1, θ_2-plane, and $\geqq \alpha$ in the other three quadrants. By continuity we have $\beta(\theta_1, 0) = \alpha$ for $\theta_1 \leqq 0$ and hence by analyticity $\beta(\theta_1, 0) = \alpha$ for all θ_1. Analogously $\beta(0, \theta_2) \equiv \alpha$. Consider now $\beta(\theta_1, \theta_2)$ for any fixed $\theta_2 > 0$ as a function of θ_1. It has a minimum at $\theta_1 = 0$ so that $\partial\beta(\theta_1, \theta_2)/\partial\theta_1 \mid_{\theta_1=0} = 0$ for all $\theta_2 \geqq 0$. Since $\partial\beta(\theta_1, \theta_2)/\partial\theta_1 \mid_{\theta_1=0}$ is again analytic in θ_2, it follows that $(\partial\beta(\theta_1, \theta_2/\partial\theta_1) \mid_{\theta_1=0}$ is identically zero. Consider now $\beta(\theta_1, \theta_2)$ for some fixed value $\theta_2 \leqq 0$. Since $\beta(\theta_1, \theta_2) \gtrless \alpha$ as $\theta_1 \gtrless 0$ and since at $\theta_1 = 0$ the derivative is zero, $\beta(\theta_1, \theta_2)$ must have a point of inflection at 0 and consequently the second derivative $\partial^2\beta(\theta_1, \theta_2)/\partial\theta_1^2 \mid_{\theta_1=0} = 0$ for all $\theta_2 < 0$ and hence for all θ_2. Since the, order of the first non-vanishing derivative $\partial^k\beta(\theta_1, \theta_2)/\partial\theta_1^k \mid_{\theta_1=0}$ must be even for $\theta_2 > 0$ and odd for $\theta_2 < 0$ we see in this manner that for any fixed θ_2 $\partial\, \beta(\theta_1, \theta_2)^k/\partial\theta_1^k \mid_{\theta_1=0} = 0$ for all $k = 1, 2, \cdots$. By analyticity it follows for each fixed θ_2 that $\beta(\theta_1, \theta_2)$ must be a constant, that is, be independent of θ_1. By symmetry it now follows that $\beta(\theta_1, \theta_2)$ must be identically constant, as was to be proved.

We digress for a moment from our search for a reasonable test of the hypothesis $\theta_1, \theta_2 \leqq 0$ to point out that there do exist non-trivial tests of H satisfying the condition of similarity

$$\beta(\theta_1, 0) = \beta(0, \theta_2) = \alpha$$

for all θ_1, θ_2. Suppose for example that X and Y are independently distributed with joint density $f_{\theta_1}(x)f_{\theta_2}(y)$ and that $\alpha = 1/m$ where m is an integer. Then we can obtain a particularly simple class of similar regions as follows. Let $S_1, \cdots, S_m$ be mutually exclusive and exhaustive sets on the real line such that

$$\int_{S_i} f_0(x)\, dx = \alpha, \qquad i = 1, \cdots, m.$$

In the x, y-plane define the set w_i to be the Cartesian product of S_i with itself, and let $w = w_1 + \cdots + w_m$. Now when $\theta_2 = 0$, X is a sufficient statistic for θ_1 and for every x we have

$$P_{\theta_1,0}((X, Y) \varepsilon W \mid x) = \alpha,$$

and hence $P_{\theta_1,0}((X, Y) \varepsilon W) \equiv \alpha$.

We now return to the original problem, and investigate the test that maximizes the minimum power over a certain class ω' of alternatives. For ω' we take the set of points (θ_1, θ_2) for which either $\theta_1 \geqq \theta_1^{**}$ or $\theta_2 \geqq \theta_2^{**}$. Let us consider first as an example the case that X and Y are independently, normally distributed, with known variances and with means θ_1 and θ_2, respectively. Then it would seem as if any reasonable test should satisfy the following two conditions:

(i) $\beta(\theta_1, \theta_2) \leqq \beta(\theta_1', \theta_2')$ whenever $\theta_1 \leqq \theta_1'$, $\theta_2 \leqq \theta_2'$, if ϕ denotes the critical function,

(ii) $\phi(x, y) \leqq \phi(x', y')$ whenever $x \leqq x'$, $y \leqq y'$. It is easily seen that (ii) implies (i); we shall now show that the test ϕ that maximizes $\inf \omega' \beta(\theta_1, \theta_2)$ does not possess property (i) and hence also not (ii), provided $\theta_1^{**} - \theta_1^*$ and $\theta_2^{**} - \theta_2^*$ are sufficiently large so that ϕ is not identically equal to α. Let β denote the power function of ϕ and suppose that $\inf \omega' \beta(\theta_1, \theta_2) = \gamma > \alpha$. Then condition (i) implies that under the hypothesis $\beta(\theta_1, \theta_2) = \alpha$ only when $\theta_1 = \theta_1^*$ and $\theta_2 = \theta_2^*$. For if $\beta(\theta_1, \theta_2) = \alpha$ also for some other point in H, it would also equal α on the line segment connecting these two points and hence by analyticity on the whole line containing this segment. But this would imply that $\beta(\theta_1, \theta_2) = \alpha$ also for points in ω' where by assumption $\beta(\theta_1, \theta_2) \geqq \gamma > \alpha$. Another consequence of condition (i) is that $\beta(\theta_1, \theta_2) > \gamma$ for all points in ω' so that the minimum point γ is never attained in ω' and is approached only as either θ_1 or θ_2 tend to $-\infty$. For if, for example, $\beta(\theta_1', \theta_2') = \gamma$ for some point with $\theta_1' \geqq \theta_2^{**}$ and finite θ_1' we would have $\beta(\theta_1, \theta_2') = \gamma$ for all $\theta_1 \leqq \theta_1'$ and hence for all θ_1. This would imply $\beta(\theta_1, \theta_2) = \gamma$ for all (θ_1, θ_2) with $\theta_1 \geqq \theta_1^{**}$, $\theta_2 \leqq \theta_2'$ and therefore $\beta(\theta_1, \theta_2) \equiv \gamma$.

From these two remarks and Theorem 3.10 of Wald's book *Statistical Decision Functions* [3], it can be shown that there exists a sequence λ_i of probability distributions over ω' with the following properties: (a) For any real number A the probability under λ_i of the intersection of ω' with the quadrant $\{\theta_1, \theta_2 \mid \theta_1, \theta_2 \geqq A\}$ tends to zero as $i \to \infty$. (b) The power of the most powerful level α test for testing $H'{:}\theta_1 = \theta_1^*$, $\theta_2 = \theta_2^*$ against the simple alternative $\int_{\omega'} p_{\theta_1,\theta_2}(x, y) d\lambda_i(\theta_1, \theta_2)$ tends to γ as $i \to \infty$. But from (a) it follows easily that as $i \to \infty$, $\int p_{\theta_1,\theta_2}(x, y) d\lambda_i(\theta_1, \theta_2)$ can be distinguished arbitrarily well from $p_{\theta_1^*,\theta_2^*}(x, y)$. This leads to the contradiction $\gamma = 1$.

We have given the proof explicitly only for the case of independent normal variables. However it applies equally well to any problem in which, in addition to the analyticity assumption of the present section, also the assumptions of Theorem 4.1 are satisfied.

4. Monotone regions. In the previous section we showed that for the hypotheses under consideration neither unbiasedness nor the minimax principle lead to desirable results. In order to arrive at a reasonable test we now impose the following preliminary conditions suggested by the negative results of the last section. We ask first that the test be nonrandomized, so that we can speak of a region w of rejection. The second restriction is one of monotonicity. Let us assume that we are concerned with two random variables X and Y whose joint distribution is given by $p_{\theta_1,\theta_2}(x, y)$. Then we shall say that the region w of rejection for the hypothesis $H: \theta_1 \leqq \theta_2^*$, $\theta_2 \leqq \theta_2^*$ is *monotone* (nondecreasing in x and y) if

$$(4.1) \qquad (x, y) \varepsilon w, x \leqq x', y \leqq y' \quad \text{imply} \quad (x', y') \varepsilon w,$$

that is, if its critical function is nondecreasing in both variables.

The restriction to monotone regions is of course suitable only in certain problems, namely, roughly speaking, when increased values of the parameters lead to higher values for the observations. To make this precise let $\theta_1 \leqq \theta_1'$, $\theta_2 \leqq \theta_2'$ and let F and G be the cumulative distribution functions of X and Y corresponding to (θ_1, θ_2) and (θ_1', θ_2'), respectively. Then we shall consider the condition of monotonicity appropriate provided for every monotone non-decreasing region w

$$(4.2) \qquad \int_w dF \leqq \int_w dG.$$

Frequently the simplest way to prove (4.2) is to establish the existence of functions $x' = f(x, y)$, $y' = g(x, y)$ with $x' \geqq x$, $y' \geqq y$ and such that when F is the cumulative distribution function of (X, Y), that of (X', Y') is G. Sometimes it is more convenient instead to prove the existence of random variables, say $Z_1, \cdots, Z_r$, and functions $X = f(Z_1, \cdots, Z_r)$, $Y = g(Z_1, \cdots, Z_r)$, $X' = f'(Z_1, \cdots, Z_r)$, $Y' = g'(Z_1, \cdots, Z_r)$ such that $X \leqq X'$, $Y \leqq Y'$ and the cdf's of (X, Y) and (X', Y') are F and G respectively. Both of these conditions clearly assure the validity of (4.2) since for any w that is nondecreasing in x and y they imply

$$(4.3) \qquad \int_w dF = P((X, Y) \epsilon w) \leqq P((X', Y') \epsilon w) = \int_w dG.$$

A remark is required also in connection with the restriction to nonrandomized tests. When dealing with discrete problems, for example binomial distributions, we must permit a certain rather trivial kind of randomization. A formal way of handling the distinction is provided by a representation of randomized tests due to M. Eudey [4]. Let X denote the number of successes in n binomial trials, and let U be uniformly distributed over [0, 1]. Then any randomized test in X is equivalent to a non-randomized test in $X + U$, and we shall consider monotone non-randomized tests in the continualized variables $X + U$, $Y + V$. Here monotonicity insures that no very heavy use is made of randomization. In fact, in the original variables X, Y randomization will occur only on the boundary of the critical region.

We shall now derive that test of the hypothesis $H: \theta_1 \leqq \theta_1^*, \theta_2 \leqq \theta_2^*$ that among all monotone regions maximizes $\inf_{\omega'} \beta(\theta_1, \theta_2)$ where ω' is the set of points (θ_1, θ_2) with $\theta_1 \geqq \theta_1^{**}$ or $\theta_2 \geqq \theta_2^{**}$. Here $\theta_1^{**} \geqq \theta_1^*$ and $\theta_2^{**} \geqq \theta_2^*$. It may be of interest to point out that if we let $\theta_1^{**} = \theta_1^*$, $\theta_2^{**} = \theta_2^*$, we get the monotone region with minimum bias.

THEOREM 4.1. *Let the joint density of X and Y be $p_{\theta_1,\theta_2}(x, y)$ where the parameter-space is a finite or infinite open rectangle $\underline{\theta}_1 < \theta_1 < \bar{\theta}_1$, $\underline{\theta}_2 < \theta_2 < \bar{\theta}_2$, and the positive sample space also is an open rectangle $\underline{x} < x < \bar{x}$, $\underline{y} < y < \bar{y}$ independent of the θ's. Suppose that (4.2) holds, that the marginal distribution of X depends only on θ_1, and that of Y only on θ_2, and that X tends in probability to $\underline{x}$ as $\theta_1 \to \underline{\theta}_1$, while Y tends to $\underline{y}$ as $\theta_2 \to \underline{\theta}_2$. Then the test that among all monotone nonrandomized tests of H maximizes the minimum power against ω' is given by the region of acceptance S:*

$$x \leqq a, \qquad y \leqq b, \tag{4.4}$$

where a and b are determined by the conditions

$$P(X \leqq a, Y \leqq b \mid \theta_1^*, \theta_2^*) = 1 - \alpha \tag{4.5}$$

and

$$P(X \leqq a \mid \theta_1^{**}) = P(Y \leqq b \mid \theta_2^{**}). \tag{4.6}$$

PROOF. We point out first that for any $x > \underline{x}$, $y > \underline{y}$

$$\lim_{\theta_1 \to \underline{\theta}_1} P(X \leqq x, Y \leqq y \mid \theta_1, \theta_2) = P(Y \leqq y \mid \theta_2) \tag{4.7}$$

$$\lim_{\theta_2 \to \underline{\theta}_2} P(X \leqq x, Y \leqq y \mid \theta_1, \theta_2) = P(X \leqq x \mid \theta_1).$$

For

$$P(X \leqq x, Y \leqq y) = P(X \leqq x) - P(X \leqq x, Y \geqq y)$$

while

$$0 \leqq P(X \leqq x, Y > y) \leqq P(Y > y)$$

and

$$\lim_{\theta_2 \to \underline{\theta}_2} P(Y > y \mid \theta_2) = 0.$$

For any monotone test the limit of the power $\beta(\theta_1, \theta_2)$ as $\theta_1 \to \underline{\theta}_1$ clearly exists and we shall denote it by $\beta(\underline{\theta}_1, \theta_2)$. The minimum power in ω' is then the smaller of the two quantities $\beta(\underline{\theta}_1, \theta_2^{**})$ and $\beta(\theta_1^{**}, \underline{\theta}_2)$. Since for the test given by (4.5) we have $\beta(\underline{\theta}_1, \theta_2^{**}) = \beta(\theta_1^{**}, \underline{\theta}_2)$ we could, if the theorem were false, increase both $\beta(\underline{\theta}_1, \theta_2^{**})$ and $\beta(\theta_1^{**}, \underline{\theta}_2)$.

Clearly any monotone test has a region of acceptance S' of the form $y \leqq g(x)$

or equivalently $x \leqq h(y)$ where g and h are non-increasing. If $S' \neq S$ is to be of size $\leqq \alpha$ we have $P(S' \mid \theta_1^*, \theta_2^*) \geqq P(S \mid \theta_1^*, \theta_2^*)$ and hence either $\lim_{x \to \underline{x}} g(x) > b$ or $\lim_{y \to \underline{y}} h(y) > a$. Suppose that the first of these conditions is satisfied and let us denote the complement of S and S' by $\bar{S}$ and $\bar{S}'$, respectively. Then there exists a constant k such that $x \geqq k$ for all points in $S \wedge \bar{S}'$, and a subset S'' of $\bar{S} \wedge S'$ such that $P(S'' \mid \theta_1, \theta_2) > 0$ for all θ_1, θ_2 and $x \leqq k$ for all points in S''. Hence by (4.7)

$$P(S \wedge \bar{S}' \mid \underline{\theta}_1, \theta_2^{**}) = 0$$

and

$$P(\bar{S} \wedge S' \mid \underline{\theta}_1, \theta_2^{**}) = P(b \leqq Y \leqq \lim_{x \to \underline{x}} g(x) \mid \theta_2^{**}) > 0,$$

which leads to

$$P(S' \mid \underline{\theta}_1, \theta_2^{**}) > P(S \mid \underline{\theta}_1, \theta_2^{**}),$$

and hence to the desired result.

The theorem becomes particularly simple if the joint density of X and Y is symmetric in its two variables when $\theta_1 = \theta_2$. For then if $\theta_1^* = \theta_2^* = \theta^*$, $\theta_1^{**} = \theta_2^{**} = \theta^{**}$, it is seen from (4.6) that $a = b$. Thus the test accepts if $\max(X, Y) \leqq a$ where a is determined by (4.5), and hence is independent of θ^{**}.

The assumptions made in Theorem 4.1 concerning the shape of the parameter and sample spaces are unnecessarily restrictive. The theorem remains valid if we assume that both the parameter space and the positive sample space are convex open sets. The proof is essentially the same, however the notation becomes considerably more complicated.

If the roles of hypothesis and class of alternatives are interchanged, we obtain

THEOREM 4.2. *For testing the hypothesis* $H': \theta_1 \geqq \theta_1^{**}$ *or* $\theta_2 \geqq \theta_2^{**}$ *against the class of alternatives* $\omega: \theta_1 \leqq \theta_1^*$, $\theta_2 \leqq \theta_2^*$, *let* S *be the region of rejection* $x \leqq c$, $y \leqq d$, *where* c *and* d *are determined by*

$$P(X \leqq c \mid \theta_1^{**}) = P(Y \leqq d \mid \theta_2^{**}) = \alpha. \tag{4.8}$$

Then under the assumptions of Theorem 4.1 S *is uniformly most powerful among all regions of rejection that are monotone non-increasing in both variables.*

PROOF. Consider any monotone region given by $x \leqq g(y)$ or $y \leqq h(x)$ with g and h non-increasing. Since the probability of rejection must be not greater than α at $(\theta_1^{**}, \underline{\theta}_2)$ and $(\underline{\theta}_1, \theta_2^{**})$ we must have

$$\lim_{x \to \underline{x}} g(x) \leqq c, \qquad \lim_{y \to \underline{y}} h(y) \leqq d. \tag{4.9}$$

But any monotone region satisfying (4.9) is contained in S and hence is uniformly less powerful than S.

5. Examples. In the present section we shall apply Theorem 4.1 to some specific problems. All other assumptions being trivially satisfied we shall in each case only check condition (4.2).

Example 5.1. Let X, Y have a multinomial distribution

$$(4.10)\quad P(X = x, Y = y) = \frac{n!}{x!y!(n - x - y)!}\theta_1^x\theta_2^y(1 - \theta_1 - \theta_2)^{n-x-y}.$$

To see that (4.2) holds let $Z_1, \cdots, Z_n$ be uniformly distributed on (0, 1) and let U, V denote the number of Z's in the intervals $(0, \theta_1)$ and $(1 - \theta_2, 1)$ respectively. If U', V' are defined analogously with respect to θ_1', θ_2' it is seen that (U, V) has the distribution (4.10) while (U', V') has the corresponding distribution for θ_1', θ_2'. Since $U \leqq U'$, $V \leqq V'$ the validity of (4.2) follows.

The same proof works of course in the case that X and Y are independently distributed, each according to a binomial distribution.

Example 5.2. Let $X_1, \cdots, X_n$ be independently and normally distributed with mean ξ and variance σ^2, and consider the hypothesis $H: \xi \leqq \xi_0, \sigma \leqq \sigma_0$. Since $\bar{X}$ and $S^2 = \sum(X_i - \bar{X})^2$ are sufficient for ξ and σ, we may restrict attention to these statistics. However, if we try to set $X = \bar{X}$, $Y = S^2$, $\theta_1 = \xi$, $\theta_2 = \sigma^2$ we encounter certain difficulties. First the distribution of X does not depend only on θ_1 as we require in Theorem 4.1. While this is not a very important condition of the theorem, a second consideration shows that it is impossible to apply the monotonicity restriction at all to the present set-up. For the joint cumulative distribution function of $\bar{X}$ and S does not satisfy condition (4.2).

This exhibits an unpleasant feature of the present approach. In a given problem it is not known a priori whether there will exist variables X, Y and a choice of the parameters θ_1, θ_2 so that (4.2) will be satisfied. On the other hand, when such variables and parameters have been found, it is not clear that these are the only possible choices. While it would of course be interesting to investigate existence and uniqueness questions, the monotonicity condition is an extraneous restriction anyway, whose suitability must be judged for each problem in terms of the choices for X, Y, θ_1, and θ_2.

In the present case we may take $X = (\bar{X} - \xi_0)/S$, $Y = S^2$, $\theta_1 = (\xi - \xi_0)/\sigma$ and $\theta_2 = \sigma^2$. To check condition (4.2) assume without loss of generality that $\xi_0 = 0$ and let $\theta_1 = \xi/\sigma < \xi'/\sigma' = \theta_1'$, $\sigma < \sigma'$. If $\sigma' = k\,\xi' = k\xi + c$, let $X_i' = kX_i + c$, so that $S' = kS$, $\bar{X}' = k\bar{X} + c$. Since $k > 1$ and $\xi < \xi'/k$ we see that $c > 0$ and $\bar{X}' > \bar{X}$, $Y' > Y$, so that (4.2) follows.

As a last problem let us consider one in which nuisance parameters are present.

Example 5.3. Let $X_1, \cdots, X_m$; $Y_1, \cdots, Y_n$ be independently normally distributed with common variance σ^2; let $E(X_i) = \xi$, $E(Y_j) = \eta$, and consider the hypothesis $H: \xi \leqq \xi_0, \eta \leqq \eta_0$. This time $\bar{X}$, $\bar{Y}$, and $S^2 = \sum(X_i - \bar{X})^2 + \sum(Y_j - \bar{Y})^2$ from a set of sufficient statistics, and again the question arises how to choose X, Y, θ_1, and θ_2. Here the principle of invariance (see [5]) leads to a solution very simply. If one rewrites the hypothesis in the form: $(\xi - \xi_0)/\sigma \leqq 0$, $(\eta - \eta_0)/\sigma \leqq 0$ it is seen that $X = (\bar{X} - \xi_0)/S$, $Y = (\bar{Y} - \eta_0)/S$ constitute a maximal invariant under a suitable group of transformations. The corresponding parameter invariants are of course $\theta_1 = (\xi - \xi_0)/\sigma$, $\theta_2 = (\eta - \eta_0)/\sigma$.

It remains, once more, to check (4.2). For this purpose let ξ, η, ξ', η', σ be numbers such that

$$\frac{\xi}{\sigma} = \theta_1, \qquad \frac{\eta}{\sigma} = \theta_2, \qquad \frac{\xi'}{\sigma} = \theta_1', \qquad \frac{\eta'}{\sigma} = \theta_2'$$

and let $U_1, \cdots, U_m$; $V_1, \cdots, V_n$ be independently normally distributed with common variance σ^2 and means $E(U_i) = \xi$, $E(V_j) = \eta$. If $T^2 = \sum(U_i - \bar{U})^2 + \sum(V_j - \bar{V})^2$, the joint distribution of $U = \bar{U}/T$, $V = \bar{V}/T$ is that corresponding to (θ_1, θ_2). Let $\xi' - \xi = c$, $\eta' - \eta = d$ and let $U_i' = U_i + c$, $V_j' = V_j + d$. Then $U' = \bar{U'}/T$, $V' = \bar{V'}/T$ have the joint distribution corresponding to (θ_1', θ_2'). Also $U' = \bar{U} + c/T > U$ and $V' = \bar{V} + c/T > V$.

Exactly analogously we can treat the hypothesis $H' : \xi/\sigma \leqq \gamma$, $\eta/\sigma \leqq \delta$, and the corresponding problems in which the two variances are not assumed to be equal.

6. A multidecision problem. As was pointed out in the introduction, some of the problems considered here really involve the choice between more than two decisions. We shall now indicate, by discussing an example, one method of treating such multidecision problems through successive reduction to problems of the simpler type.

Let us once more consider the hypothesis H : $\theta_1, \theta_2 \leq 0$ and let us assume that in case of rejection we wish to decide whether $\theta_2 > 0 \geq \theta_1$, $\theta_1 > 0 \geq \theta_2$ or whether θ_1 and θ_2 are both > 0[2]. Let us denote these three regions of the parameter space by ω_1, ω_2, and ω_3 and the associated decisions by d_1, d_2, d_3. The set $\theta_1, \theta_2 \leq 0$ will be denoted by ω_0 and the associated decision of accepting H by d_0.

We shall assume that each of the four pairs of random variables $(\pm X, \pm Y)$ is monotone with respect to the corresponding pair of parameters $(\pm\theta_1, \pm\theta_2)$ and that $p_{\theta,\theta}(x, y)$ is symmetric in x and y so that the region of acceptance is given by

$$\max(x, y) \leqq a. \tag{6.1}$$

We must now consider how to divide up the complementary region between d_1, d_2, and d_3. Here we again impose the natural monotonicity restrictions. We ask that the region for d_1 be monotone non-increasing in x and non-decreasing in y, and that the analogous conditions be satisfied by the regions for d_2 and d_3. Suppose the problem concerns a standard treatment and two new ones, where θ_1 and θ_2 measure in some way the differences between the new treatments and the standard. The circumstances are such that the most serious error consists in incorrectly rejecting the standard treatment in favor of one of the others. By proper choice of a this probability is controlled so that it is not greater than α for all $(\theta_1, \theta_2) \varepsilon \omega_0$.

Next in importance seems to come the possibility of reaching decision d_2 in

[2] A similar multidecision problem has recently been treated by Paulson [6] from a somewhat different point of view.

ω_1 or d_1 in ω_2, but we shall set these aside for the moment. The next important error presumably consists in deciding on d_3 when the parameter point lies either in ω_1 or ω_2. This error can be controlled in the usual manner by making the d_3-region sufficiently small. Thus we may select a number $0 < \beta < 1$ and impose bounds

$$P(d_3 \mid \omega_1) \leqq \beta, \qquad P(d_3 \mid \omega_2) \leqq \beta. \tag{6.2}$$

Subject to these conditions we wish to maximize $P(d_3)$ in ω_3. Let us now restrict attention to d_3-regions of the type $y \geqq g(x)$ or $x \geqq h(y)$ with g and h non-increasing. Then it is seen by the argument used to prove Theorem 4.2 that among all monotone d_3-regions satisfying (6.2) there exists one that uniformly maximizes $P(d_3)$ over ω_3. If $P(X > a \mid \theta_1 = 0) \leqq \beta$ it consists of the points satisfying either $x \geqq a$, $y \geqq b$ or $x \geqq b$, $y \geqq a$ where a, b are determined by

$$P(X \leqq a,\ Y \leqq a \mid \theta_1 = \theta_2 = 0) = 1 - \alpha,$$

$$P(X > b \mid \theta_1 = 0) = \beta.$$

If on the other hand $P(X > a \mid \theta_1 = 0) \geqq \beta$ the optimum d_3-region is given by $x \geqq b$, $y \geqq b$.

Let the remainder of the sample space be divided up symmetrically in the obvious manner between d_1 and d_2. It then follows from monotonicity that $P(d_1 \mid \omega_2)$ and $P(d_2 \mid \omega_1)$ both take on their maximum value at $\theta_1 = \theta_2 = 0$. Hence $P(d_1 \mid \omega_2) \leqq \frac{1}{2}\alpha$, $P(d_2 \mid \omega_1) \leqq \frac{1}{2}\alpha$, so that these errors also are controlled in a satisfactory manner.

7. Convex regions.[3] If we try to apply the results of Section 4 to specific examples, we occasionally find an obstacle in the condition that X and Y should tend in probability to $\underline{x}$ and $\underline{y}$ as $\theta_1 \to \underline{\theta}_1$ and $\theta_2 \to \underline{\theta}_2$, respectively. We shall now show that by restricting the acceptance region to be convex as well as monotone we can prove a result analogous to Theorem 4.1 without assuming degeneracy of the distribution at θ_1 and θ_2.

Let us consider again the joint density $p_{\theta_1,\theta_2}(x, y)$ satisfying (4.2), the hypothesis $H: \theta_1 \leqq \theta_1^*$, $\theta_2 \leqq \theta_2^*$ and the set of alternatives $\theta_1 \geqq \theta_1^{**}$ or $\theta_2 \geqq \theta_2^{**}$. Putting

$$r_\pi(x, y) = \frac{p_{\theta_1^*,\theta_2^*}(x, y)}{\pi p_{\theta_1,\theta_2^{**}}(x, y) + (1 - \pi) p_{\theta_1^{**},\theta_2}(x, y)},$$

we shall assume:

(i) For any $0 < \pi < 1$ and any C the region

$$r_\pi(x, y) \geqq C \tag{7.1}$$

is convex and non-decreasing in x and y.

---

[3] I am indebted to the referee for several valuable suggestions with regard to this section.

(ii) Given any two points (x', y') and (x'', y'') with $x' < x''$, $y' > y''$ there exists $0 < \pi < 1$ and C such that both points lie on the boundary of the set of points (x, y) satisfying (7.1).

(iii) $p_{\theta_1, \theta_2}(x, y) > 0$ for all θ_1, θ_2, x, y under consideration.

Consider now the set of alternatives $\omega' : \theta_1 \geqq \underline{\theta}_1$, $\theta_2 \geqq \theta_2^{**}$ or $\theta_2 \geqq \underline{\theta}_2$, $\theta_1 \geqq \theta_1^{**}$ where $\underline{\theta}_1$ and $\underline{\theta}_2$ may now be any numbers less than θ_1^* and θ_2^*, respectively. Let a and b be determined by

$$P(X \leqq a, Y \leqq b \mid \theta_1^*, \theta_2^*) = 1 - \alpha,$$

$$P(X \leqq a, Y \leqq b \mid \underline{\theta}_1, \theta_2^{**}) = P(X \leqq a, Y \leqq b \mid \theta_1^{**}, \underline{\theta}_2).$$

Then the acceptance region S: $x \leqq a$, $y \leqq b$ maximizes $\inf_{\omega'} \beta(\theta_1, \theta_2)$ among all monotone and convex level α tests.

For let S' be any other acceptance region satisfying the conditions that have been imposed. The boundary curves of S and S' have in common either one or

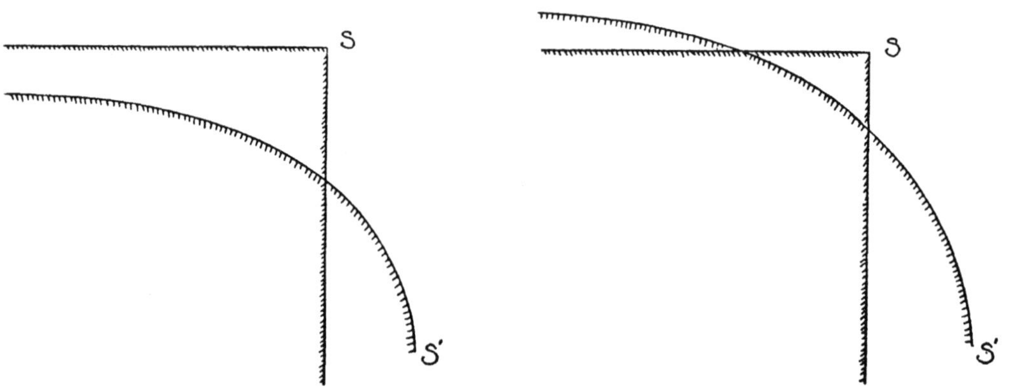

FIG. 1.

two points or an interval. Let us consider the case of two points, say (x', y') and (x'', y''). We may then assume $x' < x''$, $y' > y''$ so that there exist π and C such that the boundary of (7.1) passes through these two points. From (i) it follows that $r(x, y) \geqq C$ for all points in $S \wedge \bar{S}'$ and $\leqq C$ in $\bar{S} \wedge S'$. Since we have $P(S') \leqq P(S) = \alpha$ when the density of X, Y is $p_{\theta_1^*, \theta_2^*}(x, y)$ it follows from the fundamental lemma of Neyman and Pearson that $P(S') > P(S)$ when the density is given by $\pi p_{\theta_1^*, \theta_2^{**}}(x, y) + (1 - \pi) p_{\theta_1^{**}, \theta_2}(x, y)$. It is therefore impossible that $P(S') \geqq P(S)$ for both $(\underline{\theta}_1, \theta_2^{**})$ and $(\theta_2^{**}, \underline{\theta}_2)$ as was to be proved.

The same argument applies also in the case that the boundaries of S and S' have an interval in common. Consider finally the case of one common point, say (a, y_0). For each n let (π_n, C_n) satisfy (ii) for $(-n, b)$ and (a, y_0). Then

$$0 \leqq C_n \leqq \max \left\{ \frac{p_{\theta_1^*, \theta_2^*}(a, y_0)}{p_{\underline{\theta}_1, \theta_2^{**}}(a, y_0)}, \frac{p_{\theta_1^*, \theta_2^*}(a, y_0)}{p_{\theta_1^{**}, \underline{\theta}_2}(a, y_0)} \right\},$$

so that there is a subsequence of $\{(\pi_n, C_n)\}$ which converges, say, $\pi_n \to \pi^*$, $C_n \to$

C^*. It is easily seen from the monotonicity and convexity of the regions $r_{\pi_n}(x, y) \geqq C_n$ that the boundary of $r_{\pi^*}(x, y) \geqq C^*$ is the line $y = y_0$. The remainder of the argument is completely analogous to the two point case.

As an example let

$$(7.2) \qquad p_{\theta_1,\theta_2}(x, y) = C(\theta_1, \theta_2)\, e^{\theta_1 x + \theta_2 y}\, h(x, y),$$

and assume that (4.2) holds. Suppose that a priori lower bounds $\underline{\theta}_1$, $\underline{\theta}_2$ are given for θ_1, θ_2 such that $p_{\underline{\theta}_1,\underline{\theta}_2}(x, y)$ is again a density. If we let $\theta_1^{**} = \theta_1^*$, $\theta^{**} = \theta_2^*$, the region (7.1) becomes

$$a\, e^{(\theta_1 - \dot{\theta}_1)x} + b\, e^{(\theta_2 - \dot{\theta}_2)y} \leqq k$$

and conditions (i), (ii), and (iii) are easily checked.

8. Two-sided problems. We shall discuss only one rather trivial two-sided problem, which is enough however to indicate that the type of result one obtains here is entirely different from what we found in the one-sided case.

Let X and Y be independently, normally distributed with unit variance and means ξ and η, respectively, and consider the hypothesis $H: \xi = \eta = 0$. We shall determine the test ϕ that maximizes $\inf_{\omega'} \beta(\xi, \eta)$ where ω' is the set of points for which either $|\xi|$ or $|\eta|$ is $\geqq \gamma$, $(\gamma > 0)$. Any reasonable test for this problem would presumably attain its minimum power over ω' at the four points $(0, \gamma)$, $(0, -\gamma)$, $(\gamma, 0)$, $(-\gamma, 0)$. We therefore expect ϕ to be the most powerful test of H against the simple alternative that assigns probability ¼ to each of these 4 points. The region of acceptance for this problem is given by S:

$$e^{\gamma x} + e^{-\gamma x} + e^{\gamma y} + e^{-\gamma y} \leqq k.$$

It is easily checked that this has the following properties:

(i) S is convex.

(ii) If $0 \leqq x' \leqq x$, $0 \leqq y' \leqq y$ and $(x, y) \varepsilon S$, the point (x', y') also lies in S.

(iii) For any fixed η_0 the probability of S decreases with $|\xi|$ and for fixed ξ_0 decreases in $|\eta|$.

From (iii) it follows that ϕ is the test we are looking for, and it seems to be entirely satisfactory. In fact, if we utilize the symmetry of the situation to reduce the variables to $|X|$, $|Y|$ and the parameters to $|\xi|$, $|\eta|$ we are faced essentially with a one-sided situation and it is seen from (i) and (ii) that the acceptance region, when interpreted in this way, is both monotone and convex.

9. A general concept of monotonicity. In Sections 4–6 we made use of the notion of monotonicity, and we shall conclude this paper by indicating how this concept may be extended to the general decision problem.

Suppose that there is defined a partial ordering $\leqq$ in the sample space and a partial ordering $\dot{\leqq}$ in the parameter-space. In analogy to condition (4.2) we shall assume that if $\mathfrak{W}$ is any monotone non-decreasing region in the sample space we have for any two parameter-points $\theta \dot{\leqq} \theta' : P_\theta(\mathfrak{W}) \leqq P_{\theta'}(\mathfrak{W})$.

Suppose that there is also defined a partial ordering in the decision space, to be denoted by $\underset{\cdot}{\leqq}$. We shall assume that the loss function W satisfies the conditions:

(i) $d_1 \underset{\cdot}{\leqq} d_2 \underset{\cdot}{\leqq} d_3$ and $W(\theta, d_1) < W(\theta, d_2)$ implies $W(\theta, d_2) \leqq W(\theta, d_3)$,

(ii) $\theta_1 \dot{\leqq} \theta_2 \dot{\leqq} \theta_3$ and $W(\theta_1, d) < W(\theta_2, d)$ implies $W(\theta_2, d) \leqq W(\theta_3, d)$. Under these assumptions it seems natural to restrict consideration to monotone decision functions, where we shall call δ monotone if $x \leqq x'$ implies $\delta(x) \underset{\cdot}{\leqq} \delta(x')$.

REFERENCES

[1] J. Neyman and E. S. Pearson, "Contributions to the theory of testing statistical hypotheses. Part III." *Statistical Research Memoirs*, Vol. 2 (1938), pp. 36–57.

[2] Stanley L. Isaacson, "On the theory of unbiased tests of simple statistical hypotheses specifying the values of two or more parameters," *Annals of Math. Stat.*, Vol. 22 (1951), pp. 217–234.

[3] A. Wald, *Statistical Decision Functions*, John Wiley and Sons, New York, 1950.

[4] Mark W. Eudey, "On the treatment of discontinuous random variables," *Technical Report No. 13, ONR Project NR-042-036*, 1949.

[5] E. L. Lehmann, "Some principles of the theory of testing hypotheses," *Annals of Math. Stat.*, Vol. 21 (1950), pp. 1–26.

[6] Edward Paulson, "On the comparison of several experimental categories with a control," *Annals of Math. Stat.*, Vol. 23 (1952), pp. 239–246.

ORDERED FAMILIES OF DISTRIBUTIONS[1]

By E. L. Lehmann

University of California, Berkeley

1. Summary and introduction. A comparison is made of several definitions of ordered sets of distributions, some of which were introduced earlier by the author [7], [8] and by Rubin [10]. These definitions attempt to make precise the intuitive notion that large values of the parameter which labels the distributions go together with large values of the random variables themselves. Of the various definitions discussed the combination of two, (B) and (C) of Section 2, appears to be statistically most meaningful. In Section 3 it is shown that this ordering implies monotonicity for the power function of sequential probability ratio tests. In Section 4 the results are applied to obtaining tests that give a certain guaranteed power with a minimum number of observations. Finally, in Section 5, certain consequences are derived regarding the comparability of experiments in the sense of Blackwell [1].

2. Some definitions of order. Let $X = (X_1, \cdots, X_n)$ be a random vector with probability distribution P_θ, depending on a real parameter θ. In the problems occurring in applications such distributions are usually ordered in the sense, roughly speaking, that large values of θ lead on the whole to large values of the X's. This intuitive notion can be given a precise mathematical meaning in various ways, some of which we shall now describe.

(A′) For any $\theta < \theta'$ there exists a vector-valued function $f = (f_1, \cdots, f_n)$, depending in general on θ and θ', such that[2]

(i) $x \leqq f(x)$,

(ii) if X has distribution P_θ, then the distribution of $(f_1(X), \cdots, f_n(X))$ is $P_{\theta'}$.

This condition, which was used by the author in [7] and [8], states that one can pass from a random vector with distribution P_θ to one with distribution $P_{\theta'}$ by a transformation which increases all of the components of the vector. An example is the case of a location parameter θ where one can then put

$$f_j(x) = x_j + \theta' - \theta.$$

For technical reasons the following slightly weaker condition, which was given in [8], is sometimes more convenient.

(A) There exists a random vector Z and functions $g = (g_1, \cdots, g_n)$, $g' = (g_1', \cdots, g_n')$ such that

(i) $g(z) \leqq g'(z)$ for all z,

(ii) the distributions of $g(Z)$ and $g'(Z)$ are P_θ and $P_{\theta'}$ respectively.

Received September 14, 1954.

[1] This paper was prepared with the partial support of the Office of Naval Research.

[2] Here, as throughout, an inequality between two vectors means that this inequality holds for all the components.

A function ϕ defined on an n-dimensional euclidean space is said to be increasing if $x \leqq x'$ implies $\phi(x) \leqq \phi(x')$; a set S is said to be *increasing* if its characteristic set function is, that is, if $x \varepsilon S$, $x \leqq x'$ implies $x' \varepsilon S$.

Condition (A') is the special case in which $Z = X$, g is the identity function, and $g' = f$. Condition (A) clearly implies:

(B) If $\theta < \theta'$, then for every increasing set[3] S

$$P_\theta(S) \leqq P_{\theta'}(S), \tag{2.1}$$

and also the seemingly stronger

(B') If $\theta < \theta'$, then for every increasing function[3] $\phi(x_1, \cdots, x_n)$

$$E_\theta \phi(X) \leqq E_{\theta'}\phi(X). \tag{2.2}$$

Actually, (B) and (B') are equivalent. To see this, assume without loss of generality that ϕ is non-negative, and consider the approximation of ϕ by a sequence of nondecreasing simple functions

$$\phi_n(x) = \begin{cases} \dfrac{i-1}{2^n} & \text{for} \quad x \varepsilon S_i^{(n)} \\ n \quad \text{for} & x \varepsilon S_N^{(n)} \end{cases}$$

where

$$S_i^{(n)} = \left\{x: \frac{i-1}{2^n} \leqq \phi(x) < \frac{i}{2^n}\right\}, \qquad i = 1, \cdots, n \cdot 2^n,$$

$$S_N^{(n)} = \{x: \phi(x) > n\} \qquad N = n \cdot 2^n + 1.$$

Then it is seen that $E_\theta \phi_n(X)$ can be written in the form

$$\sum_{i=1}^{N} a_i P_\theta(S_i^{(n)} + S_{i+1}^{(n)} + \cdots + S_N^{(n)})$$

where the a_i are $\geqq 0$. But each set $S_i^{(n)} + \cdots + S_N^{(n)}$ is increasing, and it follows from condition (B) that $E_\theta \phi_n(X) \leqq E_{\theta'}\phi_n(X)$ and hence $E_\theta \phi(X) \leqq E_{\theta'}\phi(X)$.

A somewhat different condition supposes that all of the distributions P_θ possess probability densities with respect to a common σ-finite measure μ.

(C) If $\theta < \theta'$, the probability ratio

$$\frac{p_{\theta'}(x)}{p_\theta(x)} \tag{2.3}$$

is increasing.[4]

---

[3] Throughout, we restrict consideration to sets and functions which are Borel measurable.

[4] Probability densities being defined only up to sets of measure zero, condition (C) and similar conditions to be considered later, for example in connection with Theorem 3, should be interpreted to mean that there exist versions of these densities satisfying the condition in question. Furthermore, the condition is not meant to carry any implication as regards the points x at which both densities vanish.

Slightly more generally it is enough to assume the existence of real-valued functions $t_1, \cdots, t_k$ such that

$$\frac{p_{\theta'}(x)}{p_\theta(x)} = \frac{f_{\theta'}(t_1(x), \cdots, t_k(x))}{f_\theta(t_1(x), \cdots, t_k(x))}.$$

Then $t_1(x), \cdots, t_k(x)$ are sufficient statistics, and without loss of generality $f_\theta(t_1, \cdots, t_k)$ may be taken to be the generalized probability density of $T = (t_1(X), \cdots, t_k(X))$. Condition (C) is therefore essentially a generalization of one investigated by H. Rubin [10] to the effect that the ratio (2.3) is a monotone function of a real-valued statistic. We note the obvious lemma:

LEMMA 1. *If for each x the density $p_\theta(x)$ is a differentiable function of θ, then a necessary and sufficient condition for* (C) *to hold is that $\partial / \partial\theta(\log p_\theta(x))$ be nondecreasing.*

It was pointed out above that (A) implies (B). The following examples show that (A) and (B) are not equivalent, and that in general (C) is not directly comparable to (A) or (B).

The situation is summarized in Table I in which the sign + or − indicates that the condition in question does or does not hold.

TABLE I

(A)	(B)	(C)	
+	+	+	Evidently possible
+	+	−	Example 2.1
+	−	+	Impossible since (A) implies (B)
+	−	−	Impossible since (A) implies (B)
−	+	+	Example 2.2
−	+	−	Example 2.3
−	−	+	Example 2.4
−	−	−	Evidently possible

EXAMPLE 2.1. Let X be a random variable having a Cauchy distribution, with density

$$p_\theta(x) = \frac{1}{\pi}\frac{1}{1 + (x - \theta)^2}.$$

Then if $\theta < \theta'$, the transformation $f(x) = x + (\theta' - \theta)$ shows that (A) holds, and hence also (B). On the other hand, the ratio $p_{\theta'}(x) / p_\theta(x) \to 1$ as $x \to \pm\infty$, and hence obviously is not monotone.

EXAMPLE 2.2. Let $n = 2$, and let the probability be concentrated on the four squares $A, \cdots, D$ indicated in Fig. 1a. The conditional distribution over each of the four squares is assumed uniform under both θ and θ'. The probabilities of the squares are given in Table II.

It is easily checked that (B) holds. Also

$$r(x_1, x_2) = \frac{P_{\theta'}(x_1, x_2)}{P_\theta(x_1, x_2)}$$

is larger in A than in either B, C, or D, so that (C) is satisfied. On the other hand, if there existed vectors $g(Z)$ and $g'(Z)$ with distributions P_θ and $P_{\theta'}$, and such that $g(z) \leqq g'(z)$ for all z, then $g'(z) \varepsilon C$ would imply $g(z) \varepsilon C$, and hence $P_{\theta'}(C) \leqq P_\theta(C)$. Thus (A) does not hold.

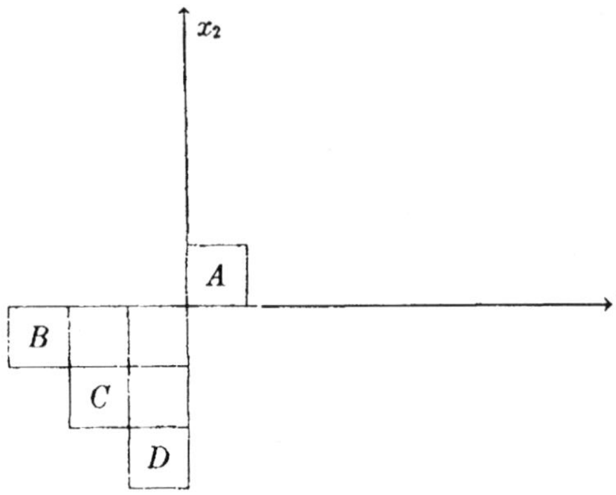

FIG. 1a

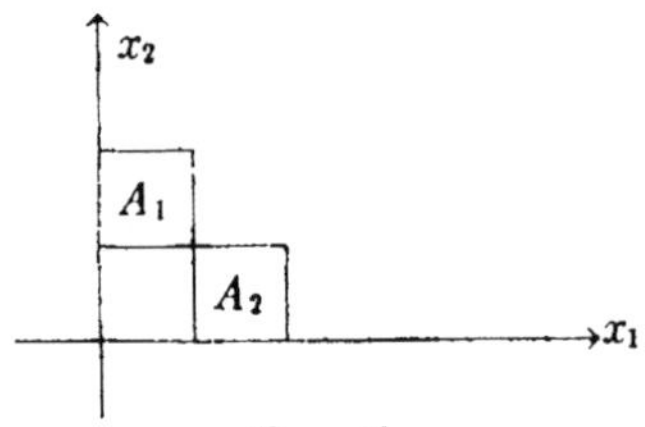

FIG. 1b

TABLE II

	P_θ	$P_{\theta'}$
A	3/16	12/16
B	6/16	1/16
C	1/16	2/16
D	6/16	1/16

Here the parameter θ takes on only two values. We obtain an example in which θ ranges over a continuum by means of the following lemma.

LEMMA 2. *Let P_0 and P_1 be two probability distributions, and let*

$$P_\theta = \theta P_1 + (1 - \theta)P_0, \qquad 0 \leqq \theta \leqq 1.$$

Then each of the conditions (A), (B), *and* (C) *holds for all* $0 \leqq \theta < \theta' \leqq 1$ *if and only if it holds for the pair* $\theta = 0$, $\theta' = 1$.

PROOF. A direct calculation shows that if (B) or (C) holds for the pair $\theta = 0$, $\theta' = 1$, it holds for all $\theta < \theta'$. To prove this for (A), suppose that $f_i(Z)$ has distribution P_i $(i = 0, 1)$ and that $f_0(z) \leqq f_1(z)$ for all z. Consider a random variable U, uniformly distributed on $[0, 1]$, and let

$$X_\theta = g(U, Z) = \begin{cases} f_0(Z) & \text{if} \quad U \leqq \theta \\ f_1(Z) & \text{if} \quad \theta < U \end{cases}$$

$$X_{\theta'} = g'(U, Z) = \begin{cases} f_0(Z) & \text{if} \quad U \leqq \theta' \\ f_1(Z) & \text{if} \quad \theta' < U. \end{cases}$$

Then X_θ and $X_{\theta'}$ have distributions P_θ and $P_{\theta'}$ respectively, and $g(u, z) \leqq g'(u, z)$ for all u and z.

The required example is now obtained by taking for P_0 and P_1 the probabilities denoted in the example by P_θ and $P_{\theta'}$, and by defining P_θ as in the lemma. This remark applies also to the examples that follow.

EXAMPLE 2.3. In Fig. 1a of Example 2.2 replace the square A by two squares A_1, A_2 as indicated in Fig. 1b. Let the probabilities $P_\theta(A) = \frac{3}{16}$ and $P_{\theta'}(A) = \frac{12}{16}$ be divided among A_1 and A_2 so that

$$P_\theta(A_1) = \tfrac{2}{16}, \qquad P_\theta(A_2) = \tfrac{1}{16}; \qquad P_{\theta'}(A_1) = \tfrac{3}{16}, \qquad P_{\theta'}(A_2) = \tfrac{9}{16}.$$

Then as before (A) does not hold and (B) does. However, (C) now also does not hold since the ratio $r(x_1, x_2)$ has the value 2 in region C but only the value $\frac{3}{2}$ in region A_1.

EXAMPLE 2.4. The (x_1, x_2)-plane is divided into 6 parts A_1, A_2, B_1, B_2, C_1, C_2 as indicated in Fig. 2. The probability ratio and the probabilities under θ and θ' of the six sets are given in Table III.

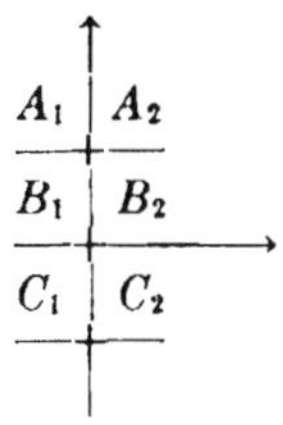

FIG. 2

TABLE III

	P_θ	$P_{\theta'}$	$r(x_1, x_2)$
A_1	.27	.45	5/3
A_2	.03	.05	
B_1	.30	.30	1
B_2	.10	.10	
C_1	.03	.01	1/3
C_2	.27	.09	

It is seen that (C) holds. On the other hand $P_\theta(A_2 + B_2 + C_2) = .40$, $P_{\theta'}(A_2 + B_2 + C_2) = .24$, so that (B) and hence (A) is not satisfied.

The rather chaotic state of things indicated by these examples is replaced by a much simpler one if the components of the vector X are independent, though not necessarily identically distributed.

THEOREM 1. *If $X_1, \cdots, X_n$ are independent, then*

$$(C) \rightarrow (B) \rightleftarrows (A).$$

PROOF. Consider first the case $n = 1$. Suppose that (B) holds, and let F_θ and $F_{\theta'}$ denote the cumulative distribution functions of the distributions P_θ and $P_{\theta'}$ respectively. If $g(z) = F_\theta^{-1}(z)$, $g'(z) = F_{\theta'}^{-1}(z)$, and Z is uniformly distributed on $[0, 1]$, then $g(Z)$ and $g'(Z)$ have the distributions F_θ and $F_{\theta'}$ respectively. That $g(z) \leqq g'(z)$ follows from the fact that $F_{\theta'}(x) \leqq F_\theta(x)$ for all x since (B) is assumed to hold.

To show that (C) implies (B) when $n = 1$, let $r(x) = p_{\theta'}(x) / p_\theta(x)$. Given any constant k there exists a number ρ between 0 and 1 such that

$$P_\theta\{X > k\} = P_\theta\{r(X) > r(k)\} + \rho P_\theta\{r(X) = r(k)\}. \tag{2.4}$$

It is then easily seen that (2.4) holds, with the same ρ, also when θ is replaced by θ'. Consider now the problem of testing θ against θ', at the level of significance α, which is the value of the probability (2.4). Then the critical function, given by

$$\phi(x) = \begin{cases} 1 & \text{if} \quad r(x) > r(k) \\ \rho & \text{if} \quad r(x) = r(k) \end{cases}$$

has size α, and is the most powerful level α test for testing θ against θ'. It follows by comparison with the test $\phi(x) \equiv \alpha$ that

$$P_\theta\{r(X) > r(k)\} + \rho P_\theta\{r(X) = r(k)\} \leqq P_{\theta'}\{r(X) > r(k)\} + \rho P_{\theta'}\{r(X) = r(k)\}$$

and hence that for each k,

$$P_\theta\{X > k\} \leqq P_{\theta'}\{X > k\}.$$

The same relation for $X \geqq k$ follows by a limiting argument.

Suppose now that $n > 1$ and that (B) holds. Then in particular

$$P_\theta\{X_i > k\} \leqq P_{\theta'}\{X_i > k\} \qquad \text{for all } k \tag{2.5}$$

and it follows from the case $n = 1$ that (A) is satisfied.

Finally let $n > 1$, $p_\theta(x_1, \cdots, x_n) = f_\theta^{(1)}(x_1) \cdots f_\theta^{(n)}(x_n)$, and assume (C) to be satisfied. Then for each i, $f_{\theta'}^{(i)}(x_i) / f_\theta^{(i)}(x_i)$ is nondecreasing in x_i as is seen by holding the other coordinates fixed. It follows from the case $n = 1$ that (2.5) holds, and the proof is complete.

We shall in the present paper be mainly concerned with families of distributions that are ordered in the sense that both conditions (B) and (C) hold. It is a consequence of Theorem 1 that this is the case in particular if $X_1, \cdots, X_n$ is a sample from a univariate distribution with density $f_\theta(x)$ where for $\theta < \theta'$ the ratio $f_{\theta'}(x) \,/\, f_\theta(x)$ is nondecreasing in x.

3. Monotonicity of the power function of some sequential tests. As a first application we consider the problem of testing sequentially the hypothesis $\theta \leqq \theta_0$ against the alternatives $\theta \geqq \theta_1$, where $\theta_0 < \theta_1$. Wald proposes as a solution the sequential probability ratio test, according to which observations are taken as long as

$$a < \sum_{i=1}^{m} \log \frac{p_{\theta_1}(x_i)}{p_{\theta_0}(x_i)} < b. \tag{3.1}$$

At the first violation of (3.1) the hypothesis is accepted or rejected according as the probability ratio is then $\leqq a$ or $\geqq b$.

Wald mentions ([12], p. 73) that in many important special cases the power function $\beta(\theta)$ of this test is an increasing function of θ. If a and b are adjusted so that $\beta(\theta_0) = \alpha$ and $\beta(\theta_1) = \beta$, this then implies that $\beta(\theta) \leqq \alpha$ for $\theta \leqq \theta_0$, and $\beta(\theta) \geqq \beta$ for $\theta \geqq \theta_1$, and hence satisfactory control of the probabilities of both kinds of error. The following result establishes such monotonicity for a large class of problems. The test treated is the *generalized probability ratio test*, where in (3.1) the constant boundaries a, b are replaced by variable boundaries, say a_m and b_m, and where some of the strong or weak inequality signs defining the test may be replaced by weak or strong ones respectively. This includes in particular the case of a single sample, or, more generally, of truncated sampling schemes if at some stage $a_m = b_m$.

THEOREM 2. *Let $X_1, X_2, \cdots$ be a sequence of random variables such that for all m the joint density $p_\theta^{(m)}(x_1, \cdots, x_m)$ of $X_1, \cdots, X_m$ satisfies* (B) *and* (C). *Then the power function $\beta(\theta)$ of any generalized probability ratio test is nondecreasing.*

PROOF. Let

$$z_m = \frac{p_{\theta_1}^{(m)}(x_1, \cdots, x_m)}{p_{\theta_0}^{(m)}(x_1, \cdots, x_m)}.$$

Then for $\theta < \theta'$ we have that for all k

$$P_\theta\{Z_m > k\} \leqq P_{\theta'}\{Z_m > k\}.$$

This follows from the fact that by (C), the set

$$\left\{(x_1, \cdots, x_m)\colon \frac{p_{\theta_1}^{(m)}(x_1, \cdots, x_m)}{p_{\theta_0}^{(m)}(x_1, \cdots, x_m)} > k\right\}$$

is increasing, and that by (B) the probability of an increasing set is monotone in θ. Since Z_m is real-valued, there exists by Theorem 1 a real-valued function f_m such that $f_m(z) \geqq z$ for all z, and the distribution of $f_m(Z_m)$ is given by

$$P_\theta\{f_m(Z_m) \leqq u\} = P_{\theta'}\{Z_m \leqq u\} \qquad \text{for all } u.$$

Consider now the points $(1, Z_1)$, $(2, Z_2)$, $\cdots$ and the path they describe in the (i, Z)-plane. With the generalized probability ratio test, observations are taken as long as this path lies within a certain prescribed band, and the hypothesis is accepted or rejected according as the path leaves the band for the first time through the upper or lower boundary. Now the path $\mathcal{C}'$ formed by the points $(1, f_1(Z_1))$, $(2, f_2(Z_2))$, $\cdots$ lies entirely above the path $\mathcal{C}$ formed by the points (i, Z_i), and hence whenever $\mathcal{C}$ leads to rejection by leaving the band through the upper boundary, so does $\mathcal{C}'$. But the probability of $\mathcal{C}$ and $\mathcal{C}'$ leading to rejection is exactly $\beta(\theta)$ and $\beta(\theta')$ respectively, which completes the proof.

It may be worth noting that use was made of condition (C) only for the pair of values (θ_0, θ_1).

Some simple applications of this theorem are to cases in which $X_1, X_2, \cdots$ are independently, identically distributed random variables, with probability density $f_\theta(x)$ for which $f_{\theta'}(x) / f_\theta(x)$ is nondecreasing in x whenever $\theta < \theta'$. In all such cases it follows from Theorems 1 and 2 that the power function of a generalized probability ratio test is nondecreasing.

EXAMPLE 3.1. Let the density of the X's be given by

$$f_\theta(x) = \theta g(x) + (1 - \theta)h(x), \qquad 0 \leqq \theta \leqq 1.$$

This is the situation in which the population under investigation is a mixture of two populations. In an experiment, for example, there may be the possibility of "gross errors" in addition to normal errors. Or it may be the problem of detecting the frequency of mutation of some gene, the effect of which is not directly observable. Since

$$\frac{f_{\theta'}(x)}{f_\theta(x)} = \frac{\theta'\left[\dfrac{g(x)}{h(x)} - 1\right] + 1}{\theta\left[\dfrac{g(x)}{h(x)} - 1\right] + 1},$$

it is seen that for $\theta < \theta'$ this ratio is increasing in x provided this is the case for $g(x) / h(x)$.

EXAMPLE 3.2. Let

$$f_\theta(x) = g(x - \theta). \tag{3.2}$$

Then (A) clearly holds without any restriction on the function g. On the other hand, (C) is exactly the condition of twice positivity of Schoenberg [11], a real-valued measurable function g being m times positive if, for every $k(=1, \cdots, m)$, $u_1 < u_2 < \cdots < u_k$, $v_1 < v_2 < \cdots < v_k$ implies that the determinant

$$\det \| g(u_i - v_j) \| \geqq 0.$$

A trivial specialization of Lemma 1 of [11] shows that a probability density g is twice positive if and only if (i) its domain of positivity is an interval (a, b), $-\infty \leqq a < b \leqq \infty$, (ii) the function $-\log g$ is convex (and hence automatically continuous) in the open interval (a, b), and if g is correctly defined at the end points, as can always be achieved.[5]

As specific examples, let

$$g_1(x) = \frac{1}{\sqrt{2\pi}} e^{-\frac{1}{2}x^2} \qquad \text{(Normal)}$$

$$g_2(x) = e^{-x-e^{-x}}$$

$$g_3(x) = e^{-x} \quad \text{for} \quad x \geqq 0 \qquad \text{(Exponential)}$$

$$g_4(x) = 1 \quad \text{for} \quad -\tfrac{1}{2} \leqq x \leqq \tfrac{1}{2} \qquad \text{(Rectangular)}$$

$$g_5(x) = \frac{e^{-x}}{(1 + e^{-x})^2} \qquad \text{(Logistic)}$$

$$g_6(x) = \tfrac{1}{2}e^{-|x|} \qquad \text{(Laplace)}$$

$$g_7(x) = \frac{1}{\pi} \frac{1}{1 + x^2} \qquad \text{(Cauchy)}$$

In the first six of these cases $-\log g$ is convex while in the last it is not. A general class of densities of form (3.2) that satisfy condition (C) is formed by the cases in which g is a Polya frequency function. This class was defined and investigated by Schoenberg (see for example [11]) who showed these functions to be totally positive (that is, k times positive for all $k = 1, 2, \cdots$) and hence in particular twice positive.

EXAMPLE 3.3. Let

$$f_\theta(x) = \frac{1}{\theta} g\left(\frac{x}{\theta}\right) \tag{3.3}$$

where g is an even function, and where, without loss of generality, one may restrict x to be nonnegative since the absolute values $|X_1|, |X_2|, \cdots$ form a set of sufficient statistics for θ. It is then seen as in the previous example, or can be deduced from it by transforming to $Y = \log X$, that (C) holds if and only if the domain of positivity of g is an interval (a, b) and $-\log g(e^x)$ is convex for $\log a < x < \log b$. This holds in the cases g_1, g_4, g_6 and g_7 of the previous example. Since the convexity of $-\log g(x)$ implies that of $-\log g(e^x)$ but not conversely, condition (C) in the case of an even function is more restrictive for a location parameter than for a scale parameter.

EXAMPLE 3.4. A well-known example, which satisfies also the stronger conditions investigated by Rubin [10], is that of an exponential family, with

$$f_\theta(x) = a(\theta)e^{xb(\theta)}h(x)$$

[5] The same condition was encountered in a slightly different context by Ruist, "Comparison for tests of nonparametric hypotheses," *Arkiv. för Mathematik*, Vol. 3 (1954), pp. 133–163. Logarithmically convex functions have also been considered by Artin in his "Einführung in die Theorie der Gammafunktion," *Hamburger Math. Einzelschriften*, No. 11, B. G. Teubner, Leipzig, (1931).

where $b(\theta)$ is a strictly increasing function of θ. This includes among others the binomial and Poisson families of distributions. It also includes the cases $f_\theta(y) = g(y - \theta)$ where g is one of the densities g_1, g_2 of Example 3.2, in the first case with $y = x$ and in the second with $y = e^{-e^{-x}}$. Still further special cases are obtained by putting $f_\theta(y) = (1/\theta)g(y/\theta)$ with g one of the functions g_1, g_3 or g_6 of Example 3.2 and $y = -x^2$, $y = -x$ and $y = -|x|$ respectively.

Without going into details we mention as further application of Theorem 2 some sequential tests of composite hypothesis, discussed among others by Wald [12], Cox [3], Johnson [4], such as the sequential t-tests or sequential analysis of variance tests. In those cases the variables $X_1, X_2, \cdots$ are dependent. That (C) holds follows from the fact that the noncentral t- and F- distributions satisfy (C) (see Section 4, Examples 4.3 and 4.4), while (B) is easily checked in all these cases.

4. Tests with guaranteed power. As another application consider the problem of testing that $\theta \leqq \theta_0$ against the alternatives $\theta \geqq \theta_1$ on the basis of $X = (X_1, \cdots, X_n)$. It is desired to find that test which, subject to

$$\beta(\theta) \leqq \alpha \qquad \text{for} \quad \theta \leqq \theta_0, \tag{4.1}$$

maximizes the minimum power over $\theta \geqq \theta_1$, that is, which gives the greatest possible guaranteed power in that range. The solution to this problem is to determine a least favorable pair of distributions λ_0, λ_1 over the sets $\omega_0 = \{\theta : \theta \leqq \theta_0\}$ and $\omega_1 = \{\theta : \theta \geqq \theta_1\}$, and to reject the hypothesis when

$$\frac{\int_{\omega_1} p_\theta(x_1, \cdots, x_n)\, d\lambda_1(\theta)}{\int_{\omega_0} p_\theta(x_1, \cdots, x_n)\, d\lambda_0(\theta)} \geq k. \tag{4.2}$$

If the family of distributions is ordered, it seems reasonable to expect that the least favorable distributions are those assigning probability 1 to the points θ_0 and θ_1 respectively, in which case (4.1) reduces to the probability ratio test

$$\frac{p_{\theta_1}(x_1, \cdots, x_n)}{p_{\theta_0}(x_1, \cdots, x_n)} > k. \tag{4.3}$$

It follows from Theorem 8.3 of [9] that (4.3) is the solution to the stated problem provided $\beta(\theta) \leqq \beta(\theta_0) = \alpha$ for $\theta \leqq \theta_0$ and $\beta(\theta) \geqq \beta(\theta_1)$ for $\theta \geqq \theta_1$. But this is certainly the case if $\beta(\theta)$ is nondecreasing. A sufficient condition for this is that both (B) and (C) hold since then the critical region (4.3) is increasing and hence its probability is a nondecreasing function of θ. (Actually, this is a special case of Theorem 2.) That (B) alone is not enough is seen, for example, in the Cauchy case. If $X_1, \cdots, X_n$ are independently, identically distributed with density $\pi^{-1} / (1 + (x - \theta)^2)$, it is seen that the region in which

$$\prod_{i=1}^{n} \frac{1 + (x_i - \theta_1)^2}{1 + (x_i - \theta_0)^2} > k$$

is a bounded set in n-space. Its probability therefore tends to zero as $\theta \to \infty$.

A limiting case, as $\theta_1 \to \theta_0$, of the property that the minimum power over $\theta \geqq \theta_1$ be a maximum is that of locally maximum power. Here one seeks the test which maximizes the derivative $\beta'(\theta_0)$ of the power function at $\theta = \theta_0$. If for any critical region w, the integral

$$\beta(\theta) = \int_w p_\theta(x)\, d\mu(x)$$

can be differentiated under the integral sign with respect to θ, the problem becomes that of maximizing

$$\beta'(\theta_0) = \int \frac{\partial}{\partial\theta} \log p_\theta(x) \bigg|_{\theta=\theta_0} p_{\theta_0}(x)\, d\mu(x)$$

subject to (4.1). If we again tentatively replace (4.1) by the side condition $\beta(\theta_0) = \alpha$, the best critical region by the Neyman-Pearson fundamental lemma is given by

$$\frac{\partial}{\partial\theta} \log p_\theta(x) \bigg|_{\theta=\theta_0} \geqq k. \tag{4.4}$$

If (C) holds, it was seen earlier that the left-hand side of (4.4) is a nondecreasing function of the x's. Hence it follows from (B) that $\beta(\theta) \leqq \beta(\theta_0) = \alpha$ for $\theta \leqq \theta_0$ and therefore that (4.4) is the desired result.

Let $X_1, \cdots, X_n$ be independently and identically distributed with density $f_\theta(x)$, which is either a mixture of two densities in proportion $\theta : 1 - \theta$, or where θ is a location or scale parameter, and suppose that the conditions of Examples 3.1–3.3 respectively are satisfied. Then the test maximizing the minimum power over $\theta \geqq \theta_0$ is given by the rejection region

$$\frac{f_{\theta_1}(x_1) \cdots f_{\theta_1}(x_n)}{f_{\theta_0}(x_1) \cdots f_{\theta_0}(x_n)} \geqq k \tag{4.5}$$

and the test maximizing the power locally by

$$\sum_{i=1}^{n} \frac{\partial}{\partial\theta} \log f_\theta(x_i) \bigg|_{\theta=\theta_0} \geqq k'. \tag{4.6}$$

A uniformly most powerful one-sided test does of course usually not exist. A notable exception is the well-known case of the exponential family of Example 3.4.

As an illustration consider the case that $f_\theta(x) = g(x - \theta)$ where g is one of the densities g_i ($i = 1, \cdots, 7$) of Example 3.2, and that $\theta_0 = 0$. For $i = 1, 2$ these are exponential families, and the test given by (4.6) is uniformly most powerful against the alternatives $\theta \geqq 0$. The same conclusion holds also for $i = 3$ since in that case $Y = \min(X_1, \cdots, X_n)$ is a sufficient statistic with density $n \exp[-n(y - \theta)]$ for $y \geqq \theta$. The case $i = 4$ is interesting in that again a uniformly most powerful one-sided test exists, although the minimal sufficient statistic is (Y, Z), with $Y = \min_i X_i$, $Z = \max_i X_i$, and hence two-dimensional. The explanation is that the statistic Y by itself is sufficient for $\theta \geqq 0$ when attention is restricted to the part of the sample space that is possible when $\theta = 0$. In the case of the logistic distribution the locally most powerful test can be written down by substituting in (4.6). It is not uniformly most powerful, but is unbiased since (C) holds. For $i = 6$, when the sample is drawn from a Laplace distribution with unknown location parameter θ, the power function of a test may not be differentiable. However, it turns out that a locally most powerful test, in the natural sense of the term, still exists and, perhaps somewhat surprisingly, is given by the sign test, as will be shown in the appendix. Finally in the case $i = 7$, that of a Cauchy distribution, (C) does not hold, and the locally most powerful test does not seem to have a simple structure even when $n = 1$.

We now turn to some applications in which the variables $X_1, \cdots, X_n$ are not independent. Dependence may for example be introduced through the elimination of nuisance parameters by the principle of invariance or because the observable variables involve some common unobservable components, and the joint density of the x's will be a mixture of densities of independent variables. We first give a sufficient condition for (C) to hold in that case.

THEOREM 3. *Let $x = (x_1, \cdots, x_n)$ and let $g_\theta(x, \xi)$ be a family of densities depending on two real parameters θ and ξ and jointly measurable in x and ξ. For each θ, let λ_θ be a measure for ξ such that for all x, the integral*

$$p_\theta(x) = \int g_\theta(x, \xi)\, d\lambda_\theta(\xi)$$

exists. Then a sufficient condition for the family of densities $p_\theta(x)$ to satisfy (C) *is that for $\theta < \theta'$ condition* (C) *holds* (i) *for $g_\theta(x, \xi)$ when ξ is fixed and θ is taken as the parameter,* (ii) *for $g_\theta(x, \xi)$ when θ is fixed and ξ is taken as the parameter,* (iii) *for $d\lambda_\theta(\xi)$.*

Here in assumption (iii) the densities $d\lambda_\theta(\xi)$ and $d\lambda_{\theta'}(\xi)$ may be computed with respect to any σ-finite measure ν that dominates both of the given measures, since only the ratio of the densities matters. In the proof that follows and later in the paper we shall therefore denote this ratio by $d\lambda_{\theta'}(\xi) / d\lambda_\theta(\xi)$. This should not be taken to imply that $\lambda_{\theta'}$ is absolutely continuous with respect to λ_θ, but should be interpreted as a shorthand notation for $(d\lambda_{\theta'} / d\nu):(d\lambda_\theta / d\nu)$.

PROOF. We must show that $x \leqq x'$ implies

$$\frac{\int g_\theta(x', \xi)\, d\lambda_\theta(\xi)}{\int g_\theta(x, \xi)\, d\lambda_\theta(\xi)} \leqq \frac{\int g_{\theta'}(x', \xi)\, d\lambda_{\theta'}(\xi)}{\int g_{\theta'}(x, \xi)\, d\lambda_{\theta'}(\xi)} \tag{4.7}$$

Let Λ and Λ' be the probability distributions given by

$$d\Lambda(\xi) = \frac{g_\theta(x, \xi)\, d\lambda_\theta(\xi)}{\int g_\theta(x, \zeta)\, d\lambda_\theta(\zeta)}, \qquad d\Lambda'(\xi) = \frac{g_{\theta'}(x, \xi)\, d\lambda_{\theta'}(\xi)}{\int g_{\theta'}(x, \zeta)\, d\lambda_{\theta'}(\zeta)}.$$

These are the *a posteriori* distributions of ξ given x, corresponding to θ and θ' respectively. Then (4.7) may be rewritten as

$$\int \frac{g_\theta(x', \xi)}{g_\theta(x, \xi)}\, d\Lambda(\xi) \leqq \int \frac{g_{\theta'}(x', \xi)}{g_{\theta'}(x, \xi)}\, d\Lambda'(\xi). \tag{4.8}$$

By assumption (i) it is enough to prove that

$$\int \frac{g_\theta(x', \xi)}{g_\theta(x, \xi)}\, [d\Lambda'(\xi) - d\Lambda(\xi)] \geqq 0. \tag{4.9}$$

By assumption (iii) the ξ-axis can be divided into two mutually exclusive and exhaustive intervals S_- and S_+ such that S_- lies to the left of S_+ and $d\Lambda'(\xi) / d\Lambda(\xi)$ is $\leqq 1$ in S_- and $\geqq 1$ in S_+. We then have that the left-hand side of (4.9) equals

$$a \int_{S_-} [d\Lambda'(\xi) - d\Lambda(\xi)] + b \int_{S_+} [d\Lambda'(\xi) - d\Lambda(\xi)] \tag{4.10}$$

where a and b are mean values of $g_\theta(x', \xi) / g_\theta(x, \xi)$ in S_- and S_+ respectively so that by assumption (ii), $a \leqq b$. Since Λ and Λ' are probability measures, (4.10) becomes

$$(b - a) \int_{S_+} [d\Lambda'(\xi) - d\Lambda(\xi)] = (b - a) \int_{S_+} \left[\frac{d\Lambda'(\xi)}{d\Lambda(\xi)} - 1\right] d\Lambda(\xi) \geqq 0,$$

and was to be proved.

COROLLARY. *Let ξ be vector-valued, $\xi = (\xi_1, \cdots, \xi_s)$ say, and let*

$$p_\theta(x) = \int g_\theta(x, \xi)\, d\Lambda_\theta(\xi).$$

Suppose that the measure Λ_θ is the product of s linear measures $\Lambda_\theta = \lambda_\theta^{(1)} \times \lambda_\theta^{(2)} \times \cdots \times \lambda_\theta^{(s)}$ each of which satisfies condition (iii) *of Theorem* 3, *and that $g_\theta(x, \xi)$ satisfies condition* (i) *of this theorem. Suppose that condition* (ii) *is replaced by*

(ii′) *for each $j = 1, \cdots, s - 1$, the ratio*

$$\frac{g_\theta(x_1', \cdots, x_n', \xi_1', \cdots, \xi_j', \xi_{j+1}, \cdots, \xi_s)}{g_\theta(x_1, \cdots, x_n, \xi_1, \cdots, \xi_j, \xi_{j+1}, \cdots, \xi_s)}$$

is nondecreasing in $\xi_{j+1}, \cdots, \xi_s$ provided $x_i \leqq x_i'$ $(i = 1, \cdots, n)$ and $\xi_i \leqq \xi_i'$ $(i = 1, \cdots, j)$.

PROOF. It is seen from Theorem 3 by induction over j that

$$g_\theta^{(j)}(x_1, \cdots, x_n, \xi_{j+1}, \cdots, \xi_s) = \int \cdots \int g_\theta(x_1, \cdots, x_n, \xi_1, \cdots, \xi_s)\, d\lambda_\theta^{(1)}(\xi_1) \cdots d\lambda_\theta^{(j)}(\xi_j)$$

satisfies conditions (i) and (ii′), and this yields the desired result.

As an application we consider:

EXAMPLE 4.1. Let $U_1, \cdots, U_s$ be a sample from an unobservable random variable with density $f_\theta(u)$. What we observe are

$$X_{ij} = U_i + V_{ij}$$

where the V's are independently normally distributed with mean zero. For the moment we shall assume the variance of the V's to be known, and hence without loss of generality to be equal to 1. A typical example is the usual simplest model II problem in which θ is a scale parameter. We shall assume that $f_\theta(u)$ is an even function of u and that for $\theta < \theta'$ the ratio $f_{\theta'}(u) / f_\theta(u)$ is an increasing function of $|u|$, and consider the problem of testing $\theta \leqq \theta_0$ against $\theta \geqq \theta_1$. The joint density of the X's is given by

$$p_\theta(x) = \prod_{i=1}^{s} \int_{-\infty}^{\infty} \frac{1}{(\sqrt{2\pi})^n} \exp\left[-\frac{1}{2}\sum_{j=1}^{n}(x_{ij} - u_i)^2\right] f_\theta(u_i)\, du_i .$$

Therefore the absolute values of the means $\bar{x}_1, \cdots, \bar{x}_s$ constitute a set of sufficient statistics for θ, and we may restrict attention to them. Putting $y_i = \sqrt{n}\bar{x}_i$ and $\xi = \sqrt{n}u$, we obtain the joint density of the Y's as

$$p_\theta(y_1, \cdots, y_s) = C \prod_{i=1}^{s} \int_{-\infty}^{\infty} \exp\left[-\tfrac{1}{2}(y_i - \xi_i)^2\right] f_\theta(\xi_i / \sqrt{n})\, d\xi_i .$$

We shall now prove condition (C) for the density of the $|Y|$'s. Since we are dealing with a sample it is enough to check this for the case $s = 1$. We have

$$p_\theta(y) = Ce^{-\frac{1}{2}y^2} \int_0^{\infty} (e^{-\xi y} + e^{\xi y}) e^{-\frac{1}{2}\xi^2} f(\xi / \sqrt{n})\, d\xi. \tag{4.11}$$

Condition (iii) of Theorem 3 is satisfied by assumption, and we need only check (i) and (ii) with

$$g_\theta(y, \xi) = e^{-\frac{1}{2}y^2}(e^{-\xi y} + e^{\xi y}) e^{-\frac{1}{2}\xi^2} \qquad \text{for} \quad \xi > 0.$$

Since this is independent of θ, assumption (i) clearly holds. Examining (ii) we have

$$\frac{g_\theta(y', \xi)}{g_\theta(y, \xi)} = e^{-\frac{1}{2}(y'^2 - y^2)} \frac{e^{-\xi y'} + e^{\xi y'}}{e^{-\xi y} + e^{\xi y}} .$$

Now if $|y| \leqq |y'|$, it is easily checked that $(e^{-\xi y'} + e^{\xi y'}) / (e^{-\xi y} + e^{\xi y})$ is an increasing function of $|\xi|$, and this completes the proof of (C). It follows that the test which rejects when

$$\frac{p_{\theta_1}(y_1) \cdots p_{\theta_1}(y_s)}{p_{\theta_0}(y_1) \cdots p_{\theta_0}(y_s)} > k$$

where $p_\theta(y)$ is given by (4.11) maximizes the minimum power for testing $\theta \leqq \theta_0$ against $\theta \geqq \theta_1$.

We next consider the following somewhat more realistic case.

EXAMPLE 4.2. Suppose that the assumptions of Example 4.1 hold but that the variance σ^2 of the X's is unknown. We assume further that the unknown parameter in the distribution of the U's is a scale parameter, say τ. The problem is to test $\tau/\sigma \leqq \theta_0$ against $\tau/\sigma \geqq \theta_1$. Putting $\theta = \tau/\sigma$ the joint probability density of the X_{ij} is

$$\prod_{i=1}^{s} \int \frac{1}{(\sqrt{2\pi}\sigma)^n} \exp\left[-\frac{1}{2\sigma^2}\sum (x_{ij} - u_i)^2\right] \frac{1}{\theta\sigma} f\left(\frac{u_i}{\theta\sigma}\right) du_i.$$

Here the statistics $V = \sum\sum (X_{ij} - \bar{X}_i)^2, \bar{X}_1, \cdots, \bar{X}_s$ are jointly sufficient. Putting $Y_i = \sqrt{n}\bar{X}_i$, $\xi_i = \sqrt{n}u_i$, the joint density of V and the Y's is given by

$$\frac{C}{\sigma^N} v^{(N-s)/2-1} \exp\left[-\frac{v}{2\sigma^2}\right] \prod_{i=1}^{s} \int \exp\left[-\frac{1}{2\sigma^2}(y_i - \xi_i)^2\right] \frac{1}{\theta\sigma^2} f\left(\frac{\xi_i}{\sqrt{n}\,\theta\sigma^2}\right) d\xi_i.$$

Now the problem of testing $\theta \leqq \theta_0$ against $\theta \geqq \theta_1$ remains invariant under multiplication of the Y's by a common positive constant a and of V by a^2, and there exists a solution to the given problem which is invariant under these transformations. We may therefore restrict attention to the maximal invariant $(z_1, \cdots, z_s)$ where $z_i = y_i / \sqrt{v}$. The joint density of the Z's is given by

$$C \int_0^\infty v^p e^{-v/2} \left[\prod_{i=1}^{s} \int_{-\infty}^{\infty} \exp\left[-\tfrac{1}{2}(z_i\sqrt{v} - \xi_i)^2\right] \frac{1}{\theta} f\left(\frac{\xi_i}{\theta\sqrt{n}}\right) d\xi_i\right] dv$$

$$= \int_0^\infty \cdots \int_0^\infty \frac{1}{\theta^s} f\left(\frac{\xi_i}{\theta\sqrt{n}}\right)\left[C \exp\left(-\tfrac{1}{2}\sum \xi_i^2\right) \int_0^\infty v^p e^{-v/2} \prod_{i=1}^{s} \left(e^{-\xi_i z_i \sqrt{v}} + e^{\xi_i z_i \sqrt{v}}\right) dv\right] d\xi_1 \cdots d\xi_s.$$

Denoting the expression in brackets by $g(z, \xi)$ we shall now show that $g(z', \xi) / g(z, \xi)$ is increasing in ξ for $z \leqq z'$, the other two conditions of Theorem 3 being satisfied as before. To prove that $g(z, \xi)$ has the desired property we apply once more Theorem 3 with z, v, ξ playing the role of θ, ξ, x in this order. The weight function for u being $d\lambda(v) = Cv^p e^{-v/2}$ independent of z, condition (iii) is satisfied. Putting

$$h_z(v, \xi) = C \exp\left[-\tfrac{1}{2}\sum \xi_i^2\right] \prod_{i=1}^{s} \left(e^{-\xi_i z_i \sqrt{v}} + e^{\xi_i z_i \sqrt{v}}\right)$$

it is enough to show that $h_{z'}(v, \xi) / h_z(v, \xi)$ is increasing in v and ξ and $h_z(v, \xi') / h_z(v, \xi)$ in ξ where the ξ's are assumed to be nonnegative and where $|z_i| \leqq |z_i'|$ for $i = 1, \cdots, s$. Now

$$\frac{h_z(v, \xi')}{h_z(v, \xi)} = \prod_i \frac{e^{-\xi_i' z_i \sqrt{v}} + e^{\xi_i' z_i \sqrt{v}}}{e^{-\xi_i z_i \sqrt{v}} + e^{\xi_i z_i \sqrt{v}}}$$

and each factor is increasing in v since $|\xi_i z_i| \leqq |\xi_i' z_i|$. Similarly $h_{z'}(v, \xi) / h_z(v, \xi)$ is increasing in v and $|\xi_i|$. Finally, condition (ii′) of the Corollary to Theorem 3 is checked in the same manner, and it therefore follows from this corollary that (C) holds for the density of the Z's.

From this and the fact that the density of the Z's is even in each of the variables it is seen that the most powerful invariant test for testing θ_0 against θ_1 has a rejection region which is increasing in $|z_1|, \cdots, |z_s|$. That the probability of such a region is increasing in θ is a consequence of the fact that condition (A) holds, θ being a scale parameter for the Z's.

As two further illustrations we prove that the noncentral t and F densities have monotone likelihood ratios, that is, satisfy (C), so that the associated tests have the minimax property discussed at the beginning of this section. The first of these results was earlier given by Kruskal [6]; the second was obtained by Rushton (personal communication) and by Meyer (in "An application of the invariance principle to the Student hypothesis," Technical Report No. 24, Department of Statistics, Stanford University, unpublished). A result containing these two as special cases was obtained about simultaneously with the present paper by Karlin ("On distributions $(p(x|\omega)$ for which $p(x|\omega_1) \cdot p(x|\omega_2)$ is monotone," Technical Report No. 26, Department of Statistics, Stanford University, unpublished), who considered densities of the form

$$g_\theta(x) = c(\theta)\phi(x) \int e^{\pm r(t)x} e^{\theta t} \, d\psi(t).$$

Example 4.3. Let $p_\theta(t)$ denote the noncentral t density with noncentrality parameter θ, (including as a particular case the central density for $\theta = 0$), that is, the density of Student's t statistic when the sample on which it is based is drawn from a normal distribution $N(\eta, \sigma^2)$. Then

$$p_\theta(t) = Ct^{-n} \int_0^\infty \exp\left[-\frac{n}{2}(w - \theta)^2\right] w^{n-1} e^{-w^2/2t^2} \, dw$$

where n is the sample size and $\theta = \eta/\sigma$. That $p_{\theta'}(t) / p_\theta(t)$ is an increasing function of t for $t \leqq 0$ follows directly from Theorem 3. For $t \geqq 0$ it can be seen by noting that Theorem 3 remains valid if the ratios considered in (ii) and (iii) are nonincreasing instead of nondecreasing, with the ratio considered in (i) remaining nondecreasing in x.

Example 4.4. The noncentral F-density with r and s degrees of freedom and noncentrality parameter θ is given by

$$p_\theta(u) = \sum_{k=0}^\infty P_\theta(k) h_{r+k,s+k}(u), \qquad u \geqq 0$$

where $h_{r+k,s+k}$ is the central F-density with $r + k$ and $s + k$ degrees of freedom, and where

$$P_\theta(k) = \theta^k e^{-\theta} / k!$$

is the Poisson probability with parameter θ. It again follows immediately from Theorem 3 that for $\theta < \theta'$ the ratio $p_{\theta'}(u) / p_\theta(u)$ is increasing in u.

We shall now mention some problems in which the conditions of Theorem 3 do not appear to be satisfied. In these situations it would be of interest to obtain basic densities f under which the probability ratio test for testing θ_0 against θ_1 maximizes the minimum power against $\theta \geqq \theta_1$. In all of these problems this is easily shown to be the case when f is the normal density.

PROBLEM 1. Let $X_1, \cdots, X_n$ be a sample from $(1/\theta)f((x - \xi) / \theta)$. Then the distribution of the differences $X_j - X_i$ depends only on θ. It follows from the Hunt-Stein theorem that for testing $\theta \leqq \theta_0$ there exists a test depending only on these differences and which maximizes the minimum power over $\theta \geqq \theta_1$. The problem mentioned then arises for the joint density of these differences, which is easily written down and which is of course independent of ξ. An elaboration of this problem is the case of two samples from densities $(1/\sigma)f((x - \xi) / \sigma)$ and $(1/\tau)f((y - \tau) / \tau)$ respectively, where $\theta = \sigma/\tau$.

PROBLEM 2. Let $X_1, \cdots, X_n$ be a sample from $(1/\sigma)f(x/\sigma - \theta)$. Here it is the ratios that play the role of the differences in Problem 1. In the two-sample version of this problem the samples came from $(1/\sigma)f((x - \xi) / \sigma)$ and $(1/\sigma)f((y - \eta) / \sigma)$, and $\theta = (\eta - \xi) / \sigma$.

PROBLEM 3. Let $X_1, \cdots, X_n$ be a sample from $f(x - \theta)$, where f is even and consider the problem of testing $|\theta| \leqq \theta_0$ against $|\theta| \geqq \theta_1$. Here one would expect the test that maximizes the minimum power to be given by the rejection region

$$\frac{f(x_1 - \theta_1) \cdots f(x_n - \theta_1) + f(x_1 + \theta_1) \cdots f(x_n + \theta_1)}{f(x_1 - \theta_0) \cdots f(x_n - \theta_0) + f(x_1 + \theta_0) \cdots f(x_n + \theta_0)} \geqq C.$$

This will be the case provided the probability of this region is an increasing function of $|\theta|$. The problem is to find conditions on f which would insure this.

5. Comparability of experiments. When a family of distributions $\{P_\theta\}$ is ordered, it seems reasonable to expect

(D). The pair of distributions (θ_0', θ_1') is more[6] informative than the pair (θ_0, θ_1) in the sense of Blackwell [1] provided $\theta_0' \leqq \theta_0 < \theta_1 \leqq \theta_1'$.

Let ϕ_α and ϕ_α' be the most powerful level α tests for testing θ_0 against θ_1 and θ_0' against θ_1' respectively. Then Blackwell showed in [2] that (θ_0', θ_1') is more informative than (θ_0, θ_1) if and only if $\beta_\alpha(\theta_1) \leqq \beta_\alpha'(\theta_1')$ for all α, where β_α and β_α' denote the power functions of ϕ_α and ϕ_α'.

A somewhat stronger property than (D) which one might also expect to hold in an ordered family is:

(E). Let $\theta_0 < \theta_1$ and let λ_0, λ_1 be any distributions over the sets $\theta \leqq \theta_0$ and $\theta \geqq \theta_1$ respectively. Then the pair of distributions $(\int p_\theta(x)\, d\lambda_0(\theta), \int p_\theta(x)\, d\lambda_1(\theta))$ is more informative than the pair $(p_{\theta_0}(x), p_{\theta_1}(x))$.

Clearly (E) is actually stronger than (D). As a trivial example, let X be normally distributed with unit variance and mean ξ and let $\theta_0 = \theta_0'$ correspond to $\xi = 0$, θ_1 to $\xi = -1$ and θ_1' to $\xi = +1$. Then (θ_0, θ_1) and (θ_0', θ_1') are equally

[6] Throughout we shall understand with Blackwell "more informative" in the weak sense of "at least as informative."

informative and both strictly more informative than $(\theta_0, \frac{1}{2}\theta_1 + \frac{1}{2}\theta_1')$. It would be interesting to know whether more natural examples of this phenomenon exist such as, for example, a family of densities $g(x - \theta)$ which satisfies (D) for all $\theta_0' \leqq \theta_0 < \theta_1 \leqq \theta'$ but for which (E) does not hold.

A condition equivalent to (E) is:

(E′). For every pair $\theta_0 < \theta_1$ and every α the power function $\beta(\theta)$ of the probability ratio test for testing θ_0 against θ_1 satisfies

$$\begin{aligned} \beta(\theta) &\leqq \beta(\theta_0) \qquad \text{for} \quad \theta \leqq \theta_0, \\ \beta(\theta) &\geqq \beta(\theta_1) \qquad \text{for} \quad \theta \geqq \theta_1. \end{aligned} \tag{5.1}$$

To see this, note that (E) states that at every level α, the pair of *a priori* distributions assigning probability 1 to θ_0 and θ_1 respectively is least favorable for testing $\theta \leqq \theta_0$ against $\theta \geqq \theta_1$. It follows from Theorem 3.10 of [13] that (E) implies (E′). The converse is also a special case of a well-known simple decision-theoretic result, or alternatively can be seen from the proof of Theorem 4. Since (E′) is a consequence of (B) + (C), so is (E). On the other hand, the following example shows that (B) is not enough to insure even (D).

EXAMPLE 5.1. Let X be uniformly distributed over the union of the two intervals $(\theta - \frac{3}{4}, \theta - \frac{1}{4})$, $(\theta + \frac{1}{4}, \theta + \frac{3}{4})$. Then (B) holds since θ is a location parameter. On the other hand, the pair of distributions $(\theta = 0, \theta = \frac{1}{2})$ is clearly strictly more informative than the pair $(\theta = 0, \theta = 1)$.

We shall finally show that (B) + (C) permit an even stronger conclusion than (E).

THEOREM 4. *Let $p_\theta(x)$ be a family of probability densities satisfying* (B) *and* (C). *Let (λ_0, λ_1) and (λ_0', λ_1') be two pairs of probability distributions for the parameter θ such that the three ratios $d\lambda_0 / d\lambda_0'$, $d\lambda_1 / d\lambda_0$, $d\lambda_1' / d\lambda_1$ are all nondecreasing. Then the experiment*

$$\left(\int p_\theta(x)\, d\lambda_0'(\theta), \int p_\theta(x)\, d\lambda_1'(\theta)\right)$$

is more informative than the experiment

$$\left(\int p_\theta(x)\, d\lambda_0(\theta), \int p_\theta(x)\, d\lambda_1(\theta)\right)$$

It is convenient to prove first the following lemma.

LEMMA 3. *Let $x = (x_1, \cdots, x_n)$, and let $p_\theta(x)$ be a family of densities satisfying conditions* (B) *and* (C). *Let λ, λ' be two probability measures for θ such that $d\lambda'(\theta) / d\lambda(\theta)$ is nondecreasing in θ. Then*

(i) $\dfrac{\int p_\theta(x)\, d\lambda'(\theta)}{\int p_\theta(x)\, d\lambda(\theta)}$ *is nondecreasing in x,*

(ii) *if $\phi(x)$ is nondecreasing in x,*

$$\int E_\theta \phi(x)\, d\lambda(\theta) \leqq \int E_\theta \phi(x)\, d\lambda'(\theta). \tag{5.1}$$

PROOF. (i) follows from Theorem 3, since $p_\theta(x') / p_\theta(x)$ is nondecreasing in x, and $d\lambda'(\theta) / d\lambda(\theta)$ is nondecreasing in θ. To see (ii), let $\psi(\theta) = E_\theta \phi(X)$. Then by (B′), $\psi(\theta)$ is nondecreasing and it is easily seen that

$$\int \psi(\theta)[d\lambda'(\theta) - d\lambda(\theta)] \geqq 0.$$

PROOF OF THEOREM 4. Let ϕ_α and ϕ'_α be the most powerful level α tests for testing $\int p_\theta(x)\, d\lambda_0(\theta)$ against $\int p_\theta(x)\, d\lambda_1(\theta)$, and $\int p_\theta(x)\, d\lambda'_0(\theta)$ against $\int p_\theta(x)\, d\lambda'_1(\theta)$ respectively. Let

$$\beta(\alpha) = \int E_\theta \phi_\alpha(x)\, d\lambda_1(\theta), \qquad \beta'(\alpha) = \int E_\theta \phi'_\alpha(x)\, d\lambda'_1(\theta)$$

denote the power of these two tests for their respective alternatives. Then the desired result follows if for all α we have $\beta(\alpha) \leqq \beta'(\alpha)$. It is seen from part (i) of Lemma 3 that the rejection functions ϕ_α and ϕ'_α are nondecreasing. Therefore, by part (ii) of the lemma

$$\int E_\theta \phi_\alpha(x)\, d\lambda'_0(\theta) \leqq \int E_\theta \phi_\alpha(x)\, d\lambda_0(\theta) = \alpha$$

so that ϕ is a level α test also for the hypothesis $\int p_\theta(x)\, d\lambda'_0(\theta)$, and

$$\int E_\theta \phi_\alpha(x)\, d\lambda'_1(\theta) \leqq \beta'(\alpha)$$

since $\beta'(\alpha)$ is the power of the most powerful level α test. Also, by part (ii) of the lemma

$$\beta(\alpha) = \int E_\theta \phi(x)\, d\lambda_1(\theta) \leqq \int E_\theta \phi(x)\, d\lambda'_1(\theta),$$

and the result follows.

In conclusion I should like to thank a referee of this paper for many very helpful suggestions.

6. Appendix. A property of the sign test. It was recently shown by Hoeffding and Rosenblatt ("The efficiency of tests," *Ann. Math. Stat.*, Vol. 26 (1955), pp. 52–63) that the sign test is asymptotically most efficient for detecting a small shift in the distribution with density $\frac{1}{2}e^{-|x|}$. We shall show below that the sign test is in fact locally most powerful for testing $H: \theta = 0$ against the alternatives $\theta > 0$ when

$$p_\theta(x_1, \cdots, x_n) = \frac{1}{2^n} e^{-\Sigma|x_i - \theta|} \tag{6.1}$$

for any fixed sample size n. In this we shall restrict ourselves to levels of significance α at which the sign test can be carried out without randomization, that is, to one of the levels

$$\alpha_m = \sum_{k=0}^{m} \binom{n}{k} \Big/ 2^n, \qquad m = 0, 1, \cdots, n-1. \tag{6.2}$$

Since the power function $\beta(\theta)$ of a test of this hypothesis may not be differentiable, we shall state the optimum property of the sign test more precisely as follows.

Let $\beta^*(\theta)$ be the power function of the sign test at one of the levels α_m and let $\beta_\phi(\theta)$ be the power function of any other test ϕ of H at the same level. Then there exists Δ such that

$$\beta_\phi(\theta) < \beta^*(\theta) \qquad \text{for} \quad 0 < \theta < \Delta. \tag{6.3}$$

To prove this, let us denote by R_k $(k = 0, \cdots, n)$ the subset of the sample space in which k is of the X's are positive and $n - k$ are negative. The proof follows easily from the following lemma.

LEMMA 4. *Let $0 \leqq k < l \leqq n$ and let S_k, S_l be subsets of R_k and R_l respectively for which*

$$P_0(S_k) = P_0(S_l). \tag{6.4}$$

Then there exists $\Delta_{k,l}$ such that

$$P_\theta(S_k) < P_\theta(S_l) \qquad \textit{for} \quad 0 < \theta < \Delta_{k,l}. \tag{6.5}$$

PROOF. We note that

$$\frac{e^{-|x-\theta|}}{e^{-|x|}} = \begin{cases} e^{-\theta} & \text{if} \quad x < 0 \\ e^{2x-\theta} & \text{if} \quad 0 < x < \theta \\ e^{\theta} & \text{if} \quad \theta < x \end{cases}$$

and that $e^{-\theta} < e^{2x-\theta} < e^{\theta}$ if $0 < x < \theta$. Let $S_{l,\theta}$ denote the subset of S_l for which the l positive x's are all $> \theta$. Then

$$P_\theta(S_l) \geqq e^{(2l-n)\theta} P_0(S_{l,\theta}) + e^{-n\theta} P_0(S_l - S_{l,\theta})$$
$$P_\theta(S_k) \leqq e^{(2k-n)\theta} P_0(S_k).$$

Putting $\eta(\theta) = P_0(S_l - S_{l,\theta})$ and denoting the common value of (6.4) by γ, we therefore have

$$P_0(S_l) - P_0(S_k) \geqq e^{(2l-n)\theta}[\gamma - \eta(\theta)] + e^{-n\theta}\eta(\theta) - \gamma e^{(2k-n)\theta}$$

This will be positive provided

$$\gamma[e^{(2l-n)\theta} - e^{(2k-n)\theta}] > \eta(\theta)[e^{-n\theta} - e^{(2k-n)\theta}].$$

Up to terms of order θ, the left- and right-hand sides are respectively $2\gamma(l-k)\theta$ and $\eta(\theta)(2k-n)\theta$. Since $\eta(\theta) \to 0$ as $\theta \to 0$, it follows that the desired inequality holds when θ is sufficiently small.

The result expressed by (6.3) is now an obvious consequence when the alternative test ϕ is nonrandomized. For consider any rejection region that does not consist of the upper tail of a sign test. Then it can be converted into a sign test of the same size by a finite number of steps, each of which consists in replacing an S_k by an S_l with $k < l$ which satisfies (6.4).

Only minor modifications of the argument are required in case the alternative test ϕ is randomized. In particular, in the lemma, the sets S_k and S_l are replaced by critical functions ϕ_k and ϕ_l over R_k and R_l respectively, such that

$$E_0 \phi_k(X_1, \cdots, X_n) = E_0 \phi_l(X_1, \cdots, X_n),$$

the conclusion being that

$$E_\theta \phi_k(X_1, \cdots, X_n) < E_\theta \phi_l(X_1, \cdots, X_n) \qquad \text{for} \quad 0 < \theta < \Delta_{k,l}.$$

It is interesting to note that the sign test, being similar for testing $H: \theta = 0$ when the density (6.1) involves an unknown scale parameter, is also locally most powerful for that problem.

It should be mentioned finally that the above proof may be modified to show that the two-sided sign test maximizes $\frac{1}{2}[\beta(\theta) + \beta(-\theta)]$ for sufficiently small θ. This test is therefore locally most powerful among all tests that are symmetric with respect to the origin.

REFERENCES

[1] D. Blackwell, "Comparison of experiments," *Proceedings of the Second Berkeley Symposium on Mathematical Statistics and Probability*, Univ. California Press, Berkeley, 1951, pp. 93–102.

[2] D. Blackwell, "Equivalent comparisons of experiments," *Ann. Math. Stat.*, Vol. 24 (1953), pp. 265–272.

[3] D. R. Cox, "Sequential tests for composite hypotheses," *Proc. Cambridge Philos. Soc.*, Vol. 48 (1952), pp. 290–299.

[4] N. L. Johnson, "Some notes on the application of sequential methods in the analysis of variance," *Ann. Math. Stat.*, Vol. 24 (1953), pp. 614–623.

[5] J. L. Hodges, Jr. and E. L. Lehmann, "Testing the approximate validity of statistical hypotheses," *J. Roy. Stat. Soc.*, to be published.

[6] W. Kruskal, "The monotonicity of the ratio of two noncentral t density functions," *Ann. Math. Stat.*, Vol. 25 (1954), pp. 162–165.

[7] E. L. Lehmann, "Consistency and unbiasedness of certain nonparametric tests," *Ann. Math. Stat.*, Vol. 22 (1951), pp. 165–179.

[8] E. L. Lehmann, "Testing multiparameter hypotheses," *Ann. Math. Stat.*, Vol. 23 (1952), pp. 541–552.

[9] E. L. Lehmann, "Some principles of the theory of testing hypotheses," *Ann. Math. Stat.*, Vol. 20 (1950), pp. 1–26.

[10] H. Rubin, "A complete class of decision procedures for distributions with monotone likelihood ratio," *Ann. Math. Stat.*, Vol. 22 (1951), p. 608 (Abstract).

[11] I. J. Schoenberg, "On Polya frequency functions. I. The totally positive functions and their Laplace transforms," *J. d'Analyse Mathematique*, Vol. 1 (1951), pp. 331–374.

[12] A. Wald, *Sequential Analysis*, John Wiley and Sons, New York, 1947.

[13] A. Wald, *Statistical Decision Functions*, John Wiley and Sons, New York, 1950.

The Annals of Statistics
1988, Vol. 16, No. 2, 521–533

COMPARING LOCATION EXPERIMENTS[1]

BY E. L. LEHMANN

University of California, Berkeley

In Sections 1–3, the classical theory of the comparison of two experiments is reviewed with particular reference to the comparison of two location experiments. It is shown that the requirement of domination of one experiment by another for *all* decision problems is too strong to provide a reasonable basis for comparison. For one-parameter problems with monotone likelihood ratio, it is therefore proposed to restrict the comparison to decision problems that are monotone in the sense of Karlin and Rubin (1956). Application of this weaker definition to the location problem is shown to give satisfactory results. A scale-free comparison of this type leads to a new tail-ordering of distributions, and this is explored in Section 6.

1. Introduction. An experiment **E** is a random quantity X and a family $\mathbf{P} = \{P_\theta,\ \theta \in \Omega\}$ of possible distributions of X. Let $\mathbf{F} = (Y, \mathbf{Q} = \{Q_\theta,\ \theta \in \Omega\})$ be another experiment, with the distributions P_θ and Q_θ corresponding to the same state of nature θ. The idea of patterning the definition of one experiment being more informative than another on the concept of sufficiency was initiated in an unpublished memorandum by Bohnenblust, Shapley and Sherman and developed into a theory by Blackwell (1951, 1953).

DEFINITION (1.1). The experiment **F** is more informative than (or sufficient for) **E** if there exists a random quantity Z with known distribution and a function h such that for all $\theta \in \Omega$,

$$Y \textit{ is distributed as } Q_\theta \Rightarrow h(Y, Z) \textit{ is distributed as } P_\theta.$$

An immediate consequence of (1.1) is:

(1.2) For any decision procedure δ based on X and any loss function $L(\theta, d)$ there exists a (possibly randomized) procedure δ' based on Y such that $R(\theta, \delta') = R(\theta, \delta)$ for all θ.

It was shown by Blackwell, and under more general conditions by Le Cam (1964) and Feldman and Ramamoorthi (1986) that typically not only does (1.1) imply (1.2) but the inverse implication also holds. In fact, in the same papers it is shown that (1.1) is implied by the following apparently even weaker statement:

(1.3) Statement (1.2) with the conclusion $R(\theta, \delta') = R(\theta, \delta)$ for all θ replaced by $R(\theta, \delta') \le R(\theta, \delta)$ for all θ.

Received September 1986; revised August 1987.

[1]Research supported by National Science Foundation Grant DMS-84-01388.

AMS 1980 *subject classification*. Primary 62C05.

Key words and phrases. Comparison of experiments, location families, monotone decision procedures, monotone likelihood ratio, spread ordering, tail ordering.

A fourth condition, which typically is equivalent to (1.1)–(1.3), is the Bayes condition that given any prior distribution Λ for θ, the Bayes risk is no larger when the experiment is based on Y than when it is based on X.

If Y is more informative in the sense of these definitions, which are equivalent in the situations to be considered in this paper, we shall write $Y \geq X$. If $Y \geq X$ and $X \geq Y$, the experiments X and Y will be said to be *equivalent* ($Y \sim X$). The experiment Y is strictly more informative than X ($Y > X$) if it is more informative than X but not equivalent to it.

The various possibilities are illustrated by the following example.

EXAMPLE 1.1 (Normal). Let $X = (X_1, \ldots, X_n)$, $Y = (Y_1, \ldots, Y_n)$, where the X_i and Y_i are independently normally distributed as $N(\xi, \sigma^2)$ and $N(\xi, \rho^2\sigma^2)$, respectively, with ρ known and $0 < \rho < 1$.

(i) $\sigma = \sigma_0$ known. Here $Y \geq X$ since $Y_i + Z_i$ has the same distribution as X_i when Z_i is $N(0, (1-\rho^2)\sigma_0^2)$. That Y is strictly more informative than X is seen by noting that the UMV unbiased estimators of ξ based on X and Y are, respectively, $\overline{X}$ with variance σ_0^2/n and $\overline{Y}$ with variance $\rho^2\sigma_0^2/n$. The latter variance cannot be matched by an unbiased estimator based on X, so that by Definition (1.1) X is not as informative as Y.

(ii) $\xi = \xi_0$ known. Assuming without loss of generality that $\xi_0 = 0$, one sees that $Y \sim X$ since the variables Y_i/ρ have the same distribution as the X_i, and the variables ρX_i the same distribution as the Y_i.

(iii) ξ and σ both unknown. The surprising fact [see Hansen and Torgersen (1974)] is that in this case X and Y are not comparable.

This example illustrates the three possibilities: strict comparability, equivalence and noncomparability, and the two principal methods used to determine comparability. If Y is more informative than X, it is typically easy to determine the function $h(Y, Z)$ required in (1.1). To prove that Y is not more informative than X, one attempts to construct a statistical task that can be performed on the basis of X, but that either cannot be performed, or at least not as well, on the basis of Y. For this latter purpose it is often most convenient to find a function $a(\theta)$ that has an unbiased estimator based on X but not on Y. One reason for looking at this particular kind of task is that unbiasedness of an estimator requires only the calculation of first moments. Another reason is a result of DeGroot (1966), Theorem 4.1, which states essentially that if $\mathbf{P}$ is complete, then Y is more informative than X if and only if for each set B for which probability is defined $a(\theta) = P_\theta(X \in B)$ has a nonnegative unbiased estimator based on Y.

2. Comparing two location experiments.

PROBLEM 1. In most of this paper, we shall be concerned with the case that $\mathbf{P}$ and $\mathbf{Q}$ are location families, i.e., they are given by

$$(2.1) \qquad P_\theta(X \leq x) = F(x - \theta) \quad \text{and} \quad Q_\theta(Y \leq y) = G(y - \theta).$$

For this case it was shown by Boll (1955) and independently by Stone (1961) that condition (1.1) greatly simplifies in that the function $h(Y, Z)$ can always be taken to be of the form $h(Y, Z) = Y + Z$, i.e., $Y \geq X$ if and only if there exists a random variable Z independent of Y such that

(2.2) $$Y \text{ has distribution } Q_\theta \Rightarrow Y + Z \text{ has distribution } P_\theta.$$

As an example, suppose that G is the uniform distribution on $(-\frac{1}{2}, \frac{1}{2})$ and F the triangular distribution on $(-1, 1)$. Then $Y \geq X$ since (2.2) holds with Z uniformly distributed on $(-\frac{1}{2}, \frac{1}{2})$.

If ϕ_X and ϕ_Y denote the characteristic functions of the distributions F and G, respectively, (2.2) is equivalent to the condition that

(2.3) $$\psi(t) = \phi_X(t)/\phi_Y(t) \text{ is a characteristic function.}$$

As an example, suppose that F is the double exponential distribution with density $\frac{1}{2}e^{-|x|}$ and G the exponential distribution with density e^{-x}, $x > 0$. Then

$$\phi_X(t) = \frac{1}{1 + t^2} \quad \text{and} \quad \phi_Y(t) = \frac{1}{1 - it},$$

so that $\psi(t) = 1/(1 + it)$, which is the characteristic function of $-Y$. Thus Y is more informative than X.

An immediate consequence of (2.3) is the surprising fact that if X is normally distributed, then Y cannot be more informative than X unless Y is also normal. This follows immediately from Cramér's theorem that if X is normal and X is the sum of two independent random variables Y and Z, then Y and Z must be normal. It is however disconcerting to learn that a normal location family

$$F(x - \theta) = \frac{1}{\sigma}\Phi\left(\frac{x - \theta}{\sigma}\right)$$

even for every large σ is never less informative than a nonnormal $G(y - \theta)$ even if the latter distribution is very tightly concentrated about θ. Of course, if

$$G(y - \theta) = \frac{1}{\tau}\Phi\left(\frac{y - \theta}{\tau}\right),$$

with $\tau < \sigma$, then Y is more informative than X by Example 1.1(i).

Given any Y with distribution $G(y - \theta)$, it is trivial to construct a less informative X simply by taking $F(x - \theta)$ to be the distribution of $Y + Z$ for any independent Z. On the other hand, any particular given F and G will be comparable only in very exceptional cases since this would require one of the characteristic functions ϕ_X, ϕ_Y to be a factor of the other.

The normal example suggests the possibility that such comparisons may be more readily available when G differs from F only by a scale factor, say.

PROBLEM 2. **P** and **Q** are given by (2.1) with

(2.4) $$G(y - \theta) = F\left(\frac{y - \theta}{\rho}\right), \qquad 0 < \rho < 1.$$

This problem has been considered by Stone (1961) and Goel and DeGroot (1979). A necessary and sufficient condition for Y given by (2.4) to be more informative than X for all ρ, $0 < \rho < 1$, is that

$$(2.5)\qquad \psi(t) = \frac{\phi_X(t)}{\phi_X(\rho t)} \text{ is a characteristic function for } all\ 0 < \rho < 1.$$

As was pointed out by Goel and DeGroot, the distributions F whose characteristic functions ϕ_X satisfy (2.5) were investigated by Lévy in a quite different context, and are called self-decomposable or belonging to class L. In particular, it follows from Lévy's work that all stable laws are self-decomposable and that, on the other hand, all self-decomposable distributions are infinitely divisible.

EXAMPLE 2.1 (Double exponential). As a simple example of a self-decomposable distribution that is not stable, consider the double exponential distribution with density $f(x) = \frac{1}{2}e^{-|x|}$ and characteristic function $\phi(t) = 1/(1 + t^2)$. To see that this is self-decomposable, note that

$$\psi(t) = \frac{1 + \rho^2 t^2}{1 + t^2} = \rho^2 \cdot 1 + (1 - \rho^2) \cdot \frac{1}{1 + t^2}.$$

Thus, $\psi(t)$ is the characteristic function of a variable which is equal to 0 with probability ρ^2 and has density $f(x)$ with probability $1 - \rho^2$.

On the other hand, let $F(x)$ be any distribution whose support is a finite interval such as the uniform or triangular distribution. Then F cannot be self-decomposable since it is not infinitely divisible.

3. The uniform case. It follows from the discussion of the preceding section that if X and Y are uniformly distributed as

$$(3.1)\qquad X\colon U\left(\theta - \tfrac{1}{2}, \theta + \tfrac{1}{2}\right), \qquad Y\colon U\left(\theta - \frac{\rho}{2}, \theta + \frac{\rho}{2}\right),$$

then very surprisingly Y is not more informative than X for all $0 < \rho < 1$. This does not rule out the possibility that it may be more informative for some ρ, and this is in fact the case.

THEOREM 3.1. *Under* (3.1), *Y is more informative than X if and only if $\rho = 1/k$ for some positive integer k.*

PROOF. (i) If $\rho = 1/k$, then (2.2) holds with Z being uniformly distributed over the points

$$-\frac{(k-1)}{2k}, -\frac{(k-3)}{2k}, \ldots, \frac{k-3}{2k}, \frac{k-1}{2k}.$$

This is easily checked and well known. That Y in this case is strictly more informative than X can be seen, for example, by considering the problem of testing H: $\theta = 0$ against the simple alternative $\theta = \theta_1$, where $1/2k < \theta_1 < \frac{1}{2}$. Then there exists a test with power 1 based on Y but not on X.

(ii) To prove that Y is not more informative than X for $\rho \neq 1/k$, we shall exhibit a statistical task which X can perform in this case but Y cannot. For this purpose, consider the estimation of

$$
\begin{aligned}
a(\theta) = P_\theta(X < 0) &= 1, && \text{if } \theta < -\tfrac{1}{2}, \\
&= \tfrac{1}{2} - \theta, && \text{if } -\tfrac{1}{2} \le \theta \le \tfrac{1}{2}, \qquad (3.2)\\
&= 0, && \text{if } \theta > \tfrac{1}{2}.
\end{aligned}
$$

The estimator

$$
\begin{aligned}
\delta_0(X) &= 1, \quad \text{if } X < 0, \\
&= 0, \quad \text{otherwise},
\end{aligned}
$$

has the following two properties:

(a) $\delta_0(X)$ is unbiased;
(b) $\operatorname{Var}_\theta[\delta_0(X)] = 0$ when $\theta < -1$ and when $\theta > 1$.

We shall now show that an estimator based on Y that shares these properties can exist only if $\rho = 1/k$. Here attention can be restricted to nonrandomized estimators $\delta(Y)$ since if $\delta'(Y, Z)$ has properties (a) and (b), where Z has a known distribution, so does $\delta(Y) = E[\delta'(Y, Z) | Y]$.

Suppose now that $\delta(Y)$ satisfies (b). Then $\delta(Y)$ must be constant (a.e.) for $y > 1 - \frac{1}{2}\rho$ and for $y < -1 + \frac{1}{2}\rho$, and these constants must be 0 and 1, respectively, if $\delta(Y)$ is to be unbiased. By differentiating the unbiasedness condition

$$
\rho^{-1} \int_{\theta - \rho/2}^{\theta + \rho/2} \delta(y)\, dy = a(\theta), \quad \text{for all } \theta,
$$

one finds further that $\delta(y)$ must satisfy (a.e.)

$$
\begin{aligned}
\delta\left(y - \frac{\rho}{2}\right) - \delta\left(y + \frac{\rho}{2}\right) &= \rho \quad \text{if } -\tfrac{1}{2} < y < \tfrac{1}{2}, \qquad (3.3)\\
&= 0, \quad \text{otherwise}.
\end{aligned}
$$

For almost all sequences $\delta(y \pm j\rho)$, $j = 0, 1, 2, \ldots$, δ must therefore decrease from its value 1 near $j = -\infty$ to its value 0 near $j = +\infty$ by steps the sizes of which are restricted to 0 and ρ. This is possible only when $\rho = 1/k$ for some positive integer k. □

It seems plausible that the difficulty in this example is caused by the insistence on δ being unbiased, and that an estimator $\delta(Y)$ with risk uniformly smaller than that of $\delta_0(X)$ does exist for most reasonable loss functions. It will be seen at the end of Section 4 that this is indeed the case.

4. Monotone decision problems. The examples discussed in the preceding sections make it clear that condition (1.1) for comparability is too strong to hold in many situations in which intuition suggests that one experiment is more informative than another.

Three approaches to weakening the requirements for comparability have been proposed.

(i) Le Cam (1964) replaces condition (1.2) by the approximate condition that each risk function based on X can be matched within ε by one based on Y.

(ii) Several authors suggest that comparisons should be made in terms of some measure(s) of information such as Fisher, Shannon or Kullback–Leibler information.

(iii) Throughout the literature on the comparison of experiments, the suggestion occurs of comparing two experiments not for all decision problems but only for some family $\mathbf{C}$ of problems. For such restricted comparisons, (1.3) is the relevant determining condition. So as to distinguish this approach from the classical one, we shall say that Y is *more effective* than X with respect to the class $\mathbf{C}$ of decision problems concerning θ if for any problem in $\mathbf{C}$ (specified by a set of possible decisions and a loss function) and any decision procedure for this problem based on X, there exists a procedure δ' based on Y such that $R(\theta, \delta') \leq R(\theta, \delta)$ for all θ. It is this last approach with which we shall be concerned in the remainder of this paper. More particularly, we shall be interested in defining a class $\mathbf{C}$ that is large enough to include most of the statistical problems of interest, but not so large that comparability becomes practically impossible.

A suitable such class was introduced by Karlin and Rubin (1956) [see also Brown, Cohen and Strawderman (1976)] for the case that θ is a real-valued parameter. This class, the class of *monotone procedures*, which we shall denote by $\mathbf{M}$, is defined in terms of the action space and the permissible loss functions. For each θ it is assumed that there is a "correct" action $a(\theta)$, and the function $a(\theta)$ is assumed to be real-valued and nondecreasing. The set of values taken on by $a(\theta)$ as θ ranges over Ω is the action space A. The loss function is assumed to satisfy $L(\theta, a(\theta)) = 0$ for all θ, with $L(\theta, d)$ nondecreasing as a function of d as d moves away from $a(\theta)$ on either side.

The basic result of Karlin and Rubin (KR) is concerned not with a comparison problem but with a single family of densities $p_\theta(x)$ with monotone likelihood ratio in $T(x)$, which without loss of generality we shall take to be x. It then states that for any monotone decision problem for such a family p_θ, the class $\mathbf{M}$ of monotone procedures is essentially complete. Here a nonrandomized procedure $\delta(x)$ is monotone if

$$x < x' \Rightarrow \delta(x) \leq \delta(x').$$

A randomized procedure δ is monotone if for any $x < x'$,

(4.1) $\delta(x)$ assigns positive probablity to $[a\ ,\ \infty)$ implies that $\delta(x')$ assigns probability 1 to $[a\ ,\ \infty)$.

[Under slightly stronger assumptions, Brown, Cohen and Strawderman (1976) show $\mathbf{M}$ to be complete rather than only essentially complete.]

For the case of a finite number of actions (or decisions) $a_1, \ldots, a_k$, a decision procedure can be described by k functions $\phi = (\phi_1, \ldots, \phi_k)$ with

$$0 \leq \phi_i \leq 1 \quad \text{and} \quad \sum_{i=1}^{k} \phi_i(x) = 1.$$

It follows from (4.1) that such a procedure ϕ is monotone provided there exist $k+1$ points $x_1 = -\infty \le x_2 \le \cdots \le x_{k+1} = +\infty$ such that $\phi_i(x) = 1$ or 0 as x lies in or outside the interval (x_i, x_{i+1}) with possible randomization on the end points. An example is provided by the case of two decisions corresponding to testing a hypothesis H: $\theta \le \theta_0$ against the alternatives $\theta > \theta_0$; or the three-decision problem in which the hypothesis H: $\theta_0 \le \theta \le \theta_1$, $\theta_0 \le \theta_1$, is to be accepted (decision d_0) or rejected in favor either of the alternatives $\theta < \theta_0$ (decision d_1) or $\theta > \theta_1$ (decision d_2) with suitable losses for wrong decisions. [This case is treated in some detail in Ferguson (1967), Section 6.1.] The infinite action case is illustrated by the estimation of a function $a(\theta)$, for example, the estimation of a continuous parameter θ with squared error loss, or the situation described by (3.2).

5. Relative effectiveness of location families. The aim of the present section is to provide necessary and sufficient conditions for Y to be more effective than X in problems 1 and 2 with respect to the class **M**. However, we begin by presenting the result in its natural, somewhat more general setting.

Let **P** and **Q** consist of the families

$$P_\theta(X \le x) = F_\theta(x) \quad \text{and} \quad Q_\theta(Y \le y) = G_\theta(y). \tag{5.1}$$

THEOREM 5.1. *Let F_θ and G_θ have densities f_θ and g_θ with respect to a common σ-finite measure μ, and suppose that the families $f_\theta(x)$ and $g_\theta(x)$ have monotone likelihood ratio in x. Then a necessary and sufficient condition for Y to be more effective than X with respect to* **M** *is that the function*

$$h_\theta(x) = G_\theta^{-1}[F_\theta(x)] \text{ is a nondecreasing function of } \theta \text{ for each } x. \tag{5.2}$$

PROOF. By the KR theorem, attention can be restricted to the ability of Y to dominate any monotone procedure δ based on X, i.e., to prove the existence of a procedure δ' based on Y with risk uniformly no greater than that of δ. Although this is not required for the proof, the construction will produce a monotone δ'.

Assuming (5.2), we begin by showing that for any θ_0 and any $0 < \alpha < 1$, given any level α test of H: $\theta \le \theta_0$ against $\theta > \theta_0$ based on X, there exists a test for the same problem based on Y that is uniformly at least as powerful for $\theta > \theta_0$ and uniformly at most as powerful for $\theta < \theta_0$.

Since there exist tests based on X and Y, respectively, that are simultaneously uniformly most powerful against $\theta > \theta_0$ and uniformly least powerful against $\theta < \theta_0$, it is enough to show that the claimed relationship holds for these two tests. For this purpose, we shall assume without loss of generality that F_{θ_0} and G_{θ_0} are continuous. This can always be achieved by the following construction. Suppose X has a discontinuity at x_0, with $P(X = x_0) = p$ when $\theta = \theta_0$. Define a new variable X^* by

$$\begin{aligned} X^* &= X, && \text{if } X < x_0, \\ &= X + pU, && \text{if } X = x_0, \\ &= X + p, && \text{if } X > x_0, \end{aligned}$$

where U is uniformly distributed on $(0,1)$. Then the distribution of X^* is continuous when $\theta = \theta_0$ and the family of distributions of X^* has monotone likelihood ratio. Furthermore, the experiments X and X^* are equivalent. This is seen from the fact that clearly X^* is equivalent to the pair (X, U), for which X is a sufficient statistic.

Assume therefore that F_{θ_0} and G_{θ_0} are both continuous. Then the optimal test based on X is given by the rejection region $X > a$ and that based on Y by $Y > b$. Then a and b satisfy $F_{\theta_0}(a) = G_{\theta_0}(b) = 1 - \alpha$ and are therefore related by

$$b = G_{\theta_0}^{-1}\left[F_{\theta_0}(a)\right]. \tag{5.3}$$

The power of the two tests against any $\theta > \theta_0$ is $1 - F_\theta(a)$ and $1 - G_\theta(b)$, respectively, so that domination of X by Y requires

$$G_\theta(b) \le F_\theta(a), \quad \text{for all } \theta > \theta_0,$$

and analogously

$$G_\theta(b) \ge F_\theta(a), \quad \text{for all } \theta < \theta_0.$$

Substitution from (5.3) shows the inequality to be equivalent to

$$G_{\theta_0}^{-1}\left[F_{\theta_0}(a)\right] \le G_\theta^{-1}\left[F_\theta(a)\right], \quad \text{for all } \theta > \theta_0,$$

with the inequality being reversed for $\theta < \theta_0$. Since this must hold for all a, it is seen that (5.2) is a necessary condition for Y to be more effective than X for all one-sided testing problems.

To complete the proof of sufficiency, one can now plug into the proof of the KR theorem. The result established so far corresponds exactly to Lemma 3 of Karlin and Rubin. To construct a procedure based on Y to dominate any given monotone procedure with finite decision space based on X, one now simply applies the arguments of their Lemma 4 and Theorem 1. Having established Y's dominance for all finite problems in $\mathbf{M}$, one then passes to the limit as in Section 7 of Karlin and Rubin to establish the result for the infinite case. □

We next specialize Theorem 5.1 to Problem 1.

Theorem 5.2. *Let the distributions of X and Y be given by* (2.1) *and suppose that $F(x-\theta)$ and $G(x-\theta)$ have densities (with respect to Lebesgue measure) that are strongly unimodal. (A density f is strongly unimodal if $-\log f$ is convex.) Then a necessary and sufficient condition for Y to be more effective than X with respect to $\mathbf{M}$ is that*

$$\frac{G^{-1}[F(b)] - G^{-1}[F(a)]}{b-a} \le 1, \quad \textit{for all } a < b. \tag{5.4}$$

Proof. If f and g are strongly unimodal, the families $f(x-\theta)$ and $g(x-\theta)$ have monotone likelihood ratio in x. The function h defined by (5.2) in the present case reduces to

$$h_\theta(x) = G^{-1}[F(x-\theta)] + \theta$$

and hence the inequality $h_\theta(x) \le h_{\theta'}(x)$ to (5.4) with $a = x - \theta'$ and $b = x - \theta$. □

Putting $F(a) = u$, $F(b) = v$, we can rewrite (5.4) as

$$G^{-1}(v) - G^{-1}(u) \le F^{-1}(v) - F^{-1}(u), \quad \text{for all } 0 < u < v < 1. \tag{5.5}$$

Condition (5.5) states that F is more spread out than G in the sense that any two quantiles are at least as far apart under F than they are under G. These are just the circumstances under which one would expect inferences about the location of F to be more difficult than those about the location of G. Restriction to monotone problems has thus replaced the original rather strange and—as it turned out—not very useful condition (2.2) with one that nicely quantifies our intuition.

Condition (5.5) was discussed in Bickel and Lehmann (1979) as the definition of F being *more spread out* than G. It was also pointed out there that if F^{-1} and G^{-1} are differentiable, (5.5) is equivalent to

$$\frac{f[F^{-1}(y)]}{g[G^{-1}(y)]} \le 1, \quad \text{for all } 0 < y < 1. \tag{5.6}$$

Still another equivalent condition is given in Theorem 1 of Bickel and Lehmann.

Let us finally specialize the preceding results to Problem 2, defined by (2.4). Since in that case $G(y) = F(y/\rho)$, $0 < \rho < 1$, we have $G^{-1}(y) = \rho F^{-1}(y)$, and it is seen that (5.5) holds for all F. We have thus proved

Theorem 5.3. *Let the distributions of X and Y be given by* (2.4) *and suppose that F has a density f which is strongly unimodal. Then Y is more effective than X relative to* **M** *for all* $0 < \rho < 1$.

Since the uniform distribution is strongly unimodal, this establishes in particular the conjecture expressed at the end of Section 3 for any loss function meeting the KR conditions of case 1 of Section 4.

6. Scale-free comparisons and a tail-ordering. Conditions (5.4)–(5.6) make it possible for any particular F and G to decide on the comparability of the two families (2.1) with respect to **M**. We shall in the present section consider a scale-free version of this comparison.

Problem 3. Given F and G, does there exist ρ sufficiently small so that $G((y - \theta)/\rho)$ is more effective than $F(x - \theta)$ with respect to **M**?

When such a ρ exists, we shall say that G has *a more effective shape* than F with respect to **M**.

It follows from (5.5) that G has a more effective shape than F if and only if there exists ρ such that

$$\frac{G^{-1}(v) - G^{-1}(u)}{F^{-1}(v) - F^{-1}(u)} \le \frac{1}{\rho}, \quad \text{for all } 0 < u < v < 1,$$

i.e., if and only if

$$\frac{G^{-1}(v) - G^{-1}(u)}{F^{-1}(v) - F^{-1}(u)} \text{ is bounded.} \tag{6.1}$$

If F^{-1} and G^{-1} are differentiable, this reduces to

$$\frac{f[F^{-1}(y)]}{g[G^{-1}(y)]} \text{ is bounded.} \tag{6.2}$$

The function $f[F^{-1}(y)]$ is studied in a different context by Parzen (1979) who calls it the density-quantile function, and considers its limiting behavior as $y \to 0$ or 1 a measure of tail weight. He also evaluates this function for a number of important distributions. Condition (6.2) is also closely related to the s-ordering of symmetric distributions introduced by van Zwet (1964), who requires that

$$\frac{f[F^{-1}(y)]}{g[G^{-1}(y)]} \text{ be } \begin{matrix} \text{decreasing for } y > \frac{1}{2}, \\ \text{increasing for } y < \frac{1}{2}. \end{matrix} \tag{6.3}$$

The ratio therefore attains its maximum at $y = \frac{1}{2}$ and if f is bounded, (6.3) implies (6.2).

As discussed by Loh (1984b), van Zwet's s-ordering and some related orderings (including Loh's t-ordering) take into account not only the heaviness of the tail but also the behavior of f (its "peakedness") at the center. In contrast, condition (6.2) provides a definition of pure tail-ordering, provided attention is restricted to densities f, g such that f is bounded and $g[G^{-1}(y)]$ is bounded away from 0 in any interval $0 < a < y < b < 1$. As an example, if G is double exponential and F Cauchy, then F is not tail-heavier according to the s-ordering, but it is strictly tail-heavier according to the ordering (6.2) since [see Parzen (1979)]

$$\begin{aligned} f[F^{-1}(u)] &= 1 - u, \quad \text{for } u > \tfrac{1}{2}, \\ &= u, \qquad\quad \text{for } u < \tfrac{1}{2}, \end{aligned}$$

when F is standard double exponential and

$$\begin{aligned} f[F^{-1}(u)] = \frac{1}{\pi}\sin^2(\pi u) &\sim \pi u^2, \qquad \text{as } u \to 0 \\ &\sim \pi(1-u)^2, \quad \text{as } u \to 1, \end{aligned}$$

when F is standard Cauchy, where $\sim$ means that the ratio tends to 1. As another example, note that according to the s-ordering, the double exponential is heavier-tailed than the logistic, whereas the two are equivalent according to the ordering (6.3). This is seen from the fact that for the logistic distribution

$$f[F^{-1}(u)] = u(1-u).$$

Let us now return to Problem 3 and provide examples of some situations in which G has a more effective shape than F by being lighter-tailed according to Definition (6.2). Suppose, for example, that G is uniform (or has any other distribution whose density g is bounded away from 0 and ∞ on its support). It then follows from (6.2) that the shape of G is more effective than that of any distribution F with bounded density, and that it is strictly more effective if the bounded density of F is not bounded away from 0, e.g., if F is triangular, normal, etc.

As another example, suppose that G is the triangular distribution with density

$$g(x) = 1 - |x|, \qquad -1 < x < 1.$$

Then $g[G^{-1}(y)] = \sqrt{2(1-y)}$, and G is lighter-tailed than the logistic and double exponential distributions but heavier-tailed than the extreme value distribution for which $f[F^{-1}(y)] = -(1-y)\log(1-y)$ (see Parzen, l.c.)

7. The case of n observations. So far, attention has been restricted to a single observation from model (2.1) or (2.4). We shall now generalize Theorem 5.1 to the case that $X_1, \cdots, X_n; Y_1, \ldots, Y_n$ are i.i.d. with distributions F_θ and G_θ, respectively, where θ continues to be real-valued. The definition of the class **M** does not require any modification since it concerns only the parameter and decision space but not the sample space.

THEOREM 7.1. *Under the assumptions of Theorem* 5.1, *condition* (5.2) *is necessary and sufficient for* $Y = (Y_1, \ldots, Y_n)$ *to be more effective than* $X = (X_1, \ldots, X_n)$ *with respect to* **M**.

PROOF. A decision procedure $\delta(x_1, \ldots, x_n)$ is said to be monotone if

$$x_i \le x_i', \quad \text{for all } i = 1, \ldots, n \text{ implies that } \delta(x) \le \delta(x'). \tag{7.1}$$

The KR theorem discussed in Section 4 was generalized to n i.i.d. variables from a MLR family by Oosterhoff (1969), Brown, Cohen and Strawderman (1976) and Van Houwelingen and Verbeek (1985), who state that as in the one-dimensional case the class of monotone procedures is essentially complete. We can therefore as in the proof of Theorem 5.1 restrict attention to the problem of dominating any monotone procedure based on X.

In analogy to the proof of Theorem 5.1, we begin by showing that for any θ_0 and $0 < \alpha < 1$, given any monotone level α test ϕ of H: $\theta \le \theta_0$ against $\theta > \theta_0$ based on $X = (X_1, \ldots, X_n)$, there exists a test for the same problem based on $Y = (Y_1, \ldots, Y_n)$, which is uniformly at least as powerful for $\theta > \theta_0$ and uniformly at most as powerful for $\theta < \theta_0$. Note, however, that there is now no longer a unique monotone level α test (which is uniformly most powerful) but a large class of such tests.

To establish a test dominating ϕ, let us denote the distributions F_{θ_0} and G_{θ_0} by F and G, respectively, and as in the proof of Theorem 5.1, assume without loss of generality that F_{θ_0} and G_{θ_0} are continuous. Let $Z_i = h(Y_i)$, where $h = F^{-1}G$. We shall show in the following discussion that for any monotone test ϕ based on X there exists a uniformly better test based on Z. Since Y is at least as informative as Z, this will prove the existence of a test based on Y, which is uniformly better than ϕ.

Note that X_i and Z_i have the same distribution $F = F_{\theta_0}$ when $\theta = \theta_0$. We shall now show that the distribution H_θ of Z_i satisfies

$$(7.2) \qquad \begin{aligned} H_\theta(z) &\le F_\theta(z) \quad \text{when } \theta > \theta_0, \\ &\ge \qquad\quad\ \ \text{when } \theta < \theta_0. \end{aligned}$$

To see this, note that

$$H_\theta(z) = P_\theta\left[F^{-1}G(Y_i) \le z\right] = G_\theta\{G^{-1}[F(z)]\}.$$

Therefore $H_\theta(z) \le F_\theta(z)$ provided

$$G_{\theta_0}^{-1}\left[F_{\theta_0}(z)\right] \le G_\theta^{-1}\left[F_\theta(z)\right],$$

which by (5.2) is the case when $\theta_0 < \theta$ with the opposite inequality holding when $\theta < \theta_0$.

Condition (7.2) states that Z_i is stochastically larger or smaller than X_i as $\theta > \theta_0$ or $< \theta_0$.

Let $\phi(X)$ be any monotone test based on $X = (X_1, \dots, X_n)$ and let $\phi^*(Y) = \phi(Z)$ be the same test based on $Z = (Z_1, \cdots, Z_n)$. Then it follows from the basic property of stochastically ordered random variables, given for example in Chapter 3, Lemma 1 of Lehmann (1986), that

$$\begin{aligned} E_\theta\phi(X) &\le E_\theta\phi(Z) \quad \text{for } \theta > \theta_0, \\ &\ge \qquad\qquad\ \ \text{for } \theta < \theta_0, \end{aligned}$$

as was to be proved.

For more general decision procedures, we can proceed as in Brown, Cohen and Strawderman (1976) or Van Houwelingen and Verbeek (1985) to complete the proof of the theorem. □

It is an immediate consequence of Theorem 7.1 that not only Theorem 5.1 but also Theorems 5.2 and 5.3 and the results of Section 6 remain valid when the single variable X is replaced by a sample $X_1, \dots, X_n$.

REFERENCES

BICKEL, P. J. and LEHMANN, E. L. (1979). Descriptive statistics for nonparametric models. IV. In *Contributions to Statistics, Jaroslav Hájek Memorial Volume* (J. Jurečková, ed.). Academia, Prague.

BLACKWELL, D. (1951). Comparison of experiments. *Proc. Second Berkeley Symp. Math. Statist. Probab.* 93–102. Univ. California Press.

BLACKWELL, D. (1953). Equivalent comparison of experiments. *Ann. Math. Statist.* **24** 265–272.

BOLL, C. H. (1955). Comparison of experiments in the infinite case and the use of invariance in establishing sufficiency. Ph.D. thesis, Stanford Univ.

BROWN, L. D., COHEN, A. and STRAWDERMAN, W. E. (1976). A complete class theorem for strict monotone likelihood ratio with applications. *Ann. Statist.* **4** 712–722.

DEGROOT, M. H. (1966). Optimal allocation of observations. *Ann. Inst. Statist. Math.* **18** 13–28.

FELDMAN, D. and RAMAMOORTHI, R. V. (1986). Equivalent comparison of experiments—another proof of Blackwell's theorem. Unpublished.

FERGUSON, T. S. (1967). *Mathematical Statistics: A Decision-Theoretic Approach.* Academic, New York.

GOEL, P. K. and DEGROOT, M. H. (1979). Comparison of experiments and information measures. *Ann. Statist.* **7** 1066–1077.

HANSEN, O. H. and TORGERSEN, E. N. (1974). Comparison of linear normal experiments. *Ann. Statist.* **2** 367–373.

KARLIN, S. and RUBIN, H. (1956). The theory of decision procedures for distributions with monotone likelihood ratio. *Ann. Math. Statist.* **27** 272–299.

LE CAM, L. (1964). Sufficiency and approximate sufficiency. *Ann. Math. Statist.* **35** 1419–1455.

LEHMANN, E. L. (1986). *Testing Statistical Hypotheses*, 2nd ed. Wiley, New York.

LOH, W.-Y. (1984a). Strong unimodality and scale mixtures. *Ann. Inst. Statist. Math.* **36** 441–449.

LOH, W.-Y. (1984b). Bounds on ARE's for restricted classes of distributions defined via tail-orderings. *Ann. Statist.* **12** 685–701.

OOSTERHOFF, J. (1969). Combination of one-sided statistical tests. *Math. Centre Tracts* **28**. Mathematical Centre, Amsterdam.

PARZEN, E. (1979). Nonparametric statistical data modeling (with discussion). *J. Amer. Statist. Assoc.* **74** 105–131.

STONE, M. (1961). Non-equivalent comparisons of experiments and their use for experiments involving location parameters. *Ann. Statist.* **32** 326–332.

VAN HOUWELINGEN, H. and VERBEEK, A. (1985). On the construction of monotone symmetric decision rules for distributions with monotone likelihood ratio. *Scand. J. Statist.* **12** 73–81.

VAN ZWET, W. R. (1964). Convex transformations of random variables. Mathematical Centre, Amsterdam.

DEPARTMENT OF STATISTICS
UNIVERSITY OF CALIFORNIA
BERKELEY, CALIFORNIA 94720

The Annals of Statistics
1992, Vol. 20, No. 4, 2100–2110

INVARIANT DIRECTIONAL ORDERINGS

BY E. L. LEHMANN[1] AND J. ROJO[2]

University of California, Berkeley, and University of Texas, El Paso

Statistical concepts of order permeate the theory and practice of statistics. The present paper is concerned with a large class of directional orderings of univariate distributions. (What do we mean by saying that a random variable Y is larger than another random variable X?) Attention is restricted to preorders that are invariant under monotone transformations; this includes orderings such as monotone likelihood ratio, hazard ordering, and stochastic ordering. Simple characterizations of these orderings are obtained in terms of a maximal invariant. It is shown how such invariant preorderings can be used to generate concepts of Y_2 being further to the right of X_2 than Y_1 is of X_1.

1. Introduction. The concept of statistical order which compares two random variables or distribution functions has many uses in both applied and theoretical statistics. In reliability theory, the concepts of increasing failure rate, increasing failure rate on the average, and new better than used, compare distribution functions F and G in terms of the function $\phi = G^{-1}F$ [Van Zwet (1964), Barlow and Proschan (1975) and Loh (1984)]. The monotonicity of the power of monotone rank tests is considered by Lehmann (1959) and Doksum (1969). Comparisons of distributions based on tail heaviness or skewness were treated respectively by Rojo (1988, 1992) and MacGillivray (1986). For some extensions to the multivariate case see, for example, Lehmann (1952, 1955), Whitt (1982), Keilson and Sumita (1983) and Karlin and Rinott (1983) and, to the comparison of stochastic processes, Pledger and Proschan (1973) and Whitt (1981).

In this paper, attention will be restricted to univariate directional orderings which are concerned with the question of whether one distribution is in some sense to the right of the other. In this connection many different ordering concepts have been proposed, of which we mention the following:

(A) Stochastic ordering. The random variable Y is said to be stochastically larger than X if their distribution functions G and F satisfy

$$F(x) \geq G(x) \quad \text{for all } x, \tag{1.1}$$

and this will be denoted by $F \leq_{\text{st}} G$.

Received November 1990; revised November 1991.
[1]Research supported by NSF Grant DMS-89-08670
[2]Research partially supported by NSF Grant DMS-86-06420.
AMS 1980 *subject classifications.* Primary 62A05, 62E10; secondary 62B15, 60E05.
Key words and phrases. Stochastic ordering, monotone likelihood ratio ordering, hazard ordering, location families, distance between distribution functions.

(B) Monotone likelihood ratio (MLR). When F and G have densities f and g with respect to some common dominating measure μ, and

$$g/f \text{ is nondecreasing,} \tag{1.2}$$

the pair (F, G) is said to have monotone likelihood ratio, which we denote by $F \leq_{\mathrm{mlr}} G$.

(C) Hazard ordering. A cdf F is said to be smaller than G in the hazard ordering ($F \leq_{\mathrm{h}} G$) if

$$\frac{g(x)}{1 - G(x)} \leq \frac{f(x)}{1 - F(x)} \quad \text{for all } x. \tag{1.3}$$

Here $f(x)/[1 - F(x)]$ is the "mortality" of a subject at time x given that it has survived to this time. If the densities f and g exist, condition (1.3) is easily seen to be equivalent to

$$\frac{1 - F(x)}{1 - G(x)} \text{ is nonincreasing.} \tag{1.4}$$

(D) Restricted definitions. It is not always appropriate to require that the comparisons (A)–(C) hold for all values of x. For example, in a comparison of lengths of lives of women and men, one may want to restrict the comparison to childhood or perhaps to ages past childbearing. We consider two cases of such relations:

(i) Definitions (A)–(C) for all x exceeding a specified x_0.
(ii) Definitions (A)–(C) for all sufficiently large x.

(E) Comparisons based on a single functional. One population of heights, ages, incomes, etc. is frequently considered to be larger than another such population if it is larger "on the average," or if the median of the first population exceeds that of the second. More generally, a rather weak comparison of two distributions can be defined in terms of some measure μ of location, that is, by seeing whether $\mu(F)$ is larger or smaller than $\mu(G)$.

Examples (A)–(E) seem to constitute a rather haphazard collection. The purpose of the present paper is to present a more systematic way of defining and studying such orderings and to obtain simple characterizations of the orderings discussed above subject only to very general restrictions on the distribution functions F and G such as continuity or being strictly increasing. Possible approaches are suggested by two well-known rather obvious characterizations of stochastic ordering, namely, that (1.1) holds if and only if

$$F^{-1}[G(x)] \leq x \quad \text{for all } x \tag{1.5}$$

or if

(1.6) $E_F\psi(X) \le E_G\psi(X)$ for all nondecreasing functions ψ, where here and below $F^{-1}(u) = \inf\{t: F(t) \ge u\}$.

The following are natural extensions of these characterizations:

1. Let $F \le G$ when $\phi = F^{-1}G$ is a member of some specified class of functions. For discussion of such an approach see, for example, Oja (1981) and Loh (1984).
2. Let $F \le G$ when

$$E_F[\psi(X)] \le E_G[\psi(X)]$$

for all ψ belonging to some specified class [see, e.g., Whitt (1980)].

Unfortunately it turns out that neither monotone likelihood ratio nor hazard ordering can be characterized in either of these ways. We develop here an alternative approach which will provide a fairly simple characterization of a large class of directional orderings including (A)–(C).

2. Invariant directional preorderings. In this section we restrict attention to the class of distributions $\mathbf{F}$ defined as follows:

(2.1) $$\mathbf{F} = \{F: F \text{ is continuous and strictly increasing on } (-\infty, \infty)\}.$$

We shall later weaken this restriction.

A preorder of the distributions $F \in \mathbf{F}$ is a set S of ordered pairs (F, G) in $\mathbf{F} \times \mathbf{F}$ satisfying

(2.2) $$(F, F) \in S \quad \text{for all } F \in \mathbf{F}$$

and

(2.3) $$(F, G) \in S \quad \text{and} \quad (G, H) \in S \quad \text{implies} \quad (F, H) \in S.$$

When $(F, G) \in S$, we say that $F \le_S G$ or $X \le_S Y$, where X and Y are random variables with distributions F and G, respectively.

Definition. The preorder S is invariant (under monotone transformations) if

(2.4) $$X \le_S Y \quad \text{implies} \quad \psi(X) \le_S \psi(Y) \quad \text{for all } \psi \in \Psi,$$

where Ψ is the class of all strictly increasing continuous functions ψ with $\psi(-\infty) = -\infty$, $\psi(+\infty) = +\infty$.

It is easy to see that (A)–(E) in Section 1 all satisfy (2.2) and (2.3). Examples (A)–(C) and (D)(ii) also satisfy (2.4) but (D)(ii) does not.

When X is distributed according to F, the distribution of $\psi(X)$ is $F\psi^{-1}$, and in terms of S condition (2.4) becomes

(2.5) $$(F, G) \in S \quad \Rightarrow \quad (F\psi^{-1}, G\psi^{-1}) \in S \quad \text{for all } \psi \in \Psi.$$

To see the simplification resulting from (2.5) consider the orbits under the group of transformations $\psi \in \Psi$ in the space $\mathbf{F} \times \mathbf{F}$ of pairs (F, G):

(2.6) $$\mathbf{O}(F, G) = \{(F\psi^{-1}, G\psi^{-1}): \psi \in \Psi\}.$$

It follows from the definition of the orbits and (2.5) that $(F, G) \in S$ if and only if the whole orbit $\mathbf{O}(F, G)$ is contained in S. Therefore, to characterize an invariant preordering S, it is sufficient to list the totality of orbits in S, and for that purpose it is convenient to have available a suitable labeling of the orbits, that is, a maximal invariant under Ψ. It is easy to see that such a labeling is provided by

$$k(u) = GF^{-1}(u), \qquad 0 \leq u \leq 1. \tag{2.7}$$

On the one hand, k is invariant under Ψ since $G\psi^{-1}(F\psi^{-1})^{-1} = GF^{-1}$; on the other hand, if $G_1F_1^{-1} = G_2F_2^{-1}$, we have $F_1^{-1}F_2 = G_1^{-1}G_2 = \psi$, and thus $F_1 = F_2\psi^{-1}$ and $G_1 = G_2\psi^{-1}$ so that (F_1, G_1) and (F_2, G_2) lie on the same orbit.

For any $F, G \in \mathbf{F}$, it is seen that $k = GF^{-1}$ has the following properties:

(2.8) $k(0) = 0$ and $k(1) = 1$, and on $(0, 1)$ k is continuous and strictly increasing.

While MLR and hazard ordering cannot be characterized in terms of $F^{-1}G$, they have very simple characterizations in terms of $k = GF^{-1}$.

THEOREM 1.

(i) $F \leq_{\mathrm{mlr}} G \Leftrightarrow k(\cdot)$ *is convex*;
(ii) $F \leq_{\mathrm{h}} G \Leftrightarrow 1 - k^{-1}(1 - \mu)$ *is star-shaped*;
(iii) $F \leq_{\mathrm{st}} G \Leftrightarrow k(u) \leq u$ *for all* $0 < u < 1$.

PROOF. The proof of (i) follows easily by noting that the derivative of k is $k' = gF^{-1}/fF^{-1}$ and recalling that k is convex if and only if k' is nondecreasing. To prove (ii), note that $(1 - k^{-1}(1 - u))$ is star-shaped if and only if $(1 - FG^{-1}(t))/(1 - t)$ is nonincreasing in t or, equivalently, if $\bar{F}/\bar{G}$ is nonincreasing, where $\bar{F}$ denotes $1 - F$. Then it is easy to verify that the latter condition is equivalent to (1.3). The proof of (iii) is trivial. □

THEOREM 2. *Inclusion relationships among* (i)–(iii) *are given in the following diagram*:

<table>
<tr><td>$k(\cdot)$ convex</td><td>$\Leftrightarrow$ MLR $\Leftrightarrow$</td><td>$1 - k^{-1}(1 - u)$ convex</td></tr>
<tr><td>‖</td><td></td><td>‖</td></tr>
<tr><td>$k(\cdot)$ star-shaped</td><td></td><td>$1 - k^{-1}(1 - u)$ star-shaped $\Leftrightarrow F \leq_h G$</td></tr>
<tr><td>‖</td><td></td><td>‖</td></tr>
<tr><td>G/F nondecreasing</td><td></td><td>$\bar{G}/\bar{F}$ nondecreasing</td></tr>
<tr><td>$\searrow$</td><td></td><td>$\swarrow$</td></tr>
<tr><td></td><td>$k(u) \leq u$</td><td></td></tr>
<tr><td></td><td>‖</td><td></td></tr>
<tr><td></td><td>$F \leq_{\mathrm{st}} G$</td><td></td></tr>
</table>

PROOF. The nontrivial aspects follow from the well-known fact that, for functions going through the origin, convexity implies star-shapedness. □

3. Characterization of sets of k-functions defining preorders. It was pointed out in Section 2 that the partial order (D)(i) is not invariant. On the other hand, for example, (D)(i) applied to (A) has a k-analogue

$$k(u) \geq u \quad \text{for all } u \geq u_0, \tag{3.1}$$

which clearly is invariant. It is easy to translate (3.1) into a condition on the pair (F, G); the condition is simply

$$G^{-1}(u) \geq F^{-1}(u) \quad \text{for all } u \geq u_0. \tag{3.2}$$

This example shows that it is sometimes convenient to start with a preorder of the k-functions and from it to derive the equivalent ordering rather than the other way around. To be able to do so, we need to know when the set

$$S = \{(F, G) : k(F, G) \in \mathbf{K}\}$$

is a preorder. Clearly the set S will satisfy (2.2) if and only if the function $k(u) = u$ for all $0 < u < 1$ is in $\mathbf{K}$. To satisfy (2.3), $\mathbf{K}$ must satisfy

$$GF^{-1} \in \mathbf{K} \quad \text{and} \quad HG^{-1} \in \mathbf{K} \quad \Rightarrow \quad HF^{-1} \in \mathbf{K}.$$

But if $GF^{-1} = k_1$ and $HG^{-1} = k_2$, then $HF^{-1} = k_2(k_1)$ so that (2.3) will hold provided $\mathbf{K}$ is closed under composition.

Thus $\mathbf{K}$ corresponds to a preorder provided it contains the identity function and is closed under composition.

As an application consider the problem of finding a preorder that is stronger than monotone likelihood ratio. In k-space, a natural strengthening of convexity is to require that

$$k^{(i)}(u) \geq 0 \quad \text{for all } 0 < u < 1 \text{ and all } i = 1, \dots, n. \tag{3.3}$$

The identity function obviously satisfies (3.3) and it is not difficult to show that the class $\mathbf{K}$ defined by (3.3) is closed under composition. The conditions on (F, G) corresponding to (3.3) do not appear to have any simple interpretation. As an illustration of (3.3), consider the following examples.

EXAMPLE 3.1. Let $k(u) = u^n$, $n \geq 3$. Then $k^{(i)}(u) \geq 0$, $0 \leq u \leq 1$, for $i = 1, 2, 3, \dots, n$. Moreover, it is easy to see that $k(u) = u^n$ implies that $G(t) = (F(t))^n$, and hence the nth order statistic $X_{(n)}$ in a random sample of size n from F satisfies (3.3) so that $X < X_{(n)}$ in the sense of (3.3).

EXAMPLE 3.2. In the normal location case, say with $F(x) = \Phi(x)$, $G(x) = \Phi(x - \theta)$, $\theta \geq 0$, we have MLR so that k satisfies (3.3) with $n = 2$; however, it does not satisfy (3.3) with $n = 3$.

A location family satisfying (3.3) for all n is given in the following example.

EXAMPLE 3.3. Consider the extreme value location family of distributions with densities given by

$$f_\theta(x) = e^{x-\theta}e^{-e^{x-\theta}}. \tag{3.4}$$

For $\theta > 0$, it is easy to see that

$$k(u) = F_\theta F_0^{-1}(u) = 1 - (1-u)^\lambda, \tag{3.5}$$

where $\lambda = e^{-\theta} < 1$. It then follows that $k(u)$ satisfies (3.3) for every positive integer n.

EXAMPLE 3.4. Consider the family of exponential distributions with density $(1/a)e^{-x/a}$. To see that this is ordered in the sense of (3.3) for all n, we need only note that if X is distributed according to (3.4), then the distribution of e^X is exponential with scale parameter $a = e^\theta$. Since any invariant order is unaffected by a monotone transformation, (3.3) follows from Example 3.3.

4. Extension to a larger class of distribution functions. At the beginning of Section 2 we restricted attention to the class $\mathbf{F}$ of distribution functions which are continuous and strictly increasing on $(-\infty, \infty)$. This is too restrictive and we shall now extend the theory of the preceding sections to the class $\mathbf{F}^*$ of distributions F whose support is an interval (which may change with F and which may be finite, semiinfinite or infinite), on which F is assumed to be continuous and strictly increasing.

If the support of $F \in \mathbf{F}^*$ is the interval (a, b), the quantile function F^{-1} is continuous and strictly increasing on the open interval $(0, 1)$ and satisfies

$$\lim_{u\to 0} F^{-1}(u) = a, \qquad \lim_{u\to 1} F^{-1}(u) = b$$

and

$$F^{-1}(0) = -\infty, \qquad F^{-1}(1) = b.$$

The nature of $k = GF^{-1}$ depends on the relative positions of the supports (a, b) of G and (c, d) of F. Seven different situations are possible:

1. $a = c$, $b = d$. In this case k increases strictly and continuously from $k(0) = 0$ to $k(1) = 1$ as it does when $a = c = -\infty$, $b = d = \infty$. No changes are needed.
2. $a < c < d < b$. k increases strictly and continuously from $(0, G(c))$ to $(1, G(d))$.
3. $c < a < b < d$. k increases strictly and continuously from $(F(a), 0)$ to $(F(b), 1)$.
4. $a < c < b < d$. k increases strictly and continuously from $(0, G(c))$ to $(F(b), 1)$.
5. $c < a < d < b$. k increases strictly and continuously from $(F(a), 0)$ to $(1, G(d))$.
6. $b \le c$. $k(u) = 1$ for $0 < u \le 1$ and $k(0) = 0$.
7. $d \le a$. $k(u) = 0$ for $0 \le u < 1$ and $k(1) = 1$.

The type remains invariant under the transformations $\psi \in \Psi$ defined in (2.4). So does the nature of the carriers C and D of F and G, that is, whether each of C and D is finite, is semiinfinite with a finite right endpoint or left endpoint or is infinite in both directions. As a result, although k remains invariant under the transformations of Ψ, it no longer is maximal invariant. Consider, for example, the pairs (F, G) with $F = G$. They all correspond to the single function $k(u) = u$ for all $0 < u < 1$, but they do not constitute a single orbit since the cases $C = D =$ finite, $C = D =$ infinite, etc. all have different orbits. Thus the maximal invariant must take account of the nature of C and D; in addition, with respect to the types (6) and (7) defined in the preceding paragraph, it must distinguish between the types

$$6(\text{a})\ \ b < c \text{ and } 6(\text{b})\ \ b = c$$

and

$$7(\text{a})\ \ d < a \text{ and } 7(\text{b})\ \ d = a,$$

which also cannot be transformed into each other.

If the assumptions made at the beginning of this section are further weakened, the complexity of the maximal invariant will increase even further.

There is an alternative way of handling a weakening of the assumptions such as that at the beginning of Section 2, and this is to enlarge the group Ψ of transformations defined in (2.4). However, this approach also runs into difficulties and we therefore consider no further extensions here.

5. Ordering the orbits. An invariant preorder S allows us to go a step further. We can say not only of certain pairs (F, G)—the pairs in S—that $F \leq G$, but also for certain quadruples $(F_1, G_1; F_2, G_2)$ that, according to the ordering $\leq_S$, G_2 is further away from F_2 than G_1 is from F_1.

To fix ideas, consider distribution functions F_1 and G_1, with $F_1 \leq_S G_1$ and define $F_2 = F_1\psi^{-1}$ and $G_2 = G_1\psi^{-1}$ for $\psi \in \Psi$. Since (F_1, G_1) and (F_2, G_2) are in the same orbit, we say F_1 and G_1 are at the same distance from each other as F_2 and G_2. That is, all pairs on the same orbit are equally distant. On the other hand, if $F_1 \leq_S G_1 \leq_S G_2$, we say that G_2 is further to the right of F_1 than G_1 is. This last situation implies that $G_1 = F_1\psi^{-1}$ and $F_1\psi^{-1} \leq_S G_2$ for some monotone ψ. Extending this argument to the case of the quadruple $(F_1, G_1; F_2, G_2)$ with $(F_i, G_i) \in S$, $i = 1, 2$, we say that G_2 is further to the right of F_2 than G_1 is of F_1 if $F_2 = F_1(\psi^{-1})$ and $G_1(\psi^{-1}) \leq_S G_2$ for some $\psi \in \Psi$. In terms of $k_1 = G_1F_1^{-1}$ and $k_2 = G_2F_2^{-1}$, this is equivalent to requiring that $k_2(k_1^{-1}) \in \mathbf{K}_S$, and this motivates the following definition.

Definition. Let $(F_1, G_1), (F_2, G_2) \in S$. Then G_2 is said to be further to the right of F_2 than G_1 is of F_1 if

$$(k_1, k_2) \in S, \tag{5.1}$$

where $k_i = G_iF_i^{-1}$, $i = 1, 2$.

Since k_1, k_2 are distribution functions, (5.1) is equivalent to the condition that $k_2(k_1^{-1}) \in \mathbf{K}_S$, that is, $k_1 \leq_S k_2$. It also follows trivially that $k_1 \leq_S k_2$ if and only if $k_2 = k_3(k_1)$ for some $k_3 \in \mathbf{K}_S$. Thus, for example, when S represents stochastic ordering, G_2 is further to the right of F_2 than G_1 is of F_1 if $G_2 F_2^{-1}(u) \leq G_1 F_1^{-1}(u)$, $0 < u < 1$. If S represents monotone likelihood ratio ordering, G_2 will be further to the right of F_2 than G_1 is of F_1 if $G_2 F_2^{-1}(u) = h(G_1 F_1^{-1}(u))$ for some convex function h. This is illustrated by the following two examples.

EXAMPLE 5.1. Consider the location family $F(x - \theta)$, and define $G_1 = F(x - \theta_1)$ and $G_2 = F(x - \theta_2)$ with $\theta_2 > \theta_1 > \theta$. Then $F \leq_{\text{st}} G_1$ and $F \leq_{\text{st}} G_2$. Moreover, $k_1 \leq_{\text{st}} k_2$, where $k_1 = G_1 F^{-1}$ and $k_2 = G_2 F^{-1}$ so that G_2 is further to the right of F than G_1 is.

EXAMPLE 5.2. Consider the family of beta distributions F_α, with densities

$$f_\alpha(x) = \alpha x^{\alpha - 1}, \qquad \alpha \geq 0, 0 < x < 1.$$

Then $F_{\alpha_1} \leq_{\text{mlr}} F_{\beta_1}$ and $F_{\alpha_2} \leq_{\text{mlr}} F_{\beta_2}$ when $\alpha_1 < \beta_1$ and $\alpha_2 < \beta_2$. Moreover, when $\beta_2/\alpha_2 > \beta_1/\alpha_1$, $F_{\beta_2} F_{\alpha_2}^{-1} F_{\alpha_1} F_{\beta_1}^{-1}$ is convex and hence F_{β_2} is further to the right of F_{α_2} than F_{β_1} is of F_{α_1}.

6. Distances between ordered distributions. The qualitative distance relation, introduced by (5.1) on pairs of distribution functions $(F, G) \in S$ by comparing their corresponding k-functions raises the question of whether there exists a metric $d(F, G)$ which is consistent with the approach using k-functions. To be consistent, such a metric must satisfy the following two conditions: Since the k-functions are invariant under monotone transformations, the distance $d(F, G)$ must also be invariant under monotone transformations. Formally, we shall require (i) that

$$d(F, G) = d(F\psi^{-1}, G\psi^{-1}) \quad \text{for all } \psi \in \Psi$$

and (ii) that if $k_i = G_i F_i^{-1} \in \mathbf{K}_S$, $i = 1, 2$ with $k_1 <_S k_2$, then

$$d_S(F_1, G_1) \leq d_S(F_2, G_2).$$

Condition (ii) immediately implies the additional desirable requirement that if the ordering S_1 is stronger than S_2, then

$$d_{S_1}(F_1, G_1) \leq d_{S_1}(F_2, G_2) \quad \Rightarrow \quad d_{S_2}(F_1, F_1) \leq d_{S_2}(F_2, G_2).$$

A metric that satisfies (i) and (ii) above when S is stochastic ordering—and hence also for any stronger ordering—is provided by the supnorm d_0. Since

$$d_0(F, G) = \sup_x |G(x) - F(x)| = \sup_{0 \leq u \leq 1} |GF^{-1}(u) - u| = \sup_{0 \leq u \leq 1} |k(u) - u|,$$

d_0 is invariant under monotone transformations. Also, if $k_1, k_2 \in \mathbf{K}_S$ with $k_1(u) \leq k_2(u)$ for all u, then

$$d_0(F_1, G_1) \leq d_0(F_2, G_2).$$

EXAMPLE 6.1. Let $G(x) = F(x - \theta)$, $\theta > 0$, so that $F \leq_{st} G$. Then

$$d_0(F,G) = \sup_{0 \leq u \leq 1} \{u - F[F^{-1}(u) - \theta]\}.$$

If F is unimodal with a density f which is symmetric about t_0, the supremum is attained at $t = t_0 + \theta/2$, and hence

$$d_0(F,G) = F\left(t_0 + \frac{\theta}{2}\right) - F\left(t_0 - \frac{\theta}{2}\right).$$

Having defined a distance function consistent with stochastic ordering, we next consider the orderings:

S_1: hazard ordering
S_2: the ordering in which $F \leq_S G$ if $k(\cdot)$ is star-shaped
S_3: MLR ordering

and for each $i = 1, 2, 3$ define a distance function d_i consistent with S_i as follows:

$$d_1(F,G) = \sup_x \left| \ln\left(\frac{\overline{G}(x)}{\overline{F}(x)}\right)\right|;$$

$$d_2(F,G) = \sup_x \left| \ln\left(\frac{G(x)}{F(x)}\right)\right|;$$

$$d_3(F,G) = \sup_x \left| \ln\left(\frac{g(x)}{f(x)}\right)\right|.$$

It is clear that d_1, d_2, and d_3 satisfy the invariance condition (i) given above. That they also satisfy condition (ii) follows from the following theorem.

THEOREM 3. *If* $k_1, k_2, k_2(k_1^{-1}) \in \mathbf{K}_{S_i}$, *then*

$$d_i(F_1, G_1) \leq d_i(F_2, G_2), \qquad i = 1,2,3.$$

PROOF. Clearly, the membership of k_1, k_2 and $k_2(k_1^{-1})$ in $\mathbf{K}_{S_i}$, $i = 1$, 2 or 3, implies that

$$k_1(u) \leq u, \qquad k_2(u) \leq u \quad \text{and} \quad k_1(u) \geq k_2(u). \tag{6.1}$$

We now proceed by cases.

(i) For the S_1 ordering, (6.1) implies that

$$d_1(F_j, G_j) = \ln \sup_x \frac{\overline{G}_j(x)}{\overline{F}_j(x)} = \ln \sup_{0 \leq u \leq 1} \frac{\overline{G}_j \overline{F}_j^{-1}(u)}{u}, \qquad j = 1,2, \tag{6.2}$$

and that $\overline{k}_2/\overline{k}_1$ is nondecreasing. This latter implication is equivalent to

$$\frac{\overline{G}_2 \overline{F}_2^{-1}(u)}{\overline{G}_1 \overline{F}_1^{-1}(u)} \text{ is nonincreasing in } u. \tag{6.3}$$

Therefore, $\bar{G}_2\bar{F}_2^{-1}(u)/u \geq \bar{G}_1\bar{F}_1^{-1}(u)/u$ and the result for S_1 follows from (6.2).

(ii) For the S_2 ordering, condition (6.1) implies that

$$d_2(F_j, G_j) = \ln \sup_{0 \leq u \leq 1} \frac{u}{G_j F_j^{-1}(u)}, \tag{6.4}$$

and the fact that $k_2(k_1^{-1}) \in \mathbf{K}_{S_2}$ implies that k_2/k_1 is nondecreasing and hence

$$\frac{u}{G_2F_2^{-1}(u)} \geq \frac{u}{G_1F_1^{-1}(u)}, \qquad 0 < u < 1. \tag{6.5}$$

The result then follows from (6.4) and (6.5).

(iii) In the case of S_3, (6.1) together with the condition that k'_j is nondecreasing for $j = 1, 2$, implies that

$$d_3(F_j, G_j) = \max\{-\ln k'_j(0^+), \ln k'_j(1^-)\}. \tag{6.6}$$

But $k_1(u) \geq k_2(u)$, together with $k_j(0) = 0$ and $k_j(1) = 1$, $j = 1, 2$, imply that $k'_1(0^+) \geq k'_2(0^+)$ and $k'_1(1^-) \leq k'_2(1^-)$, and now the result follows from (6.6). □

EXAMPLE 6.2. Let $\bar{G}(x) = e^{-x}$ and $\bar{F} = [e^{-4x} + e^{-x}]/2$, $x > 0$. Then $F \leq_{\text{mlr}} G$ and we have

$$d_1(F, G) = \sup_x \left| \ln \frac{\bar{G}(x)}{\bar{F}(x)} \right| = \ln 2$$

and

$$d_3(F, G) = \sup_x \left| \ln \frac{g(x)}{f(x)} \right| = \ln\left(\frac{5}{2}\right).$$

REFERENCES

BARLOW, R. E. and PROSCHAN, F. (1975). *Statistical Theory of Reliability and Life Testing*. Holt, New York.

DOKSUM, K. (1969). Starshaped transformations and the power of rank test. *Ann. Math. Statist.* **40** 1167–1176.

KARLIN, S. and RINOTT, Y. (1983). Unbiasedness in the sense of Lehmann in n-action decision problems. In *A Festschrift for Erich L. Lehmann in Honor of His Sixty-fifth Birthday* (P. J. Bickel, K. A. Doksum and J. L. Hodges, Jr., eds.) 261–285. Wadsworth, Monterey, Calif.

KEILSON, J. and SUMITA, U. (1983). Uniform stochastic ordering and related inequalities. *Canad. J. Statist.* **10** 181–198.

LEHMANN, E. L. (1952). Testing multiparameter hypotheses. *Ann. Math. Statist.* **23** 541–552.

LEHMANN, E. L. (1955). Ordered families of distributions. *Ann. Math. Statist.* **26** 399–419.

LEHMANN, E. L. (1959). *Testing Statistical Hypotheses*. Wiley, New York.

LOH, W. Y. (1984). Bounds on ARE's for restricted classes of distributions defined via tail-orderings. *Ann. Statist.* **12** 685–701.

MACGILLIVRAY, H. L. (1986). Skewness and asymmetry: Measure and orderings. *Ann. Statist.* **14** 994–1011.

OJA, H. (1981). On location, scale, skewness and kurtosis of univariate distributions. *Scand. J. Statist.* **8** 154–168.

PLEDGER, G. and PROSCHAN, F. (1973). Stochastic comparisons of random processes, with applications to reliability. *J. Appl. Probab.* **10** 572–585.

ROJO, J. (1988). On the concept of tail-heaviness. Technical Report 175, Dept. Statistics, Univ. California, Berkeley.

ROJO, J. (1992). A pure tail ordering based on the ratio of the quantile functions. *Ann. Statist.* **20** 570–579.

VAN ZWET, W. R. (1964). *Convex Transformations of Random Variables. CWI Tract* **7**. Math. Centrum, Amsterdam.

WHITT, W. (1980). Uniform conditional stochastic order. *J. Appl. Probab.* **17** 112–123.

WHITT, W. (1981). Comparing counting processes and queues. *Adv. in Appl. Probab.* **13** 207–220.

WHITT, W. (1982). Multivariate monotone likelihood ratio and uniform conditional stochastic order. *J. Appl. Probab.* **19** 695–701.

DEPARTMENT OF STATISTICS
STATISTICAL LABORATORY
UNIVERSITY OF CALIFORNIA
BERKELEY, CALIFORNIA 94720

MATHEMATICAL SCIENCES DEPARTMENT
UNIVERSITY OF TEXAS
EL PASO, TEXAS 79968

Part IX
Probability Theory

The Impact of Lehmann's Work in Applied Probability

Walter W. Piegorsch and Moshe Shaked

Lehmann's works in applied probability crossed a broad spectrum of topics, and many of his papers had substantial impact. We comment on two such examples below. The first [19] provided a valuable beginning for the study and application of notions of positive and negative dependence. The second [20] also touched on issues of probability theory and the nature of random variables—in particular, on the distributional aspects of the reciprocal of a random variable. The contributions each represented are a study in contrasts, however: the former was a substantial work from which many wide-ranging contributions developed, while the second was a shorter, later-appearing piece whose impact(s) are still evolving.

1 Comments on Lehmann [19]

Erich Lehmann introduced and studied in [19] three concepts of positive dependence of two random variables. These are the concepts of positive quadrant dependence (PQD), positive regression dependence (PRD), and positive likelihood ratio dependence (PLRD). In the same paper Lehmann also introduced and studied the negative dependence analogs of these concepts. Since the publication of [19], these concepts have been used and applied in diverse areas of applied probability and statistics. They can be found in some abstract probabilistic papers, as well as in practical studies in medicine, operations research, management science, biology, sociology, economics, and physics. For example, a search in the ISI Web of Knowledge on May 11, 2009, yielded more than 500 papers that cite [19]. Apart from mainstream journals in statistics and probability, the references to [19] included papers in journals such as *Monthly Weather Review*, *Control and Cybernetics*, *Pharmaceutical Statistics*,

W.W. Piegorsch and M. Shaked
Department of Mathematics, University of Arizona, Tucson, Arizona 85721, USA
e-mail: piegorsch@math.arizona.edu; shaked@math.arizona.edu

J. Rojo (ed.), *Selected Works of E. L. Lehmann*, Selected Works in Probability and Statistics, DOI 10.1007/978-1-4614-1412-4_63, © Springer Science+Business Media, LLC 2012

Journal of Hydrologic Engineering, Stochastic Hydrology and Hydraulics, Journal of Geophysical Research, Progress in Natural Science, Psychometrika, Multivariate Behavioral Research, Journal of Mathematical Psychology, Applied Psychological Measurement, and *Stochastic Environmental Research and Risk Assessment*, to name a few.

The study in [19] was a main starting point for later studies of concepts of stochastic positive and negative dependence. These studies continue up to the present time. Two main routes leading from [19] were used by researchers in the area of dependence concepts. One type consists of studies that introduce further notions of *bivariate* dependence. For example, Harris [10] studied the notion of right corner set increasing dependence (that, in fact, was already mentioned in [19], but was not thoroughly studied there). Yanagimoto [31] refined the notions of Lehmann [19] and of Harris [10], and thus introduced some families of positive dependence notions. A more systematic extension of all the above bivariate notions of dependence was made in Shaked [24]. Another type involves studies that extend the concepts of [19] to the *multivariate* case. A basic paper in this vein is the paper by Esary, Proschan, and Walkup [9] in which the notion of association of a set (not necessarily a pair) of random variables is introduced. The multivariate ideas in [9], that built on [19], led to a large number of further studies. Perhaps it is worthwhile to mention here the work of Lee [17] which extended the positive dependence notions of Shaked [24] from the bivariate to the multivariate case. In another paper of Lee [18] a parallel development of negative dependence notions is given. Other studies of positive and negative dependence concepts, that start from [19] and extend it in various directions, can be found in the papers by Block, Savits, and Shaked [3,4].

It is intriguing to speculate how familiar Lehmann was with the relationship between some of his concepts of positive dependence in [19] and the terminology and basic results of the theory of total positivity; see Karlin [12]. For example, the notion of PLRD is actually the same as requiring the joint density function of the two underlying random variables to be totally positive of order 2 (TP_2). By noticing the total positivity condition above, the relation of the notions of Lehmann [19] to notions in statistical physics become apparent, and new routes open for their extensions to the multivariate case. For example, the PLRD notion can be extended to the multivariate case by requiring the joint density function of the underlying random variables to be TP_2 in pairs. Another multivariate extension of the PLRD notion is obtained by requiring the joint density function of the underlying random variables to satisfy another condition that is an extension of the bivariate TP_2 condition; see, for example, Karlin and Rinott [13]. Block and Ting [5] studied these kinds of extensions.

The three main concepts of positive dependence in [19] are related as follows

$$\text{PLRD} \Longrightarrow \text{PRD} \Longrightarrow \text{PQD}. \tag{1.1}$$

It follows that every property that PQD random vectors have, is also a property of PLRD and PRD random vectors. For example, Lehmann [19] proved that if the covariance of a PQD (and hence also of a PRD or a PLRD) random vector exists then it must be nonnegative. Lehmann [19] also showed that other measures of association, such as Kendall's τ, Spearman's ρ, and Blomqvist's q, are all nonnegative for PQD random vectors.

An important and useful closure property of the PQD concepts that is proven in [19] is the following. Suppose that the independent random vectors $(Y_1, V_1), (Y_2, V_2), \ldots, (Y_n, V_n)$ are all PQD. Let r and s be n-dimensional real functions that are nondecreasing in each of their n variables when the other variables are held fixed. Then

$$r(Y_1, Y_2, \ldots, Y_n) \quad \text{and} \quad s(V_1, V_2, \ldots, V_n)$$

are PQD. In particular the PQD concept is closed under nondecreasing transformations of the underlying random variables.

By looking at the formal definition of the PQD concept in [19] (which is also given in (1.2) below) it is easy to see that for every random variable X we have that (X, X) is PQD. From the closure property that is described in the previous paragraph it follows that $(X_1, X_2) \stackrel{\text{def}}{=} (X, s(X))$ is PQD whenever s is a nondecreasing function. The random variables X_1 and X_2 described above are called comonotone in the recent actuary and insurance literature; see, for instance, Ludkovski and Rüschendorf [21] and references therein. From the above discussion it follows that comonotone random variables are positively dependent in the sense of all the notions in [19].

An elegant and useful result that is proven in [19] shows that if a PQD random vector (X_1, X_2) has a zero covariance then X_1 and X_2 must be independent. This property has been known for bivariate normal random variables – the above result extends this surprising property to a host of jointly distributed pairs of random variables.

The concepts of positive and negative dependence in [19] (and also their later extensions) have been applied in many areas of statistics and probability. Lehmann himself, in [19], applied them to some hypotheses testing situations. One application involves a multiple comparisons setting. Suppose that we have a number of pairs of hypotheses that are tested using the same sample. Then the test statistics are not independent, and it is often of interest to control not the individual significance levels (that is, error probabilities), but the experimentwise error rate. Lehmann [19] noticed that sometimes the test statistics are positively or negatively dependent, and as a consequence he obtained tight bounds on the experimentwise error rate.

Another application of the concepts of positive dependence in statistics, that Lehmann described in [19], is concerned with testing the null hypothesis H_0 of independence versus the alternative H_a of positive dependence. One benefit of the various definitions of positive dependence in [19] is that they enable the statistician to formally state H_a. Furthermore, by identifying some quantity that must be positive under H_a (that is, under the assumption of positive dependence), but 0 under H_0, one can use the empirical sample estimator of that quantity as a test statistic. Lehmann [19] showed that such testing procedures yield unbiased tests. This very idea has been used extensively again and again in later literature. For example, see Shetty and Pandit [27] and references therein.

Other applications of the concepts of positive and negative dependence are based on various inequalities involving useful probabilistic quantities that follow from these concepts. For example, Barlow and Proschan [1] obtained useful bounds on reliability systems using the notions of [19], whereas various useful inequalities in queueing theory have been derived using the theory of positive dependence; see, for instance, Daduna and Szekli [7] and references therein.

Positive or negative dependence concepts are a useful and convenient tool for relaxing the assumption of independence in various studies of limit theorems. Many papers have adopted this tool. We cannot list them all here, but we will just mention a recent paper, by Ko, Kim, and Han [16], in which the reader may find further references for the use of positive and negative dependence in obtaining limit theorems.

For the purpose of motivating the following discussion, recall formally from [19] that the two random variables X_1 and X_2 are PQD if

$$P\{X_1 \le x_1, X_2 \le x_2\} \ge P\{X_1 \le x_1\}P\{X_2 \le x_2\} \quad \text{for all } (x_1, x_2) \in \mathbb{R}^2. \tag{1.2}$$

Notice that the inequality (1.2) compares the probability that (X_1, X_2) is in a lower quadrant, with the probability of being in the same lower quadrant if X_1 and X_2 "were independent". This observation led several researchers to introduce and consider a stochastic order between bivariate random vectors with the same marginals, (X_1, X_2) and (Y_1, Y_2), say, through the definition

$$P\{X_1 \le x_1, X_2 \le x_2\} \le P\{Y_1 \le x_1, Y_2 \le x_2\} \quad \text{for all } (x_1, x_2) \in \mathbb{R}^2; \tag{1.3}$$

we may denote (1.3) as $(X_1, X_2) \le_{\text{PQD}} (Y_1, Y_2)$, and call it the PQD positive dependence stochastic order. Intuitively, $(X_1, X_2) \le_{\text{PQD}} (Y_1, Y_2)$ indicates that Y_1 and Y_2 are "more positively dependent" than X_1 and X_2. Tchen [28], among others, studied this order.

For the purpose of comparing multivariate (rather than just bivariate) random vectors in the sense of positive dependence, one may consider extending the defining inequality (1.3) to the multivariate case. It turns out that there is more than one way of doing this. Some authors considered defining a positive dependence comparison of random vectors by comparing probabilities of lower or upper orthants in $\mathbb{R}^n$; see, for example, Joe [11]. Others, such as Bäuerle [2] and Shaked and Shanthikumar [25], studied a stronger integral stochastic order, which compares the expectations of supermodular functions of the compared random vectors.

The above discussion illustrates how the PQD concept in [19] led to the introduction and study of corresponding positive dependence stochastic orders. A similar discussion applies to the other two positive dependence concepts in (1.1). For example, a stochastic order that is based on the PRD notion has been introduced and studied in Yanagimoto and Sibuya [32], whereas a stochastic order that is based on the PLRD notion was considered in Kimeldorf and Sampson [15]. If we denote the corresponding orders by $\le_{\text{PRD}}$ and $\le_{\text{PLRD}}$, then, motivated by (1.1), we should expect the following relationships among the three stochastic orders above:

$$\le_{\text{PLRD}} \Longrightarrow \le_{\text{PRD}} \Longrightarrow \le_{\text{PQD}} .$$

Generally speaking, this is indeed the case, but the interested reader is advised to consult Chapter 9 in Shaked and Shanthikumar [26] where a detailed study of this issue is given.

2 Comments on Lehmann and Shaffer [20]

In contrast to the far-reaching contributions made by Lehmann's work [19] on positive dependence, the article [20] by Lehmann and Shaffer on inverted distributions has to date seen lesser development. The two papers do share a common feature, however: their primary focus is on distributional attributes of random variables rather than new statistical methodology. This leads us to speculate that the future, potential impact of [20] is still untapped.

The questions dealt with in this article [20] are, on the surface, somewhat straightforward: how can the relationship between the probability density function f_X of a continuous random variable X, and that of its inverted distribution—the probability density function f_Y of $Y = 1/X$—be characterized? Lehmann and Shaffer [20] motivate the question from previous discussions on the nature of the inverse moment $E[1/X]$ (also called the negative moment or the reciprocal moment by some authors; see, e.g., Khuri and Casella [14], Piegorsch and Casella [23], Wooff [30], or Cressie, Davis, Folks, and Policello [6]). Indeed, their discussion concerns only continuous random variables, since clearly, if a discrete probability mass function has non-zero mass at $X = 0$, $E[1/X]$ does not exist. This makes the issue of inverted distributions for discrete random variables somewhat less interesting.

Expanding on results in [23], Lehmann and Shaffer [20] observe that the nature of $f_X(x)$ near $x = 0$ helps characterize the nature of the inverse probability density function f_Y, and further that the behavior of $f_X(x)$ as $x \to \infty$ can tell us something about the behavior of $f_Y(y)$ near $y = 0$. This is not surprising—as the authors themselves note [20, page 192], "it is difficult to imagine that these results are new, but we have not been able to find them in the literature." Indeed, the basic mathematical finding simply reflects the relationship between the behavior of $f_X(x)$ near $x = 0$ and the stability of the integral $\int x^{-1} f_X(x)dx$. Nonetheless, some fascinating distributional results emerge from the effort. Lehmann and Shaffer [20] show that if

$$x^2 f_X(x) \to 0 \text{ as } |x| \to \infty, \tag{2.4}$$

then $f_Y(y) \to 0$ as $y \to 0$. To do so, they characterize how heavy the tail behavior of f_X appears, related to what the authors call *Cauchy tails*. Specifically, they find that if $f_X(0+) = \lim_{x \to 0+} f_X(x)$, then the ratio of $f_Y(y)$ to $1/(1 + y^2)$ tends to $f_X(0+)$ as $y \to \infty$. Thus if $f_X(0+) = 0$ the (right) tail behavior of f_Y is lighter than that of a Cauchy probability density function, if $f_X(0+)$ is positive and finite the tail behavior of f_Y is similar to the Cauchy, and if $f_X(0+)$ is infinite the tail behavior is heavier than Cauchy. (Analogous left-tail results are obtained by studying $x \to 0-$ and $y \to -\infty$.)

Lehmann and Shaffer [20] also use these tail-behavior results to describe the modal nature of f_Y. In particular, and rather intriguingly, they elicit that if (2.4) holds such that $f_X(x) > 0$ for both positive and negative values of x, then f_Y will be multi-modal, with at least one mode at some value of $y < 0$ and one mode at some value of $y > 0$. (In most cases, the inverted probability density function is exactly bimodal.) They expand upon these results via some clever limiting operations, showing, e.g., that the strange bi- or multi-modal behavior of an appropriately standardized version of $1/X$ can actually disappear as certain location and scale parameters are driven to their extremes. (See Sections 3 and 4 of the Lehmann and Shaffer [20] for more details.) The authors refer to this spectrum of behavior for $1/X$ as "bizarre," a term difficult to argue with, but in doing so they also yield insights into the fascinating and sometimes unusual nature of inverted distributions.

As we note above, the impact of the Lehmann and Shaffer [20] article has been limited, perhaps due to its focus on these curious aspects of (inverted) distributional theory. A few applications have appeared for selected econometric problems—see, e.g., Diebold and Lamb [8] and ten Raa and Steel [29]—but these have generally been confined to calculations on inverse moments and not to the more general features of the inverted distribution. An interesting, hierarchical modeling effort is given by Peng [22], who estimates an inverse moment via Bayesian methods in an engineering reliability analysis. Indeed, we speculate that when working with a posterior probability density function $\pi(\mu|X)$, characterization of the larger distributional features of a reciprocal parameter such as $\psi = 1/\mu$, beyond just the inverse moment $E[\psi|X]$, may have unexploited value. The Lehmann and Shaffer [20] results available for the tail and modal behavior of an inverted (posterior) distribution $\pi(\psi|X)$ might be important considerations in future Bayesian studies with inverse or reciprocal quantities such as $\psi = 1/\mu$.

References

[1] Barlow, R. E. and Proschan, F. (1975). *Statistical Theory of Reliability and Life Testing, Probability Models*. Holt, Rinehart, and Winston, New York.

[2] Bäuerle, N. (1997). Inequalities for stochastic models via supermodular orderings. *Communications in Statistics - Stochastic Models* **13**, 181–201.

[3] Block, H. W., Savits, T. H., and Shaked, M. (1982). Some concepts of negative dependence. *Annals of Probability* **10**, 765–772.

[4] Block, H. W., Savits, T. H., and Shaked, M. (1985). A concept of negative dependence using stochastic ordering. *Statistics and Probability Letters* **3**, 81–86.

[5] Block, H. W. and Ting, M.-L. (1981). Some concepts of multivariate dependence. *Communications in Statistics A - Theory and Methods* **10**, 749–762.

[6] Cressie, N., Davis, A. S., Folks, J. L., and Policello, G. E., II (1981). The moment-generating function and negative integer moments. *American Statistician* **35**, 148–150.

[7] Daduna, H. and Szekli, R. (2004). On the correlation structure of closed queueing networks. *Stochastic Models* **20**, 1–29.

[8] Diebold, F. X. and Lamb, R. L. (1997). Why are estimates of agricultural supply response so variable? *Journal of Econometrics* **76**, 357–373.

[9] Esary, J. D., Proschan, F., and Walkup, D. W. (1967). Association of random variables, with applications. *Annals of Mathematical Statistics* **38**, 1466–1474.
[10] Harris, R. (1970). A multivariate definition for increasing hazard rate distributions. *Annals of Mathematical Statistics* **41**, 713–717.
[11] Joe, H. (1997). *Multivariate Models and Dependence Concepts*, Chapman and Hall, London.
[12] Karlin, S. (1968). *Total Positivity*, Stanford University Press, Stanford, California.
[13] Karlin, S. and Rinott, Y. (1980). Classes of orderings of measures and related correlation inequalities. I. Multivariate totally positive distributions. *Journal of Multivariate Analysis* **10**, 467–498.
[14] Khuri, A. and Casella, G. (2002). The existence of the first negative moment revisited. *American Statistician* **56**, 44–47.
[15] Kimeldorf, G. and Sampson, A. R. (1987). Positive dependence orderings. *Annals of the Institute of Statistical Mathematics* **39**, 113–128.
[16] Ko, M.-H., Kim, T.-S., and Han, K.-H. (2009). A note on the almost sure convergence for dependent random variables in a Hilbert space. *Journal of Theoretical Probability* **22**, 506–513.
[17] Lee, M.-L. T. (1985). Dependence by total positivity. *Annals of Probability* **13**, 572–582.
[18] Lee, M.-L. T. (1985). Dependence by reverse regular rule. *Annals of Probability* **13**, 583–591.
[19] Lehmann, E. L. (1966). Some concepts of dependence. *Annals of Mathematical Statistics* **37**, 1137–1153.
[20] Lehmann, E. L. and Shaffer, J. P. (1988). Inverted distributions. *American Statistician* **42**, 191–194.
[21] Ludkovski, M. and Rüschendorf, L. (2008). On comonotonicity of Pareto optimal risk sharing. *Statistics and Probability Letters* **78**, 1181–1188.
[22] Peng, C. Y. (2008). The first negative moment in the sense of the Cauchy principal value. *Statistics and Probability Letters* **78**, 1765–1774.
[23] Piegorsch, W. W. and Casella, G. (1985). The existence of the first negative moment. *American Statistician* **39**, 60–62.
[24] Shaked, M. (1977). A family of concepts of dependence for bivariate distributions. *Journal of the American Statistical Association* **72**, 642–650.
[25] Shaked, M. and Shanthikumar, J. G. (1997). Supermodular stochastic orders and positive dependence of random vectors. *Journal of Multivariate Analysis* **61**, 86–101.
[26] Shaked, M. and Shanthikumar, J. G. (2007). *Stochastic Orders*, Springer, New York.
[27] Shetty, I. D. and Pandit, P. V. (2003). Distribution-free tests for independence against positive quadrant dependence: A generalization. *Statistical Methods and Applications* **12**, 5–17.
[28] Tchen, A. H. (1980). Inequalities for distributions with given marginals. *Annals of Probability* **8**, 814–827.
[29] ten Raa, T. and Steel, M. F. J. (1994). Revised stochastic-analysis of an input-output model. *Regional Science and Urban Economics* **24**, 361–371.
[30] Wooff, D. A. (1985). Bounds on reciprocal moments with applications and developments in Stein estimation and post-stratification. *Journal of the Royal Statistical Society, Series B* **47**, 362–371.
[31] Yanagimoto, T. (1972). Families of positively dependent random variables. *Annals of the Institute of Statistical mathematics* **24**, 559–573.
[32] Yanagimoto, T. and Sibuya, M. (1972). Stochastically larger component of a random vector. *Annals of the Institute of Statistical Mathematics* **24**, 259–269.

SOME CONCEPTS OF DEPENDENCE[1]

E. L. LEHMANN

University of California, Berkeley

1. Summary and introduction. Problems involving dependent pairs of variables (X, Y) have been studied most intensively in the case of bivariate normal distributions and of 2×2 tables. This is due primarily to the importance of these cases but perhaps partly also to the fact that they exhibit only a particularly simple form of dependence. (See Examples 9(i) and 10 in Section 7.) Studies involving the general case center mainly around two problems: (i) tests of independence; (ii) definition and estimation of measures of association. In most treatments of these problems, there occurs implicitly a concept which is of importance also in other contexts (for example, the evaluation of the performance of certain multiple decision procedures), the concept of *positive* (or *negative*) dependence or association. Tests of independence, for example those based on rank correlation, Kendall's t-statistic, or normal scores, are usually not omnibus tests (for a discussion of such tests see [4], [15] and [17], but designed to detect rather specific types of alternatives, namely those for which large values of Y tend to be associated with large values of X and small values of Y with small values of X (positive dependence) or the opposite case of negative dependence in which large values of one variable tend to be associated with small values of the other. Similarly, measures of association are typically designed to measure the degree of this kind of association.

The purpose of the present paper is to give three successively stronger definitions of positive dependence, to investigate their consequences, explore the strength of each definition through a number of examples, and to give some statistical applications.

2. Quadrant dependence. For a first definition, we compare the probability of any quadrant $X \leqq x$, $Y \leqq y$ under the distribution F of (X, Y) with the corresponding probability in the case of independence. We say that the pair (X, Y) or its distribution F is *positively quadrant dependent* if

$$(2.1) \qquad P(X \leqq x, Y \leqq y) \geqq P(X \leqq x)P(Y \leqq y) \qquad \text{for all } x, y.$$

The dependence is strict if inequality holds for at least some pair (x, y). The family of all distributions F satisfying (2.1) will be denoted by $\mathfrak{F}_1$. Similarly, (X, Y) or F is *negatively quadrant dependent* if (2.1) holds with the inequality sign reversed, and the totality of negatively quadrant dependent distributions

Received 3 September 1965; revised 2 February 1966.

[1] Prepared with the partial support of the Office of Naval Research, Contract Nonr-222-43. This paper in whole or in part may be reproduced for any purpose of the United States Government.

will be denoted by $\mathcal{G}_1$. To simplify the notation we shall write $(X, Y) \varepsilon \mathcal{F}$ to mean that the distribution of (X, Y) belongs to $\mathcal{F}$.

LEMMA 1.

(i) $(X, X) \varepsilon \mathcal{F}_1$ *for all* X

(ii) $(X, Y) \varepsilon \mathcal{F}_1 \Leftrightarrow (X, -Y) \varepsilon \mathcal{G}_1$

(iii) $(X, Y) \varepsilon \mathcal{F}_1$ *implies* $(r(X), s(Y)) \varepsilon \mathcal{F}_1$ *for all non-decreasing functions* r *and* s. *The concept of positive quadrant dependence is thus invariant under non-decreasing transformations (and similarly under nonincreasing transformations) of both variables.*

(iv) *The set of inequalities* (2.1) *is equivalent to that obtained by replacing one or both of the inequalities* $X \leqq x$ *or* $Y \leqq y$ *by the corresponding* $X < x$ *or* $Y < y$.

(v) *The set of inequalities* (2.1) *is equivalent to each of the following, where again the equality signs inside the probabilities are optional:*

$$P(X \leqq x, Y \geqq y) \leqq P(X \leqq x)\, P(Y \geqq y) \tag{2.1$'$}$$

$$P(X \geqq x, Y \leqq y) \leqq P(X \geqq x)\, P(Y \leqq y) \tag{2.1$''$}$$

$$P(X \geqq x, Y \geqq y) \geqq P(X \geqq x)\, P(Y \geqq y). \tag{2.1$'''$}$$

PROOF. (i), (ii) and (iii) are obvious.

(iv) To see that (2.1) implies the corresponding inequalities with one or both of the equal signs omitted, replace x and/or y by $x - 1/n$ and/or $y - 1/n$. To go in the other direction replace x and/or y by $x + 1/n$ and/or $y + 1/n$.

(v) The equivalence of (2.1) and (2.1$'$) follows from (iv) and the fact that $(X \leqq x, Y \geqq y) = (X \leqq x) - (X \leqq x, Y < y)$. The equivalence with (2.1$''$) follows analogously, and the equivalence with (2.1$'''$) from the fact that the sets involved in the left hand sides of the four inequalities (with some of the equality signs omitted) add up to the whole space.

An important class of examples of distributions with quadrant dependence is furnished by Theorem 1 below. Before stating this theorem, it is convenient to introduce the following definition. We shall say that two real-valued functions r and s of n arguments are *concordant* for the ith coordinate if, considered as functions of the ith coordinate (with all other coordinates held fixed), they are monotone in the same direction, i.e. either both non-decreasing or both non-increasing. Similarly r and s will be called *discordant* for the ith coordinate if they are monotone in opposite directions.

THEOREM 1. *Let* $(X_1, Y_1), \cdots, (X_n, Y_n)$ *be independent pairs of random variables with joint distributions* $F_1, \cdots, F_n$. *Let* r *and* s *be functions of* n *variables and let*

$$X = r(X_1, \cdots, X_n), \qquad Y = s(Y_1, \cdots, Y_n). \tag{2.2}$$

Then (i) $(X, Y) \varepsilon \mathcal{F}_1$ *if for each* i *one (not necessarily the same for all* i*) of the following conditions hold:* (a) $F_i \varepsilon \mathcal{F}_1$ *and* r, s *are concordant for the* i*th coordinate or* (b) $F_i \varepsilon \mathcal{G}_1$ *and* r, s *are discordant for the* i*th coordinate;*

(ii) *similarly* $(X, Y) \varepsilon \mathcal{G}_1$ *if for each* i *either* $F_i \varepsilon \mathcal{F}_1$ *and* r, s *are discordant, or* $F_i \varepsilon \mathcal{G}_1$ *and* r, s *are concordant.*

(iii) *Let* U, V *be independent and independent of* $(X_1, Y_1), \cdots, (X_n, Y_n)$ *and let*

$$X = r(U, X_1, \cdots, X_n), \qquad Y = s(V, Y_1, \cdots, Y_n). \tag{2.3}$$

Then the conclusions of (i) *and* (ii) *continue to hold under the assumptions made there without any assumptions concerning the behaviour of* r *and* s *as functions of* U *and* V *respectively.*

PROOF. This will be proved as a consequence of Theorem 2 in the next section.

EXAMPLE 1. The following are some pairs of random variables (X, Y) with positive quadrant dependence; the property in each case follows from Theorem 1 and the fact that $(X, X) \varepsilon \mathcal{F}_1$ for all X.

(i) $X, Y = s(X)$ for any random variable X and any non-decreasing function s.

(ii) $X = U + aZ$, $Y = V + bZ$ for any independent random variables U, V, Z if a and b have the same sign.

(iii) $X, Y = X + V$ for any independent X, V

(iv) $X = r(U, Z)$, $Y = s(V, Z)$ where U, V, Z are independent and r and s are non-decreasing in Z but otherwise arbitrary.

Part (ii) of the example can be used to show that any bivariate normal distribution with positive correlation coefficient is in $\mathcal{F}_1$. We shall later give different proofs of this result in Example 5 (Section 5) and Example 10 (i) (Section 8).

3. A property of quadrant dependent distributions. An important property of distributions in $\mathcal{F}_1$ is given by Theorem 2 below, which is a generalization of an inequality of Chebyshev (cf. Hardy, Littlewood and Polya (1943), p. 43). I am grateful to Professors W. J. Hall and W. Hoeffding for providing me with a much simpler proof of the basic inequality (3.2) on which this theorem rests than my original proof. They point out that it is, in fact, an immediate consequence of the following Lemma 2, due to Hoeffding (1940), the proof of which was also communicated to me by Professor Hoeffding.

LEMMA 2 (Hoeffding). *If* F *denotes the joint and* F_X *and* F_Y *the marginal distributions of* X *and* Y, *then*

$$E(XY) - E(X)E(Y) = \int_{-\infty}^{\infty} \int_{-\infty}^{\infty} [F(x, y) - F_X(x)F_Y(y)]\, dxdy \tag{3.1}$$

provided the expectations on the left hand side exist.

PROOF. Let (X_1, Y_1), (X_2, Y_2) be independent, each distributed according to F. Then

$$\begin{aligned} 2[E(X_1Y_1) - E(X_1)\, E(Y_1)] &= E[(X_1 - X_2)(Y_1 - Y_2)] \\ &= E \int_{-\infty}^{\infty} \int_{-\infty}^{\infty} [I(u, X_1) - I(u, X_2)][I(v, Y_1) - I(v, Y_2)]\, dudv \end{aligned}$$

where $I(u, x) = 1$ if $u \leqq x$ and $= 0$ otherwise. Since $E|XY|$, $E|X|$ and $E|Y|$ are assumed finite, we can take the expectation under the integral sign, and the resulting expression is seen to reduce to twice the right hand side of (3.1). This completes the proof.

A simple consequence is the following

LEMMA 3. *If* $(X, Y) \varepsilon \mathfrak{F}_1$ *and if the expectations in* (3.2) *exist, then*

$$E(XY) \geqq E(X)\, E(Y) \tag{3.2}$$

with equality holding if and only if X *and* Y *are independent.*

PROOF. That (3.2) holds is obvious from (3.1). Suppose now that (2.1) holds and that equality holds in (3.2). Then $F(x, y) = F_X(x)\, F_Y(y)$ except possibly on a set of Lebesgue measure zero. From the fact that cumulative distribution functions are continuous on the right, it is easily seen that if two distributions agree almost everywhere with respect to Lebesgue measure, they must agree everywhere. Thus X and Y are independent, and this completes the proof.

Lemma 3 is a special case of the following theorem.

THEOREM 2. *Throughout Theorem* 1, *the conclusion* $(X, Y) \varepsilon \mathfrak{F}_1$ *can be replaced by* (3.2), *and the conclusion* $(X, Y) \varepsilon \mathfrak{G}_1$ *by* (3.2) *with the inequality sign reversed, provided the expectations in* (3.2) *exist.*

PROOF. (i)

1. We begin by proving the theorem for the case that X and Y are given by (2.2) and that $n = 1$. Suppose that r and s are both non-decreasing and that (X_1, Y_1) is in $\mathfrak{F}_1$. Then $(r(X_1), s(Y_1))$ is in $\mathfrak{F}_1$ by Lemma 1 (iii), and the result follows from Lemma 3.

If instead of non-decreasing, the functions r and s are non-increasing, the result follows if we replace them by $-r$ and $-s$ respectively. If one of them, say r, is increasing and the other decreasing and (X_1, Y_1) is in $\mathfrak{G}_1$ we put $s'(y_1) = s(-y_1)$ and apply the result already proved to r and s'. This completes the proof of (i) for the case $n = 1$.

2. To prove (i) for general n we proceed by induction. From the result for $n = 1$ it follows that for all fixed $x_2, \cdots, x_n$ and $y_2, \cdots, y_n$,

$$E[r(X_1, x_2, \cdots, x_n)\, s(Y_1, y_2, \cdots, y_n)] \geqq E[r(X_1, x_2, \cdots, x_n)]\, E[s(Y_1, y_2, \cdots, y_n)].$$

Taking expectation of both sides we get

$$E[r(X_1, X_2, \cdots, X_n)\, s(Y_1, Y_2, \cdots, Y_n)) \geqq E[r^*(X_2, \cdots, X_n)\, s^*(Y_2, \cdots, Y_n)]$$

where $r^*(x_2, \cdots, x_n) = E\, r(X_1, x_2, \cdots, x_n)$ and s^* is defined correspondingly. Now r^* and s^* have the same monotonicity properties in $x_2, \cdots, x_n$ and $y_2, \cdots, y_n$ as do the functions r and s, and the result therefore follows by induction.

(ii) This is seen by putting $s'(y_1, \cdots, y_n) = s(-y_1, \cdots, -y_n)$ and applying (i) to r and s'.

(iii) This is proved by applying (i) or (ii) to $r(u, X_1, \cdots, X_n)$ and

$s(v, Y_1, \cdots, Y_n)$ for fixed u and v and then taking expectations on both sides, and using the independence of U and V.

We are now in a position to prove Theorem 1.

PROOF OF THEOREM 1. Apply Theorem 2 with (X, Y) replaced by the variables (X', Y'): $X' = I(U, X_1, \cdots, X_n)$, $Y' = J(V, Y_1, \cdots, Y_n)$, where I and J indicate the events

$$r(U, X_1, \cdots, X_n) \leqq x \quad \text{and} \quad s(V, Y_1, \cdots, Y_n) \leqq y$$

respectively. If r and s satisfy the assumptions of Theorem 1, so do I and J, and this completes the proof.

For $n = 1$, Part (i) of Theorem 2 shows that if $(X, Y) \varepsilon \mathfrak{F}_1$ and r and s are non-decreasing, then

$$E\, r(X)\, s(Y) \geqq E\, r(X)\, E\, s(Y). \tag{3.3}$$

Equality in (3.3) may hold even when X and Y are not independent, since it requires only the independence of $r(X)$ and $s(Y)$. As an example, let $X = I_1 T$, $Y = I_2 T$ where I_1, I_2, T are independent, I_1 and I_2 take on the values ± 1 with probability $\frac{1}{2}$ each, and T is any positive random variable. Then $(X, Y) \varepsilon \mathfrak{F}_1$ by Example 1 (iv), and X and Y are dependent. However, if $r(x) = s(x) = \text{sgn}\,(x)$, it follows that $r(X) = I_1$ and $s(Y) = I_2$ are independent, so that equality holds in (3.3).

Lemma 3 states that if $(X, Y) \varepsilon \mathfrak{F}_1$ and the covariance of X and Y exists, then $\text{Cov}\,(X, Y) \geqq 0$. Covariance is only one of a number of measures of association that have been proposed in the literature. (For a discussion of such measures cf. [14] and [16].) We shall now prove that several other such measures are non-negative for all distributions in $\mathfrak{F}_1$.

COROLLARY 1. *If F is positively quadrant dependent, then Kendall's τ, Spearman's ρ_s and the quadrant measure q discussed by Blomqvist* (1950) *are all non-negative.*

PROOF. (i) Since Kendall's τ is the covariance of $X = \text{sgn}\,(X_2 - X_1)$ and $Y = \text{sgn}\,(Y_2 - Y_1)$ it is by Lemma 2 enough to show that $(X, Y) \varepsilon \mathfrak{F}_1$. But this follows from Theorem 1 (i).

(ii) Since $\rho_s/3$ is the covariance of $X = \text{sgn}\,(X_2 - X_1)$ and $Y = \text{sgn}\,(Y_3 - Y_1)$ the result follows from Lemma 2 and Theorem 1 (i) by putting $n = 3$, $r(x_1, x_2, x_3) = \text{sgn}\,(x_2 - x_1)$ and $s(y_1, y_2, y_3) = \text{sgn}\,(y_3 - y_1)$.

(iii) Let μ and ν denote the medians of the marginal distributions of X and Y, and let $r(X)$ and $s(Y)$ indicate the events $X > \mu$ and $Y > \nu$ respectively. Then

$$\begin{aligned} q &= E[rs + (1 - r)(1 - s) - r(1 - s) - s(1 - r)] \\ &= E[1 + 4rs - 2r - 2s]. \end{aligned}$$

By Theorem 2(i), $E(rs) \geqq E(r)\, E(s)$ and hence

$$q \geqq [1 - 2E(r)]\, [1 - 2E(s)] \geqq 0.$$

4. Application to slippage problems. The basic idea of this section is due to Cochran (1941). It was further developed by Paulson (1952), Halperin, Greenhouse, Cornfield and Zalokar (1955), Hartley (1955), H. A. David (1956). Doornbos (1956) and by Doornbos and Prins (1956 and 1958).

Consider the testing of a hypothesis H against a number of different sets of alternatives K_i $(i = 1, \cdots, s)$. Suppose H is rejected in favor of K_i if

$$T_i \geqq C_i \tag{4.1}$$

at level α_i and that the tests are similar so that

$$P_H(T_i \geqq C_i) = \alpha_i . \tag{4.2}$$

Typical examples are the so-called slippage problems in which it is assumed of s parameters $\theta_1, \cdots, \theta_s$ that they are either all equal (H) or that exactly one of them has "slipped", i.e. is different from the others, and where then K_i represents the possibility that θ_i has slipped. In such situations, rather than controlling the individual error probabilities (4.2), it is frequently of interest to control the experimentwise error rate, that is, the probability P of falsely rejecting H in favor of any of the alternatives K_i. Applying Bonferroni's inequalities to the events (4.1), we obtain for P the inequalities

$$\sum \alpha_i - \sum_{i<j} P(T_i \geqq C_i, T_j \geqq C_j) \leqq P \leqq \sum \alpha_i . \tag{4.3}$$

Suppose now that every pair (T_i, T_j) is negatively quadrant dependent, so that

$$P(T_i \geqq C_i, T_j \geqq C_j) \leqq \alpha_i\alpha_j .$$

Then it follows from (4.3) that

$$\sum \alpha_i - \sum_{i<j} \alpha_i\alpha_j \leqq P \leqq \sum \alpha_i . \tag{4.4}$$

To see how close these bounds are, note that if $\alpha = \sum \alpha_i$, then $\sum_{i<j} \alpha_i\alpha_j \leqq \frac{1}{2}(1 - s^{-1})\alpha^2$ so that

$$\alpha - \tfrac{1}{2}\alpha^2(1 - s^{-1}) \leqq P \leqq \alpha . \tag{4.5}$$

This shows α to be an excellent approximation for P whenever α is small. In the following examples, we shall consider some cases of negative quadrant dependence that have arisen in this context.

Example 2. *The problem of m rankings.* A number of objects, say n, are ranked independently by m observers. The hypothesis H to be tested is that there is no preference among the objects, so that each observer ranks them at random. Under the alternative K_i, the ith object is typically preferred but there is no preference among the remaining objects.

Let R_{ik} be the rank assigned to the ith object by the kth observer, and let $R_i = \sum R_{ik}$ denote the rank sum for the ith object. Then H is rejected in favor of K_i if R_i is sufficiently large. In [8], it is shown by induction over m that R_i, R_j are negatively quadrant dependent, the result being obvious for $m = 1$ from the formula

$$P(R_i \leqq x, R_j \leqq y) = [xy - \min(x, y)]/n(n-1). \tag{4.7}$$

Instead of proving the result by induction, we here only need note that $R_i = \sum R_{ik}$, $R_j = \sum R_{jk}$, the pairs (R_{ik}, R_{jk}) $k = 1, \cdots, n$ are independent, and each pair is in $\mathcal{G}_1$ by (4.7). The result now follows from Theorem 1.

Example 3. In connection with the Poisson slippage problem, Doornbos and Prins (1958) proved that $(U_i, U_j) \varepsilon \mathcal{G}_1$ where U_i is the number of trials, in a sequence of n multinomial trials with s possible outcomes, resulting in outcome i. Their method is closely related to that of the next section. We note here that if $U_{ik} = 1$ when the kth trial results in outcome i and $U_{ik} = 0$ otherwise, a simple calculation shows that $(U_{ik}, U_{jk}) \varepsilon \mathcal{G}_1$ and since $U_i = \sum_{k=1}^{n} U_{ik}$, $U_j = \sum_{k=1}^{n} U_{jk}$, it follows from Theorem 1 that also $(U_i, U_j) \varepsilon \mathcal{G}_1$.

Example 4. Consider testing the nonparametric hypothesis of the equality of s distributions $H: F_1 = \cdots = F_s$ against the alternatives K_i that F_i has slipped to the right, on the basis of samples X_{ik} $(k = 1, \cdots, n_i)$ from F_i. If the $N = \sum n_i$ observations are ranked, and R_{ik} denotes the rank of X_{ik} and $R_i = \sum R_{ik}$, a distribution free test rejects H in favor of K_i when R_i is sufficiently large. In this connection Doornbos and Prins [8] give a proof by Kesten of the fact that $(R_i, R_j) \varepsilon \mathcal{G}_1$. An examination of the proof shows that it carries over to the case in which an arbitrary set of numbers $v_1, \cdots, v_N$ (rather than the set of integers $1, \cdots, N$) is divided at random into s groups of $n_1, \cdots, n_s$ elements respectively, with R_i denoting the sum of the v-values in the ith group. (Kesten's proof is by induction over N. The population size is reduced by conditioning on which of the groups contains the integer N. If in the extension the largest v is not unique, the proof applies if one conditions on the position of any one of the largest v's.) Another interesting special case of the generalized result is that in which all v's are zero or one. This proves negative quadrant dependence for the components of a multiple hypergeometric distribution, which also follows from Example 10 (iii) of Section 8.

5. Regression dependence. Definition (2.1) can be rewritten as

$$P(Y \leqq y \mid X \leqq x) \geqq P(Y \leqq y) \tag{5.1}$$

and in this form clearly expresses the fact that knowledge of X being small increases the probability of Y being small. It may be felt that the intuitive concept of positive dependence is better represented by the stronger condition.

$$P(Y \leqq y \mid X \leqq x) \geqq P(Y \leqq y \mid X \leqq x') \quad \text{for all} \quad x < x' \quad \text{and all} \quad y. \tag{5.2}$$

Rather than (5.2), we shall here consider the still stronger condition

$$P(Y \leqq y \mid X = x) \quad \text{is non-increasing in } x, \tag{5.3}$$

which was discussed earlier by Tukey (1958) and Lehmann (1959). If (5.3) holds, we shall say that Y is *positively regression dependent* on X; the family of all distributions F of (X, Y) for which (5.3) holds will be denoted by $\mathcal{F}_2$. Similarly, Y is *negatively regression dependent* on X if $P(Y \leqq y \mid X = x)$ is non-decreasing in x; the associated family of distributions F will be denoted by $\mathcal{G}_2$.

EXAMPLE 5. Let $Y = \alpha + \beta X + U$, where X and U are independent. Then Y is positively or negatively regression dependent on X as $\beta \geqq 0$ or $\leqq 0$. This is obvious since the conditional distribution of Y given $X = x$ is that of $\alpha + \beta x + U$ and hence is clearly stochastically increasing in x if $\beta > 0$. In particular, it follows from this example that the components of a bivariate normal distribution are positively or negatively regression dependent as $\rho \geqq 0$ or $\rho \leqq 0$.

Before discussing further examples, we state formally the relationship between the different definitions of positive dependence given up to this point.

LEMMA 4. *The definitions* (5.1) *to* (5.3) *are connected by the implications* (5.3) $\Rightarrow$ (5.2) $\Rightarrow$ (5.1) *and hence* $\mathfrak{F}_1 \subset \mathfrak{F}_2$.

PROOF. That (5.2) $\Rightarrow$ (5.1) is obvious. To show that (5.3) $\Rightarrow$ (5.2), let $h(u) = P(Y \leqq y \mid X = u)$, so that

$$P(Y \leqq y, X \leqq x) = \int_{-\infty}^{x} h(u)\, dF_X(u)$$

where F_X denotes the marginal distribution of X. Under the assumption that h is non-increasing, we must therefore show that

$$\int_{(-\infty, x]} h(u)\, dF_X(u)/P(X \leqq x) \geqq \int_{(-\infty, x']} h(u)\, dF_X(u)/P(X \leqq x') \qquad \text{for} \quad x < x',$$

which is obviously true.

Lemma 4 frequently provides a convenient method of proving membership in $\mathfrak{F}_1$. As a first example, let us consider once more the dependence of two components from a multinomial distribution.

EXAMPLE 3 (continued). If $(U_1, \cdots, U_s)$ have a multinomial distribution corresponding to n trials and success probabilities $(p_1, \cdots, p_s)$, the conditional distribution of U_j given $U_i = x$ is a binomial distribution with success probability $p_i/(1 - p_j)$ and corresponding to $n - x$ trials. Since $P(U_j \leqq y \mid U_i = x)$ is thus a decreasing function of x, it follows that (U_i, U_j) is in $\mathcal{G}_2$ and therefore in $\mathcal{G}_1$. (The proof of quadrant dependence for this case in [8] uses related ideas.) The case of a multiple hypergeometric distribution mentioned in Example 4 can be treated similarly.

EXAMPLE 6. An interesting example in which negative quadrant dependence was first stated by Cochran (1941) and proved by Doornbos (1956) permits a very simple treatment by the present method. The problem is that of testing H: $\sigma_1 = \cdots = \sigma_s$ in the case of s normal distributions $N(\xi_i, \sigma_i^2)$, on the basis of samples of equal size, say n. The hypothesis is rejected in favor of the alternative K_i that the ith variance has slipped to the right if

$$A_i = S_i^2/\sum_{k=1}^{s} S_k^2 > C$$

where $S_k^2 = \sum (X_{kl} - X_{k\cdot})^2/(n - 1)$. Cochran showed that the joint density of A_i and A_j is proportional to

$$(5.4) \qquad a_1^{f_1} a_2^{f_2} (1 - a_1 - a_2)^{f_3} \quad \text{for} \quad 0 \leqq a_1, a_2; \quad a_1 + a_2 \leqq 1$$

with $f_1 = f_2 = \frac{1}{2}(n - 1) - 1$ and $f_3 = \frac{1}{2}(s - 2)(n - 1) - 1$. The conditional distribution of A_2 given $A_1 = a_1$ under (5.4) is that of $(1 - a_1)B$ where B has a beta distribution with density $C\, b^{f_2} (1 - b)^{f_3}$. Since $(1 - a_1)\, B$ is clearly stochastically decreasing as a_1 increases, this shows that $(A_1, A_2) \,\varepsilon\, \mathcal{G}_2$ for the general case (5.4).

While in many standard examples of positive dependence, the distributions satisfy not only (5.1) but also (5.3), the second of these conditions is much stronger than the first. The following are some examples, in which there is a strong intuitive feeling of positive or negative dependence, where there is quadrant dependence, but where the stronger condition of regression dependence is not satisfied.

EXAMPLE 4 (continued). As was stated in Section 4, the variables (R_i, R_j) of Example 4 are in $\mathcal{G}_1$ for all values $v_1, \cdots, v_N$. While they are also in $\mathcal{G}_2$ for the special case that the v's are all zero or one, this is not true in general, as is seen by putting $v_1 = 3$, $v_2 = 3$, $v_3 = 5$, $v_4 = 15$; $n_1 = 2$, $n_2 = 1$; and $s = 2$. It is easily checked that in this case $P(R_2 \leqq 1 \mid R_1 \leqq x)$ is not increasing in x, and that therefore $X = R_1$, $Y = R_2$ do not even satisfy (5.2).

This example also demonstrates the asymmetry in x, y of Definitions (5.2) and (5.3). For while $P\{R_2 \leqq y \mid R_1 \leqq x\}$ and $P\{R_2 \leqq y \mid R_1 = x\}$ are not increasing functions of x for all y, $P\{R_1 \leqq x \mid R_2 \leqq y\}$ and $P\{R_1 \leqq x \mid R_2 = y\}$ are nondecreasing functions of y for all x. In fact, consider any $v_1, \cdots, v_N$ from which we draw first a sample of size n_1 with sample sum R_1 and then a sample of size $n_2 = 1$ with sample-value R_2. Then $P\{R_1 \leqq x \mid R_2 = y\}$ is an increasing function of y. This is easily seen by constructing from a random variable with the distribution of R_1 given y, a variable with the distribution of R_1 given $y' > y$, which is always at least as large as the first variable and with positive probability is actually larger.

As a definition of positive dependence one might wish to symmetrize (5.2) or (5.3) by adding to it the dual condition in which X and Y are interchanged. On the other hand, the asymmetric form is preferable as a tool for proving quadrant dependence since it is then enough to prove either (5.2) (or (5.3)) or the dual condition.

It may seem surprising that a pair of random variables, which intuitively appears to exhibit such strong negative dependence as that of the preceding example, does not belong to $\mathcal{G}_2$. However, the following example exhibits the same phenomenon.

EXAMPLE 7. If U, V are independent, then $X = U + V$, $Y = U$ belong to $\mathcal{F}_1$ by Example 1 (iii). That they do not necessarily satisfy (5.2) or (5.3) is shown by the case in which both variables take on the values 0, 2, 3 with probabilities p, q, r, $(p + q + r = 1)$. For it is then easily checked that

$$P(Y \leqq 2 \mid X \leqq 3) < P(Y \leqq 2 \mid X \leqq 4).$$

Again, the example also exhibits asymmetry. Although (X, Y) does not even satisfy (5.2), we see that (Y, X) satisfies (5.3) since $P(X \leqq x \mid Y = y) = P(V \leqq x - y)$ which decreases as y increases.

It was shown by Efron (1965) that $(U + V, U) \varepsilon \mathfrak{F}_2$ if U, V are independently distributed according to Pólya frequency functions of order 2. A simple proof of this result will be given in Example 12 (Section 8).

EXAMPLE 8 (Tukey). Let $X_1, \cdots, X_n$ be independently distributed with distribution F, let the ordered sample be denoted by $X^{(1)} < \cdots < X^{(n)}$, and suppose that F is subexponential to the right, i.e. that $[1 - F(x + a)]/[1 - F(x)]$ is an increasing function of x for each fixed $a > 0$. Then it was shown by Tukey (1958) that $X^{(s)} - X^{(r)}$ is negatively regression dependent on $X^{(r)}$ if $r < s$.

For the sake of completeness, we conclude this section with an example in which (X, Y) satisfies (5.2) but not (5.3).

EXAMPLE 9. Let the distribution of (X, Y) be given by the following 3×3 table

		y		
		1	2	3
	1	p	0	0
x	2	0	q	0
	3	s	0	r

Then it is easily checked that (X, Y) satisfies (5.2) provided

$$qs \leqq pr; \tag{5.5}$$

on the other hand (X, Y) satisfies (5.3) only if $s = 0$.

6. Unbiasedness of tests of independence. Let $(X_1, Y_1), \cdots, (X_n, Y_n)$ be independently distributed with a common distribution F. Suppose that the marginal distributions of F are continuous (so that the X's and the Y's will be distinct with probability one) and consider the problem of testing the hypothesis H that X_i and Y_i are independent (without any further specification of F) against alternatives of positive dependence. Then any similar test (and hence any unbiased test) $\varphi(x_1, y_1; \cdots; x_n, y_n)$ is a permutation test. (See [18], Chap. 5). Suppose without loss of generality that in addition φ is invariant under permutations of the pairs $(x_1, y_1), \cdots, (x_n, y_n)$.

THEOREM 3. *If under the above assumptions and for any real-valued non-decreasing functions* $f_1, \cdots, f_n$ *satisfying*

$$f_1(v) \leqq f_2(v) \leqq \cdots \leqq f_n(v) \tag{6.1}$$

it is true that

$$x_1 < \cdots < x_n \quad \textit{and} \quad y_i' = f_i(y_i) \qquad \textit{for} \quad i = 1, \cdots, n \tag{6.2}$$

implies

$$\varphi(x_1, y_1; \cdots; x_n, y_n) \leqq \varphi(x_1, y_1'; \cdots; x_n, y_n') \tag{6.3}$$

then φ is unbiased against all alternatives F in $\mathfrak{F}_2$.

PROOF. If α is the size of the test, φ satisfies

$$\alpha = E_H(\varphi(X_1, Y_1; \cdots; X_n, Y_n) \mid x_1, \cdots, x_n). \tag{6.4}$$

This implies that for any fixed $x_1, \cdots, x_n$

$$\alpha = E\,\varphi(x_1, Y_1; \cdots; x_n, Y_n)$$

whenever $Y_1, \cdots, Y_n$ are identically, independently distributed. Consider now any alternative $F \varepsilon \mathfrak{F}_2$. Since then the conditional distributions of Y given x_i are stochastically increasing in i, there exist (see [18], p. 73) functions f_i satisfying (6.1) and independent, identically distributed random variables V_i such that $Y_i' = f_i(V_i)$ has the conditional distribution of Y given x_i. Since α is the conditional expectation of φ given $x_1, \cdots, x_n$ when $Y_i = V_i$ independent of X_i, it follows—if $\beta(x_1, \cdots, x_n)$ denotes the conditional expectation of φ under F—that

$$\beta(x_1, \cdots, x_n) = E\,\varphi(x_1, Y_1'; \cdots, x_n, Y_n' \mid x_1, \cdots, x_n)$$

and hence by (6.1) that

$$\alpha \leqq \beta(x_1, \cdots, x_n) \qquad \text{for all} \quad x_1, \cdots, x_n.$$

If β denotes the unconditional power of the test against φ, then

$$\beta = E_F\,\beta(X_1, \cdots X_n)$$

and hence $\alpha \leqq \beta$ as was to be proved.

COROLLARY 1. *The theorem remains valid if φ satisfies* (6.3) *for all pairs of points $(x_1, y_1; \cdots; x_n, y_n)$ and $(x_1, y_1'; \cdots; x_n, y_n')$ satisfying*

$$x_1 < \cdots < x_n, \qquad i < j, \qquad y_i < y_j \Rightarrow y_i' < y_j'. \tag{6.5}$$

PROOF. It is easily seen that (6.2) and (6.5) are equivalent.

COROLLARY 2. *The following one-sided tests of independence are unbiased against all alternatives belonging to $\mathfrak{F}_2$. (For a discussion of these tests see for example* [14].)

(i) *The difference sign covariance test based on*

$$t = \textstyle\sum_{i \neq j} \operatorname{sgn}(x_i - x_j) \operatorname{sgn}(y_i - y_j)/n(n-1)$$

or equivalently on the number of concordant pairs (x_i, y_i), (x_j, y_j) (i.e. *pairs satisfying the condition $(x_i - x_j)(y_i - y_j) > 0$*).

(ii) *The test based on the unbiased grade correlation*

$$R' = \sum \operatorname{sgn}(x_i - x_j) \operatorname{sgn}(y_i - y_k)/n(n-1)(n-2)$$

where the summation extends over all different subscripts i, j, k, or equivalently on the number of triplets (x_i, y_i), (x_j, y_j), (x_k, y_k) with at least two concordant pairs.

(iii) *The test based on the rank correlation coefficient*

$$R = [(n-2)R' + 3t]/(n+1).$$

(iv) *The quadrant test discussed by Blomqvist* (1950), *which is based on the number Q of points among $(x_1, y_1), \cdots, (x_n, y_n)$ lying in quadrants* I *or* III *when the origin is taken as (m_X, m_Y) where m_X is the median of the X's and m_Y the median of the Y's (so that Q is the number of pairs (x_i, y_i) concodrant with the pair (m_X, m_Y)).*

PROOF. (i) to (iii). Since (6.5) states that

$$(6.6) \quad (x_i, y_i), (x_j, y_j) \text{ concordant} \Rightarrow (x_i, y_i'), (x_j, y_j') \text{ concordant}$$

it is obvious that (6.5) implies $t(x, y) \leqq t(x, y')$ and $R'(x, y) \leqq R'(x, y')$ and hence also $R(x, y) \leqq R(x, y')$ where (x, y) stands for $(x_1, y_1), \cdots, (x_n, y_n)$. Thus in all three cases (6.5) implies (6.3) as was to be proved.

(iv) We shall consider only the case $n = 2m$. Let $(x_1, y_1), \cdots, (x_n, y_n)$ and $(x_1, y_1'), \cdots, (x_n, y_n')$ satisfy (6.5), let m_Y' denote the median of the y_i', and let k be the number of points among $(x_1, y_1), \cdots, (x_n, y_n)$ that are in quadrant I, so that there are also k points in III. Suppose that in moving from the unprimed to the primed y's, at least one point (x_i, y_i) moved from III to II$'$. Then $y_i' > m_Y'$ and hence the y' coordinates of all points originally in I are greater than m_Y' (since they are greater than y_i'). Thus all k points which originally were in I will be in I$'$. Suppose now that exactly $r \geqq 1$ points have risen from III to II$'$. By definition of m_Y' there must still be m points below m_Y'. Since none of the points from I have dropped to IV$'$, at least r must have dropped from II to III$'$. Thus the number of points in III$'$ is $\geqq$ the number of points in III, and hence $Q(x, y') \geqq Q(x, y)$.

7. Unbiasedness of tests of independence based on normal and other scores. In the present section we shall apply Theorem 3 to prove unbiasedness, against alternatives in $\mathfrak{F}_2$ with continuous marginals, of a class of rank tests for the hypothesis H of Section 6 proposed by Bhuchongkul (1964). These tests are based on statistics of the form $\sum A(r_i) B(s_i)$, where $(r_1, \cdots, r_n)$ denote the ranks of $(X_1, \cdots, X_n)$ and $(s_1, \cdots, s_n)$ the ranks of $(Y_1, \cdots, Y_n)$. Let $(t_1, \cdots, t_n)$ be the ranks of $(Y_{i_1}, \cdots, Y_{i_n})$ where $(i_1, \cdots i_n)$ is the permutation for which $X_{i_1} < \cdots < X_{i_n}$ so that the ranks of $(X_{i_1}, \cdots, X_{i_n})$ are $(1, \cdots, n)$. Then the test statistics can be written equivalently as

$$(7.1) \qquad T_n = \textstyle\sum_{i=1}^n A(i) B(t_i).$$

The purpose of the present section is to prove the following theorem.

THEOREM 4. *The tests with rejection region $T_n \geqq C$, where T_n is given by* (7.1), *are unbiased against all alternatives belonging to $\mathfrak{F}_2$ provided the functions A and B are non-decreasing.*

Some important special cases are

(i) $A(i) = B(i) = i$, in which case the test reduces to the test based on the rank correlation coefficient considered in Corollary 2 (iii) of the preceding section.

(ii) $A(i) = B(i) = E(U^{(i)})$, where $U^{(1)} < \cdots < U^{(n)}$ denotes an ordered sample from a normal distribution. This leads to the normal scores test studied by Bhuchongkul.

(iii) More generally $A(i) = E[V^{(i)}]$, $B(i) = E[W^{(i)}]$, where $V^{(1)} < \cdots < V^{(n)}$ and $W^{(1)} < \cdots < W^{(n)}$ are ordered samples from any two continuous distributions J and K.

To prove Theorem 4, we shall define the following partial ordering of permutations and establish some properties of this ordering. Let

$$\{i\} = (i_1, \cdots, i_n) \quad \text{and} \quad \{j\} = (j_1, \cdots, j_n)$$

be two permutations of $(1, \cdots, n)$. We shall say that $\{i\}$ is better ordered than $\{j\}$ if for all a, b

$$a < b \quad \text{and} \quad j_a < j_b \Rightarrow i_a < i_b . \tag{7.2}$$

A more usual comparison defines $\{i\}$ to be better ordered than $\{j\}$ if it is possible to transform $\{j\}$ into $\{i\}$ by a number of steps, each of which consists in correcting an inversion. This definition neither implies nor is it implied by (7.2) as is seen by the following two examples:

(a) (132) is better ordered than (312) according to the second definition but not according to (7.2).

(b) (1324) is better ordered than (2413) according to (7.2) but not according to the second definition.

We shall here restrict use of the term "better ordered" to (7.2). It is clear that (7.2) does define a partial ordering among the $n!$ permutations of $(1, \cdots, n)$. The basic property required is contained in the following theorem.

THEOREM 5. *If* $(i_1, \cdots, i_n)$ *is better ordered than* $(j_1, \cdots, j_n)$ *and if for any* $m \leqq n$

$$i'_{1m} < \cdots < i'_{mm} \quad \text{and} \quad j'_{1m} < \cdots < j'_{mm} \tag{7.3}$$

denote the ordered m-tuples $(i_1, \cdots, i_m)$ *and* $(j_1, \cdots, j_m)$, *then*

$$i'_{km} \leqq j'_{km} \qquad \text{for all} \quad 1 \leqq k \leqq m \quad \text{and all} \quad 1 \leqq m \leqq n. \tag{7.4}$$

PROOF. The proof will be by induction over m.

We note first the following. Let the rank of i_k among $(i_1, \cdots, i_m)$ be r and that of j_k among $(j_1, \cdots, j_m)$ be s. (For simplicity we suppress the subscripts km on r and s.) Then the number of i's to the right of i_m and less than i_k is $(i_k - 1) - (r - 1) = i_k - r$; the number of j's to the right of j_m and less than j_k is $j_k - s$. If $(i_1, \cdots, i_n)$ is better ordered than $(j_1, \cdots, j_n)$, it follows that

$$i_k - r \leqq j_k - s. \tag{7.5}$$

Applying (7.5) to $m = k = 1$ (so that also $r = s = 1$) it follows that $i_1 \leqq j_1$ and hence $i'_{11} \leqq j'_{11}$, so that the theorem is correct for $m = 1$.

Suppose now that the result holds for $m = 1, \cdots, a - 1$ and consider the situation for $m = a$. Let r and s denote the ranks of i_a and j_a among $(i_1, \cdots, i_a)$ and $(j_1, \cdots, j_a)$ respectively. Suppose first that $r \leqq s$. Then it is an easy consequence of the induction hypothesis that $i'_{ka} \leqq j'_{ka}$ for all $k = 1, \cdots, a$. Consider therefore the case $r > s$. Then it again follows easily from the induction hypothesis that $i'_{ka} \leqq j'_{ka}$ for $k < s$ and $k > r$. It only remains to consider the case $s \leqq k \leqq r$. Since $i_a = i'_{rm}$, it follows that i_a must exceed i'_{km} by at least $r - k$ so that $i_a \geqq i'_{km} + (r - k)$ and analogously $j_a \leqq j'_{km} + (k - s)$. Hence,

$$j'_{km} - i'_{km} \geqq [j_a + (k - s)] - [i_a - (r - k)]$$
$$= (j_a - s) - (i_a - r).$$

The right hand side is $\geqq 0$ by (7.5), and this completes the proof.

COROLLARY 1. *If $\{i\}$ is better ordered than $\{j\}$ and if h is non-decreasing, then*

$$\sum_{k=1}^{m} h(i_k) \leqq \sum_{k=1}^{m} h(j_k) \qquad \textit{for all} \quad 1 \leqq m \leqq n. \tag{7.6}$$

PROOF. This is obvious since

$$\sum_{k=1}^{m} h(i_k) = \sum_{k=1}^{m} h(i'_{km}) \leqq \sum_{k=1}^{m} h(j'_{km}) = \sum_{k=1}^{m} h(j_k).$$

(Although we do not need it here, it is in fact easily seen that (7.3) is equivalent to (7.6) holding for all non-decreasing functions h.)

COROLLARY 2. *If $\{i\}$ is better ordered than $\{j\}$, if $a_1 < \cdots < a_n$ and h is non-decreasing, then*

$$\sum_{k=1}^{n} a_k h(i_k) \geqq \sum_{k=1}^{n} a_k h(j_k). \tag{7.7}$$

PROOF. Since the inequality is not affected by the addition of a constant to h, we may assume without loss of generality that h is positive, and then further that $\sum_{i=1}^{n} h(i) = 1$.

Let us now interpret $h(i_1), \cdots, h(i_n)$ and $h(j_1), \cdots, h(j_n)$ as probability distributions over the integers $1, \cdots, n$. Then a random variable with the first of these distributions is, by Corollary 1, stochastically larger than one with the second distribution, and the result follows from a well-known property of stochastically ordered distributions. (See, for example, [18], Chapter 3, Lemma 2 and Problem 11.)

PROOF OF THEOREM 4. Consider the application of Corollary 1 of Theorem 3 to a rank test. If $(t_1', \cdots, t_n')$ and $(t_1'', \cdots, t_n'')$ denote the ranks of $(y_1', \cdots, y_n')$ and $(y_1'', \cdots, y_n'')$, condition (6.5) states that $(t_1', \cdots, t_n')$ is better ordered than $(t_1'', \cdots, t_n'')$. Thus Theorem 4 now follows from Theorem 5 and Corollary 1 of Theorem 3.

8. Likelihood ratio dependence. Condition (5.3) which defined regression dependence requires the conditional variable Y given x to be stochastically increasing. An even stronger condition is obtained by requiring the conditional density of Y given x to have monotone likelihood ratio. (For a discussion of the relation between these two conditions, see [18], p. 74). We shall then say that (X, Y) or its distribution F shows *positive likelihood ratio dependence*. The family of all distributions F satisfying this condition will be denoted by $\mathfrak{F}_{\cdot}$. If $f(x, y)$ is the joint density of X and Y with respect to some product measure, the condition may be written formally as

$$f(x, y') f(x', y) \leqq f(x, y) f(x', y') \qquad \text{for all} \quad x < x', \quad y < y'. \tag{8.1}$$

If the inequality is reversed, F is negatively likelihood ratio dependent and belongs to $\mathcal{G}_3$. Like conditions (2.1) but unlike (5.2) and (5.3), condition (8.1) is symmetric in X and Y. Checking (8.1) is frequently the easiest way of proving positive dependence.

EXAMPLE 10. The following are some bivariate densities with monotone likelihood ratio dependence:

(i) a bivariate normal density is in $\mathfrak{F}_3$ or $\mathcal{G}_3$ as $\rho \geqq 0$ or $\rho \leqq 0$;

(ii) any two components of a multinomial distribution (considered earlier in Example 3) are in $\mathcal{G}_3$;

(iii) the Dirichlet distribution with density (5.4) is in $\mathcal{G}_3$. In each of these cases it is easy to verify that the density satisfies the required inequality.

EXAMPLE 11. Consider the joint distribution F of any two dependent indicator variables I_0 and I_1 with, say, $p_{ij} = P(I_0 = i, I_1 = j)$, $i = 0, 1; j = 0, 1$. (Such a distribution defines the random structure of a two by two table if I_0 and I_1 indicate the occurrence of the two characteristics in question). Then it is seen that F is in $\mathfrak{F}_3$ (and hence in $\mathfrak{F}_1$ and $\mathfrak{F}_2$) if

$$p_{00}p_{11} \geqq p_{01}\, p_{10} \tag{8.2}$$

and in $\mathcal{G}_3$ if the inequality is reversed. As in the normal case, all distributions in this family therefore show likelihood ratio dependence.

EXAMPLE 12. Let U, V be independently distributed with densities g and h respectively, and let $X = U$, $Y = U + V$. Then we saw earlier in Example 7 that Y is always positively regression dependent on X but not necessarily X on Y. Now the joint density of X and Y is $g(x)\, h(y - x)$ and Condition (8.1) therefore reduces to

$$h(y - x')/h(y - x) \leqq h(y' - x')/h(y' - x). \tag{8.3}$$

This condition is satisfied provided $-\log h$ is convex (see [18], p. 330). For such densities (X, Y) belongs to $\mathfrak{F}_3$ and hence not only (X, Y) but also (Y, X) belongs to $\mathfrak{F}_2$. This last result was proved earlier by Efron (1965).

EXAMPLE 13. Let $Z_1, \cdots, Z_n$ be independently distributed according to a univariate distribution G, and let X and Y be two order statistics of the Z's, say $X = Z^{(r)}$ and $Y = Z^{(s)}$ with $r < s$. Then the joint distribution F of X and Y has density

$$f(x, s) = [G(x)]^{r-1}\, [G(y) - G(x)]^{s-r-1}\, [1 - G(y)]^{n-s}, \qquad x < y$$

with respect to the product measure $G \times G$. This density satisfies (8.1) so that $F \,\varepsilon\, \mathfrak{F}_3$. That $F \,\varepsilon\, \mathfrak{F}_2$ and hence Cov $(Z^{(r)}, Z^{(s)}) \geqq 0$ was proved by Bickel (1965).

EXAMPLE 14. Let $f_\theta(x)$, $g_\theta(y)$ be two families of univariate densities, each with monotone (increasing) likelihood ratio. Then for any mixing distribution λ, the distribution with density.

$$h(x, y) = \int f_\theta(x)\, g_\theta(y)\, d\lambda(\theta)$$

belongs to $\mathfrak{F}_3$.

PROOF. The inequality (8.1) in the present case may be written as

$$\iint f_\theta(x)\, g_\theta(y) f_\eta(x')\, g_\eta(y')\, d\lambda(\theta)\, d\lambda(\eta)$$
$$\geqq \iint f_\theta(x)\, g_\theta(y') f_\eta(x')\, g_\eta(y)\, d\lambda(\theta)\, d\lambda(\eta).$$

Since the integrands on the two sides are equal for $\theta = \eta$, it is enough to integrate over the region $\theta \neq \eta$. Consider separately the contributions for $\theta < \eta$ and $\theta > \eta$. In the latter integrals interchange the variables of integration θ and η, so that all integrals extend over the region $\theta < \eta$. On combining all four integrals into a single integral on the left hand side, the integrand is seen to factor into

$$[f_\theta(x)f_\eta(x') - f_\eta(x)f_\theta(x')][g_\theta(y)g_\eta(y') - g_\eta(y)g_\theta(y')]$$

which is non-negative. This completes the proof.

REFERENCES

[1] BHUCHONGKUL, S. (1964). A class of nonparametric tests for independence in bivariate populations. *Ann. Math. Statist.* **35** 138–149.

[2] BICKEL, PETER J. (1965). Some contributions to the theory of order statistics. Submitted.

[3] BLOMQVIST, NILS (1950). On a measure of dependence between two random variables. *Ann. Math. Statist.* **21** 593–600.

[4] BLUM, J. R., KIEFER, J. AND ROSENBLATT, M. (1961). "Distribution free tests of independence based on the sample distribution," *Ann. Math. Statist.* **32** 485–498.

[5] COCHRAN, W. G. (1941). The distribution of the largest of a set of estimated variances as a fraction of their total. *Ann. Eugenics* **11** 47–52.

[6] DAVID, H. A. (1956). On the application to statistics of an elementary theorem in probability. *Biometrika* **43** 85–91.

[7] DOORNBOS, R. (1956). Significance of the smallest of a set of estimated normal variances. *Statistica Neerlandica*, **10** 17–26.

[8] DOORNBOS, R. AND PRINS, H. J. (1958). On slippage tests. *Indag. Math.* **20** 38–55 and 438–447.

[9] EFRON, BRADLEY (1965). Increasing properties of Pólya frequency functions. *Ann. Math. Statist.* **36** 272–279.

[10] HALPERIN, M., GREENHOUSE, S. W., CORNFIELD, J. AND ZALOKAR, JULIA (1955). Tables of percentage points for the studentifed maximum absolute deviate in normal samples. *J. Amer. Statist. Assoc.* **50** 185–195.

[11] HARDY, G. H., LITTLEWOOD, J. E. AND PÓLYA, G. (1934). *Inequalities*. Cambridge Univ. Press.

[12] HARTLEY, H. O. (1955). Some recent developments in analysis of variance. *Comm. Pure Appl. Math.* **8** 47–72.

[13] HÖFFDING, WASSILY (1940). Masstabinvariante Korrelations-theorie. *Schriften Math. Inst. Univ. Berlin* **5** 181–233.

[14] HÖEFFDING, WASSILY (1948). A class of statistics with asymptotically normal distributions. *Ann. Math. Statist.* **19** 293–375.

[15] HÖEFFDING, WASSILY (1948). A nonparametric test of independence. *Ann. Math. Statist.* **19** 546–557.

[16] Kruskal, William H. (1958). Ordinal measures of association. *J. Amer. Statist. Assoc.* **53** 814–861.

[17] Lehmann, E. L. (1951). Consistency and unbiasedness of certain nonparametric tests. *Ann. Math. Statist.* **22** 165–179.

[18] Lehmann, E. L. (1959). *Testing statistical hypotheses*. Wiley, New York.

[19] Paulson, Edward (1952). On the comparison of several experimental categories with a control. *Ann. Math. Statist.* **23** 239–246.

[20] Paulson, Edward (1952). An optimum solution to the k-sample slippage problem for the normal distribution. *Ann. Math. Statist.* **23** 610–616.

[21] Tukey, J. W. (1958). A problem of Berkson, and minimum variance orderly estimators. *Ann. Math. Statist.* **29** 588–592.

Inverted Distributions

E. L. LEHMANN and JULIET POPPER SHAFFER*

By the inverted distribution of a random variable X we mean the distribution of its reciprocal $1/X$. Every distribution G that assigns probability 0 to the value 0 is the inverted distribution of some random variable, namely of $X = 1/Y$, where Y is distributed according to G. Thus it is impossible to assert any special properties of inverted distributions in general. As we show in Section 2, however, very weak assumptions concerning the distribution of X permit quite strong conclusions concerning G, such as Cauchy tails and multiple modes. Sections 3 and 4 discuss certain approximations and limit results concerning the inverted distributions considered in Section 2.

KEY WORDS: Bimodality; Cauchy tails; Limit distributions; Multimodality; Reciprocals.

1. INTRODUCTION

Piegorsch and Casella (1985) reminded us of the fact (or, possibly, surprised us with it) that the random variable $1/X$ does not possess an expectation if X has a continuous density f with

$$f(0) > 0, \tag{1.1}$$

for example if it is normally distributed with any mean and variance. The result and its generalizations are of statistical interest although [as is pointed out by Piegorsch and Casella and by Hannan (1985)] from an analytic point of view they simply reflect the relation between the integral $\int (1/x)f(x)\,dx$ and the behavior of f at 0. Other aspects of reciprocal moments were discussed, for example, by Wooff (1985). For a general discussion of inverted distributions, see Folks (1983); the normal case was studied by Zellner (1978, appendix).

In the present article, we show that the assumptions of Piegorsch and Casella permit the even stronger conclusion that the distribution of $1/X$ has a Cauchy tail if $0 < f(0) < \infty$ [and an even heavier tail if $f(0) = \infty$]. This can be interpreted as a statistical explanation of the nonexistence of $E(1/X)$.

The aforementioned result suggests that, conversely, the behavior of f at ∞ can tell us something about the behavior of the density g of $Y = 1/X$ at 0. In fact, we shall show that if the density f satisfies

$$x^2 f(x) \to 0 \quad \text{as} \quad x \to +\infty \quad \text{or} \quad -\infty, \tag{1.2}$$

then

$$g(y) \to 0 \quad \text{as} \quad y \to 0 \tag{1.3}$$

from the right or left, respectively.

This second result has a surprising consequence. Suppose that f satisfies (1.2) and, in addition, $f(x)$ is positive for both positive and negative values of x. Then the density g of its inverted distribution must be at least (and typically is exactly) bimodal.

2. THE SHAPE OF INVERTED DISTRIBUTIONS

In this section, we shall be concerned with the effect of the behavior of the density f of a random variable X at ± 0 and $\pm\infty$ on that of the density g of $Y = 1/X$ at $\pm\infty$ and ± 0, respectively. For the sake of simplicity, we shall assume throughout this and later sections that the support of X is an interval with endpoints $a < b$ (finite or infinite) and f is continuous in the open interval (a, b), which we shall call the open support of X.

Theorem 2.1. Assume that $f(x)$ tends to a limit $f(0+)$ (finite or infinite) as $x \to 0+$, that is, as x tends to 0 from the right. Then the density g of $Y = 1/X$ satisfies

$$\frac{g(y)}{1/(1 + y^2)} \to f(0+) \quad \text{as} \quad y \to \infty. \tag{2.1}$$

Proof. By making the transformation $y = 1/x$, one sees that the density of $Y = 1/X$ is given by

$$g(y) = (1/y^2)f(1/y), \tag{2.2}$$

and the result follows.

When the left side of (2.1) tends to a finite positive constant, we shall say for short that g has a right Cauchy tail; when it tends to 0 or ∞, we shall say that the right tail of g is lighter or heavier, respectively, than Cauchy.

Under the assumptions of Theorem 2.1 it follows that if $f(0+) = 0$, the right tail of g is lighter than Cauchy; if $0 < f(0+) < \infty$, the right tail of g is Cauchy; and if $f(0+) = \infty$, the right tail of g is heavier than Cauchy.

Note that the analogous results are obtained for the left tail of g by letting $y \to -\infty$, $x \to 0-$, and replacing $f(0+)$ by $f(0-)$.

This simple theorem establishes Cauchy tails for the reciprocals of most of the random variables commonly encountered. Some examples are: (a) positive and negative Cauchy tails for the inverted distributions of variables with normal, t, logistic, and double-exponential distributions, or uniform distributions on (a, b) with $a < 0 < b$; and (b) positive Cauchy tails for many positive random variables, including those with uniform distributions on $(0, a)$ and with exponential distributions, including the χ^2 distribution with $n = 2$ degrees of freedom. (For χ^2 with $n = 1$ and $n > 2$ the tail of G is heavier and lighter, respectively, than Cauchy.)

Theorem 2.2. If X has a positive tail lighter than Cauchy, that is, a density satisfying

*E. L. Lehmann is Professor and Juliet Popper Shaffer is Senior Lecturer, both at the Department of Statistics, University of California, Berkeley, California 94720. Research was supported by National Science Foundation Grant DMS84-01388. The authors thank a referee for supplying Formula (4.6).

J. Rojo (ed.), *Selected Works of E. L. Lehmann*, Selected Works in Probability and Statistics, DOI 10.1007/978-1-4614-1412-4_65, © Springer Science+Business Media, LLC 2012

$$\frac{f(x)}{1/(1+x^2)} \to 0 \quad \text{as} \quad x \to \infty \tag{2.3}$$

or, equivalently,

$$x^2 f(x) \to 0 \quad \text{as} \quad x \to \infty, \tag{2.4}$$

then

$$g(y) \to 0 \quad \text{as} \quad y \to 0+. \tag{2.5}$$

Proof. The proof follows directly from (2.2).

The analogous result holds as $y \to 0-$. If f is continuous and has both positive and negative tails lighter than Cauchy such as in the normal, t_ν with $\nu \geq 2$, logistic, or double-exponential distributions, this theorem shows that $g(y) \to 0$ as $y \to 0$ and hence that g is continuous for all y provided $g(0)$ is defined to be 0.

Theorem 2.3. Suppose that

$f(x)$ is positive for some positive

and some negative values of x, (2.6)

and (1.2) holds. Then $g(y)$ has at least one positive and one negative mode.

Proof. Assume without loss of generality that (2.4) holds. It follows from Theorem 2.2 that $g(y) \to 0$ as y tends to $0+$, and it follows from (2.6) that there exist $y_1 < 0 < y_2$ such that both $g(y_1)$ and $g(y_2)$ are positive. On the other hand, g is a density and, therefore, there exist values $y_1' < y_1$ and $y_2' > y_2$ such that $g(y_1') < g(y_1)$ and $g(y_2') < g(y_2)$. It follows that g must have at least one positive and one negative mode.

Note from the proof that, in the statement of Theorem 2.3, Assumption (1.2) can be replaced by the weaker assumption:

there exists a sequence $x_n \to +\infty$ or $-\infty$

for which $x_n^2 f(x_n)$ tends to 0. (2.7)

This result turns out to be useful for an application to be made in Section 4.

It is difficult to imagine that these results are new, but we have not been able to find them in the literature. Bimodality of g, when f is normal, was discussed by Zellner (1978).

As an illustration, Figure 1 shows the density g of $1/X$ when X has a t distribution with 1 df (Cauchy), 2 df, and infinite df (normal). It is seen that g is unimodal in the Cauchy case but bimodal in the other two cases in which the tail of f is lighter than Cauchy. The figure also gives a hint of the heavy tails, which we know by Theorem 2.1 that all three distributions have.

As a second example, Figure 2 shows g when f is a normal distribution with unit variance and means 0, .5, and 1.0, respectively. The bimodality is clearly seen in all three cases. The figure also illustrates, however, the decreasing importance of the negative mode as the mean increases. A limit result of Section 3 will show that as the mean tends to infinity, the shape of a suitably normalized version of g tends to that of the standard normal density.

3. A LIMITING RESULT

In the preceding section, we showed that for typical f, the shape of the density g of $1/X$ tends to be rather bizarre. We shall now consider a limiting case for which this bizarre behavior disappears so that one can expect it to be muted as the limiting situation is approached.

Consider as the density of X a location-scale family, say

$$f_{\theta,\sigma}(x) = \frac{1}{\sigma} f^*\left(\frac{x-\theta}{\sigma}\right). \tag{3.1}$$

As $\theta \to \infty$ (σ remaining fixed), X tends to infinity in probability and hence $Y = 1/X$ tends to 0 in probability. A more useful limit result is obtained by considering (for $\theta \neq 0$) the distribution of the linear transform of $1/X$,

$$Z = \frac{\theta^2}{\sigma}\left(\frac{1}{X} - \frac{1}{\theta}\right). \tag{3.2}$$

The distribution of Z becomes particularly simple when it is expressed in terms of the standardized variable

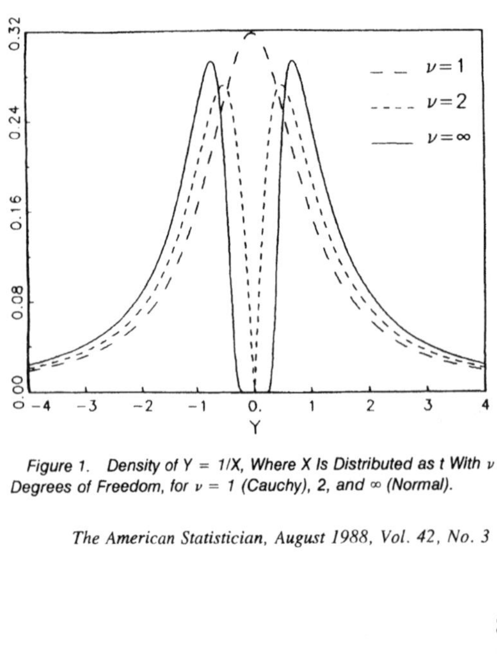

Figure 1. Density of $Y = 1/X$, Where X Is Distributed as t With ν Degrees of Freedom, for $\nu = 1$ (Cauchy), 2, and ∞ (Normal).

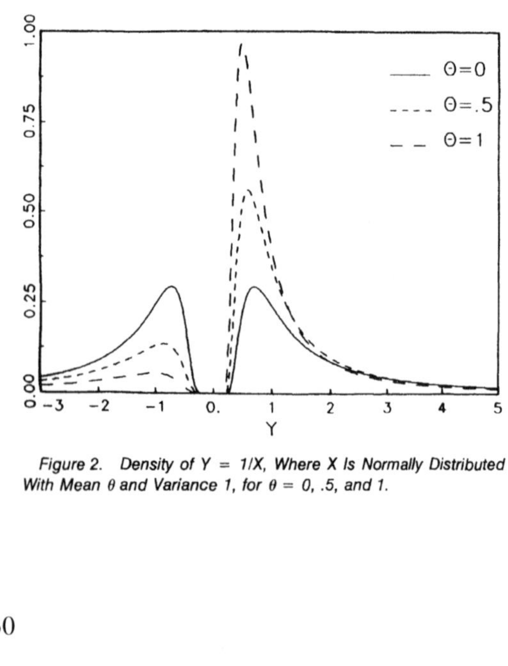

Figure 2. Density of $Y = 1/X$, Where X Is Normally Distributed With Mean θ and Variance 1, for $\theta = 0$, .5, and 1.

$$X^* = \frac{X - \theta}{\sigma} = \frac{-Z}{1 + \delta Z}, \quad (3.3)$$

where $\delta = \sigma/\theta$.

Theorem 3.1. The density of Z is given by

$$g^*(z) = \frac{1}{(1 + \delta z)^2} f^*\left(\frac{-z}{1 + \delta z}\right). \quad (3.4)$$

Proof. The proof follows directly from (3.1) and (3.2).

It follows that for any fixed z

$$g^*(z) \to f^*(-z) \quad \text{as} \quad \delta \to 0. \quad (3.5)$$

Hence, if f^* is unimodal, so is the limit of $g^*(z) = g_\delta^*(z)$ as $\delta \to 0$, although for any fixed δ, $g_\delta^*(z)$ may be bimodal.

Formula (3.5) provides approximations for two different situations:

1. *For fixed θ and small σ.* This can be viewed as a generalization of the normal approximation for $(1/\bar{X} - 1/\theta)$, where $\bar{X}$ is the mean of n iid variables and $f_{\theta,\sigma}$ is the density of $\bar{X}$.
2. *For fixed σ and large θ.* This is of course obvious, since g^* depends only on δ.

4. TWO KINDS OF LIMITS

Let $X_1, \ldots, X_n$ be iid with mean $\theta \neq 0$, variance $\sigma^2 < \infty$, and density f. If $\bar{X} = (X_1 + \cdots + X_n)/n$, it follows from standard asymptotic theory that the distribution of

$$Z_n = (\sqrt{n}\,\theta^2/\sigma)(1/\bar{X} - 1/\theta)$$

tends to the standard normal distribution. On the other hand, it was seen in the preceding sections that for each n the distribution of Z_n has a right Cauchy tail provided $f_n(0+)$ is defined and $0 < f_n(0+) < \infty$, where f_n is the density of $X_1 + \cdots + X_n$. The apparent inconsistency results from the fact that two different limit processes are applied. In this section, the probability $\Pr(Z_n \geq z)$ is considered for fixed z as $n \to \infty$; in the preceding sections, the probability $\Pr(Z_n \geq z)$ was considered for fixed n as $z \to \infty$.

Related is the fact that the limit distribution of Z_n as $n \to \infty$ has finite moments of all orders, whereas for any fixed n not even the first moment of Z_n exists. The possibility of such a discrepancy between the moments of the limit distribution and the limit of the actual moments is usually illustrated by artificial examples constructed for the purpose. In the present case, the discrepancy arises in a natural setting.

A Cauchy (or heavier) right tail for the distribution of Z_n is derived under the assumption that $f_n(0+)$ exists and is positive, and the proof of bimodality is based on Assumptions (2.6) and (1.2) or (2.7). This raises the question of when these conditions are satisfied. For this purpose, it is necessary to distinguish two cases.

Case 1: The Random Variables Take on Both Positive and Negative Values. Suppose that f is bounded and satisfies the continuity conditions assumed at the beginning of Section 2.

(a) Cauchy tails: Assume that $a < 0 < b$ and hence f is positive in some interval around 0. Then $f_n(0) > 0$ for all n. To see this, suppose that X and X' are independent with densities f and f', both of which are positive in $(-\varepsilon/2, \varepsilon/2)$. Then it is easily seen that the density of $W = X + X'$ given by

$$h(w) = \int f(w - x) f'(x)\,dx \quad (4.1)$$

again satisfies the required continuity conditions and $h(w) > 0$ for all $|w| < \varepsilon$ and hence, in particular, for $w = 0$. The result now follows by induction over n.

(b) Bimodality: Suppose that, instead of (2.7), f satisfies the stronger assumption

$$E(X_i^2) < \infty. \quad (4.2)$$

Then f_n satisfies (2.7), and the density of $1/\bar{X}$ is, therefore, at least bimodal. To see this, note that (4.2) implies

$$E[(X_1 + \cdots + X_n)^2] < \infty, \quad (4.3)$$

since $(\Sigma X_i)^2 < n \Sigma X_i^2$. In turn, (4.3) implies that the density of $X_1 + \cdots + X_n$ must satisfy (2.7).

Case 2: The Random Variables X_i Are Positive so That $f(x) = 0$ for $x < 0$. For this case we shall show that, typically, even if $f(0+) > 0$, $f_n(0+) = 0$ for all $n \geq 2$ (so the distribution of $1/\bar{X}$ is not necessarily heavy-tailed). At the other extreme, we shall give an example with $f(0+) = \infty$ for which $f_n(0+) = \infty$ for all n.

Lemma 4.1. If $X_1, \ldots, X_n$ are iid positive random variables with cdf F and density f, and if $f_n(0+)$ exists, then a necessary and sufficient condition for $f_n(0+) = 0$ is that

$$[F(\Delta)]^n/\Delta \to 0 \quad \text{as} \quad \Delta \to 0+. \quad (4.4)$$

Proof. Note that

$$[F(\Delta/n)]^n \leq \Pr[X_1 + \cdots + X_n \leq \Delta] = F_n(\Delta) \leq [F(\Delta)]^n, \quad (4.5)$$

where F_n denotes the cdf of $X_1 + \cdots + X_n$. On dividing all sides by Δ, it is seen that the middle term tends to $f_n(0+)$. That (4.4) is necessary and sufficient for $f_n(0+) = 0$ then follows from the left and right inequalities, respectively.

Theorem 4.1. If $f(0+) < \infty$, $n \geq 2$, and $f_n(0+)$ exists, then $f_n(0+) = 0$.

Proof. By l'Hospital's rule

$$\lim_{\Delta \to 0+} \frac{F^n(\Delta)}{\Delta} = \lim_{\Delta \to 0+} [nF^{n-1}(\Delta) \cdot f(\Delta)]$$

and the right side is 0, since $F(\Delta) \to 0$ as $\Delta \to 0+$. The result now follows.

If $f(0+) = \infty$, the conclusion of this theorem no longer holds.

Example 4.1. Let $F(x) = x^\alpha$ $(0 \leq x \leq 1)$, and suppose that $0 < \alpha < 1$, so $f(0+) = \infty$. It is easily verified by induction over n that

$$F_n(\Delta) = \Delta^{\alpha n} F_n(1) \quad \text{for} \quad 0 \leq \Delta \leq 1 \quad (4.6)$$

and hence $f_n(0+)$ exists and satisfies

$$f_n(0+) = \infty \quad \text{for} \quad n < 1/\alpha$$
$$= 0 \quad \text{for} \quad n > 1/\alpha$$

and is positive and finite for $n = 1/\alpha$.

In Example 4.1, $f_n(0+) \neq 0$ for small n but $f_n(0+) = 0$ for n sufficiently large. In the following example, $f_n(0+) = \infty$ for all n.

Example 4.2. Let

$$F(x) = \frac{1}{\log(1/x)}, \quad f(x) = \frac{1}{x \log^2 x} \quad \text{for} \quad 0 < x < 1/e, \tag{4.7}$$

and $f(x) = 0$ elsewhere. This is the one-sided version of a distribution considered in another context by Gnedenko and Kolmogorov (1954, p. 223). Then

$$f_n(\Delta) > \frac{c_n}{|\Delta \log^{n+1} \Delta|}, \quad c_n > 0, \quad \Delta < 1/e^2, \tag{4.8}$$

and hence $f_n(0+) = \infty$. The result (4.8) was given for the two-sided version by Gnedenko and Kolmogorov, and their proof also applies to the present one-sided case. The proof uses the facts that

$$f_n(\Delta) = \int_0^{\Delta} f(\Delta - x) f_{n-1}(x)\, dx \quad \text{for} \quad 0 \leq x \leq \Delta, \quad \Delta < 1/e \tag{4.9}$$

and f attains its minimum over $(0, \Delta)$ at Δ when $\Delta < 1/e^2$. [The requirement $\Delta < 1/e^2$ (rather than $\Delta < 1/e$ as stated by Gnedenko and Kolmogorov) was pointed out by a referee. In any case, the result is only needed in the neighborhood of $\Delta = 0$.] This shows that

$$f_n(\Delta) \geq f(\Delta) F_{n-1}(\Delta), \quad \Delta < 1/e^2, \tag{4.10}$$

and (4.8) now follows by induction over n.

[Received July 1986. Revised February 1988.]

REFERENCES

Folks, J. Leroy (1983), "Inverse Distributions," in *Encyclopedia of Statistical Sciences* (Vol. 4), eds. S. Kotz and N. L. Johnson, New York: John Wiley, pp. 244–249.

Gnedenko, B. V., and Kolmogorov, A. N. (1954), *Limit Distributions for Sums of Independent Random Variables*, Cambridge, MA: Addison-Wesley.

Hannan, James (1985), "Sufficient and Necessary Conditions for Finite Negative Moment," *The American Statistician*, 39, 326.

Piegorsch, Walter W., and Casella, George (1985), "The Existence of the First Negative Moment," *The American Statistician*, 39, 60–62.

Wooff, David A. (1985), "Bounds on Reciprocal Moments With Applications and Developments in Stein Estimation and Post-stratification," *Journal of the Royal Statistical Society*, Ser. B, 47, 362–371.

Zellner, A. (1978), "Estimation of Functions of Population Means and Regression Coefficients Including Structural Coefficients," *Journal of Econometrics*, 8, 127–158.

Part X
Biographical Work

Erich L. Lehmann's Biographical Work

Javier Rojo

Erich L. Lehmann received his Ph.D. degree from the University of California at Berkeley in 1946. Although he was not the first of Jerzy Neyman's students, he was the first of Neyman's students to be hired by the mathematics department at Berkeley in 1947. His retirement in 1988 marked forty one years of service, all of which – except for visits to Columbia, Princeton, and Stanford – were spent in Berkeley. His professional activities did not diminish after his retirement nor did his social involvement in the department and in the broader statistical community. Having arrived in Berkeley on January 2nd, 1941, Erich spent close to 68 years in Berkeley. The family's departure from Germany in 1933 was accompanied by very difficult circumstances. The experience of becoming a refugee had a strong impact on Erich's future life. In Lehmann (2008), he writes:

I had lost my country, my language, and my friends.

In addition, he had to cope with the death of his sibling. In one of my visits to Erich and Julie, I asked him about a bust of a young man that sat on his working desk. He explained that the bust was of his brother who had died young. It was clear that, even after more than sixty years, the experience was still painful. It is then easy to understand Erich's desire to belong to a community. In Lehmann (2008), he writes:

What I most longed for is indicated by the title of an early draft of this book, "In Search of Community," a community to which I could belong.

He found two communities, his family and the professional community, both of which he embraced with gusto. The former community has been documented in his *Family Memoirs* which he left as a legacy to his family. The latter community has been fascinatingly described in his reminiscences published in Lehmann (2008).

The experiences that led to this desire of belonging, coupled with his arrival in Berkeley to observe first hand, and be such an important part of, the birth and

J. Rojo
Professor of Statistics, Department of Statistics, Rice University, Houston, TX 77005
e-mail: jrojo@rice.edu

J. Rojo (ed.), *Selected Works of E. L. Lehmann*, Selected Works in Probability and Statistics, DOI 10.1007/978-1-4614-1412-4_66, © Springer Science+Business Media, LLC 2012

development of the Statistics Department, together with his substantial involvement in the social life of his professional community, provided Erich with a wealth of information from which to draw from. Throughout his life, Erich felt grateful to the many individuals that impacted his career along the way. In his early years, there were Edmund Landau, Rolf Noskwith, Richard Courant, Griffith C. Evans, and Jerzy Neyman. During his Ph.D. dissertation stage, Pao-Lu Hsu and George Polya played very important roles. The life-long friendships developed during the years as a graduate student and as a young faculty member in the department, included Joseph L. Hodges, Evelyn Fix, Charles M. Stein, Henry Scheffé, and many others. In addition, there were his 43+ Ph.D. students.

A large portion of Erich's social life revolved around his professional life. In DeGroot (1986), Erich states that

My social life, and this is partly the result of being at Berkeley, is practically confined to statisticians.

He clearly enjoyed the social interactions with his colleagues and students. Given his interests in becoming a writer, it is only natural that he would be involved in writing about many of the members of this professional community. The papers reproduced in this section represent a small collection of Erich's writings about those who impacted his professional development. Knowing how strongly Wald's work intertwined with his own, and knowing the appreciation that Erich had for Wald having interacted with him both professionally and socially, it is perhaps puzzling that Erich co-authored a single-page summary on Wald's work after Wald's death. And, although Erich did not have much interaction with Harold Hotelling, Erich co-authored a twelve page biographical memoir. A possible explanation for the former is that Erich was a very polite individual. As Wald spent all his professional life at Columbia, and since Wolfowitz (1952), a close collaborator of Wald's, had already written a full biography, Erich most likely felt that it was the prerogative of the Statistics Department at Columbia to write about Wald. In the case of Hotelling, Erich writes in Lehmann (2008) that, despite the fact that he had very little interaction with Hotelling, he found himself in the position of Hotelling's biographer thirty years after his death. This was a consequence of the National Academy's custom to honor their deceased members with a memoir. Being asked by the National Academy to find someone to write the Hotelling's memoirs, and not being able to find anyone, Erich decided to write it in collaboration with Kenneth Arrow. As it regards to Wald, Erich took advantage of the opportunity to write extensively about Wald and the impact of his work, forty three and fifty eight years later – respectively, in Lehmann (1993) and Lehmann (2008).

Besides the articles reproduced here, there is, of course, Lehmann (2008). In it, Erich described, masterfully, the work of, and the collaborations and social interactions with, more than seventy members of his professional community. In addition, the set of references below provides several other biographical writings. Together, they provide fascinating accounts of, and many insights about, Erich's colleagues and the development of Statistics.

References

[1] T. W. Anderson, H. Cramer, J. L. Hodges, Jr. H. A. Freeman, E. L. Lehmann, A. M. Mood, C. Stein. The life of Abraham Wald. In *Selected Papers in Statistics and Probability*, pp 1. New York: McGraw-Hill, 1955.

[2] T. W. Anderson, K. L. Chung, and E. L. Lehmann. Pao-Lu Hsu, 1909-1970. *Ann. Statist.*, Vol. 7, pp. 467-470, 1979.

[3] K. J. Arrow and E. L. Lehmann. Harold Hotelling. In *Biographical Memoirs*, National Academy of Sciences, Vol. 87, pp. 221-233. The National Academies Press, 2005.

[4] P. J. Bickel and E. L. Lehmann. Joseph Lawson Hodges, Jr. *IMS Bulletin*, Vol. 4 (2000), pp. 411-412, 2000. Also in *ISI Newsletter*, Vol. 24, 10-11, 2000.

[5] D. Blackwell, L. Le Cam, E. L. Lehmann, H. Lewy, E. L. Scott. Jerzy Neyman, Statistics: Berkeley 1894-1981. *University of California: In Memoriam*, pp 313-315, 1985. http://content.cdlib.org/view?docId=hb4d5nb20m&chunk.id=div00120&brand=calisphere&doc.view=entire_text

[6] D. Blackwell, R. N. Bellah, P. J. Bickel, W. W. Borah, E. L. Lehmann. Edward William Barankin, Statistics: Berkeley 1920-1985. *University of California: In Memoriam*, pp. 8-9, 1986. http://content.cdlib.org/view?docId=hb767nb3z6&chunk.id=div00005&brand=calisphere&doc.view=entire_text

[7] D. Blackwell, E. Colson, S. Ervin-Tripp, L. Le Cam, E. L. Lehmann, L. Nader. Elizabeth Leonard Scott, Statistics: Berkeley 1917-1988. *University of California: In Memoriam*, pp 186-190, 1991. http://content.cdlib.org/view?docId=hb4t1nb2bd&chunk.id=div00061&brand=calisphere&doc.view=entire_text

[8] D. Blackwell, A. H. Bowker, and E. L. Lehmann. Joseph L. Hodges, Jr. *University of California: In Memoriam*, pp. 83-85, 2000.

[9] C. Daniel and E. L. Lehmann. Henry Scheffé, 1907-1977. *Ann. Statist.*, Vol. 7, pp. 1149-1161, 1979.

[10] M. H. DeGroot. A Conversation with Erich L. Lehmann. *Statist. Sci.*, Vol. I, 243-258, 1986.

[11] E. L. Lehmann. Hsu's work on inference. *Ann. Statist.*, Vol. 7, pp. 471-473, 1979.

[12] E. L. Lehmann, D. Blackwell, J. L. Hodges. Henry Scheffé, Statistics; Engineering: Berkeley and Los Angeles, 1907-1977. *University of California: In Memoriam*, September, pp 96-98, 1979. http://content.cdlib.org/view?docId=hb4q2nb2px&chunk.id=div00039&brand=calisphere&doc.view=entire_text

[13] E. L. Lehmann, D. R. Brillinger, and I. M. Singer. Jack Carl Kiefer, Statistics and Mathematics: Berkeley, 1924-1981. *University of California: In Memoriam*, pp. 215-216, 1985. http://content.cdlib.org/viewdocId=hb4d5nb20m&chunk.id=div00081&brand=calisphere&doc.view=entire_text

[14] E. L. Lehmann. Mentors and Early Collaborators: Reminiscences from the Years 1940-1956 with an Epilogue. *Statist. Sci.*, Vol. 8, No. 3, pp. 331-341, 1993.

[15] E. L. Lehmann. Jerzy Neyman (1894-1981). In *Biographical Memoirs of the Nat. Acad. of Sci.*, Vol. 63, pp. 395-420. The National Academies Press, 1994.

[16] E. L. Lehmann. Le Cam at Berkeley. In *Festschrift for Lucien Le Cam*, pp. 297-304, Springer-Verlag, 1997.

[17] E. L. Lehmann. *Reminiscences of a Statistician: The Company I Kept*, Springer, 2008.

[18] J. Neyman, J. L. Hodges, Jr., L. Le Cam, E. L. Lehmann, E. L. Scott. Evelyn Fix, Statistics: Berkeley *University of California: In Memoriam*, June pp 36-38, 1967. http://content.cdlib.org/view?docId=hb629006vt&chunk.id=div00009&brand=calisphere&doc.view=entire_text

[19] J. Wolfowitz. Abraham Wald, 1902-1950. *Ann. Math. Statist.*, Volume 23, Number 1, pp. 1-13, 1952.

THE LIFE OF ABRAHAM WALD

Abraham Wald was born October 31, 1902, in Cluj, Rumania. His entire elementary and secondary school education was obtained at home, mainly under the direction of an older brother. After graduating from the local university in Cluj, he went to Vienna, entering the University of Vienna in 1927. Here he studied mathematics, particularly geometry, and contributed to the colloquium led by Karl Menger. After receiving his Ph.D. in 1931, Wald could not find an academic post because of religious prejudice; he began work in econometrics first with Karl Schlesinger, a banker and economist, and then with Oskar Morgenstern at the Business Cycle Research Institute.

By 1938 the rise of Nazism made life in Vienna untenable for Wald. In the summer of that year he came to the United States as a fellow of the Cowles Commission for Research in Economics, and in the fall he came to Columbia University on a fellowship of the Carnegie Corporation. He began intensive study of modern statistical inference under the tutelage of Harold Hotelling. It was not long before Wald began research in statistics; in fact, his path-breaking paper in decision theory {37} was written during this academic year. In the following academic year (1939–1940) Wald lectured in statistics while Hotelling was absent on sabbatical leave; this was the beginning of Wald's pedagogical career at Columbia. In 1941 he was made an assistant professor of economics, in 1943 an associate professor, and in 1944 a professor. In 1946 when Hotelling was called to the University of North Carolina, Columbia met the possibility of Wald also leaving by setting up a Department of Mathematical Statistics with Wald as executive officer.

In 1941 Wald married Lucille Lang. A daughter, Betty, was born in 1943, and a son, Robert, was born in 1947. In the fall of 1950, Wald and his wife traveled to India for a lecture tour. It was on one leg of this trip that they met their deaths in a plane crash.

A fuller biography of Wald is given in papers by J. Wolfowitz and Karl Menger in the March issue of the 1952 (Wald Memorial) volume of the *Annals of Mathematical Statistics.*[1] These papers also survey Wald's work in statistics and geometry; another paper in that issue by G. Tintner[2] surveys Wald's contributions to econometrics.

[1] J. Wolfowitz, "Abraham Wald, 1902–1950," *Annals of Math. Stat.*, Vol. 23 (1952), pp. 1–13.

K. Menger, "The formative years of Abraham Wald and his work in geometry," *Annals of Math. Stat.*, Vol. 23 (1952), pp. 14–20.

[2] G. Tintner, "Abraham Wald's contributions to econometrics," *Annals of Math. Stat.*, Vol. 23 (1952), pp. 21–28.

J. Rojo (ed.), *Selected Works of E. L. Lehmann*, Selected Works in Probability and Statistics, DOI 10.1007/978-1-4614-1412-4_67, © Springer Science+Business Media, LLC 2012

The Annals of Statistics
1979, Vol. 7, No. 3, 467–470

PAO-LU HSU 1909–1970

BY T. W. ANDERSON, K. L. CHUNG AND E. L. LEHMANN
Stanford University and University of California, Berkeley

Pao-Lu Hsu was born in Peking in 1909 and obtained his B.Sc. degree from Tsing Hua University in 1933. He spent the years 1936–1940 at University College, London where he received a Ph.D. in 1938 and an D.Sc. in 1940. From London he returned to China to accept a professorship in the Department of Mathematics at Peking University.

The war years were difficult (letters from 1943–44 to Professor Neyman in Berkeley mention starvation), but Hsu continued his research. Efforts to bring him to the United States finally succeeded, and in 1945 he arrived just in time for the First Berkeley Symposium on Probability and Statistics. A semester of teaching at the University of California was followed by a semester at Columbia. This was the year in which Hotelling moved from Columbia to North Carolina, and he offered Hsu an associate professorship in the department he was creating there. Thus Hsu spent 1946–47 in Chapel Hill from where—in spite of many efforts to keep him in America—he returned to his professorship at Peking University.

Professor H. F. Tuan of Peking University informs us that Hsu died in his home on the campus of Peking University on December 18, 1970, from chronic tuberculosis. There was a memorial meeting in his honor.

Hsu was born into a mandarin family from the famed lake city of Hangchow, but was brought up in Peking. Perhaps because of this background he spoke the "common" (or "official") dialect with an interesting soft overtone. In appearance and carriage he was a Chinese scholar in the classic mold. His British education formed his taste in mathematics; he preferred the hard and concrete to the general and abstract. Setting even higher standards for his own work than for others, he would temper his critical sense with a gentle mocking humor. He could work feverishly on research for spells, but used to lament that life's diverse interests conflicted with a single-minded devotion to science. He enjoyed the company of men and women of the other culture and was versed in traditional Chinese letters. A particular hobby was chanting the musical drama of the Yuan dynasty with a small group of connoisseurs accompanied by ancient instruments. Partly owing to fragile health he never married but apparently came close to it. He returned to China in the summer of 1947, shortly after the Communist victory, despite offers of positions in the U.S. by A. Wald and others. He was looking forward to being part of the emerging new society in his homeland. In his last years he was too ill to get around but continued to teach from his room. The Chinese scientific community

Received June 1978.
AMS 1970 *subject classifications.* Primary 01A70; secondary 00A15.
Key words and phrases. Obituary, bibliography.

gathered to pay homage to him at his funeral. Apart from his published papers and a few remarks given us by an old friend, we are unable to obtain further information about Hsu's life and work in the twenty some years he lived in Peking.

Hsu is affectionately remembered by many students and colleagues as a gentle, shy and modest man who was reticent about his personal life, but had a strong influence as a teacher and model of a scientist. Isadore Blumen, who was a student of Hsu during his year at Chapel Hill, writes: "It was Hsu's insistence on simplicity combined with depth of understanding, clarity without avoidance of difficulty, and above all a deep and obvious but unspoken commitment to the highest goals and standards of scholarship which attracted us to him". Ralph Bradley remembers Hsu's lectures as "models for the future", and Herbert Robbins says "he was unforgettable, and I fear irreplaceable". While Hsu may not have been the formal supervisor of any Ph.D. students he strongly influenced a number, among them Kai-Lai Chung, Sen-Ming Len and Shou-Jen Wang before he came to the West, and Isadore Blumen, Albert Bowker, Erich Lehmann and Ingram Olkin in the U.S.

Hsu's statistical work was concerned primarily with inference in univariate and multivariate linear models and with the associated distribution theory, both exact and asymptotic. A more detailed discussion of this and his probabilistic work follows.

We gratefully acknowledge the assistance of the following persons, who provided us with their recollections of Hsu, with information about his life, and with suggestions for and corrections of our account: I. Blumen, R. Bradley, S. S. Chern, F. N. David, C. Eisenhart, J. Neyman, I. Olkin, E. S. Pearson, H. Robbins, S. Stigler, H. F. Tuan.

Pao-Lu Hsu: Publications

⟨1935⟩

1. On the limit of a sequence of point sets. *Bull. Amer. Math. Soc.* **41** 502–504.

⟨1938⟩

2. Contributions to the two-sample problem and the theory of the "Student's" T-test. *Statist. Res. Mem.* **2** 1–24.
3. On the best quadratic estimate of the variance. *Statist. Res. Mem.* **2** 91–104.
4. Notes on Hotelling's generalized *T*. *Ann. Math. Statist.* **9** 231–243.

⟨1939⟩

5. On the distribution of roots of certain determinantal equations. *Ann. Eugenics* **9** 250–258.

⟨1940⟩

6. A new proof of the joint product moment distribution. *Proc. Cambridge Philos. Soc.* **(2) 6** 185–189.
7. On iterated limits. *J. Chinese Math. Soc.* **2** 40–63.
8. An algebraic derivation of the distribution of rectangular coordinates. *Proc. Edinburgh Math. Soc.* **(2) 6** 185–189.
9. On generalized analysis of variance. *Biometrika* **31** 221–237.

⟨1941⟩

10. On the limiting distribution of roots of a determinantal equation. *J. London Math. Soc.* **16** 183–194.
11. The limiting distribution of the canonical correlations. *Biometrika* **32** 38–45.
12. Analysis of variance from the power function standpoint. *Biometrika* **32** 62–69.

13. Canonical reduction of the general regression problem. *Ann. Eugenics* **11** 42–46.
14. On the problem of rank and the limiting distribution of Fisher's test function. *Ann. Eugenics* **11** 39–41.

⟨1942⟩

15. The limiting distribution of a general class of statistics. *Acad. Sinica Science Record* **1** 37–41.

⟨1943⟩

16. Some simple facts about the separation of degrees of freedom in factorial experiments. *Sankhyā* **6** 253–254.

⟨1945⟩

17. The approximate distribution of the mean and variance of a sample of independent variables. *Ann. Math. Statist.* **16** 1–29.
18. On the approximate distribution of ratios. *Ann. Math. Statist.* **16** 204–210.
19. On the power functions for the *E*-test and the *T*-test. *Ann. Math. Statist.* **16** 278–286.

⟨1946⟩

20. On a factorization of pseudo-orthogonal matrices. *Quart. J. Math. Oxford Ser.* **17** 162–165.
21. Sur un théorème de probabilité dénombrables. *C. R. Acad. Sci. Paris* (with Kai-Lai Chung) 467–469.
22. On the asymptotic distribution of certain statistics used in testing the independence between successive observations from a normal population. *Ann. Math. Statist.* **17** 350–354.

⟨1947⟩

23. Complete convergence and the law of large numbers. (with H. Robbins) *Proc. Nat. Acad. Sci. U.S.A.* **33** 25–31.

⟨1949⟩

24. The limiting distributions of functions of sample means and application to testing hypotheses. *Proc. Berkeley Symp. Math. Statist. Probability* 359–402. Univ. of California Press, Berkeley and Los Angeles.

⟨1951⟩

25. Absolute moments and characteristic function. *J. Chinese Math. Soc. (New Series)* **1** 257–280. (In English with abstract in Chinese).
26. A lemma on the coefficient of reduction of a sum of independent variables. *Acad. Sinica Science Record* **4** 197–200.

⟨1953⟩

27. On symmetric, orthogonal, and skew-symmetric matrices. *Proc. Edinburgh Math. Soc.* Ser. **2** 37–44.

⟨1954⟩

28. On characteristic functions which coincide in the neighborhood of zero. *Acta Math. Sinica* (Succeeding *J. Chinese Math. Soc. (New Series))* **4** 21–32. (In Chinese with abstract in English).

⟨1955⟩

29. On a kind of transformations of matrices. *Acta Math. Sinica* **5** 333–346. (In Chinese with abstract in English).
30. On a kind of transformations of matric pairs. *Acta Sci. Natur. Univ. Pekinensis* **1** 1–16. (In Chinese with abstract in English).

⟨1957⟩

31. On the simultaneous transformation of a hermitian matrix and a symmetric or skew-symmetric matrix. *Acta Sci. Natur. Univ. Pekinensis* **3** 167–210. (In Chinese with abstract in English).

⟨1958⟩

32. The differentiability of the probability transition function of a purely discontinuous stationary Markhoff process on the euclidean space. *Acta Sci. Natur. Univ. Pekinensis* **4** 257–270. (In Chinese with abstract in English).
33. The absolute continuity of the distribution function in the class *L*. *Acta Sci. Natur. Univ. Pekinensis* **4** 145–150. (In Chinese with abstract in English).

⟨1964⟩

34. An association scheme $M_3(6)$ which is not a L_3-scheme. *Acta Math. Sinica* **14** 177–178. (In Chinese).

⟨1968⟩

35. Appendix III in *Limit Distributions of Sums of Independent Random Variables*, revised edition (Eds. B. V. Gnedenko and A. N. Kolmogorov) (translated by K. L. Chung, with Appendices by J. L. Doob and P. L. Hsu), Addison-Wesley.

· **Papers by Bann Cheng as representative for a seminar conducted by Pao-Lu Hsu.**

⟨1964⟩

1. The limiting distribution of order statistics. *Acta Math. Sinica* **14** 694–714. (In Chinese).
2. Partially balanced incomplete block designs. *Prog. Math.* **7** 240–281. (In Chinese).

Department of Statistics
Stanford University
Stanford, California 94305

Department of Mathematics
Stanford University
Stanford, California 94305

Department of Statistics
University of California
Berkeley, California 94720

The Annals of Statistics
1979, Vol. 7, No. 3, 471–473

HSU'S WORK ON INFERENCE

BY E. L. LEHMANN
University of California, Berkeley

Hsu spent four years (1936-1940) at University College, London, where E. S. Pearson had recently succeeded his father in the chair of statistics and where, during the first two years, Neyman was a Reader in the Statistics Department. During this period Hsu wrote a remarkable series of papers on statistical inference which show the strong influence of the Neyman-Pearson point of view.

In 1938 Hsu's first two statistical papers appeared in Vol. II of the Neyman-Pearson edited *Statistical Research Memoirs*. The first of these [2] is concerned with what today is called the Behrens-Fisher problem. If X_i and $Y_j (i = 1, \cdots, m; j = 1, \ldots, n)$ denote samples from normal distributions $N(\xi, \sigma^2)$ and $N(\eta, \tau^2)$, Hsu considers the class of statistics $u = (\bar{Y} - \bar{X})^2/(A_1 S_x^2 + A_2 S_Y^2)$ where $S_X^2 = \Sigma(X_i - \bar{X})^2$ and $S_Y^2 = \Sigma(Y_j - \bar{Y})^2$. This reduces to u_1, Student's t, for $A_1 = A_2 = N/mn(N - 2)$ where $N = m + n$ and to the Behrens-Fisher statistic u_2 for $A_1 = 1/m(m - 1)$, $A_2 = 1/n(n - 1)$.

Hsu finds a series expansion for the density of u, and utilizes this to study the power function of the rejection regions $u \geqslant C$ in terms of the parameters $\theta = \tau^2/\sigma^2$ and $\lambda = (\eta - \xi)^2/(\frac{\sigma^2}{m} + \frac{\tau^2}{n})$. It is an exact (not asymptotic) analysis, described by Scheffé (1970) as "a model of mathematical rigor". In the process, he obtains stochastic bounds for u_2 which were later taken up independently and generalized by Hájek, Lawton and others (cf. Eaton and Olshen (1972)). Hsu's main conclusion, obtained by a combination of his analytical study with some numerical work, is that for $\lambda = 0$ and varying θ neither u_1 nor u_2 control the rejection probability at all well (except when $m = n$) although of the two, u_2 is less sensitive to variation of θ.

In the second paper [3], Hsu treats the question of optimal estimators of the variance σ^2 in the Gauss-Markov model. In the spirit of the Gauss-Markov theorem, he considers estimators Q which are (a) quadratic and (b) unbiased. In addition he imposes the restriction (c) that the variance of Q be independent of the unknown means. (This is a forerunner of the condition he imposed in [12] for the power function of analysis of variance tests).

Hsu then obtains a necessary and sufficient condition for the usual unbiased estimate S^2 of σ^2 to have uniformly minimum variance within this class of estimators. He illustrates the condition on a number of examples and, in particular, shows that S^2 has the desired property in the one-sample case. The problem was

Received June 1978.
AMS 1970 *subject classifications.* Primary 01A70; secondary 62F05, 62F10.
Key words and phrases. Obituary, hypothesis testing, estimation.

taken up again by Rao (1952) who replaced (c) by the condition that Q be definite, while Seely (1971) instead considered an equivalent invariance condition. Hsu's paper may be regarded as the point of origin of the recent large literature on best quadratic estimates of variances and variance components.

The remaining papers in this series devoted to small-sample inference are concerned with the problem of testing univariate and multivariate linear hypotheses, particularly with power properties of the tests of these hypotheses. In the first of these papers [4], Hsu obtains the power of Hotelling's T^2-test, showing that the nonnull distribution of T^2 is that of the ratio of a noncentral χ^2 to an independent central χ^2 variable. He also proves that the test is in a certain sense locally most powerful. In addition, he points out some new applications of T^2.

From a study of T^2 it was a natural step to a multivariate extension of the general linear (univariate) hypothesis discussed by Kolodzieczyk (1935). Generalizing some examples considered earlier in the literature, Hsu in [9] formulates the general multivariate linear hypothesis in its canonical form. He obtains the nonnull distribution of the likelihood ratio test statistic when the covariance matrix is known. For the case of unknown covariance matrix, if θ_i denote the nonvanishing roots of the associated determinantal equation, Hsu considers the test statistics $W = \Pi(1 - \theta_i)$ and $V = \Sigma\theta_i/(1 - \theta_i)$, and proves that their asymptotic power agrees as the sample size tends to infinity. In a succeeding paper [13], Hsu shows how the general multivariate regression problem can be reduced to the earlier canonical form so that the theory of [9] becomes applicable.

Perhaps the most important paper in this series is [12], where Hsu obtains the first optimum property for the likelihood ratio test of the univariate linear hypothesis, in fact essentially the first nonlocal optimum property for any hypothesis specifying the value of more than one parameter. It had been shown by Kolodzieczyk that a uniformly most powerful similar test does not exist for linear hypotheses with more than one constraint. If λ denotes the noncentrality parameter, which is the only parameter on which the nonnull distribution of the likelihood ratio test for this problem depends, Hsu proves that this test is uniformly most powerful among all tests with power function depending only on λ. It was later realized that this condition is equivalent to invariance of the power function under the natural class of transformations. The resulting optimum property is essentially also equivalent to the fact that the test is uniformly most powerful invariant.

This paper started two lines of development. On the one hand, Hsu's formulation was applied to multivariate problems (Hotelling's T^2 and the multiple correlation coefficient) by his student Simaika (1941) and modifications of the formulation were considered by Wald (1942), Hsu (1945) [19], Wolfowitz (1949), Lehmann (1959) and others. On the other hand, in the course of this paper Hsu provided a new method for obtaining all similar tests. At Hsu's suggestion this method was applied to other problems by Simaika (1941) and Lehmann (1947), and was formalized by means of the concept of completeness by Lehmann and Scheffé (1950).

REFERENCES

EATON, MORRIS L. and OLSHEN, RICHARD A. (1972). Random quotients and the Behrens-Fisher problem. *Ann. Math. Statist.* **43** 1852–1860.

HSU, P. L. (1945). On the power functions for the E^2-test and the T^2-test. *Ann. Math. Statist.* **16** 278–286.

KOLODZIECZYK, S. (1935). On an important class of statistical hypotheses. *Biometrika* **21** 161-190.

LEHMANN, E. L. (1947). On optimum tests of composite hypotheses with one constraint. *Ann. Math. Statist.* **18** 473–494.

LEHMANN, E. L. (1959). Optimum invariant tests. *Ann. Math. Statist.* **30** 881–884.

LEHMANN, E. L. and SCHEFFÉ, HENRY (1950). Completeness, similar regions and unbiased estimation, *Sankhyā* **10** 305–340.

RAO, C. RADHAKRISHNA (1952). Some theorems on minimum variance estimation. *Sankhyā* **12** 27–44.

SCHEFFÉ, HENRY (1970). Practical solutions of the Behrens-Fisher problem. *J. Amer. Statist. Assoc.* **65** 1501–1508.

SEELY, JUSTUS (1971). Quadratic subspaces and completeness. *Ann. Math. Statist.* **42** 710–721.

SIMAIKA, J. B. (1941). On an optimum property of two important statistical tests. *Biometrika* **32** 70–80.

WALD, ABRAHAM (1942). On the power function of the analysis of variance test. *Ann. Math. Statist.* **13** 434–439.

WOLFOWITZ, J. (1949). The power of the classical tests associated with the normal distribution. *Ann. Math. Statist.* **20** 540–551.

DEPARTMENT OF STATISTICS
UNIVERSITY OF CALIFORNIA
BERKELEY, CALIFORNIA 94720

The Annals of Statistics
1979, Vol. 7, No. 6, 1149–1161

HENRY SCHEFFÉ 1907–1977

BY C. DANIEL AND E. L. LEHMANN

Rhinebeck, New York, and University of California, Berkeley

Henry Scheffé was born on April 11, 1907, in New York City. His father and mother were German, originally from Alsace. He went to elementary school in New York and graduated from high school in Islip, Long Island in 1924. Following graduation, he entered the Cooper Union Free Night School for the Advancement of Science and Art to study electrical engineering, and in 1925 became a student at the Polytechnic Institute of Brooklyn. There his grades were all A's except for a D in mechanical engineering which on reexamination was changed to an E! During this time he also worked as a technical assistant at the Bell Telephone Laboratories and took a training course there.

In 1928 he was admitted to the University of Wisconsin with advanced standing to study mathematics. The record shows but one course related to statistics. It was "Theories of probabilities (sic) and methods of least squares" given by Warren Weaver. Scheffé was an intercollegiate wrestler at Wisconsin. (Contrary to widespread rumor, he acquired his broken nose at the age of three, not in wrestling.) He received his B.A. with high honors in 1931 and remained at Wisconsin for graduate study, obtaining his Ph.D. in 1935. His thesis, written under R. E. Langer, on asymptotic solutions of certain differential equations, was published in 1936 [S1][1].

During the years 1935-1938, Scheffé taught mathematics at the University of Wisconsin, at Oregon State University, and at Reed College. His first statistical paper [S3] was written at Oregon State and appeared in the American Mathematical Monthly in 1942.

At some time in 1940-1941, he decided that the field of mathematical statistics promised more interesting opportunities for research than analysis. (Harry Goheen remembers that when Scheffé found that parts of his thesis subject had already been worked on by Gauss he felt that he should move to less well-travelled paths.) This resulted in his going to Princeton in 1941 to teach (as Instructor from 1941-1943 and as Lecturer during 1943-1944) and to do research with the statistics group in the Mathematics Department. In the company of Ted Anderson, George Brown, Bill Cochran, Will Dixon, Alex Mood, Fred Mosteller, John Tukey, Sam Wilks and Charlie Winsor it is easy to imagine his growing attachment to statistics. From 1943 to 1946, he worked as consultant and Senior Mathematics Officer at the Office of Scientific Research and Development, under a contract of this office with

Received October 1978.

[1]The numbers [S1], . . . refer to the list of Scheffé's publications at the end of the article.

Princeton University. In this capacity, he wrote a number of reports, which we have not seen but which describe work carried out under the general heading "Effects of impact and explosion."

In 1944–45 Scheffé taught statistics at Syracuse University and from 1946-1948 served as Associate Professor of Engineering at UCLA. Actually, the first year he was on leave as a Guggenheim Fellow, part of it in Berkeley where he began his joint work with E. Lehmann. In 1948 Scheffé moved once more, this time to Columbia University, where he remained as Associate Professor of Mathematical Statistics until 1953, serving as executive officer of the department for 1951-1953.

Scheffé joined the faulty of the University of California at Berkeley as Professor of Statistics and Assistant Director of the Statistical Laboratory in 1953 and remained there until his retirement in 1974. In 1954 he was President of the Institute of Mathematical Statistics of which he had been elected Fellow ten years earlier, and from 1954 to 1956 he was a Vice President of the American Statistical Association. The year 1962-1963 was spent at the University of London on a Fulbright Research Award. From 1965 to 1968 he was Chairman of the Berkeley department. It was a period of great unrest at the University, which put a heavy strain on him. But in spite of violently conflicting attitudes by different groups of faculty and students, he managed to hold the department together and keep the atmosphere within the department pleasant. His fairmindedness was greatly valued by all members of the department.

Scheffé enjoyed teaching and would recall the better students in his courses for many years. He preferred not only to correct his own examinations, even when the classes were rather large, but also to conduct the laboratory sections of his courses himself without the aid of teaching assistants. He was equally conscientious and sympathetic as a referee for several journals. An associate editor of one of these (Technometrics) tells us that of all the refereeing he has seen, Henry's was the most patient and helpful.

Along with his teaching and research, Scheffé managed a schedule of daily bicycling and swimming and he engaged in frequent summer snorkeling and backpacking. He was a dedicated tourist, especially to Mexico and France, who often returned with small works of art chosen with a sure and strongly individual taste. He was sensitive to the beauties of nature and had a particular enthusiam for desert country. He loved music and as an adult learned to play the recorder and treasured the opportunity this brought of playing chamber music with friends. A few months before his death, he had just finished reading all of Trollope's novels.

In 1974 Scheffé retired from Berkeley and accepted a three-year appoinment as Professor of Mathematics at the University of Indiana in Bloomington. In June of 1977 he returned to Berkeley, which he considered home and where he was planning to prepare a new edition of his book. This was not to be—he died on July 5, 1977 from injuries sustained in a bicycle accident earlier that day.

He is survived by his wife, Miriam, by his daughter Molly, now a mathematician working near Boston, and by his son Michael, a commericial artist and designer in Los Angeles. Friends and colleagues all over the world join them in mourning his loss.

Work in Statistics. Scheffé's first major statistical investigation was concerned with two-sided tests and confidence intervals for the ratio of two normal variances. However, in studying this problem he found that determination of the optimum procedures he was seeking required first an extension of the Neyman-Pearson theory of optimum tests. In [9], these authors had developed a theory of uniformly most powerful (UMP) and (in the presence of nuisance parameters) UMP similar tests. This theory was successful in dealing with an important class of one-sided problems but typically was not applicable to two-sided alternatives. To overcome this difficulty, Neyman and Pearson in [10] introduced the notion of unbiased tests, but only for one-parameter situations, that is, when there are no nuisance parameters. The main result of Scheffé's paper [S4] fills in this gap by providing conditions for UMP unbiased tests in the presence of nuisance parameters.

These results were applied to the problem of obtaining optimum tests for the hypothesis of equality of two normal variances σ^2 and τ^2 and corresponding confidence intervals for their ratio $\theta = \tau^2/\sigma^2$ in the succeeding statistical publication [S6]. In the first part of this paper it is taken for granted that the hypothesis will be rejected when the appropiate F-statistic T is either too small or too large, so that the associated confidence intervals have the form $T/B < \theta < T/A$, and a number of choices for A and B are considered: (i) minimizing $E[\log(T/A) - \log(T/B)]$ or equivalently B/A to obtain the "logarithmically shortest" confidence intervals; (ii) the reciprocal limits, for which $A = 1/B$; (iii) the limits corresponding to the likeihood ratio test; (iv) those determined by the equal-tails test. In particular, there is a numerical comparison of (i) with the more convenient limits (iv).

The second part of the paper uses the results of [S4] to show that the test (i) is in fact UMP among all similar tests that are unbiased. This optimum test is now part of the classical literature. It is a sign of the times that the paper does not mention the disastrous sensitivity of the test to the assumption of normality. Many years later, Scheffé devoted considerable attention to the problem of robustness in his book [S29].

Having "disposed" of the problem of the equality of variances in the presence of unknown means, it was natural for Scheffé to turn next to the dual problem of testing the equality of two normal means when both variances are unknown, the so-called Behrens-Fisher problem. Unfortunately, the conditions under which he had derived his optimum tests did not apply in the new situation. Even more seriously, there was now no obvious statistic corresponding to the F-statistic T in the earlier case, which would provide a class of "natural" similar tests of the hypothesis. In [S7] Scheffé set himself the task of providing such a class. His starting point was a test mentioned by Neyman and attributed by him to Bartlett, for the case of equal sample sizes. If the two samples are then denoted by $X_1, \cdots, X_n$ and $Y_1, \cdots, Y_n$ and their expectations by ξ and η, the differences $Z_i = Y_i - X_i$ constitute a sample of n from a normal distribution with mean $\zeta = \eta - \xi$ and variance $\sigma^2 + \tau^2$. Bartlett's suggestion was to use Student's one-sample t-statistic calculated form the Z's to test the hypothesis $\zeta = 0$, i.e., $\eta = \xi$. When the sample sizes are unequal, $m < n$ say, a statistic with a t-distribution can of course be obtained by discarding n-m of the Y's. Since this is clearly wasteful, Scheffé considered a more general class of statistics which under the hypothesis $\eta = \xi$ have a t-distribution and determined the member of this class which leads to the confidence intervals for $\eta = \xi$ with smallest expected length. He compared his solution with the optimum confidence intervals when the value of $\theta = \tau^2/\sigma^2$ is known, and showed that there is very little loss of efficiency.

Scheffé's solution of the Behrens-Fisher problem received some criticism because of its lack of symmetry, more precisely, because it is not invariant under permutations of the X's among themselves and the Y's among themselves. In the case $m = n$ for example, the denominator of the t-statistic is proportional to $\Sigma(Y_i - X_i)^2$, which is clearly changed by such permutations. Scheffé reacted to this criticism in a note [S12] in the succeeding volume of the Annals by showing that there exists no symmetric t-statistic with the properties he had postulated.

This note is however not his last word on the subject. In 1970, more than twenty-five years later, he returned once more to the Behrens-Fisher problem in [S35] to join his earlier critics and disavow his t-solution, which he discussed in a section entitled "An impractical solution." Referring to [S7] and [S12], he wrote: "These articles were written before I had much consulting experience, and since then I have never recommended the solution in practice. The reason is that the estimate s_d [the estimate of the standard deviation in the denominator of the t-statistic] requires putting in random order the elements of the larger sample, and the value of s_d and hence the length of the interval depends very much on the result of this randomization of the data. The effect of this in practice would be deplorable." He then proceeded to discuss some alternative solutions of the problem, in particular the Welch-Aspin test and Welch's t-test with estimated degrees of freedom, neither of which has the similarity property, which earlier had seemed so essential. The question of the existence of a similar test of reasonable structure, which was initiated by Scheffé in [S7] and [S12], was answered negatively in the 1960's by Linnik and his students. (An account of this work is given by Linnik: "Lecons sur les problèmes de statistique analytique." Gauthier-Villars, Paris, 1967.)

The problem of similar two-sample tests is also the subject of the paper following [S7], this time in a nonparametric context. If $X_1, \cdots, X_m$ and $Y_1, \cdots, Y_n$ are samples from F and G, Scheffé sought in [S8] to determine the totality of level α tests of the hypothesis $F = G$ when this common distribution is completely unknown, except possibly restricted to be continuous or to have a density. Expressed in modern terminology, his necessary and sufficient condition is essentially that the conditional rejection probability given the $m + n$ order statistics $z_{(1)} < \cdots < z_{(m+n)}$ of the combined sample is α a.e.

[S8] was Scheffé's first contribution to nonparametric statistical inference which, particularly as the result of the recent work of Wald and Wolfowitz, was beginning to become a "hot" topic. Scheffé himself contributed significantly to the heating up process by following [S8] by the first comprehensive review of the new field [S9].

The longest section of his review paper is again devoted to the problem of similar tests. It is pointed out how the structure of these tests, which had been uncovered in [S8], extends to other nonparametric problems and encompasses the two important special classes of nonparametric tests to be found in the literature; randomization tests and rank tests. The paper then goes on to describe the tests in these classes which had been proposed for problem of randomness, the two-sample problem, the hypothesis of independence, and the analysis of variance.

The second part of the paper deals with estimation. There is a brief section on point estimation, which formulates the problem as that of estimating a functional defined over the nonparametric class of distributions in question, and points out how the concepts of unbiasedness and consistency continue to apply in this new setting. The survey then turns to confidence sets and discusses confidence intervals for a population median or the difference of two medians, and confidence bands for an unknown distribution function. This second part of the paper concludes with an account of the work of Wald and Wilks on nonparametric tolerance limits.

The third and last part of the paper is entitled, "Toward a general theory." It discusses the consistency criterion and the modified likelihood principle of Wolfowitz and contains a short description of Wald's decision theoretic approach to statistics.

In his introduction to this survey, Scheffé wrote: "Only a very small fraction of the extensive literature of mathematical statistics is devoted to the nonparametric case, and most of this is of the last decade. We may expect this branch to be rapidly explored however: The prospects of a theory freed from specific assumptions about the form of the population distribution should excite both the theoretician and the practitioner since such a theory might combine elegance of structure with wide applicability. "

The paper made an important contribution to the burgeoning new field by providing it with a solid foundation for the spectacular development which Scheffé had so accurately predicted, but it leaves the present-day reader with the question of why the author himself did not participate in this development. He saw the attraction of the new methods, and by detailing what had been accomplished made clear some of the outstanding gaps. Why did he not try to fill some of them? A partial answer may be found in his view of the principal task that lay ahead, which is outlined in the concluding sentences of his paper: " . . . we have in the parametric case a large body of constructive theory which may be applied in particular examples to yield the optimum tests or estimates; thus we have the Fisher theory of maximum likelihood statistics for point estimation, and the constructive theorems of the Neyman-Pearson theory for the existence of critical regions of types A, A_1, B, B_1 and the related types of "best" confidence intervals. The contrasting lack of any constructive general methods challenges us in the nonparametric theory". This program probably was not feasible then and in fact has not yet been carried out. Scheffé himself later came to a quite different view of the needs of statistics. By that time he no longer thought so highly of nonparametric methods and hence did not return to that subject.

During the next few years Scheffé wrote only a few shorter notes, some of which may have grown out of his war work. Two of the these were his first directed to applied statisticians. The first of these [S14] shows some of the relations between control charts, the analysis of variance, and chi-square tests. The latter [S17] combines an imaginative understanding of the problems of quality control engineers with his typical thoroughness in expounding the operating characteristics of the usual mean and range charts.

In spite of these more applied concerns, his heart at that time was still in optimum theory and the problem of similar tests. In these interests he was joined in 1946 by one of us (EL); the collaboration led to a preliminary note [S16] and then to the papers [S18] and [S24]. This work is concerned with two statistical problems, determining the totality of similar tests for a given family of distributions and finding unbiased estimators with uniformly minimum variance. A test represented by a critical function ϕ is similar at level α with respect to a family $\mathcal{P}$ of distribution P if

$$EP\phi(X) = \alpha \quad \text{for all } P \in \mathcal{P},$$

while δ is an unbiased estimator of a real-valued functional g defined over $\mathcal{P}$ if

$$EP\delta(X) = g(P) \quad \text{for all } P \in \mathcal{P}.$$

The tool which provides an easy key to the two problems in a large class of cases is that of a complete family of distributions. If T is a sufficient statistic for $\mathcal{P}$, then T or more accurately the induced family $\mathcal{P}^T$ of distributions of T, is said to be complete if

$$EPf(T) = 0 \quad \text{for all } P \in \mathcal{P} \quad \text{implies } f(t) = 0 \quad (\text{a.e. } \mathcal{P}^T).$$

In the presence of such a complete sufficient statistic one easily obtains the results:

(i) A necessary and sufficient condition for ϕ to be similar at level α is that

$$E[\phi(X)|t] = \alpha \quad (\text{a.e. } \mathcal{P}^T).$$

and

(ii) Any functional g which has an unbiased estimator, has an unbiased estimator with uniformly minimum variance, namely the unique unbiased estimator depending only on T.

These results, together with various extensions and ramifications, provide simple and unified proofs of most of the special results that had been obtained earlier by various authors on these two problems. It must however be emphasized that the basic ideas underlying (i) and (ii) were not really new. In particular, the method (i) of obtaining all similar tests had been used by Hsu in [5] while the method underlying (ii) had been employed by Rao in [12], [13] and [14]. The main contribution appears to be the isolation of the concept of *completeness*, which plays an important role in mathematical statistics even beyond its impact through the results (i) and (ii). Another concept which was introduced and investigated in the course of this work is that of *minimal* sufficient statistic. This concept was introduced independently in the year following the publication of [S18] by Dynkin [3]. The two concepts are related by the fact that a sufficient statistic cannot be complete unless it is minimal.

With this series of papers, Scheffé's concern with similar tests and optimality theory, which had started with his first major statistical paper in 1942, essentially came to an end. However, before abandoning the subject, he wrote one final paper related to the problem of optimum tests, which grew out of the thesis of one of his Ph.D. students, Stanley Isaacson. Scheffé's interest had been primarily in the problem indicated in the title of [S4], testing hypotheses with one constraint, that is, which specify the value of a single real-valued parameter. Isaacson [6], at Scheffé's suggestion, had studied hypotheses specifying the values of two or more parameters. For this purpose, he required an extension of the fundamental lemma of Neyman and Pearson, and in [6] established sufficient conditions for a solution to his optimization problem. In [S20], Chernoff and Scheffé generalized Isaacson's result, and proved that their condition is not only sufficient but also necessary.

The work described so far, which constitutes the first phase of Scheffé's research, was dominated by issues belonging to the domain of mathematical statistics. The second phase begins with [S21]. It is devoted primarily to problems relating to the analysis of variance and is concerned with issues belonging to the domain of statistical methodology. The extent of his change of attitude is indicated by the fact that in his comprehensive and rigorous book on *The Analysis of Variance*, the optimality properties are compressed into five pages. The shift in point of view is seen perhaps most clearly by an entry in the subject index. Under optimality properties we find the sub-entry: "Of lesser importance than robustness". The page to which this entry refers contains the following footnote: "When I became aware that the nominal probability of type 1 error for the standard test of the equality of variances of two populations is invalidated by non-normality to the same order of magnitude as found in Table 10.2.1, I found little consolation in the optimum properties someone once established for that test (Scheffé, 1942)."

This development was not due to any relaxation in the intensity, depth or rigor of his work. It reflected, rather , the growth of a well trained but extremely modest mathematician into a more self-confident—though still modest—scientist who wanted to be useful to a wider circle of research workers.

A short expository account for chemical engineers on the background of statistical methods [S19] was his first effort in the new direction. As a consultant for

Consumer's Union he became involved in the problem of "organoleptic" preference tests. The resulting paper was on an analysis of variance for paired comparisons [S21] which extends the general methods of the analysis of variance to situations in which only subjective degrees of preference between pairs of samples are recorded. This becomes an analysis of *differences*, which is not transformable into the more familiar partitioning of direct measurements. The designs and methods of analysis produced have been widely used in many areas ever since.

The next paper, on judging all contrasts in the analysis of variance [S22], was motivated in large part by the problems of his colleagues in applied statistics. There are of course many settings in which experimenters—and hence statisticians—are willing to commit themselves, before the results are in, to a specified set of comparisons among the results. But there are at least as many other settings in which the experimenter rightly refuses this restriction. He must see all that the data tell him, and this may well require him to make comparisons that were not foreseen.

The problem of multiple comparisons has been a source of concern to statisticians and to others for a long time. It was J. W. Tukey who first devised a method of simultaneously estimating by confidence intervals *all* contrasts among a set of means like those usually produced in a planned experiment, all being independent with the same variance. The method he chose—called the T-Method by Scheffé—gives priority to the $K(K-1)/2$ pairwise comparisons among K means, using the critical point of the Studentized range as a multiplier of the standard error of each mean, to guarantee with predetermined confidence coefficient that all intervals cover their corresponding parametric values. These intervals are the shortest possible (in expectation) for the differences between pairs of population means. But the T-method pays a price for this in requiring larger intervals than need be for more complex groupings, for example, for the comparison of the average of two means with the average of a different set of three.

Scheffé set himself the problem of making allowances, which would control the probability of no errors of coverage at preassigned confidence level, for the experimenter who reserves the right to make *any* comparison among the means, even those suggested by long mulling of data. He saw and proved that the confidence ellipsoid in $(K-1)$-dimensional space is equivalent to the intersection of the sets between all pairs of parallel tangent hyperplanes that support the ellipsoid. This proved that the confidence ellipsoid provides a confidence set (at the same level) for all contrasts among the means (from one line of an analysis of variance table). The size of the ellipsoid depends only on the value of the F-statistic at the chosen level of confidence, and on K. Scheffé showed that the (infinite) set of confidence interval statements given by the S-method for a given set of K means is equivalent to the usual F-test in the sense that at least one confidence interval on a contrast among the means fails to cover zero if and only if the F-test is significant at the α-level. (He later discovered a forerunner of this result in a paper by Working and Hotelling 1929.)

The allowance for any observed contrast $\hat{\psi} = \Sigma_{i=1}^{K} c_i X_i$ ($\Sigma\ c_i = 0$) is the interval from $\hat{\psi} - A$ to $\hat{\psi} + A$ where A is a constant times the estimated standard error of $\hat{\psi}$. The constant is calculated from the F-distribution with $K - 1$ and ν (the degrees of freedom for error) degrees of freedom. For the case $K = 6$, $\nu = 18$ and $\alpha = .05$, for example, the constant is 3.72. This (typical) value is disturbingly large to statisticians accustomed to using a tabular t-value of, say, 2.101 for such cases. The paper (in Tables 5 and 6) gives some numerical consequences of the latter practice. Scheffé often said that he consoled himself with the thought that users who did not mind using the F-test for overall tests of null hypotheses about a set of K means, should not mind the values produced by the S-method since they were equivalent to the F-test. He also suggested some relaxation of the size of the critical region by setting $\alpha = .1$, which of course results in some shortening of the expected values of the intervals. In the example just given, the effect is to reduce the constant from 3.72 to 3.32.

One of us (CD) takes pride in the fact that his copy of this paper is inscribed "To the guy who hounded me into this." This required listening to long disquistions throughout more than a year, on how messy the distribution problem was, climaxed one evening at 11:30 with "Guess what; it's the F-distribution."

Extensions, limitations, alternatives continue to be matters of active research and controversy. R. G. Miller's book [7] and recent paper [8] give an excellent overview and bibliography up to 1977. Judging from the frequency of its citation, the S-method is one of the most widely used devices in interpreting F-ratios found in the analysis of variance and in regression.

An objection to the S-method, or at least to one interpretation of it, was raised by Olshen [11]. Suppose that the S-intervals are reported only when a preliminary F-test rejects the overall null hypothesis H. Olshen showed that the conditional coverage probability (given that H has been rejected) which seems then to be the probability of interest, is at least in some situations always smaller than the stated confidence probability. The issue is further discussed in a note by Scheffé [S37] together with a comment by Olshen and a rejoinder by Scheffé.

In this note Scheffé distinguished between two situations. In one only a test of H is to be carried out, with no further statements in case of rejection. (This is basically the case in which the parameters being tested are nuisance parameters, which one hopes to discard). In the other, in which the parameters being tested are of primary interest, rejection of H is to be followed by some confidence statements for significant contrasts. Scheffé pointed out that in the latter type of situation, the S-procedure (interpreted as consisting of an initial statement that H is accepted or rejected, supplemented in the latter case by more detailed confidence statements) has an overall probability of a correct statement equal to that claimed by his method.

Finally, in his rejoinder following Olshen's comment, Scheffé noted that the difficulty is shared by many standard statistical procedures, since in fact studies are usually published only when they report statistically significant results. The debate thus joins an important on-going discussion of differences between statistical theory and practice.

The paper [S22] on the S-method was followed in 1954 by a study of partially hierarchal (crossed and nested) models in the analysis of variance [S23]. Its purpose was to describe as compactly as possible the effects of several process variables and several sources of random variation on the corrosion resistance of sheet steel used in making tin cans. A companion paper by the experimenter and the applied statisticians Vaurio and Daniel [16] gave the details of the system and the experiment. In a later letter to the same journal (see [S23]) Scheffé withdrew the model he had used in favor of that of Wilk and Kempthorne [17].

His work on mixed and alternative models for analysis of variance resulted in two papers [S25, S26], both published in 1956. These reflected his efforts to develop "Model II" which he considered less well understood than "Model I", the so-called fixed effects model. They constituted part of his program for the book *The Analysis of Variance* that was taking shape during this period.

The paper on fitting straight lines when one variable is controlled [S27] contains first a detailed explication of Berkson's celebrated 1950 paper [1] and then a generalization to sets of straight lines with possibly different parameters. The paper solved an industrial consulting problem which he believed to be of frequent occurrence.

The two papers on experiments with mixtures [S28, S32] were responses to real problems in chemical and petrochemical industry. The former appears to be the fundamental work in the field. It is extended, discussed and applied in a large number of later papers by other authors. Scheffé devised symmetrical designs[2] for q mixture components, x_i, each at $(m + 1)$ levels, called $\{q, m\}$ lattices, and gave symmetrical polynomial fitting equations of degrees 1, 2, and 3. The linear equation (which contains no constant term) is:

$$\eta = \Sigma_1^q \beta_i x_i, \qquad i = 1, \ldots, q.$$

The quadratic equation (which contains no constant or squared terms) is:

$$\eta = \Sigma\beta_i x_i + \Sigma\beta_{ij} x_i x_j, \qquad \text{with } 1 \leqslant i < j \leqslant q.$$

A canonical cubic equation is also given.

For each fitting equation, simple expressions for the least-squares estimates of the coefficients are derived in terms of the observed responses, as well as formulas for their variances. (The coefficient $-4/3$ in his equation (4.8) should be -12; the -3 should be $+3$.) Extensions to equations of the fourth order, to correspond to gasoline blending problems, were made by Gorman and Hinman [4]. They also included a useful explanation of the equations of lower order.

Scheffé's second paper in mixture designs [S32] was also written in response to applied problems. It is more difficult to read, even though as G. A. Barnard

[2]Optimality properties of these designs were established by Kiefer in *Ann. Math. Statist.* **32** (1961) 298-325.

complained in the discussion following its presentation, the paper does not conform to Royal Statistical Society usage, since "it leaves too few loose ends for discussants to jump on." The title does not indicate that the paper handles experiments with some mixture-variables (constrained by $\Sigma x_i = 1$) and some "process variables (unconstrained)." Perhaps it still awaits a more elementary exposition to complete the bridge to physical application.

The Analysis of Variance [S29] is Scheffé's magnum opus. It has established itself as the basic work on the theory and application of that form of analysis. The review in the Journal of the Royal Statistical Society [2] by D. R. Cox starts: "The first adjective that comes to mind to describe this book is *professional.*" The review ends:

> Altogether this is a most important book, deserving to be widely read. Statisticians working in fields where the analysis of variance is used extensively are likely to find the book extremely valuable in consolidating their knowledge of the theoretical side of the subject.

L. J. Savage, writing in Mathematical Reviews [15], concluded his review:

> Whatever may be said in criticism is overshadowed by the merits of this book which is unique in its field and will be indispensible to all who seriously do, study, or teach statistics in connection with experimentation.

The long life of the first edition has moreover justified the judgments of the reviewers. This must be attributed to its combination of thoroughness and generality (e.g., in Chapters 1 and 2) with its intuitive and practical insights (especially in Chapter 10).

The note on separation of variables [S34] extends a result on transformability to additivity from the two-variable case given in his book (page 95 ff.) to any number of independent (continuous) variables. It includes a number of examples, some of them with surprising outcomes.

In his last major paper "A statistical theory of calibration" Scheffé attacked a problem of great practical importance. The result was a thirty-four page treatment of characteristic thoroughness but with no numerical example. (He was working at the end on an expository paper that was to include several sets of real calibration.) One of the inhibiting factors to its wide use must be the eleven-page set of tables (of over 13,000 entries) giving values needed for actual use in calibration. Only two values from the tables are needed for a single calibration problem. The tables are extraordinarily condensed in that all entries are one- or two-digit numbers, less than 18 in the first and less than 7 in the second. He gives a procedure that for given α and δ, and for every possible sequence of later observations (after the calibration curve is made), guarantees that the probability is $\geqslant (1 - \delta)$ that the proportion of true statements about coverage that are made is in the long run $\geqslant (1 - \alpha)$.

Scheffé made path-breaking contributions both to statistical theory and to methodology. Again and again he showed us that it is possible to solve problems posed by statistical practice on the firm basis of carefully stated assumptions. He thus served as a model for younger statisticians, and his influence will long be felt in the statistical world.

We gratefully acknowledge the assistance of the following persons who provided us with information, and with comments and suggestions about earlier drafts of this obituary: D. Brunk, C. Eisenhart, K. Green, W. Kruskal, D. Lowry, F. Mosteller, I. Olkin, M. Scheffé, S. Stigler, J. Tukey, G. Watson.

The publications of Henry Scheffé

[1] Asymptotic solutions of certain linear differential equations in which the coefficient of the parameter may have a zero. *Trans. Amer. Math. Soc.* **40** 127–154 (1936).

[2] At what level of rigor should advanced calculus for undergraduates be taught. *Amer. Math. Monthly* **47** 635–640 (1940).

[3] An inverse problem in correlation theory. *Amer. Math. Monthly* **49** 99–104 (1942).

[4] On the theory of testing composite hypotheses with one constraint. *Ann. Math. Statist.* **13** 280–293 (1942).

[5] Linear differential equations with two-term recurrence formulas. *J. Math. Physics* **21** 240–249 (1942).

[6] On the ratio of the variances of two normal populations. *Ann. Math. Statist.* **13** 371–388 (1942).

[7] On solutions of the Behrens-Fisher problem, based on the *t*-distribution. *Ann. Math. Statist.* **41** 35–44 (1943).

[8] On a measure problem arising in the theory of non-parametric tests. *Ann. Math. Statist.* **14** 227–233 (1943).

[9] Statistical inference in the non-parametric case. *Ann. Math. Statist.* **14** 305–332 (1943).

[10] A formula for sample sizes for population tolerance limits. (with J. W. Tukey) *Ann. Math. Statist.* **15** 217 (1944).

[11] Note on the use of the tables of percentage points of the incomplete beta function to calculate small sample confidence intervals for a binomial *p*. *Biometrika* **33** 181 (1944).

[12] A note on the Behrens-Fisher problem. *Ann. Math. Statist.* **15** 430–431 (1944).

[13] Non-parametric estimation. I. Validation of order statistics. (with J. W. Tukey) *Ann. Math. Statist.* **16** 187–192 (1945).

[14] The relation of control charts to analysis of variance and chi-square tests. *J. Amer. Statist. Assoc.* **42** 425–431 (1947).

[15] A useful convergence theorem for probability distributions. *Ann. Math. Statist.* **18** 434–438 (1947).

[16] On the problem of similar regions. (with E. L. Lehmann) *Proc. Natl. Acad. Sci. USA* **33** 382–386 (1947).

[17] (Book review). Operating characteristics of average and range charts. *Indust. Quality Control* **5** 13–18 (May 1949).

[18] Completeness, similar regions, and unbiased estimation. Part I. (with E. L. Lehmann) *Sankhyā* **10** 305–340 (1950). *Corrigendum, Sankhyā* **17** 250 (1950).

[19] Theoretical backgrounds of the statistical methods. *Indust. Engin. Chem.* **43** 1292–1294 (1951).

[20] A generalization of the Neyman-Pearson fundamental lemma. (with H. Chernoff) *Ann. Math. Statist.* **23** 213–225 (1952).

[21] An analysis of variance for paired comparisons. *J. Amer. Statist. Assoc.* **47** 381–400 (1952).

[22] A method for judging all contrasts in the analysis of variance. *Biometrika* **40** 87–104 (1953).

[23] Statistical methods for evaluation of several sets of constant and several sources of variability. *Chem. Engin. Progress* **50** 200–205 (1954). Correction, March 1955, same journal.

[24] Completeness, similar regions, and unbiased estimation. Part II. (with E. L. Lehmann) *Sankhyā* **15** 219–236 (1955).

[25] A mixed model for the analysis of variance. *Ann. Math. Statist.* **27** 23–36 (1956).

[26] Alternative models for the analysis of variance. *Ann. Math. Statist.* **27** 251–271 (1956).

[27] Fitting straight lines when one variable is controlled. *J. Amer. Statist. Assoc.* **53** 106–117 (1958).

[28] Experiments with mixtures. *J. Roy Statist. Soc. B* **20** 344–360 (1958).

[29] *The Analysis of Variance*. Wiley, New York, 1959.

[30] Simultaneous interval estimates of linear functions of parameters. *Bull. Int. Statist. Inst.* **38** 245–253 (1961).

[31] Reply to Mr. Quenouille's comments about my paper on mixtures. *J. Roy Statist. Soc. B* **23** 171–172 (1961).

[32] The simplex-centroid design for experiments with mixtures. (including discussion) *J. Roy. Statist. Soc. B* **25** 235–263 (1963).

[33] Multiple testing versus multiple estimation. Improper confidence sets. Estimation of directions and ratios. *Ann. Math. Statist.* **41** 1–29 (1970).

[34] A note on separation of variables. *Techometrics* **12** 388–393 (1970).

[35] Practical solutions of the Behrens-Fisher problem. *J. Amer. Statist. Assoc.* **65** 1501–1508 (1970).

[36] Statistical theory of calibration. *Ann. Math. Statist.* **1** 1–37 (1973).

[37] A note on a reformulation of the *S*-method of multiple comparison. *J. Amer. Statist. Assoc.* **72** 143–146 (1977).

REFERENCES

[1] Berkson, J. (1950). Are there two regressions? *J. Amer. Statist. Assoc.* **45** 164–180.

[2] Cox, D. R. (1960). Scheffé: The analysis of variance. *J. Roy. Statist. Soc. A* **123** 482.

[3] Dynkin, E. B. (1951). Necessary and sufficient statistics for a family of probability distributions. *Uspekhi Mat. Nauk* (*N. S.*) **6** (41), 68–90. (Also Selected Transl. Math. Statist. Prob. 17–40).

[4] Gorman, J. W. and Hinman, J. E. (1962). Simplex lattice designs for multicomponent systems. *Techometrics* **4** 462–487.

[5] Hsu, P. L. (1941). Analysis of variance from the power function standpoint. *Biometrika* **32** 62–69.

[6] Isaacson, S. L. (1951). On the theory of unbiased tests of simple statistical hypotheses specifying the values of two or more parameters. *Ann. Math. Statist.* **22** 217–234.

[7] Miller, R. G., Jr. (1966). *Simultaneous Statistical Inference*. McGraw Hill, New York.

[8] Miller, R. G., Jr. (1977). Developments in multiple comparisons. *J. Amer. Statist. Assoc.* **72** 779–788.

[9] Neyman, J. and Pearson, E. S. (1933). On the problem of the most efficient tests of statistical hypotheses. *Phil. Trans. Roy. Soc. Ser. A* **231** 289–337.

[10] Neyman, J. and Pearson, E. S. (1936, 1938). Contributions to the theory of testing statistical hypotheses. *Statist. Res. Memoirs* **1** 1–37 and **2** 25–57.

[11] Olshen, R. A. (1973). The conditional level of the *F*-test. *J. Amer. Statist. Assoc.* **68** 692–698.

[12] Rao, C. R. (1945). Information and accuracy attainable in the estimation of statistical parameters. *Bull. Calcutta Math. Soc.* **37** 81–91.

[13] Rao, C. R. (1947). Minimum variance and the estimation of several parameters. *Proc. Camb. Phil. Soc.* **43** 280–283.

[14] Rao, C. R. (1949). Sufficient statistics and minimum variance estimates. *Proc. Camb. Phil. Soc.* **45** 213–218.

[15] Savage, L. J. (1961). Scheffé: The analysis of variance. *Math. Rev.* **22** 661–7152.

[16] Vaurio, V. W. and Daniel, C. (1954). Evaluation of several sets of constants and several sources of variability. *Chem. Engin. Progress* **50** 81–86.

[17] Wilk, M. B. and Kempthorne, O. (1955). Fixed, mixed, and random models. *J. Amer. Statist. Assoc.* **50** 1144–1167.

[18] Working, H. and Hotelling, H. (1929). Application of the theory of error to the interpretation of trends. *J. Amer. Statist. Assoc.* (March Suppl.) 73–85.

Box 313, R.D. 2 Rhinebeck
New York 12572

Department of Statistics
University of California
Berkeley CA. 94720

JERZY NEYMAN

April 16, 1894–August 5, 1981

BY E. L. LEHMANN

DURING THE 1930s Jerzy Neyman developed a new paradigm for theoretical statistics, which derives optimal statistical procedures as solutions to clearly stated mathematical problems. He applied these ideas to the theories of hypothesis testing, estimation by confidence intervals, and survey sampling. During the following decades this became the dominant approach to theoretical statistics. In addition to his scientific work, Neyman was a far-seeing and highly efficient administrator who in the decade 1945–55 created in Berkeley a substantial Department of Statistics of international stature. Starting in 1945 he also established a Berkeley series of Symposia on Mathematical Statistics and Probability, meeting at five-year intervals, which for the next twenty-five years became the principal series of international meetings in statistics.

Neyman's long life was dominated by his work, of which he took a comprehensive view encompassing its academic, administrative, and social aspects.

Adapted from a biographical article written for the *Dictionary of Scientific Biography*, Supplement II. I am grateful to the American Council of Learned Societies for permitting the use of this material.

J. Rojo (ed.), *Selected Works of E. L. Lehmann*, Selected Works in Probability and Statistics, DOI 10.1007/978-1-4614-1412-4_71, © Springer Science+Business Media, LLC 2012

EARLY YEARS

Jerzy Neyman was born in Bendery (Russia) to parents of Polish ancestry. His full name with title—Splawa-Neyman—the first part of which he dropped at age thirty, reflects membership in the polish nobility. Neyman's father Czeslaw, who died when Jerzy was twelve, was a lawyer and later judge and an enthusiastic amateur archeologist. Since the family had been prohibited by the Russian authorities from living in Central Poland, then under Russian domination, Neyman grew up in Russia: in Kherson, Melitopol, Simferopol, and (after his father's death) Kharkov, where in 1912 he entered the university.

At Kharkov, Neyman was first interested in physics, but because of his clumsiness in the laboratory he abandoned it in favor of mathematics. On reading Lebesgue's "Leçons sur L'integration et la Recherche des Fonctions Primitives," he later wrote (in a Festschrift in honor of Herman Wold [1970]): "I became emotionally involved. I spent the summer of 1915 at a country estate, coaching the son of the owner. There were three summer houses on the estate, filled with young people, including girls whom I found beautiful and most attractive. However, the involvement with sets, measure and integration proved stronger than the charms of young ladies and most of my time was spent either in my room or on the adjacent balcony either on study or on my first efforts at new results, intended to fill in a few gaps that I found in Lebesgue." A manuscript on Lebesgue integration (500 pp., handwritten) that Neyman submitted to a prize competition won a gold medal.

One of his mentors at Kharkov was Serge Bernstein who lectured on probability theory and statistics (including application of the latter to agriculture), subjects that did not particularly interest Neyman. Nevertheless, he later acknowledged the influence of Bernstein from whom he "tried to acquire his tendency of concentrating on some 'big problem'." It was also Bernstein who introduced him to Karl Pearson's *Grammar of Science,* which made a deep impression. After the first World War, Poland regained its independence but soon became embroiled in a war with Russia over borders. Neyman, still in Kharkov, was jailed as an enemy alien. In 1921, in an exchange of prisoners, he finally went to Poland for the first time at the age of twenty-seven.

In Warsaw he established contact with Sierpinski, one of the founders of the journal *Fundamenta Mathematicae,* which published one of Neyman's gold medal results (1923, vol. 5, pp. 328–30). Although Neyman's heart was in pure mathematics, the statistics he had learned from Bernstein was more marketable and enabled him to obtain a position as (the only) statistician at the Agricultural Institute in Bydgoszcz (formerly Bromberg). There, during 1921–22, he produced several papers on the application of probabilistic ideas to agricultural experimentation. In light of Neyman's later development, this work is of interest because of its introduction of probability models for the phenomena being studied, particularly a randomization model for the case of a completely randomized experiment. (A key section was translated and published with an introduction by Speed and comments by Rubin in *Statistical Science,* vol. 5, 1990, pp. 463–80.) He had learned the philosophy of such an approach from Karl Pearson's book, where great stress is laid on models as mental constructs the formulation of which constitutes the essence of science.[1]

In December 1922 Neyman gave up his job in Bydgoszcz to take charge of equipment and observations at the State Meteorological Institute, a change that enabled him to move to Warsaw. He did not like the work and soon left to become an assistant at the University of Warsaw and Special Lecturer in mathematics and statistics at the Central College of Agriculture; he also gave regular lectures at the University of Krakow. In 1924 he obtained his doctorate from the University of Warsaw with a thesis based on the papers he had written at Bydgoszcz.

Since no one in Poland was able to gauge the importance of his statistical work (he was "sui generis," as he later described himself), the Polish authorities provided an opportunity for him to establish his credibility through publication in British journals. For this purpose they gave him a fellowship to work with Karl Pearson in London. He did publish three papers in *Biometrika* (based in part on his earlier work), but scientifically the academic year (1925–26) spent in Pearson's laboratory was a disappointment. Neyman found the work of the laboratory old-fashioned and Pearson himself surprisingly ignorant of modern mathematics. (The fact that Pearson did not understand the difference between independence and lack of correlation led to a misunderstanding that nearly terminated Neyman's stay at the laboratory.) So when, with the help of Pearson and Sierpinski, Neyman received a Rockefeller fellowship that made it possible for him to stay in the West for another year, he decided to spend it in Paris rather than in London.

There he attended the lectures of Lebesgue and a seminar of Hadamard. "I felt that this was real mathematics worth studying" he wrote later, "and, were it not for Egon Pearson, I would have probably drifted to my earlier passion for sets, measure and integration, and returned to Poland as a faithful member of the Warsaw school and a steady contributor to *Fundamenta Mathematicae*."

THE NEWMAN–PEARSON THEORY

What pulled Neyman back into statistics was a letter he received in the fall of 1926 from Egon Pearson, Karl Pearson's son, with whom Neyman had had only little contact in London. Egon had begun to question the rationale underlying some of the current work in statistics, and the letter outlined his concerns. Correspondence developed and, reinforced by occasional joint holidays, continued even after the end of the Rockefeller year when Neyman returned to a hectic and difficult life in Warsaw. He continued to lecture at the university (as docent after his habilitation in 1928), at the Central College of Agriculture, and at the University of Krakow. In addition, he founded a small statistical laboratory at the Nencki Institute for Experimental Biology. To supplement his meager academic income, and to provide financial support for the students and young co-workers in his laboratory, he took on a variety of consulting jobs. These involved different areas of application, with the majority coming from agriculture and from the Institute for Social Problems, the latter work being concerned with Polish census data.

Neyman felt harassed, and his financial situation was always precarious. The bright spot in this difficult period was his work with the younger Pearson. Trying to find a unifying, logical basis that would lead systematically to the various statistical tests that had been proposed by Student and Fisher was a "big problem" of the kind for which he had hoped since his student days with Bernstein.

In 1933 Karl Pearson retired from his chair at University College, London, and his position was divided between R. A. Fisher and Egon Pearson. The latter lost no time, and, as soon as it became available, in the spring of 1934 offered Neyman a temporary position in his laboratory. Neyman was enthusiastic. This would greatly facilitate their joint work and bring relief to his Warsaw difficulties.

The set of issues addressed in the joint work of Neyman and Pearson between 1926 and 1933 turned out indeed to be a "big problem," and their treatment of it established a new paradigm that changed the statistical landscape. What concerned Pearson when he first approached Neyman in 1926 was the ad hoc nature of the small sample tests being studied by Fisher and Student. In his search for a general principle from which such tests could be derived, he had written to Student. In his reply Student suggested that one would be inclined to reject a hypothesis under which the observed sample is very improbable, "if there is an alternative hypothesis which will explain the occurrence of the sample with a more reasonable probability" (E. S. Pearson in *Research Papers in Statistics,* F. N. David, ed., 1966). This comment led Pearson to propose to Neyman the likelihood ratio criterion, in which the maximum likelihood of the observed sample under the alternatives under consideration is compared to its value under the hypothesis. During the next year Neyman and Pearson studied this and other approaches, and worked out likelihood ratio tests for some important examples. They published their results in 1928 in a fundamental two-part paper, "On the Use and Interpretation of Certain Test Criteria for Purposes of Statistical Inference." The paper contained many of the basic concepts of what was to become the Neyman-Pearson Theory of Hypothesis Testing, such as the two types of error, the idea of power, and the distinction between simple and composite hypotheses.

Although Pearson felt that the likelihood ratio provided the unified approach for which he had been looking, Neyman was not yet satisfied. It seemed to him that the likelihood principle itself was somewhat ad hoc and had no fully logical basis. However, in February 1930 he was able to write Pearson that he had found "a rigorous argument in favour of the likelihood method." His new approach consisted of maximizing the power of the test, subject to the condition that under the hypothesis (assumed to be simple) the rejection probability has a preassigned value (the level of the test). He reassured Pearson that in all cases he had examined so far this logically convincing test coincided with the likelihood ratio test. A month later Neyman announced to Pearson that he now had a general solution of the problem of testing a simple hypothesis against a simple alternative. The result in question is the "Fundamental Lemma," which plays such a crucial role in the Neyman-Pearson theory.

The next step was to realize that in the case of more than one alternative there might exist a uniformly most powerful test that would simultaneously maximize the power for all of them. If such a test exists, Neyman found, it coincides with the likelihood ratio test, but in the contrary case—alas—the likelihood ratio test may be biased. These results, together with many examples and elaborations, were published in 1933 under the title, "On the Problem of the Most Efficient Tests of Statistical Hypotheses." While in the 1928 paper the initiative and insights had been those of Pearson, who had to explain to Neyman what he was doing, the situation was now reversed, with Neyman as leader and Pearson a somewhat reluctant follower.

The 1933 paper is the fundamental paper in the theory of hypothesis testing. It established a framework for this theory[2] and states the problem of finding the best test as a clearly formulated, logically convincing mathematical problem that one can then proceed to solve. Its importance transcends the theory of hypothesis testing since it also provided the inspiration for Wald's later, much more general, statistical decision theory.

SURVEY SAMPLING AND CONFIDENCE ESTIMATION

The following year Neyman published another landmark paper. An elaboration of work on survey sampling he had done earlier for the Warsaw Institute for Social Problems, it was directed toward bringing clarity into a somewhat muddled discussion about the relative merits of two different sampling methods. His treatment, described by Fisher as "luminous," introduced many important concepts and results and may be said to have initiated the modern theory of survey sampling.

The year 1935 brought two noteworthy events. The first was Neyman's appointment to a permanent position as reader (associate professor) in Pearson's department. Although at the time he was still hoping to eventually return to Poland, he in fact never did except for brief visits. The second event was the presentation at a meeting of the Royal Statistical Society of an important paper on agricultural experimentation in which he raised some questions concerning Fisher's Latin square design. This caused a break in their hitherto friendly relationship and was the beginning of lifelong disputes. (Fisher, who opened the discussion of the paper, stated that "he had hoped that Dr. Neyman's paper would be on a subject with which the author was fully acquainted, and on which he could speak with authority, as in the case of his address to the Society last summer. Since seeing the paper, he had come to the conclusion that Dr. Neyman had been somewhat unwise in his choice of topics.")

Neyman was to remain in England for four years (1934–38). During this time he continued his collaboration with

Egon Pearson on the theory and applications of optimal tests, efforts that also included contributions from graduate and postdoctoral students. To facilitate publication and to emphasize the unified point of view underlying this work, Neyman and Pearson set up a series, "Statistical Research Memoirs," published by University College and restricted to work done in the Department of Statistics. A first volume appeared in 1936 and a second in 1938.

Another central problem occupying Neyman during his London years was the theory of estimation: not point estimation in which a parameter is estimated by a unique number, but estimation by means of an interval or more general set in which the unknown parameter can be said to lie with specified confidence (i.e., probability). Such confidence sets are easily obtained under the Bayesian assumption that the parameter is itself random with a known probability distribution, but Neyman's aim was to dispense with such an assumption that he considered arbitrary and unwarranted.

Neyman published brief accounts of his solution to this problem in 1934 and 1935 and the theory in full generality in 1937 in "Outline of a Theory of Statistical Estimation Based on the Classical Theory of Probability." He had first submitted this paper to *Biometrika*, then being edited by Egon Pearson, but to his great disappointment Pearson rejected it as too long and too mathematical.

Neyman's approach was based on the idea of obtaining confidence sets $S(X)$ for a parameter θ from acceptance regions for the hypotheses that $\theta = \theta_0$ by taking for $S(X)$ the set of all parameter values θ_0 that would be accepted at the given level. This formulation established an equivalence between confidence sets and families of tests and enabled him to transfer the test theory of the 1933 paper lock, stock, and barrel to the theory of estimation. (Unbeknownst to Neyman, the idea of obtaining confidence sets by inverting an acceptance rule had already been used in special cases by Laplace, in a large-sample binomial setting, and by Hotelling).

In his paper on survey sampling, Neyman had referred to the relationship of his confidence intervals to Fisher's fiducial limits, which appeared to give the same results, although derived from a somewhat different point of view. In the discussion of the paper, Fisher welcomed Neyman as an ally in the effort to free statistics from unwarranted Bayesian assumptions. He went on to say that "Dr. Neyman claimed to have generalized the argument of fiducial probability, and he had every reason to be proud of the line of argument he had developed for its perfect clarity. The generalization was a wide and very handsome one, but it had been erected at considerable expense. . . ." He then proceeded to indicate the disadvantages he saw in Neyman's formulation, of which perhaps the most important one (revealing the wide difference between their interpretations) was nonuniqueness. The debate between the two men over their respective approaches continued for many years, usually in less friendly terms; it is reviewed by Neyman in "Silver Jubilee of My Dispute with Fisher" (*Journal of the Operation Research Society of Japan*, 1961, pp. 145-54).

STATISTICAL PHILOSOPHY

During the period of his work with Pearson, Neyman's attitude toward probability and hypothesis testing gradually underwent a radical change. In 1926 he tended to favor a Bayesian approach[3] in the belief that any theory would have to involve statements about the probabilities of various alternative hypotheses and hence an assumption of prior probabilities. In the face of Pearson's (and perhaps also Fisher's) strongly anti-Bayesian position, he became less certain, and in his papers of the late 1920s (both alone and jointly with Pearson) he presented Bayesian and non-Bayesian approaches side by side. A decisive influence was von Mises's book *Wahrscheinlichkeit, Statistik und Wahrheit* (1928) about which he later wrote (in his Author's Note to *A Selection of Early Statistical Papers of J. Neyman*) that it "confirmed him as a radical 'frequentist' intent on probability as a mathematical idealization of relative frequency." He remained an avowed frequentist and opposed any subjective approach to science for the rest of his life.

A second basic aspect of Neyman's work from the 1930s on is a point of view, that he formulated clearly in the closing pages of his presentation at the 1937 Geneva Conference ("L'estimation statistique traitée comme un problème classique de probabilité," *Actualites Scientifiques et Industrielles*, no. 739, pp. 25–57). He states that his approach is not based on inductive reasoning but on the concept of "comportement inductif" or inductive behavior. That is, statistics is to be used not to extract from experience "beliefs" but as a guide to appropriate action.

He summarized his views in a paragraph in a paper presented in 1949 to the International Congress on the Philosophy of Science ("Foundations of the General Theory of Estimation," *Actualites Scientifiques et Industreilles*, 1951, No. 1146, p. 85).

"Why abandon the phrase 'inductive reasoning' in favor of 'inductive behavior'?" As explained in 1937,[4] the term inductive reasoning does not seem appropriate to describe the new method of estimation because all the reasoning behind this method is clearly deductive. Starting with whatever is known about the distribution of the observable variables X, we deduce the general form of the functions f(X) and g(X) which have the properties of confidence limits. Once a class of such pairs of functions is found, we formulate some properties of these functions which may be considered desirable and deduce either the existence or non-existence of an "optimum" pair, etc. Once the various possibilities are investigated we may decide to use a particular pair of confidence limits for purposes of statistical estimation. This decision, however, is not 'reasoning'. This is an act of will just as the decision to buy insurance is an act of will. Thus, the mental processes behind the new method of estimation consist of deductive reasoning and of an act of will. In these circumstances the term 'inductive reasoning' is out of place and, if one wants to keep the adjective 'inductive', it seems most appropriate to attach to it the noun 'behavior'.

In other writings (e.g., *Reviews of the International Statistical Institute,* 1957, vol. 25, pp. 7–22), Neyman acknowledges that a very similar point of view was advocated by Gauss and Laplace. It is, of course, also that of Wald's later general statistical decision theory. On the other hand, this view was strongly attacked by Fisher (e.g., JRSS (B), 1955, vol. 17, pp. 69-78) who maintained that decision making has no role in scientific inference and that his fiducial argument provides exactly the mechanism required for this purpose.

MOVING TO AMERICA

By 1937 Neyman's work was becoming known not only in England and Poland but also in other parts of Europe and in America. He gave an invited talk about the theory of estimation at the International Congress of Probability in 1937 in Geneva, and in the spring of 1937 he spent six weeks in the United States on a lecture tour organized by S. S. Wilks. The visit included a week at the Graduate School of the Department of Agriculture in Washington arranged by Edward Deming. There he gave three lectures and six conferences on the relevance of probability theory to statistics and on his work in hypothesis testing, estimation, sampling, and agricultural experimentation as illustrations of this approach. These "Lectures and Conferences on Mathematical Statistics," which provided a coherent statement of the new paradigm he had developed and exhibited its successful application to a number of substantive problems, were a tremendous success. A mimeographed version appeared in 1938 and was soon sold out. Neyman published an augmented second edition in 1952.

After his return from the United States, Neyman debated whether to remain in England, where he had a permanent position but little prospect of promotion and independence, or to return to Poland. Then completely unexpectedly, in the fall of 1937, he received a letter from an unknown, G. C. Evans, chairman of the Mathematics Department at Berkeley, offering him a position in the department. Neyman hesitated for some time. California and its university were completely unknown quantities to him, while the situation in England, although not ideal, was reasonably satisfactory and stable. An attractive aspect of the Berkeley offer was the nonexistence there of any systematic program in statistics, so that he would be free to follow his own ideas. What finally tipped the balance in favor of Berkeley was the threat of war in Europe. Thus, in April 1938 Neyman decided to accept the Berkeley offer and emigrate to America, with his wife Olga (from whom he later separated) and his two-year-old son Michael. He had just turned forty-four, and he would remain in Berkeley for the rest of his life.

THE BERKELEY DEPARTMENT OF STATISTICS

Neyman's top priority after his arrival in Berkeley was the development of a statistics program, that is, a systematic set of courses and a faculty to teach them. He quickly organized a number of core courses and began to train some graduate students and one temporary instructor in his own approach to statistics. Administratively, he set up a statistical laboratory as a semiautonomous unit within the Mathematics Department. However, soon America's entry into World War II in 1941 put all further academic development on hold. Neyman took on war work, and for the next several years this became the laboratory's central and all-consuming activity. The work dealt with quite specific military problems and did not produce results of lasting interest.

The building of a faculty began in earnest after the war, and by 1956 Neyman had established a permanent staff of twelve members, many his own students, but also including three senior appointments from outside (Loève, Scheffé, and Blackwell). Development of a substantial faculty, with the attendant problems of space, clerical staff, summer support, and so on, represented a major sustained administrative effort. A crucial issue in the growth of the program concerned the course offerings in basic statistics by other departments. Although these involved major vested interests, Neyman gradually concentrated the teaching of statistics within his program, at least at the lower division level. This was an important achievement both in establishing the identity of the program and in obtaining the student base that alone could justify the ongoing expansion of the faculty. In his negotiations with the administration, Neyman was strengthened by the growing international reputation of his laboratory and by the increasing postwar importance of the field of statistics itself.

An important factor in the laboratory's reputation was the series of international "Symposia on Mathematical Statistics and Probability" that Neyman organized at five-year intervals during 1945–70, and the subsequent publication of their proceedings. The first symposium was held in

August 1945 to celebrate the end of the war and "the return to theoretical research" after years of war work. The meeting, although rather modest compared to the later symposia, was such a success that Neyman soon began to plan another one for 1950. In later years the symposia grew in size, scope, and importance and did much to establish Berkeley as a major statistical center.

The spectacular growth Neyman achieved for his group required a constant struggle with various administrative authorities, including those of the Mathematics Department. To decrease the number of obstacles, and also to provide greater visibility for the statistics program, Neyman, soon after his arrival in Berkeley, began a long effort to obtain independent status for his group as a Department of Statistics. A separate department finally became a reality in 1955, with Neyman as its chair. He resigned the chairmanship the following year (but retained the directorship of the laboratory to the end of his life). He wrote in his letter of resignation that "the transformation of the old Statistical Laboratory into a Department of Statistics closed a period of development . . . and opened a new phase." In these circumstances he stated that "it is only natural to have a new and younger man take over."

There was perhaps another reason. Much of Neyman's energy during his nearly twenty years in Berkeley had gone into administration. His efforts had been enormously successful—a first-rate department, the symposia, and a large number of grants providing summer support for faculty and students. It was a great accomplishment and his personal creation, but now it was time to get back more fully into research.

APPLIED STATISTICS

Neyman's theoretical research in Berkeley was largely motivated by his consulting work, one of the purposes for which the university had appointed him and through which he made himself useful to the campus at large. Problems in astronomy, for example, led to the interesting insight (1948, with E. L. Scott) that maximum likelihood estimates may cease to be consistent if the number of nuisance parameters tends to infinity with increasing sample size. Also, to simplify maximum likelihood computations, which in applications frequently became very cumbersome, he developed linearized, asymptotically equivalent methods—his BAN (best asymptotically normal) estimates (1949), which have proved enormously useful.

His major research efforts in Berkeley were devoted to several large-scale applied projects. These included questions regarding competition of species (with T. Park), accident proneness (with G. Bates), the distribution of galaxies and the expansion of the universe (with C. D. Shane and particularly Elizabeth Scott, who became a steady collaborator and close companion), the effectiveness of cloud seeding, and a model for carcinogenesis. Of these perhaps the most important was the work in astronomy, where the introduction of the Neyman-Scott clustering model brought new methods into the field that "were remarkable and perhaps have not yet been fully appreciated and exploited." (*Peeble: Large Scale Structure of the Universe,* Princeton Univeristy Press, Princeton, N.J., 1979).

Neyman's applicational work, although it extends over many different areas, exhibits certain common features, which he made explicit in some of his writings and which combine into a philosophy for applied statistics. The following are some of the principal aspects:

1. *The studies are indeterministic.* Neyman pointed out that the distinction between deterministic and indeterministic studies lies not so much in the nature of the phenomena as in the treatment accorded to them ("Indeterminism in Science and New Demands on Statisticians," *Journal of the American Statistical Assocication,* 1960, vol. 55, pp. 625–39). In fact, many subjects that traditionally were treated as deterministic are now being viewed stochastically. Neyman himself contributed to this change in several areas.

2. *An indeterministic study of a scientific phenomenon involves the construction of a stochastic model.* In this connection, Neyman introduced the important distinction between models that are interpolatory devices and those that embody genuine explanatory theories. The latter he describes as "a set of reasonable assumptions regarding the mechanism of the phenomena studied," while the former "by contrast consist of the selection of a relatively ad hoc family of functions, not deduced from underlying assumptions, and indexed by a set of parameters" ("Stochastic Models and Their Application to Social Phenomena," coauthored by W. Kruskal and presented at a joint session of the IMS, ASA, and the American Sociological Society, Sept. 1956). The distinction was discussed earlier, and again later, in Neyman's paper in *Annals of Mathematical Statistics* (1939, vol. 10, pp. 372–73) and in *A View of Biometry.* Siam. (Philadelphia, 1974, pp. 185–201).

Neyman's favorite example of the distinction between the two types was the family of Pearson curves, which for a time was very popular as an interpolatory model that could be fitted to many different data sets, and Mendel's model for heredity. "At the time of its invention," he wrote about the latter in the *Journal of the American Statistical Association* (1960, vol. 55, pp. 625–39), "this was little more than an interpolatory procedure invented to summarize Mendel's experiments with peas," but eventually it turned out to satisfy Neyman's two criteria for a model of genuine scientific value, namely (i) broad applicability (i.e., in Mendel's case not restricted to peas) and (ii) identifiability of details (the genes as entities with identifiable location on the chromosomes).

Most actual modeling, Neyman pointed out, is intermediate between these two extremes, often exhibiting features of both kinds. Related is the realization that investigators will tend to use as building blocks models that "partly through experience and partly through imagination, appear to us familiar and, therefore, simple." (with E. L. Scott, "Stochastic Models of Population Dynamics," *Science* 1959, vol. 130, pp. 303–8).

3. *To develop a "genuine explanatory theory" requires substantial knowledge of the scientific background of the problem.* When the investigation concerns a branch of science with which the statistician is unfamiliar, this may require a considerable amount of work. For his collaboration with Scott in astronomy, Neyman studied the astrophysical literature, joined the American Astronomical Society, and became a member of the Commission on Galaxies of the International Astronomical Union. When he developed an interest in carcinogenesis, he spent three months at the National Institutes of Health to learn more about the biological background of the problem.

Neyman summarized his own attitude toward this kind of research in the closing sentence of his paper, "A Glance of My Personal Experiences in the Process of Research" (in *Scientists at Work*, Almquist and Wicksell, Uppsala, Sweden, 1970): "The elements that are common . . . seem to be my susceptibility to becoming emotionally involved in other individuals' interests and enthusiasm, whether these individuals are sympathetic or not, and, particularly in the more recent decades, the delight I experience in trying to fathom the chance mechanisms of phenomena in the empirical world."

An avenue for learning about the state of the art in a field and bringing together diverse points of view, which Neyman enjoyed and of which he made repeated use, was to arrange a conference. Two of these (on weather modification in 1965 and on molecular biology in 1970) became parts of the then-current symposia. In addition, in 1961, jointly with Scott, he arranged a "Conference on the Instability of Systems of Galaxies." In 1973 he edited for the National Academy of Sciences a volume in celebration of the five-hundredth anniversary of the birth of Nicholas Copernicus: *The Heritage of Copernicus: Theories Pleasing to the Mind.* The volume presents revolutionary changes that occurred, respectively, in astronomy and cosmology, biology, chemistry and physics, mathematics, technology, and in the thinking in mathematics and science brought about by the introduction of chance mechanisms. Finally, in July 1981, jointly with Le Cam and on very short notice, Neyman arranged an interdisciplinary cancer conference.

EPILOGUE

A month after the cancer conference, Neyman died of heart failure at age eighty-seven on August 5, 1981. He had been in reasonable health up to two weeks earlier and on the day before his death was still working in the hospital on a book on weather modification.

Neyman is recognized as one of the founders of the modern theory of statistics, whose work on hypothesis testing, confidence intervals, and survey sampling has revolutionized both theory and practice. His enormous influence on the development of statistics is further greatly enhanced through the large number of his Ph.D. students. These are:

From Poland: Kolodziejczyk, Iwaszkiewicz, Wilenski, Pytkowski.

From London: Hsu, David, Johnson, Eisenhart, Sukhatme, Beall, Sato, Shanawany, Jackson, Tang.

From Berkeley: Chen, Dantzig, Lehmann, Massey, Fix, Chapman, Eudey, Gurland, Hodges, Seiden, Hughes, Taylor, Jeeves, Le Cam, Tate, Chiang, Agarwal, Sane, Read, Singh, Borgman, Marcus, Buehler, Kulkarni, Davies, Clifford, Samuels, Oyelese, Ray, Grieg, Green, Tsiatis, Singh, Javitz, Darden.

Neyman was completely and enthusiastically dedicated to his work, which filled his life—there was no time for hobbies. Work, however, included not only research and teaching but also their social aspects, such as traveling to meetings and organizing conferences. Pleasing his guests was an avocation; his hospitality had an international reputation. In his laboratory he created a family atmosphere that included students, colleagues, and visitors, with himself as paterfamilias.

As administrator, Neyman was indomitable. He would not take "no" for an answer and was quite capable of resorting to unilateral actions. He firmly believed in the righteousness of his causes and found it difficult to understand how a reasonable person could disagree with him. At the same time, he had great charm that often was hard to resist.

High on Neyman's list of values were his opposition to injustice and his sympathy for the underdog. Two illustrations must suffice. In 1946, as a member of the allied mission to supervise the Greek elections, his feeling for fairness and justice led him to disobey orders of his superiors, with the consequence of his services being abruptly terminated. The details are given by Constance Reid (*Neyman—from Life*, Springer, New York, 1982, pp. 201–8). This book also describes (pp. 262–68) how deeply Neyman was affected, when in 1963 as a visiting lecturer for the Mathematical Association of America, he came into firsthand contact with segregation in the South. This led him to various efforts that culminated in the establishment at Berkeley of a special scholarship program to help prepare talented underprivileged young people for a university education.

On a personal level, the characteristic that perhaps remains above all in the minds of his friends and associates is his generosity—furthering the careers of his students, giving credit and doing more than his share in collaboration, and extending his help (including financial assistance out of his own pocket) to anyone who needed it.

NOTES

1. "One of my favorite ideas," Neyman later wrote in 1957 (*Rev. Int. Stat. Inst.* 25:8), "learned from Mach via Karl Pearson's 'Grammar of Science,' is that scientific theories are no more than models of natural phenomena."

2. Its role in today's statistical climate is discussed by Lehmann, "The Neyman-Pearson theory after 50 Years" (In *Proceedings of the Berkeley Conference in Honor of Jerzy Neyman and Jack Kiefer*, vol. 1 [Wadsworth, 1985]:1–14).

3. In "Frequentist Probability and Frequentist Statistics" (*Synthese* 36[1977]:97–131, Neyman states, "I began as a quasi-Bayesian. My assumption was that the estimated parameter (just one!) is a particular value of a random variable having an unknown prior distribution."

4. Loc. cit.

HONORS AND DISTINCTIONS

HONORARY DEGREES

1959 D.Sc., University of Chicago
1963 LL.D., University of California, Berkeley
1964 Ph.D., University of Stockholm, Sweden
1974 D.Sc., University of Warsaw, Poland
1974 D.Sc., Indian Statistical Institute

MEMBERSHIPS

1963 U.S. National Academy of Sciences
1963 Royal Swedish Academy
1966 Polish National Academy
1979 Royal Society (London)

AWARDS

1958 AAAS Newcomb Cleveland Prize
1966 U.K. Royal Statistical Society Guy Medal in Gold
1969 U.S. National Medal of Science
1973 Copernicus Society of America, Medal and Award

SELECTED BIBLIOGRAPHY

I. Original Works

A complete bibliography of Neyman's work is given at the end of David Kendall's memoir (*Biographical Memoirs of Fellows of the Royal Society*, vol. 28, 1982).

Some of the early papers are reprinted in the two volumes, *A Selection of Early Statistical Papers of J. Neyman* and *Joint Statistical Papers of J. Neyman and E. S. Pearson* (University of California Press).

Neyman's letters to E. S. Pearson from 1926 to 1933 (but not Pearson's replies) are preserved in Pearson's estate.

An overall impression of Neyman's ideas and style can be gained from his book, *Lectures and Conferences on Mathematical Statistics and Probability*, Second Revised and Enlarged Edition (Graduate School, U.S. Department of Agriculture, Washington, D.C., 1952). The following partial list provides a more detailed view of his major theoretical contributions.

A. Paradigmatic Papers

1928

With E. S. Pearson. On the use and interpretation of certain test criteria for purposes of statistical inference. *Biometrika* 20A:175–240, 263–94.

1933

With E. S. Pearson. On the problem of the most efficient tests of statistical hypotheses. *Philos. Trans. R. Soc. Lond. Ser. A* 231:289–337.

1934

On the two different aspects of the representative method. *J. R. Stat. Soc.* 97:558–625 (A Spanish version of this paper appeared in *Estad. J. Inter-Am. Stat. Inst.* 17:587–651.)

1937

Outline of a theory of statistical estimation based on the classical theory of probability. *Philos. Trans. R. Soc. Lond. Ser. A* 236:333–80.

B. Some other theoretical contributions

1937

'Smooth' test for goodness of fit. *Skand. Aktuar. Tidskr.* 20:149–99.

1939

On a new class of 'contagious' distributions, applicable in entomology and bacteriology. *Ann. Math. Stat.* 10:35–57.

1948

With E. L. Scott. Consistent estimates based on partially consistent observations. *Econometrica* 16:1–32.

1949

Contribution to the theory of the chi–square test. In *Proceedings of the Berkeley Symposium on Mathematical Statistical Probability*, ed. J. Neyman, pp. 239–73. Berkeley: University of California Press.

Optimal asymptotic tests of composite statistical hypotheses. In *Probability and Statistics*, Harald Cramér Volume, ed. U. Granander, pp. 213–34. Uppsala, Sweden: Almquist & Wiksells.

Neyman's position regarding the role of statistics in science can be obtained from the following more philosophical and sometimes autobiographical articles.

1951

Foundation of the general theory of statistical estimation. *Actual. Sci. Ind.* 1146:83–95.

1955

The problem of inductive inference. *Commun. Pure Appl. Math.* VIII:13–45.

1957

'Inductive behavior' as a basic concept of philosophy of science. *Rev. Inst. Int. Stat.* 25:7–22.

1959

With E. L. Scott. Stochastic models of population dynamics. *Science N.Y.* 130:303–8.

1970

A glance at some of my personal experiences in the process of research. In *Scientists at Work*, ed. T. Dalenius, G. Karlsson and S. Malmquist, pp. 148–64. Sweden: Almquist & Wiksells.

1971

With E. L. Scott. Outlier proneness of phenomena and of related distributions. In *Optimizing Methods in Statistics*, ed. J. S. Rustagi, pp. 413–30. New York: Academic Press.

1977

Frequentist probability and frequentist statistics. *Synthèse* 36:97–131.

II. Secondary Literature

The most important source for Neyman's life and personality is Constance Reid's *Neyman—from Life* (Springer, New York, 1982), which is based on Neyman's own recollections (obtained during weekly meetings over a period of more than a year) and those of his colleagues and former students and on many original documents. A useful account of his collaboration with E. S. Pearson was written by Pearson for the Neyman Festschrift, *Research Papers in Statistics* (F. N. David, ed., Wiley, New York, 1966). Additional accounts of his life and work are provided by the following papers.

1974

L. Le Cam and E. L. Lehmann. J. Neyman—on the occasion of his 80th birthday. *Ann. Stat.* 2:vii–xiii.

1982

D. G. Kendall, M. S. Bartlett, and T. L. Page. Jerzy Neman, 1894–1981. In *Biographical Memoirs of Fellows of the Royal Society*, vol. 28, pp. 379–412.

1982

E. L. Lehmann and Constance Reid. In memoriam – Jerzy Neyman, 1894–1981. *Am. Stat.* 36:61–62.

1985

E. L. Scott. Neyman, Jerzy. *Encycl. Stat. Sci.* 6:214–23.

18

Le Cam at Berkeley

E. L. Lehmann[1]

Written in appreciation of the pleasure and many benefits I have received from over forty years of friendship and collegiality with Lucien Le Cam.

18.1 Early Life

In 1950 a young Frenchman came to Berkeley for a one-year visit. To his great surprise he stayed on and has remained in Berkeley ever since (though he is still a French citizen).

Le Cam was born in 1924 and spent his early life on a farm leased by his family in Creuse, Central France.

After a somewhat unconventional education, he obtained his License en Sciences (roughly equivalent to a BA) at the University of Paris in 1945. For reasons of convenience he selected as one of his examination topics Statistics,a subject in which he had no background and for which he prepared himself in less than a week by reading the lecture notes of the examiner, George Darmois. With the help of Darmois, he obtained a position as statisticianat what was evolving into the Electricite de France. This job enabled him occasionally to attend seminars at the university, and at one of these he met Neyman. The meeting, together with recommendations by Charles Stein (who had spent the preceding year in Paris) and Maurice Frechet, led to an invitation to spend a year in Berkeley as Lecturer in Statistics.

At Berkeley, he found that his background did not quite match what was expected either in mathematics or statistics. On the one hand, his statistics was applied and included little theory. On the other, the only modern mathematics he knew came from reading the books of Bourbaki. So he accepted a suggestion by Neyman to stay on for another year and complete his training by getting a California Ph.D.

Throughout his school career, Le Cam had problems with examinations, occasionally even managing to fail them. An explanation is offered by a story he tells of himself. In an early intelligence test, he was asked to give the next entry in the sequence 3,5,7, ... His answer was 11, and his score on the test dismal. Now, in Berkeley, he managed to fail the Ph.D. Qual-

[1]University of California at Berkeley

J. Rojo (ed.), *Selected Works of E. L. Lehmann*, Selected Works in Probability and Statistics, DOI 10.1007/978-1-4614-1412-4_72, © Springer Science+Business Media, LLC 2012

ifying Examination. The assignment was to give an hour lecture on fixed point theorems and Le Cam decided to present the latest and most general version he was able to find in the literature. This required first providing a considerable amount of preparatory algebraic topology, and before he knew it the hour was over-he had never reached the main result. Repeating the examination a few months later, he took a less abstract approach. However, he included a new proof of his own which was vigorously (and, as it later turned out, incorrectly) disputed by one of the examiners. This time he passed, but only barely.

Le Cam obtained his doctorate in 1952, but this was preceded by another event, his marriage earlier in the year to Louise Romig. In an autobiograhical sketch (Albers, Alexanderson & Reid 1990)[2] he reports that when he told Neyman about the impending marriage, "he flew into a rage. He called my future father-in-law, Harry Romig, who was also a statistician, and told him it was 'illegal' for me to get married because I had not finished my thesis. Actually it was finished except for a few pages still to be typed. I went ahead and got married anyway in April 1952. I was awarded my Ph.D. that June".

The thesis (Le Cam 1953), "On Some Asymptotic Properties of Maximum Likelihood Estimates and Related Bayes Estimates", has since became a classic. It grew out of the discovery in the preceding year by Joe Hodges of the phenomenon of superefficiency, which had confounded the universally held belief in the efficiency of maximum likelihood estimators. One of the principal results ofLe Cam's thesis now provided damage control by showing that superefficiency can be achieved only on a set of measure zero.

On completing his degree, Le Cam accepted Neyman's offer of a ladder appointment as Instructor. He obtained tenure in 1958 and the Professorship in 1960. The following year he succeeded David Blackwell as chair of the Statistics Department.

18.2 University and Scientific Administration

Le Cam's term of office was marked by a phenomenal growth of the Department. In 1961 the Faculty consisted of twelve members; by the end of his term, four years later, the number had increased to twenty. In large measure this success was due to his vigorous and untiring efforts. They were helped by the general popularity of science in the post-Sputnik era, and by continuing growth of enrollment in statistics courses and of the field of statistics in the world at large.

Both as a member of the Department and as its chair Le Cam held

[2]This sketch provides much additional detail on Le Cam's early life.

opinions which he liked to bring to the attention of his colleagues and the University administration in long, provocative and amusing letters. These were enjoyed by the Faculty but were less popular with the Deans and Chancellors.

Following are excerpts from one such communication (written in 1958) regarding a proposal by his colleague Loeve to strengthen the probability part of a one-year introduction to probability and statistics at the senior/first year graduate level. Some of the statisticians in the Department had protested that such additional material was more than what students at this level and in the time available could handle, and that the change would infringe on the statistical content of the course. Here are some excerpts from Le Cam's "Comment on 'Comments on Loeve's proposed revision' ".

> The proposed revision seems to arouse such violent opposition that the inarticulate undersigned feels obligated to communicate his feelings on the subject by the present memo.
>
> Loeve's proposal seemed at first so innocuous and so gentle that I did not expect much reaction to it from any side, except possibly that practical people (are statisticians supposed to be practical?) might desire even more emphasis on applicable probability theory. As of now I consider the differences of opinion ariSIng in our midst are irreconcilable and due to such basic differences of understanding of what is and what should be probability theory and how it should be taught that no useful purpose will be derived by arguing any further in the same direction. Therefore the present memo will not propose any solutions but only try to generate some heat on controversial matters.
>
> Our teaching of Probability and consequently Statistics is hope-Lessly inadequate. As far as I can tell, the people who struggle through [our basic courses] do not even get a knowledge of probability remotely comparable to the knowledge imparted by J. Dubourdieu in his evening lectures to candidates to the Actuarial examination in Paris.
>
> In case one should object that Probability Theory per se is not a worthy subject of study, I shall make my next point: The statisticians who are going to make the headlines in the future are not those who can in a twinkling give an analysis of variance for a Latin Square or a three-way classification. Statistics is basically much more complicated than Probability, would it be only because where the Probabilist has only one measure to cope with, the statistician has a family of measures. There is of course plenty to do in statistics without using complicated machinery. It just requires brains. This last commodity we cannot dispense to the students, but we can give them tools.

> If we would give our students a good background in mathematics and probability, we might be able to teach them more, not, less, statistics in a shorter time. Otherwise we are bound to produce Ph.D.'s in Statistics who cannot even read the statistical papers of our Symposia, not to mention the sundry applied papers to be 'found in the same Symposia.
> If this memo seems unreasonable to you, I will gladly restore it to its original strength and triple the emphasis on the necessity of teaching probability theory as a major part of our task and not as an undignified accessory to some statistical arguments.

Talking with chairman Le Cam, some time in 1962, Neyman mused on the fact that the following year marked not only the 150th anniversary of Laplace's "TMorie Analytique des Probabilites" but also the 200th anniversary of the posthumous publication of Bayes' "Essay Toward Solving a Problem in the Doctrine of Chances" and the 250th of Jacob Bernoulli's "Ars Conjectandi". To celebrate this remarkable triple anniversary Neyman proposed to Le Cam an international research seminar to be held at Berkeley. It was difficult to obtain financial support for such a meeting in the short time available, and the seminar was replaced by a series of individuallectures over a period of time during 1963, which were published two years later (Neyman & Le Cam 1965).

After plans had been announced and invitations sent, the organizers realized that the first edition of Laplace's book had appeared not in 1813 as claimed but in 1812, and the second edition containing the famous "Essai Philosophique sur les Probabilites" in 1814. The editors got around this difficulty by stating in their foreword that this Essai "must have been written in 1813, 150 years before the Berkeley Seminar of 1963".

The collaboration of Neyman and Le Gam continued with the Fifth Berkeley Symposium (Le Cam & Neyman 1967) scheduled for 1965. The first four symposia, which had taken place at five-year intervals since 1945 had been Neyman's sole responsibility. He now asked Le Cam to chair the Organizing Committee and to serve as Coeditor of the Proceedings which by then had grown to five substantial volumes. Five years later (Le Cam, Neyman & Scott 1973), Le Cam also coedited the Proceedings of the sixth (and last) symposium.

A last conference on which the two men collaborated was devoted to an interdisciplinary meeting on cancer models. Since coming to Berkeley, Le Cam had concentrated on theoretical aspects of statistics, but this changed in 1974 as a result of his daughter Linda's being diagnosed with osteosarooma. At that time he began to collaborate[3] with Vera Byers and her

[3]Some comments on this collaboration by Vera Byers are given in Albers et aI. (1990,p.171).

husband Al Levin who successfully treated Linda with the immunotherapy they were developing in their laboratory.

Neyman had became interested in models for the mechanism of carcinogenesis in the 1950's, and in 1958 had spent three months at the National Institutes of Health to learn more about the biological background of the problem and to participate in the work being conducted there. In the Spring of 1981, Le Cam told Neyman of experimental work being done at Berkeley that was relevant to his models. In response, Neyman proposed a conference that would bring together researchers approaching this problem from different points of view, and asked Le Cam to undertake its organization. He stressed the urgency of the problem, and the arrangements were in fact completed in less than a month. The conference took place in July 1981, only a few weeks before Neyman's death of heart failure, and the Proceedings-dedicated to his memory-were published the following year (Le Cam & Neyman 1982).

The week following Neyman's death another member of the Statistics Department, Jack Kiefer, who had experienced no previous symptoms, suffered a fatal heart attack at the young age of only 57. Stunned by these two deaths within a few days of each other, the Department hesitated for some time about a suitable commemoration. Eventually, Le Cam and Elizabeth Scott decided to organize a "Conference in Honor of Jerzy Neyman and Jack Kiefer". The conference took place in June 1983 and the Proceedings (Le Cam & Olshen 1985) were published in two volumes.

18.3 Research

Beginning with his thesis, Le Cam's statistical research has centered on large-sample theory. This has included very general treatments of maximum likelihood and Bayes estimation, introduction of contiguity and local asymptotic normality, superefficiency, and his general theory of the (approximate) comparison of experiments. Additional contributions have been made by many of his close to 40 Ph.D. students, and this body of work-which has been referred to (Feigin 1987) as the Le Cam school of asymptotic methods-was given a comprehensive exposition in Le Cam's 1986 book.

A review of this book (Shiryaev & Koshevnik 1989) states that "the endeavor for generality distinguishes Professor Le Cam himself as well as the book under review". Le Cam strongly disclaims any efforts at generality for its own sake, and justifies it as providing better organization of the material[4]. "It was just more convenient", he writes in the preface to his 'book, "to rewrite the definitions in such a way that the desired theorems are always true, instead of imposing restrictive conditions on a more stan-

[4]Personal communication

dard system". As a result, his work is a powerful resource that takes care of the many special nonregular cases not covered by the simpler classical approaches.

Another reviewer (Dacunha-Castelle 1988) characterizes the book [my translation from the French original] as "the work of a mathematician. It so happens that he is talking about statistics. This is due to fortuitous circumstances, but it is a stroke of good luck for statistics".

Without doubt, Le Cam's work has brought statistics closer to becoming part of mathematics. By the same token, as a consequence of the generality and abstraction of his treatment, it presents considerable difficulty to students and colleagues.[5] His course in large-sample theory was notoriously hard, and students frequently audited it once or even twice before daring to take it for credit. The somewhat forbidding nature of his work has been greatly alleviated locally by his friendliness and patience. He kept his office door open and never seemed to tire of helping his students (and anyone else) with their problems.

Although the construction of his own general theory of experiments occupies the center of Le Cam's work, it by no means exhausts it. There are for example several papers dealing with the Central Limit Theorem; three expository papers with a historical or philosophical slant; a number of applied papers, particularly on immunotherapy and other aspects of cancer research; and papers on occasional special topics.

Le Cam's asymptotic work has been enormously influential. His theory of contiguity is a principal topic of books by Roussas (1972) and Greenwood & Shiryaev (1985), and it plays a major role in Hajek & Sidak (1967). His general approach to large-sample theory and the comparison of experiments forms the central topic of books by Heyer (1982) and Torgersen (1991).

In acknowledgement of his work, Le Cam received in 1957 a prestigious Sloan Fellowship which enabled him to spend a year in Paris in order "to renew contacts with the Bourbaki group of French mathematicians" and in 1971 a Miller Research Professorship at Berkeley. During 1972/73 he served as Director of the Centre de Recherches MatMmatiques of the University of Montreal. The fear that he might accept this as a permanent position was relieved when at the end of the year he decided to return to Berkeley. In 1976 Le Cam was elected to the American Academy of Arts and Sciences. He took early retirement from Berkeley in 1991, but he continues to come to his office daily, with his door remaining open as before.

18.4 References

Albers, D., Alexanderson, J. & Reid, C. (1990), *More Mathematical*

[5]To alleviate this difficulty Le Cam & Yang (1990) provide a more accessible introduction to essential parts of the theory. In this connection see also Strasser (1985).

People, H!trcourt, Brace, Jovanovich, Boston.

Dacunha-Castelle, D. (1988), 'A propos du livre de Lucien Le Cam: Asymptotic Methods in Statistical Decision Theory', *Metrika* p. 254. (Review of Le Cam 1986.).

Feigin, P. (1987), 'Review of Greenwood and Shiryaev (1985)', *Journal of the American Statistical Association* 82, 1195. .

Greenwood, P. & Shiryaev, A. (1985), *Contiguity and the Statistical Invariance Principle,* Gordan and Breach. New York.

Hajek, J. & Sidak, Z. (1967), *Theory of Rank Tests,* Academic Press. (Also published by Academia, the Publishing House of the Czechoslavak Academy of Sciences, Prague.).

Heyer, H. (1982), *Theory of Statistical Experiments,* Springer-Verlag, New York. (Second edition of *Mathematische Theorie Statischer Experimente,* Springer-Verlag 1973.).

Le Cam, L. (1953), 'On some asymptotic properties of maximum likelihood estimates and related Bayes estimates', *University of California Publications in Statistics* 1, 277-330.

Le Cam, L. & Neyman, J., eds (1967), *Proceedings of the Fifth Berkeley Symposium on Mathematical Statistics and Probability,* University of California Press, Berkeley.

Le Cam, L. & Neyman, J., eds (1982), *Probability Models and Cancer,* North Holland. Amsterdam.

Le Cam, L. & Olshen, R., eds (1985), *Proc. Berkeley Confer. in Honor of Jerzy Neyman and Jack Kiefer,* Wadsworth. Monterey.

Le Cam, L. & Yang, G. L. (1990), *Asymptotics in Statistics: Some Basic Concepts,* Springer-Verlag.

Le Cam, L., Neyman, J. & Scott, E. L., eds (1973), *Proceedings of the Sixth Berkeley Symposium on Mathematical Statistics and Probability,* University of California Press, Berkeley.

Neyman, J. & Le Cam, L., eds (1965), *Bernoulli, Bayes, Laplace,* Springer-Verlag, Berlin.

Roussas, G. (1972), *Contiguity of Probability Measures,* Cambridge University Press.

Shiryaev, A. N. & Koshevnik, Y. A. (1989), 'Review of Le Cam (1986)', *Bulletin of the Amererican Mathematical Society* 20,280-285.

Strasser, H. (1985), *Mathematical Theory of Statistics: Statistical Experiments and Asymptotic Decision Theory,* De Gruyter, Berlin.

Torgersen, E. (1991), *Comparison of Statistical Experiments*, Cambridge University Press.

University of California: In Memoriam, 2000

Joseph L. Hodges, Jr., Statistics: Berkeley

1922-2000
Professor Emeritus

Joseph L. Hodges, Jr. had a long distinguished career at Berkeley as a researcher, gifted teacher, and active participant in the University Senate. He held important positions in professional societies and in both campus and university administration--and was widely known on campus as a lunchtime fixture at the Faculty Club.

Born in Shreveport, Louisiana, on April 10, 1922, Joe entered the University of California at the age of sixteen and earned his B.A. in mathematics in 1942. He remained at Berkeley as a graduate student of mathematics, attracted by Professor Tarski, but soon found himself in a group started by Professor Jerzy Neyman to deal with statistical problems for the military, particularly bombing patterns. This led to his going to Guam as an operations analyst with LeMay's Twentieth Air Force. Another member of that group was Erich Lehmann (now Professor Emeritus of Statistics) who became a lifelong friend and frequent scientific collaborator.

After the war he continued this work for another year in Washington, D.C. then returned to Berkeley to continue graduate work. While in Washington, he met Theodora Jane Long (Teddy), and they married in 1947. He obtained his Ph.D. in 1949 under the supervision of Professor Neyman and joined the statistics faculty (then part of the Mathematics Department). He remained at Berkeley for the rest of his life, except for a year's leave at the University of Chicago (1951-52), where he rejected a permanent move, and a sabbatical year in Sweden (1956-57).

Joe was superb and inspiring as a teacher and well liked as a major advisor. He was particularly concerned with undergradu

— 84 —

ate teaching. In connection with this teaching he co-authored the elementary texts *Basic Concepts of Probability and Statistics* (1964; 2nd ed. 1970) jointly with Lehmann (which was translated into Danish, Hebrew, Italian and Farsi), and *Stat Lab--An Empirical Approach* (1975) jointly with the psychologists Krech and Crutchfield.

In statistics, Joe is best known for his work in nonparametric inference. In a technical report (with Fix), written in 1951 but published only in 1989, he pioneered nonparametric density estimation. He is the inventor of the Hodges bivariate-sign test and (with Lehmann) of the Hodges-Lehmann estimator. One of his most striking discoveries was the phenomenon of superefficiency, which exploded long-held beliefs about maximum likelihood estimation and had a profound effect on asymptotic theory. His bibliography lists a steady stream of publications, at least one, but usually more, per year until the 1970s when administrative duties began to consume his energy.

He served on the governing boards of both the International Biometrics Society and the Institute of Mathematical Statistics, and perhaps his major service to the statistical profession was his term as Editor of the *Annals of Mathematical Statistics*, the leading journal of theoretical statistics. During his term the size of the journal increased from 1500 to 2000 pages a year, but he worked so efficiently that he was praised for reducing the average time from submission of a paper to publication. At first he handled both the editing and publishing of the journal; however, it became clear during his tenure that the workload was becoming unmanageable. He convinced the Institute to divide the two functions and found and established a suitable Managing Editor.

While he was still editor of the *Annals* he was invited by the University to become a member of the Committee on Budget and Interdepartmental Relations of the Academic Senate. Against the advice of his colleagues, Hodges felt he could not turn down the invitation to serve on the Budget Committee (surely the most important Senate activity); at the same time, he felt he had made a commitment to the Institute that he could

J. Rojo (ed.), *Selected Works of E. L. Lehmann*, Selected Works in Probability and Statistics, DOI 10.1007/978-1-4614-1412-4_73, © Springer Science+Business Media, LLC 2012

not go back on. His effectiveness is indicated by the fact that after his sabbatical he was appointed Chairman of the Budget Committee. Then began a long period of University service, focused on academic matters.

He became Assistant to the Chancellor for Academic Affairs, member and then Chair of the Universitywide Budget Committee, and Academic Assistant to the Vice President for Academic Affairs. Known for his devotion to quality, he was considered impeccable in his fairness and objectivity. Other service included stints on the Library Committee, the Committee on Year-Round Operation, the Board of the Faculty Club, the Executive Committee of the Miller Institute, and various departmental chores.

Joe joined the Faculty Club as soon as he became a member of the faculty and was widely known for his conversations, most of the time at a set table in the Great Hall but on Fridays with the group known as the Little Thinkers. At six feet four inches, Joe was an imposing presence, and not just physically. A reliable source of University and campus gossip, Joe kept the conversation lively. Being a contrarian, he would often exaggerate his innate conservatism for the sake of argument. For example, he opposed the admission of women to the Faculty Club and praised Ronald Reagan. He delighted in strong advocacy of faculty research that was considered controversial or not accepted by the establishment.

Those who had the opportunity to visit his home on Campus Drive know how much he enjoyed his family. Even here, the University was important. Teddy taught on campus for many years, and four of their five children attended UC campuses. Later, his eleven grandchildren gave him further delight. On March 1, 2000, he died unexpectedly from a heart attack.

His vivid presence will be missed by all who knew him.

David Blackwell Albert H. Bowker Erich L. Lehmann

HAROLD HOTELLING

September 29, 1895–December 26, 1973

BY K. J. ARROW AND E. L. LEHMANN

HAROLD HOTELLING WAS A man of many interests and talents. After majoring in journalism at the University of Washington and obtaining his B.A in that field in 1919, he did his graduate work in mathematics at Princeton, where he received his Ph.D. in 1924 with a thesis on topology. Upon leaving Princeton, he took a position as research associate at the Food Research Institute of Stanford University, from where he moved to the Stanford Mathematics Department as an associate professor in 1927. It was during his Stanford period that he began to focus on the two fields—statistics and economics—in which he would do his life's work. He was one of the few Americans who in the 1920s realized the revolution that R. A. Fisher had brought about in statistics and he spent six months in 1929 at the Rothamstead (United Kingdom) agricultural research station to work with Fisher.

In 1931 Hotelling accepted a professorship in the Economics Department of Columbia University. He taught a course in mathematical economics, but most of his energy during his 15 years there was spent developing the first program in the modern (Fisherian) theory of statistics. With the aid of a grant from the Carnegie Corporation, he was able to appoint a research associate who would also teach courses; among those who held this position were Abraham Wald and Henry B. Mann. He had as graduate students Samuel S. Wilks, W. Allen Wallis, Jacob Wolfowitz, Albert Bowker, and Meyer A. Girshick, who in turn set up similar programs at Princeton, Stanford, Chicago, and Cornell.

With the aid of the research associates the statistics program was indeed first-rate; it had a separate listing in the catalogue but there was no department and no degree associated with it. His intellectual leadership, kindness, and generosity to his students were legendary among them, and his house in Mountain Lakes, New Jersey, witnessed his and his wife's monthly open houses for them and other statisticians. He had none of the prejudices then still common; refugees from Europe and students from India could count on his warm support and very practical help.

His human and liberty-loving sympathies made him a bitter opponent of Hitler and an early advocate of intervention in World War II. When we entered the war, he persuaded the U.S. military to create a statistical research group at Columbia, of which he was made director. The group worked on many problems of quality control for munitions and other statistical issues relevant to the war effort.

The Columbia administration resisted his efforts to create an independent department of statistics with permanent faculty. It persisted with its refusal even in 1946, when Hotelling received an exceptionally attractive offer from the University of North Carolina to start a statistics program with strong external financial backing. As a result, Hotelling left Columbia for Chapel Hill, where he quickly built up a strong Department of Mathematical Statistics.

Hotelling married Floy Tracy in 1920, and they had two children. After her early death in 1934, he married Susanna Edmundson, with whom he had five sons. He retired in

1966 but remained at Chapel Hill, where he died in 1973 after a long period of ill health.

CONTRIBUTIONS TO ECONOMICS

(For some of the information on Hotelling's career as an economist, we draw on the work of Adrian Darnell [1990].)

Harold Hotelling wrote six major papers on economics. Four of them have had a profound effect on the field, though frequently only after a long period of time. The other two are equally ingenious but deal with matters on which others were also working. Hotelling was interested in economics from his undergraduate days. Though his undergraduate degree was in journalism, he did take some courses in economics and in later years seemed to regard them as equally important to his journalism. His Ph.D. thesis in mathematics concerned topology, but his economic interests led to a paper, published almost contemporaneously with one based on his dissertation, on the true economic meaning of depreciation (1925). (Incidentally and possibly irrelevantly, the uncle of his thesis advisor, Oswald Veblen, was a very well-known though highly unorthodox economist, Thorstein Veblen.) The topic of depreciation is somewhat specialized, but Hotelling's treatment is at once highly original and fully in accord with standard economic reasoning. This paper became the standard for all subsequent work in the field. It did require a reasonable amount of mathematics, including calculus of variations and solving an integral equation in the most general case. Although there had already been a tradition of the use of mathematics in economics, indeed with predecessors in the late eighteenth century the use of mathematics had not spread far at this point. The continuous use of mathematics and economics really begins with Hotelling and some European contemporaries.

Another paper, whose inception appears to date from the same period but was published somewhat later (1929) in a major economic journal, dealt with a topic that he introduced, that of spatial competition. He recognized that when establishments are spatially separated, the cost to the consumer includes not only the price but also the transportation costs. As a result, he concluded that firms will tend to be concentrated in the middle. He noted that spatial differentiation was not only interesting in itself but could also be regarded as a metaphor for quality differentiation in products. His method of analysis was to find the Nash equilibrium point of a two-stage game; in the first state the players locate themselves; in the second they choose prices. (This was, of course, long before game theory was defined.) He noted that the analysis could also be applied to competition between political parties; they would both tend toward the center. Thirty years later and with additional contributions by Howard Bowen (Bowen, 1943), Duncan Black (Black, 1948), and Anthony Downs (Downs, 1957), the idea entered political science as the median voter theorem and has remained a staple of formal political theory to this day.

Unfortunately, Hotelling's work on the economics of spatial location was vitiated by an error (he used local conditions for a maximum, which were not sufficient in this case), as first pointed out 50 years later (d'Aspremont et al., 1979), an error that, however, did not affect the political implications.

The most influential of all his papers, at least in recent years, was that on the economics of exhaustible resources (e.g., minerals, such as oil) (1931). He gave a formal argument that the price of a commodity (more precisely, the excess of the price over the cost of extraction) would have to rise over time at the same rate as the rate of interest. The argument, like some parts of the paper on depreciation, used the calculus of variations, well beyond the capacity of the great bulk of economists then, and the paper was rejected by one journal before being published. With the post-World War II concern over resource exhaustion, the paper became acknowledged and forms to this day an important component of the analysis of the future of resource scarcity.

The two papers on the theories of the firm and of the consumer (1932, 1935) were the best expositions of their subjects. He connected them soundly to the general mathematical theory of unconstrained and constrained maxima and put especial emphasis on the symmetries in the cross elasticities of demand and supply implied by the first-order conditions and on the implications of the second-order conditions. The first paper was a systematic treatment for any number of commodities of standard results derived graphically for two commodities. The second broke ground new to the general economics profession, but it turned out that its basic ideas had been anticipated by Evgenii Slutsky, the Russian probability theorist, in an Italian actuarial journal 21 years earlier. These papers were the basis of his Columbia course in mathematical economics, at least in 1941-1942.

The paper that received the most immediate recognition was his presidential address to the Econometric Society on the economic criteria for judging whether a policy change represents an economic improvement (1938). Though the general idea had been adumbrated over the history of economics, the precise statement and interpretation of the welfare criteria had never been stated clearly. He not only clarified the interpretation and gave a precise and simple proof but raised new ideas for measurement of welfare improvements, especially in a world of many commodities.

This paper and some remarks in earlier papers also made clear Hotelling's inclinations to some kind of market socialism. Despite the fact that his work was based on standard economic principles, he pointed out that they did not necessarily lead to *laissez-faire* implications.

Finally, one of the most used of Hotelling's economic insights is contained not in a paper but in a letter to the director of the National Park Service (Hotelling, 1947). The director had asked him and others how to measure the economic value of national parks. Since the entrance free is relatively small, it is clear that people are probably getting satisfaction from the park well in excess of the fee charged. Hotelling noted that people usually have to travel considerable distances and thereby incur significant money costs. If we find the largest distance traveled, then those individuals can be thought of as getting zero net satisfaction (the value of the park less the transportation costs). Then all nearer individuals get a surplus that can be computed. By integrating over the population, we can get a total measure of satisfaction. This method, called the "transport cost method," has been used numerous times to evaluate the social gain to parks and similar geographically defined sites.

CONTRIBUTIONS TO STATISTICS

Hotelling's statistical work falls into two categories. On the one hand, he made deep and highly original research contributions; on the other, he was in the broadest sense the teacher of a generation of American statisticians.

In one of his earliest statistical papers, written with Holbrook Working (1929), Hotelling obtained confidence bands for regression curves several years before Neyman developed his theory of confidence sets. His best-known and most influential paper (1931) extended Student's t-test (in both the one- and the two-sample cases) from univariate to multivariate distributions. The resulting test is known as Hotelling's T^2-test. As in the earlier paper, he points out that the tests can be converted into confidence statements, this time for the unknown multivariate mean. He also introduces invariance considerations to simplify determining the null distribution of the test statistic. This is an idea that entered the mainstream only much later. Hotelling continued his contributions to multivariate analysis with, among others, two basic papers on principal components (1933) and canonical correlations (1936).

A paper pointing in quite a different direction was "Rank Correlation and Tests of Significance Involving No Assumption of Normality" (1936). Of it, Richard Savage, in his 1953 "Bibliography of Nonparametric Statistics," writes that "papers related to nonparametric problems were published in the nineteenth century, but the true beginning of the subject may be taken as 1936, the year in which Hotelling and Pabst published their paper on rank correlation."

Again, invariance considerations are central and are used to motivate the reduction to the ranks of the observations.

It may seem surprising that Hotelling, having conceived of such basic ideas as confidence sets and invariance and having used them more than once, did not take the additional step of formulating them abstractly and developing their properties. But that was not his style. Mathematicians distinguish between problem solvers and system builders. Hotelling (at least in his statistical work) was emphatically the former. He took on a difficult problem and after having solved it, moved on to tackle the next one. As a result, confidence sets are credited to Neyman, who discovered the idea independently a few years later and developed it into a general theory; a general concept of invariance was first formulated in the late 1940s by Hunt and Stein.

Hotelling's role as an educator may be said to have begun when—it seems on his own initiative—he reviewed the first edition of Fisher's *Statistical Methods for Research Workers,* published in 1925, for the *Journal of the American Statistical*

Association (JASA). His review ended by noting that "the author's work is revolutionary and should be far better known in this country." Hotelling considered the task of making American statisticians aware of Fisher's work so important that he went on to review the next six editions of the book as well as the first two editions of Fisher's *The Design of Experiments.*

In the same vein, after spending six months with Fisher in Rothamstead, Hotelling published two survey papers in *JASA*: "British Statistics and Statisticians Today" (1930) and "Recent Improvements in Statistical Inference" (1931). They summarized what he had learned in England.

During the 1930s, knowledge of the new methodology developed by Fisher and expanded by his successors spread in America into many different fields of application, and led to the setting up of statistics courses in the corresponding university departments. However, these courses were not taught by statisticians, since that profession did not yet exist. Instead, they were typically assigned to the youngest, most recently hired member of the department, who often knew nothing of the subject and had to learn it one lecture ahead of the students.

The situation dismayed Hotelling, and in 1940 he analyzed it in a paper (1940,2), "The Teaching of Statistics," that raised a question basic to the future of the discipline.

> The growing need, demand and opportunity have confronted the educational system of the country with a series of problems regarding the teaching of statistics. Should statistics be taught in the department of agriculture, anthropology, astronomy, biology, business, economics, education, engineering, medicine, physics, political science, psychology or sociology, or in all these departments? Should its teaching be entrusted to the department of mathematics, or a separate department of statistics, and in either of these cases should other departments be prohibited from offering duplicating courses in statistics, as they are often inclined to do?

Hotelling continued this discussion five years later with a talk on the slightly broader topic—"The Place of Statistics in the University" (1949). He concluded the paper by recommending that "organization of the teaching of statistical methods should be centralized and should provide also for the joint functions of research and advice and service needed by others in the institution and possibly outside it, regarding the statistical aspects of their problems of designing experiments and interpreting observations."

In the following decades, the issues raised by these two papers were hotly debated in universities across the country, but in most cases were gradually settled along the lines suggested by Hotelling, with independent departments of statistics offering not only centralized teaching of the subject but also consulting services to fill the needs of researchers in other departments.

The most central aspect of Hotelling's educational activities was, of course, his development of two outstanding statistics groups at Columbia and Chapel Hill and the training of statisticians he carried out at those institutions. It is no exaggeration to state that during the 1930s and early 1940s, Hotelling nearly single-handedly brought American statistics into the modern age and laid the foundation for the extraordinary development of the subject after the Second World War.

REFERENCES

d'Aspremont, P., J.-J. Gabszewicz, and J.-T. Thisse. 1979. On Hotelling's "Stability in Competition." *Econometrica* 47:1145-1156.

Black, D. 1948. On the rationale of group decision-making. *J. Polit. Econ.* 56:23-34.

Bowen, H. 1943. The interpretation of voting in the allocation of economic resources. *Q. J. Econ.* 58:27-48.

Darnell, A. C. 1990. The life and economic thought of Harold Hotelling. In *The Collected Economics Articles of Harold Hotelling*, ed. A. C. Darnell, pp. 1-28. New York: Springer-Verlag.

Downs, A. 1957. *An Economic Theory of Democracy*. New York: Harper.

Hotelling, H. 1947. Letter of June 18, 1947, to Newton B. Drury. Included in the report *The Economics of Public Recreation: An Economic Study of the Monetary Evaluation of Recreation in the National Parks*, 1949. Mimeographed. Washington, D.C.: Land and Recreational Planning Division, National Park Service.

Olkin, I., S. Ghurye, W. Hoeffding, W. Madow, and H. Mann, eds. 1960. *Contributions to Probability and Statistics: Essays in Honor of Harold Hotelling*. Stanford, Calif.: Stanford University Press.

SELECTED BIBLIOGRAPHY

For a more complete bibliography, see Olkin et al. (1960).

1925

A general mathematical theory of depreciation. *J. Am. Stat. Assoc.* 20:340-353.

1926

Multiple-sheeted spaces and manifolds of states of motion. *Trans. Am. Math. Soc.* 28:479-490.

1927

An application of analysis situs to statistics. *Bull. Am. Math. Soc.* 33:467-476.

1929

With H. Working. Applications of the theory of error to the interpretation of trends. *J. Am. Stat. Assoc.* 24(Mar. suppl.):73-85.

Stability in competition. *Econ. J.* 39:41-57.

1930

British statistics and statisticians today. *J. Am. Stat. Assoc.* 25:186-190.

The consistency and ultimate distribution of optimum statistics. *Trans. Am. Math. Soc.* 32:847-859.

1931

Recent improvements in statistical inference. *J. Am. Stat. Assoc.* 28(Mar. suppl.):79-89.

The economics of exhaustible resources. *J. Polit. Econ.* 89:137:175.

The generalization of students' ratio. *Ann. Math. Stat.* 2:360-378.

1932

Edgeworth's taxation paradox and the nature of demand and supply functions. *J. Polit. Econ.* 40:577-616.

1933

Analysis of a complex of statistical variables into principal components. *J. Educ. Psychol.* 24:417-441, 498-520.

1935

Demand functions with limited budgets. *Econometrica* 3:66-78.

1936

With M. R. Pabst. Rank correlation and tests of significance involving no assumption of normality. *Ann. Math. Stat.* 7:29-43.

Relations between two sets of variates. *Biometrika* 28:321-377.

Curtailing production is anti-social. *Columbia Alumni News* 28(4):3, 16.

1938

With L. R. Frankel. The transformation of statistics to simplify their distribution. *Ann. Math. Stat.* 9:87-96.

The general welfare in relation to problems of taxation and of railway and utility rates. (Presidential address to the Econometric society.) *Econometrica* 6:242-269.

1939

Tubes and spheres in *n*-spaces and a class of statistical problems. *Am. J. Math.* 61:440-460.

The relation of prices to marginal cost in an optimum system. *Econometrica* 7:151-155.

1940

The selection of variates for use in prediction with some comments on the general problem of nuisance parameters. *Ann. Math. Stat.* 11:271-283.

The teaching of statistics. *Ann. Math. Stat.* 11:457-470.

1941

Experimental determination of the maximum of a function. *Ann. Math. Stat.* 12:20-45.

1942

Problems of prediction. *Am. J. Sociol.* 48:61-76.

1943

Income-tax revision as proposed by Irving Fisher. *Econometrica* 11:83-87.

1944

Some improvements in weighing and other experimental techniques. *Ann. Math. Stat.* 15:297-306.

1951

A generalized *T* test and measure of multivariate dispersion. In *Proceedings of the Second Berkeley Symposium on Mathematical Statistics and Probability*, ed. J. Newman, pp. 23-42. Berkeley: University of California Press.

1953

New light on the correlation coefficient and its transform. *J. R. Stat. Soc. B* 15:193-225 (with discussion, pp. 225-232).

1957

The relations of the newer multivariate statistical methods to factor analysis. *Brit. J. Stat. Psychol.* 10(2):69-79.

Part XI
Historical Work

Erich L. Lehmann's Work on the History of Classical Statistics

Javier Rojo

The papers in this chapter present a fascinating collection of insights and historical facts about the gestation and early periods of classical statistics. Erich, having played a very important role in these developments, is able to masterfully interweave his memoirs with information obtained trough careful readings of the correspondence between some of the other major players in these developments. Although all the papers are brought together by common themes, they can roughly be classified into two groups. One group of papers go to the root of some of the important breakthrough ideas and concepts in classical statistics; their origin, evolution, and impact. In the second group of papers, Erich in a somewhat more leisurely style, describes the personalities and the history of events at the end of the 19th, and beginning of the 20th, century that led to the development of statistics.

The earliest paper in the first group, Lehmann (1958), is being published here for the first time. It discusses the interpretation of confidence statements beginning with Lapace (1812). It is here that Erich found the first expression that may be interpreted as a confidence statement:

$$P\left\{\frac{-T\sqrt{2i(n-1)}}{n\sqrt{n}} \leq p \leq \frac{T\sqrt{2i(n-1)}}{n\sqrt{n}}\right\} = \frac{2}{\sqrt{\pi}}\int_0^T e^{-t^2}dt, \qquad (1)$$

where i, in Laplace's notation, indicates the number of successes in n independent and identical Bernoulli trials with probability of success p. Erich includes the original paragraphs in French and states that, although Laplace did not provide an interpretation for (1), the use of this expression by him and others indicated that the interpretation was that of p falling in the interval defined by the fixed left and right limits, and with the probability given, in (1). Furthermore, Lehmann (1958) states, Poisson incorporated (1) into his own work, Poisson (1837), and developed a fiducial distribution for p.

J. Rojo
Professor of Statistics, Department of Statistics, Rice University, Houston, TX 77005
e-mail: jrojo@rice.edu

J. Rojo (ed.), *Selected Works of E. L. Lehmann*, Selected Works in Probability and Statistics, DOI 10.1007/978-1-4614-1412-4_75, © Springer Science+Business Media, LLC 2012

Lehmann (1958) presents Poisson's original discussion in French, with some changes in notation and some minor annotations. Several other authors discussed (1), but it is not until 1927 that Erich discovered the first proper (frequentist) interpretation of (1) by E. B. Wilson.

Extensions to the estimation of the difference of two binomial probabilities are also discussed and Erich concludes the paper with the following paragraph:

> *It is the achievement of Laplace and Gauss, and of their followers throughout the* 19th *century, to have recognized the importance of obtaining probability limits for unknown parameters. They also developed a method for determining such limits, which is free of a priori assumptions regarding these parameters: The inversion of direct probabilities statements based on large sample approximations. However, it appears to have been understood only in the present century that in the resulting statements the parameter is constant and the probability refers to the variation of the limits.*

Lehmann (1990) addresses the question: *Where do models come from?* Examining what Fisher and Neyman had to say in this regard, Erich concludes that Fisher did not envision a general theory of model selection, preferring to leave this task to others ("the practical statisticians"), and adds that Fisher later offered some general advice: (i) use those models that are familiar ("and know how to handle") while (ii) understanding that the parsimony of the selected model will depend on the "volume of data". By contrast, Neyman had a constant interest in the theory of modeling. Among the important Neyman's contributions to this area, Erich writes, was his recognition of the existence of two classes of models: "explanatory" and "interpolating". The former type needs to be informed by the available knowledge about the scientific question at hand, while the latter type is more concerned with selecting models that provide a good fit to the data. Addressing objections from the applied statistician camp that stated "that they know exactly where their models come from", Erich asks the provocative question:

> *Is applied statistics, and more particularly model building, an art with each new case having to be treated from scratch (although even artistic endeavors require techniques which can be systematized and learned), completely on its own merits, or does theory have a contribution to make to this process?*

Erich then proceeds to discuss three general ways in which theory can contribute in this regard: (i) by providing a "reservoir of models"; (ii) by developing techniques for choosing a model from a restricted class; (iii) by classifying models, e.g. explanatory vs. empirical, that can lead to a better understanding of its applicability. The paper ends with some comments on areas that, it is stated, required further work. For example, the assumption of independence. It is also suggested that ideas external to statistics, e. g. from artificial intelligence, may contribute to model selection.

In Lehmann (1993a), Erich examines the question of who should be credited for inspiring the development of the Neyman-Pearson theory. Through his incisive writing, Erich takes the reader to the end of the 19th, and beginning of the 20th, century to a debate between Bertrand and Borel on a probabilistic approach to testing. At play, are the conflicting versions of Neyman and Pearson as to the source of their inspiration. It is pointed out by Erich that, while Neyman credited Borel's debate with Bertrand, and in particular the thoughts expressed by Borel, for his inspiration, Pearson credited

Gosset. In trying to resolve the issue, Erich points out that a letter from Gosset to Pearson included a *crucial suggestion:*

> *If there is an alternative hypothesis which will explain the occurrence of the sample with more reasonable probability ... you will be very much more inclined to consider that the original hypothesis is not true.*

After a careful review of the correspondence from Neyman to Pearson, Erich concludes:

> *... there appears to be no mention of Borel in this or other letters of Neyman's to Pearson during this period. This combination of circumstances lends some support to Pearson's recollections.*

The motivation behind Lehmann (1999) is the following statement by F. N. David, as discussed in Laird (1989), that Erich found very intriguing:

> *I think he [Gosset] was really the big influence in statistics .... He asked the questions and Pearson and Fisher put them into statistical language and then Neyman came to work with the mathematics. But I think most of it stems from Gosset.*

Erich then goes on to state "The aim of this paper is to consider to what extent David's conclusion is justified." The paper examines three different aspects of Gosset's contributions to small-sample methodology: (i) Distributional problems, (ii) Robustness against departures from normality, (iii) The choice of a test. Through the journey of Erich's analysis of the correspondence of Fisher, Gosset, and Pearson, a completely new image of Gosset comes alive. Although Gosset, by his own admission was unable to solve some important problems due to his lack of mathematical preparation, he is now seen in his new role of pointing both Fisher and Pearson in the direction of important statistical issues that required attention. Gosset is not afraid to push Fisher on these issues and does not back down even when Fisher pushes back.

Erich concludes by agreeing with F. N. David's assessment regarding Pearson. Regarding Fisher, although Gosset exerted influence on some of Fisher's work, Fisher's contributions were numerous and many did not benefit from Gosset's insights. Thus F. N. David's assessment does not apply to Fisher. Finally, Neyman, through Pearson – see Lehmann (1993a), was influenced by Gosset in the development of the Neyman-Pearson theory, although as discussed earlier, Neyman credited Borel for his inspiration.

In Lehmann (2008a), Erich undertakes a thorough critique of the underlying assumptions of the linear model. The three aspects of the model that are the subjects of this scrutiny are the assumptions of: (i) normality; (ii) independence; and (iii) the linear structure in the model. Whereas assumption (i) has received more than its share of attention, (ii) and (iii) have not. Furthermore, Erich writes, the types of data that arise in the various areas of research, (e.g. physical, biological, and social sciences), are quite different and present different challenges. As the paper was published in a volume in honor of David A. Freedman, Erich concludes with a discussion of Freedman's strong criticism of the misuse of the linear regression methodology.

In the last paper in the first group, Lehmann (2009), Erich examines the history of optimality. From Laplace's and Gauss' work on best linear estimates of parameters, through Fisher's work on maximum likelihood estimation and the Neyman-Pearson developments, Wald's decision theory, and the Hunt-Stein theorem and its applications, Erich's clear and concise narrative summarizes most of the important developments in optimality. After discussing some extensions of the Neyman-Pearson theory to – for example – sequential analysis, robustness, and multiple testing, and after briefly discussing Kiefer's work on optimal experimental design, Erich briefly touches on some criticisms of optimality (e.g. Tukey's). The paper concludes by stating that, as it regards to the state of optimality today,

> *Ad hoc solutions are proposed and tried out on a few examples. This is a natural first step, but eventually we will want to justify the solutions at which we have arrived intuitively and by trial and error. A theoretical underpinning will be provided and the conditions will be found under which these solutions are the best possible.*

The second group of papers includes a brief account of young Erich's immediate family and the tragic circumstances that led the family to leave Germany and settle in Switzerland. After this passage, in a conversation masterfully orchestrated by DeGrot (1986), Erich provides a more leisurely expedition into his career that started in Switzerland after his father suggested mathematics as a field of study instead of German literature which is what Erich felt a passion for. After five years in Switzerland, Erich attended Trinity College in Cambridge for two years and then went to New York where he was ushered by Courant to Berkeley where he spent close to sixty eight years. Together with Lehmann (1993b), Lehmann (1996), and Lehmann (2008b), these fascinating accounts of Erich's professional career immerse the reader in the developments of statistics in The United States during the 20th, and early part of the 21st, century. The portraits of the major personalities in these developments and in Erich's career come to life through Erich's pen. The origin and development of the Statistics Department in Berkeley is wonderfully chronicled as is the broader picture of the development of statistics. In addition, Lehmann (1997) succeeds in conveying well the excitement, success and tribulations of being a successful textbook writer. Finally, Lehmann (2004) which appeared in a volume of papers presented at the inaugural Lehmann Symposium, discusses the connection between optimality and symposia embodied in the person of Jerzy Neyman, who was the originator of the Berkeley Symposia and played a major role in the development of optimality theory. This paper contains a facsimile of a letter from Neyman to Pearson that indicates that Neyman had not, at that point, yet reached the final solution to the problem. An excerpt of a letter dated two weeks later, however, confirms that the solution is finally, for all practical purposes, at hand. The origin, life, strong legacy, and death of the Berkeley Symposium is also discussed. The paper ends with Erich's blessing of the "new" Lehmann Symposium:

> *We may consider Neyman to be the godparent of this symposium, both through his founding of optimality theory and the inspiration provided by his symposia.*

The fourth Lehmann Symposium recently took place at Rice University during May of 2011. Plans are currently under way for a Lehmann Centennial in 2017.

References

[1] M. H. DeGroot. A Conversation with Erich L. Lehmann. *Statist. Sci.*, Vol. I, 243-258, 1986.

[2] P. S. Laplace. *Théorie analytique des probabilités*. Paris. 1812.

[3] E. L. Lehmann. *Some Early Instances of Confidence Statements*. Statistical Laboratory, University of California, Berkeley. ONR 5 Technical Report to the Office of Naval Research. Unpublished. September 1958.

[4] E. L. Lehmann. Model specification: The views of Fisher and Neyman, and later developments. *Statist. Sci.*, Vol. 5, pp. 160-168, 1990.

[5] E. L. Lehmann. The Bertrand-Borel Debate and the origins of the Neyman-Pearson theory. In *Statistics and Probability. A Raghu Raj Bahadur Festschrift*, (Ghosh *et al* Eds.), pp. 371-380. Wiley Eastern Ltd. New Delhi, 1993a.

[6] E. L. Lehmann. Mentors and early collaborators: reminiscences from the years 1940-1956 with an epilogue. *Statist. Sci.*, Vol. 8, pp. 331-341, 1993b.

[7] E. L. Lehmann. The creation and early history of the Berkeley statistics department. In *Festschrift for David Blackwell*, IMS Lecture Notes - Monograph Series, Vol. 30, (T. Ferguson, Ed.), pp. 139-146, 1996.

[8] E. L. Lehmann. Testing statistical hypotheses: The story of a book. *Statist. Sci.* Vol. 12, pp. 48-52, 1997.

[9] E. L. Lehmann. 'Student' and small-sample theory. *Statist. Sci.*, Vol. 14, pp. 418-426, 1999.

[10] E. L. Lehmann. Optimality and symposia: Some history. In *The First Erich L. Lehmann Symposium - Optimality*. (V. Perez-Abreu and J. Rojo, Eds.), IMS Lecture Notes-Monograph Series, Vol. 44, pp. 1-10, 2004.

[11] E. L. Lehmann. On the history and use of some standard statistical models. In *Probability and Statistics: Essays in Honor of David A. Freedman*, (D. Nolan and T. Speed, Eds), IMS Collections Vol. 2, pp. 114-126, 2008a.

[12] E. L. Lehmann. *Reminiscences of a Statistician: The Company I Kept*, Springer, 2008b.

[13] E. L. Lehmann. Some history of optimality. In *The Third Erich L. Lehmann Symposium – Optimality*, (J. Rojo, Ed.) IMS Lecture Notes-Monograph Series, Vol. 57, pp. 10-17, 2009.

[14] S. D. Poisson. *Recherches sur la probabilité des jugements*. Paris. 1837.

[15] E. B. Wilson. Probable inference, the law of succession, and statistical inference. *J. Amer. Statist. Assoc.*, Vol 27, pp. 209-212, 1927.

SOME EARLY INSTANCES OF CONFIDENCE STATEMENTS

By E. L. Lehmann

Statistical Laboratory, University of California, Berkeley,
ONR 5 Technical Report to the Office of Naval Research
prepared under Contract Nonr-222(43) (NR 042-036)

September, 1958

SOME EARLY INSTANCES OF CONFIDENCE STATEMENTS

By E. L. Lehmann[1]

University of California, Berkeley

A general theory for obtaining limits within which an unknown parameter can be asserted to lie with known probability was developed by Fisher and Neyman in the 1930's.[2] The basic device through which such limits are obtained consists in "solving" a probability statement which involves an unknown parameter θ, for this parameter. The resulting statement concerning θ is then equivalent to the original statement and thus holds with the same known probability.

The above inversion procedure seems to have been employed for the first time by Laplace. In his Théorie Analytique des Probabilités (p. 287),[3] he applies it to the approximate equation

$$P\left\{\left|\frac{i}{n} - p\right| \leqq T\sqrt{2p(1-p)/n}\right\} = \frac{2}{\sqrt{\pi}} \int_0^T e^{-t^2} dt \tag{1}$$

$$+ \text{ a correction term of order } n^{-1/2}$$

where i/n denotes the frequency of success in n binomial trials with probability p of success. Replacing $p(1-p)$ by $i(n-i)/n^2$ and solving for p, he thus obtains the approximate equation

$$P\left\{- \frac{T\sqrt{2i(n-i)}}{n\sqrt{n}} \leqq p \leqq \frac{T\sqrt{2i(n-i)}}{n\sqrt{n}}\right\} = \frac{2}{\sqrt{\pi}} \int_0^T e^{-t^2} dt. \tag{2}$$

This appears to be the first instance of what, at least formally, can be considered a confidence statement.

Since the passage is short, it may be worth giving it

verbatim; however, in the formulae I shall suppress all terms of order less than that of the leading term; in particular, I have replaced Laplace's expression x, which is equal to np plus a term less than 1, by np:

Si l'on connaît le nombre de fois que sur n coups l'événement a[4] est arrivé, la formule (o)[5] donnera la probabilité que sa facilité p, supposée inconnue, sera comprise dans les limites données. En effet si l'on nomme i ce nombre de fois, on aura, par ce qui précède, la probabilité que la différence $\frac{i}{n} - p$ sera comprise dans les limites $\pm T\sqrt{2p(1-p)}/n\sqrt{n}$; par conséquent, on aura la probabilité que p sera compris dans les limites

$$\frac{i}{n} \mp \frac{T\sqrt{2p(1-p)}}{n\sqrt{n}}.$$

La fonction $T\sqrt{2p(1-p)}/n\sqrt{n}$ étant de l'ordre $1/\sqrt{n}$, on peut, en négligeant les quantités de l'ordre 1/n, y substituer i au lieu de np et n-i au lieu de n(1-p); les limites précédentes deviennent ainsi, en négligeant les termes de l'ordre 1/n,

$$\frac{i}{n} \mp \frac{T\sqrt{2i(n-i)}}{n\sqrt{n}}$$

et la probabilité que la facilité de l'événement a est contenue dans cas limites est égale à

$$\frac{2}{\sqrt{\pi}} \int dt\, e^{-t^2}.$$

On voit ainsi que, à mesure que les événements se multiplient, l'intervalle des limites se resserre de plus en plus, et la probabilité que la valeur de p tombe dans ces limites approche de plus en plus de l'unité ou de la certitude. C'est ainsi que les événements, en se developpant, font connaître leurs probabilités respectives.

Laplace does not comment directly on the interpretation of the probability statement (2). However, from the way this and related results are applied it seems clear that both he and his successors thought of p as the variable which with the stated probability falls in the fixed interval $\pm Ti(n-i)/n\sqrt{n}$. This is indicated also by Laplace's alternative derivation of essentially the same formula (pp. 377/378). In this second derivation he uses a Bayesian argument, assuming a uniform a priori distribution for p, and he clearly considers the results obtained by these two methods as completely equivalent.

The first derivation of (2) was taken over by Poisson into his Recherches sur la Probabilité des Jugements (1837). It is interesting that Poisson utilizes the result to obtain a fiducial distribution for p (p. 211). He gives the one-sided limits for $q = 1 - p$ corresponding to (2). Treating the totality of these limits as a cumulative distribution function, he differentiates it to obtain a formal probability density for q. In quoting the passage, I have changed the notation to make it correspond to that of Laplace:

La quantité Q, déterminée par la première formule (18) [that is, given up to terms of first order by $(1/\sqrt{\pi})\int e^{-t^2}dt$], étant la probabilité que le nombre $n-i$ est inférieur a $nq - T\sqrt{2npq}$, elle est également la probabilité que la chance inconnue q de l'événement F, arrivé $n-i$ fois dans n epreuves, est supérieure à $(n-i)/n + T\sqrt{2pq/n}$, ou bien à $(n-i)/n + (T/n)\sqrt{2i(n-i)/n}$, en substituant dans le second terme de cette limite, à la place de p et q, leur valeurs approchées i/n et $(n-i)/n$. Si l'on met $T-dT$ au lieu de T dans cette formule, et que l'on conserve seulement les inifiment petits du premier ordre, $Q - \frac{dQ}{dT}\,dT$ sera donc aussi la probabilite que q surpasse

$$\frac{i}{n} + \frac{T}{n}\sqrt{\frac{2i(n-i)}{n}} - \sqrt{\frac{2i(n-i)}{n}}\,\frac{dT}{n}\ ;$$

par conséquent, $-dQ/dT$ exprimera la probabilité infiniment petite que l'on a précisément

$$q = \frac{i}{n} + \frac{T}{n}\sqrt{\frac{2i(n-i)}{n}}$$

pour toutes las valeurs de T positives et très petites par rapport à $\sqrt{n}$, comme le suppose l'expression de Q.

Poisson's work was the basis for certain other writers, particularly Gavarret (1840) and Lexis (1875), who quote his statement of (2) which is obtained by Laplace's first method. However, generally Laplace's second (Bayesian) derivation was preferred as being on a firmer foundation. In fact, the first derivation was attacked by De Morgan (1837) as illegitimate; an attempt was made by Monro (1874) to legitimize it by showing it to be equivalent to the Bayesian derivation. Todhunter discusses both derivations in his History of the Theory of Probability

(1865, pp. 554-558). His own attitude is indicated by his remark: "As Poisson in his latest discussion of the question adopted the inversion of James Bernoulli's theorem,[6] we may perhaps infer that he considered the amount of assumption thus involved to be no greater than that which is required in the use of Bayes' theorem." Both derivations are also given, without comment regarding any differences between them, by Czuber in his report on probability theory to the Deutsche Mathematiker Vereinigung in 1899 (pp. 82-84) and again in the third edition of his textbook on the theory of probability (1914, pp. 203-206).

For the first time as far as I have been able to determine, the proper interpretation of (2) is given by E. B. Wilson in 1927.[7] He describes two procedures as the usual methods for assigning limits to a binomial probability p, one based on Tchebycheff's inequality, the other equivalent to (2). He then criticises this approach in the following passage in which p_0 stands for the frequency previously denoted by i/n: "Strictly speaking, the usual statement of probable inference as given above is elliptical. Really, the chance that the true probability p lies outside a specified range is either 0 or 1; for p actually lies within that range or does not. It is the observed rate p_0 which has a greater or less chance of lying within a certain interval of the true rate p. If the observer has the hard luck to have observed a relatively rare event and to have based his inference thereon, he may be fairly wide of the mark."

On the basis of Poisson's work, an extension of (2) was given by Lexis (1875, p. 106) to the case of two binomial probabilities.

After first testing the hypothesis that the two probabilities p and p' are equal, he wishes to estimate the difference between these probabilities. With

$$F_u = \frac{2}{\sqrt{\pi}} \int_0^u e^{-t^2} dt,$$

he obtains the approximate probability $F_3 = 99998$ that the difference between the frequency of successes i/n in the first and i'/n' in the second sequence of trials falls in the interval

$$p - p' \pm 3\sqrt{\frac{2i(n-i)}{n^3} + \frac{2i'(n'-i'}{n'^3}}$$

and hence that $p-p'$ falls in the interval

$$\frac{i}{n} - \frac{i'}{n'} \pm 3\sqrt{\frac{2i(n-i)}{n^3} \quad \frac{2i'(n'-i'}{n'^3}} .$$

There is another line of development, which utilizes the inversion of probability statements to obtain special cases of formally correct confidence intervals. In the theory of errors of observations, probable errors are frequently used by 19th century authors to obtain intervals within which the true value of a parameter lies with probability 1/2, that is confidence intervals with confidence coefficient 1/2. This application is made by Gauss in the first publication (1816) on the subject, in a paper concerned with the "determination of the accuracy of observations." It is clear that he suggests the value 1/2 purely as a convention; at the beginning of the paper, he gives the limits which a normally distributed error will not exceed with probability 1/2, but also those corresponding to the probabilities 1/10 and 1/100.

In a similar context, an even more explicit statement is given by Fourier (1826). Having previously discussed the concept of probable error, and having defined the quantities $A = (a+b+\cdots+n)/m$, $B = (a^2+b^2+\cdots+n^2)/m$, $g = \sqrt{2(B-A^2/m}$, where $a, b, \cdots, n$ are m observations of the "true" quantity H, he writes (Section XI):

"Pour compléter cette discussion, il faut déterminer quelle probabilité il y a que la quantité cherchée H est comprise entre des limites proposees A + D et A - D. A est le resultat moyen qu'on a trouve, H est la valeur fixe que donnerait un nombre infini d'observations, et D est une quantité proposée que l'on ajoute a la valeur A ou que l'on retranche. La Table suivante fait connaître la probabilité P d'une erreur positive ou negative plus grande que D, et cette quantite D est le produit de g par un facteur proposee ∂

∂	P
0.47708	1/2
1.38591	1/20
1.98495	1/200
2.46130	1/2000
2.86783	1/20000

Chacun des nombres de la colonne P fait connaître quelle probabilité il y a que la valeur exacte H, qui est l'objet de la recherche, est comprise entre les limites $A+g\partial$ et $A-g\partial$. ..."[10]

Although most later authors restrict their attention to the conventional value $\frac{1}{2}$, they sometimes mention in passing that the methods apply equally well to other values. Thus for example,

Jevons in his Principles of Science writes (2nd edition, 1883, p. 387):

"The <u>probable error</u> is taken by mathematicians to mean the limits within which it is as likely as not that the truth will fall. ... Any other limits might have been selected at will. We might calculate the limits within which it was one hundred or one thousand to one that the truth would fall; but there is a convention to take the even odds one to one, as the quantity of probability of which the limits are to be estimated."

The connection between the idea of probable error and statements of the type (2) is recognized by Lexis who advocates the use of a higher probability of covering the true value thereby coming closer to the modern point of view. In his discussion of (2) he writes (1875, pp. 99/100):[9]

...so ist mit der Wahrscheinlichkeit F_u die Unbekannte p hoechstens um $\pm T/\sqrt{n}\sqrt{2i(n-i)}$ von i/n verschieden, oder es besteht mit der bezeichneten Wahrscheinlichkeit die Grenzgleichung

$$p = \frac{i}{n} (\pm \frac{T}{\sqrt{n}} \sqrt{2i(n-i)}$$

wenn wir durch die einseitige Klammer andeuten, dass p irgend einen Wert in den angegebenen Grenzen haben kann. Setzt man z.B. $u = .476936$, so ist $F_u = 1/2$, und man kann also 1 gegen 1 wetten dass die Abweichung der empirischen Groesse i/n von p hoechstens

$$\pm \frac{.476936}{\sqrt{n}} \sqrt{2i(n-i)}$$

betraegt. Diese letztere Groesse wuerde die sogennante

"wahrscheinliche" Abweichung des Beobachtungsresultates darstellen. Fuer die Zwecke der Statistik duerfte es jedoch gaeigneter sein, fer u eine so grosse Zahl zu nehmen, dass F_u der Einheit sehr nahe kommt, also eine Wahrscheinlichkeit ausdrueckt, die in der Praxis der Gewissheit gleich geachtet werden kann.[10] Es genuegt, wie oben, u gleich 3 zu setzen, und man erhaalt aladann die Wahrscheinlichkeit $F_3 = .999978$ fuer die Grenzgleichung:

$$p = \frac{i}{n} \left(\pm \frac{3}{\sqrt{n}} \sqrt{2i(n-i)} \right).$$

Die mit dieser Wahrscheinlichkeit zulaessige positive und negative Abweichung koennte man als die "aeusserste" Abweichung bezeichnen."

It is the achievement of Laplace and Gauss, and of their followers throughout the 19th century, to have recognized the importance of obtaining probability limits for unknown parameters. They also developed a method for determining such limits, which is free of *a priori* assumptions regarding these parameters: The inversion of direct probability statements based on large sample approximations. However, it appears to have been understood only in the present century that in the resulting statements the parameter is constant and the probability refers to the variation of the limits.

REFERENCES

Czuber, Emanuel (1899). Jahresbericht der Deutschen Mathematiker vereinigung, **7**, No. 2.

Czuber, Emanuel (1914). Wahrscheinlichkeitsrechung. Dritte Auflage. Leipzig.

Fourier, J. B. (1826). Recherches statistiques sur la ville de Paris et le departement de la Seine, **3**.

Gauss, C. F. (1816). Zeitschrift fuer Astronomie und verwandte Wissenschaften, **1**, 185. (Reprinted in Gauss: Werke, **4**, 109-119, Goettingen, 1873).

De Morgan, A. (1837). Cambridge Philos. Trans., **6**, 422-

Hotelling, Harold (1931). Ann. Math. Stat., **2**, 360.

Jevons, W. Stanley (1873). The principles of science. MacMillan, London.

Laplace, P. S. (1812). Théorie analytique des probabilités. Paris. (The 3rd edition of 1820 is reprinted as Vol. 7 of Laplace's collected works.)

Lexis, W. (1875). Einleitung in die Theorie der Bevoelkerungs-statistik. Strassburg.

Monro, C. J. (1874). Proc. London Math. Soc., **5**, 74.

Neyman, J. (1952). Lectures and conferences on mathematical statistics and probability. 2nd edition. Grad. School U.S. Dept. of Agriculture, Washington.

Poisson, S. D. (1837). Recherches sur la probabilite des jugements. Paris.

Todhunter, I. (1865). A history of the mathematical theory of probability. MacMillan. Cambridge.

Wilson, E. B. (1927). J. Am. Stat. Assoc., **27**, 209.

Working, Halbrook and Hotelling, Harold (1929). J. Am. Stat. Assoc. Suppl. **24**, 73.

Translation of the Lexis passage

...then with probability F_u, the unknown quantity p differs from i/n by at most $\pm T/\sqrt{n}\sqrt{2i(n-i)}$, or there holds with this probability the relation $p = i/n\ (\pm T/\sqrt{n}\sqrt{2i(n-i)}$ if we indicate by a one-sided parenthesis that p can take on any value within the stated limits. If for example u = .476936 then $F_u = 1/2$ and one can bet 1 against 1 that the deviation of the observed quantity i/n differs from p by at most $\pm .476936\sqrt{2i(n-i)/n}$. This latter quantity would represent the so-called "probable" deviation of the observed frequency. For the purposes of statistics it would seem to be more appropriate to take for u a number so large that F_u comes very close to unity, and thus represents a probability, which for all practical purposes can be considered equal to certainty. It is sufficient to put, as above, u equal to 3, and one then obtains the probability $F_3 = .999978$ for the relation: $P = i/n\ (\pm 3\sqrt{2i(u-i)/n}$ The positive and negative deviation which is permitted with this probability could be called the "extreme" deviation.

FOOTNOTES

[1]This paper was prepared with the partial support of the Office of Naval Research (Nonr-222-43). This paper in whole or in part may be reproduced for any purpose of the United States Government.

[2]Special cases were treated by Working and Hotelling (1929) and by Hotelling (1931).

[3]The page references here are to the 3rd edition of 1820.

[4]The event whose probability is p.

[5]Essentially our formula (1).

[6]That is, the first derivation.

[7]Attention was recently drawn to this paper by J. Neyman (1952, p. 222).

[8]I have here changed the notation so that it agrees with that used above.

[9]The notation has again been changed to agree with that used earlier.

[10]A similar definition is given by Fourier who attaches to the quantity $3g$ the name "la plus grande erreur."

Statistical Science
1986, Vol. 1, No. 2, 243–258

A Conversation with Erich L. Lehmann

Morris H. DeGroot

Erich L. Lehmann was born in Strasbourg, France, on November 20, 1917, and came to the United States in 1940. He received an M.A. in mathematics from the University of California, Berkeley, in 1942 and a Ph.D. in mathematics from the same university in 1946. He has been a faculty member there since that time. He was the Editor of *The Annals of Mathematical Statistics* from 1953–1955, and President of the Institute of Mathematical Statistics in 1961. He was awarded a Guggenheim Fellowship in 1955, 1966, and 1980. He was elected to membership in the American Academy of Arts and Sciences in 1975 and to membership in the National Academy of Sciences in 1978. In February 1985 he received an honorary doctorate from the University of Leiden.

The following conversation took place in his home in Berkeley one afternoon in October 1984.

"THE ONE THING IN WHICH I WAS REALLY INTERESTED WAS GERMAN LITERATURE"

DeGroot: Tell me how you got interested in statistics and how you came to the field of statistics.

Lehmann: It's a longish story because I have to start way back. I was raised in Germany and I am of Jewish ancestry. In 1933 when the Nazis came into power, my father decided that we had better leave. That was very early April 1933. After some wandering, because I had a brother who was dying and needed to be cared for first, we settled in Switzerland and I went to high school there. At the beginning of my senior year my father asked me "What do you plan to do after you graduate from high school?" The answer was totally obvious. He knew what I was going to say because the one thing in which I was really interested was German literature. Poetry, novels, short stories—I was reading, I was writing. You know the way it is at 17 or 18. That was what excited me, and that was what I wanted to do with my life. So I said, "Well, of course, study German literature." And he said, "I don't know if that's such a good idea because you can't go to Germany, and studying German literature outside of Germany doesn't make very much sense. You've always had an interest in mathematics. Why don't you study mathematics." This might seem hard for somebody raised in this country to believe, but I said, "Well, if you think, so, o.k." [Laughs] It didn't even cause me terrible pangs. I don't know whether I really believed it or not, but anyway I started supposedly studying mathematics in Zurich.

But you know how European universities are. There are no exams or anything, and nobody controls what you do. So I actually went to lectures in literature and music and history and things like that. I occasionally went to one of the calculus courses, but I really didn't do very much. After a year and a half or two years of that, it looked very much like war. This was 1938, and my father was worried. There was a considerable possibility that the Germans would march through Switzerland, which was not a very comfortable position to be in. So he suggested that I go to England. I got enrolled in Trinity College, Cambridge, and there I really had to start studying more seriously. But there was another very bad obstacle. In those days in Cambridge, when you studied mathematics you also had to study physics and astronomy in at least 50% of your program. That was what the schedule covered.

DeGroot: There was no combination of mathematics with other disciplines?

Lehmann: No, there was a fixed curriculum and that's what you took. Physics has always been my worst subject in school. I *hated* it. I had the same teacher for mathematics and physics and he said, "You know, if I didn't see the results of your exams, I wouldn't believe it." I was always the best student in mathematics and I was among the very worst students in physics. Therefore, at Cambridge they told me that I'd better start concentrating on the physics because the mathematics I could sort of take on the wing. The result was I didn't do terribly well in either. I spent all my time studying something that I did not enjoy at all.

DeGroot: Was it the physics laboratory that you didn't like?

Lehmann: No, it was the mechanics and electricity and that kind of stuff. I guess they didn't call it physics, they called it applied mathematics. I really disliked it intensely. And then somehow at one point

J. Rojo (ed.), *Selected Works of E. L. Lehmann*, Selected Works in Probability and Statistics, DOI 10.1007/978-1-4614-1412-4_77, © Springer Science+Business Media, LLC 2012

in 1940 the idea of coming to the United States came up. By then the war had started in Europe. For some reason I had never thought about coming to the United States as a serious possibility, partly because I hadn't realized that I could come in on the French quota.

DeGroot: How did that happen?

Lehmann: Well, I was born during the first World War in Strasbourg. At the time, that was part of Germany and my father was stationed there as an officer on the German general staff. He was drafted, essentially, like all German males were. So I was born there, and then after 2 months we moved back to Frankfurt where my family had lived before. But because of the fact that Strasbourg is in France today, I was able to come in on the French quota. So in 1940 I came to this country, and I arrived in New York. The only two universities I had ever heard of in America were Harvard and Princeton, but I had a letter of recommendation to Courant.

DeGroot: Had you applied to any of these places in advance?

Lehmann: No, I hadn't applied to anything. I just came. It was very hard to get here because of visa problems and all kinds of problems. But somehow I managed. My mother's closest girlhood friend was the wife of Edmund Landau, the number theorist. She was a close friend of Courant, so when my mother asked her if she had any suggestions, she said she would drop Courant a note. So I got to see Courant and I said, "Where should I study? I want to study mathematics in this country." And he said, "First tell me. Do you want to live in New York or in the United States of America?" [Laughs] There were all these ethnic pockets and people could live as German refugees or whatever, without Americanization, if they stayed in New York. I said I had no particular interest in New York. So he said "Well, why don't you go to Berkeley, the University of California. It's up and coming. It's a young university and it will be a very good place, and I think you'll like it." An *amazing* piece of advice. I said to Courant, "Wonderful, don't tell me anything else." California. I didn't know about Northern California and Southern California. I had visions of a Garden of Eden with palm trees and figs and so on. What could be better. So I came here at the end of 1940 and started studying mathematics.

I didn't have a B.A., which was a little peculiar, and so I was a probationary graduate student. I was allowed to do this because of Evans, who was the chairman of the Math Department—a very nice man of whom I became very fond. He felt that students at European universities were about 2 years ahead of American students at the same level, because the high schools covered so much more. So he thought I could roughly have the equivalent of a B.A. and he was willing to give me this probationary graduate student appointment. After half a year, I got a teaching assistantship and everything was wonderful. And then in late '41 or late '42, he called me in and said, "You know, with this war, pure mathematics isn't very useful right now. I think you should do something that's more useful to the war effort." I said, "Well, how do I do that?" He said, "There are two possibilities. Either you go in the direction of physics—that's very useful—or you go in the direction of statistics." Well, the answer to me was clear. I was not going into physics again. I had just escaped. I didn't know anything about statistics—I had never heard of it—but if I should do one or the other of these two things . . .

This was again a push from the outside, you see; nothing from inside. If that's what I was supposed to do, all right. In German education you have this tremendous respect for authority. I *still* find it very hard to shake it off. So I said, "All right, I'll give statistics a try." He said, "Well, see Professor Neyman." I saw Neyman and he enrolled me in courses. After about a year, I decided this isn't for me. I don't like this one bit. This isn't really mathematics. This is *messy* stuff. Just writing down that normal density they were always talking about was messy; it was practically impossible to remember, it was so complicated. Number theory was what I liked, prime numbers and things like that.

DeGroot: Were these statistics courses particularly applied?

Lehmann: No, no, they were not applied. It was just the kind of mathematics that I didn't like in connection with physics in the first place. I decided that despite Evans and Neyman, I was going to get out of it again and go back and work with Tarski, who did the kind of mathematics that I really enjoyed. But it took me a little time to get up my courage to face these two rather formidable people and tell them I wasn't going to do what they wanted me to do. And before I had a chance to do that, Neyman called me in and said, "I'm in an awkward position. Dorothy Bernstein (whom Neyman had hired half a year before as a probabilist) has just come to me and said she doesn't like the kind of stuff that we do because it's not really nice mathematics. She wants to leave and I'm going to be short-handed. Would you be interested in teaching statistics?" And there were overtones, I don't know how strong, that it might develop into a more permanent position if things worked out. Here I was, a not-very-far-along graduate student—I didn't even have a thesis topic yet—a foreigner with no backer, no connections, no nothing, and somebody offers me what comes close to being a job. I said to Neyman, "That sounds wounderful."

DeGroot: How long had you been at Berkeley at that point?

Lehmann: About 2 years. So three times: First, German literature—my father says do mathematics. Second, I try to do mathematics and Evans says do something applied. The third time, I want to go back to pure mathematics and Neyman says here's a job for you. I gave up. [Laughs] One of the things that I disliked about physics originally, and about statistics when I wanted to get out of it, was the applied flavor, their connection with the real world, instead of their being this ideal abstract stuff. I always had the feeling that whatever abilities I had were more in the abstract direction. But the curious thing is that over the years, I have gotten to like the applied aspect of statistics. It hasn't gotten so far that I actually do applied work. Julie [his wife Juliet Shaffer] calls me an "armchair" applied statistician. [Laughs] But I like to think about statistics in connection with real situations, not totally in the abstract. So my career in statistics has actually worked out better than I had any reason to expect. I think you will find generally that in my generation everybody came to statistics in a peculiar way, because the subject didn't really exist.

"I HAD A TRIPLE OF Ph.D. SUPERVISORS"

DeGroot: Did your study of statistics prove to be helpful in the war effort?

Lehmann: No, not at all. First, Neyman put me on a calculator, a Monroe or Friden, that sort of machine. They worked on very complicated problems and I hadn't the faintest notion of what it was all about. I must have totally repressed that. I just don't remember anything about it. It has always been the kind of thing that I disliked more than practically anything. I was very inaccurate, I would make mistakes, and I can't imagine that I could have turned up anything that was of the slightest use. It didn't last very long because one day again I was called in and was told that they were forming operations research groups and they needed somebody, and they thought maybe I would be interested. This was again one of those times where you really couldn't say no.

I had no idea what was involved. I couldn't imagine that it could involve much statistics because I didn't know any. I had had a year or a year-and-a-half course, or something like that. But I did go. I had no idea where we were going until we got to San Francisco, and there we were told we were going to Guam. I was not sent to the European theater of operations because of my European background. They were worried that I might have divided sympathies. How they could imagine that, I don't know; but these things happen. So I spent a year in Guam sharing a tent with Joe Hodges and George Nicholson.

I actually did mostly photo interpretation. What I was supposed to do, Jack Youden had started with the Eighth Air Force in England: bombing accuracy studies, particularly using a record on the accuracies to pit the various groups against each other, and so on. You know, it's hard to imagine that people who fly and are in danger of their lives daily would worry about which group has achieved the best record. But that was the kind of thing I was supposed to do. Anyway, the first requirement was to keep pretty good records. So that meant that you studied the photographs of each bombing run and plotted the impacts. Sometimes you couldn't see them, but you could extrapolate and then keep records from that. That's really what I mainly did for a year. Not activity at a very high intellectual level.

There were some statistical problems. Joe and I were once asked what was potentially a very important question, but I don't think that anything we said had any impact on it: The B29's had been flying at substantial altitudes and the bombing accuracy was not very good. The question was to determine by extrapolation what would happen in terms of the greater losses and the gain in accuracy if they flew somewhat lower. But then what they actually did was fly at an altitude that was very, very low; these conditions were so totally different, there was no possibility of extrapolating to them.

DeGroot: Did Hodges go with you from Berkeley at that time?

Lehmann: No. He had been in Berkeley, and we knew each other quite well. But he left a little before me and said he couldn't tell me where he was going; it was very secret stuff. We met in Colorado Springs at the training session. [Laughs]

DeGroot: Did you return to Berkeley after the war?

Lehmann: Yes, I returned when the Japanese war ended. I can tell you amusing stories about my military service, but I don't think that's really what you wanted to talk about. So I came back in August 1945. And then I got a thesis topic, but it didn't work out. That is to say, by the time I had the results and was about to write them up, I found out that it was all in Markov and other places. It was some aspect of the moment problem. So I had to get a new problem and Hsu gave me a problem that he had already started a little bit on, and that was a very helpful thing. Then Hsu left for North Carolina, and subsequently went back to China. Then Neyman left and went to Greece to help supervise the Greek elections. Have you read the Neyman book? [Constance Reid, *Neyman—from Life*, Springer-Verlag, New York, 1982].

DeGroot: Yes, I have.

Lehmann: Then you know about these things. And so I was left a little stranded, but he arranged for Polya to supervise me. So I had a triple of Ph.D. supervisors. The problem was given by Hsu; most of the thing was supervised by Polya at Stanford, in the sense that I went up there once a week or once every 2 weeks and told him what I had done; and then Neyman came back just in time for the final exam.

DeGroot: What was the topic of your thesis?

Harald Cramér, Erich Lehmann, and George Polya, 1951

Lehmann: What today we would call uniformly most powerful unbiased tests, although they weren't called that then—Type A regions or Type A_1 regions, that kind of thing. Polya did something for me which I have always remembered. One day he said, "Well, I think that's enough. Write it up into a nice form, and that will be your thesis." And I said, "Really? It doesn't seem to me that we have done all the interesting things." And he said "Well, that's my problem not yours. You can do the good work after you get your degree. It's enough for a degree." Since then, that's been my attitude toward my own students. And then I was in fact kept on at Berkeley, so I have been in Berkeley since 1940.

"NEYMAN GOT RUMORS THAT I WAS NOT FOLLOWING THE SCRIPT CLOSELY ENOUGH"

DeGroot: You've had Guggenheim Fellowships on three different occasions. That must be close to a record. Tell me about the projects you worked on at those times.

Lehmann: I think they were basically all trying to work on the books. [*Testing Statistical Hypotheses*, John Wiley and Sons, New York, 1959; *Theory of Point Estimation*, John Wiley and Sons, New York, 1983] The books have been sort of an interesting thing in my life. They grew out of notes that Colin Blyth took in the late 1940s. It is surprising to me today that a year after my Ph.D. I basically had the pattern that I've worked on for the last 40 years. Although of course the books today look a lot different from those notes, the basic organization and the basic approach has been kept.

DeGroot: Were those some of the first courses that you taught after your Ph.D.?

Lehmann: I think they were, yes. I think they were two summer courses. I don't know whether I also taught them as regular graduate courses. And then I got into this trouble which Constance Reid describes. Neyman got rumors that I was not following the script closely enough. It was pretty painful. I still feel that of all of Neyman's students, I probably am the one who followed his approach most closely. I developed it a little bit further, but I didn't really do anything terribly innovative. And yet he found even these small innovations so bothersome.

DeGroot: For example?

Lehmann: Well, I used invariance, and that he hadn't used. I don't know just what bothered him. We never had a real conversation about it. But he said, "I've heard that you're not doing it the way I did it. I understand you're writing a book. I want to see the manuscript." I said, "Well, I gave you the set of notes that Blyth took, and that's really the best thing I have." I considered it quite bothersome that he would ask me for the manuscript. And then he barred me from teaching the course. I never taught it again until he resigned the chairmanship.

DeGroot: This was the course in statistical theory?

Lehmann: It still is, and I still teach it quite often. I'm teaching it this semester. But it seemed to me out of character for Neyman to do that because he was a very generous man in many ways and a very liberal person. But we had lots of troubles.

DeGroot: It certainly was ironic for him to object on those grounds, given his own history with Fisher.

Lehmann: It's very strange. It went from Karl Pearson to Fisher, from Fisher to Neyman, and in a different way, from Neyman to me. I must say though that despite the fact that we had these conflicts, he was always tremendously good in furthering my career. He always got me promotions, he got me better salary increases, he got me invited to conferences, and all of that kind of thing. That's very different from Karl Pearson who tried to squash Fisher, tried to prevent him from getting a job, tried to prevent him from publishing. I don't know how generous Fisher was in these respects. So Neyman and I had our differences, but he was always very good about these things. I was very appreciative of that.

DeGroot: I gather you feel that he was a major influence on your career.

Lehmann: Well, certainly in the sense that I've worked on Neyman-Pearson theory in one aspect or another all my life.

DeGroot: Are there other people that you regard as major influences on your career?

Lehmann: Yes. When I first started teaching on the faculty, there were four of us sharing a little office with four desks. Two desks on one side and two desks on the other side forming a block with just enough space that one could walk around. That led to a lot of joint work. Joe Hodges, Evelyn Fix, Charles Stein, and I. So I got to learn a lot from Charles about invariance among other things. We wrote quite a number of papers together. He was a very important influence on me. You know, all the people with whom you collaborate are important influences. Joe Hodges, with whom I collaborated for a long, long time, was important. We wrote many papers together and a book; I also collaborated with Henry Scheffé, and learned a lot from both of them. But Neyman really set me on the course, as I said earlier. In a way I rebelled against it, but in a way I also accepted it because I really stayed pretty much on the course.

Here's another example of how difficult Neyman found it to accept something that was a little different. The first thing I wrote was as a graduate student, and there was a little note which is in the *Annals* on what is today called minimal complete families. ["On families of admissible tests," *Ann. Math. Statist.* **18** (1947), 97–105] I set out the concept and applied it to a very simple hypothesis-testing situation. I wrote it up—it was just a few pages—and said to Neyman that I would like to publish it. He essentially said, "It's junk. Don't bother." But I sent it in to Wilks anyway. [Laughs] And this essentially gave me the friendship of Wald and Wolfowitz, because Wald seized on this and it became one of the two fundamental concepts of his decision theory. Minimax is one, and complete families is the other one.

DeGroot: So your note really introduced the concept of minimal complete families?

Lehmann: Yes, quite explicitly, without the terminology and without having any realization that this could be much broadened, that there was such a thing as decision theory, that there was a connection with Bayes solutions, or any of that. This has always interested me in connection with priority assignments. I think very many things come about in two stages. First, somebody has a germ of an idea in a rather specific context and perhaps uses it to solve a specific problem, but without developing any of its consequences or seeing any of its logic or contextual possibilities. Then somebody else comes along and develops it into a general theory. Now I've been on both sides of this. In general, I'm more on the system building side than on the initiating side. But on this particular topic I was on the initiating, and that's rare.

With the concept of the completeness of a sufficient statistic I was on the other side, because that idea is basically apparent in some of Rao's papers, you can find it in one of the Wolfowitz papers, and there are one or two other places where you can find it. In the Rao-Blackwell theorem, if the estimator is unique then you're finished. So it had already been applied. I think the main contribution of these big *Sankhyā* papers by Henry (Scheffé) and me is that we isolated this concept and then explored its general theory. ["Completeness, similar regions, and unbiased estimation, Parts I and II" *Sankhyā*, **10** (1950), 305–340, and **15** (1955), 219–236]. In this case other people had the idea and we expanded it and made a big thing of it, while with the notion of complete families it was the other way around. It seems to me that typically it's the second people who get the credit, I think a little unfairly.

FAVORITE PAPERS

DeGroot: You have published about 90 papers in statistics, as well as your famous books. Let's get to the books in a moment. Are there some favorites of yours among the papers that you liked doing or you think were particularly influential?

Lehmann: I enjoyed doing them all. You know,

it's fun. But yes, there are probably some that I like better than others. The .864 paper with Joe (Hodges) I like very much. ["The efficiency of some nonparametric competitors of the t test," *Ann. Math. Statist.* **27** (1956), 324–335]. I think that was influential in the sense that it dispelled the belief that while nonparametric techniques are very convenient because you don't have many assumptions, they have so little power that they are no good.

DeGroot: Was that really the first paper that did an explicit efficiency calculation?

Lehmann: No, Pitman was the first to do explicit calculations, and those are in his mimeographed nonparametric notes which students at Columbia published. There he calculated it for the normal, $3/\pi$. That was already very surprising that you would get such high efficiency in the normal case. But that for *all* possible distributions you would get at least .864, that I think is really a great surprise. I also had a paper on the power of nonparametric rank tests which has played a rather peculiar role in the literature. ["The power of rank tests," *Ann. Math. Statist.* **24** (1953), 23–43] What I noted there was that the difficulty of calculating power for rank tests against the usual types of alternatives is that the power depends on the following: Suppose you have a two-sample problem. One distribution is F and the other is G. Then the power depends on the way G is a function of F. So if $G = g(F)$, then it depends on this function g and doesn't depend on F at all. This function g is very complicated for the usual types of problems. So I said well, one way of getting approximate power is to approximate the true g by, say, a polynomial for which it is very easy to calculate the exact power. And the very simplest is of course if g is just a power.

Erich Lehmann, 1968

DeGroot: The Lehmann alternatives . . .

Lehmann: Yes, that's gotten the term the Lehmann alternatives. Because it is so very convenient to calculate, it has been used quite a bit. That is a paper which Neyman for some reason appreciated. I remember that I was at Columbia when I did that, and I wanted to visit Berkeley. He invited me to give a seminar talk and he was very pleased with that. It's the only paper that I wrote that I remember he was pleased with.

DeGroot: When were you at Columbia?

Lehmann: In 1950. You see there was a policy really in Berkeley that you didn't keep your own students. In statistics that policy was practically impossible because there were so few people being trained. Hotelling was training, Wilks was training, Neyman was training, and for theoretical statistics I think that was about it in the 1940's. So unless you were willing to keep your own students, you couldn't build up. So Neyman kept a lot of his students—Hodges, myself, Le Cam—But there was sort of an understanding that we should leave at one point or another for a year or two to see another place and get more experience. I found that the ideal time to do that was when that horrible loyalty oath was introduced here, which caused Charles (Stein) to leave. In the beginning I was here and the atmosphere was just so poisoned that you couldn't work. You just went to faculty meetings of one kind or another practically every day. Some people were not speaking to other people. It was just awful. So I thought it was a good time to take the opportunity of leaving. I arranged for one year away, 1950–1951, and I spent one semester at Columbia and one semester at Princeton. And the second year it was still going on and I didn't want to go back, so I went to Stanford and spent a year there. Then after that I came back here and settled down.

DeGroot: Was that period at Berkeley at all comparable, in terms of the political turmoil, to the 1960's and the period of the Vietnam War?

Lehmann: Well, the students, of course, weren't involved and so there were no mass movements.

DeGroot: Do you recall the faculty, for example, being divided during the Vietnam period?

Lehmann: Oh, the faculty certainly was divided. I had a very funny experience. I had always refused to chair the department for a variety of reasons. And then in 1970, Betty (Scott) was department chair and she said she was going to take off in the spring, and she had run through all the people whom she could reasonably ask to do this kind of thing, and would I be acting chair for this one semester. She said it wouldn't involve any work because she had prepared everything, and all that would happen was that from time to time the administrative assistant would ask me to sign something. Well, that was the Cambodia spring of 1970, when the university exploded. When they stopped having graduation ceremonies, so I suddenly was saddled with organizing a departmental graduation ceremony. When we had department meetings around the clock, with the faculty varying from the people who said they would not teach on campus to the people who said they wanted it in their records that they taught each class at the assigned time at the assigned place. [Laughs] It was a pretty wild period.

DeGroot: So you paid your dues for having avoided it earlier.

Lehmann: I paid my dues, but I found that I could do it and I was willing to take it on not too long after that on a regular basis for 3 years. But you asked me about papers and books. I want to mention a third category because it's one of the things that I enjoy the most and also it seems to me one of the most important ones, and that is Ph.D. students. I've had over 50 Ph.D. students who have spread all over the world, many of whom continue to be close friends. I occasionally collaborate with one or the other now. I have found that of all the teaching activities, the one that I enjoy the most is working one on one and getting to know people as people. Also, because we have this very international student body, I've had students from Norway, Switzerland, Germany, Thailand, Nigeria, Singapore, Israel, you name it. I have found that to be really one of the nicest and most appealing aspects of university teaching. But both the book writing and the supervision of students are, in my opinion, vastly underrated by university administrators and by organizations like the National Science Foundation. We don't get teaching credit for supervising Ph.D. students. On one occasion, I simultaneously had 10 Ph.D. students; it just happened to build up like that. Now I have usually one or two. And I didn't get any credit for that. It's an enormously time- and energy-consuming activity, and it's an extension of your research activities.

On the other hand, book writing is not considered research in our profession, which I think is totally crazy. I will admit that writing an elementary text like Joe (Hodges) and I did [*Basic Concepts in Probability and Statistics*, 2nd ed., Holden-Day, San Francisco, 1970] is not research, and yet there could be considerable intellectual activity even in that, as you very well know. But an enormous amount of research goes into writing advanced books. You know, in the humanities that *is* research and there book writing is the thing that you have to do if you want to be recognized or advanced. But I was told at one point by an NSF program director in statistics that if I mentioned that I was working on the *Estimation* book in my research proposal, I wouldn't get a grant. It's totally crazy because I think a book like that can have a much greater impact than just writing another paper.

DeGroot: There's no question that your books have had much impact on the field and serve a vital research purpose.

Lehmann: I have never understood why the mathematical sciences take this dim view of book writing. To try to synthesize things in some way seems to me a very important activity.

"I WANTED TO BECOME A WRITER"

DeGroot: You tell me that you are now involved in preparing a new edition of the *Testing* book. Is that occupying much of your time?

Lehmann: Yes, most of my time, whatever free time I have. Actually, I didn't realize it until fairly recently, but in a way the book writing ties in with what I originally wanted to do. I wanted to become a writer. It's true, I thought of a different kind of writing from *this* ... And, of course, in many ways they are very different, but they also have many similarities.

DeGroot: There is certainly good writing and bad writing in technical fields just as in other fields. And your books are beautifully written.

Lehmann: Well, trying to write a book like that has many points of contact with any other kind of writing. And so in this very back door way, I have finally realized that I am doing something that was my life's ambition when I started out and that I thought I had given up. So I do it with much happier feelings today than when I originally did it.

DeGroot: What kind of changes can we look for in the new edition?

Lehmann: First of all there is of course an enormous amount of updating. The book is 25 years old, so a lot has happened since then. I have practically no robustness considerations in the book, and in those days I didn't really understand that optimality wasn't everything. [Laughs] Today I see that optimality is only one side of a rather complicated picture and so I have a lot of robustness in it now. I struggled with whether to put in Bayesian things. I think I will not, partially because Bayesians don't believe in hypothesis testing anyway. [Laughs] Why worry about it. But there is another criticism which I think in some sense is the most legitimate if you take the Neyman-Pearson theory more or less on its own terms. There is a very legitimate attack on testing that has been mounted by several people. John Pratt has been very effective that way. That is, you sometimes get totally unreasonable things, like optimal confidence sets which turn out to be intervals for σ^2 which go from −100 to −7. It's a little embarrassing to say the least. [Laughs] There are other related things which really raise the question, in the Neyman-Pearson theory, "What is the relevant frame of reference?" That brings up the question of determining the proper conditioning, so I will have a last chapter in the book which is a very tentative, speculative chapter on conditional inference.

DeGroot: Will this be in the spirit of Kiefer's papers? [J. Kiefer, "Admissibility of conditional confidence procedures," *Ann. Statist.* **4** (1976), 836–865; and "Conditional confidence statements and confidence estimators" (with discussion), *J. Amer. Statist. Assoc.* **72** (1977), 789–827].

Lehmann: Well, Kiefer's ideas will be mentioned at the very end. He died too early to push them all the way through. With regard to conditioning on ancillaries, Cox, I think, is the person who saw this all so clearly. Fisher saw everything, too, but not very clearly. You can find a lot of it there. And then there's relevant subsets. I don't know whether you follow that literature. It actually has some Bayesian connections. That's a different type of conditioning. When conditioning on ancillaries, the conditioning event has a known probability. With relevant subsets it has an unknown probability. So the unknown parameter is involved not only in the conditional distribution, but also in the probability of the conditioning event. Which makes it more suspect, and some people don't believe in it. There is an Australian, G. K. Robinson, who has done some very interesting work on this. And Kiefer brought some new ideas. So I will give an introduction to that area, with a paragraph or two summarizing Kiefer's approach. That will be the most novel feature, I think.

DeGroot: So some of your revisions have been carried out with an eye on John Pratt's review of the first edition? [*J. Amer. Statist. Assoc.* **56** (1961), 163–167]

Lehmann: Right. [Laughs]

DeGroot: How is the *Estimation* book doing?

Lehmann: I think that it is being used. It's been out about a year and a half, and it's just going into the third printing. The first two printings together were 3000 copies. So it has sold about 3000 copies in a year and a half. Those are not huge numbers, and as you know you will never get rich on it. But on the other hand, for a book at that level I don't think that's too bad. At the moment the book suffers from the fact that the *Estimation* and the *Testing* books don't fit together. The *Testing* book is just obsolete. I hope that when the two books are together they will really form a more or less cohesive, natural one-year course at the graduate level.

DeGroot: That will be wonderful. You are a victim of having scooped yourself in a sense, since the original notes were used by so many other writers before your own books came out. So you have to compete with yourself now.

What courses do you like to teach? You tell me that you still enjoy teaching the statistical inference course.

Lehmann: Yes, I think that's the one I enjoy most. Then I typically teach a nonparametric course at some level or other. I have a book on that topic.

Erich Lehmann, Elizabeth Scott, and Jerzy Neyman

[*Nonparametrics: Statistical Methods Based on Ranks*, Holden-Day, San Francisco, 1975] I try to teach only courses where I can use my own books. Then I don't get into that horrible dilemma: Do you fight the text or do you give in? Nobody else's text ever satisfies anybody who teaches. I teach sometimes from *Basic Concepts*; last year I taught a freshman course which, if I don't do it too often, I quite enjoy. And then I'm developing a new course. I call it "Asymptotics for Applied Statistics," or something like that. It's for people who want to understand the applicational side of statistics; for example, robustness. Why does the t test work reasonably well while the F test for variances is absolutely disastrous? You cannot understand that unless you understand a little bit about asymptotics. You don't need very much. You need the central limit theorem, the theorem that people sometimes call Slutsky's theorem, and one or two other little things like that which one can explain pretty well. And there are all kinds of other asymptotic things. I've given the course twice now and I hope to give it again next year. I'll give a course which requires a year of calculus, preferably a year and a half, and some exposure to probability and statistics.

DeGroot: But it's not mainly for statistics majors?

Lehmann: Well, the people at the M.A. level who cannot take the serious probability course would get something out of it. I think they all need to understand a little of this. I start out with $O(\ )$ and $o(\ )$ and Taylor's theorem, and I get applications of all of these things to statistics. I find one can explain this to people who don't have much background so that at least they understand what the intuitive ideas are with practically no proofs.

DeGroot: Is the main purpose of teaching the background that you just described to get to these robustness results? Or are there also other kinds of applied results?

Lehmann: There are all kinds of things. For example, to understand what heavy tails means, you have to understand $O(\ )$ and $o(\ )$ and rates of convergence and things like that. That's a different type of asymptotics. That's going out to infinity on the real line, rather than the sample size going to infinity. And to understand consistency. Or asymptotic relative efficiency; that's a tremendously useful concept. I do some of that so that I can discuss the efficiency of nonparametric tests relative to t tests. So there's quite a body of applications.

DeGroot: That's an interesting idea—to take a topic like asymptotic theory and make it tie in with the Master's program.

Lehmann: I don't think there's anything like that available at the moment.

"THE PENDULUM WILL SWING BACK AGAIN"

DeGroot: I would be interested in your views about the current state of statistics as a field and where you think it's going. I've heard some bleak predictions. Computer science seems to be sweeping up students and cornering the market on interesting topics, and I've heard people say they don't even see a future for statistics.

Lehmann: No I don't see it that way, but you know I'm not at the forefront of what goes on today. I'm not really a good person to look into the future. I find it interesting to see the swing of the pendulum, and at the moment we are swinging away from the kind of stuff that I was interested in. You know, Student and Fisher, and then following them, Neyman-Pearson, really brought about a revolution and created this theory of mathematical statistics which is very strongly model-oriented. I would agree with all the critics who say it's *too* strongly model-oriented; we don't know these models. So they are now interested in adaptive and nonparametric methods. They are all trying to get away from narrow models but they are still perhaps too bound by assumptions.

Well, now one of the big swings is in the direction of data analysis which, if you take it seriously, is just as silly as the model business, if you take *it* seriously. They say ideally we don't make any assumptions whatever; we just treat data on their own terms. Well, if you make believe that you have absolutely no knowledge of how the data were generated and where they come from, then you are throwing away a lot of very useful things. The big thing today is data analysis and nonparametric procedures—regression analysis where you assume nothing about the regression function and nothing about the distribution of the errors, and so on. But it seems to me clear that after a while the pendulum will partially swing back again, because this direction is chaos. If you have no way of evaluating how well you're doing and of comparing different techniques, I don't think you're going to be satisfied.

DeGroot: It used to be that the role of statistics was *precisely* to help you evaluate how you were doing.

Lehmann: Consider optimality theory, which is not very alive at the moment. I never took those optimality results very seriously. Well, when I was young I probably did take them seriously, but not for terribly long. But I think they provide what Tukey calls "bench marks" and as such they are quite useful. I was recently at a talk given by Brad Efron where he said that for one of the problems he was talking about, he was in real trouble because there was no optimal procedure with which to compare the various results. So I think this kind of approach does have some use. It's not as overriding as people maybe thought when they first worked on it. And I'm not saying that all of us will necessarily go back to optimality theory, although people even now talk about optimal convergence rates. They talk about that in computer science, too. The smallest number of steps, exponential versus polynomial times, and so on. So maybe in this slightly looser sense, people will be interested. But certainly they will be interested in looking at some kind of performance characteristics. Of course, model assumptions get very hard when you have really large data sets because every model assumption is going to be proved false if you have enough data. There is the interesting approach due to Huber where you use neighborhoods of parametric models, but it hasn't really caught on.

DeGroot: Those are nonparametric neighborhoods of parametric models?

Lehmann: Yes. I thought that was an awfully interesting idea, and I hoped that people would do a little more with it. Another area in which there has been surprisingly little progress is inference in stochastic processes. People are just beginning to come to grips with it now. Practically all of the standard theory of estimation and testing is for Euclidean spaces, where the observations are random variables, preferably i.i.d. I remember telling Wald that I thought an awfully interesting problem in sequential analysis would be to decide at which time points you are going to observe a stochastic process. This would be an extension of sequential analysis to design-type problems. He encouraged me to work on that problem but I said I didn't think I knew enough mathematics. The whole area is just wide open.

DeGroot: Was Wald at Columbia when you were there?

Lehmann: Wald died while I was at Columbia. It was a horrible thing. They wouldn't release the information. We were first approached in the department by newspapers, and everybody was terribly worried that the family would first hear it that way. We made telephone calls and sent telegrams to India trying to find out what happened, but we couldn't get contact. The government didn't want to release any information.

DeGroot: The Indian government?

Lehmann: The Indian airline and the Indian government. And I actually took over three of Wald's students. So in addition to the students that I supervised here, I had three students from Columbia that I worked with. Allan Birnbaum was one of them. They finished there while I was back at Berkeley.

DeGroot: Had you gotten to know Wald fairly well?

Lehmann: Yes. Wald had spent a year here prior to that. He finished his decision theory book here. He and Charles and Joe and I went on a hiking trip to Yosemite with all kinds of interesting events. [Laughs] So I got to know him fairly well. Wald was fun. I remember one thing he loved to do. He looked up at a mountain that we were going to climb and he estimated—he loved to estimate—that it would be a climb of 1800 feet. And then a couple of hours later, when we were on top, he would look down and say, "It was 1900 feet. A very good estimate."

DeGroot: Do you still hike?

Lehmann: Yes, Julie and I love to hike. We go up and do some parts of that loop trail in Yosemite occasionally. Two years ago we went to Europe and hiked in the Alps. It's very nice hiking out of Berkeley, except that we have both been so busy that we haven't been doing much.

"I DON'T THINK EXACT NONPARAMETRICS WILL PROVE TO BE VERY VIABLE IN THE LONG RUN"

DeGroot: What do you think have been some of the major breakthroughs in statistics and some of the major influences on the field?

Lehmann: Well, let's say since Fisher. I'm amazed how often, when I try to write something up and then look back, I find that Fisher really understood it very well. One of the most amazing examples to me is that in my estimation book I took as my two basic types of models, the exponential families and group families. And I found that there's a 1934 paper by Fisher in which he basically defines these two types of families, and for exactly the right reasons. Namely, that in one way or another, they reduce the dimensionality of the problem. In one case by having a low-dimensional sufficient statistic, and in the other case by having a high-dimensional ancillary which has a low-dimensional conditional distribution. So certainly Fisher, with the analysis of variance and the design of experiments . . .

DeGroot: Maximum likelihood . . .

Lehmann: Well, actually Edgeworth had quite a bit on maximum likelihood. That's a very nebulous thing. I don't know whether you read Jimmie Savage and John Pratt on that. [L. J. Savage, "On rereading R. A. Fisher," *Ann. Statist.* 4 (1976), 441–483 (edited for publication by J. W. Pratt)] The priority situation as to what Fisher did or did not know about this is nebulous. Edgeworth is impossible to read, so you can't blame anybody for not really having understood him. One doesn't really know just how much Edgeworth understood . . . The 1922 paper by Fisher on sufficient statistics . . . Then the Neyman-Pearson theory, which was quite precise and modeled optimality theory. Fisher had no small-sample optimality, although he was interested in large-sample optimality . . . Certainly the Bayesian movement was a major

Harald Cramér, Kim Cramér, Erich Lehmann, Charles Stein, Joseph L. Hodges, and Juliet P. Shaffer, 1983

development. I guess, as far as we are concerned, Jimmie Savage is the hero of this, although he credited de Finetti. I have never gone back to read de Finetti, but I've read Jimmie quite a bit.

I think that the whole nonparametric development is interesting, but personally I don't think the exact nonparametric rank-based approach will prove to be very viable in the long run. It's too restricted. For example, you can't treat linear models in a really exact nonparametric fashion with rank type procedures. It only works in rather simple situations. The robust approach seems to work much better and it is much more general. I don't know whether you've looked at Hettmansperger's book that recently came out [T. P. Hettmansperger, *Statistical Inference Based on Ranks*, John Wiley and Sons, New York, 1984], but it does some of that. And Julie and I jointly supervised a student, David Draper, who tried to develop robust nonparametric analysis of variance and linear models types of procedures which are pretty close to being operational now. The tests of Wilcoxon type will probably continue to be around just because they are so neat. However, on the whole, there are better procedures. Recently, adaptive procedures are being developed. That goes back to Charles in 1956 or thereabouts ... Certainly Tukey was very important in the robust direction, and I guess everybody agrees that he was also the one who revived data analysis.

DeGroot: You mentioned Charles Stein in connection with adaptive procedures. What was that connection?

Lehmann: He conjectured that, for example, in the two-sample problem you can get tests which are asymptotically fully efficient. He also gave an intuitive, heuristic argument which isn't quite rigorous but is fairly close to rigorous. Suppose you have $F(x - \xi)$ and $F(y - \eta)$, and you want to test $\xi = \eta$. You can get fully efficient tests without making any assumptions about F. Today it can be done in a more sophisticated way, but the way he did it was this: First you estimate F from part of the sample, say $n^{1/2}$ observations, and then you use the asymptotically best test for that F. And that works; asymptotically you do as well as if you knew F all along.

DeGroot: When you think about asymptotic theory, do you distinguish between asymptotic asymptotic theory and reasonable-sample-size asymptotic theory?

Lehmann: Well, the asymptotic theory that most people do consists of limit theorems. You can embed the true situation in lots of different sequences, and try to find one which gives you a good approximation. You use the limit result as an approximation. But when I write or talk about these things, I always say that you have to supplement this method with spot checks of how well it works for a number of different distributions and a number of different sample sizes. And then you have this blind faith, for which there is no analytic argument that I know of and which doesn't always hold precisely, that as you increase the sample size the approximation will improve. I mean, it could dip or go wildly, but in practice it typically doesn't. So, if you know that it works fairly well for $n = 30$ and a distribution which looks this way or that way, then you say that for $n = 50$ or 80 or 100 it will work even better.

DeGroot: That's the statistician's religion.

Lehmann: Right. You take that on faith. But I don't know whether that answers your question or not.

DeGroot: Well, I was just thinking that there are some asymptotic results that are truly very large sample results . . .

Lehmann: Yes, I think that one wants to warn against those. In fact, at first Charles didn't pursue his adaptive stuff further, and people thought it would take hundreds of thousands of observations. It turns out that (Charles) Stone has a procedure for a problem of that type where you can do quite well with 40 observations.

Well, to continue . . . Certainly Wald's decision theory was a very major step, but I think it has been disappointing in its influence. If you look at theorems about decision theory, there haven't been very many since Wald. However, it has freed people from necessarily sticking to hypothesis testing, point estimation, or confidence intervals. One of Wald's other contributions, sequential analysis, is important but it hasn't really blossomed the way I would have expected either.

DeGroot: Well, it may have been that the wartime applications were really the right ones for sequential analysis.

Lehmann: That's true. Quality control and that type of application. Now let me ask *you* something. What do you think of Jim Berger's book? [*Statistical Decision Theory: Foundations, Concepts, and Methods,* Springer-Verlag, New York, 1980].

DeGroot: I like it a lot. Especially the Bayesian parts. I think Jim would be the first to admit, especially now that he's in the process of revising the book, that despite his statement in the Preface that he is a rabid Bayesian, there is some non-Bayesian material in there.

Lehmann: Yes, that's of course what I liked. I feel that it's a very peace-making book. As a method for generating procedures, of course, anybody with my kind of persuasion likes Bayesian procedures.

DeGroot: As a technical device.

Lehmann: Well, even as a way of thinking about a problem. This would be the right procedure if you thought that this was approximately how θ behaves. But I do like the fact that he talks about risk functions.

DeGroot: Sure. He doesn't immediately integrate out over θ.

Lehmann: And that of course in a certain sense makes him a nonrabid Bayesian.

DeGroot: I've been kidding him ever since his book came out that it's not enough simply to say you're Bayesian. If you want to get into the Bayesian group, you have to prove it in your writing. And he says that he is revising the book and it is becoming more Bayesian. So you better hold on to that first edition. [Laughs]

"WHY DO PEOPLE USE .05?"

Lehmann: But Morrie, where does the movement away from narrow models leave the Bayesian? You are even worse off than I am.

DeGroot: In many ways that is true. Just as we were saying about Neyman-Pearson theory, taking the Bayesian approach strictly on its merits is the extreme case of optimality theory. In many problems it's impossible to find an optimal procedure, and where does that leave the Bayesian? I think the answer is that the Bayesian view is useful in helping to *think* about the problem.

Lehmann: No question about it. Stein estimation is a wonderful example of that. Lindley came up with a Bayesian interpretation which led essentially to the Stein-type procedure. I think that's great.

DeGroot: Nowadays we often think about large data sets and multiparameter problems. Even though Bayesians don't have nice distributions to handle them, since nobody knows how to put down prior distributions on a high-dimensional space . . .

Lehmann: But before you can even have the prior distributions, you have to have distributions with parameters on which you can put those priors.

DeGroot: Yes, you have to be able to think about parameters. What is needed and has to be developed, in the same spirit as all of statistics, is some reasonable approximation theory. Nobody is going to get a distribution that precisely represents the experimenter's or the statistician's beliefs about some complicated setup. But just like asymptotic theory in the rest of statistics, if you can't find the Bayes decision or the optimal decision with some model, you try to get a model that works reasonably well and you try to find procedures that work reasonably well.

Lehmann: Doesn't that go against Bayesian principles? I would certainly agree with what you have just said, and I would push it slightly further and say that statistical problems don't have unique answers. There are lots of different ways of formulating problems and analyzing them, and there are different aspects that one can emphasize. Personally, I very much like an eclectic point of view where you say let's try two or three different things. For example, let's try two or three different prior distributions and see how they compare and how I like what I get with them. But I thought that most Bayesians would not buy that.

DeGroot: Well, I can't speak for other Bayesians but certainly I would accept it. The idea is that I could spend much time specifying a prior that really represents my information and my beliefs, but instead I might simply try a few standard distributions. I agree with you. Given a problem I am working on, I will try various techniques. And if the answers are more or less the same then I don't have to worry about it.

Lehmann: And if they are not, then that also guides you as to how to think about the problem. You can see *why* you get different answers.

DeGroot: Especially if you are doing this work for some client who is going to be using your analysis. You can point out that different approaches, different prior distributions, or even different methods of analysis would yield different answers. So even within the Bayesian framework, it never hurts to try a few different prior distributions, a few different likelihood functions, a few different loss functions.

Lehmann: Then as far as what we would do in practice, I think that there's practically no difference between you and me and most other statisticians. The philosophy behind it may be different, but the funny thing is that it doesn't really matter when you actually do it. The same kind of question comes up in a different context. I read over and over again that hypothesis testing is dead as a door nail, that nobody does hypothesis testing. I talk to Julie and she says that in the behaviorial sciences, hypothesis testing is what they do the most. All my statistical life, I have been interested in three different types of things: testing, point estimation, and confidence-interval estimation. There is not a year that somebody doesn't tell me that two of them are total nonsense and only the third one makes sense. But which one they pick changes from year to year. [Laughs] I think these views result from the fact that different people work primarily in different areas of application. In some areas of application, one type of inference is particularly appropriate and in other areas another one is particularly appropriate.

DeGroot: That's right. There's no doubt that testing is widely used in applications. The question is, *why* is it, and should it be? I think that, in effect, we statisticians have been very successful. They have now learned this methodology, it is appealing to them, and it's being widely used.

Lehmann: It's really surprising that testing has been so successful, because it's a very involuted way of thinking.

DeGroot: It is, but it's also easy to carry out the procedures. Take the simple example of a χ^2 test. Why is it universally used? Well, it's a simple procedure. You calculate it, and you don't have to think about it.

Lehmann: That leads to an interesting topic. This whole question of standardization, with its tremendous drawbacks. There are people who have to do statistics who can't possibly understand it at a research level or at an innovative level. It is necessary for them to have more or less cut-and-dried procedures to apply. And this is not going to get any better with the computer. We now have these packages, and people will understand even less of what they are doing than they did before.

DeGroot: And they will be dealing with much more complex situations.

Lehmann: They will just push buttons and they won't have the faintest notion of what's going on. But there is another aspect to it. Why do people use .05? I always try to talk to my class about that, because it's an interesting question. It's obviously a silly thing to do. You should take into account what power you can get. And yet there are very interesting studies that show that people use testing in situations where they have so little power that they might as well forget about carrying out the experiment, because there is almost no chance of discovering the kind of effect that they are interested in. They really needed a larger sample size. But short of that, they would do better to carry out the test at a somewhat higher significance level where they would have a better balance between the two kinds of error. But besides the fact that people like to be told to do things in a fixed way, and editors like to apply a fixed rule such as not to accept a paper unless the result is significant at the .05 level, there is also the advantage that if you use procedures in a standardized way, you get used to them and you understand them at that level. You talk about the same thing and it creates a universal language in a way. People can communicate much better than they would if you didn't have that standardization. So although I am basically strongly opposed to it, I can see that there are also some positive values to it.

DeGroot: Yes, that's a very good point. To come back to your other point, that everybody criticizes two out of your three topics: It has always amazed me about statistics that we argue among ourselves about which of our basic techniques are of practical value. It seems to me that in other areas one can argue about whether a methodology is going to prove to be useful, but people would agree whether a technique is useful in practice. But in statistics, as you say, some people believe that confidence intervals are the only procedures that make any sense on practical grounds, and others think they have no practical value whatsoever. I find it kind of spooky to be in such a field.

Lehmann: After a while you get used to it. If

somebody attacks one of these, I just know that next year I'm going to get one who will be on the other side.

DeGroot: Nonparametrics, your other area, suffers from the same criticism. Some people think that's the only way to go because parametric models are never correct, and as a Bayesian I would say that, strictly speaking, there's no such thing as a nonparametric problem. Maybe that's what keeps the field interesting.

Lehmann: I think it's interesting both on a technical level and on an intellectual level. There is a lot of turmoil, and the computer has changed the landscape quite considerably. As we were saying a minute ago, the problems that these packages are going to pose for the practice of statistics are enormous. I think they will make robust techniques—techniques that are not very sensitive to errors in the data—particularly important, because people may not even notice that one of the numbers is wrong by an order of magnitude. If you do a calculation by hand or on a calculator and you have 1,000,000 where you should have 100, you are likely to notice it. But if the data are just fed into the computer and all you see is a final summary number, it may be totally off without anybody noticing it if you don't worry about this. We teach statistics courses to about 5,000 students each year at Berkeley. We will have to worry about what kinds of things we should teach them so that they can make the best use of these techniques. I think it continues to be an absolutely fascinating field.

"STATISTICS TOTALLY DOMINATES MY SOCIAL LIFE"

DeGroot: What relationship do you see between your professional work and your social life? Does one influence the other?

Lehmann: My social life, and this is partly the result of being at Berkeley, is practically entirely confined to statisticans. We have so many visitors and so many interesting people coming through whom Julie and I both enjoy seeing. In fact, we find that there is more of this than we can handle. My closest friends are people in the department that I have known for a long time. I find one of the great pleasures of academic life, particularly of statistics, is that the statistical community is a worldwide community of people that one enjoys being with. So statistics totally dominates my social life.

But the most important impact of statistics on my social life was of course that it brought Julie into my life. She was at the time on the psychology faculty at the University of Kansas and had obtained a fellowship to strengthen her statistics. Having seen some of my papers, she thought she would like to work with me and asked me to sponsor her year in Berkeley, assuring me that she would not take up much of my time—a prediction that turned out to be rather inaccurate. The fact that we are now in the same discipline and department, that we do joint work, that each of us reads in draft whatever the other writes, and that I rely very heavily on her judgment and advice, means of course that professional concerns spill over into our private lives much more than would normally be the case.

Erich Lehmann in Leiden, 1985, at a party celebrating his honorary degree

A very specific impact Julie has had on my work is connected with the estimation book. After the publication of *Hypothesis Testing*, it would have been natural also to turn the estimation notes into a book. However, I never felt as comfortable with the theory of estimation, partly because of the dependence on the choice of loss function, partly because the subject was growing so much that the job seemed daunting. It was Julie's influence that eventually caused me to tackle this project.

In general, I find university life just absolutely wonderful. I don't know whether it's going to continue to be that wonderful in the future. I worry, for example, about what unionization may do to academic life. But the thing that I have appreciated so much about it is that you get the best of both worlds: You get the security of a salaried job with tenure, and at the same time you are to a very large extent your own agent. You can shape your activities in your own way. You can vary and emphasize differently the teaching, the research, the writing, and the administrative aspects. I chaired the department, I edited the *Annals*—things like that. Usually if you want to shape your own activities like that, you have to take more risks.

But coming back to statistics: The second way in which it influences you very strongly, and I don't see how anybody can be a statistician and not be influenced by it, is that it gets you used to stochastic thinking. Almost every newpaper article you read makes you say, "My God, they claim this or that, but don't they see that the real reason may not be that at all—that there is no causal relationship there." Probability and stochastic things just dominate our lives. I *wish* I understood what probability is. You Bayesians, of course, think you know. [Laughs]

DeGroot: Do you think about probability at that foundational level?

Lehmann: Not really; just occasionally. The fact is that after you do all of the technical and professional things you have to do, there isn't that much time left. And I like to read recreational things. I just finished reading the wonderful autobiography by Iris Origo, *Images and Shadows*. I am now reading this book [*Refugee Scholars in America: Their Impact and Their Experiences* by Lewis A. Coser] on a topic that interests me very much, namely the effect that the immigration from Europe during the Nazi period had on the development of American art and science. This book is particularly aimed at the social sciences, but I have other books that deal more with the arts and physical sciences. These books are not about my generation. They're about people who had already established themselves in Europe, and then came over because they were kicked out or because they had to flee from the Nazis in the 1930's and '40's, or because of some other difficulties.

I think that's how this university came to be so great. They brought over people like Tarski and Neyman. It was just an incredible achievement that Evans, an American traditional mathematician and a fairly conservative man in some ways, would have the foresight to bring *two* people like that, each of whom essentially created a new field and had a tremendous impact on its development. Neyman and Tarski were politically as far apart as people can be. Tarski was quite conservative and Neyman was a radical in a way. Neither of them were traditional mathematicians.

DeGroot: Did Tarski come at about the same time as Neyman?

Lehmann: No, Tarski was later. Neyman opposed Tarski's coming because he wanted Zygmund to come. There is a bit about that in Constance Reid's book. But to come back to the question you raised—I don't really think very much about foundational questions. After I've done my stint at teaching and so-called paper writing and research, and talked with students, then I really enjoy reading other types of books. I always read three or four books simultaneously. And I listen to a lot of music and go to concerts. The day only has 24 hours and my energy is not unlimited.

"COMING TO AMERICA WAS A VERY GOOD THING FOR ME"

DeGroot: Did you ever get around to writing, other than your technical writing?

Lehmann: Well, about 10 years ago I really thought I'd get out of statistics. I thought I would sort of retire. And I started translating German literature. There's a nineteenth-century German writer whom I was particularly fond of and who had not yet been translated. I translated one of his stories and then two by other writers. So, I have a set of three stories which I thought I would like to publish. Three love stories, by three very different writers of the same period who took three very different approaches, and that needed very different translation techniques because the language was so different in the three. But then I never polished it and never got it into publishable shape. I also started translating a very baroque nineteenth-century novel. Julie is still mad at me because it's a mystery story in a way, and about halfway through I stopped and she never found out how it ends. [Laughs] It's by an author who was essentially untranslated and very hard to translate. So then I got quite interested in translation and read a lot of things about it.

I've also done a little autobiographical writing for my children and grandchildren. Since I am the generation that came over from Europe, all of this European stuff will be lost if I don't put some of it down. And so I wrote a little bit about my family. I can trace my mother's family directly back to about 1500. These German Jews kept very good records. There were these ghettos with about 200 families, and they recorded whenever a house was sold, the new family that took it over, what professions these people had, and so on. I have a book. [Getting the book, *Stammbuch der Frankfurter Juden*, published in 1907] It's like a book of family trees, and it traces the Jews of Frankfurt back to the 1500's.

DeGroot: Did your mother's family remain in Frankfurt all through those years?

Lehmann: Yes. My father's family came to Frankfurt in the early ninetheenth century, around 1830, from Hamburg. I've always wondered whether there exists a similar book about Hamburg where I can trace his family back. But I haven't made a real effort to try and find it.

DeGroot: Are you still writing your memoirs?

Lehmann: Not at the moment because I want to finish the revision of the *Testing* book. After that I might get back to it. Anyway, I'm approaching retirement. I'll be 67 next month. So at worst or at best, I'm not quite sure, I'll continue working at the university for another 3 years. After that, I'll have more time.

DeGroot: What do you see yourself doing over the next 3 years?

Lehmann: Well, the book hopefully will be finished by January or February (1985). And I mentioned to you that I have an idea for this asymptotics project; that's one possibility. And Peter Bickel, Bill van Zwet, and I have been talking about an advanced asymptotics book. I don't know whether I can really play a part, but that's another possibility. Julie and I are talking about various projects. And occasionally I think that maybe I should see if I can write a book that will *sell*; a popular book that represents a point of view that I like and that might be useable. But I think it's probably a little late for that.

DeGroot: But the Hodges and Lehmann book must have sold quite well.

Lehmann: Well, it sold something like 50,000 copies over the years, which is respectable. It would have done much better if we had been willing to revise it every few years, because you know what happens: After 5 years there are stacks of second-hand copies. So even if the book is still used in classes, few new copies are sold.

DeGroot: And after retirement?

Lehmann: Spend more time with music. I play the piano a little bit and I will be able to spend more time with that. Go to more concerts, read more, try to write more about the family, all those things. Coming to America was a very good thing for me. My life would have been totally different if I had stayed in Europe. I can't tell you how glad I am that I came. I had a chance to go back to Zurich in the early 1950's. They wanted to start a chair in mathematical statistics at ETH. But I have no real roots in Europe. In Germany obviously there are no roots of any kind left. Everyone in my family there either emigrated or died. I have some kind of attachment to Switzerland because I spent some of my formative years there, although they were pretty unhappy years. But I don't want to live in Europe. Things are so much narrower and so much more restricted. There's much less freedom and everything is much more prescribed.

America is really great. Despite the fact that people are very much worried about the various ways in which the government interferes with our lives, this is still an amazingly free country. Berkeley is, of course, particularly free. Whenever our children accuse Julie and me of being squares, I tell them that Berkeley is a place where even squares can feel comfortable. [Laughs] There are many things that I would enjoy doing if I had enough energy, but as you get older the main drawback is that you become tired more easily. Otherwise I haven't found it too bad.

DeGroot: Thank you, Erich.

Statistical Science
1990, Vol. 5, No. 2, 160–168

Model Specification: The Views of Fisher and Neyman, and Later Developments

E. L. Lehmann

Abstract. Since Fisher's formulation in 1922 of a framework for theoretical statistics, statistical theory has been concerned primarily with the derivation and properties of suitable statistical procedures on the basis of an assumed statistical model (including sensitivity to deviations from this model). Until relatively recently, the theory has paid little attention to the question of how such a model should be chosen. In the present paper, we consider first what Fisher and Neyman had to say about this problem and in Section 2 survey some contributions statistical theory has made to it. In Section 3 we study a distinction between two types of models (empirical and explanatory) which has been discussed by Neyman, Box, and others. A concluding section considers some lines of further work.

Key words and phrases: Empirical model, explanatory model, model selection, reservoir of models, binomial model, technology, revolutionary and normal science.

Where do probability models come from? To judge by the resounding silence over this question on the part of most statisticians, it seems highly embarrassing. In general, the theoretician is happy to accept that his abstract probability triple (**Ω**, **A**, **P**) was found under a gooseberry bush, while the applied statistician's model "just growed".

A. P. Dawid (1982)

1. THE VIEWS OF FISHER AND NEYMAN

A general framework for theoretical statistics was proposed by Fisher (1922) in his fundamental paper "On the mathematical foundations of theoretical statistics." In it, Fisher defines the principal task of statistics to be "the reduction of data," and states that "this object is accomplished by constructing a hypothetical infinite population, of which the actual data are regarded as constituting a sample. The law of distribution of this hypothetical population is specified by relatively few parameters, ..." In other words, Fisher states that the first step is the construction of a low-dimensional parametric model.

On this basis, Fisher divides the problem of statistics into three types: (1) *problems of specification* (that is, the problem of specifying the parametric model); (2) *problems of estimation* (this terminology for the second stage is explained by the fact that Fisher in this paper is concerned only with point estimation; more generally, this type of problem could be described as the derivation of a suitable statistical procedure; and (3) *problems of distribution* (this refers to the distribution of the estimator derived in (2); a more general description of this last stage would be assessing the performance of the procedure obtained in (2)).

The present paper is concerned only with the first of these stages, and it is with considerable interest that one wonders what Fisher has to say about it. Disappointingly, his discussion is confined to a single paragraph which is dominated by the first sentence: "As regards problems of specification, these are entirely a matter for the practical statistician, ..."

This statement constitutes Fisher's answer to the question raised in the heading of Section 2 of the present paper and the quotation at its beginning. Fisher's statement implies that in his view there can be no theory of modeling, no general modeling strategies, but that instead each problem must be considered entirely on its own merits. He does not appear to have revised this opinion later; the index to the five-volume collection of his papers (published by the University of Adelaide) has only one entry under "Specification, problems of"—the 1922 statement cited in the preceding paragraph.

Actually, following this uncompromisingly negative statement, Fisher unbends slightly and offers two general suggestions concerning model building: (a) "We must confine ourselves to those forms which we know how to handle," an (b) "More or less elaborate forms will be suitable according to the volume

E. L. Lehmann is Professor of Statistics, Emeritus, at the University of California, Berkeley. His mailing address is Department of Statistics, University of California, Berkeley, California 94720.

of the data." We shall return to both of these suggestions later.

To Neyman, not only the practice but also the theory of modeling was a central concern. In 1959 he introduced a course in the subject into the Berkeley curriculum and taught it with few interruptions until the time of his death in 1981. We shall here discuss three comments of his on modeling.

1. Models of complex phenomena are constructed by combining simple building blocks which, "partly through experience and partly through imagination, appear to us familiar, and, therefore, simple." Although not making exactly the same point, this comment is somewhat reminiscent of Fisher's suggestion that we should restrict attention to models we know how to handle.

2. An important contribution to the theory of modeling is Neyman's distinction between two types of models: "interpolatory formulae" on the one hand and "explanatory models" on the other. The latter try to provide an explanation of the mechanism underlying the observed phenomena; Mendelian inheritance was Neyman's favorite example. On the other hand an interpolatory formula is based on a convenient and flexible family of distributions or models given a priori, for example the Pearson curves, one of which is selected as providing the best fit to the data. (Neyman, 1939). The same distinction is emphasized in the writings of George Box, who uses the terms "empirical model" and "theoretical" or "mechanistic" model for the two concepts (mechanistic since it identifies the underlying mechanism).

3. The last comment of Neyman's we mention here is that to develop a "genuine explanatory theory" requires substantial knowledge of the scientific background of the problem. The same general idea is expressed by John Stuart Mill who writes in his *System of Logic*: "The guesses which serve to give mental unity and wholeness to a chaos of scattered particulars, are accidents which rarely occur to any minds but those abounding in knowledge and disciplined in intellectual combinations" (Mill, 1879, page 344). This requirement is agreed on by all serious statisticians but it constitutes of course an obstacle to any general theory of modeling, and is likely a principal reason for Fisher's negative feeling concerning the possibility of such a theory.

2. WHERE DO MODELS COME FROM?

Several statisticians with extensive applied experience have objected to this question, stating that they know exactly where their models come from. However, it seems difficult to find explicit statements of how models are obtained, which raises the question in a different form: Is applied statistics, and more particularly model building, an art, with each new case having to be treated from scratch (although even artistic endeavors require techniques which can be systematized and learned), completely on its own merits, or does theory have a contribution to make to this process?

2.1 A Reservoir of Models

One crucial (although somewhat indirect) contribution of theory is indicated by Fisher's and Neyman's reference to "those forms which we know how to handle," and to building blocks which "appear to us familiar and, therefore, simple." These references presuppose the existence of a reservoir of models which are well understood and whose properties we know. Probability theory and statistics have provided us with a rich collection of such models. One need only think of the various families of univariate and multivariate distributions, of the different kinds of stochastic processes, of linear and generalized linear models, and so on. The list seems inexhaustible and furnishes the modeler with an indispensable tool.

One aspect of models that is of particular importance to realistic modeling is the way they are described or characterized. (For some references to the literature on characterization of distributions, see, for example, Galambos, 1982a.)

EXAMPLE 2.1. As an illustration consider the model that is traditionally described by

(2.1) $X_1, \ldots, X_n$ are iid with normal distribution $N(0, \sigma^2)$.

If asked to justify the normality assumption, a statistician might refer to the central limit theorem and for corroboration might cite previous experience with similar data. There is however an alternative approach which may be more convincing. The model (2.1) with $n > 1$ can be characterized by the two conditions

(2.2a) the X's are independent

and

(2.2b) the joint density of the X's is spherically symmetric, i.e., the density is the same at all points equidistant from the origin.

(For a discussion of the equivalence of (2.2) and (2.1) see, for example, Feller, 1966, pages 77–78, and Letac, 1981.)

The meaning of condition (2.2b) is much easier to grasp than that of normality. As an example, suppose that X_1 and X_2 are the coordinates of the impact of a shot fired at a circular target, where we shall assume that the expected point of impact is the bull's eye which is taken to be the origin. If the method of aiming is asymmetric in the horizontal and vertical directions (as is typically the case), the assumption of circular symmetry will usually not be justified; on the other hand, one can imagine automated computer methods of aiming for which this symmetry does hold.

EXAMPLE 2.2. As a second example, consider the assumption that a variable X is distributed according to an exponential distribution. This family of distributions is characterized by the property of "no aging," i.e., the distribution of the lifetime remaining at time t (given that life has not terminated prior to t) is the same as the distribution of the lifetime X at birth. In many situations, this property clearly will not hold and an exponential distribution therefore be unsuitable. Situations in which the assumption may be reasonable are described for example in Mann, Schafer and Singpurwalla (1974).

EXAMPLE 2.3. Consider finally the Poisson distribution. In many applications, it arises as the distribution of the number of events occurring in a fixed interval, where the events are generated by a Poisson process. The latter is characterized by two assumptions: (a) the numbers of events occurring in non-overlapping intervals are independent, and (b) the probability of more than one event occurring in a very short interval is of smaller order than the probability of a single occurrence.

Simple characterizations such as those of Examples 2.1–2.3 not only provide relatively clear criteria for the applicability of a model in a given situation but they may also suggest, when the assumptions are not satisfied, in what way the assumptions are violated and on this basis lead to ideas as to how the model should be modified. For example, a Poisson model may not be suitable because of the presence of "multiple births," and this possibility can be incorporated into the model by specifying the distribution of the "litter size." (For a discussion of this idea from a modeling point of view, see, for example, Neyman and Scott, 1959, and Cox and Isham, 1980; some related ideas are considered by Feller, 1943.)

An assumption that makes an appearance in both Examples 2.1 and 2.3 and which generally enjoys great popularity is the assumption of independence. As has recently been emphasized by Kruskal (1988), this assumption is often made rather casually. It is, for example, frequently taken for granted that successive observations by the same observer are independent. That this in fact tends not to be the case was noted by Pearson (1902) who carried out some experiments for this purpose. The issue is discussed by Student (1927) and Cochran (1968). Further references can be found in Kruskal's paper. Unfortunately, the independence assumption is not only very seductive but the resulting references are liable to serious error when the assumption is not justified.

Classes of models for dependent observations taken in series are treated in the theory of time series. Observations which, while not independent, retain the complete symmetry of the iid assumption are called *exchangeable*. The theory of exchangeable events was initiated by de Finetti; for references and recent developments, see, for example, Kingman (1978), Koch and Spizzichino (1981), Galambos (1982b) and Diaconis and Freedman (1984).

2.2 Model Selection

Procedures for choosing a model not from the vast storehouse mentioned in (2.1) but from a much more narrowly defined class of models are discussed in the theory of *model selection*. A typical example is the choice of a regression model, for example of the best dimension in a nested class of such models. The best fitting model is of course always the one that includes the largest number of variables. However, this difficulty can be overcome in a number of ways, for example by selecting the dimension k which minimizes the expected squared prediction error, i.e., the expected squared difference between the next observation and its best prediction from the model.

This measure has two components:

$$\begin{aligned} &E(\text{squared prediction error}) \\ &\quad = (\text{squared bias}) + (\text{variance}). \end{aligned} \tag{2.3}$$

As the dimension k of the model increases, the bias will decrease. At the same time the variance will tend to increase since the need to estimate a larger number of parameters will increase the variability of the estimator. Typically there will exist a minimizing value k_0 which then provides the desired solution. The value of k_0 will tend to increase as the number of observations increases and its determination thus constitutes an implementation of Fisher's suggestion that "more or less elaborate forms will be suitable according to the volume of the data." (In practice, one minimizes not the expected squared prediction error which depends on unknown parameters but a suitable estimator of this expected value.)

Areas in which model selection has been successfully employed include, besides regression (Breiman and Freedman, 1983), the choice of an appropriate ARMA model (Poskitt, 1987); choosing the order of a Markov chain (Katz, 1981); or the maximal order of the interactions to be included in factorial and log-linear models (Linhart and Zucchini, 1986; Edwards and Havranek, 1987).

As described above, this approach appears to make the choice of model automatically, solely on the basis of the data. However, this view of model selection ignores a preliminary step: the specification of the class of models from which the selection is to be made. This specification often will also be quite routine along the line: we are modeling a smooth surface, so let's approximate it by a polynomial regression equation including all terms up to a given degree. In choosing the variables to include in the list of possibilities, we may be quite liberal and let the procedure make the choice from this list (as well as the degree of the polynomial) from the data. However, in other cases, the choice of class of models may be strongly informed by subject matter considerations.

2.3 Modeling a Sequence of Dichotomous Trials

The great difficulty of model specification is well illustrated by the history of models for sequences of dichotomous trials such as tosses with a coin, births classified by sex, defective and nondefective items on an assembly line, clinical trials or free throws in basket ball.

In his 1710 paper, "An argument for divine providence, taken from the constant regularity observ'd in the births of both sexes," Arbuthnot assumed a binomial distribution, and on this basis tested (and rejected) the hypothesis that $p = ½$. The binomial model of course is the result of two assumptions: (1) the constancy of the success probability p; and (2) the independence of the trials. Heyde and Seneta (1977) and Stigler (1986) discuss the history of the struggle with these assumptions which went on throughout the 19th century, and which led to the development of more general alternative models and of tests of the binomial hypothesis.

A curious debate sprang up in the early 20th century in publications by Marbe (1899, 1916, 1919), Sterzinger (1911) and Kammerer (1919), in which these authors claimed that the results of probability theory contradict the workings of the real world and are therefore not applicable to reality. These beliefs were based on their theories of the "bunching" of events (Knäuelungstheorie) of Sterzinger, the "uniformity" (Gleichförmigkeit) of the world (Marbe) or Kammerer's principle of seriality, bolstered by inconsistencies which the authors thought they had discovered between the observed numbers of runs in various dichotomous sequences such as coin tosses or births and the probabilities of such runs calculated on a binomial model. The arguments show that what is really being put in question is the assumption of independence. The flaws in some of this work are analyzed by Bortkiewicz (1917) and Czuber (1923). (For a recent study of related issues, see Gilovich, Vallone and Tversky, 1985; Tversky and Gilovich, 1989a, b; Larkey, Smith and Kadane, 1989; and Hooker, 1989).

3. TWO TYPES OF MODELS

The distinction between the two types of models mentioned in Section 2, *empirical* (or interpolatory) on the one hand, and *explanatory* (or mechanistic) on the other, has been noted in one form or another by many writers and tends to make an appearance in most general discussions of modeling. Models of these two types differ in many respects and the distinction throws light on some aspects of the modeling process.

3.1 Purpose

As pointed out by Box (for example, in Box and Hunter, 1965, and Box and Draper, 1987) and others, the two kinds of models differ in their basic purpose. Empirical models are used as a guide to *action*, often based on forecasts of what to expect from future observations (e.g., the number of college applications, demand for goods, or stock market performance). In contrast, explanatory models embody the search for the basic mechanism underlying the process being studied; they constitute an effort to achieve *understanding*.

The following description of the distinction is from a paper dealing with the role of models in ecology (Goldstein, 1977).

> Within a given ecological research program, modeling can be a valuable procedure in helping to address a number of frequently occurring research objectives. A basic research objective is increased fundamental understanding of the system being studied. This need not be an objective of all research programs. Oftentimes, there is a desire to produce a specific output from a given system. In many circumstances, this goal can be achieved through a well-designed manipulation of the system's inputs, without any attempt to derive a basic understanding as to how the system functions. This type of approach is frequently referred to as an "input-output" analysis and the system is described as a "black box". Mathematical modeling techniques can be very helpful in this type of analysis as well.

3.2 Environment

The two kinds of aims and attitudes described in 3.1 tend to arise in somewhat different environments. This distinction is discussed for example in a paper by Healy (1978) who contrasts technology with science, or applied with theoretical science, as follows: "I merely want to propose that much of what is commonly described as science comes more appropriately under the heading of technology. I could soften the blow . . . if I substituted for "technology" the term "applied science."" Asking what distinguishes the two, Healy goes on to state that: "I hold that, in contrast to the scientist, the technologist is not concerned with truth at all. . . . The mark of the technologist is that he must act; everything that he does has some sort of deadline. He has to manage therefore, with as much truth as is available to him, with the scientific theories current in his time."

It is interesting to contrast this with the following passage from Box and Hill (1967).

> It should be noted that the objective we are presently considering is that of finding out 'how a system works'. The reason for this may be no more than scientific curiosity. However, if we know how the system works and can describe it by a mathematical model, then we can use this knowledge for practical aims such as predicting the behavior of the process under various experimental conditions and, in particular, in finding optimum operating conditions. This fact leads to some confusion because if all we need to do is either to estimate the behavior of the process under various experimental conditions or to find optimum operating conditions, we do not necessarily need a mechanistic model. In some circumstances, an attempt to discover a mechanism merely to develop an operable system would be needlessly time consuming.

Both passages are somewhat defensive, but they defend against attacks from opposite directions. In the expectation that his audience will feel insulted by having their work described as technological rather than scientific, Healy says that he "could soften the blow" of calling what much of statistics is concerned with technology rather than science by using a less offensive terminology. Box and Hill, on the other hand, apologize for being needlessly scientific in a situation which calls for a solution to a practical problem. They point out that developing an explanatory model rather than being satisfied with an empirical one might be justified by more than mere "scientific curiosity": the knowledge gained in this way might actually be utilized "for practical aims."

3.3 Ad Hoc Nature versus Broad Applicability

The difference in attitudes just described leads quite naturally to different positions concerning the appropriate level of generality of the models and the conclusions derived from them. A scientific model describing the structure of the underlying mechanism and the laws governing it will strive for the most general formulation possible. (A heated debate concerning the status of the logistic curve as a general law of population growth is discussed in Kingsland, 1985.) This typically requires abstraction and idealization in order to eliminate the specific circumstances of particular situations. General physical laws explaining the motion of bodies or biological laws describing the genetic mechanisms of inheritance apply without regard to the nature of particular bodies and to very general classes of biological organisms.

On the other hand, a technological model intended to provide guidance in a particular situation at hand will want to make full use of all the special circumstances of that situation. In particular, it needs to provide a good approximation only over the range of values of interest. Thus a linear regression may be perfectly adequate for the problem under consideration even when it is clear that a linear approximation can provide a reasonable fit only over a very limited range.

3.4 The Role of Subject Matter

An explanatory model, as is clear from the very nature of such models, requires detailed knowledge and understanding of the substantive situation that the model is to represent. On the other hand, an empirical model may be obtained from a family of models selected largely for convenience, on the basis solely of the data without much input from the underlying situation. (Examples of both situations can be found in the writings of Box and his coworkers, for example in Box and Hunter, 1965; Box and Hill, 1967; Box, Hunter and Hunter, 1978; and Box and Draper, 1987.)

It is interesting to examine Kepler's elliptical orbits and Mendel's laws of inheritance from this point of view. A first look at Kepler's work may lead to the conclusion that he just tried to fit a simple curve to the available data and found that an ellipse was both mathematically simple and provided an excellent fit. However, closer study reveals that his discovery came about as the result of a great deal of previous thinking and theorizing about the subject matter. (For a detailed discussion of this point, see, for example, Aiton, 1975, and Wilson, 1975.)

Similarly it is tempting to believe that Mendel, often depicted as working in monastic isolation, came to his startling innovative laws solely by contemplating the results of his experiments with peas. Again it emerges from a more detailed study that his conclusions and explanations were by no means based only on his data. (See, for example, Orel, 1984, and Olby, 1985.)

3.5 Validating the Model

The difference in the aims and nature of the two types of models implies very different attitudes toward checking their validity. Techniques such as goodness of fit tests or cross validation serve the needs of checking an empirical model by determining whether the model provides an adequate fit for the data. Many different models could pass such a test, which reflects the fact that there is not a unique correct empirical model. On the other hand, ideally there is only one model which at the given level of abstraction and generality describes the mechanism or process in question. To check its accuracy requires identification of the details of the model and their functions and interrelations with the corresponding details of the real situation. And the explanation must provide a detailed fit not only for the particular phenomenon at hand but must also fit appropriately into the existing scientific fabric of related phenomena in which it is to become embedded.

It is interesting to note that the distinction discussed in this section was realized, and considered important, long before the modern scientific age. Following are three quotes, one each from the 16th, 17th and 18th centuries, as given in Popper (1965).

The first is from Osiander's "Preface to Copernicus' 'De Revolutionibus'", putting a "spin" on this work of which Copernicus would hardly have approved had he lived to see it: "There is no need for these hypotheses to be true, or even to be at all like the truth; rather one thing is sufficient for them—that they should yield calculations which agree with the observations." For additional discussion of Osiander's Preface, the complete text of which is given for example in Rosen (1971), see Toulmin (1961) and particularly Blumenberg (1987). Note that throughout this paper it is tacitly assumed that there exists an underlying "true" situation which one is attempting to model. This is an attitude much discussed by philosophers of science. Some recent references are Cartwright (1983), Hacking (1983), Jardine (1986) and Giere (1988).

Osiander was a Protestant Theologian. The issue became more heated in the following century in the dispute between Galileo and the Catholic Church. The position of the latter as stated by Cardinal Bellarmino in 1615 was that the church would raise no objections if Galileo stated his theory as a mathematical hypothesis, "invented and assumed in order to abbreviate and ease the calculations," provided he did not claim it to be a true description of the world.

As a last expression of this thought, here is a quote from Bishop Berkeley (1734). "A mathematical hypothesis may be described as a procedure for calculating certain results. It is a mere formalism ... judged by its efficiency.... The question of truth of a mathematical hypothesis does not arise—only that of its use as a calculating tool." Although these passages of course refer to deterministic rather than stochastic models, the idea of empirical modeling could hardly be expressed more clearly today.

4. A SPECTRUM OF POSSIBILITIES

The discussion in the preceding section of empirical and explanatory models ignores the fact that many, perhaps most, actual modeling situations have an intermediate character, exhibiting some features of each of the two types. This was clearly recognized by Kruskal and Neyman (1956) who wrote:

> Although the distinction between true theory and interpolation is sometimes quite sharp in specific cases, the two models of analysis really represent somewhat extreme points of a continuous spectrum.... The continuum is by no means precise, nor is it meant to be so....

(The authors then proceed, as an amusing pastime, to place various kinds of models on a "fanciful scale" meant to represent this imprecise continuum. To factor analysis, for example, on a scale from 0 (pure interpolation) to 10 (fully explanatory) they assign a score of 4.)

The corresponding remark applies to the discussion of science versus technology. In this connection, it is helpful to recall the distinction made by Kuhn (1970) between normal and revolutionary science. As suggested by Nelder (1986), normal science occupies a position somewhere between technology and revolutionary science, and again there exists a whole spectrum of intermediate shadings. Concerning the role of statistics in these differents environments Nelder writes:

> I doubt if statistics has much to offer to revolutionary science.... The position is very different, however, with both normal science and technology, though some justification may be needed for grouping them together here. I would argue that there are major differences between normal science and technology, for example in the relevance of cost-benefit assessments to their progress, their long-term objectives, and their attitudes to theory (Healy, personal communication), it is nonetheless true that, on the scale of day to day activity, the procedures of (normal) scientist and technologist will be found to be very much alike. They will both be working within a given theoretical framework and be concerned, for instance, with estimating quantities defined within that framework, with confirming the estimates of others, and with relating their estimates to those predicted by theory.

The construction of a satisfactory "revolutionary" model is essentially the problem of scientific discovery. Where the ideas for such discoveries come from is one of the central problems in the philosophy of science which has been discussed by many scientists and philosophers. Most scientists agree with Nelder's implication that there can be no systematic aid to discovery, that it is a matter of imagination and inspiration. One of the most influential philosophers of science, Karl Popper, sums this up by saying (Popper, 1965, page 192) that scientific hypotheses are "the free creations of our own minds, the result of an almost poetic intuition."

Yet even in this extreme case, the situation is not as clearcut as it seems at first. Examination of some of the classical examples of revolutionary science shows that the eventual explanatory model is often reached in stages, and that in the earlier efforts one may find models that are descriptive rather than fully explanatory. (Such descriptive models should perhaps be considered as a distinct third type intermediate between empirical and explanatory models. A division of models into these three types by Berkeley is discussed by Popper, 1965, page 169.) This is, for example, true of Kepler whose descriptive model (laws) of planetary motion precede Newton's explanatory model. To some extent, this remark also applies to Mendel's revolutionary model for inheritance. His theory does offer an explanation for his laws of segregation and independent assortment in terms of factors responsible for the transmission of genetic material. However, he was not in a position to identify these factors and so obtain the required identification between his model and biological reality.

5. CONCLUSIONS

This paper has been concerned with the question of what contribution statistical theory has to make to model specification or construction. Three areas of such contributions turned out to be the following.

1. *A reservoir of models.* Here particular emphasis is placed on transparent characterizations or descriptions of the models that would facilitate the understanding of when a given model is appropriate. (A special problem in this area which requires additional work is a study of the circumstances under which independence is or is not a suitable assumption.)

2. *Model selection.* This is a body of techniques for selecting a particular model (or parametric subfamily of models) from a rather narrowly specified family of models. It seems likely that this approach will be developed much further, particularly with the help of techniques from artificial intelligence. The use of expert systems will make it easier to inject subject matter information into this process. On the other hand, it is difficult to see how this approach can break out of a current paradigm and thus lead to revolutionary scientific discoveries. (An interesting and thoughtful "progress report" on the possible role of artificial intelligence in scientific discovery is provided by Langley, Simon, Bradshaw and Zytkow (1987).

3. *Classification of models.* The distinction between explanatory and empirical models provides only one of the many ways in which models can be classified. A better understanding of the modeling process can perhaps be obtained by distinguishing various other types of situations which require different kinds of models. Fundamental differences between the needs and possibilities in the social and physical sciences, for example, are discussed by Lieberson (1985) and in the debates initiated by Freedman (1985, 1987). The corresponding issues regarding biology and physics (with extensive reference to the literature) are considered for example in Rosenberg (1985).

Another important distinction is that between deterministic and stochastic models. Of course any statistical model has a stochastic component. This may, however, enter either only through the errors in the measurements of an essentially deterministic situation (as in Kepler's astronomical data) or also through the basic stochastic nature of the underlying phenomenon (for example, in Mendel's theory).

As a last example, we mention the distinction between models which do or do not employ what psychologists have called "hypothetical constructs" (see, for example, MacCorquodale and Meehl, 1948). These are entities whose existence is required by the theory but has not actually been established, although it is hoped that observation will eventually change their status. Mendel's genes or some of the elementary particles are cases in which eventual verification did occur; structures required for certain mental abilities, for example color vision, provide another illustration. On the other hand, the postulated existence of ether and phlogiston in the end had to be abandoned.

In addition to these different types of models, it is useful to distinguish between two aspects, both of which are typically present in the same model: the subject matter part of the model and the part played by "error." Here the latter term is meant to include not only measurement error but impurities in the material, changes in temperature or time of day, in fact all the contributions to the observations of the various environmental and observer effects that are extraneous to the subject matter.

ACKNOWLEDGMENTS

This paper greatly benefited from discussions with Persi Diaconis, Lucien Le Cam and Juliet Shaffer. This article is a modified version of the 1988 Fisher Lecture delivered on August 24, 1988 at the Joint Statistical Meetings in New Orleans. The research was supported by NSF Grant No. DMS-89-08670.

REFERENCES

AITON, E. J. (1975). The elliptical orbit and the area law. In *Kepler. Four Hundred Years* (Proceedings of conferences held in honour of Johannes Kepler) (A. Beer and P. Beer, eds.). Pergamon, Oxford.

ARBUTHNOT, J. (1710). An argument for Divine Providence, taken from the constant regularity observ'd in the births of both sexes. *Philos. Trans. Roy. Soc. London* **27** 186–190.

BLUMENBERG, H. (1987). *The Genesis of the Copernican World.* MIT Press, Cambridge, Mass.

BORTKIEWICZ, L. VON (1917). *Die Iterationen.* Springer, Berlin.

BOX, G. E. P. and DRAPER, N. (1987). *Empirical Model-Building and Response Surfaces.* Wiley, New York.

BOX, G. E. P. and HILL, W. J. (1967). Discrimination among mechanistic models. *Technometrics* **9** 57–71.

BOX, G. E. P. and HUNTER, W. G. (1965). The experimental study of physical mechanisms. *Technometrics* **7** 23–42.

BOX, G. E. P., HUNTER, W. G. and HUNTER, J. S. (1978). *Statistics for Experimenters.* Wiley, New York.

BREIMAN, L. and FREEDMAN, D. (1983). How many variables should be entered into a regression equation? *J. Amer. Statist. Assoc.* **78** 131–136.

CARTWRIGHT, N. (1983). *How the Laws of Physics Lie.* Oxford Univ. Press, New York.

COCHRAN, W. G. (1968). Errors of measurement in statistics. *Technometrics* **10** 637–666.

COX, D. R. and ISHAM, V. (1980). *Point Processes.* Chapman and Hall, London.

CZUBER, E. (1923). *Die Philosophischen Grundlagen der Wahrscheinlichkeitsrechnung.* Teubner, Leipzig.

DAWID, A. P. (1982). Intersubjective statistical models. In *Exchangeability in Probability and Statistics* (G. Koch and F. Spizzichino, eds.) 217–232. North-Holland, Amsterdam.

DIACONIS, P. and FREEDMAN, D. (1984). Partial exchangeability and sufficiency. In *Statistics: Applications and New Directions* (J. K. Ghosh and J. Roy, eds.) 205–236. Indian Statistical Institute, Calcutta.

EDWARDS, D. and HAVRANEK, T. (1987). A fast model selection procedure for large families of models. *J. Amer. Statist. Assoc.* **82,** 205–213.

FELLER, W. (1943). On a general class of 'contagious' distributions. *Ann. Math. Statist.* **14** 389–400.

FELLER, W. (1966). *An Introduction to Probability Theory and Its Applications* **2** 77–78. Wiley, New York.

FISHER, R. A. (1922). On the mathematical foundations of theoretical statistics. *Philos. Trans. Roy. Soc. London Ser. A* **222** 309–368.

FREEDMAN, D. (1985). Statistics and the scientific method. In *Cohort Analysis in Social Research: Beyond the Identification Problem* (W. M. Mason and S. E. Fienberg, eds.) 343–359. Springer, New York.

FREEDMAN, D. (1987). As others see us: A case study in path analysis (with discussion). *J. Educ. Statist.* **12** 101–223.

GALAMBOS, J. (1982a). Characterizations of distributions. *Encyclopedia of Statistical Sciences* **1** 422–428. Wiley, New York.

GALAMBOS, J. (1982b). Exchangeability. *Encyclopedia of Statistical Sciences* **2** 573–577. Wiley, New York.

GALAMBOS, J. (1982c). Exponential distribution. *Encyclopedia of Statistical Sciences* **2** 583–587. Wiley, New York.

GIERE, R. N. (1988). *Explaining Science.* Univ. Chicago Press, Chicago.

GILOVICH, T., VALLONE, R. and TVERSKY, A. (1985). The hot hand in basketball: On the misperception of random sequences. *Cognitive Psychology* **17** 295–314.

GOLDSTEIN, R. A. (1977). Reality and models. In *Mathematical Models in Biological Discovery* (D. L. Solomon and C. Walter, eds.) 207–215. Springer, New York.

HACKING, I. (1983). *Representing and Intervening.* Cambridge Univ. Press, Cambridge.

HEALY, M. J. R. (1978). Is statistics a science? *J. Roy. Statist. Soc. Ser. A* **141** 385–393.

HEYDE, C. C. and SENETA, E. (1977). *I. J. Bienaymé.* Springer, New York.

HOOKER, R. (1989). Basketball, baseball, and the null hypothesis. *Chance* **2** (4) 35–37.

JARDINE, N. (1986). *The Fortunes of Inquiry.* Clarendon Press, Oxford.

KAMMERER, P. (1919). *Das Gesetz der Serie.* Deutsche Verlags-Anstalt, Stuttgart.

KATZ, R. W. (1981). On some criteria for estimating the order of a Markov chain. *Technometrics* **23** 243–249.

KINGMAN, J. F. C. (1978). Uses of exchangeability. *Ann. Probab.* **6** 183–197.

KINGSLAND, S. E. (1985). *Modeling Nature.* Univ. Chicago Press, Chicago.

KOCH, G. and SPIZZICHINO, F. (eds.) (1981). *Exchangeability in Probability and Statistics.* North-Holland, Amsterdam.

KRUSKAL, W. (1988). Miracles and statistics. The casual assumption of independence. *J. Amer. Statist. Assoc.* **83** 929–940.

KRUSKAL, W. and NEYMAN, J. (1956). Stochastic models and their applications to social phenomena. Unpublished lecture presented at joint session of IMS, ASA and the Amer. Social Soc., Detroit, Sept. 1956.

KUHN, T. S. (1970). *The Structure of Scientific Revolutions,* 2nd ed. Univ. Chicago Press, Chicago.

LANGLEY, P., SIMON, H., BRADSHAW, G. and ZYTKOW, J. (1987). *Scientific Discovery: Computational Explorations of the Creative Processes.* MIT Press, Cambridge, Mass.

LAPLACE, P. S. (1917). *A Philosophical Essay on Probabilities.* Wiley, New York. (English translation.)

LARKEY, P., SMITH, R. and KADANE, J. (1989). It's okay to believe in the "hot hand." *Chance* **2** (4) 22–30.

LETAC, G. (1981). Isotropy and sphericity: Some characterizations of the normal distribution. *Ann. Statist.* **9** 408–417.

LIEBERSON, S. (1985). *Making it Count.* Univ. California Press, Berkeley.

LINHART, H. and ZUCCHINI, W. (1986). *Model Selection.* Wiley, New York.

LUCE, A. D. (1986). *Response Times. Their Role in Inferring Elementary Mental Organization.* Oxford Univ. Press, New York.

MACCORQUODALE, K. and MEEHL, P. (1948). On a distinction between hypothetical constructs and intervening variables. *Psych. Rev.* **55** 95–107.

MAKRIDAKIS, S., ANDERSEN, A., CARBONE, R., FILDES, R., HIBON, M., LEWANDOWSKI R., NEWTON, J., PARZEN, E. and WINKLER, R. (1984). *Forecasting Accuracy of Major Time Series Methods.* Wiley, New York.

MANN, N. R., SCHAFER, R. E. and SINGPURWALLA, N. D. (1974). *Methods for Statistical Analysis of Reliability and Life Data.* Wiley, New York.

MARBE, K. (1899). *Naturphilosophische Untersuchungen zur Wahrscheinlichkeitstheorie.* Leipzig.

MARBE, K. (1916, 1919). *Die Gleichförmigkeit in der Welt* **1, 2.** C. H. Beck, München.

MILL, J. S. (1879). *System of Logic* **1,** 10th ed.

NELDER, J. A. (1986). Statistics, science and technology. *J. Roy. Statist. Soc. Ser. A* **149** 109–121.

NEYMAN, J. (1939). On a new class of 'contagious' distributions, applicable in entomology and bacteriology. *Ann. Math. Statist.* **10** 35–57.

NEYMAN, J. and SCOTT, E. L. (1959). Stochastic models of population dynamics. *Science* **130** 303–308.

OLBY, R. (1985). *Origins of Mendelism,* 2nd ed. Univ. Chicago Press, Chicago.

OREL, V. (1984). *Mendel.* Oxford Univ. Press, New York.

PEARSON, K. (1902). On the mathematical theory of errors of judgment. *Philos. Trans. Roy. Soc. London Ser. A* **198** 235–299.

POPPER, K. (1965). *Conjectures and Refutations: The Growth of Scientific Knowledge*, 2nd ed. Basic Books, New York.

POSKITT, D. S. (1987). A modified Hannan-Rissanen strategy for mixed autoregressive-moving average order determination. *Biometrika* **74** 781–790.

ROSEN, E. (1971). *Three Copernican Treatises*, 3rd ed. Octagon Books, New York.

ROSENBERG, A. (1985). *The Structure of Biological Science*. Cambridge Univ. Press, Cambridge.

SHIBATA, R. (1981). An optimum selection of regression variables. *Biomerika* **68** 45–54.

STERZINGER, O. (1911). *Zur Logik und Naturphilosophie der Wahrscheinlichkeitslehre*. Doctoral thesis, Leipzig.

STIGLER, S. M. (1986). *The History of Statistics: The Measurement of Uncertainty Before 1990*. Belknap, Cambridge, Mass.

STUDENT (1927). Errors of routine analysis. *Biometrika* **19** 151–164.

THOMPSON, J. R. and BROWN, B. W. (eds.) (1987). *Cancer Modeling*. Dekker, New York.

TONG, H. (1975). Determination of the order of a Markov chain by Akaike's information criteria. *J. Appl. Probab.* **12** 488–497.

TOULMIN, S. (1961). *Foresight and Understanding*. Indiana Univ. Press, Bloomington, Ind.

TVERSKY, A. and GILOVICH, T. (1989a). The cold facts about the "hot hand" in basketball. *Chance* **2** (1) 16–21.

TVERSKY, A. and GILOVICH, T. (1989b). The "hot hand": Statistical reality or cognitive illusion? *Chance* **2** (4) 31–34.

VAN FRAASSEN, B. (1989). *Laws and Symmetries*. Oxford Univ. Press, New York.

WILSON, C. A. (1975). Kepler's ellipse and area rule: Their derivation from fact and conjecture. In *Kepler. Four Hundred Years* (Proceedings of conferences held in honour of Johannes Kepler) (A. Beer and P. Beer, eds.). Pergamon, Oxford.

J.K. Ghosh, S.K. Mitra, K.R. Parthasarathy and B.L.S. Prakasa Rao (*eds.*),
Statistics and Probability : A Raghu Raj Bahadur Festschrift
Published by Wiley Eastern Limited, Publishers, (1993), 371 – 380

The Bertrand-Borel Debate and the Origins of the Neyman-Pearson Theory[1]

E.L. LEHMANN
University of California, Berkeley

Abstract

The views of the mathematicians Bertrand and Borel on hypothesis testing and the usefulness of probability models are reviewed as background to the Neyman-Pearson theory. There is a discussion of Neyman's claim that these views had an influence on his work with Pearson, and of Pearson's denial of this claim.

1 Introduction

Throughout the 19th Century, hypothesis testing was carried out in various different areas of application along the lines laid down in the "Théorie Analytique des Probabilités" by Laplace (1812, 1814, 1820) and in Poisson's "Recherches sur la Probabilité des Jugements" (1837). Nearly all of this work was asymptotic, based on De Moivre's normal approximation to the

[1] Research supported by NSF Grant No. DMS-8908670.
AMS Subject Classifications : Primary 62-03; secondary 62 F 04
Key words and Phrases: Hypothesis Testing, Neyman-Pearson Theory, Probability Model.

binomial and Laplace's central limit theorem. Toward the end of the century, the approach was summarized by Edgeworth (1895) and extended in an important new direction by Karl Pearson (1900).

A new phase, small-sample theory, was ushered in by Student (1908) and was developed by Fisher in a series of papers culminating in his book "Statistical Methods for Research Workers" (1925). In all the work up to this point the test was taken as given (on intuitive grounds) and the emphasis was on the calculation of the p-value or, when a Bayesian approach was used, on the probability of the hypothesis being true.

The final step, asking how to choose the test, was taken by Neyman and E.S. Pearson (1928, 1933) who formulated the problem and to a large extent solved it. These authors give somewhat different accounts of the main influences that inspired them in this work. Pearson (1939, 1966) attributes the suggestion that the appropriate test must depend on the alternatives against which the hypothesis is to be tested to a letter of Student while Neyman cites the influence of a debate between Bertrand and Borel.

The Bertrand-Borel debate[2] is concerned not only with the problem of how to choose a test, but also with the more fundamental question of whether calculations of p-values are appropriate at all since they are based on probability models that at best are rough approximations of a much more complex real situation. In presenting the views of Bertrand and Borel in sections 2 and 3, we shall be concerned with both these aspects of the debate. Sections 4 and 5 discuss its relation to the Neyman-Pearson theory.

2 Bertrand's views

Neyman's source for Bertrand's views was Bertrand's book "Calcul des Probabilités (1889, 1897, 1907). The author explains the purpose of his book in a short preface. Probability theory is one of the most attractive branches of the mathematical sciences, he says, and yet one of the most neglected. He suggests that one of the reasons for this state of affairs may be Laplace's book (the Théorie Analytique), which is indispensable for the study of probability theory, but requires a great deal of mathematical preparation. Bertrand's book is to remedy this situation by providing an account of the most useful and best known results with the simplest possible proofs.

Emulating Laplace, Bertrand begins his book with a nontechnical essay, "Les lois du hasard" [The laws of chance], which gives an introductory overview.[3] Besides the essay, the relevant chapter for the present purpose is chapter 7 on the probability of causes. Much of this chapter is devoted

[2] This seems the best way to describe the situation although it was not a debate in the sense of a formal discussion. Bertrand stated certain views in his probability text of 1889. He was no longer alive when Borel criticized them rather sharply in a popular book on probability in 1914.

[3] However, in contrast to Laplace's ebullience and optimism, Bertrand's attitude toward applications is rather negative.

to an attack on hypothesis testing on the grounds that the plausibility of a hypothesis depends on the prior probability of it being true, which is rarely known.

Then, speaking about Michell's famous problem[4] of whether the closeness to each other of the six (visible to the naked eye) stars forming the Pleiades can be attributed to chance, he changes the direction of attack. He again first refers to the unavailability of prior probabilities, but then asks: "How can we decide on the unusual results that chance is incapable of producing?" and continues (p.166)[5]

"The Pleiades appear closer to each other than one would naturally expect. This statement deserves thinking about; but when one wants to translate the phenomenon into numbers, the necessary ingredients are lacking. In order to make the vague idea of closeness more precise, should we look for the smallest circle that contains the group? the largest of the angular distances? the sum of squares of all the distances? the area of the spherical polygon of which some of the stars are the vertices and which contains the others in its interior? Each of these quantities is smaller for the group of the Pleiades than seems plausible. Which of them should provide the measure of implausibility? If three of the stars form an equilateral triangle, do we have to add this circumstance, which is certainly very unlikely apriori, to those that point to a cause?"

What Bertrand seems to consider here is the choice between a number of possible test statistics, small values of which are significant. However, in the last sentence he addresses an additional issue, namely what features of the available material to include in the observations to be analysed.

Beyond these particular questions, Bertrand here also raises doubts about the scientific validity of such a probabilistic approach to testing. In the introductory essay he goes much further. After discussing Michell's own analysis, he states (p.xviii): "The application of such calculations to questions of this kind is a delusion and an abuse."

Quite generally, Bertrand opposes probability models, particularly in the social sciences, as quite unrealistic,[6] and this aspect of the debate continues today. His statement concerning Laplace's law of succession: "If time is to confirm and increase [the conviction that the sun will rise tomorrow], it will be by the discovery of the laws of astronomy and not by the repeated success of an analogous game of chance", is echoed in recent writing of David Freedman (for example, Freedman (1985) and (1987)).

---

[4]For recent reviews of some of the extensive literature on this problem see Gower (1982) and Dale (1991).

[5]Since copies of the book by Bertrand (and that of Borel) cited here are easily available, I have translated the passages to be quoted for easier reading. The page references are to the edition of 1907.

[6]For further discussion of his views see for example Stigler (1986) and Daston (1988).

3 Borel's reply

A very different position is taken by Borel in his book "Le Hasard" (1914); the passages given here in translation are from the revised edition of (1948). His treatment of these issues occurs in chapter 4 which, like Bertrand's corresponding chapter, is entitled "probability of causes". After discussing Bayes' theorem, Borel mentions the example of the frequency of male births and restates it as an urn problem. After some discussion he points out that "this example shows the particular form that problems of causes often take on. This form is the following: Is such and such a result due to chance or does it have a cause? It has often been observed how much this statement lacks in precision. Bertrand has strongly emphasized this point.[7] But whatever objection one can raise from a logical point of view cannot prevent the preceding question from arising in many situations: the theory of probability cannot refuse to examine it and to give an answer; the precision of the response will naturally be limited by the lack of precision in the question; but to refuse to answer under the pretext that the answer cannot be absolutely precise, is to place oneself on purely abstract grounds and to misunderstand the essential nature of the application of mathematics. True, the calculations provide an exact answer to every exact question; but in practice the questions are never exact: the given experimental facts necessarily permit à certain amount of play; the same lack of precision then affects the result of the calculations and the would-be absolute theoretical precision of these calculations is pure delusion."

Borel next turns to problems in astronomy and after some introductory remarks quotes the passage about the Pleiades from Bertrand that was given in the preceding section. Referring to Bertrand's equilateral triangle, he writes: "If one has observed a number such as that quoted above [1.342517] as evaluation of an angle in tenths of seconds (in the decimal system), one would not think of asking to know the probability that this angle is exactly $13^G\ 42'\ 51''\ 7$; because one would never have asked that precise question before having measured the angle...

"The question is whether one has the same reservations in the case in which one states that one of the angles of the triangle formed by three stars has "une valeur *remarquable*" [a striking or noteworthy value], and is for example equal to the angle of the equilateral triangle ($66^G\ 66'\ 66''$),[8] or to a half right angle: 50 decimal degrees, or 5.000.000 tenths of a second. Here is what one can say on this subject:

"One should carefully guard against the tendency to consider as striking an event that one has not specified *beforehand*, because the number of

[7] Borel here refers to Bertrand's discussion of the different interpretations that can be given to the idea of randomness, as illustrated by his well-known treatment of Buffon's needle problem; cf Bertrand (1889, ch.1).

[8] The old-fashioned french angular measure "grade" is a decimal measure; a right angle is 100 grades.

such events that may appear striking, from different points of view, is very substantial.

"In contrast to this, the question of grouping [i.e. clustering] in space is one that it is natural to ask apriori; the theory of probability does not enable one to resolve it with certainty but makes precise the conclusions one can draw from observations.

"This precision is quite useful, because it often effectively replaces inferences at which one arrives too quickly by 'common sense' ... Between the two extreme situations [those in which the results clearly are or are not significant] there are many others, namely those in which the probability lies between for example one hundredth and one hundred-thousandth: the value [the probability] has between these limits is not at all useless and what one can learn from it may be very valuable, no matter what Bertrand thinks".

In summary, Bertrand raises two questions: (i) on what grounds to prefer one test statistic to another; (ii) whether the calculation of a p-value, once the test statistic has been chosen, is at all useful in view of the shaky assumptions underlying the typical probability model.

To the first point Borel replies with the idea that the test statistic should represent some striking feature of the situation; on the second he asserts his unshaken faith in the value of probabilistic calculations even though they are only approximations to the true values.

It should be noted that neither of these authors offers any direct suggestion that the test should be based on more plausible alternatives although both Paul Lévy and Neyman were later to interpret Borel's comments in this way.

4 Circonstances remarquables

The first mention of the views of Bertrand and Borel in Neyman's papers occur in Neyman (1929) where he refers to Bertrand's critique of the Bayesian approach to hypothesis testing and then mentions:[9]

"M. Borel who also discusses analogous questions, observes that the correct judgment concerning a statistical hypothesis depends on the choice of the aspect of the observed fact $f(E)$ which in some way must be 'remarquable'. This idea of a remarkable aspect which is a bit vague can be made precise if one agrees to consider also alternative hypotheses."

The debate between the two french mathematicians is discussed at greater length in Neyman and Pearson (1933), and Neyman returns to it again in Neyman (1935) where he writes:[9]

"... the origin of the general theory of testing hypotheses is linked to most interesting remarks of french mathematicians, particularly of J.

[9] My translation of the original french.

Bertrand (1889) and M.E. Borel (1932). They considered the process of hypothesis testing which is basically the one employed by the english school...

"It is known that Bertrand was skeptical about the scientific value of the results of such a process. On the other hand, M. Borel insisted that such a result can be valid provided the feature $f(E)$ of the observed facts that serve as basis for testing N, be 'en quelque sorte remarquable'.

"This last remark preceded a series of works [by Neyman Pearson, and others] which basically strove to give a precise meaning to Borel's somewhat vague words..."

How much importance Neyman attached to the influence of Borel's remark can be seen from the fact that late in life he returned to it again in Neyman (1977), an account of his views on statistics. He mentions that the methodology for testing hypotheses "was the subject of a lively discussion by Borel and others", and goes on to say that "Borel was optimistic but insisted that: (a) the criterion to test a hypothesis ... must be selected *not after the examination of the results of observation*, but before, and (b) this criterion should be a function of the observations 'en quelque sorte remarquable'. It is these remarks of Borel that served as an inspiration to Egon S. Pearson and myself in our effort to build a frequentist theory of testing hypotheses".

Neyman sent a preprint of this paper to E.S. Pearson who expressed himself[10] delighted with it ("so delighted that I have started devouring it, although only received five hours ago!" he wrote) and who attached to his reply to Neyman a large number of comments. In one of the first of these he takes exception to Neyman's attribution of inspiration cited above. He writes:

"I remember that you produced this quotation when we began to get our [1933] paper into shape and write it up. This must have been during the period 1930/31, or even as early as my summer visit to Poland of 1929. I had not heard of the Borel quotation until you started to draft the introductory section of our paper. The words may have for a year or two been acting as an inspiration to you, but while I at once felt the words relevant and important, my approach was a different one, and was not I think *altered* by inspiration from the words."

Pearson goes on to say that in 1925 or 1926 he had begun to question the basis of the t-test and to ask: Why use that particular statistic? and that this led to the question: 'How choose the rejection region?' He then continues:

"The above stages led up to Borel's requirement of finding ... a criterion which was 'a function of the observations 'en quelque sorte remarquable'.

"Now my point is that you and I (perhaps my first leading) had our-

[10] In an unpublished letter of September 28, 1976, which is contained in the Neyman Collection of the Bancroft Library, University of California, Berkeley.

selves reached the Borel requirement independently of Borel, because we were serious humane thinkers; Borel's expression neatly capped our own requirement."

As mentioned in the introduction, a strong outside influence had also affected Pearson's thinking. This was a letter[11] by Student written in an answer to a question by Pearson. This letter contains the crucial suggestion: "If there is an alternative hypothesis which will explain the occurrence of the sample with a more reasonable probability ... you will be very much more inclined to consider that the original hypothesis is not true."

Pearson must very soon have communicated this idea to Neyman because a letter of Neyman's to Pearson, dated December 9, 1926 begins: "I think that to have the possibility of testing it is necessary to adopt such a principle as Student's, ..." (Reid (1982), p.70). On the other hand, there appears to be no mention of Borel in this or other letters of Neyman's to Pearson of this period. This combination of circumstances lends some support to Pearson's recollections.[12]

It is interesting to note that Neyman was not the only one on whom Borel's discussion of tests of randomness made a deep impression. Paul Lévy in his "Calcul des Probabilités (1925) treats this topic in a section which he acknowledges to be largely borrowed from Borel. He there also mentions the role played by "propriétés remarquables" and goes on to analyse the meaning of this phrase. And he comes to the conclusion which — by whatever common or separate routes — Neyman and Pearson were also to reach. "Une propriété remarquable", he writes, "est une propriété facile à expliquer par une cause autre que le hasard."[13] (p.93). [A striking property is a property that is easy to explain by a cause other than pure chance.]

5 The Bertrand-Borel debate and the Neyman-Pearson theory

The basic concepts of the Neyman-Pearson theory: rejection region, type 1 error, alternatives to the hypothesis, and power, have became so embedded in our statistical language and thought that it is difficult for us to see the situation as it was in the first quarter of the century. The debate of the French probabilists of that period provides a window into their thinking.

What strikes the modern reader is the vagueness of the discussion which is of course inevitable when no conceptual framework and suitable terminology are available. As a partial explanation of "the obscurity which envelops the theoretical bases of statistical methods" Fisher in his fundamental paper, "On the mathematical foundations of theoretical statistics" (1922) suggests that "it appears to be widely thought, or rather felt, that in

[11] This letter of May 11, 1926 was published in part in Pearson (1939).

[12] The Neyman-Pearson story is told from Pearson's point of view in Pearson (1966) and from Neyman's in Reid (1982).

[13] Neyman's only reference to Lévy in this general context is in Neyman (1932) where be attributes the concept of "remarquable" characters to "E. Borel and P. Lévy" in the books quoted here.

a subject in which all results are liable to greater or smaller errors, precise definition of ideas or concepts is, if not impossible, at least not a practical necessity."

In this paper Fisher did in fact create the needed theoretical framework in the context of estimation. He formulated his theory in terms of a parametric family of distributions (of the observations), posed the problem of choosing a suitable estimator, defined a measure of accuracy, and introduced both relative and absolute efficiency. However, Fisher did not extend the theory to hypothesis testing, which — like his predecessors — he discussed only in the context of specific examples or classes of examples.

It remained for Neyman and Pearson to formulate a general theory of hypothesis testing. Fisher's parametric families provided a way to specify both hypothesis and alternatives. Instead of speaking, as Bertrand and Borel had done, of choosing among vaguely indicated test criteria, the class of available tests is now clearly defined as the class of all possible level α rejection regions. Furthermore the concept of power made possible a comparison of the advantages and disadvantages of different tests and led to maximal power as a principle for chosing a "best" test.

There is a world of difference between the vague musings of Bertrand and Borel, and the clearly defined structure of the Neyman-Pearson theory. In particular, the latter of course permits quantitative evaluations and comparisons.

It is just this clarity and precision, at the time only an unrealized idea, that is at the heart of the debate between Bertrand and Borel. Bertrand would undoubtedly object to many of the uses to which tests are put (and he would not be alone in this). It is true, in particular, that the models from which the analysis is derived often have a rather flimsy basis and even at best are only approximations. The theory has tried to meet these and related objections by studying the robustness of existing tests, developing new more robust alternative tests, by providing simultaneous test procedures to deal with the problem of multiplicity, and so on, all within the general Neyman-Pearson framework.

To the many objections that remain, Borel has an answer in the passage quoted in Section 3 which can be summarized as follows. "The theory of probability cannot refuse to examine" the situation and to come up with an answer. Of course, the real situation is much more complex than the model allows for; the precise answers that the calculations provide for hypothetically precise questions are only rough approximate solutions of the much less precise real problems. We should therefore not take the analysis too literally, put too much stock in the precise numerical answers, but — we should not give up on the enterprise. It provides invaluable guidance, a check on undisciplined speculation, on "inferences at which one arrives too quickly by 'common sense'."

REFERENCES

Bertrand, J. (1889, 1897, 1907) Calcul des Probabilités Gauthier-Villars, Paris.

Borel, E. (1914, 1920, 1932; New Revised Edition 1948) Le Hasard. Presses Universaires de France, Paris.

Dale, A.I. (1991) A History of Inverse Probability: From Thomas Bayes to Karl Pearson, Springer Verlag, New York.

Daston, L. (1988) Classical Probability in the Enlightenment, Princeton Univ. Press.

Edgeworth, F.Y. (1895) Methods of Statistics, Jubilee Volume of the Statist. Soc., 181-217.

Fisher, R.A. (1925) *Statistical Methods for Research Workers*, Oliver and Boyd, Edinburgh.

Gower, B. (1982) Astronomy and probability. Forbes versus Michell on the distribution of the stars. *Annals of Science*, **39**, 145-160.

Freedman, D. (1985) Statistics and the scientific method, Cohort Anal., 343-359.

Freedman, D. (1987) As others see us: A case stüdy in path analysis, *J. Educ. Statist.*, **12**, 101-128.

Hochberg, Y. and Tamhane, A. (1987) *Multiple Comparison Procedures*, Wiley, New York.

Laplace, P.S. (1812, 1814, 1820) *Théorie Analytique des Probabilités*, Courcier, Paris.

Lévy, P. (1925) *Calcul des Probabilités*, Gauthier-Villars. 1925.

Neyman, J. (1929) Méthodes nonvelles de vérification des hypothèses. Comptes Rendus du 1 Congrès de Mathématiciens des Pays Slaves. Warsaw, 355-366.

Neyman, J. (1932) On methods of testing hypotheses, *Atti Congr. Internat. Matematici*, **6**, 35-41.

Neyman, J. (1935) Sur la vérification des hypothèses statistiques composées, *Bull. Soc. Math. de France*, **63**, 346-366.

Neyman, J. and Pearson, E.S. (1928) On the use and interpretation of certain test criteria, *Biometrika*, **20A**, 175-240 and 263-294.

Neyman, J. and Pearson, E.S. (1933) On the problem of the most efficient tests of statistical hypotheses, *Phil. Trans. Roy. Soc. (A)*, **231**, 289-337.

Pearson, E.S. (1939) "Student" as a statistician, *Biometrika*, **30**, 205-250.

Pearson, E.S. (1966) *The Neyman-Pearson story, 1926-34*, In Research Papers in Statistics (Festschrift for J. Neyman), F.N. David (Ed). Wiley. New York.

Pearson, K. (1900) On the criterion that a given system of deviations from the probable in the case of a correlated system of variables is such that it can be reasonably supposed to have arisen from random sampling, *Phil. Mag.* *5*, Vol. **50**, 157-175.

Poisson, S.D. (1837) Recherches sur la Probabilité des Jugements, Bachelier, Paris.

Reid, C. (1982) *Neyman from Life*, Springer-Verlag, New York.

Shaffer, J. (1988) *Simultaneous testing*, In Encycl. Statist. Sci. Vol. 8, 484-490. Wiley, New York.

Stigler, S. (1986) *The History of Statistics*, Belknap Press, Cambridge.

Student (1908) On the probable error of a mean, *Biometrika*, **6**, 1-25.

Department of Statistics
367 Evans Hall
University of California
Berkeley, California 94720
U. S. A.

Statistical Science
1993, Vol. 8, No. 3, 331–341

Mentors and Early Collaborators: Reminiscences from the Years 1940–1956 with an Epilogue

E. L. Lehmann

Abstract. These reminiscences extend from the year 1940, in which I arrived in Berkeley, to 1956, the year in which Neyman resigned from the chairmanship of the Berkeley Statistics Department and handed its leadership over to the next generation. They sketch my experiences with six scientists who have influenced me as mentors or collaborators: Evans, Neyman, Wald, Scheffé, Stein and Hodges. The origin of these recollections was a conversation with Agnes Herzberg and Persi Diaconis, which was videotaped under the sponsorship of Pfizer Central Research and the American Statistical Association on April 28, 1992 by the Department of Statistics at the University of Connecticut under the direction of Harry Posten. Although the conversation went in a somewhat different direction and the overlap is moderate, it got me thinking about the people who influenced me in my 20's and 30's and thus led to the present paper.

1. INTRODUCTION

It was exactly fifty years ago that I first heard of statistics as a possible subject of study, and at the urging of Professor Griffith Evans agreed to give it a try. However, the applied nature of the material did not appeal to me. After a semester, I decided to return to pure mathematics but just then was offered a teaching job in the statistics program that I could not refuse. Gradually, I got to like the work and even found the relevance of probabilistic and statistical ideas to so many different aspects of the real world of great interest. So when recently at a meeting with a group of students at the University of Connecticut I was asked to assess my career, my heartfelt response was: "It turned out better than expected."

That it turned out well was partly due to the situation of the field at the time I entered it. The seminal work of Fisher and Neyman-Pearson had laid the foundations; building a superstructure on this basis was a much easier task for which students of Neyman were well prepared. But there was another factor. This was the cohesiveness of the developing profession, the friendship, collegiality and mutual support that the small but growing group of theoretical statisticians extended to one another. The Institute of Mathematical Statistics, which was my scientific home, had fewer than 500 members when I joined it in 1943; *The Annals of Mathematical Statistics*, which published many of my papers and on the editorial board of which I served for many years, in 1943 ran to about 450 pages. As a result, the atmosphere was encouraging rather than intimidating, and many opportunities were available even to a beginner.

During my career, I had the good fortune of enjoying the acquaintance and friendship of many remarkable people who shared their ideas with me and gave me their support. In the early stages, it was my teachers who influenced me; later, I learned from my coworkers; and finally, more recently, it was my students who struck out in new directions and required me to follow them as well as I could. The scientific contributions of many of the figures whom I encountered in the early part of my career have become part of the history of our subject. It therefore seemed that it might be of interest to record the following personal recollections.

E. L. Lehmann is Professor Emeritus, Department of Statistics, University of California at Berkeley, 367 Evans Hall, Berkeley, California 94720.

2. GRIFFITH C. EVANS (1887–1973)

Evans, one of the most distinguished American mathematicians of his generation, made major contributions to functional analysis (the subject of his Colloquium Lectures to the American Mathematical Society in 1916) and potential theory. In recognition of his work, he received many honors, including membership in the National Academy, the American Academy of Arts and Sciences and the American Philosophical Society. He was on the Faculty of Rice University when in 1934 he was asked by the University of California to take over and revitalize the moribund Berkeley Mathematics Department. He was given some new positions and was able to persuade some of the existing faculty to retire. As a result, it was possible for him to hire a group of active young mathematicians: Alfred Foster, Derrick Lehmer, Hans Lewy, Charles Morrey, Anthony Morse and Raphael Robinson. (For a more detailed account of this early history of the Berkeley Mathematics Department, see Rider, 1989. An account of Evans' work can be found in Morrey, 1983.)

Evans, one of whose fields of specialization was mathematical economics, took a broad view of mathematics. He had become interested in statistics through discussions in the summer of 1931 with R. A. Fisher, and "as early as 1935 [had] envisaged California as the place for a really outstanding statistician, if possible of the level of R. A. Fisher himself" (Reid, 1982, p. 142). However, Fisher's visit to Berkeley in 1936 to give the prestigious Hitchcock lectures was not a success, and Evans' choice eventually fell on the Polish statistician Jerzy Neyman.

Neyman who by that time had done (partly in collaboration with E. S. Pearson) his fundamental work on hypothesis testing, confidence intervals and survey sampling, came to Evans' attention through the series of lectures and conferences he gave in Washington, D.C. in 1937. After considerable hesitation and some negotiations, Neyman accepted Evans' offer of a Professorship in the Berkeley Mathematics Department, and at age 44 started work in Berkeley in 1938.

I first met Evans in late December 1940 when on my arrival in Berkeley I presented myself at the office of the Mathematics Department (the staff at that time consisted of a half-time secretary) to see whether I could enroll as a student. Although I had no appointment, I was ushered in to see Evans, a gentle, soft-spoken man who not only on the spot accepted me as a student but suggested that I should start, on a probationary basis, as a graduate student since he thought my studies in Europe had given me the equivalent of an American B.A.

I thus greatly benefited from Evans' informal and personal approach to administrative problems, of which the following story is another example. In his first year at Berkeley, Hans Lewy, who tended to work late into the night, had a morning class for which he sometimes came late and for which occasionally he overslept altogether. One morning when this had happened again, Lewy was awakened by the ringing of his doorbell. In his pajamas and still half asleep, he opened the door to find his chairman, Evans, standing there to remind him that his class was waiting for him. I am told that there was no need for Evans to intervene again.

After my first term had gone satisfactorily, Evans offered me a teaching assistantship. The seven or eight TA's of the Mathematics Department at that time taught their own courses independently, and this included the making up and correcting of the examinations and the assigning of grades. My first assignment was a course in analytic geometry, an ideal choice. It is more advanced and had better students than remedial algebra or trigonometry, which are beset by pedagogical problems; on the other hand, it is a much easier subject to teach than calculus. I believe that Evans chose the course deliberately to make it easy for a student who was new to the program and to the country. It is only one example of his thoughtfulness and his concern for me and the other graduate students. He cared for us and was always available for advice and help when we needed it.

In the summer of 1942, half a year after Pearl Harbor and Germany's declaration of war against the U.S., Evans expressed to me his feelings that I could be more useful to the war effort if I left pure mathematics to work in either physics or statistics. Since I greatly respected his judgment and had a very negative attitude toward physics (about which he knew and which he ascribed to a mental block), I agreed to try statistics, a subject of which I had never heard before. He told me to see Professor Neyman, whom I had not met. Thus, as a result of Evans' intervention, I began in the Fall of 1942 what would become a lifelong career in statistics.

Soon after this switch, I became part of Neyman's group and no longer had much contact with Evans. From these later years, only one memory of him stands out. The story is characteristic both of his originality and his self-effacing modesty. When a new building was planned in 1969 which would house the mathematics and statistics departments and their libraries, it was decided to name it in honor of Evans. Even before the cornerstone was laid, a painting of Evans by Erle Loran was unveiled which was intended for the lobby of the new building and which now hangs in the mathematics–statistics library. The unveiling and naming of the building was celebrated by a dinner at which Morrey and others spoke of Evans' accomplishments. When he rose to reply, his opening sentence was rather surprising: "Who was Bacon?" He then proceeded to go through the Bacons listed in the encyclopedia and concluded his short talk by returning to his original question, now slightly amplified: "But who was the Bacon for whom Bacon Hall was named?"[1]

3. JERZY NEYMAN (1894–1981)

Neyman was 44 when in 1938 he accepted the offer to start a statistics program in the Berkeley Mathematics

[1] Bacon Hall was a building on the Berkeley Campus named after Henry Douglas Bacon (1813–1893), a businessman primarily in banking and real estate. He gave his art collection and large library to the University of California, together with some money for a building which later became Bacon Hall (Slutes, 1947).

Department. In Poland, he had headed a small statistical laboratory; in London, he had been a member of Egon Pearson's Department of Statistics at University College. Now he quickly recruited a small staff of assistants from the students in his graduate course to help with the laboratories that were part of his courses and with some of the applied work he was undertaking for faculty members in other departments.

When in 1942 I accepted Evans' suggestion to study statistics, the statistics program consisted essentially of three courses: a one-semester lower division course, and a one-year course each at the upper division and graduate level. In addition, there was a graduate course in probability theory. To take over the teaching of the latter, Neyman had in the Fall of 1942 obtained the services of a young mathematician, Dorothy Bernstein, who was also giving the lower division course.

A more ambitious development of courses and faculty to teach them had to wait until the end of the war but eventually led to the creation during the decade 1946–1956 of the Berkeley Statistics Department, which became one of Neyman's principal American achievements. For many years, it was the leading department of theoretical statistics in the county. Its curriculum set the standards that were followed by many others; it trained hundreds of Ph.D. students from all over the world. In addition, the Berkeley Symposia on Mathematical Statistics and Probability, international meetings that Neyman organized at 5-year intervals from 1945 to 1970, made Berkeley an international center of the first magnitude.

In accordance with Evans' suggestion, I enrolled in Neyman's upper division course but after one semester decided that I did not like statistics. It seemed messy, and the assumptions often appeared to be quite arbitrary. However, before I was able to tell Evans and Neyman of my intention to return to pure mathematics, something happened that changed my mind. Dorothy Bernstein had come to the same conclusion as I and had requested to be released from her contract. Caught in a bind, Neyman asked me whether I would like to take over some of her duties; I was to receive a promotion and a corresponding increase in salary. The offer was too good to refuse.

With this change of status, I became a member of Neyman's group, the small group of budding statisticians who assisted Neyman in the various aspects of his work. This meant in particular the ever present possibility of being drafted for some of the tasks required by Neyman's activities, such as the Symposia and his applied projects. An exciting, but rather scary, example occurred very quickly when Neyman told me that he was leaving for three weeks and wanted me to take over the lectures of the upper division course (my first statistics course!) in which I was a student. He outlined the material to be covered, suggested Uspensky's (1937) book on probability theory as a reference, and then I was on my own.

It turned out not to be too difficult since the course provided only an introduction to probability theory and least squares estimation, roughly corresponding to the first 14 chapters of F. N. David's book *Probability Theory for Statistical Methods* (1949). In hindsight, it is surprising that it contained nothing about testing or confidence intervals. These topics, which Neyman must have considered as still too nonstandard for inclusion in the undergraduate program, were treated along the lines of his own work in the graduate course.

Neyman had an interesting way of teaching. He liked to call students to the board, giving preference to the women in the class ("Ladies first"), and would try to get them to derive the new results under his guidance, which often would essentially turn into dictation. It was somewhat of an ordeal for the hapless victim, and I recall Neyman once coming to class saying that he had just received a letter concerning this practice accusing him of sadism. He clearly was at a complete loss – "Sadism?" – but announced that any student feeling this way could be excused from coming to the board. To the best of my knowledge, no one took advantage of this offer.

In 1944, Neyman recommended me for the Operations Analysis Section that was being formed by the 20th Air Force. Thus, I spent the year 1944–1945 on Guam, where I found myself sharing a tent with my fellow student Joe Hodges and with the statistician George Nicholson.

I returned to Berkeley just in time to participate in the Symposium on Mathematical Statistics and Probability that Neyman organized in 1945 to "mark the end of the war and to stimulate the return to theoretical research." It was a grand event, which gave us graduate students a chance to meet some of the great names in the field. I drew a particularly nice assignment: to be the driver of the probabilists Feller and Doob. I recall a memorable drive to Stanford in which my two passengers entertained each other and me by playing various games such as asking how big a salary it would take for each of them to accept a position in Berkeley. The figure went up and down with the quality of the view and the surroundings. When we reached the Pacific, Feller lowered his figure substantially, but in response Doob raised his since he might be tempted to swim in the ocean and could drown.

The symposium was such a success that Neyman soon started organizing a second one, which took place in 1950. From then on, the Berkeley Symposia became a regular event every fifth year, the sixth and last being held in 1970. In addition, there were many visitors who came to give summer courses or to substitute for a faculty member on leave. This gave us a chance to get to know such luminaries as Cramér, H. B. Mann,

Cochran, Wald, the mathematicians Besicovich and Bochner, and others.

Pleasing his guests was an avocation for Neyman; his hospitality had an international reputation. One of the toasts he would propose at all social events was "To the international intellectual community!" I now realize – although I didn't at the time – how close to Neyman's heart this was. Like many emigrants, he was not completely at home anywhere; the international intellectual community had become his true home, and the toast celebrated this fact.

During the year 1945–1946, I wrote my thesis (on a problem in the Neyman–Pearson theory suggested to me by P. L. Hsu), and the next academic year was appointed Instructor, the first regular faculty member in statistics after Neyman. In the succeeding years, faculty appointments were extended to other students as they completed their degrees: Hodges, Barankin, Scott, Fix, Le Cam. Retention of Berkeley Ph.D.'s on the faculty generally was frowned upon, but an exception was made for Neyman since no one else was providing the training he wanted for his faculty. He did, however, gradually also make some important outside appointments: Stein, Loève, Scheffé and Blackwell, who strengthened and broadened the group enormously.

The year 1947 brought Neyman two great victories. From the beginning, he had envisaged an independent statistics department, separate from the mathematics department. Another aim was a journal under his control that would provide a means of publication for himself and his students and associates. He was in a strong position to confront the administration with these demands when he received an offer from Abraham Wald to join him in the new statistics department that Wald was then forming at Columbia.

Evans adamantly opposed a separate department since he believed in a "greater mathematics department" that would include all the mathematical sciences. A compromise was now reached that left statistics within the mathematics department but with a separate budget that no longer required Evans' approval, and with the right to make its own research appointments although Evans would still have a say on teaching appointments.

A journal of his own had been of great importance to Neyman since in England he had seen the efforts of Karl Pearson to block publication of Fisher's papers, and even more so after the painful experience of having his own fundamental paper on confidence intervals rejected for *Biometrika* by his friend and close collaborator Egon Pearson. It turned out that a journal was not feasible, but Neyman was given something nearly as good: the series *University of California Publications in Statistics*, which for many years stayed essentially under his control.

Neyman's dislike of editorial restrictions was illustrated by an occasion which also brought out his somewhat puckish sense of humor. While working on the proofs of the second edition of his *Lectures and Conferences on Mathematical Statistics* (1952), a very engaging book which is still enjoyable today, he wanted to add an insert of several pages. Since this would have violated the editorial instructions, he cut the proofsheet in question and glued the insert onto the upper and lower part of the sheet, folding it so that when opened it flowed accordion-like to the floor. Since I was going to Washington at the time, he asked me to deliver these proofs to Ed Deming, who was in charge of the project. I left them at Deming's office without waiting for his reaction.

Neyman's wish for a completely independent Department of Statistics had to wait a few more years. When Evans retired in 1954, his successor Charles Morrey had no desire to keep a substantial group of statisticians in the Mathematics Department against their wishes and recommended the creation of a separate statistics department. Thus, in 1955 Neyman finally obtained his own department which consisted of seven tenured faculty members (Barankin, Blackwell, Hodges, Lehmann, Loève, Neyman, Scheffé), three tenure-track assistant professors (Fix, Le Cam, Scott) and several lecturers and visitors. There was a substantial course program including at the graduate level such specialized courses as "Stochastic Processes," "Nonparametric Inference," "Decision Theory," "Experimental Design" and "Large-Sample Theory."

In achieving this expansion in the short span of ten years, Neyman was helped by the fact that in this, as in other campaigns, he was always convinced of the righteousness of his cause. This justified for him the relentless pursuit of his goal. Each compromise reached, each partial success, provided the basis from which to launch his next demand. No wonder that one of his Deans was once heard to complain that he would rather grant whatever Neyman wanted than have to discuss it with him.

When the Berkeley Chancellor finally recommended to President Sproul the creation of a Department of Statistics, Sproul asked an assistant to summarize the history of the issue. As part of his summary, the assistant wrote: "Here a willful, persistent and distinguished director has succeeded, step by step, . . . against the original wish of his department chairman and dean in converting a small 'laboratory' or institute into, in terms of numbers of students taught, an enormously expensive unit; and he then argues that the unit should be renamed a "department" because no additional expense will be incurred. . . . How, in the future, can a bureau, laboratory, or institute director be restrained from enlarging his empire against the judgement of his administrative superiors?" (quoted from Reid, 1982, p. 239).

What Sproul's assistant left out of the account, but

which provided the basis for Neyman's vision, was the enormous developing growth of statistics which would soon make it into an extensive scientific field of world-wide importance. By establishing his program immediately after the war, at the beginning of this development, Neyman was able to obtain for Berkeley a leading position that it was to maintain for many years. The particular charge of inordinate expense in terms of the numbers of students taught, although true at the time it was made, lost its validity as gradually the new department took over the teaching of all lower division statistics courses and as a result soon regularly taught statistics to almost 5000 students a year.

The creation of the Department was followed by an unexpected development. Disputes between Neyman and some of his faculty (including myself) that had been simmering for some time now broke into the open – disputes which he himself later characterized as father–son conflicts. Neyman rightly considered the Department his creation. He wanted to continue to run it as a family enterprise of which he was the father and head, a benevolent dictator who would consult with other members but who in the end had sole responsibility and – with the interests of the group in mind – would make the final decisions.

However, even Neyman's own students on the faculty were by now in their mid and later 30's, had national and international reputations, and had their own ideas about appointments and courses which did not always agree with his. He once told me about his reaction to this situation. There is a bird, he said: if someone touches the eggs in its nest, it will no longer have anything to do with them. Feeling the same way, he took a step that came as as complete surprise to all of us. Only a year after the creation of the department for which he had worked so hard, he resigned as its chair and resisted all our efforts and those of the administration to change his mind. For himself he kept the directorship of the statistical laboratory with a small staff, and the university administration agreed to his request that during his lifetime the laboratory would be independent of the Department.

4. ABRAHAM WALD (1902–1950)

The structure that formed the framework for my work in theoretical statistics is due primarily to Jerzy Neyman and Abraham Wald. Although Wald's ideas, unlike Neyman's, came to me not directly from him but via his student Charles Stein, I did have some personal contacts with Wald during the short period between 1947, when I first met him, and his untimely death in an airplane accident three years later.

Wald made revolutionary contributions to sequential analysis, large-sample theory and decision theory. Of these, the last was most relevant to my own interests. In fact, it was in connection with this work that I first met him in 1947. The *Annals* of that year contained a short paper of mine in which I introduced in terms of a very simple example – testing a simple hypothesis in a one-parameter exponential family – a formulation of the testing problem that differed from the traditional Neyman–Pearson approach (Lehmann, 1947). Instead of trying to obtain a single optimum test, I pursued a more modest aim and worked out the smallest family of all possible candidates for this honor, suggesting that beyond that applied considerations would be needed to narrow down the choice. When shortly thereafter I happened to be in New York, I was informed that Wald wanted to meet me. He then told me that he had found my idea applicable to the general decision theory he was developing. He called such families minimal complete and under very general assumptions showed that they consisted exactly of all Bayes procedures together with certain of their limits.

During 1948–1949, Wald spent part of a sabbatical leave in Berkeley, giving a summer course and completing his book on statistical decision theory. Toward the end of that year, Joe Hodges, Charles Stein and I had planned a four-day hiking trip in Yosemite, and we asked Wald whether he would like to join us. Although he was considerably older than we and the hiking promised to be fairly strenuous (the first day we had to cover a distance of 20 miles and climb close to 6000 feet), he was game and turned out to be a good sport and fun to be with. We talked a great deal about statistics, but he was also very much interested in our surroundings. He particularly loved to estimate distances, speeds, heights, etc. I recall how at the bottom of a mountain we were about to climb, he estimated the rise to be 1500 feet. When we arrived at the top he looked down and declared himself satisfied: "Actually it was 1450 feet," he proclaimed, "I wasn't too far off."

Despite his accomplishments and fame, Wald was completely unpretentious, easy going and good natured. He was far less mercurial and competitive than his collaborator Jack Wolfowitz. (Since in statistics it has been traditional to list joint authors alphabetically, it was a standard joke in the profession that Wald had searched a long time before finding a suitable collaborator.) Mathematicians are often classified as being either problem solvers or system builders. Wald excelled at both.

Since I wanted to take a leave from Berkeley during 1950–1951, Wald arranged a visiting appointment at Columbia for me for the Fall semester; I had already made arrangements to spend the spring in Princeton. With Ted Anderson, Howard Levene, Henry Scheffé and Wald and Wolfowitz on the staff, Columbia was an interesting place – together with Berkeley, the strongest group in theoretical statistics in the country. Toward the end of the semester, Wald and his wife left for India, where he was to lecture on his new decision theory. Shortly before Christmas, we began to hear rumors of an airplane accident in which he might have been involved; it took two or three days before the Indian Government confirmed that indeed both Wald and his wife had died in the crash of a local plane that was to have taken them to Nepal. (For further information on Abraham Wald, see Wolfowitz, 1952.)

The death of Wald at age 48 was not only a great personal and scientific loss but also spelled the end of the group he was still in the process of building. Within a few years, Wolfowitz had moved to Cornell, Anderson to Stanford and Scheffé to Berkeley. It was a blow from which the Department still has not recovered, more than 40 years later.

One of the many problems caused by Wald's death was the fate of the students who had been working with him or were planning to do so. Two of the latter group approached me to ask whether I would take them on. So it came about that two Columbia Ph.D.'s, Allan Birnbaum and Jack Laderman, obtained their degrees under the supervision of a Berkeley faculty member.

It has been a great privilege to have known both Neyman and Wald. I only regret that I never met the third of the great founders of modern statistical theory, R. A. Fisher.

5. HENRY SCHEFFÉ (1907–1977)

Besides Evans, Neyman and (indirectly) Wald, the people who had the greatest influence on my development and career were three colleagues with whom I began to collaborate in the years immediately after completing my Ph.D.

The first of these was Henry Scheffé. Having started in analysis with a thesis on asymptotic solutions of certain differential equations, he had switched fields in his mid 30's since he believed statistics to be a more promising area of research. In 1941, to learn the new subject, he joined the famous group of students and associates (containing among others Ted Anderson, Bill Cochran, Fred Mosteller and John Tukey) whom Wilks had gathered at Princeton.

I first met Scheffé in 1946, when he came to spend a year at Berkeley on a Guggenheim Fellowship. Despite the image of a prizefighter conjured up by his picture, Henry turned out to be a very nonbelligerent, sensitive, rather shy person with interests in art, music and literature. He had worked on hypothesis testing problems similar to those considered in my thesis, and his papers had constituted some of my principal references. So we had many common interests and soon became friends.

Although we had dealt with very similar problems, Henry had used the original approach of Neyman and Pearson, while I had employed a method due to P. L. Hsu, who had given me my thesis problem. Discussing these different approaches to the problem of similar regions (i.e., of characterizing the totality of similar tests), we came to understand the common feature that lay behind both methods. It was the existence of sufficient statistics T having a property that we called completeness. For all cases in which such a statistic T exists, we then had a complete solution of the problem of similar regions. The result seemed exciting not only to us but also to Neyman, who arranged for quick publication of our results in the *Proceedings of the National Academy of Sciences* for 1947.

At the end of the year, Henry left for UCLA, from where the following year he moved to join Wald's Department at Columbia. In the meantime, our joint work continued by correspondence and through occasional visits of mine to Los Angeles and New York and of Henry's to Berkeley. An impetus for a fuller development of our completeness concept came from the realization that it played a crucial role also in the theory of unbiased estimation, where the existence of a complete sufficient statistic T assures that any function of the parameters that has an unbiased estimator has one with uniformly minimum variance, namely the unique unbiased estimator that is a function of T. This is an immediate consequence of the Rao-Blackwell theorem.

Since completeness has come to play such a central role in mathematical statistics, it should be pointed out that we deserve only limited credit for this concept. The basic idea was not original with us but was contained in the applications made in testing by Neyman-Pearson and more explicitly by Hsu, and in unbiased estimation in work by Blackwell, Halmos, Rao and Wolfowitz. Our contribution was to isolate the common feature of these various applications and to provide it with an identity by giving it a name and investigating its properties.

In view of its importance, we decided to explore the properties of the completeness concept (and some of its variants such as bounded and strong completeness) rather fully. The result was a massive two-part paper that we published in the Indian statistical journal, *Sankhyā* (Lehmann and Scheffé, 1950/1955/1956), which we thought would give us the space needed for the detailed discussion and the many examples we wanted to present. The paper also introduced the concept of minimal sufficient statistic, minimality being a necessary condition for completeness. (We found later that a theory of minimal sufficient statistics was being developed at the same time by Dynkin in the Soviet Union.)

In 1953, Neyman brought Henry to Berkeley on a permanent basis as Professor and Assistant Director of the Statistical Laboratory. Both he and I had hoped that this move would lead to further joint work, but to my great regret, although we remained close friends, we never again collaborated on a scientific investigation.

6. CHARLES STEIN (1920-)

When in 1945 I opened the September issue of *The Annals of Mathematical Statistics*, I was excited to see a paper containing a highly original solution of a problem that had long been of interest to Neyman and his students. In it, Charles Stein, in a way that was extremely elegant as well as effective, constructed a two-stage test of the hypothesis specifying the value of a normal mean that achieves a given power, independent of the variance, against an alternative value of the mean. I brought the paper to Neyman's attention, and he was as impressed as I was. The next year when he was visiting Columbia, he sought out Charles (who was then still a graduate student) with the result that in the Fall of 1947 Charles joined the Berkeley Statistics faculty.

Neyman's group was expanding, but the space assigned to it was limited; so Charles, Joe Hodges, Evelyn Fix (who had been Neyman's principal assistant during the war years) and I shared an office barely large enough to accommodate our four desks, which were arranged as a square block in the middle of the room. This enforced proximity of four congenial people with common interests led to many discussions and to joint work of mine with both Charles and Joe.

While Henry and I had been interested in the same kinds of problems, Charles, having studied with Wald, had a more decision theoretic background in which admissibility, minimaxity and least favorable distributions were the central concepts. On the other hand, Charles' two-stage paper of 1945 that had so impressed me was concerned with a problem in hypothesis testing. So it was not difficult to interest him in some of the problems in this theory that concerned me at the time. There was one gap that particularly bothered me. Neyman and Pearson had shown that the t-test was UMP among all similar tests. Was it in fact most powerful among all tests at the given level α?[2] This turned out to be essentially a minimax problem to which Wald's theory of least favorable distributions is applicable. In the t-problem, we were able to determine the least favorable distribution and to show that for $\alpha < 1/2$ (i.e., all levels of interest), the most powerful test does depend on the alternative, so that a UMP test does not exist. (The t-test is UMP for $\alpha \geq 1/2$.) By the same method, we were able to solve the corresponding problem for a number of other hypotheses concerning normal distributions. We had expected to extend the results, published as "Most Powerful Tests of Composite Hypotheses. I. Normal Distributions" (Lehmann and Stein, 1948), to exponential and perhaps other distributions but became more interested in other problems, so that the planned part II was never written.

Our next paper was motivated in part by a survey paper by Scheffé (1943), which called attention to "the need for constructive methods of obtaining 'good' and 'best' tests in the nonparametric case." This was the problem on which Charles and I decided to work next, and it resulted in our paper "On the Theory of Some Nonparametric Hypotheses" (1949). The optimal tests for which we obtained a "constructive method" turned out to be generalizations of Fisher's permutation version of the t-test.

Charles and I wrote two more papers together. One was a short note on completeness in the sequential case (1950); the other (1953) took up a problem raised by our first paper. The fact that at the usual levels the t-test is not UMP poses the question whether it is admissible. Since it is UMP among all similar tests, its admissibility follows immediately in a rather trivial local sense. We now proved that it is also admissible in a much stronger sense, namely against the class of alternatives specifying any fixed value of the standardized mean.

That our collaboration did not continue was mainly due to the fact that after only two years Charles left Berkeley for Chicago and later for Stanford. The reason was an anti-Communist loyalty oath that the Regents of the University of California imposed on the Faculty in 1949. After interminable discussions, the great majority of the Faculty signed, but 31 members refused to do so and were dismissed. A few years later, the California Supreme Court ordered their reinstatement. Some returned, while others remained in the positions they had found in the meantime.

For the statistics group, the controversy resulted in the loss of Charles who left the University rather than waiting to be fired. The rest of us, including Neyman who declared himself to be "color blind" on this issue, had found the oath unpalatable but in itself of no great importance and, though disgusted, had signed it without much difficulty.

In the course of our joint work, I learned from Charles about techniques for proving minimaxity including, in particular, the concept of invariance and the Hunt-Stein theorem. This theorem was included in the lecture notes of my course in hypothesis testing recorded by Colin Blyth (Lehmann, 1948–49), which led an existence as an underground text until the official publication of a much expanded version in 1959 (Lehmann, 1959). The notes refer to the Hunt–Stein paper

[2] I was teaching a graduate course on hypothesis testing, and it seemed to me only natural that some student would ask this question. In a similar way, much of my research arose from questions that came up in preparing my lectures. In contrast, Henry Scheffé told me that most of his research came out of his consulting practice.

"Most Stringent Tests of Statistical Hypotheses" as "to be published." However, by 1959, the paper still had not appeared, so that with Charles' permission I presented the theorem and its proof in my book. The reason for the postponement of publication of the Hunt–Stein theorem (told to me by Charles) is interesting. The theorem requires that the group in question satisfy a certain condition (essentially what is now called amenability). This condition holds for the group of translations and the group of orthogonal transformations which are the groups required for (univariate) analysis of variance.

Charles expected it to hold also for the full linear group, required for Hotelling's T^2-test, and felt uncomfortable publishing the paper without it. By 1959, he had constructed a simple counterexample, but by that time the theorem was pretty widely known, so that separate journal publications no longer held much interest for him.

For Charles, invariance was of interest as a condition that may insure stringency or other minimax properties of tests. For the exposition on which I was working for my course and later the book, it played an additional role. Since UMP tests exist only rarely, I was interested in reasonable conditions of "impartiality" which might lead to tests that are UMP within the class restricted in this way. Unbiasedness, introduced by Neyman and Pearson, was one such condition; invariance now provided another. In this way, the results on UMP invariant tests that I had learned from Charles in connection with most stringent tests became important in their own right and played a central role in my account of hypothesis testing. During the three years in which we worked together, Charles thus became an important influence on my development as a statistician and on my career. (For further information on Charles Stein see DeGroot, 1986.)

7. JOSEPH L. HODGES, JR. (1922-)

The third, last and most extensive of my early collaborations was with my fellow student and Guam tentmate Joe Hodges. Joe had come to Berkeley in 1938 as a beginning undergraduate and had stayed on until he obtained his Ph.D. under Neyman in 1948 with a two-part thesis: "I. Initial Sample Size in the Stein Procedure" and "II. Stringency in Acceptance Sampling." The first part he never published. The second part was the first paper in the series of UC Publications in Statistics (Hodges, 1949). (I think it might have had more of an impact had it appeared in a regular statistical journal.) Joe, like myself two years before, was retained by Neyman for his Faculty.

After we both had learned about Wald's decision theory from our officemate Charles Stein, Joe and I were struck by the dearth of concrete examples for which minimax solutions had been worked out, and we decided to tackle a simple case: estimating binomial p with squared error loss. Theory suggested that as a first step we should look for a Bayes estimator with constant risk, since any such estimator would automatically be minimax. The only priors for which we could see how to obtain the Bayes estimator and its risk explicitly were the beta-distributions, and beginners' luck was with us. There was a beta prior which led to a constant risk estimator, and so we had solved the problem on the first try.

But while our solution was a theoretical success, it was from a practical point of view a disaster. Since p is much easier to estimate accurately when it is close to 0 or 1, a constant risk estimator is typically not what one wants and, except for very small sample sizes, the standard unbiased estimator has much smaller risk than our minimax estimator over most of the range of p. What we had thus obtained, unwittingly, was a counterexample showing how unsatisfactory a minimax estimator can be.

Following this 1950 paper, we published during the next fifteen years roughly a paper a year. In the second paper, we presented a new method of proving admissibility by solving an appropriate differential inequality, an approach that has proved more widely applicable than we would have expected. One of my favorites is the next paper, "The Use of Previous Experience in Reaching Statistical Decisions" (1952), which proposes a compromise between the Bayes and minimax approach, namely to minimize the Bayes risk subject to a bound (somewhat larger than the minimax risk) on the maximum risk. Although the principle seems appealing and has found a number of applications, unfortunately such restricted Bayes solutions tend to be rather messy.

By the mid 1950's, Joe's and my interests had shifted from decision theory to the relatively new methodology of nonparametric inference. We were intrigued by Pitman's surprising result that the asymptotic relative efficiency (ARE) of the Wilcoxon to the t-test is $3/\pi = .955$ in the normal case. Computation of the efficiency in some other cases gave even higher values, which raised the question of how low this efficiency can get. The answer given in our paper "The efficiency of some nonparametric competitors of the t-test" (1956) came as a great surprise. The sharp lower bound for this ARE turned out to be .864. These high values gave a boost to nonparametric tests such as the Wilcoxon which generally had been thought to be inefficient.

What made our collaboration so pleasant and effective was the way we complemented each other. Joe was an outstanding problem solver, who enjoyed tackling a problem from scratch and inventing and developing whatever methods were needed to solve it. For me, it was more useful and important to see how the question fitted into the whole fabric of related results and if possible to be guided to a solution by the context. And I then liked to see what the solution contributed to the whole pattern and to consider how far it could be extended. To this difference between us one could add that Joe tended to think more geometrically and I more in terms of formal manipulation.

Joe and I did much of our joint work on walks with or without trails, in the Berkeley hills. Conversation on these walks was not confined to statistics; we talked about music, literature, politics and worked on composing limericks. Here is one of Joe's finest, in which I had no part, and which I quote with his permission.

The music of Johannes Brahms
Has strange, ineluctable charms.
And sometimes it seems
It might lapse into themes,
But alas, they are all false alarms.

Our most memorable hike took place in Summer 1950. We had planned a four-day hike with Charles Stein in Yosemite. Since Wald was in Berkeley at the time we asked him to join us and, as reported in Section 4, he agreed. On the second day of this walking tour, Charles fell sick, and we decided that we had to return. When our trail crossed the road late that afternoon, Joe and I were delegated to hitch a ride down to the valley where our car was parked and then to return to pick up the other two. Joe and I, neither of us experienced hitchhikers, had no luck and decided to try one at a time. Eventually a car stopped for me, and the two women in it opened the door for me to get in. At that moment Joe, with his big frame and height of 6 feet 4 came lumbering out of the bushes. The women gave a scream, slammed the door shut and sped off. We gave up the effort for that night, which was lucky. The next morning Charles felt all right, and the second part of the hike (over Vogelsang Pass) turned out to be the most scenic one.

8. EPILOGUE

The epilogue to a story deals with the future of its characters. The present section will do this for the three subjects of the present account who continued as members of the Berkeley Statistics Department after 1956: Neyman, Scheffé and Hodges.

Neyman

Neyman's resignation as chair constituted a major crisis for the new Department. The statistical community at large expected open warfare and the end of the Berkeley statistics group. Neyman was, after all, the dominant figure; in addition, as project and laboratory director, he controlled all research funds, the total of which was about equal to the departmental budget. There was thus plenty of opportunity for conflict, but the dire predictions did not come true.

That the transition proceeded peacefully was in the first place due to David Blackwell, who, as first post-Neyman chair of the Department, was able to smooth over potential conflicts and to maintain his close and friendly relationship with Neyman as well as with other members of the Department. However, a great deal of credit is also due to Neyman himself, who neither tried to dominate nor to wash his hands of the new departmental organization. He participated in staff meetings like any other member of the faculty and went out of his way to acknowledge the authority of the chair. This found a curious symbolic expression when nearly 20 years later I became department chair. As a student I, like Neyman's other students, had always called him Mr. Neyman, while he addressed us by our first names. This did not change after I joined the faculty; he never encouraged me to use his first name. Now he started signing his notes to me "Jerry."

Neyman had had such a fundamental influence on the development of statistics through his scientific and organizational work and his personal international efforts that it seemed an account of his life and work—the two are largely synonymous—would be of great value. When in 1978 I read Constance Reid's biography of Courant, it appeared to me that she would be the ideal author for such a project. When I approached her with this suggestion, her first reaction was negative, but a few months later she expressed a cautious interest. We agreed that I would sound Neyman out about his willingness to cooperate in such an undertaking. Neyman had often indicated to me that he had no interest in contemplating the past but wanted to think only about the future. It was therefore with some trepidation that I went to see him, with Reid's (1970) Hilbert and (1976) Courant biographies under my arm. Nevertheless, I was taken aback by his reply to my question of how he felt about such a project. "It's a free country," he said curtly.

I made it clear that this was not a basis on which we could proceed. There was no intention of going against his wishes, and in any case his cooperation would be required. "How much cooperation?", he wanted to know. When I told him that Constance needed to meet with him once or twice for an hour or two, he replied that he always enjoyed talking to young ladies. These initial conversations must have gone well, for a pattern soon developed of a meeting between them every Saturday morning, followed by lunch and a drink, and then more conversation. When after a year, Constance told Neyman that she now had enough information and would instead have to start writing, he was quite disappointed. The book, which came out in 1982, a few months after Neyman's death, not only tells the story of his life and accomplishments, but also gives a very lifelike picture of him as a person.

Scheffé

Blackwell was succeeded as chair of the Statistics Department by Le Cam, and he in turn by Scheffé, who chaired the department from 1965 to 1968. It was a period of great unrest in Berkeley, the time of the Free Speech Movement. Student strikes in particular caused difficult problems for the chair. But in spite of violently conflicting attitudes within both faculty and student body, Scheffé managed to hold the Department together and to keep the atmosphere within the Department pleasant. His fair mindedness was greatly valued by all members of the Department.

Scientifically, soon after our *Sankhyā* paper, Henry's main interests turned from optimality theory to more methodologically oriented work, of which his famous paper on the *S*-method for judging all contrasts in analysis of variance was the first important accomplishment. He brought to this new orientation the rigor of his mathematical origins. This is particularly noticable in his emphasis on clearly defined and justified models and on the consequences of violations of these model assumptions which is such a striking feature and one of the great strengths of his pioneering book *The Analysis of Variance* (1959). He had planned to revise this work after his retirement, but he died in 1977 from injuries sustained in a bicycle accident without completing the revision. It is a pity that he did not live to see the beautiful optimality property of his *S*-method, established by Wijsman in 1979. I believe it would have given him great pleasure. (For further information on Henry Scheffé, see the obituary by Daniel and Lehmann, 1979.)

Hodges

After 1956, Joe and I continued our joint work particularly in the area of nonparametric inference. Of this later work, I mention only one paper, "Estimates of Location Based on Rank Tests" (1963), which brought the pairing of our names into the statistical terminology. Here we showed how the rank methods that had proved so unexpectedly successful for hypothesis testing could be transferred to point estimation. The class of estimators defined in this way, the so-called *R*-estimators, includes in particular the estimator now known as the Hodges-Lehmann estimator, which shares the good efficiency properties of the Wilcoxon test.

Joe and I collaborated not only on research but also joined to write an elementary text, *Basic Concepts of Probability and Statistics* (1964; second edition 1970), which we dedicated to Neyman and which was later translated into Danish, Hebrew, Italian and recently into Farsi. It gave a rigorous development of probability and statistical inference in finite sample spaces without calculus. While the probability part was fairly conventional, I believe it was the first elementary book discussing such topics as the Neyman–Pearson Lemma and optimal design. The lower division course ("Stat. 1") which used it as a text was very different from the more cookbookish methods course "Stat. 2." Henry Scheffé once explained the difference by saying that "Stat. 1" was intended for students who wanted to understand statistics but were not planning to use it, while the reverse was true for "Stat. 2." [Since then "Stat. 2" has changed radically under the influence of our colleagues Freedman, Pisani and Purves, whose text *Statistics* (1978; second edition 1991) has made this into an attractive and intellectually stimulating course.]

Some years later, Joe wrote another elementary text jointly with the psychologists Krech and Crutchfield. Called *Stat Lab*, it was based on a single large data set which provided the values of a large number of variables on 6^4 families. A third book for a first year upper division course that Joe taught for many years contains many interesting ideas but unfortunately has remained unpublished. The reason is Joe's dislike of the editorial process, which he shared with Neyman. One of the victims of this dislike was a pioneering paper, joint with Fix, on density estimation (Fix and Hodges, 1951). Because of its historical importance, this paper was published with Joe's permission in the *ISI Review* by Silverman and Jones (1989), who also provided an introduction.

After 1970, our joint work decreased since by then much of Joe's efforts had gone into higher administration, as member and then chair of the Budget Committee both at the campus and state level, and also as advisor on academic personnel matters to both the Chancellor at Berkeley and the President of the whole University system.

Joe retired from the University in 1991, taking advantage of a "golden handshake," a special early retirement offer, which – he claims – netted him an extra 32 cents a month.

REFERENCES

Daniel, C. and Lehmann, E. L. (1979). Henry Scheffé 1907–1977. *Ann. Statist.* 7 1149–1161.

David, F. N. (1949). *Probability Theory for Statistical Methods.* Cambridge Univ. Press.

DeGroot, M. (1986). A conversation with Charles Stein. *Statist. Sci.* 1 454–462.

Fix, E. and Hodges, J. L., Jr. (1951). Discriminatory analysis – nonparametric discrimination: consistency properties. Unpublished report.

Freedman, D., Pisani, R. and Purves, R. (1978). *Statistics.* Norton, New York. [2nd ed., 1991, with Adhikari.]

Hodges, J. L., Jr. (1949). The choice of inspection stringency in acceptance sampling by attributes. *Univ. Calif. Publ. Statist.* 1 1–14.

Hodges, J. L., Jr., Krech, D. and Crutchfield, R. S. (1975). *Stat Lab.* McGraw Hill, New York.

Hodges, J. L., Jr. and Lehmann, E. L. (1950). Some problems in minimax point estimation. *Ann. Math. Statist.* 21 182–197.

Hodges, J. L., Jr. and Lehmann, E. L. (1951). Some applications of the Cramér-Rao inequality. *Proc. Second Berkeley Symp. Math. Statist. Probab.* 13–22. Univ. California Press, Berkeley.

Hodges, J. L., Jr. and Lehmann, E. L. (1952). The use of previous experience in reaching statistical decisions. *Ann. Math. Statist.* 23 396–407.

Hodges, J. L., Jr. and Lehmann, E. L. (1956). The efficiency of some nonparametric competitors of the *t*-test. *Ann. Math. Statist.* 27 324–335.

Hodges, J. L., Jr. and Lehmann, E. L. (1963). Estimates of location based on rank tests. *Ann. Math. Statist.* 34 598–611.

Hodges, J. L., Jr. and Lehmann, E. L. (1964). *Basic Concepts of Probability and Statistics.* Holden-Day, San Francisco. [2nd ed., 1970.]

Lehmann, E. L. (1947). On families of admissible tests. *Ann. Math. Statist.* 18 97–104.

Lehmann, E. L. (1948–49). *Theory of Testing Hypotheses.* Notes recorded by Colin Blyth.

Lehmann, E. L. (1959). *Testing Statistical Hypotheses.* Wiley, New York. [2nd ed., 1986.]

Lehmann, E. L. and Scheffé, H. (1947). On the problem of similar regions. *Proc. Nat. Acad. Sci. U.S.A.* 33 382–386.

Lehmann, E. L. and Scheffé, H. (1950/1955/1956). Completeness, similar regions, and unbiased estimation. *Sankhyā* 10 305–340; 15 219–236; 17 250 (a correction).

Lehmann, E. L. and Stein, C. (1948). Most powerful tests of composite hypotheses. I. Normal distributions. *Ann. Math. Statist.* 19 495–516.

Lehmann, E. L. and Stein, C. (1949). On the theory of some nonparametric hypotheses. *Ann. Math. Statist.* 20 28–45.

Lehmann, E. L. and Stein, C. (1950). Completeness in the sequential case. *Ann. Math. Statist.* 21 376–385.

Lehmann, E. L. and Stein, C. (1953). The admissibility of certain invariant statistical tests involving a translation parameter. *Ann. Math. Statist.* 24 473–479.

Morrey, C. B. (1983). *Jerzy Neyman (1894–1981).* Biographical Memoirs. National Academy of Sciences, Washington, DC.

Neyman, J. (1952). *Lectures and Conferences on Mathematical Statistics*, 2nd ed. Graduate School, U.S. Dept. Agriculture, Washington. [1st ed., 1938.]

Reid, C. (1970). *Hilbert.* Springer, New York.

Reid, C. (1976). *Courant in Göttingen and New York.* Springer, New York.

Reid, C. (1982). *Neyman from Life.* Springer, New York.

Rider, R. (1989). An opportune time: Griffith C. Evans and Mathematics at Berkeley. In *A Century of Mathematics in America, Part 2* (P. Duren, ed.) 283–302. Amer. Math. Soc., Providence, RI.

Scheffé, H. (1943). Statistical inference in the nonparametric case. *Ann. Math. Statist.* 14 305–332.

Scheffé, H. (1959). *The Analysis of Variance.* Wiley, New York.

Silverman, B. W. and Jones, M.C. (1989). E. Fix and J. L. Hodges (1951): An important contribution to nonparametric discriminant analysis and density estimation. *Internat. Statist. Rev.* 57 233–238.

Shutes, M. H. (1947). Henry Douglas Bacon (1813–1893). *California Historical Quarterly* 26 193–200.

Uspensky, J. V. (1937). *Introduction to Mathematical Probability.* McGraw Hill, New York.

Wald, A. (1950). *Statistical Decision Functions.* Wiley, New York.

Wijsman, R. A. (1979). Constructing all smallest simultaneous confidence sets in a given class, with applications to MANOVA. *Ann. Statist.* 7 1003–1018.

Wolfowitz, J. (1952). Abraham Wald, 1902–1950. *Ann. Math. Statist.* 23 1–13.

Statistics, Probability and Game Theory
IMS Lecture Notes – Monograph Series (1996) Volume 30

THE CREATION AND EARLY HISTORY OF THE BERKELEY STATISTICS DEPARTMENT

E. L. LEHMANN
Educational Testing Service and University of California, Berkeley

Abstract

This paper sketches the development of the Berkeley statistics group from Neyman's appointment as Professor of Mathematics in 1938 to its status as an independent Department of Statistics in 1955. It ends with a brief description of Neyman's resignation as chair of the Department in 1956 and its transition to a new era under the chairmanship of David Blackwell.

1. The Beginning. In her Neyman biography,[1] Constance Reid describes how in 1934 the University of California brought Griffith Evans[2] to Berkeley to revitalize its Mathematics Department. She mentions Evans' early interest in statistics, and tells of his search for a suitable statistician, which ended with his offering a position to Jerzy Neyman.[3] Evans took some risk with this offer since he had never met Neyman. But I believe Neyman fulfilled his expectations as a serious and original scientist, interested in both the mathematical theory of statistics, and the broad spectrum of its applications. He also turned out to be an ambitious and dynamic administrator, sometimes perhaps more so than Evans had bargained for.

On Neyman's part, the risk was of course also very great. He had never been to California and knew little about its University except the—to him very attractive—fact that it had as yet no statistics program. However, he quickly came to like both Berkeley and the University, and he does not seem to have had any later regrets.

Neyman arrived in the summer of 1938 to take up a position as Professor of Mathematics, with the specific responsibility of developing a statistics program and of assisting Berkeley faculty members with the statistical problems they might encounter in their work.

Over the next three years, he set up a skeleton teaching program and acquired a few graduate students who, as Teaching Assistants, helped with the laboratories attached to the courses. He also engaged in a steadily increasing amount of consulting work with various faculty colleagues, which he enjoyed and which was much appreciated. For the Assistants, his secretary,

J. Rojo (ed.), *Selected Works of E. L. Lehmann*, Selected Works in Probability and Statistics, DOI 10.1007/978-1-4614-1412-4_81, © Springer Science+Business Media, LLC 2012

and the calculators that were needed for teaching and research, he was provided with a number of rooms. He dubbed his organization the Statistical Laboratory ("The Lab") and soon obtained official status for it, with himself as Director.

Further progress was halted with the coming of World War II during which all his energy was absorbed by war work. As a result, when the war ended in 1945, the statistical faculty still consisted of just one person: Neyman as Professor of Mathematics and Director of the Statistical Laboratory. Ten years later it had grown to ten tenured or tenure track faculty members who formed a separate Department of Statistics, independent of the Mathematics Department. How was it possible in so short a time to achieve such a spectacular development?

2. A Period of Growth. Building a substantial program required attention to many different aspects: Faculty, students, support staff, equipment, space, and—very importantly—the reputation of the group both within and outside the University. Each of these presented its own problem.

(i) *Faculty.* One question Neyman had to face if he wanted to assemble a teaching staff, was from where to recruit his faculty. A natural pool—his own students who were trained in his still quite novel approach to statistics—was problematical because the University strongly discouraged departments from hiring their own Ph.D.students. Neyman overcame this difficulty by persuading the administration that this was the principal source available to him since few other universities were turning out students he considered satisfactory.

As a result, during the next several year he appointed to the faculty, as they completed their degrees: Lehmann (1946), Barankin (1947), who was enrolled in the statistics program but wrote his thesis in algebra, Fix and Scott (1950), Hodges (1951), and Le Cam (1952). In addition, he made some superb appointments from the outside: Stein (1947) who came with a fresh degree from Columbia and was familiar with Wald's ideas, Loève (1948) to develop the probability program, Scheffé (1953), and Blackwell (1955).

A faculty, half of which has been trained by its senior member, presents two dangers: That of inbreeding, which may result in a certain narrowness, and—in the other direction—difficulties as former students become independent colleagues. Although in this case the first of these dangers was somewhat mitigated by the very strong outside appointments, a certain narrowness did result. There was a lack of attention to any Fisherian or Bayesian aspects, and in general the program reflected the Neyman-Pearson-Wald point of view so strongly that Lindley (1990), in his Wald lectures, referred to this approach simply as "Berkeley". This weakness was at the same time a source of strength. It resulted in a congenial group which shared a basic point of view and which engaged in much collaborative work.[4] As we shall see later, the second difficulty mentioned above, eventually also caused some problems.

(ii) *Students and Courses.* The expansion of the teaching staff made possible a corresponding broadening of the offerings of both theoretical and applied courses. The resulting increase in enrollments, in turn, justified further faculty appointments. Although the statistics group never obtained a monopoly in the teaching of statistics, Neyman cooperated closely with the applied departments that sent him students, and gradually persuaded other departments to relinquish their lower division courses so that, eventually, nearly all lower division teaching of statistics, involving several thousand students, was done by the members of the Statistical Laboratory.

(iii) *The Symposia.* As the war ended in 1945, Neyman was elated at the prospect of returning to free scientific research. To celebrate the occasion, he proposed a meeting "to mark the end of the war and to stimulate the return to theoretical research". He issued invitations to statisticians all over the country, and the result was the Berkeley Symposium on Mathematical Statistics and Probability, August 13–18, 1945. The majority of the speakers came from the various campuses of the University of California, but the participants also included such luminaries as Doob, Feller, Hotelling,Hsu, Pólya, and Wolfowitz. The papers were published in a Proceedings Volume which ran to 501 pages.

Not only Neyman, but also Evans and the university administration considered the symposium a great success and agreed to his request for financial support for a second symposium to be held in 1950. This second meeting had an even broader cast, including some speakers from abroad, among them Cramér, De Finetti, Dvoretzky, Erdös, Fortet and Paul Lévy. The table of contents of the Proceedings volume, reflecting Neyman's wide interests, was divided by subject matter: Mathematical statistics, probability, astronomy, biometry, econometrics, physics, traffic engineering, and wave analysis. At the conclusion of the Preface Neyman looked "forward to a third symposium within a few years."

Thus started a tradition of symposia, which were held every five years (until 1970). They became the principal international statistical meetings during this period, and were an important factor in establishing an international reputation for Berkeley as an important statistical center.

These various aspects of the program were driven by Neyman's energy and involved an unending flurry of activities on his part. Providing his staff with space (as the group grew we were constantly on the move), equipment, summer support, promotions; dealing with a stream of consulting problems; organizing not only the symposia and the publication of their proceedings, but also developing a serial publication (*The University of California Publications in Statistics*); arranging summer sessions with distinguished visitors—one gets dizzy just enumerating them and wonders at the prodigious amount of energy Neyman was able to muster. But this effort achieved its end; the group grew steadily, and Berkeley acquired an international reputation as a first-rate center of mathematical statistics.

Yet, even Neyman's energy and talent could not have generated this sustained development without the support of an outside factor: the growing importance of the field of statistics itself. Statistics had proved its usefulness during the war and now was expanding in ever widening circles in universities, industry and government. The nature of the subject was changing from a primarily descriptive to an inferential mode in which the new approaches developed by Fisher, Neyman and Pearson played a central role. Correspondingly, there was increased demand for statistical services (both teaching and consulting) throughout the University, which greatly strengthened Neyman's hand in his negotiations with the administration.

3. From Lab to Department. In the efforts on behalf of his laboratory, Neyman usually had the support of Evans, but on one issue they strongly disagreed, and the resulting conflict was a cause on continuing tension. From the start, Neyman had envisioned a Department of Statistics which would be independent of the Mathematics Department. In contrast, Evans was trying to create a broad-based Mathematics Department that would contain all the mathematical sciences: Not only pure and applied mathematics, but also areas such as statistics, logic and actuarial and computer science.

One reason for Neyman's drive for independence was undoubtedly his strong personal desire for freedom from outside controls. He liked to be the master of his fate. For publishing his papers, in order to avoid the interference of editors who might not approve of his ideas or his style, he instituted a series that would be under his control. Similarly, he did not want to have to clear his appointments or his budget with a chairman who might have different views.

There was, however, a second more objective reason. Although Neyman had played a central role in the mathematification of statistics, he felt that statistical theory—unlike mathematics—needed to be motivated by its importance and usefulness for applications, and that this applied orientation led to attitudes and value judgments different from those held by mathematicians. He thus considered statistics to be a distinct subject, and believed that it could develop more naturally in a separate department. In addition, the departmental status would provide the emerging subject with greater visibility and recognition. This conclusion came to be shared by other statisticians and eventually led to the establishment of separate departments of statistics in most American universities.

Convinced of the righteousness and importance of his cause, Neyman pressed relentlessly for greater independence for the Lab. Occasional disagreements with Evans about specific issues, offers from other universities, all were used to obtain, step by step, official recognition of the Lab, a separate budget, greater authority in hiring, etc. until in the end, the connection with the mathematics department became little more than a formality.

Under these circumstances, when Evans retired from the chairmanship in 1949, his successor (Charles Morrey) no longer opposed the separation of the two groups, and recommended the formation of a separate department of statistics. But in Berkeley the administrative wheels grind slowly, and it took another five years until, in 1954, the Berkeley Chancellor (Clark Kerr) and the President of the university (Gordon Sproul), and finally in 1955 the Board of Regents, approved the new arrangement.

4. Two Crises. The steady progress from the appointment of Neyman in 1938 as Professor of Mathematics to the establishment of a separate department, the faculty of which would eventually attain a size of over 20, was interrupted by two crises, each of which had the potential to destroy much of what had been achieved by then.

The University of California is governed by an independent Board of Regents. However, it depends on funds appropriated by the state government in Sacramento, and in addition, most of the regents are appointed by the Governor. In politically turbulent times, Sacramento, therefore, can make its influence felt. Such an occasion arose during the MacCarthy period. In order to preempt legislation proposed by the State Unamerican Activities Committee, which would have deprived the university of some of its authority, in 1950 the Regents voted to require from university faculty and staff, an anticommunist oath. Many faculty members considered such a special oath offensive and demeaning, and a number (in the end about 30) refused to comply. After an extended, bitter struggle, the regents voted to dismiss the non-signers.[5] Most members of the statistics group (including Neyman) signed with distaste. However, Charles Stein decided instead to leave the university. (After some years in Chicago, he returned to California, but to Stanford rather than Berkeley.)

The conflict deeply split the faculty of the university and poisoned the atmosphere. To escape the turmoil, Hodges and Lehmann took extended leaves. Several members of the group received attractive offers from other universities; Scheffé whom Neyman tried to bring to Berkeley, refused his offer. The outlook was bleak and Neyman so pessimistic that he tried (unsuccessfully) to move the whole group to another university.

By mid-1951, the Legislature had removed the motive for the University oath by requiring an anticommunist oath of all state employees. In addition, changes had occurred on the Board of Regents, and the Board rescinded the special oath. With the exception of Stein, no member of the group had left, and Scheffé accepted a renewed offer. The cohesiveness and loyalty that Neyman had built, together with the attractiveness of Berkeley, had been strong enough for the Lab to survive.

The second crisis occurred during the year following the creation of the department when Neyman, completely unexpectedly, announced his resignation as chair of the new department. Chancellor Kerr persuaded him to withdraw his resignation, but a few months later he renewed it and this time remained firm. It must have been a difficult decision for him, and his reasons undoubtedly were complex, but two stand out.

One motive was a wish to return to the kind of work he had enjoyed so much as a younger man. The twenty-year period since then had been devoted to the building of the Berkeley statistics group and had left little time and energy for his research. He was now 61, and if he still wanted to do some major research, this was the time to do it.

A second reason for relinquishing the responsibility for the department was a gradual change that had developed in his relations with the group. His former students had grown up, had acquired new responsibilities and their own views on issues, which did not always agree with his. He was still the "grand old man", but conflicts and tensions had arisen which Neyman found bothersome. Under these circumstances he preferred to keep as his domain, the laboratory, separated from the department, and completely under his control, and to leave the running of the department to his younger colleagues.

Complete separation of the two operations was, of course, not possible. Neyman's laboratory was in charge of all grants, with a research budget that was about equal in size to the budget of the department. Research and teaching assistants came from a common pool, and there were many other points of contact and possible conflict. With authority thus divided, it was essential that the new chair get along well with both Neyman and the rest of the group. The obvious person for the position was David Blackwell who had spent 1954/55 in Berkeley as a visitor, and had joined the permanent faculty the following year.[6] Blackwell was respected and liked by everyone. He had a good way of dealing with little difficulties by not taking them too seriously. Perhaps it was also an advantage that as a relative newcomer, he had not been involved in earlier disagreements. In any case, it turned out to be a great choice. Blackwell managed the transition in a smooth and gracious way, and steered the statistics department onto its new course without a serious break of continuity, thus confounding the predictions of many professional colleagues who expected the department to disintegrate after Neyman's resignation.

David has now been an admired and respected colleague for over forty years, and it is a great pleasure to dedicate to him these memories of earlier times in friendship and with affection.

---

[1]Constance Reid's indispensable book, *Neyman—from life,* tells the story of Neyman's life and of the department he created. I have repeatedly drawn on her account to check, bolster, and sometimes correct my own memories.

[2]Griffith Conrad Evans (1887-1973) was one of the most distinguished mathematicians of his generation. (Among other honors he was Colloquium Lecturer of the *American Mathematical Society* (1916), was elected to the National Academy of Sciences (1933), and served as President of the *American Mathematical Society* (1938-1940).) He came to the University of California, Berkeley in 1934 to chair its mathematics department, which at the time was in the doldrums. During the 15 years of his chairmanship, he developed the department into a first-rate institution. Evans retired in 1955 and when later the university decided on a new building to house the mathematics and statistics departments, it was named Evans Hall.

For a more detailed account of Evan's life and accomplishments, see Rider (1990). Some of my own memories of Evans can be found in Lehmann (1993).

[3]For more details on Neyman's life, see Reid (1982) and Lehmann (1990).

[4]At one time or another, there were joint papers by Blackwell and Hodges; Fix and Hodges; Fix, Hodges, and Lehmann; Hodges and Le Cam, and Lehmann and Stein. In addition, there were long collaborations by Neymann and Scott and by Hodges and Lehmann.

[5]Eventually the State Supreme Court ruled against the Regents. A detailed account of the very complicated history of this struggle, is provided by Gardner (1967).

[6]Neyman had in fact tried to bring Blackwell to Berkeley once before. In 1942 he was looking for a young faculty member and Doob had recommended Blackwell as an outstanding student. Neyman had obtained the agreement of Evans who, however, shortly thereafter told Neyman that Blackwell's race would cause difficulties in the department. His own wife, a southerner, had said that she would not feel able to invite Blackwell to her house. Neyman, not wanting to cause dissension in the department that had been so hospitable to him, and administratively dependent on Evans, backed down. I am happy to be able to add that David recently told me that later, after he joined the statistics department, Mr. and Mrs. Evans did invite him and his wife Ann to their house and that it was a very pleasant evening.

REFERENCES

GARDNER, D. (1967). *The California Oath Controversy.* University of California Press, Berkeley, CA.

LEHMANN, E. L. (1990). Jerzy Neyman, 1894–1981. *Dictionary of Scientific Biography,* **18** 669–674.

LEHMANN, E. L. (1993). Mentors and early collaboration. *Statistical Science* **8** 331–341.

LINDLEY, D. V. (1990). The 1988 Wald Memorial Lectures: The present position in Bayesian statistics. *Statistical Science* **5** 44–65.

REID, C.(1982). *Neyman—from Life*. Springer, New York.

RIDER, R. E. (1990). Griffith Conrad Evans, 1887–1973. *Dictionary of Scientific Biography*, **17** 270–272.

E. L. LEHMANN
DEPARTMENT OF STATISTICS
UNIVERSITY OF CALIFORNIA
BERKELEY, CA 94702

Statistical Science
1997, Vol. 12, No. 1, 48–52

Testing Statistical Hypotheses: The Story of a Book

E. L. Lehmann

Abstract. This is an account of the life of the author's book *Testing Statistical Hypotheses*, its genesis, philosophy, reception and publishing history. There is also some discussion of the position of hypothesis testing and the Neyman–Pearson theory in the wider context of statistical methodology and theory.

Key words and phrases: Testing statistical hypotheses, Neyman–Pearson theory, textbook publishing.

1. INTRODUCTION

As author of several textbooks, I am often asked about various aspects of book writing—from the severely practical to the quite personal. These questions suggest a certain curiosity about the writing and publishing of such texts; the following is the account of the life history of one of them.

The book I shall write about, *Testing Statistical Hypotheses* (*TSH*), was my first and has now been in print for almost 40 years. Before describing its history, let me get one question out of the way which is frequently hinted at although less often asked outright: "What was the financial reward, how much money did you make from the book?" I am not being coy in saying that I do not really know since I have not kept adequate records. However, I have a rough idea in terms of a scale suggested by the number theorist Edmund Landau, who stated that from the payments for one of his books he built his house in Göttingen. I can say categorically that from the earnings of *TSH* I could not have built a house, not even a modest one, much less Landau's mansion in Göttingen. On the other hand, the proceeds would have enabled me to buy a car—in fact, a somewhat fancier one than the compact we drive. These upper and lower bounds are a bit crude, but then mathematical bounds often are.

TSH gives an account of the small-sample theory and methodology of hypothesis testing. The idea of a small-sample approach based on an assumed model was introduced by Student in 1908 (Student, 1908) and greatly extended by R. A. Fisher in the 1920s. Fisher's 1925 book, *Statistical Methods for Research Workers*, brought the new methodology to the attention of users of statistics. A theoretical framework for testing was formulated by Neyman and Pearson (1933), who showed that the Student–Fisher tests had some optimal properties within the normal models on which they were based.

When in 1942 I became a student of Neyman, I learned this theory (with the basic ingredients of level, power, similarity, unbiasedness and optimality) at the source, and wrote my thesis (1946) on its application to a special case suggested to me by P. L. Hsu.

With this background I was very receptive to a second group of ideas, which had been strongly influenced by Neyman's approach to statistics. These were the concepts of minimaxity and admissibility of Wald's decision theory and the related invariance considerations which I learned from Charles Stein when in 1947 he became a colleague, friend and collaborator.

2. THE BLYTH NOTES

These two sets of ideas easily combined into an integrated whole which formed the basis of the graduate course on hypothesis testing which I gave in 1948. Among the students attending it was Colin Blyth, to whom the testing book to a large extent owes its existence. In a letter to me of April 1981 he describes how this came about:

> I attended your course 260A in Testing in the Fall 1948 and wrote up careful notes. (After each lecture I went home and wrote up the rough notes I'd taken

E. L. Lehmann is a senior scholar, Educational Testing Service, 15-T Rosedale Road, Princeton, New Jersey 08541, and Professor Emeritus, Department of Statistics, University of California, Berkeley, California 94720.

> in class, putting them into readable form, and settling to my own satisfaction any details omitted or points I hadn't been clear about in class.) Sometime in the Spring of 1949 you saw these notes (I don't remember how this happened, and can think of no reason that it might have) and suggested mimeographing them for other students. I preferred to attend your course again in the summer of 1949 and put them in a form more suitable for distribution and this we did. I wrote them up, you read them carefully and made numerous suggestions, and after a final revision they went to the typist.

The notes were mimeographed and sold at cost, first by the Berkeley Statistical Laboratory and later by the University Bookstore. As our students graduated and took up positions at other universities, they recommended them or used them in their courses. As a result, orders started arriving from other colleges. In the absence of any other systematic treatment of this still fairly novel material, the notes gradually became something of an underground text. They formed a slight paper-covered package of 163 pages, which provided a skeletal account of the theory and contained only the simplest applications.

3. THE CONCEPT OF THE BOOK

An increasing demand, and references to the material in books by colleagues, gradually led to the idea of expanding the notes into a book. In the intervening years I had occasionally been teaching a graduate course along the lines of the notes, and came across questions concerning the theory which I tried to answer in my research. In particular, this included the joint work with Henry Scheffé on the completeness of sufficient statistics. In the course of this work I also became familiar with the theory of exponential families, and this provided an important unifying idea.

I mentioned earlier that the notes combined threads from two different lines of work: the Neyman–Pearson treatment of hypothesis testing and Wald's general decision theory. These provided the underlying structure. However, I did not think of the book as an account of the theory for its own sake, but rather as a means of justifying the many standard tests whose optimum properties the theory had established. Presenting these tests and their properties was my central concern. The theory makes it possible to do so in a systematic fashion. Without such a unifying structure they would form an unmotivated, disorganized collection of ad hoc procedures (see Note 1), dependent on the ability of their originators to intuit which particular test statistics would prove effective.

NOTE 1. This statement is perhaps too strong. The likelihood ratio principle, for example, does provide a systematic approach which frequently leads to the standard tests. However, it can be justified only asymptotically; there is no reason to expect the resulting tests to be satisfactory for small samples.

A question that caused me some difficulty concerned the mathematical level at which to present the material. The natural mathematical framework is provided by measure theory; however, such an advanced level threatened to erect a formidable barrier for many potential readers. This would be particularly regrettable since the measure theoretic considerations contribute little to the understanding of the statistical ideas and results. On the other hand, a clean, rigorous treatment is not possible without them.

In the end I decided to include a brief introduction to measure theory, giving the principal definitions and results, and then using them in the rest of the book where needed. Russian colleagues later told me of a surprising effect of this inclusion when a translation appeared in their country. According to these reports, it helped to legitimize statistics in the eyes of some Russian mathematicians. If it is based on measure theory, they felt, there may be something to it!

4. FROM NOTES TO BOOK

The next problem was to obtain a publisher. I am sometimes asked about this aspect, but at the time the answer was easy. The statistical textbook literature was dominated by the "Wiley Publications in Statistics," a series that contained among others such distinguished volumes as Feller's *Probability Theory*, Cochran's *Sampling Techniques*, Doob's *Stochastic Processes*, Wald's *Decision Theory*, Rao's *Advanced Statistical Methods*, Blackwell and Girshick's *Theory of Games and Statistical Decisions*, Savage's *Foundations of Statistics*, Scheffé's *Analysis of Variance* and Anderson's *Multivariate Analysis*. Furthermore, a Wiley representative regularly visited the Department and had expressed an interest in the book. Thus, it appeared in 1959 under the Wiley imprint.

The notes had been issued in 1949, the book came out 10 years later. What took it so long? I can think of various excuses. The book was about three times as long as the notes; I was also doing other things: teaching, research, a three-year stint as Editor of the *Annals* and so on. However, the main reason can be summed up in one word: rewriting! There is no page, section or chapter that I wrote, which looking at it again a few months later I could not improve, and a few months after that improve again. This process might have gone on forever had Neyman not told me one day that I had dawdled long enough; get off the dime! So I finally took the plunge, but with great trepidation. Publication is so irreversible.

5. ERRORS

After publishing a book one waits with some anxiety for the reviews. But at least in my case another ordeal arrives much sooner: the discovery of errors, big and little, subtle and obvious—how could one have been so foolish and so careless! Over the years I have learned how to interpret letters starting with, "I have difficulty following your proof..." or "I was unable to solve problem...". These are polite ways of telling me: "You botched it again!" Only days after publication I found that one of the figures was seriously in error. And of all the figures in the book this was the one the publisher had chosen to put on the dustjacket. Most embarrassing!

The biggest source of errors was the more than 200 problems. The difficulty often resulted from some fine points or special cases that I had overlooked and that naturally caused readers much trouble when they struggled with them. Letters asking for clarification were not only painful reminders of my ineptitude but they could also take quite a bit of time and effort to answer at a time when I was no longer working in this area. I was saved from this bondage to my past errors when a heroic group of 15 Dutch statisticians decided to work through the whole collection systematically and in 1984 with Wiley's permission published the solutions as a 310-page book (Kallenberg et al., 1984). One member of the group told me later that this was the most painful job he had ever undertaken. However, from then on I was able to answer queries about the problems with a simple reference.

6. THE BOOK AS TEXT

Shortly after the book appeared, I was scheduled to teach the graduate course for which it was intended as a text. But, it was not clear to me how best to utilize it. There seemed little point in rephrasing in lectures what I had labored hard to express as clearly as possible in print, and so I announced that I would proceed differently than in the past. I would outline and highlight a section in class; the students were then to read it; and I would answer questions at the following class meeting. If that left any time I would use it to talk about auxiliary material that was of interest but not covered in the book. After a week of this program, three students appeared in my office. They told me that they were a delegation from the class which had met to discuss the situation. They felt unanimously that my new approach was unacceptable since the text was much too terse for independent reading, and that the whole class would drop the course unless I returned to the usual method of lecturing.

While the experiment was thus a failure, the availability of the book did provide a new flexibility to the course. When I taught it to an audience with diverse background, it was possible to omit some material (e.g., the measure theoretic aspects or some particular applications) and ask interested students to work this through on their own.

7. REVIEWS

It typically takes a year or so after publication before the first reviews begin to appear. Since serious critical reviews require a careful reading of the book for which reviewers often have neither time nor inclination (remember, these reviews are unpaid!), many reviews of technical books consist of a sketch of the contents and the level of difficulty, with comments about omissions and relevance for different potential groups of readers. In the present case there was one exception: a serious, highly critical review by John Pratt (Pratt, 1961).

After some friendly praise for the writing and mathematical exposition, John expressed his strong reservations concerning the Neyman–Pearson theory which constituted the subject matter of the book. His criticisms had various strands, but in the end many of them had their origin in his Bayesian orientation with its emphasis on coherence and logical necessity. In contrast, the approach of *TSH* is pragmatic. It looks at a variety of properties (unbiasedness, invariance, minimaxity and Bayes averages) which are not always compatible. Each of them is taken as far as it will carry and is illustrated on situations to which it is applicable. This is not the place to revisit this old dispute except to repeat my long-held belief that neither approach is right or wrong, that each has its place and the choice must depend on the situation. However, although I did not agree with his position, I have always been grateful to John for giving the book such serious attention.

Another discussion of the book appeared in 1970 as a section of a paper by Jack Wolfowitz on the future of mathematical statistics (Wolfowitz, 1970). His main reason for including this critique was to warn potential readers against the book's pernicious influence. He characterized it as being both bad statistics (it treats the wrong problem) and bad mathematics (it lacks deep and difficult theorems), and he concluded his review with the thought that it would be "a disaster for statistics if this book should determine the direction for research for any appreciable period of time."

8. SECOND EDITION

In fact, Wolfowitz need not have worried: by 1970 the heyday of small-sample theory was past. It had become clear by then that concepts such as unbiasedness and invariance which are so central to the book, while important and applicable to a large class of basic problems, are nevertheless severely limited. They provide results in exponential and transformation families but do not extend much further. (An exception is the minimax approach, which applies in great generality. It is computationally difficult but is finding renewed interest; see, e.g., Brown, 1994, and Donoho and Johnstone, 1995.) Thus, someone faced with a new and complex problem is not likely to resort to small-sample theory but will use maximum likelihood or one of its variants, or perhaps a Bayes procedure with a noninformative prior.

Despite the waning interest in small-sample theory as a research area, *TSH* (which had been translated into Russian, Polish and Japanese) continued to be used as a text and in 1983 was joined by a parallel volume on point estimation. However, it was now 25 years old and showing its age. It required a general updating; in addition, certain issues that had not been addressed in the original version deserved attention. One of these was conditional inference, the importance of which had been pointed out in Pratt's review. Another was multiple comparisons and other simultaneous inference procedures, a methodology dealing more realistically with problems presented traditionally as tests of hypotheses. This is often an oversimplification, which had justifiably been criticized among others by Wolfowitz.

As a result, in 1986 I brought out a second edition. In this new incarnation, the book had grown from 400 to 600 pages, with a new chapter on conditional inference and an exposition of Wijsman's beautiful theory of optimal simultaneous confidence procedures. Some time later, on a visit to China, I brought a copy of the new edition for Zheng Zhongguo, who had made many of the arrangements for our trip. He told me that he would accept it only if in exchange he could give me his copy of the pirated (English) version of the first edition. When I agreed, he signed it in a reversal of the traditional inscription, "To the author of the book with best wishes, Z.G."

9. AN ODYSSEY

For many years, both the testing and estimation books had fared well with Wiley, and relations between publisher and author had been cordial. I was therefore shocked when in 1991 I received a notification from an unknown Wiley official informing me that they were letting one of the books go out of print (the title was garbled, so that I was not sure which one). The tone was peremptory and did not invite discussion. So I accepted the verdict, requesting only that the copyright of both books should be returned to me.

A natural new home for them at this point seemed to be the Wadsworth Series in Probability and Statistics, of which my friend John Kimmel was the Sponsoring Editor and which had previously published a Festschrift for my 65th birthday. John agreed, and both books were reissued by Wadsworth with new covers but otherwise unchanged.

What I did not know was that Wadsworth was one of a number of publishing companies owned by the Thomson Corporation. Soon after the reissue it was decided (for some internal reasons) to transfer the series that Kimmel had built up over the years to a different Thomson Company, Van Nostrand Reinhold (VNR). There it lingered without an editor or much activity for nine months, and was then moved to still another Thomson Company, Chapman and Hall, New York, where John in the meantime had become Statistics Editor. In 1994 history repeated itself. The statistics program for Chapman and Hall, New York, was terminated and the list of books sent to Chapman and Hall, London. And John Kimmel also moved again, from Chapman and Hall to Springer-Verlag.

While these transfers were going on, I had begun work on a second edition of *Point Estimation*, jointly with George Casella. We both believed that the books were not likely to do well under the new arrangement and, with some effort, persuaded Chapman and Hall to relinquish the two copyrights. They are now slated to go to Springer, and will thus over a period of five years have had six different publishers! It is discouraging to see books treated as purely commercial property, and to realize how little power authors have to control the fate of their work.

10. FUTURE PROSPECTS

At the time of Blyth's notes, nearly 50 years ago, the theory of hypothesis testing was at the center of theoretical research. In the intervening years the field of statistics has expanded enormously and the once "hot" Neyman–Pearson theory has become classical. It is still struggling with important and interesting issues, for example, the combinatorial problems arising in the exact treatment of randomization tests (see, e.g., Diaconis and Holmes, 1994) and the still unresolved question of when and how far to condition. However, in the wider context of statistics as a whole these problems no longer hold center stage.

Interest in the theory of testing is threatened also from another direction: attacks on the practice of hypothesis testing itself. These attacks are partially a reaction to a period in which testing was greatly overused and misused (with some editors urging significance at conventional levels as a precondition for publication). Estimation by confidence intervals is an alternative preferred by many of the opponents of testing. (For a balanced discussion of this issue see, e.g., Bailar and Mosteller, 1992.) In the 19th century, particularly in the work of Laplace, fixed level tests, p-values and confidence intervals were used very flexibly, often in conjunction. This seems sensible and is advocated, for example, in Levin (1993).

The theory described in *TSH* should not be viewed as contradicting or inhibiting such an approach. In fact, *TSH* includes a treatment of confidence estimation and discusses the relation of the two theories. Most of the material could have been written in terms of confidence sets. The distinction between the two approaches is largely one of reporting and interpretation. Much of the theory can also be presented in terms of p-values (see Schweder, 1988).

Hypothesis testing has been around for a long time and appears to provide a solution to a type of question that seems natural in many inquiries. Thus I believe that, singly or as a component of more comprehensive strategies, testing will continue to remain an important part of statistical methodology. If this is the case, the associated theory, which together with its application is the subject of *TSH*, will continue to be of interest.

REFERENCES

BAILAR, J. C. and MOSTELLER, F., eds. (1992). *Medical Uses of Statistics*, 2nd ed. NEJM Books, Boston.

BROWN, L. D. (1994). Minimaxity, more or less. In *Statistical Decision Theory and Related Topics V* (S. S. Gupta and J. O. Berger, eds.) 1–18. Springer, New York.

DIACONIS, P. and HOLMES, S. (1994). Gray codes for randomization procedures. *Statistics and Computing* **4** 287–302.

DONOHO, D. L. and JOHNSTONE, I. M. (1995). Wavelet shrinkage: asymptopia? *J. Roy. Statist. Soc. Ser. B* **57** 301–369.

KALLENBERG, W. C. M. et al. (1984). *Testing Statistical Hypotheses: Worked Solutions*. Centrum voor Wiskunde en Informatica, Amsterdam.

LEVIN, J. R. (1993). Statistical significance testing from three perspectives. *J. Exper. Educ.* **61** 378–382.

NEYMAN, J. and PEARSON, E. S. (1993). On the problem of the most efficient tests of statistical hypotheses. *Philos. Trans. Roy. Soc. London Ser. A* 289–337.

PRATT, J. W. (1961). Book review. *J. Amer. Statist. Assoc.* **56** 163–167.

SCHWEDER, T. (1988). A significance version of the basic Neyman–Pearson theory for scientific hypothesis testing. *Scand. J. Statist.* **15** 225–235.

STUDENT (W. S. GOSSET) (1908). On the probable error of a mean. *Biometrika* **6** 1–25.

THOMPSON, B., ed. (1993). Statistical significance testing in practice. *J. Exper. Educ.* **61** 285–396.

WOLFOWITZ, J. (1970). Reflections on the future of mathematical statistics. In *Essays in Probability and Statistics* (R. C. Bose, ed.) 443–453. Univ. California Press, Berkeley.

Statistical Science
1999, Vol. 14, No. 4, 418–426

"Student" and Small-Sample Theory

E. L. Lehmann

Abstract. The paper discusses the contributions Student (W. S. Gosset) made to the three stages in which small-sample methodology was established in the period 1908–1933: (i) the distributions of the test-statistics under the assumption of normality, (ii) the robustness of these distributions against nonnormality, (iii) the optimal choice of test statistics. The conclusions are based on a careful reading of the correspondence of Gosset with Fisher and E. S. Pearson.

Key words and phrases: History of statistics, "exact" distribution theory, assumption of normality, robustness, hypothesis testing, Neyman–Pearson theory.

1. INTRODUCTION

In an interview published in *Statistical Science* [Laird (1989)], F. N. David talks about statistics in the 1920s and 30s as developed by Gosset (Student), Fisher, Egon Pearson and Neyman. [In the remainder of this paper we shall usually refer to E(gon) S. Pearson simply as Pearson and to his father as Karl Pearson or occasionally as K. P.] She describes herself as a contemporary observer who saw "all the protagonists from a worm's eye point of view." (For anyone like myself who knew the feisty David later in her life it is hard to imagine her ever playing the role of a "worm.") Her surprising assessment:

> I think he [Gosset] was really the big influence in statistics.... He asked the questions and Pearson or Fisher put them into statistical language and then Neyman came to work with the mathematics. But I think most of it stems from Gosset. (1.1)

(Note: Numbers have been added to quotations for easier cross-reference.)

Here she is of course not talking about all of statistics in this period, but of the development of the new small-sample approach. Nevertheless, her claim is surprising because Gosset is mainly known for only one, although a pathbreaking, contribution: Student's t-test. The aim of this paper is to consider to what extent David's conclusion is justified.

The basis for the new methodology was established in three stages. Stage 1 (Student–Fisher) determined the distributions of the statistics used to test means, variances, correlation and regression coefficients under the assumption of normality. At the second stage (Pearson), the robustness of these distributions under nonnormality was investigated. Finally, at the last stage Neyman and Pearson laid the foundation for a rational choice of test statistics not only in the normal case but quite generally. In the following sections we shall consider the contributions Gosset made to each of these stages.

An author writing about this period is fortunate to have available a wealth of material that frequently makes it possible to trace mutual influences and the development of ideas in considerable detail. The principal sources I have used are acknowledged at the end of the paper.

2. THE NEW METHODOLOGY

2.1 Gosset's 1908 Papers

The event that with little fanfare and no particular enthusiasm on the part of the Editor ushered in a new era in statistics was the *Biometrika* publication in 1908 of a paper "The probable error of the mean" by "Student," the pseudonym of William Sealy Gosset. The reason for the pseudonym was a policy by Gosset's employer, the brewery Arthur Guinness Sons and Co., against work done for the firm being made public. Allowing Gosset to publish under a pseudonym was a concession that resulted in the birth of the statistician "Student," the creator of Student's t-test.

Today the pathbreaking nature of this paper is generally recognized and has been widely commented upon, among others by Pearson (1939), Fisher (1939), Welch (1958), Mosteller and Tukey (1977, Section B), Box (1981), Tankard (1984),

E. L. Lehmann is Professor Emeritus, Department of Statistics, University of California, Berkeley, California 94720.

Pearson (1990), Lehmann (1993) and Hald (1998, Section 27.5).

The core of the paper consists of the derivation of the distribution of

$$z = (\bar{X} - \mu)/S \tag{2.1}$$

where $X_1, \dots, X_n$ are i.i.d. with normal distribution $N(\mu, \sigma^2)$ and where $S^2 = \Sigma(X_i - \bar{X})^2/n$. (The definitions of z and S^2 do not agree with current usage but are those used by Student.) This derivation was a difficult task for Gosset who was a chemist, not a mathematician, and although he obtained the correct answer, he was not able to give a rigorous proof.

However, this distributional result was not the reason for the enormous influence of the paper. The principal contribution was that it brought a new point of view. It stated the need for methods dealing with small samples, for which the normal approximations of the theory of errors were not adequate, and it brought the crucial insight that exact results can be obtained by making an additional assumption: that the form of the distribution of the observations is known. Concerning this assumption, Student wrote in the Introduction to his paper [Student (1908a)]:

> ...the sample is not sufficiently large to determine what is the law of distribution of individuals. It is usual, however, to assume a normal distribution... : since some law of distribution must be assumed it is better to work with a curve whose area and ordinates are tabled, and whose properties are well known. This assumption is accordingly made in the present paper, so that its conclusions are not strictly applicable to populations known not to be normally distributed; yet it appears probable that the deviation from normality must be very extreme to lead to serious error. (2.2)

What did Student mean when he wrote that "it is usual to assume a normal distribution"? He learned statistics by reading two books: Airy's *Theory of Errors of Observations* (1879) and Merriman's *Method of Least Squares* (1884). Both emphasize that errors are typically sums of a large number of independent small components and hence are approximately normally distributed. In fact, the normal distribution is called the Law of Probability of Errors. Merriman says about it (page 33), "Whatever may be thought of the theoretical deductions of the law of probability of error, there can be no doubt but that its practical demonstration by experience is entirely satisfactory." Airy (page 24) is slightly more cautious because he warns, "It must always be borne in mind that the law of frequency of errors does not exactly hold except the number of errors (i.e., components of error) is indefinitely great. With a limited number of errors, the law will be imperfectly followed; and the deductions, made on the supposition that the law is strictly followed, will be or may be inaccurate or inconsistent."

Despite these reservations, Student's assumption of normality is thus grounded in a well-established tradition.

Student illustrated the use of his distribution by three examples, including one of a paired comparison experiment which he reduces to the one-sample situation by taking differences. Finally, he gave a table of the z-distribution for sample sizes 4 to 10. He later extended it to sample sizes 2 to 30 [Student, (1917)]. The change from z to $t = z\sqrt{n-1}$ which is now called Student's t, is due to Fisher (1925a, b) and is discussed in Eisenhart (1979). Student provided tables for t in 1925.

The paper on z was followed by another paper (1908b) in which Student tackled the small-sample distribution of the sample correlation coefficient in the normal case when the population correlation coefficient is 0. Because a mathematical derivation was beyond his powers, he decided to "fit a Pearson curve" and, using some elementary properties of correlation coefficients, he "guessed" (his own word) the correct form.

For the case that the population correlation coefficient is different from 0, he came to the conclusion that it "probably cannot be represented by any of Professor Pearson's types of frequency curves" and admits that he "cannot suggest an equation which will accord with the facts."

Gosset wrote no further papers on small-sample distributions (except for providing tables of z and t). An obvious explanation is that he had a full-time job as brewer. However, he himself denies that this was the reason, explaining to Fisher (July 14, 1924): "By the way it is not time but ability which has prevented me following up my work by more on your lines." Gosset (1970) His 1908 papers did not receive much attention and his ideas might have continued to go unnoticed had they not acquired a new champion of exceptional brilliance and enormous energy.

2.2 Fisher's Proof

In 1912 R. A. Fisher, then 22 years old and a Cambridge undergraduate, was put into contact with Gosset through Fisher's teacher, the astronomer F. J. M. Stratton. As a result, Gosset received from Fisher a proof of the z-distribution and asked Karl Pearson to look at it, admitting that he could not follow the argument (which was based on n-dimensional geometry) and suggesting, "It seemed to me that if it's alright perhaps you might like to put the proof in a note [in *Biometrika* of which K. P. was the Editor]. It's so nice and mathematical that it might appeal to some people. In any case I should be glad of your opinion of it . . . ."

Pearson was not impressed. "I do not follow Mr. Fisher's proof and it is not the kind of proof which appeals to me," he replied. Pearson (1990, page 47) As a result, the proof was only published in 1915 together with the corresponding proof for the distribution of the correlation coefficient that Student had conjectured in his second 1908 paper. In the correlation case, the n pairs of observations are considered as the coordinates of a point in $2n$-dimensional space, in which the two sample means, two sample variances, and the sample covariance have, as Fisher writes, "a beautiful interpretation," [Fisher (1915)] from which the desired density can be obtained.

During the next few years, Fisher did no further work on such distributional problems, but he was pulled back to them when undertaking an investigation of the difference between the inter- and intraclass correlation coefficients. The distribution of the latter was still missing, and Fisher derived it by the same geometrical method he had used previously (Fisher, 1921).

A clue to Fisher's thinking about such problems at the time can be gleaned from his fundamental paper, "On the mathematical foundations of theoretical statistics" (1922a), that was submitted in June 1921 and read in November of that year. As the principal problems in statistics, he mentions specification, estimation and distributions. He lists the work on χ^2 by Karl Pearson and himself, the papers by Student and his own papers of 1915 and 1921 as "solving the problem of distribution" for the cases that they cover.

He continues,

> The brevity of this list is emphasized by the absence of investigation of other important statistics, such as the regression coefficients, multiple correlations, and the correlation ratio. [Fisher (1922a)] (2.3)

He takes up this theme again in the Summary of the paper where he states,

> In problems of Distribution relatively little progress has hitherto been made, these problems still affording a field for valuable enquiry for highly trained mathematicians. [Fisher (1922a)] (2.4)

2.3 Extensions

These two passages suggest that Fisher thought the outstanding distributional problems were difficult and also that he had no plans to work on them himself. However, in April 1922 he received two letters from Gosset that apparently changed his mind. In the first letter Gosset urged,

> But seriously I want to know what is the frequency distribution of $r\sigma_X/\sigma_Y$ [the regression coefficient] for small samples, in my work I want that more than the r distribution [the correlation coefficient] now happily solved. (2.5) Gosset (1970)

In his later summaries of the correspondence, Fisher comments on this letter: ". . . enquiry about the distribution of an estimated regression coefficient (a problem to which he [Gosset] presumably received the solution by return)."

This solution (together with that of the two-sample problem) appeared in *JRSS* (1922b). The paper is primarily concerned with a different problem, that of testing the goodness of fit of regression lines. At the end, Fisher appends a section which, in view of the dates, must have been added at the last moment and of which he later states in his Author's Note Fisher (1971–1974): "Section 6 takes up a second topic, connected with the first only by arising also in regression data." He introduces this second topic by explaining that

> . . . an exact solution of the distribution of regression coefficients . . . has been outstanding for many years; but the need for its solution was recently brought home to the writer by correspondence with 'Student' whose brilliant researches in 1908 form the basis of the exact solution. (2.6)

A comparison of (2.6) with the statement (2.4) of a year earlier indicates the change of mind brought about by Gosset's letter. The earlier statement suggests that Fisher thought the problem was hard and that he had no intention of working on it himself. After reading Gosset's letter, he must have looked at the problem again and realized that it easily yielded to the geometric method he had used earlier; in fact, so easily that he was able to send Gosset the solution "by return [mail]." Gosset (1970)

This seems to be the point at which Fisher realized the full power of his method, and the opportunity to apply this new-found confidence arose immediately. For within days there followed another request from Gosset (April 12): Gosset (1970)

> I forgot to put another problem to you in my last letter, that of the probable error of partial $\left\{\begin{matrix}\text{correlation}\\ \text{regression}\end{matrix}\right\}$ coefficients for small samples.

And Fisher in his later summary comments, "... this also was probably quickly answered for on May 5 he [Gosset] refers to the solution." Gosset (1970) [The result was published in Fisher (1924a).]

This series of papers by Fisher on "exact" small-sample distributions culminates in two summary papers in 1924 and 1925. In the first of these, (a lecture delivered in 1924b but only published in 1928), he introduces the distribution of the F-statistic for testing the equality of two normal variances and points out how χ^2 and t^2 are special cases of F. He also shows how this distribution can be used for an "analysis of variance." In the second paper he surveys the many uses to which the t-distribution can be put, and points out that this distribution applies whenever one is dealing with the ratio of two independent variables of which the numerator is normally distributed (with mean 0) and the denominator as the square root of a χ^2-variable divided by its degrees of freedom. He also gives for the first time an algebraic proof of the t-distribution.

However, this is still not quite the end, for in 1928 Fisher obtained the distribution of the multiple correlation coefficient, the last "of the problems of the exact distribution of statistics in common use to have resisted solution," [Fisher (1928)], as he writes in the opening sentence. The paper is remarkable in that it obtains the distribution not only under the null hypothesis but also in the noncentral case (a term Fisher introduced here) and as a result also the nonnull distribution for the analysis of variance and all the other statistics treated earlier.

The solutions of the indicated distributional problems led Fisher to some further developments, in particular the analysis of variance and the design of experiments [for details, see Pearson (1939)]. In these, Gosset participated but no longer in the earlier role of catalyst and we shall therefore not discuss them here.

3. ROBUSTNESS

3.1 Student's Questions

The small-sample "exact" methodology discussed in the preceding section is based on the assumption of normally distributed observations. This was emphasized by Student in his first 1908 paper where he stated that

> ... the conclusions are not strictly applicable to populations known not to be normally distributed; yet it appears probable that the deviation from normality must be very extreme to lead to serious error. [Student (1908b)] (3.1)

He was naturally curious about the effect of nonnormality and in 1923 (July 3) wrote to Fisher:

> What I should like you to do is to find a solution for some other population than a normal one. It seems to me you might assume some sort of an equation for the frequency distribution of x which would lend itself to treatment besides the Gaussian. I tried $y = a$ [i.e. the rectangular distribution] once but soon got tied up.
>
> I had hoped to go on to a (right-) triangular distribution ◺: but having been defeated by ▭ I hadn't the heart to try.

Later that year (October 1), he acknowledges a reply by Fisher, unfortunately lost: "I like the result for z in the case of that horrible curve you are so fond of. I take it that in skew curves the distribution of z is skew in the opposite direction." (I have made various inquiries about what this distribution might have been, but without any definite answers. George Barnard has suggested the log normal as a possibility.)

At that time no further discussion between Gosset and Fisher on robustness is recorded. However, the problem continues to concern Gosset, so he raises it again, this time with E. S. Pearson. In response to a letter in which Pearson inquires about a different matter (which will be taken up in the next section). Gosset writes (May 11, 1926): Gosset (1970)

> I'm more troubled really by the assumption of normality and have tried from time to time to see what happens with other population distributions, but I understand that you get correlation between s and m [the denominator and numerator of t] with *any* other population distribution.
>
> Still I wish you'd tell me what happens with the even chance population ▭ [rectangular] or such as △ [symmetrical triangular]: it's beyond my analysis. (3.2)

Pearson describes his reaction to this appeal in Pearson (1990, page 90):

> The existence of these random numbers [i.e., Tippet's table of random numbers (1927)] opened out the possibility scarcely dreamed of before, of carrying out a great variety of experimental programmes, particularly of answering in considerable depth and breadth the kind of questions about the robustness of the 'normal theory' tests based on z (or t), s^2, r and χ^2 *raised by Gosset* [my italics] in his letter to me

> of 11 May 1926. This programme I started on in 1927 and results began to appear, as they became available, in *Biometrika* papers published between 1928 and 1931. (3.3)

3.2 A Fisher – Gosset Debate

While this robustness work of Pearson and his coworkers was progressing from 1926 to 1931, a heated argument broke out between Pearson and Fisher. Trying to mediate, Gosset was drawn into a lengthy debate with Fisher that is of interest here because it produced several statements that clarify the views of the three participants on some aspects of the robustness question. [For additional material on this debate, see Pearson (1990).] The dispute was sparked by a critical remark that Pearson made in his 1929 review in *Nature* of the second edition (1928) of Fisher's book *Statistical Methods for Research Workers* and involved two issues:

(i) Whether Fisher's writing had been misleading.
(ii) How robust the normal theory tests actually are under nonnormality.

Fisher, in a letter to Gosset of June 20, 1924, made it clear that he was concerned only with the first of these, but Gosset was less interested in bruised egos than in the validity of the new methods, and replied,

> The really important point is, not your misunderstanding of Pearson, or, if there was one, his of you, but the crying practical problem of *How much does it matter*? And in fact *that* is *your* business: none of the rest of us have the slightest chance of *solving* the problem: we can play about with samples [i.e., perform simulation studies], I am not belittling E. S. P.'s work, but it is up to you to get us a proper solution. (3.4)

This is a remarkable statement, particularly from Gosset, whose statistical concerns are practical and who himself was a pioneer in the use of simulation. Here he makes clear the inadequacy of simulation alone and the need to supplement it by theory.

But Fisher will have none of it. In a long reply on June 27 he brushes off Gosset's suggestion:

> I do not think what you are doing with nonnormal distributions is at all my business, and I doubt if it is the right approach. What I think is my business is the detailed examination of the data, and of the methods of collection, to determine what information they are fit to give, and how they should be improved to give more or other information. In this job it has never been my experience to want to make the variation more normal; I believe even in extreme cases a change of variate [i.e., a transformation] will do all that is wanted, but that of course depends on the limitation of my own experience. I have fairly often applied a z-test to crude values, and to log values, even when the translation is a severe strain, . . . , but have never found it to make an important difference.
>
> Where I differ from you, I suppose, is in regarding normality as only a part of the difficulty of getting good data; viewed in this collection of difficulties I think you will see that it is one of the least important.
>
> You want to regard it as a part of the mathematical problem, which I do not, because a mathematical problem must start with precise data, and data other than normally [sic] are either not precise or very uninteresting.

To bring greater clarity to the issue, Gosset in his next letter makes use of a distinction that Fisher (in a letter of June 27) had introduced with respect to an assistant of Gosset's and which Gosset now turns around and applies to Fisher himself (June 28): Gosset (1970)

> I think you must for the moment consent to be analysed into α-Fisher the eminent mathematician and β-Fisher the humble applier of his formulae.
>
> Now it's α-Fisher's business, or I think it is, to supply the world's needs in statistical formulae: true β-Fisher doesn't think the particular ones that I want are necessary but between ourselves that's just his cussedness. In any case I quite agree that what *we* are doing with nonnormal distributions is no business of either of them; it is merely empirical whereas α-Fisher is interested in the theoretical side and β-Fisher in whatever seems good to him. But when β-Fisher says that the detailed examination of the data is his business and proceds to examine them by means of tables which are strictly true only for normally distributed variables I think I'm entitled to ask him what difference it makes if in fact the samples are not taken from this class of variables.

As a result of Gosset's intervention, Fisher did not publish his planned (apparently rather intemperate) rebuttal (which has not been preserved) to Pearson's review. Instead, at Fisher's suggestion, Gosset submitted a diplomatic response which was published in *Nature* (Student, 1929). Much of it was concerned with Pearson's comment that

Fisher's book could be misleading, but he also addressed the substantive issue of the robustness of the t-test:

> The question of the applicability of normal theory to non-normal material is, however, of considerable importance and merits attention both from the mathematician and from those of us whose province it is to apply the results of his labours to practical work. Personally, I have always believed, without perhaps any very definite grounds for this belief, that in point of fact 'Student's' distribution will be found to be very little affected by the sort of small departures from normality which obtain in most biological and experimental work, and recent work on small samples confirms me in this belief. We should all of us, however, be grateful to Dr. Fisher if he would show us elsewhere on theoretical grounds what *sort* of modification of his tables we require to make when the samples with which we are working are drawn from populations which are neither symmetrical nor mesokurtic [i.e., whose coefficient of kurtosis is not zero].

(Gosset's "indicate to us" was replaced by the editor's "show us elsewhere" that greatly annoyed Fisher.) Fisher was not willing to leave Gosset's challenge unanswered and published a reply in *Nature* (Fisher, 1929) which, as he mentions in a letter to Gosset "seems free from Billingsgate (i.e., abusive language) and may even help members and others to understand better where we stand." Gosset (1970) In this reply, Fisher does not address the question of robustness of the tests, but in response to the last sentence of Gosset's letter considers alternatives that would avoid Pearson's criticism. Two comments in this reply are of particular interest.

In the first of these he considers what would happen if it were possible to generalize the normal-theory distributions and discusses the criticisms to which such an extension would be exposed. The most interesting of these is

> ...that the particular statistics, means and mean squares entering into these tests are only efficient for normal distributions, and that for Pearson curves quite other statistics are required, and not only revised distributions of the familiar statistics appropriate to normal material.

This statement is of course correct and interesting in light of the later Neyman–Pearson theory.

Later in the letter Fisher makes another interesting suggestion:

> Beyond all questions of metrical variates there are, largely undeveloped, a system of tests which depend only on frequency and on order of magnitude. Examples exist in 'Student's' writings, and in my own. They are free from all taints of normality, but are too insensitive to be very useful; still, their development would be of more interest than the programme of research first considered.

These two comments show the enormous breadth of Fisher's vision. They foreshadow two of the most significant later developments of the small-sample approach: the Neyman–Pearson theory and the nonparametric methodology of rank tests.

A referee has suggested that Fisher might have been referring to his randomization tests rather than to rank tests. However, randomization tests depend on the values of the observations, not just on frequencies and order of magnitude. Any remaining doubt is removed by Fisher's comment that such tests are "too insensitive to be useful."

While the Fisher–Gosset debate concerning the robustness (against nonnormality) of the t-test and the tests Fisher had developed in its wake brought no meeting of minds, some clarification was achieved by Pearson's simulation studies. They indicated that the t-test and those of the model I analysis of variance are fairly insensitive under nonnormality, but that this is not true for the F-test for variances or some of the tests for variance components. These suggestions were later confirmed by theoretical results of George Box and others as well as by additional simulation work. Since the vulnerable F-tests were included in Fisher's *Statistical Methods* without any warning about their unreliability, Gosset's insistence on verification seems justified.

4. CHOICE OF TEST

On the robustness question Gosset clearly was the driving force. It was his suggestion that led to Pearson's empirical investigations, and he tried repeatedly, though unsuccessfully, to get Fisher to study the issue theoretically.

At the next stage, which led to the Neyman–Pearson theory, Gosset's role was quite different. As Pearson recalled the origin of this development in his obituary of Gosset (Pearson, 1939):

> I had been trying to discover some principle beyond that of practical expediency which would justify the use of "Student's" z.

He addressed some of his questions in a letter to Gosset and

> Gosset's reply had a tremendous influence on the direction of my subsequent work, for the first paragraph contains the germ of that idea which has formed the basis of all the later joint researches of Neyman and myself. It is the simple suggestion that the only valid reason for rejecting a statistical hypothesis is that some alternative explains the observed events with a greater degree of probability. (4.1)

A referee has pointed out that the consideration of alternative hypotheses was not new. For example, Jevons (1873, 1877, Chapter 12) discusses the problem of choosing among a number of alternative hypotheses.

Pearson goes on to quote the relevant paragraph of Gosset's letter of May 11, 1926. The crucial point that was to have such far reaching consequences is contained in a single sentence. Speaking about the observation of a very unlikely event Gosset writes,

> ... even if the chance is very small, say .00001, that doesn't in itself necessarily prove that the sample is not drawn randomly from the population [specified by the hypothesis]: what it does is to show that if there is any alternative hypothesis which will explain the occurrence of the sample with a more reasonable probability, say .05 (such as that it belongs to a different population or that the sample wasn't random or whatever will do the trick) you will be very much more inclined to consider that the original hypothesis is not true. (Pearson, 1939) (4.2)

Pearson passed the suggestion on to Neyman who was spending the year in Paris and who acknowledged it in a letter of December 6, 1926, agreeing that "to have the possibility of testing, it is necessary to adopt such a principle as Student's" [quoted in Reid (1982), page 70].

Gosset's suggestion led Pearson to the idea of likelihood ratio tests as a reasonable method of test construction, and the result was a pair of joint papers by Neyman and Pearson in the 1928 volume of *Biometrika*, "On the use and interpretation of certain test criteria for purposes of statistical inference," which together took up 98 pages. In it the authors introduced not only the likelihood ratio principle, but also the concept of first and second kinds of errors. The formulation of both ideas required not only the hypothesis H but also a class of alternatives to H.

Neyman and Pearson followed the likelihood ratio paper with an attack from first principles on how to choose a test not only in the normal case but quite generally. This work resulted in their 1933 paper "On the problem of the most efficient tests of statistical hypothesis," which formed the basis of the theory of hypothesis testing as we now know it. The approach made use not only of the class of alternatives suggested by Student, but also of an innovation introduced by Fisher in his *Statistical Methods* (1925b), namely to define significance in terms of a preassigned level instead of reporting p-values. This proposal (that later was overused and as a result attracted strong opposition) has a curious connection with Gosset.

A few years after starting to work for Guinness and learning the statistical methods he needed from Airy and Merriman, Gosset in 1905 (before he became "Student") made an appointment to consult Karl Pearson about some questions he was unable to resolve. In a report to the brewery (Pearson, 1939) he states as one of these questions that

> none of our books mentions the odds, which are conveniently accepted as being sufficient to establish any conclusion.

Had Gosset addressed this question twenty years later to Fisher, we might credit (or blame) him for having suggested the idea of fixed levels, a concept that constituted a crucial element of the Neyman–Pearson formulation.

5. CONCLUSION

Let us now return to the question posed at the beginning of this paper regarding the influence of Gosset on Fisher, Pearson and Neyman.

5.1 Gosset and Pearson

Pearson's contributions to small-sample theory are twofold. They consist on the one hand of his simulation studies of robustness culminating in his 1931 paper "The analysis of variance in cases of non-normal variation." The other strand is his joint work with Neyman in which they developed what is now called the Neyman–Pearson theory. For both aspects, the crucial ideas came from Student. As Pearson himself acknowledges (Pearson, 1990, page 82) in commenting on Student's letter from 1926,

> His letter left me with two fundamental ideas:
>
> (a) The rational human mind did not discard a hypothesis unless it could conceive at least one plausible *alternative* hypothesis.

manuscript of E. S. Pearson's by R. L. Plackett with the assistance of G. A. Barnard. This book contains many additional references.

For Fisher, Pearson and Neyman we again have available convenient collections of their papers: the five volumes of Fisher's papers, many with helpful later "Author's Notes," edited by Bennett and published by the University of Adelaide (1971–1974), and three volumes, one each, of papers by Pearson (1966a), Neyman (1967), and Neyman–Pearson (1967), published by the University of California Press. Of the secondary literature, we mention only the biographies of Fisher by his daughter Joan Fisher Box (1978) and of Neyman by Constance Reid (1982).

REFERENCES

AIRY, G. B. (1879). *On the Algebraic and Numerical Theory of Errors of Observations and the Combination of Observations.* 3rd Ed. MacMillan, London.

BOX, J. F. (1978). *R. A. Fisher: The Life of a Scientist.* Wiley, New York.

BOX, J. F. (1981). Gosset, Fisher and the t-distribution. *Amer. Statist.* **35** 61–66.

EISENHART, C. (1979). On the transition from "Student's" z to "Student's" t. *Amer. Statist.* **33** 6–10.

FISHER, R. A. (1915). Frequency distribution of the values of the correlation coefficient in samples from an indefinitely large population. *Biometrika* **10** 507–521.

FISHER, R. A. (1921). On the "probable error" of a coefficient of correlation deduced from a small sample. *Metron* **1** 3–32.

FISHER, R. A. (1922a). On the mathematical foundations of theoretical statistics. *Philos. Trans. Roy. Soc. London Ser. A* **222** 309–368.

FISHER, R. A. (1922b). The goodness of fit of regression formulae, and the distribution of regression coefficients. *J. Roy. Statist. Soc.* **85** 597–612.

FISHER, R. A. (1924a). The distribution of the partial correlation coefficient. *Metron* **3** 329–332.

FISHER, R. A. (1924b). On a distribution yielding the error functions of several well known statistics. In *Proc. Internat. Congress of Mathematics, Toronto* **2** 805–813.

FISHER, R. A. (1925a). Applications of 'Student's' distribution. *Metron* **5** 90–104.

FISHER, R. A. (1925b). *Statistical Methods for Research Workers,* 2nd ed. (1928) Oliver and Boyd, Edinburgh.

FISHER, R. A. (1928). The general sampling distribution of the multiple correlation coefficient. *Proc. Roy. Soc. London Ser. A* **121** 654–673.

FISHER, R. A. (1929). Statistics and biological research. *Nature* **124** 266–267.

FISHER, R. A. (1939). Student. *Ann. Engenics* **9** 1–9.

FISHER, R. A. (1971–1974). *Collected Papers of R. A. Fisher* (J. H. Bennett, ed.). Univ. Adelaide.

GOSSET, W. S. (1970). *Letters from W. S. Gosset to R. A. Fisher, 1915–1936 With Summaries by R. A. Fisher and a Foreword by L. McMullen.* Printed for private circulation.

HALD, A. (1998). *A History of Mathematical Statistics (from 1750–1930).* Wiley, New York.

JEVONS, W. S. (1874, 1877). *The Principles of Science.* McMillan, London.

LAIRD, N. M. (1989). A conversation with F. N. David. *Statist. Sci.* **4** 235–246.

LEHMANN, E. L. (1993). The Bertrand–Borel debate and the origins of the Neyman–Pearson theory. In *Statistics and Probability: A Raghu Raj Bahadur Festschrift* (J. K. Ghosh, S. K. Mitra, K. R. Parthasarathy and B. L. S. Prakasa Rao, eds.). Wiley Eastern Limited, New Delhi.

MCMULLEN, L. (1939). 'Student' as a man. *Biometrika* **30** 205–210.

MERRIMAN, M. (1884). *A Textbook on the Method of Least Squares.* Wiley, New York.

MOSTELLER, R. and TUKEY, J. W. (1977). *Data Analysis and Regression.* Addison-Wesley, Reading, MA.

NEYMAN, J. (1929). Méthodes nouvelles de vérification des hypothèses. In *Comptes Rendus du I Congrès de Mathématiciens des Pays Slaves, Warsaw* 355–366.

NEYMAN, J. (1967). *A Selection of Early Statistical Papers of J. Neyman.* Univ. California Press, Berkeley.

NEYMAN, J. (1977). Frequentist probability and frequentist statistics. *Synthèse* **36** 97–131.

NEYMAN, J. and PEARSON, E. S. (1928). On the use and interpretation of certain test criteria for purposes of statistical inference. *Biometrika* **20A** 175–240 and 263–294.

NEYMAN, J. and PEARSON, E. S. (1933). On the problem of the most efficient tests of statistical hypotheses. *Philos. Trans. Roy. Soc. Ser. A* **231** 289–337.

NEYMAN, J. and PEARSON, E. S. (1967). *Joint Statistical Papers of J. Neyman and E. S. Pearson.* Univ. California Press, Berkeley.

PEARSON, E. S. (1929a). Review of *R. A. Fisher: Statistical Methods for Research Workers, 2nd ed. Nature.*

PEARSON, E. S. (1931). The analysis of variance in cases of nonnormal variation. *Biometrika* **23** 114–133.

PEARSON, E. S. (1939). 'Student' as a statistician. *Biometrika* **30** 210–250.

PEARSON, E. S. (1966a). *The Selected Papers of E. S. Pearson.* Univ. California Press, Berkeley.

PEARSON, E. S. (1967). Some reflexions on continuity in the development of mathematical statistics, 1885–1920. *Biometrika* **54** 341–355.

PEARSON, E. S. (1968). Some early correspondence between W. S Gosset, R. A. Fisher and Karl Pearson with notes and comments. *Biometrika* **55** 445–457.

PEARSON, E. S. (1990). *'Student'—A Statistical Biography of William Sealy Gosset* (R. L. Plackett and G. A. Barnard, eds.). Clarendon Press, Oxford.

REID, C. (1982). *Neyman—From Life.* Springer, New York.

STUDENT (1908a). The probable error of a mean. *Biometrika* **6** 1–25.

STUDENT (1908b). Probable error of a correlation coefficient. *Biometrika* **6** 302–310.

STUDENT (1917). Tables for estimating the probability that the mean of a unique sample of observations lies between $-\infty$ and any given distance of the mean of the population from which the sample is drawn. *Biometrika* **6** 414–417.

STUDENT (1925). New table for testing the significance of observations. *Metron* **5** 105–108.

STUDENT (1929). Statistics in biological research. *Nature* **124** 93.

STUDENT'S COLLECTED PAPERS (1942). E. S. Pearson and J. Wishart, eds. Cambridge Univ. Press.

TANKARD, J. W., JR. (1984). *The Statistical Pioneers.* Schenkman, Cambridge.

TIPPETT, L. H. C. (1927). *Random Sampling Numbers.* Cambridge Univ. Press.

WELCH, B. L. (1958). 'Student' and small sample theory. *J. Amer. Statist. Assoc.* **53** 777–788.

OPTIMALITY AND SYMPOSIA: SOME HISTORY

ERICH L. LEHMANN

1. Introduction

As a contribution to a symposium focusing on optimality, this paper will consider the origin of the concept of optimality in statistics and the history of the earliest statistical symposia. These two very disparate subjects are linked by the fact that they originated in the imagination of the same person, Jerzy Neyman. A brief account of his life will therefore be useful as a framework for the two main subjects.

Neyman was born in 1894 to Polish parents, but he grew up in Russia and saw Poland for the first time when he moved there in 1921 at age 27. In Warsaw he became associated with the Polish School of Mathematics and published a paper in its journal, Fundamenta Mathematicae. However, mathematical positions were in short supply and his knowledge of statistics (which he had learned from Sergei Berstein) was more marketable. So he worked as a statistician in agriculture and meterology until he found university positions in Warsaw and Krakow, where he lectured in both mathematics and statistics.

The Polish authorities were interested in Neyman's statistical work but felt unable to evaluate it. To obtain an assessment of its worth, they provided him with a fellowship for the academic year 1925/26 at Karl Pearson's Laboratory in London, which was then the center of the statistical world. His success or failure there, they hoped, would furnish them with a basis for judgment.

His year in London was a great disappointment for Neyman, since the mathematical level at Pearson's Laboratory was much lower than he had expected. To salvage something from his stay in the West, he obtained a fellowship for a second year of leave which he spent in Paris. There the lectures of Borel and Lebesgue and Hadamard's seminar drew him back into pure mathematics.

At this crucial moment Neyman's life was changed by a letter he received in late 1926 from Karl Person's son Egon. Continuing a conversation they had started in

J. Rojo (ed.), *Selected Works of E. L. Lehmann*, Selected Works in Probability and Statistics, DOI 10.1007/978-1-4614-1412-4_84, © Springer Science+Business Media, LLC 2012

Jerzy Neyman and Egon Pearson

London, Egon Pearson's letter proposed that they work jointly on a problem that had been occupying him for some time.

The problem had its origin in Fisher's work during the preceding decade, which culminated in his enormously successful book of 1925, "Statistical Methods for Research Workers". The book presented Fisher's new tests based on the t, F and χ^2 distributions and his analysis of variance. However, these methods were put forth without any justification, backed only by Fisher's authority.

The question Pearson now asked himself, and then raised with Neyman, was: Why these tests rather than any of the many others that could be proposed? The problem captured Neyman's imagination and resulted in a collaboration carried out mainly through correspondence (with Neyman in Poland and Pearson in England) supplemented by occasional visits.

Neyman's letters to Pearson have been preserved and, even without Pearson's letters which are lost, make it possible to trace the progress of the work step by step. Their study shows two clearly distinguishable phases. During the early years Pearson is the leader and originator. He explains what he is doing to Neyman to whom all this is quite new and who often misunderstands. Neyman's main contribution at this stage is to force Pearson to greater clarity and to help working out the more complicated examples.

This first part of the work resulted in a long 1928 paper on the likelihood ratio method of hypothesis testing. This approach had been proposed by Pearson at the start of their work. In a large number of examples it was now seen to give the answers previously suggested by intuition. In particular, under the assumption of normality it led to Fisher's tests. Pearson was confirmed in his belief that it provided the general answer he had been looking for.

2. The birth of optimality theory

However, Neyman was not satisfied. He agreed that likelihood ratio was an appealing principle but felt that it was lacking a logically convincing justification. The tables were now turned. Neyman took the lead and gradually arrived at a radically new formulation which he first mentions to Pearson in a letter of February 20, 1930. To understand this and the following letter some explanations are necessary.

Neyman and Pearson define a test by a contour $\phi(X_1, \ldots, X_m) = c$. The hypothesis is rejected if the sample point falls inside this contour and accepted when it falls outside. (Today, the reverse of this definition would seem more natural). The contours given by the likelihood ratio criterion the authors call the λ- contours. It should also be mentioned that Neyman's English at the time was still far from perfect. In particular, there are many spelling errors which in the following quotations have been retained.

From the letter of Feb. 20, 1930:

"At present I am working on a variation calculus problem connected [with the] likelihood method. The results already obtained are a vigorous argument in favour of the likelihood method. I considerably forget the variation calculus and until the present time I have only results for samples of two. But in all cases considered I have found the following: We test a simple hypothesis concerning the value of some caracter $a = a_0$, and wish to find contours $\phi(X_1, \ldots, X_n) = c$ such that 1) the probability P(a) of a sample lying inside the contour (which probability is determined by the hypothesis H) is equal

$$P(a_0) = \epsilon$$

where ϵ is a certain fixed value, say .01. (This is for controlling the errors in rejecting a true hypothesis) and 2) that the probability $P(a_1)$ determined by some other hypothesis H' that $a = a_1 \neq a_0$ of sample lying inside the same contour be maximum. Using such contours and rejecting H when Σ (the sample point) is inside $\phi = const.$ we are sure that a true hypothesis is rejected with a frequency less than ϵ , and that if H (the hypothesis tested) is false and the true hypothesis is say H', then most often the observed sample will be inside $\phi = const$ and hence the hypothesis H will be rejected.

I feel you start to be angry as you think I am attacking the likelihood method! Be quiet! In all cases I have considered the $\phi = const.$ contours are the λ contours!" (A facsimile of the first part of this passage is shown on the next page.)

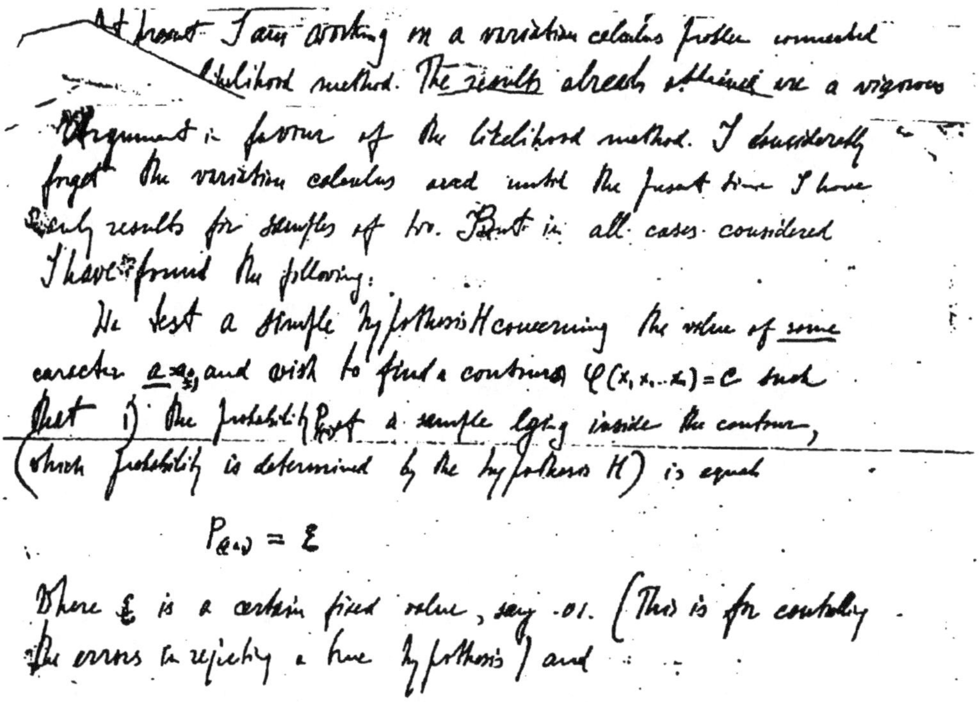

At present I am working on a variation calculus problem connected [with the] likelihood method. The results already attained are a vigorous argument in favour of the likelihood method. I considerably forgot the variation calculus and until the present time I have only results for samples of two. But in all cases considered I have found the following:

We test a simple hypothesis H concerning the value of some character $a = a_0$ and wish to find a contour $\varphi(x_1, x_2, \ldots x_n) = C$ such that 1) the probability of a sample lying inside the contour, (which probability is determined by the hypothesis H) is equal

$$P_{(a_0)} = \varepsilon$$

where ε is a certain fixed value, say .01. (This is for controlling the errors in rejecting a true hypothesis) and

Letter of February 20, 1930

In this letter Neyman is still struggling with the problem of testing a simple hypothesis against a simple alternative. Two weeks later, he essentially has the solution. In this letter he outlines a number of problems on which the authors want to work. One of these problems is:

From the letter of March 8, 1930

"f) To finish what I started to do with the variation calculus. You will understand it in a moment. To reduce to a given level the errors of rejecting a true hypothesis, we may use any test. [Today we might say, any test statistic]. Now we want to find a test which would 1) reduce the probability of rejecting a true hypothesis to the level $\leq \epsilon$ and 2) such that the probability of accepting a false hypothesis should be minimum. – We find that if such a best test exists, then it is the λ-test. I am now shure [sic] that in a few days I shall be ready. This will show that the "λ-principle" is not only a principle, but that there are arguments to prove that it is really "the best test"."

The result which Neyman announces here is of course what is now known as the Neyman-Pearson lemma. The formulation he sets forth in these two letters of 1930 constitute the birth of exact optimality theory.

Neyman realizes that this solution only settles the case of a simple alternative and begins to worry about the more practical situation in which there are many alternatives. In a letter of June 3, 1930 he considers testing $H : \sigma = \sigma_0$ against $\sigma \neq \sigma_0$ in the normal case with zero mean. He states that "In this case also the [best test] is independent of the alternative, 'but not intirely [sic]'"; it depends on whether σ is greater or less than σ_0.

He returns to this problem a year later and in a letter of August 17, 1931 suggests that the test should reject $H : \sigma = \sigma_0$ when ΣX_i^2 is $< c_1$ or $> c_2$. But how to choose c_1 and c_2? He mentions equal tails as a possibility but then interrupts himself and exclaims:

"I think I have got the point!"

And he goes on to argue that the probability of rejection under an alternative should never be less than under H; what we now call unbiasedness.

Their work on optimum tests of hypotheses was published by Neyman and Pearson in 1933. It was followed in 1934 by Neyman's paper establishing a rigorous theory of survey sampling and his 1937 paper on confidence intervals. In these three papers Neyman established a new point of view and in the process achieved a transformation that is sometimes referred to, not always favorably, as the mathematization of statistics.

3. The move to America

Neyman's work now was becoming known, and he received various invitations to lecture. Of particular importance was a 1937 American lecture tour ending in Washington, D.C. where he gave a series of lectures and conferences which were a huge success.

Shortly after returning from this tour to London, he received a letter from a person of whom he had never heard, Griffith Evans, chairman of the mathematics department of the University of California at Berkeley. Completely unexpectedly, the letter offered him a position in Evans' department. After considerable hesitation and some negotiations Neyman decided to accept the offer and in 1938 moved to Berkeley. He was 44, at the halfway point of his life. The great achievement of the first (European) half had been his entirely new approach to inference. The principal achievement of the second (American) half was to be the creation of an institutional base for the dissemination of this new approach.

Although Evans held Neyman in great esteem and gave him strong support, the two men had very different visions of Neyman's role which inevitably led to conflicts. Evans considered statistics a branch of mathematics such as algebra or differential equations, and thought of Neyman as a Professor of Mathematics whose specialty was statistics. For Neyman, on the other hand, statistics was a separate discipline, which should be represented by a separate department. He fought unrelentingly for independence and gradually obtained ever increasing autonomy. However, it took nearly 20 years before he reached his goal of a separate department of statistics, a department that was to grow to a faculty of over 20 and by the end of the century had graduated more than 400 Ph.D.'s.

But this development started slowly. Between 1938 and 1941, Neyman established a skeleton course program and a small statistical laboratory with a small temporary staff. After the outbreak of war in December 1941 growth of the academic program came to a halt as Neyman's efforts and those of his fledgling laboratory were concentrated on war work. It did not resume until the end of the European war in the spring of 1945.

4. The Berkeley Symposia

In June 1945, with the end of the war in sight Neyman returned to science and the development of his laboratory. As a first step, he proposed a symposium on mathematical statistics and probability, "to mark the end of the war and stimulate the return to theoretical research." The meeting took place in August 1945, with a sequel in January 1946 for those who had been unable to come at such short notice. Evans reported to the Provost that the symposium

"was an outstanding success and the success was due to Mr. Neyman's foresight in seeing its possibility at this time, and to his initiative and resourcefulness in planning it. It constitutes a significant page in the history of the University of California.

But what is of even more importance, it was convincing evidence of the growing importance of the relation of statistics to experimental work in many branches of science and of the service which the statistical laboratory is rendering to the university and the public."

The Proceedings of the Symposium, a volume of 500 pages consisting of 30 papers, appeared after some delays in 1949.

It seems likely that this first symposium was not part of a long range plan, but was conceived on the spur of the moment. However it was so successful, and Neyman enjoyed so much acting as host on his own turf, that by 1948 he began planning for a second symposium, which took place in 1950. By the third, in 1955, he was clearly viewing the symposia as an ongoing series.

The third symposium was on a much grander scale than the earlier two. The proceedings took up five volumes totaling over 1000 pages. The titles of the volumes:

(1) Theory of Statistics
(2) Probability Theory
(3) Astronomy and Physics
(4) Biology and Problems of Health
(5) Econometrics, Industrial Research, and Psychometry

reflect Neyman's broad range of interests.

At 5-year intervals Neyman organized three more symposia of increasing size and complexity. The proceedings of the sixth came to 6 volumes with a total of close to 3000 pages. By 1975, time for the seventh, Neyman was tired and the enterprise had become a burden to the department. And so the sixth symposium turned out to be the last.

During their lifetime the symposia constituted the most important international statistical meetings. All the leading probabilists and statisticians participated in one or more of them, as did many of the greatest physicists, astronomers, mathematicians, economists, ... The proceedings contained many seminal papers; for example Stein estimation, Stein's idea of adaptive inference, and Robbins' Empirical Bayes approach all were introduced in the Proceedings of the 3rd Symposium. Thus the symposia played an important role in establishing Berkeley as an internationally known center of statistics.

In an extensive review of the second symposium in the British journal Nature, M.G. Kendall predicted that "the Berkeley symposia will take a high place among the statistical literature of the world." The third symposium received an even more interesting review. It appeared in Science under the title: "Statistics – Servant of all Sciences", and its author–surprisingly–was Jerzy Neyman.

What gives this review its special interest is that it contains a succinct account of Neyman's statistical philosophy. He stresses the role of indeterminism–and hence of probability and statistics–in modern science. "As a result", he writes, "the field of application is literally limitless. Briefly, their role in scientific research consists
i) in providing tentative stochastic models of given classes of phenomena,
and
ii) in developing optimal methods of dealing with the observations in order either to supplement the original models with certain details such as the various constants they involve [i.e. estimation] or to decide for or against the further retention of these models [hypothesis testing]."

Point (ii), optimal methods of analysis, was of course his program of the 1930's. To this he now adds that statistical considerations must be preceded by construction of appropriate probabilistic models, and that this is a task for the statistician (in collaboration with a subject matter expert).

Modeling as an explicit programmatic requirement was new at the time and added an important component to Neyman's statistical philosophy. It also became a central aspect of his Berkeley research, particularly his work on some large-scale projects in several fields: competition of species, accident proneness, distribution of galaxies, effectiveness of cloud seeding, and the mechanisms of carcinogenesis. He was working on the proceedings of a conference on the last of these subjects during the last days in the hospital before his death on August 5, 1981.

We may consider Neyman to be the godparent of this symposium, both through his founding of optimality theory and the inspiration provided by his symposia. A detailed account of his life is given in Constance Reid's book "Neyman–From Life".

I should like to thank Mrs. Reid for sharing with me Neyman's letters to Pearson and other documents she collected while working on her book.

References

[1] Le Cam, L. and Neyman, J. (Eds) (1982). Probability Models and Cancer. North Holland. Amsterdam.

[2] Neyman, J. (1934). On two different aspects of the representative method: The method of stratified sampling and the method of purposive selection. J. Roy. Statist. Soc. 97, 558-625.

[3] Neyman, J. (1938). Lectures and Conferences on Mathematical Statistics. Graduate School of the U.S. Department of Agriculture. Washington. [2nd expanded Ed. 1952].

[4] Neyman, J. and Pearson. E.S. (1928). On the use and interpretation of certain test criteria. Biometrika 20A, 175-240, 263-294.

[5] Neyman, J. and Pearson. E.S. (1933). On the problem of the most efficient tests of statistical hypotheses. Phil. Trans. Roy. Soc. Ser. A. 231, 289-337.

[6] Pearson, E.S. (1966). The Neyman-Pearson story. In Research Papers in Statistics (F.N.David, Ed). John Wiley. New York.

[7] Reid, C. (1982). Neyman - from Life. Springer, New York.

Erich L. Lehmann
Department of Statistics
367 Evans Hall
University of California
Berkeley, CA 94720

IMS Collections
Probability and Statistics: Essays in Honor of David A. Freedman
Vol. 2 (2008) 114–126

DOI: 10.1214/193940307000000419

On the history and use of some standard statistical models

E. L. Lehmann[1]

University of California, Berkeley

Abstract: This paper tries to tell the story of the general linear model, which saw the light of day 200 years ago, and the assumptions underlying it. We distinguish three principal stages (ignoring earlier more isolated instances). The model was first proposed in the context of astronomical and geodesic observations, where the main source of variation was observational error. This was the main use of the model during the 19th century.

In the 1920's it was developed in a new direction by R.A. Fisher whose principal applications were in agriculture and biology. Finally, beginning in the 1930's and 40's it became an important tool for the social sciences. As new areas of applications were added, the assumptions underlying the model tended to become more questionable, and the resulting statistical techniques more prone to misuse.

1. Introduction

It was 200 years ago, in 1805, that Legendre first published the method of least squares and a vague formulation of what has come to be known as the standard linear model [27]. This model has played a central role in the statistical methodology used in the physical, biological and social sciences, and it is the aim of the present paper to sketch this role. I am not trying to write a history of the linear model[2], but am mainly concerned with the role the underlying assumptions have played in these three areas of application.

The model (defined in Section 6) assumes that each observation is the sum of two components: a deterministic term (which is a linear combination of the relevant explanatory variables) and a random term representing error or other "disturbances". The error terms are assumed to be independently, identically distributed (i.i.d.) according to a common normal distribution, and Sections 2–5 are therefore concerned with the role of these assumptions in the special case of the simple normal i.i.d. model. This model is of interest also in its own right as the standard model for the one-sample problem. Sections 6 and 7 take up issues concerning the non-random term of the linear model.

To conclude this introduction, let me briefly consider the general nature of mathematical, and particularly statistical, models.

[1]University of California, Department of Statistics, 367 Evans Hall #3860, Berkeley, CA 94720-3860, USA, e-mail: shaffer@stat.berkeley.edu

AMS 2000 subject classifications: 62A01, 62-03, 62J05.

Keywords and phrases: assumptions, independence, least squares, linear model, normality, observational studies.

[2]For accounts of this history see Seal [41] and Plackett [38], and in a broader context Stigler [45] and Hald [20].

The first use of the term "model" by a statistician that I have found occurs in Karl Pearson's Grammar of Science [33]. In this book he emphasizes the distinction between the real phenomenon and the model, which he usually calls "conception", but to which on a few occasions he refers as "models". For example he writes [33], p. 206:

> "The scientist postulates nothing in the world beyond sense [i.e. sense perceptions]; for him the atom and the ether are–like the geometric surface–models by which he resumes the world of sense.

The role and construction of mathematical models played a central role in the work of the mathematician Richard von Mises (1883-1953) who developed such models for a number of disciplines, among them aerodynamics, hydrodynamics, and plasticity. Besides undertaking these efforts in the physical sciences, von Mises felt that there was a great need for a similar treatment of probability theory. Instead of relying on Laplace's inadequate definition in terms of equally likely cases, he wanted to build a model for probability that would represent the physical reality underlying this concept.

In his fundamental paper of 1919 on the foundations of probability theory [48], he describes his approach as follows [my translation from the German]:

> "The present treatment is based on the assumption that probability theory is a *natural science, of the same kind as geometry or theoretical mechanics.* It has the aim to present the relations and dependencies of specific observable phenomenon, not as a faithful description of reality, but as its abstraction and idealization."

The observable phenomenon for which the paper builds a model is the stability of the long-run frequency of an outcome in a long sequence of repeated random events, for example, the frequency of heads in a long sequence of tosses of a coin. The principal constituents of the model are infinite sequences of trials with random outcomes, of which it is assumed that the frequency of a given outcome tends to a limit. To define randomness, von Mises requires that the same limit should obtain in any predetermined subsequence which can be chosen in the light of the observations up to this point.

It turned out that this formulation was too complicated; it was also too narrow in that it only applied to situations which allowed a large number of repetitions. As a formulation of probability theory it was replaced by a quite different approach proposed by Kolmogorov [23]. Kolmogorov did not construct a model for probability theory. Instead, he stated a small number of very simple axioms which any ideal model should satisfy. In this system probability itself was left undefined, subject only to these axioms. In this way the system could be interpreted and fleshed out not only by the frequency concept of probability but also by the idea of probability as degree of belief (The background of Kolmogorov's formulation is discussed in Shafer and Vovk [42]).

Although von Mises' modeling approach was not the one that was ultimately adopted by the profession, it nevertheless exerted great influence, particularly on Kolmogorov who cited the effect it had on his own formulation and who later took it as a starting point of a renewed effort to get a grip on the crucial (and difficult) concept of randomness[3].

[3]For a more detailed review of these foundational issues, see von Plato [49].

2. The assumption of normality in the 19th century

The most widely used statistical model for a sequence of repeated measurements has been the i.i.d. normal model according to which the observations are independently distributed with a common normal distribution. The first person to write down the formula for what today is called the normal density was De Moivre [10] who derived it as the limit of the binomial but did not consider it as a probability density in its own right. It was only in the beginning of the 19th Century through the combined insights of Laplace and Gauss that the central role of the normal distribution was realized. The normal distribution became the acknowledged model for the distribution of errors of physical (particularly astronomical) measurements and was called the Law of Errors.[4] It had a theoretical basis in the so called Law of Elementary Errors which assumed that an observational error is the sum of a large number of small independent errors and is therefore approximately normally distributed by the Central Limit Theorem. This argument was reinforced by extensive experience which showed good approximate agreement with the normal form.

The many textbooks on the subject all agreed on this point. For example, Brunt [6, 7], after citing theoretical arguments in favor of the normal distribution, states that "the final justification of Gauss' error curve rests upon the fact that it works well in practice and yields curves which in very many cases agree very closely with the observed frequency curves. The normal law is to be regarded as *proved by experience* and *explained* by Hagen's hypothesis [i.e. the law of elementary errors.]"

In a German text, Helmert [21, 22] explains [my translation]: "The form of the distribution of errors can only be determined through observation.... According to experience the [norma] law usually provides a close approximation."

Similar statements can be found in other British, German and French texts.

The history of the Theory of Errors is recounted in meticulous detail in a book on the subject by Czuber [9]. It was realized of course that the normal law could only be an approximation since in practice the observations were discrete and bounded. Nevertheless, as Czuber writes [my translation]:

> "An essential support of Gauss' Law of Errors is provided by the agreement which exists between its consequences and the results of observations that have really been obtained. It has led to the general acceptance by observers despite the concerns that can be raised against the various theoretical arguments in its favor."

Czuber devotes 14 pages to empirical comparisons of the law of errors with experience, and another section to the "elimination of contradictory observations" [i.e. gross errors], because it is only after such "doubtful observations" have been removed that the law applies.

He analyzes seven data sets from astronomy and geodesy, and in addition some experimental results of measurements carried out specifically to test the Gaussian law. His method in these examples is to compare the numbers of observations in various intervals with those predicted by theory. His conclusion is that the agreement is "satisfactory".

It should be noted that when Czuber mentions the general acceptance of the law, this statement is implicitly restricted to the principal areas of application he considers: the physical sciences and particularly astronomy and geodesy.

The normal distribution lost its exclusive position toward the end of the 19th Century when a strong interest developed in systematic application of statistical

[4]The term "normal distribution" was first suggested by Peirce [37], Lexis [28] and Galton [16] and began to take hold in the 1880's (see Stigler [46], Chapter 22).

methods to biological, sociological and economic investigations. There the distributions encountered were often far from normal and frequently asymmetric. As a result, families of models were developed that include skewed and heavy-tailed distributions. The most influential of these at the time was the system proposed by Karl Pearson [34]. Stigler [45] (p. 335) explains Pearson's success:

> "In thirty pages of detailed examples he demonstrated the successful flexibility and practicality of the system with a force that bludgeoned any potential skeptic into submission. The examples ranged from Venn's barometric pressures to the heights of St. Louis school girls to Weldon's crabs to statistics on pauperism .... Not only was his work more general than others, it was practical and came to public view with a record of proven accomplishment."

3. Fisher and the assumption of normality

Early in the 20th Century the normal distribution regained its ascendancy through the small sample work of R.A. Fisher. As Geary [17] summarized the situation in a historical overview:

> "Our historian will find a significant change of attitude a quarter of a century ago following the brilliant work of R.A. Fisher who showed that when universal normality could be assumed, inferences of the widest practical usefulness could be drawn. Prejudice in favor of normality returned in full force and interest in non-normality receded in the background."

But could universal normality be assumed? Fisher developed his new methods not in the context of astronomical measurements but for application to biological and agricultural data, and here the assumption is on much shakier ground.

In the earlier applications, the principal source of variation in the observations was observational error. Now to this is added the variability of the subjects (trees, agricultural plots, farms, ...) being measured. These subjects Fisher modeled as random samples from an infinite population, and he based his derivations on the assumption that the population distribution of the characteristic being measured is normal. But this distribution depends on the nature of the population, and the basis for such an assumption in these circumstances often is weak.

Fisher's treatment of the assumptions of normality in his enormously successful 1925 book, "Statistical Methods for Research Workers" (SMRW) [11], gave rise to a heated controversy between Fisher and E.S. Pearson. The conflict had its origin in Pearson's 1929 review of the 2nd Edition of SMRW for the journal "Nature". On the whole the review was favorable, but it contained the following critical paragraph.

> "There is one criticism however which must be made from the statistical point of view. A large number of these tests are based ... on the assumption that the population sampled is of the 'normal' form. That this is the case can be gathered from a careful reading of the test, but the point is not sufficiently emphasized. It does not appear reasonable to lay stress on the 'exactness' of the tests when no means whatever are given of appreciating how rapidly they became inexact as the population sampled diverges from normality. That the tests, for example, connected with the analysis of variance [here he is referring to the F-test for variances] are far more dependent on normality than those involved Student's z (or t) distribution is almost certain, but no clear indication of the need for caution in their application is given."

To make things worse, the review also contained the sentence: "It would seem wiser in the long-run even in a textbook, to admit the incompleteness of theory in this direction, rather than risk giving the reader the impression that the solution of all his problems has been achieved."

In defense of his criticism,[5] Pearson (in a letter to Gosset) cited a paper by the American economist Tolley [47] which showed that its author indeed had been misled by Fisher's book. Tolley wrote:

> "Recently the English School of Statisticians has developed formulas and probability tables to accompany them which, they state, are applicable regardless of the form of the frequency distribution. These formulas are given, most of them without proof, in Fisher's book [11] .... If we accept the statements of those who have developed these newer formulas, skew frequency distributions and small samples need cause no further difficulty as far as measurement of error is concerned."

Fisher was shaken by Tolley's misunderstanding and in a letter to Gosset (June 27, 1929) admitted some culpability:

> "The claim of exactness for the solutions and tests given was wrong, although a careful reader would find that I had kept within the letter of the law by hidden allusions to normality."

A careful reading of SMWR concerning its treatment of normality in fact shows the following.

1. For some of the results, the assumption is stated (though never emphasized).
2. More often, results are stated without any qualifications.
3. The treatment of examples is very deficient in this respect. As Gosset pointed out to Fisher (June 24, 1929):

   > "Although when you think about it you agree that 'exactness' or even appropriate use depends on normality, in practice you don't consider the question at all when you apply your tables to your examples: *not one word.*"

4. In the chapter on distributions (which introduces the normal, Poisson and binomial and no others), Fisher states that it is important to know "the experimental conditions upon which they occur." However, neither this nor any later chapter contains any discussion of the conditions under which data can be expected to be normally distributed (not one word, as Gosset might say.)
5. Another important omission (not even taken up in later editions after it had been pointed out by Pearson) is the great sensitivity to the assumption of normality in tests of variances (rather than means), which make these tests essentially useless.

This criticism must of course be seen against the background of the enormous achievements and novelty of the book. In the process of making accessible Fisher's research of the preceding decade, it established a new paradigm and revolutionized statistical methodology.

The book provides little information about Fisher's own attitude toward the assumption of normality, which forms the basis of his work on the analysis of variance, covariance and regression. The clearest statement of his position that I have been able to find is in a letter of 1929 to "Nature" entitled "Statistics and biological research", which was the concluding document in the controversy with Pearson. Basing his defense of the assumption on experience rather than theory, Fisher claims:

> "On the practical side there is little enough room for anxiety, especially among biologists, who are used to checking the adequacy of their methods by control experiments. The difficulty of obtaining decisive results often flows from heterogeneity of materials, often from causes of bias, often, too, from the difficulty of setting up an experiment in such a way as to obtain a valid estimate of errors. I have never known difficulty to

[5]For more details on this controversy see Pearson [32] and the Fisher-Gosset correspondence (Gosset [18]).

> arise in biological work from imperfect normality of the variation, often though I have examined data for this particular cause of difficulty; nor is there, I believe, any case to contrary in the literature. This is not to say that the deviation from 'Student's' t-distribution found by Shewhart and Winters [43], for samples from rectangular and triangular distributions, may not have a real application in some technological work, but rather that such deviations have not been found, and are scarcely to be looked for, in biological research and ordinarily conducted."

Fisher is not so naive as to claim universal normality even for the kind of biological data with which he is dealing. What he claims instead is that his methods are insensitive to departures from normality, in modern terminology that they are fairly robust against non-normality. This turned out, as Pearson had already discovered through simulation studies, to be true for tests of means but not for tests of variances. (However, even for means a theorem of Bahadur and Savage [3] suggest the need for some caution. They showed that for any given sample size, no matter how large, there exist distributions for which the size of the t-test is arbitrarily close to 1.)

4. The role of normality after Fisher

Fisher's "Statistical Methods" was enormously influential. Together with the work on which it was based it was the most important instrument for the movement from 19th Century large-sample to 20th Century small-sample statistics. The volume sold well, particularly considering how small the statistical community was at the time. A second edition became necessary in 1928 and a third in 1930, with the first three editions selling 1050, 1250 and 1500 copies respectively. This flow continued throughout Fisher's life with new editions appearing every 2 to 3 years.

However, in the 1930's other texts also began to bring the new methodology to ever widening audiences. By far the most successful of these was George Snedecor's "Statistical Methods" [44], which was published in 1937, and the seven editions of which sold the unparalleled number of 237,000 copies. For several years it was one of the most cited publications in the Science Citation Index, and in 1995 still had nearly 2000 entries.[6]

Snedecor's book is essentially a more clearly written, simpler, very user-friendly, version of Fisher's SMRW. It also contains a large number of numerical examples, mainly from agriculture and biology. The book avoids complications and pays practically no attention to the assumptions underlying the recommended procedures including the assumption of normality. This is in line with Snedecor's philosophy which he explains in the Preface:

> "To the mathematical statistician must be delegated the task of developing the theory and devising the methods, *accompanying these latter by adequate statements of the limitations of their use* (my italics)."

He adds that "None but the biologist can decide whether the conditions are fulfilled in his experiments." This seems begging the question because how can the biologists or other users make this decision when they are not clearly told what are the conditions.

This lack of attention to assumptions is shared by most of the many texts on statistics in the 40's and 50's that followed those of Fisher and Snedecor. A fairly comprehensive look at the role of the assumptions underlying Fisher's small-sample methods was provided in 1959 by Scheffé in his book "The Analysis of Variance"

[6] Cited from Carriquiri and David [8].

[40], in a justly famous chapter of nearly 40 pages entitled "The effects of departures from the underlying assumptions."

Scheffé measures the departure of a distribution from normality in terms of the coefficients γ_1 of skewness and γ_2 of kurtosis, the standardized 3rd and 4th moments. For any symmetric distribution $\gamma_1 = 0$, and for the normal distribution also $\gamma_2 = 0$. In the introduction to the chapter, he provides an idea of the range of (γ_1, γ_2) in engineering data, in routine chemical analyses, in ages of marriage, in barometric heights, and in the length and breadth of beans.

Scheffé points out that for large n the distribution of Student's t is approximately normal and hence "the inferences about the mean which are valid in the case of normality must be correct for large n regardless of the form of the population." This is followed by a discussion of the effect of non-normality on the χ^2-test for variance, where it is shown that non-normality causes serious errors. Thus, Scheffé concludes that

> "the effect of violation of the normality assumption is slight on inferences about the mean but dangerous on inferences about variances."

He points out that these results had already been noted by E. S. Pearson [31], and their reason by Box [4].

However, Scheffé's concern is an exception. The standard textbook attitude toward the assumption of normality is well summarized by Brownlee [5] (p. 179):

> "It is presumably on this foundation [that means are approximately normal by the Central Limit Theorem] that applied statisticians have found empirically that usually there is no great need to fuss about the normality assumption. After a statistician has analyzed several quite widely differing transformations of a variable in a fair number of specific instances and found that the conclusions reached are substantially identical for all the transformations, then he ceases to worry unduly about the normality assumption in most situations."

5. The assumption of independence

The standard model for the one-sample problem, i.e. for a number of repeated measurements or other observations of a common quantity, assumed not only that the observations are normally distributed but also that they are independent. This assumption has received much less attention than the assumption of normality. As Kruskal [26] states: "An almost universal assumption in statistical models for repeated measurements of real-word quantities is that these measurements are independent, yet we know that such independence is fragile."

As an explanation of the casual assumption of independence he suggests:

> "One answer is ignorance.... Far more important than simple ignorance, is seductive simplicity: It is so easy to multiply marginal probabilities, formulas simplify, and manipulation is relatively smooth, so the investigator neglects dependence, or hopes that it makes little difference. Sometimes the hope is realized, but more often dependence can make a tremendous difference."

In many situations, the assumption of independence seems natural, even obvious, because of the absence of any direct influence of one observation on another. However, a "spurious" (but nevertheless very real) dependence may be caused by the presence of a common factor. This phenomenon was investigated (and the term "spurious correlation" coined) by Karl Pearson in two papers of 1897 and 1902 [35, 36], in which he reports extensive experiments on measurements carried out by himself.

A difficulty for any serious investigation of the affect of dependence is the great variety of forms that dependence can take. Even Scheffé, who includes a discussion of dependence in his chapter on failures of assumptions (mentioned in the preceding section) cannot do more than examine two or three special cases. His conclusion: "The effect of correlation in the observations can be very serious on inferences about means."

Following Scheffé, many of the later books on linear models and regression analysis included discussion of the effects of non-normality, dependence, and inequality of variances. Particularly noteworthy in this regard is Miller [30] who provides a careful treatment of these issues as an integral part of each chapter of his book.

There was also a trickle-down effect on some of the more general introductions to statistical methods. Thus, the seventh (1980) edition of Snedecor's text (Snedecor had died in 1974 and Cochran had become a co-author) contained a substantial chapter on "Failure in the assumptions". Its treatment of dependence closely paralleled that of Scheffé and in particular also warned about the effect of dependence on the t-test.

6. The linear model

The previous sections have been concerned with the i.i.d. normal model for the one-sample problem, i.e. for repeated measurements or observations of the same quantity. In the remainder of the paper we shall consider the general linear model where the same issues arise as well as some new ones.

The model is given by

$$Y_i = \sum_{j=1}^{n} x_{ij}\beta_j + \varepsilon_i \quad (i = 1, \dots, n)$$

where the Y's are the observed values, the x's are known constants, the β's unknown parameters, and the ε's the errors. Of the ε's it is typically assumed that they are independently normally distributed with mean 0 and common variance σ^2.

This model and the proposal to estimate the β's by means of least squares, are due to Gauss and Legendre at the beginning of the 19th Century. Known as the theory of combinations of observations, the resulting methodology, applied primarily to astronomy and geodesy, became the principal statistical activity throughout much of the 19th Century, with many textbooks devoted to it. (When Gosset in 1904 needed statistical methods for his brewery work, his main sources of information were Airy's [1] (3rd Ed) book with the ponderous title "On the Algebraic and Numerical Theory of Errors of Observations" and Merriman's "A Textbook on Least Squares" [29]. As Stigler [45] (p.11) puts it: "The method of least squares was the dominant theme – the leitmotif – of nineteenth-century mathematical statistics."

In this theory, probability calculations such as that of determining the probable error of the estimates were carried out assuming large samples. In this way the variance of the normal error distribution could be assumed known, and the estimates (which were linear functions of the observations) could therefore be assumed to be (approximately) normally distributed with known variance.

The linear model became a much more flexible instrument through Fisher's introduction of analysis of variance and covariance, and regression analysis. At the same time he extended its application to the more complicated data of biology with their complex sources of variation mentioned in Section 4. Fisher was able to overcome many of the difficulties arising in this extension through the use of randomization and other aspects of experimental design set forth in his 1935 book on this subject [12].

After Fisher's biological applications, the linear model still faced one important challenge: the application of regression models to the social sciences, particularly to economics. Here in addition to the assumptions concerning the random terms ε_i (normality and independence), questions concerning the adequacy of the structural part $\Sigma x_{ij}\beta_j$ became particularly important.

Freedman [15] (p. 9) compares the situation with that in astronomy where

> "The relevant variables were known from Newtonian mechanics, and so were the functional forms of the equations connecting them. Measurement could be done with high precision. Much was known about the errors in the measurements and the equations."

Regarding the corresponding issues in the social sciences, I shall here restrict attention to economics[7] where after some early isolated instances, statistical model building was pursued systematically (beginning in the 1930's) by writers such as Tinbergen, Haavelmo [19] and Koopmans. It was developed further in the 1940's by the work of the Cowles Commission, which was then the center of economic research. About these efforts Koopmans [24] points out that

> "this theory has been widely applied to data obtained from agricultural experiments or from measurements in biological populations. There are some essential differences between data of this kind and those usually encountered in economic problems.
>
> In agricultural experiments some of the determining variables can be completely controlled by the experimenter .... Other determining variables less under his control are usually by their nature subject to adequate independent variation. In that respect they bear a resemblance to the variables representing measurable characteristics of individuals of a biological population which is usually conceived as random drawings from a *stable* (my italics) probability distribution.
>
> In economic analysis variables at the control of an experimenting institution are exceptional. Further only a few types of variables ... are so erratic in nature that they could reasonably be regarded as drawings from any stable distribution ....
>
> Further the relations between the variables studied in this type of analysis are themselves subject to gradual or abrupt changes, according to institutional or technical changes in society .... Therefore the number of observations from which the regression coefficients have to be estimated is limited by the very nature of the problem."

Commenting on the relative roles of the economist and the statistician in this work, Koopmans states:

> "The economist – or, in general, the expert in the field to which the dependent variable belongs – should by economic reasoning and general economic experience... devise a set of determining variables which he expects to be a complete set [i.e. such that the effect of any additional variables can safely be absorbed into the error term]."

Koopmans makes it clear that the determination of the appropriate structural part is not the task of the statistician but of the subject matter expert. This point is also made by Arrow [2] when he states that

> "The method of scientific investigation indicated in the preceding paragraphs calls then for intensive a priori thinking to formulate a model, followed by the selection of a best-fitting structure from that model by appropriate statistical techniques."

Once the model has become formulated, statistical techniques make it possible to deduce far-reaching conclusions. However the reliability of these conclusions is of course limited by the reliability of the model. The difficulty of verifying the model assumptions on the one hand, and the power of the statistical machinery to deliver

[7]For applications to psychology see Krüger et al. [25] (Chapters 2 and 3).

results on the other, have provided a fertile ground for misuse of the regression methodology. The chief critic of this misuse has been David Freedman who over the past 25 years has called attention to it, both in a series of close to 50 substantial papers, and as a statistical consultant and expert witness in more than 100 cases.

As an example of Freedman's criticism consider his detailed analysis of a book by Hope, which deals with the effects of education on class mobility (Freedman [13]). In this paper Freedman points out that

> "one problem noticeable to a statistician is that investigators do not pay attention to the stochastic assumptions behind the models. It does not seem possible to derive these assumptions from current theory, nor are they easily validated empirically on a case-by-case basis"

The paper ends with the devastating conclusion:

> "My opinion is that investigators need to think more about the underlying process, and look more closely at the data, without the distorting prism of conventional (and largely irrelevant) stochastic models. Estimating nonexistent parameters cannot be very fruitful. And it must be equally a waste of time to test theories on the basis of statistical hypotheses that are rooted neither in prior theory nor in fact, even if the algorithms are recited in every statistics text without caveat."

The opening sentence of this conclusion suggests an alternative to the methodology Freedman is criticizing: better data, more substantive knowledge and input, multiple studies under varying conditions as had been carried out for example in establishing smoking as a major cause of lung cancer and other diseases. These requirements had been foreshadowed in the passages from Koopmans and Arrow cited earlier in this section.

Freedman capped his long involvements in these issues as scientist, consultant and expert witness with a text: "Statistical Models–Theory and Practice" [15]. As the title suggests, the book provides an account not only of the theory of statistical modeling but also–through the careful examination of many real-life examples–a guide to how such modeling should and should not be used.

An additional valuable feature of the book is a review of the literature on statistical modeling which is both a resource and encouragement for further study.

7. Conclusions

This paper has been concerned with the assumptions underlying the linear model and the special case of the normal i.i.d. model. It considered three assumptions: normality, independence and in the more general case, the linear structure of the deterministic part. Of these three, the assumption of normality has received by far the most attention in the literature. For the i.i.d. case and a few simple linear models it has led to an alternative nonparametric methodology which has been developed to avoid it. Unfortunately these nonparametric methods are no help with respect to dependence (or the inequality of variances, a topic we have not discussed here.)

A point emphasized by this survey is that different fields of study involve different kinds of data and have very different modeling situations. Thus, when the only source of variation is observational error, as often is the case in the physical sciences, the situation is much simpler than in the biological sciences where one is dealing with samples from a somewhat heterogeneous (but stable) population. And these situations, in turn, are easier to handle than those of observational studies in the social sciences where the populations are less stable and where random sampling typically is not possible.

This subject matter dependence creates a difficulty for general texts on statistical methods but it does not absolve them from responsibility. As a minimum, such texts should provide warnings against the unsubstantiated use of standard models. A useful cautionary example might be the F-test of variances with its strong dependence on the assumption of normality. Even with the t-test which is fairly robust against non-normality one should not tempt students into Tolley's error (discussed in Section 4).

A warning that should be included in such texts is the danger of the too facile assumption of independence. Another distinction that would be worth mentioning is that between observational and experimental studies, perhaps with references to the literature (for example the book by Rosenbaum [39]) treating the special methods developed for the former.

Though general statistics texts could, and should include, this kind of material, they can only go so far. As mentioned by Koopmans, the statistician and the subject matter specialist each have a role to play. A general text cannot provide the subject matter knowledge and the special features that are needed for successful modeling in specific cases. Experience with similar data is required, knowledge of theory and, as Freedman points out: shoe leather.

Acknowledgments. I am grateful to Persi Diaconis and Juliet Shaffer for many helpful critical comments.

References

[1] Airy, G. (1861, 1879). *On the Algebraical and Numerical Theory of Errors of Observations*, 3rd ed. Macmillan, London.

[2] Arrow, K. (1951). Mathematical models in the social sciences. In *The Policy Sciences* (Lerner and Lasswell, eds.). Stanford Univ. Press.

[3] Bahadur, R. and Savage, L. (1956). The nonexistence of certain statistical procedures in nonparametric problems. *Ann. Math. Statist.* **27** 1115–1122.

[4] Box, G. (1953). Non-normality and tests of variances. *Biometrika* **40** 318–335.

[5] Brownlee, K. (1960). *Statistical Theory and Methodology in Science and Engineering.* Wiley, New York.

[6] Brunt, D. (1917). *The Combination of Observations.* Cambridge Univ. Press.

[7] Brunt, D. (1931). *The Combination of Observations.* Cambridge Univ. Press.

[8] Carriquiry, A. and David, H. (2001). George Waddel Snedecor. In *Statisticians of the Centuries* (Heyde and Seneta, eds.). Springer, New York.

[9] Czuber, E. (1891). *Theorie der Beobachtungsfehler.* Teubner, Leipzig.

[10] De Moivre, A. (1733). Approximatio ad Summam Terminorum Binomii $(a+b)^n$ in Seriem Expansi. Printed for private circulation.

[11] Fisher, R. (1925). *Statistical Methods for Research Workers.* Oliver and Boyd, Edinburgh.

[12] Fisher, R. (1935). *The Design of Experiments.* Oliver and Boyd, Edinburgh.

[13] Freedman, D. (1987). As others see us (with discussion). *J. Educ. Statist.* **12** 101–223. Reprinted in J. Shaffer, ed. The Role of Models in Nonexperimental Social Science AERA/ASA Washington, D.C. (1997).

[14] Freedman, D. (1991). Statistical models and shoe leather. In *Sociological Methodology* (P. Marsden, ed.). Amer. Social Assoc., Washington, D.C.

[15] Freedman, D. (2005). *Statistical Models: Theory and Practice.* Cambridge Univ. Press.

[16] GALTON, F. (1877). Typical laws of heredity. *Nature* **15** 492–495, 512–514, 532–533.

[17] GEARY, R. (1947). Testing for normality. *Biometrika* **34** 209–242.

[18] GOSSET, W. (1970). Letters from W. S. Gosset to R. A. Fisher, 1915–1936 with summaries by R. A. Fisher and a Foreword by L. McMullen. Printed for private circulation.

[19] HAAVELMO, T. (1944). The probability approach in econometrics. *Econometrica* **12** Supplement.

[20] HALD, A. (1998). *A History of Mathematical Statistics from 1750 to 1930.* Wiley, New York.

[21] HELMERT, F. (1872). *Die Ausgleichsrechnung nach der Methode der Kleinsten Quadrate.* Teubner, Leipzig.

[22] HELMERT, F. (1907). *Die Ausgleichsrechnung nach der Methode der Kleinsten Quadrate.* Teubner, Leipzig.

[23] KOLMOGOROV, A. (1933). *Grundbegriffe der Wahrscheinlichkeitsrechnung.* Springer, Berlin.

[24] KOOPMANS, T. (1937). *Linear Regression Analysis of Economic Time Series.* Netherlands Economic Institute, Haarlem.

[25] KRÜGER, L., GIGERENZER, G. AND MORGAN, M., EDS. (1987). *The Probabilistic Revolution* **2**. MIT Press, Cambridge, MA.

[26] KRUSKAL, W. (1988). Miracles and statistics: The casual assumption of independence. *J. Amer. Statist. Assoc.* **83** 929–940.

[27] LEGENDRE, A. (1805). *Nouvelles Méthodes pour la Détermination des Orbites des Comètes.* Courcier, Paris.

[28] LEXIS, W. (1877). *Theorie der Massenerscheinungen in der Menschlichen Gesellschaft.* Wagner, Freiburg.

[29] MERRIMAN, M. (1884). A textbook on the method of least squares, 8th ed. 1900.

[30] MILLER, R. (1986). *Beyond ANOVA, Basics of Applied Statistics.* Wiley, New York.

[31] PEARSON, E. (1931). The analysis of variance in cases of non-normal variation. *Biometrika* **23** 114–133.

[32] PEARSON, E. (1990). *Student.* Clarendon Press, Oxford.

[33] PEARSON, K. (1892). *Grammar of Science.* Walter Scott, London. 3rd ed. of 1911 reprinted by Meridian Books, 1957.

[34] PEARSON, K. (1895). Contributions to the mathematical theory of evolution, II. Skew variation in homogeneous material. *Phil. Trans. Roy. Soc. London* **186** 343–414.

[35] PEARSON, K. (1897). On a form of spurious correlation which may arise when indices are used in the measurement of organs. *Proc. Roy. Soc.* **60** 489–497.

[36] PEARSON, K. (1902). On the mathematical theory of errors of judgement, with special reference to the personal equation. *Phil. Trans. Roy. Soc. London* **198** 235–299.

[37] PEIRCE, C. (1873). On the theory of errors of observations. Appendix 21 to the Report of the Superintendent of the U.S. Coast Survey for the year ending June 1870. 200–224. Reprinted in *Stigler: American Contributions to Mathematical Statistics in the Nineteenth Century* **2** (1980). Arno Press, New York.

[38] PLACKETT, R. (1972). The discovery of the method of least squares. *Biometrika* **59** 239–251.

[39] ROSENBAUM, P. (1995, 2002). *Observational Studies.* Springer, New York.

[40] SCHEFFÉ, H. (1959). *The Analysis of Variance.* Wiley, New York.

[41] SEAL, H. (1967). The historical developmnent of the gauss linear model. *Biometrika* **54** 1–24.

[42] SHAFER, G. AND VOVK, V. (2006). The sources of kolmogorovs *Grundbegriffe*. *Statist. Sci.* **21** 70–98.

[43] SHEWHART, W. AND WINTERS, F. (1928). Small samples – new experimental results. *J. Amer. Statist. Assoc.* **23** 144–153.

[44] SNEDECOR, G. (1937). *Statistial Methods*. The Iowa State College Press, Ames, Iowa.

[45] STIGLER, S. (1986). *The History of Statistics*. Belknap Press, Cambridge, MA.

[46] STIGLER, S. (1999). *Statistics on the Table*. Harvard Univ. Press, Cambridge, MA.

[47] TOLLEY, H. (1929). Economic data from the sampling point of view. *J. Amer. Statist. Assoc.* **24** 69–72.

[48] VON MISES, R. (1919). Grundlagen der Wahrscheinlichkeitsrechnung. *Math. Zeitschrift* **5** 52–99.

[49] VON PLATO, J. (1994). *Creating Modern Probability*. Cambridge Univ. Press.

IMS Lecture Notes–Monograph Series
Optimality: The Third Erich L. Lehmann Symposium
Vol. 57 (2009) 11–17

DOI: 10.1214/09-LNMS5703

Some History of Optimality

Erich L. Lehmann

University of California, Berkeley

Abstract: This paper gives an account of the history of both small- and large-sample optimality for both estimation and testing. It covers this history from the beginnings by Laplace and Gauss to the ongoing research on optimal multiple comparison procedures, and also includes the optimality of designs.

Contents

1. Combination of Observations

The earliest optimality considerations appear to be those of Laplace and Gauss at the beginning of the 19th Century concerned with determining the best linear estimates of parameters in linear structures. Laplace calls these optimal estimates "the most advantageous" while Gauss refers to them as "the most plausible values." Various aspects of this problem were discussed throughout the 19th Century under the heading "Combination of Observations." The version of the principal result generally accepted today is the so-called Gauss-Markov theorem. It states (in modern language) that the least squares estimates are the linear unbiased estimates with minimum variance. While restricted to linear estimates, the result is nonparametric in that it makes no assumptions about the distribution of the errors. For an account of this work see, for example, Stigler [23], Hald [7], or Chatterjee [3].

2. Maximum Likelihood Estimation

Optimality next played an important part in Fisher's fundamental paper of 1922. In this paper (followed by a clarifying paper in 1925), Fisher considers estimation

Keywords and phrases: optimality, estimation, testing, design, maximum likelihood, Neyman-Pearson lemma.

in parametric models, proposes the maximum likelihood estimator (MLE) as a generally applicable solution, and claims (but does not prove) that such estimators are consistent and asymptotically efficient efficient (i.e., that they minimize the asymptotic variance). Note that unlike the Gauss-Markov theorem, maximum likelihood estimation does assume that the distribution of the variables belongs to a given parametric family.

Maximum likelihood has become the most widely used method of estimation, and there has been an enormous amount of work connected with Fisher's claims concerning it. It has led to the discovery of superefficiency on the one hand and of second order efficiency on the other. Many counterexamples have been found (even to consistency), and a number of modifications (bias correction and replacement of the MLE by a consistent root of the likelihood equation, for example) have been proposed. The situation is complex, but under suitable restrictions Fisher's conjecture is essentially correct when the likelihood equation has a unique root. For a more precise statement, see Shao [22], and for further discussion, for example, Efron [4], Le Cam [15], and Barndorff-Nielsen and Cox [1].

3. The Neyman-Pearson Program

Least squares and maximum likelihood were first proposed on intuitive grounds and then justified by showing that they possessed certain optimality properties. Optimality as a deliberate program for determining good procedures was introduced in 1933 by Neyman and Pearson in a paper (on testing rather than estimation) appropriately called, "On the problem of the most efficient tests of statistical hypotheses." As they explain in the introduction:

> The main purpose of the present paper is to find a general method of determining tests which ... would be most efficient" [in the sense of minimizing the probability of erroneous conclusions].

In a certain sense this is the true start of optimality theory.

4. The Neyman-Pearson Theory of Hypothesis Testing

Neyman and Pearson [17] implemented the above program by seeking, for any given situation, the test which, among all those controlling the probability of false rejection at a given level α, has the maximum power (and hence the minimum probability of false acceptance).

For testing a simple hypothesis against a simple alternative, they found the solution to this problem to be the likelihood ratio test. This result, which is mathematically quite elementary but has crucial statistical consequences, is known as the Neyman-Pearson Lemma.

It turns out that in some (very rare) cases the same test is most powerful against all alternatives under consideration. Such a uniformly most powerful (UMP) test is then the optimal solution to the given testing problem. Where a UMP test does not exist, additional criteria must be invoked.

For example, when nuisance parameters are present, Neyman and Pearson require that under the hypothesis the rejection probability be α for all values of the nuisance parameters. They call such rejection regions similar regions. As an important example, they show that the one-sided t-test is UMP among all similar regions.

For two-sided alternatives, one would not expect UMP tests to exist even without nuisance parameters. For such cases, Neyman and Pearson then impose the additional condition of unbiasedness, i.e., that the power of the test is $\geq \alpha$ for all alternatives. In follow-ups to their 1933 paper (1936 and 1938), they show, for example, that the two-sided t-test is UMP among all unbiased tests.

UMP similar or unbiased tests exist for important classes of testing problems concerning a single real-valued parameter (in the presence or not of nuisance parameters) but not for hypotheses such as

$$H : \theta_1 = \cdots = \theta_s$$

concerning several parameters.

A different condition, the principle of invariance (suggested by Hunt and Stein, unpublished), is successful in a number of important such multiparameter situations. If both the hypothesis and the class of alternatives remain invariant under a group G of transformations of the sample space, there does in these cases exist a UMP test among all tests invariant under G.

5. Wald's Optimality Criteria

A quite-different approach to optimality was initiated by Wald in his 1939 paper, "Contributions to the theory of statistical estimation and testing hypotheses," and was then developed further in a series of publications culminating in his 1950 book, "Statistical Decision Functions."

This approach was part of Wald's formulation of a general theory of decision procedures. Instead of seeking procedures that are uniformly optimal among some suitably restricted class of decision procedures, Wald proposes to minimize some global feature of the performance.

Specialized to hypothesis testing, these proposals reduce to:

i Maximize the average power, averaged with respect to some suitable weight function over the alternatives. For obvious reasons, Wald called such maximizing procedures Bayes solutions.

ii Maximize the minimum power over alternatives bounded away from the hypothesis.

6. The Hunt-Stein Theorem

An important connection between the earlier invariance approach and that of Wald is the Hunt-Stein theorem. It states that if a UMP invariant test exists under a group G satisfying certain conditions (called amenability), then this test also maximizes the minimum power over any invariant set of alternatives.[1]

To illustrate this theorem, consider the univariate linear hypothesis. Under the combined action of the groups of location, scale and orthogonal transformation, a UMP invariant test exists. Since these groups satisfy the Hunt-Stein conditions, the resulting test therefore maximizes the minimum power against all invariant sets of alternatives.

[1] The 1946 paper by Hunt and Stein, "Most stringent tests of statistical hypotheses," containing this theorem was never published. The theorem appeared in print for the first time in Lehmann, "Testing Statistical Hypotheses," John Wiley, 1959.

As a second example, consider the multivariate one-sample problem. Hotelling's T^2-test is UMP among all tests that are invariant under the group G of all non-singular linear transformations. Since G is not amenable, the Hunt-Stein theorem does not apply. The resulting maximin problem poses considerable difficulties.

7. Some Extension of the Neyman-Pearson Theory

The Neyman-Pearson theory has been extended in a number of directions. The following are two extensions of the Neyman-Pearson Lemma, which is so basic to this theory.

(i) Sequential Analysis

During World War II, Wald proposed sequential procedures, as a way of obtaining good power with fewer observations. In particular, he suggested the probability ratio test for testing a simple hypothesis against a simple alternative. This test continues observation as long as the likelihood ratio remains between two fixed limits and takes the indicated decision (acceptance or rejection) as soon as it falls outside these limits.

In 1948, Wald and Wolfowitz proved the remarkable result that for testing a simple hypothesis against a simple alternative, the sequential probability ratio test, among all tests with the same (or smaller) probabilities of error, minimizes the expected number of observations both under the hypothesis and the alternative.

(ii) Robust Inference

All the work reported so far (except for the Gauss-Markov theorem) was carried out under the assumption of an underlying parametric model. In practice, such an assumption can be expected to hold at best approximately. As a more realistic formulation, Huber [8] suggested replacing the assumption of a parametric model by that of a neighborhood of such a model.

In the following year, he obtained the analog of the Neyman-Pearson Lemma. For testing the neighborhood of a distribution P_0, the test maximizing the minimum power over the neighborhood of an alternative P_1 is a censored version of the likelihood ratio test of P_0 against P_1.

(iii) Multiple Testing

A very different extension of the Neyman-Pearson theory that is of great practical importance deals with the situation in which a number of hypotheses (sometimes a very large number) are being tested rather than just one. Unlike the work discussed so far, which is classical, the theory of multiple testing is an area of very active ongoing research.

The first problem here (before optimization) is to find a suitable generalization of the concept of significance level that would provide satisfactory control of the probability of false rejections. After this, maximin tests have been obtained, which require, however, not only unbiasedness and invariance but also a condition of monotonicity. Surveys of the current state of this work are provided by Shaffer [20, 21].

8. Large Sample Optimality of Testing

A small-sample theory of optimal estimation parallels that of optimal testing sketched in Sections 4-7, with concepts such as unbiasedness, equivariance (instead of invariance), and minimax variance (instead of maximin power), and will not be discussed here. Asymptotic optimality for estimation goes back to Fisher (1922), as mentioned in Section 2, and is defined as minimum asymptotic variance.

For testing, asymptotic optimality is considerably more complex, both conceptually and technically. It was first studied by Wald in 1941. Consider testing a simple hypothesis $\theta = \theta_0$ against a simple alternative $\theta = \theta_1$. If we keep both θ_0 and θ_1 fixed, and carry out the tests at a fixed level α, the power of any reasonable test sequence will tend to 1. Thus any such test sequence will in a trivial sense be asymptotically UMP.

A more useful approach is obtained by considering a sequence of alternatives

$$\theta_n = \theta_0 + h/\sqrt{n}. \tag{8.1}$$

For a sequence with fixed h, the power will typically tend to a limit between 0 and 1 as $n \to \infty$. As h varies from 0 to ∞, the limiting power will be an increasing function of h, going from α to 1; we shall call this the asymptotic power function. A sequence of tests can then be defined as being asymptotically most powerful (AUMP) if it maximizes the asymptotic power for all h.

Unlike the finite sample situation where UMP tests exist only rarely and then are unique, it turns out that AUMP tests exist under very weak assumptions, and that in fact many different AUMP tests exist for the same situation, among them the likelihood ratio test, the Wald test, and the locally most powerful (Rao) test. To distinguish them, one must resort to higher order asymptotics. (See, for example, Barndorff-Nielsen and Cox [1]). An exposition of the first order theory due to Le Cam can be found in Lehmann and Romano [16].

9. Optimal Design

In generalization of minimizing the variance of an estimator, optimal design theory is concerned with determining the design that minimizes some function of the covariance matrix of the best linear estimators of the parameters in question. In particular, D-optimality minimizes the determinant, E-optimality is a minimax criterion, and so on. After some isolated earlier efforts, the problem of optimal design was studied systematically by Kiefer in more than 40 papers between 1958 and his early death in 1981. They make up vol. III of his collected papers.

10. A Culture Clash

Not everyone was enthusiastic about optimality as a guiding principle. Criticism was highlighted at a 1958 meeting of the Royal Statistical Society at which Kiefer presented a survey talk on "Optimum Experimental Designs." It was a discussion paper, and the reaction was nearly uniformly negative. The core of the disagreement is stated clearly when Barnard quotes Kiefer as saying of procedures proposed in a paper by Box and Wilson [2] that they "often [are] not even well-defined rules of operation." Barnard's reply:

> In the field of practical human activity, rules of operation which are not well defined may be preferable to rules which are.

The conflict is discussed by Henry Wynn in his introduction to a reprint of Kiefer's paper in "Breakthroughs in Statistics," vol. I (Kotz and Johnson, Eds.). Wynn calls it "a clash of statistical cultures."

This clash is between the Wald school of abstraction and optimality on the one hand and what Tocher, in his discussion of Kiefer's paper, calls "the school of British experimental design – a very practical people," on the other.

11. Tukey's Criticism

Criticism of optimality was not confined to England. An outspoken American critic questioning the value and importance of optimization was John Tukey. His attitude is indicated by the titles of two philosophical papers in 1961 and 1962, which are titled respectively, "The tyranny of the best" and "Dangers of optimization."

Tukey's concern with optimality had its origin in the fact that at the time optimization had become the dominant interest of mathematical statistics. In the 1962 paper, he writes:

> Some [statisticians] seem to equate [optimization] to statistics an attitude which, if widely adopted, is guaranteed to produce a dried-up, encysted field with little chance of real growth.

12. Conclusion

That optimality considerations had become so dominant is explained by the historical situation. Periods of great innovation are followed by periods of consolidation, in which the new work is given a final (or, as Tukey says, encysted) form. Such a dichotomy is also discussed by Huber [10]. Thus, the revolutionary work of Student and Fisher was followed by the optimization approach of Neyman, Pearson, Wald and their students described in this paper.

Today we are faced with the opposite situation. We are confronted with new problems arising from immense data sets and involving problems of great complexity. Ad hoc solutions are proposed and tried out on a few examples. This is a natural first step, but eventually we will want to justify the solutions at which we have arrived intuitively and by trial and error. A theoretical underpinning will be provided and the conditions will be found under which these solutions are the best possible.

References

[1] BARNDORFF-NIELSEN, O. E. and COX, D. R. (1994). *Inference and Asymptotics.* Chapman & Hall, London.

[2] BOX, G. E. P. and WILSON, K. B. (1951). On the experimental attainment of optimum conditions. *J. Roy. Statist. Soc. Ser. B* **13** 1–45.

[3] CHATTERJEE, S. K. (2003). *Statistical Thought.* Oxford Univ. Press.

[4] EFRON, B. (1982). Maximum likelihood and decision theory. *Ann. Statist.* **10** 309–368.

[5] FISHER, R. A. (1922). On the mathematical foundations of theoretical statistics. *Phil. Trans. Roy. Soc. London Ser. A* **222** 401–415.

[6] FISHER, R. A. (1925). Theory of statistical estimation. *Cambridge Philos. Soc.* **22** 700–725.

[7] HALD, A. (1998). *A History of Mathematical Statistics - from 1750 to 1930.* Wiley, New York.

[8] HUBER, P. (1964). Robust estimation of a location parameter. *Ann. Math. Statist.* **35** 73–101.

[9] HUBER, P. (1965). A robust version of the probability ratio test. *Ann. Math. Statist.* **36** 1753–1758.

[10] HUBER, P. (1975). Applications vs. abstraction: The selling out of mathematical statistics? In *Proc. Conference on Directions of Mathematical Statistics. Suppl. Adv. Prob.* **7** 84–89.

[11] KIEFER, J. C. (1959). Optimum experimental designs (with discussion). *J. Roy. Statist. Soc. Ser. B* **21** 273–319. (Reprinted in Kotz and Johnson (1992). *Breakthroughs in Statistics, Vol. 1.* Springer.)

[12] KIMBALL, G. E. (1958). A critique of operations research. *J. Wash. Acad. Sci.* **48** 33–37.

[13] KOTZ, S. and JOHNSON, N. L. (Eds. 1992, 1997). *Breakthroughs in Statistics* **1**. Springer, New York.

[14] LE CAM, L. (1953). On some asymptotic properties of maximum likelihood estimates and related Bayes' estimates. *Univ. of Calif. Publ. in Statist.* **1** 277–330.

[15] LE CAM, L. (1990). Maximum likelihood - an introduction. *ISI Review* **58** 153–171.

[16] LEHMANN, E. L. and ROMANO, J. P. (2005). *Testing Statistical Hypotheses,* 3rd ed. Springer, New York.

[17] NEYMAN, J. and PEARSON, E. S. (1933). On the problem of the most efficient tests of statistical hypotheses. *Phil. Trans. Roy. Soc. Ser. A* **231** 289–337.

[18] NEYMAN, J. and PEARSON, E. S. (1936, 1938). Contributions to the theory of testing statistical hypotheses. *Statist. Res. Memoirs* **1** 1–37; **2** 25–57.

[19] PEARSON, E. S. (1939). "Student" as a statistician. *Biometrika* **30** 210–250.

[20] SHAFFER, J. P. (2004). Optimality results in multiple hypothesis testing. In *Proceedings of the First Lehmann Symposium* (J. Rojo and V. Pérez-Abreu, eds.). *IMS Lecture Notes Monogr. Ser.* **44** 11–35. Inst. Math. Statist., Beachwood, OH.

[21] SHAFFER, J. P. (2006). Recent developments towards optimality in multiple hypothesis testing. In *Proceedings of the Second Lehmann Symposium* (J. Rojo, ed.). *IMS Lecture Notes Monogr. Ser.* **49** 16–32. Inst. Math. Statist., Beachwood, OH.

[22] SHAO, J. (1999). *Mathematical Statistics.* Springer, New York.

[23] STIGLER, S. (1986). *The History of Statistics.* Harvard Univ. Press, Cambridge, MA.

[24] TUKEY, J. W. (1961). Statistical and quantitative methodology. In *Trends in Social Science* (D. P. Ray, ed.). Philosophical Library, New York. (Reprinted in Vol. III of Tukey's Collected Works.)

[25] TUKEY, J. W. (1962). The future of data analysis. *Ann. Math. Statist.* (Reprinted in Vol. III of Tukey's Collected Works.)

[26] WALD, A. (1943). Tests of statistical hypotheses concerning several parameters when the number of observations is large. *Trans. Amer. Math. Soc.* **54** 426–482.

[27] WALD, A. (1950). *Statistical Decision Functions.* Wiley, New York.

[28] WALD, A. and WOLFOWITZ, J. (1948). Optimum character of the sequential probability ratio test. *Ann. Math. Statist.* **19** 326–339.

Part XII
Philosophical Work

Erich L. Lehmann's Work on the Philosophy of Statistics

Javier Rojo

Erich Lehmann's work on the foundations of probability and statistics is discussed in this chapter but his contributions in this area are dispersed over other chapters in this volume, particularly the chapter on his work on the history of classical statistics. The books Theory of Point Estimation (TPE) and Testing Statistical Hypotheses (TSH) also contain implicit and explicit ideas on the foundations of statistics. As expected, the Neyman-Pearson-Wald school is dominant in these works. In addition, Lehmann (2008) and the soon-to-be-published *Fisher, Neyman, and the Creation of Classical Statistics*, contain many of Erich's thoughts on philosophical issues and the historical breakthroughs that have impacted and provided directions for the future of the discipline.

Erich commented in various occasions that he did not involve himself too much in the debate of the foundational issues of statistics. Thus, in DeGroot (1986), after having been asked if he thought about the foundations of probability, he stated: "I don't really think very much about foundational questions." In Lehmann (2008), responding to the self-posed question *What is my outlook?* Erich writes:

> *Foundational issues have not been an active interest of mine, so I find it easiest to describe my attitude with a number of (rather superficial) comments ...*

His work, however, demonstrates otherwise. Erich believed in the frequentist interpretation of probability but also understood that such a perspective is not applicable in some situations (see, e.g. page 188, Lehmann (2008)), and in Lehmann (2001) he discusses some of the philosophical shortcomings of such an interpretation.

Erich was a strong believer in the Neyman-Pearson-Wald school of optimality, while recognizing its limitations. Thus, for example, in Lehmann (1985), Erich writes:

> *Complete reliance on optimality, on the other hand, suffers from some drawbacks.*

J. Rojo
Professor of Statistics, Department of Statistics, Rice University, Houston, TX 77005
e-mail: jrojo@rice.edu

J. Rojo (ed.), *Selected Works of E. L. Lehmann*, Selected Works in Probability and Statistics, DOI 10.1007/978-1-4614-1412-4_87, © Springer Science+Business Media, LLC 2012

In particular, optimality considerations achieve solutions but only in simple situations, and such solutions may lack robustness and other desirable properties.

There are at least three recurring themes in his work on foundational issues: (i) model selection; (ii) frequentist statistical inference; (iii) Bayesian statistical inference; and (iv) exploratory data analysis. Model selection has been discussed in Erich's contributions to the history of classical statistics. Therefore, attention will be focused on (ii), (iii), and (iv).

It is interesting that the likelihood principle is not discussed in any of Erich's writings, as far as I have been able to ascertain, except for a brief reference in Lehmann (2008) to Berger and Wolpert (1984). This is a bit puzzling in light of the fact that one of his former Ph.D. students at Columbia University published, see Birnbaum (1962), a long treatise on the likelihood principle. The reader must surmise that Erich, rather than expressing his disagreement with the likelihood principle, decided not to discuss it.

Erich viewed the trichotomy (ii), (iii), and (iv) as being ordered by the level of model assumptions made. Thus, (iv) does not make any assumptions of an underlying probability model, and lets the data speak for itself. The frequentist approach, on the other hand, relies on a probability model to carry out an evaluation of the statistical procedures under consideration, while the Bayesian approach introduces the additional component of the prior distribution. None of these three approaches is perfect, however. Exploratory data analysis, except in very limited cases, cannot determine if the observed effects in the analysis are indeed real without the use of a probability model, and the Bayesian paradigm has to deal with the selection of the prior distribution and the sensitivity of the solution to such a choice of prior. On the other hand, the frequentist Neyman-Pearson-Wald approach runs into problems in its search for a procedure that minimizes the risk uniformly (in the parameter). Since this is not possible, then attention has to be restricted to a class of procedures satisfying some additional principles none of which are always satisfactory. Against this background of lack of satisfaction with any given approach, Lehmann (1985) and Lehmann (1995) develop ideas that bridge the chasm that has been created by the heated philosophical debates. For example, Lehmann (1985) discusses how the Neyman-Pearson can contribute to the exploration of underlying data structure and its relation with Bayesian inference. In Lehmann (1995), Erich continues with this line of thought and writes that:

> *In practice, the three approaches can often fruitfully interact, with each benefiting from considerations of the other points of view.*
> *It seems clear that model-free data analysis, frequentist and Bayesian model-based inference and decision making each has its place. The question appears not to be – as it is often phrased – which is the correct approach but in what circumstances each is most appropriate.*

The reader learns to appreciate Erich's balanced view on foundational issues – he discusses both the benefits and the shortcomings of the various approaches to inference and interpretations of probability.

Erich did not like confrontation and, given the tone of some of the published work on the philosophy of statistics, must have preferred to express his views in a low-key and conciliatory manner – but see Lehmann (1990) for an exception to this statement.

Erich writes, in Lehmann (2008), that his original position was solidly in the frequentist camp. With time, however, his position was somewhat influenced by classical Bayesian ideas, but remained unconvinced by the radical Bayesian position. Thus, he writes, in Lehmann (1995), that *bridge building to the "radical" [Bayesian] position is more difficult*. Erich never defined what he meant by a radical Bayesian position, but the reader may surmise that he was referring to a Bayesian paradigm that insists on the elicitation of a prior distribution at all costs. Thus, in Lehmann (2008), he writes:

> *However, it seems to me that the strength of these beliefs tends to be rather fuzzy, and not sufficiently well defined and stable to assign a definite numerical value to it. If, with considerable effort, such a value is elicited, it is about as trustworthy as a confession extracted through torture.*

In the end, Erich's work reflects a philosophy that indicates that no single approach can be totally satisfactory. Rather than debating their differences, Erich proposed that a fruitful approach can be obtained from the bringing together of ideas from (ii), (iii), and (iv) – with (iii) serving as a bridge that connects all three.

References

[1] J. O. Berger and R. L. Wolpert. *The Likelihood Principle*. Lecture Notes–Monograph Series, Volume 6, (S. S. Gupta, Ed.), Institute of Mathematical Statistics, 1984.

[2] P. J. Bickel and E. L. Lehmann. Frequentist interpretation of probability. In *International Encyclopedia of the Social and Behavioral Sciences*, N. J. Smelser and Paul B. Baltes (Eds.), pp. 5796-5798. Pergamon, Oxford, 2001.

[3] A. Birnbaum. On the Foundations of Statistical Inference. *J. Am. Statist. Assoc.* Vol. 57, No. 298, pp. 269-306, 1962.

[4] M. H. DeGroot. A Conversation with Erich L. Lehmann. *Statist. Sci.*, Vol. I, 243-258, 1986.

[5] E. L. Lehmann. The Neyman-Pearson Theory After Fifty Years. In *Proceedings of the Berkeley Conference in Honor of Jerzy Neyman and Jack Kiefer*, Vol. I. (L. Le Cam and R. A. Olshen, Eds), pp. 1-14, Wadsworth 1985.

[6] E. L. Lehmann. Comment. [The 1988 Wald Memorial Lectures: The Present Position in Bayesian Statistics] *Statist. Sci.*, Volume 5, Number 1, pp. 82-83, 1990.

[7] E. L. Lehmann. Foundational issues in statistics: theory and practice. *Foundations of Science*, Vol. 1, pp. 45-49, 1995.

[8] E. L. Lehmann. *Reminiscences of a Statistician: The Company I Kept*, Springer, 2008.

[9] E. L. Lehmann. *Fisher, Neyman, and the Creation of Classical Statistics*, Springer, 2011.

THE NEYMAN-PEARSON THEORY AFTER FIFTY YEARS

E. L. LEHMANN University of California, Berkeley

ABSTRACT: To commemorate the 50th anniversary of the Neyman-Pearson paradigm, this paper sketches some aspects of its development during the past half-century. In particular, the relevance of this approach to data analysis and Bayesian statistics is discussed.

1. INTRODUCTION

The Neyman-Pearson theory of hypothesis testing was developed by its two authors during the years 1926-1934.[1] However, its official birthday may be assigned to 1933, which saw the publication of the seminal paper "On the Problem of the Most Efficient Tests of Statistical Hypotheses." In this paper, the authors provided both the first complete statement of their new approach and some of its most important applications.

Research partially supported by National Science Foundation Grant MCS82-01498.

From *Proceedings of the Berkeley Conference in Honor of Jerzy Neyman and Jack Kiefer*, Volume I, Lucien M. Le Cam and Richard A. Olshen, eds.,

[1]Accounts of the collaboration can be found in Pearson's recollections (1966) and in the Neyman biography by Reid (1983).

The theory contains three principal ingredients:

(i) a *parametric model* specifying the assumed family of distributions of the observable random variables;

(ii) a *criterion* measuring the performance of a given statistical procedure, namely, the power function of the test in question;

(iii) the advocacy of *optimality*: that the most desirable test is obtained by maximizing the power.

Despite the fact that these ideas had their forerunners in the work of Fisher,[2] the Neyman-Pearson paradigm formulated for the first time a clear program and provided a completely novel approach to hypothesis testing, the first "exact" small-sample theory of its kind.

It is the purpose of this paper to commemorate the 50th anniversary of the NP theory by considering some aspects of its development during the past half-century and to inquire into its relevance for statistics today. A particular concern will be the contribution the NP approach can make to data analysis and Bayesian statistics. All three of these paradigms have had extensive developments, generalizations, and elaborations during the period in question.

Data analysis had its origin in descriptive statistics, which dealt with the description and summary of data without making any assumptions, in particular, without assuming an underlying stochastic model. Under the leadership of Tukey (see, for example, his paper of 1962), many new techniques were developed and new tasks added such as the discovery of underlying structure and of transformations to simplify such structure. Recent treatments of the subject can be found in the books by Tukey (1977), Mosteller and Tukey (1977), and Hoaglin, Mosteller, and Tukey (1983). For critiques, see, for example, Dempster (1983) and Mallows (1983).

The Neyman-Pearson theory, which was originally concerned only with hypothesis testing, was adapted by Neyman (1937) to deal also

[2]Particularly in his fundamental paper of 1922, which dealt with point estimation but also emphasized parametric models and the fact that maximum likelihood estimators minimize the asymptotic variance. Fisher was preceded by Edgeworth.

with estimation by confidence sets and was generalized by Wald to his general decision theory, in which the power function was replaced by a general risk function, new optimality criteria were introduced, and the restriction to parametric models was jettisoned (although in Wald's final exposition of 1950 the examples were all parametric).

Bayesian statistics had come to a first flowering in the work of Laplace. It had been pushed aside in deliberate efforts by Fisher, and Neyman and Pearson, to build an "objective" theory and was reawakened by L. J. Savage[3] (particularly in his 1954 book) and earlier, from a somewhat different point of view, by Jeffreys (1939). One of its principal features is the assumption that the parameters of the parametric model should themselves be viewed as random variables whose distribution is assumed to be known.

The needs that lie behind the development of data analysis and Bayesian statistics had their impact also on the NP theory. In particular, the desire to free statistical analysis from the stringent and unreliable assumptions of parametric models moved the NP theory in the direction of data analysis, and the need to incorporate prior information and subjective impressions brought it closer to Bayesian statistics. The two trends will be sketched in Sections 2-4.

2. ROBUSTNESS AND ADAPTATION

A typical example of the classical NP approach is the normal two-sample problem. If one assumes that $X_1, \ldots, X_m$ and $Y_1, \ldots, Y_n$ are independent with normal distributions

$$\begin{aligned} X_1, \ldots, X_m &: N(\xi, \sigma^2) \\ Y_1, \ldots, Y_n &: N(\eta, \sigma^2), \end{aligned} \tag{2.1}$$

then the optimal test of the hypothesis $\eta = \xi$ against the one-sided alternatives $\eta > \xi$ under a variety of optimality criteria (UMP

[3]Savage acknowledged the profound influence on his thinking of de Finetti, as presented, for example, in de Finetti (1937).

unbiased, UMP invariant, most stringent, and so forth) is the two-sample t-test.

However, the validity of this model is frequently doubtful: the distribution of the errors may not be normal, the two variances not be equal, the observations not be independent. We shall here consider only the effect of non-normality, and hence the model

$$\begin{aligned} &X_1, \ldots, X_m : F(x - \xi) \\ &Y_1, \ldots, Y_n : F(y - \eta), \end{aligned} \tag{2.2}$$

where F is unknown. Before embarking on this program, we should mention a type of model that is intermediate between (2.1) and (2.2). In this formulation, it is assumed that the distribution F in (2.2) is approximately normal, for example, that

$$F(x) = (1 - \varepsilon)\Phi(x) + \varepsilon H(x), \tag{2.3}$$

where ε is given and H is an arbitrary distribution function. Such models, which have been studied by Huber [see, for example, Huber (1981)], will not be considered in the present paper.

The first question that was asked about the robustness of the t-test concerned the actual level $\alpha_{m,n}(F)$ of the t-test carried out at nominal level α when the model is given by (2.2). This question was raised quite early; for a review of some of the relevant literature, see Tan (1982). It immediately follows from the central limit theorem and Slutsky's theorem that

$$\alpha_{m,n}(F) \to \alpha \quad \text{as } m, n \to \infty \tag{2.4}$$

for any F with finite second moment. The same type of argument shows further that, for large m and n, not only the level but also the power of the t-test is approximately equal to the power predicted under the normal model.

If the value of the significance level is taken very seriously, the approximate nature of the level under (2.2) may be considered inadequate. The difficulty can be overcome by a modification of the t-test, the *permutation* (or *randomization*) t test due to R. A. Fisher (1935). In this test the t-statistic is compared not with all the values it takes on as the X's and Y's range over the sample space but only with the values obtained by permuting the $m + n$ observed values $(x_1, \ldots, x_m, y_1, \ldots, y_n)$. The resulting test has a level that is independent of F in the model (2.2). In addition, it has asymptotically the same power as the t-test (Hoeffding, 1952). It thus appears to be an ideal solution of the robustness problem.

Unfortunately, this conclusion overlooks the fact that, if the normal model is wrong, the t-test loses its optimality properties and may actually not be a particularly desirable test to use. The distinction is that between robustness of validity (or, more generally, performance robustness) and robustness of efficiency, introduced by Tukey and McLaughlin (1963). It is closely related to the concepts of criterion and inference robustness proposed by Box and Tiao (1964). The t-test and its permutation version are performance-robust in that their performance is approximately that asserted by the normal theory even under model (2.2). However, since the efficiency of these tests does not continue to hold even approximately in (2.2) when F is not normal, performance robustness is not as strong a recommendation as appears at first sight.

There exist, in fact, tests (based on the rank of the observations) whose significance level is independent of F and whose power for many realistic distributions F is much higher than that of the t-test. A particularly striking example is the normal-scores test N, whose asymptotic relative efficiency (ARE) $e_{N,t}(F)$ with respect to the t-test satisfies (Chernoff and Savage, 1958)

$$e_{N,t}(F) \geqslant 1 \quad \text{for all } F, \tag{2.5}$$

with equality holding only when F is normal, and with $e_{N,t}(F)$ becoming arbitrarily large when the tail of F is sufficiently heavy. The test N has the asymptotic optimality property of maximizing $\inf e_{N,t}(F)$ in model (2.2).

In this solution to the robustness problem the original model (2.1) still plays a central (though implicit) role because it sets up the t-test as the standard with which other tests are being compared. If we want to treat (2.2) without any special relevance to the normal case, a more natural efficiency measure of a test is its *absolute* efficiency $e(F)$, that is, its efficiency relative to the best parametric test of $H: \Delta = \eta - \xi = 0$ when F is known (for example, the UMP invariant test). Stein (1956) suggested the existence of a test that would satisfy

$$e(F) = 1 \tag{2.6}$$

for all sufficiently smooth F. The program of working out such *adaptive* procedures (which adapt themselves to the unknown true F) has been carried out for a number of problems. For the estimation of the center of a symmetric distribution on the basis of a sample $X_1, \ldots, X_n$, for example, fully efficient adaptive estimators were obtained by Beran (1974) and Stone (1975). The latter shows that the small-sample efficiency of his estimator is of the order of .9 for a variety of distributions of quite different shapes (such as normal, Cauchy, and double-exponential) for sample sizes as small as $n = 40$. The conditions under which adaptation is possible have been investigated by Bickel (1982).

3. RELATION WITH DATA ANALYSIS

For removing the assumption of normality from model (2.1), adaptive tests and estimators provide the ultimate answer. However, they seem to be far from the methods and concerns of data analysis as an approach to statistics without any modeling. In fact, they are closer than appears at first sight.

Adaptive procedures begin by estimating the unknown distribution F or its density f; they then apply the procedure that would be best if the estimated f were the true one. However, the problem of estimating an unknown density f is quite close to that of describing a data set by means, say, of a histogram. Exactly the same issues arise in both (interval width, smoothing, etc.); see, for example, Scott (1979), Freedman and Diaconis (1981), and Hoaglin, Mosteller, and Tukey (1983).

There is, of course, a difference between the two methods of attack. The modeling approach can measure and compare the performance of different procedures; it can ask how close the estimated density is likely to be to the true one and obtain the interval width that provides the best convergence rate. However, this requires making some assumptions: for example, that the observations are n independent realizations of a random variable whose distribution has a smooth density.

The Neyman-Pearson program as represented by (ii) and (iii) of the Introduction can be applied not only to data description but also to other aspects of data analysis. We shall briefly consider its contribution to two of those: data summaries and exploration of underlying structure.

To summarize a simple one-dimensional batch of data, one calculates statistics that indicate, for example, the center and the spread of the batch. If one is willing to assume that the observations $X_1, \ldots, X_n$ are a sample from a distribution F, one will be concerned instead with measures $\mu(F)$ of the center and spread of F (or of skewness, Kurtosis, etc.), and the summary statistics become estimators of the unknown $\mu(F)$.[4] Some of the concerns about such summaries can be shared by the two approaches; resistance to outliers is an important example. However, the assumption of an underlying probability model provides new possibilities. In particular, one can now again measure and compare the performance of different estimators. The choice of suitable measures $\mu(F)$ (and their estimators) from this point of view is discussed in Bickel and Lehmann (1975/1976).

Methods for discovering underlying structure (including the choice of transformations to simplify the structure) can again be strengthened by making some assumptions. As an example, we mention recent work concerned with model selection that assumes a class of possible models of varying complexity; see, for example, Stone (1981).

[4]Typically $\mu(F)$ is estimated by $\mu(F_n)$, the functional μ evaluated at the empirical distribution function F_n.

The data are used to choose a suitable model within this class. An interesting new feature of this application of the Neyman-Pearson-Wald approach is that the loss function must include a component representing the complexity of the model.

It thus appears that many of the problems considered in data analysis can also be studied from the NPW point of view. The latter will always require some (although possibly rather weak) assumptions, and this can never quite match the freedom of a totally unrestrained examination of the data. On the other hand, it has the advantage of allowing evaluation and comparison of different methods. In addition, it may permit the extrapolation of findings to other data sets, whereas a purely data-analytic approach is bound to the data set under consideration.

4. RELATION WITH BAYESIAN INFERENCE

The move toward data analysis has as its aim a weakening of assumptions that are frequently over-detailed and unrealistic. In the other direction, the Bayesian approach tries to take account of additional information, not represented in the classical models, that affects our expectations regarding the parameter values likely to obtain. This is achieved by postulating that the parameters themselves are random variables (though unobservable) with a known probability distribution.

Non-Bayesians and most Bayesians appear to be in agreement on two basic facts:

(i) The procedures one wants to use are Bayes' procedures with respect to some prior distribution (or at least limits of such procedures). For Bayesians this follows from first principles. From a decision theoretic point of view it is a consequence of the fact that only Bayes' procedures and their limits are admissible.

(ii) Obtaining the appropriate prior distribution is usually a difficult task. Different Bayesian ideas of how to go about this are represented, for example, by the books of Raiffa and Schlaifer (1961), Box and Tiao (1964), and Lindley (1965). In the present section, we shall briefly discuss the contribution of decision theory to this problem.

Since a large-sample approach often brings great simplification, let us begin by considering the asymptotic situation of a large number of i.i.d. random variables having density $f(x, \theta)$, where θ is real-valued and has prior density $\pi(\theta)$ with $\pi(\theta) > 0$ for all θ and $E|\theta| < \infty$. Then, under suitable regularity conditions on $f(x, \theta)$, the Bayes' estimator $\delta_{\pi,n}$ of θ with respect to squared error loss satisfies[5]

$$\sqrt{n}(\delta_{\pi,n} - \theta) \to N(0, 1/I(\theta)) \quad \text{in law,} \tag{4.1}$$

where $I(\theta)$ is the amount of information each X_i contains about θ. See, for example, Lehmann (1983), Section 6.7.

Thus, to the accuracy of this approximation, the Bayes' procedures share the asymptotic optimality of maximum likelihood estimators: they are efficient. However, it is a consequence of this result that the theory cannot distinguish between them. In fact, the Bayes' estimators corresponding to two different prior densities π and π' satisfy

$$\sqrt{n}(\delta_{\pi,n} - \delta_{\pi',n}) \to 0 \quad \text{in probability.} \tag{4.2}$$

Thus, large-sample theory does indeed greatly simplify the situation. Unfortunately it simplifies it too much.

A crucial assumption in these considerations is that $\pi(\theta) > 0$ for all θ, because a Bayes' procedure pays no attention to parameter values that are impossible a priori. This suggests that optimal risk can be attained most rapidly (in a minimax sense) if π is not too uneven, i.e., if there are no θ-values for which $\pi(\theta)$ is relatively very low. Actually, such an argument should also take into account the amount of information the sample contains about different values of θ, which then points to an uninformative prior, and results in an approach that some authors call *objective* Bayesian. Furthermore, in problems in which invariance considerations are applicable, the decision theoretic invariance principle leads directly, even for small

[5]However, see Diaconis and Freedman (1983) and Freedman and Diaconis (1983).

samples, to the Bayes' solution corresponding to an invariant (and hence noninformative) prior.

If, based on previous experience and perhaps other less tangible and more subjective impressions, one has a preferred prior distribution, decision theory suggests an alternative approach. Compare the behavior of procedures corresponding to a number of representative priors, including the prior of one's choice, and adopt the latter if its performance is satisfactory. Essentially the same strategy is recommended in a book, written from a Bayesian point of view, by J. Berger (1980). In the contrary case, one may want to tone down one's prior opinions in the direction of a noninformative prior.

A compromise of this kind is offered by the theory of restricted Bayes solutions. Suppose one has arrived at a preferred prior Λ, which leads to very high values of the risk function for some parameter values and about which one does not feel very confident. Consider for a moment the minimax solution as a competitor that has a much smaller maximum risk M but which corresponds to a prior very different from Λ. As a compromise between these alternatives, let us limit the allowable maximum risk not to its minimum value M but to $M(1 + \varepsilon)$, so that the risk function is required to satisfy

$$R(\theta, \delta) \leqslant M(1 + \varepsilon) \quad \text{for all } \theta. \tag{4.3}$$

The *restricted Bayes' procedure* $\delta_{\Lambda, \varepsilon}$ is then obtained by minimizing the Bayes' risk $\int R(\theta, \delta)d\Lambda(\theta)$ subject to (4.3); see Hodges and Lehmann (1953). The choice of ε in (4.3) will reflect both the confidence one places on Λ and the cost of excessively high risk. Applications of this approach have been developed by Efron and Morris (1971), Bickel (1983), and Jelihovschi (1984).

5. CONCLUSIONS AND ACKNOWLEDGMENTS

The Neyman-Pearson theory makes two principal recommendations: Evaluate the behavior of statistical procedures in terms of some performance criterion, and then opt for the procedure that performs best. The application of these recommendations requires a probabilistic model for the observations, although such models can be much broader than those originally envisaged by Neyman and Pearson.

As has been indicated in the preceding sections, the first of these recommendations can be usefully applied to a wide spectrum of statistical techniques ranging from procedures inspired by data analysis to those obtained from Bayesian principles. It is difficult to imagine statistics as a subject in which this idea would not play a central role.

Complete reliance on optimality, on the other hand, suffers from some drawbacks. In particular,

(i) procedures that minimize the risk may be quite unsatisfactory in other respects such as robustness or interpretability;

(ii) most optimality criteria lead to solutions only in fairly simple problems.

As a consequence, an optimal solution (if it exists) should not be considered the ultimate answer, but its performance should be studied, just as that of any other proposed procedure, from all relevant points of view. On the other hand, a procedure that is best in some way is of value as a benchmark with which competing procedures can be compared. It shows how well one can do in this particular direction and how much is lost when other criteria have to be taken into account.

It is suggested in Sections 3 and 4 that the procedures obtained from a decision theoretic point of view may not differ as much from those reached via a data analytic or Bayesian philosophy as is frequently claimed and that all three ways of treating problems can have contributions to make. This point has been made by other authors.

The relationship between decision theory and data analysis is explored by Thisted (1981) and in the book by Hoaglin, Mosteller, and Tukey (1983). Regarding the uses of decision theory in a Bayesian setting, the work of Berger (1980) has already been mentioned. A different combination of Bayesian and Neyman-Pearson theory has been recommended by Box (1980), who proposes a Bayesian approach to the estimation of the parameters of the model and a test based on sampling considerations for testing the model.

Finally, it was shown how all three lines of attack can be used in a substantive study by Mosteller and Wallace (1964), and a plea for "ecumenism" in statistics is put forth by Box (1983).

REFERENCES

Andrews, D. F. (1978). Article on "Data Analysis, Exploratory." In *Internat. Encycl. of Statist.* Kruskal and Tanur, eds. Free Press, New York.

Beran, R. (1974). Asymptotically efficient adaptive rank estimates in location models. *Ann. Statist. 2*, 63-74.

Berger, J. (1980). *Statistical Decision Theory.* Springer, New York.

Bickel, P. J. (1981). Minimax estimation of the mean of a normal distribution when the parameter space is restricted. *Ann. Statist. 9*, 1301-1309.

Bickel, P. J. (1982). On adaptive estimation. *Ann. Statist. 10*, 647-671.

Bickel, P. J. (1984). Parametric robustness. *Ann. Statist. 12*.

Bickel, P. J., and Lehmann, E. L. (1975/1976). Descriptive statistics for nonparametric models. *Ann. Statist. 3*, 1038-1069; *4*, 1139-1158.

Box, G. E. P. (1980). Sampling and Bayes' inference in scientific modelling and robustness. *J. Roy. Statist. Soc.* (A) *143*, 383-430.

Box, G. E. P. (1983). An apology for ecumenism in statistics. In *Scientific Inference, Data Analysis, and Robustness.* Box, Leonard, and Wu, eds. John Wiley, New York.

Box, G. E. P., and Tiao, G. C. (1964). A note on criterion robustness and inference robustness. *Biometrika 51*, 169-173.

Box, G. E. P., and Tiao, G. C. (1973). *Bayesian Inference in Statistical Analysis.* Addison-Wesley, Reading.

Chernoff, H., and Savage, I. R. (1958). Asymptotic normality and efficiency of some nonparametric competitors of the *t*-test. *Ann. Math. Statist. 29*, 972-994.

de Finetti, B. (1937). La prévision: ses lois logiques, ses sources sujectives. *Ann. de l'Inst. Henri Poincaré 7*, 1-68. [Trans. in Kyburg, H., and Smokler, H. (1964), *Studies in Subjective Probability*, John Wiley, New York.]

Dempster, A. P. (1983). Purposes and limitations of data Analysis. In *Scientific Inference, Data Analysis, and Robustness*. Box, Leonard, and Wu, eds. John Wiley, New York.

Diaconis, P., and Freedman, D. (1983). Frequency properties of Bayes rules. In *Scientific Inference, Data Analysis, and Robustness*. Box, Leonard, and Wu, eds. John Wiley, New York.

Efron, B., and Morris, C. (1973). Stein's estimation rule and its competitors—an empirical Bayes approach. *J. Amer. Statist. Assoc. 68*, 117-130.

Fisher, R. A. (1922). On the mathematical foundations of theoretical statistics. *Phil. Trans. Roy. Soc.* (A) *222*, 309-368.

Fisher, R. A. (1935). *The Design of Experiments*. Oliver and Boyd, Edinburgh.

Freedman, D., and Diaconis, P. (1981). On the histogram as a density estimator: L_2 theory. *Zeitsch. Wahrsch. 57*, 453-476.

Freedman, D., and Diaconis, P. (1983). On inconsistent Bayes estimates in the discrete case. *Ann. Statist. 11*, 1109-1118.

Hoaglin, D. C., Mosteller, F., and Tukey, J. W. (1983). *Understanding Robust and Exploratory Data Analysis*. John Wiley, New York.

Hodges, J. L., Jr., and Lehmann, E. L. (1953). The use of previous experience in reaching statistical decisions. *Ann. Math. Statist. 23*, 396-407.

Hoeffding, W. (1952). The large-sample power of tests based on permutations of observations. *Ann. Math. Statist. 23*, 169-192.

Huber, P. J. (1981). *Robust Statistics*. John Wiley, New York.

Jeffreys, H. (1939). *The Theory of Probability*. Oxford University Press, Oxford.

Jelihovschi, E. (1984). Estimation of Poisson parameters, subject to constraint. Ph.D. dissertation, University of California, Berkeley.

Lehmann, E. L. (1983). *Theory of Point Estimation*. John Wiley, New York.

Lindley, D. (1965). *Probability and Statistics*. Vol. 2: *Inference*. Cambridge University Press, Cambridge.

Mallows, C. L. (1983). Data description. In *Scientific Inference, Data Analysis, and Robustness*. Box, Leonard, and Wu, eds. John Wiley, New York.

Mosteller, F., and Tukey, J. W. (1977). *Data Analysis and Regression*. John Wiley, New York.

Mosteller, F., and Wallace, D. (1964). *Inference and Disputed Authorship: The Federalist*. Addison-Wesley, Reading.

Neyman, J. (1937). Outline of a theory of statistical estimation based on the classical theory of probability. *Phil. Trans. Roy. Soc.* (A) *236*, 333-380.

Neyman, J., and Pearson, E. S. (1933). On the problem of the most efficient tests of statistical hypotheses. *Phil. Trans. Roy. Soc.* (A) *231*, 289-337.

Pearson, E. S. (1966). The Neyman-Pearson story: 1926-1934. In *Research Papers in Statistics: Festschrift for J. Neyman.* F. N. David, ed. John Wiley, New York.

Raiffa, H., and Schlaifer, R. (1961). *Applied Statistical Decision Theory.* Harvard University Pres, Cambridge.

Reid, C. (1982). *Neyman—From Life.* Springer, New York.

Savage, L. J. (1954). *Foundations of Statistics.* John Wiley, New York. 2nd revised ed., Dover (1972).

Scott, D. W. (1979). On optimal and data-based histograms. *Biometrika 66*, 605-610.

Stein, C. (1956). Efficient nonparametric testing and estimation. *Proc. Third Berkeley Symp. Math. Statist. and Prob. 1*, 187-196.

Stone, C. J. (1975). Adaptive maximum likelihood estimators of a location parameter. *Ann. Statist. 3*, 267-284.

Stone, C. J. (1981). Admissible selection of an accurate and parsimonious normal linear regression model. *Ann. Statist. 9*, 475-485.

Student [W. S. Gosset] (1908). On the probable error of a mean. *Biometrika 6*, 1-25.

Tan, W. Y. (1982). Sampling distributions and robustness of *t*, *F* and variance-ratio in two samples and ANOVA models with respect to departure from normality. *Communications in Statistics, Theory and Methods 11* (No. 22), 2485-2511.

Thisbed, R. (1982). Decision theoretic regression diagnostics. In *Statistical Decision Theory and Related Topics.* Gupta and Berger, eds. Academic Press, New York. Vol. II, pp. 363-382.

Tukey, J. W., and McLaughlin, D. H. (1963). Less vulnerable confidence and significance procedures for location based on a single sample: Trimming/Winsorization 1. *Sankhya 25*, 331-352.

Tukey, J. W. (1977). *Exploratory Data Analysis.* Addison-Wesley, Reading.

Wald, A. (1950). *Statistical Decision Functions.* John Wiley, New York.

E. L. Lehmann
Department of Statistics
University of California
Berkeley, USA

FOUNDATIONAL ISSUES IN STATISTICS: THEORY AND PRACTICE

Abstract. The foundational issues discussed by David Freedman are examined from the point of view of an investigator faced with the choices they imply. It turns out that in practice the boundaries between the various philosophies are less sharp than might be expected from the passion with which their differences have been argued for a long time. There is considerable common ground, and each approach gains by borrowing tools from the other.

1. Introduction

David Freedman's paper consists of two parts. The first describes the frequentist and Bayesian approaches to probability and statistics, the second is concerned with the validity of statistical models. Both emphasize the importance of putting realism above convenience and dogma. I shall here consider these issues from the point of view of an investigator faced with the choices implied by Freedman's distinctions, namely between

- data analysis without a model;
- a frequentist model involving unknown parameters;
- A Bayesian model which assigns a known distribution to these parameters. (I am restricting attention here to what Freedman calls the "classical" Bayesian position; bridge building to the "radical" position is more difficult).

2. Model-free data analysis

At this stage, a data set will typically be examined, analyzed, and organized in many different ways in an attempt to bring out its salient features and to pinpoint meaningful effects. However, the question then arises whether these effects are real. Unfortunately, except in very special circumstances (for an example see Freedman and Lane (1983)), this question cannot be answered without a probability model. The difficulty is greatly exacerbated by the multiple facets of such an approach. The suggested effects of interest, for example, will tend to be those that appear most significant out of large numbers that were examined, some in detail, some

J. Rojo (ed.), *Selected Works of E. L. Lehmann*, Selected Works in Probability and Statistics, DOI 10.1007/978-1-4614-1412-4_89, © Springer Science+Business Media, LLC 2012

only out of the corner of an eye, and many of which – in a phrase of Cournot (1843) – "have left no traces". The appearance of their significance may thus be greatly exaggerated. Cournot thought that the problem posed by this distortion was insoluble. However, often the situation can be saved by a second stage with new data at which a probability model is used to test the much smaller and more clearly defined set of effects found interesting at the first stage.

3. The modeling decision

The choice between an analysis with or without model depends largely on the extent of our knowledge concerning the underlying situation. Do we know enough from past experience and/or a theoretical understanding of the situation to be clear about the questions to be asked and to be able to formulate an adequately realistic model? Or do we have to examine the data to see what the problems are and to get an idea of the kind of model that might be appropriate?

In this connection it is important to realize that models used for different purposes need to meet different standards of realism. A well known classification (see for example Box, Hunter and Hunter (1978), Cox (1990) and Lehmann (1990)) distinguishes between on the one hand explanatory or substantive models which try to portray in detail the principal features of the situation and to explain their interrelations, and on the other empirical or off-the-shelf models, which employ the best fitting model from a predetermined standard family. The latter have little explanatory power but nevertheless can be very useful for limited practical purposes such as predicting school enrollments or forecasting elections. (Early astronomers were able to make very accurate predictions based on elaborate models [1] that bore little resemblance to reality. Their insistence on circular motions is reminiscent of the ubiquitous assumption of linearity in the kind of regression model discussed by Freedman in Section 5 of his paper.) As Freedman points out, the success of such predictions could and should be routinely monitored. A danger of empirical models is that once they have entered the literature, their shaky origin and limited purpose have a tendency to be forgotten.

4. Model-based inference

Suppose now that a model has been selected which we believe to be adequate, and that we know exactly which questions we want to ask. Let us assume that the problem has been formulated as a choice among a set of possible decisions and that we can assess the loss $L(\theta, d)$ that would result from taking decision d when θ is the true parameter value. The observations are represented by a random quantity X which has a probability distribution depending on θ. A statistical procedure is

[1] A recent account of the accomplishments and ultimate failures of these models can be found in North (1994).

a rule (or function) δ which specifies the decision $d = \delta(x)$ to be taken when X takes the value x. Finally, the performance of the decision rule δ is measured by the average loss resulting from its use,

$$R(\theta, \delta) = EL[\theta, \delta(X)] \tag{1}$$

This expected loss is called the risk function of δ.

A classical Bayesian has a prior distribution W for θ; for the sake of simplicity we suppose that W has a density $w(\theta)$. Within this Bayesian framework, the quantity of interest is the expected risk

$$r_\delta = \int R(\theta, \delta) w(\theta) d\theta \tag{2}$$

and the most desirable procedure is that for which r_δ is as small as possible. Determining this optimum procedure is in principle straightforward, although it may be computationally difficult.

The hardest part often is the choice of W. In practice, prior distributions, just like many probability models, frequently are of the off-the-shelf kind, that is, they are chosen from a convenient standard family which is sufficiently flexible to permit modeling the rough qualitative features desired. A popular choice is an uninformative prior that corresponds to a state of ignorance concerning θ.

Frequentists do not face the issue of determining a prior, but they have to contend with another difficulty. Rather than minimizing (2), they seek a procedure that minimizes $R(\theta, \delta)$ for all θ. Since, unfortunately, such a procedure does not exist, various devices are used to work around this problem, none of them really compelling. We mention here only one such criterion: minimizing the maximum risk. A procedure achieving this is said to be minimax.

5. Common ground

The distinctions between the three approaches: (i) model-free data analysis, (ii) frequentist and (iii) Bayesian model-based inference have been heatedly debated for a long time. However, in practice there is more contact – and the lines separating the three modes are less sharp – than this debate suggests.

For example, an exploratory data analysis may perform informal tests with ad hoc models, for guidance on which hypothesis suggested by the data to pursue; conversely, the subsequent formal-inference stage may utilize the first stage to search for the models it needs.

We shall in the remainder of this section restrict attention to some of the relationships between (ii) and (iii). Additional discussion of these and other points of contact between the two approaches can be found in Diaconis and Freedman (1986) and Lehmann (1985).

(a) Previous experience. For a large class of important situations in areas such as medicine, agriculture, business and education, a great body of related earlier experience is available to draw on. This experience provides an observed frequency distribution of θ's of which we have reason to believe that the present (unknown) θ-value is a typical member. A Bayesian is likely to accept this distribution as an appropriate prior (although possibly making slight modifications to allow for special features in which the present situation might differ from the past). A frequentist, even without a clearly defined random mechanism for producing the θ's, may analogously work with the empirical distribution as if it were the true frequency distribution of a random quantity. Although they will think very differently both of this distribution W and of the assumed distribution of X given θ (along the lines indicated in Sections 2 and 3 of Freedman's paper), both will be led to the procedure δ minimizing (2) and thus to the same decision.

(b) A fundamental connection. Even when no prior information is available, Bayes procedures are of importance to frequentists. This is a consequence of a fundamental theorem of Wald, which states that the only sensible (from a frequentist point of view) procedures are those that are Bayes solutions for some prior W or limits of such solutions. In the light of this result, it is often useful to examine a frequentist procedure from a Bayesian point of view to see whether it is Bayes with respect to some prior and if so, whether this prior seems at all plausible. In this connection it is interesting to note that minimax procedures are often Bayes solutions corresponding to an uninformative prior.

(c) Performance. In the opposite direction from (b), a Bayesian, after having computed the Bayes procedure δ corresponding to a preferred prior, can learn much about the procedure by studying its risk function $R(\theta, \delta)$. This latter step, although inconsistent with a Bayesian philosophy, is now acknowledged as useful by many Bayesians. (For a discussion from a contrary viewpoint, see Berger (1986).)

5. Conclusions

1. It seems clear that model-free data analysis, frequentist and Bayesian model-based inference and decision making each has its place. The question appears not to be – as it so often is phrased – which is the correct approach but in what circumstances each is most appropriate.

2. In practice, the three approaches can often fruitfully interact, with each benefiting from consideration of the other points of view.

3. Although the debates between adherents of Bayesian and frequentist philosophies have been carried on with much heat for a long time, in practice neither side very often lives up to its principles, since both models and priors frequently are of the off-the-shelf variety.

References

Berger, J. ,1986, Discussion of Diaconis and Freedman, *Annals of Statistics* **14**, 30-37.

Box, G. E. P., Hunter, W. G., and Hunter, J. S., 1978, *Statistics for Experiments.* New York, Wiley.

Cournot, A. A., 1843, *Exposition de la Théorie des Chances et des Probabilités.* Paris, Hachette

Cox, D. R., 1990, Models in Statistical Analysis, *Statistical Science* **5**, 169-174.

Diaconis, P. and Freedman, D., 1986, On the Consistency of Bayes Estimates, *Annals of Statistics* **14**, 1-67, (with discussion).

Freedman, D. A. and Lane, D., 1983, Significance Testing in a Nonstochastic Setting, pp.185-208 in P. J. Bickel, K. A. Doksum, and J. L. Hodges, Jr. (eds), *A Festschrift for Erich L. Lehmann in Honor of his Sixty-Fifth Birthday.* Belmont, Calif., Wadsworth.

Lehmann, E. L., 1985, The Neyman-Pearson Theory After Fifty Years, pp.1-14 in L. LeCam and R. Olshen (eds), *Proceedings of the Berkeley Conference in Honor of Jerzy Neyman and Jack Kiefer.* Vol. 1. Monterey, Calif., Wadsworth.

Lehmann, E. L., 1990, Model Specification, *Statistical Science* **5**, 160-168.

North, J., 1994, *Astronomy and Cosmology.* London, Fontana.

PROBABILITY
AND
MATHEMATICAL STATISTICS
Vol. 15 (1995), pp. 29–36

NEYMAN'S STATISTICAL PHILOSOPHY

BY

E. L. LEHMANN (BERKELEY, CALIFORNIA)

1. Inductive behavior vs. inductive inference: Neyman and Fisher. A dramatics meeting of the Royal Statistical Society occurred on December 18, 1934, when R. A. Fisher – already at the height of his fame although not universally popular – presented a paper on *The logic of inference.* He stated that it was concerned with the problem of drawing inferences from the particular to the general or, in statistical language, from the sample to the population. He pointed out that deductive arguments based on probability were not adequate for such inferences, "for reasoning of a genuinely inductive kind," unless one used the theory of inverse probability [i.e. Bayesian inference] thereby "making it a deduction from the general to the particular." He then advanced the ambitious claim that truly inductive reasoning could be based on his concept of likelihood, which appeared "to take its place as a measure of rational belief when we are reasoning from the sample to the population," and proceeded to develop some of its properties.

The paper was attacked ferociously in the discussion following Fisher's presentation, and further objections were raised in later written contributions to the discussion. The last of these was by Neyman who started out in a highly complimentary vein. He then went on: "The psychology of some readers lead them to develop a theory already started along the way indicated. They probably think: 'What an interesting problem is raised! how could I develop it further?'... I personally seem to have another kind of psychology and can't help thinking: 'What an interesting way of asking and answering questions, but can't I do it differently?'"

Neyman points out that the properties Fisher establishes in his paper such as the asymptotic normality and efficiency of the maximum likelihood estimator are obtained on the basis of probability calculations and "not because we believe blindly in some magic property of the [likelihood] function."

He then turns to his principal theme, "whether it is possible to construct a theory of mathematical statistics independent of the conception of likeli-

presentation, and an elaboration of Neyman's earlier discussion of it. The difference between these two points of view was thought by both Fisher and Neyman to be at the heart of their lifelong dispute.

The possibility and nature of induction is a classical problem of philosophy, and it seems of interest to inquire into the reception of the ideas of Neyman and Fisher by philosophers of science. Of course, it takes time for one discipline to absorb, acknowledge, or even become aware of the concepts and ideas from another. A bridge from the Neyman–Pearson theory of hypothesis testing and Neyman's theory of confidence intervals to the philosophy of science was built by two highly influential books, which examined these theories from a philosophical point of view: Braithwaite's *Scientific Explanation* (1953) and Hacking's *Logic of Statistical Inference* (1965). These were followed by a number of other monographs and conference proceedings devoted entirely or at least in part to philosophical interpretations of various theories of statistical inference, among them Gillies (1973), Stegmüller (1973), Kyburg (1974), Harper and Hooker, Eds. (1974), Maxwell and Anderson, Eds. (1975), Rosenkrantz (1977), Asquith and Kyburg, Eds. (1979), Seidenfeld (1979), Fetzer (1981), and Howson and Urbach (1989).

In its treatment of induction, this literature mirrors the Fisher–Neyman conflict, dividing the authors into 'inductionists,' who discuss how inductive inference is or should be carried out, and 'deductionists,' who believe that only deductive arguments are legitimate.

The former group subdivides further according to the medium through which induction is to be effected: Fisher's likelihood, his fiducial method, the posterior Bayesian probabilities of Jeffreys, De Finetti or Savage, or possibly some new approach of their own. For these philosophers, Neyman's ideas are either irrelevant or serve only as a foil against which to set forth their preferred alternative approach.

2. Deductivism: Neyman and Popper. The deductivists also speak with more than one voice. Here Neyman has a rival in the person of Karl Popper. The views of these two men with respect to induction have important similarities, as can be seen from Popper's first book of 1934, which appeared in an English translation in 1959 under the title *The Logic of Scientific Discovery.*

After a brief introductory paragraph, Chapter 1 begins with a discussion of induction: "According to a widely accepted view – to be opposed in this book – the empirical sciences can be characterized by the fact that they use 'inductive methods,' as they are called." A few pages later, Popper contrasts this inductivist view with his own approach: "The theory to be developed in the following pages stands directly opposed to all attempts to operate with the ideas of inductive logic. It might be described as the *deductive method of testing*, or as the view that a hypothesis can only be empirically tested – and only after it has been advanced." Actually, at the time Popper was concerned only with deterministic situations. It was only in later editions that he extended his considerations to a stochastic setting. And in his autobiography of 1976 he refers (p. 79) to a "nonexistent 'method of induction,' of which nobody has ever given a sensible description." One could think that one hears Neyman speaking. Not only the terms they use but also the times at which they first put forth their ideas are very close. The appearance of Popper's book in 1934(1) is sandwiched between the first mention in Neyman–Pearson (1933) of statistical tests as 'rules of behavior,' and Neyman's proposal to replace Fisher's inductive approach by a purely deductive one in his 1935 discussion of Fisher's paper to the Royal Statistical Society.

Since the difficulties with induction had been known for a long time, it is remarkable that independently and nearly simultaneously Neyman and Popper found a revolutionary way to finesse the issue by replacing inductive reasoning with a deductive process of hypothesis testing. They then proceeded to develop this shared central idea in different directions, with Popper pursuing it philosophically while Neyman (in his joint work with Pearson) showed how to implement it in scientific practice.

The difference in these two developments is illustrated by their respective attitudes toward the outcome of a test. For Popper, the meaning of this outcome is central to his approach. Rejection of hypotheses, which he calls *falsification*, is for him the principal tool of the scientific method. On the other hand, he never permits hypotheses to be accepted. At best, they receive some support from the test, which he calls *corroboration*.

In contrast, Neyman and Pearson use the terms *acceptance* and *rejection* rather casually, with Pearson later (1955) admitting that some of their "wording may have been chosen inadequately." And Neyman strongly denies that they have any cognitive significance. According to him (Neyman, 1957), the outcome of a test is "an act of will or decision to take a particular action,"... and again in the next paragraph: "These processes are certainly not any sort of 'reasoning'...; they are acts of will."

He states his position even more strongly in his text (Neyman, 1950). A test is a choice between two actions A and B which is "interpreted as the *adoption* or the *acceptance* of one of the hypotheses H or $\bar{H}$ and the *rejection* of the other." He goes on to warn the reader: "The terms 'accepting' and 'rejecting' a hypothesis are very convenient and are well established. It is important, however, to keep their exact meaning in mind and to discard various additional implications which may be suggested by intuition. Thus, to accept a hypothesis H means only to decide to take action A rather than action B. This does not mean that we necessarily believe that the hypothesis is true. Also, if the application of a rule of inductive behavior "rejects H, this means only that the rule prescribes action B and does not imply that we believe that H is false."

(1) It appeared in the fall of 1934 but bore the imprint of 1935.

However, statisticians tend not to analyze terms as carefully as philosophers do. The line between actions, decisions and acts of will on the one hand, and assertions, statements and judgements on the other, is often rather thin. So it should not come as a surprise that Neyman's own applied writing sometimes crosses this line which he has drawn so firmly. In a paper on the expansion of galaxies, for example (Lovasich, Mayall, Neyman, and Scott, 1961), a hypothesis of stability is being tested. Among the conclusions we read: "If we admit [certain assumptions], then, with outliers included, there is strong evidence that at least some of the clusters (and this would include Coma) are unstable. In this case, the chance of error in asserting instability of at least some of the clusters is less than one in a hundred. If the outliers are excluded, no such evidence exists; the chance of error in asserting instability may be as large as three in four." Similar language is found, for example, in Neyman and Scott (1961).

In denying that the outcome of a test has any cognitive impact, Neyman thus seems to have carried his argument too far: it contradicts the reality even of his own statistical practice. On the other hand, scientific experiments typically are followed by decisions – if only whether or not to make a public statement. A test then offers a substitute for the unattainable, but so desperately desired, ideal of induction. On the basis of the incomplete information provided by the observations, it furnishes a guide to which action to take and thus – Neyman's evocative phrase – to appropriate 'inductive behavior.'

Despite the close connection of their ideas, Neyman and Popper – in stark contrast to the lifelong feuding of Neyman and Fisher – assiduously ignored each other. I can find no reference to Popper in Neyman's writings, and the few times Popper mentions Neyman, it is as translator of von Mises or relates to Neyman's frequentists interpretation of probability. This can, of course, partly be explained by the fact that the two men worked in different disciplines. It is nevertheless surprising.

3. The impact of Neyman's work. Neyman's ideas on testing and estimation channelled statistical theory into new directions which culminated in Wald's general decision theory. How strongly Wald was influenced by Neyman's behavioristic point of view can be seen by the explanation Wald (1950) offers when formulating his general decision problem: "The adoption of a particular decision function by the experimenter may be termed 'inductive behavior,' since it determines uniquely the procedure for carrying out the experiment and for making a terminal decision. Thus, the above decision problem may be called the problem of inductive behavior." (In a footnote he attributes the term to Neyman (1938).)

But Neyman thought of his behavioristic formulation (as Fisher did of his likelihood approach) as an important contribution not only to statistics but also to the philosophical problem of induction. It is of interest to consider the reaction of philosophers of science to this claim. In assessing the reception of Neyman's ideas by the philosophical community, it is necessary to consider more generally the influence of the whole of the modern statistical enterprise based on the work of Fisher, Neyman, Pearson and Wald, and also of the Bayesians, particularly Jeffreys, De Finetti and Savage. This body of work has affected the relevant parts of the philosophy of science in two important ways: By changing the emphasis from a deterministic to a stochastic setting([2]), and by developing methods that permit the implementation of the abstract formulations of the philosophy of science.

The approaches of Fisher, Neyman–Pearson and the Bayesians each have their proponents among philosophers of science. But even the expositions of inductivists and other opponents of Neyman typically find room for a critical discussion of the Neyman–Pearson theory. Thus, Gillies (1973) explains that although the Neyman–Pearson theory "in certain circumstances gives results which directly contradict our own approach," nevertheless "an examination of the Neyman–Pearson theory is not for us an option but a most pressing necessity" because "the Neyman–Pearson theory is still the generally accepted theory of testing statistical hypotheses."

Similarly, Kyburg (1974) justifies devoting a chapter to the "classical" theory of Neyman and Pearson by reminding the reader: "By the 1940's, the view, primarily associated with the name of Neyman that the fundamental form of statistical inference was the choice between statistical hypotheses, had come to dominate other views in the English speaking world. It merely dominated other views among statisticians; but it utterly overwhelmed other views among those whose interest in statistics was primarily practical."

These and the other authors listed in Section 2 discuss Neyman's approach not because of their liking for it but because of its prominence. However, there are also philosophers who strongly support Neyman's point of view. Prominent among these is Giere. In his essay (Giere, 1975), for example, he states that the approaches of Fisher, Neyman–Pearson, Neyman–Wald, and of "subjective Bayesians like Savage" have developed "prototype inductive logics which can deal with interesting scientific cases." — "Of these possibilities," he says, "the Neyman–Pearson–Wald tradition seems to me most promising." In a later section he writes: "The logic adopted here is a version of the theory of statistical hypothesis testing developed by Neyman and Pearson... This choice is based partly on my personal conviction that the Neyman–Pearson viewpoint is basically correct. But it is also justified by the tremendous influence this work has had for nearly half a century. The Neyman–Pearson approach must count as one of the three or four major candidates for a comprehensive inductive logic."

([2]) It has been pointed out to me by Brian Skyrms that historically much of the impetus for this change came from physics.

Another philosopher who strongly supports Neyman's ideas is Hacking who writes in 1980: "In my opinion there have been two important contributions to the study of rationality in this century, both negative. One is by Jerzy Neyman, the other by Imre Lakatos." And later on: "if you have firm beliefs about the future, then you can have detailed beliefs about the future, and you can engage in inductive behaviour such as avoiding betting. That is what Neyman teaches, and of course, I endorse it. There is no such thing as inductive inference, nor a logic of statistical inference or whatever..."

The supporters of Neyman, however, are in the minority. As Giere (1975) points out "in recent years, the most active area of interest for philosophers of science has been the subjective Bayesian approach to inference and decision making. Whatever the causes and reasons for this interest, it is a fact that, of all philosophers seriously concerned with the foundations of probability and statistics, more prefer some form of subjective account than any other approach."

Had he known about it, Neyman would have been interested in the attention philosophers have paid to his ideas, and in the importance statistics has acquired in the discussion of induction. He would no doubt have been disappointed to learn that his approach has not found more acceptance among philosophers. I believe he would have reacted the way he usually did when hearing unwelcome news. He would have shrugged his shoulders and said, "too bad."

REFERENCES

Asquith, P., and Kyburg, H., Eds. (1979), *Current Research in Philosophy of Science*, Philosophy of Science Association, East Lansing.

Braithwaite, R. B. (1953), *Scientific Explanation*, Cambridge University Press.

Fetzer, J. (1981), *Scientific Knowledge*, Reidel, Dordrecht.

Fisher, R. A. (1935), *The logic of inference* (with discussion), J. Roy. Statist. Soc.

Fisher, R. A. (1935), *The Design of Experiments*, Oliver and Boyd, Edinburgh.

Fisher, R. A. (1955), *Statistical methods and scientific induction*, J. Roy. Statist. Soc. (B) 17, pp. 69–78.

Giere, R. (1975), *The epistemological roots of scientific knowledge*, in: *Induction, Probability and Confirmation*, W. L. Harper and C. A. Hooker (Eds.), University of Minnesota Press, Minneapolis.

Gillies, D. A. (1973), *An Objective Theory of Probability*, Methuen, London.

Hacking, I. (1965), *Logic of Statistical Inference*, Cambridge University Press.

Hacking, I. (1980), in: *Scientific Discovery, Logic and Rationality*, T. Nickles (Ed.), Reidel, Dordrecht.

Harper, W. L., and Hooker, C. A. (Eds.) (1974), *Foundations of Probability Theory, Statistical Inference, and Statistical Theories of Science*, Vol. 2: *Foundations and Philosophy of Statistical Inference*, Reidel, Dordrecht.

Howson, C., and Urbach, P. (1989), *Scientific Reasoning: The Bayesian Approach*, Open Court, La Salle.

Kyburg, H. (1974), *The Logical Foundations of Statistical Inference*, Reidel, Dordrecht.

Lovasich, J. L., Mayall, N. U., Neyman, J., and Scott, E. L. (1961), *The Expansion of Clusters of Galaxies*, in: Proc. 4th Berkeley Symp. Math. Statist. Probability, Vol. 3, University of California Press, pp. 187–227.

Maxwell, G., and Anderson, R. M., Jr. (1975), *Induction, Probability, and Confirmation*, University of Minnesota Press, Minneapolis.

Neyman, J. (1938), *L'estimation statistique traitée comme un problème classique de probabilité*, Actualités Sci. Indust., No. 739, pp. 25–57.

Neyman, J. (1947), *Raisonnement inductif ou comportement inductif*, Proc. Internat. Statist. Conference 3, pp. 423–433.

Neyman, J. (1950), *First Course in Probability and Statistics*, Holt, New York.

Neyman, J. (1951), *Foundations of the general theory of statistical estimation*, Actualités Sci. Indust., No. 1146, pp. 83–95.

Neyman, J. (1952), *Lectures and Conferences on Mathematical Statistics*, 2nd ed., Graduate School, U.S. Department of Agriculture, Washington, D. C.

Neyman, J. (1955), *The problem of inductive inference*, Commun. Pure Appl. Math. 8, pp. 13–45.

Neyman, J. (1957), *'Inductive behavior' as a basic concept of philosophy of science*, Rev. Inst. Internat. Statist. 25, pp. 7–22.

Neyman, J. (1966), *Behavioristic Points of View on Mathematical Statistics in Political Economy and Econometrics — Essays in Honour of Oskar Lange*, PWN — Polish Scientific Publishers, Warszawa.

Neyman, J. (1971), *Foundations of behavioral statistics*, in: *Foundations of Statistical Inference*, V. P. Godambe and D. A. Sprott (Eds.), Holt, Rinehart and Winston, Toronto.

Neyman, J. (1977), *Frequentist probability and frequentist statistics*, Synthèse 36, pp. 97–131.

Neyman, J., and Pearson, E. S. (1933), *On the problem of the most efficient tests of statistical hypotheses*, Philos. Trans. Roy. Statist. Soc. Sec. A. 231, pp. 289–337.

Neyman, J., and Scott, E. L. (1961), *Magnitude — redshift relation of galaxies in clusters in the presence of instability and absorption*, Astron. J. 66, pp. 581–589.

Pearson, E. S. (1955), *Statistical concepts in their relation to reality*, J. Roy. Statist. Soc. (B) 17, pp. 204–207.

Popper, K. (1935), *Die Logik der Forschung*, Springer, Wien. English translation: *The Logic of Scientific Discovery* (1959), Basic Books, New York.

Popper, K. (1976), *Unended Quest*, Open Court, La Salle.

Rosenkrantz, R. D. (1977), *Inference, Method and Decision*, Reidel, Dordrecht.

Seidenfeld, T. (1979), *Philosophical Problems of Statistical Inference*, Reidel, Dordrecht.

Stegmüller, W. (1973), *Personelle und statistische Wahrscheinlichkeit*, Vol. 2: *Statistisches Schliessen, statistische Begründung, statistische Analyse*, Springer, Berlin.

Wald, A. (1950), *Statistical Decision Functions*, Wiley, New York.

Department of Statistics
University of California
Berkeley, CA 94720, U.S.A.

Received on 8.3.1994

the French Revolution. Yet even these enfranchisements revealed telling fault lines in the revolutionary conception of citizenship; the National Assembly, for instance, insisted that the Jews must give up their particular identity and corporate privileges as Jews in order to become French citizens. Since discussions of free black and slave rights inevitably involved the slave colonies, and since France's colony of Saint Domingue (San Domingo) was home to the first successful slave revolt in history beginning in 1791, the interest in citizenship has also stimulated study of France's slave colonies. Because the French Revolution put both revolution and modernity on the agenda, it continually offers rich possibilities for historical and theoretical debate.

See also: Bourgeoisie/Middle Classes, History of; Citizenship, Historical Development of; Democracy; Democratic Theory; Enlightenment; Freedom/Liberty: Impact on the Social Sciences; Ideology: History of the Concept; Nationalism: General; Political Representation; Representation: History of the Problem; Rousseau, Jean-Jacques (1712–78); Social Science, the Idea of; Socialism: Historical Aspects; Tocqueville, Alexis de (1805–59)

Bibliography

Arendt H 1963 *On Revolution*. Viking Press, New York

Barnave A 1988 *Introduction à la Révolution française. De la Révolution et de la Constitution*. Presses Universitaires de Grenoble, Grenoble, France

Buonarotti P 1828 *Conspiration pour l'égalité dite de Babeuf suivie du procès auquel elle donna lieu, et des pièces justificatives*. A la Libraire Romantique, Brussels, 2 Vols.

Burke E 1790 *Reflections on the Revolution in France, and on the Proceedings in Certain Societies in London Relative to that Event*, 3rd edn. J. Dodsley, London

Furet F 1978 *Penser la Révolution française*. Gallimard, Paris

Furet F 1986 *Marx et la Révolution française*. Flammarion, Paris

Godineau D 1988 *Citoyennes tricoteuses: les femmes du peuple à Paris pendant la Révolution française*. Alinéa, Aix-en-Provence, France

Hunt L A 1984 *Politics, Culture, and Class in the French Revolution*. University of California Press, Berkeley, CA

James C L R 1938 *The Black Jacobins: Toussaint L'Ouverture and the San Domingo Revolution*. Secker & Warburg, London

Landes J B 1988 *Women and the Public Sphere in the Age of the French Revolution*. Cornell University Press, Ithaca, NY

Taine H 1885–88 *La Révolution*. Hachette, Paris, 3 Vols.

Tocqueville A de 1856 *L'ancien régime et la Révolution*, 2nd edn. Michel Lévy frères, Paris

Tocqueville A de 1977 Oeuvres complètes. In: Jardin A, Lesourd J-A (eds.) *Correspondance Tocqueville-Louis de Kergolay*. Gallimard, Paris, Vol. 13

L. Hunt

Frequentist Inference

1. Data, Models, and Inference

The starting point of a statistical analysis is a set of data, for example, of counts or measurements. One aim may be simply to study what these data have to tell us. If they consist of a set of real numbers we might want to see, for example, whether they are small or large; tightly concentrated or spread out, whether they are stable or tend to increase with time, etc. If they are points in the plane we can get an idea of the shape of this set of points, for example, whether they cluster about a line. The branch of statistics dealing with this kind of investigation used to be called descriptive statistics, but now goes by the name data analysis or, more precisely, exploratory data analysis (EDA), a term introduced by Tukey (see *Exploratory Data Analysis: Univariate Methods*. Instead we consider how we can quantify the conclusions or decisions drawn from an analysis. Frequentist inference requires that any quantifying measure be interpretable in terms of frequentist probability of events (see *Frequentist Interpretation of Probability*) . That is, we assume the data are random quantities produced by some probability distribution and that something is known about this distribution. For example, if we have a set of n measurements of some quantity, we may assume that these measurements are independently and identically distributed. We may stop there or may go further and make some assumptions about this common distribution.

Statistical analyses based on such assumptions were common in the nineteenth century, and isolated instances can be found in the eighteenth century; but the first general framework was proposed by R. A. Fisher (1922) (see *Fisher, Ronald A (1890–1962)*). He states:

> The object of statistical methods is the reduction of data. A quantity of data which usually by its mere bulk is incapable of entering the mind, is to be replaced by relatively few quantities which shall adequately represent the whole.
> This object is accomplished by constructing a hypothetical infinite population of which the actual data are regarded as constituting a random sample. The law of distribution of this hypothetical population is specified by relatively few parameters, which are sufficient to describe it exhaustively in respect of all quantities under discussion.

In modern terminology we might paraphrase this proposal by saying that we construct a mathematical model, according to which the data are produced by a probability distribution assumed to belong to some specified parametric family of distributions.

What are the statistical methods based on these models supposed to achieve? Fisher called the aim 'inductive inference' or 'inductive reasoning,' and described it variously as 'learning by experience' and as 'drawing inferences from the particular to the general, from consequences to causes' or—using statistical language—'from the sample to the population.' 'The purpose of inductive reasoning based on empirical observations,' he wrote 'is to improve our understanding of the systems from which these observations are drawn.'

This interpretation was criticized by Neyman (1961) who objected that 'After a conscientious effort to find the exact meaning of this term [inductive reasoning] I came to the conclusion that, at least in the sense of Fisher, the term is empty.' Neyman proposed instead that the purpose of statistical analysis is to serve as guide to appropriate action; and, in contrast to Fisher's inductive reasoning, named it inductive behavior (see also *Neyman, Jerzy (1894–1981)*).

Neyman's point of view was taken up by Wald who used it to construct a comprehensive framework for statistical decision making (Wald 1950). It involved three principal elements

(a) A family of probability distributions representing the various possible true situations.

(b) A set of actions from which it is the statistician's task to choose the most appropriate one.

(c) A loss function which measures the loss resulting from any action taken for any of the possible true situations.

Wald's formulation was castigated by Fisher as completely inappropriate for scientific work. 'It is important,' he wrote (Fisher 1973, p. 106) 'that the scientific worker introduces no cost functions for faulty decisions. To do so would imply that the purposes to which new knowledge was to be put were known and capable of evaluation. As workers in Science we aim at methods of inference which shall be equally convincing to all freely reasoning minds, entirely independently of any intentions that might be furthered by utilizing the knowledge inferred.'

While it is difficult not to sympathize with Fisher's stress on scientific aims as an alternative to immediate utilitarian use, his statements are distressingly vague. An effort to be more specific, which still seems rather vague, was made by Tukey (1960). For the outcome of a nonaction-oriented statistical analysis he proposed the term 'conclusion.' Concerning the difference he writes: 'Conclusions are established with careful regard to evidence but without regard to consequences of specific actions in specific circumstances. Conclusions are withheld until adequate evidence has been accumulated.'

The distinction between Fisher's point of view and that of Neyman and Wald, between conclusions and decisions, concerns the interpretation of the results of a statistical analysis. How those results are reached from the given model by means of frequentist inference f orms the subject of this article. (For the frequentist meaning of probability see *Frequentist Interpretation of Probability*.) It includes the derivation of appropriate procedures, the comparison of different procedures, investigating the sensitivity of a procedure to departures from the assumed model, and the checking of the suitability of the model. Another topic of interest, considered in Sect. 3.6, is the selection of an appropriate model.

There is an alternative approach to inference which can embody both the decision theory aspects of the Neyman–Wald approach and Fisher's views on learning by experience. This is the Bayes approach which requires not only the frequentist's specification of a family of possible distributions for the observation, but also that the distribution of the observation has itself been drawn from the specified model according to a known probability mechanism. This is coupled with the subjective point of view of probability in which the chance of an event represents a quantitative measure of the observers' odds on the occurrence or nonoccurrence of the event. This point of view, though attractive, puts a burden on the observer to specify his or her state of mind with possibly unreasonable precision and, of course, leaves open the issue of the generalizability of the conclusions by scientists at large (see also *Bayesian Statistics*).

The problems of inference have been the subject of much discourse among philosophers as well as scientists whose primary interest was not statistics. We refer to Keynes (1921), von Mises (1928) and Jeffreys (1939) among others.

A problem in discussing frequentist inference is that, while distinctions such as those between Fisher, Neyman, and Wald are required in order to formulate the goals of theory, they are artificial in practice. Even in scientific inquiry the cost of initiating extensive further investigation on the basis of weak evidence has to be kept in mind. On the other hand, the most interesting consequences may not be foreseen and the costs of actions difficult to quantify. Is foisting a slightly better-than-average drug with serious side effects on most patients better than doing nothing? What if the drug turns out to have major benefits for a small group of patients not represented in the original clinical trial of the drug? A chemist, E. B. Wilson (1952), after considering the work of the thinkers we mention, pleads eloquently, 'There is a great need for further work on the subject of scientific inference. To be fruitful it should be carried out by critical original minds who are not only well-versed in philosophy but also familiar with the way scientists actually work (and not just with the way some of them say they work).'

Wilson concludes pessimistically: 'Unfortunately the practical nonexistence of such people almost suggests that the qualities of mind required by a good philosopher and those needed by a working scientist are incompatible.'

In what follows we

(a) begin, for illustrative purposes, by describing a few simple models that have arisen from a frequentist point of view (Sect. 2), and

(b) discuss by example the principal frequentist inference procedures (hypothesis testing, point estimation, confidence regions, prediction, and model selection) (Sect. 3).

We use the term frequentist inference if all probabilities involved (such as significance level, power, and coefficient) are interpreted as frequentist probabilities in the sense of the article (*Frequentist Interpretation of Probability*). In particular, the assertion that the value of such a probability is p then means that in a large number of independent cases with probability p of some outcome, we expect the outcome to occur with a frequency close to p. In this connection it is important to realize that there is no need for these situations to be at all like each other, as long as the probability p is the same in each (see Neyman 1937).

2. *Some Simple Models*

Frequentist models are characterized by two principal features.

(a) The variability of the data. If the observations were repeated, different values would be obtained. This variability is represented in the model by postulating that the data are generated by a probability distribution.

(b) The unknown aspect of the situation which the statistical investigation is to elucidate. This unknown feature is represented in the model by the fact that the probability distribution of the observations is assumed to be only partially known.

These two features are seen clearly in the following model, which was mentioned briefly in the preceding section.

Example 1. (Error-measurement model) If $X_1, \ldots, X_n$ denote n measurements of an unknown quantity θ, this model can be written as

$$X_i = \theta + \varepsilon_i. \tag{1}$$

Here θ represents the unknown quantity we wish to estimate or test. The errors ε_i are the source of the variability of the observations. This frequency model assumes that, in a replication of the situation, the unknown value of θ would be the same but that the ε's would take on different values. The most common assumption concerning their variability is that they are independent random variables with a common distribution with mean zero. This distribution may be assumed to be known. More usually it too embodies some unknown features; for example, its form may be known but it may contain an unknown scale parameter.

Simpson and others toward the middle of the eighteenth century considered various possible forms for this distribution; but by the early nineteenth century it was agreed generally that the most suitable distribution of such measurement errors, in most cases, was the normal. The basis for this belief was the hypothesis of 'elementary errors,' which assumes that observational errors are the sum of a large number of small, elementary errors; and, by the Central Limit Theorem, are therefore approximately normally distributed.

If the ε's in (1) are assumed to be independently normally distributed with the mean 0 and variance σ^2, the model (1) is equivalent to assuming that

$$X_1, \ldots, X_n \text{ are independent } N(\theta, \sigma^2). \tag{2}$$

Example 2 (Normal one-sample model) Model (2) arises also in contexts quite different from the measurement situation described above. Suppose X is some numerical characteristic of a person such as height, weight, blood pressure, intelligence, etc., or of an animal, plant or manufactured product. However, this time we are not taking several observations X_i on the same subject but one observation each on different subjects drawn at random from a population which, in the model, we shall assume to be infinitely large. In this setting θ is the mean value of X in the population, and (2) is called the normal one-sample model. (The measurement situation can be viewed as a special case if one considers the n measurements actually taken as a sample from an essentially infinite number of measurements that could be taken.)

In these applications the assumption of normality is frequently made even though it is often not suitable and each situation ought to be considered in its own right. Often normality can be brought closer by applying it not to the observations X_i themselves but to some transformation $T(X_i)$. In other cases the normal model (2) is replaced by the more general model

$$X_1, \ldots, X_n \text{ are independently distributed according to } F(x-\theta) \tag{3}$$

where F has mean zero and otherwise is arbitrary, or where it is assumed to be an arbitrary distribution that is symmetric about zero. Some smoothness conditions may also be imposed; for example, that F has a density f.

Example 3. (Linear Model) Large areas of statistical methodology are based on the following extension of (1), due to Gauss (1809) and known as the linear model

$$X_i = \sum_{j=1}^{p} \beta_j z_{ij} + \varepsilon_i \quad (i = 1, \ldots, n) \tag{4}$$

where the z's are known constants (sometimes called explanatory variables) and the β's unknown parameters. The ε's are the observational errors as before and are assumed to have expectation zero.

Examples of (4) abound. Here are a few.

(a) The measurement model (1) is the special case of (4) with $p = 1$, $z_{i2} = 1$ and $\beta_1 = \theta$.

(b) The two-sample problem with $n_1 + n_2$ independent variables $X_1, \ldots, X_{n_1} : N(\theta_1, \sigma^2)$ and $X_{n_1+1}, \ldots, X_{n_1+n_2} : N(\theta_2, \sigma^2)$ corresponds to $p = 2$,

$$z_{11} = \cdots = z_{1_{n1}} = 1, \quad z_{1n_1+1} = \cdots = z_{1n_1+n_2} = 0, \quad \beta_1 = \theta_1 \tag{5}$$

and

$$z_{21} = \cdots = z_{2n_1} = 0, \quad z_{2n_1+1} = \cdots = z_{2n_1+n_2} = 1, \quad \beta_2 = \theta_2 \tag{6}$$

(c) The k-sample model is defined analogously. The two-way layout without interactions assumes that the variables X_{ijk} $(k = 1, \ldots, n_{ij}; i = 1, \ldots, a; j = 1, \ldots, b)$ are independent normal with common variance σ^2 and with means

$$E(X_{ijk}) = \mu + \alpha_i + \beta_j (\Sigma\alpha_i = \Sigma\beta_j = 0). \tag{7}$$

The k-sample model and the two-way layout are the simplest cases of analysis of variance models.

Another important class of examples of (4) are models for regression. The simplest case is

(d) Simple linear regression, given by

$$X_i = \alpha + \beta t_i + \varepsilon_i \tag{8}$$

where the t's are known constants, and α and β unknown parameters.

This is a special case of

(e) Polynomial regression, with

$$X_i = f(t_i) + \varepsilon_i \tag{9}$$

where f is a polynomial (for example, quadratic) of known degree and with unknown coefficients.

Other functions f, of course, are also possible.

A modification of (4) that is sometimes suitable is to have some of the z's random. For example (7) in some situations is replaced by

$$X_{ijk} = \mu + A_i + \beta_j + \varepsilon_{ijk} (\Sigma\beta_j = 0) \tag{10}$$

where the A's are unobservable random variables assumed to be independent $N(0, \sigma_A^2)$ and independent of ε. For more, see *Linear Hypothesis*.

When the distribution of the observational errors the ε's in (1) and (4) is specified, the models considered so far are parametric, that is, they can be smoothly parameterized by Euclidean labels. Models such as (3) with the distributional form of errors unspecified up to symmetry are semiparametric. Nonparametric models are ones in which as little is assumed as possible.

Example 4. (Nonparametric regression) This model is suitable in situations where, on each member of a sample of size n from some population, we observe not only a response Y but also observe characteristics (covariates) $\mathbf{Z} = (Z_1, \ldots, Z_k)$. For example, Y might be the income of a randomly sampled individual, Z_1 might be age, Z_2 educational level, etc. Of principal interest here is the average relation between Y and $\mathbf{Z}$

$$E(Y|Z_1 = z_1, \ldots, Z_k = z_k) \equiv m(z_1, \ldots, z_k). \tag{11}$$

If we assume nothing about the form of m, then subject to the assumptions that $E|Y| < \infty$ and that $(Y_1, \mathbf{Z}_1), \ldots, (Y_n, \mathbf{Z}_n)$ are independent and identically distributed, this model is completely general since we can always write

$$Y_i = m(Z_{i1}, \ldots, Z_{ik}) + \varepsilon_i \tag{12}$$

where $E(\varepsilon_i|Z_i) = 0$. Note that if we assume

$$m(z_1, \ldots, z_k) = \sum_{j=1}^{k} \beta_j z_j,$$

then, given $\mathbf{Z}_i = z_i$ $(i = 1, \ldots, n)$ we are back in the linear model (4).

In order to identify m in (12) we need to make some assumptions, for instance that m is continuous.

An important method for constructing more complex models from simpler ones is that of hierarchical models. As exemplified in (10), such models often contain unobservable random variables which are themselves of interest.

From this point of view Bayes models can be considered as two-stage hierarchical models in which the first stage is a frequentist model involving unobservable parameters for which the Bayes model specifies a distribution as a second stage. If the frequentist model is itself hierarchical, a last stage is added when the model is Bayesianized. To this difference—that the unknown parameters of the frequentist model become unobservable random variables in the corresponding Bayes model of course, is added the difference in interpretation of the probabilities involved as frequentist or subjective.

3. *Inference Methods*

As indicated in Sect. 1 linked closely with the frequentist paradigm is that of decision theory. The principal types of problems considered in that framework are

(a) hypothesis testing,
(b) point estimation,
(c) confidence regions,

(d) prediction,
(e) model selection.
We shall discuss these in the above order.

3.1 Hypothesis Testing

Wald's decision theory grew out of the Neyman–Pearson theory of hypothesis testing (Neyman and Pearson 1933) which the authors formulated as choosing between two decisions, whether a prespecified statement (the hypothesis H) about the distribution P generating the data is or is not correct. If $\theta \in \Theta$ parameterizes the model, this is equivalent to determining whether $H: \theta \in \Theta_0$ or the alternative $K: \theta \in \Theta_1$ is true.

This formulation is satisfactory in many practical contexts such as occur, for example, in medicine, agriculture, or industry. For instance, in Example 3(b) the two samples may correspond to control and treatment groups of patients. In Example 3(b) the assumption of normality is made, and implicitly that of an additive treatment effect measured by the difference $\theta_2 - \theta_1$. The hypothesis to be tested is usually formulated as $H: \theta_2 \leqslant \theta_1$; i.e., that the treatment has no beneficial effect. The alternative $K: \theta_2 > \theta_1$ claims the existence of a beneficial effect.

The test for this problem is a test function δ which takes on the values $\delta(x) = 0$ or $\delta(x) = 1$ for the sample points x for which we decide to accept or reject H. The performance of such a test is measured by the probabilities of

(a) Type I errors—rejecting H when H is true, and
(b) Type II errors—accepting H when H is false,
that is, by the probabilities

$$P_\theta[\delta(x) = 1] \text{ when } \theta \in \Theta_0 \quad \text{and}$$
$$P_\theta[\delta(x) = 0] \text{ when } \theta \in \Theta_1. \tag{13}$$

The test δ is required to satisfy the condition

$$P_\theta(\text{Type I error}) \leqslant \alpha \quad \text{for all } \theta \in \Theta_0 \tag{14}$$

where the level α of the test is preassigned (usually as 0.05 or 0.01) as the maximum probability of false rejection to be tolerated. Subject to (14) we look for tests δ which make P_θ (Type II error) as small as possible.

This formulation is meaningful if it is possible to distinguish sharply between hypothesis and alternatives, between those θ-values for which rejection of H is clearly desirable or undesirable. In Example 3(b), for instance, we can specify that only beneficial effects exceeding $\varepsilon\sigma$ (for some given $\varepsilon > 0$) are of interest. Such a distinction is often possible in the contexts we have mentioned.

The Neyman–Pearson formulation of hypothesis testing in terms of the two types of error works well when we have clear definitions of both the hypothesis and the alternatives. In scientific contexts it is often the case that the hypothesis is clear but the alternatives are vague. It is then the unlikelihood of observations that would be surprising if the hypothesis were true (but less so if the vaguely entertained alternatives were valid) which takes on particular significance. It is still necessary to define a test statistic T and to suppose the hypothesis is rejected when its values are sufficiently large, since such values would be surprising if H were true but not if H were false. Then the probability of observing values greater than or equal to the observed value of T is the p-value or significance probability of the hypothesis. If it is sufficiently small, it draws attention to anomalies not expected under H. The calculation of such probabilities is central to science. However, whether such attention should be followed by action as assumed in the Neyman–Pearson theory depends on extraneous considerations.

The Bayesian approach to testing assigns prior probabilities to hypothesis and alternative and calculates the ratio of posterior probabilities, the Bayes factor, as a measure of evidence in favor of or against the hypothesis rather than the significant probability. This approach is very attractive in providing exactly what everyone wants: the probability of the hypothesis being true, given the data. However, this probability depends on the prior probability distribution assigned to the parameters; and these priors can overwhelm strong opposite data evidence. Disparate priors can result in conflicting conclusions based on the same data by analysts holding different prior opinions (see also *Hypothesis Testing in Statistics*).

3.2 Point Estimation

The decision-oriented point of view of Neyman and Pearson led to the full-fledged decision theory of Wald, based on assigning losses to the consequences of inappropriate decisions. Perhaps the most successful example of this approach is point estimation (for which the idea of loss functions had already been discussed by Gauss and Laplace).

The prototypical estimation problem is that of estimating an unknown physical constant subject to measurement error discussed in Example 1. The loss function most commonly considered is squared error leading to its expected value, the mean squared error (MSE), as its risk. The standard assumption of independent normal errors with mean zero and common variance implies that the sample mean $\bar{X}$ is the appropriate estimator according to a number of different criteria.

(a) Minimax. This is a worst case analysis in which the maximum risk is minimized.

(b) Uniform minimum variance unbiased. An estimator of θ is said to be unbiased if its expectation is equal to θ. For unbiased estimators the MSE is equal to the variance; and, in the present case, $\bar{X}$ minimizes the variance among all unbiased estimators.

(c) Maximum likelihood (ML). This is a non-decision-theoretic approach based on the likelihood the density of the data viewed as a function of the parameter. Fisher considered the likelihood of an observation as a measure of support for the different θ-values that might have generated it and proposed the θ-value maximizing it as an appropriate estimator of θ. When a uniform prior distribution can be placed on the parameter, the ML estimator is just the mode of the posterior distribution. However, when θ ranges over the infinite line (as is the case in the measurement model and many others), no uniform probability distribution for θ exists; and only an 'improper Bayes' interpretation is then possible.

The ML principle is popular for a number of reasons

(a) general applicability,

(b) invariance under reparameterization,

(c) in cases such as Example 2, where meaning can be attached to 'large sample sizes,' the method is approximately optimal,

(d) in many standard cases it leads to explicit solutions.

However, there are examples in which the resulting estimators are completely misleading, (LeCam 1990, Ferguson 1996). In particular, ML is either not applicable or can lead to very poor results in many non- and semiparametric models such as Example 4. For a general discussion of different approaches to estimation see, for example, Bickel and Doksum (2000), Lechmann and Casella (1998).

The assumption of joint normality of the data, for instance in Examples 1–3, is often seriously in doubt. In such cases robustness criteria are invoked and lead to alternatives to the linear procedures appropriate for normally distributed data. For example, the median rather than the mean (or some intermediate trimmed mean) might be used in Example 1, and minimum L_1 (least absolute values) rather than L_2 (least squares) estimators in Example 3. For an extended discussion of robustness see Hampel et al. (1986) and Staudte and Sheather (1990).

3.3 Bayes Estimation

Still another principle is that of Bayes estimation, i.e., minimizing the expected risk under a prior distribution π of the parameter $\boldsymbol{\theta}$. Since this is the same as minimizing the average risk, averaged with respect to the weight-function π, such estimators are of interest both from a Bayesian and a frequentist point of view. For estimating a real-valued parameter $g(\boldsymbol{\theta})$ with data X and squared error loss, the Bayes estimator is

$$E[g(\boldsymbol{\theta})|X] \tag{15}$$

the mean of the posterior distribution of $g(\boldsymbol{\theta})$ given X.

Bayes estimators are essentially never unbiased (Girshick and Savage 1951) since they naturally 'pull' the estimator toward one's prior opinion, the prior mean of $g(\boldsymbol{\theta})$. For example, in the one-sample model of Example 2. If π is normal (μ, τ^2), the Bayes estimator is

$$\left(\frac{n\bar{X}}{\sigma^2}+\frac{\mu}{\tau^2}\right)\Big/\left(\frac{n}{\sigma^2}+\frac{1}{\tau^2}\right). \tag{16}$$

Frequentist and Bayes inference generally merge for large sample sizes since the data then wash out the influence of the prior distribution (provided the prior does not exclude parts of the sample space); see, for example, Blackwell and Dubins (1962).

In estimation even more can be said. If the prior distribution is locally uniform, Bayes estimators are asymptotically optimal in a frequentist sense by the Bernstein–von Mises theorem (see, for example, Lehmann and Casella 1998, Bickel and Doksum 2000). This is, however, a limit result; in practice, the effect of a strong prior often persists even for substantial sample sizes.

Both aspects can be illustrated by the Bayes estimator (16). For fixed μ, the difference of this estimator from $\bar{X}$ is of the order $1/n\tau^2$. For fixed τ^2, it is accordingly of the order $1/n$, while the difference of both estimators from θ is of the order $1/\sqrt{n}$. On the other hand, the convergence of (16) can become arbitrarily slow if $\tau^2 \to 0$ as $n \to \infty$. The basic message is that optimal frequentist estimators are appropriate for Bayesian problems and vice versa.

For the sake of simplicity, we have here focused attention to the estimation of real-valued parameters. Of course, estimation problems arise also for multivariate parameters (for instance, in Example 3) and for function-valued parameters such as the function m in Example 4 or in the estimation of an unknown density. Squared error loss is then replaced by summed or integrated MSEs. However, the univariate and multivariate situations present unexpected differences. For example, the typically unique minimax estimator of the univariate case is replaced by an infinite set of minimax estimators; and among these the natural extension of the classical univariate estimator is no longer optimal (Stein 1956).

Point estimates by themselves are hardly ever sufficient. Typically one also requires an idea of the error committed in using the estimator; that is, an estimate of the MSE. This is again the problem of estimating a parameter, and there is little to add. However, interpretation of the estimated measure of error is often difficult. It is clarified by the concept of confidence estimation, which we take up next.

3.4 Confidence Regions

When estimating θ in the measurement error model of Example 1, it is customary to indicate the reliability of the estimate $\bar{X}$ by attaching to it the error estimate $\pm S/\sqrt{n}$ with $S^2 = \Sigma(X_i - \bar{X})^2/(n-1)$ as a proxy for $\pm\sigma/\sqrt{n}$. In the case of normal errors with mean zero.

$$P_\theta\left[\bar{X} - \frac{\sigma}{\sqrt{n}} \leqslant \theta \leqslant \bar{X} + \frac{\sigma}{\sqrt{n}}\right] = 0.67 \quad \text{for all } \theta. \tag{17}$$

If σ is known, (17) provides intervals of fixed length (and with random midpoints) which, in many repetitions of the experiment of taking n measurements, will 'cover' the true θ 67 percent of the time whatever are the true values of θ and σ. The intervals (17) are confidence intervals for θ with confidence coefficient 0.67.

In practice, σ is usually unknown. We can then obtain confidence intervals for θ by replacing σ by its estimator S. If t_α is the $100(1-\alpha/2)$ percent point of the t-distribution with $n-1$ degrees of freedom, the intervals

$$\bar{X} - t_\alpha \frac{S}{\sqrt{n}} \leqslant \theta \leqslant \bar{X} + t_\alpha \frac{S}{\sqrt{n}} \tag{18}$$

cover the true θ with probability $1-\alpha$ so that the statement (18) is correct about $100(1-\alpha)$ percent of the time. The length of the intervals is no longer fixed but random, its length tending to increase as the accuracy of the observations (the inverse of which is measured by σ and estimated by S) decreases.

In general, in estimating a parameter θ taking values in an arbitrary space Θ, a $100\ (1-\alpha)$ percent confidence region for θ is a random subset $C(X)$ of Θ, depending on the data with the property that

$$P[\theta \in C(X)] \geqslant 1-\alpha \tag{19}$$

for all probability distribution P in the model. Here the region $C(X)$ can, for example, be an interval as above, an ellipse as in the Scheffé regions for the parameters of the linear model in Example 3, or a 'band' about an unknown distribution function.

Confidence regions can be viewed as simultaneously making statements about the acceptability of a family of hypotheses (see Lehmann 1986). In particular, they provide a measure of the acceptability of the alternatives. A 95 percent confidence interval for the treatment effect of Example 3(b), for instance, which includes zero, tells us not only that the null hypothesis is accepted at the 5 percent level but also specifies all alternatives to the null hypothesis which are accepted at that level. An interval which does not contain zero but to which zero is close suggests that although we have seen something that is surprising under the null hypothesis, the data are consistent with alternatives that do not differ materially from zero.

The interpretation of confidence regions is conceptually difficult because the probability statements refer not to any random variation of the parameter, which, although unknown, is considered fixed, but to that of the data-dependent region. In contrast, a Bayesian analysis views θ as random and constructs credible regions $C^*(X)$ such that the posterior probability of θ falling into $C^*(X)$ is $\geqslant 1-\alpha$. This is just the kind of conclusion one would like to make; but, as in the case of testing, tends to be strongly influenced by assumptions concerning the prior.

Again, as in point estimation, Bayes regions derived on the basis of optimality criteria of size (length, volume, etc.) for large sample sizes agree approximately with frequentist confidence regions satisfying minimax size properties subject to fixed probability of coverage (see also *Estimation: Point and Interval*).

3.5 Prediction

This is the part of frequentist inference that is closest to the Bayesian approach and that fits least well into the decision-theoretic framework. Typically we are given a sample from a population of the form $(\mathbf{Z}_i, Y_i)$, $i = 1, \ldots, n$ where $\mathbf{Z}_i$ is a vector of predictors. Using this sample (which provides information concerning the relationship of $\mathbf{Z}$ and Y) and a new $\mathbf{Z}_{n+1}$ we wish to predict the unobserved Y_{n+1}. The predictor of Y_{n+1} is a function

$$\delta(\mathbf{Z}; \mathbf{Z}_1, Y_1, \ldots, \mathbf{Z}_n, Y_n) = \hat{\delta}(\mathbf{Z})$$

such that $\hat{\delta}(\mathbf{Z}_{n+1})$ is used to predict Y_{n+1}. In the continuous case (regression), the classical measure of loss resulting from incorrect prediction is squared error,

$$[Y_{n+1} - \hat{\delta}(Z_{n+1})]^2.$$

On the other hand, in the case of categorical variables taking on values $\{1, \ldots, k\}$ (classification), the standard loss function takes on the value 0 if $Y_{n+1} = \hat{\delta}(\mathbf{Z}_{n+1})$ and 1 otherwise. Prediction has the attractive feature that it allows nonparametric estimates of error since for any procedure we can compute its performance by applying it to the training sample on which it was built, for instance, $(1/n)\Sigma_{i=1}^{n}(\hat{\delta}(Z_i) - Y_i)^2$. This measure typically underestimates error, but techniques such as cross validation can be used to adjust it.

3.6 Model Selection

An important aspect of a frequentist analysis is the specification of a model. As an illustration, consider Example 3(e) where the regression function is assumed to be well approximated by a polynomial. Suppose the errors are normally distributed with zero mean and common variance. To complete the specification of the model, it is then necessary to decide on the degree d of the polynomial. With classical (least squares) prediction methods, the variance of the prediction error increases and its bias decreases with d. On a frequentist

basis the data can then be used to select d so as to obtain a desirable balance between these two aspects. Criteria for selecting d (and making analogous choices within a family of proposed models in other problems) have been developed by Akaike, Mallows, Schwarz, and others (for more details, see, for example, Linhart and Zucchini 1986):

Bayesian solutions to this problem differ in a number of ways, reflecting whether we assume that the true model belongs to the hypothesized class of models (e.g., is really a polynomial) or can merely be approximated arbitrarily closely by such models. For more on this topic see Shao (1997).

See also: Estimation: Point and Interval; Robustness in Statistics.

Bibliography

Bickel P, Doksum K 2000 *Mathematical Statistics*, 2nd edn. Prentice Hall, New York, Vol. 1

Blackwell D, Dubins L 1962 Merging of opinions with increasing information. *Annals of Mathematical Statistics* **33**: 882–6

Ferguson T S 1996 *A Course in Large Sample Theory*. Chapman & Hall, London

Fisher R A 1922 On the mathematical foundations of theoretical statistics. *Philosophical Transactions of the Royal Society of London, Series A* **222**: 309–68

Fisher R A 1973 *Statistical Methods and Scientific Inference*, 3rd edn. Hafner Press, New York

Gauss C F 1809 *Theoria Motus Corporum Celestium*. Perthes, Hamburg, Germany

Girshick M A, Savage L J 1951 *Bayes and Minimax Estimates for Quadratic Loss Functions*. *Proceedings of the 2nd Berkeley Symposium of Mathematics, Statistics, and Probability*. University of California Press, Berkeley, CA

Hampel F, Ronchetti E, Rousseeuw P, Stahel W 1986 *Robust Statistics*. Wiley, New York

Jeffreys H 1939 *Theory of Probability*. Clarendon Press, Oxford, UK

Keynes J M 1921 *A Treatise on Probability*. Macmillan, London

LeCam L 1990 Maximum likelihood: An introduction. *International Statistical Review* **58**: 153–71

Lehmann E L 1986 *Testing Statistical Hypotheses*, 2nd edn. Springer, New York

Lehmann E L, Casella G 1998 *Theory of Point Estimation*, 2nd edn. Springer, New York

Linhart H, Zucchini W 1986 *Model Selection*. Wiley, New York

Neyman J 1937 Outline of a theory of statistical estimation based on the classical theory of probability. *Philosophical Transactions of the Royal Society Series A* **236**: 333–80

Neyman J 1961 Silver Jubilee of my dispute with Fisher. *Journal of the Operations Research Society of Japan* **3**: 145–54

Neyman J, Pearson E S 1933 On the problem of the most efficient tests of statistical hypothesis. *Philosophical Transactions of the Royal Society Series A* **231**: 289–337

Shao J 1997 An asymptotic theory for linear model selection (with discussion). *Statistica Sinica* **7**: 221–66

Staudte R G, Sheather S J 1990 *Robust Estimation and Testing*. Wiley, New York

Stein C 1956 Inadmissibility of the usual estimator for the mean of a multivariate distribution. *Proceedings of the 3rd Berkeley Symposium of Mathematics, Statistics, and Probability*. University of California Press, Vol. 1, pp. 187–95

Tukey J W 1960 Conclusions versus decisions. *Technology* **2**: 423–33

von Mises R 1928 *Wahrscheinlichkeit, Statistik and Wahrheit*. Springer, Wien, Austria

Wald A 1950 *Statistical Decision Functions*. Wiley, New York

Wilson E B 1952 *An Introduction to Scientific Research*. McGraw Hill, New York

P. J. Bickel and E. L. Lehmann

Frequentist Interpretation of Probability

If the outcome of an event is observed in a large number of independent repetitions of the event under roughly the same conditions, then it is a fact of the real world that the frequency of the outcome stabilizes as the number of repetitions increases. If another long sequence of such events is observed, the frequency of the outcome will typically be approximately the same as it was in the first sequence.

Unfortunately, the real world is not very tidy. For this reason it was necessary in the above statement to insert several weasel words. The use of 'roughly the same,' 'typically,' 'approximately,' and 'long sequence' make it clear that the stability phenomenon being described cannot be stated very precisely. A much clearer statement is possible within a mathematical model of this phenomenon. This discovery is due to Jacob Bernoulli who raised the following question.

It is well known, Bernoulli says in Part IV of his book *Ars Conjectandi* (published posthumously in 1713), that the degree to which the frequency of an observed event varies about the probability of the event decreases as the number of events increases. He goes on to say that an important question that has never been asked before concerns the behavior of this variability as the number n of events increases indefinitely. He envisages two possibilities.

(a) As n gets larger and larger, the variability eventually shrinks to zero, so that for sufficiently large n the frequency will essentially pinpoint the probability p of the outcome.

(b) Alternatively, it is conceivable that there is a positive lower bound below which the vari ability can never fall so that p will always be surrounded by a cloud of uncertainty, no matter how large a number of events we observe.

Bernoulli then proceeds to prove the law of large numbers, which shows that it is (a) rather than (b) that pertains. More precisely, he proves that for any $a > 0$

$$\Pr\left(\left|\frac{X}{n}-p\right| < a\right) \to 1 \quad \text{as} \quad n \to \infty \qquad (1)$$

J. Rojo (ed.), *Selected Works of E. L. Lehmann*, Selected Works in Probability and Statistics, DOI 10.1007/978-1-4614-1412-4_92, © Springer Science+Business Media, LLC 2012

where X/n is the frequency under consideration.

It is easy to be misled into the belief that this theorem proves something about the behavior of frequencies in the real world. It does not. The result is only concerned with properties of the mathematical model. What it does show is that the behavior of frequencies in the model is mirrored in a way that is much neater and more precise, the very imprecise stability phenomenon stated in the first paragraph of this article.

In fact, a result for the model much more precise than (1) was obtained by De Moivre (1733). It gives the normal approximation

$$\Pr\left(\left|\frac{X}{n}-p\right| < c/\sqrt{n}\right) \to \int_{-c/\sqrt{pq}}^{c/\sqrt{pq}} \varphi(x)\,dx \qquad (2)$$

where φ denotes the standard normal density. This is again a theorem in the model. The approximate corresponding real-world phenomenon can be seen, for example, by observing a quincunx, a mechanical device using balls falling through 'random' paths to generate a histogram.

De Moivre's result was given a far reaching generalization by Laplace (1810) in the central limit theorem (CLT) concerning the behavior of the average $\bar{X}$ of n identically, independently distributed random variable $X_1, \ldots, X_n$ with mean ξ and finite variance σ^2. It shows that

$$\Pr\left(|\bar{X}-\xi| < c/\sqrt{n} \right) \to \int_{-c/\sigma}^{c/\sigma} \varphi(x)\,dx. \qquad (3)$$

This reduces to (2) when X takes on the values of 1 and 0 with probabilities p and q, respectively. The CLT formed the basis of most frequentist inference throughout the nineteenth century.

The first systematic discussion of the frequentist approach was given by Venn (1866), and an axiomatization based on frequencies in infinite random sequences (*Kollectives*) was attempted by von Mises (1928). Because of technical difficulties his concept of a random sequence was modified by Solomonoff (1964), Martin-Löf (1966), and Kolmogorov (1968), with the introduction of computational complexity. (An entirely different axiomatization based on events and their probabilities rather than random sequences was put forward by Kolmogorov in 1933, and has successfully served as a basis for both frequentist and subjective interpretations of probability). For more details on these different approaches see Barnett (1982).

The frequentist concept of probability described so far has met considerable criticism. One of the main objections is that it is not applicable to many situations to which one might want to apply probability assessments. To see this, consider the following three possibilities.

(a) An actual sequence of repetitions may be available; for example, a sequence of coin tosses or a sequence of independent measurements of the same quantity.

(b) A sequence of repetitions may be available in principle but not likely to be carried out in practice; for example, the polio experiment of 1954 involving a sample of over a million children.

(c) A unique event which by its very nature can never be replicated, such as the outcome of a particular historical event; for example, whether a particular president will survive an impeachment trial. The conditions of this experiment cannot be duplicated.

The frequentist concept of probability can be applied in cases (a) and (b) but not in the third situation. An alternative approach to probability which is applicable in all cases is the notion of probability as degree of belief; i.e., of a state of mind (for a discussion of this approach, see Robert (1994)). The inference methods based on these two interpretations of the meaning of probability are called frequentist and Bayesian, respectively.

Although frequentist probability is considered objective, it has the following subjective feature. Its impact on a particular person will differ from one person to another. One patient facing a surgical procedure with a 1 percent mortality rate will consider this a dire prospect and emphasize the possibility of a fatal outcome. Another will shrug it off as so rare as not to be worth worrying about.

There exists a class of situations in which both approaches will lead to the same probability assessment. Suppose there is complete symmetry between the various outcomes; for example, in random sampling which is performed so that the drawing favors no sample over another. Then we expect the frequencies of the various outcomes to be roughly the same and will also, in our beliefs, assign the same probability to each of them.

Let us now turn to a second criticism of frequentist probability. This concerns the difficulty of specifying what is meant by a repetition in the first sentence of this section. Consider once more the surgical procedure with 1 percent fatalities. This figure may represent the experience of thousands of cases, with the operation performed by different surgeons in different hospitals and—of course—on different patients. The rate of fatalities may vary from one hospital or surgeon to another and may, in particular, vary drastically with the condition, for example, the age and general health, of the patient.

Suppose a young woman requires this operation although her general health is very good. The frequency of a fatal outcome with patients sharing these characteristics may be much lower, and the 1 percent figure in that sense would be quite misleading for her. And yet she might be considered to have been obtained under 'roughly the same conditions,' namely to be drawn at random from the total population of persons requiring this surgery. To obtain the most useful figures one should identify the most important variables, classify the cases accordingly (for example, young, middle-aged, old; male, female; etc.) and then provide the frequency for each class. They will, of course, be meaningful only for the classes which contain a reasonable number of cases.

1. A Terminology Note

A source of much confusion in the discussion of probability is the fact that 'probability' is used both as a mathematical term, i.e., as a concept in the mathematical model and also in everyday language when talking about real events. When reading about probability, it is important to be aware of these two meanings and to keep them distinct.

See also: Frequentist Inference; Probability: Formal; Probability: Interpretations; Statistical Methods, History of: Post-1900; Statistical Methods, History of: Pre-1900

Bibliography

Barnett V 1982 *Comparative Statistical Inference*, 2nd edn. Wiley, New York

De Moivre A 1733 1738, 1756. In: *Doctrine of Chance*, Millar, London, Reprinted by Chelsea Publishers, New York (1967), pp. 235–43 (1738), 243–54 (1754) 1878–1912

Kolmogorov A 1933 *Grundbegriffe der Wahrscheinlichkeitsrechnung*. Ergebnisse der Mathematik, Vol. 2. Springer, Berlin

Kolmogorov A 1968 Logical basis for information theory and probability theory. *IEEE Transactions on Information Theory* **14**: 662–4

Laplace P S *Oevres complètes de Laplace*. Gauthier-Villars, Paris

Martin-Löf P 1966 The definition of random sequences. *Information and Control* **9**: 602–19

Robert C P 1994 *The Bayesian Choice*. Springer, New York

Solomonoff R J 1964 The formal theory of inductive inference. *Information and Control* **7**: 224–54

Venn J 1866 *The Logic of Change*. Reprinted by Chelsea Publishers, New York, (1967)

von Mises R 1957 *Probability, Statistics and Truth*. Macmillan, New York

P. J. Bickel and E. L. Lehmann

Freud, Sigmund (1856–1939)

1. Family Background and Early Life

Sigmund Freud, neuroscientist turned founder of psychoanalysis, was born on May 6, 1856, in Freiberg, Moravia, now part of the Czech Republic. His father, Jakob (1815–96), a wool merchant, came from Galicia, and married Amalie (1835–1930, neé Nathansohn), 20 years younger, when he was 40. Jakob had two sons by a previous marriage, Emanuel and Philipp. Jakob was warm and affectionate, with a strong sense of humor, and Freud evidently admired and respected him. Amalie was a lively and attractive woman who was proud of Sigmund, her first-born.

The family multiplied quickly: five daughters and two more sons were born within 10 years. The first of these, Julius, died when 8 months old and Sigmund was aged 19 months. Both parents were Jews, with no religious affiliations, though a devout Catholic Nanny took Sigmund to Church. She left when he was two and a half. A nephew, John, a year older than Freud was very close to him: the two fought and played together. That childhood ambivalence, Freud reflected years later, must have had a profound effect on his character and on his ability to defend himself (Freud 1900).

Financially, times were hard. In 1859 the family moved to Leipzig, and in the following year to Vienna. Freud missed the beauty of the countryside round Freiberg. Although the children were well looked after, poverty was not assuaged by rapid family growth. By 1875 family fortunes improved: Amelie's family were helping, making possible a move to a larger house. Freud had a room of his own that served as a study.

Freud's memories of his childhood as revealed in his writings are without early detail, but one unforgettable recollection was of urinating in his parents' bedroom while they were present. Reprimanding him, his father said bluntly: 'The boy will come to nothing!' Freud's wounded ambitions were reflected recurrently in dreams in which he listed his achievements as if to say: 'You see! I have come to something!' (Freud 1900; see also Shengold 1993).

2. Education and Early Interests

Freud's schooling began with his mother, until his father took over before sending him to a private school. Freud learned rapidly, and was reading Shakespeare (whom he loved all his life) from the age of eight years. When nine years of age, he won a place at the Sperl Gymnasium, where he was a distinguished pupil, passing out, at 17, with the distinction '*summa cum laude*.' He became an accomplished linguist, with a firm foundation in Latin and Greek, a sound knowledge of Hebrew, fluency in English and French, and good Italian. He corresponded with an adolescent friend in Spanish. He became one of the greatest stylists in the German language (Kaufmann 1980). He studied at every opportunity. He loved the Arts, but took an early dislike to music, which he never felt able to appreciate. He was fond of walking, swimming, and skating. He liked travel, and made his first visit to England at the age of 19.

Part XIII
Books

The Books

Javier Rojo

As a young man, Eric Leo Lehmann had a desire to become a writer. In Lehmann (2008), he writes that "*My passion was German literature, my dream to become a writer, perhaps another Thomas Mann or Gottfried Keller.*" But the Nazis had come to power in 1933, and his father had decided to take the family to Zürich. Now, two years later, Erich's father advises him to consider other options as it will be impossible to return to Germany. Perhaps mathematics, with a universal character, would be a good choice. It happened that Edmund Landau's wife was a good friend of Erich's mother. Erich's father thought that Landau should be consulted on Erich's mathematical ability. For this purpose, Landau was invited to visit in his next trip to Zürich. After a two hour session with Erich, Landau recommended to his parents that mathematics could be an option. Realizing that Switzerland was not a good place for a German refugee, and seeking and then following Landau's advice once more, Erich went to England and enrolled in Trinity College. After two years, and with a letter of introduction for Courant from Landau's wife – Landau had passed away earlier from a heart attach, Erich arrived in New York. After being asked if he wanted to live in the United States or in New York, and having preferred the former, Courant recommends Berkeley as "an up-and-coming university" that will be a good place for Erich. And so it happened that Erich Leo Lehmann arrived in Berkeley in 1941.

Griffith C. Evans, who had been recruited from the Rice Institute – now Rice University –, was the chair of the Mathematics Department at Berkeley. Evans moved to Berkeley to build up the Mathematics Department which was in a bit of a disarray. Evans had a broad vision of Mathematics that included Statistics. Fisher was invited to visit Berkeley as a way to preview a possible association with Berkeley. Reid (1982) writes that Evans "... had become personally acquainted with Fisher during summer weeks they had spent together at the University of Michigan in 1931". In addition, Reid (1982) continues, in a set of notes, "A Brief History of the Mathematics

J. Rojo
Professor of Statistics, Department of Statistics, Rice University, Houston, TX 77005
e-mail: jrojo@rice.edu

J. Rojo (ed.), *Selected Works of E. L. Lehmann*, Selected Works in Probability and Statistics, DOI 10.1007/978-1-4614-1412-4_93, © Springer Science+Business Media, LLC 2012

Department", Evans writes that he considered California "as the place for a really outstanding statistician, if possible of the level of R. A. Fisher himself". Fisher's visit, however, did not go well. But Evans persevered in his efforts to build up statistics within the mathematics department. Although he did not know Neyman, word got to him about the success of Neyman's rounds of visits to several universities (e.g., Columbia, Princeton, Chicago) that culminated with a very successful visit to the U.S. Department of Agriculture. Without knowing him, Evans offered Neyman a position. These details as well as Erich's interactions with Evans and Neyman have been fascinatingly described in Lehmann (2008) and Reid (1982).

But it was, very likely, the early passion for writing that led Erich to write several successful and influential books. The list includes:

1. **Testing statistical hypotheses**. First edition, 1959. Second edition, 1986. Third edition, with Joseph P. Romano, 2005. The first edition has been translated into Russian (1964), Polish (1968), and Japanese.
2. **Basic Concepts of Probability and Statistics**, with Joseph L. Hodges. First edition 1964. Reprinted in 2005 as part of the SIAM series Classics in Applied Mathematics. Second edition 1970. This book has been translated into Hebrew (1972), Farsi (1994), Italian (1971), and Danish (1969).
3. **Elements of Finite Probability**, with Joseph L. Hodges. First edition 1965. Second edition 1970.
4. **Nonparametrics: Statistical Methods Based on Ranks**, with the assistance of H. J. M. D'Abrera. First published as a hardcover by Holden-Day in 1975. In paperback by Prentice-Hall, Inc., Simon & Schuster in 1998, and then by Springer Science + Business Media, LLC in 2006. The book has been translated into Japanese (1998).
5. **Theory of Point Estimation**. First edition, 1983. Second edition, with George Casella, 1998. The first edition was translated into Russian (1991), and the second into Chinese (2004).
6. **Elements of Large-Sample Theory**, 1999.
7. **Reminiscences of a Statistician: The Company I Kept**, 2008.

In addition, Lehmann was a *Special Editor* for the second edition of *Statistics: A Guide to the Unknown*, edited by Judith M. Tanur. There were two printings of this second edition: the original edition was published in 1978 and it has been translated into Chinese (1980) and Spanish (1992). The second printing was published in 1985. For the third edition, published in 1989, Judith M. Tanur was the editor while Lehmann continued as a co-editor. He was joined by Frederick Mosteller, William H. Kruskal, Richard F. Link, Richard S. Pieters and Gerald R. Rising. Two more works, *Statistics: a guide to the study of the biological and health sciences* and *Statistics: a guide to political and social issues*, both published in 1977, were a result of this group's collaborative efforts. Lehmann was a *Special Editor* for the latter and a co-editor for the former.

At the time of his death, Erich was working on a manuscript for a new book soon to be published by Springer-Verlag. The title of the book is *Fisher, Neyman, and the Creation of Classical Statistics*. After his death, Juliet Shaffer has been working diligently to bring the book to its final form for publication. In addition, Lehmann was collaborating with Fritz Scholz on a revision of the *Nonparametrics: Statistical Methods Based on Ranks* book that will incorporate the usage of R. At the time of this writing, Fritz continues his indefatigable efforts on the revision and plans to have the book out in approximately two years.

Testing Statistical Hypotheses (TSH) and Theory of Point Estimation (TPE) have been the most influential of Erich's books. These books have been a mainstay in many Ph.D. programs, helping to educate future generations of statisticians. The genesis of the first edition of TSH has been discussed in Lehmann (1997), and the preface of the first edition of TPE gives a glimpse of how the book evolved. In both cases, Colin Blyth played an important role in facilitating the gestation of these books.

Several reviews of these books have appeared in the literature. For a review of Lehmann (1959) see, e.g. Pratt (1961); for a review of Lehmann and Romano (2005) see, e.g. Steinebach (2006). For a review of Hodges and Lehmann (1964), see Hedayat (1970); Grum (2006) reviewed the 2nd edition. Gastwirth (1977) reviewed Lehmann (1975). Rao (1984), Bickel (1984), and Berger (1984), reviewed TPE, 1st edition. Strawderman (2000) reviewed TPE, 2nd edition. Finally, Bellhouse (2008) reviewed Lehmann (2008).

The review by Speed (2008) on *Reminiscences of a Statistician: The Company I Kept*, and the review by Holmes (2000) on *Elements of Large-Sample Theory* are reproduced in this chapter.

In addition, George Casella and Joseph Romano provide fascinating accounts on their collaboration with Erich on the second and third editions of TPE and TSH, respectively. Finally, Doug Wolfe reviews the impact that *Nonparametrics: Statistical Methods Based on Ranks* had, and continues to have, on the discipline and on him personally.

References

[1] Bellhouse, D. (2008). Review of Reminiscences of a Statistician: The Company I Kept by Erich L. Lehmann. *Mathematical reviews*, MR2367933 (2008k:01021).

[2] Berger, J. O. (1984). Review of Theory of Point Estimation, by E. Lehmann. *Journal of the American Statistical Association*, **79**, 941-942.

[3] Bickel, P. J. (1984). Review of Theory of Point Estimation, by E. Lehmann. *Metrika*, **31**, Number 1, 253-258.

[4] Bose, A. (1999) Review of Elements of Large-Sample Theory. *Sankhyā A, The Indian Journal of Statistics*, **61**, No. 3, 453-455.

[5] Bose, A. (2005). Review of Testing Statistical Hypothesis 3rd edition. *Sankhyā A. The Indian Journal of Statistics*, **67**, No. 4, 795-796.

[6] Gastwirth, J. L. (1977). Review of E. L. Lehmann, Nonparametrics: Statistical methods based on ranks. *Bull. Amer. Math. Soc.*, *83*, 967-973.

[7] Grum, J. (2006). Review of Basic Concepts of Probability and Statistics, by J.L. Hodges, Jr. and E.L. Lehmann. 2nd Ed. SIAM series Classics in Applied Mathematics. *International Journal of Materials and Product Technology*, **27**, 295 - 295.

[8] Hedayat, S. (1970). Book Review: Basic Concepts of Probability and Statistics, 2nd edition, by J. L. Hodges, Jr. and E. L. Lehman, *Biometrics*, **26**, 589-590.

[9] Holmes, S. P. (2000). Review of Elements of Large-Sample Theory. *Journal of the American Statistical Association, 95*, 328-329.

[10] Lehmann, E. L. (1959). *Testing Statistical Hypotheses*. John Wiley & Sons, New York.

[11] Lehmann, E. L., and Hodges, J. L. (1964). *Basic Concepts of Probability and Statistics*. Holden-Day, Inc., San Francisco, London, Amsterdam. Reprinted in 2005 as part of the SIAM series Classics in Applied Mathematics.

[12] Lehmann, E. L. (1975). *Nonparametrics: Statistical Methods Based on Ranks*, with the assistance of H. J. M. D'Abrera. First published as a hardcover by Holden-Day in 1975. In paperback by Prentice-Hall, Inc., Simon & Schuster in 1998, and then by Springer Science + Business Media, LLC in 2006.

[13] J. M. Tanur, F. Mosteller, W. H. Kruskal, E. L. Lehmann, R. F. Link, R. S. Pieters, and G. R. Rising, Co-editors (1977). *Statistics: a guide to the study of the biological and health sciences*. Holden-Day, Inc., Publishers, San Francisco, 140 pp.

[14] J. M. Tanur (Editor), E. L. Lehmann (Special Editor), F. Mosteller, W. H. Kruskal, R. F. Link, R. S. Pieters, and G. R. Rising, Co-editors (1977). *Statistics: a guide to political and social issues*. Holden-Day, Inc., Publishers, San Francisco, 141 pp.

[15] J. M. Tanur (Editor), E. L. Lehmann (Special Editor), F. Mosteller, W. H. Kruskal, R. F. Link, R. S. Pieters, and G. R. Rising, Co-editors (1978). *Statistics: A Guide to the Unknown, Second Edition*, Holden-Day, San Francisco.

[16] Lehmann, E. L. (1983). *Theory of Point Estimation*. Wiley & Sons, New York. Transferred to Wadsworth & Brooks/Cole 1991.

[17] J. M. Tanur (Editor), E. L. Lehmann (Special Editor), F. Mosteller, W. H. Kruskal, R. F. Link, R. S. Pieters, and G. R. Rising, Co-editors (1985). Reprint of *Statistics: A Guide to the Unknown, Second Edition*, Wadsworth & Brooks/Cole.

[18] J. M. Tanur, F. Mosteller, W. H. Krusal, E. L. Lehmann, R. F. Link, R. S. Pieters, G. R. Rising, Co-editors. *Statistics: A Guide to the Unknown, Third Edition*. Wadsworth, 1989.

[19] J. M. Tanur (Editor), E. L. Lehmann (Special Editor), F. Mosteller, W. H. Kruskal, R. F. Link, R. S. Pieters, and G. R. Rising, Co-editors. **Chinese** Translation of *Statistics: A Guide to the Unknown, Second Edition*, 1990.

[20] J. M. Tanur (Editor), E. L. Lehmann (Special Editor), F. Mosteller, W. H. Kruskal, R. F. Link, R. S. Pieters, and G. R. Rising, Co-editors. **Spanish** translation of *Statistics: A Guide to the Unknown, Second Edition*. Alianza Editorial, S. A., Spain, ISBN 842069651X, 1992.

[21] Lehmann, E. L. (1986). *Testing Statistical Hypotheses*, second edition. John Wiley & Sons, New York. Transferred to Wadsworth & Brooks/Cole, 1991.

[22] Lehmann, E. L. (1997). Testing Statistical Hypotheses: The Story of a Book. *Statistical Science*, **12**, No. 1, 48-52.

[23] Lehmann, E. L., and Romano, J. P. (2005). *Testing Statistical Hypotheses,* third edition. Springer-Verlag, New York.

[24] Lehmann, E. L. (2008). *Reminiscences of a Statistician: The Company I Kept*. Springer-Verlag, New York.

[25] Pratt, J. W. (1961). Review of Testing Statistical Hypotheses by E. L. Lehmann. *J. Amer. Statist. Ass.*, **56**, 163-167.

[26] Rao, C. R. (1984). Review of Theory of Point Estimation, by E. Lehmann. *Journal of the American Statistical Association*, **79**, 942.

[27] Reid, C. (1982). *Neyman from Life*. Springer-Verlag, New York.

[28] Speed, T. (2008). Review of Reminiscences of a Statistician: The Company I Kept by Erich L. Lehmann. *International Statistical Review*, **76**, 153.

[29] Steinebach, J. (2006). Book Review: E. L. Lehmann, J. P. Romano: Testing Statistical Hypotheses Springer New York 2005 (3rd edition), corr. 2nd printing 2006, XIV, 786 pages. *Metrika*, **64**, 255-256.

[30] Strawderman, W. E. (2000). Review of Theory of Point Estimation (2nd ed.). E. L. Lehmann and George Casella. *Journal of the American Statistical Association*, **95**, 329.

On the Third Edition of *Testing Statistical Hypotheses*

Joseph P. Romano

Abstract In this article, the history of the third edition of *Testing Statistical Hypotheses* will be discussed.

Sometime during the fall of 1998, Erich Lehmann and Julie Shaffer visited the Statistics Department at Stanford during one of our Tuesday joint colloquia with the Berkeley Statistics Department. While having coffee prior to the seminar, Erich asked me if we could have a quick chat. During this seemingly impromptu conversation, Erich asked if I would be willing to work on the revision of his testing book. Coming from a man I so truly admire, I was genuinely taken aback by this offer. After all, Erich's testing book was rightly regarded as one of the most important books in statistics, and it has had an enormous and enduring impact since the first edition appeared in 1959. Overwhelmed by the offer, I told Erich I would soon get back to him.

My immediate thoughts were that this was an extraordinary opportunity for me to work with Erich on the book, but it also was an enormous responsibility to live up to the high standard of excellence that the first and second editions already held. There was also the matter that I was already working on a book on subsampling with Dimitris Politis and Michael Wolf, and I never imagined I would undertake another book project, at least not so soon.

Over the years at Stanford, I had taught estimation and testing from Erich's texts. But, I wondered why Erich chose to ask me to work on this project. As a student at Berkeley, I took Erich's first-year graduate sequence, Statistics 210, with Javier Rojo and Dorota Dabrowska as the teaching assistants for the course that year. I was quite proud that I got perfect scores on all six exams which I took from Erich during my first year at Berkeley. But after graduating in 1986, I had not had much contact with Erich,

J.P. Romano
Professor of Statistics, Departments of Statistics and Economics, Stanford University, Stanford, CA 94305-4065
e-mail: romano@stanford.edu

J. Rojo (ed.), *Selected Works of E. L. Lehmann*, Selected Works in Probability and Statistics, DOI 10.1007/978-1-4614-1412-4_94, © Springer Science+Business Media, LLC 2012

aside from seeing him at the joint colloquia. In any case, I called Erich the next day or so, and told him that I was indeed quite excited about the project, but that I could not begin until the subsampling book was complete, which I expected to be in early 1999. Erich was fine with the delay, and we began in earnest during the spring of 1999.

Our first meeting took place at Erich's home in Berkeley. Erich made a couple of points that have stayed with me. First, one of the nice things about working on hypothesis testing is that the notion of power is so compelling as a measure of performance that one need not bring in arbitrary loss functions. Second, even though the first and second editions are written in such a way where optimal tests are derived as examples of a general approach (or theorem), in fact almost all the tests considered in the text were already proposed prior to any optimality treatment. A prime example of this is the t-test, dating back to Gossett's publication in 1908 under the name Student, which was invented without regard to such properties as uniformly most powerful among invariant, or unbiased, tests. One of Erich's main goals was to be able to find some optimality properties or objective justification for ad hoc procedures.

We initially discussed various topics that could be incorporated into the new edition, such as foundational issues and the Bayesian criticism of the Neyman-Pearson approach, new developments in multiple testing, and the emerging literature of tests claiming superiority over the likelihood ratio, as discussed in Perlman and Wu (1999). Erich even pointed out sections from the second edition which he thought could be improved. But for the most part, I felt the topic which needed the most attention was large-sample theory, and Erich agreed. Starting with large-sample theory had several advantages from my point of view. First, I was very reluctant to rewrite anything that Erich had already published, given the book's success and influence. Why mess with a good thing? Second, based on my experience with teaching hypothesis testing, I would already supplement Erich's book with more modern asymptotic approaches. (The second edition did include a little asymptotics; for example, the robustness of the t-test was discussed using the Central Limit Theorem.) The drawback to finite-sample theory is that it offers amenable solutions in only a restricted class of models. On the other hand, a large-sample approach applies much more generally, and easily offers solutions to such classic problems as the Behrens-Fisher problem and the optimality of the Chi-squared goodness of fit test, as well as many others where finite-sample considerations are less fruitful.

Of course, a reasonably general theory of asymptotic optimality requires the development of new machinery, such as contiguity, the Hellinger metric, quadratic mean differentiable families, and convergence to a normal limiting experiment. Such concepts, no doubt inspired by Wald, were developed by Hájek, Le Cam, and others. In fact, such ideas were no longer that recent, and it was important to bring in the new technology. Naturally, the development of large-sample theory would require a substantial addition to the book. We mistakenly thought the project would take about two years.

Basically, the plan was that I would start writing some sections, and Erich would read, criticize, and sometimes rewrite them. I would visit Erich every couple of weeks in Berkeley, usually on a Thursday morning. Without fail, as I approached Erich's building, I would see Erich watching from his balcony at the appointed time, so that he could meet me downstairs and let me in. (Fortunately, I am punctual.) Then, we would work on the book until we were hungry for some lunch. Usually, Julie would join us for lunch, and we would discuss statistics, but also politics and music. Erich would often lend me some of his audio discs, so that I might better appreciate some of his favorite composers as well as fill some gaps in my musical knowledge. Despite my being a self-proclaimed amateur musician, I greatly appreciated being exposed to the Shostakovich string quartets and some of the works of Schnittke.

It was absolutely clear that Erich was actually quite concerned with pedagogical issues. He really tried to view the writing from a student's perspective, and was concerned with clarity, motivation, and having sufficiently many examples. He was also concerned that we not raise the level of the text, so that the book remains accessible to first year students. For example, is almost sure convergence a concept that is really needed, or does convergence in probability suffice for statistical applications? Do quadratic mean differentiable models greatly increase our ability to provide methods for important models? Do bootstrap and resampling methods really fit in a book concerned largely with optimality?

At some point, it was nice to realize that not only was I learning from Erich, but that I had something to offer as well. On such occasions, Erich would always ask relevant, insightful questions, and it became obvious that any new material would be treated in an honest and scholarly manner. If I could not convince Erich of the value of a section, it risked being abandoned. For the most part, our philosophical views were in agreement, and it was just a matter of reaching the point where our writing styles were in harmony.

Erich's concerns were evident in the pages of critique he would mail to me or hand me on one of my visits. (I still have a file full of them.) Some sections required several rounds of rewriting. Of course, we made some compromises, but we tried to empathize with the reader. As an example, Erich convinced me to include some material on Edgeworth expansions, because he thought a few introductory pages on the subject could be of enormous benefit in exposing students to a topic which they might not otherwise see.

Other areas of the book required major reworking. After including such tests as the Kolmogorov-Smirnov test and the Chi-squared goodness of fit test as examples in various chapters of the manuscript, Erich thought it might be a good idea to have an entire chapter on goodness of fit. This chapter was probably the most challenging (and least satisfying), because the available results are not completely satisfactory. Even so, Erich viewed goodness of fit as such a fundamentally important problem that, by laying out some key considerations and discussing some currently used tests, such a chapter could be quite valuable.

Soon, another problem emerged. The length of the manuscript was getting to be quite long. At one point, we decided there should be two volumes, with the first volume devoted to finite-sample theory and the second to large-sample theory. We even signed a new contract with Springer, and we were fortunate enough to work with a patient editor, John Kimmel. But, we then became concerned that some departments might only adopt the first volume. After some back and forth, we decided upon one volume if we could limit the number of pages to 800, which we barely succeeded in doing.

(It was particularly surprising to see how many pages were saved by moving all the references to the end of the book, rather than to the end of each chapter.)

Another change to the third edition was the addition of a separate chapter on multiple testing and simultaneous inference. In fact, the book became an impetus for me to direct some of my research effort on multiple testing. Our discussions led to a couple publications in the *Annals of Statistics* in 2005, one of which was coauthored with Julie Shaffer as well. I am extremely proud that this joint work appeared, extending the number of consecutive decades where Erich has published in the *Annals* to an incredible 7, dating back to the 1940s (when the original name of the journal was the *Annals of Mathematical Statistics*).

Toward the end of the project, Erich proposed that we dedicate the book to two of the great pioneers in our field, Lucien Le Cam and John Tukey. The reasons were clear. Tukey inspired much of the work on multiple testing, due to his 1953 mimeographed volume on the subject, which was only officially published in 1994 as Volume VIII in *The Collected Works of John W. Tukey*. Le Cam's groundbreaking work on large-sample optimality culminated in his 1986 volume entitled *Asymptotic Methods in Statistical Decision Theory*. Given my own personal memorable interactions with Tukey as an undergraduate at Princeton and Le Cam as a graduate student at Berkeley, I thought Erich's idea to dedicate the book to their memory was perfect.

By the time the third edition appeared in 2005, the original plan of a two year project had turned into six. But, I view the time as well spent. Working on this project with Erich was certainly one of the most rewarding academic experiences I have undertaken. I would like to think the end result is a success, with the whole exceeding the sum of the individual parts.

References

[1] Le Cam, L. (1986). *Asymptotic Methods in Statistical Decision Theory*, Springer-Verlag, New York.

[2] Lehmann, E. and Romano, J. (2005). Generalizations of the familywise error rate. *Annals of Statistics* **33**, 1138-1154.

[3] Lehmann, E., Romano, J. and Shaffer, J. (2005). On optimality of stepdown and stepup procedures. *Annals of Statistics* **33**, 1084-1108.

[4] Perlman, M. and Wu, L. (1999). The Emperor's new tests. *Statistical Science* **14**, 355-381.

[5] Student. (1908). The probable error of a mean. *Biometrika* **6**, 1-25.

[6] Tukey, J. (1953). The problem of multiple comparisons. Published in *The Collected Works of John W. Tukey: Multiple Comparisons, Volume VIII. (1999)*. Edited by H. Braun, CRC Press, Boca Raton, Florida.

Writing TPE2

George Casella

As I write this memory of working with Erich Lehmann on the second edition of *Theory of Point Estimation* (TPE2), I am still trying to process the news from Bin Yu about Erich's death. The passing of Erich Lehmann not only signals the passing of a great statistician and a great man, but also the end of an era. Post World War II statistics, from approximately the late 1940s until the middle 1970s, was an era of "mathematical statistics". It was marked by the development of much of the underlying theory, but also marred by a marginalization - some thought that statistics was merely a branch of applied mathematics.

Not Erich. Although his work (some with Scheffé and Hodges) was very mathematical, he strove to solve real statistical problems, and create the underlying theory of statistics: *theoretical statistics*, not *mathematical statistics*. He was never concerned with elegant proofs, or math for math's sake, but to provide sound and understandable solutions to real problems. As an example, take a look at Erich's 1983 JASA paper "Estimation with Inadequate Information", that shows how (and why) unbiasedness can sometimes produce an absurd result.

But to the Story of TPE2. In 1994 I got an email from John Kimmel, Editor of the Spring-Verlag Statistics Textbook Series, asking if I would be interested in working with Erich Lehmann on a revision of *Theory of Point Estimation* (TPE1), which was originally published in 1983. As you might expect, this was an offer that no one could refuse! I talked with Erich on the phone about this (he was at Berkeley and I was at Cornell), and we agreed that I would come to Berkeley to discuss how to proceed. I worked for about two months putting together a plan for what new material to put in, and for what old material to take out. I was working under the assumption that John Kimmel had suggested this collaboration to Erich, and that I would have to convince him to take me on.

G. Casella
Distinguished Professor, Department of Statistics, University of Florida, Gainesville, FL 32611
e-mail: casella@stat.ufl.edu

J. Rojo (ed.), *Selected Works of E. L. Lehmann*, Selected Works in Probability and Statistics, DOI 10.1007/978-1-4614-1412-4_95, © Springer Science+Business Media, LLC 2012

After meeting with Erich the first time (I took me over a year to get up enough nerve to address him as "Erich"), as I was in the mode of trying to convince him to work with me, he stopped and said, "George, you don't understand. Kimmel didn't suggest you to me, I suggested you to him. So just relax." It turned out that Erich was a fan of *Statistical Inference* (Casella and Berger) and decided that he wanted to have me work on TPE2. Wow.

So now the "experienced" text book writer (me- I had already done two books) really learned how to write a textbook. Erich was adamant that whatever was included in the text had to be done right. That is, the entire story must be told - introduce the problem, show how it can be solved, explain the scope and limitations of the solutions, and provide examples and exercises. A text book must have the whole story. We don't say, "for details see some paper in the literature", we give, and explain, the details. It was Erich's insistence that any good qualified teacher, who may not be totally familiar with the topic, could pick up the textbook and teach a course out of it.

A number of people had suggestions for the second edition. Persi Diaconis told me that one of the charms of TPE1 was that it was not an enormous tome - Erich had done a wonderful job of distilling the important concepts. Persi urged me to not make the second edition too big, or else it would lose its charm. Although that was a little difficult (the second edition grew by 100 pages) another suggestion, by Larry Brown, caused me a bit more difficulty. Larry did not like the way the Information Inequality was stated, and wanted me to completely revise the treatment. The problem, Larry said, is that in the statement of the Information Inequality in TPE1, the conditions were on the estimator (it had to allow differentiation under the integral sign), and this restricted the class. Larry wanted the conditions on the density. I liked this idea, and proposed it to Erich. Erich felt that changing the assumptions made the presentation a bit too difficult, and did not want to do it. So there I was caught in the middle, between Lehmann and Brown. The final result? See Section 2.5, where there are two information inequalities.

In other places the material changed, with perhaps the most significant change being the addition of Chapter 5: Average Risk Optimality, which is all about Bayesian estimation. Erich always wanted the main chapters to reflect optimality, hence the title. Here is one interesting anecdote about this chapter. I wrote a first draft in Ithaca, and sent it to Erich in Berkeley. After a week or so he called me to discuss what I had written, and in his gentle way he informed me that I had gotten it all wrong! What happened was that I had structured the chapter according to distributions, rather than types of estimators. Erich told me that as I rewrote and rearranged everything I would see how things would flow better and, of course, he was right.

Other new material included much on Stein estimation and the accompanying calculations, the addition of complete class theorems, and some material on EM and Gibbs. Erich was not initially in favor of the computational material, but I convinced him that it was necessary so that we could tell the entire story!

In any second edition it is necessary to not only add new material, but also to delete some material from the first edition. Of course, one could just add, but a good textbook is not an encyclopedia, but rather a filter of what the authors believe comprises a thorough treatment of the topic. One major deletion was material on robust estimation, not because it is not important but because with the appearance of some excellent textbooks on the subject, we could defer this material and make room for the Bayes chapter - which we both wanted to have in the book. Other topics, or sections, were judiciously edited (for example, Section 5.1 in TPE1 is Section 1.8 in TPE 2), since hindsight always allows for a more coherent flow of material.

We worked together for four years, 1994-1998, during which I made about two trips per year to Berkeley, where I always stayed with Erich and his wife Juliet Shaffer, and was treated to good food and good conversation (and hard work). Julie was on sabbatical one year at Princeton, and during that year I could drive from Ithaca to Princeton for more frequent meetings, and one time Erich and Julie came and stayed at my house in Ithaca. (Much later, after I moved to Florida, I convinced Erich and Julie to visit, and Erich gave a talk at the department. As a reward, I took them on a hike to a nature preserve where we saw about fifty alligators.)

TPE2 is still, as I understand, a mainstay in many graduate programs, and has been translated into Chinese. For the most part, the material still fits well into modern graduate programs because it is core material. To do good statistics, whether theory or applications (with the latter always needing some theory) it is necessary to have a solid theoretical core. TPE2 fills that role. Will there be a TPE3? I expect so. As we monitor the change in our field, we see trends that sometimes lead to necessary changes in the core curriculum. When that happens, TPE3 will be there.

Nonparametrics: Statistical Methods Based on Ranks and Its Impact on the Field of Nonparametric Statistics

Douglas A. Wolfe

As I was doing background reading in the mid-1960s in preparation for my dissertation in the area of nonparametric statistics, it seemed that everywhere I turned there was yet another important paper on rank-based statistical methodology by Erich Lehmann, including the article in which he proposed the use of structured nonparametric alternatives (that bear his name) for approximate power calculations (1953), the joint work with Joe Hodges on the surprising efficiency of rank tests (1956), the series of papers on various aspects of rank-based approaches to linear models (1963a, 1963b, 1964), and his pair of definitive articles on nonparametric confidence intervals based on rank procedures (1963c) and estimates of location associated with rank tests (1963), the latter once again being joint work with Hodges. Therefore, it was certainly no surprise when Lehmann published Nonparametrics: Statistical Methods Based on Ranks (NSMBR) in 1975 and it turned out to be one of the most influential textbooks in the field.

Lehmann's text was not the first on the topic of nonparametric statistics. Hájek and Šidák (1967), Puri and Sen (1971), and Gibbons (1971) had previously published books that emphasized the theory associated with nonparametric rank-based procedures, while Conover (1971) and Hollander and Wolfe (1973) had provided applications-oriented texts on the subject. Even so, Nonparametrics: Statistical Methods Based on Ranks was somewhat unique at the time, as it carefully addressed the theoretical underpinnings of rank-based procedures, while at the same time showcasing their naturally appealing, intuitive flavor to motivate their importance for data analysis. Lehmann also chose to use interesting, real data examples and problems throughout the text that helped convince the reader of the applicability and relevance of rank-based approaches to statistical inference.

NSMBR appeared on the scene at a time when research on statistical procedures based on sample rankings was near its peak and they were experiencing widespread

D.A. Wolfe
Professor of Statistics, Department of Statistics, The Ohio State University, Columbus, OH 43210
e-mail: daw@stat.osu.edu

J. Rojo (ed.), *Selected Works of E. L. Lehmann*, Selected Works in Probability and Statistics, DOI 10.1007/978-1-4614-1412-4_96, © Springer Science+Business Media, LLC 2012

popularity in data analysis. As a result, the response to NSMBR was immediate and distinctly positive. Reviewers at the time were generous with their praise. In a review for the Bulletin of the American Mathematical Society, Gastwirth (1977) comments, "Professor Lehmann's text will undoubtedly become the standard to which other mathematical introductions to nonparametric inference will be compared. It has an excellent selection of problems, ranging from routine calculations to challenging problems for the best students, and an extensive bibliography." In a review for the Journal of the Royal Statistical Society, Series A, Haigh (1977) states, "The reader learns not only non-parametric methods, but a healthy attitude towards statistics in general – to think about the underlying experimental situation, to consider the effect of possible errors in the data, to see whether one-sided or two-sided tests are appropriate." He adds, "The book is rich in detail, well-organized and produced and easy to read. It persuasively points out the high power and efficiency that many non-parametric tests have against their parametric alternatives, and makes out a strong case for non-parametric methods per se." Finally, in a review for JASA, Berengut (1978) listed a number of qualities of the book that he felt were "particularly outstanding", including "the careful delineation of subtle but important distinctions between models", "the explicit description of possible limitations on inference in a particular model", and "the lucid explanation of point estimation and confidence procedures". Personally, I find one of the biggest plusses for the text to be Lehmann's ability to dissect and present conceptually complicated concepts (such as small sample power and asymptotic efficiency for rank-based procedures) in such a way that the reader immediately understands both the particulars and the generalities. This is especially true (both in his original papers and in NSMBR) of his development and discussion of point estimators and confidence procedures associated with rank-based tests.

While NSMBR has not had the same dominance that Lehmann's testing hypotheses and estimation books have had on the entire statistics profession, it has certainly had significant impact within the nonparametric community. It has been translated into an impressive number of languages and sold extensively in every part of the world. It provided important guidance to many researchers in the field, including myself, during the 1970s and 80s and helped introduce an entire generation of statistical practitioners to the novelty and importance of rank-based procedures and their relevance for data analysis and interpretation of results.

In my opinion, it still serves today as perhaps the best single source for anyone who wants to understand not only the philosophy and motivation behind the development of rank-based statistical procedures but also the rationale for their practical usage. All of us who have worked in the arena of rank-based inference at some time in our careers owe a signal sense of appreciation to Erich Lehmann not only for his seminal and novel research contributions to this field but also for his commitment to disseminating the accumulated knowledge through the publication of NSMBR. Thank you Erich for being so involved in both.

References

[1] Berengut, D. (1978). Review of Nonparametrics: Statistical Methods Based on Ranks by E. L. Lehmann. *Journal of the American Statistical Association*. **73**: 892-893.
[2] Conover, W. J. (1971). *Practical Nonparametric Statistics*. New York: John Wiley.
[3] Gastwirth, J. L. (1977). Review of Nonparametrics: Statistical Methods Based on Ranks by E. L. Lehmann. *Bulletin of the American Mathematical Society*. **83**: 967-973.
[4] Gibbons, J. D. (1971). *Nonparametric Statistical Inference*. New York: McGraw-Hill.
[5] Haigh, J. (1977). Review of Nonparametrics: Statistical Methods Based on Ranks by E. L. Lehmann. *Journal of the Royal Statistical Society, Series A*. **140**: 108-109.
[6] Hájek, J., and Z. Šidák (1967). *Theory of Rank Tests*. New York: Academic Press.
[7] Hodges, J. L., Jr., and E. L. Lehmann (1956). The efficiency of some nonparametric competitors of the t-test. *Annals of Mathematical Statistics*. **27**: 324-355.
[8] Hodges, J. L., Jr., and E. L. Lehmann (1963). Estimates of location based on rank tests. *Annals of Mathematical Statistics*. **34**: 598-611.
[9] Hollander, M., and D. A. Wolfe (1973). *Nonparametric Statistical Methods*. New York: John Wiley.
[10] Lehmann, E. L. (1953). The power of rank tests. *Annals of Mathematical Statistics*. **24**: 23-43.
[11] Lehmann, E. L. (1963a). Robust estimation in analysis of variance. *Annals of Mathematical Statistics*. **34**: 957-966.
[12] Lehmann, E. L. (1963b). Asymptotically nonparametric inference: An alternative approach to linear models. *Annals of Mathematical Statistics*. **34**: 1494-1506.
[13] Lehmann, E. L. (1963c). Nonparametric confidence intervals for a shift parameter. *Annals of Mathematical Statistics*. **34**: 1507-1512.
[14] Lehmann, E. L. (1964). Asymptotically nonparametric inference in some linear models with one observation per cell. *Annals of Mathematical Statistics*. **35**: 726-734.
[15] Lehmann, E. L. (1975). *Nonparametrics: Statistical Methods Based on Ranks*. San Francisco: Holden-Day.
[16] Puri, M. L., and P. K. Sen (1971). *Nonparametric Methods in Multivariate Analysis*. New York: John Wiley.

consisting of modification of some of those already mentioned, are briefly stated with citation of references. Chapter 1 ends with an interesting discussion of criteria for the comparison of different algorithms.

The statement of the translation editor that this is the only book in English which deals extensively with the approximate solution of integral equations (with Cauchy kernels) may still be true. At least this reviewer could find no other. However, for the case of convolution equations there is the excellent and extensive work by Gohberg and Fel'dman [7], which forms an effective complement to the present book, since the approach and the methods in the two monographs are quite distinct for the most part.

One feature of the book which aids the reader is to set off passages which can be omitted on a first reading or are suitable "for a reader with an advanced mathematical training", by a vertical line within the margin of the text. The book is "closely written", in the sense of the theorem-proof style, sometimes proceeding without clear motivation from the readers point of view. However this style has the advantage of the presentation of a large amount of material in a short compass, and of clearly setting out what is known and not known. For a book on numerical analysis there are remarkably few numbers included. The only example noticed where any details of calculation were cited was the computer solution of the system which arose in dispersion theory (p.263)

References

1. N. I. Mushelišvili, *Singular integral equations*, 2nd ed., Noordhoff, Groningen, 1953. MR **15**, 434.

2. I. N. Vekua, *On the theory of singular integral equations*, Soobšč. Akaa. Nauk Gruzin. SSR **3**(1942), 869–876. MR **5**, 268.

3. B. V. Hvedelidze, *Regularization problem in the theory of integral equations with Cauchy kernel*, Dokl. Akad. Nauk SSSR **140** (1961), 66–68 = Soviet Math. Dokl. **2** (1961), 1169. MR **26** #563.

4. A. N. Tihonov, *Solution of incorrectly formulated problems and the regularization method*, Dokl. Akad. Nauk SSSR **151** (1963), 501–504 = Soviet Math. Dokl. **4** (1963), 1035–1038. MR **28** #5576.

5. ______, *Regularization of incorrectly posed problems*, Dokl Akad. Nauk SSSR **153** (1963), 49–52 = Soviet Math. Dokl. **4**(1963), 1624–1627. MR **28** #5577.

6. J. N. Franklin, *On Tikhonov's method for ill-posed problems*, Math. Comp. **28** (1974), 889–907. MR **51** #12007.

7. I. C. Gohberg and I. A. Fel'dman, *Convolution equations and projection methods for their solution*, "Nauka", Moscow, 1971; English transl., Transl. Math. Monographs, vol. 41, Amer. Math. Soc., Providence, R. I., 1974. MR **50** #8148; #8149.

D. H. Hyers

BULLETIN OF THE
AMERICAN MATHEMATICAL SOCIETY
Volume 83, Number 5, September 1977

Nonparametrics: *Statistical methods based on ranks*, by E. L. Lehmann, Holden-Day, Inc., San Francisco, and McGraw-Hill International Book Company, Düsseldorf, Johannesburg, London, Mexico, New York, Panama, São Paulo, Singapore, Sydney, Toronto, 1975, xvi + 457 pp.

The methods taught in the standard basic statistics course assume that the observations come from a normal (Gaussian) distribution. Students are taught that the best estimate of the average or typical value of a population is the average (mean) of a sample. In some applications, however, the median describes the typical member of a population more accurately than the mean. For example, in summarizing the income level of a town, a few rich families have a far greater effect on the value of the mean than they do on the median.

Practical statisticians have long noticed that a few extremely deviant observations sometimes occur even in samples from populations which look "nearly normal" and that these observations have an unduly large impact on classical statistical analyses. Professor Tukey's 1961 article **[11]** stimulated a great deal of research in the development of *robust* statistical estimators and tests which are relatively unaffected by deviations from the usual assumptions underlying statistical methods. Nonparametric tests are the most robust of all procedures, as they are valid for samples from any population. Rank tests, the class of nonparametric methods discussed by Professor Lehmann, are also of interest because every rank test generates a unique estimator which is often robust.

Rather than attempt to summarize the entire text, let us concentrate on the "two-sample problem" which arises in testing whether two populations differ or are essentially the same. For example, suppose one is interested in deciding whether two groups (e.g., white and Mexican Americans) have the same distribution of heights. Thus, we are testing that the height distribution of Mexicans $G(x)$ equals that of the whites $F(x)$. As an *alternative* one might consider $G(x) = F(x - \Delta)$, i.e., the height distribution of whites is shifted to the right, i.e., for any fixed height, h, the proportion of Mexicans (x's) who are less than h is greater than the proportion of whites (y's) who are less than h. In addition, we are interested in estimating the size of the shift Δ required to make the distributions identical.

To resolve the issue, we take random samples $x_1, \ldots, x_m$ from the first group and $y_1, \ldots, y_n$ from the second. Consider the combined sample of $N = m + n$ observations. Under the null hypothesis that the two distributions are the same, i.e., $F(x) = G(x)$, it can be shown that each of the original observations has probability $1/N$ of being the kth largest in the pooled sample. Thus, the ranks in the combined sample that the n y's will have, can be considered as a random sample of n integers chosen from $1, \ldots, N$, irrespective of the form of the distribution function $F(x)$. Any test which is solely a function of the ranks that one group of observations has in the combined sample is called a rank test. If we let R_i be the rank y_i has in the ordered combined sample of N, then the Wilcoxon test is defined as $W = \Sigma R_i$, and its distribution is that of the sum of n randomly selected integers from $1, 2, \ldots, N$. Since the average of the first N integers is $(N + 1)/2$, under the null hypothesis the expected value of W is $n(N + 1)/2$. If the actual value of W is much larger than that, statisticians reject the hypothesis that the x's and y's have the same distribution in favor of the alternative and conclude that the distribution of the y's is shifted to the right, i.e., $\Delta > 0$. From the distribution of W we compute the probability of obtaining a value at least as distant from its expected value as the one actually obtained. Usually when this significance probability level is less than 5% we reject the null hypothesis.

At this juncture probability theory is required to calculate the exact distribution of the statistic W and show that in large samples it is approximately normally distributed. Under the null hypothesis the ranks of the y's are a random sample from $1, \ldots, N$. Thus, they are negatively correlated and the ordinary Central Limit Theorem does not apply. Lehmann's appendix provides an elegant treatment of asymptotic normality based on Hájek's approach to samples from finite populations. Fortunately, the asymptotic approximation to the distribution of rank statistics is quite accurate even in relatively small samples ($m,n \geqslant 10$).

So far we have limited our attention to insuring that the probability of rejecting the hypothesis that both populations are the same when it is indeed true, is small (5% or less). The strength of nonparametric statistics is that this calculation is the same regardless of the form of the density function of the population. On the other hand, we also desire to reject the null hypothesis when the two populations truly differ. If c is the critical point of a test of size α (often .05), i.e., the probability, when the null hypothesis is true, of obtaining a value $> c$, is $\leqslant \alpha$, written $P_0[W > c] \leqslant \alpha$, then the power of the test is the probability of $[W > c]$ calculated under the alternative assumption (e.g., $\Delta = 1$). While the size (α) of the Wilcoxon and all rank tests *does not* depend on the form of the underlying density function, their power does. The remarkable fact about the Wilcoxon test is that it is about 95% as powerful as the usual t-test for normal data. Hence, one pays a rather small price in terms of loss of power for guaranteeing that the Type I error (size) is not affected by the form of the density.

By now the reader is probably asking if can we develop a test that is as powerful as the t-test on normal data and still has the property that its size (α) is unaffected by the shape of the density. For practical purposes the answer is yes, although the result is strictly true only for large samples. The idea is to use statistics of the form $\Sigma a(R_i)$, where R_i is the rank of y_i and $a(R_i)$ is specified by $a(R_i/(N+1))$, where $a(u)$ is a function on $(0,1)$. The fundamental result, due to Chernoff and Savage [2] is the following:

Let $x_1, \ldots, x_m; y_1, \ldots, y_n$ be two independent samples from the distributions $F(x)$ and $G(x) = F(x - \Delta)$, respectively, and assume that $f = F'$ has finite Fisher information, i.e., $I = \int (f'/f)^2 f\, dx < \infty$. The asymptotically most powerful rank test of H_0: $\Delta = 0$ against $\Delta \neq 0$ is based on the function

$$(1) \qquad a(u) = I^{-1/2} f'\left[F^{-1}(u)\right]/f\left[F^{-1}(u)\right], \qquad 0 < u < 1,$$

and is asymptotically as powerful as the best parametric (maximum likelihood) procedure. When the population sampled is the logistic,

$$f(x) = e^{-x}/(1 + e^{-x})^2,$$

then the function $a(u) = 2\sqrt{3}\,(u - \frac{1}{2})$ is a straight line which corresponds to the Wilcoxon test normalized to have mean 0 and variance 1. Since the functional form of the logistic density is similar to the normal, it is not surprising that the rank test which is best for the logistic is quite powerful on normal data.

Mathematicians looking at formula (1) will notice that the functions $a(u)$

are in $L^2(0,1)$ and realize that there should be a connection between the inner product $\int a_1(u)a_2(u)\,du$ and the closeness of the two densities f_1,f_2 generating the test scores a_1 and a_2.

The connection arises from the consideration of Pitman efficiency, a criterion enabling us to compare tests for large samples. A special criterion is needed here because almost any reasonable test will have power approaching 1 as the sample sizes (m,n) increase. The idea underlying Pitman efficiency is to consider testing H_0: $\Delta = 0$ against a sequence of alternatives H_1: $\Delta = \delta/\sqrt{N}$. In this case the power will not approach 1 as N increases, because the difference between the null and alternative hypotheses will decrease as N increases. It can be shown that the limiting power of a test T_1 (based on a_1) is given by

$$\Phi(c_1\delta - u_\alpha) \tag{2}$$

where $\Phi(-u_\alpha) = \alpha$ the level of the test. Notice that the larger c_1 is, the more powerful the test is. The constant c_1 is called the efficacy of the test and the Pitman efficiency of a test T_1 relative to another test T_2 is

$$e(T_1,T_2) = (c_1/c_2)^2. \tag{3}$$

The quantity $e(T_1,T_2)$ can also be interpreted as the limiting ratio of the sample sizes required by the tests T_1 and T_2 to achieve the same limiting power π against the same sequence of alternatives. For example, if the Pitman efficiency of test T_1 relative to T_2 $(e(T_1,T_2)) = \frac{1}{2}$, then the test T_2 results require approximately half as many observations as the test T_1 to achieve the same large sample power for critical regions of the same size α. Besides its statistical interpretations, the Pitman efficiency can be computed readily. It can be shown that the efficiency of test T_1 (based on $a_1(u)$) relative to the best test, T_2, on data from the density f_2 is given by

$$\langle a_1,a_2\rangle = \int a_1(u)a_2(u)\,du, \tag{4}$$

where a_2 is obtained from (1). By the symmetry of the inner product T_2 has this same efficiency relative to the best test, T_1, for data from the density f_1.

The fact that the functions $a(u)$ generating the most powerful rank tests for a wide family of densities are in $L^2(0,1)$ yields insight into other problems as well. If one truly knew the form of the density, it would not make sense to use a rank test instead of the usual maximum likelihood test. Suppose one knew something about the density, e.g., it was either f_1 or f_2 (normal or double exponential): Is there a reasonably powerful rank test for this problem? Considering the functions a_1, a_2 as vectors in $L^2(0,1)$, it is clear that a test corresponding to the angle bisector will maximize the minimum efficiency when data comes from either density. This problem is discussed in **[1]** and **[6]**. Again statisticians lose relatively little power against either density when they insure that the rank test is reasonably powerful for samples from a family of densities.

Another problem arises when data is censored or missing. This often occurs in reliability and clinical trials when we observe lifetimes (survivors) over time and must end the experiment early. Essentially the data allows us only to order the x's whose value was less than t and to know that $m - m_1$ x's were greater than t. Similarly, we have the ordered $y_1, \ldots, y_k$ $(k = n - n_1)$ and the knowledge that $n - n_1$ y's were greater than t. Thus all the $n + m - m_1 - n_1$ observations can be considered as missing or tied. If the test based on $a(u)$ is appropriate for complete samples we desire the closest L^2 approximation to it which is constant on $(1 - r,1)$ where $r = (m_1 + n_1)/(m + n)$, the proportion of observations which are larger than t. That approximation **[5]**, **[10]** is clearly the function equaling $a(u)$ on $(0,1 - r)$ and the average $(1 - r)^{-1}\int_{1-r}^{1} a(u)\, du$ on $(1 - r,1)$. Thus, we score the ranks of the true observations with the same scores we would use in the complete data situation and score the censored observations by their "average rank." Although this formula is derived from asymptotic theory, the result agrees with common sense and yields tests whose statistical properties are close to the special small sample optimal procedures that have been derived in the literature.

Mathematicians realize that Pitman efficiency is only one of several approximations that can be used to compare the power of tests in large samples. Cochran considered the rate at which the level (rejection probability under H_0) decreases to zero when the alternative is a fixed distance from the null hypothesis and the power is also fixed (< 1). A different formulation which often yields equivalent results is due to Bahadur. Recently Hájek **[7]** showed that in terms of Bahadur efficiency, rank tests are as powerful as any function of the observations against a fixed alternative. Formally, consider the null hypothesis H_0: $F = G$. The statistic $\psi(x_1, \ldots, x_m, y_1, \ldots, y_n)$ which rejects H_0 when its value is $\geqslant \psi_0$ has level

$$L(\psi_0) = P_0\left[\psi(x_1, \ldots, x_m, y_1, \ldots, y_n) \geqslant \psi_0\right].$$

The Bahadur efficiency is essentially the rate at which the *observed level* tends to zero at a fixed alternative. For example, consider a fixed alternative $F = F^*$, $A = A^*$ (e.g., $F^* = N(0,1)$, $G = N(\Delta,1)$, $\Delta > 0$). Bahadur showed that for almost all sample sequences $N^{-1}\log L(\psi(x_1, \ldots, x_n; y_1, \ldots, y_m))$ tends to a limit, $-K(\psi)$. The larger $K(\psi)$ the faster the level (calculated under the alternative) approaches zero. Since smaller levels lead to rejection of the null hypothesis, this shows that tests with large values of K are better. Hájek showed that the Max $K(\psi)$ over all tests ψ of H_0 against any specific alternative, $H_A(F^*,G^*)$, is achieved by a rank test of the form $\Sigma a(R_i)$. Thus, no power is lost asymptotically by restricting attention to rank tests.

While the book briefly discusses some of the alternative notions of efficiency and refers to the recent work of Bickel, the reviewer would have appreciated both a more comprehensive analysis of when and why they give the same or different results and Professor Lehmann's expert opinion as to which notion of efficiency is most appropriate to each problem treated. In particular, Rubin and Sethuraman **[9]** developed the concept of Bayes risk efficiency which often agrees with Pitman's notion. In view of the "heated discussion" between Bayesians and non-Bayesians **[3]** in the literature on the foundations of statistics, even a short section by Professor Lehmann could aid in converting the "heat" to light.

So far our discussion has focused on a somewhat artificial problem, namely testing the hypothesis that the two populations underlying our samples are identical. In the real world the problem of interest is how to estimate the shift (difference), Δ. Statisticians have shown that for every rank test there is a corresponding estimator, Δ, and its efficiency relative to the best parametric estimator is the same as the Pitman efficiency of the test to the a.m.p.r.t. Since the fundamental paper in this area is a joint work of the author **[8]**, it is not surprising that the sections on estimating the treatment effect, illustrated by estimates derived from the Wilcoxon test, are both clear and comprehensive.

Due to space limitations we have illustrated the fundamental notions of nonparametric inference on just one problem. The text covers a much wider class of situations in which rank tests have been used, including paired comparisons (used in matching studies), the comparison of several treatments and randomized complete blocks. This last topic is quite important in biological research as it provides a method for dividing the population into homogeneous subgroups and testing to see if there is a treatment effect across all subgroups. These methods take advantage of the fact that "background" variation within each subgroup is normally much smaller than the variation in the entire population.

The mathematician reading the text will be exposed both to a wide variety of applications of nonparametric methods and to the probabilistic tools which underly the theory. Indeed, the mathematical appendix (approximately 80 pages long) is superb. Although the book discusses a variety of real problems, from a practical statistician's viewpoint it spends too much time on the problem of ties (observations with the same value) and not enough time on "heuristic" or common sense methods. For example, in the discussion of testing whether a sequence of numbers is a sample from the same distribution *or* has an increasing trend, i.e., the mean of the x_i's increases with i, the standard test due to Mann is discussed. If one draws a picture of samples coming from both the null and alternative hypotheses, one notices that their major difference is highlighted at the ends. This suggests that a reasonably powerful test might be developed by considering the lowest third of the observations as x's and the upper third as y's and testing whether the y's are larger than the x's. In the reviewer's opinion it is unfortunate that students generally are not taught the reasoning behind this type of "quick" and relatively efficient statistical procedure **[4]**.

In summary, Professor Lehmann's text will undoubtedly become the standard to which other mathematical introductions to nonparametric inference will be compared. It has an excellent selection of problems, ranging from routine calculations to challenging problems for the best students, and an extensive bibliography. Although students may not find the book particularly easy to read, Professor Lehmann has provided a very well organized guide to the field of rank tests. Mathematically inclined students should be encouraged to and able to pursue the study of Kolmogoroff-Smirnov and other "goodness of fit" tests which stimulated a significant amount of research in the properties of tied-down Wiener processes. The reviewer has no doubt that a student with initiative will learn a great deal about both the mathematical foundations and use of rank tests from this book.

References

1. A. Birnbaum and E. Laska, *Efficiency robust two-sample rank tests*, J. Amer. Statist. Assoc. **62** (1967), 1241–1251. MR **36** #6083.

2. H. Chernoff and I. R. Savage, *Asymptotic normality and efficiency of certain nonparametric test statistics*, Ann. Math. Statist. **29** (1958), 972–994. MR **20** #6755.

3. J. Cornfield, *The Bayesian outlook and its application*, Biometrics **25** (1969), 617–642. MR **41** #2833.

4. D. R. Cox and A. Stuart, *Some quick sign tests for trend in location and dispersion*, Biometrika **42** (1955), 80–95. MR **16**, 842.

5. J. L. Gastwirth, *Asymptotically most powerful rank tests for the two-sample problem with censored data*, Ann. Math. Statist. **36** (1965), 1243–1247. MR **31** #852.

6. ______, *On robust procedures*, J. Amer. Statist. Assoc. **61** (1966), 929–948. MR **34** #5226.

7. J. Hájek, *Asymptotic sufficiency of the vector of ranks in the Bahadur sense*, Ann. Statist. **2** (1974), 75–83. MR **50** #8825.

8. J. L. Hodges, Jr. and E. L. Lehmann, *Estimates of location based on rank tests*, Ann. Math. Statist. **34** (1963), 598–611. MR **27** #2050.

9. H. Rubin and J. Sethuraman, *Bayes risk efficiency*, Sankhyā Ser. A **27** (1965), 347–356. MR **34** #6928.

10. P. K. Sen, *Asymptotically most powerful rank order tests for grouped data*, Ann. Math. Statist. **38** (1967), 1229–1239. MR **35** #3820.

11. J. W. Tukey, *The future of data analysis*, Ann. Math. Statist. **33** (1962), 1–67. MR **24** #A3761.

J. L. Gastwirth

BULLETIN OF THE
AMERICAN MATHEMATICAL SOCIETY
Volume 83, Number 5, September 1977

Representations of real numbers by infinite series, János Galambos, Lecture Notes in Math., vol. 502, Springer-Verlag, Berlin, Hiedelberg, New York, 1976, vi + 146 pp., $7.40.

During the last thirty years a large amount of research has been devoted to the study of various algorithms for the representation of real numbers by means of sequences of integers. In addition to the two "classical" algorithms, i.e. digit expansions and continued fractions, and to several types of additional, slightly less well-known ones like Cantor, Lüroth, Engel and Oppenheim series, more general classes of such algorithms have been defined and investigated, particularly by F. Schweiger [6] and also by the author himself [3].

The book under review introduces the reader to some of the most important features of these developments. The exposition is based on what the author calls an (α,γ)-expansion $y(x)$ of a real number x. For each j, two strictly decreasing sequences $\alpha_j(n)$ and $\gamma_j(n)$ of positive real numbers are given, satisfying the condition

$$\alpha_j(n-1) - \alpha_j(n) \leqslant \gamma_j(n) \qquad (n = 2,3,\dots).$$

In order to define the algorithm for a given number $x \in (0,1]$, an auxiliary sequence $d_j(x)$ of integers is defined in such a way that the infinite series

$$y(x) = \alpha_1(d_1) + \gamma_1(d_1)\alpha_2(d_2) + \gamma_1(d_1)\gamma_1(d_2)\alpha_3(d_3) + \cdots$$

is always convergent and has, under fairly general assumptions, the limit x. A

Elements of Large-Sample Theory.

Erich L. LEHMANN. New York: Springer, 1999. ISBN 0-387-98595-6. xii + 631 pp. \$79.95.

This introductory book on the most useful parts of large-sample theory is designed to be accessible to scientists outside statistics and certainly to master's-level statistics students who ignore most of measure theory. According to the author, "the subject of this book, first-order large-sample theory, constitutes a coherent body of concepts and results that are central to both theoretical and applied statistics." All of the other existing books published on the subject over the last 20 years, from Ibragimov and Has'minskii in 1979 to the most recent by Van der Waart in 1998 have a common prerequisite in mathematical sophistication (measure theory in particular) that do not make the concepts available to a wide audience.

The idea of a much more elementary book is a brilliant one (which, in the Preface, the author attributes to his wife, Juliet Schaffer). There is a real need for scientists from other fields to understand and use such notions as efficiency, consistency, robustness, nonparametric estimation and goodness of fit. Many examples of reinventions of these notions may be found by browsing biological, computer science, physics, economics, and chemistry journals. The fault for ignorance of the more complex features of statistics is not to be lightly assigned to the creative *rediscoverers*, but rather to statisticians. Most classical mathematical statisticians' articles and books treating these subjects are unreadable for the lay public. Unfortunately, these successive reincarnations build communication gaps between the different communities and statisticians themselves. This book should help curb this trend, and I am happy to recommend it to my non-mathematically trained colleagues.

The table of contents gives a good idea of the book's structure: Mathematical Background, Convergence in Probability and in Law, Performance of Statistical Tests, Estimation, Multivariate Extensions, Nonparametric Estimation, and Efficient Estimators and Tests.

The book starts with all of the essentials that a nonmathematically trained scientist often lacks. I have already successfully sent Figure 2.61 in Chapter 2, of the difference between uniform and nonuniform convergence, to several master students in statistics and Ph.D students from other disciplines.

This book's self-explanatory style makes it ideal for isolated scholars from other disciplines. The professorial voice is present throughout the text in a most friendly way. Many exercises are given to illustrate each chapter, with useful hints provided when necessary.

The author is well known for his scholarship, but the reader will not be overwhelmed by too many extensions or research articles. In most cases, a theorem is credited in the text to its author, and each chapter ends in a complete bibliographical review of relevant material.

This book clearly reveals what a "gentle man" the author really is in the literal sense—each notion is gently introduced, first in its simplest form, then generalized. For instance, on page 488 an example showing the efficiency of a Bayes estimator for the binomial with a beta prior is developed, then the reader is led to the general asymptotic efficiency result for Bayes estimators. This introduction *by example* is most helpful in leading to the general theory painlessly, providing both motivation and illustration.

I can only find one slight lapse in the author's discernment in his choice of presentation and material. Chapter 2, Section 2.7 includes the presentation of limit theorems in the cases when the random variables are not independent or are not identically distributed.

The book combines a delightfully rich perspective as the author spans so much of modern statistics' development. Take, for instance, Von Mises' "plug-in" (A term, I am told, is from Ned Glick in 1972) principle. It's current incarnation is Efron's bootstrap and the clarity with which the author develops this picture makes exposing it to master level students a real pleasure. Actually the entire Chapter 6, which is dedicated to nonparametric estimation, should be a suggested reading for many biologists/economists/philosophers who assimilate statistics with parametric statistics, believing the other type to be simply *heuristics*. This chapter opens with the simplest plug-in estimates $h(\hat{F}_n)$, which can be written as expectations and that can be estimated by U and V statistics.

The more general case is developed by using linearization techniques, the delta method, its generalizations to statistical functionals, and the influence function. This provides real insight into much of what asymptotics is about. (There are even statisticians who claim that the only really useful theorem in mathematical statistics is Taylor's theorem!)

Finally, the author shows his vision of the bootstrap as a simple extension of the plug-in principle to include the case where the functional depends both on the distribution and the sample size. Again, this way of leading the reader step by step through simple examples all the way to being able to grasp the proof of the bootstrap's consistency properties is a real achievement.

Of course, Chapter 7 on efficient estimation can be prescribed to the many researchers now wedded to the maximum likelihood estimator. It provides the methods of comparing the various choices available for estimators, although the author does overemphasize his opinion that the first property expected of a good estimator in the iid case is that it be consistent. At this stage I would have also included a sharp recall to those who use maximum likelihood estimation that such a conclusion is moot as soon as the assumptions of the model's validity are dropped, and robustness must always be considered concurrently. The brevity with which the concept of identifiability was treated is to be regretted, as there are few books that clearly illustrate and define this notion.

Asymptotic theory demands a growing number of observations. The book uses the classical fiction of triangular arrays to make this rigorous. This embeds $X_1, \ldots, X_n$ into an array $X_{ni}, 1 \leq n \leq \infty, 1 \leq i \leq n$. The distributions in different rows can be arbitrarily dependent, or even unspecified.

This sometimes seems natural; for example, in the classical matching problem, a Poisson limit is proved for $S_n = X_1 + X_2 + \cdots + X_n$ with X_i 1 or 0, depending on whether the ith place of the permutation is fixed or not. These $X_1, \ldots, X_n$ cannot be extended in any symmetric fashion, and triangular arrays seem natural. One recent advance in asymptotic theory replaces the hypothetical triangular array by an approximation theorem with error for the actual variables under consideration. For instance, in the permutation example one proves, for n fixed, $\|\mathcal{L}(S_n) - Po(1)\| \leq 2/n!$.

Such tools as Stein's method (Stein 1986) and coupling inequalities offer close to automated ways of deriving such approximations with explicit errors. It would have been nice to incorporate these ideas into statistical applications, as the methods do not use measure theory or any mathematics beyond the prerequisites of this book.

The book has some epidermic defects that will disappear with future editions with regards to some of the mathematical copy-editing, which does not seem quite on a par with the high quality of all the other aspects of the book. For instance, the characters on the Figure 7.1.1 (p. 454) are set in an unreadable way; after a long decryption, one realizes that the font is too large and that the expressions $\theta - c(\theta)$ and $\theta + c(\theta)$ were supposed to lie on the same line. For a TEX "æsthete," the parentheses are a little painful to follow; opening left parentheses of a certain size like to be closed with the same-sized right counterpart, and when two quantities are going to become infinitely similar, as in $h(\hat{F}_n) \to h(F)$, why shouldn't the size of their brackets be the same, too?

Equation (6.5.14) has a $\to$ that leads nowhere special, and the brackets for the normal distribution's parameters are left unclosed in the middle of page 389. There are also a few author's names consistently mispelled in the references and author index, (e.g., W. Hoeffding, O. Zeitouni, and H. Jeffrey's). But I will stop quibbling and linger rather on what this book provides.

Elements of Large-Sample Theory will give many of us the opportunity to create the intermediate courses that we lack, between the elementary level and the specialists' courses that we teach from the authors' other books (Lehmann 1983, 1986; Lehmann and Casella 1998). These books stress small-sample results, which are valid in precisely specified parametric models. In a personal communication, Professor Lehmann has suggested that if he had to choose all over again, he would have developed the kind of useful approximations developed in the present book.

We can only be thankful for the time that Lehmann has set aside to write carefully a book containing a simple picture of the intricacies of the best in mathematical statistics, instead of striving, as many of our colleagues, to add a few more technically beautiful but practically useless articles to our poor "unreadable" journals.

Susan P. HOLMES
Stanford University

REFERENCES

Casella, G., and Berger, R. L. (1990), *Statistical Inference*, Pacific Grove, CA: Wadsworth.

Ferguson, T. S. (1996), *A Course in Large-Sample Theory*, London: Chapman and Hall.

Glick, N. (1972), "Sample-Based Classification Procedures Derived From Density Estimators," *Journal of the American Statistical Association*, 67, 116–122.

Ibragimov, I. A., and Has'minskii, R. Z. (1981), *Statistical Estimation*, New York: Springer-Verlag.

Lehmann, E. L. (1975), *Nonparametrics*, San Francisco: Holden-Day.

——— (1983), *Theory of Point Estimation*, London: Chapman and Hall.

——— (1986), *Testing Statistical Hypotheses* (2nd ed.), London: Chapman and Hall.

Lehmann, E. L., and Casella, G. (1998), *Theory of Point Estimation* (2nd ed.), New York: Springer-Verlag.

Van der Vaart, A. (1998), *Asymptotic Statistics*, Cambridge, U.K.: Cambridge University Press.

Theory of Point Estimation (2nd ed.).

E. L. LEHMANN and George CASELLA. New York: Springer-Verlag, 1998. ISBN 0-387-98502-6. xxvi + 589 pp. $74.95.

This second edition of Erich Lehmann's 1983 classic text is a most welcome addition to the statistical literature. George Casella has joined in the writing of the new edition and both the (more) senior and (more) junior author have clearly left the stamps of their individual interests and styles on this excellent book.

The new edition is approximately 80 pages thicker and includes roughly 1,000 references, of which about 250 are post-1983. The original edition is clearly visible in the revision and many sections have been carried over with minimal changes.

The first edition was reviewed in JASA by Berger (1984) and Rao (1984). The reader is encouraged to read these reviews, as much of what they say (especially Berger) remains relevant to the second edition.

The major changes are splitting Chapter 4 in the first edition on "Global Properties" into two chapters, one (Chapter 4) on "Average Risk Optimality" (Bayesian Estimation) and one (Chapter 5) on "Minimality and Admissibility"; and concatenating (and reducing) the material in Chapter 5 (Large-Sample Theory) and Chapter 6 (Asymptotic Optimality) into a new Chapter 6 (Asymptotic Optimality).

As a result of these changes, the new edition includes quite a bit more material on Bayesian estimation and quite a bit less material on robust estimators (e.g., medians, trimmed means, and M, L, and R estimates). No doubt space considerations played a role in the decision to omit much of the discussion of robust alternatives to the sample mean. I would have preferred that at least some of this discussion had been retained; it added substantially to the utility and balance of the first edition as a graduate text.

In a somewhat minor change in organization, the discussion of "Convergence in Probability and in Law" is moved to Chapter 1.

Typographically, the new edition is not as attractive as the first. This seems largely due to the increased bulk of material included, which has resulted in a smaller type size and smaller section headings. I find the latter significantly more annoying than the former. A return to larger section headings would be most welcome and helpful without adding greatly to the page count.

One major improvement over the first edition is the placement of the bibliography at the end of the text as opposed to a separate bibliography at the end of each chapter. This excellent 44-page bibliography is extensive and up to date. It is a major addition and should be useful to and welcomed by purchasers of the text. It is in itself almost worth the price of the book.

The "Reference" section at the end of each chapter has been replaced by a "Notes" section. This section contains extensions of the discussions of selected topics, historical comments, and indications of other possible references. These "Notes" sections are typically expansions on mostly historical comments at the beginning of the "Reference" section of the first edition. They are interesting, informative, and well done.

The problem sets are an excellent source of additional information and practice, and they give good opportunities for testing and honing of skills (for students and teachers!). The problem sets are extensive (15–30 pages per chapter), well chosen, and of varying difficulty. The error rate seems to be quite low. One amusing lapse is in Problem 6.24 in Chapter 1, where one is to show that a particular matrix is inevitable (instead of invertible).

The new edition continues the tradition of the first in presenting the material at a non–measure-theoretic level. The careful scholarship and readability of the first edition is admirably continued. The density of typos seems quite small and that of technical errors extremely small. In the spirit of Berger's review, here are some minor errors and one slightly more major one to indicate that I have indeed read the book:

- Example 1.3 on page 229 gives a Bayes estimator for a Poisson sample (of size n) and gamma prior, but the expression given is for the case $n = 1$.
- In Example 3.2 on page 169, the scale-equivariant estimator should be $B|X|^r$ if $X < 0$.
- In Example 4.5 on page 116, $calF_0'$ should be $\mathcal{F}_0'$.
- In Problem 6.9 on page 295, the definition of the n-fold convolution is incorrect (convoluted?).
- In Equation 1.11 on page 88, the exponent of q should be $n - t$, not $n - 1$.
- The slightly more serious error is in the statement of Corollary 1.6 on page 311. The corollary claims that δ_Λ is minimax if and only if $\Lambda(\omega_\Lambda) = 1$, where ω_Λ is the set of θ for which the risk of the Bayes estimator δ_Λ takes on its supremum. The condition is sufficient but not necessary. A class of counterexample is provided by any proper Bayes minimax estimator of a p-variate normal mean for $p \geq 5$ with respect to a prior whose support is all of R^p. For all such procedures, the Bayes risk is strictly less than the supremum (minimax) risk.

This revised and updated edition of *Theory of Point Estimation* is an excellent addition to the literature. It is well written, up to date, well organized, and at an appropriate level. It should remain a standard graduate-level text and reference work for many years. The first edition has been my favorite statistics book for some time, and the new edition is a worthy successor.

William E. STRAWDERMAN
Rutgers University

REFERENCES

Berger, J. O. (1984), Review of *Theory of Point Estimation*, by E. Lehmann *Journal of the American Statistical Association*, 79, 941–942.

Rao, C. R. (1984), Review of *Theory of Point Estimation*, by E. Lehmann *Journal of the American Statistical Association*, 79, 942.

Mathematical Statistics.

Jun SHAO. New York: Springer Verlag, 1999. ISBN 0-387-98674-X. xiv + 529 pp. $74.95.

This text is a nice addition to the collection of graduate textbooks in theoretical statistics at the level of Lehmann's (1986) *Testing Statistical Hypothesis* and Lehmann and Casella's (1998) *Theory of Point Estimation*. It is intended as a first- or second-year graduate text for a two-semester course in mathematical statistics. Students should have a good knowledge of advanced calculus. In fact some knowledge of probability and, to a lesser extent, measure theory is a practical necessity, although the Chapter 1 gives a nice review/overview of measure theory and probability, albeit largely without proofs but with some nice examples. The book is generally quite rigorous and well written, with numerous good examples. It should be a good main or supplementary text for its intended audience.

Two aspects of the book separate it from its principal competitors. The first is the treatment of asymptotic aspects of statistics, which is integrated into the discussion of each topic as opposed to being separated out in a separate chapter. The second distinguishing feature is the treatment in a

Reminiscences of a Statistician: The Company I Kept
Erich L. Lehmann
Springer, 2008, xii ї 309 pages, 34.94 / US$ 44.95, softcover
ISBN: 978-0-387-71596-4

Table of contents

Readership: For anyone interested in the people and ideas from the academic side of our profession over the last 75 years.

This book is a gem, a must-read. It is a thoroughly enjoyable mixture of autobiographical and broader historical material, presented as mini-biographies of over 60 mathematicians and statisticians connected to the author, these being grouped by time, place or a scientific or personal theme. Lehmann is extremely well known in the statistical world for his research contributions, and his books on testing, estimation, nonparametrics, and introductory statistics, which have been widely used for decades, and translated into many languages other than their original English. The wonderful writing style which makes his texts so popular shines out in this one. He suggests early on in the book that as a teenager his love was German literature, and that perhaps he would have become a writer, had circumstances in his native Germany not dictated otherwise. Instead he became a mathematician, and later a statistician. A reader of this volume will come away with the distinct impression that he realized this early ambition, though perhaps not quite as he originally imagined.

There are so many fascinating pieces in this book that it is probably unwise to single any out, but I cannot resist doing so. The portrait of the German mathematician Edmund Landau is a delight, showing a side to him and recounting his final, sad years in a way that cannot fail to amuse and move the reader. The book contains many anecdotes relating to what we might call career development, implicitly emphasizing the role of chance in life. It contains much of interest on academic research, administration, teaching, the writing of papers and textbooks, collaboration, advising and, most prominently, friendship. It is also full of photos of those who people the book.

Terry Speed: terry@stat.berkeley.edu
Department of Statistics, 367 Evans Hall #3860
University of California, Berkeley, CA 94720-3860, USA

International Statistical Review (2008), **76**, 1, 140–156